ADME-ENABLING TECHNOLOGIES IN DRUG DESIGN AND DEVELOPMENT

ADME-ENABLING TECHNOLOGIES IN DRUG DESIGN AND DEVELOPMENT

EDITED BY

DONGLU ZHANG
SEKHAR SURAPANENI

WILEY

A JOHN WILEY & SONS, INC., PUBLICATION

Published by John Wiley & Sons, Inc., Hoboken, New Jersey.
Published simultaneously in Canada.

For general information on our other products and services or for technical support, please contact our Customer Care Department within the United States at (800) 762-2974, outside the United States at (317) 572-3993 or fax (317) 572-4002.

Wiley also publishes its books in a variety of electronic formats. Some content that appears in print may not be available in electronic formats. For more information about Wiley products, visit our web site at www.wiley.com.

Library of Congress Cataloging-in-Publication Data:
ADME-enabling technologies in drug design and development / edited by Donglu
Zhang, Sekhar Surapaneni.
    p. ; cm.
  Includes bibliographical references and index.
  ISBN 978-0-470-54278-1 (cloth)
  I. Zhang, Donglu. II. Surapaneni, Sekhar.
  [DNLM: 1. Drug Design. 2. Drug Evaluation, Preclinical. 3. Pharmaceutical
Preparations–metabolism. 4. Pharmacokinetics. 5. Technology,
Pharmaceutical–methods. QV 744]
  LC-classification not assigned
  615.1'9–dc23
                                                        2011030352

Printed in the United States of America.

ISBN: 9780470542781

10 9 8 7 6 5 4 3 2 1

CONTENTS

FOREWORD

The discovery, design, and development of drugs is a complex endeavor of optimizing on three axes: efficacy, safety, and druggability or drug-likeness. Each of these axes is a potential cause of attrition as a new molecular entity progresses through the many phases of drug development. Out of the 5000–10,000 compounds evaluated in discovery efforts, only 250 enter preclinical testing, 5 enter clinical trials, and only 1 is granted approval by the Food and Drug Administration at a cost that is estimated between US$1.3–1.6 billion [1]. Efforts to increase innovation, decrease attrition, and lower the cost of drug development are the focus of the pharmaceutical industry and regulatory agencies alike. Advances have been made in some disciplines such as drug metabolism and pharmacokinetics (PK), particularly in the area of absorption, distribution, metabolism, and excretion (ADME) studies. For example, a root cause analysis of clinical attrition [2] showed that unacceptable PK or bioavailability accounted for 40% of clinical attrition in the 1990s but within a decade had been reduced to less than 10%, in large part by the identification and mitigation of risks associated with ADME/PK properties earlier in the drug discovery process. This was enabled by the introduction of automated high- and medium-throughput screening of lead optimization candidates in the discovery space. While impressive, this improvement alone is not sufficient to reverse the rising costs and long development cycle times. It is, however, a step in the right direction. As the pharmaceutical industry has evolved, the focus of ADME studies has shifted from studies conducted primarily in support of regulatory submissions to playing a significant role in the earliest stages of the discovery phase of drug development. The engagement of ADME scientists in the discovery space has allowed drug candidates to progress in the development pipeline to the next milestone with greater probability of success because desirable characteristics, such as good aqueous solubility for absorption, high bioavailability, and balanced clearance, have been engineered into the molecules, and liabilities such as high first-pass metabolism and unacceptable drug–drug interactions potential have been engineered out.

The history of the discipline of drug metabolism and PK and ADME studies, with its roots in organic chemistry and pharmacology, has been well chronicled [3–8]. The rapid advancement of the discipline over the past 50 years is clearly linked to the development of ever-increasingly sophisticated analytical tools and the growth of the pharmaceutical industry. The vast number of tools at the disposal of drug metabolism scientists has transformed the study of xenobiotics from descriptive to quantitative, in vivo to the molecular levels, and from simply characterizing to predicting ADME properties.

It would be beyond the scope of this introduction to provide a historical accounting of the numerous advances of technology that have shaped the field. There are, however, three noteworthy milestones in the evolution of the discipline that merit mention: the use of radioisotopes in metabolism and distribution studies; the discovery of the superfamily of drug metabolizing enzymes, the cytochrome P450s; and the revolutionizing impact of mass spectrometry as both a qualitative and quantitative tool.

With the discovery of a new radioisotope of carbon, ^{14}C, by Martin and Ruben [9], this powerful analytical tool enabled the first radiolabeled studies that elucidated the metabolic pathways and the disposition of xenobiotics in rats [10, 11]. The use of radiotracers went on to become an indispensable tool in biochemical pathway elucidation and in drug disposition studies. While ^{14}C-labeled compounds are predominantly used in *in vivo* studies to fulfill regulatory requirement, the development of new reagents and techniques in tritium labeling now have allowed stereo- and site-selective synthesis with high specific activity, making these labeled molecule readily available for use in the earliest phases of drug discovery [12, 13].

The discovery of the cytochrome P450s and their role in the metabolism of endo- and xenobiotics opened a field of science that continues to grow and have a tremendous impact on the development of drugs and the practice of medicine. The pioneering research in this field has been well documented by Estabrook, a key contributor to our current understanding of this super-family of enzymes [14]. The magnitude of research on the cytochrome P450s has exploded since 2003 (from greater than 2000 literature references to over 67,000 citations, as reflected by searching the PubMed database in 2011) The expanding knowledge of the cytochrome P450s has impacted early discovery efforts via assays for metabolic stability, species comparison in the selection of the most relevant species for toxicology studies, identification of the primary enzymes involved in the metabolism of a candidate drug, and potential polymorphic or drug–drug interaction liabilities of a candidate drug. The influence of the research on the cytochrome P450s also reaches into the clinical realm of drug development in the need for and design of clinical drug–drug interaction trials as well as in the regulatory guidance on drug interactions [15, 16].

No single analytical technique has had a more powerful effect on drug development than mass spectrometry, with an impact on multiple disciplines, such as chemistry, biology, and ADME [17]. An excellent review of mass spectrometry and its applications in drug metabolism and PK has recently been published [18] Mass spectrometry moved from the being a specialized tool largely used in structure identification to a "routine," but albeit powerful, analytical technology used across the pharmaceutical industry and academia alike. The selectivity, sensitivity, and speed of mass spectrometry enabled much of the success seen with high-throughput screening and advances in bioanalytical analysis in a multitude of biological matrices in both PK and biotransformation studies.

The ADME scientist of today is fortunate to have an arsenal of tools at his or her disposal, many of which will be expanded upon in this book. The advances in technologies often have implications in adjacent technologies that further the discipline of drug metabolism and PK and allow an integrated approach to solving problems and advancing drug candidates through the phases of drug development.

LISA A. SHIPLEY

REFERENCES

1. Burrill & Company. *Analysis for Pharmaceutical Research and Manufacturers of America; and Pharmaceutical Research and Manufacturers of America, PhRMA Annual Member Survey (Washington, DC: PhRMA, 2010).* Citations at http://www.phrma.org/research/infographics, 2010.

2. Kola I, Landis J (2004) Can the pharmaceutical industry reduce attrition rates? *Nature Reviews. Drug Discovery* 3:711–715.

3. Conti A, Bickel MH (1977) History of drug metabolism: Discoveries of the major pathways in the 19th century. *Drug Metabolism Reviews* 6(1):1–50.

4. Bachmann C, Bickel MH (1985) History of drug metabolism: The first half of the 20th century. *Drug Metabolism Reviews* 16(3):185–253.

5. Murphy PJ (2001) Xenobiotic metabolism: A look from the past to the future. *Drug Metabolism and Disposition* 29:779–780.

6. Murphy PJ (2008) The development of drug metabolism research as expressed in the publications of ASPET: Part 1, 1909–1958. *Drug Metabolism and Disposition* 36:1–5.

7. Murphy PJ (2008) The development of drug metabolism research as expressed in the publications of ASPET: Part 2, 1959–1983. *Drug Metabolism and Disposition* 36:981–985.

8. Murphy PJ (2008) The development of drug metabolism research as expressed in the publications of ASPET: Part 3, 1984–2008. *Drug Metabolism and Disposition* 36:1977–1982.

9. Ruben S, Kamen MD (1941) Long-lived radioactive carbon: C14. *Physical Review* 59:349–354.

10. Elliott HW, Chang FNH, Abdou IA, Anderson HH (1949) The distribution of radioactivity in rats after administration of C14 labeled methadone. *The Journal of Pharmacology and Experimental Therapeutics* 95:494–501.

11. Morris HP, Weisburger JH, Weisburger EK (1950) The distribution of radioactivity following the feeding of carbon 14-labeled 2-acetylaminofluorene in rats. *Cancer Research* 10:620–634.

12. Saljoughian M (2002) Synthetic tritium labeling: Reagents and methodologies. *Synthesis* 13:1781–1801.

13. Voges R, Heys JR, Moenius T *Preparation of Compounds Labeled with Tritium and Carbon -14,* Chichester, U.K.: John Wiley and Sons, 2009.

14. Estabrook RW (2003) A passion for P450's (remembrances of the early history of research on cytochrome P450). *Drug Metabolism and Disposition* 31:1461–1473.

15. *Guideline on the Investigation of Drug Interactions (EMA/CHMP/EWP/125211/2010).* (2010) http://www.ema.europa.eu/ema/pages/includes/document/open_document.jsp?webContentId=WC500090112.

16. *Guidance for Industry: In Vivo Drug Metabolism/Drug Interaction Studies-Study Design, Data Analysis, and Recommendations for Dosing and Labeling.* (1999) http://www.fda.gov/cder/guidance/index.htm.

17. Ackermann BL, Berna MJ, Eckstein JA, Ott LW, Chadhary AK (2008) Current applications of liquid chromatography/mass spectrometry in pharmaceutical discovery after a decade of innovation. *Annual Review Of Analytical Chemistry* 1:357–396.

18. Ramanathan R, ed. *Mass Spectrometry in Drug Metabolism and Pharmacokinetics.* Hoboken, NJ: John Wiley and Sons, 2009.

PREFACE

Understanding and characterizing absorption, metabolism, distribution, and excretion (ADME) properties of new chemical entities and drug candidates is an integral part of drug design and development. ADME is the discipline that is involved in the entire process of drug development, right from discovery, lead optimization, and clinical drug candidate selection through drug development and regulatory process. The complexity of ADME studies in drug discovery and development requires a drug metabolism scientist to know all available technologies in order to choose the right experimental approach and technology for solving the problems in a timely manner. During the last decade, tremendous progress has been made in wide array of technologies including mass spectrometry and molecular biology tools, and these enabling technologies are widely employed by ADME scientists. The generation of ADME data to support discovery and development teams is a gated process and timely generation of data to make right decisions is of paramount importance. Given the complexity of the drug discovery and development process, right techniques and tools should be used to generate timely data that is useful for decision making and regulatory filing. This requires an understanding of not only the breadth and depth of ADME technologies but also their limitation and pitfalls so scientists can make appropriate choices in employing these tools. A book on integrated enabling technologies will not only be useful to drug metabolism scientists but also could be a very helpful reference for scientists from the fields of pharmacology, medicinal chemistry, pharmaceutics, toxicology, and bioanalytical sciences in academia and industry.

This book is divided into four main sections. Part A provides the reader with an overview of ADME concepts and current topics including ADME and transporter studies in drug discovery and development, active and toxic metabolites, modeling and simulation, and developing biologics and individual medicines. Part B describes the ADME systems and methods; these include ADME screening technologies, permeability and transporter studies, distribution across specialized barriers such as blood–brain barrier (BBB) or placenta, cytochrome P450 (CYP) inhibition, induction, phenotyping, animal models for studying metabolism and transporters, and bile collection. Part C of the book discusses analytical tools including liquid chromatography-mass spectrometry (LC-MS) technologies for quantitation, metabolite identification and profiling, accelerator mass spectrometry (AMS) and radioprofiling, nuclear magnetic resonance (NMR), supercritical fluid chromatography (SFC) and other separation techniques, mass spectrometric imaging, and quantitative whole-body autoradiography (QWBA) tissue distribution techniques. Part D presents new and evolving technologies such as stem cells, genetically modified animal models, and siRNA techniques in ADME studies. Other techniques included in this section are target imaging technologies, radiosynthesis, formulation, and testing of cardiovascular toxicity potential.

We would like to thank our colleagues who are the experts and leading practitioners of the techniques described in the book for their contributions. We hope that this book is useful and serves as a quick reference to all drug hunters and to all those who are new to the discipline of ADME.

DONGLU ZHANG
SEKHAR SURAPANENI

CONTRIBUTORS

Suresh K. Balani, DMPK/NCDS, Millennium: The Takeda Oncology Company, Cambridge, MA, USA

Praveen V. Balimane, Bristol-Myers Squibb, Princeton, NJ, USA

Vanessa N. Barth, Translational Sciences, Eli Lilly and Company, Indianapolis, IN, USA

Leslie Bell, Novartis Institutes for BioMedical Research, Cambridge, MA, USA

Rajinder Bhardwaj, DMPK, Chemical Sciences and Pharmacokinetics, Lundbeck Research USA, Paramus, NJ, USA

Catherine L. Booth-Genthe, Respiratory Therapeutic Area Unit, GlaxoSmithKline, King of Prussia, PA, USA

Hong Cai, Bristol-Myers Squibb, Pennington, NJ, USA

Gamini Chandrasena, DMPK, Chemical Sciences and Pharmacokinetics, Lundbeck Research USA, Paramus, NJ, USA

Jiwen Chen, Bristol-Myers Squibb, Pennington, NJ, USA

Saeho Chong, College of Pharmacy, Seoul National University, Seoul, Korea

Lisa J. Christopher, Bristol-Myers Squibb, Princeton, NJ, USA

Jun Dai, Bristol-Myers Squibb, Princeton, NJ, USA

Li Di, Pfizer Global Research and Development, Groton, CT, USA

Ashok Dongre, Bristol-Myers Squibb, Pennington, NJ, USA

Dieter M. Drexler, Bristol-Myers Squibb, Wallingford, CT, USA

Richard W. Edom, Janssen Pharmaceutical Companies of Johnson & Johnson, Raritan, NJ, USA

Charles S. Elmore, Radiochemistry, AstraZeneca, Mölndal, Sweden

Adrian J. Fretland, Nonclinical Safety, Early ADME Department, Roche, Nutley, NJ, USA

Timothy J. Garrett, Clinical and Translational Science Institute, University of Florida, Gainesville, FL, USA

Lingling Guan, Ricerca Biosciences, Concord, OH, USA

Anshul Gupta, Drug Metabolism and Pharmacokinetics, AstraZeneca, Waltham, MA, USA

Yong-Hae Han, Bristol-Myers Squibb, Princeton, NJ, USA

Imad Hanna, Drug Metabolism and Pharmacokinetics, Novartis Institutes for BioMedical Research, East Hanover, NJ, USA

David C. Hay, MRC Centre for Regenerative Medicine, Edinburgh, UK

Haizheng Hong, College of Oceanography and Environmental Sciences, Xiamen University, Fujian, China

Cornelis E.C.A. Hop, Department of Drug Metabolism and Pharmacokinetics, Genentech, South San Francisco, CA, USA

Matthew Hoffmann, Celgene Corporation, Summit, NJ, USA

Stella Huang, Bristol-Myers Squibb, Wallingford, CT, USA

W. Griffith Humphreys, Bristol-Myers Squibb, Princeton, NJ, USA

Wenying Jian, Johnson & Johnson Pharmaceutical Research & Development, Raritan, NJ, USA

Xi-Ling Jiang, Department of Pharmaceutical Sciences, School of Pharmacy and Pharmaceutical Sciences, University at Buffalo, The State University of New York, Buffalo, NY, USA

Kim A. Johnson, Bristol-Myers Squibb, Wallingford, CT, USA

Janan Jona, Small Molecule Process and Product Development/Preformulation, Amgen Inc., Thousand Oaks, CA, USA

Elizabeth M. Joshi, Department of Drug Disposition, Lilly Research Laboratories, Indianapolis, IN, USA

Nataraj Kalyanaraman, Pharmacokinetics and Drug Metabolism, Amgen Inc., Thousand Oaks, CA, USA

Jiesheng Kang, Sanofi-Aventis U.S. Inc., Bridgewater, NJ, USA

Edward H. Kerns, Therapeutics for Rare and Neglected Diseases, NIH Center for Translational Therapeutics, Rockville, MD, USA

Yuan-Hon Kiang, Small Molecular Process and Product Development/Preformulation, Amgen Inc., Thousand Oaks, CA, USA

Wing Wah Lam, Janssen Pharmaceutical Companies of Johnson & Johnson, Raritan, NJ, USA

Chun Li, Metabolism and Pharmacokinetics, Genomics Institute of the Novartis Research Foundation, San Diego, CA, USA

Mingxiang Liao, DMPK/NCDS, Millennium: The Takeda Oncology Company, Cambridge, MA, USA

Heng-Keang Lim, Janssen Pharmaceutical Companies of Johnson & Johnson, Raritan, NJ, USA

Zhongping (John) Lin, Frontage Laboratories, Inc. Malvern, PA, USA

Chang-Xiao Liu, State Key Laboratory of Drug Technology and Pharmacokinetics, Tianjin Institute of Pharmaceutical Research, Tianjin, China

Tom Lloyd, Worldwide Clinical Trials Drug Development Solutions Bioanalytical Sciences, Austin, TX, USA

Anthony Y.H. Lu, Department of Chemical Biology, Ernest Mario School of Pharmacy, Rutgers University, Piscataway, NJ, USA

Qiang Ma, Receptor Biology Laboratory, Toxicology and Molecular Biology Branch, National Institute for Occupational Safety and Health, Centers for Disease Control and Prevention, Morgantown, WV, USA

Daniel P. Magparangalan, Covidien, St. Louis, MO, USA

Brad D. Maxwell, Bristol-Myers Squibb, Princeton, NJ, USA

Kaushik Mitra, Merck & Co. Inc., Rahway, NJ, USA

Voon Ong, San Diego, CA, USA

Ryan M. Pelis, Department of Pharmacology, Dalhousie University, Halifax, Nova Scotia, Canada

Natalia Penner, Department of Drug Metabolism and Pharmacokinetics, Biogen Idec, Cambridge, MA, USA

Chandra Prakash, Department of Drug Metabolism and Pharmacokinetics, Biogen Idec, Cambridge, MA, USA

Darren L. Reid, Small Molecular Process and Product Development/Preformulation, Amgen Inc., Thousand Oaks, CA, USA

Kevin L. Salyers, Pharmacokinetics and Drug Metabolism, Amgen Inc., Thousand Oaks, CA, USA

Mark Seymour, Xceleron, Heslington, York, UK

Adam Shilling, Incyte Corp, Wilmington, DE, USA

Lisa A. Shipley, Drug Metabolism and Pharmacokinetics, Merck & Co., Inc., West Point, PA, USA

Yue-Zhong Shu, Bristol-Myers Squibb, Princeton, NJ, USA

Jose Silva, Janssen Pharmaceutical Companies of Johnson & Johnson, Raritan, NJ, USA

Matthew D. Silva, Amgen Inc., Thousand Oaks, CA, USA

Sekhar Surapaneni, Drug Metabolism and Pharmacokinetics, Celgene Corporation, Summit, NJ, USA

Adrienne A. Tymiak, Bristol-Myers Squibb, Princeton, NJ, USA

Jianling Wang, Novartis Institutes for BioMedical Research, Cambridge, MA, USA

Lifei Wang, Bristol-Myers Squibb, Princeton, NJ, USA

Xiaomin Wang, Celgene Corporation, Summit, NJ, USA

David B. Wang-Iverson, Bristol-Myers Squibb, Princeton, NJ, USA

Naidong Weng, Janssen Pharmaceutical Companies of Johnson & Johnson Raritan, NJ, USA

Caroline Woodward, Department of Drug Metabolism and Pharmacokinetics, Biogen Idec, Cambridge, MA, USA

Cindy Q. Xia, Biotransformation/DMPK, Drug Safety and Disposition, Millennium: The Takeda Oncology Company, Cambridge, MA, USA

Yang Xu, Pharmacokinetics and Drug Metabolism, Amgen Inc., Thousand Oaks, CA, USA

Ming Yao, Bristol-Myers Squibb, Princeton, NJ, USA

Richard A. Yost, Department of Chemistry, University of Florida, Gainesville, FL, USA

Ai-Ming Yu, Department of Pharmaceutical Sciences, School of Pharmacy and Pharmaceutical Sciences, University at Buffalo, The State University of New York, Buffalo, NY, USA

Haoyu Zeng, Safety Assessment, Merck Research Laboratories, West Point, PA, USA

Donglu Zhang, Bristol-Myers Squibb, Princeton, NJ, USA

Yingru Zhang, Bristol-Myers Squibb, Princeton, NJ, USA

Zhoupeng Zhang, Merck & Co. Inc., Rahway, NJ, USA

Mingshe Zhu, Bristol-Myers Squibb, Princeton, NJ, USA

Peijuan Zhu, Respiratory Therapeutic Area Unit, GlaxoSmithKline, King of Prussia, PA, USA

PART A

ADME: OVERVIEW AND CURRENT TOPICS

1

REGULATORY DRUG DISPOSITION AND NDA PACKAGE INCLUDING MIST

SEKHAR SURAPANENI

1.1 INTRODUCTION

Drug metabolism and pharmacokinetics (DMPK) plays an important and integral part in drug discovery and development. In drug discovery, during lead optimization and drug candidate identification, metabolism studies are conducted to screen a large number of compounds with potential liabilities. The emphasis is on generating data efficiently and in a timely fashion related to a compound's absorption, distribution, metabolism, and excretion (ADME) characteristics. Advances in analytical technologies, drug metabolism, and transporter biology has enabled drug metabolism scientists develop *in vitro* and *in vivo* tools to screen a large number of compounds efficiently and incorporate ADME information in lead optimization and identification. This early characterization for metabolic and pharmacokinetic (PK) properties is an essential element of lead optimization and candidate selection [1, 2]. Metabolism studies in the early drug discovery stages often involve evaluating a series of compounds to help identify and select a candidate for further development. The main purposes of these studies are to see if the compound has adequate metabolic stability, has low potential for any drug–drug interactions due to cytochrome P450 (CYP) induction or inhibition, and is not metabolized by polymorphic enzymes (such as CYP2D6) or exclusively by a single enzyme. In addition, the metabolites are characterized to assess if any reactive

metabolites of safety concern are generated. Recent survey suggests that the DMPK role in early optimization has resulted in reduced attrition due to PK-related issues from 40% (1990s) to <10% (2000s) [3]. Once the compound is selected for clinical development, detailed PK/ADME studies are conducted to characterize the bioavailability, metabolic properties, distribution, and excretion and elimination of the drug. These studies provide information to assess safety and provide data for registration. The focus of this chapter is to describe the various DMPK studies performed at various stages of drug development and how they are captured when filing a new drug application (NDA). Figure 1.1 shows the various DMPK studies conducted at different stages of drug development. The schematic is intended as an illustration, but the precise timing of the studies depends not only on the drug and its properties but also on the intended therapeutic benefit and target population. There are a number of regulatory guidance documents issued by regulatory authorities (U.S. Food and Drug Administration [FDA], European Medicines Agency [EMEA], etc.), and these guidelines are expected to be adhered to during the conduct of *in vitro* and *in vivo* metabolism studies. In addition, DMPK provides bioanalytical support for safety (toxicology and first-in-human) and efficacy (proof of concept or pivotal clinical) studies. The method development, validation, and sample analysis are expected to be conducted according to the guidelines issued by regulatory

ADME-Enabling Technologies in Drug Design and Development, First Edition. Edited by Donglu Zhang and Sekhar Surapaneni.
© 2012 John Wiley & Sons, Inc. Published 2012 by John Wiley & Sons, Inc.

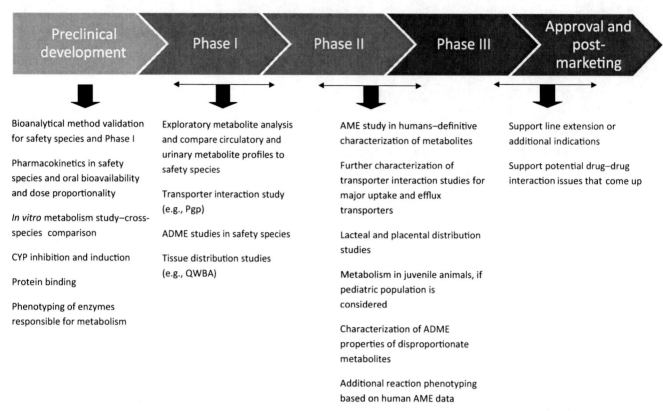

Preclinical development	Phase I	Phase II	Phase III	Approval and post-marketing

Bioanalytical method validation for safety species and Phase I

Pharmacokinetics in safety species and oral bioavailability and dose proportionality

In vitro metabolism study–cross-species comparison

CYP inhibition and induction

Protein binding

Phenotyping of enzymes responsible for metabolism

Exploratory metabolite analysis and compare circulatory and urinary metabolite profiles to safety species

Transporter interaction study (e.g., Pgp)

ADME studies in safety species

Tissue distribution studies (e.g., QWBA)

AME study in humans–definitive characterization of metabolites

Further characterization of transporter interaction studies for major uptake and efflux transporters

Lacteal and placental distribution studies

Metabolism in juvenile animals, if pediatric population is considered

Characterization of ADME properties of disproportionate metabolites

Additional reaction phenotyping based on human AME data

Support line extension or additional indications

Support potential drug–drug interaction issues that come up

FIGURE 1.1. Various DMPK studies conducted at different stages of drug development.

agencies worldwide, and any sample analysis conducted under good laboratory practice (GLP) guidelines is expected to meet the standards set forth under these guidelines. The metabolism data generated at various stages of development need to be integrated and summarized for regulatory filing and approval. The agreement to assemble all the quality, safety, and efficacy information in a common format (Common Technical Document or CTD) and the technical requirements for the registration of pharmaceuticals has been harmonized by the International Committee on Harmonization (ICH) process. This has revolutionized the regulatory review process and harmonized electronic submissions. Table 1.1 shows the table of contents for the drug metabolism contributions in a CTD for an NDA. The objectives of this chapter are to discuss DMPK studies conducted during development and to reference the regulatory guidance documents that apply to various studies. The purposes are to integrate the information across studies and species and to present information related to safety and efficacy in an unambiguous and transparent manner to help regulators evaluate the content and main features of the drug.

TABLE 1.1. DMPK Summaries in the Sections of the Common Technical Document (CTD)

2.4 Nonclinical overview
2.6 Nonclinical written/tabulated summaries
   2.6.4 Pharmacokinetic written summary
    • Brief summary
    • Methods of analysis
    • Absorption
    • Distribution
    • Metabolism (including interspecies comparison)
    • Excretion
    • Pharmacokinetic drug interactions
    • Other pharmacokinetic studies
   2.6.5 Pharmacokinetic tabulated summary
Module 4 (Reports of studies summarized in Sections 2.6.4 and 2.6.5)
   4.2.2.1 Analytical methods and validation reports
   4.2.2.2 Absorption
   4.2.2.3 Distribution
   4.2.2.4 Metabolism
   4.2.2.5 Excretion
   4.2.2.6 Pharmacokinetic drug interactions
   4.2.2.7 Other pharmacokinetic studies
4.3 Literature references (single PDF/reference)

1.2 NONCLINICAL OVERVIEW

Nonclinical overview presents the summary of information related to pharmacology, PK/ADME, and toxicology. Nonclinical PK information in this section is not intended to contain a summary of each study conducted. The purposes are to integrate drug metabolism information generated across studies and to provide a summary of findings that may have safety and efficacy implications. Taking the pharmacology, PK, and toxicology results into account, the implications of the nonclinical findings for safe human use of the pharmaceutical should be discussed. It should briefly describe the PK/toxicokinetic (TK) data in safety species, metabolism characteristics *in vitro* and *in vivo* and any interspecies differences, tissue distribution properties, and relevance to safety or efficacy. In addition, any inconsistencies and limitations of the data should be discussed.

1.3 PK

Section 2.2 of Module 2 describes the nonclinical PK information. This section gives a brief list of all ADME studies and methods of analysis. It also gives a brief summary of principal findings. The information is summarized in approximately three to four pages.

The "Methods of Analysis" section contains a brief summary of analytical methods employed for analysis of biological samples. Bioanalytical laboratories support a range of regulatory studies ranging from early toxicology studies through clinical studies (Phases I, II, and III) to definitive bioequivalence and bioavailability studies in support of the marketed product. PK/TK data aid the interpretation of efficacy and safety information, and therefore the quality of PK/TK data is directly related to the quality of bioanalytical data and the methods employed to generate such data. Therefore, the method validation and sample analysis should be carried out as per GLP regulatory guidelines [4, 5]. Although bioanalytical methods are validated to ensure that they function and perform as intended, the actual study or "incurred" samples from animals and human subjects may differ in composition and could potentially differ in their behavior compared with standard or quality control samples. This has resulted in a recommendation to analyze a fraction of study samples for reproducibility. The current regulatory expectations is such that many companies routinely conduct incurred sample reanalysis (ISR) to ensure the reproducibility and quality of study sample data. The detailed discussion of regulatory requirements is not within the scope of this chapter but the reader is referred to recent reviews in this area [6–8]. The bioanalytical methods used in toxi-cology studies should describe the species, detection and quantitation limits, and validation and stability of data.

1.4 ABSORPTION

Early in drug development, PK studies are conducted in rodent (rats and mice) and nonrodent (dogs or monkeys) species to assess absorption (extent and rate of absorption), bioavailability ($\%F$), dose proportionality, and kinetic parameters (C_{max}, AUC, and $t_{1/2}$). The choice of species is dependent on the metabolic similarities between animal species and humans based on *in vitro* and *in vivo* metabolism studies. It is important to ensure that metabolites generated in humans are covered at least in one of the species used in toxicology testing. In addition, it is important to ensure that adequate exposures are obtained in the dose range tested for safety assessment. Any nonlinearity in PK should be assessed and should aid in dose selection for safety studies. Typically, absorption and exposure assessment is done in male and female animals to evaluate any gender differences. This information should be taken into consideration when selecting doses for toxicology studies.

The route of administration should be based on the intended route of delivery for humans. Although many small-molecule drugs are orally delivered, intravenous PK are also determined to characterize the absolute bioavailability, clearance, and volume of distribution of molecule for better understanding of oral kinetics. It is important to carry out PK characterization with adequately characterized compound for stability and purity. The salt and crystalline form of the compound should be the same as the one that is used for safety studies. The dose selection should cover the dose range where the compound was found effective in pharmacology models to 30- to 50-fold of the efficacious dose for safety studies. The selected doses would generate exposure data across the dose range for dose proportionality assessment and help to assess any metabolism issues at higher doses. It is important that PK studies employ dose volumes and acceptable formulations used in toxicology studies so the data obtained in PK studies can be useful for carrying out safety studies. Formulation strategies and approaches used in preclinical setting are discussed in detail in Chapter 31. Although nonclinical PK studies are not required to be conducted as per GLP, they are often carried out under the spirit of GLP guidelines. The bioanalysis of study samples employs robust and reproducible methods. The PK analysis typically employs noncompartmental methods and describes PK information such as C_{max}, T_{max}, AUC, CL, V_{ss}, $t_{1/2}$, and $\%F$. It is important to determine half-life accurately so that any potential accumulation could be understood before repeat dose studies are conducted.

The absorption data are summarized in Section 2.6.4. The written summaries describe the test system (species, gender, number of replicated animals), the route of administration, formulations used, and analytical and data analysis methodologies used. The summary of findings should describe the absorption (rate and extent of absorption), percent oral availability, linearity of kinetics, and gender differences, if any.

1.5 DISTRIBUTION

Distribution is a process by which drug and its metabolites partition in and out of various cells and body tissues. Distribution of a compound into tissues requires the compound to permeate cell membranes, and this primarily occurs via passive diffusion. However, there are specialized barriers such as the blood–brain barrier (BBB) that express transporter proteins (e.g., P-glycoprotein [Pgp]) to prevent or minimize access to tissue. Drug distribution into tissues depends on the physicochemical properties of the compound (log P, pK_a, molecular weight, etc.), protein binding, permeability, and transporter activity [9, 10].

1.5.1 Plasma Protein Binding

Plasma protein binding dictates the rate and extent of distribution of drugs in the body. It is generally thought that the free drug is considered pharmacologically and toxicologically relevant [11, 12]. It is the free fraction and not the bound fraction that is available for distribution into tissues. Drugs bind to plasma protein albumin, alpha-acid glycoprotein (AGP), and lipoproteins. Binding and transport of exogenous and endogenous are important functions of plasma proteins. Drugs that have high lipophilicity tend to bind to a greater extent to plasma proteins. A more in-depth discussion of techniques and issues related to plasma protein binding are presented in Chapters 9 and 12.

As outlined above, the plasma protein binding is an important determinant for the pharmacodynamics (PD)/toxicodynamics of drugs, and the free fraction dictates the drug distribution and also influences clearance and half-life. Therefore, understanding the extent of protein binding across species is an important consideration for interpretation of efficacy and safety. It is important to take into consideration any difference in protein binding across species in explaining the effects observed in pharmacology or toxicology studies. In addition, range of concentrations that are observed in efficacy and safety studies across various species need to be evaluated to determine if the protein binding is linear or saturable. This may have an important role in explaining the effects of drugs at higher concentrations if the binding is saturable. Furthermore, it is known that disease conditions, age, and pregnancy alter the levels of plasma proteins, thereby influencing the free drug concentrations, clearance, and distribution [13]. For example, AGP levels are known to increase during pregnancy and this needs to be taken into consideration for drugs that are highly bound to this plasma protein [13]. Although protein binding is determined *ex vivo* using plasma from healthy animals and humans, sometimes the plasma protein binding is also determined in disease populations to ensure that protein binding is similar.

In addition to plasma protein binding, blood-to-plasma ratio is also determined for drugs. This is particularly important for drugs that are distributed into blood cells and bind to cellular components [14]. In such cases where blood-to-plasma partitioning is much greater than one, the plasma PK profile alone will not adequately capture the PK of the drugs. Therefore, it is important to understand the blood portioning of the compound early on so that PK will be assessed in relevant matrix.

Plasma protein and blood-to-plasma partitioning is captured under "Distribution" in Section 2.6.4 of the written summaries. The species, matrix, concentration range, and techniques used are described. The results should describe the extent of binding across species and highlight any species differences. In addition, it will discuss the observation of any saturable binding and the relevance of this to the interpretation of PK/TK. There are no specific guidances that address plasma protein binding alone, although it is referred in some guidances.

1.5.2 Tissue Distribution

Tissue distribution studies are essential in understanding the distribution of a compound and its metabolites into tissues and any potential for accumulation. Typically, tissue distribution studies are conducted as single-dose studies with radiolabeled compounds (e.g., ^{14}C) via the intended route of administration [15]. Upon administration of the compound, the tissues of interest are sampled over a time course (up to 1–2 weeks) depending on PK properties. The samples are analyzed by a combination of techniques using tissue homogenization, combustion of tissue homogenate, and determination of radioactivity using liquid scintillation counting. This traditional method is labor intensive and permits limited understanding of distribution as not all tissues are sampled for practical consideration. Increasingly, the quantitative whole-body autoradiography (QWBA) technique is used for determination of tissue distribution for its ease and comprehensive nature [16]. Once the radiolabeled drug is administered to an animal, the animal is sacrificed at specified time points, frozen and embedded in a matrix, and the carcass sliced into thin sections for determination of radioactivity by an imaging technique. This allows to capture concentrations virtually in all tissues and the sections are available for any further investigations if questions arise later during the drug development. A limitation of QWBA studies is that only small animals can be accommodated and any studies with large animals may have to use a more traditional method described above. Another limitation is that the concentration data obtained from these studies are based on total radioactivity and not on the parent compound. Therefore, care must be exercised when interpreting concentration data and should be combined with other PK information before making any inferences.

The purposes of conducting tissue distribution are to understand the extent of distribution of drug-related material (parent and metabolites), to determine the potential for accumulation in any tissues, to understand distribution into tissues with special barriers (e.g., CNS) and binding to melanin, and to provide dosimetry analysis and guidance for conducting radiolabeled human absorption, metabolism, and excretion (AME) studies.

1.5.3 Lacteal and Placental Distribution Studies

In addition to the tissue distribution studies, drug distribution into milk and placenta is assessed during development to understand the potential exposure and risk for breastfeeding infant and fetus, respectively. These studies are carried out much later in the drug development as a part of the registration package and typically conducted in species used for reproductive and developmental toxicology studies. There is some guidance regarding placental and lacteal transfer studies. Both FDA and EMEA guidances on reproductive and developmental toxicity make statements that imply that the distribution of drug and/or its metabolites into milk and placenta (fetus) be evaluated [17–19]. For example, the EMEA draft guidance ("Guideline on Risk Assessment of Medicinal Products on Human Reproduction and Lactation," 2006) states that "information about the excretion into milk of the active substance and/or metabolites should be available." In addition, the guidance also states that "the exposure in pregnant animals measured by plasma concentrations of the compound and/or metabolites should be assessed." Similarly, ICH guidance (ICH S3A) refers to the need for assessment of exposure in newborns, dams, or fetuses and states that "secretion in milk may be assessed to define its role in the exposure to newborns" [20]. In addition to the studies stated above, specialized distribution studies such as distribution into semen may need to be considered. For some drugs, there is a possibility of drug eliciting potential effects in the female subjects when the drug is administered to male due to excretion into semen and subsequent exposure to females. This is not routinely assessed for many drugs and there is no regulatory guidance specifically requiring a study. However, many drugs have been shown to be excreted into semen to some extent and if there is concern for reproductive developmental toxicity, then it may be prudent to evaluate this proactively. The compound properties, techniques used for evaluation, and *in vitro* and *in vivo* models used for determination of lacteal and placental transfer are discussed in great detail in Chapter 17.

The milk and placental information is summarized under "Distribution" in Section 2.6.4 and the summary data is captured in the tabulated summaries under Section 2.6.5. The summary details the species, strain, the analytical techniques and methods, exposure or concentration data, and milk-to-plasma or placenta- or fetus-to-plasma concentration ratio. This data is useful for interpretation and risk assessment for infants or fetuses.

1.6 METABOLISM

Majority of the drugs are biotransformed to metabolites before being excreted in urine and feces. Early on in drug discovery, compounds are characterized and optimized for their metabolic properties leading to a compound with desired PK properties. When elimination of a drug occurs via metabolism, the rate and routes of metabolism can significantly affect a drug's safety and efficacy. Therefore, a number of *in vitro* and *in vivo* metabolism studies are conducted to fully understand the metabolic profile of a compound during drug development. A thorough understanding is required to address potential metabolism-based issues such as reactive metabolites of safety concern, potential drug–drug interactions due to CYP induction or inhibition, PK variability due to polymorphism, and potential active metabolites influencing PK-PD understanding and intellectual property rights. The information generated from these *in vitro* and *in vivo* metabolism studies constitutes an integral part of an NDA/marketing authorization application (MAA).

1.6.1 *In vitro* Metabolism Studies

In vitro drug metabolism studies are conducted to identify the metabolic rates and routes and to help understand the potential safety and efficacy issues related to drug or metabolites. The purposes of these studies are as follows: (1) to identify and characterize metabolites in safety species and humans in order to identify and select relevant species for safety assessment; (2) to identify any potential for drug–drug interactions; (3) to identify the enzymes responsible and to determine any genetic polymorphisms that may influence the PK, PD, and safety of the drug; and (4) to identify any potential reactive or genotoxic structural alerts that require further assessment in safety studies. Liver subcellular fractions such as microsomes or S9 and hepatocytes are predominantly used for *in vitro* metabolism studies. For most drugs that are metabolized by the CYP system, the liver microsomes are the test system of choice. However, the hepatocyte system provides a full complementary of metabolic enzymes. Regardless of the system used, it is important to use appropriate controls to ensure the viability of the system; employing known marker activity controls and sensitive analytical systems to detect and profile metabolites. The choice of metabolic system and the advantages and disadvantages of each system are discussed in greater detail in Chapter 9. Chapters 19 and 20 describe various mass spectrometry methods and strategies employed in drug metabolism studies.

Regulatory agencies have issued guidance documents regarding the conduct of *in vitro* metabolism and drug–drug interaction studies during development [21–24]. Both FDA and Health Canada documents outline the various model systems, probes, and considerations for experiments (choice of concentration and time course). It is apparent from these guidelines that the test system needs to be shown as viable and is well characterized to investigate new chemical entities or drug products. The choice of test system should be justified based on the metabolic pathways of the drug product from preliminary experimentation. Recombinant test systems may be used where applicable to further characterize if polymorphic enzymes are involved in the metabolism. The guidance document states that *in vitro* studies should be conducted at concentrations similar to those seen *in vivo*. This is particularly important if a metabolic pathway is saturable. Depending on the concentration used, the metabolic rate and relative abundance of metabolites may be different. These guidances emphasize that the *in vitro* studies should be confirmed with *in vivo* studies and cannot replace *in vivo* studies. They should be used as guidance and be confirmed with *in vivo* data.

The description of metabolite profiling, identification, and cross-species comparison, and major metabolic pathways should be captured in Module 2.6.4 under the "Metabolism" section. The data should be tabulated under the "Metabolism" section in Section 2.6.5. The summaries should describe the experimental conditions used (test system, concentration, time of incubation, detection system, etc.), relative rates of metabolism,

major metabolic pathways and identification, and cross-species comparison. Summaries should highlight if there is any unique metabolism in a particular species. The study reports pertaining to metabolism studies are submitted in Module 4, Section 4.2 under "Metabolism."

1.6.2 Drug–Drug Interaction Studies

1.6.2.1 CYP Inhibition Metabolism is the major route of clearance for most drugs, and CYP plays a major role in metabolism with approximately 75% of the drugs metabolized by this family of enzymes [25]. The major CYP enzymes involved in drug metabolism are CYP3A4, CYP2D6, CYP2C9, CYP2C19, CYP1A2, CYP2E1, CYP2B6, and CYP2A6. Inhibition of these enzymes can result in altered clearance and PK of drugs leading to adverse effects. This is particularly important considering that the aging population is often on multiple drugs (polypharmacy) for various ailments. Therefore, evaluation of drug interactions is part of drug effectiveness and safety. A detailed discussion of methods, conditions for incubation, probe substrates, and liquid chromatography-mass spectrometry (LC-MS) technologies are presented in Chapters 9 and 14.

1.6.2.2 CYP Induction Induction of drug metabolism enzymes refers to a process where the activity of enzymes increase upon administration of a compound through increased expression and synthesis or by stabilization of enzyme. Most oxidative and conjugative enzymes and drug transporters are inducible to a varying degree. However, the induction of CYP enzymes is of most concern during drug development due to their prominent role in the metabolism of drugs. There are three principle nuclear receptors, namely, the pregnane X receptor (PXR), the constitutive androstane receptor (CAR), and the aryl hydrocarbon receptor (AhR), that regulate the induction of CYP enzymes [26]. Drug binding to receptors leads to a series of molecular events resulting in increased mRNA expression and synthesis of enzyme. Increased levels of enzyme increases clearance and reduces exposure of drugs that are metabolized by the induced enzyme(s). Enzyme induction leads to reduced PD activity due to lowered exposures, although there is potential for manifestation of toxicity due to increased levels of metabolite(s) that are of safety concern.

The detailed discussion of the mechanisms of induction is not within the scope of this chapter. However, Chapter 15 provides a detailed account of *in vitro* and *in vivo* techniques and their merit, strategies, modeling and simulation, and risk assessment.

Regulatory agencies have issued guidance for the assessment of induction potential for drugs during drug development, and it is expected that this information is available at the time of NDA filing. Many companies routinely conduct this evaluation prior to going into Phase III due to a large number of patients involved in a clinical trial and potential for drug–drug interactions due to induction. The FDA guidance recommends using a positive control inducer in the experiments to account for variability between individual hepatocyte preparations. The positive control inducers should produce at least >2-fold induction at the recommended concentrations. The guidance provides recommended substrates for conducting induction studies. The guidance recommends using three concentrations spanning maximal concentration observed at therapeutic dose and an order of magnitude higher than average plasma concentrations. The hepatocyte preparations should be treated for at least 2–3 days for induction to occur. Following treatment, enzymatic activity of CYP3A, CYP2B6, and CYP1A2 should be evaluated using recommended probe substrates. Although enzymatic activity determination is most reliable, other means of induction evaluation such as immunoquantitation of enzymes, mRNA determination, and reporter gene assays are also acceptable. However, recent survey by Pharmaceutical Research and Manufacturers of America (PhRMA) recommends enzymatic as well as mRNA determination as the most reliable [27]. If a drug produces a change in catalytic activity or mRNA that is >40% of the positive control, then it is considered as an inducer and further evaluation in clinic is warranted.

1.6.2.3 Transporter Interaction Studies

Transporters can be major determinants for absorption, distribution, and disposition of drugs. Transporters are expressed at key physiological barriers to limit the distribution of drugs or facilitate excretion. Recent advances in biochemical and molecular biological techniques have led not only to identify many transporters but also to clone and characterize function and distribution in tissues. Transporters are broadly classified into solute carriers (SLC) or ATP binding cassette (ABC) transporters. Chapter 3 describes in detail the transporter family, their role in drug transport and disposition, and techniques used to characterize and understand the role of transporters in drug absorption and disposition. Modulation of the function of these transporters can result in drug–drug interactions and it is well documented in the literature (e.g., statins, digoxin, and cephalosporin antibiotics) [28, 29].

Regulatory agencies worldwide recognized the issue of transporter-based drug interactions and addressed this in their respective documents. For example, the FDA draft guidance issued in 2006 lists some of the major human transporters and known substrates, inhibitors, and inducers. The guidance states that Pgp is the best studied of all the transporters and it is appropriate to evaluate during drug development. Pgp is expressed at the intestinal barrier limiting the absorption of drugs.

It is also expressed in the kidney and liver playing a critical role in excretion of drugs and their metabolites. Pgp also plays a critical role in limiting distribution of drugs into the CNS and fetus due to its expression in these barriers [29]. Any interference in the function of Pgp can lead to altered levels of drugs in circulation and tissues and compromise the safety and efficacy of drugs. Therefore, studies should be conducted to determine if a drug candidate is a substrate and inhibitor of Pgp. This can be accomplished with Caco-2 cells or other engineered cell lines that overexpress Pgp. Irrespective of test system used, the experimentation should include known substrates and inhibitors to demonstrate the suitability of test systems. Bidirectional transport measurements are preferred and net flux should be calculated for interpretation of results. If the compound has a net flux ratio of >2, then it is considered a substrate and further evaluation should be carried out. If the test compound has an efflux ratio <2, then it is not a Pgp substrate; and follow-up interaction studies are not needed. The range of concentrations (e.g., 1, 10, and 100 μM) should be considered during evaluation and the selection depends on maximal concentrations observed in the plasma and at the intestinal barrier and on the solubility limitation of the drug. A test compound should also be tested for its inhibition, irrespective of whether or not it is a substrate. Again, a wide range of concentrations should be considered to generate inhibitory data (IC_{50} or Ki). The draft guidance provides a list of acceptable probe substrate and positive control inhibitors. If the test compound has an $I/IC_{50} > 0.1$, then it is an inhibitor of Pgp and an in vivo interaction study should be conducted with digoxin as a probe. If the ratio is <0.1, then the test compound is a weak inhibitor and an in vivo interaction study is not needed. The potential for Pgp induction is also discussed in the guidance document and it is recommended that if the drug is shown not to induce CYP3A in vivo, then no further test of Pgp induction in vivo is necessary. However, if the in vivo CYP3A induction test is positive, then an additional study of the investigation drug's effect on a Pgp probe substrate is recommended. Similar guidance is also provided by Japan's Ministry of Health, Labor, and Welfare (MHLW) under the name of "Methods for Drug Interaction Studies." This document discusses in some parts the guidance for assessing transporter-mediated interactions at the intestinal barrier, tissue distribution, and elimination into urine and bile.

1.6.2.4 Identification of Enzymes Responsible for Metabolism

Most drugs are metabolized and cleared from the body and it has been reported that 75% of the marketed drugs are primarily metabolized by the CYP system. Although many CYP enzymes exhibit overlapping substrate specificity, in most cases a single isoform seem to contribute to the majority of the metabolism. Therefore, the goal early on in the development is to identify if the drug is metabolized by a single enzyme or multiple enzymes or different family of enzymes. It is ideal to have equal contribution from multiple enzymes to make less susceptible for drug interactions when coadministered with a potent inhibitor or inducer of that particular enzyme. Another goal of phenotyping is to identify if a polymorphic enzyme is involved in metabolism. Recent FDA guidance addresses the reaction phenotyping in detail and states that if human *in vivo* data indicate that CYP enzymes contribute >25% of a drug's clearance, studies to identify drug metabolizing CYP enzymes *in vitro* should be conducted [22]. This recommendation includes cases in which oxidative metabolism is followed by transferase reactions because a drug–drug interaction that inhibits oxidation of the parent compound can result in elevated levels of the parent compound. The guidance suggests that clinically relevant concentrations of the drug should be considered for *in vitro* experiments. Preliminary experiments should be conducted to assess linearity of metabolite formation and protein concentrations, and time course should be optimized. Reliable and robust analytical methods should be used to monitor the formation of metabolites, and analytical methods should have acceptable sensitivity to measure the percent of inhibition over the range of concentrations tested. The guidance outlines three principal ways, namely, use of CYP isoform chemical inhibitors or selective antibodies, recombinant enzyme systems, and correlation analysis with a bank of microsomes whose activities were characterized with known isoform selective probe substrates. Chapter 13 discusses in detail various aspects of CYP and non-CYP enzymes involved in drug metabolism.

Identification of enzymes responsible for metabolism should be summarized under the "Metabolism" section in Section 2.6.4. The summary should describe the test systems used, conditions of incubations, analytical methods, and the results. In summary, the data should also speak about the consequences of the findings, for example, state that if exclusively metabolized by particular enzymes, potential for interaction exists if coadministered with strong inhibitors of that particular enzymes. It is also important to convey a consistent message across documents such as written summaries, overviews, and investigator brochures.

1.6.3 *In vivo* Metabolism (ADME) Studies

Biodisposition studies or ADME of drugs play a critical role in its initial selection as well as in its subsequent clinical development. *In vivo* ADME studies provide important information regarding the absorption of the drug, distribution of the drug-related material into the target tissues, important metabolic pathways, and the eventual excretion of the drug-related material from the body [30, 31]. The information generated from these studies is helpful in understanding the outcomes of the safety studies and efficacy of pharmacology studies [32]. Therefore, *in vivo* metabolism studies with radiolabeled tracers have become a crucial component of the drug development package for regulatory submissions.

Typically, *in vivo* drug metabolism and disposition studies are conducted with a radiolabel (14C or 3H) material to provide quantitative information on the rate and extent of metabolism, routes of excretion for parent compound, and its metabolites and circulating metabolites. The nonclinical ADME studies are conducted in two species, rodents (rats or mice) and nonrodents (dogs or monkeys), and the choice is based on the species used in toxicology studies. The selection of dose, route of administration, and formulations should mimic the safety studies. These studies are typically single-dose studies with sample collection up to a week; however, longer duration of sample collection may be needed if the drug has a long half-life. Plasma, urine, and fecal samples are collected during the duration of the study to analyze for radioactivity and for metabolite profiling. Tissue samples may be collected if there is a particular issue to address safety or efficacy concerns. Metabolic profiles of plasma, urine, and fecal samples are generated by a combination of techniques (e.g., high-performance liquid chromatography [HPLC] radioactivity detection, liquid chromatography [LC] fractionation followed by scintillation counting, and LC-MS coupled to radioactivity detection). The reader is referred to Chapter 22 for a detailed description of the radioactivity profiling techniques and recent advances. Also of interest are Chapter 19 and Chapter 20, which describe mass spectrometry methods for metabolite identification.

Metabolic profiles of urine and fecal samples generated in ADME studies provide information about the extent of metabolism and the routes of excretion for the parent and metabolites. This will enable one to assess the important metabolic pathways, the importance of organs in the elimination, the total amount of drug substance absorbed, and if any metabolites of safety or activity concern are generated. Metabolic profiles of plasma samples collected over the time course enable one to calculate exposure for metabolites, understanding major circulatory metabolites, and the half-life of metabolites that could potentially result in accumulation upon repeat administration. In addition, it provides information about metabolites that would enable comparison between toxicology species and humans.

1.6.3.1 Metabolites in Safety Testing (MIST) One of the most debated and controversial guidances in the area of metabolism was issued by the FDA in 2005 related to MIST [33]. This guidance made recommendations on when to identify and characterize metabolites. This guidance followed an earlier publication by the pharmaceutical industry on the role of metabolites as potential mediators of adverse effects, which itself was discussed extensively in the literature [34]. Following an extensive discussion at scientific meetings, workshops, and in the literature, the FDA has issued a final guidance in 2008 [35]. While there were some differences between the draft and the final guidance, the final guidance stated that the disproportionate metabolites, present at >10% of the parent exposure, need to be considered for safety assessment. The guidance recommends conducting drug metabolite profiling and identification early in development such that any metabolites of safety concern (disproportionate or unique human metabolites) are addressed before exposing a large number of patients in pivotal clinical studies. The guidance states that coverage of disproportionate metabolites observed in humans must be demonstrated in at least one of the species used for toxicology studies. It is noteworthy that the guidance does not state sensitive species but rather one of the species. The guidance had great impact on the generation of quantitative and qualitative profiles early on in development so that comparisons of metabolic profiles can be made in human and safety species. This has led to conducting and completing radiolabeled studies so that all necessary information is available before initiation of Phase III studies. There are a number of publications on the timing and the different approaches for this data early on in development with and without the use of radiolabeled compound. For any disproportionate metabolites observed in humans but not present in any of the safety species, further testing needs to be done with the metabolite. The FDA recommends conducting further studies. These include (1) general toxicology studies with the metabolite at the exposure equivalent to those obtained in humans; (2) *in vitro* genotoxicity testing in an acceptable test; (3) embryo-fetal developmental studies if the drug is intended for women with child-bearing potential; and (4) carcinogenicity studies with the metabolite. The metabolic characterization and issues are complex and the guidance speaks the need to consider issues case by case. The guidance also states that certain conjugative metabolites (O-glucuronides and sulfate conjugates) other than acyl glucuronides are pharmacologically inactive and not of safety concern even when they occur at greater than the 10% cutoff. Although the FDA guidance considers >10% of parent exposure as disproportionate metabolite, the ICH guidance differs by stating that major metabolites are >10%

drug-related exposure [36]. This may have great impact for drugs that extensively metabolized and may reduce the burden to monitor a number of metabolites if based on total drug-related material rather than parent. Precisely for these reasons, the FDA guidance provides examples of case studies and highlights the complexity of issues and the need for considering metabolism in safety testing on a case-by-case basis.

1.7 EXCRETION

As stated in previous sections, ADME studies provide important information regarding metabolism and excretion of compounds into urine and feces. The purpose of these studies is to understand not only the metabolism but also how the parent and metabolites are eliminated. The information generated from these studies is useful in evaluating if the kidney or liver is an important organ of elimination and if there is any safety concern in hepatic or renal-impaired populations. In addition, the excretion data also sheds light on the role of transporters. For example, if the renal excretion of parent or metabolite is greater than the glomerular filtration, then it is likely that an active secretion may be occurring and further studies need to be conducted to evaluate potential issues. In nonclinical studies early in the development, excretion studies are conducted in intact and bile duct-cannulated rodent and nonrodent species to understand the excretion of drug and metabolites. This enables to understand the metabolic pathways and rate of excretion. The reader is referred to Chapter 18 for a detailed discussion of bile collection in animals and humans. Also, Chapter 16 discusses various animal models used to study drug metabolism and transporters.

The information generated regarding excretion should be captured in Module 2.6.4 and tabulated summaries should contain summary data in Module 2.6.5 under the "Excretion" section. The written summary should discuss if the parent is excreted intact or metabolized, the relative contribution of liver and kidney for excretion, and the species comparison of excretory profiles of parent and metabolites.

1.8 IMPACT OF METABOLISM INFORMATION ON LABELING

As described in the previous sections, drug metabolism and disposition data is generated to fully understand the metabolic pathways, clearance, drug–drug interaction potential, and changes in exposure in diseased or special populations. The results from the studies described above should be integrated and included in the NDA. This integration of data not only helps reviewers evaluate drug for approval but also summarizes the key aspects of drug in the prescription label for prescribes and patients. The FDA also recommended that when the information has important implications for the safe and effective use of the drug and the drug metabolism information results in recommendations for dosage adjustments, contraindications, or warnings, this information should be included in the appropriate sections, such as boxed warning, dosage and administration, contraindications, and drug interactions. The drug metabolism information is captured in the "Clinical Pharmacology" section of the prescription label under the subsections of Pharmacokinetics, Absorption, Distribution, Metabolism, and Excretion. The information is brief and conveys important information to the prescriber to either adjust the dose or avoid prescribing the medicine when the patient is taking concomitant interacting medications.

1.9   CONCLUSIONS

Drug metabolism plays a central role from drug discovery through drug development and approval. A number of studies are conducted early in lead optimization and candidate selection. With advances in analytical technologies and *in vitro* and *in vivo* metabolism tools, rapid screening methods can be employed to speed up discovery. However, once a compound is nominated, detailed metabolism studies need to be conducted in a timely fashion and staged according to the development of a product candidate. Most importantly, drug metabolism scientists should be aware of the regulatory guidances and expectations for the conduct of these studies so that best practices can be followed in generation of data.

REFERENCES

1. Alavijeh MS, Palmer AM (2004) The pivotal role of drug metabolism and pharmacokinetics in the discovery and development of new medicines. *Drugs* 7:755–763.
2. Davis AM, Riley RJ (2004) Predictive ADMET studies, the challenges and the opportunities. *Curr Opin Chem Biol* 8:378–386.
3. Kola J, Landis J (2004) Opinion: Can the pharmaceutical industry reduce attrition rates? *Nat Rev Drug Discov* 3:711–716.
4. U.S. Department of Health and Human Services, Food and Drug Administration (FDA), Center for Drug Evaluation and Research (CDER) and Center for Veterinary Medicine (CVM) (2001) Guidance for industry: Bioanalytical method validation.
5. International Conference on Harmonization (ICH) of Technical Requirements for the Registration of Pharmaceuticals for Human Use (1995 and 1996), ICH Q2A—Text on Validation of Analytical Procedures, and ICH Q2B—Validation of Analytical Procedures—Methodology.
6. Viswanathan CT, Bansal S, DeSilva B, et al. (2007) Quantitative bioanalytical method validation and implementation: Best practices for chromatographic and ligand binding assays. *AAPS J* 9(1):E30–E42.
7. Bansal S, DeStefano A (2007) Key elements of bioanalytical method validation for small molecules. *AAPS J* 9(1):E109–E114.
8. Rocci ML, Devanarayanan V, Haughey DB, et al. (2007) Confirmatory reanalysis of incurred bioanalytical samples. *AAPS J* 9(1):E336–E343.
9. Schmidt S, Gonzalez D, Derendorf H (2009) Significance of protein binding in pharmacokinetics and pharmacodynamics. *J Pharm Sci* 99:1107–1122.
10. Pacifici G, Viani A (1992) Methods of determining plasma and tissue binding of drugs. *Clin Pharmacol* 23:449–468.
11. Rowland M, Tozer TN. *Clinical Pharmacokinetics: Concepts and Applications*, 3rd ed., Philadelphia, PA: Williams &Wilkins, 1995.
12. MacKichan JJ. Influence of protein binding and use of unbound (free) drug concentration. In *Applied Pharmacokinetics: Principles of Therapeutic Drug Monitoring*, 3rd ed., Evans WE, Schentag JJ, Jusko WJ, eds. Vancouver, WA: Applied Therapeutics, 1992:5-1–5-48.
13. Rolan PE (1994) Plasma protein binding displacement interactions: Why are they still regarded as clinically important? *Br J Clin Pharmacol* 37:125–128.
14. Fagerholm U (2007) Prediction of human pharmacokinetics: Evaluation of methods for prediction of volume of distribution. *J Pharm Pharmacol* 59:1181–1190.
15. Marathe PH, Shyu WC, Humphreys WG (2004) The use of radiolabeled compounds for ADME studies in discovery and exploratory development. *Curr Pharm Des* 10:2991–3008.
16. Solon EG, Kraus L (2001) Quantitative whole-body autoradiography in the pharmaceutical industry: Survey results on study design, methods, and regulatory compliance. *J Pharmacol Toxicol Methods* 46:73–81.
17. FDA Guidance for Industry (2005) Clinical lactation studies: Study design, data analysis, and recommendations for labeling.
18. FDA Reviewer Guidance (2001) Integration of study results to assess concerns about human reproductive and developmental toxicities.
19. EMEA Guidance (2006) Guideline on risk assessment of medicinal products on human reproduction and lactation.
20. ICH Guidance for Industry (1995) Toxicokinetics: The assessment of systemic exposure in toxicity studies.
21. FDA Guidance for Industry (1997) Drug metabolism/drug interaction studies in the drug development process: Studies *in vitro*.

22. FDA Draft Guidance (2006) Drug interaction studies: Study design, data analysis, and implications for dosing and labeling.

23. Health Canada (2000) Drug-drug interactions: Studies *in vitro* and *in vivo*.

24. EMEA Guidance (2010) Guideline on the investigation of drug interactions (draft).

25. Williams J, Bauman JN, Cai H, Conlon K, Hansel S, Hurst S, Sadagopan N, Tugnait M, Zhang L, Sahi J (2005) *In vitro* ADME phenotyping in drug discovery: Current challenges and future solutions. *Curr Opin Drug Discov Devel* 8:78–88.

26. Lin JH (2006) CYP induction-mediated drug interactions: *In vitro* assessment and clinical implications. *Pharm Res* 23(6):1089–1116.

27. Chu V, Einolf HJ, Evers R, Kumar G, Moore D, Ripp S, Silva J, Sinha V, Sinz M, Skerjanec A (2009) *In vitro* and *in vivo* induction of cytochrome p450: A survey of the current practices and recommendations: A pharmaceutical research and manufacturers of America perspective. *Drug Metab Dispos* 37(7):1339–1354.

28. Xia CQ, Milton MN, Gan LS (2007) Evaluation of drug-transporter interactions using *in vitro* and *in vivo* models. *Curr Drug Metab* 8(4):341–363.

29. Giacomini KM, Huang SM, Tweedie DJ, Benet LZ, Brouwer KLR, Chu X, Dahlin A, Evers R, Fischer V, Hillgren KM, Hoffmaster KA, Ishikawa T, Keppler D, Kim RB, Lee CA, Niemi M, Polli JW, Sugiyama Y, Swaan PW, Ware JA, Wright SH, Wah Yee S, Zamek-Gliszczynski MJ, Zhang L (2010) Membrane transporters in drug development. *Nat Rev Drug Discov* 9(3):215–236.

30. Caldwell J, Gardner I, Swales N (1995) An introduction to drug disposition: The basic principle of absorption, distribution, metabolism and excretion. *Toxicol Pathol* 23:102–114.

31. Campbell DB (1994) Are we doing too many animal biodisposition investigations before Phase I studies in man? A re-evaluation of the timing and extent of ADME studies. *Eur J Drug Metab Pharmacokinet* 19:283–293.

32. Bischoff H, Angerbauer R, Boberg M, Petzinna D, Schmidt D, Steinke W, Thomas G (1998) Preclinical review of cerivastatin sodium—A step forward in HMG-CoA reductase inhibition. *Atherosclerosis* 139(Suppl 1):S7–S13.

33. Center for Drug Evaluation and Research, Food and Drug Administration (2005) FDA guidance (Draft) for industry: Safety testing of drug metabolites.

34. Baillie TA, Cayen MN, Fouda H, et al. (2002) Drug metabolites in safety testing. *Toxicol Appl Pharmacol* 182:188–196. (MIST—Metabolites in Safety Testing).

35. U.S. Food and Drug Administration (FDA) (2008) Guidance for industry: Safety testing of drug metabolites (Washington, DC).

36. ICH Guidance (2009) Non-clinical safety studies for the conduct of human clinical trials and marketing authorization for pharmaceuticals.

2

OPTIMAL ADME PROPERTIES FOR CLINICAL CANDIDATE AND INVESTIGATIONAL NEW DRUG (IND) PACKAGE

Rajinder Bhardwaj and Gamini Chandrasena

2.1 INTRODUCTION

The clinical success of a potential drug candidate depends on how closely it achieves an overall balance of "optimal" drug-like properties between efficacy, safety, and pharmacokinetics (PK) (Sugiyama, 2005; Segall et al., 2006). To achieve "balanced" properties successfully and efficiently, a significant consideration has to be given to the ideal profile of the therapeutic target and indication from the exploratory stages of drug discovery. Although potency is an important requirement of a potential drug candidate, the PK profile invariably dictates its effectiveness as a successful drug in the therapeutic use. A series of preclinical studies on the new chemical entity (NCE) must be performed in discovery and development to characterize the compound for a suitable combination of potency, selectivity, PK, and safety before it can be administered to humans (Kennedy, 1997). Drug discovery process involves various pharmacological mechanistic, functional, or efficacy studies as dictated by a specific disease or biology area to offer the greatest promise for therapeutic effectiveness (DiMasi, 1994; Kennedy, 1997). While the "optimal" binding properties of an NCE to the therapeutic target of choice is very crucial, nevertheless it is important that the NCE reaches the target site in desired concentrations to engage and modulate the target in a way to therapeutically change or alleviate the disease state. Thus, the interplay between PK and phar-

macodynamics (PD) is critical for effective drug therapy (Gabrielsson et al., 2009). The PK profile of a compound is defined by determining its absorption, distribution, metabolism, and excretion (ADME) properties. It is now recognized that unacceptable physicochemical, ADME, and toxicity properties could result in preventing a drug candidate from advancing into NCE status. Therefore, it has been widely accepted that early ADME profiling of drug candidates is one of the crucial factors in determining the potential success of an NCE in drug development (DiMasi, 1994, 2001; Kennedy, 1997). A typical path for a compound's journey from the chemist's shelf to become a therapeutic drug is summarized in Table 2.1.

Unlike traditional preclinical operations providing data in support of regulatory filings in recent years, areas of PK, drug metabolism, drug absorption, and related disciplines such as biopharmaceutics have developed and utilized state-of-the-art tools in preclinical testing that add value, mitigating risk in reducing the attrition of a potential NCE in clinical development (DiMasi, 2001; Reichel, 2006; Gabrielsson et al., 2009). The advancement of automation technology, miniaturization, and the advent of new ADME screens as well as mechanism-based assays with enhanced efficiency and the utilization of predictive computational tools (Kirchmair et al., 2008; Mayr and Bojanic, 2009) and data mining capabilities enable scientists to move drug discovery leads effectively and swiftly to development

ADME-Enabling Technologies in Drug Design and Development, First Edition. Edited by Donglu Zhang and Sekhar Surapaneni.
© 2012 John Wiley & Sons, Inc. Published 2012 by John Wiley & Sons, Inc.

TABLE 2.1. Various Stages and Testing Schemes in a Typical Pharmaceutical Preclinical and Clinical Development Program (Modified from Kramer et al., 2007)

Drug Discovery and Development	Stages	Testing Scheme
Preclinical IND filing	Hit identification	Target identification and validation, *in vitro* efficacy assays, physical-chemistry assays
	Lead optimization	*In vivo* efficacy assays (preclinical proof of concept), *in vitro* ADME assays, and preliminary PK and toxicology studies
	Preclinical development	PK/PD modeling, salt form selection, crystal form assessment; good laboratory practice (GLP) toxicology studies: genetic toxicology, safety pharmacology, *in vivo* toxicology in two species
Clinical	Phase 1	First–in-human dose, safety and tolerability in normal healthy volunteers
	Phase 2	Safety and tolerability in patients, early clinical proof of principle
	Phase 3	Definitive clinical proof of principle
	Phase 4/marketing license	Postmarketing safety review

candidates (DiMasi, 2001; Reichel, 2006). Given the challenging timelines with greater competitive landscape, the industry has adopted a more multidimensional lead optimization approach to address critical issues in drug discovery and development, taking advantage of the new technology rather than the traditional sequential approaches. It has been observed in the past 10 years that implementation of a balanced drug metabolism and pharmacokinetic (DMPK) strategy has reduced the compound attrition rate in the clinical development phase due to ADME and PK issues (Kennedy, 1997; Frank and Hargreaves, 2003). This chapter attempts to provide an overview of general ADME and PK characteristics and currently available approaches one utilizes in advancing a lead compound to the clinical development. The ensuing chapters will provide more comprehensive details on each of the specific ADME processes and the available technologies that enable such processes.

2.2   NCE AND INVESTIGATIONAL NEW DRUG (IND) PACKAGE

Filing an IND application is the initial step of early preclinical development program with the primary objectives of determining whether or not (1) the NCE is reasonably safe to administer in humans and (2) it exhibits pharmacological activity that warrants further commercial development in order to secure an approval for investigating in humans. The title 21 of the federal register (21 CFR 312) contains the regulations regarding the IND applications. The U.S. Food and Drug Administration's (FDA) Center for Drug Evaluation and Research IND Applications website page provides the relevant information and links about the application process and applicable regulatory requirements to assemble an IND package (http://www.fda.gov/Drugs/DevelopmentApprovalProcess/HowDrugsare DevelopedandApproved/ApprovalApplications/InvestigationalNewDrugINDApplication/default.htm).

In general, the IND package submitted to the FDA should include key elements such as general safety pharmacology and toxicology in preclinical animal species, the manufacturing information, and clinical protocols and investigator information. The FDA's primary objective in reviewing an IND is to ensure that the relevant data provided on the NCE is adequate to ensure its effectiveness and safety to further development in phase studies. From an ADME perspective, a typical IND-enabling package should contain the information ranges from a number of *in vitro* and *in vivo* ADME assays (Figure 2.1), which could suggest that the compound has minimal PK and drug–drug interaction (DDI) liability in moving forward in phase studies. The *in vitro* information generally includes data obtained from metabolism, permeability, and protein/blood partitioning of an NCE, whereas, under *in vivo*, single- and multiple-dose PK in rodent and nonrodent species as well as the role of different routes of administration on PK and bioanalytical methods information is obtained (DiMasi, 1994; Zhang et al., 2009). Relevant mechanistic studies such as achieving mass balance by evaluating routes of elimination, metabolite profiling, and tissue distribution in both

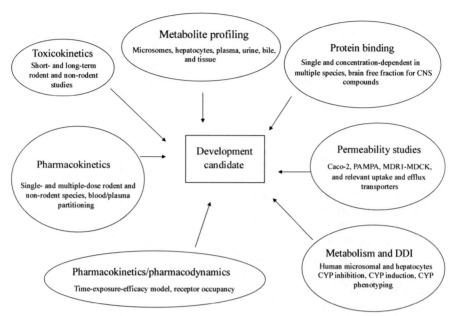

FIGURE 2.1. A schematic presentation of relevant ADME assays that could be utilized for screening an NCE.

rodent and/or nonrodent species should be integrated, for example, incorporating quantitative whole-body autoradiography (Luffer-Atlas, 2008; Zhang et al., 2009). In addition, performance of toxicokinetic analysis in relevant animal species of critical importance is needed to be included in the IND package since it can define a therapeutic window with respect to a specific toxicity outcome that may potentially be encountered in humans (Cayen, 1995). Moreover, additional metabolite structural identification and its relevant distribution profile will help in providing necessary supportive information to augment the interpretation of any toxicology findings (Cayen, 1995). Once the IND is submitted, the sponsor will have 30-day stay prior to initiating a trial, allowing the regulatory authority time to review the application for safety to assure minimal risk on the study subjects. The FDA would expect to respond requesting additional information if there are issues with the submission, or place clinical phase studies on hold until the FDA-mandated requirements are met.

2.3 ADME OPTIMIZATION

During the design phase of an NCE, considerable efforts should be given to optimize ADME properties. It has been demonstrated that the ADME properties of an NCE can be predicted to some extent based on physicochemical parameters (Lipinski et al., 1997). Thus, ideal physiochemical properties would be the first approach that one should look in a lead candidate for develop-

ment. Optimal molecular weight (MW) (<450), lipophilicity (log P < 5), and total hydrogen bond acceptors and donors (<10) are some of the general properties identified by Lipinski (Lipinski et al., 1997) that are associated with good oral bioavailability. Over the past decade, many additional physiochemical properties such as polar surface area (PSA), free rotatable bonds, molar refractivity, and number of heavy atoms (Bajpai and Adkison, 2000; Veber et al., 2002) have gained attention so that in early drug discovery, one should look for "lead-like" properties on compounds in addition to "drug-like" properties. In a recent study, physicochemical properties of 309 commercially available drugs were compared against intestinal absorption (Fa), intestinal extraction (Fg), and liver extraction (Fh) after oral administration (Varma et al., 2010). The study suggested that higher MW (>500) significantly impacted Fa, while Fg and Fh decreased with increasing lipophilicity. In addition, free rotatable bonds in a molecule has a negative effect on first-pass metabolism suggesting that compact molecules are constrained to fewer conformations and therefore would be more restrictive in binding to drug metabolizing enzymes (Varma et al., 2010). In a nutshell, there is a plethora of information available in the literature on the effect of physicochemical parameters on ADME properties; however, it is important to note that optimal ADME properties require a careful "balancing act" among several physicochemical parameters, which increases the odds of becoming a potential drug.

Generation of *in silico* data provides a rationale for establishing qualitative, semiquantitative, or quantitative structure–activity relationship (QSAR) models (Bidault, 2006; Subramanian et al., 2006; Singh et al., 2007a,b). This machine learning enables one to increase predictability of a set of molecular interactions in guiding synthesis beginning from the hit-to-lead process. Moreover, *in silico* predictive models offer an advantage in evaluating the ADME and physical-chemical properties of molecules even prior to chemical synthesis, enabling virtual screens with significant savings in costs as well as in time. The commercially available predictive software packages such as ClogP 4.0 (BioByte Corp., Claremont, CA), ADMET Predictor and ClassPharmer (Simulations Plus, Lancaster, CA), ACD/Structure Design Suite (Advanced Chemistry Development Inc., Toronto, ON), META (Multicase Inc., Cleveland, OH), and MetaDrug™ (Genego, Saint Joseph, MI) work on a set of programmed rules annotated from both the literature and proprietary sources to predict the ADME properties of compounds and/or their metabolites (Greene et al., 1999; Testa et al., 2005). Software packages such as MetaDrug/ToxHunter™ (Genego) are designed to take advantage of known pathways to link parent and/or predicted metabolite structures with potential targets and/or toxicity. In a related example, early screening hits with *in silico* methodology has been shown to reasonably predict oral absorption and permeability utilizing molecular descriptors such as PSA, molecular size, flexibility, and hydrogen bonding capacity from QSAR-type models (Tetko and Oprea, 2008). However, despite the cost and time involved in synthesizing, purifying, and characterizing compounds, an ADME optimization process with only virtual models are no substitution for the utilization of relevant *in vitro* and *in vivo* assays in making key decisions for selecting development candidates.

The ideal ADME properties that one should look for in a development candidate are (1) good oral bioavailability, (2) low to moderate blood clearance, (3) ideal volume of distribution (beyond plasma volume), (4) low potential for DDI, (5) low metabolism-dependent toxicity relevant to humans, and (6) a desirable projected human half-life that enables dosing regimen consistent with patient compliance. However, not many or all potential NCEs are able to adhere to an ideal package with all the desired properties described above, suggesting that oftentimes a decision to move forward a compound is based on the balance of the risk-benefit ratio in terms of efficacy, potency, and PK characteristics. The PK properties can be optimized by utilizing *in vitro* and *in vivo* ADME screening as a selection process from chemical libraries or one-off synthesis of few analogs in a rational drug design paradigm. The *in vitro* and *in vivo* assays can be tiered as per project decision requirements to align with relevant ADME issues and tailor

screening cascades according to available resources and assay capacity needs. The assay decision tree should be logically used in making rational decisions to narrow the number of leads from hit-to-lead to lead optimization rather than using a check box approach. A typical project decision tree the authors have utilized in a project at lead optimization stage for central nervous system (CNS) drug discovery is shown in Figure 2.2.

2.3.1 Absorption

An overwhelming majority of new drugs are being developed with the intention of being administered orally and the expectation that NCEs will have good oral bioavailability. Oral bioavailability can be determined in animal models, though it is not the most optimal approach to screen large set of compounds in the discovery mode. A number of *in vitro* models utilizing cell monolayer or tissue segments have been developed to facilitate the assessment of absorption across intestinal membrane. The most commonly used approach in the pharmaceutical industry is the cell culture-based studies utilizing cell lines such as Caco-2, MDCKII, or LLC-PK (Artursson, 1991; Artursson and Karlsson, 1991; Rubas et al., 1996). In the absence of any potential dissolution issues, the gut luminal solubility of a compound plays a critical role in determining the absorption across the intestinal epithelium (Avdeef, 2003; Yalkowsky and He, 2003). The solubility of an NCE can be determined by either leveraging a thermodynamic approach utilizing the shake-flask method (Avdeef, 2003; Yalkowsky and He, 2003) or by measuring kinetic solubility utilizing the nephelometric-type assays. The more resource- and compound-intensive thermodynamic solubility assays are being utilized for establishing stable PK and PK and/or PD study formulations (Hageman, 2006; Zhou et al., 2007; Bard et al., 2008; Yamashita et al., 2008), whereas kinetic solubility assays are being leveraged for higher-throughput ADME screens (Yamashita et al., 2008).

The Caco-2, a human colon carcinoma cell line is the most commonly used cell-based model in determining the permeability of drugs in the industry (Artursson, 1991; Artursson and Karlsson, 1991; Rubas et al., 1996). The Caco-2 cells form a confluent monolayer on semipermeable filter membrane. Under suitable culture conditions, Caco-2 cells undergo enterocytic-like differentiation and become polarized with well-established tight junctions, resembling human intestinal epithelium (Artursson and Karlsson, 1991). The rate of permeation of compounds across the Caco-2 monolayer can be used to determine a permeability coefficient (Pc), which can be related to the *in vivo* absorption of a compound (Artursson and Karlsson, 1991). These cell lines have

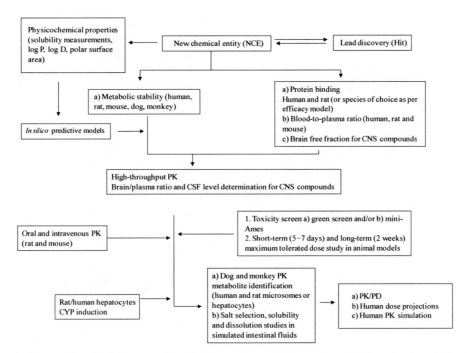

FIGURE 2.2. A typical screening decision tree for triaging compounds from hit-to-lead to lead optimization for identifying potential drug candidates.

also shown to express many efflux and influx transporters and drug metabolism enzymes (Sun et al., 2002), but the functional contribution of many of them remain to be fully elucidated. The Caco-2 cells can be useful in performing mechanistic studies such as a surrogate measurement of prodrug permeability across the intestinal epithelium or engineering cells to overexpress transporters or knockdown of a specific transporter. The FDA guideline has also listed Caco-2 cell line model system as an acceptable method to evaluate the permeability potential of a drug substance. The bidirectional permeability data generated in a Caco-2 monolayer can be binned into high, medium, and low permeability categories consistent with the Biopharmaceutics Classification System (BCS) classification (Artursson et al., 2001), if recoveries and efflux ratios are known. In general, the compounds with high permeability and minimal or no P-glycoprotein (Pgp) activity are recommended to carry forward; however, to advance a compound with low to medium permeability with or without Pgp efflux liability, a better understanding of the potency, efficacy, and metabolism properties of a molecule is needed.

Parallel artificial membrane permeability assay (PAMPA) data could also be utilized in making a rational choice for cherry-picking discovery compounds with good membrane permeability. The PAMPA is less resource-and time-intensive and could be run as a high-throughput assay in contrast to cell culture-based assays (Avdeef, 2005). The rate of permeation across phospho-

lipids has shown to correlate well with the extent of drug absorption in humans with the caveat that it may not predict accurately for the compounds that interact with drug transporters (Avdeef, 2005). PAMPA and unidirectional Caco-2 cell models can also be utilized in combination to evaluate the permeability properties (Balimane et al., 2006). A high PAMPA and Caco-2 permeability (apical to basal) would indicate high intestinal permeability in humans, whereas high PAMPA permeability coupled with low Caco-2 cell permeability (apical to basal) would suggest an involvement of an efflux transporter. A compound showing both low permeability in PAMPA and Caco-2 cell permeability (apical to basal) should be deprioritized; hence, low intrinsic permeability could result in poor absorption in humans (Balimane et al., 2006). In addition, intestinal pH varies from the acidic to basic range and could affect the permeability characteristics of drugs due to different degrees of ionization. Both PAMPA and Caco-2 assays can be performed at different pH to capture the effect on solubility and ionization to predict intestinal absorption in different regions of the intestine (Avdeef, 2005; Balimane et al., 2006).

Beside cell line models, everted intestinal rings and brush-border membrane vesicles (BBMVs) are also commonly used systems for assessing membrane permeability (Alcorn et al., 1993; Barthe et al., 1999). Both everted intestinal rings and BBMVs are most useful in determining the rate of drug uptake rather than measuring the transepithelial flux. However, both of these assays are not really amenable to high-throughput iterative assay format as the preparation of vesicles or everted guts are somewhat labor intensive. The rat *in situ* intestinal perfusion is another method, where an intestinal segment of an anesthetized rat is exposed to a drug solution via a single-pass perfusion or recirculation through the lumen (Fagerholm et al., 1996). The rate of permeability is estimated from the rate of disappearance of the drug from collected perfusate (or measured in mesentry blood), and the resulting permeability rates have been demonstrated to correlate fairly well with the permeability determined in the human jejunum (Fagerholm et al., 1996). In addition, Lennernäs (1998) has suggested that *in situ* perfusion of rat jejenum is a useful technique to classify compounds according to the BCS. In the discovery phase, the utilization of either everted intestinal rings or *in situ* intestinal perfusion models can help understand the regional differences as well as the mechanism of drug absorption across the intestine, which may be a valuable tool for de-risking the potential formulation strategy of potent leads with less than optimal properties (Barthe et al., 1998, 1999).

Drug transporters, most importantly efflux transporters, have been identified as playing a prominent role in the oral bioavailability of drugs. It is generally accepted that coadministration of drugs that interact with transporters, similar to what would be in cytochrome P450 (CYP) enzymes (as a substrate, inhibitor, or inducer), can potentially lead to DDIs (Giacomini et al., 2010). In addition, the modulation of transporter proteins can also affect the biliary and/or renal clearance, as well as drug permeability across the blood–brain barrier (BBB). Among various efflux transporters, Pgp plays an important role in drug discovery. The Pgp interaction, whether a compound is a substrate or inhibitor, can be determined by bidirectional transport measurements such as in the mdr-1-transfected MDCKI or MDCKII cell lines, which is considered to be the most definitive assay in early drug discovery (Feng et al., 2008). A bidirectional efflux of >2 is generally considered as a Pgp substrate, whereas the digoxin efflux inhibition in the presence of the NCE is used for determining Pgp inhibition effect in the same model system. However, other higher-throughput assays such as Pgp binding and ATPase activity have been used in the discovery phase to determine Pgp affinity and to rank order compounds for Pgp substrate specificity. The Pgp knockout mouse model has been utilized very effectively to de-risk efflux potential both at the intestinal and the BBB levels. Also, one should note that the impact of Pgp or any other efflux transporter for that matter may not be significant to the development of a compound as long as the compound

shows high transcellular permeability. Thus, the transcellular permeability component in conjunction with the efflux ratio should also be considered when interpreting the bidirectional permeability assay, as the low permeability compounds may confound the Pgp effect. Follow-up mechanistic studies can be conducted in the presence of a very selective inhibitor to further confirm the involvement of transporters (viz, transporter phenotyping) using MDCK or LLC-PK1 cell lines stably transfected with a specific efflux transporter (e.g., MDR1, MRP2, or BCRP) to tease out potential transporter-mediated DDI. Moreover, it has been shown that pregnane X receptor (PXR) has the propensity to induce Pgp expression similar to CYP3A4 induction (Collett et al., 2004), and it would be recommended to screen leads for PXR up-regulation prior to nominating for potential development candidates. Beside induction potential, the efflux transporters may well be susceptible to genetic variability as observed for drug metabolism enzymes, and therefore the impact of pharmacogenetics of efflux transporters in drug disposition should also be noted (Cascorbi, 2006). The relevance of uptake transporters in drug disposition should also be taken into consideration, especially transporters located in the two major clearance organs—liver and kidney. However, there is much work to be done to evaluate uptake transporters and their importance to overall drug clearance in the body (Kusuhara and Sugiyama, 2009).

2.3.2 Metabolism

The rate of *in vivo* biotransformation of a compound plays an important role in delineating the systemic availability and, ultimately, the therapeutic efficacy of a drug candidate. It has been established that, typically, drug biotransformation processes results into more water-soluble product(s) for enhancing the potential elimination of the drug from the body, and thus the overall structure of the compound would dictate the metabolic stability of a compound. It is recommended to analyze the chemical structure of the NCE, which could provide certain physicochemical and safety attributes. For example, furan, thiophene, and methylene dioxide ring systems, and aromatic nitro and amine groups should be evaluated for the formation of reactive intermediates with potential toxicity consequences (Kalgutkar and Didiuk, 2009; Walsh and Miwa, 2011). In this endeavor, the glutathione conjugation system could be utilized for screening compounds for generating electrophilic species (Korfmacher, 2009). In addition, acyl glucuronide is another type of active metabolites where it have the potential to form covalent adducts with proteins and thus lead to idiosyncratic-type drug reactions in some instances (Korfmacher, 2009). There are various *in silico* tools available such as MetaDrug, META, and METOR (Lhasa Ltd., Leeds, U.K.) that can be used to predict potential metabolism-prone soft spots, namely, oxidative, reductive, as well as conjugative, with the possibility of predicting and identifying formation of reactive intermediates.

Discovery teams utilize number of *in vitro* assay screens to test the metabolic stability of compounds. Among the most popular and widely utilized systems are liver microsomes and hepatocytes generated from different species. The metabolic stability assay using human liver microsome (HLM) preparation, which is amenable to higher-throughput screening, is widely used in the discovery phase for the prediction of human intrinsic clearance and considered to be the gold standard in industry (Shou et al., 2005; Fonsi et al., 2008). Isolated hepatocytes, either fresh or cryopreserved, have demonstrated to retain a broader spectrum of drug metabolism enzymatic activities, including not only reticular systems but also cytosolic and mitochondrial enzymes (Caldwell et al., 1999; Li, 1999). In addition, liver slices, consisting of more than one layer of hepatocytes, have also been shown to have metabolism activity and has gained popularity over time. Both systems have the capability in assessing metabolism enzyme induction *in vitro*. Most compounds are rank ordered on the basis of liver microsomal data on the rate of disappearance of the parent drug, and the usual strategy is to bin compounds into high, intermediate, and low in the screening cascade for selection purposes. Oftentimes, it has been observed that microsomes may not very well predict *in vivo* clearance of an NCE which could be due to involvement of Phase II metabolism. Thus, other *in vitro* metabolism assay systems such as liver slices (Ekins et al., 2001; Vanhulle et al., 2001), hepatic sandwich cultures (Swift et al., 2010), hepatic membrane subcellular fractions (Umehara et al., 2000), and purified P450 enzyme preparations (Crespi, 1999) could be utilized in capturing Phase II metabolism in addition to Phase I, and have more accurate predictions of *in vivo* clearance. The CYP reaction phenotyping is also another component in drug metabolism that is important when compounds advance to the lead optimization phase and beyond into development. Information generated from CYP phenotype studies is especially useful in determining the individual CYP contribution in drug clearance as well as the potential clinical implication of the known genetic polymorphisms (Zhang et al., 2007) of the human CYP enzymes in the predicted clearance. The most common models available are liver microsomes, heterologous cDNA expressed human CYP isoenzymes in yeast (*Saccharomyces cerevisiae*), bacteria (*Escherichia coli*), or mammalian B-lymphoblastoids cells (Ohgiya et al., 1989; Crespi et al.,

1991). In addition, these enzymes scaled up in bioreactors could also be very useful in generating usable amounts of metabolite products, which otherwise could be technically challenging to chemically synthesize.

The DDI potential of a compound can be assessed by *in vitro* inhibition studies using either HLMs or hepatocytes (Obach, 2008; Shou and Dai, 2008). Usually, in early screening cascades, the CYP inhibition is measured utilizing recombinant CYP isoenzymes and candidate drugs are incubated with specific probe substrates with the parent molecule, or the major metabolite of substrates is quantified by either fluorescence (Crespi, 1999) or liquid chromatography coupled with tandem mass spectrometry (LC-MS/MS) techniques (Smith et al., 2007). Many recombinant human CYP enzyme systems as described in the previous paragraph could be utilized in CYP inhibition studies (Crespi, 1999), and generally, early in the screening cascade, CYP inhibition is performed using a single concentration (5–10-μM range) and binned compounds into low, moderate, or strong inhibitors based on percent inhibition. The determination of an apparent IC_{50} of a candidate drug toward a specific probe substrate is pursued in the lead optimization phase as a de-risk strategy (Yan et al., 2002), and usually, compounds are binned as weak ($IC_{50} > 10\ \mu M$), moderate ($1\ \mu M < IC_{50} < 10\ \mu M$), or strong ($IC_{50} < 1\ \mu M$) inhibitors for ranking purposes. The moderate and strong inhibitors may be further subjected to mechanism-based inhibition studies (Yan et al., 2002) to differentiate reversible from either quasi-irreversible or irreversible inhibitors (Fowler and Zhang, 2008). The characteristic mechanism-based inhibitors require metabolism, and it is time and concentration dependent and is experimentally assessed using microsomes with a preincubation step followed by dilution steps in determining residual enzymatic activity (Di et al., 2007). An observed decreased in IC_{50} value obtained in the coincubation study following a preincubation step is a good indication of potential irreversible inhibition. However, estimation of Ki values along with partition ratios will provide the extent of irreversible inhibition potential, and thus relative clinical DDI risk could be evaluated using a ratio (I/Ki) between the projected maximum human plasma concentration (I) and the Ki of the inhibitor (Blanchard et al., 2004). In addition, these *in vitro* parameters are being integrated into physiologically based pharmacokinetics (PBPK) models such as SimCyp (Grime et al., 2009) to simulate DDI potential using human PK parameters obtained from clinically relevant compounds. The advancement of LC-MS/MS analytical techniques has enabled the recent development of a "CYP cocktail inhibition" assay (Smith et al., 2007; Bacolod et al., 2009) that utilizes clinically relevant human CYP enzyme probe substrates to simultaneously measure both reversible and irreversible inhibition. This is a significant improvement in assessing clinical DDI potential early in the preclinical phase enabling to closely mimic the clinical outcome.

The induction of human CYPs, CYP1A2, 2B6, 2C9, 2C19, and 3A4, has been implicated in clinical DDI and thus important to evaluate the clinical risk in preclinical development of an NCE with such induction potential (Lewis, 1996). The human pregnane X receptor (hPXR) (Cui et al., 2008) and human aryl hydrocarbon receptor (hAhR) (Flaveny et al., 2009) have shown to be key regulators of CYP3A4 and CYP1A2 expression, respectively. In the context of polypharmacy, an induction of a CYP-isozyme responsible for clearance of a particular drug could result in decreased plasma levels that may potentially alter the therapeutic impact; otherwise the drug is given alone. Although there are number of binding and transfected cell-based *in vitro* assays amenable to high-throughput hPXR or hAhR screening available (Cui et al., 2008), the human primary hepatocyte monolayer model with capability to execute in a high-throughput mode has been the primary assay of choice in industry (Flaveny et al., 2009). It is often being used as a mechanistic model for determination of human CYP induction potential via the measurement of both mRNA and enzymatic activity of CYP1A2 and 3A4/5 with corresponding aryl hydrocarbon receptor (Ahr) and PXR receptor inductions by omeprazole and rifampicin, respectively (Yueh et al., 2005; Fahmi et al., 2008). Induction potential of CYP3A4 is normalized against the 40% induction of the clinically relevant CYP3A inducer; rifampicin at 10 μM to generate an inducer index with above 40% is considered a potential risk by the regulatory authorities. The LC-MS/MS is utilized for quantification of the parent and/or metabolite responsible for induction potential (Li, 1999). The advantage of hepatocyte model over PXR assay is to measure both CYP mRNA and activity, thus identifying the relative contribution from a parent and a potential metabolite with respect to CYP induction.

2.3.3 PK

The animal PK studies are an important component in drug discovery through development as these studies are generally designed to provide whole-body drug disposition kinetics, subsequently being scaled to potential human dose projections. Not all *in vitro* metabolism data will be able to successfully predict *in vivo* plasma or blood exposure profiles of both rodent and nonrodent species, thus creating certain uncertainty *in vitro–in vivo* correlations (IVIVC) for human scaling. Nevertheless, the two key animal PK parameters, blood or plasma clearance and volume of distribution, have been rou-

tinely utilized to perform allometry-type scaling to predict human PK parameters and to estimate terminal half-life of compounds (Mahmood et al., 2006; Fagerholm, 2007). In addition, animal oral bioavailability measurements are evaluated to estimate potential human systemic availability of drugs upon oral administration. Therefore, considerable efforts are given in the design and execution of animal PK studies since these PK parameters are being utilized to simulate human dose and dosing regimens to support the IND filing in first-in-human (FIH) studies.

In the early stages of discovery, it is not as critical to determine the complete animal PK profile of compounds. Most often, organizations follow a quick rank ordering strategy to help mitigate rapid deprioritization of those compounds with less than optimal plasma exposure to work on fewer leads with desirable PK profiles in the animal species to move them along the discovery value chain. Various *in vivo* screen methods such as cassette dosing (Bayliss and Frick, 1999; White and Manitpisitkul, 2001; Manitpisitkul and White, 2004) or cassette-accelerated rapid rat (CARR) (Korfmacher et al., 2001) method have been employed to select compounds with desirable plasma exposure profiles. Cassette dosing has been used by a number of groups to rank order compounds on the basis of oral (PO) plasma exposure or to estimate systemic clearance after an intravenous dose. Although this method can be utilized in a relatively high-throughput manner with reduced animal usage than the conventional PK studies, cassette dosing has not gained sufficient traction within the industry due to its potential DDI risk with the anticipated increased in false positive/negative outcomes. In the rapid rat screen method (Korfmacher et al., 2001), the compounds are ranked on the basis of area under curve (AUC) determination. In brief, each compound at a fixed dose is administered in two rats and the blood samples are withdrawn at 1-h time intervals over a 6-h period. All blood samples that correspond to each time point is then pooled and processed for cassette bioanalysis for plasma concentration determination. The area under the concentration–time curve (AUC_{0-6hr}) is estimated and AUC_{0-6hr} below 500 hr*ng/mL is deprioritized. Similar to cassette dosing, the CARR method also do not provide reliable half-life estimates or full PK profile of the compound. Korfmacher et al. (2001) have reported an evaluation of screen results of more than 5000 compounds suggesting CARR method had a filtering efficiency of only 50%, and suggested that this method provides a useful decision criterion at the early stage gates of drug discovery to weed out nonviable candidates with minimal use of *in vivo* resources.

The major limitations in performing PK studies are the ethical nature of the routine use of animals, and the time and the intensive labor it would take to execute the in-life portion, in addition to needing significant bioanalytical resources for analysis of a large set of plasma/blood samples for establishing drug levels. The PK characterization is therefore considered in the latter part of the screening cascade where a number of potential leads have been narrowed with respect to druggability characteristics. If the PK input is required in early discovery decisions, more efficient methods such as a single time point plasma exposure analysis or a cassette-type dosing as eluted in the previous paragraph could be initiated. A semisimultaneous bioavailability (F) estimation is a method that has been successfully utilized in drug discovery (Bredberg and Karlsson, 1991), where an intravenous dose is administered first and at a suitable postintravenous dose (i.e., postdistributional) an oral dose is administered to the same animal. The key PK parameters such as CL and V_{ss} can be estimated from either noncompartmental or compartmental modeling of intravenous plasma concentration–time curve. Oral PK parameters, C_{max}, T_{max} and F, are subsequently determined from the PO and combined intravenous and PO plasma concentration–time profiles, respectively (Bredberg and Karlsson, 1991). Once the potential leads for further development are identified, the PK studies can be performed on both rodent and nonrodent species with a sufficient number of animals to generate relevant PK parameters with appropriate statistical model fit. This would also facilitate the identification of *in vivo* circulating metabolites, which are quite difficult to generate in the early discovery setting, especially with the cassette dosing approach. Additional PK studies with multiple species (both rodent and nonrodent) including nonhuman primates will enable to support allometry scaling for human PK projections, while multiple-dose PK studies in relevant species will enable to support toxicokinetics in estimating the therapeutic index (TI) of potential leads.

Drug distribution with respect to target tissue exposure is an important component that is needed to be evaluated in context with PK for a PD effect to build a solid PK/PD relationship (Jusko and Gretch, 1976; Tillement et al., 1978). The use of free fraction either in blood and/or plasma and tissue has gained attention because for numerous drugs, it has been observed that therapeutic response correlates better with the concentration of a diffusible, unbound drug than with the total drug concentration (Rowland, 1980). For protein binding, the most common techniques used are either equilibrium dialysis, or ultrafiltration or determination of compound binding affinity to human serum albumin (HSA) and/or alpha-acid glycoprotein (AGP) proteins immobilized on high-performance liquid chromatography (HPLC) columns. Equilibrium dialysis has been

considered to be the gold standard for protein binding as it has low potential for experimental artifacts and helps determine the entire spectrum of protein binding. On the other hand, ultrafiltration has been a popular alternative as this approach eliminates extensive setup time as needed in the equilibrium dialysis and enables generating biological samples for concentration analysis in a short time frame and provides a means to test metabolically and/or chemically unstable compounds (Banker and Clark, 2008). In drug discovery, the protein binding measurements are usually performed in humans and species of choice (rat, mouse, dog, or monkey). The high protein bound (>99%) compounds are the one that needs most attention experimentally since a slight change in binding affinity can result in a greater-fold change in the free fraction. On the other hand, a compound with high free fraction (<95%) could also be problem with respect to toxicity outcome, if the compound has a narrow therapeutic window. However, it has been also suggested that plasma protein binding can have an important influence on key PK parameters, whereas the changes in plasma protein binding may not significantly influence the clinical PK (Benet and Hoener, 2002). In drug discovery, the decision to take forward or deprioritize a compound on the basis of protein binding profile should be considered in consort with other ADME, PK, and PD parameters (e.g., potency or site of action).

2.4 ADME OPTIMIZATION FOR CNS DRUGS

Crossing the BBB for a drug is a very important ADME characteristic that one would look for, especially when the compound is being targeted in the CNS (Smith, 1992). These compounds should have CNS "likeness" properties such as low MW (<400), moderate lipophilicity (clog P < 3), and number hydrogen bond donors and acceptors <4 and <8, respectively, for enabling a compound to cross the BBB (Ajay and Aurcko, 1999; Reichel, 2006) with increasing odds. There are several assays available to be used that can be interpreted alone or in consort to make rational decisions on the ADME behavior of compounds being developed for CNS programs (Di et al., 2009). In one of the assays, the permeability across BBB is determined *in vivo* using *in situ* brain perfusion technique (Dagenais et al., 2000; Bhattacharjee et al., 2001), where the drug is perfused into the brain from the carotid artery or by using brain microdialysis (Dagenais et al., 2000; Sawchuk and Elmquist, 2000; Mano et al., 2002; Cano-Cebrián et al., 2005). However, this method has not been extensively used as a higher-throughput model within discovery because of the limitation of using a large number of animals and the amount of labor it would take to perform the studies on routine basis. In terms of *in vitro* cell culture model, primary cultured porcine brain endothelial cells have been generally used in academic research; however, the isolation and cultivation of primary brain endothelial cells is a long-drawn laborious process and not conducive for routine application (Gumbleton and Audus, 2001). However, additionally, Kusch-Poddar et al. (2005) have suggested the use of immortalized cell models of endothelial origin to predict BBB permeability.

Interestingly, MDCK cell line (dog kidney epithelial cells) transfected with MDR1 gene has been widely used in the pharmaceutical industry for predicting BBB permeability as well as determining Pgp efflux potential (Doan et al., 2002). The binning of compounds in terms of permeability and efflux ratio (basal-apical/apical-basal) in MDR1-MDCK cells (MDCK I or II) has been an attractive tool for rank ordering compounds according to permeability, which in turn allows establishing structure–activity relationships (SARs) driving medicinal chemistry. In early discovery, the data generated from the MDR1-MDCK assay can also be used as a surrogate measure for intrinsic CNS penetrability of compounds (Doan et al., 2002; Carrara et al., 2007). However, predicting interactions with other BBB efflux transporters such as multi drug resistance protein (MRP) or breast cancer resistance protein (BCRP) may be challenging, and thus engineered cell lines with individual transfected transporters similar to MDR1-MDCK cell line may be needed as some in industry has already begun to evaluate the efflux potential of these novel transporters (Jin and Di, 2008; Nies et al., 2008; Colabufo et al., 2009).

The *in vivo* assays in conjunction with *in vitro* assays can be utilized to investigate BBB permeability. It has been reported that a good balance between the rate and the extent of brain penetration is an important factor in designing optimal CNS PK (Liu and Chen, 2005). The determination of *in vivo* brain/plasma ratio, either in rat or in mice, has been observed to be the most widespread accepted practice in industry, which relates to the extent of brain penetration. However, the selection of the time points to measure brain/plasma ratio is very crucial for the interpretation of brain/plasma ratio (Doran et al., 2005). It is generally considered that a compound with a brain/plasma ratio >0.3 would be considered to have sufficient access to the CNS tissue, whereas compounds with a value greater than 1 are expected to freely cross the BBB. However, high brain/plasma ratio may confound the results as a consequence of high binding to lipids, membranes, proteins, or perhaps lysosomal trapping. Doran et al. (2005) have shown that brain/plasma ratio could be affected by the nature of the compounds such as acidic or neutral, where compounds showed a

ratio between 0.5 and 1. However, basic compounds have demonstrated brain/plasma ratio as high as 6. Thus, interpreting brain-to-plasma ratio is an important component of CNS penetration; the efficacy should also be considered prior to deprioritizing compounds with low brain/plasma ratio. It is possible that a compound with low intrinsic BBB permeability properties can show relevant pharmacological effect due to good target tissue binding and or engagement, whereas a compound with high permeability may not achieve high brain levels due to extensive plasma protein binding negating brain tissue penetration or engaged in rapid disposition from the CNS compartment. Besides the determination of compound concentrations in brain and plasma, cerebrospinal fluid (CF) concentrations are often used as surrogate measures for free concentration in the brain or even for CNS penetration per se (Martin, 2004). In addition, the ratio of free brain and free plasma concentration utilizing an *in vitro* experiment such as equilibrium dialysis has been used to determine the extent of brain penetration (Kalvass et al., 2007). The free brain concentration may also be estimated from CSF drug concentration, assuming there is no drug transporter interaction (Kalvass et al., 2007). Shen et al. (2004) has shown that CSF concentration approximates free brain concentration for moderate- to high-permeability compounds; however, CSF concentration may not represent free brain concentration for low-permeability compounds and/or compounds with transporters interaction.

2.5 SUMMARY

In the world of polypharmacy, one of the biggest challenges in ADME/PK would be to identify the potential role of drug metabolizing enzymes and transporters (both uptake and efflux) in DDI interplay, and enabling an optimization process in discovery to provide lead candidates success in FIH studies. The utilization of PBPK, PK/PD tools, as well as PK simulation modeling enable better prediction of human outcomes to position drug candidates with improved drug properties to be successful clinical candidates. A recent white paper on transporters has emphasized several major uptake and efflux transporters that could be relevant in terms of the potential clinical outcome (Giacomini et al., 2010) and the challenge to the industry, especially in drug discovery, to develop both relevant screens and mechanistic models that are capable of evaluating the potential interplay between key transporters and drug metabolizing enzymes to mitigate efficacy, safety, and potential genetic polymorphic issues in the clinic.

In a nutshell, a complete understanding of all facets of chemical structure (SAR) and the impact of that structure on the ADME/PK and safety properties would be necessary for advancing a compound to a lead candidate status. Most often, it is not feasible to attain all the desired ADME profiles for a drug candidate in discovery, but this should not prevent the candidate from further development, should an unmet need or a good business case were to be made. Ultimately, it is the pharmacological activity coupled with the drug candidate's superior PK profile that is considered pivotal for hedging success in the clinical development programs. Thus, the development of early lead candidates should be judged on a case-by-case basis, addressing the risk-to-benefit ratio while having a well-designed risk mitigation plan to evaluate the perceived risk in development to make a commercially viable drug. In this chapter, the authors have attempted to provide a general overview of what it would take in the discovery ADME/PK strategy to position a clinically viable drug candidate.

REFERENCES

Ajay BGW, Aurcko MA (1999) Designing libraries with CNS activity. *J Med Chem* 42(24):4942–4951.

Alcorn CJ, Simpson RJ, Leahy DE, Peters TJ (1993) Partition and distribution coefficients of solutes and drugs in brush border membrane vesicles. *Biochem Pharmacol* 45(9):1775–1782.

Artursson P (1991) Cell cultures as models for drug absorption across the intestinal mucosa. *Crit Rev Ther Drug Carrier Syst* 8:305–330.

Artursson P, Karlsson J (1991) Correlation between oral drug absorption in humans and apparent drug permeability coefficients in human intestinal epithelia (Caco-2) cells. *Biochem Biophys Res Commun* 175:880–890.

Artursson P, Palm K, Luthman K (2001) Caco-2 monolayers in experimental and theoretical predictions of drug transport. *Adv Drug Deliv Rev* 46(1–3):27–43.

Avdeef A (2003) *Absorption and Drug Development: Solubility, Permeability and Charge State*. Wiley-Interscience, New York.

Avdeef A (2005) The rise of PAMPA. *Expert Opin Drug Metab Toxicol* 1(2):325–342.

Bacolod M, Nguyen K, Chandrasena G (2009) Development and validation of a higher throughput *in vitro* cocktail CYP inhibition assay to simultaneously assess the reversible and time dependent inhibitory potencies on new chemical entity. Poster presented at the ISSX 2009, Baltimore, MD.

Bajpai M, Adkison KK (2000) High-throughput screening for lead optimization: A rational approach. *Curr Opin Drug Discov Devel* 3(1):63–71.

Balimane PV, Han YH, Chong S (2006) Current industrial practices of assessing permeability and P-glycoprotein interaction. *AAPS J* 8(1):E1–13.

Banker MJ, Clark TH (2008) Plasma/serum protein binding determinations. *Curr Drug Metab* 9:854–859.

Bard B, Martel S, Carrupt PA (2008) High throughput UV method for the estimation of thermodynamic solubility and the determination of the solubility in biorelevant media. *Eur J Pharm Sci* 33(3):230–240.

Barthe L, Bessouet M, Woodley JF, Houin G (1998) The improved everted gut sac: A simple method to study intestinal P-glycoprotein. *Int J Pharm* 173:255–258.

Barthe L, Woodley JF, Houin G (1999) Gastrointestinal absorption of drugs: Methods and studies. *Fundam Clin Pharmacol* 13:154–168.

Bayliss MK, Frick LW (1999) High-throughput pharmacokinetics: Cassette dosing. *Curr Opin Drug Discov Devel* 2(1):20–25.

Benet LZ, Hoener B (2002) Changes in plasma protein binding have little clinical relevance. *Clin Pharmacol Ther* 71:115–121.

Bhattacharjee AK, Nagashima T, Kondoh T, Tamaki N (2001) Quantification of early blood-brain barrier disruption by in situ brain perfusion technique. *Brain Res Brain Res Protoc* 8(2):126–131.

Bidault Y (2006) A flexible approach for optimising *in silico* ADME/Tox characterization of lead candidates. *Expert Opin Drug Metab Toxicol* 2(1):157–168.

Blanchard N, Richert L, Coassolo P, Lavé T (2004) Qualitative and quantitative assessment of drug-drug interaction potential in man, based on Ki, IC50 and inhibitor concentration. *Curr Drug Metab* 5(2):147–156.

Bredberg U, Karlsson MO (1991) *In vivo* evaluation of the semi-simultaneous method for bioavailability estimation using controlled intravenous infusion as an "extravascular" route of administration. *Biopharm Drug Dispos* 12(8):583–597.

Caldwell GW, Masucci JA, Chacon E (1999) High throughput liquid chromatography-mass spectrometry assessment of the metabolic activity of commercially available hepatocytes from 96-well plates. *Comb Chem High Throughput Screen* 2(1):39–51.

Cano-Cebrián MJ, Zornoza T, Polache A, Granero L (2005) Quantitative *in vivo* microdialysis in pharmacokinetic studies: Some reminders. *Curr Drug Metab* 6(2):83–90.

Carrara S, Reali V, Misiano P, Dondio G, Bigogno C (2007) Evaluation of *in vitro* brain penetration: Optimized PAMPA and MDCKII-MDR1 assay comparison. *Int J Pharm* 345(1–2):125–133.

Cascorbi I (2006) Role of pharmacogenetics of ATP-binding cassette transporters in the pharmacokinetics of drugs. *Pharmacol Ther* 112(2):457–473.

Cayen MN (1995) Considerations in the design of toxicokinetic programs. *Toxicol Pathol* 23:148–157.

Colabufo NA, Berardi F, Contino M, Niso M, Perrone R (2009) ABC pumps and their role in active drug transport. *Curr Top Med Chem* 9(2):119–129.

Collett A, Tanianis-Hughes J, Warhurst G (2004) Rapid induction of P-glycoprotein expression by high permeability compounds in colonic cells *in vitro*: A possible source of transporter mediated drug interactions? *Biochem Pharmacol* 68(4):783–790.

Crespi CL (1999) Higher-throughput screening with human cytochromes P450. *Curr Opin Drug Discov Devel* 2(1):15–19.

Crespi CL, Gonzalez FJ, Steimel DT, Turner TR, Gelboin HV, Penman BW, Langenbach R (1991) A metabolically competent human cell line expressing five cDNAs encoding procarcinogen-activating enzymes: Application to mutagenicity testing. *Chem Res Toxicol* 4:566–572.

Cui X, Thomas A, Gerlach V, White RE, Morrison RA, Cheng KC (2008) Application and interpretation of hPXR screening data: Validation of reporter signal requirements for prediction of clinically relevant CYP3A4 inducers. *Biochem Pharmacol* 76(5):680–689.

Dagenais C, Rousselle C, Pollack GM, Scherrmann JM (2000) Development of an *in situ* mouse brain perfusion model and its application to mdr1a P-glycoprotein-deficient mice. *J Cereb Blood Flow Metab* 20(2):381–386.

Di L, Kerns EH, Li SQ, Carter GT (2007) Comparison of cytochrome P450 inhibition assays for drug discovery using human liver microsomes with LC-MS, rhCYP450 isozymes with fluorescence, and double cocktail with LC-MS. *Int J Pharm* 335(1–2):1–11.

Di L, Kerns EH, Bezar IF, Petusky SL, Huang Y (2009) Comparison of blood-brain barrier permeability assays: *In situ* brain perfusion, MDR1-MDCKII and PAMPA-BBB. *J Pharm Sci* 98(6):1980–1991.

DiMasi JA (1994) Risks, regulation, and rewards in new drug development in the United States. *Regul Toxicol Pharmacol* 19:228–235.

DiMasi JA (2001) New drug development in the United States from 1963 to 1999. *Clin Pharmacol Ther* 69(5):286–296.

Doan KM, Humphreys JE, Webster LO, Wring SA, Shampine LJ, Serabjit-Singh CJ, Adkison KK, Polli JW (2002) Passive permeability and P-glycoprotein-mediated efflux differentiate central nervous system (CNS) and non-CNS marketed drugs. *J Pharmacol Exp Ther* 303(3):1029–1037.

Doran A, Obach RS, Smith BJ, Hosea NA, Becker S, Callegari E, Chen C, Chen X, Choo E, Cianfrogna J, Cox LM, Gibbs JP, Gibbs MA, Hatch H, Hop CE, Kasman IN, Laperle J, Liu J, Liu X, Logman M, Maclin D, Nedza FM, Nelson F, Olson E, Rahematpura S, Raunig D, Rogers S, Schmidt K, Spracklin DK, Szewc M, Troutman M, Tseng E, Tu M, Van Deusen JW, Venkatakrishnan K, Walens G, Wang EQ, Wong D, Yasgar AS, Zhang C (2005) The impact of P-glycoprotein on the disposition of drugs targeted for indications of the central nervous system: Evaluation using the MDR1A/1B knockout mouse model. *Drug Deliv Rev* 33(1):165–174.

Ekins S, Ring BJ, Grace J, McRobie-Belle DJ, Wrighton SA (2001) Present and future *in vitro* approaches for drug metabolism. *J Pharmacol Toxicol Methods* 44(1):313–324.

Fagerholm U (2007) Prediction of human pharmacokinetics-evaluation of methods for prediction of volume of distribution. *J Pharm Pharmacol* 59(9):1181–1190.

Fagerholm U, Johansson M, Lennernas H (1996) Comparison between permeability coefficients in rat and human jejunum. *Pharm Res* 13:1336–1342.

Fahmi OA, Boldt S, Kish M, Obach RS, Tremaine LM (2008) Prediction of drug-drug interactions from *in vitro* induction data: Application of the relative induction score approach using cryopreserved human hepatocytes. *Drug Metab Dispos* 36:1971–1974.

Feng B, Mills JB, Davidson RE, Mireles RJ, Janiszewski JS, Troutman MD, de Morais SM (2008) *In vitro* P-glycoprotein assays to predict the *in vivo* interactions of P-glycoprotein with drugs in the central nervous system. *Drug Metab Dispos* 36(2):268–275.

Flaveny CA, Murray IA, Chiaro CR, Perdew GH (2009) Ligand selectivity and gene regulation by the human aryl hydrocarbon receptor in transgenic mice. *Mol Pharmacol* 75(6):1412–1420.

Fonsi M, Orsale MV, Monteagudo E (2008) High-throughput microsomal stability assay for screening new chemical entities in drug discovery. *J Biomol Screen* 13(9):862–869.

Fowler S, Zhang H (2008) *In vitro* evaluation of reversible and irreversible cytochrome P450 inhibition: Current status on methodologies and their utility for predicting drug-drug interactions. *AAPS J* 10(2):410–424.

Frank R, Hargreaves R (2003) Clinical biomarkers in drug discovery and development. *Nat Rev Drug Discov* 2(7): 566–580.

Gabrielsson J, Dolgos H, Gillberg PG, Bredberg U, Benthem B, Duker G (2009) Early integration of pharmacokinetic and dynamic reasoning is essential for optimal development of lead compounds: Strategic considerations. *Drug Discov Today* 14(7–8):358–372.

Giacomini KM, Huang S-M, Tweedie DJ, Benet LZ, Brouwer KLR, Chu X, Dahlin A, Evers R, Fischer V, Hillgren KM, Hoffmaster KA, Ishikawa T, Keppler D, Kim RB, Lee CA, Niemi M, Polli JW, Sugiyama Y, Swaan PW, Ware JA, Wright SH, Yee SW, Zamek-Gliszczynski MJ, Zhang L, for The International Transporter Consortium (2010) Membrane transporters in drug development. *Nat Rev Drug Discov* 9:215–236.

Greene N, Judson PN, Langowski JJ, Marchant CA (1999) Knowledge-based expert systems for toxicity and metabolism prediction: DEREK, StAR and METEOR. *SAR QSAR Environ Res* 10(2–3):299–314.

Grime KH, Bird J, Ferguson D, Riley RJ (2009) Mechanism-based inhibition of cytochrome P450 enzymes: An evaluation of early decision making *in vitro* approaches and drug-drug interaction prediction methods. *Eur J Pharm Sci* 36(2–3):175–191.

Gumbleton M, Audus KL (2001) Progress and limitations in the use of *in vitro* cell cultures to serve as a permeability screen for the blood-brain barrier. *J Pharm Sci* 90(11): 1681–1698.

Hageman MJ (2006) Solubility, solubilization and dissolution in drug delivery during lead optimization. *Biotechnol Pharm Aspects* 4:99–130.

Jin H, Di L (2008) Permeability—*in vitro* assays for assessing drug transporter activity. *Curr Drug Metab* 9(9):911–920.

Jusko WJ, Gretch M (1976) Plasma and tissue protein binding of drugs in pharmacokinetics. *Drug Metab Rev* 5:43–140.

Kalgutkar AS, Didiuk MT (2009) Structural alerts, reactive metabolites, and protein covalent binding: How reliable are these attributes as predictors of drug toxicity? *Chem Biodivers* 6(11):2115–2137.

Kalvass JC, Maurer TS, Pollack GM (2007) Use of plasma and brain unbound fractions to assess the extent of brain distribution of 34 drugs: Comparison of unbound concentration ratios to *in vivo* p-glycoprotein efflux ratios. *Drug Metab Dispos* 35(4):660–666.

Kennedy T (1997) Managing the drug discovery/development interface. *Drug Discov Today* 2:436–444.

Kirchmair J, Distinto S, Schuster D, Spitzer G, Langer T, Wolber G (2008) Enhancing drug discovery through *in silico* screening: Strategies to increase true positives retrieval rates. *Curr Med Chem* 15(20):2040–2053.

Korfmacher WA (2009) Advances in the integration of drug metabolism into the lead optimization paradigm. *Mini Rev Med Chem* 9(6):703–716.

Korfmacher WA, Cox KA, Ng KJ, Veals J, Hsieh Y, Wainhaus S, Broske L, Prelusky D, Nomeir A, White RE (2001) Cassette-accelerated rapid rat screen: A systematic procedure for the dosing and liquid chromatography/atmospheric pressure ionization tandem mass spectrometric analysis of new chemical entities as part of new drug discovery. *Rapid Commun Mass Spectrom* 15(5):335–340.

Kramer JA, Sagartz JE, Morris DL (2007) The application of discovery toxicology and pathology towards the design of safer pharmaceutical lead candidates. *Nat Rev Drug Discov* 6(8):636–649.

Kusch-Poddar M, Drewe J, Fux I, Gutmann H (2005) Evaluation of the immortalized human brain capillary endothelial cell line BB19 as a human cell culture model for the blood-brain barrier. *Brain Res* 1064(1–2):21–31.

Kusuhara H and Sugiyama Y (2009) *In vitro-in vivo* extrapolation of transporter-mediated clearance in the liver and kidney. *Drug Metab Pharmacokinet* 24(1):37–52.

Lennernäs H (1998) Human intestinal permeability. *J Pharm Sci* 87:403–410.

Lewis DF (1996) *Cytochromes P450.* Taylor & Francis, Bristol, PA.

Li AP (1999) Overview: Hepatocytes and cryopreservation—A personal historical perspective. *Chem Biol Interact* 121(1):1–5.

Lipinski CA, Lombardo F, Dominy BW, Feeney PJ (1997) Experimental and computational approaches to estimate solubility and permeability in drug discovery and development settings. *Adv Drug Deliv Rev* 23:3–25.

Liu X, Chen C (2005) Strategies to optimize brain penetration in drug discovery. *Curr Opin Drug Discov Devel* 8(4):505–512.

Luffer-Atlas D (2008) Unique/major human metabolites: Why, how, and when to test for safety in animals. *Drug Metab Rev* 40(3):447–463.

Mahmood I, Martinez M, Hunter RP (2006) Interspecies allometric scaling. Part I: Prediction of clearance in large animals. *J Vet Pharmacol Ther* 29(5):415–423.

Manitpisitkul P, White RE (2004) Whatever happened to cassette-dosing pharmacokinetics? *Drug Discov Today* 9(15):652–658.

Mano Y, Higuchi S, Kamimura H (2002) Investigation of the high partition of YM992, a novel antidepressant, in rat brain—*In vitro* and *in vivo* evidence for the high binding in brain and the high permeability at the BBB. *Biopharm Drug Dispos* 23(9):351–360.

Martin I (2004) Prediction of blood-brain barrier penetration: Are we missing the point? *Drug Discov Today* 9(4):161–162.

Mayr LM, Bojanic D (2009) Novel trends in high-throughput screening. *Curr Opin Pharmacol* 9(5):580–588.

Nies AT, Schwab M, Keppler D (2008) Interplay of conjugating enzymes with OATP uptake transporters and ABCC/MRP efflux pumps in the elimination of drugs. *Expert Opin Drug Metab Toxicol* 4(5):545–568.

Obach S (2008) Inhibition of drug-metabolizing enzymes and drug–drug interactions in drug discovery and development. *In Drug-Drug Interactions in Pharmaceutical Development.* Li, AP, ed. Hoboken, NJ: John Wiley & Sons, Inc.; 75–93.

Ohgiya S, Komori M, Fujitani T, Miura T, Shinriki N, Kamataki T (1989) Cloning of human cytochrome P450 cDNA and its expression in *Saccharomyces cerevisiae. Biochem Int* 18:429–438.

Reichel A (2006) The role of blood-brain barrier studies in the pharmaceutical industry. *Curr Drug Metab* 7:183–203.

Rowland M (1980) Plasma protein binding and therapeutic drug monitoring. *Ther Drug Monit* 2:29–37.

Rubas W, Cromwell ME, Shahrokh Z, Villagran J, Nguyen TN, Wellton M, Nguyen TH, Mrsny RJ (1996) Flux measurements across Caco-2 monolayers may predict transport in human large intestinal tissue. *J Pharm Sci* 85:165–169.

Sawchuk RJ, Elmquist WF (2000) Microdialysis in the study of drug transporters in the CNS. *Adv Drug Deliv Rev* 45(2–3):295–307.

Segall MD, Beresford AP, Gola JM, Hawksley D, Tarbit MH (2006) Focus on success: Using a probabilistic approach to achieve an optimal balance of compound properties in drug discovery. *Expert Opin Drug Metab Toxicol* 2(2):325–337.

Shen DD, Artru AA, Adkison KK (2004) Principles and applicability of CSF sampling for the assessment of CNS drug delivery and pharmacodynamics. *Adv Drug Deliv Rev* Oct 14;56(12):1825–1857.

Shou M, Dai R (2008) Analysis of *in vitro* cytochrome P450 inhibition in drug discovery and development. In *Drug Metabolism in Drug Design and Development: Basic Concepts and Practice*, Zhang D, Humphreys WG, eds. Hoboken, NJ: John Wiley & Sons; 513–544.

Shou WZ, Magis L, Li AC, Weng N, Bryant MS (2005) A novel approach to perform metabolite screening during the quantitative LC-MS/MS analyses of *in vitro* metabolic stability samples using a hybrid triple-quadruple linear ion trap mass spectrometer. *J Mass Spectrom* 40(10):1347–1356.

Singh B, Dhake AS, Sethi D, Paul Y (2007a) *In silico* ADME predictions using quantitative structure pharmacokinetic relationships. Part I: Fundamental aspects. *The Pharma Review* August–September: 93–100.

Singh B, Parle M, Paul Y, Khurana L (2007b) *In silico* ADME predictions using quantitative structure pharmacokinetic relationships. Part II: Descriptors. *The Pharma Review* October–November: 63–68.

Smith D, Sadagopan N, Zientek M, Reddy A, Cohen L (2007) Analytical approaches to determine cytochrome P450 inhibitory potential of new chemical entities in drug discovery. *J Chromatogr B Analyt Technol Biomed Life Sci* 850(1–2):455–463.

Smith QR (1992) In *Physiology and Pharmacology of the Blood-Brain Barrier*, Bradbury MWB, ed., pp. 23–52. Springer, Berlin.

Subramanian G, Mjalli AM, Kutz ME (2006) Integrated approaches to perform *in silico* drug discovery. *Curr Drug Discov Technol* 3(3):189–197.

Sugiyama Y (2005) Druggability: Selecting optimized drug candidates. *Drug Discov Today* 10(23–24):1577–1579.

Sun D, Lennernas H, Welage LS, Barnett JL, Landowski CP, Foster D, Fleisher D, Lee KD, Amidon GL (2002) Comparison of human duodenum and Caco-2 gene expression profiles for 12,000 gene sequences tags and correlation with permeability of 26 drugs. *Pharm Res* 19(10):1400–1416.

Swift B, Pfeifer ND, Brouwer KL (2010) Sandwich-cultured hepatocytes: An *in vitro* model to evaluate hepatobiliary transporter-based drug interactions and hepatotoxicity. *Drug Metab Rev* 42(3):446–471.

Testa B, Balmat AL, Long A, Judson P (2005) Predicting drug metabolism—An evaluation of the expert system METEOR. *Chem Biodivers* 2(7):872–885.

Tetko IV, Oprea TI (2008) Early ADME/T predictions: Toy or tool? In *Chemoinformatics Approaches to Virtual Screening*, Varnel A and Tropsha A, Eds. Cambridge, UK. Royal Society of Chemistry; pp. 240–267.

Tillement JP, Lhoste F, Giudicelli JF (1978) Diseases and drug protein binding. *Clin Pharmacokinet* 3:144–154.

Umehara K, Kudo S, Hirao Y, Morita S, Uchida M, Odomi M, Miyamoto G (2000) Oxidative cleavage of the octyl side chain of 1- (3,4-dichlorobenzyl)-5-octylbiguanide (OPB-

2045) in rat and dog liver preparations. *Drug Metab Dispos* 28(8):887–894.

Vanhulle VP, Martiat GA, Verbeeck RK, Horsmans Y, Calderon PB, Eeckhoudt SL, Taper HS, Delzenne N (2001) Cryopreservation of rat precision-cut liver slices by ultrarapid freezing: Influence on phase I and II metabolism and on cell viability upon incubation for 24 hours. *Life Sci* 68(21):2391–2403.

Varma MV, Obach RS, Rotter C, Miller HR, Chang G, Steyn SJ, El-Kattan A, Troutman MD (2010) Physicochemical space for optimum oral bioavailability: Contribution of human intestinal absorption and first-pass elimination. *J Med Chem* 53(3):1098–1108.

Veber DF, Johnson SR, Cheng HY, Smith BR, Ward KW, Kopple KD (2002) Molecular properties that influence the oral bioavailability of drug candidates. *J Med Chem* 45(12):2615–2623.

Walsh JS, Miwa GT (2011) Bioactivation of drugs: Risk and drug design. *Annu Rev Pharmacol Toxicol* 51:145–167.

White RE, Manitpisitkul P (2001) Pharmacokinetic theory of cassette dosing in drug discovery screening. *Drug Metab Dispos* 29(7):957–966.

Yalkowsky SH, He Y (2003) *Handbook of Aqueous Solubility Data*. CRC Press, Boca Raton, FL.

Yamashita T, Dohta Y, Nakamura T, Fukami T (2008) High-speed solubility screening assay using ultra-performance liquid chromatography/mass spectrometry in drug discovery. *J Chromatogr A* 1182(1):72–76.

Yan Z, Rafferty B, Caldwell GW, Masucci JA (2002) Rapidly distinguishing reversible and irreversible CYP450 inhibitors by using fluorometric kinetic analyses. *Eur J Drug Metab Pharmacokinet* 27(4):281–287.

Yueh MF, Kawahara M, Raucy J (2005) Cell-based high-throughput bioassays to assess induction and inhibition of CYP1A enzymes. *Toxicol In Vitro* 19(2):275–287.

Zhang H, Davis CD, Sinz MW, Rodrigues AD (2007) Cytochrome P450 reaction-phenotyping: An industrial perspective. *Expert Opin Drug Metab Toxicol* 3(5):667–687.

Zhang L, Zhang YD, Zhao P, Huang SM (2009) Predicting drug-drug interactions: An FDA perspective. *AAPS J* 11(2):300–306.

Zhou L, Yang L, Tilton S, Wang J (2007) Development of a high throughput equilibrium solubility assay using miniaturized shake-flask method in early drug discovery. *J Pharm Sci* 96(11):3052–3071.

3

DRUG TRANSPORTERS IN DRUG INTERACTIONS AND DISPOSITION

Imad Hanna and Ryan M. Pelis

3.1 INTRODUCTION

Drug transporters are gaining greater appreciation in the drug development industry due to their ability to influence drug absorption, distribution, metabolism, and excretion (ADME), and their potential involvement in adverse drug reactions and drug–drug interactions (DDIs). Drug transporters are membrane-bound proteins that physically transport relatively small organic molecules (i.e., nonbiologics), including those of physiological, pharmacological, and toxicological significance, across cell membranes. Although these transporters are expressed in many polarized epithelia, and some nonpolarized cell types, the intestine, liver, kidney, and capillary endothelium of the brain (blood–brain barrier) are sites where drug transporters are largely expected to influence drug disposition. For example, the expression of drug transporters in the apical membranes of intestinal enterocytes and endothelial cells of the brain are suspected of slowing the oral absorption and brain penetration of drugs, respectively. Drug transporters in liver hepatocytes and renal tubule cells are positioned to facilitate drug excretion from the body. Drug transporters expressed in liver hepatocytes may also control intracellular drug concentrations, which could influence the amount of drug that is available for metabolism by drug metabolizing enzymes. Although the expression of drug transporters in barrier (intestine and blood–brain barrier) and excretory (liver and kidney) tissues can

protect the body's organs from potential toxicity due to drug exposure, they can also be a major limiting factor in drug efficacy.

The drug transporters are contained within two distinct superfamilies, the ATP-binding cassette (ABC) and solute carrier (SLC) families. Whereas ABC transporters always efflux their substrates out of cells, SLC family members comprise both uptake and efflux transporters. Although there are approximately 50 known members of ABC transport proteins, only P-glycoprotein (Pgp or MDR1; ABCB1), multidrug resistance-associated protein 2 (MRP2; ABCC2), and the breast cancer resistance protein (BCRP; ABCG2) will be discussed, as there is mounting data in the literature suggesting that they play an important role in drug disposition. Similarly, there are many SLC family members that have the capacity to transport therapeutics, but the primary focus will be on the organic cation transporters 1 (OCT1; SLC22A1) and 2 (OCT2; SLC22A2); the multidrug and toxin extrusion transporters 1 (MATE1; SLC47A1) and 2K (MATE2K; SLC47A2); organic anion transporters 1 (OAT1; SLC22A6) and 3 (OAT3; SLC22A8); and the organic anion transporting polypeptides 1B1 (OATP1B1; SLCO1B1), 1B3 (OATP1B3; SLCO1B3), and 2B1 (OATP2B1; SLCO2B1). Figure 3.1 shows the cellular and subcellular localization of the aforementioned drug transporters in the kidney, liver, intestine, and brain. Importantly, despite their omission from Figure 3.1,

ADME-Enabling Technologies in Drug Design and Development, First Edition. Edited by Donglu Zhang and Sekhar Surapaneni.
© 2012 John Wiley & Sons, Inc. Published 2012 by John Wiley & Sons, Inc.

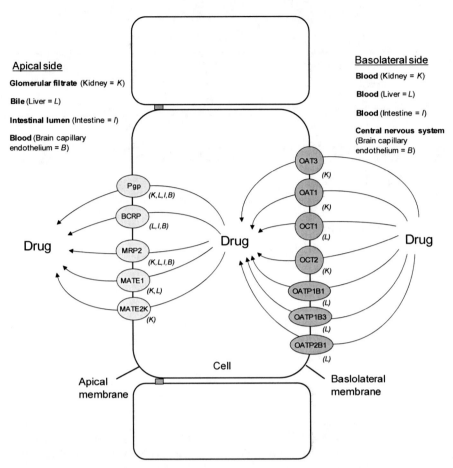

FIGURE 3.1. Model depicting the role of uptake and efflux transport proteins in drug transport by kidney proximal tubules, liver hepatocytes, intestinal enterocytes, and brain capillary endothelial cells. As noted in the text, this is not an exhaustive list of drug transporters that localize to these barrier tissues, but instead, the ones that are thought to be most relevant to drug disposition in humans. Uptake transporters localize to basolateral membranes, whereas efflux transporters localize to apical membranes, and the cellular localization of the individual transporters is noted (K, kidney proximal tubules; L, liver hepatocytes; I, intestinal enterocytes; and B, brain capillary endothelial cells). The arrows indicate the direction of net transport. The substrate specificity of the individual transporters is discussed in the text.

other drug transporters may be as/or more important to drug disposition and drug interaction potential, and thus their involvement should not necessarily be overlooked in drug development.

A hallmark feature of the drug transporters is their ability to transport compounds with diverse physicochemical properties (multispecificity), and thus they have a tendency for overlapping substrate specificities. For example, the human orthologs of OCT1, OCT2, MATE1, and MATE2K all transport the H_2 histamine receptor antagonist cimetidine and the antidiabetic drug metformin. The multispecific behavior of the drug transporters, their tendency for overlapping substrate specificities, and in some cases, their coexpression in the same cell type often makes it challenging to determine the contributions of individual transporters to drug disposition, adverse drug reactions, and DDIs.

A major goal of preclinical drug development activities is to determine whether a drug is safe for effective use on the market. Toward this end, it may be important to understand what transporters transport a drug of interest. This information, along with knowing where the transporters are expressed, may help anticipate how a drug will distribute throughout the body and where the majority of excretion will occur. Since comedications are commonplace in the clinic, it may be necessary to identify whether a particular drug is an inhibitor of uptake and/or efflux transporters, which may influence the comedication's pharmacokinetics and/or pharmacodynamics. For these reasons, this chapter will provide background information on the drug transporters, that is, where they are expressed, their transport mechanism/energetics, and their substrate/inhibitor specificities. Additionally, because of its relevance to the drug development industry, clinical cases where transporters have been shown to influence drug disposition or are thought to contribute to DDIs will be highlighted. The current methodologies used to study drug transporters in drug development will also be discussed. The current view of which drug transporters are most relevant to drug disposition and drug interactions and how to most appropriately study their transport activity in drug development was the focus of a recent review (Giacomini et al., 2010).

3.2 ABC TRANSPORTERS

Pgp, MRPs, and BCRP are ATP-dependent efflux transporters that were first identified in tumor cells where they were found to confer resistance to numerous chemotherapeutic agents. ABC transporters are ATP-dependent transporters, and the hydrolysis of ATP is linked to ABC transporter-mediated drug efflux out of cells. Because ABC proteins are efflux transporters, there are challenging issues associated with assessing the interaction of drugs as substrates/inhibitors with these transporters, which will be addressed later in this chapter.

3.2.1 Pgp (MDR1, ABCB1)

3.2.1.1 Tissue Expression Pgp was the first of the ABC transport proteins to be cloned, and thus it is the most well studied of the drug transporters. Pgp is expressed in numerous barrier and excretory epithelial tissues including intestinal enterocytes, hepatocytes, renal tubule cells, and capillaries of the brain, where it is expressed exclusively in apical membranes (Schinkel and Jonker, 2003). Subsequently, Pgp is involved in the efflux of drug substrates into the intestinal lumen (enterocytes), bile (hepatocytes), urine (tubule cells), and bloodstream (brain endothelium). Accordingly, Pgp plays a predominant role in limiting the absorption of orally administered drugs, the removal of drug substances from the body, as well as limiting the brain penetration of drugs.

3.2.1.2 Substrate/Inhibitor Specificity In general, Pgp substrates are large-molecular-weight (>400 Da) amphipathic organic molecules that are neutral or positively charged, although Pgp also transports some anionic compounds, albeit at low rates (Schinkel and Jonker, 2003). Examples of Pgp substrates include paclitaxel, etoposide, and methotrexate (anticancer), digoxin (cardiac glycoside), cyclosporin A (immunosuppressant), erythromycin (antibiotic), and verapamil (calcium

channel blocker) (Schinkel and Jonker, 2003). PSC833 (valspodar), GF120918 (elacridar), and LY335979 (zosusquidar) are often used as *in vitro* and *in vivo* inhibitors of Pgp (Thomas and Coley, 2003; Nobili et al., 2006; Mayur et al., 2009). However, both PSC833 and GF120918 have also been shown to inhibit BCRP; LY335979 does not appear to inhibit BCRP (Shepard et al., 2003; Mahringer et al., 2009). HM30181 is a recently identified Pgp inhibitor that appears to be somewhat specific and extremely potent toward Pgp (IC_{50}<100 nM) (Kwak et al., 2010).

3.2.1.3 Animal Models The relative importance of Pgp in drug disposition has been noted from *in vivo* studies with mice deficient in Pgp. Whereas humans only have one Pgp isoform (MDR1), mice have two functional Pgp isoforms (Mdr1a and Mdr1b). Plasma levels of the Pgp substrate digoxin were significantly higher in Pgp double knockout mice ($Mdr1^{a/b-/-}$) compared with wild-type mice following intravenous injection, and this occurred concomitantly with a reduction in biliary excretion, suggesting an important role for Pgp in biliary digoxin excretion (Schinkel et al., 1997). Brain levels of digoxin were also elevated (27-fold) in Pgp null mice, indicating that loss of Pgp activity can compromise the drug barrier function of the blood–brain barrier (Schinkel et al., 1997).

3.2.1.4 Clinical Studies There are numerous synonymous and nonsynonymous mutations that have been identified in the gene encoding Pgp (Maeda and Sugiyama, 2008). Most notably is the 3435C>T mutation in the coding region of Pgp, which does not result in an amino acid change (synonymous mutation), but appears to impact protein function, an effect likely manifested through impairment of normal protein folding (Komar, 2007; Kimchi-Sarfaty et al., 2007a,b). Plasma levels of digoxin were lower in Japanese patients carrying the 3435C>T mutation following oral digoxin administration, but the exact mechanism by which this occurred is unclear (Sakaeda et al., 2001). The absorption of digoxin across the immortalized human cell line Caco-2, which expresses Pgp and is a model of intestinal drug absorption, was increased following shRNA ablation of Pgp (Watanabe et al., 2005).

DDIs at the level of Pgp have been cited. A two- to threefold increase in the plasma digoxin area under the curve (AUC) ratio in healthy human volunteers following coadministration of the Pgp inhibitor PSC833 was attributed to inhibition of Pgp-mediated digoxin clearance at the kidney and liver (Kovarik et al., 1999). Following intravenous administration, the level of ^{11}C-verapamil, determined by positron emission tomography, in the brains of healthy human volunteers was significantly elevated when the Pgp inhibitor tariquidar (XR9576) was present, implicating an important role for Pgp in blood–brain barrier function (Wagner et al., 2009).

3.2.2 BCRP (ABCG2)

3.2.2.1 Tissue Expression
BCRP was first cloned in 1998 from MCF-7 human breast cancer carcinoma cells that were resistant to mitoxantrone, doxorubicin, and daunorubicin (Doyle et al., 1998). Like Pgp, BCRP is expressed in apical membranes of intestinal enterocytes, hepatocytes, renal tubule cells, and brain capillary endothelial cells (Robey et al., 2009) (Figure 3.1). Thus, BCRP is expected to play an important role in limiting drug absorption across the intestine and blood–brain barrier, and in biliary and renal drug excretion.

3.2.2.2 Substrate/Inhibitor Specificity
The substrate/inhibitor specificity of BCRP is broad, interacting with small to large (200–700 Da) organic molecules that carry a negative or positive charge or those that are neutral. Examples of BCRP substrates include lamivudine and zidovudine (antivirals), erythromycin and norfloxacin (antibiotics), topotecan, imatinib, and methotrexate (anticancer), nitrendipine and dipyridamole (calcium channel blockers), and rosuvastatin and pravastatin (HMG-CoA reductase inhibitors) (Staud and Pavek, 2005; Ieiri et al., 2009; Robey et al., 2009). Fumitremorgin C, Ko143, and novobiocin are inhibitors of BCRP commonly used to identify the contribution of BCRP activity to drug transport in various primary and immortalized cells (Nicolle et al., 2009). Ko143 is also a relatively potent MRP2 inhibitor (Matsson et al., 2009).

3.2.2.3 Animal Models
Studies with Bcrp knockout mice (Bcrp$^{-/-}$) have demonstrated an important role for BCRP in the disposition of not only therapeutic agents but also environmental toxins, such as pheophorbide A (Jonker et al., 2002). The oral bioavailability of topotecan was elevated (~6-fold) in Bcrp$^{-/-}$ mice, highlighting the importance of Bcrp activity in limiting intestinal topotecan absorption (Jonker et al., 2002). In another study, the tissue-to-plasma ratios of topotecan were elevated 1.5-fold in Mdr1$^{a/b-/-}$ mice and 1.6-fold in Bcrp$^{-/-}$ mice, but were 12-fold higher in Mdr1$^{a/b-/-}$Bcrp$^{-/-}$ knockout mice (de Vries et al., 2007). These data illustrate that Pgp and BCRP can play a redundant role in limiting the brain penetration of drugs. Compared with wild-type mice, the AUC of sulfasalazine after either an oral or intravenous dose was considerably higher in Bcrp$^{-/-}$ mice, but not in Mdr1$^{a/b-/-}$ mice, suggesting that

sulfasalazine may be a suitable probe for assessing the DDI potential of drug candidates as BCRP inhibitors in humans (Zaher et al., 2006).

3.2.2.4 Clinical Studies
The BCRP gene is highly polymorphic, and in many cases single-nucleotide polymorphisms have been shown *in vitro* to influence protein levels, membrane expression, and/or BCRP transport activity (Mizuarai et al., 2004). The most well-studied single-nucleotide polymorphism in the gene encoding BCRP is the nonsynonymous Gln141Lys mutation, which occurs at high frequencies in Caucasians (10–15%) and Asians (25–35%) (Niemi, 2010). Plasma levels of the BCRP substrate diflometacan were elevated ~3-fold following its intravenous injection in patients with the Gln141Lys mutation, likely due to reduced BCRP-mediated renal and/or hepatic elimination of the drug (Sparreboom et al., 2004). Following oral administration, the mean plasma AUC of rosuvastatin and atorvastatin were substantially higher in individuals with the Gln141Lys mutation compared with individuals without the mutation (Keskitalo et al., 2009). In a retrospective study that analyzed data from 291 children with acute lymphoblastic leukemia that had received chemotherapy, patients with the Pgp 3435C>T mutation and patients with the BCRP Gln141Lys mutation had more frequent encephalopathy episodes than patients carrying the wild-type alleles (Erdilyi et al., 2008). Although not direct evidence, these data are consistent with increased exposure of the central nervous system of these patients to chemotherapeutics, and thus increased risk for central nervous system toxicity. Patients carrying both mutations had even more adverse events than patients with no or only one of the predisposing genotypes (Erdilyi et al., 2008). Like the mouse knockout studies, these data show that BCRP and Pgp can often compensate for either one's absence due to an overlap in substrate specificity and tissue-level expression.

3.2.3 MRP2 (ABCC2)

3.2.3.1 Tissue Expression
MRP2 was initially characterized in human and rat hepatocytes, and the human ortholog of MRP2 was cloned shortly thereafter from a cisplatin-resistant head and neck cancer KB cell line (Mayer et al., 1995; Taniguchi et al., 1996). MRP2 localizes to apical membranes of intestinal enterocytes, hepatocytes, and renal tubule cells (Jedlitschky et al., 2006) (Figure 3.1). Expression of MRP2 was not detected in isolated rat or healthy human brain microvessels (Kubota et al., 2006; Yousif et al., 2007). However, MRP2 expression was detected in brain microvessels isolated from human patients with epilepsy and rats with experimentally induced epilepsy

(van Vliet et al., 2005; Kubota et al., 2006). The up-regulation of MRP2 in the blood–brain barrier in patients with epilepsy may have implications toward treatment of epilepsy with pharmacotherapy.

3.2.3.2 Substrate/Inhibitor Specificity

MRP2 transports a wide range of endobiotic and xenobiotic compounds including glutathione, glucuronide, and sulfate conjugates (Konig et al., 1999). MRP2 substrates are largely anionic; however, MRP2 can also transport neutral and weakly basic drugs. Therapeutic substrates of MRP2 include doxorubicin, paclitaxel and methotrexate (anticancer), pravastatin (HMG-CoA reductase inhibitor), cefodizime and ironotecan (antibiotics), saquinavir, ritonavir, and indinavir (HIV protease inhibitors), and valsartan and olmesartan (angiotensin II receptor antagonists) (Huisman et al., 2002, 2005; Jedlitschky et al., 2006; Nakagomi-Hagihara et al., 2006; Yamashiro et al., 2006). MK571 is a commonly used inhibitor of MRP2, although it has also been shown to inhibit Pgp and BCRP with near equal potencies (IC_{50s} within one log unit) (Matsson et al., 2009).

3.2.3.3 Animal Models

Mutant transport-deficient (TR-) and Esai hyperbilirubinemic (EHBR) rats are deficient in Mrp2 and have been used extensively as a model of Dubin–Johnson syndrome, a human hereditary condition in which MRP2 is deficient. Numerous studies with these rats have helped assign a role for MRP2 in drug disposition and in particular its role in hepatic and renal drug elimination; several examples are mentioned below. EHBR rats showed reduced biliary excretion of the Mrp2 substrates olmesartan and methotrexate compared with control rats, signifying the importance of Mrp2 in hepatic elimination of these drugs (Chen et al., 2003; Takayanagi et al., 2005). The urinary and fecal excretion of methylmercury was lower in TR- rats compared with controls following intravenous administration of the heavy metal chelators 2,3-dimercapto-1-propanesulfonic acid and *meso*-2,3-dimercaptosuccinic acid, suggesting that Mrp2 transports methylmercury conjugates of the respective chelators and is relevant to their renal excretion (Zalups and Bridges, 2009).

In addition to the TR- and EHBR rats, Mrp2 knock-out mice (Abcc2$^{-/-}$) have been generated and have proven useful for examining Mrp2 transport activity *in vivo* (Chu et al., 2006). In the initial characterization of these mice, the biliary excretion of the organic anion and diagnostic of liver function bromosulfophthalein was suppressed in the knockout animals. With respect to dissecting contributions of the individual transporters, the Abcc2$^{-/-}$ mice have been particularly valuable when used in combination with knockout mice for other

ABC transporters, such as Pgp (see review by Kruh et al., 2007). For example, both Pgp and MRP2 transport the anticancer agent doxorubicin, and the biliary excretion of doxorubicin following its intravenous administration was reduced ~2-fold in Abcc2$^{-/-}$ mice, ~10-fold in Mdr1$^{a/b-/-}$ mice, and ~54-fold in mice lacking both proteins (Vlaming et al., 2006). Importantly, drug disposition data generated from knockout animal models must be interpreted with caution, as expression of other transporters and metabolizing enzymes with overlapping substrate specificities have been shown to be altered (Chen et al., 2005; Johnson et al., 2006; Vlaming et al., 2006).

3.2.3.4 Clinical Studies

Compared with Pgp and BCRP, relatively little is known about the role of MRP2 in drug disposition in humans, and most of what is known comes from observations in individuals with mutations in the gene encoding the transport protein. Dubin–Johnson syndrome is an autosomal recessive disorder caused by loss of function mutations in the MRP2 gene. The disorder is characterized by reduced biliary excretion of conjugated bilirubin resulting in hyperbilirubinemia. Dubin–Johnson syndrome has also been linked to impaired plasma clearance of other organic anions, including methotrexate and bromosulfophthalein (Abe and Okuda, 1975) (Hulot et al., 2005). In addition to the mutations in MRP2 associated with Dubin–Johnson syndrome, numerous single-nucleotide polymorphisms with varying degrees of effects on transport activity have been noted (Ieiri et al., 2009). Most notably is a 24C>T mutation in the 5′ untranslated region of MRP2. This mutation was associated with a twofold higher mean plasma methotrexate AUC in female pediatric patients with acute lymphoblastic leukemia receiving intravenous methotrexate therapy (Rau et al., 2006). In another study, individuals with the 24C>T MRP2 variant developed diclofenac-related hepatotoxicity at higher frequencies, likely due to impaired MRP2-mediated efflux of diclofenac acyl glucuronide (Daly et al., 2007).

3.3 SLC TRANSPORTERS

Drug transporters in the SLC family comprise both uptake and efflux transport proteins. Unlike the ABC efflux transporters, transport mediated by the SLC family members is not linked directly to ATP hydrolysis. Among these transport proteins include those that operate via electrogenic facilitated diffusion and exchange. The energetics driving drug uptake or efflux mediated by these transporters will be discussed below, as it is relevant to the role these transporters play in drug disposition.

3.3.1 OCT1 (SLC22A1) and OCT2 (SLC22A2)

3.3.1.1 Tissue Expression and Transport Mechanism

OCT1 was the first of the OCT transporters to be cloned, and in the human OCT1 is largely expressed in the sinusoidal membrane of hepatocytes, with little expression in the kidney (Figure 3.1); however, species differences in OCT1 distribution have been described (see below) (Gründemann et al., 1994; Gorboulev et al., 1997; Meyer-Wentrup et al., 1998; Motohashi et al., 2002). Conversely, OCT2 expression appears to be specific to renal tubule cells where it is localized in basolateral membranes (Motohashi et al., 2002).

The OCTs are electrogenic facilitated diffusion transport proteins that transport relatively small monovalent organic cations. The cell interior is electronegative relative to the outside primarily due to the activity of Na^+,K^+-ATPase and K^+ channels. OCTs are uptake transporters, driven by the inside negative membrane potential. At a typical transmembrane potential (-60 to $-70\,mV$), the intracellular concentration of a cationic drug would be 10–15-fold higher inside the cell due to the activity of OCTs.

3.3.1.2 Substrate/Inhibitor Specificity

As already alluded to, the OCTs transport relatively small (<500 MW) organic cations, including fixed cations as well as weak bases; these have historically been referred to as Type I organic cations (Wright, 2005). The interaction of weak bases as inhibitors and substrates of OCT2 has been shown to be dependent on their degree of ionization; weak bases are transported and/or inhibit more efficiently at lower pH values (Barendt and Wright, 2002). The substrate/inhibitor specificity of the human orthologs of OCT1 and OCT2 has been described in detail elsewhere (see Koepsell, 2004; Koepsell et al., 2007; Nies et al., 2011). Drugs of therapeutic relevance that have been shown to interact as substrates/inhibitors of OCT1 and OCT2 include quinine (antimalarial), procainamide (Na^+ channel blocker), metformin (antidiabetic), cisplatin (anticancer), cimetidine, and clonidine (α-adrenergic agonist). Commonly used inhibitors of OCT1 and OCT2 include decynium 22, 1-methyl-4-phenylpyridinium, tetraethylammonium, and tetrapentylammonium.

3.3.1.3 Animal Models

Mice deficient in Oct1 (Oct1$^{-/-}$), Oct2 (Oct2$^{-/-}$), and both genes (Oct1/2$^{-/-}$) have been generated and shown to be useful for examining the role of OCTs in hepatic and renal elimination of cationic drugs. Levels of the model OCT substrate tetraethylammonium in the livers of wild-type mice were ~11-fold higher than plasma, but this fold difference was only ~2 in Oct1$^{-/-}$ mice (Jonker et al., 2001). Concomitant with the reduction in hepatic tetraethylammonium accumulation was a decrease in its biliary clearance in Oct1$^{-/-}$ mice. Whereas OCT1 is not expressed at appreciable levels in human kidney, its expression in the kidneys of rodent species is functionally significant. For example, the renal excretion of the OCT substrate cisplatin was significantly reduced in Oct1/2$^{-/-}$ mice, but not in mice deficient in either Oct1 (Oct1$^{-/-}$) or Oct2 (Oct2$^{-/-}$), and animals deficient in both transporters were less susceptible to cisplatin-induced nephrotoxicity, highlighting the importance of OCTs in the renal uptake of cisplatin and its subsequent elimination (Filipski et al., 2009). In another study, the pharmacokinetics of tetraethylammonium in Oct2$^{-/-}$ mice was not altered compared with wild-type mice, but the renal clearance of tetraethylammonium was reduced ~2.5-fold in Oct1/2$^{-/-}$ mice (Jonker et al., 2003). Oct1 has also been suggested to be expressed at the basolateral membrane of mouse intestinal enterocytes, and accordingly, direct intestinal secretion of tetraethylammonium was reduced ~2-fold in Oct1$^{-/-}$ mice (Jonker et al., 2001). The abundance of OCT1 in the gastrointestinal tract of humans appears to be relatively low, and thus it is not clear to what extent intestinally expressed OCT1 may influence drug disposition in humans (Hilgendorf et al., 2007).

3.3.1.4 Clinical Studies

Numerous nonsynonymous mutations in the genes encoding human OCT1 and OCT2 have been identified, and many have been shown in vitro to alter transport function (Takane et al., 2008; Zair et al., 2008). Metformin pharmacokinetics, most notably the plasma AUC and maximal plasma clearance, were significantly higher in individuals carrying a reduced function OCT1 allele, an effect likely caused by impaired hepatic metformin uptake, and thus its hepatobiliary excretion (Shu et al., 2008). Importantly, the glucose-lowering effects of metformin were blunted in individuals carrying reduced-function OCT1 alleles, linking the transporter to metformin pharmacodynamics (Shu et al., 2007). The nonsynonymous mutation, Ala270Ser in OCT2, which has a high allelic frequency (~10%), was found to result in enhanced transport activity when expressed heterologously in Madin–Darby canine kidney (MDCK) cells. Consistent with the in vitro results, the renal clearance of metformin, and in particular the renal tubular secretion of the drug, was greater in healthy volunteers of Caucasian or African-American ancestry that were heterozygous for the variant OCT2 allele (808G/T) compared with individuals homozygous for the reference allele (808G/G) (Chen et al., 2009a). For reasons that are unclear, studies in healthy volunteers from Asian populations with the heterozygous variant showed reduced, as opposed to enhanced, clearance of metformin (Song et al., 2008; Wang et al., 2008). Consistent with the 808G/T mutation resulting in reduced OCT2 function *in vivo*, the mutation was associated with reduced cisplatin-induced nephrotoxicity in patients (Filipski et al., 2009).

3.3.2 MATE1 (SLC47A1) and MATE2K (SLC47A2)

3.3.2.1 Tissue Expression and Transport Mechanism

Experiments largely using brush-border membrane vesicles isolated from hepatocytes and renal tubule cells had identified electroneutral organic cation/H^+ exchange as the dominant transport process suspected of mediating the final step in hepatic and renal tubular secretion of Type I organic cations (Oude Elferink et al., 1995; Pritchard and Miller, 1996; Wright and Dantzler, 2004). The molecular identity of the apically located organic cation/H^+ exchanger was recently identified when the human orthologs of MATEs were cloned by screening the human genome for sequences similar to MATE-type transporters from bacteria (Otsuka et al., 2005). Whereas MATE1 localizes to apical membranes of hepatocytes and renal tubule cells, MATE2K expression is restricted to the apical membranes of renal tubule cells (Otsuka et al., 2005; Masuda et al., 2006) (Figure 3.1). Thus, as the earlier studies with native tissues had demonstrated, the MATEs are suspected of playing an important role in the final step in hepatic and renal tubular secretion of Type I organic cations, efflux across apical membranes for final elimination into bile and urine, respectively. Because they are electroneutral transporters, the transmembrane electrical potential does not directly impact their transport activity. In contrast, transport mediated by the MATEs is driven by the substrate (i.e., organic cation) and/or pH gradients across the plasma membrane. Because the MATEs were only recently identified at the molecular level, relatively little is known regarding their substrate/inhibitor specificity and involvement in drug disposition.

3.3.2.2 Substrate/Inhibitor Specificity

The MATEs interact with relatively small (<500 MW) organic cations (Type I), although some organic anions such as estrone-3-sulfate have been identified as substrates (Tanihara et al., 2007). Like the OCTs, tetraethylammonium and 1-methyl-4-phenylpyridinium appear to be very good substrates of MATE1 and MATE2K, and are obvious choices for assessing their activities *in vitro* (Tanihara et al., 2007). Clinically relevant substrates of MATE1 and MATE2K include cimetidine, metformin, procainamide (Na^+ channel blocker), and cisplatin and carboplatin (anticancer) (Yonezawa et al., 2006; Tanihara et al., 2007). Creatinine, an endogenous metabolite of creatine, which is used as a marker of glomerular filtration

rate in humans, is also a substrate of MATE1 and MATE2K, as well as OCT2 (Urakami et al., 2004; Tanihara et al., 2007). Active tubular secretion of creatinine mediated by OCTs and MATEs likely explains why renal creatinine clearance can overestimate true glomerular filtration rates in humans (Shannon, 1935; Miller and Winkler, 1938; Miller et al., 1952). Inhibiting active tubular secretion of creatinine with OCT2 and/or MATE inhibitors, such as cimetidine, would likely result in a more reliable estimate of glomerular filtration rate. Pyrimethamine is a recently identified potent and relatively specific inhibitor of MATE1 ($K_i = 145$ nM) when compared with OCTs (Ito et al., 2010).

3.3.2.3 Animal Models

Consistent with the involvement of MATE1 in the renal handling of creatinine, blood creatinine levels were significantly elevated in Mate1$^{-/-}$ mice compared with wild-type controls (Tsuda et al., 2009). The mRNA levels of Oct1 and Oct2 in the kidneys and liver of Mate1$^{-/-}$ mice was not different from wild-type mice, whereas the accumulation of metformin in the kidney and liver following a single intravenous administration was significantly higher in Mate1$^{-/-}$ mice. The plasma AUC of metformin was higher and the amount of metformin excreted in the urine via tubular secretion was lower in Mate1$^{-/-}$ mice, suggesting an important role for the transport protein in the renal elimination of the antidiabetic drug.

3.3.2.4 Clinical Studies

A number of nonsynonymous single-nucleotide polymorphisms in the coding region of MATE1 and MATE2K have been identified, with most of the polymorphisms causing a reduction to complete loss of transport activity *in vitro* (Kajiwara et al., 2009; Chen et al., 2009b). For example, the MATE1 Gly64Asp and Val480Met variants, and MATE2K Gly-211Val variant, were found to be nonfunctional, an effect largely attributed to loss of membrane expression (Kajiwara et al., 2009; Chen et al., 2009b). Individuals with the rs2289669 (G>C) single-nucleotide polymorphism in an intronic region on MATE1 responded better to the glucose-lowering effects of metformin, suggesting a reduction in activity of the MATE1 transport protein; although the mechanism behind the supposed altered transport activity was not clear (Becker et al., 2009). Selective inhibition of MATE1 in cells doubly transfected with MATE1 and OCT2 causes reduced transepithelial transport and cellular accumulation of OCT2/MATE1 substrates, and demonstrates how inhibition of efflux *in vivo* could lead to a DDI, that is, reduced hepatic and/or renal elimination and increased hepatic and/or renal accumulation of potentially cytotoxic drugs (Meyer zu Schwabedissen et al., 2010).

3.3.3   OAT1 (SLC22A6) and OAT3 (SLC22A8)

3.3.3.1   Tissue Expression and Transport Mechanism

OAT1 and OAT3 are organic anion transporters predominantly expressed in the basolateral membranes of renal tubule cells (Figure 3.1). OAT1 and OAT3 are anion exchangers, and the primary driving force suspected of mediating drug uptake into tubule cells via these transport proteins is an outwardly directed gradient for α-ketoglutarate, a Kreb's cycle intermediate (Wright and Dantzler, 2004). The outwardly directed α-ketoglutarate gradient is maintained in part by the activity of basolateral (NaDC1; SLC13A2) and apical (NaDC3; SLC13A3) Na$^+$-dependent dicarboxylate cotransporters that get their energy (in part) from the Na$^+$ gradient established by Na$^+$,K$^+$-ATPase. OAT1 and OAT3 likely operate with a 1:1 stoichiometry, and thus depending on the substrate being transported, the transport process could either be electrogenic (monovalent organic anions) or electroneutral (divalent organic anions) (Aslamkhan et al., 2003). The electrogenic exchange of intracellular α-ketoglutarate (net charge of −2) for an extracellular monovalent organic anion (net charge of −1) via OAT1 or OAT3 would be further stimulated by the inside negative membrane potential of the cell.

3.3.3.2   Substrate/Inhibitor Specificity

OAT1 and OAT3 transport relatively small (<500 MW) organic anions, although they can also transport a limited number of neutral drugs (Wright and Dantzler, 2004; Rizwan and Burckhardt, 2007). The mouse orthologs of Oat1 and Oat3 have recently been shown to be inhibited by several organic cations, and mouse Oat3 was additionally shown to transport the fixed cation 1-methyl-3-phenylpyridinium (Ahn et al., 2009). Due to the efficiency with which it is cleared at the kidney, in part due to active secretion by renal tubule cells, *para*-aminohippurate is used as a diagnostic marker of renal plasma flow, and has served as a prototypic substrate/inhibitor of OAT1; *para*-aminohippurate is only weakly transported by OAT3 (Wright and Dantzler, 2004). OAT1 substrates of clinical significance include many antiviral drugs such as cidofovir, adefovir, acyclovir and zidovudine (AZT), the H$_2$ histamine receptor antagonists cimetidine and ranitidine, and the loop diuretics furosemide and bumetanide (Rizwan and Burckhardt, 2007). Several nonsteroidal anti-inflammatory drugs (NSAIDs), such as ibuprofen and indomethacin, have also been identified as OAT1 substrates *in vitro*. However, these drugs are sufficiently hydrophobic, and thus OAT1 activity is unlikely to be a rate-limiting step in their renal clearance. In addition to many drugs in the above classes being substrates of OAT1, they can also act as inhibitors. Estrone-3-sulfate, a metabolite of the hormone estrone, is a prototypic substrate/inhibitor of OAT3; estrone-3-sulfate is only weakly transported by OAT1 (Wright and Dantzler, 2004). OAT3 also has the capacity to interact with antivirals, H$_2$ histamine receptor antagonists, loop diuretics, and NSAIDs, as well as the β-lactam antibiotic, benzylpenicillin (Rizwan and Burckhardt, 2007). The uricosuric probenecid is a common inhibitor of OAT1 and OAT3 and has been used *in vivo* to prevent nephrotoxic substrates of OATs from gaining access into the cytosol of renal tubule cells (see below). Probenecid is also the inhibitor of choice to use for clinical DDI studies aimed at assessing the impact of OAT1 and OAT3 on the renal elimination of drug candidates (Giacomini et al., 2010).

3.3.3.3   Animal Models

Oat1$^{-/-}$ mice exhibited diminished renal excretion of *para*-aminohippurate and the loop diuretic furosemide compared with wild-type mice (Eraly et al., 2006). In this study, the renal clearance of *para*-aminohippurate in Oat1$^{-/-}$ mice was near equivalent to the renal clearance of inulin, a marker of glomerular filtration rate, signifying the essential role Oat1 plays in the renal tubular secretion of *para*-aminohippurate. The diuretic response to furosemide was blunted in Oat1$^{-/-}$ mice, presumably due to reduced tubular secretion of the drug, demonstrating the importance of Oat1 transport activity to the pharmacodynamic effect of the diuretic (Eraly et al., 2006). The plasma clearance of benzylpenicillin and methotrexate, both known substrates of OAT3, were reduced in Oat3$^{-/-}$ mice (Vanwert et al., 2007; Vanwert and Sweet, 2007).

3.3.3.4   Clinical Studies

Several nonsynonymous single-nucleotide polymorphisms in the genes encoding OAT1 and OAT3 have been identified in various ethnic groups (Bleasby et al., 2005; Fujita et al., 2005; Erdman et al., 2006). However, it is not clear what impact, if any, these mutations have on the disposition of OAT substrates. Notable mutations in OAT3 resulting in a complete loss of transport function *in vitro* include Arg-149Ser, Gly239Stop and Iso260Arg, and the Iso305Phe mutation, which is found in 3.5% of Asian-Americans, exhibits an altered substrate specificity (Erdman et al., 2006). The Arg454Gln mutation in OAT1 was shown to be nonfunctional in *Xenopus laevis* oocytes, but despite this observation, the renal secretory clearance (total clearance minus glomerular filtration) of the antiviral drug and OAT1 substrate adefovir was unchanged in individuals heterozygous for the mutation (Fujita et al., 2005). Possible reasons suggested for the lack of change in the renal secretory clearance of adefovir were that (1) sufficient OAT1 activity may have been present in the heterozygous individuals since they only had one nonfunctional allele, (2) the apical efflux of adefovir may be the rate-limiting step in tubular secretion, and/or (3) other transporters may compensate for loss/reduced OAT1 activity (Fujita et al., 2005). However, OAT3 is not expected to play a compensatory role in the renal tubular secretion of adefovir since it is a relatively poor substrate of this transport protein (Fujita et al., 2005).

OAT1 and OAT3 have also been implicated in adverse events and DDIs at the level of renal tubular secretion. The antiviral drugs adefovir and cidofovir are transported by OAT1, and heterologous expression of OAT1 renders cells more susceptible to cytotoxicity from these drugs (Cihlar et al., 1999). Indeed, nephrotoxicity is the dose-limiting toxicity for adefovir and cidofovir in the clinic, presumably through the uptake activity of OAT1 at the basolateral membrane of renal tubule cells (Lalezari et al., 1997; Cundy, 1999). Moreover, the organic anion transport inhibitor probenecid has been used clinically to reduce the potential for adefovir and cidofovir to cause nephrotoxicity (Cundy, 1999). Concomitant methotrexate and NSAID administration can cause a pharmacokinetic drug interaction leading to increased plasma levels of methotrexate and an increased potential for toxicity (Johnson et al., 1993). A good correlation between NSAID inhibition of OAT3-mediated methotrexate uptake in heterologous cells *in vitro* and a reduction in renal methotrexate clearance *in vivo* provides support for the contention that the drug interaction between NSAIDs and methotrexate occurs at the level of OAT3-mediated transport (Maeda et al., 2008).

3.3.4 OATP1B1 (SLCO1B1, SLC21A6), OATP1B3 (SLCO1B3, SLC21A8), and OATP2B1 (SLCO2B1, SLC21A9)

3.3.4.1 Tissue Expression and Transport Mechanism
The OATP family is composed of 11 members. OATP1B1, OATP2B1, and OATP1B3 are expressed in sinusoidal (basolateral) membranes of human hepatocytes, and at least OATP1B1 and OATP1B3 are considered relevant to hepatic elimination of therapeutic drugs (Smith et al., 2005; Niemi, 2007; Hagenbuch and Gui, 2008; Kalliokoski and Niemi, 2009; Fahrmayr et al., 2010; Giacomini et al., 2010). While expression of OATP1B1 and OATP1B3 appear to be liver specific, OATP2B1 is also expressed in apical membranes of intestinal enterocytes and brain capillary endothelial cells (Kobayashi et al., 2003; Bronger et al., 2005). The most widely investigated of the three proteins is OATP1B1, which is evident from the number of studies demonstrating the ability of OATP1B1 to transport various xenobiotics and endobiotics. OATP1B1 is expressed in both basal and lateral membranes of hepatocytes, and using immunofluorescence microscopy,

Konig et al. (2000) showed that OATP1B1 staining was more abundant in hepatocytes near the central vein versus those close to the portal vein. Similarly, OATP1B3 expression was highest in hepatocytes around the central vein when assessed by immunohistochemistry (Abe et al., 2001).

The energetic mechanism of OATP-mediated transport is not well understood, but appears to occur via Na^+-independent anion exchange (Kullak-Ublick et al., 1995; Noe et al., 1997; Abe et al., 1999; Walters et al., 2000). When expressed in oocytes, OATP1B1- and OATP1B3-mediated uptake of taurocholate (OATP1B3) and estrone-3-sulfate (OATP1B1) were *trans*-stimulated by preloading the oocytes with several organic anions, including estradiol 17β-*D*-glucuronide, taurocholate, and estrone-3-sulfate (Mahagita et al., 2007). However, sulfate, bicarbonate, and glutathione failed to *trans*-stimulate uptake activity of OATP1B1 and OATP1B3 (Mahagita et al., 2007). Uptake activity of OATP1B1 and OATP1B3 was not altered by manipulation of membrane potential (by changing extracellular potassium levels), suggesting that the transport mechanism is electroneutral (Mahagita et al., 2007). However, using oocytes expressing OATP1B1 and OATP1B3 under voltage-clamped conditions, Martinez-Becerra et al. (2010) showed that an outward current was generated when substrate was added to the extracellular buffer, which is consistent with electrogenic transport. Several studies have shown that manipulation of pH can influence OATP transport activity, but the mechanism by which this occurs is currently unknown (Kobayashi et al., 2003; Mahagita et al., 2007; Leuthold et al., 2009; Martinez-Becerra et al., 2010). Importantly, the endogenous organic anion exiting hepatocytes in exchange for extracellular OATP substrates remains elusive.

3.3.4.2 Substrate/Inhibitor Specificity
In general, the OATPs transport a wide variety of endogenous and exogenous amphipathic organic molecules, many of which are organic anions at physiological pH. Endogenous substrates of OATPs include hormones and hormone conjugates, such as thyroxine, estrone-3-sulfate, dehydroepiandrosterone sulfate and estradiol 17β-*D*-glucuronide, eicosanoids (leukotriene C4), bile acids (taurocholate), as well as bilirubin glucuronide (Abe et al., 1999; Hagenbuch and Meier, 2004; Smith et al., 2005; Hagenbuch and Gui, 2008; Hagenbuch, 2010). Therapeutic substrates of OATPs occur in many different classes, including HMG-CoA reductase inhibitors (atorvastatin, rosuvastatin, pravastatin, etc.), antibiotics (rifampicin), angiotensin II receptor antagonists (valsartan, olmesartan, and telmisartan), anticancer drugs (methotrexate, paclitaxel, and SN-38), antihistamines (fexofenadine), and antidiabetic drugs (repaglinide)

(Smith et al., 2005; Niemi, 2007; Kalliokoski and Niemi, 2009; Fahrmayr et al., 2010). Compared with OATP1B1 and OATP1B3, the substrate specificity of OATP2B1 has been suggested to be more limited (Hagenbuch and Meier, 2004).

Due to the overlapping substrate specificity of OATP1B1 and OATP1B3, and in some cases OATP2B1, and their colocalization in sinusoidal membranes of human hepatocytes, a challenge faced by the pharmaceutical industry is how to determine the respective contribution of each transporter to hepatic drug uptake. Indeed, efforts to identify substrates and inhibitors that exclusively interact with a single hepatic OATP have met limited success. For example, the OATP1B1 substrates estrone-3-sulfate and estradiol 17β-*D*-glucuronide are also transported by and inhibit OATP2B1 and OATP1B3, respectively, and bromosulfophthalein is a substrate of all three hepatic OATPs (Kullak-Ublick et al., 2001; Smith et al., 2005). However, OATP1B3 does appear to display a preference for several substrates, including digoxin and cholecystokinin 8 (Ismair et al., 2001; Kullak-Ublick et al., 2001). Using a 96-well fluorescence drug-based plate assay with fluorescein-methotrexate as a probe substrate, Gui et al. (2010) showed that OATP1B1 and OATP1B3 display differential selectivity toward select compounds. For example, estropipate selectively inhibited OATP1B1 (IC_{50} = 0.06 μM vs. 19.3 μM for OATP1B3), while ursolic acid was selective against OATP1B3 (IC_{50} = 2.3 μM vs. 12.5 μM for OATP1B1). Recent studies examining the inhibitory interaction of green tea catechins with OATPs highlight the complexity of OATP substrate binding surface (Roth et al., 2011). In these studies, epicatechin gallate and epigallocatechin gallate inhibited OATP2B1 and OATP1B1 in a concentration-dependent manner, whereas their effect on OATP1B3 was substrate dependent; epigallocatechin gallate stimulated OATP1B3-mediated estrone-3-sulfate transport at low concentrations, did not alter estradiol 17β-*D*-glucuronide transport, and noncompetitively inhibited fluo-3 transport. These data suggest that the OATP1B3 (and potentially the other OATPs) have multiple substrate binding sites. These observations have clear implications for programs that use only one probe substrate for assessing inhibition of OATPs.

3.3.4.3 Animal Models

Species differences in OATPs are apparent, questioning the use of preclinical species for predicting the role of OATPs in the hepatic disposition of therapeutics in humans. In humans, the OATP1B subfamily contains two members, OATP1B1 and OATP1B3, whereas rodents contain a single ortholog (Oatp1b2) that has functional characteristics common to both human orthologs; gene duplication in the OATP1B subfamily occurred after divergence from rodents (Hagenbuch and Meier, 2004). Human OATP2B1 has a single ortholog in rodents (Oatp2b1), with the human and rat orthologs sharing 77% sequence identity. However, rat Oatp2b1 appears to have a broader substrate specificity compared with human OATP2B1 (Hagenbuch and Meier, 2003).

Despite species differences in hepatic OATPs, mice deficient in Oatp1b2 have been generated, and given that Oatp1b2 has a substrate specificity common to both OATP1B1 and OATP1B3, they may be useful for predicting the role of OATP1B1 and OATP1B3 in drug disposition in humans (Chen et al., 2008; Zaher et al., 2008). Intravenous administration of rifampicin (a substrate common to Oatp1b2, OATP1B3, and OATP1B1) led to a 1.7-fold increase in the plasma AUC with a concomitant decrease in the liver-to-plasma ratio in Oatp1b2$^{(-/-)}$ mice compared with wild-type mice (Zaher et al., 2008). In addition, a transgenic mouse that expresses human OATP1B1 has been generated, and was used to examine the disposition of methotrexate, a substrate common to OATP1B1 and OATP1B3 (van de Steeg et al., 2009). The plasma AUC of methotrexate was lower and liver-to-plasma concentration ratio higher in the transgenic mice compared with wild-type mice, suggesting that OATP1B1 plays an important role in hepatic methotrexate uptake for subsequent metabolism and/or biliary clearance in humans.

3.3.4.4 Clinical Studies

A number of mutations, with varying functional consequences *in vitro*, have been indentified in OATP1B1, OATP1B3, and OATP2B1 (Tamai et al., 2000; Iida et al., 2001; Tirona et al., 2001; Michalski et al., 2002; Nozawa et al., 2002; Letschert et al., 2004). However, *in vivo* consequences of these mutations regarding drug disposition is only conclusive for OATP1B1. The impact of a number of OATP1B1 polymorphisms on drug pharmacokinetics and pharmacodynamics has been reviewed in detail elsewhere (Fahrmayr et al., 2010). In particular, the single-nucleotide polymorphisms Asn130Asp (OATP1B1*B) and Val174Ala (OATP1B1*5) occur with high allelic frequencies (Caucasian, African-American, and Japanese populations were assessed), and can occur either alone or in combination (the two alleles combined is denoted by OATP1B1*15) (Romaine et al., 2010). With the OATP1B1*15 mutation, the *5 allele appears to have a dominant effect on transport function. When expressed in heterologous expression systems, the OATP1B1*5 mutant showed reduced transport function that was caused by a reduced maximal rate of transport (J_{max}), and consistent with the reduction in J_{max}, plasma membrane expression of the *5 mutant was reduced (Kameyama et al., 2005). The AUC of repaglinide was significantly higher in individuals carrying both *5 alleles compared with individuals carrying either one *5 allele or individuals with two reference alleles (Niemi et al., 2005). Additionally, the blood glucose-lowering effect of repaglinide was more pronounced in individuals carrying both *5 alleles (Niemi et al., 2005). Systemic exposure to simvastatin acid is also demonstrably higher in individuals with both *5 alleles, and these individuals are at greater risk of simvastatin-induced myopathy (Link et al., 2008; Niemi, 2010). The importance of OATP1B1 to hepatic statin uptake and clearance, and the prevalence and functional consequence of the *5 allele, has led to statin dosage recommendations (U.S. Food and Drug Administration) based on OATP1B1 genotype (Niemi, 2007).

A number of studies have also suggested DDIs to occur at the level of OATP1B1 inhibition of hepatic drug uptake (as reviewed by (Smith et al., 2005; Niemi, 2007; Fahrmayr et al., 2010). In these cases, administration of known OATP inhibitors, such as cyclosporine, gemfibrozil, and rifampicin have been shown to increase the plasma exposure of OATP substrates (e.g., statins) in humans. However, since many of the OATP inhibitors and substrates examined also interact (as substrates/inhibitors) with hepatic drug metabolizing enzymes, it is difficult to determine the contribution of inhibition of transport versus metabolism to the observed increase in drug exposure.

3.4 *IN VITRO* ASSAYS IN DRUG DEVELOPMENT

The involvement of drug transporters in adverse reactions and DDIs, as well as the observation that polymorphisms in drug transporters can influence drug disposition, makes it important to identify the specific transporter(s) involved in the disposition of a drug candidate. In this regard, it is not only important to assess whether the drug candidate is a substrate of a specific transporter, but also whether it can act as an inhibitor. Importantly, as the number of transporters involved in a drug's disposition increases, the likelihood for transporter polymorphisms or inhibition to influence its overall disposition decreases. Given the accumulating data suggesting their involvement in drug disposition, the aforementioned transporters should be considered when developing drug candidates. The assays described here are not an exhaustive list of assays available for studying drug transporters, but rather, the *in vitro* assays most often used in later-stage drug development when radiolabeled compounds are available. The assays described are conducted to assist in transitioning studies from preclinical species to first-in-human. Additionally,

these assays are used to support the clinical plan for candidate drugs, in particular, recommendations for *in vivo* DDI studies.

A number of *in vitro* models have been developed to study drug transporters, and the ones most often used in drug development are discussed below. The human colorectal carcinoma cell line, Caco-2, expresses several drug transporters that localize to luminal membranes of human intestinal enterocytes, and has been used as a model system for predicting drug permeability across the human intestine, including the influence of transporters on this process (Sun et al., 2008). Primary hepatocytes in several configurations (plated, suspension, and sandwich culture) have been used for assessing hepatic drug transport. Currently, there is a lack of an equivalent primary culture model for assessing renal drug transport. That is, it is comparatively difficult to isolate and/or maintain differentiated human renal proximal tubule cells in primary culture. However, renal transporters, as well as the other drug transporters, can be cloned and studied in heterologous expression systems. Each one of the systems described below has advantages and disadvantages, with no one system proving to be superior in all aspects. Clearly, the use of multiple assay formats and matrices is a contributor to the inability to clearly associate transport processes and their perturbation with clinical outcomes. Efforts to standardize the conduct of these studies by pharmaceutical practitioners are likely to aid in such associations.

3.4.1 Considerations for Assessing Candidate Drugs as Inhibitors

There are several issues to consider when assessing the inhibition potential of a candidate drug. Foremost, the specific transporters or tissues assessed should be determined by the clearance pathway of the coadministered drug. For example, if the coadministered drug is mainly excreted in the urine via active transport (renal clearance > glomerular filtration) then an assessment of the potential for the candidate drug to inhibit renal drug transporters should be a priority, whereas the potential for inhibition of hepatic transporters is likely inconsequential on the pharmacokinetics of the victim comedication. Since it is not known a priori which marketed drugs a candidate drug will be coadministered with, it is necessary to comprehensively assess the inhibitory potential of a drug candidate against the transporters expected to be most clinically relevant.

3.4.2 Considerations for Assessing Candidate Drugs as Substrates

The physicochemical properties of a candidate drug, the administration route, and the pathway(s) by which it is cleared should "ideally" drive the *in vitro* studies that are conducted to assess its interaction as a substrate of drug transporters (transporter phenotyping). For example, intestinal transporters are most relevant to orally administered drugs, and drug transporters in the intestine, and possibly the liver inlet, are unlikely to influence the overall flux of compounds that are both highly permeable and highly soluble (Biopharmaceutical Classification System Class I). Similarly, intestinal transporters are unlikely to influence the absorption of highly permeable compounds with limited solubility (Class II) provided that the free drug concentration is high enough to fully saturate the substrate binding site. In contrast, drug transporters are much more likely to influence the disposition of candidate drugs that are characterized as Biopharmaceutical Classification System Classes III (low permeability/high solubility) and IV (low permeability/high solubility) drugs. Finally, studies to assess transporters are likely to be warranted when the majority of an administered drug is found to be excreted as parent (nonmetabolized) or when transport activity can influence the amount of drug available for clearance by intracellular drug metabolizing enzymes (Benet, 2009).

3.4.3 Assay Systems

3.4.3.1 Caco-2
Caco-2 cells have been used in drug development to predict intestinal permeability of drug candidates and the fraction of an oral dose absorbed. Caco-2 cells have also been used to determine the potential for drug candidates to be substrates and inhibitors of drug transporters. For these studies, Caco-2 cells are seeded onto a permeable support membrane and cultured for a period of time ranging from 7 to 28 days to allow for the formation of a polarized monolayer with tight junctions and the expression of relevant transporters (Hidalgo et al., 1989; Hilgers et al., 1990). Pgp, MRP2, and BCRP are efflux transporters common to apical membranes of Caco-2 cells and intestinal enterocytes (Hunter et al., 1993; Gutmann et al., 1999; Xia et al., 2005) (see Figure 3.1). Compounds of interest are applied to either side of the cell monolayer in an apparatus that allows for the measurement of the unidirectional fluxes (basolateral-to-apical flux and apical-to-basolateral flux) across the monolayer. Due to the expression of efflux transporters in the apical membrane, the flux of an efflux transporter substrate would be higher in the basolateral-to-apical direction compared with that in the opposite direction. Inhibition of the efflux transporter by drug candidates or specific transport protein inhibitors would result in the individual unidirectional fluxes approaching unity. The reader is directed to a review by Hubatsch et al. (2007)

for a detailed description of the assay design. Similar assays using other cell lines such as MDCK cells and the porcine kidney cell line (LLC-PK1) are used and have an advantage since they can heterologously express a single efflux transporter, as opposed to the suite of transporters found in Caco-2 cells. Data generated from Caco-2 studies have been used to accurately predict the extent of intestinal absorption of some drug candidates, especially those that have a high apparent passive permeability, where transporters and diffusion across tight junctions are relatively unimportant to total flux (Sun et al., 2008). However, for drug candidates whose net flux across the monolayer is influenced to a large extent by passive diffusion across tight junctions and the involvement of active transport, variability in tight junctional resistance and the expression level of drug transporters/metabolizing enzymes in Caco-2 cell cultures can lead to poor predictions of fraction absorbed (Sun et al., 2008).

3.4.3.2 Primary Hepatocytes in Suspension
Hepatocytes have a tendency to lose expression of drug transporters after they have been plated on standard two-dimensional supports, such as plastic (Kukongviriyapan and Stacey, 1989; Luttringer et al., 2002; Richert et al., 2006). Conversely, Richert et al. (2006) have found that the expression of select hepatic uptake transporters in freshly isolated and cryopreserved hepatocytes is representative of their expression in the native tissue. Once isolated, however, hepatocytes lose their polarity and the efflux transporters that are typically expressed in apical membranes of hepatocytes *in vivo* are rapidly internalized (Bow et al., 2008). Accordingly, hepatocytes in suspension are most suited for transport studies assessing transporter (i.e., OATP1B1, OATP1B3, OATP2B1, and OCT1) involvement in candidate drug uptake across sinusoidal membranes, or alternatively, inhibition of sinusoidal uptake transporters by candidate drugs.

A methodology using a centrifugation step through a layer of mineral oil has been developed for hepatocytes in suspension, and the reader is directed to a review by Maeda and Sugiyama (2010) for details on assay design and applications. In this assay, uptake is initiated by mixing the hepatocytes with a transport solution containing a compound of interest in the absence or presence of transport protein inhibitors for a predetermined amount of time. After the uptake period, the entire mixture is transferred to a microcentrifuge tube containing mineral oil layered over NaOH. After layering the transport solution containing the hepatocytes on top of the mineral oil, the entire contents are rapidly centrifuged such that the hepatocytes move through the oil layer (leaving behind the transport buffer) into the

NaOH, where they are lysed. The microcentrifuge tubes are then snap-frozen and the tubes cut at the oil/NaOH interface, and the lysed cells (in the NaOH) are analyzed for the amount of compound that was taken up into the hepatocytes. Transport experiments are routinely conducted at 37°C versus 4°C in order to determine the amount of uptake due to active transport versus passive diffusion (including nonspecific binding); however, passive diffusion can be altered by temperature (Poirier et al., 2008). A cocktail of inhibitors designed to completely inhibit all possible uptake transporters involved may be an alternative approach for determining active versus passive uptake into hepatocytes in suspension. When assessing the potential for a drug candidate to inhibit a specific hepatic uptake transporter, the transport solution would contain a specific probe substrate of a hepatic uptake transporter (e.g., 1-methyl-4-phenylpyridinium as a probe substrate of OCT1; see Table 3.2) in the presence of increasing concentrations of a drug candidate. When assessing the potential for a drug candidate to be a substrate of a specific hepatic uptake transporter, the transport solution would contain the drug candidate along with a specific inhibitor of the transporter of interest. Caution should be used when interpreting transporter phenotyping and inhibition data from these experiments since suspension hepatocytes express a suite of transporters with a tendency for overlapping substrate/inhibitor specificities. Regardless, hepatocytes in suspension have been used to accurately predict uptake transporter involvement in hepatic clearance of drugs (Watanabe et al., 2009, 2010).

3.4.3.3 Primary Hepatocytes in a Sandwich Configuration

Unlike hepatocytes grown on traditional two-dimensional supports, hepatocytes grown in a three-dimensional matrix between collagen maintain polarity, reorganize to form canalicular networks, and retain many physiological and pharmacological functions of hepatocytes *in vivo* (Dunn et al., 1989; Berthiaume et al., 1996). A number of drug transporters, including Pgp, MRP2, OATP1B1, and OATP1B3, traffic to the appropriate membrane in sandwich-cultured human (and rat) hepatocytes, and are retained over several days in culture (Swift et al., 2010). Sandwich-cultured hepatocytes have been used to predict hepatobiliary drug clearance and the effect on drug clearance of transporter inhibition (Swift et al., 2010). A drawback of hepatocytes in culture is the apparent reduction of cytochrome P450 expression (Boess et al., 2003). However, drug metabolizing enzyme expression can be induced, and thus sandwich-cultured hepatocytes are a potentially useful system for examining the interplay between metabolism and transport in hepatic drug handling (Swift et al., 2010).

Sandwich-cultured hepatocytes allow not only for substrate accumulation into hepatocytes to be measured (sinusoidal uptake) but also for determination of canalicular and sinusoidal efflux (Swift et al., 2010). The general methodology for measuring sinusoidal uptake (uptake into the cells from the bath) and canalicular efflux (efflux from the cell into the canaliculi) involves incubating the hepatocytes with a compound diluted in a buffer supplemented with a physiological level of calcium (standard buffer). After the incubation period, the hepatocytes are either washed in the standard buffer or a buffer lacking calcium (Ca^{2+}-free). The amount of compound accumulating in the sandwich-cultured hepatocytes washed with the standard buffer represents both sinusoidal uptake and canalicular efflux. Since washing the hepatocytes in Ca^{2+}-free buffer disrupts the tight junctions, and hence allows the substrate in the canalicular compartment to be washed away, the amount of compound accumulating in the hepatocytes washed with Ca^{2+}-free buffer is a measure of sinusoidal uptake only. The amount of compound effluxed into the canalicular space can be determined by subtracting the amount of compound accumulating after washing the hepatocytes in Ca^{2+}-free buffer from the amount of substrate accumulating after washing the hepatocytes with the standard buffer. Transport assays conducted at 4°C or in the presence of transport protein inhibitor cocktails can be used to determine active versus passive transport (or nonspecific binding). Since sandwich-cultured hepatocytes express a suite of transporters, using them to determine if a candidate drug is a substrate of a specific transport protein requires the use of transporter-specific inhibitors. Alternatively, assessing if a candidate drug is an inhibitor of a specific transporter requires the use of transporter specific substrates in the absence or presence of the drug candidate. As with the suspension hepatocytes, caution should be used when interpreting transporter phenotyping and inhibition data generated from sandwich-cultured hepatocytes since the transporters expressed in this system have a tendency for overlapping substrate and inhibitor specificities.

3.4.3.4 Cloned Transporters in Heterologous Expression Systems

All of the transporters discussed earlier in this chapter have been expressed heterologously in various cell types. Compared with primary cells or immortalized cell lines such as Caco-2, the use of heterologous expression systems allows for the individual transporters to be studied in isolation; this is particularly useful since drug transporters have a tendency for overlapping substrate/inhibitor specificities. Tumor cells expressing select ABC transporters have been established by selection with chemotherapeutics. For example, increased levels of BCRP expression have been established in the IGROV-1 ovarian cancer cell line by selection with the BCRP substrate topotecan (Ma et al., 1998). Cell lines stably expressing drug transporters can also be established by transfection of cDNA into various cell types. Cells most often used for transient or stable expression of drug transporters are the nonpolarized cells, human embryonic kidney (HEK), and Chinese hamster ovary (CHO). *Xenopus* oocytes have also been used for heterologous expression of drug transporters. Polarized cells that have been used for transient or stable expression of drug transporters include MDCK and LLC-PK1 cells. More recently, polarized cells have been used for expressing uptake and efflux transporters in the same cell in order to examine the coordination of uptake and efflux transporters in mediating vectorial transport, that is, transepithelial transport (Shitara et al., 2006). Inside-out membrane vesicles isolated from cells heterologously expressing ABC transporters are particularly useful for studying the interaction of relatively hydrophilic drugs with these ATP-dependent efflux transporters.

3.4.3.5 Transporter Phenotyping with Cell Lines Expressing a Single Transporter

To determine if a candidate drug is a substrate of a specific uptake transporter, uptake of the candidate drug can be determined in cells expressing the individual uptake transporter and compared with uptake of the candidate drug into the parental control cells, that is, those that do not express appreciable levels of the transporter. If the candidate drug is a substrate of the transporter, its accumulation should be considerably higher in the cells heterologously expressing the transport protein. When a candidate drug is deemed to be a substrate, then it may be appropriate to determine the Michaelis constant (K_m) and maximal rate of transport (J_{max}). Before determining transport kinetics, a time course of candidate drug uptake into the cells expressing the transporter should be conducted in order to establish an initial rate time point (i.e., a time point at which uptake is linear with time) at which to conduct the kinetic studies. Performing the kinetic studies at initial rate will ensure that the process examined is primarily specific to the transporter, and not other processes (e.g., passive diffusion). For kinetic experiments, the uptake of increasing concentrations of a candidate drug into parental control cells is subtracted from its uptake into cells expressing the uptake transporter at each concentration of candidate drug tested. It is important to subtract uptake into the parental control cells in order to account for endogenous transport activity (if present) as well as nonspecific binding of the drug candidate to either the outside of the cell or the transport apparatus.

Transporter phenotyping studies are inherently difficult to interpret with ABC transporters expressed in living cells since it requires that the candidate drug access the substrate binding site, which resides either in the inner leaflet of the lipid bilayer in the case of Pgp or at the intracellular aspect of the transport protein; access will be limited for hydrophilic drug candidates. Regardless, accumulation studies can be conducted whereby the candidate drug is placed in the extracellular buffer and its accumulation is measured in cells expressing the ABC efflux transporter and compared with accumulation in parental control cells. If the candidate drug is a substrate, its accumulation in the cells expressing the ABC efflux transporter is expected to be lower than in control cells. Several concentrations (low to high) of drug candidate should be used when phenotyping to ensure that the substrate binding site of the transporter is not saturated, and hence efflux activity potentially overlooked (false negative). Kinetic analysis of transporter-mediated efflux using living cells would result in an apparent K_m value since the exact concentration of candidate drug at the active site of the ABC efflux transporter would be unknown, or at best could only be estimated based on several assumptions, including cell volume and the fraction of the drug candidate that is free for interaction with the substrate binding site.

3.4.3.6 Transporter Inhibition with Cell Lines Expressing a Single Transporter

When assessing inhibition of uptake transporters using cells expressing an individual transport protein, a probe substrate is added to the extracellular buffer and its intracellular accumulation is determined in the absence or presence of a drug candidate. Fluorescent (Table 3.1) and radiolabeled (Table 3.2) probe substrates of drug transporters are often used for transporter inhibition studies. If the drug candidate is an inhibitor of the uptake transporter, the intracellular accumulation of the probe substrate should be considerably lower when it is present. Probe substrate uptake can be conducted in control parental cells in parallel in order to determine the percent inhibition of transport activity caused by the drug candidate; this is particularly important to do if there is endogenous transporter-mediated uptake of the probe substrate in the heterologous cell being used, as the endogenous transport activity could be influenced by the drug candidate. If transport protein inhibition is observed, it may be appropriate to determine the kinetics of inhibition, that is, either K_i (inhibitory constant) or IC_{50} (concentration to produce half maximal transport). K_i determinations are labor intensive and require the K_m value of the specific transporter toward a probe substrate to be determined in the absence and presence of increasing concentrations of a candidate drug. Determination of an IC_{50} value can be done relatively quickly as it requires that inhibition of a single concentration of

TABLE 3.1. Examples of Fluorescent Substrates of Human Drug Transporters That Can Be Used for *in vitro* Inhibition Studies

Transporter	Fluorescent Substrate	References
Pgp	Rhodamine 123	Weaver et al. (1991)
	Doxorubicin	Weaver et al. (1991)
	Daunorubicin	Weaver et al. (1991)
	Calcein	Hollo et al. (1994)
BCRP	Mitoxantrone	Litman et al. (2000); Robey et al. (2001); Minderman et al. (2002)
	boron-dipyrromethene-prazosin (BODIPY-prazosin)	Robey et al. (2001)
MRP2	5-(6)-carboxy-2',7'-dichlorofluorescein	Pratt et al. (2006); Lechner et al. (2010)
	Glutathione methylfluorescein	Forster et al. (2008)
	Calcein	Masereeuw et al. (2003)
	Fluo-3	Masereeuw et al. (2003)
	Cholyl-L-lysyl-fluorescein	de Waart et al. (2010)
OAT1	6-carboxyfluorescein	Cihlar and Ho (2000)
OAT3	6-carboxyfluorescein	Rodiger et al. (2010)
OATP1B1	8-fluorescein-cyclic adenosine monophosphate	Bednarczyk (2010)
	Fluorescein-methotrexate	Gui et al. (2010)
	Chenodeoxycholate-NBD	Yamaguchi et al. (2006)
OATP1B3	8-fluorescein-cyclic adenosine monophosphate	Bednarczyk (2010)
	Fluorescein-methotrexate	Gui et al. (2010)
	Chenodeoxycholate-NBD	Yamaguchi et al. (2006)
	Fluo-3	Baldes et al. (2006)
	Cholyl-L-lysyl-fluorescein	de Waart et al. (2010)
OATP2B1	Unknown	
OCT1	4-[4-(dimethylamino)styryl]-N-methylpyridinium (ASP⁺)	Ciarimboli et al. (2004)
OCT2	4-[4-(dimethylamino)styryl]-N-methylpyridinium (ASP+)	Cetinkaya et al. (2003)
MATE1	4',6-diamidino-2-phenylindole (DAPI)	Yasujima et al. (2010)
MATE2K	4',6-diamidino-2-phenylindole (DAPI)	Yasujima et al. (2010)

probe substrate be determined in the presence of increasing concentrations of a candidate drug. Moreover, if the probe substrate is used at a concentration well below its K_m value, the IC_{50} obtained will closely approximate the K_i value, as the K_i is equal to $K_i(1 + [S]/K_m)$.

Again, compared with uptake transporters, assessing inhibition of ABC transporters in living cell systems is less straightforward given that their substrate binding site is intracellular (or within the inner leaflet of the lipid bilayer) and they only operate in an efflux mode. When using cells overexpressing a single ABC transporter, the effect of a candidate drug on the intracellular accumulation of a probe substrate (Tables 3.1 and 3.2) can be examined. If the candidate drug is an effective inhibitor, an increase in the intracellular accumulation of the probe substrate should be observed. In parallel, probe substrate uptake can be conducted in control parental cells and the percent inhibition of transport activity caused by the candidate drug determined. There

are a few considerations when conducting inhibition studies with ABC transporters: (1) as noted above, the intracellular concentration of probe substrate and candidate drug is unknown, which makes it difficult to get an accurate estimate of the K_i or IC_{50} value; (2) inhibition of transport is usually conducted once the probe substrate and inhibitor have reached some steady-state intracellular concentration; (3) the probe substrate concentration should be low enough as to not saturate the substrate binding site (this can be established early on in method development); and (4) the physicochemical properties will influence the ability of the drug candidate to access the substrate binding site, and thus can greatly influence inhibitory potential; that is, even the most potent inhibitor will not demonstrate inhibition if it cannot get into the cell.

3.4.3.7 Cell Lines Expressing Two or More Transporters Polarized cells, such as MDCK cells, can be transiently or stably transfected with uptake and

TABLE 3.2. Examples of Radiolabeled Substrates of Human Drug Transporters That Can Be Used for *in vitro* Inhibition Studies

Transporter	Substrate	References
Pgp	*N*-methylquinidine	Karlsson et al. (2010)
	Digoxin	de Lannoy and Silverman (1992); Tanigawara et al. (1992)
BCRP	Estrone-3-sulfate	Karlsson et al. (2010)
	Methotrexate	Volk and Schneider (2003)
MRP2	Estradiol 17β-*D*-glucuronide	Karlsson et al. (2010)
	Methotrexate	Bakos et al. (2000)
OAT1	*Para*-aminohippurate	Lu et al. (1999)
	Adefovir	Ho et al. (2000); Mulato et al. (2000)
OAT3	Estrone-3-sulfate	Takeda et al. (2002)
OATP1B1	Estradiol 17β-*D*-glucuronide	Abe et al. (1999)
	Taurocholate	Abe et al. (1999)
	Estrone-3-sulfate	Abe et al. (1999)
OATP1B3	Estrone-3-sulfate	Kullak-Ublick et al. (2001)
	Methotrexate	Abe et al. (2001)
	Estradiol 17β-*D*-glucuronide	Konig et al. (2000)
	Taurocholate	Abe et al. (2001)
OATP2B1	Estrone-3-sulfate	Tamai et al. (2001)
OCT1	Tetraethylammonium	Sakata et al. (2004)
	1-methyl-4-phenylpyridinium	Kerb et al. (2002)
OCT2	Tetraethylammonium	Barendt and Wright (2002)
	1-methyl-4-phenylpyridinium	Barendt and Wright (2002)
	Metformin	Kimura et al. (2005)
MATE1	Tetraethylammonium	Tanihara et al. (2007)
	1-methyl-4-phenylpyridinium	Tanihara et al. (2007)
	Metformin	Tanihara et al. (2007)
MATE2K	Tetraethylammonium	Tanihara et al. (2007)
	1-methyl-4-phenylpyridinium	Tanihara et al. (2007)
	Metformin	Tanihara et al. (2007)

The substrates indicated are commercially available with either ^{3}H or ^{14}C labels.

efflux transporters to examine their coordinated roles in transepithelial drug transport as well as inhibition of this process. In these studies, the polarized cells are seeded onto permeable supports such that the basolateral and apical compartments are accessible. Transport studies can be conducted in a manner similar to what was described for the Caco-2 system. The intracellular concentration of a drug candidate, its movement from the basolateral-to-apical compartment, and its movement from the apical-to-basolateral compartment can be determined in (1) control parental cells, (2) cells expressing an uptake transporter only, (3) cells expressing an efflux transporter only, and (4) cells expressing both uptake and efflux transporters (see the following studies for examples of experimental design: Cui et al., 2001; Sasaki et al., 2002; Mita et al., 2005; Liu et al., 2006; Nies et al., 2008). If there is coordination of uptake and efflux in mediating transepithelial transport, the vectorial

movement of a drug candidate should be highest in cells expressing both uptake and efflux transporters. For this strategy to be effective, however, requires that once transfected into the cells, the uptake and efflux transporters target to the appropriate membrane, which can be assessed using several strategies, including functional assays, immunocytochemistry, and cell surface biotinylation followed by Western blotting. MDCK cells expressing select hepatic and renal uptake transporters have been generated to examine their potential role in hepatobiliary and renal tubular drug clearance *in vivo*, respectively. For example, Sasaki et al. (2004) used MDCK cells doubly transfected with rat Oatp1b2 and Mrp2 to show that several organic anions, including estradiol 17β-*D*-glucuronide, pravastatin, lithocholate-3-sulfate, BQ123, and leukotriene C4, are preferentially transported in the basolateral-to-apical direction, and after accounting for differences in transporter expression *in vitro* to *in vivo*, the *in vitro* transepithelial transport of the organic anions correlated well with their *in vivo* biliary clearance. However, caution should be used when inferring data from doubly transfected cells to the *in vivo* situation since the heterologous system does not express the entire suite of drug transporters/metabolizing enzymes that occur in the native tissue.

3.4.3.8 *Inside-Out Membrane Vesicles*

Membrane vesicles in an inside-out orientation can be used for studying the interaction of candidate drugs as substrates and inhibitors of ABC transporters. Inside-out membrane vesicles have been prepared from both nonmammalian and mammalian cells that heterologously express ABC transporters, such as Sf9 insect and HEK293 cells. The inside-out orientation renders the substrate binding surface of the ABC transporter accessible from the extracellular buffer. Thus, unlike studying ABC transporters in living cells, inside-out membrane vesicles allows for an accurate determination of the K_m, K_i, or IC_{50} value. The reader is directed to the following reviews on the use of inside-out membrane vesicles for studying ABC transporters (Glavinas et al., 2004, 2008).

There are two major assay types for studying ABC-mediated transport with membrane vesicles, the ATPase assay (indirect method) and the vesicular transport assay (direct method). The ATPase assay does not directly measure transport, but monitors the amount of inorganic phosphate produced due to ATP hydrolysis, which is linked to the transport cycle (Sarkadi et al., 1992). Sarkadi et al. (1992) found that for a small set of substrates tested, the level to which they stimulated Pgp ATPase activity correlated with their affinity for the transporter. The ATPase assay is particularly amenable for high-throughput screening and when assessing the interaction of hydrophobic drugs (with a high passive permeability) as substrates. However, substrate binding may be overlooked (false negative) when transporter turnover is low, as these conditions can result in low levels of ATP that are undetectable (Glavinas et al., 2008). For example, cyclosporin A is a known Pgp substrate, but results in a false negative in the ATPase assay (Glavinas et al., 2008).

In contrast to the ATPase assay, translocation of substrate across the lipid bilayer of the vesicle is directly measured in the vesicular transport assay. In this assay, a compound is added to the extravesicular buffer and its accumulation in the intravesicular space is determined in the absence versus presence of ATP. Transport is "stopped" by a rapid filtration method in which the vesicles are trapped on a nitrocellulose membrane, and the amount of compound accumulating inside the vesicles can be determined using the appropriate analytical method. If the compound is a substrate, its accumulation in the intravesicular space should be greater in the presence of ATP versus its absence. If a drug candidate is identified as a substrate, it may be appropriate to determine the J_{max} and K_m values associated with the transport activity. Since ABC transporters appear to have multiple binding sites, the increase in substrate uptake with increasing concentrations may not follow simple Michaelis–Menten kinetics, but instead may be sigmoidal (indicating positive cooperativity); this has been noted for MRP2 (Borst et al., 2006). Complete inhibition of ABC transporter-mediated uptake of a probe substrate by a drug candidate would result in levels of substrate in the intravesicular space that are comparable both in the presence versus the absence of ATP. Assessing a drug candidate as an inhibitor requires the use of a broad concentration range (of drug candidate) since ABC transporters have multiple sites of interaction, which can result in probe substrate uptake exhibiting a "bell-shaped" curve with increasing concentrations of drug candidate. For example, MRP2-mediated transport of estradiol 17β-*D*-glucuronide is stimulated by relatively low concentrations of furosemide, but then inhibited by higher concentrations (Zelcer et al., 2003). A disadvantage of the vesicular transport assay is that compounds with relatively high passive permeability are not retained inside the vesicles (Glavinas et al., 2008).

3.5 CONCLUSIONS AND PERSPECTIVES

It is becoming increasingly clear that select drug transporters are important for the ADME properties of many drugs on the market as well as drug candidates in the pipeline. Additionally, drug transporters are a potential site of DDIs and adverse drug reactions, and genetic variation in the genes encoding drug transporters can influence a drug's pharmacokinetics. A number of drug transporters in the SLC (OATs, OCTs, OATPs, and MATEs) and ABC (Pgp, BCRP, and MRP2) transporter families are beginning to receive considerable attention in the course of drug development, as there is evidence that they are clinically relevant. Currently, industry scientists have multiple tools for assessing the interaction of drug candidates as substrates and inhibitors with these and other drug transporters, and are able to generate potentially useful kinetic data. Still in its infancy, however, is the ability to accurately predict the pharmacokinetics of a drug from *in vitro* transporter data. What is missing is the ability to scale transporter activity from *in vitro* to *in vivo*. Efforts are currently underway using either quantitative Western blotting or mass spectrometry to determine expression factors of select transporters both *in vitro* as well as *in vivo*, which would allow for scaling of *in vitro* transporter activity (namely J_{max}).

Also, computer models that incorporate *in vitro* transporter data for predicting drug pharmacokinetics are in the early stages of development and testing. Scaling factors are currently available for enzyme activity of select cytochrome P450 enzymes, and with the appropriate *in vitro* data, computer models can accurately predict pharmacokinetics of drug candidates that are largely cleared by cytochrome P450-mediated metabolism. Also lacking on the transporter front is the ability to predict the fold change in plasma exposure of a drug following an alteration in its transporter-mediated clearance, for example, due to a DDI or genetic variation. The current recommendation for predicting DDI potential is to use the I/K_i (or I/IC_{50}) or I_2/K_i (or I_2/IC_{50}) approach for inhibition of systemic transporters (those in the kidney, liver, and brain) and intestinal efflux transporters, respectively; where I is the unbound maximal plasma concentration of the inhibitor, I_2 is the oral dose of the inhibitor/250 mL of water, and K_i (or IC_{50}) is the inhibitor constant generated from *in vitro* data; the K_i or IC_{50} value should be corrected for plasma protein binding when assessing potential for inhibition of systemic transporters (Giacomini et al., 2010). Clinical DDI trials should be at least considered when $I/K_i \geq 0.1$ or when $I_2/K_i \geq 10$ (Giacomini et al., 2010). An additional *in vitro–in vivo* extrapolation approach, which takes into consideration the maximal unbound drug concentration at the liver inlet, is available for specifically assessing inhibition of hepatic uptake transporters, such as OATPs (Giacomini et al., 2010). While these approaches are available, there is currently not enough data showing a correlation between *in vitro* inhibition and clinical outcome. A future goal should be to further develop/refine our *in vitro* assays and to determine how best to incorporate this data into models in order to better predict the influence of drug transporters on the disposition of therapeutic drugs.

REFERENCES

Abe H, Okuda K (1975) Biliary excretion of conjugated sulfobromophthalein (BSP) in constitutional conjugated hyperbilirubinemias. *Digestion* 13:272–283.

Abe T, Kakyo M, Tokui T, Nakagomi R, Nishio T, Nakai D, Nomura H, Unno M, Suzuki M, Naitoh T, Matsuno S, Yawo H (1999) Identification of a novel gene family encoding human liver-specific organic anion transporter LST-1. *J Biol Chem* 274:17159–17163.

Abe T, Unno M, Onogawa T, Tokui T, Kondo TN, Nakagomi R, Adachi H, Fujiwara K, Okabe M, Suzuki T, Nunoki K, Sato E, Kakyo M, Nishio T, Sugita J, Asano N, Tanemoto M, Seki M, Date F, Ono K, Kondo Y, Shiiba K, Suzuki M, Ohtani H, Shimosegawa T, Iinuma K, Nagura H, Ito S, Matsuno S (2001) LST-2, a human liver-specific organic

anion transporter, determines methotrexate sensitivity in gastrointestinal cancers. *Gastroenterology* 120:1689–1699.

Ahn SY, Eraly SA, Tsigelny I, Nigam SK (2009) Interaction of organic cations with organic anion transporters. *J Biol Chem* 284:31422–31430.

Aslamkhan A, Han YH, Walden R, Sweet DH, Pritchard JB (2003) Stoichiometry of organic anion/dicarboxylate exchange in membrane vesicles from rat renal cortex and hOAT1-expressing cells. *Am J Physiol Renal Physiol* 285:F775–F783.

Bakos E, Evers R, Sinko E, Varadi A, Borst P, Sarkadi B (2000) Interactions of the human multidrug resistance proteins MRP1 and MRP2 with organic anions. *Mol Pharmacol* 57:760–768.

Baldes C, Koenig P, Neumann D, Lenhof HP, Kohlbacher O, Lehr CM (2006) Development of a fluorescence-based assay for screening of modulators of human organic anion transporter 1B3 (OATP1B3). *Eur J Pharm Biopharm* 62:39–43.

Barendt WM, Wright SH (2002) The human organic cation transporter (hOCT2) recognizes the degree of substrate ionization. *J Biol Chem* 277:22491–22496.

Becker ML, Visser LE, van Schaik RH, Hofman A, Uitterlinden AG, Stricker BH (2009) Genetic variation in the multidrug and toxin extrusion 1 transporter protein influences the glucose-lowering effect of metformin in patients with diabetes: A preliminary study. *Diabetes* 58:745–749.

Bednarczyk D (2010) Fluorescence-based assays for the assessment of drug interaction with the human transporters OATP1B1 and OATP1B3. *Anal Biochem* 405:50–58.

Benet LZ (2009) The drug transporter-metabolism alliance: Uncovering and defining the interplay. *Mol Pharm* 6:1631–1643.

Berthiaume F, Moghe PV, Toner M, Yarmush ML (1996) Effect of extracellular matrix topology on cell structure, function, and physiological responsiveness: Hepatocytes cultured in a sandwich configuration. *FASEB J* 10:1471–1484.

Bleasby K, Hall LA, Perry JL, Mohrenweiser HW, Pritchard JB (2005) Functional consequences of single nucleotide polymorphisms in the human organic anion transporter hOAT1 (SLC22A6). *J Pharmacol Exp Ther* 314:923–931.

Boess F, Kamber M, Romer S, Gasser R, Muller D, Albertini S, Suter L (2003) Gene expression in two hepatic cell lines, cultured primary hepatocytes, and liver slices compared to the *in vivo* liver gene expression in rats: Possible implications for toxicogenomics use of *in vitro* systems. *Toxicol Sci* 73:386–402.

Borst P, Zelcer N, van de Wetering K, Poolman B (2006) On the putative co-transport of drugs by multidrug resistance proteins. *FEBS Lett* 580:1085–1093.

Bow DA, Perry JL, Miller DS, Pritchard JB, Brouwer KL (2008) Localization of P-gp (Abcb1) and Mrp2 (Abcc2) in freshly isolated rat hepatocytes. *Drug Metab Dispos* 36:198–202.

Bronger H, Konig J, Kopplow K, Steiner HH, Ahmadi R, Herold-Mende C, Keppler D, Nies AT (2005) ABCC drug efflux pumps and organic anion uptake transporters in human gliomas and the blood-tumor barrier. *Cancer Res* 65:11419–11428.

Cetinkaya I, Ciarimboli G, Yalcinkaya G, Mehrens T, Velic A, Hirsch JR, Gorboulev V, Koepsell H, Schlatter E (2003) Regulation of human organic cation transporter hOCT2 by PKA, PI3K, and calmodulin-dependent kinases. *Am J Physiol Renal Physiol* 284:F293–F302.

Chen C, Scott D, Hanson E, Franco J, Berryman E, Volberg M, Liu X (2003) Impact of Mrp2 on the biliary excretion and intestinal absorption of furosemide, probenecid, and methotrexate using Eisai hyperbilirubinemic rats. *Pharm Res* 20:31–37.

Chen C, Slitt AL, Dieter MZ, Tanaka Y, Scheffer GL, Klaassen CD (2005) Up-regulation of Mrp4 expression in kidney of Mrp2-deficient TR- rats. *Biochem Pharmacol* 70:1088–1095.

Chen C, Stock JL, Liu X, Shi J, Van Deusen JW, DiMattia DA, Dullea RG, de Morais SM (2008) Utility of a novel Oatp1b2 knockout mouse model for evaluating the role of Oatp1b2 in the hepatic uptake of model compounds. *Drug Metab Dispos* 36:1840–1845.

Chen Y, Li S, Brown C, Cheatham S, Castro RA, Leabman MK, Urban TJ, Chen L, Yee SW, Choi JH, Huang Y, Brett CM, Burchard EG, Giacomini KM (2009a) Effect of genetic variation in the organic cation transporter 2 on the renal elimination of metformin. *Pharmacogenet Genomics* 19:497–504.

Chen Y, Teranishi K, Li S, Yee SW, Hesselson S, Stryke D, Johns SJ, Ferrin TE, Kwok P, Giacomini KM (2009b) Genetic variants in multidrug and toxic compound extrusion-1, hMATE1, alter transport function. *Pharmacogenomics J* 9:127–136.

Chu X-Y, Strauss JR, Mariano MA, Li J, Newton DJ, Cai X, Wang RW, Yabut J, Hartley DP, Evans DC, Evers R (2006) Characterization of mice lacking the multidrug resistance protein Mrp2 (Abcc2). *J Pharmacol Exp Ther* 317:579–589.

Ciarimboli G, Struwe K, Arndt P, Gorboulev V, Koepsell H, Schlatter E, Hirsch JR (2004) Regulation of the human organic cation transporter hOCT1. *J Cell Physiol* 201:420–428.

Cihlar T, Ho ES (2000) Fluorescence-based assay for the interaction of small molecules with the human renal organic anion transporter 1. *Anal Biochem* 283:49–55.

Cihlar T, Lin DC, Pritchard JB, Fuller MD, Mendel DB, Sweet DH (1999) The antiviral nucleotide analogs cidofovir and adefovir are novel substrates for human and rat renal organic anion transporter 1. *Mol Pharmacol* 56:570–580.

Cui Y, König J, Keppler D (2001) Vectorial transport by double-transfected cells expressing the human uptake transporter SLC21A8 and the apical export pump ABCC2. *Mol Pharmacol* 60:934–943.

Cundy KC (1999) Clinical pharmacokinetics of the antiviral nucleotide analogues cidofovir and adefovir. *Clin Pharmacokinet* 36:127–143.

Daly AK, Aithal GP, Leathart JB, Swainsbury RA, Dang TS, Day CP (2007) Genetic susceptibility to diclofenac-induced hepatotoxicity: Contribution of UGT2B7, CYP2C8, and ABCC2 genotypes. *Gastroenterology* 132:272–281.

de Lannoy IA, Silverman M (1992) The MDR1 gene product, P-glycoprotein, mediates the transport of the cardiac glycoside, digoxin. *Biochem Biophys Res Commun* 189:551–557.

de Vries NA, Zhao J, Kroon E, Buckle T, Beijnen JH, van Tellingen O (2007) P-glycoprotein and breast cancer resistance protein: Two dominant transporters working together in limiting the brain penetration of topotecan. *Clin Cancer Res* 13:6440–6449.

de Waart DR, Hausler S, Vlaming ML, Kunne C, Hanggi E, Gruss HJ, Oude Elferink RP, Stieger B (2010) Hepatic transport mechanisms of cholyl-L-lysyl-fluorescein. *J Pharmacol Exp Ther* 334:78–86.

Doyle LA, Yang W, Abruzzo LV, Krogmann T, Gao Y, Rishi AK, Ross DD (1998) A multidrug resistance transporter from human MCF-7 breast cancer cells. *Proc Natl Acad Sci U S A* 95:15665–15670.

Dunn JC, Yarmush ML, Koebe HG, Tompkins RG (1989) Hepatocyte function and extracellular matrix geometry: Long-term culture in a sandwich configuration. *FASEB J* 3:174–177.

Eraly SA, Vallon V, Vaughn DA, Gangoiti JA, Nigam SK, Nagle M, Monte JC, Rieg T, Truong DM, Long JM, Barshop BA, Kaler G, Nigam SK (2006) Decreased renal organic anion secretion and plasma accumulation of endogenous organic anions in OAT1 knock-out mice. *J Biol Chem* 281:5072–5083.

Erdilyi DJ, Kamory E, Csokay B, Andrikovics H, Tordai A, Kiss C, Filni-Semsei A, Janszky I, Zalka A, Fekete G, Falus A, Kovacs GT, Szalai C (2008) Synergistic interaction of ABCB1 and ABCG2 polymorphisms predicts the prevalence of toxic encephalopathy during anticancer chemotherapy. *Pharmacogenomics J* 8:321–327.

Erdman AR, Mangravite LM, Urban TJ, Lagpacan LL, Castro RA, de la Cruz M, Chan W, Huang CC, Johns SJ, Kawamoto M, Stryke D, Taylor TR, Carlson EJ, Ferrin TE, Brett CM, Burchard EG, Giacomini KM (2006) The human organic anion transporter 3 (OAT3; SLC22A8): Genetic variation and functional genomics. *Am J Physiol Renal Physiol* 290:F905–F912.

Fahrmayr C, Fromm MF, Konig J (2010) Hepatic OATP and OCT uptake transporters: Their role for drug-drug interactions and pharmacogenetic aspects. *Drug Metab Rev* 42:380–401.

Filipski KK, Mathijssen RH, Mikkelsen TS, Schinkel AH, Sparreboom A (2009) Contribution of organic cation transporter 2 (OCT2) to cisplatin-induced nephrotoxicity. *Clin Pharmacol Ther* 86:396–402.

Forster F, Volz A, Fricker G (2008) Compound profiling for ABCC2 (MRP2) using a fluorescent microplate assay system. *Eur J Pharm Biopharm* 69:396–403.

Fujita T, Brown C, Carlson EJ, Taylor T, de la Cruz M, Johns SJ, Stryke D, Kawamoto M, Fujita K, Castro R, Chen CW,

Lin ET, Brett CM, Burchard EG, Ferrin TE, Huang CC, Leabman MK, Giacomini KM (2005) Functional analysis of polymorphisms in the organic anion transporter, SLC22A6 (OAT1). *Pharmacogenet Genomics* 15:201–209.

Giacomini KM, Huang SM, Tweedie DJ, Benet LZ, Brouwer KL, Chu X, Dahlin A, Evers R, Fischer V, Hillgren KM, Hoffmaster KA, Ishikawa T, Keppler D, Kim RB, Lee CA, Niemi M, Polli JW, Sugiyama Y, Swaan PW, Ware JA, Wright SH, Yee SW, Zamek-Gliszczynski MJ, Zhang L (2010) Membrane transporters in drug development. *Nat Rev Drug Discov* 9:215–236.

Glavinas H, Krajcsi P, Cserepes J, Sarkadi B (2004) The role of ABC transporters in drug resistance, metabolism and toxicity. *Curr Drug Deliv* 1:27–42.

Glavinas H, Mehn D, Jani M, Oosterhuis B, Heredi-Szabo K, Krajcsi P (2008) Utilization of membrane vesicle preparations to study drug-ABC transporter interactions. *Expert Opin Drug Metab Toxicol* 4:721–732.

Gorboulev V, Ulzheimer JC, Akhoundova A, Ulzheimer-Teuber I, Karbach U, Quester S, Baumann C, Lang F, Busch AE, Koepsell H (1997) Cloning and characterization of two human polyspecific organic cation transporters. *DNA Cell Biol* 16:871–881.

Gründemann D, Gorboulev V, Gambaryan S, Veyhl M, Koepsell H (1994) Drug excretion mediated by a new prototype of polyspecific transporter. *Nature* 372:549–552.

Gui C, Obaidat A, Chaguturu R, Hagenbuch B (2010) Development of a cell-based high-throughput assay to screen for inhibitors of organic anion transporting polypeptides 1B1 and 1B3. *Curr Chem Genomics* 4:1–8.

Gutmann H, Fricker G, Torok M, Michael S, Beglinger C, Drewe J (1999) Evidence for different ABC-transporters in Caco-2 cells modulating drug uptake. *Pharm Res* 16:402–407.

Hagenbuch B (2010) Drug uptake systems in liver and kidney: A historic perspective. *Clin Pharmacol Ther* 87:39–47.

Hagenbuch B, Gui C (2008) Xenobiotic transporters of the human organic anion transporting polypeptides (OATP) family. *Xenobiotica* 38:778–801.

Hagenbuch B, Meier PJ (2003) The superfamily of organic anion transporting polypeptides. *Biochim Biophys Acta* 1609:1–18.

Hagenbuch B, Meier PJ (2004) Organic anion transporting polypeptides of the OATP/ SLC21 family: Phylogenetic classification as OATP/ SLCO superfamily, new nomenclature and molecular/functional properties. *Pflugers Arch* 447:653–665.

Hidalgo IJ, Raub TJ, Borchardt RT (1989) Characterization of the human colon carcinoma cell line (Caco-2) as a model system for intestinal epithelial permeability. *Gastroenterology* 96:736–749.

Hilgendorf C, Ahlin G, Seithel A, Artursson P, Ungell AL, Karlsson J (2007) Expression of thirty-six drug transporter genes in human intestine, liver, kidney, and organotypic cell lines. *Drug Metab Dispos* 35:1333–1340.

Hilgers AR, Conradi RA, Burton PS (1990) Caco-2 cell monolayers as a model for drug transport across the intestinal mucosa. *Pharm Res* 7:902–910.

Ho ES, Lin DC, Mendel DB, Cihlar T (2000) Cytotoxicity of antiviral nucleotides adefovir and cidofovir is induced by the expression of human renal organic anion transporter 1. *J Am Soc Nephrol* 11:383–393.

Hollo Z, Homolya L, Davis CW, Sarkadi B (1994) Calcein accumulation as a fluorometric functional assay of the multidrug transporter. *Biochim Biophys Acta* 1191:384–388.

Hubatsch I, Ragnarsson EG, Artursson P (2007) Determination of drug permeability and prediction of drug absorption in Caco-2 monolayers. *Nat Protoc* 2:2111–2119.

Huisman MT, Smit JW, Crommentuyn KM, Zelcer N, Wiltshire HR, Beijnen JH, Schinkel AH (2002) Multidrug resistance protein 2 (MRP2) transports HIV protease inhibitors, and transport can be enhanced by other drugs. *AIDS* 16:2295–2301.

Huisman MT, Chhatta AA, van Tellingen O, Beijnen JH, Schinkel AH (2005) MRP2 (ABCC2) transports taxanes and confers paclitaxel resistance and both processes are stimulated by probenecid. *Int J Cancer* 116:824–829.

Hulot JS, Villard E, Maguy A, Morel V, Mir L, Tostivint I, William-Faltaos D, Fernandez C, Hatem S, Deray G, Komajda M, Leblond V, Lechat P (2005) A mutation in the drug transporter gene ABCC2 associated with impaired methotrexate elimination. *Pharmacogenet Genomics* 15:277–285.

Hunter J, Jepson MA, Tsuruo T, Simmons NL, Hirst BH (1993) Functional expression of P-glycoprotein in apical membranes of human intestinal Caco-2 cells. Kinetics of vinblastine secretion and interaction with modulators. *J Biol Chem* 268:14991–14997.

Ieiri I, Higuchi S, Sugiyama Y (2009) Genetic polymorphisms of uptake (OATP1B1, 1B3) and efflux (MRP2, BCRP) transporters: Implications for inter-individual differences in the pharmacokinetics and pharmacodynamics of statins and other clinically relevant drugs. *Expert Opin Drug Metab Toxicol* 5:703–729.

Iida A, Saito S, Sekine A, Mishima C, Kondo K, Kitamura Y, Harigae S, Osawa S, Nakamura Y (2001) Catalog of 258 single-nucleotide polymorphisms (SNPs) in genes encoding three organic anion transporters, three organic anion-transporting polypeptides, and three NADH:ubiquinone oxidoreductase flavoproteins. *J Hum Genet* 46:668–683.

Ismair MG, Stieger B, Cattori V, Hagenbuch B, Fried M, Meier PJ, Kullak-Ublick GA (2001) Hepatic uptake of cholecystokinin octapeptide by organic anion-transporting polypeptides OATP4 and OATP8 of rat and human liver. *Gastroenterology* 121:1185–1190.

Ito S, Kusuhara H, Kuroiwa Y, Wu C, Moriyama Y, Inoue K, Kondo T, Yuasa H, Nakayama H, Horita S, Sugiyama Y (2010) Potent and specific inhibition of mMate1-mediated efflux of type I organic cations in the liver and kidney by pyrimethamine. *J Pharmacol Exp Ther* 333:341–350.

Jedlitschky G, Hoffmann U, Kroemer HK (2006) Structure and function of the MRP2 (ABCC2) protein and its role in drug disposition. *Expert Opin Drug Metab Toxicol* 2:351–366.

Johnson AG, Seideman P, Day RO (1993) Adverse drug interactions with nonsteroidal anti-inflammatory drugs (NSAIDs). Recognition, management and avoidance. *Drug Saf* 8:99–127.

Johnson BM, Zhang P, Schuetz JD, Brouwer KL (2006) Characterization of transport protein expression in multidrug resistance-associated protein (Mrp) 2-deficient rats. *Drug Metab Dispos* 34:556–562.

Jonker JW, Wagenaar E, Mol CA, Buitelaar M, Koepsell H, Smit JW, Schinkel AH (2001) Reduced hepatic uptake and intestinal excretion of organic cations in mice with a targeted disruption of the organic cation transporter 1 (Oct1 [Slc22a1]) gene. *Mol Cell Biol* 21:5471–5477.

Jonker JW, Buitelaar M, Wagenaar E, Van Der Valk MA, Scheffer GL, Scheper RJ, Plosch T, Kuipers F, Elferink RP, Rosing H, Beijnen JH, Schinkel AH (2002) The breast cancer resistance protein protects against a major chlorophyll-derived dietary phototoxin and protoporphyria. *Proc Natl Acad Sci U S A* 99:15649–15654.

Jonker JW, Wagenaar E, Van Eijl S, Schinkel AH (2003) Deficiency in the organic cation transporters 1 and 2 (Oct1/Oct2 [Slc22a1/Slc22a2]) in mice abolishes renal secretion of organic cations. *Mol Cell Biol* 23:7902–7908.

Kajiwara M, Terada T, Ogasawara K, Iwano J, Katsura T, Fukatsu A, Doi T, Inui K (2009) Identification of multidrug and toxin extrusion (MATE1 and MATE2-K) variants with complete loss of transport activity. *J Hum Genet* 54:40–46.

Kalliokoski A, Niemi M (2009) Impact of OATP transporters on pharmacokinetics. *Br J Pharmacol* 158:693–705.

Kameyama Y, Yamashita K, Kobayashi K, Hosokawa M, Chiba K (2005) Functional characterization of SLCO1B1 (OATP-C) variants, SLCO1B1*5, SLCO1B1*15 and SLCO1B1*15+C1007G, by using transient expression systems of HeLa and HEK293 cells. *Pharmacogenet Genomics* 15:513–522.

Karlsson JE, Heddle C, Rozkov A, Rotticci-Mulder J, Tuvesson O, Hilgendorf C, Andersson TB (2010) High-activity P-glycoprotein, multidrug resistance protein 2, and breast cancer resistance protein membrane vesicles prepared from transiently transfected human embryonic kidney 293-Epstein-Barr virus nuclear antigen cells. *Drug Metab Dispos* 38:705–714.

Kerb R, Brinkmann U, Chatskaia N, Gorbunov D, Gorboulev V, Mornhinweg E, Keil A, Eichelbaum M, Koepsell H (2002) Identification of genetic variations of the human organic cation transporter hOCT1 and their functional consequences. *Pharmacogenetics* 12:591–595.

Keskitalo JE, Zolk O, Fromm MF, Kurkinen KJ, Neuvonen PJ, Niemi M (2009) ABCG2 polymorphism markedly affects the pharmacokinetics of atorvastatin and rosuvastatin. *Clin Pharmacol Ther* 86:197–203.

Kimchi-Sarfaty C, Marple AH, Shinar S, Kimchi AM, Scavo D, Roma MI, Kim IW, Jones A, Arora M, Gribar J, Gurwitz D, Gottesman MM (2007a) Ethnicity-related polymorphisms and haplotypes in the human ABCB1 gene. *Pharmacogenomics* 8:29–39.

Kimchi-Sarfaty C, Oh JM, Kim IW, Sauna ZE, Calcagno AM, Ambudkar SV, Gottesman MM (2007b) A "silent" polymorphism in the MDR1 gene changes substrate specificity. *Science* 315:525–528.

Kimura N, Masuda S, Tanihara Y, Ueo H, Okuda M, Katsura T, Inui K (2005) Metformin is a superior substrate for renal organic cation transporter OCT2 rather than hepatic OCT1. *Drug Metab Pharmacokinet* 20:379–386.

Kobayashi D, Nozawa T, Imai K, Nezu J, Tsuji A, Tamai I (2003) Involvement of human organic anion transporting polypeptide OATP-B (SLC21A9) in pH-dependent transport across intestinal apical membrane. *J Pharmacol Exp Ther* 306:703–708.

Koepsell H (2004) Polyspecific organic cation transporters: Their functions and interactions with drugs. *Trends Pharmacol Sci* 25:375–381.

Koepsell H, Lips K, Volk C (2007) Polyspecific organic cation transporters: Structure, function, physiological roles, and biopharmaceutical implications. *Pharm Res* 24:1227–1251.

Komar AA (2007) Silent SNPs: Impact on gene function and phenotype. *Pharmacogenomics* 8:1075–1080.

Konig J, Nies AT, Cui Y, Leier I, Keppler D (1999) Conjugate export pumps of the multidrug resistance protein (MRP) family: Localization, substrate specificity, and MRP2-mediated drug resistance. *Biochim Biophys Acta* 1461:377–394.

Konig J, Cui Y, Nies AT, Keppler D (2000) Localization and genomic organization of a new hepatocellular organic anion transporting polypeptide. *J Biol Chem* 275: 23161–23168.

Kovarik JM, Rigaudy L, Guerret M, Gerbeau C, Rost KL (1999) Longitudinal assessment of a P-glycoprotein-mediated drug interaction of valspodar on digoxin. *Clin Pharmacol Ther* 66:391–400.

Kruh GD, Belinsky MG, Gallo JM, Lee K (2007) Physiological and pharmacological functions of Mrp2, Mrp3 and Mrp4 as determined from recent studies on gene-disrupted mice. *Cancer Metastasis Rev* 26:5–14.

Kubota H, Ishihara H, Langmann T, Schmitz G, Stieger B, Wieser HG, Yonekawa Y, Frei K (2006) Distribution and functional activity of P-glycoprotein and multidrug resistance-associated proteins in human brain microvascular endothelial cells in hippocampal sclerosis. *Epilepsy Res* 68:213–228.

Kukongviriyapan V, Stacey NH (1989) Comparison of uptake kinetics in freshly isolated suspensions and short-term primary cultures of rat hepatocytes. *J Cell Physiol* 140: 491–497.

Kullak-Ublick GA, Hagenbuch B, Stieger B, Schteingart CD, Hofmann AF, Wolkoff AW, Meier PJ (1995) Molecular and functional characterization of an organic anion transporting polypeptide cloned from human liver. *Gastroenterology* 109:1274–1282.

Kullak-Ublick GA, Ismair MG, Stieger B, Landmann L, Huber R, Pizzagalli F, Fattinger K, Meier PJ, Hagenbuch B (2001) Organic anion-transporting polypeptide B (OATP-B) and its functional comparison with three other OATPs of human liver. *Gastroenterology* 120:525–533.

Kwak JO, Lee SH, Lee GS, Kim MS, Ahn YG, Lee JH, Kim SW, Kim KH, Lee MG (2010) Selective inhibition of MDR1 (ABCB1) by HM30181 increases oral bioavailability and therapeutic efficacy of paclitaxel. *Eur J Pharmacol* 627:92–98.

Lalezari JP, Stagg RJ, Kuppermann BD, Holland GN, Kramer F, Ives DV, Youle M, Robinson MR, Drew WL, Jaffe HS (1997) Intravenous cidofovir for peripheral cytomegalovirus retinitis in patients with AIDS. A randomized, controlled trial. *Ann Intern Med* 126:257–263.

Lechner C, Reichel V, Moenning U, Reichel A, Fricker G (2010) Development of a fluorescence-based assay for drug interactions with human multidrug resistance related protein (MRP2; ABCC2) in MDCKII-MRP2 membrane vesicles. *Eur J Pharm Biopharm* 75:284–290.

Letschert K, Keppler D, Konig J (2004) Mutations in the SLCO1B3 gene affecting the substrate specificity of the hepatocellular uptake transporter OATP1B3 (OATP8). *Pharmacogenetics* 14:441–452.

Leuthold S, Hagenbuch B, Mohebbi N, Wagner CA, Meier PJ, Stieger B (2009) Mechanisms of pH-gradient driven transport mediated by organic anion polypeptide transporters. *Am J Physiol Cell Physiol* 296:C570–C582.

Link E, Parish S, Armitage J, Bowman L, Heath S, Matsuda F, Gut I, Lathrop M, Collins R (2008) SLCO1B1 variants and statin-induced myopathy—A genomewide study. *N Engl J Med* 359:789–799.

Litman T, Brangi M, Hudson E, Fetsch P, Abati A, Ross DD, Miyake K, Resau JH, Bates SE (2000) The multidrug-resistant phenotype associated with overexpression of the new ABC half-transporter, MXR (ABCG2). *J Cell Sci* 113:2011–2021.

Liu L, Cui Y, Chung AY, Shitara Y, Sugiyama Y, Keppler D, Pang KS (2006) Vectorial transport of enalapril by Oatp1a1/Mrp2 and OATP1B1 and OATP1B3/MRP2 in rat and human livers. *J Pharmacol Exp Ther* 318:395–402.

Lu R, Chan BS, Schuster VL (1999) Cloning of the human kidney PAH transporter: Narrow substrate specificity and regulation by protein kinase C. *Am J Physiol* 276:F295–F303.

Luttringer O, Theil FP, Lave T, Wernli-Kuratli K, Guentert TW, de Saizieu A (2002) Influence of isolation procedure, extracellular matrix and dexamethasone on the regulation of membrane transporters gene expression in rat hepatocytes. *Biochem Pharmacol* 64:1637–1650.

Ma J, Maliepaard M, Nooter K, Loos WJ, Kolker HJ, Verweij J, Stoter G, Schellens JH (1998) Reduced cellular accumulation of topotecan: A novel mechanism of resistance in a human ovarian cancer cell line. *Br J Cancer* 77:1645–1652.

Maeda A, Tsuruoka S, Kanai Y, Endou H, Saito K, Miyamoto E, Fujimura A (2008) Evaluation of the interaction between nonsteroidal anti-inflammatory drugs and methotrexate using human organic anion transporter 3-transfected cells. *Eur J Pharmacol* 596:166–172.

Maeda K, Sugiyama Y (2008) Impact of genetic polymorphisms of transporters on the pharmacokinetic, pharmacodynamic and toxicological properties of anionic drugs. *Drug Metab Pharmacokinet* 23:223–235.

Maeda K, Sugiyama Y (2010) The use of hepatocytes to investigate drug uptake transporters. *Methods Mol Biol* 640:327–353.

Mahagita C, Grassl SM, Piyachaturawat P, Ballatori N (2007) Human organic anion transporter 1B1 and 1B3 function as bidirectional carriers and do not mediate GSH-bile acid cotransport. *Am J Physiol Gastrointest Liver Physiol* 293:G271–G278.

Mahringer A, Delzer J, Fricker G (2009) A fluorescence-based *in vitro* assay for drug interactions with breast cancer resistance protein (BCRP, ABCG2). *Eur J Pharm Biopharm* 72:605–613.

Martinez-Becerra P, Briz O, Romero MR, Macias RI, Perez MJ, Sancho-Mateo C, Lostao MP, Fernandez-Abalos JM, Marin JJ (2011) Further characterization of the electrogenicity and pH-sensitivity of the human organic anion-transporting polypeptides OATP1B1 and OATP1B3. *Mol Pharmacol* 79:596–607.

Masereeuw R, Notenboom S, Smeets PH, Wouterse AC, Russel FG (2003) Impaired renal secretion of substrates for the multidrug resistance protein 2 in mutant transport-deficient (TR(-)) rats. *J Am Soc Nephrol* 14:2741–2749.

Masuda S, Terada T, Yonezawa A, Tanihara Y, Kishimoto K, Katsura T, Ogawa O, Inui K (2006) Identification and functional characterization of a new human kidney-specific H+/organic cation antiporter, kidney-specific multidrug and toxin extrusion 2. *J Am Soc Nephrol* 17:2127–2135.

Matsson P, Pedersen JM, Norinder U, Bergstrom CA, Artursson P (2009) Identification of novel specific and general inhibitors of the three major human ATP-binding cassette transporters P-gp, BCRP and MRP2 among registered drugs. *Pharm Res* 26:1816–1831.

Mayer R, Kartenbeck J, Buchler M, Jedlitschky G, Leier I, Keppler D (1995) Expression of the MRP gene-encoded conjugate export pump in liver and its selective absence from the canalicular membrane in transport-deficient mutant hepatocytes. *J Cell Biol* 131:137–150.

Mayur YC, Peters GJ, Prasad VV, Lemo C, Sathish NK (2009) Design of new drug molecules to be used in reversing multidrug resistance in cancer cells. *Curr Cancer Drug Targets* 9:298–306.

Meyer zu Schwabedissen HE, Verstuyft C, Kroemer HK, Becquemont L, Kim RB (2010) Human multidrug and toxin extrusion 1 (MATE1/SLC47A1) transporter: Functional characterization, interaction with OCT2 (SLC22A2), and single nucleotide polymorphisms. *Am J Physiol Renal Physiol* 298:F997–F1005.

Meyer-Wentrup F, Karbach U, Gorboulev V, Arndt P, Koepsell H (1998) Membrane localization of the electrogenic cation transporter rOCT1 in rat liver. *Biochem Biophys Res Commun* 248:673–678.

Michalski C, Cui Y, Nies AT, Nuessler AK, Neuhaus P, Zanger UM, Klein K, Eichelbaum M, Keppler D, Konig J (2002) A naturally occurring mutation in the SLC21A6 gene causing impaired membrane localization of the hepatocyte uptake transporter. *J Biol Chem* 277:43058–43063.

Miller BF, Winkler AW (1938) The renal excretion of endogenous creatinine in man. Comparison with exogenous creatinine and inulin. *J Clin Invest* 17:31–40.

Miller BF, Leaf A, Mamby AR, Miller Z (1952) Validity of the endogenous creatinine clearance as a measure of glomerular filtration rate in the diseased human kidney. *J Clin Invest* 31:309–313.

Minderman H, Suvannasankha A, O'Loughlin KL, Scheffer GL, Scheper RJ, Robey RW, Baer MR (2002) Flow cytometric analysis of breast cancer resistance protein expression and function. *Cytometry* 48:59–65.

Mita S, Suzuki H, Akita H, Stieger B, Meier PJ, Hofmann AF, Sugiyama Y (2005) Vectorial transport of bile salts across MDCK cells expressing both rat Na^+-taurocholate cotransporting polypeptide and rat bile salt export pump. *Am J Physiol Gastrointest Liver Physiol* 288:G159–G167.

Mizuarai S, Aozasa N, Kotani H (2004) Single nucleotide polymorphisms result in impaired membrane localization and reduced atpase activity in multidrug transporter ABCG2. *Int J Cancer* 109:238–246.

Motohashi H, Sakurai Y, Saito H, Masuda S, Urakami Y, Goto M, Fukatsu A, Ogawa O, Inui K-I (2002) Gene expression levels and immunolocalization of organic ion transporters in the human kidney. *J Am Soc Nephrol* 13:866–874.

Mulato AS, Ho ES, Cihlar T (2000) Nonsteroidal anti-inflammatory drugs efficiently reduce the transport and cytotoxicity of adefovir mediated by the human renal organic anion transporter 1. *J Pharmacol Exp Ther* 295:10–15.

Nakagomi-Hagihara R, Nakai D, Kawai K, Yoshigae Y, Tokui T, Abe T, Ikeda T (2006) OATP1B1, OATP1B3, and mrp2 are involved in hepatobiliary transport of olmesartan, a novel angiotensin II blocker. *Drug Metab Dispos* 34:862–869.

Nicolle E, Boumendjel A, Macalou S, Genoux E, Ahmed-Belkacem A, Carrupt PA, Di Pietro A (2009) QSAR analysis and molecular modeling of ABCG2-specific inhibitors. *Adv Drug Deliv Rev* 61:34–46.

Niemi M (2007) Role of OATP transporters in the disposition of drugs. *Pharmacogenomics* 8:787–802.

Niemi M (2010) Transporter pharmacogenetics and statin toxicity. *Clin Pharmacol Ther* 87:130–133.

Niemi M, Backman JT, Kajosaari LI, Leathart JB, Neuvonen M, Daly AK, Eichelbaum M, Kivisto KT, Neuvonen PJ (2005) Polymorphic organic anion transporting polypeptide 1B1 is a major determinant of repaglinide pharmacokinetics. *Clin Pharmacol Ther* 77:468–478.

Nies AT, Herrmann E, Brom M, Keppler D (2008) Vectorial transport of the plant alkaloid berberine by double-transfected cells expressing the human organic cation transporter 1 (OCT1, SLC22A1) and the efflux pump MDR1 P-glycoprotein (ABCB1). *Naunyn Schmiedebergs Arch Pharmacol* 376:449–461.

Nies AT, Koepsell H, Damme K, Schwab M (2011) Organic cation transporters (OCTs, MATEs), *in vitro* and *in vivo* evidence for the importance in drug therapy. *Handb Exp Pharmacol* 201:105–167.

Nobili S, Landini I, Giglioni B, Mini E (2006) Pharmacological strategies for overcoming multidrug resistance. *Curr Drug Targets* 7:861–879.

Noe B, Hagenbuch B, Stieger B, Meier PJ (1997) Isolation of a multispecific organic anion and cardiac glycoside transporter from rat brain. *Proc Natl Acad Sci U S A* 94:10346–10350.

Nozawa T, Nakajima M, Tamai I, Noda K, Nezu J, Sai Y, Tsuji A, Yokoi T (2002) Genetic polymorphisms of human organic anion transporters OATP-C (SLC21A6) and OATP-B (SLC21A9): Allele frequencies in the Japanese population and functional analysis. *J Pharmacol Exp Ther* 302:804–813.

Otsuka M, Matsumoto T, Morimoto R, Arioka S, Omote H, Moriyama Y (2005) A human transporter protein that mediates the final excretion step for toxic organic cations. *Proc Natl Acad Sci U S A* 102:17923–17928.

Oude Elferink RP, Meijer DK, Kuipers F, Jansen PL, Groen AK, Groothuis GM (1995) Hepatobiliary secretion of organic compounds; molecular mechanisms of membrane transport. *Biochim Biophys Acta* 1241:215–268.

Poirier A, Lave T, Portmann R, Brun ME, Senner F, Kansy M, Grimm HP, Funk C (2008) Design, data analysis, and simulation of *in vitro* drug transport kinetic experiments using a mechanistic *in vitro* model. *Drug Metab Dispos* 36:2434–2444.

Pratt S, Chen V, Perry WI III, Starling JJ, Dantzig AH (2006) Kinetic validation of the use of carboxydichlorofluorescein as a drug surrogate for MRP5-mediated transport. *Eur J Pharm Sci* 27:524–532.

Pritchard JB, Miller DS (1996) Renal secretion of organic anions and cations. *Kidney Int* 49:1649–1654.

Rau T, Erney B, Gores R, Eschenhagen T, Beck J, Langer T (2006) High-dose methotrexate in pediatric acute lymphoblastic leukemia: Impact of ABCC2 polymorphisms on plasma concentrations. *Clin Pharmacol Ther* 80:468–476.

Richert L, Liguori MJ, Abadie C, Heyd B, Mantion G, Halkic N, Waring JF (2006) Gene expression in human hepatocytes in suspension after isolation is similar to the liver of origin, is not affected by hepatocyte cold storage and cryopreservation, but is strongly changed after hepatocyte plating. *Drug Metab Dispos* 34:870–879.

Rizwan AN, Burckhardt G (2007) Organic anion transporters of the SLC22 family: Biopharmaceutical, physiological, and pathological roles. *Pharm Res* 24:450–470.

Robey RW, Honjo Y, van de Laar A, Miyake K, Regis JT, Litman T, Bates SE (2001) A functional assay for detection of the mitoxantrone resistance protein, MXR (ABCG2). *Biochim Biophys Acta* 1512:171–182.

Robey RW, To KK, Polgar O, Dohse M, Fetsch P, Dean M, Bates SE (2009) ABCG2: A perspective. *Adv Drug Deliv Rev* 61:3–13.

Rodiger M, Zhang X, Ugele B, Gersdorff N, Wright SH, Burckhardt G, Bahn A (2010) Organic anion transporter 3 (OAT3) and renal transport of the metal chelator 2,3-dimercapto-1-propanesulfonic acid (DMPS). *Can J Physiol Pharmacol* 88:141–146.

Romaine SP, Bailey KM, Hall AS, Balmforth AJ (2010) The influence of SLCO1B1 (OATP1B1) gene polymorphisms on response to statin therapy. *Pharmacogenomics J* 10:1–11.

Roth M, Timmermann BN, Hagenbuch B (2011) Interactions of green tea catechins with organic anion transporting polypeptides. *Drug Metab Dispos* 39:920–96.

Sakaeda T, Nakamura T, Horinouchi M, Kakumoto M, Ohmoto N, Sakai T, Morita Y, Tamura T, Aoyama N, Hirai M, Kasuga M, Okumura K (2001) MDR1 genotype-related pharmacokinetics of digoxin after single oral administration in healthy Japanese subjects. *Pharm Res* 18:1400–1404.

Sakata T, Anzai N, Shin HJ, Noshiro R, Hirata T, Yokoyama H, Kanai Y, Endou H (2004) Novel single nucleotide polymorphisms of organic cation transporter 1 (SLC22A1) affecting transport functions. *Biochem Biophys Res Commun* 313:789–793.

Sarkadi B, Price EM, Boucher RC, Germann UA, Scarborough GA (1992) Expression of the human multidrug resistance cDNA in insect cells generates a high activity drug-stimulated membrane ATPase. *J Biol Chem* 267:4854–4858.

Sasaki M, Suzuki H, Ito K, Abe T, Sugiyama Y (2002) Transcellular transport of organic anions across a double-transfected Madin-Darby canine kidney II cell monolayer expressing both human organic anion-transporting polypeptide (OATP2/SLC21A6) and Multidrug resistance-associated protein 2 (MRP2/ABCC2). *J Biol Chem* 277:6497–6503.

Sasaki M, Suzuki H, Aoki J, Ito K, Meier PJ, Sugiyama Y (2004) Prediction of *in vivo* biliary clearance from the *in vitro* transcellular transport of organic anions across a double-transfected Madin-Darby canine kidney II monolayer expressing both rat organic anion transporting polypeptide 4 and multidrug resistance associated protein 2. *Mol Pharmacol* 66:450–459.

Schinkel AH, Jonker JW (2003) Mammalian drug efflux transporters of the ATP binding cassette (ABC) family: An overview. *Adv Drug Deliv Rev* 55:3–29.

Schinkel AH, Mayer U, Wagenaar E, Mol CA, van Deemter L, Smit JJ, Van Der Valk MA, Voordouw AC, Spits H, van Tellingen O, Zijlmans JM, Fibbe WE, Borst P (1997) Normal viability and altered pharmacokinetics in mice lacking mdr1-type (drug-transporting) P-glycoproteins. *Proc Natl Acad Sci U S A* 94:4028–4033.

Shannon JA (1935) The renal excretion of creatinine in man. *J Clin Invest* 14:403–410.

Shepard RL, Cao J, Starling JJ, Dantzig AH (2003) Modulation of P-glycoprotein but not MRP1- or BCRP-mediated drug resistance by LY335979. *Int J Cancer* 103:121–125.

Shitara Y, Horie T, Sugiyama Y (2006) Transporters as a determinant of drug clearance and tissue distribution. *Eur J Pharm Sci* 27:425–446.

Shu Y, Sheardown SA, Brown C, Owen RP, Zhang S, Castro RA, Ianculescu AG, Yue L, Lo JC, Burchard EG, Brett CM, Giacomini KM (2007) Effect of genetic variation in the organic cation transporter 1 (OCT1) on metformin action. *J Clin Invest* 117:1422–1431.

Shu Y, Brown C, Castro R, Shi R, Lin E, Owen R, Sheardown S, Yue L, Burchard E, Brett C, Giacomini K (2008) Effect of genetic variation in the organic cation transporter 1, OCT1, on metformin pharmacokinetics. *Mol Ther* 83:273–280.

Smith NF, Figg WD, Sparreboom A (2005) Role of the liver-specific transporters OATP1B1 and OATP1B3 in governing drug elimination. *Expert Opin Drug Metab Toxicol* 1:429–445.

Song IS, Shin HJ, Shim EJ, Jung IS, Kim WY, Shon JH, Shin JG (2008) Genetic variants of the organic cation transporter 2 influence the disposition of metformin. *Clin Pharmacol Ther* 84:559–562.

Sparreboom A, Gelderblom H, Marsh S, Ahluwalia R, Obach R, Principe P, Twelves C, Verweij J, McLeod HL (2004) Diflomotecan pharmacokinetics in relation to ABCG2 421C>A genotype. *Clin Pharmacol Ther* 76:38–44.

Staud F, Pavek P (2005) Breast cancer resistance protein (BCRP/ABCG2). *Int J Biochem Cell Biol* 37:720–725.

Sun H, Chow EC, Liu S, Du Y, Pang KS (2008) The Caco-2 cell monolayer: Usefulness and limitations. *Expert Opin Drug Metab Toxicol* 4:395–411.

Swift B, Pfeifer ND, Brouwer KL (2010) Sandwich-cultured hepatocytes: An *in vitro* model to evaluate hepatobiliary transporter-based drug interactions and hepatotoxicity. *Drug Metab Rev* 42:446–471.

Takane H, Shikata E, Otsubo K, Higuchi S, Ieiri I (2008) Polymorphism in human organic cation transporters and metformin action. *Pharmacogenomics* 9:415–422.

Takayanagi M, Sano N, Takikawa H (2005) Biliary excretion of olmesartan, an anigotensin II receptor antagonist, in the rat. *J Gastroenterol Hepatol* 20:784–788.

Takeda M, Babu E, Narikawa S, Endou H (2002) Interaction of human organic anion transporters with various cephalosporin antibiotics. *Eur J Pharmacol* 438:137–142.

Tamai I, Nezu J, Uchino H, Sai Y, Oku A, Shimane M, Tsuji A (2000) Molecular identification and characterization of novel members of the human organic anion transporter (OATP) family. *Biochem Biophys Res Commun* 273:251–260.

Tamai I, Nozawa T, Koshida M, Nezu J, Sai Y, Tsuji A (2001) Functional characterization of human organic anion transporting polypeptide B (OATP-B) in comparison with liver-specific OATP-C. *Pharm Res* 18:1262–1269.

Tanigawara Y, Okamura N, Hirai M, Yasuhara M, Ueda K, Kioka N, Komano T, Hori R (1992) Transport of digoxin by human P-glycoprotein expressed in a porcine kidney epithelial cell line (LLC-PK1). *J Pharmacol Exp Ther* 263:840–845.

Taniguchi K, Wada M, Kohno K, Nakamura T, Kawabe T, Kawakami M, Kagotani K, Okumura K, Akiyama S, Kuwano M (1996) A human canalicular multispecific organic anion transporter (cMOAT) gene is overexpressed in cisplatin-resistant human cancer cell lines with decreased drug accumulation. *Cancer Res* 56:4124–4129.

Tanihara Y, Masuda S, Sato T, Katsura T, Ogawa O, Inui K (2007) Substrate specificity of MATE1 and MATE2-K, human multidrug and toxin extrusions/H(+)-organic cation antiporters. *Biochem Pharmacol* 74:359–371.

Thomas H, Coley HM (2003) Overcoming multidrug resistance in cancer: An update on the clinical strategy of inhibiting p-glycoprotein. *Cancer Control* 10:159–165.

Tirona RG, Leake BF, Merino G, Kim RB (2001) Polymorphisms in OATP-C: Identification of multiple allelic variants associated with altered transport activity among European- and African-Americans. *J Biol Chem* 276: 35669–35675.

Tsuda M, Terada T, Mizuno T, Katsura T, Shimakura J, Inui K (2009) Targeted disruption of the multidrug and toxin extrusion 1 (mate1) gene in mice reduces renal secretion of metformin. *Mol Pharmacol* 75:1280–1286.

Urakami Y, Kimura N, Okuda M, Inui K (2004) Creatinine transport by basolateral organic cation transporter hOCT2 in the human kidney. *Pharm Res* 21:976–981.

van de Steeg E, van der Kruijssen CM, Wagenaar E, Burggraaff JE, Mesman E, Kenworthy KE, Schinkel AH (2009) Methotrexate pharmacokinetics in transgenic mice with liver-specific expression of human organic anion-transporting polypeptide 1B1 (SLCO1B1). *Drug Metab Dispos* 37:277–281.

van Vliet EA, Redeker S, Aronica E, Edelbroek PM, Gorter JA (2005) Expression of multidrug transporters MRP1, MRP2, and BCRP shortly after status epilepticus, during the latent period, and in chronic epileptic rats. *Epilepsia* 46:1569–1580.

Vanwert AL, Sweet DH (2008) Impaired clearance of methotrexate in organic anion transporter 3 (Slc22a8) knockout mice: A gender specific impact of reduced folates. *Pharm Res* 25:453–462.

Vanwert AL, Bailey RM, Sweet DH (2007) Organic anion transporter 3 (Oat3/Slc22a8) knockout mice exhibit altered clearance and distribution of penicillin G. *Am J Physiol Renal Physiol* 293:F1332–F1341.

Vlaming ML, Mohrmann K, Wagenaar E, de Waart DR, Elferink RP, Lagas JS, van Tellingen O, Vainchtein LD, Rosing H, Beijnen JH, Schellens JH, Schinkel AH (2006) Carcinogen and anticancer drug transport by Mrp2 *in vivo*: Studies using Mrp2 (Abcc2) knockout mice. *J Pharmacol Exp Ther* 318:319–327.

Volk EL, Schneider E (2003) Wild-type breast cancer resistance protein (BCRP/ABCG2) is a methotrexate polyglutamate transporter. *Cancer Res* 63:5538–5543.

Wagner CC, Bauer M, Karch R, Feurstein T, Kopp S, Chiba P, Kletter K, Loscher W, Muller M, Zeitlinger M, Langer O (2009) A pilot study to assess the efficacy of tariquidar to inhibit P-glycoprotein at the human blood-brain barrier with (R)-11C-verapamil and PET. *J Nucl Med* 50: 1954–1961.

Walters HC, Craddock AL, Fusegawa H, Willingham MC, Dawson PA (2000) Expression, transport properties, and chromosomal location of organic anion transporter subtype 3. *Am J Physiol Gastrointest Liver Physiol* 279: G1188–G1200.

Wang ZJ, Yin OQ, Tomlinson B, Chow MS (2008) OCT2 polymorphisms and *in-vivo* renal functional consequence: Studies with metformin and cimetidine. *Pharmacogenet Genomics* 18:637–645.

Watanabe T, Onuki R, Yamashita S, Taira K, Sugiyama Y (2005) Construction of a functional transporter analysis system using MDR1 knockdown Caco-2 cells. *Pharm Res* 22:1287–1293.

Watanabe T, Maeda K, Kondo T, Nakayama H, Horita S, Kusuhara H, Sugiyama Y (2009) Prediction of the hepatic and renal clearance of transporter substrates in rats using *in vitro* uptake experiments. *Drug Metab Dispos* 37: 1471–1479.

Watanabe T, Kusuhara H, Maeda K, Kanamaru H, Saito Y, Hu Z, Sugiyama Y (2010) Investigation of the rate-determining process in the hepatic elimination of HMG-CoA reductase inhibitors in rats and humans. *Drug Metab Dispos* 38:215–222.

Weaver JL, Pine PS, Aszalos A, Schoenlein PV, Currier SJ, Padmanabhan R, Gottesman MM (1991) Laser scanning and confocal microscopy of daunorubicin, doxorubicin, and rhodamine 123 in multidrug-resistant cells. *Exp Cell Res* 196:323–329.

Wright SH (2005) Role of organic cation transporters in the renal handling of therapeutic agents and xenobiotics. *Toxicol Appl Pharmacol* 204:309–319.

Wright SH, Dantzler WH (2004) Molecular and cellular physiology of renal organic cation and anion transport. *Physiol Rev* 84:987–1049.

Xia CQ, Liu N, Yang D, Miwa G, Gan LS (2005) Expression, localization, and functional characteristics of breast cancer resistance protein in Caco-2 cells. *Drug Metab Dispos* 33:637–643.

Yamaguchi H, Okada M, Akitaya S, Ohara H, Mikkaichi T, Ishikawa H, Sato M, Matsuura M, Saga T, Unno M, Abe T, Mano N, Hishinuma T, Goto J (2006) Transport of fluorescent chenodeoxycholic acid via the human organic anion transporters OATP1B1 and OATP1B3. *J Lipid Res* 47:1196–1202.

Yamashiro W, Maeda K, Hirouchi M, Adachi Y, Hu Z, Sugiyama Y (2006) Involvement of transporters in the hepatic uptake and biliary excretion of valsartan, a selective antagonist of the angiotensin II AT1-receptor, in humans. *Drug Metab Dispos* 34:1247–1254.

Yasujima T, Ohta KY, Inoue K, Ishimaru M, Yuasa H (2010) Evaluation of 4′,6-diamidino-2-phenylindole as a fluorescent probe substrate for rapid assays of the functionality of human multidrug and toxin extrusion proteins. *Drug Metab Dispos* 38:715–721.

Yonezawa A, Masuda S, Yokoo S, Katsura T, Inui K (2006) Cisplatin and oxaliplatin, but not carboplatin and nedaplatin, are substrates for human organic cation transporters (SLC22A1-3 and multidrug and toxin extrusion family). *J Pharmacol Exp Ther* 319:879–886.

Yousif S, Marie-Claire C, Roux F, Scherrmann JM, Decleves X (2007) Expression of drug transporters at the blood-brain barrier using an optimized isolated rat brain microvessel strategy. *Brain Res* 1134:1–11.

Zaher H, Khan AA, Palandra J, Brayman TG, Yu L, Ware JA (2006) Breast cancer resistance protein (Bcrp/abcg2) is a major determinant of sulfasalazine absorption and elimination in the mouse. *Mol Pharm* 3:55–61.

Zaher H, Meyer zu Schwabedissen HE, Tirona RG, Cox ML, Obert LA, Agrawal N, Palandra J, Stock JL, Kim RB, Ware JA (2008) Targeted disruption of murine organic anion-transporting polypeptide 1b2 (Oatp1b2/Slco1b2) significantly alters disposition of prototypical drug substrates pravastatin and rifampin. *Mol Pharmacol* 74:320–329.

Zair ZM, Eloranta JJ, Stieger B, Kullak-Ublick GA (2008) Pharmacogenetics of OATP (SLC21/SLCO), OAT and OCT (SLC22) and PEPT (SLC15) transporters in the intestine, liver and kidney. *Pharmacogenomics* 9:597–624.

Zalups RK, Bridges CC (2009) MRP2 involvement in renal proximal tubular elimination of methylmercury mediated by DMPS or DMSA. *Toxicol Appl Pharmacol* 235:10–17.

Zelcer N, Huisman MT, Reid G, Wielinga P, Breedveld P, Kuil A, Knipscheer P, Schellens JHM, Schinkel AH, Borst P (2003) Evidence for two interacting ligand binding sites in human multidrug resistance protein 2 (ATP binding cassette C2). *J Biol Chem* 278:23538–23544.

4

PHARMACOLOGICAL AND TOXICOLOGICAL ACTIVITY OF DRUG METABOLITES

W. Griffith Humphreys

4.1 INTRODUCTION

The understanding that drug metabolites often play a critical role in the efficacy and side effect profile of drugs has propelled drug metabolism research to the stage of being an integral part of the lead optimization and development phases of modern drug research. The importance of determination of metabolite profiles has been followed hand-in-hand by dramatic developments in the tools necessary to perform this research.

Work in the area can be divided into four phases that follow a drug's development path and have slightly different goals. The phases can be described as follows: (1) design and optimization phase to aid in discovering the best molecule; (2) initial characterization and prediction phase to aid in compound selection and early development planning; (3) descriptive phase where full metabolite profile of the compound is determined and used to optimize the clinical and safety plans; and (4) retrospective studies designed to help understand unexpected clinical or toxicological findings.

The goal of the work in the discovery phase is to attempt to optimize the metabolism properties of clinical candidates in parallel with the optimization of the potency and efficacy. Drugs that have rapid metabolic clearance are likely to have a high degree of inter- and intrapatient variability (Hellriegel et al., 1996) and be more likely to suffer from drug–drug interactions. Also, drugs with rapid clearance and low %F are likely to require suboptimal dosing regimes, that is, b.i.d. or t.i.d.,

and/or relatively high doses. The former leads to poor patient compliance, suboptimal efficacy, and marketing issues, and the latter can lead to unanticipated toxicities due to a large flux of drug and drug metabolites. Other benefits to understanding and optimizing the metabolism of new chemical entities (NCEs) are as follows:

1. A decreased clearance may translate into a lower overall dose
2. Lower rates of formation and overall amounts of reactive intermediates, which may mediate acute or idiosyncratic toxicities
3. Increased pharmacokinetic (PK) half-life, which will hopefully translate into a longer duration of action, less frequent dosing, and better patient compliance
4. Better understanding of the extrapolation of animal data to humans, making human dose projections more reliable and reducing risk upon entry into clinical development
5. Lower risk of drug–drug interactions as low-clearance drugs are less susceptible to drug–drug interaction caused by the coadministration of inhibitor
6. Lower risk of drug–food interactions due to the reduced dose
7. Decreased formation of metabolites that may have pharmacological activity against the target or may have significant off-target activity

ADME-Enabling Technologies in Drug Design and Development, First Edition. Edited by Donglu Zhang and Sekhar Surapaneni.
© 2012 John Wiley & Sons, Inc. Published 2012 by John Wiley & Sons, Inc.

These considerations make it important to understand the metabolism characteristics of candidate drug molecules, and to optimize these characteristics preclinically, the last item on the list is the subject of this chapter. Drug metabolism plays a central role in modern drug discovery and candidate optimization, and recent reviews have detailed how metabolism has impacted the discovery and design process and challenges that the field faces in the future (Baillie, 2006; Caldwell et al., 2009; Tang and Lu, 2009; Zhang et al., 2009; Sun and Scott, 2010).

Work in the late discovery phase after candidate selection involves the further characterization of the selected drug to allow prediction of clinical PK and metabolism properties, which may include prediction of efficacious human dose (Huang et al., 2008). These properties are important in the design of regulatory toxicology and first-in-human studies and can affect things such as choice of toxicology species, dose selection, and important metabolites to monitor in early studies (Baillie, 2009; Walker et al., 2009; Zhu et al., 2009).

The development phases shifts from a predictive, prospective nature trying to influence drug design to a descriptive study of the complete metabolic profile of the candidate drug. Work done in this phase has significant impact on both the safety plan as well as the clinical pharmacology plan. Additional characterization of the metabolism of the drug after all the metabolic pathways have been determined can yield important information on potential for drug–drug interactions and variability in special populations.

This chapter will give an overview of topics relating to metabolites with the potential for pharmacological or toxicological activity that are often incorporated into discovery and development efforts attempting to advance the optimal drug candidate into clinical investigations and then fully characterize it as part of development.

4.2 ASSESSMENT OF POTENTIAL FOR ACTIVE METABOLITES

In most cases, the metabolism of drugs leads to pharmacological inactivation through biotransformation to therapeutically inactive molecules. However, drug metabolism can also result in the generation of pharmacologically active metabolites. Although formation of pharmacologically active metabolites can be mediated by all types of biotransformation reactions, biotransformation resulting from oxidative metabolism mediated by cytochrome P450 (CYP) enzymes is the more common pathway leading to active metabolites.

Active metabolites may have superior pharmacological, PK, and safety profiles compared with their respec-

tive parent molecules (Fura et al., 2004; Fura, 2006). As a result, a number of active metabolites have been developed and marketed as drugs with improved profiles relative to their parent molecules. Examples of active metabolites of marketed drugs that have been developed as drugs include acetaminophen, oxyphenbutazone, oxazepam, desloratadine (Clarinex: Merck Corp., Whitehouse Station, NJ), cetirizine (Zyrtec: McNeil Corp., Ft. Washington, PA), fexofenadine (Allegra: Sanofi Corp., Bridgewater, NJ), and fesoterodine (Toviaz: Pfizer Corp., New York, NY) (Figure 4.1). Each of these drugs provides a specific benefit over the parent molecule and is superior in one or more property important for the drug's action.

During lead optimization, drug candidates are routinely screened for metabolic stability or in vivo systemic exposure and rank ordered according to the rate and extent of metabolism or systemic exposure level. In the case of metabolic screening, this is usually performed in vitro after incubations of the drug candidates with subcellular fractions such as liver microsomes or intact cellular systems (e.g., hepatocytes) containing a full complement of drug metabolizing enzymes. Compounds with low metabolic stability are then excluded from further consideration because most therapeutic targets require compounds with an extended PK half-life. The same is true with in vivo exposure studies, where high-clearance compounds are discarded. In these early screens, the concentration of the parent compound is typically the only measurement made. Consequently, there is no information on the number, identity, and pharmacological significance of metabolites that may have been formed. Even when metabolic profiling is completed and metabolites are identified, the information is typically used to direct synthesis of analogs with improved metabolic stability through the modification of metabolic soft/hot spots. Thus, the information is rarely used for the purpose of searching for pharmacologically active products as new analogs. However, rapid metabolism of parent compounds could lead to the formation of pharmacologically active metabolites that may have comparatively superior developability characteristics. As a result, metabolic instability, which otherwise may be considered a liability, can become advantageous as a method of drug design.

There are a number of advantages for screening drug candidates for active metabolites during drug discovery. The primary reason is that the process could lead to the discovery of a drug candidate with superior drug developability attributes such as

1. Improved pharmacodynamics (PD)
2. Improved PK
3. Lower probability for drug–drug interactions

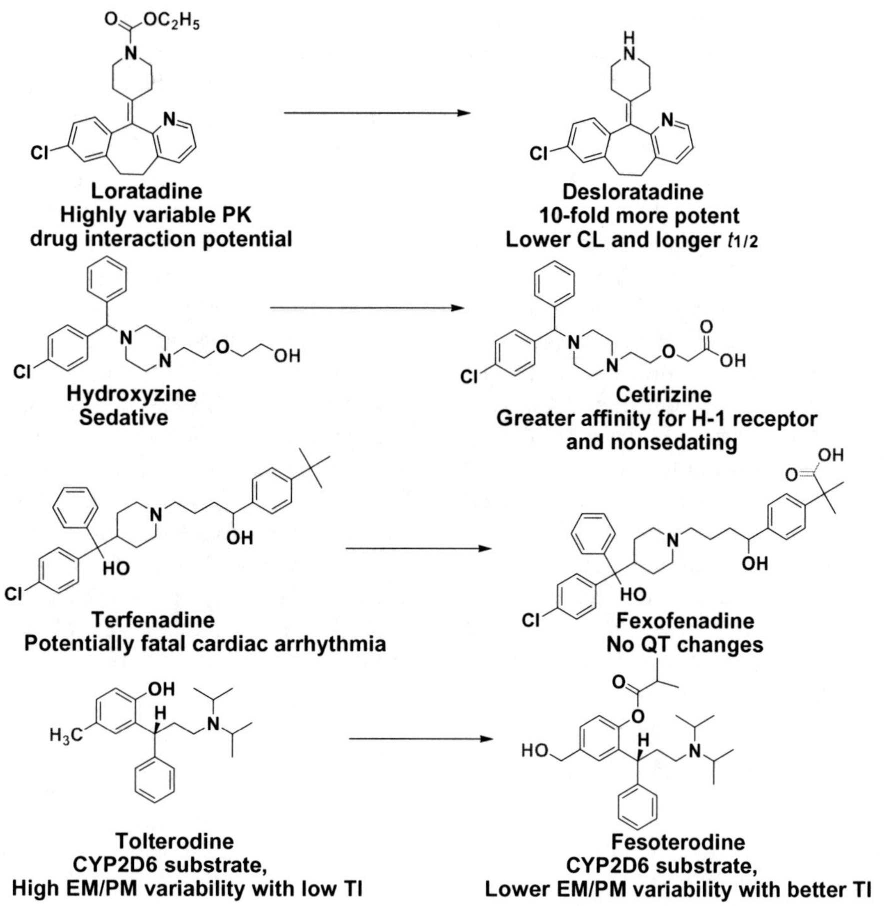

FIGURE 4.1. Metabolites of marketed drugs that have become marketed drugs. EM/PM: extensive metabolizer/poor metabolizer; CL: clearance; CYP2D6: cytochrome P450 2D6; TI: therapeutic index; H-1: nothing needed.

4. Less variable PK and/or PD

5. Improved overall safety profile

6. Improved physicochemical properties (e.g., solubility)

Other advantages of early screening for active metabolites include the potential for modifications of the entire chemical class (chemotype) to improve overall characteristics (Clader, 2004; Fura, 2004). Additionally, tracking active metabolites at the drug discovery stage will allow for the correct interpretation of the pharmacological effect observed in preclinical species in relation to a predicted effect in humans. In other words, if an active metabolite is responsible for significant activity in a species used for preclinical efficacy determination, there is a significant risk that the effect will be dramatically different in humans unless similar levels of metabolite can be expected in humans. Lastly, the early discovery of active metabolites will allow for more complete patent protection of the parent molecule.

An active metabolite may have a low potential for off-target toxicity as it, in most cases, leads to the formation of fewer number of metabolites compared with the parent compound. Moreover, most active metabolites are products of functionalization reactions, and as such are more susceptible to conjugation reactions. Conjugation reactions result in the formation of secondary metabolites that, in general, are safely cleared from the body. For example, phenacetin is metabolized to a number of metabolites, of which the O-deethylation pathway leads to the formation of acetaminophen, a more analgesic agent, whereas N-hydroxylation of phenacetin leads to the formation of a toxic metabolite. On the other hand, the corresponding active metabolite, acetaminophen, is predominantly cleared via conjugation reactions (sulfation and glucuronidation) and has a greater margin of safety relative to phenacetin.

In general, drug metabolism reactions convert lipophilic compounds to more hydrophilic, more water-soluble products. An improvement in the solubility profile is an added advantage, particularly in the current drug discovery paradigm where many drug candidates generated during lead optimization have poor aqueous solubility.

The discovery of an active metabolite can serve as a modified lead compound around which new structure–activity relationships can be examined during the lead optimization stage of drug discovery. For example, this approach was used in the discovery of ezetimibe, a cholesterol absorption inhibitor (Van Heek et al., 1997; Clader, 2004). In these studies, a lead candidate (SCH48461) gave rise to a pharmacologically active biliary metabolite that upon oral administration to rats was approximately 30-fold more potent than the parent molecule. Further optimization of the metabolite through structural modification led to the discovery of ezetimibe, a molecule that was approximately 400-fold more potent than the initial lead candidate.

In summary, tracking active metabolites at the drug discovery stage is not only important to correctly interpret the pharmacological effects in preclinical species but may also lead to the discovery of a lead candidate with superior drug developability characteristics.

4.2.1   Detection of Active Metabolites during Drug Discovery

The exploration of the potential for formation of active metabolites can be carried out with varying degrees of direction from information gathered through metabolism, PK, and biological/pharmacological assays. An example of undirected screening of active metabolites would be the modification of chemical libraries by subjecting them to metabolizing systems and subsequently using these modified libraries for high-throughput screens, either against the intended target or more broadly. This example is a way to generate increased molecular diversity from a given chemical library. However, this approach requires significant "deconvolution" efforts when activity is found in mixtures. To increase the success rate and decrease the number of compounds screened to a manageable size, the search for active metabolites could be limited to those compounds/chemotypes showing high clearance rates in in vitro metabolic stability or in in vivo exposure screens.

Activity assays in vivo may serve as a more rational approach to the exploration of active metabolites. This is most often and most effectively done in the setting of an efficacy experiment that allows for both PD and PK information to be gathered. Analysis of the relationship between the PD endpoint and the PK profile will sometimes demonstrate an apparent disconnect between the two data sets and point to the possibility that an active metabolite is responsible for some of the activity. These disconnects can serve as clear trigger points for the initiation of active metabolite searches.

For example, Van Heek et al. (1997) observed a lead candidate that underwent extensive first-pass metabolism and yet elicited a significant level of pharmacological activity. To evaluate the biological activity of the in vivo biotransformation products, they collected samples of bile from rats dosed with a lead compound and directly administered the samples to bile duct-cannulated rats via an intraduodenal cannula. As a control study, the parent compound prepared in blank bile was dosed in a similar fashion to the recipient rats. The results indicated that the in vivo activity elicited by the bile samples was higher than the parent control sample, clearly indicating the presence of an active metabolite(s) that was more potent than the parent compound. To identify the active component, the bile sample was then fractionated and each fraction tested for biological activity. The structure of the metabolite was then established following the detection of the active fraction. Further modification of the active metabolite led to the discovery of ezetimibe.

Although a lack of correlation between PK and PD data is the clearest trigger point for pursuing the possibility of metabolite contributions to the observed pharmacology, there are several other potential triggers that can be used that include (1) the observation of a greater pharmacological effect upon extravascular administration of a compound relative to parenteral administration; (2) a reduced pharmacological effect upon coadministration in vivo/in vitro with compounds that inhibit metabolism (e.g., aminobenzotriazole, ketoconazole); and (3) a prolonged PD effect despite rapid in vitro metabolism. Examples of the utilization of metabolite structural information in drug design can be found in several reviews (Fura et al., 2004; Fura, 2006).

4.2.2   Methods for Assessing and Evaluating the Biological Activity of Metabolite Mixtures

In order to assess the biological activity, and hence usefulness, of metabolic products, several approaches can be used. The most straightforward approach is to take samples of interest (i.e., microsomal incubations, plasma samples, etc.) and isolate and purify any metabolites present, after which the structure of the metabolites is determined and each is tested for biological activity. An alternative approach is to use bioassay-guided methods where biological samples containing biotransformation products are first evaluated for their pharmacological activity without any effort to isolate or structurally characterize the metabolites. The bioassay methods may be based on the assessment of the pharmacological activity using in vitro ligand binding (Soldner et al., 1998; Lim et al., 1999), cell-based assays, or *in vivo* pharmacological assays (Van Heek et al., 1997). Metabolites can be generated by multiple *in vitro* and *in vivo* methods, some of which are discussed in the following section. Biological activity in the sample mixture can then be evaluated as is or after fractionation of the sample mixture by using chromatographic techniques. The structural identity of the active metabolite can then be determined and its *in vitro* and *in vivo* activity confirmed after isolation and/or after further biological or chemical synthesis.

A systematic approach to profiling active metabolites using a 96-well plate format has been described (Shu et al., 2002). Drug metabolite mixtures are separated and fractions collected into microtiter plates such as 96-well plates. The fractions are then subjected to one or more relevant activity (e.g., receptor ligand binding) assays.

4.2.3 Methods for Generation of Metabolites

There are a number of *in vitro* and *in vivo* biotransformation techniques available to generate metabolites. The *in vitro* techniques include the use of subcellular fractions prepared from cells that mediate drug metabolism, intact cell-based systems, intact organs, and isolated expressed enzymes. *In vivo* methods involve the use of biological fluids (plasma, bile, urine, etc.) obtained from laboratory animals or humans dosed with the parent molecule. Microbial methods and biomimetic systems based on metalloporphyrin chemistry can also be used as bioreactors to produce metabolites.

4.3 ASSESSMENT OF THE POTENTIAL TOXICOLOGY OF METABOLITES

The metabolism of drugs to reactive intermediates followed by covalent binding to cellular components is generally considered to be the basis for the idiosyncratic toxicities caused by some drugs (Walgren et al., 2005; Amacher, 2006; Baillie, 2008; Uetrecht, 2009; Adams et al., 2010; Guengerich, 2010; Park et al., 2010). It is also thought to be related to acute or long-term toxicity seen in animal testing protocols (Guengerich and MacDonald, 2007). The testing of new drug candidates for their potential to form reactive metabolites and the challenges associated with data interpretation from those experiments has been reviewed (Guengerich and Mac-Donald, 2007; Baillie, 2008; Kalgutkar, 2010; Park et al., 2011). Many pharmaceutical companies examine new drug candidates for the potential to form reactive metabolites and if present make attempts to design the property out through targeted structural modification (Doss and Baillie, 2006; Kumar et al., 2010). Reactive

intermediates are most commonly thought to arise through the generation of high-energy intermediates during the oxidation of drugs by CYP or other enzymes (Guengerich, 2005, 2008; Kalgutkar et al., 2005). Examples of these intermediates are epoxides, oxirenes, arene oxides, and quinoid species. Reactive esters formed by the conjugation of carboxylic acids with glucuronic acid or acyl coenzyme A are also thought to be a source of reactive metabolites (Grillo, 2010; Regan et al., 2010; Sawamura et al., 2010).

The mechanisms by which reactive intermediates produce observed toxicity remains controversial (Adams et al., 2010; Guengerich, 2010; Park et al., 2010). The proposed downstream impacts of reactive metabolite formation are generally proposed as (1) damage through generalized oxidative stress or (2) damage through alteration of specific protein function or formation of specific protein adducts that lead to neoantigen formation. The two general mechanisms may work in combination as well.

Multiple recent publications have linked genetic polymorphisms in genes for immune system components to drug-induced toxicity (DIT) (Daly and Day, 2009). These associations have been seen with abacavir (Hughes et al., 2009), flucloxacillin (Daly et al., 2009), ximelgatran (Kindmark et al., 2008), and even acetaminophen (Harrill et al., 2009), and underscore the fact the while covalent binding may serve as an initiating event for idiosyncratic toxicity, the downstream responses are very complicated and difficult to predict based on early steps in the process.

This manuscript will give an overview of how data can be generated and strategies for using that data as part of optimization and design efforts in drug discovery.

The concept that reactive metabolites are the basis for many drug-induced toxicities has been built over many years of research on drugs that demonstrated some type of toxicity, either in laboratory animals or in humans. The positive link of compounds that caused toxicity also being compounds that formed reactive intermediates led to the conclusion that minimization of the formation of reactive metabolites would also minimize the risk of toxicity. While that conclusion is still valid, the need to reduce the concept to practice has resulted in several recent publications that have described the protein covalent binding (PCB) properties of drugs known to cause clinical DIT and have importantly included data on drugs that are not associated with clinical toxicity. This type of data on "safe drugs" has not been available previously and is essential in judging how to use information from reactive metabolite studies. The results from the three major reports from the Pfizer (New York, NY), Daiichi (Tokyo, Japan), and Dianippon (Osaka, Japan)

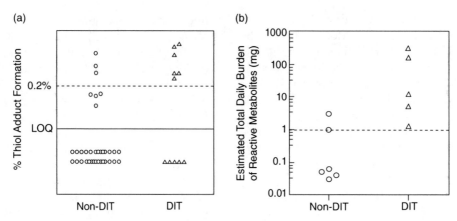

FIGURE 4.2. Scatter plot of % dGSH adduct formation (a) and estimated total daily burden (b) in the DIT and non-DIT groups. The open circles and triangles represent drugs not associated and associated with DIT, respectively. LOQ: limit of quantitation. For illustrational purposes, a horizontal dotted line is plotted at the 0.2% adduct level in panel (a), and another is plotted at the 1 mg level in panel (b). (Results reprinted from Gan et al., 2009; with permission.)

groups were similar. They all conclude that (1) some compounds not associated with DIT do form measurable amounts of reactive intermediate; (2) there is a weak correlation between reactive intermediates and whether a compound is considered generally safe or associated with DIT; (3) there are multiple factors that complicate the simple extrapolation of PCB/ glutathione (GSH) adduct data to a prediction of toxicology liabilities; and (4) there is a large "gray zone" between positives and negatives (Takakusa et al., 2008; Bauman et al., 2009; Nakayama et al., 2009; Usui et al., 2009). A study with a broad range of positive and negative compounds that measured thiol adduct levels after metabolic activation also reached similar conclusions (Gan et al., 2009; Figure 4.2). Although, the conclusions of these studies certainly make the use of reactive metabolite data in a prospective manner more challenging, they do not alter the fact that central to all hypotheses regarding DIT is the initial formation of reactive species.

An important consideration in the discovery arena is that while the studies listed above did measure the reactive metabolite potential of "positives," that is, drugs that caused DIT, the dynamic range of the response was not as great as might be seen in a typical discovery setting. This is possibly because the drugs selected as positives had all progressed through clinical development with associated safety studies in animals and humans. Drug candidates in the discovery setting can be much more proficient at generating reactive species in *in vitro* experiments and the extent of the formation of adducts can easily exceed 50% of the total drug (Evans and Baillie, 2005). The most proficient true positives studied in the *in vitro* experiments described above produced PCB values equivalent to conversion of ca. 10–20% of total drug, while the majority of compounds

studied gave values in the range of 1–5% of total drug converted to trapped reactive intermediate. Whether there is any better correlation between observed toxicity and compounds predicted to produce a very high flux of reactive metabolites is unknown. Perhaps compounds that form very high levels of reactive intermediates *in vitro* are underrepresented in the true positive compound libraries because these types of compound fail during development.

Failure in long-term toxicology studies is another potential risk in advancing a compound that forms reactive metabolites. Whether in a rodent carcinogenicity bioassay or a multiple-dose study in rodent or higher species, many toxicity-related failures are likely linked to downstream events subsequent to the formation reactive metabolites (Guengerich and MacDonald, 2007). The prospective nature of reactive metabolite data has many limitations in the prediction of toxicology failures, similar to those pointed out above for the prediction of idiosyncratic toxicities (Park et al., 2011). Findings in these exploratory toxicity studies are often complicated by uncertainties around the target pharmacology and whether the observed toxicity could be somehow related to the pharmacological target. In these cases, it may be beneficial to try to delink the toxicity from the target by demonstrating that the toxicity is or is not metabolism related. Metabolism-related toxicity is more likely to be compound specific and not related to the target.

4.3.1   Methods to Study the Formation of Reactive Metabolites

The study of the interaction of drugs with cellular components generally involves either measurement of covalent binding after reaction with nucleophilic sites on cellular macromolecules or studies with small-molecule nucleophiles. Newer assays attempt to use generic endpoint readouts to measure the impact of reactive metabolite formation. Finally, mechanistic studies seek to modulate formation of reactive species to attempt to alter observed toxicology.

4.3.2 Reactive Metabolite Studies: *In vitro*

Trapping experiments are most often done with unlabeled drug and glutathione with detection by liquid chromatography-tandem mass spectrometry (LC-MS/MS). Because unlabeled drug is used, these experiments can be done earlier in the discovery cycle than the covalent binding experiments (Shu et al., 2008; Kumar et al., 2010). The detection of glutathione adducts is aided by the characteristic fragmentation pattern of glutathione. Other analytical strategies have been developed to provide qualitative and/or quantitative determination of reactive metabolite formation at the screening stage (Argoti et al., 2005; Dieckhaus et al., 2005; Gan et al., 2005; Mutlib et al., 2005; Yan et al., 2005). These assays provide only qualitative information or utilize a non-physiological trapping agent and are thus somewhat limited in utility. The approach most often employed to circumvent this problem is through the use of radiolabeled trapping agents, typically tritiated glutathione.

PCB studies require radiolabeled drug and are thus usually carried out late in the discovery phase or in early development. A typical experiment is done either with labeled drug in microsomes or by administering the labeled drug to rodents. Both types of studies involve isolation of cellular proteins through precipitation, followed by extensive washing steps to remove noncovalently bound drug. Residual radioactivity bound to proteins is then determined by scintillation counting. The use of this type of data has received much recent attention as a potential predictor of toxicity, especially of idiosyncratic toxicities, although this remains a controversial subject. As described above, multiple recent publications have determined the PCB properties of drugs known to cause clinical DIT and have importantly included data on drugs that are not associated with clinical toxicity.

Many additional types of screens for reactive metabolite generation have been suggested in the literature but have not gained wide acceptance. Most notably are cell-based screens that utilize toxicity readouts in metabolically competent cells (Tuschl et al., 2008) or alternatively measure activation of cellular stress signals (Simmons et al., 2009).

4.3.3 Reactive Metabolite Studies: *In vivo*

Questions regarding the formation of metabolites that covalently bind proteins can be addressed *in vivo* with experiments similar in nature to the experiments described in the *in vitro* section. Animal studies are often focused on the liver proteins as a target for drug binding, while human studies are usually focused on blood components due to obvious ethical limitations. The determination of whether an observed toxicology is associated with reactive metabolites can often be an important question in the interpretation of toxicology results. The experimental approaches utilized often revolve around the modulation of metabolism thought to be involved in bioactivation. This can be done with a number of tools including chemical inhibitors or inducers, and knockout or humanized mouse models. Results from the studies could potentially yield data that would feed into drug design efforts.

4.3.4 Reactive Metabolite Data Interpretation

The information generated from the screening approaches outlined above can be utilized in a number of ways: (1) after determination of the formation of adduct and/or quantitation of adduct level, the adduct structure can be characterized and led to medicinal chemistry approaches to design new analogs with the goal of limiting the amount of adduct formed; (2) as a trigger to perform more advanced covalent binding studies; and (3) as a trigger to do toxicology studies in a cell-based assay or in animals. The first way of utilizing data outlined above is directed solely at improving a chemotype through designing out the liability, while the second and third make an attempt to contextualize the liability for the compound in question. There are multiple literature examples of the use of reactive metabolite data in aiding in the design of drug candidates (Kumar et al., 2010). These studies follow a minimization paradigm and have been successful in many cases in finding structure-based solutions that greatly reduce the reactive metabolite formation of compounds from a given chemotype.

Drug design efforts that seek to minimize the property of reactive metabolite formation naturally lead to questions such as, "how low is low enough?" and "is there a threshold limit that needs to be achieved?" There are several strategies that can be employed to aid in decision making: (1) set a threshold for acceptance—this approach would include setting the threshold at the detection limits of the assay, in other words a zero tolerance policy; (2) minimization of the property relative to a set of standards (should include known positives and negatives); and (3) minimization relative to a local reference standard, that is, a compound from the chemotype being examined. All of these approaches, except for the zero tolerance approach, require some type of quantitation of reactive metabolite formation.

While trying to put reactive metabolite data into perspective, it is important to understand the limitations. As mentioned above, the recent studies examining positives and negatives in PCB and GSH trapping experiments found a very loose correlation of reactive metabolite formation and toxicity endpoint. All studies conclude that it is important to consider the total adduct burden rather than simply the level of adduct formed in the *in vitro* incubation, as the correlation of covalent binding measurement with DIT improved in all cases. The important parameters to consider for calculation of adduct burden are total dose and some measure of the fraction of drug predicted to proceed down the pathway to reactive metabolite formation (e.g., Total burden of adduct = Dose × Fa × Fm × Fadduct, where Fa is fraction absorbed, Fm is fraction metabolized, and Fadduct is the ratio of covalent adduct/total metabolite). Compounds where these considerations may be especially important are compounds that are slowly metabolized *in vitro* yet predicted based on animal studies to be completely cleared via metabolism *in vivo*. Although the experiments are very challenging in these cases, the results may be important as a simple determination of *in vitro* rate will underpredict that total burden of adduct seen *in vivo*.

Another important conclusion from these studies was the fact that there were both false positives and false negatives found in each of the studies. The false positives are not that surprising in that it is very difficult to find highly metabolized compounds that are completely devoid of covalent binding. False negatives in the experiments can be explained by bioactivation by alternate pathways in some but not all cases.

4.3.5   Metabolite Contribution to Off-Target Toxicities

The biotransformation of parent molecules often produces metabolites with relatively minor structural alterations (addition of a hydroxyl group, demethylation, etc.). As these changes may not be dramatic enough to depart from the parent structure–activity relationship, these metabolites may have potent interactions with the same pharmacological target as the parent. Although it is rare for metabolites to have greater potency than parent, a loss of less that one order of magnitude is fairly common and the metabolites may be potent enough to contribute to the overall pharmacological response.

Although it is fairly common for metabolites to potently interact with the same pharmacological target as the parent, it is fairly uncommon for metabolites to have potent interactions with other "off-target" receptors that are not already affected by parent. Of course, this relationship would likely break down as the dose

and plasma concentration of drug and metabolites increase. This relationship between dose and consideration of metabolite toxicity forms the basis of arguments put forth that the metabolites of low-dose drugs do not need to be characterized as extensively as high-dose drugs (Smith and Obach, 2005). Duration of dosing administration is also an important consideration (Smith et al., 2009). The chances for new pharmacology would be expected to increase for metabolites that result from major structural alterations, but even in these cases there are relatively few examples of this type of behavior. The chances for off-target pharmacology of metabolites would also be expected to increase when the target is a member of a receptor family with multiple related members (e.g., peroxisome proliferator-activated receptors [PPARs], kinase families) (Humphreys and Unger, 2006).

4.4   SAFETY TESTING OF DRUG METABOLITES

The topic of safety testing of drug metabolites, the process through which sponsors ensure for the suitability of the species used for toxicological investigations, has received a great deal of recent attention. Much of this is because of the publication of the "Safety Testing of Drug Metabolites Guidance" by the U.S. Food and Drug Administration (FDA) in 2008. This guidance, along with the even more recent International Committee on Harmonization (ICH) guidance ("Non-Clinical Safety Studies for the Conduct of Human Clinical Trials for Pharmaceuticals"), provides a framework to ensure proper exposure of human metabolites in the species used for toxicological investigations and also a path forward in the event the toxicology species are judged to be not adequate to model human metabolite exposure.

As discussed above, the most significant toxicological concern related to metabolite formation is generally thought to be the generation of reactive metabolites. However, the monitoring of plasma has major limitations with regard to detection of reactive metabolite formation. The species detected in plasma are almost always downstream products formed from the reactive metabolites and not the metabolites themselves, but these species can still be a useful signal and possible biomarker of reactive metabolite formation. The absence of circulating reactive metabolites or downstream products of reactive metabolites does not rule out the formation of reactive species that still could be formed at high levels but be undetectable in the plasma. The property that is much better covered with the monitoring of plasma is the potential for metabolites to mediate either on-target (active metabolites) or off-target pharmacology. The potential for metabolites to mediate on-target pharmacology is relatively high, as discussed above many drugs have metabolites with significant activity. However, the potential for a metabolite to mediate receptor-based pharmacology not already displayed by the parent is relatively low (Humphreys and Unger, 2006). The potential for this type of toxicity becomes even lower for drugs that are administered at low dose; this point has been made in several recent reviews (Humphreys and Unger, 2006; Smith and Obach, 2009). As well, it is the basis for the special treatment of drugs administered at less than 10 mg with regard to metabolite characterization in the ICH M3 guidance on nonclinical toxicity studies.

To allow for better early decision making and more complete early development, metabolite profiling has been moved forward in the development programs employed by many pharmaceutical companies. This is an activity that traditionally was not done until well into a development program and was initiated through the conduct of a clinical ADME study with a radiolabeled drug. Many companies now profile human plasma obtained in early clinical trials and detect metabolites through the use of mass spectrometry, often with advanced tools such as accurate mass-mass spectrometry and data mining techniques (Tiller et al., 2008). There are various methods and strategies that allow comparisons to the metabolite levels found in animals. The information gathered in these studies can be used to trigger characterization activities for important human metabolites and can potentially be used to prioritize or deprioritize the conduct of the human ADME study. There has been recent suggestions that the incorporation of trace levels of radioactive material in first-in-human studies (done with a pharmacologically active dose level of the parent) along with accelerator mass spectrometry would allow very early quantitative plasma profiling and potentially full elimination profiling. This type of study could potentially be done with multiple doses of trace-labeled material during the multiple ascending dose protocol to allow metabolite profiling at or close to steady state. Overall, it is important to have a strategy in place that allows a thorough characterization of metabolites (Baillie, 2009; Walker et al., 2009; Zhu et al., 2009).

4.5 SUMMARY

Drug metabolites can play an important role in both the efficacy and toxicity of new drugs, and as such must be fully understood before the full profile of the drug can be described. The types of metabolism information needed along the discovery and development path varies, from medium- to high-throughput screens that allow the metabolic properties of new compounds to be optimized during the discovery phase, to the detailed metabolite profiling from human samples completed during development.

An understanding of the profile and structural characteristics of pharmacologically active metabolites early in the discovery/development process can lead to design changes such as chemotype structural alterations or selection of an alternate lead compound as well as allowing for better patent coverage. These changes have the potential to lead to the selection of an improved candidate, demonstration of this being the large number of successful drugs that have been developed only after finding them as an active metabolite of a marketed drug. As well, the heavy reliance on model-based drug development important for many clinical programs relies on a complete understanding of the PK-PD relationships that drives efficacy, including the contribution of active metabolites.

The relationship of drug metabolites to toxicity will remain a major challenge in the years to come. Even with the full consideration of all factors that may be involved in the prediction of toxicity based on reactive metabolite formation as discussed above, it must be understood that there are large gaps in the basic understanding of the science involved that add a great deal of uncertainty to the prediction process. While a paradigm of screening for evidence of reactive species and modifying structure based on the results is fairly straightforward to implement, it comes with a great deal of risk that the property being optimized is not a true liability. For this reason, the degree with which reactive metabolite information is used in drug design and decision making in early development must be carefully considered on a program-by-program and compound-by-compound basis.

These considerations make drug development an increasingly challenging endeavor as all of the characteristics of a candidate need to be synthesized in a complete picture of the molecule's metabolism and disposition. An important part of this picture is the role metabolites have in the PD and safety profile of a new drug and this ensures that the field of drug metabolism will continue to play an important role in the endeavor to provide high-quality drug candidates that will become safe, highly efficacious medicines.

REFERENCES

Adams DH, Ju C, et al. (2010) Mechanisms of immune-mediated liver injury. *Toxicol Sci* 115(2):307–321.

Amacher DE (2006) Reactive intermediates and the pathogenesis of adverse drug reactions: The toxicology perspective. *Curr Drug Metab* 7(3):219–229.

Argoti D, Liang L, et al. (2005) Cyanide trapping of iminium ion reactive intermediates followed by detection and structure identification using liquid chromatography-tandem mass spectrometry (LC-MS/MS). *Chem Res Toxicol* 18(10):1537–1544.

Baillie TA (2006) Future of toxicology-metabolic activation and drug design: Challenges and opportunities in chemical toxicology. *Chem Res Toxicol* 19(7):889–893.

Baillie TA (2008) Metabolism and toxicity of drugs. Two decades of progress in industrial drug metabolism. *Chem Res Toxicol* 21(1):129–137.

Baillie TA (2009) Approaches to the assessment of stable and chemically reactive drug metabolites in early clinical trials. *Chem Res Toxicol* 22(2):263–266.

Bauman JN, Kelly JM, et al. (2009) Can *in vitro* metabolism-dependent covalent binding data distinguish hepatotoxic from nonhepatotoxic drugs? An analysis using human hepatocytes and liver S-9 fraction. *Chem Res Toxicol* 22(2):332–340.

Caldwell GW, Yan Z, et al. (2009) ADME optimization and toxicity assessment in early- and late-phase drug discovery. *Curr Top Med Chem* 9(11):965–980.

Clader JW (2004) The discovery of ezetimibe: A view from outside the receptor. *J Med Chem* 47(1):1–9.

Daly AK, Day CP (2009) Genetic association studies in drug-induced liver injury. *Semin Liver Dis* 29(4):400–411.

Daly AK, Donaldson PT, et al. (2009) HLA-B*5701 genotype is a major determinant of drug-induced liver injury due to flucloxacillin. *Nat Genet* 41(7):816–819.

Dieckhaus CM, Fernandez-Metzler CL, et al. (2005) Negative ion tandem mass spectrometry for the detection of glutathione conjugates. *Chem Res Toxicol* 18(4):630–638.

Doss GA, Baillie TA (2006) Addressing metabolic activation as an integral component of drug design. *Drug Metab Rev* 38(4):641–649.

Evans DC, Baillie TA (2005) Minimizing the potential for metabolic activation as an integral part of drug design. *Curr Opin Drug Discov Devel* 8(1):44–50.

Fura A (2006) Role of pharmacologically active metabolites in drug discovery and development. *Drug Discov Today* 11(3–4):133–142.

Fura A, Shu YZ, et al. (2004) Discovering drugs through biological transformation: Role of pharmacologically active metabolites in drug discovery. *J Med Chem* 47(18):4339–4351.

Gan J, Harper TW, et al. (2005) Dansyl glutathione as a trapping agent for the quantitative estimation and identification of reactive metabolites. *Chem Res Toxicol* 18(5):896–903.

Gan J, Ruan Q, et al. (2009) *In vitro* screening of 50 highly prescribed drugs for thiol adduct formation—comparison of potential for drug-induced toxicity and extent of adduct formation. *Chem Res Toxicol* 22(4):690–698.

Grillo MP (2010) Drug-S-acyl-glutathione thioesters: Synthesis, bioanalytical properties, chemical reactivity, biological formation and degradation. *Curr Drug Metab* 2011 Mar; 12(3):229–244.

Guengerich FP (2005) Generation of reactive intermediates. *J Biochem Mol Toxicol* 19(3):173–174.

Guengerich FP (2008) Cytochrome p450 and chemical toxicology. *Chem Res Toxicol* 21(1):70–83.

Guengerich FP (2010) Mechanisms of drug toxicity and relevance to pharmaceutical development. *Drug Metab Pharmacokinet* 2011;26(1):3–14.

Guengerich FP, MacDonald JS (2007) Applying mechanisms of chemical toxicity to predict drug safety. *Chem Res Toxicol* 20(3):344–369.

Harrill AH, Watkins PB, et al. (2009) Mouse population-guided resequencing reveals that variants in CD44 contribute to acetaminophen-induced liver injury in humans. *Genome Res* 19(9):1507–1515.

Hellriegel ET, Bjornsson TD, et al. (1996) Interpatient variability in bioavailability is related to the extent of absorption: Implications for bioavailability and bioequivalence studies. *Clin Pharmacol Ther* 60(6):601–607.

Huang C, Zheng M, et al. (2008) Projection of exposure and efficacious dose prior to first-in-human studies: How successful have we been? *Pharm Res* 25(4):713–726.

Hughes AR, Brothers CH, et al. (2009) Genetic association studies to detect adverse drug reactions: Abacavir hypersensitivity as an example. *Pharmacogenomics* 10(2):225–233.

Humphreys WG, Unger SE (2006) Safety assessment of drug metabolites: Characterization of chemically stable metabolites. *Chem Res Toxicol* 19(12):1564–1569.

Kalgutkar AS (2010) Handling reactive metabolite positives in drug discovery: What has retrospective structure-toxicity analyses taught us? *Chem Biol Interact* [Epub].

Kalgutkar AS, Gardner I, et al. (2005) A comprehensive listing of bioactivation pathways of organic functional groups. *Curr Drug Metab* 6(3):161–225.

Kindmark A, Jawaid A, et al. (2008) Genome-wide pharmacogenetic investigation of a hepatic adverse event without clinical signs of immunopathology suggests an underlying immune pathogenesis. *Pharmacogenomics J* 8(3):186–195.

Kumar S, Mitra K, et al. (2010) Approaches for minimizing metabolic activation of new drug candidates in drug discovery. *Handb Exp Pharmacol* 196:511–544.

Lim HK, Stellingweif S, et al. (1999) Rapid drug metabolite profiling using fast liquid chromatography, automated multiple-stage mass spectrometry and receptor-binding. *J Chromatogr A* 831(2):227–241.

Mutlib A, Lam W, et al. (2005) Application of stable isotope labeled glutathione and rapid scanning mass spectrometers in detecting and characterizing reactive metabolites. *Rapid Commun Mass Spectrom* 19(23):3482–3492.

Nakayama S, Atsumi R, et al. (2009) A zone classification system for risk assessment of idiosyncratic drug toxicity using daily dose and covalent binding. *Drug Metab Dispos* 37(9):1970–1977.

Park BK, Boobis A, et al. (2011) Managing the challenge of chemically reactive metabolites in drug development. *Nat Rev Drug Discov* 10(4):292–306.

Park BK, Laverty H, et al. (2010) Drug bioactivation and protein adduct formation in the pathogenesis of drug-induced toxicity. *Chem Biol Interact* [Epub].

Regan SL, Maggs JL, et al. (2010) Acyl glucuronides: The good, the bad and the ugly. *Biopharm Drug Dispos* 31(7):367–395.

Sawamura R, Okudaira N, et al. (2010) Predictability of idiosyncratic drug toxicity risk for carboxylic acid-containing drugs based on the chemical stability of acyl glucuronide. *Drug Metab Dispos* 38(10):1857–1864.

Shu YZ, Johnson BM, et al. (2008) Role of biotransformation studies in minimizing metabolism-related liabilities in drug discovery. *AAPS J* 10(1):178–192.

Shu YZ, Li WY, et al. (2002) Biogram enabled evaluation of active metabolites: An exploratory approach for detecting and characterizing active/toxic drug metabolites. *Drug Metab Rev* 34(Suppl 1):1.

Simmons SO, Fan CY, et al. (2009) Cellular stress response pathway system as a sentinel ensemble in toxicological screening. *Toxicol Sci* 111(2):202–225.

Smith DA, Obach RS (2005) Seeing through the mist: Abundance versus percentage. Commentary on metabolites in safety testing. *Drug Metab Dispos* 33(10):1409–1417.

Smith DA, Obach RS (2009) Metabolites in safety testing (MIST): Considerations of mechanisms of toxicity with dose, abundance, and duration of treatment. *Chem Res Toxicol* 22(2):267–279.

Smith DA, Obach RS, et al. (2009) Clearing the MIST (metabolites in safety testing) of time: The impact of duration of administration on drug metabolite toxicity. *Chem Biol Interact* 179(1):60–67.

Soldner A, Spahn-Langguth H, et al. (1998) A radioreceptor assay for the analysis of AT1-receptor antagonists. Correlation with complementary LC data reveals a potential contribution of active metabolites. *J Pharm Biomed Anal* 17(1):111–124.

Sun H, Scott DO (2010) Structure-based drug metabolism predictions for drug design. *Chem Biol Drug Des* 75(1):3–17.

Takakusa H, Masumoto H, et al. (2008) Covalent binding and tissue distribution/retention assessment of drugs associated with idiosyncratic drug toxicity. *Drug Metab Dispos* 36(9):1770–1779.

Tang W, Lu AY (2009) Drug metabolism and pharmacokinetics in support of drug design. *Curr Pharm Des* 15(19): 2170–2183.

Tiller PR, Yu S, et al. (2008) Fractional mass filtering as a means to assess circulating metabolites in early human clinical studies. *Rapid Commun Mass Spectrom* 22(22): 3510–3516.

Tuschl G, Lauer B, et al. (2008) Primary hepatocytes as a model to analyze species-specific toxicity and drug metabolism. *Expert Opin Drug Metab Toxicol* 4(7):855–870.

Uetrecht J (2009) Immune-mediated adverse drug reactions. *Chem Res Toxicol* 22(1):24–34.

Usui T, Mise M, et al. (2009) Evaluation of the potential for drug-induced liver injury based on *in vitro* covalent binding to human liver proteins. *Drug Metab Dispos* 37(12): 2383–2392.

Van Heek M, France CF, et al. (1997) *In vivo* metabolism-based discovery of a potent cholesterol absorption inhibitor, SCH58235, in the rat and rhesus monkey through the identification of the active metabolites of SCH48461. *J Pharmacol Exp Ther* 283(1):157–163.

Walgren JL, Mitchell MD, et al. (2005) Role of metabolism in drug-induced idiosyncratic hepatotoxicity. *Crit Rev Toxicol* 35(4):325–361.

Walker D, Brady J, et al. (2009) A holistic strategy for characterizing the safety of metabolites through drug discovery and development. *Chem Res Toxicol* 22(10):1653–1662.

Yan Z, Maher N, et al. (2005) Rapid detection and characterization of minor reactive metabolites using stable-isotope trapping in combination with tandem mass spectrometry. *Rapid Commun Mass Spectrom* 19(22):3322–3330.

Zhang Z, Zhu M, et al. (2009) Metabolite identification and profiling in drug design: Current practice and future directions. *Curr Pharm Des* 15(19):2220–2235.

Zhu M, Zhang D, et al. (2009) Integrated strategies for assessment of metabolite exposure in humans during drug development: Analytical challenges and clinical development considerations. *Biopharm Drug Dispos* 30(4):163–184.

5

IMPROVING THE PHARMACEUTICAL PROPERTIES OF BIOLOGICS IN DRUG DISCOVERY: UNIQUE CHALLENGES AND ENABLING SOLUTIONS

JIWEN CHEN AND ASHOK DONGRE

5.1 INTRODUCTION

The past three decades have seen remarkable growth of biopharmaceuticals for the treatment of serious human diseases. From its humble beginnings in 1976, the biopharmaceutical industry has grown to over $75 billion in revenue, and now there are more than 20 biologics with annual sales over $1 billion (IMS Health, June 17, 2008). In 2009, 6 of the 25 new drug approvals by the U.S. Food and Drug Administration (FDA) were biologics (Hughes, 2010). The extremely robust growth of biologics is expected to continue in the foreseeable future, and will likely make up 40% of the product portfolio at major pharmaceutical companies by 2013, as the entire pharmaceutical industry has reallocated significantly more resources to biologics research and development (Goodman, 2009).

Although the primary drivers for the growth of biologics have been their clinical and commercial success, many other factors played an important role as well. Protein therapeutics are often highly specific, with little off-target toxicity and are therefore less likely to fail in clinical development. Many biologics are very stable and have long half-life, allowing for less frequent dosing. In addition, they can block biological targets that are not amenable to small-molecule inhibitors, such as tumor necrosis factor-alpha (TNFα) (Palladino et al., 2003). Furthermore, biopharmaceuticals are often heterogeneous, and sometimes even a small change in the manufacturing process can result in large perturbations in the efficacy and safety profile of the product. Therefore, the barrier of entry for generics is significantly higher for biologics, making their effective patent life substantially longer than small-molecule drugs.

In the early days of the biotechnology industry, major products were dominated by variants of autologous proteins manufactured by recombinant DNA technology. Examples of these products include insulin, interferon alpha and beta, erythropoetin, and granulocyte colony-stimulating factor. More recently, monoclonal antibodies (mAbs) have emerged as effective and targeted therapies for serious disease conditions such as cancer and rheumatoid arthritis with significant clinical benefits. There are 23 mAbs in the U.S. market (Table 5.1), and their number is expected to grow substantially in the coming decade as over 150 mAbs are in various stages of clinical development (Reichert and Dewitz, 2006).

Twenty years ago, almost half of the drug candidates failed in the clinic due to poor pharmaceutical properties, such as pharmacokinetics (Kola and Landis, 2004). As a result of the early involvement of drug metabolism and pharmacokinetic assessment in drug discovery in virtually all pharmaceutical companies, by the year 2000 less than 10% of the clinical candidates were discontinued because of pharmacokinetic and bioavailability reasons. Although the overall attrition rate is still painfully high, and bringing a drug to the market remains a daunting task for drug developers, optimization of drug-like properties in drug discovery is one way of maximizing success in clinical trials. This chapter focuses on

ADME-Enabling Technologies in Drug Design and Development, First Edition. Edited by Donglu Zhang and Sekhar Surapaneni.

TABLE 5.1. Licensed Monocolonal Antibody Therapeutics in the United States as of August 2010

Generic Name	Target	First Approved Indication	Date of FDA Approval
Muromonab-CD3	CD3	Transplant rejection	June 19, 1986
Abciximab	GP IIb/IIIa	Prevention of cardiac ischemic complications	December 22, 1994
Rituximab	CD20	Non-Hodgkin's lymphoma	November 26, 1997
Daclizumab	IL-2Rα	Transplant rejection	December 10, 1997
Basiliximab	IL-2Rα	Transplant rejection	May 12, 1998
Palivizumab	Respiratory syncytial virus	Respiratory syncytial virus infection	June 19, 1998
Infliximab	TNFα	Rheumatoid arthritis	August 24, 1998
Trastuzumab	HER2	Breast cancer	September 25, 1998
Alemtuzumab	CD52	Chronic lymphocytic leukemia	May 7, 2001
Adalimumab	TNFα	Rheumatoid arthritis	December 31, 2002
Omalizumab	IgE	Asthma	June 20, 2003
Tositumomab; I131	CD20	Non-Hodgkin's lymphoma	June 27, 2003
Cetuximab	EGFR	Colorectal cancer	February 12, 2004
Bevacizumab	VEGF	Colorectal cancer	February 26, 2004
Natalizumab	α4-intergrin	Multiple sclerosis	November 23, 2004
Ranibizumab	VEGF-A	Neovascular age-related macular degeneration	June 30, 2006
Panitumumab	EGFR	Colorectal cancer	September 27, 2006
Eculizumab	CP C5	Paroxysmal nocturnal hemoglobinuria	March 16, 2007
Certolizumab pegol	TNFα	Rheumatoid arthritis	April 18, 2008
Golimumab	TNFα	Rheumatoid arthritis	April 25, 2009
Ustekinumab	IL-12/IL-23	Plaque psoriasis	September 25, 2009
Tocilizumab	IL-6 receptor	Rheumatoid arthritis	January 8, 2010
Denosumab	RANK ligand	Osteoporosis	June 1, 2010

This list does not include drugs that have been withdrawn from the market, such as gemtuzumab and efalizumab. The suffix in the generic name indicates the type of antibody: "o" in murine antibody (e.g., muromonab), "xi" in chimeric antibody (e.g., abciximab), "zu" in humanized antibody (e.g., daclizumab), and "u" in fully human antibody (e.g., adalimumab). CD: cluster of differentiation; CP C5: complement protein C5; EGFR: epidermal growth factor receptor; GP IIb/IIIa: glycoprotein IIb/IIIa; HER2: human epidermal growth factor receptor 2; IgE: immunoglobulin E; IL: interleukin; IL-2Rα: 2 receptor α; RANK ligand: receptor activator of nuclear factor kappa-B ligand; TNFα: tumor-necrosis factor α; VEGF: vascular endothelial growth factor.

novel and significant technologies for assessing the pharmaceutical properties of biologics. These are areas of active research and exciting developments, and will likely have positive impacts on the continued growth of protein therapeutics in addressing significant yet unmet medical needs.

5.2   PHARMACOKINETICS

The vast majority of biologics are administered by intravenous and subcutaneous injections. As a result, bioavailability is typically high. Residence time of peptide and protein therapeutics *in vivo* is often correlated with their molecular sizes. An endogenous peptide with molecular weight below 4 kDa may be degraded in the body in minutes, while larger peptides have half-life in hours (Lin, 2009). Proteins with molecular weight greater than 50 kDa can last for days or weeks before they are cleared. Based on this observation, a common approach of prolonging the plasma half-life of peptide and small protein therapeutics is to conjugate polyethylene glycol polymers to the active molecule, a process termed PEGylation (Jevsevar et al., 2010). PEGylation increases the molecular weight, and perhaps more importantly, the hydrodynamic radius of the peptide drug. It is expected to remain an important technology as engineered protein scaffolds are emerging as next-generation antibody therapeutics, such as adnectins and domain antibodies (Gebauer and Skerra, 2009). These antibody fragments have attractive properties, yet their smaller size compared with a conventional antibody requires structural modification to achieve optimal pharmacokinetic properties.

Plasma concentration of protein drugs is usually determined by sandwich enzyme-linked immunoassay (ELISA) (Damen et al., 2009). In this technique, an antibody specific for the analyte is immobilized on the wells of a microtiter plate. Plasma samples, often prediluted, are then applied to the plate and incubated for a period of time. After washes to remove nonspecific binding, a secondary antibody linked to an enzyme is added. In the final step, a chromogenic substrate is applied, which is converted to a colored product by the enzyme. The amount of colored product is determined by the amount of enzyme present, and in turn is determined by the amount of analyte present. Other immunoassays, such as immunofluorescence assay (Washburn et al., 2006) and chemiluminescent assay (Zhu et al., 2009), are based on similar principles, but use different detection methods. ELISA is rapid and highly sensitive, with some assays having a limit of quantification of 1 pM (Ezan et al., 2009).

Even though an overwhelming majority of pharmacokinetic evaluations for protein therapeutics are performed by ELISA, the success of an assay largely depends on the availability of a highly specific antibody, which could take many months to produce. This long lead time may not be an issue when the drug is in clinical development, but could be problematic when the drug is in the discovery stage. In addition, it is often difficult to compare results on the same sample when different capture antibodies are used (Damen et al., 2009). Arguably the most significant advance in the quantification of biologics in complex matrices is immunoaffinity capture followed by mass spectrometry (ImmunoMS) (Ackermann and Berna, 2007; Ezan et al., 2009). Its principle is similar to a sandwich ELISA assay, and instead of using a secondary antibody, a mass spectrometer is employed as a universal detector. As the antibody provides significant selective enrichment of the analyte, ImmunoMS is more sensitive than a traditional mass spectrometry (MS) assay. Likewise, as different molecular entities typically yield different signals that can be differentiated in the mass spectrometer, ImmunoMS is more selective than an ELISA assay. Therefore, ImmunoMS combines the sensitivity of ELISA with the specificity of the MS. Using this approach, Erbitux (ImClone LLC, Bridgewater, NJ, and Bristol-Myers Squibb Company, New York, NY), an mAb for the treatment of colorectal cancer and squamous cell carcinoma, was quantified in human plasma (Dubois et al., 2008). The limit of quantification was similar to ELISA, whereas the assay variability was lower. In another study (Wolf et al., 2004), both the active and inactive forms of the incretin hormones glucose-dependent insulinotropic peptide (GIP) and glucagon-like peptide 1 (GLP-1) were simultaneously measured by ImmunoMS in human plasma with limits of quantification of 5 and 11 pM, respectively. The active and inactive hormones differ by only two residues in the N-terminus, and obtaining antibodies specific for each with little cross-reactivity was rather challenging. Besides enhanced selectivity, robust

quantification from ImmunoMS enables comparison of results from different labs, even though the capture antibodies were different. Furthermore, the capture antibodies used can be polyclonal and less specific than the ones used for a high-quality ELISA assay. This new approach is therefore uniquely suited for drug discovery studies. ImmunoMS could also be a viable alternative to ELISA in clinical development when specificity or data variability becomes a concern.

Although promising, ImmunoMS is significantly more expensive than ELISA. Additionally, appropriate internal standards for MS quantification may not always be available. For recombinant proteins, obtaining an isotopically labeled internal standard is a not a trivial task. Variability of ImmunoMS appears to come primarily from proteolytic digestion and the lack of a suitable internal standard (Damen et al., 2009). Both issues may potentially be overcome by detecting the intact protein using inductively coupled plasma mass spectrometry (ICPMS). In a recent study (Yan et al., 2010), insulin was quantitatively labeled with a reagent containing an elemental tag (^{151}Eu), and detected by isotope dilution high-performance liquid chromatography (HPLC)-ICPMS. ^{153}Eu was used as a universal standard that is not analyte specific. Excellent precision was obtained at concentrations as low as 10 pM. Compared with electrospray MS, ICPMS is attractive with a much larger dynamic range, and a uniform signal response in the MS detector independent of the sequence or size of the protein of interest. With appropriate labeling, ICPMS may become the method of choice for absolute protein quantification.

For peptide drugs with molecular weight below 5 kDa, direct quantification without proteolytic digestion can often be reliably accomplished by liquid chromatography-tandem mass spectrometry (LC-MS/MS). In contrast to ELISA, metabolites of these peptides seldom interfere with the assay. The strategy is similar to that for small-molecule drugs, with some modifications. For instance, the human immunodeficiency virus (HIV) fusion inhibitor enfuvirtide is a 36-amino acid peptide with a molecular weight of 4492 Da. During the method development for its quantification in human plasma, it was found that using three or four volumes of acetonitrile for protein precipitation (a common protocol for small molecule extraction) resulted in poor recovery of the analyte (Chang et al., 2005). This is presumably because the peptide, having significant secondary structure in aqueous solutions, was precipitated to some extent in high organic solvents. When the amount of acetonitrile was reduced to two volumes, excellent recovery of enfuvirtide was obtained. Similar to the sample clean-up for small-molecule drugs, liquid-liquid extraction and solid-phase extraction are commonly employed. An additional approach not typical for small-molecule compounds is ultrafiltration (Cho et al., 2010) using a semipermeable membrane with a certain molecular cutoff. Ultrafiltration separates peptides from proteins with higher molecular weight, thereby providing an alternative to protein precipitation if the recovery is unsatisfactory.

Many biologics are glycoproteins. The glycan (carbohydrate) content in the therapeutic could have a significant impact on its pharmacokinetics. For example, the half-life of recombinant erythropoietin increases with increasing sialic acid content (Egrie et al., 2003). This observation led to the development and commercialization of the hyperglycosylated darbepoetin alfa, which has two extra sialic acid chains and threefold longer residence time than erythropoietin alpha. On the other hand, the mannose receptor was found to play an important role in the clearance of tissue plasminogen activator (tPA) (Biessen et al., 1997). Preadministration of a high-affinity mannose receptor ligand reduced the clearance of tPA by approximately 60% in rats.

Glycan analysis is a challenging analytical problem. It typically involves cleavage of the glycans from the protein, and subsequent separation, identification, and quantification of the released glycans. The procedures are labor intensive and time consuming. An exciting new development in this area is the increasing popularity of "lab-on-a-chip" (Figeys and Pinto, 2000) that com-bines sample preparation, LC separation, and MS detection in one integrated unit. In one of these setups (Bynum et al., 2009), three microfluidic chips staggered on top of each other were employed for N-glycan analysis (Figure 5.1). The first chip had a reaction chamber packed with immobilized PNGase F for deglycosylation. Due to the very high ratio of enzyme to substrate, cleavage of the glycans was accomplished in a matter of seconds, compared with hours using traditional methods. The protein and the glycans then passed through the second chip with a C8 column. Proteins were trapped while the glycans flowed through. The third layer had an enrichment column and an LC column for the separation of the glycans. The end of the LC column was interfaced directed to the mass spectrometer. This integrated design minimized extra-column band broadening and sample consumption, resulting in complete N-glycan profiling of an mAb in as little as 10 minutes using as little as 100 ng of starting material. Unlike N-glycans, release of O-glycans from glycoproteins is more challenging and typically requires chemical hydrolysis. Microwave radiation can be employed to shorten the reaction time considerably (Maniatis et al., 2010).

5.3 METABOLISM AND DISPOSITION

Similar to their pharmacokinetics, metabolism and clearance of peptide and protein drugs to a large extent

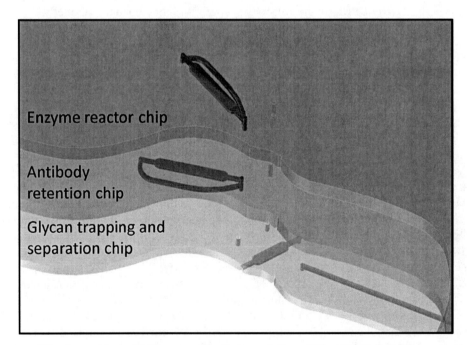

FIGURE 5.1. Diagram of the three chips used for enzymatic deglycosylation, deglycosylated protein removal and glycan concentration, and glycan separation. From top to bottom are the PNGase F enzyme chip, C8 chip, and HPLC chip, respectively. Sample and solvent are automatically introduced through the capillary flow pump and the nanoflow pump.

depends on their molecular sizes (Lin, 2009). The major clearance pathway for peptide therapeutics is metabolism by various proteases. Many endogenous peptides are hormones or growth factors with potent biological activities. Their functions are regulated in part by their rapid degradation in plasma. Through structural modifications such as substitution with D-amino acids, therapeutic peptides have longer residence time, yet in many cases their total clearance still exceeds the glomerular filtration rate, and *in vitro* metabolic stability studies point to the important role metabolism plays in their clearance. Small proteins with a size less than the effective pore size of the glomerular filtration (8–10 nm) are primarily cleared by the renal route. By contrast, for large proteins such as mAbs, renal and biliary excretions are not quantitatively important (Mascelli et al., 2007). They are eliminated mainly by endocytosis followed by proteolytic hydrolysis. All antibodies have two regions: the antigen-binding Fab region, and the "crystallizable" Fc region. Receptor-mediated endocytosis through interactions between the Fab and the corresponding receptor is specific, typically low capacity but rapid (Lin, 2009). Receptors on the cell surface are limited in number and saturable, giving rise to nonlinear kinetics observed for many mAbs. Fc receptor-mediated endocytosis, on the other hand, is nonspecific, high capacity, and relatively slow (Presta, 2008).

When the structure of the circulating metabolite is similar to the protein drug, detailed characterization of the metabolite is possible using ImmunoMS. The power of this approach was first illustrated by the quantitative determination of *in vivo* deamidation for a humanized mAb (Huang et al., 2005). In this work, the mAb was administered to monkeys, and plasma samples were collected and immunoprecipitated. Both the parent drug and the deamidated metabolite at Asn55 were captured simultaneously by this affinity enrichment step. The proteins were digested and the resulting peptides were analyzed by liquid chromatography-mass spectrometry (LC-MS). The rate of deamidation was measured by comparing the relative signal intensity of the deamidated and the native peptide from the digest that contains Asn55. Deamidation can change the biological function of an mAb by the introduction of negative charge. Importantly, Asn55 is located at the antigen binding region, and when it is replaced by Asp55 recombinantly, the binding affinity decreased by more than 10-fold. If the ELISA assay in the pharmacokinetic study measures both the parent and the deamidated metabolite, identification and quantification of the deamidated metabolite would allow for better correlation of pharmacokinetics and pharmacodynamics.

The success of ImmunoMS for metabolite profiling depends on the ability of the capture antibody to bind to the metabolites. The unique structure of peptibodies enables the identification of most major metabolites in preclinical species. Peptibodies are fusion proteins that combine a biologically active peptide with the Fc portion of the antibody to prolong the half-life of the peptide. In a recent study (Hall et al., 2010), *in vivo* metabolism of three thrombopoietin-mimic peptibodies in rats was examined in considerable detail. Following dosing of the drugs to rats, plasma samples were collected and enriched using an antibody specific for human Fc. The captured proteins were subsequently analyzed by LC-MS. Five proteolytic points were identified in one peptibody (Figure 5.2), while the other two peptibodies were more metabolically stable. This information was used to rationally design drug candidates with enhanced stability *in vivo*.

While drug–drug interaction (DDI) remains a critical concern in the discovery and development of small-molecule therapeutics, the risks for DDI with biologics are relatively low. Compared with cytochrome P450 enzymes, proteases responsible for the degradation of biologics are extremely abundant with overlapping specificity. As a result, the clearance capacity is not easily saturated (P. Theil, AAPS National Biotechnology Conference, June 2008, Seattle, WA) and DDI potential for the coadministration of two biologics (uncommon in current clinical practice) is rather small. The DDI risk is not significant either when a biologic is administered concomitantly with a small-molecule drug, as they are metabolized by two distinct groups of enzymes. A protein therapeutic can, however, impact the pharmacokinetics of a small-molecule drug indirectly. For example, interferons have been shown to inhibit CYP1A2 production (Williams et al., 1987). In one of the most comprehensive DDI clinical studies involving a biologic, treatment with PEGylated interferon-α 2a once weekly for 4 weeks in healthy volunteers was associated with a modest increase in the area under the curve (AUC) of theophylline, a substrate for CYP1A2, while no detectable effects were observed with probe substrates for CYP2C9, CYP2C19, CYP2D6, and CYP3A4 (PEGASYS [Hoffmann-La Roche, Basel, Switzerland] package insert).

5.4 IMMUNOGENICITY

Toxicity concerns for biologics are quite distinct from that of small-molecule drugs. Protein therapeutics are typically more specific to their targets, therefore off-target toxicities are less common. On the other hand, while immune response of small-molecule drugs is largely restricted to compounds that form reactive metabolites, immunogenicity is a common concern for

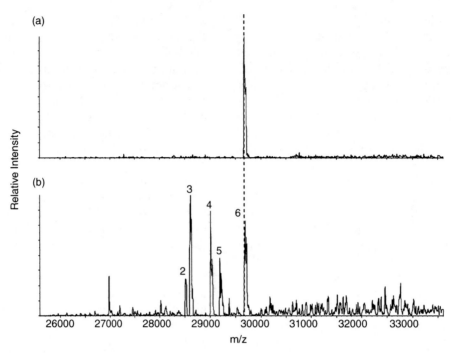

FIGURE 5.2. Reconstructed mass spectra of romiplostim and its metabolites in plasma samples from rats following an intravenous dose of 10 mg/kg: (a) 30 min postdose, showing only one peak corresponding to intact romiplostim; and (b) 24 h postdose. Five metabolites corresponding to proteolytic fragments of romiplostim were identified. (Reprinted from Hall et al., 2010; with permission.)

biologics. Immunogenicity is the capacity of a biologic to illicit an unwanted immune response after administration of the drug. Acute immune responses range from skin reactions at the injection site, to potentially fatal general immune effects such as anaphylaxis and cytokine release syndrome. A hallmark of longer-term immunogenicity is the development of neutralizing anti-drug antibodies (ADAs) in patients. ADAs alter the pharmacokinetics of the drug, and may reduce its efficacy. If the ADAs are directed against recombinant proteins that are endogenous to the patient, life-threatening autoimmune complications may occur. For example, due to a change in the manufacturing of the drug product, ADAs were detected in some patients receiving Eprex (Johnson and Johnson, New Brunswick, NJ) (erythropoietin alpha) (Bennett et al., 2004). These ADAs cross-react with endogenous epoetin, causing a severe type of anemia in these patients. In light of the potential for serious consequences, assessment of immunogenicity is a crucial part of the development plan for biologics.

The first mAb on the market, muromonab, is murine in origin. mAbs approved in the 1990s are either chimeric or humanized. Chimeric and humanized mAbs retain approximately 30% and 5% of the murine sequence, respectively (Strohl, 2009). The first fully human mAb,

adalimumab, was introduced in 2003. With the advancement of antibody engineering technology, mouse components of the antibody have been reduced progressively in the past two decades, with a concurrent reduction in immunogenicity (Hwang and Foote, 2005). Therefore, antibody humanization has resulted in biologics that are better tolerated. Many other factors also affect the immunogenicity of a therapeutic protein (Scott and De Groot, 2010). Protein aggregates, for example, are highly immunogenic. Degradation products of the protein therapeutic can induce immune response as well. These observations have made chemistry, manufacturing, and control (CMC) attributes important contributors to the safety profile of a biologic. The route of administration is another determinant of immunogenicity. Subcutaneous administration tends to be more immunogenic than intravenous injection of the same biopharmaceutical (Ponce et al., 2009), as skin tissues have abundant dendritic cells, which can present the administered protein therapeutics as antigens to trigger an immune response.

Both the FDA and the European Medicines Agency (EMEA) have issued guidance documents for immunogenicity assessment of biologics before submission of clinical trial application and new drug application (FDA and EMEA websites), highlighting concerns from the regulatory agencies. While rigorous immunogenicity testing is a crucial part of drug development, it is equally important to minimize the immunogenic potential of clinical candidates during drug discovery. Immunogenicity of biologics can be screened in three sequential stages, with decreasing throughput and increasing predictive power: *in silico*, *in vitro*, and *in vivo*.

In the context of immune response to biologics, the therapeutic protein is taken up by antigen presenting cells (APCs), processed into a large number of peptide fragments. Only a small number of these fragments have the right sequence to bind to the major histocompatibility complex (MHC) molecules on the surface of the APCs and be presented to the T cells (De Groot and Martin, 2009). There are currently over 10 software tools to predict the ability of peptide fragments to bind to the MHC. All of them can analyze a large amount of protein sequences in a short period of time, and many are able to screen for high-affinity binders (Bryson et al., 2010). A common basis for these different prediction algorithms is a collection of peptides that have been experimentally determined to bind to the MHC. The public Immune Epitope Database, for example, now contains over 100,000 such T-cell epitopes (Peters and Sette, 2007).

Peptides predicted *in silico* to contain T-cell epitopes can be confirmed by *in vitro* human leukocyte antigen (HLA, i.e., human MHC) binding assays in a high-throughput format (Steere et al., 2006). These assays measure the competition of the peptides to a radiolabeled ligand to specific HLA molecules, so a large collection of different HLA types are required. A common limitation of both the *in silico* prediction and *in vitro* binding assay is that they tend to be overpredictive, as many peptides capable of binding to HLA will not be processed from their parent protein during antigen processing, a necessary step for T-cell-mediated immune response (Baker and Jones, 2007). Human T-cell assays (De Groot and Moise, 2007), on the other hand, have fewer false positives as they involve both antigen processing and T-cell activation. In this technique, T cells isolated from the blood of donors are stimulated with either the peptide or the whole protein, and cytokine secretion, a marker of T-cell activation, is measured by ELISpot (Tangri et al., 2005). ELISpot is similar to ELISA, but here the capture antibody is precoated onto the assay plate. Direct correlation between T-cell assays and observed immunogenicity of therapeutic proteins in the clinic has been demonstrated (Figure 5.3).

Although *in vivo* assessment of immunogenicity has the advantage of examining the impact of a biologic in an intact immune system, the challenge is that almost all therapeutic proteins are foreign to wild-type laboratory animals. MHCs have different peptide binding properties in humans and nonhuman primates, and very few of the same peptides are presented by both the

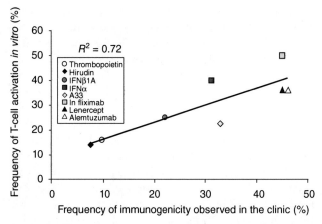

FIGURE 5.3. Correlation between the reported frequency of the development of antitherapeutic antibodies in the clinic (average of literature values) and the frequency of T-cell activation in human blood in an *in vitro* T-cell assay. IFN, interferon; A33, humanized monoclonal antibody A33. (Adapted from Baker and Jones, 2007; with permission.)

human and primate MHCs. To overcome species differences in immune response, transgenic mice with human MHCs have been created. The humanized mice carry common human MHC alleles, including HLA A, B, and DR. Several studies demonstrated significant correlation between immunogenicity in these mice and in humans (Man et al., 1995; Shirai et al., 1995), suggesting that HLA transgenic mice are among the best available proxies for predicting the immunogenic potential of biologics in humans. An alternative approach is to implant human hematopoietic stem cells into newborn immunodeficient mice (Traggiai et al., 2004; Ishikawa et al., 2005). These mice have been shown to produce functional human immune responses.

Through *in silico*, *in vitro*, and *in vivo* studies, drug candidates with high risk of immunogenicity in human patients can be screened out before they reach the clinic. Once the immunogenic epitopes have been identified, residues within the epitopes can be mutated to decrease its binding affinity to MHC molecules. Such a rational drug design approach may lead to new drug candidates with reduced immunogenicity; however, in certain cases, binding affinity to the target may also be compromised. A new trend in "deimmunizing" therapeutic proteins is to incorporate structural elements in the drug to induce tolerance to the immune system. The rationale for this approach is based on our increasing appreciation of the central role regulatory T cells (Tregs) play in maintaining immune homeostasis (Brusko et al., 2008). Tregs are dedicated suppressive T cells that limit the activation of the immune response, and have been implicated in a variety of autoimmune diseases. A significant body of work in animal models has demonstrated the immuosuppressive properties of Tregs to induce tolerance, suggesting that they have the potential to treat autoimmune diseases and facilitate transplantation tolerance. Recent studies have identified a unique class of eptitopes in the Fc fragment of immunoglobulin G (IgG) that activate Tregs, and suppress effector T-cell reactions both *in vitro* and *in vivo* (De Groot et al., 2008). These results also provided a possible mechanism for the immunosuppressive properties of intravenous IgG infusion, an efficacious therapy for inflammatory and autoimmune diseases for several decades in the clinic. Therefore, engineering Tregs activating epitopes in biologics represents a novel approach for reducing the immunogenicity of these therapeutics (Scott and De Groot, 2010).

5.5 TOXICITY AND PRECLINICAL ASSESSMENT

With the exception of the propensity to elicit immune reactions, biologics are generally well tolerated in patients. Although their safety track records have been encouraging, a number of side effects including infection and cancer have been well documented (Hansel et al., 2010). In several cases, toxicity of biologics was well publicized due to their fatal consequences. In early 2005, marketing of natalizumab, an mAb for the treatment of relapsing multiple sclerosis, was suspended only 3 months after it was approved. The suspension was due to the report of three cases of progressive multifocal leukoencephalopathy (PML) in patients taking natalizumab. PML is an opportunistic, usually fatal brain infection that is caused by the John Cunningham virus (JCV). Although JCV is common in the general population, reactivation of the virus leading to PML occurs only in patients with compromised immune functions. The incidence of PML in patients receiving natalizumab treatment ranges from 0.5% to 1.3%, with a positive correlation between the infection risk and the number of infusions the patient receives (FDA Drug Safety Communication, February 2010). It is not clear how the mAb contributes to the tendency to develop PML, but natalizumab administration appears to be associated with CD4+ T lymphopenia and altered trafficking of T lymphocytes into the central nerve system (Carson et al., 2009). Not surprisingly, efalizumab and rituximab, two other mAbs that have been reported to cause PML, also modulate immune functions. Due to the risk of PML, efalizumab was subsequently withdrawn from the market (Pugashetti and Koo, 2009).

Another mAb that caused severe adverse events in the clinic is an experimental drug called TGN1412. This humanized antibody is a CD28 agonist that activates T cells. In the first-in-human trial in early 2006, all six healthy volunteers developed systemic inflammatory response, resulting in serious organ damages such as renal failure, and two of the trial participants required intensive critical care for more than 1 week (Suntharalingam et al., 2006). Symptoms of the trial participants were consistent with cytokine release syndrome, or "cytokine storm," characterized by uncontrolled release of a number of cytokines leading to multiple organ damages that could be life-threatening. The nearly fatal reaction to TGN1412 in humans was unexpected, as doses 500-fold higher than that used in the clinic resulted in no adverse reactions in monkeys during preclinical toxicology testing (Stebbings et al., 2009). This incident highlights the challenges of conducting investigational new drug (IND)-enabling toxicology studies to accurately predict the safety of biologics with novel modes of action in humans. Recent data (Eastwood et al., 2010) indicated that in contrast to other mAbs such as alemtuzumab and rituximab, TGN1412 triggered cytokine release through a unique mechanism involving the activation of CD4+ effector memory T cells. Interestingly, CD28 is expressed in CD4+ effector memory T cells in humans, but not in any of the animal species tested in the TGN1412 toxicology studies. This subtle immunological difference in receptor expression may explain species difference in toxicological response.

In light of the potential for serious adverse events, several *in vitro* tests have been developed to predict the tendency of a biologic to cause cytokine storm. In one of these systems (Findlay et al., 2010), the mAb of interest was immobilized onto 96-well plates by dry-coating. After incubation of the mobilized mAb with human peripheral blood mononuclear cells (PBMCs) overnight, accumulation of cytokines such as TNFα, interleukin-6 (IL-6) and interleukin-8 (IL-8) was quantified by ELISA. Profound stimulation from TGN1412 was observed, while Avastin (Hoffmann-La Roche, Basel, Switzerland), an mAb rarely associated with infusion reactions, resulted in minimal cytokine release. Depending on the mechanism of action of the mAb, whole blood or cocultures such as PBMCs with endothelial cells may be used. It should be noted that at present the predictive power of cytokine release assays is still limited, and they should best be considered as hazard identification tools rather than risk quantification tools for go/no go decision making (Vidal et al., 2010).

5.6 COMPARABILITY

Due to the heterogeneity of recombinant proteins, changes made in the CMC process, from upstream cell culture and fermentation, to downstream protein purification and formulation, may have profound impacts on the safety and efficacy of the clinical candidate (Lynch et al., 2009). As a result, demonstrating comparability at the molecular level is a crucial part of the development program if such changes do occur. Even though this is by no means a trivial task, thanks to recent advances in analytical technologies, especially mass spectrometry, complete chemical characterization of many therapeutic proteins is now routinely obtainable. Primary structure of the protein, including sequence variants and posttranslational modifications, can be determined by orthogonal peptide mapping using ultraviolet (UV) and MS detection. A recent study (Xie et al., 2010) illustrates how comparability can be rapidly examined using advanced LC-MS technologies. In this work, molecular similarity between trastuzumab and a candidate biosimilar was first examined by accurate mass measurement of the intact proteins, and a mass deference of 64 Da was observed. Subsequently, the proteins were digested by trypsin and the resulting peptides were mapped by tandem mass spectrometry (MS/MS) using rapidly alternating low and high collision energy. Unlike traditional data dependent acquisition, this "broadband" MS/MS scan mode is free from biases from the user and the precursor ion signal, allowing for the sequencing of low-abundance peptides. Finally, peptide sequences and observed posttranslational modifications such as glycosylation and deamidation were assigned by the software with minimal user intervention. The mass difference from the intact protein measurement was attributed to sequence difference at two residues in the heavy chain.

Once comparability of the primary structure is confirmed, higher-order structure of the protein can be confirmed by activity assays, circular dichroism (CD) spectroscopy, and nuclear magnetic resonance (NMR) spectroscopy (Locatelli and Roger, 2006; Skrlin et al., 2010). NMR is generally applicable to small proteins. For larger proteins such as mAbs, a promising approach is hydrogen-deuterium exchange coupled with mass spectrometry detection (Chen et al., 2001; Engen, 2009).

5.7 CONCLUSIONS

In drug discovery, improving the drug-like properties of a biologic maximizes the chances of its success in the clinic. Compared with small molecules, a scientist working in this area is faced with a different set of challenges. As a result of the high specificity of many biologics, off-target toxicity is less common. As a result of the high capacity of common clearance pathways for protein therapeutics, DDI is typically not a concern. As a result of the structural similarity of fully human antibodies, their pharmacokinetics in humans is often predictable.

On the hand, many biologics are immunogenic in humans to various extents, thereby necessitating vigorous assessments of the immunogenicity of drug candidates before they reach the clinic. Thanks to advances in protein engineering and our understanding of T-cell-mediated immune reactions, immunogenic potential of drug candidates can often be minimized through rational design. In addition to immunogenicity, the spectacular failure of preclinical toxicology studies in predicting the safety of TGN1412 in the first-in-human trial has underscored the risks associated with bringing biologics to the clinic, some of which act on completely novel targets that are not amenable for small-molecule interventions. Furthermore, unlike small molecules, biologics are produced in living cells with considerable amount of heterogeneity. State-of-the-art analytical technologies, especially those utilizing mass spectrometry, have enabled scientists to establish comparability of two biologics from different manufacturing processes with unprecedented speed and accuracy. We anticipate that novel and improved methods for measuring pharmaceutical properties of biologics will help bring better treatment options to millions of patients more efficiently.

REFERENCES

Ackermann BL, Berna MJ (2007) Coupling immunoaffinity techniques with MS for quantitative analysis of low-abundance protein biomarkers. *Expert Rev Proteomics* 4:175–186.

Baker MP, Jones TD (2007) Identification and removal of immunogenicity in therapeutic proteins. *Curr Opin Drug Discov Devel* 10:219–227.

Bennett CL, Luminari S, Nissenson AR, Tallman MS, Klinge SA, McWilliams N, McKoy JM, Kim B, Lyons EA, Trifilio SM, Raisch DW, Evens AM, Kuzel TM, Schumock GT, Belknap SM, Locatelli F, Rossert J, Casadevall N (2004) Pure red-cell aplasia and epoetin therapy. *N Engl J Med* 351:1403–1408.

Biessen EA, van Teijlingen M, Vietsch H, Barrett-Bergshoeff MM, Bijsterbosch MK, Rijken DC, van Berkel TJ, Kuiper J (1997) Antagonists of the mannose receptor and the LDL receptor-related protein dramatically delay the clearance of tissue plasminogen activator. *Circulation* 95:46–52.

Brusko TM, Putnam AL, Bluestone JA (2008) Human regulatory T cells: Role in autoimmune disease and therapeutic opportunities. *Immunol Rev* 223:371–390.

Bryson CJ, Jones TD, Baker MP (2010) Prediction of immunogenicity of therapeutic proteins: Validity of computational tools. *BioDrugs* 24:1–8.

Bynum MA, Yin H, Felts K, Lee YM, Monell CR, Killeen K (2009) Characterization of IgG N-glycans employing a microfluidic chip that integrates glycan cleavage, sample purification, LC separation, and MS detection. *Anal Chem* 81:8818–8825.

Carson KR, Focosi D, Major EO, Petrini M, Richey EA, West DP, Bennett CL (2009) Monoclonal antibody-associated progressive multifocal leucoencephalopathy in patients treated with rituximab, natalizumab, and efalizumab: A review from the Research on Adverse Drug Events and Reports (RADAR) Project. *Lancet Oncol* 10:816–824.

Chang D, Kolis SJ, Linderholm KH, Julian TF, Nachi R, Dzerk AM, Lin PP, Lee JW, Bansal SK (2005) Bioanalytical method development and validation for a large peptide HIV fusion inhibitor (Enfuvirtide, T-20) and its metabolite in human plasma using LC-MS/MS. *J Pharm Biomed Anal* 38:487–496.

Chen J, Walter S, Horwich AL, Smith DL (2001) Folding of malate dehydrogenase inside the GroEL-GroES cavity. *Nat Struct Biol* 8:721–728.

Cho SY, Xu M, Roboz J, Lu M, Mascarenhas J, Hoffman R (2010) The effect of CXCL12 processing on CD34+ cell migration in myeloproliferative neoplasms. *Cancer Res* 70:3402–3410.

Damen CW, Schellens JH, Beijnen JH (2009) Bioanalytical methods for the quantification of therapeutic monoclonal antibodies and their application in clinical pharmacokinetic studies. *Hum Antibodies* 18:47–73.

De Groot AS, Martin W (2009) Reducing risk, improving outcomes: Bioengineering less immunogenic protein therapeutics. *Clin Immunol* 131:189–201.

De Groot AS, Moise L (2007) Prediction of immunogenicity for therapeutic proteins: State of the art. *Curr Opin Drug Discov Devel* 10:332–340.

De Groot AS, Moise L, McMurry JA, Wambre E, Van Overtvelt L, Moingeon P, Scott DW, Martin W (2008) Activation of natural regulatory T cells by IgG Fc-derived peptide "Tregitopes." *Blood* 112:3303–3311.

Dubois M, Fenaille F, Clement G, Lechmann M, Tabet JC, Ezan E, Becher F (2008) Immunopurification and mass spectrometric quantification of the active form of a chimeric therapeutic antibody in human serum. *Anal Chem* 80:1737–1745.

Eastwood D, Findlay L, Poole S, Bird C, Wadhwa M, Moore M, Burns C, Thorpe R, Stebbings R (2010) Monoclonal antibody TGN1412 trial failure explained by species differences in CD28 expression on CD4+ effector memory T-cells. *Br J Pharmacol* 161:512–526.

Egrie JC, Dwyer E, Browne JK, Hitz A, Lykos MA (2003) Darbepoetin alfa has a longer circulating half-life and greater *in vivo* potency than recombinant human erythropoietin. *Exp Hematol* 31:290–299.

Engen JR (2009) Analysis of protein conformation and dynamics by hydrogen/deuterium exchange MS. *Anal Chem* 81:7870–7875.

Ezan E, Dubois M, Becher F (2009) Bioanalysis of recombinant proteins and antibodies by mass spectrometry. *Analyst* 134:825–834.

Figeys D, Pinto D (2000) Lab-on-a-chip: A revolution in biological and medical sciences. *Anal Chem* 72:330A–335A.

Findlay L, Eastwood D, Stebbings R, Sharp G, Mistry Y, Ball C, Hood J, Thorpe R, Poole S (2010) Improved *in vitro* methods to predict the *in vivo* toxicity in man of therapeutic monoclonal antibodies including TGN1412. *J Immunol Methods* 352:1–12.

Gebauer M, Skerra A (2009) Engineered protein scaffolds as next-generation antibody therapeutics. *Curr Opin Chem Biol* 13:245–255.

Goodman M (2009) Sales of biologics to show robust growth through to 2013. *Nat Rev Drug Discov* 8:837.

Hall MP, Gegg C, Walker K, Spahr C, Ortiz R, Patel V, Yu S, Zhang L, Lu H, DeSilva B, Lee JW (2010) Ligand-binding mass spectrometry to study biotransformation of fusion protein drugs and guide immunoassay development: Strategic approach and application to peptibodies targeting the thrombopoietin receptor. *AAPS J* 12:576–585.

Hansel TT, Kropshofer H, Singer T, Mitchell JA, George AJ (2010) The safety and side effects of monoclonal antibodies. *Nat Rev Drug Discov* 9:325–338.

Huang L, Lu J, Wroblewski VJ, Beals JM, Riggin RM (2005) *In vivo* deamidation characterization of monoclonal antibody by LC/MS/MS. *Anal Chem* 77:1432–1439.

Hughes B (2010) 2009 FDA drug approvals. *Nat Rev Drug Discov* 9:89–92.

Hwang WY, Foote J (2005) Immunogenicity of engineered antibodies. *Methods* 36:3–10.

Ishikawa F, Yasukawa M, Lyons B, Yoshida S, Miyamoto T, Yoshimoto G, Watanabe T, Akashi K, Shultz LD, Harada M (2005) Development of functional human blood and immune systems in NOD/SCID/IL2 receptor {gamma} chain(null) mice. *Blood* 106:1565–1573.

Jevsevar S, Kunstelj M, Porekar VG (2010) PEGylation of therapeutic proteins. *Biotechnol J* 5:113–128.

Kola I, Landis J (2004) Can the pharmaceutical industry reduce attrition rates? *Nat Rev Drug Discov* 3:711–715.

Lin JH (2009) Pharmacokinetics of biotech drugs: Peptides, proteins and monoclonal antibodies. *Curr Drug Metab* 10:661–691.

Locatelli F, Roger S (2006) Comparative testing and pharmacovigilance of biosimilars. *Nephrol Dial Transplant* 21(Suppl 5):v13–v16.

Lynch CM, Hart BW, Grewal IS (2009) Practical considerations for nonclinical safety evaluation of therapeutic monoclonal antibodies. *MAbs* 1:2–11.

Man S, Newberg MH, Crotzer VL, Luckey CJ, Williams NS, Chen Y, Huczko EL, Ridge JP, Engelhard VH (1995) Definition of a human T cell epitope from influenza A nonstructural protein 1 using HLA-A2.1 transgenic mice. *Int Immunol* 7:597–605.

Maniatis S, Zhou H, Reinhold V (2010) Rapid de-O-glycosylation concomitant with peptide labeling using microwave radiation and an alkyl amine base. *Anal Chem* 82:2421–2425.

Mascelli MA, Zhou H, Sweet R, Getsy J, Davis HM, Graham M, Abernethy D (2007) Molecular, biologic, and pharmacokinetic properties of monoclonal antibodies: Impact of these parameters on early clinical development. *J Clin Pharmacol* 47:553–565.

Palladino MA, Bahjat FR, Theodorakis EA, Moldawer LL (2003) Anti-TNF-alpha therapies: The next generation. *Nat Rev Drug Discov* 2:736–746.

Peters B, Sette A (2007) Integrating epitope data into the emerging web of biomedical knowledge resources. *Nat Rev Immunol* 7:485–490.

Ponce R, Abad L, Amaravadi L, Gelzleichter T, Gore E, Green J, Gupta S, Herzyk D, Hurst C, Ivens IA, Kawabata T, Maier C, Mounho B, Rup B, Shankar G, Smith H, Thomas P, Wierda D (2009) Immunogenicity of biologically-derived therapeutics: Assessment and interpretation of nonclinical safety studies. *Regul Toxicol Pharmacol* 54:164–182.

Presta LG (2008) Molecular engineering and design of therapeutic antibodies. *Curr Opin Immunol* 20:460–470.

Pugashetti R, Koo J (2009) Efalizumab discontinuation: A practical strategy. *J Dermatolog Treat* 20:132–136.

Reichert JM, Dewitz MC (2006) Anti-infective monoclonal antibodies: Perils and promise of development. *Nat Rev Drug Discov* 5:191–195.

Scott DW, De Groot AS (2010) Can we prevent immunogenicity of human protein drugs? *Ann Rheum Dis* 69(Suppl 1):i72–i76.

Shirai M, Arichi T, Nishioka M, Nomura T, Ikeda K, Kawanishi K, Engelhard VH, Feinstone SM, Berzofsky JA (1995) CTL responses of HLA-A2.1-transgenic mice specific for hepatitis C viral peptides predict epitopes for CTL of humans carrying HLA-A2.1. *J Immunol* 154:2733–2742.

Skrlin A, Radic I, Vuletic M, Schwinke D, Runac D, Kusalic T, Paskvan I, Krsic M, Bratos M, Marinc S (2010) Comparison of the physicochemical properties of a biosimilar filgrastim with those of reference filgrastim. *Biologicals* 38:557–566.

Stebbings R, Poole S, Thorpe R (2009) Safety of biologics, lessons learnt from TGN1412. *Curr Opin Biotechnol* 20:673–677.

Steere AC, Klitz W, Drouin EE, Falk BA, Kwok WW, Nepom GT, Baxter-Lowe LA (2006) Antibiotic-refractory Lyme arthritis is associated with HLA-DR molecules that bind a *Borrelia burgdorferi* peptide. *J Exp Med* 203:961–971.

Strohl WR (2009) Therapeutic monoclonal antibodies: Past, present, and future. In *Therapeutic Monoclonal Antibodies: From Bench to Clinic*, An Z, ed. John Wiley & Sons, Inc., Hoboken, NJ.

Suntharalingam G, Perry MR, Ward S, Brett SJ, Castello-Cortes A, Brunner MD, Panoskaltsis N (2006) Cytokine storm in a phase 1 trial of the anti-CD28 monoclonal antibody TGN1412. *N Engl J Med* 355:1018–1028.

Tangri S, Mothe BR, Eisenbraun J, Sidney J, Southwood S, Briggs K, Zinckgraf J, Bilsel P, Newman M, Chesnut R, Licalsi C, Sette A (2005) Rationally engineered therapeutic proteins with reduced immunogenicity. *J Immunol* 174:3187–3196.

Traggiai E, Chicha L, Mazzucchelli L, Bronz L, Piffaretti JC, Lanzavecchia A, Manz MG (2004) Development of a human adaptive immune system in cord blood cell-transplanted mice. *Science* 304:104–107.

Vidal JM, Kawabata TT, Thorpe R, Silva-Lima B, Cederbrant K, Poole S, Mueller-Berghaus J, Pallardy M, Van der Laan JW (2010) *In vitro* cytokine release assays for predicting cytokine release syndrome: The current state-of-the-science. Report of a European Medicines Agency Workshop. *Cytokine* 51:213–215.

Washburn WK, Teperman LW, Heffron TG, Douglas DD, Gay S, Katz E, Klintmalm GB (2006) A novel three-dose regimen of daclizumab in liver transplant recipients with hepatitis C: A pharmacokinetic and pharmacodynamic study. *Liver Transpl* 12:585–591.

Williams SJ, Baird-Lambert JA, Farrell GC (1987) Inhibition of theophylline metabolism by interferon. *Lancet* 2:939–941.

Wolf R, Hoffmann T, Rosche F, Demuth HU (2004) Simultaneous determination of incretin hormones and their truncated forms from human plasma by immunoprecipitation and liquid chromatography-mass spectrometry. *J Chromatogr B Analyt Technol Biomed Life Sci* 803:91–99.

Xie H, Chakraborty A, Ahn J, Yu YQ, Dakshinamoorthy DP, Gilar M, Chen W, Skilton SJ, Mazzeo JR (2010) Rapid comparison of a candidate biosimilar to an innovator monoclonal antibody with advanced liquid chromatography and mass spectrometry technologies. *MAbs* 2:379–394.

Yan X, Xu M, Yang L, Wang Q (2010) Absolute quantification of intact proteins via 1,4,7,10-tetraazacyclododecane-1,4,7-trisacetic acid-10-maleimidoethylacetamide-europium labeling and HPLC coupled with species-unspecific isotope dilution ICPMS. *Anal Chem* 82:1261–1269.

Zhu Y, Hu C, Lu M, Liao S, Marini JC, Yohrling J, Yeilding N, Davis HM, Zhou H (2009) Population pharmacokinetic modeling of ustekinumab, a human monoclonal antibody targeting IL-12/23p40, in patients with moderate to severe plaque psoriasis. *J Clin Pharmacol* 49:162–175.

6

CLINICAL DOSE ESTIMATION USING PHARMACOKINETIC/PHARMACODYNAMIC MODELING AND SIMULATION

LINGLING GUAN

6.1 INTRODUCTION

Research for a new drug can be divided in two different stages: discovery and development. Drug development can be divided into preclinical and clinical phases. Knowledge accumulated from the preclinical *in vitro* and *in vivo* animal pharmacology and toxicology data is used to enable the drug to be administered to humans with minimal risk. Increased costs and reduced productivity of pharmaceutical research and development have been growing concerns in recent years. One potential reason could be that applied sciences have not kept pace with the advances of basic sciences. However, there are ways to improve the process that may make it more cost-effective and get new drugs to the market more quickly and safely.

The quality and quantity of preclinical data that support the development of a new drug have considerably improved in recent years. The incredible advances in discovery technology and lead optimization have generated more compounds than ever in preclinical phase to be tested in clinical trials. One of the major benefits of pharmacokinetic (PK)/pharmacodynamic (PD) modeling in this stage is to offer assistance with the identification of potential surrogates or biomarkers for a particular mechanism of action. This can expedite the decision making to initiate the clinical trials and also allow an early read on the drug candidate's performance in the clinical drug development, especially when the clinical effects of a drug are not easily measurable or are slow to develop.

The information-gathering activities for clinical drug development (Phases 1–4) begin when a lead compound is first introduced into human and end when the accumulated information is summarized and presented to a regulatory agency for a market-access decision. Phase 1 typically starts with dose escalation studies in healthy volunteers with dense sampling. The main objective is to investigate the PK profile of drug candidate. An initial dose–concentration–effect relationship may be established to predict and assess initial safety and tolerability, and to assess the PD that may provide an early read of activity/efficacy and to use biomarkers to establish a probable dose range. This is generally viewed as a learning phase of clinical development, but it can also be a confirming phase for the preclinical development, especially with regard to PK. The knowledge gained here is then communicated back to the preclinical in order to facilitate a more informed selection of the next generation of drug candidates.

Traditional paradigm in clinical drug development is an established and accepted approach in the development of novel therapeutic products based on an empirical linear model (Rooney et al., 2001). A few pharmaceutical companies have adapted the learn-and-confirm drug development paradigm. Drug development is an information-gathering process of two successive cycles (Sheiner and Steimer, 2000). The first cycle (Phase 1 and Phase 2a) addresses the question of whether benefit over existing therapies can reasonably be expected in terms of efficacy/safety. It involves Phase 1 to learn the largest short-term dose in humans without

ADME-Enabling Technologies in Drug Design and Development, First Edition. Edited by Donglu Zhang and Sekhar Surapaneni.
© 2012 John Wiley & Sons, Inc. Published 2012 by John Wiley & Sons, Inc.

harm, and then Phase 2a to test whether that dose induces some measurable short-term therapeutic benefit in intended patients. The second cycle (Phase 2b and Phase 3) attempts to learn (Phase 2b) what is a good drug regimen to achieve acceptable benefit/risk ratio and ends with just one formal Phase 3 trial of that regimen versus a comparator.

The conducting Phases 2 and 3 clinical trials constitute a significant proportion of the development costs for any new drug. Phase 2 studies test the dose range hypothesis and are typically divided into two stages, Phase 2a and Phase 2b. The objective of Phase 2a is to test the efficacy hypothesis of a drug candidate, demonstrating the proof of concept. This stage ends the first learning–confirming cycle according to Sheiner (1997). One might view learning as an exercise in confirming; it is far more natural to view learning as the stage of constructing a model of the input–outcome relationship itself. The confirmed positive data from this stage allows the drug candidate to advance to the next cycle. One of the major objectives of Phase 2b is to develop the concentration–response relationship in efficacy and safety by exploring a large range of doses in the target patient population. The PK/PD relationship that has evolved from the preclinical phase up to Phase 2b is used to assist in designing the Phase 3 trial.

Phase 3 studies aim to demonstrate efficacy in a statistically robust manner. The objective of Phase 3 studies is to provide confirmatory evidence that demonstrates an acceptable benefit/risk in a large target patient population. This stage provides the condition for final characterization of the PK/PD relationship and for explanation of the sources of interindividual variability in patient response using population PK (PopPK)/PD approaches. This approach relies on empirical decision making, that is, decisions made in a semiqualitative and semiquantitative manner from a limited series of study outputs. It can lead to suboptimal development strategies because it does not fully quantitatively assess the likelihood of success and risk associated with particular decisions.

Because failure of a drug in Phase 3 trials is increasingly being viewed as commercially unacceptable, a novel paradigm using quantitative analysis has been utilized (Rooney et al., 2001). As a means of improving decision making in product development, the quantitative analysis can be implemented at various stages throughout the drug development to maximize the likelihood of success for those compounds entering Phase 3. Model-based drug development (MBDD) has been recognized as a promising tool to address some of the related challenges. This approach includes use of adaptive trial designs, extensive use of biomarkers, development of personalized medicines, and pharmacometrics,

which have been addressed on regulatory approval. The novel paradigm can be used to test various "what if" scenarios that are related to drug, disease characteristics, and trial design.

The learning phase of drug development preferred model-dependent methods, thus a model must interpolate between and extrapolate beyond the conditions of the actual study design from which they are defined. The types of models must be predictive, and most importantly, mechanistic as opposed to empirical. Model-independent approaches are generally suited in the confirmatory stage. This approach sometimes actually means that the models are independent from assumptions and is achieved from a simple statistic whose distribution depends on controllable study design, but not on the "model" for the data. By using this technique, the development team can predict and evaluate a range of potential results any given trial might generate before dosing any patient in a clinical study. This chapter will focus on some novel paradigms to predict first–in–human dose and to apply PK/PD modeling and simulation in clinical drug development.

6.2   BIOMARKERS IN PK AND PD

Biomarkers have the potential to improve the efficiency of clinical trials and drug development because they can be reliable substitutes for clinical responses. Several recent review articles discuss the potential for PK/PD modeling and simulation in drug discovery and development using biomarkers, which create value throughout the drug development process by providing input to help early decision making.

6.2.1   PK

PK, concerned with what the body does to the drug, is the study of a drug and/or its metabolite kinetics in the body, from which drug concentration–time profiles are generated. The body is a very complex system and a drug undergoes many steps as it is being absorbed, distributed, metabolized, and excreted (ADME). PK has been broadly divided into absorption and disposition (Ruiz-Garcia et al., 2008). Disposition is further subdivided into distribution and elimination, and the term elimination includes metabolism and excretion as illustrated in Figure 6.1.

PK is a discipline that uses mathematical models to describe and predict the time course of a drug and its metabolite concentrations in serum, plasma, or whole blood, tissue, and organs over time. PK is the relationship between drug inflow (a more general view than dose) and drug concentration(s) at various body sites,

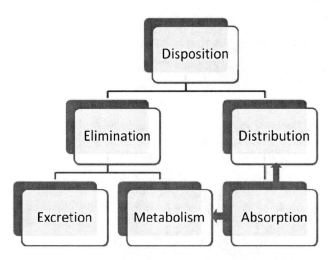

FIGURE 6.1. Schematic relationship of drug absorption, distribution, metabolism, and excretion (ADME).

also called biophase(s) of drug action, and for which subprocesses (submodels) for ADME determine the relationship. From the PK perspective, the drug has been eliminated when it is no longer in its original chemical structure and when biotransformation of the parent compound takes place.

Different methodologies or mathematical approaches have been described, and we can classify PK strategies under two categories, namely noncompartmental analysis (NCA) and modeling. The NCA method uses algebraic equations and provides a descriptive knowledge of the test compound. Very minimal assumptions are made (the principal one is the first-order exponential terminal phase), which minimizes bias by eliminating the assumption required in any modeling (Ruiz-Garcia et al., 2008). It is the method of choice for bioequivalence (BE) studies and the only analysis required by the Food and Drug Administration (FDA) for a new drug application (NDA). However, no simulations can be performed from NCA. Under modeling, the analyses include the classical compartmental analysis, physiologically based pharmacokinetic (PBPK) analysis, and hybrid models where the classical compartmental analysis and physiologically based models are mixed.

6.2.2 PD

PD, concerned with what the drug does to the body, is how the body responds to the presence of drug, for example, changes in blood pressure, from which response–time profiles are generated. PD is the relationship between drug concentrations and pharmacological effects (sometimes called surrogate effects, but

more properly called bioresponses), and in turn, the relationship of these responses to clinical outcomes. PD can also include the probability of experiencing a side effect at a given dose of drug.

There may be other factors in PD that affect how well the drug works.

Both the PK and the PD of drugs can be described using mathematical and statistical functions. Dose–exposure–effect relationship, one in which "exposure" can be the concentration–versus-time profile, or area under the concentration curve or C_{max}, and "effect" may be a pharmacological marker, an index of efficacy, or a measure of safety (Sheiner and Steimer, 2000). In many cases, the distribution of PK and PD behavior resembles a bell curve or normal distribution. A typical patient is perhaps mid-aged, reasonably healthy, and not taking other medications, and the PK for this patient would be at the center of the distribution.

The difference in the PK of the patients from the typical provides a means of dose adjustments to ensure that all patients have the same exposure to the drug. This is the essence of PopPK/PD modeling, in which functions are tested to see how well they describe data collected from clinical trials and trends are examined to determine how patient factors such as demography, disease status, progression, and comedications might affect exposure to the drug and their subsequent response in these patients.

6.2.3 Biomarkers

Biomarkers serve a wide range of purposes for therapeutic assessment strategies in clinical drug development, by providing a basis for the selection of lead candidates for further clinical trials, and contribute to the understanding of the pharmacology of drug candidates. Biomarkers may provide information on mechanism of action, on guidance in dosing, and for characterization of the subtypes of disease for which a therapeutic intervention is most appropriate. Biomarkers can also be used as diagnostic tools, or as a tool for staging disease, and for monitoring of clinical response to an intervention. Figure 6.2 defines biomarkers, surrogate endpoints, and clinical endpoints (Frank and Hargreaves, 2003; Matfin, 2007; Wagner, 2008).

A well-defined clinical endpoint such as death or stroke is the most reliable way to assess the clinical impact of a therapeutic intervention (drug or device). However, this assessment may be impractical for the evaluation of some chronic disease therapies with increased cost and complexity because long study periods are usually required to achieve clinical endpoints and large numbers of patients are needed in trials for evaluation. Through evaluation, biomarkers can

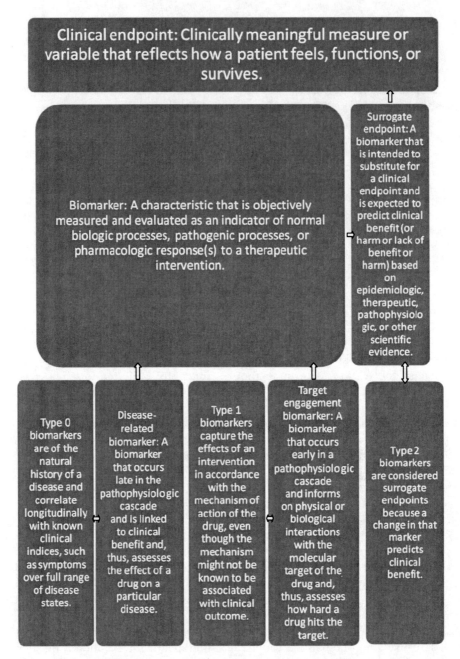

FIGURE 6.2. The biomarker concepts and definitions.

progress to become surrogate endpoints (blood pressure) for "hard" clinical endpoints (death, stroke, endstage renal disease). This process of progression of a biomarker requires rigorous statistical criteria for assessing correlation of biomarkers or surrogate endpoints with clinical endpoints or outcomes.

Assessments of biomarkers should not be redundant, nor should a vast array of biomarkers be measured without good scientific or clinical rationale to avoid false claims or misinterpretations. The biomarkers for a given study should represent both efficacy and safety because assessment of benefit and risk must be the primary goal of the clinical development plan for all therapeutic interventions. It is very difficult to capture both benefit and risk within a single measure. Therefore, a range of biomarkers is often utilized. The techniques used in discovery of biomarkers include genomics, proteomics, transcriptomics, metabolomics, biostatistics, and thorough understanding of disease mechanism by systems biology.

Biomarkers in clinical trials and drug development are invaluable. Proof of concept is achieved when it is established that a drug candidate works to improve a disease condition in a way predicted by the proposed mechanism of action. A biomarker may be measured from a biosample, a physical examination or an imaging test, more easily or frequently and with higher precision. Qualification is the fit-for-purpose evidentiary process of linking a biomarker with biological processes and clinical endpoints. Validation is the fit-for-purpose process of assessing the assay and its measurement performance characteristics, determining the range of conditions under which the assay will give reproducible and accurate data. Developing and using biomarkers can replace a distal clinical endpoint with a more proximal one, which can be measured earlier.

Use of biomarkers and surrogate endpoints can contribute to better understanding of the mechanism of drug action, justification of selected dose range, and giving an estimate of the benefit/risk ratio. Biomarkers can indicate that the agent is working even when the patient does not feel subjective changes, which can result in making faster "go or no-go" decisions and reduced sample size requirements for clinical studies. Surrogate endpoints can support drug approval by regulatory authorities (FDA) on the basis of adequate and well-controlled clinical trials establishing that the drug has an effect on a surrogate endpoint that can reasonably predict clinical benefit. However, further Phase 4 postmarketing studies may be required to confirm clinical benefit.

The next-generation biomarkers include biomarkers for pharmacogenomics and pharmacogenetics (Frank and Hargreaves, 2003). Pharmacogenomics is an assay intended to study interindividual variations in whole-genome or candidate gene single-nucleotide polymorphism (SNP) maps, haplotype markers, and alterations in gene expression or inactivation that may be correlated with pharmacological function and therapeutic response. Pharmacogenetics is an assay intended to study interindividual variations in deoxyribonucleic acid (DNA) sequence related to drug absorption and disposition (PK) or drug action (PD) including polymorphic variation in the genes that encode the functions of transporters, metabolizing enzymes, receptors, and other proteins.

Biomarkers are receiving increasing use as indicators of disease presence, severity, activity, prognosis, and therapeutic efficacy because they can provide reliable substitutes for clinical responses. An ideal surrogate endpoint should be practicable, low cost, acceptable to patient, consistent with the known mechanism of drug action, and detectable and quantifiable (with acceptable variability) in all patients. They should also be consistent with the pathogenesis of the disease, biologically plausible and valid, and linked to the ultimate clinical outcome to be accepted as proof of effectiveness. Through validation, biomarkers can be used as surrogate endpoints for "hard" clinical endpoints. Suitable biomarkers and surrogate endpoints for assessing cardiometabolic risk in clinical trials may be identified while long-term clinical trials for various agents are ongoing.

6.3 MODEL-BASED CLINICAL DRUG DEVELOPMENT

Under conventional drug development paradigms, the details of qualification and assumptions are often lost as more data become available. This poor attrition in the recent development provides an incentive for a different approach. MBDD is a more integrated approach to organize and construct quantitative relationships of complex data from discrete experiments conducted along the drug development pathway. The relationships are ultimately reviewed by the global regulatory agencies at critical development stages to provide a more rational explanation for decisions regarding a more focused clinical development program and lead to optimal decision making.

Biomarkers are characteristics that are measured and evaluated as indicators of normal biological processes, pathogenic processes, or pharmacological responses. Modeling is often very useful for prediction of PK/PD exposure following multiple-dose regimens based on single-dose data, usually the earliest and one of the most common extrapolations. Modeling is invaluable for certain situations, where the only way to resolve the problem is to model. PD modeling is used to relate exposure to effect, where effects are measurements of safety, biomarkers, or clinical response. MBDD uses biomarkers and drug and disease models to integrate knowledge of the drug effect(s), disease progression, dose response, relevant covariates, and safety or toxicity over time.

6.3.1 Modeling

The term "modeling" is data-driven (exploratory) analysis based on appropriate mathematical and statistical functions that describe how study inputs and outputs relate to each other, and how the models can be used to predict study outputs under a variety of different conditions. Inputs may be descriptive data, such as dose, patient demographics, or compliance information.

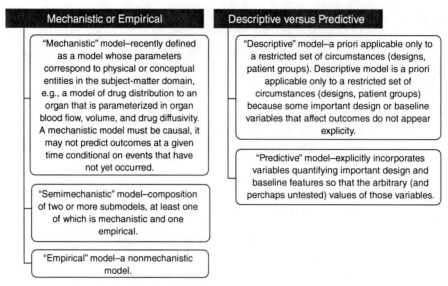

FIGURE 6.3. Different types of models.

Outputs are the results obtained from the conduct of the study, such as concentration–time data or PD and efficacy results. The model cannot be specified in the protocol prior to the study, and may be developed or refined based on the results, as "driven" by the collected "data." The definitions of the different models are listed in Figure 6.3 (Sheiner and Steimer, 2000; Aarons et al., 2001).

Models can be classified as drug, disease, or trial models, and can be generated by biological, pharmacological, or statistical modeling. Biological modeling models pathways of disease as targets of intervention. Pharmacological modeling determines relationships for dose, exposure, and response. Statistical modeling assesses development strategies and trial designs in populations. Drug models describe the relationship between exposure (PK) and response (PD) for both desired and undesired effects. Disease models describe the time course of disease and placebo effects. Trial models describe variations from the nominal trial protocol due to things such as patient dropout and lack of adherence to the dosing regimen. Disease and drug models are used to understand the relationship between treatment, biomarker changes, and clinical outcomes.

Model for the data is a differential equation to gain insight and to specify the probability distribution of random variables representing observations, drug concentration(s), and effect(s) in a subject matter domain. The model may be the composition of several submodels (such as PK or PD), including ones for the distribution of unobservable conceptual entities, for example, drug clearance in the population. The model can be totally empirical where the compartments are black boxes (one, two, or three-compartment models) or have some physiological meaning assigned to the compartments defined in the PBPK models. PK models can be for parent drug and active metabolites. PD models are for surrogate and clinical endpoints and side effects, characterizing the interaction between disease progression, placebo, and drug effect. Covariate models relate patient characteristics and inter- and intrapatient variability to PK and PD parameters. Modeling can also capture patient compliance and dropout, and uncertainty.

6.3.2 Simulation

Simulation is applied modeling bounded by assumed input limits and their relationships to the outcomes to predict the outputs. In some regards, simulation is the opposite of modeling. Modeling starts with data and ends with a model with parameters; simulation starts with a model and parameters and end up with simulated outputs. Simulation is used to try out various doses before the trials, thus reducing the number of dose regimens that are given in the patients. Simulation can be based on resampling procedures, where new data sets are produced by random sampling from empirical databases. The "bootstrap" (Efron and Tibshirani, 1993) is such a resampling technique increasingly popular for determination of uncertainty on estimated or predicted quantities (confidence intervals).

Simulation encompasses two types of situations where random variables are absent (deterministic simulations) and present (stochastic simulations) (Aarons et al., 2001; Sale, 2004). An example of deterministic simulations is when the multiple-dose profile is predicted from the typical (mean) parameters as estimated from data from a single-dose study. Deterministic simulation does not intend to reflect reality since it fails to incorporate the random variability in the calculation of outputs and only addresses the outcome for the mean of population (Sale, 2004). An example of stochastic simulations is when a possible outcome for all individuals of a future trial is simulated based on individual parameter values sampled from a probability distribution (Aarons et al., 2001). Scholastic simulation incorporates the random sources of variability into the model using a technique called Monte Carlo simulation.

The simulation model built using preclinical and early clinical data must be viewed as a "working model" that will become refined and improved as more data and information becomes available. After a model is developed, it should be tested rigorously to ensure that the predictions from this model are reasonable. This model may be used to simulate the PK and PD of a new study and then compare the simulated results to the actual study data. PK and PD models can in turn be tested to simulate different dose regimens and treatment options. This is part of the "learn and confirm" cycle where information from each study is modeled, and the model is then used to help design the next study, making the study more informative and robust.

6.3.3 Population Modeling

The main purpose of a PK analysis is to obtain a set of parameters that describe the kinetic behavior of the drug in the body after administration. We tend to extrapolate the PK analysis results from one study to the potential target population for the study drug. The information obtained from any clinical trial is based on the scientific signal obtained out of random noise. "Signal" is the variability that can be explained and predicted by the computerized model, while "noise" in a clinical trial is any source of unexplained variability. Population modeling is a way of making the signals easier to measure by reducing noise, and thus is particularly useful to identify trends in the data. It is important to identify the factors for the variability in patients' response to a drug, because PK and/or PD of this drug can be different for each patient.

FDA defines the term PopPK as "the study of the sources and correlates of variability in drug concentrations among individuals who are the target patient population receiving clinically relevant doses of a drug of interest." PopPK involves the simultaneous analysis of all data from a group (population) of individuals to provide information about the variability of the

models' parameters. The richness of this information depends on the mathematical approach performed and the data set available. The phrase "population analysis" not only is used to refer to onstage analysis but also is equally applicable to naive pooled data (NPD), naive averaged data (NAD), standard two-stage (S2S) approach, three-stage (3S) analysis, Bayesian estimation, and one-stage analysis (linear and nonlinear mixed effects modeling).

Population modeling uses a smaller number of samples per individual, which is ethically desirable, and handles imbalance data, which can occur in dose escalation studies or when some samples are below the limit of quantification, in a proper fashion. Population modeling performs combined analyses and evaluates complex models when data from many subjects are analyzed together (Aarons et al., 2001) to obtain a direct estimate of the population characteristics of PK/PD model parameters. PopPK/PD analysis can detect deviation from dose proportionality and determine the factors involved, and can allow more rapid identification of appropriate dose regimens. It also can provide better insights to patient tolerance and response to a drug, and optimize studies to shorten development time via the conduct of informative trials. The ability to make timely and well-informed development decisions translates to major cost savings.

6.3.4 Quantitative Pharmacology (QP) and Pharmacometrics

QP is used as an organized approach in MBDD to integrate individuals from different disciplines in a combined effort to develop quantitative models in order to solve specific, complex, and multivariate problems in drug development. QP translates the relationships between disease, drug action, and variability to improve patient outcomes by enabling quantitative decision-making processes from discovery to development. QP results in greater efficiency in drug development by encouraging more transparent and objective study designs and more data-driven risk-taking to optimize timelines, analyses, and decision making. This in turn will help better understand the mechanism of drug–disease interactions, identify molecular targets with high probability of impacting disease, select the ideal drug candidate in early drug development, optimize clinical trials through modeling and clinical trial simulations, and identify optimal patients and treatment regimens for particular drugs.

There is a growing interest in using PK/PD modeling and simulation to support drug discovery/development and to improve decision making in clinical drug development and regulatory review. Pharmacometrics is defined as the science that uses models based on pharmacology, physiology, and disease for quantitative analysis of interactions between drugs and patients to aid efficient drug development, regulatory decisions, and rational drug treatment in patients. A major focus of pharmacometrics is to understand variability in drug response, which involves PK, PD, and disease progression with a focus on populations and variability. Variability may be predictable (differences in body weight or kidney function) or apparently unpredictable (a reflection of current lack of knowledge).

PK/PD modeling and simulation can be used in all phases of drug development, from preclinical to clinical phases. The development of biomarker/surrogate models and the data-driven (exploratory) analysis should ideally progress throughout drug development, including preclinical and Phase 1. The dominant paradigm for drug development is to conduct empirical clinical trials and to test against the null hypothesis as the primary method for evaluating the performance of new medicines. The Phase 2 trials are to assess the particular dose, while Phase 2b has a broader objective, that is, to predict the best dose for each patient type. The modeling and simulation of patient data can often be more efficiently and reliably undertaken if models have previously been developed from rich, high-quality Phase 1 data.

During the learning phase of drug development, a model is constructed to explanatorily analyze and estimate the quantitative relationship between inputs and outputs in some mechanistic view. Inputs are drug dose and timing, patient characteristics, disease stage, and so on, and the outputs are observable clinical results, such as occurrence of an adverse event. The crucial in the learning phase is that drug action is being modeled and the sought model is a predictive (not descriptive) one, thus value of the modeling becomes greater as more data are available. The useful learning models must be mechanistic as opposed to empirical to extrapolate beyond the bounds of the design on which they are defined (models must explicitly express the values of those bounds), and provide credible extrapolations (models must incorporate the current scientific understanding of their subject matter field).

The model-based approach becomes the basis of clinical trial simulation, which has received increasing attention and is being used to guide the design of patient studies based on models derived from early clinical data. Using modeling and simulation to try out various study designs is a good way of ensuring that clinical trials are able to detect a drug effect. Modeling the probability of a desirable response and experiencing side effects at a given dose helps justify the risk-to-benefit ratio of a drug to a patient. More importantly,

modeling can show the relative risks and benefits by changing the dose in certain patients, thus to provide a better and more rational dose for patients that improves their response and reduces the side effects. Optimal use of PK/PD modeling and simulation may ultimately be used for registration support.

6.4 FIRST-IN-HUMAN DOSE

FDA requires that the "first-in-human" dose be based on preclinical animal safety and efficacy to ensure safe Phase 1 clinical trials. Preclinical animal models include mouse and nonhuman primates. Although interspecies scaling of PK parameters is not optimal, it is important for predicting doses in human. It is necessary to understand the drug disposition in different species through *in vitro* and *in vivo* experiments so that the appropriate species for human estimates can be selected.

6.4.1 Drug Classification Systems as Tools for Development

Patients prefer oral dose in compliance with the drug therapy, and development effort has been focused on a good understanding of bioavailability (F) from drug physicochemical properties. F has been defined by the FDA as the rate and extent to which the active ingredient or active moiety is absorbed from a drug product and becomes available at the site of action. The Biopharmaceutics Classification System (BCS) divides drugs into four groups based on the values of their *in vitro* solubility and *in vivo* permeability (Ruiz-Garcia et al., 2008). The Biopharmaceutical Drug Disposition and Classification System (BDDCS) has been suggested to divide compounds into four classes according to their solubility, transport and metabolism activities, and the food effects to be predicted (Wu and Benet, 2005), as shown in Figure 6.4.

Class 1 (High Solubility, High Permeability): The high permeability and high solubility of such compounds allow high concentrations in the gut to saturate any transporter, both efflux and absorptive. Class 1 compounds may be substrates for both uptake and efflux transporters *in vitro*, but transporter effects on absorption will not be clinically important. However, efflux transporters may have measurable effect on the penetration of the compounds through the blood–brain barrier (BBB). If the systemic concentration of the compounds is lower, transporters may overcome the effect of the high passive permeability. These compounds can also be involved in transporter-

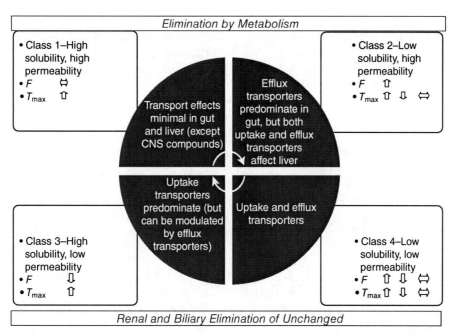

FIGURE 6.4. Biopharmaceutical Drug Disposition and Classification System (BDDCS). CNS, central nervous system.

mediated drug–drug interactions. Because of the complete absorption from high solubility and high permeability of Class 1 compounds, high-fat meals will have no significant effect on F but may increase the time to maximum plasma concentration (T_{max}) due to delayed stomach emptying.

Class 2 (Low Solubility, High Permeability): These high-permeability compounds will pass through the gut membranes and uptake transporters will have no effect on absorption. However, the low solubility will limit the concentration at the enterocytes to prevent saturation of the efflux transporters. Consequently, efflux transporters will affect the extent of oral F and the rate of absorption of Class 2 compounds. High-fat meals will increase F of Class 2 drugs due to inhibition of efflux transporters in the gut and additional solubilization of drug in the intestinal lumen (e.g., micelle formation). High-fat meals will decrease T_{max} if there is inhibition of efflux cycles or increase T_{max} if there is slowing of stomach emptying. The delayed emptying will dominate, especially when membrane permeation is passive. However, high-fat meals may have little effect on F but increase T_{max} due to delayed gastric emptying if high permeability of a Class 2 compound results from uptake transporters. Food effects on F and other transporter-related drug interactions will decrease or eliminate if formulation changes could increase the solubility of Class 2 compounds and enable them to function like Class 1 drugs.

Class 3 (High Solubility, Low Permeability): For Class 3 compounds, the drug availability will be sufficient in the gut lumen due to good solubility, but an uptake transporter will be necessary to overcome the poor permeability of these compounds. Apical efflux transporters may also be important for the absorption of such compounds when sufficient penetration is achieved via an uptake transporter. High-fat meals will decrease F for Class 3 compounds due to inhibition of uptake transporters in the intestine. Some Class 3 compounds can be substrates for intestinal efflux transporters, the modulation on efflux transporters could result in an increase or no effect on F. However, T_{max} would increase from a combination of slower absorption and delayed stomach emptying.

Class 4 (Low Solubility, Low Permeability): Due to the low permeability and low solubility of these compounds, both uptake and efflux transporters play an important role in the oral F of Class 4 compounds. It is difficult to predict the food effects for Class 4 compounds, but F may increase because of increased solubilization of drug in the intestine and inhibition of efflux transporters by high-fat meals.

6.4.2 Interspecies and Allometric Scaling

A few interspecies scaling methods, such as maximum life span potential (MLP), brain weight, body surface area (BSA), physiologically based models, and allometric scaling, have been used to predict drug PK parameters in human from preclinical data. Varying degrees of success have been achieved from these techniques for different drugs over the years. However, it still remains controversial which method gives the best prediction. Although it was able to correlate BSA directly to the kidney weight of different mammalian species (rat, dog, and human) (Pinkel, 1958), and BSA may work well within a small size range for some parameters; however, it may not be linear for all physiological parameters in all species. Physiochemical parameters such as log D and pK_a values have been proposed to predict volume of distribution of the drug in humans (Lombardo et al., 2002), in which the predicted result was within two- to threefold of the observed value.

Two approaches of interspecies scaling have been applied extensively over the past decade, PBPK models and allometric scaling. Both methods have some advantages and some disadvantages and vary significantly in their ability to predict. PBPK models employ mechanistic-based evaluation of drug disposition, and are mathematically complex and time consuming. Allometric scaling assumes body weight to scale PK parameters of a drug across several species. It is an interspecies scaling method that utilizes log transformation of body weight and PK parameters extrapolated from animals to humans by the equation $Y = aWb$. Y is the physiological parameter, such as clearance (CL), half-life ($t_{1/2}$), or volume of distribution, where W is the body weight, a is the allometric coefficient, and b is the allometric exponent. The equation can also be expressed as $Y = \log a + b \log W$, in which a is the y-intercept and b is the slope.

Many physical and physiological parameters have been shown to correlate with body weight in animals as well as humans (Dedrick et al., 1970; Weiss et al., 1977; Boxenbaum, 1982). Therefore, it is assumed that PK parameters such as clearance and half-lives would also correlate with body weight. In a situation when drug metabolism occurs only in the liver and drug clearance is high and dependent on hepatic blood flow, blood flow correlates with body weight to the 0.75 power. The basal metabolic rate has been suggested to scale to the two-third or three-fourth power of body mass (West et al., 1997; Dodds et al., 2001). The exponents of the allometric equations for blood flow, enzyme activities, and renal excretion parameters are around 0.7 (body weight to the 0.7 power).

This body weight rule, also called rule of exponents, is useful to make reasonable interspecies predictions even if the elimination mechanism is not known. However, improvements are needed to obtain qualitative, accurate extrapolations because metabolism is frequently not limited by hepatic blood flow, and species difference in absorption, plasma protein binding, biliary excretion,

glomerular filtration, and intestinal flora also contributes to the complexity of scaling. The use of various correction factors including brain weight, MLP, and glomerular filtration was reported (Nagilla and Ward 2004), in which it was concluded that allometric scaling with or without correction factors was suboptimal in predicting human clearance from *in vivo* preclinical data.

6.4.3   Animal Species, Plasma Protein Binding, and *in vivo–in vitro* Correlation

The efficiency of using two or three species for allometric scaling of drug CL and volume of distribution in humans was investigated (Mahmood and Balian, 1996; Tang et al., 2007). Tang and colleagues developed CLhuman = 0.152·CLrat or 0.410·CLdog or 0.407·CLmonkey for one-species prediction, and CLhuman = a·Whuman^b, where $b = 0.628$ (rat/dog) and 0.650 (rat/monkey), as a two-species method. Mahmood (2009) pointed out that human CL prediction from rule of exponents remains more accurate than the one- or two-species methods. Often it is assumed that the more species used in the scaling, the more reliable the prediction (Mahmood and Balian, 1996). It is important to include a wider range of body weights for allometric scaling for a better defined slope of extrapolation. However, the improvement in prediction by including more species needs to be carefully evaluated for additional time, resources, and cost.

Protein binding is a reversible process that maintains equilibrium of bound and unbound forms of drug in the blood. Only the unbound fraction exhibits pharmacological effects and may be metabolized and/or excreted. Human serum albumin, lipoprotein, glycoprotein, and α, β, and γ globulins are the common blood proteins that drug binds to in human. Since albumin is basic, acidic, and neutral drugs will primarily bind to albumin. If albumin becomes saturated, then these drugs will bind to lipoprotein. Basic drugs bind to acidic alpha-1-acid glycoprotein. Various medical conditions may affect the levels of these drug binding proteins. The affinity to albumin is increased for large molecules, particularly those containing aromatic rings, amino acids, and linkers. Drugs that require a long duration of action can be achieved by increasing plasma protein binding. The high protein binding leads drug to a lower water solubility and a longer half-life.

Generally, drugs bind to albumin to higher extent in human than in other species. The differences may be explained by the varying amounts of albumin in each species or simply by structural differences in albumin from different species. A correction on plasma protein binding is needed for allometric scaling for high protein binding drugs, but may not make a difference if protein binding is low. It may be appropriate to scale unbound drug if protein binding is high or varies between species (Mordenti, 1985). However, Mahmood (2000) concluded that unbound clearance cannot be predicted more accurately than total clearance using the allometric approach for 20 randomly selected drugs in his studies. A reasonable qualitative estimate in human was obtained by dividing clearance, volume of distribution, and terminal half-life by 40, 200, and 4, respectively, from allometric scaling of *in vivo* rat data alone (Caldwell et al., 2004).

It is important to understand the drug *in vivo* metabolic pathway, as the metabolic stability of drug can vary among species and influence the drug pharmacological effects, toxicity, and clearance. Drug metabolism is classified into two phases. Phase 1 reactions are oxidation, reduction, and hydrolyses, which form products that can proceed to Phase 2 reactions. Phase 2 reactions are conjugations (i.e., glucuronidation) that ultimately make the metabolites more polar expediting renal excretion. Drugs metabolized by both Phase 1 and 2 reactions may be excreted as parent drug, metabolites from Phase 1, and conjugation products from Phase 2 reactions. Interspecies differences in drug metabolism may result from differences in the amount of an enzyme, an inducer or inhibitor of the enzyme, the activity of an enzyme, or the extent of competing reactions. The focus should be on the qualitative and quantitative differences in the activity and specificity of the enzymes catalyzing both Phase 1 and 2 reactions among different species.

6.5 EXAMPLES

6.5.1 First-in-Human Dose

The single doses to be administered in first clinical trial were selected based on the knowledge about the compound at the investigational new drug (IND) stage. The *in situ* permeability data classified this drug as a Class 2 compound (high permeability and low solubility) according to the BCS and BDDCS. It was neither a P-glycoprotein (Pgp) substrate nor an inhibitor, but a transporter could possibly have mediated its Caco-2 permeability, and the *in vitro* metabolic studies showed that P450 was the major enzyme involved. The absorption of this drug should be primarily passive as a function of lipophilicity because intestinal uptake transporters generally would have no effect on its absorption due to rapid permeation into the gut membranes. However, its low solubility would limit the drug entering into the enterocytes, and prevent saturation of the apical efflux transporters and the intestinal metabolic enzymes such as CYP3A4.

Efflux transporters should predominate for this compound to affect the extent of oral *F* and the rate of

absorption, and transporter–enzyme interplay in the gut would be important for this drug. The *in vitro–in vivo* scaling with a correction for human plasma protein binding and allometric scaling from preclinical pharmacological and toxicological models (rodent and nonrodent animal species) were used. The average of predicted human clearance through body weight allometric scaling and using unbound *in vitro* intrinsic human CL corrected for plasma protein binding was estimated to be moderate at 0.13 L/h/kg. The drug was predicted to have a moderate volume of distribution and a long $t_{1/2}$ in human.

Because *in vitro* studies indicated that this drug was extensively metabolized and both transporters and enzymes may play a role in its absorption, *F* in human for this drug was estimated to be about 10%, much lower than the observed *F* in animals (more than 50%), although the latter values appeared to agree with the *in vitro–in vivo* estimate. The exposure to this drug after a single oral dose of 1 mg in an average human being was predicted to be about 13 ng·h/mL. The mean area under the plasma concentration–time curve (AUC) in the relevant animal efficacy model was around 1.2 µg·h/mL. If this animal pharmacology model truly represented human therapeutic indication, an oral dose of 90 mg would achieve an "efficacious" exposure in an average human (e.g., 60 kg).

High-fat meals were expected to increase *F* for Class 2 compounds due to inhibition of efflux transporters in the intestine and additional solubilization of drug in the intestinal lumen. T_{max} was expected to decrease due to inhibition of efflux cycling or to increase due to slowing of stomach emptying. The calculated PK data from the human single ascending dose (SAD) study was in agreement with a drug of high clearance and long half-life. The mean observed AUC was about 0.9 µg·h/mL at 60 mg, and CL/F was 1.2 L/h/kg, and $t_{1/2}$ was 40 h. C_{max} and AUC increased less than dose proportionally after single oral dose administrations and reached plateaus at high doses. A twofold in food effect was also observed, the extent of *F* (AUC and C_{max}) doubled and the rate of absorption decreased (T_{max} increased from 2.4 to 4.3 h).

The dominated effect of delayed gastric emptying on its absorption indicated that membrane permeation of this drug was most likely passive. Two major metabolites identified in human metabolic profiling samples, a *N*-descarbonyl acid primarily in plasma and a *N*-descarbonyl glucuronide dominantly in urine, had been previously detected *in vivo* in animals and in *in vitro* studies of metabolic stability. The clinical data confirmed that the metabolites were the dominant species in circulation, in agreement with that very minimal amount of parent drug was present in collected urine excretion samples. It is important to identify species differences with respect to metabolic stability or transport activity in order to successfully predict a reasonable dose and in turn a target exposure in first-in-human study.

6.5.2 Pediatric Dose

There has been increasing interest in pediatric exclusivity and requirements to conduct clinical studies in children. Regulatory authorities recommend that a clinical study in children be initiated when a product is developed for a disease or a condition in adults and the product is anticipated to be used in children. According to the International Conference on Harmonization E11 guidelines (Abernethy and Burckart, 2010), efficacy can be extrapolated from adults to pediatric population if the drug is intended for the same indication, and the disease process is similar and the outcome of therapy is likely to be comparable in children and adults. How to find a safe and efficacious dose of a drug in children becomes the focus of pediatric clinical pharmacology.

Interspecies scaling widely used in adult humans was also tried to predict pediatric PK parameters. It has been suggested that the drug clearance in children can be predicted as $CL_{child} = CL_{adult} \cdot (\text{Weight of child}/70)^b$. Mahmood (2007) developed allometric equations using double log plots of clearance versus children body weight or age. The results showed that an exponent of 0.75, 0.80, and 0.85 provided the same degree of accuracy for children (Mahmood, 2006), and an allometric equation is drug specific to predict clearance in children with reasonable accuracy because there was no single method suitable for all drugs or for all ages.

A recent study (Mahmood, 2010) indicated that interspecies scaling of adult rat, dog, and human clearance could be useful to predict drug clearance in children at different ages. Four methods, for example, simple allometry, MLP, or MLP with an empirical correction factor and a fixed exponent of 0.75, have also been used to estimate drug clearance in children. The results indicated that simple allometry would overpredict and use of MLP would underpredict drug clearance in children. However, an empirical correction factor to MLP substantially improved the prediction.

Now there are fewer limitations in the study of pediatric clinical pharmacology with improvements in PK/PD modeling and simulation methodology. Allometric scaling has inherent limitations and may not make optimal use of the available drug-specific information, a combination of allometric and pharmacometric approaches may be a promising tool to predict children clearance from adult data. The same consideration hold in the selection of study design and analysis methods in children as in adults in order to obtain adequate PK, efficacy, and safety information about the drug.

Pediatric Phase 2 and 3 clinical trials were planned for a drug to treat children and adolescents. The dose estimation for pediatric population was a challenge, although the dose range and PK/PD of this drug had been well characterized in adults. The Phase 2 pediatric study examined PK, safety, and efficacy in children (7–11 years) and adolescents (12–17 years). Each group received four different doses. The mean body weights for children and adolescents are 45 and 60 kg, respectively. There was no correlation of CL/F with either dose or age. CL/F values are approximately similar between adolescents and adults. Similar to adults, linear increases in AUC were observed with increasing doses. The effect of body weight on dose-normalized AUC could be described by an exponential equation as $AUC/D = 430 \cdot W^{(-0.55)}$ as shown in Figure 6.5.

The predicted AUC from this power equation was plotted along with the observed AUCs from this study as shown in Figure 6.6. The predicted values were similar to the observed parameters for pediatric patients with similar body weights receiving equivalent doses. Because body weight alone provided an adequate predictor for exposure, AUC values could be controlled based on maximum doses allowed for lower body weights from the relation between exposure to drug and body weight. Typical exposures in adults for therapeutic doses would provide a reasonable starting point for dose selection in younger age groups to test this drug in Phase 3 studies. Moreover, a fixed-dose design was feasible to provide a range of exposures in children and adolescents overlapping the exposures in adults administered a safe and effective dose.

6.6 DISCUSSION AND CONCLUSION

Current drug discovery and development programs are under growing scrutiny for low productivity and escalating costs. The concept and evolving role of MBDD and the associated quantitative pharmacology-based iterative "learn and confirm" paradigm provide concise information for rational decision making. The lack of critical infrastructure elements is a major impediment in the successful deployment of modeling and simulations in drug development. Currently, modeling and simulations play a supportive role to set the design of empirical clinical trials, and models may become the primary outcome of a development program in a fully realized MBDD paradigm.

A series of questions (Colburn, 2003) based on the available preclinical and clinical information should be asked before PK/PD modeling and simulation. The analysis plan is the key component in a clinical trial analysis. It should include the specific objectives, the

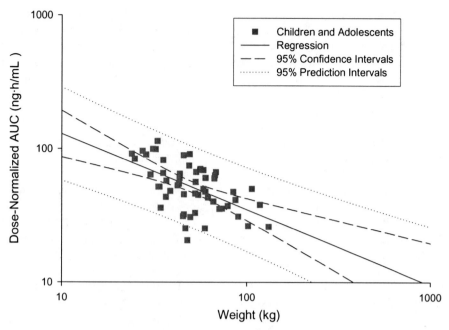

FIGURE 6.5. Effect of weight on dose-normalized AUC in children and adolescents following a single oral dose.

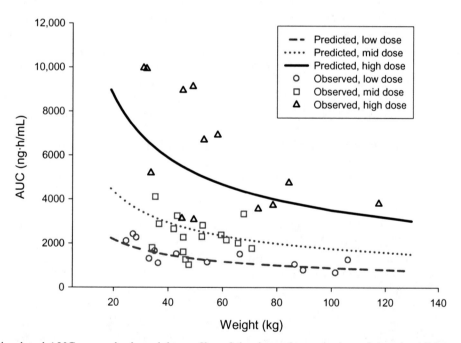

FIGURE 6.6. Simulated AUC versus body weight profiles of the drug after a single oral dose in children and adolescents.

sensitivity analysis, and the model validation. PK/PD models are used to better understand this relationship and how it might change as a function of drug input and other variables. But biomarker responses do not always parallel drug and/or active metabolite concentrations, and there is generally a delay between concentrations and effects. The final analysis report should have the assumptions made, and the extrapolation degree and dimension as well as the data available to support the model.

No single method has been identified to date to offer the "perfect" solution (Mahmood, 2005) for first–in-human dose prediction, although weight-based allometric scaling adjusted for plasma protein binding and metabolic stability can clearly be useful but not optimal (Bonate and Howard, 2000). Studies for dose prediction are generally conducted with low doses in animal efficacy models and high doses in toxicity models. This wide range of doses provides a handful of information on saturation of enzymes and/or receptor bindings, which may affect the drug PK/PD. Selection of appropriate preclinical animal models that will represent the human is a challenge for robust and reliable estimates. A relevant animal efficacy model needs to be defined for each human disease. This may involve testing preclinical models with the disease or applying known variations caused by the disease to normal animal data.

Allometric scaling from animals to human combined with PK/PD modeling and simulation should increase the utility and capability of first-in-human dose estimates. The parent and metabolites should be measured over the entire blood concentration-versus-time profile, and drug–drug interactions on protein and receptor binding should be investigated because PK/PD mechanisms may vary among species and affect success in scaling. Having only one active moiety rather than several or one that dominates drug efficacious and toxic effects makes PK/PD analysis simpler. If drug plasma binding is linear in the therapeutic and toxic dose range, unbound and total drug concentrations are just simple ratios. However, nonlinear plasma binding requires unbound drug concentrations to be measured because unbound concentrations increase greater than total drug concentrations.

Competitive endogenous ligands complicate the development and validation of PK/PD assays and thus inputs for PK/PD models. The competitive endogenous substances should be identified and used as biomarkers in drug development programs because they may compete with drug for receptor binding, or stimulate or inhibit the pharmacological activity of drug. It is important to determine the concentrations of the competing ligands in both active and placebo groups so baseline activity associated with endogenous competing in the absence of drug can be comprehensively built into the PK/PD models. PK/PD models should be simple or unique, and should never be more complicated than required to fit current data or predict future data. Models should only contain the rate-limiting, mechanism-based components that are required to describe the data. Sensitivity analysis has been used to identify robust and unique models when transitioning the models from modeling to simulation.

A good PK analysis will provide a mathematical model that is able to fit the data, simulate, and predict various case scenarios with a certain degree of confidence. This will not necessarily be the best model, but it should be the simplest model that successfully describes the data. The disease component of the PK/PD model can be better understood and applied if disease progression is modeled without therapeutic intervention. The best model is selected by some statistical criterion along with diagnostic plots. A good and useful model is the one that can characterize and include most of the important features of the data to understand the system and make accurate and precise predictions. It should be as simple as possible but robust, logically plausible, and effective to serve its purposes with appropriate precision, detail, and flexibility for extrapolation. The successful mathematical PK/PD relationship verifies PK and PD data, drug mechanism of action, and is further utilized for simulation and biomarker qualification and clinical validation.

Modeling and simulation should carry forward one basic repository of summary information from preclinical through clinical phases. The knowledge about the drug should be continuously updated and formally incorporated into the model when data is accumulated during drug development. Appropriate dose regimens can be identified early and dose adjustments for special populations can be inferred with fewer studies using modeling and simulation. Furthermore, fewer ineffective or toxic doses will be tested as the dose adjustments will be based on modeled and simulated data obtained from earlier studies rather than those used for related drugs. Mathematical modeling can expedite the selection of an appropriate clinical dose, even for drugs that have a narrow therapeutic index. PK/PD analysis can also help understand and modeling "placebo response," a phenomenon where patients on placebo without taking the drug improve in a clinical trial. The evaluation of placebo response can help distinguish drug effect from placebo effect, which can be marked because the study may not be long enough to determine the drug benefit for therapeutic indications such as Alzheimer's disease or depression.

In conclusion, PK/PD modeling and simulation is not only essential in designing clinical trials and for labeling during drug development but also very effective in government regulation for approval and policy guidance. The use of PK/PD modeling and simulation for regulatory submission should be considered and discussed as early as possible or feasible. Although modeling and simulations cannot replace an experimental study yet when confirmatory clinical data are needed, it helps to plan it properly. Optimal use of modeling and simulation will lead to fewer study failures due to poor study design, more successful compounds in the late stage of development, and smaller number of studies needed to support the drug registration. PK/PD modeling and simulation can be an invaluable tool for drug development in making critical decisions, such as compound selection, dose selection, study design, and patient population, all of which can lead to a considerable reduction in the cost of development.

REFERENCES

Aarons L, Karlsson MO, Mentré F, Rombout F, Steimer JL, van Peer A, COST B15 Experts (2001) Role of modeling and simulation in Phase I drug development. *Eur J Pharm Sci* 13(2):115–122.

Abernethy DR, Burckart GJ (2010) Pediatric dose selection. *Clin Pharmacol Ther* 87(3):270–271.

Bonate PL, Howard D (2000) Prospective Allometric scaling: Does the emperor have clothes? *J Clin Pharmacol* 40(6):665–670; discussion 671–6.

Boxenbaum H (1982) Interspecies scaling, allometry, physiological time, and the ground plan of pharmacokinetics. *J Pharmacokinet Biopharm* 10(2):201–227.

Caldwell GW, Masucci JA, Yan Z, Hageman W (2004) Allometric scaling of pharmacokinetic parameters in drug discovery: Can human CL, V_{dss} and $t_{1/2}$ be predicted from in-vivo rat data? *Eur J Drug Metab Pharmacokinet* 29(2):133–143.

Colburn WA (2003) Biomarkers in drug discovery and development: From target Identification through drug marketing. *J Clin Pharmacol* 43(4):329–341.

Dedrick R, Bischoff KB, Zaharko DS (1970) Interspecies correlation of plasma concentration history of methotrexate (NSC-740). *Cancer Chemother Rep* 54(2):95–101.

Dodds PS, Rothman DH, Weitz JS (2001) Re-examination of the "3/4-law" of metabolism. *J Theor Biol* 209(1):9–27.

Efron B, Tibshirani TJ (1993) *An Introduction to the Bootstrap*. Chapman and Hall, New York.

Frank R, Hargreaves R (2003) Clinical biomarkers in drug discovery and development. *Nat Rev Drug Discov* 2(7):566–580.

Lombardo F, Obach RS, Shalaeva MY, Gao F (2002) Prediction of volume of distribution values in humans for neutral and basic drugs using physicochemical measurements and plasma protein binding data. *J Med Chem* 45(13):2867–2876.

Mahmood I (2000) Interspecies scaling: Role of protein binding in the prediction of clearance from animals to humans. *J Clin Pharmacol* 40(12 Pt 2):1439–1446.

Mahmood I (2005) The correction factors do help in improving the prediction of human clearance from animal data. *J Pharm Sci* 94(5):940–945; author reply 946–7.

Mahmood I (2006) Prediction of drug clearance in children from adults: A comparison of several allometric methods. *Br J Clin Pharmacol* 61(5):545–557.

Mahmood I (2007) Prediction of drug clearance in children: Impact of allometric exponents, body weight, and age. *Ther Drug Monit* 29(3):271–278.

Mahmood I (2009) Role of fixed coefficients and exponents in the prediction of human drug clearance: How accurate are the predictions from one or two species? *J Pharm Sci* 98(7):2472–2493.

Mahmood I (2010) Interspecies scaling for the prediction of drug clearance in children: Application of maximum lifespan potential and an empirical correction factor. *Clin Pharmacokinet* 49(7):479–492.

Mahmood I, Balian JD (1996) Interspecies scaling: A comparative study for the prediction of clearance and volume using two or more than two species. *Life Sci* 59(7):579–585.

Matfin G (2007) Biomarkers in clinical trials and drug development: Measurement of cardiometabolic risk. *Br J Diabetes Vasc Dis* 7(3):101–106.

Mordenti J (1985) Pharmacokinetic scale-up: Accurate prediction of human pharmacokinetic profiles from animal data. *J Pharm Sci* 74(10):1097–1099.

Nagilla R, Ward KW (2004) A comprehensive analysis of the role of correction factors in the allometric predictivity of clearance from rat, dog, and monkey to humans. *J Pharm Sci* 93(10):2522–2534.

Pinkel D (1958) The use of body surface area as a criterion of drug dosage in cancer chemotherapy. *Cancer Res* 18:853–856.

Rooney KF, Snoeck E, Watson PH (2001) Modeling and simulation in clinical drug development. *Drug Discov Today* 6(15):802–806.

Ruiz-Garcia A, Bermejo M, Moss A, Casabo VG (2008) Pharmacokinetics in drug discovery. *J Pharm Sci* 97(2):654–690.

Sale M (2004) *Clinical Trial Simulation in Pharmacokinetics in Drug Development: Clinical Study Design and Analysis*. Springer, New York.

Sheiner LB (1997) Learning versus confirming in clinical drug development. *Clin Pharmacol Ther* 61(3):275–291.

Sheiner LB, Steimer J-L (2000) Pharmacokinetic/pharmacodynamic modeling in drug development. *Annu Rev Pharmacol Toxicol* 40:67–95.

Tang H, Hussain A, Leal M, Mayersohn M, Fluhler E (2007) Interspecies prediction of human drug clearance based on scaling data from one or two animal species. *Drug Metab Dispos* 35(10):1886–1893.

Wagner JA (2008) Strategic approach to fit-for-purpose biomarkers in drug development. *Annu Rev Pharmacol Toxicol* 48:631–651.

Weiss M, Sziegoleit W, Förster W (1977) Dependence of pharmacokinetic parameters on the body weight. *Int J Clin Pharmacol Biopharm* 15(12):572–575.

West GB, Brown JH, Enquist BJ (1997) A general model for the origin of allometric scaling laws in biology. *Science* 276(5309):122–126.

Wu C-Y, Benet LZ (2005) Predicting drug disposition via application of BCS: Transport/absorption/elimination interplay and development of a biopharmaceutics drug disposition classification system. *Pharm Res* 22(1):11–23.

7

PHARMACOGENOMICS AND INDIVIDUALIZED MEDICINE

ANTHONY Y.H. LU AND QIANG MA

7.1 INTRODUCTION

It is well established that in a large patient population, individuals may respond to the same medication differently in drug efficacy, drug safety, or both. One measurable parameter of variability in drug response is the plasma drug level that sometimes varies as much as several hundredfold among patients. Therefore, even with the best drug available, the standard daily dose that is proven to be efficacious and safe for the majority of patients may be ineffective or even harmful for a small number of patients.

Although numerous factors can influence the outcome of a drug treatment in individual patients, it is generally accepted that genetic variations in humans play a major role in determining the variability of disease phenotypes, drug efficacy, and drug side effect (Lu, 1998; Meyer, 2000; Evans and McLeod, 2003; Weinshilboum, 2003a; Evans and Relling, 2004; Eichelbaum et al., 2006; Lin, 2007). The human genome sequence provides a special record of human evolution. This sequence varies among populations and individuals. As the complete sequence of the human genome became available, the impact of the variability of the human genome on the pathogenesis of important diseases and the responses to drug therapy in humans can be readily analyzed. Parallel to the rapid accumulation of the knowledge on the genome–disease and genome–drug interactions, there arises a high hope that individualized medicine will soon become a reality. In this chapter, we will examine the origins of individual variability in

drug treatment; the role of drug targets, drug metabolizing enzymes, and drug transporters in determining the individual variability in drug therapy; and the many challenges we face in reaching the goal of individualized medicine.

7.2 INDIVIDUAL VARIABILITY IN DRUG THERAPY

It has been known for years that the optimum doses required for many therapeutic agents can vary to a significant extent from patient to patient. For example, the required daily dose of warfarin for inhibition of thrombosis and embolism in many disease conditions can vary up to 20–30-fold among individual patients, thus necessitating frequent blood coagulation testing to ensure effective and safe anticoagulation in patients.

The study by Davidson et al. (1997) on simvastatin, a member of the statin class of 3-hydroxy-3-methyl-glutaryl-coenzyme A (HMG-CoA) reductase inhibitors and cholesterol-lowering agents, provides a clear example of large individual variability in drug response and drug safety. In a population of 156 healthy men and women, a 40-mg daily dose of simvastatin reduced the low-density lipoprotein (LDL) cholesterol levels by 41% on average in 6 weeks, whereas the 80- and 160-mg doses resulted in median reductions of 47% and 53%, respectively, demonstrating that simvastatin is highly effective in reducing LDL cholesterol levels for the majority of the population. However, a small number of

ADME-Enabling Technologies in Drug Design and Development, First Edition. Edited by Donglu Zhang and Sekhar Surapaneni.
© 2012 John Wiley & Sons, Inc. Published 2012 by John Wiley & Sons, Inc.

individuals (approximately 5%) showed little or no reduction in their LDL cholesterol levels even at the high dose of 160 mg per day. In addition, a few individuals (<2%) had slight elevations of plasma hepatic transaminase activities and signs of myopathy. The reason for such variability is currently unknown but genetic variations are potential contributing factors. Chasman et al. (2004) reported that individuals carrying an HMG-CoA reductase genetic variant allele may exhibit significantly smaller reductions in the cholesterol levels when treated with pravastatin, another member of the statin cholesterol-lowering drugs. In a separate study, The SEARCH Collaborative Group (2008) identified several common variants of the *SLCO1B1* gene, which encodes the solute carrier organic anion transporter family member 1B1 protein; moreover, these variants are strongly associated with an increased risk of statin-induced myopathy.

7.3   WE ARE ALL HUMAN VARIANTS

Despite the overwhelming similarity in the nucleotide sequences of the human genome, there exist millions of points of DNA variations between any two randomly selected individuals. Genetic variations can result from a single-nucleotide polymorphism (SNP), nucleotide repeats, insertions, or deletions in DNA nucleotide sequences. Such changes can alter the amino acid sequence of a coded protein or the transcriptional expression of the gene. SNP is likely the most common gene variation. More than 1.42 million SNPs have been identified in the human genome, among which more than 60,000 SNPs are found in the coding regions of the genes (Sachidanandam et al., 2001). Most human genes (>90%) contain at least one SNP and nearly every human gene is marked by a sequence variation. Therefore, at the best approximation, we are all mutants or variants and we are all "flawed" to some extents at the genomic level.

The majority of SNPs appear to have no apparent effect on gene function. However, certain SNPs do have profound impact on the function of associated genes, whether the SNPs occur in the coding regions or at a significant distance from the transcription starting sites of the genes. Some SNPs are known to be associated with significant changes in drug efficacy and drug disposition (Evans and Relling, 1999; McLeod and Evans, 2001; Eichelbaum et al., 2006; Roden et al., 2006). It is becoming increasingly clear that an identification of a single SNP may not be sufficient to relate the variation of a target protein to a disease or a drug response. Therefore, new techniques are being developed to integrate sets of SNPs across the entire genome to identify genetic loci that exist in linkage disequilibrium, thereby identifying new disease susceptibility genes and pathways. In this respect, a haplotype defines a set of genetic variants that are inherited together in linkage disequilibrium, and thus, is particularly useful in genome-phenotype analyses.

Variations in genome sequences among individuals may well underlie the differences in the human susceptibility to diseases, the onset and severity of illnesses, and the way humans respond to a drug treatment. Unraveling the role of genetic variations in the pathogenesis of a disease could lead to the identification of disease genes. Understanding the role of genetic variations in responses to therapeutic agents could streamline the clinical development of new drugs by customizing drug interventions to specific drug target genotypes of patients; it could also improve the drug efficacy and safety profile by optimizing a drug dose according to the genotypes of drug metabolism and pharmacokinetics of individuals.

7.4   ORIGINS OF INDIVIDUAL VARIABILITY IN DRUG THERAPY

Multiple factors, either genetic or environmental, contribute to individual variability in drug therapy (Table 7.1). Genetic polymorphisms of the proteins involved in drug targeting and drug disposition are likely the most important source of individual variations in drug response and drug safety. Genetic variations can change the protein structure via mutations in the coding region or the amount of the protein expressed by modulating gene regulation, both of which alter the function of the protein, or the rate and kinetic constants in the case of an enzyme. Structural changes of receptors or enzymes can profoundly impact on the interaction between drugs and intended targets, and consequently, the drug response. Genetic polymorphisms of drug metabolizing enzymes and transporters are known to affect the absorption, distribution, metabolism, and elimination of many drugs. Alterations of DNA repair enzymes may weaken the ability of cells to defend against mutations and other toxic effects induced by many alkylating anticancer agents. Glutathione (GSH) plays an important role in protecting cells from oxidative stress and reactive intermediates generated from drugs. Structural alterations of enzymes involved in the biosynthesis of GSH may reduce the cellular GSH content to a low level at which cells are susceptible to attacks by reactive species, resulting in cell damage and death.

Whereas genetic factors commonly cause permanent changes in protein structures and individual responses to drug therapy, the effect of environmental factors on

TABLE 7.1. Origins of Individual Variability in Drug Therapy

Factors	Consequences[a]
Genetic factors	
Drug targets	Alter drug efficacy
Drug metabolizing enzymes	Alter drug metabolism
Drug transporters	Alter drug absorption, distribution, and elimination
DNA repair enzymes, GSH level	Affect drug safety
Environmental factors	
P450 induction	Decrease drug efficacy
P450 inhibition	Cause potential drug–drug interactions
Physiological factors	
Age, gender, disease, inflammatory mediators, and so on	Affect drug absorption, distribution, metabolism, and elimination

[a] Stated consequences regarding cytochrome P450 induction and inhibition do not apply to prodrugs that require P450-catalyzed activation steps to form an active metabolite, or drugs that are converted by P450 to biologically active metabolites. In the former case, that is, prodrugs, only the metabolites are active, whereas in the latter case, both the parent drug and its metabolites can be active. One type of prodrugs takes the form of esters to improve drug availability; in this scenario, activation of the prodrugs generally involves hydrolysis by esterases.

TABLE 7.2. Mutations in *VKORC1* Coding Region Leading to Warfarin Resistance

Amino Acid Change	Daily Warfarin Dose (mg)	Resistance Phenotype	Reference
Wild type	4–6	–	
Ala41Ser	16	Moderate	Rieder et al. (2005)
Arg58Gly	34	Major	Rost et al. (2004)
Leu128Arg	>45	Severe	Rost et al. (2004); Bodin et al. (2005)

drug response may be more transient in nature. Dietary constituents, environmental chemicals, and multiple drug use are known to induce or inhibit drug metabolizing enzymes, particularly cytochrome P450s, resulting in drug levels that are either too low or too high for a proper drug response. In these scenarios, the drug response may return to a normal level after the environmental factors are removed from the cells. A large individual variability in the induction and inhibition of human cytochrome P450 enzymes has been established (Lin and Lu, 2001). Physiological factors, such as age and disease state, also contribute significantly to pharmacokinetic variability and drug response in patients.

7.5 GENETIC POLYMORPHISM OF DRUG TARGETS

Genetic variations in drug targets can have a profound effect on drug efficacy. For instance, vitamin K epoxide reductase complex 1 (*VKORC1*) is the target of warfarin in the treatment and prevention of thromboembolic diseases. Mutations in the *VKORC1* coding region lead to warfarin resistance. The effective daily dose of warfarin is 4–6 mg for individuals carrying the wild-type alleles (Table 7.2). A number of *VKORC1* mutations (Ala41Ser, Arg58Gly, and Leu128Arg) have been identified (Rost et al., 2004; Bodin et al., 2005; Rieder et al., 2005; Rettie and Tai, 2006). Although the frequencies of these mutations in human populations are rather low, all three variants exhibit a warfarin resistance phenotype. In particular, individuals carrying the Leu128Arg variant require very high doses of warfarin (>45 mg per day) to achieve effective treatment of a thrombotic event.

Mutations of the β_2-adrenoreceptor, which is encoded by *ADRB2*, may alter the airway response to β-agonist, such as albuterol. Lima et al. (1999) showed that albuterol evokes a larger and more rapid bronchodilation response in Arg16/Arg16 homozygotes (wild type) than in carriers of the Gly16 variant (Arg16/Gly16 and Gly16/Gly16). The maximal percent increase in albuterol-evoked forced expiratory volume in 1 s (FEV_1) is 18% for individuals carrying the Arg16/Arg16, but only 5% for those carrying the Gly16 variants after an oral dose of 8 mg of albuterol.

Chemokine receptor 2 (CCR 2) and chemokine receptor 5 (CCR 5) are two cofactors that are essential for the infection of human immunodeficiency virus (HIV) in humans. The CCR 2 Val64Ile polymorphism is common in Caucasians and African-Americans with an allele frequency reaching 10% of the populations. Individuals carrying the Ile allele progress to acquired immune deficiency syndrome (AIDS) 2–4 years later than those carrying the wild-type receptor (Smith et al., 1997). One important variant of CCR 5 carries a 32 base pair deletion (Δ32) (Samson et al., 1996). About 9% of Caucasians carry this allele, but the polymorphism is generally not found in Africans. Individuals carrying the Δ32 deletion are protected from the transmission of HIV. These examples reveal that genetic polymorphisms of CCR 2 and CCR 5 can have significant impact on HIV infection and AIDS onset.

Individualized therapy is particularly important for cancer patients, owing in part to the complexity of the disease, the severe toxicity of many anticancer agents, and the existence of many different genotypes of the same disease in patients. Understanding the crucial molecular abnormalities in cancer is an essential step in the design of an effective anticancer drug. Once this is established, the drug can be targeted to patients who have a particular molecular abnormality. The key for the success in this targeted therapy is to separate the nonresponders from the responders in clinical practice. Success has been achieved with this approach for a number of malignancies. For example, imatinib has been used to specifically inhibit the tyrosine kinase activity of *bcr-abl* in tumor cells of chronic myeloid leukemia (Druker and Lydon, 2000); gefitinib selectively targets the overexpressed EGFR mutant proteins in malignant cells for the treatment of nonsmall-cell lung cancer (Sordella et al., 2004); and trastuzumab is used for the treatment of breast cancer patients who have overexpressed HER2 receptor in cancer cells (Hudis, 2007).

7.6   GENETIC POLYMORPHISM OF CYTOCHROME P450s

Since the 1960s—long before the concept of individualized medicine was established—individual variability of drug levels and genetic polymorphisms of drug metabolizing enzymes have been studied vigorously by pioneers including A. Conney, W. Evans, M. Eichelbaum, W. Kalow, R. Smith, E. Vesell, W. Weber, and R. Weinshilboum. Exemplary studies that address drug dosing, efficacy, and toxicity issues include the phenotyping of thiopurine *S*-methyltransferase (TPMT) to identify cancer patients with a low activity in methylating toxic cancer drugs (an inactivation pathway) (Evans and McLeod, 2003; Weinshilboum, 2003a; Evans and Relling, 2004); the identification of "slow acetylator" (*N*-acetyltransferase 2; NAT2) for isoniazid acetylation in the treatment of tuberculosis (Weber, 1987); and the phenotyping of "poor metabolizer" (PM) of CYP2D6-mediated debrisoquine hydroxylation (Eichelbaum et al., 2006). All these studies followed a classical approach: identifying individual phenotypes by measuring drug levels in the urine or plasma before the genetic mechanism was known; establishing the pharmacokinetics of drugs in "normal or extensive metabolizers" versus "poor or slow metabolizers" and their impact on drug efficacy and safety; and finally, establishing the molecular mechanism of the genetic defect to explain the low or lack of the enzyme activity years later. Although this research process appears slow and tedious, the outcome of the genetic polymorphism studies of drug metabolizing enzymes is clinically significant and meaningful.

Since the completion of the human genome project, numerous variants of drug metabolizing enzymes, particularly those of cytochrome P450s, have been identified and categorized. At the present, most of the variants were only evaluated for enzyme activities. With a few exceptions, the significance of genetic variations of these variants in pharmacokinetics and clinical outcome of drug therapy has not been established. Thus, even though the DNA sequence-based analysis of variants is much faster than the classical approach used in past research for identifying new variants, the impact of many of the newly identified genetic variants on drug therapy remains unclear.

The CYP2D6 polymorphism is one of the best studied among cytochrome P450s (Eichelbaum et al., 2006; Zhou et al., 2008). The *CYP2D6* variant alleles are classified on the basis of enzymatic activities (Table 7.3). The frequency and genetic basis of major CYP2D6 variants are well documented. In addition, methods for rapid and effective clinical testing of these variants are now available. The CYP2D6 phenotypes are particularly important in antidepressant therapy, as many of the available antidepressants are metabolized predominantly by this enzyme. If CYP2D6 is mainly responsible for the blood level of a drug in humans, and the genetic polymorphism of the drug target is not an issue, knowing the CYP2D6 phenotype of an individual patient would allow physicians to prescribe a safe and effective dose of the drug to the patient.

CYP2C19 catalyzes the metabolism of many commonly used drugs, including *S*-mephenytoin (anticonvulsant), omeprazole (antiulcer), and diazepam (antianxiety). To date, more than 20 variants of *CYP2C19* have been identified (Zhou et al., 2008). *CYP2C19*2* and *3* are null alleles that result in a total loss of the enzyme activity. The majority of PMs of CYP2C19 are due to these two variant alleles. About 15–25% of the Chinese, Japanese, and Korean populations are PMs of *S*-mephenytoin. On the other hand, the frequency of PMs in Caucasians is much lower (<5%).

CYP2C19 plays a very important role in the proton pump inhibitor therapy for peptic ulcer and gastroesophageal reflux diseases. Furuta et al. (1999) showed that the effect of omeprazole on the intragastric pH value largely depends on the individual *CYP2C19* genotype. At a single dose (20 mg) of omeprazole, the plasma drug area under the curve (AUC) is the highest among PM subjects, the lowest among extensive metabolizer (EM) subjects, and medium in heterozygous EMs. These *CYP2C19* genotype-based differences in AUC translate into differences in the extent and duration of inhibition of gastric acid secretion by omeprazole. The pharmaco-

TABLE 7.3. CYP2D6 Genetic Polymorphism: Characteristics and Clinical Consequences

Phenotype	Characteristics	Consequences
Poor metabolizer (PM)	Frequency: 5–10% Caucasians; 1–2% Chinese and Japanese	High plasma drug level
	Major variants: 2D6*3, *4, *5, *6	Risk of drug-related side effects
	Enzyme inactive	Use of reduced drug dose
Ultrarapid metabolizer (UM)[a]	Frequency: 1–2% Caucasian; 30% Ethiopian	Very low plasma drug level
	Multiple copy of CYP2D6 gene	Loss of drug efficacy
	Very high enzyme activity	Use of higher drug dose
Intermediate metabolizer (IM)	Major variants: 2D6*9, *10, *41	Lower dose for some patients
	Chinese: very high frequency of *10	
	Low residual enzyme activity	
Extensive metabolizer (EM)	Remaining population	Standard dose for most individuals
	Normal rate of metabolism	
	Not a uniform group of individuals	

[a] Does not apply to prodrugs or drugs with active metabolites; see Table 7.1 for more explanation.

kinetic data are in a good agreement with the observations that the intragastric pH is 4.5 for the PM subjects, 3.3 for the heterozygous EM subjects, and 2.1 for the EM subjects. Schwab et al. (2004) have shown that the rate of eradication of *Helicobacter pylori* by a combination therapy with lansoprazole and antibiotics is highly dependent on *CYP2C19* in white patients who take a standard dose of lansoprazole (30 mg twice a day). Besides resistance to antibiotics, CYP2C19 polymorphisms were identified as the most important factor that affects the success of *H. pylori* eradication. Since individuals with the EM phenotype have lower serum concentrations of lansoprazole and lower rates of *H. pylori* eradication, the patients would benefit from a higher dose of the proton pump inhibitor than the patients with the PM phenotype.

CYP2C9 is involved in the metabolism of many clinically important drugs including tobutamide (hypoglycemic agent), glipizide (hypoglycemic agent), phenytoin (anticonvulsant), S-warfarin (anticoagulant), and flubiprofen (anti-inflammatory agent). At the present, more than 30 variants of *CYP2C9* have been identified. Two of the most common allelic variants are *CYP2C9*2* and *CYP2C9*3*. The variants exhibit largely reduced enzymatic activities; the extent of such reduction is substrate dependent. There is a significant difference in the frequency of *CYP2C9* variants among different ethnic groups (Lee et al., 2002). Among Caucasians, about 1% are the *CYP2C9*2* homozygous carriers and 0.4% the *CYP2C9*3* homozygous carriers. In Chinese and Japanese populations, homozygous *2, homozygous *3, and heterozygous *1/*2 carriers are very rare, whereas the heterozygous *1/*3 accounts for 4%. The impaired metabolism of a low therapeutic index drug, such as warfarin, by CYP2C9 variants has very important clinical implications, which will be discussed in more detail in Section 7.11.

CYP3A4 is the most abundant P450 enzyme in the human liver and is responsible for the metabolism of more than 50% of the drugs used clinically. More than twenty *CYP3A4* variants have been identified. Many of these variants have altered enzyme activities varying from a modest to significant loss in catalytic efficiency; the extent of the activity loss is often dependent on the substrate used (Miyazaki et al., 2008; Zhou et al., 2008). There is also a large difference across ethnic groups in the frequency of *CYP3A4* variants. For example, there is a high frequency of *CYP3A4*2* and *7* in Caucasian and a high frequency of *CYP3A4*16* and *18* in Asian populations (Sata et al., 2000; Lamba et al., 2002).

Although large individual variations in the CYP3A4 activity have been identified in human populations, the clinical significance of the *CYP3A4* variant alleles for many drugs that are metabolized by CYP3A4 remains uncertain. Based on current data, it appears that the clinical impact of *CYP3A4* alleles is only minimal to moderate. These coding variants are unlikely to account for the more-than-10-fold differences in the CYP3A4 activity observed *in vivo* because the alleles produce only small changes in the enzyme activity and many of the alleles exist in low frequencies (Lamba et al., 2002). One factor that may contribute to the complexity of the CYP3A4 puzzle is CYP3A5, which is another member of the CYP3A family. Virtually all CYP3A4 substrates (with a few exceptions) are also metabolized by CYP3A5. Although CYP3A5 metabolizes these drugs at slower rates in most cases, some drugs can be metabolized by CYP3A5 at equal or greater rates than by the CYP3A4 enzyme. Therefore, the metabolic rates of CYP3A4 drugs measured *in vivo* may not be just a measurement of the activity of CYP3A4, but also that of CYP3A5. Since about 25% of whites and 50% of blacks express functional CYP3A5 (Kuehl et al., 2001), this dual pathway potentially obscures the clinical effects of *CYP3A4* variants in human studies. The prediction of CYP3A4 phenotypes is also complicated by the fact that the effect of polymorphisms of the CYP3A4 gene on the activity of a variant enzyme is often substrate dependent. Additionally, unlike the phenotyping of CYP2D6 and CYP2C19, clinically meaningful phenotyping of CYP3A4 by using probe drug substrates in a population has not been successfully established.

7.7 GENETIC POLYMORPHISM OF OTHER DRUG METABOLIZING ENZYMES

In addition to cytochrome P450s, many other drug metabolizing enzymes also play an important role in the metabolism of drugs of diverse chemical structures. Genetic polymorphisms of these enzymes are often associated with various human diseases and drug efficacy and toxicity issues.

The cytoplasmic TPMT catalyzes the S-methylation of a number of thiopurine drugs, such as 6-mercaptopurine, azathioprine, and thioguanine, that are commonly used for the treatment of leukemia and autoimmune diseases. The efficacy and safety of these agents lie in the balance of two metabolic pathways: (1) activation of the drugs to 6-thioguanine nucleotides followed by incorporation of the nucleotides into nucleic acids that cause the death of leukemia cells; and (2) inactivation of the drugs to inactive metabolites (Evans and Johnson, 2001; Eichelbaum et al., 2006). Since S-methylation by TPMT is the predominant inactivation pathway of thiopurine drugs, individuals carrying defective *TPMT* variants accumulate higher levels of cytotoxic thioguanine nucleotides than those with the wild-type alleles after receiving a standard dose of the drugs, thereby leading to severe hematological toxicity. More than 20 variant alleles of the *TPMT* gene have been documented (Zhou et al., 2008). In Caucasians, about 90% of the population inherit a high enzyme activity, 10% with an intermediate activity (heterozygous), and 0.3% with a low or no activity. Individuals carrying defective *TPMT* alleles, such as *TPMT*2*, *TPMT*3A*, and *TPMT*3C*, have poor enzymatic activities. They are at risk of developing hematological toxicity if given the normal dose; therefore, a reduced drug dose should be prescribed to these patients.

N-acetyltransferases (NATs) catalyze the acetylation of aromatic amines and hydrazines. Human variability in drug acetylation was discovered more than 50 years ago during the initial clinical trials of isoniazid as an antituberculosis drug (Evans et al., 1960). Although isoniazid is a remarkably effective drug, a high percentage of patients treated with isoniazid developed devastating nerve toxicity due to high blood levels of the drug. Two major genetically distinct phenotypes were identified and were referred to as "rapid acetylators" and "slow acetylators"; the phenotypes were later attributed to differences in the enzymatic activities of NAT1 and NAT2 (Weber, 1987). To date, more than 15 *NAT2* alleles have been identified in humans. Individuals carrying *NAT2*5A*, *NAT2*6A*, and *NAT2*7A* are associated with the slow acetylator phenotype (Zhou et al., 2008). *NAT2* polymorphisms are also associated with cancer susceptibility to certain industrial chemicals. A poor acetylator phenotype leads to increased risks of lung, bladder, and gastric cancers if the individuals are exposed to carcinogenic arylamines for a long period of time.

Uridine 5'-diphosphate-glucuronosyltransferase (UGT) 1A1 plays an important role in the glucuronidation of many commonly used drugs as well as certain endogenous substrates, such as bilirubin (Tukey and Strassburg, 2000). More than 100 *UGT1A1* variant alleles have been identified so far. There is a significant difference among ethnic groups in the frequency of *UGT1A1* variants (Zhou et al., 2008). For instance, the frequency of the *UGT1A1*6* mutation is high in Japanese and Chinese (16–23%) but low in Caucasian populations (<1%). Because UGT1A1 is mainly responsible for the glucuronidation of bilirubin, the high frequency of *UGT1A1*6* variant appears to contribute to the high incidence of neonatal hyperbilirubinemia in Asian children. Three inherited forms of unconjugated hyperbilirubinemia are known to occur in humans (Kadakol et al., 2000; Tukey and Strassburg, 2000): the Crigler–Najjar syndrome Type I and Type II are caused by variant alleles in the *UGT1A1* coding region and the Gilbert's syndrome by polymorphisms in the *UGT1A1* promoter. In the Crigler–Najjar syndrome Type I, bilirubin glucuronidation in patients is completely lacking (e.g., UGT1A1*6), which leads to very high serum levels of unconjugated bilirubin and early childhood death. Patients with the Crigler–Najjar syndrome Type II show markedly reduced activities of bilirubin glucuronidation (10–30% of normal). A genetic polymorphism in the promoter region of the UGT1A1 gene (e.g., UGT1A1*28) results in a reduced expression of the UGT enzyme, leading to the Gilbert's syndrome. The significance of UGT1A1*28 variant in the toxicity of the anticancer drug irinotecan will be discussed in Section 7.9.

7.8 GENETIC POLYMORPHISM OF TRANSPORTERS

Influx and efflux transporters play important roles in the absorption, distribution, and elimination of therapeutic agents. Genetic polymorphisms of transporters can profoundly affect drug disposition, drug response, and drug safety. P-glycoprotein (Pgp, MDR1, or ABCB1) encoded by *ABCB1* has received much attention because many clinically important drugs are substrates of Pgp. The *ABCB1* gene is highly polymorphic with numerous documented variants. Ethnic-dependent frequencies of some allelic variants are well known. Among many of the naturally occurring genetic variants of *ABCB1*, SNP C3435T is of particular interest due to its high frequency (20–60%) in many populations (Zhou et al., 2008). The functional significance of C3435T has been studied on the disposition of digoxin and other Pgp substrates. Conflicting results have been reported: One study found lower serum digoxin concentrations in individuals carrying the variant T alleles than in wild-type subjects, whereas another study showed higher plasma digoxin levels in mutant carriers than in carriers of the wild-type gene (Sakaeda et al., 2001; Verstuyft et al., 2003). The discrepancy between these studies is in part due to the fact that C3435T may not be the only polymorphism that affects the Pgp expression level. In this respect, the expression level of Pgp may be determined by polygenic traits rather than a monogenic one. Thus, the influence of the C3435T polymorphism on the pharmacokinetics and pharmacodynamics of Pgp substrates remains to be defined.

The human organic anion transporting polypeptide-C (OATP-C) coded by *SLC21A6* is a liver-specific transporter and is important for the hepatic uptake of a variety of endogenous compounds and therapeutic agents. Tirona et al. (2001) characterized 16 *OATP-C* alleles *in vitro* and found several variants (e.g., *OATP-C*5*, *OATP-C*9*) that show reduced uptake of OATP-C substrates, such as estrone sulfate and estradiol 17β-D-glucuronide. Since the genotypic frequency of the *OATP-C*5* variant is 14% in European-Americans and the frequency of *OATP-C*9* is 9% in African-Americans, these variants may have significant effects on the disposition of OATP-C drug substrates. Indeed, individuals carrying *OATP-C*5* have been found to have high plasma levels of the cholesterol-lowering drug pravastatin (Nishizato et al., 2003; Mwinyi et al., 2004) and the antidiabetic drug repaglinide (Niemi et al., 2005), both of which are OATP-C substrates.

The breast cancer resistance protein (BCRP/ABCG2) plays an important role in regulating the intestinal absorption and biliary secretion of drugs, drug metabolites, and toxic xenobiotics (Gradhand and Kim, 2008). The *ABCG2* C421A (Q141K) genotype is widely present in ethnic groups (30–60% in Asians and 5–10% in Caucasians and African-Americans). Sparreboom et al. (2004) found that heterozygous patients for Q141K have 300% higher plasma levels of diflomotecan, an anticancer agent, after an intravenous drug administration of the drug. These findings indicate that the *ABCG2*

alleles may alter drug exposure by affecting the biliary secretion of drugs and metabolites. Therefore, *ABCG2* genotypes clearly influence the disposition, efficacy, and safety of drugs.

7.9 PHARMACOGENOMICS AND DRUG SAFETY

One of the most challenging issues in drug safety is related to drug toxicities that are idiosyncratic in nature. Idiosyncratic adverse drug reactions, which are characterized by their rare occurrence and the requirement of multiple exposures, represent the most extreme cases of individual variability in drug safety. Because the number of individuals affected by this type of toxicity is rather small, a major challenge is to identify the gene or genes that are responsible for the toxic events and the individuals that are prone to the injury caused by a specific drug. It is by no means an easy task, but the large number of genetic variants known to be present in human populations may aid in addressing this difficult safety issue.

Statins, such as simvastatin, atorvastatin, and pravastatin, are HMG-CoA reductase inhibitors. Statin therapies result in a large decrease in LDL cholesterol levels and a reduction in cardiovascular events. In rare cases, however, statins cause myopathy (muscle pain or weakness) that occasionally leads to rhabdomyolysis (muscle breakdown and myoglobin release), which in turn causes renal failure and even death, particularly at high doses of statins. In a large SEARCH (Study of the Effectiveness of Additional Reduction in Cholesterol and Homocysteine) trial involving 12,064 subjects, 96 participants from the 80-mg-simvastatin-daily-dose group known to have developed myopathy during treatment were selected for a genome-wide association study using 316,184 SNPs in comparison with 96 control subjects (no documented myopathy) from the 80-mg-daily-dose group (The SEARCH Collaborative Group, 2008). A single strong association of myopathy was established with the rs4363657 SNP marker within the *SLCO1B1* gene on chromosome 12. No SNPs in any other region of the chromosome were clearly associated with myopathy. *SLCO1B1* encodes the organic anion transporting polypeptide OATP1B1 that is known to regulate the hepatic uptake of statins. It appears that the *SLCO1B1* variants diminish the hepatic uptake of statins leading to higher drug concentrations in the circulation to cause myopathy. Therefore, genotyping of *SLCO1B1* may help screen out those individuals with abnormal SLCO1B1 activities, and thereby achieve the benefits of statin therapy safely and effectively. This study illustrates the power of genome-wide associations in relating genetic variants to drug response and drug toxicity, particularly if a single gene is involved in such events.

Irinotecan is a potent DNA topoisomerase I inhibitor used in the treatment of colorectal and lung cancers (Tukey et al., 2002). Irinotecan is a prodrug that is converted by liver carboxylesterases to SN-38, the active topoisomerase inhibitor. SN-38 is glucuronidated by UGT1A1 and then eliminated from the body. High levels of SN-38 are associated with bone marrow toxicity (leucopenia) and gastrointestinal (GI) toxicity (severe diarrhea). Thus, the benefit of irinotecan therapy lies in a balance between the inhibition of topoisomerase I in cancer cells and the glucuronidation of SN-38 by UGT1A1 to minimize toxicity. Individuals carrying UGT1A1 variants could face severe toxicity problems in irinotecan therapy due to the accumulation of toxic SN-38. The *UGT1A1*28* polymorphism is such an example. The presence of seven, instead of six, TA repeats in the *UGT1A1* promoter region results in a decreased expression of the UGT1A1 protein and low glucuronidation of SN-38 (Kadakol et al., 2000; Zhang et al., 2007). Patients that are homozygous or heterozygous for the UGT1A1*28 polymorphism have elevated levels of SN-38 due to a decrease in the glucuronidation activity, and therefore are more susceptible to bone marrow and GI toxicities if given an irinotecan treatment. The FDA has recommended that patients be genotyped for the *UGT1A1*28* polymorphism and dose adjustments be made before an irinotecan treatment.

Many drug-related toxicity issues remain unsolved. A key to address the issues is to understand the mechanism of toxicity (drug related versus drug target related), determine the gene or genes responsible for the toxic events, and develop reliable biomarkers for screening. The two examples discussed above illustrate the utility of these approaches and bring hope that many of the toxicity problems can eventually be resolved and measures can be taken, for instance, by adjusting drug dose or using alternative drugs, to minimize the injury to patients according to individual genotypes and phenotypes.

7.10 WARFARIN PHARMACOGENOMICS: A MODEL FOR INDIVIDUALIZED MEDICINE

The anticoagulant agent warfarin is widely used for the treatment and prevention of thromboembolic diseases. Due to its narrow therapeutic index and wide individual variability in drug response, warfarin has been used as a model drug for studying individualized medicine (Rettie and Tai, 2006). The standard daily dose of warfarin is between 4 and 6 mg; however, in a large patient population, the effective daily dose of warfarin can vary from 0.5 to over 30 mg. Patients who receive an insufficient dose of warfarin will be at risk of failing blood clot control; on the other hand, patients given too high a dose can experience excessive and uncontrolled bleeding. Due to high rates of adverse events associated with warfarin therapy, frequent monitoring is necessary and is commonly used clinically to achieve appropriate anticoagulation. For years, physicians have used the "prothrombin time," a blood test for measuring the time required for blood to clot after medication, to adjust the warfarin dose and obtain a desired balance between the clinical benefit and bleeding risk. Prothrombin time has been standardized to the "international normalized ratio" (INR). The current practice involves the use of a standard 4–6-mg daily dose of warfarin or an initial dose estimated according to the patient's clinical characteristics; prothrombin time is then measured and adjustment is made based on the INR response with a goal to maintain INR in a range between 2 and 3. The time period required to reach this INR range can differ significantly among individuals, ranging from several days to several months. Nevertheless, the test does provide a simple, rapid, and relatively cheap way for clinicians to identify warfarin doses at which individual patients can be treated safely and effectively.

The pharmacological target of warfarin therapy is *VKORC1*, a gene encoding a subunit of the vitamin K epoxide reductase complex (Rettie and Tai, 2006; Limdi and Veenstra, 2008; Kim et al., 2009). The vitamin K reductase catalyzes the conversion of vitamin K to reduced vitamin K, a component required for the carboxylation of hypofunctional blood clotting factors including factors II, VII, IX, X, and other regulatory proteins to activated factors by γ-glutamyl carboxylase (GGCX). Blocking the generation of the reduced form of vitamin K by warfarin interferes with the formation of functional clotting factors through GGCX. Oral warfarin is well absorbed, and the S-enantiomer of warfarin is responsible for most of the VKORC1 inhibition. In the human liver, the S-warfarin is metabolized primarily to the inactive metabolite 7-hydroxywarfarin by CYP2C9.

In patients carrying the wild-type *CYP2C9*1* allele, S-warfarin is cleared from the body normally, and the standard daily dose results in a modest elevation of INR. PMs who carry the *CYP2C9*2* and/or *CYP2C9*3* alleles have impaired capacity of metabolizing S-warfarin, and thus require a reduced daily dose of warfarin. These patients have a two- to threefold higher risk of an adverse event than those with the wild-type allele when warfarin therapy is initiated. The most common polymorphisms of the *VKORC1* gene are within the noncoding region, which alter the expression of the enzyme (Rettie and Tai, 2006; Limdi and Veenstra, 2008). For example, individuals carrying the 1173T/T allele require about half of the warfarin daily dose compared with patients with the wild-type 1173C/C allele. In other cases, a higher-than-standard daily dose is required for individuals carrying certain *VKORC1* noncoding SNPs. Polymorphisms in the coding regions of the *VKORC1* gene often lead to varying degrees of warfarin resistance (Table 7.2). The number of resistant patients examined to date is relatively small.

Because of the large variation and difficulties in achieving effective and safe anticoagulation associated with warfarin therapy, it is apparently preferable to have a more accurate prediction of the initial dose based on the pharmacogenetics of each individual to improve anticoagulation control. It is believed that genotyping of patients prior to a warfarin therapy potentially reduces warfarin adverse events and facilitates achieving stable INR efficiently in patients with genetic variations in *CYP2C9* and *VKORC1* (Gage and Lesko, 2008; Kim et al., 2009). In August 2007, FDA revised the warfarin labeling to include pharmacogenetic information in the labeling. FDA has also approved four genetic tests for warfarin use, one of which provides fast testing in less than an hour. However, shortly after the new warfarin labeling was initiated, FDA issued a brief press release, apparently due to the negative response from the medical community, indicating that the new label is not a mandatory requirement for clinicians to genotype patients before the initiation of a warfarin therapy. In a special article, Limdi and Veenstra (2008) stated that, although the influence of *CYP2C9* and *VKORC1* genotypes on warfarin dose requirements has been consistently demonstrated in observational studies and randomized clinical trials, evidence to date from prospective, controlled studies has not demonstrated an added benefit of incorporating genotype-guided therapy in improving anticoagulation control or in preventing or reducing adverse effects. Therefore, they concluded that the routine use of *CYP2C9* and *VKORC1* genotyping in the general patient population who begin warfarin therapy is not supported by evidence currently available.

Variability in warfarin therapy is apparently a complex issue involving more than just the genotypes of *CYP2C9* and *VKORC1* and several other known physiological factors (Table 7.4). Although the variable metabolism of S-warfarin by CYP2C9 was identified initially as a major contributor to variable responses of the drug, it is now recognized that the contribution of the variable metabolism by CYP2C9 is rather small, estimated to be about 10% of warfarin dose variations. The contribution of *VKORC1* genotypes to dose variations is approximately 25%, and clinical factors such as age, gender, diet drugs, and body mass index another 20%. Therefore, the identified factors that contribute to variable warfarin responses are roughly 50%, whereas

TABLE 7.4. Contributions of Genetic and Nongenetic Factors to Population Variance in Warfarin Dose

Factors	% Contribution to Dose Variance
Vitamin K epoxide reductase complex 1 (VKORC1)	25
CYP2C9	10
γ-Glutamyl carboxylase	2
Clinical factors (age, gender, drugs, diet, body mass index)	20
Unknown	43

Adapted from data in Au and Rettie (2008).

other, yet-to-be identified factors reach almost 50%. Perhaps this is the major reason that many clinicians do not believe that the time is ripe for incorporating genotyping testing in warfarin therapy, particularly when the "prothrombin time" test, though not perfect, is available in clinics and provides reasonable information for dose adjustment. In an editorial titled "Genetic testing for warfarin dosing? Not yet ready for prime time," Bussey et al. (2008) stated that "some experienced clinicians question whether genetic testing adds significantly to the information one may discern by carefully monitoring the INR and by taking into consideration the numerous patient-specific factors that influence warfarin dosing requirements, such as age, underlying disease states, and concomitant drugs."

The International Warfarin Pharmacogenetics Consortium (2009) has used clinical data (age, height, weight, and race) and genetic data (*CYP2C9*1, *2, and *3, and *VKORC1* variants) of 4043 patients from nine countries to estimate the warfarin doses to achieve targeted INR. The pharmacogenetic algorithm accurately identified 50% of the patients who require 3 mg of warfarin or less per day and 25% who require 7 mg or more per day to achieve the target dose. Thus, despite impressive progress made in identifying *CYP2C9* and *VKORC1* variants, the successful prediction of a target clinical warfarin daily dose based on clinical and genetic data is only 25–50%, which is consistent with the fact that almost 50% of the factors responsible for variable warfarin responses remain unknown.

The clinical use of pharmacogenetic testing, such as in the case of warfarin therapy, is severely limited by a lack of prospective clinical trials that demonstrate that incorporating genetic testing can indeed benefit the selection of appropriate therapeutic agent and drug dose for individual patients to improve therapeutic response and reduce adverse drug effects. Research efforts designed to evaluate the effectiveness of genotype-guided warfarin therapy in improving therapeutic outcomes are under way. Definitive clinical outcomes in such studies would certainly be welcome news for the new era of individualized medicine.

7.11 CAN INDIVIDUALIZED DRUG THERAPY BE ACHIEVED?

The answer to this question varies from optimistic to pessimistic, in part depending on to what extent one considers it a success in achieving the goal of individualized medicine. When asked about the status of pharmacogenomics and individualized medicine in drug therapy, Richard Weinshilboum (2003b) replied, "The future is here! In psychiatry, where many of the drugs are metabolized by CYP2D6, our psychiatrists at Mayo Clinic have begun to request that the CYP2D6 genotype information be made available before drug therapy." He was, of course, referring to the fact that individuals carrying *CYP2D6*3*, *4*, *5*, or *6* variants should be given reduced doses of antidepressants to avoid or reduce drug side effects. In this regard, individualized medicine could be achieved when the metabolism of a drug is the major factor affecting drug response and safety. Of course, this is the simplest case in individualized medicine in which the variable factor is simple and well defined. However, in most cases, variable drug responses involve multiple factors, such as in the case of warfarin. Therefore, achieving individualized medicine for many diseases requires one to understand the pathogenesis of the disease, identify the gene(s) responsible for the disease, and establish the role of genetic polymorphisms of drug targets, transporters, and drug metabolizing enzymes involved in the drug therapy. It is particularly challenging in achieving these goals if disease genes are unidentified.

On the other hand, Nerbert and coworkers (Nebert and Vesell, 2006; Nebert et al., 2008) questioned whether personalized drug therapy can ever be achieved by means of DNA testing alone. At the present time, the major goal of clinical pharmacology and pharmacogenomics is to establish phenotype–genotype associations through genetic tests that reveal genetic predispositions to a disease and drug toxicity. The practical purpose is to identify patients who are drug responders and patients who are prone to drug toxicity. Although some success has been achieved in recent years in establishing such phenotype–genotype associations for monogenic disorders, it was argued that, because of the complexity of the genome, this task is far more challenging than originally anticipated. Based on their analysis of literature data, the authors concluded that, for complex diseases involving multiple genes, it would be very difficult to determine unequivocally either an exact phenotype or

genotype. Therefore, it remains unclear whether individualized drug therapy will ever be achievable by means of DNA testing. It was suggested that, perhaps in combination with proteomics, metabonomics might complement genomics in achieving individualized drug therapy.

7.12 CONCLUSIONS

The ultimate goal of individualized medicine is for physicians to prescribe an appropriate medication for the right target of the disease and at the right dose for individual patients to achieve maximal efficacy with minimal adverse effects. Although the concept is very attractive and the goal is noble, good clinical data to support the use of genetic testing for drug treatment for most diseases are still not yet available. Take warfarin dosing as an example: Despite excellent research and high hopes, the target dose can be successfully predicted based on genetic and clinical information for only 25–50% of the patients. Factors responsible for the remaining 50% of drug variations remain unknown.

The challenge for achieving individualized drug therapy is enormous. It is perceivable that a long and rough road lies ahead. The achievements so far have been limited. Perhaps, instead of asking for a full package to address all questions involved in individualized medicine, we may deal with individual issues, one at a time, at different stages in the quest for answers. From this prospective, several important achievements can be highlighted at what is the starting point in this long journey. Target therapies with imatinib, gefitinib, and trastuzumab for certain genotype cancer patients have pointed to the direction for future cancer therapy. The doses for antidepressant drugs can be adjusted in patients based on their CYP2D6 phenotypes to minimize drug toxicity. Genotyping of *UGT1A1*28* in colon cancer patients before irinotecan treatment is an important measure to decrease the GI and bone marrow toxicities of irinotecan. Genotyping of *SLCO1B1* in patients undergoing a statin therapy potentially helps achieve the benefit of cholesterol-lowering drugs more safely and effectively.

Based on these successes, it is believed that basic research will continue to be essential to understanding the mechanism of the pathogenesis of diseases and the roles of genetic variations of disease gene(s), drug targets, and all proteins important for drug disposition in determining drug response variations. The integration of genomics, proteomics, and metabonomics in genome-wide association studies will facilitate the identification of predisposing genetic factors associated with multifactorial diseases and drug response. Finally, prospective clinical trials that evaluate the utility and effectiveness of genotyping and individualized medicine are critical in guiding the research and clinical practice of individualized medicine in the future.

DISCLAIMER

The findings and conclusions in this chapter are those of the authors and do not necessarily represent the views of the National Institute for Occupational Safety and Health.

CONTACT INFORMATION

Q. Ma: Receptor Biology Laboratory, Toxicology and Molecular Biology Branch, Health Effects Laboratory Division, National Institute for Occupational Safety and Health, Centers for Disease Control and Prevention, 1095 Willowdale Rd., Morgantown, WV, 26505, USA; Telephone: (304) 285-6241; Email: qam1@cdc.gov.

A.Y.H. Lu: Department of Chemical Biology, Ernest Mario School of Pharmacy, Rutgers University, Piscataway, NJ, 08854 USA; Telephone: (732) 445-3400; Email: antylu@rci.rutgers.edu

REFERENCES

Au N, Rettie AE (2008) Pharmacogenomics of 4-hydroxycoumarin anticoagulants. *Drug Metab Rev* 40: 355–375.

Bodin L, Horellou MH, Flaujac C, Loriot MA, Samama MM (2005) A vitamin K epoxide reductase complex subunit-1 (VKORC1) mutation in a patient with vitamin K antagonist resistance. *J Thromb Haemost* 3:1533–1535.

Bussey HI, Wittkowsky AK, Hylek EM, Walker MB (2008) Editorial: Genetic testing for warfarin dosing? Not yet ready for prime time. *Pharmacotherapy* 28:141–143.

Chasman DI, Posada D, Subramanyun L, Cook NR, Stanton VP, Ridker PM (2004) Pharmacogenetic study of statin therapy and cholesterol reduction. *JAMA* 291:2821–2827.

Davidson MH, Stein EA, Dujovne CA, Hunninghaka DB, Weiss SR, Knopp RH, Illingworth DR, Mitchel YB, Melino MR, Zupkis RV, Dobrinska MR, Amin RD, Tobert JA (1997) The efficacy and six-week tolerability of simvastatin 80 and 160 mg/day. *Am J Cardiol* 79:38–42.

Druker BJ, Lydon NB (2000) Lessons learned from the development of an Abl tyrosine kinase inhibitor for chronic myelogenous leukemia. *J Clin Invest* 105:3–7.

Eichelbaum M, Ingelman-Sundberg M, Evans WE (2006) Pharmacogenomics and individualized drug therapy. *Annu Rev Med* 57:119–137.

Evans DAP, Manley KA, McKusick VA (1960) Genetic control of isoniazid metabolism in man. *Br Med J* 2:485–491.

Evans WE, Johnson JA (2001) Pharmacogenomics: The inherited basis for interindividual differences in drug response. *Annu Rev Genomics Hum Genet* 2:9–39.

Evans WE, McLeod HL (2003) Pharmacogenomics: Drug disposition, drug targets and side effects. *N Engl J Med* 348:538–549.

Evans WE, Relling MV (1999) Pharmacogenomics: Translating functional genomics into rational therapeutics. *Science* 286:487–491.

Evans WE, Relling MV (2004) Moving towards individualized medicine with pharmacogenomics. *Nature* 429:464–468.

Furuta T, Ohashi K, Kosuge K, Zhao X, Takashima M, Kimura M, Nishimoto M, Hanai H, Kaneko E, Ishizaki T (1999) CYP2C19 genotype status and effect of omeprazole on intragastric pH in humans. *Clin Pharmacol Ther* 65: 552–561.

Gage BF, Lesko LJ (2008) Pharmacogenetics of warfarin: Regulatory, scientific and clinical issues. *J Thromb Thrombolysis* 25:45–51.

Gradhand U, Kim RB (2008) Pharmacogenomics of MRP transporters (ABCC1-5) and BCRP (ABCG2). *Drug Metab Rev* 40:317–354.

Hudis CA (2007) Trastuzumab: Mechanism of action and use in clinical practice. *N Engl J Med* 357:39–51.

Kadakol A, Ghosh SS, Sappal BS, Sharma G, Chowdhury JR, Chowdhury NR (2000) Genetic lesions of bilirubin uridine-diphosphoglucuronate glucuronosyltransferase (UGT1A1) causing Crigler-Najjar and Gilbert syndromes: Correlation of genotype and phenotype. *Hum Mutat* 16:297–306.

Kim MJ, Huang SM, Meyer UA, Rahman A, Lesko LJ (2009) A regulatory science perspective on warfarin therapy: A pharmacogenetic opportunity. *J Clin Pharmacol* 49: 138–146.

Kuehl P, Zhang J, Lin Y, Lamba J, Assem M, Schuetz J, Watkins PB, Daly A, Wrighton SA, Hall SD, Maurel P, Relling M, Brimer C, Yasuda K, Venkataramanan R, Strom S, Thummel K, Boguski MS, Schuetz E (2001) Sequence diversity in CYP3A promoters and characterization of the genetic basis of polymorphic CYP3A5 expression. *Nat Genet* 27:383–391.

Lamba JK, Lin YS, Thummel K, Daly A, Watkins PB, Strom S, Zhang J, Schuetz EG (2002) Common allelic variants of cytochrome P4503A4 and their prevalence in different populations. *Pharmacogenetics* 12:121–132.

Lee CR, Goldstein JA, Pieper JA (2002) Cytochrome P450 2C9 polymorphisms: A comprehensive review of the *in vitro* and human data. *Pharmacogenetics* 12:251–263.

Lima JJ, Thomason DB, Mohamed MH, Eberle LV, Self TH, Johnson JA (1999) Impact of genetic polymorphisms of the beta2-adrenergic receptor on albuterol bronchodilator pharmacodynamics. *Clin Pharmacol Ther* 65:519–525.

Limdi NA, Veenstra DL (2008) Warfarin pharmacogenetics. *Pharmacotherapy* 28:1084–1097.

Lin JH (2007) Pharmacokinetic and pharmacodynamic variability: A daunting challenge in drug therapy. *Curr Drug Metab* 8:109–136.

Lin JH, Lu AYH (2001) Interindividual variability in inhibition and induction of cytochrome P450 enzymes. *Annu Rev Pharmacol Toxicol* 41:535–567.

Lu AYH (1998) Drug-metabolism research challenges in the new millennium: Individual variability in drug therapy and drug safety. *Drug Metab Dispos* 26:1217–1222.

McLeod HL, Evans WE (2001) Pharmacogenomics: Unlocking the human genome for better drug therapy. *Annu Rev Pharmacol Toxicol* 41:101–121.

Meyer UA (2000) Pharmacogenetics and adverse drug reactions. *Lancet* 356:1667–1671.

Miyazaki M, Nakamura K, Fujita Y, Guengerich FP, Horiuchi R, Yamamoto K (2008) Defective activity of recombinant cytochromes P450 3A4.2 and 3A4.16 in oxidation of midazolam, nifedipine and testosterone. *Drug Metab Dispos* 36:2287–2291.

Mwinyi J, Johne A, Bauer S, Roots I, Gerloff T (2004) Evidence for inverse effects of OATP-C (SLC21A6) *5 and *1b haplotypes on pravastatin kinetics. *Clin Pharmacol Ther* 75:415–421.

Nebert DW, Vesell ES (2006) Can personalized drug therapy be achieved? A closer look at pharmaco-metabonomics. *Trends Pharmacol Sci* 27:580–586.

Nebert DW, Zhang G, Vesell ES (2008) From human genetics and genomics to pharmacogenetics and pharmacogenomics: Past lessons, future directions. *Drug Metab Rev* 40:187–224.

Niemi M, Backman JT, Kajosaari LI, Leathart JB, Neuvonen M, Daly AK, Eichelbaum M, Kivisto KT, Neuvonen PJ (2005) Polymorphic organic anion transporting polypeptide 1B1 is a major determinant of repaglinide pharmacokinetics. *Clin Pharmacol Ther* 77:468–478.

Nishizato Y, Ieiri I, Suzuki H, Kimura M, Kawabata K, Hirota T, Takane H, Irie S, Kusuhara H, Urasaki Y, Urae A, Higuchi S, Otsubo K, Sugiyama Y (2003) Polymorphisms of OATP-C (SLC21A6) and OAT3 (SLC22A8) genes: Consequences for pravastatin pharmacokinetics. *Clin Pharmacol Ther* 73:554–565.

Rettie AE, Tai G (2006) The pharmacogenomics of warfarin: Closing in on personalized medicine. *Mol Interv* 6:223–227.

Rieder MJ, Reiner AP, Gage BF, Nickerson DA, Eby CS, McLeod HL, Blough DK, Thummel KE, Veewnstra DL, Rettie AE (2005) Effect of VKORC1 haplotypes on transcriptional regulation and warfarin dose. *N Engl J Med* 352:2285–2293.

Roden DM, Altman RB, Benowitz NL, et al (2006) Pharmacogenomics: Challenges and opportunities. *Ann Intern Med* 145:749–757.

Rost S, Fregin A, Ivaskevicius V, Conzelmann E, Hortnagel K, Pelz HJ, Lappegard K, Seifried E, Scharrer I, Tuddenham EGD, Muller CR, Strom TM, Oldenburg J (2004) Mutations in VKORC1 cause warfarin resistance and multiple coagulation factor deficiency type 2. *Nature* 427:537–541.

Sachidanandam R, Weissman D, Schmidt SC, et al (2001) A map of human genome sequence variation containing 1.42 million single nucleotide polymorphisms. *Nature* 409:928–933.

Sakaeda T, Nakamura T, Horinouchi M, Kakumoto M, Ohmoto N, Sakai T, Morita Y, Tamura T, Aoyama N, Hirai M, Kasuga M, Okumura K (2001) MDR1 genotype-related pharmacokinetics of digoxin after single oral administration in healthy Japanese subjects. *Pharm Res* 18:1400–1404.

Samson M, Libert F, Doranz BJ, et al (1996) Resistance to HIV-1 infection in Caucasian individuals bearing mutant alleles of the CCR-5 chemokine receptor gene. *Nature* 382:722–725.

Sata F, Sapone A, Elizondo G, Stocker P, Miller VP, Zheng W, Raunio H, Crespi CL, Gonzalez FJ (2000) CYP3A4 allelic variants with amino acid substitutions in exons 7 and 12: Evidence for an allelic variant with altered catalytic activity. *Clin Pharmacol Ther* 67:48–56.

Schwab M, Schaeffeler E, Klotz U, Treiber G (2004) CYP2C19 polymorphism is a major predictor of treatment failure in white patients by use of lansoprazole-based quadruple therapy for eradication of *Helicobacter pylori*. *Clin Pharmacol Ther* 76:201–209.

Smith MW, Dean M, Carrington M, et al (1997) Contrasting genetic influence of CCR2 and CCR5 variants on HIV-1 infection and disease progression. *Science* 382:959–965.

Sordella R, Bell DW, Haber DA, Settleman J (2004) Gefitinib-sensitizing EGFR mutations in lung cancer activate anti-apoptotic pathways. *Science* 305:1163–1167.

Sparreboom A, Gelderblom H, Marsh S, Ahluwalia R, Obach R, Principe P, Twelves C, Verweij J, McLeod HL (2004) Diflomotecan pharmacokinetics in relation to ABCG2 421C>A genotype. *Clin Pharmacol Ther* 76:38–44.

The International Warfarin Pharmacogenetics Consortium (2009) Estimation of the warfarin dose with clinical and pharmacogenetic data. *N Engl J Med* 360:753–764.

The SEARCH Collaborative Group (2008) SLCO1B1 variants and statin-induced myopathy: A genomewide study. *N Engl J Med* 359:789–799.

Tirona RG, Leake BF, Merino G, Kim RB (2001) Polymorphisms in OATP-C: Identification of multiple allelic variants associated with altered transport activity among European- and African-Americans. *J Biol Chem* 276:35669–35675.

Tukey RH, Strassburg CP (2000) Human UDP-glucuronosyltransferases: Metabolism, expression and disease. *Annu Rev Pharmacol Toxicol* 40:581–616.

Tukey RH, Strassburg CP, MacKenzie PI (2002) Pharmacogenomics of human UDP-glucuronosyltransferases and irinotecan toxicity. *Mol Pharmacol* 62:446–450.

Verstuyft C, Schwab M, Schaeffeler E, Kerb R, Brinkman U, Jaillon P, Funck-Brentano C, Becquemont L (2003) Digoxin pharmacokinetics and MDR1 genetic polymorphisms. *Eur J Clin Pharmacol* 58:809–812.

Weber WW (1987) *The Acetylator Genes and Drug Response.* Oxford University Press, New York.

Weinshilboum R (2003a) Inheritance and drug response. *N Engl J Med* 348:529–537.

Weinshilboum R (2003b) Pharmacogenetics: The future is here. *Mol Interv* 3:118–122.

Zhang D, Zhang D, Cui D, Gambardella J, Ma L, Barros A, Wang L, Fu Y, Rahematpura S, Nielsen J, Donegan M,

Zhang H, Humphreys WG (2007) Characterization of the UDP glucuronosyltransferase activity of human liver microsomes genotyped for the UGT1A1*28 polymorphism. *Drug Metab Dispos* 35:2270–2280.

Zhou S, Di YM, Chan E, Du Y, Chou VD, Xue CC, Lai X, Wang J, Li CG, Tian M, Duan W (2008) Clinical pharmacogenetics and potential application in personalized medicine. *Curr Drug Metab* 9:738–784.

8

OVERVIEW OF DRUG METABOLISM AND PHARMACOKINETICS WITH APPLICATIONS IN DRUG DISCOVERY AND DEVELOPMENT IN CHINA

CHANG-XIAO LIU

8.1 INTRODUCTION

Studies on drug metabolism and pharmacokinetics (PK) have been conducted in China since the 1960s. Early studies, mainly in Beijing, Shanghai, and Tianjin, were involved in investigating the disposition and biotransformation of several drugs in experimental animals. In the 1980s, the studies developed greatly. The pharmacology departments of many research institutes and universities have been devoted to studies of the disposition and metabolism of new drugs, drug metabolizing enzymes, mechanisms of drug action, PK models, and pharmacodynamic (PD)–PK analysis. Some industrial institutes and company laboratories extended PK studies to include absorption, distribution, and clinical pharmacology, drug level monitoring, dosage schedule, toxicity, and safety. Quantitative determination of xenobiotic concentrations in body fluids and tissues needs extremely sensitive and selective methods. Some laboratories have done pioneering work in developing methods utilizing gas chromatography (GC), high-performance liquid chromatography (HPLC), and in combination with mass spectrometry (MS).

In the last decade, Chinese researchers have made contributions to drug metabolism and PK. New analytical methods and new instrumentation have been developed for separation and identification of drugs and their metabolites. The theories and technologies have been applied to drug metabolism and PK of traditional Chinese medicines (TCMs), including quantitatively describing the kinetic changes of absorption, distribution, metabolism, and elimination/efficacy (ADME/E) of complex system of the TCM products. Drug metabolic pathways, enzymes, and enzyme complexes responsible for metabolism of drugs and xenobiotics have been elucidated. Genetic variations in metabolic pathways or rate of biotransformation have been explored, since most drugs are biotransformed, which have important clinical implication.

8.2 PK–PD TRANSLATION RESEARCH IN NEW DRUG RESEARCH AND DEVELOPMENT

In the "critical path research" for new drugs, the implementation of translational medicine or translational research will not only affect the pharmaceutical enterprises of the new product development process but also impact the mechanism of innovative drug research. In particular, to improve the efficiency of research and development, it is very important to grasp the three elements of translational research: scientific support systems, and new drug research and development. Scientific support systems include technical standards, research tools, registration policies, and registration of scientific criteria for promoting the realization of the invention. Translational research plays an extremely important role for PK as a main line of absorption,

ADME-Enabling Technologies in Drug Design and Development, First Edition. Edited by Donglu Zhang and Sekhar Surapaneni.
© 2012 John Wiley & Sons, Inc. Published 2012 by John Wiley & Sons, Inc.

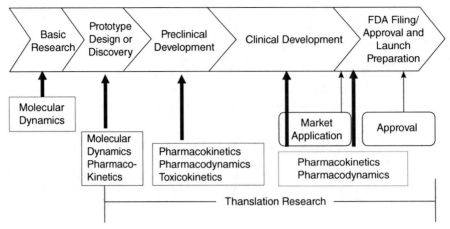

FIGURE 8.1. PK/PD/TK translation research in new drug research and development. FDA: Food and Drug Administration of China.

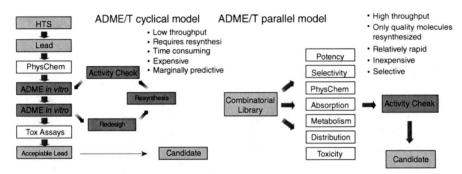

FIGURE 8.2. ADME/T models at early stage of drug discovery: ADME/T cyclical model and parallel model (Selick et al., 2002). HTS: high throughput screen.

distribution, metabolism, elimination/efficacy, and toxicity (ADME/E/T) research (Liu, 2010) as shown in Figure 8.1. PK and PD as well as toxicokinetics (TK) are important parts of ADME/E/T translation research.

8.3 ABSORPTION, DISTRIBUTION, METABOLISM, EXCRETION, AND TOXICITY (ADME/T) STUDIES IN DRUG DISCOVERY AND EARLY STAGE OF DEVELOPMENT

ADME/T studies are widely used in drug discovery to optimize the balance of properties necessary to convert leads into good medicines. However, throughput using traditional methods is now too low to support recent development in combinatorial and library chemistry, which have generated many more molecules of interest. To absorption, distribution, metabolism, and excretion (ADME) scientists, this situation is generating both the problem and the solution: An opportunity is now forming to use higher-throughput ADME screens and computational models to access this wide chemical diversity and to dissect out the properties that dictate

PK or metabolic profiles. In the future, we could see ADME properties designed-in from the first principles in drug design (Selick et al., 2002) with the cyclical and parallel models as shown in Figure 8.2. According to our experiences in using the models, the parallel model has produced a higher throughput.

Evaluation of drug metabolism and PK data plays an important role in drug discovery and development. Recently, higher-throughput *in vitro* ADME/T screening facilities have been established in order to evaluate an appreciable fraction of synthesized compounds. The ADME/T screening process can be divided into five distinct steps: (1) plate management of compounds in need of *in vitro* ADME data; (2) optimization of the mass spectrometry (MS/MS) method; (3) *in vitro* ADME experiments and sample cleanup; (4) collection and reduction of raw liquid chromatography mass spectrometry (LC-MS/MS) data; and (5) archiving of the processed ADME data (Hop et al., 2008).

For the early stage of drug discovery, many novel techniques with humanized tissues or human-derived preparations combined with high-throughput techniques have been recently developed. Applications of screening and evaluation of drug-like properties led to the optimization of suitable ADME properties and drug safety. In addition, the in vitro ADME/T screening can generate large quantities of data obtained under identical conditions to allow building of reliable in silico models (Selick et al., 2002).

Over the years, various in vitro methodologies have been developed to predict drug interaction potential in vivo; and in vitro study has become a critical first step in the assessment of drug interactions. Well-executed in vitro studies can be used as a screening tool for the need for further in vivo assessment and can provide the basis for the design of subsequent in vivo drug interaction studies. Besides in vitro experiments, in silico modeling and simulation may also assist in the prediction of drug interactions (Zhang et al., 2009a). Many computational drug design methods have been developed and are being applied to study the interactions between drug candidates and cytochrome P450 enzymes. P450 assays offer a challenge for computational methods because of the ambiguities of the catalytic cycle and the significant flexibility of the active site. Different computational methods display different limitations, which is crucial to take into account when choosing the method appropriate to each application. P450 assays offer an additional challenge for methodology development that can be applied automatically to all the cytochromes and can be used by chemists to detect metabolically labile positions that should be protected and to evaluate new tools in early ADME/T assays.

In addition, the in silico modeling can also predict the site of metabolism. For example, the regioselective metabolism of paclitaxel and cephalomannine at C3- is determined by the ligand and receptor interaction. Substituents with a lower C-H bond energy value are more susceptible to P450 oxidation because it needs lower activation energy. The bond energy of C-H at C4- is lower than that at 3-p-phenyl; thus, C4- instead of 3-p-phenyl in cephalomannine is hydroxylated. Molecular docking by AutoDock (Scripps, CA, USA) suggested that cephalomannine adopted an orientation in favor of 4-hydroxylation, whereas paclitaxel adopted an orientation favoring 3-p-hydroxylation. Kinetic studies showed that CYP3A4 catalyzed cephalomannine more efficiently than paclitaxel due to an increased V_{max}. The results obtained by Zhang et al. (2008) demonstrate that a relatively minor modification of taxane at C3- has major consequence on the metabolism. Actually, the main metabolites of paclitaxel varied substantially among different species, but for cephalomannine, 4-hydroxycephalomannine was the main metabolite in human liver microsomes (HLMs), rat liver microsomes (RLMs), and minipig liver microsomes (PLMs). In RLMs and PLMs, the activity of cephalomannine

4-hydroxylation was strongly inhibited by the CYP3A inhibitors troleandomycin and ketoconazole. These observations suggested that human CYP3A4 orthologs in minipig (CYP3A29) and rat (CYP3A1/2) might play a major role in cephalomannine metabolism; however, detailed examinations with recombinant rat and minipig P450s were needed to ascribe the reaction to particular animal P450 enzymes (Zhang et al., 2008).

For certain P450 enzymes, enzyme activity is lacking or is greatly reduced in some population. Based on these differences, the population can be divided into poor (slow) or extensive (rapid) metabolizers. The drug metabolizing enzyme systems, which are localized principally in the liver, are usually divided into two groups responsible for Phase I and Phase II metabolic reactions. During Phase I metabolism, one or more polar groups are introduced into the hydrophobic parent molecule, thus creating a handle, or position, for Phase II conjugating enzymes. The conjugated products are sufficiently polar that these detoxified chemicals are now excreted from the cell and the body (Zhou, 2003, 2004).

The drug metabolizing enzymes, transporters, and receptors were known to exhibit genetic polymorphisms with clinical consequences and contribute to interindividual variability in drug response. However, the pharmacological effects of drugs that are determined by several genes encoding proteins involved in drug disposition and efficacy are more difficult to elucidate in clinical studies. The predictive genotyping for drug metabolizing enzymes, transporters, and receptors in the clinic has not yet become routine at present. Since the knowledge about the benefits of predictive genotyping for a more effective therapy is increasing remarkably, the predictive genotyping for drug metabolizing enzymes, transporters, and receptors will become routine for the development of some specific drugs in the future.

8.4 DRUG TRANSPORTERS IN NEW DRUG RESEARCH AND DEVELOPMENT

During the last decade, a greater focus has been given to the impact of genetic variation of membrane transporters on the PK and toxicity of numerous therapeutic drugs. While the majority of transporter-related pharmacogenetic research has been related to classic genes encoding the outward-directed ATP-binding cassette (ABC) transporters, such as ABCB1 (P-glycoprotein [Pgp]), ABCC2 (multidrug resistance protein 2 [MRP2]), and ABCG2 (breast cancer resistance protein [BCRP]), more studies have been conducted in recent years evaluating genes encoding solute carriers (SLCs) that mediate the cellular uptake of drugs, such as SLCO1B1 (OATP1B1) and SLC22A1 (OCT1). The distribution of

ABC and SLC transporters in tissues key to PK, such as intestine (absorption), blood–brain barrier (BBB) (distribution), liver (metabolism), and kidneys (excretion), strongly suggests that genetic variation associated with changes in protein expression or function of these transporters may have a substantial impact on systemic drug exposure and toxicity (Franke et al., 2010). These studies are developed and applied for drug metabolism and PK in China.

The *in vivo* metabolic clearance in human has been successfully predicted by using *in vitro* data of metabolic stability in cryopreserved preparations of human hepatocytes. In the prediction by human hepatocytes, the systematic underpredictions of *in vivo* clearance have been commonly observed among different data sets. The regression-based scaling factor for the *in vitro*-to-*in vivo* extrapolation has mitigated discrepancy between *in vitro* prediction and *in vivo* observations. In addition to the elimination by metabolic degradation, the important roles of transporter-mediated hepatic uptake and canalicular excretion have been increasingly recognized as a rate-determining step in hepatic clearance. Therefore, it is proposed that the *in vitro* assessment should allow the evaluation of clearances for both transporter(s)-mediated uptake/excretion and metabolic degradation. It highlights the advantages of cryopreserved human hepatocytes as one of the versatile *in vitro* systems for the prediction of *in vivo* metabolic clearance in human at the early development stage. The following section discusses the mechanisms underlying the systematic underprediction of *in vivo* intrinsic clearance by hepatocytes. Glutathione is a tripeptide composed by L-glutamate, L-cysteine, and glycine that serves antioxygenation and detoxication functions within the cell. Recent study in China has found that glutathione is the main driving force for bile salt-independent bile flow; impaired biliary excretion of glutathione can lead to cholestasis. Based on the evidence of choleretic effect of glutathione, enhancement of biliary excretion of glutathione may be a good strategy for prevention and treatment of cholestasis (Zhang et al., 2009b).

Liu et al. (2007, 2008) found that no significant difference of the extracted berbemine (EB) concentration in the brain cortex was found between diabetic rats and control rats. Electron microscope examination of the brain cortex did not show a clear damage to the endothelial cells of microvessel in the diabetic rats. In addition, the protein level of Pgp in the brains of the diabetic rats examined was significantly lower than that of the control rats. These results suggested that the function and expression of Pgp might be impaired in the BBB of streptozotocin (STZ)-induced diabetic rats (Liu et al., 2006). Their study showed that insulin restored impaired function and expression of Pgp in diabetic BBB, and

further study showed that insulin up-regulated Pgp expression and function in normal BBB, so insulin might be one of the factors that regulated the function and expression of Pgp in the BBB in diabetes. In this study, the intracellular pathways where insulin regulated the Pgp were investigated using primarily cultured rat brain microvessel endothelial cells model. These results indicated that insulin regulated Pgp function and expression through signal transduction pathways involving activation of the PKC/NF-κB but not PI3K/Akt pathway (Liu et al., 2009).

Liu et al. used a model of the brain of pentylenetetrazole (PTZ)-kindled rats to investigate whether Pgp is overexpressed and the effects of Pgp up-regulation on the distribution of phenobarbital (PB) in the brain and its antiepileptic effects. Coadministration of cyclosporine (CsA) reversed the decrease of PB concentration in the brain without affecting the PB level in plasma and significantly potentiated the anticonvulsive effects of PB. The study demonstrated that chronic PTZ-kindling might increase Pgp expression and function in the brain of rats, resulting in the decrease of rhodamine 123 and PB levels in brain tissues. Coadministration of CsA increased PB levels in the brain and enhanced the anticonvulsive effects of PB by inhibiting Pgp function (Liu et al., 2007). A study was carried out by Liu et al. to investigate whether repetitive/temporal hypoxia up-regulated Pgp in cultured rat brain microvascular endothelial cells (rBMECs). Cultured rBMECs were used as *in vitro* BBB model. Cells that reached confluence were subjected to temporal hypoxic exposure. In this study, it was found that 8-temporal hypoxic exposure induced a 1.6-fold increase of Pgp level in cells, accompanied by a decrease of cellular accumulation of rhodamine 123. Cellular accumulation of PB was also decreased. These findings indicated that repetitive/temporal hypoxia may be one of the factors resulting in Pgp overexpression in refractory epilepsy (Liu et al., 2008).

Previous studies demonstrated the antidiabetic effects of berberine (BBR). However, the facts that BBR had low bioavailability and poor absorption through the gut wall indicated that BBR might exert its antihyperglycemic effect in the intestinal tract before absorption. The purpose of this study was to investigate whether BBR attenuates disaccharidase activities and β-glucuronidase activity in the small intestine of STZ-induced diabetic rats. Two groups of STZ-induced diabetic rats were treated with protamine zinc insulin (10 U/kg) subcutaneously twice daily and BBR (100 mg/kg) orally once daily for 4 weeks, respectively. Both age-matched normal rats and diabetic control rats received physiological saline only. Fasting blood glucose levels, body weight, intestinal disaccharidase, and β-glucuronidase activities in the duodenum, jejunum, and ileum were assessed for changes. Their findings suggested that BBR treatment significantly decreases the activities of intestinal disaccharidases and β-glucuronidase in STZ-induced diabetic rats. The results demonstrated that the inhibitory effect on intestinal disaccharidases and β-glucuronidase of BBR might be one of the mechanisms for BBR as an antihyperglycemic agent (Liu et al., 2008).

8.5 DRUG METABOLISM AND PK STUDIES FOR NEW DRUG RESEARCH AND DEVELOPMENT

8.5.1 Technical Guidelines for PK Studies in China

Increasingly improved ADME/T preclinical studies with improved methods and technology are able to ensure the development of PK laboratories to meet international standards. Biotechnology and drug research are also a combination of multimethods studies Chinese. PK experts drafted and revised the technical guidelines to make them consistent with international standards.

Regulatory guidelines direct trends of drug metabolism and PK research in the pharmaceutical industry to meet the requirements of drug metabolism data for drug development and registrations in China, under the lead of the State Drug and Food Administration of China. Chinese scientists discussed and drafted four regulatory guidelines from 1990 to 2005 on drug metabolism and PK research including bioanalytical methods, preclinical PK, clinical PK, and bioavailability. The regulatory requirements on drug–drug interactions (DDIs) including screening of cytochrome P450 (CYP) enzyme inhibition and induction, and reaction phenotyping of metabolizing enzymes were a focus for new molecule entries. The Guidance for Industry on Safety Testing, Drug Metabolites, and Toxicokinetics was emphasized for animal and human ADME studies and for validation of preclinical species for safety evaluation of drug candidates. To avoid delays in the drug development process due to the discovery of unique human metabolites in the late stages of clinical trials, many pharmaceutical companies are currently developing new strategies for early assessment of metabolite exposure in humans related to preclinical species. Those include plasma metabolite profiling and quantification in the first-in-human study and early ADME study. Regulatory guidelines also provide detailed recommendations on study design, analytical method validation, and compliance and data interpretation in bioanalytical and DDI studies. During the drug development and registration processes, the State Food and Drug Administration of China (SFDA) regulations require a sponsor to submit investigational new drug applications (INAs), carcino-

genicity study protocol, and new drug applications (NDAs), in which certain types of drug metabolism data are included. SFDA accepted the Technical Requirements for Registration of Pharmaceuticals for Human Use of the International Conference on Harmonization (ICH). Common drug metabolism and disposition studies in support of clinical development and regulatory submissions are described.

According to "The Provisions for New Drug Approval in China," applicants must carry out animal PK or clinical PK studies before investigation exemption for new drugs (INDs) or NDAs. The rate and extent of absorption, distribution, and retention in important organs and tissues, as well as the rate and degree of excretion, should be investigated.

Experiments on absorption, distribution, and excretion of the drug can be done with rats, rabbits, or mice. The amount of the drug excreted into the urine, feces, and bile should be at least examined. There should be at least five animals in each dose group. The influences of high, medium, and low doses on absorption and excretion should be studied in a selected species of animals. Selected species should be used for studying the effects of high, medium, and low doses of the test drug on the PK parameters. Metabolism study should be carried out for new drugs. The study must obtain the main pathways of biotransformation *in vivo*. In Phase I and II clinical trials, attempt should be made to establish conditions for using sensitive detecting techniques to measure drug concentrations in studying the PK of single dose administration. Clinical PK study is of special significance in guiding the safe and effective use of drugs in clinic. If possible, PK study should be conducted by experienced clinical pharmacologists or researchers to study the correlation of drug level in body fluid with efficacy and toxicity. Bioavailability and bioequivalence of pharmaceutical products should be defined clinically and these studies are also used to control the quantity of drug preparations.

In order to practice studies on preclinical and clinical PK, the State Drug Administration of China (SDA) and Chinese scientists edited four guidelines: preclinical and clinical PK, PK of biotechnical products, and bioavailability/bioequivalence.

At present, multicomponent analyses are carried out most conveniently by HPLC. This method provides profiles of drugs and metabolites that can be used for quantification and identification of the major and minor metabolites. MS is a relatively old field of science. Its development has been marked by intermittent surges of high activity in the design of new types of instruments, and these periods of activity have led to new uses of MS and to gaining new knowledge in each area of application. The development of the gas chromatography-mass spectrometry (GC-MS) and the high-performance liquid chromatography-mass spectrometry (LC-MS), especially LC-MS and LC-MS/MS, resulted in important changes in drug metabolism and PK studies. LC-MS/MS can obtain more structural information on metabolites. Atmospheric pressure chemical ionization (APCI), thermospray ionization (TSI), electrospray ionization (ESI), sonic spray ionization (SSI), and atmospheric pressure ionization (API) have been developed for use as an interface in LC-MS. Those techniques are able to make LC-MS/MS a very powerful tool for analyses of drugs and metabolites. LC-MS/MS analytical systems are now used for qualitative and quantitative analyses of drugs and metabolites in many institutes, universities, and hospitals in China.

Development and validation of analytical methods are improved gradually with international standards. New analytical methods and new instrumentation have been developed for the separation and identification of drugs and their metabolites (Xia, 2001; Liu et al., 2001a,b).

In order to develop standards for preclinical and clinical PK, the standard studies on preclinical PK of prodrug and biotech products were listed in the National Key Scientific Technical Plan (863 Plan) in 1998 and carried out by the author at the Tianjin Institute of Pharmaceutical Research. In 2003, the preclinical PK standard studies of new drugs were listed in the National 863 Plan and carried out by China Pharmaceutical University, Military Medical Academy, Tianjin Institute of Pharmaceutical Research, and Shenyang Pharmaceutical University. In the National Base Research Plan (973 Plan), Shanghai Institute of Materia Medica, Chinese Academy of Sciences and Tianjin Institute of Pharmaceutical Research carried out ADME/Tox projects to establish a rapid screen ADME/Tox system.

Drug metabolism and PK study has internationally become a pivotal and integrated part of drug research and development. However, the previous research capacity and level of preclinical drug metabolism and PK study in China was very low, mainly because of lacking powerful technologies and research systems, which had become a main obstacle for initiating the novel drug research and development in China. Under this background, a project of key technologies and systems for preclinical PK and drug metabolism was initiated and presided. After 10 years of development and innovation, the bottlenecks of most key technologies relating to drug metabolism and PK studies have been broken through, and a complete platform and the researching systems for preclinical drug metabolism and PK studies have been essentially established in China. The main achievements of this project can be summarized as follows: (1) great advancement in the high-sensitivity and high-throughput biological sample quantification technology based on various hyphenated MS, and especially that for the simultaneous determination of multicomponents from herbal drugs; (2) original development of various cellular, subcellular, and molecular models *in vitro* for high-throughput screening and evaluating of drug absorption, metabolism, and distribution properties and mechanisms; (3) creative extension of PK/PD combined research strategy, especially in the cellular level into disclosing the pharmacological targets and mechanisms of effective components contained in herbal drugs; and (4) creative development of series novel strategies and methods, including "multicomponents integrated PK," "metabonomics-based global PK," and dynamical biofingerprint strategy for the PK-related study of the complicated multicomponents characterizing TCM systems (Liu, 2008). Through these technological breakthroughs, and in light of the international guidelines, this project also contributed to the establishment of a well-organized platform that can satisfy the requirements of high-throughput PK screening of lead compounds, complete evaluation of PK properties of drug candidates, and the innovative PK study of herbal medicines from a global viewpoint (Hong Kong Medical Publisher, 2008).

Based on the well-developed technologies and systems, the preclinical PK evaluation of 20 innovative drugs had been fulfilled complying with the international guidelines; the PK properties and underlying mechanisms of over 70 effective components of herbal drugs had been investigated in depth. This project led to 285 local and international scientific publications and 23 patents of invention. In general, the achievements of this project would be a great promotion for drug research and development and the modernization of herbal medicines in China (Hong Kong Medical Publisher, 2008).

8.5.2 Studies on New Molecular Entity (NME) Drugs

In the last 5 years, the following four NME drugs have finished preclinical studies and Phase I, II, III, or IV clinical trials.

8.5.2.1 Salvicine Salvicine (Figure 8.3) is a novel diterpenoid quinone compound obtained by structural modification of a natural product isolated from a Chinese herb with potent growth inhibitory activity against a wide spectrum of human tumor cells *in vitro* and in mice bearing human tumor xenografts. Salvicine has also been found to have a profound cytotoxic effect on multidrug-resistant (MDR) cells. Moreover, salvicine significantly reduced the lung metastatic foci of MDA-

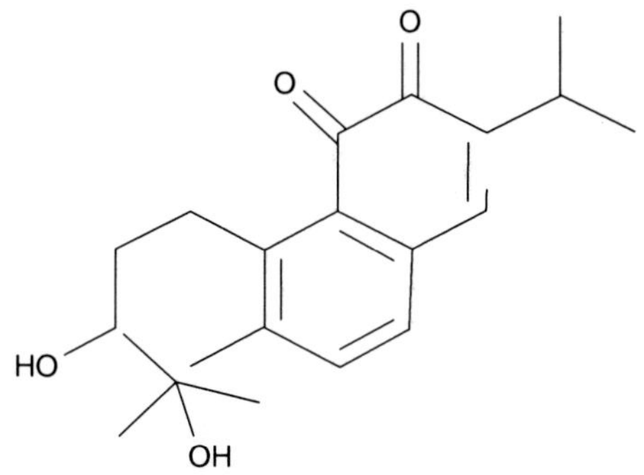

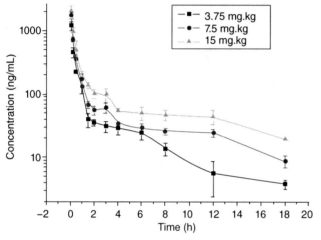

FIGURE 8.3. The chemical structure of salvicine.

FIGURE 8.4. The concentration–time curves after intravenous injection of salvicine to rats at doses of 3.75, 7.5, and 15 mg/kg.

MB-435 orthotopic xenograft. Recent studies demonstrated that salvicine is a novel nonintercalative topoisomerase II (Topo II) poison by binding to the ATPase domain, promoting DNA-Topo II binding and inhibiting Topo II-mediated DNA relegation and ATP hydrolysis. Further studies have indicated that salvicine-elicited reactive oxygen species (ROS) plays a central role in salvicine-induced cellular response including Topo II inhibition, DNA damage, circumventing MDR, and tumor cell adhesion inhibition. At doses of 3.75, 7.5, and 15 mg/kg to rats, the PK parameters were $t_{1/2\beta}$ 3.40, 4.77, and 5.75 h, and areas under concentration-time curves (AUCs) 658.4, 1097.2, and 1713.6 ng·h/mL, respectively. The result suggests that the drug possesses linear kinetic characteristics (Figure 8.4) (Meng and Ding, 2007; Meng et al., 2007).

8.5.2.2 Butylphthalide
S-(-)-3-n-Butylphthalide [S-(-)-NBP] (Figure 8.5) was extracted as a pure component from seeds of *Apium graveolens* Linn. Then,

(±)-NBP was synthesized and developed as an anticerebral ischemic agent. In 2002, (±)-n-butylphthalide soft capsule was approved by the SFDA to be marketed in China.

A rapid, sensitive, and specific reversed-phase high-performance liquid chromatographic method was developed for the determination of 3-n-butylphthalide, a drug currently being developed for treatment of stroke, in rabbit plasma. Fluorescence detection at an excitation wavelength of 280 nm and an emission wavelength of 304 nm was used for quantification of 3-n-butylphthalide (Zhao et al., 2003). In 2008, a rapid, sensitive, and specific high-performance liquid chromatography-electrospray ionization tandem quadrupole mass spectrometry (HPLC-MS/MS) method was developed and validated for the determination of 3-n-butylphthalide in rat plasma. The total chromatographic running time was 2.5 min. The method was linear over the concentration range of 11.14–3480.00 ng/mL, using as little as 100 μL of plasma. The lower limit of quantification (LLOQ) was 5.57 ng/mL. The method was successfully used to support a preclinical PK study of 3-n-butylphthalide in rats following intravenous administration (Niu et al., 2008).

8.5.2.3 Schiprizine
Schiprizine (ZT-1), N-(2-hydroxy-3-methoxy-5-chlorobenziliidene) huperzine A (Hup A) is a novel potent chlonesterase (ChE) inhibitor, which is rapidly transformed into the active metabolite Hup A. Originally isolated form Chinese club moss by Shanghai Institute of Materia Medica, Chinese Academy of Sciences, Hup A was demonstrated to have neuroprotective activities. ZT-1 was selected out of over 100 Hup A derivatives identified in the same institute. Figure 8.6 shows the chemical structure of ZT-1 and Hup A. *In vitro* pharmacological tests showed a marked concentration-dependent inhibition of acetylchlonesterase (AchE). *In vivo* investigations conducted in mice, rats, and monkeys showed that ZT-1 is equipotent to Hup A and more potent than donepezil and tacrine. The bioanalytical method was established by LC-MS (Li et al., 2004; Wei et al., 2006a). Preclinical PK was studied in animals (Wei et al., 2006b). Study showed that ZT-1 is a prodrug of Hup A. After oral administration of ZT-1 (1, 2.5, 5.0, and 10.0 mg/kg to rats, the PK parameters are listed in Table 8.1. Tissue distribution results showed that Hup A was rapidly distributed in lung, liver, kidney, and digestive tissues after oral 5 mg/kg of ZT-1 to rats. The drug levels in most tissues were much higher at 15 min than those at 2 or 6 h after dosing. The parent drug (ZT-1) was not found in urine and feces during 0–48 h after oral dosing of 5 mg/kg to rats. The total excretion of metabolite Hup A from feces, urine, and bile amounted to 3.28%, 20.5%, and 0.27% of the dose.

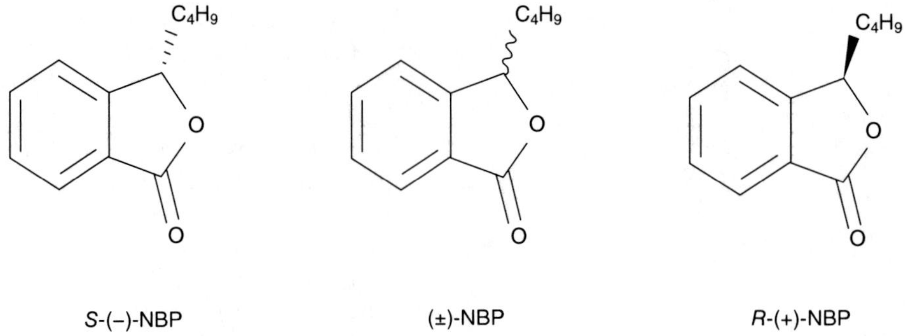

FIGURE 8.5. The chemical structures of *n*-butylphthalide.

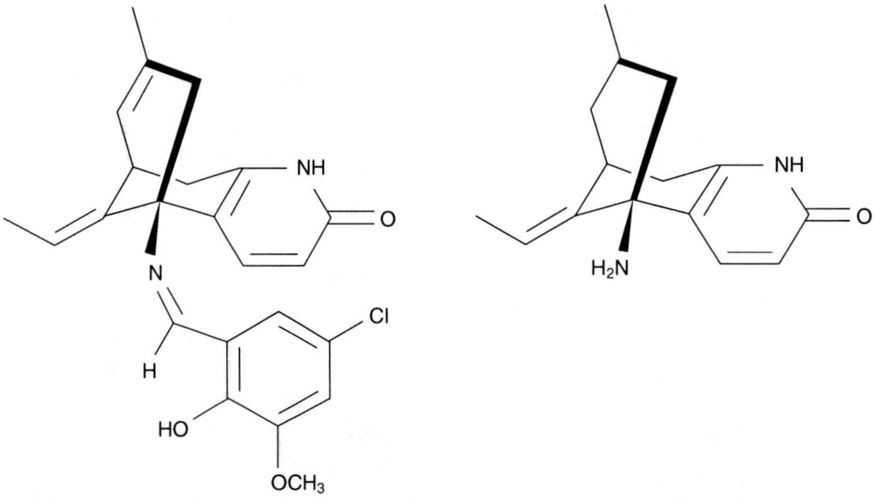

FIGURE 8.6. The chemical structures of schiprizine (ZT-1) and huperzine A.

TABLE 8.1. Pharmacokinetic Parameters of Hup A after Oral Administration of 1, 2.5, 5.0, and 10.0 mg/kg of ZT-1 to Rats

Dose (mg/kg)	T_{peak} (h)	$t_{1/2\beta}$ (h)	C_{max} (nmol/mL)	AUC (nmol·h/mL)
1.0	0.25	6.45	0.59	2.62
2.5	0.25	6.19	1.45	6.19
5.0	0.25	6.36	2.29	12.88
10.0	0.25	7.41	3.37	22.29

After oral administration of ZT-1 at 2.5 mg/kg to dogs T_{peak}, $t_{1/2\beta}$, C_{max}, and AUC were 1–3 h, 5.11–7.14, 2.58–3.44 nmol/mL, and 19.40–25.15 nmol·h/mL for Hup A, respectively. Preclinical studies have been finished in China and Phase I clinical trial has also been completed

in Switzerland and China. In 2004, a Phase II clinical trial for treating Alzheimer's disease (AD) is being conducted in patients with mild to moderate AD.

8.5.2.4 Bromotetrandrine *Stephania tetrandra* S. Moore, containing bisbenzylisoquinoline alkaloids is commonly used as anti-inflammatory and analgesic medicine in China. Recent studies show that the alkaloids enhanced the cytotoxicity of anticancer drugs in Pgp-dependent tumor cells. Synthesized derivatives of tetradrine including bromotetrandrine (Figure 8.7) reversed some MDR *in vitro* and *in vivo* (Wang et al., 2005). Its activity may be related to the inhibition of Pgp overexpression and the increase in intracellular accumulation of anticancer drugs. Bromotetrandrine may be a promising MDR modulator for eventual assessment in the clinic (Jin et al., 2005). Now the preclinical studies of bromotetrandrine including PD, PK, pharmacology

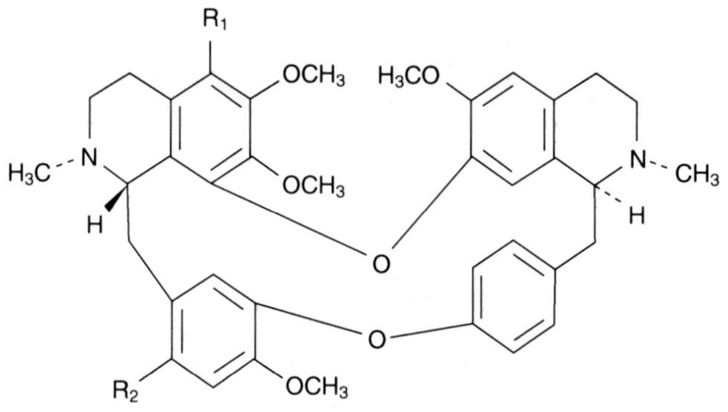

FIGURE 8.7. The chemical structures of bromotetrandrine and tetradrine: bromotetrandrine (R_1 = Br, R_2 = H); tetradrine (R_1 = H, R_2 = H).

and toxicology, quality control tests, and Phase I clinical trials have been completed in China (Xiao et al., 2004, 2005), suggesting that bromotetrandrine may be good modifiers of MDR in cancer chemotherapy.

8.5.3 PK Calculation Program

At present, the computer program is widely applied to calculate PK parameters. Many PK programs were complied and used for treatment of the concentration–time data. A practical PK program (3p87) is one of these programs. It is a complied BASIC program operating on IBM-PC-compatible micros. The program offers different optimization algorithms including Maquardt, Hartley, and Simplex algorithms to get the best-fit parameter estimates. An automatic step-size Meson differential equation solver handles Michaelis–Menten PK, and a splint function method deals with statistical moment parameters. The features of the program are as follows: (1) a convenient data management system—it is easy to input, modify, append, or delete dosage groups or concentration–time data, and recomputation or output of computed results can be executed whenever it is necessary; (2) automated discrimination of linear or nonlinear kinetics, automated initial estimation of parameters, automated changing of algorithms, and convergent precision and models to met the requirements of optimization; (3) objective comparison and choice of models or weighting by comprehensive statistical criteria and graphic displays; (4) automatic processing of batch or dosage group data and giving the mean and standard deviation of the primary and second parameters of different dosage groups; (4) the user can select model, initial values, algorithms, weighting, or convergent precision if necessary; and (5) table display or printout with high-quality graphic output including parameters, various statistical criteria of goodness of fit, linear or semilog scale C–T curves, error scale-gram, and linear regression display of observed and calculated C–T data.

At present, more than 40 institutions, including research institutes, medical colleges, universities and hospitals, have used the 3p87 program to study PK. Some scholars from the United States, Sweden, and New Zealand were interested in this program. The 3p97, new edition, was modified based on the 3p87 program. It added PK parameters calculated from urine drug data, and calculation of bioavailability and bioequivalence of drug preparations. The 3p87/3p97 is now widely used for PK calculation. More than 80% of research papers on PK for experimental and clinical studies applied the program for NDAs in China.

8.6 STUDIES ON THE PK OF BIOTECHNOLOGICAL PRODUCTS

Biotech drugs, including peptides, proteins and antibodies, oligonucleotides, and DNA, are projected to cover a substantial market share in the future health care systems. For their widespread applications in pharmacotherapy, respective drug development programs of biotech products should be successfully completed in a rapid, cost-efficient, and goal-oriented manner.

Research and development of biotechnical products is a hot point in new drug research and development in recent years in China. In PK studies, radiation determination (RA), HPLC, or HPLC-RA, enzyme-linked immunosorbent assay (ELISA), and bioassay methods are used for the studies on the PK of protein or peptide drugs. There are two laboratories at the Institute of Radiation Medicine, Beijing, and Tianjin Institute of Pharmaceutical Research in China. They carried out the PK of some new recombined biotechnical products.

Model-based drug development utilizing PK/PD concepts (Figure 8.8) (Meibohm, 2006) including exposure–response correlations has repeatedly been promoted by industry, academia, and regulatory authorities for all preclinical and clinical phases of drug development, and is believed to result in a scientifically driven, evidence-based, and more focused and accelerated drug product development process. Thus, PK/PD concepts are likely to continue expanding their role as a cornerstone in the successful development of biotech drug products in the future.

Recombinant human granulocyte colony-stimulating factor (rhG-CSF) is a hematopoietic growth factor that selectively stimulates granulopoietic cells of the neutrophilic lineage, and is widely used in clinical practice, particularly in the management of cancer chemotherapy-induced neutropenia. Studies have shown that modification of various proteins by the chemical addition of

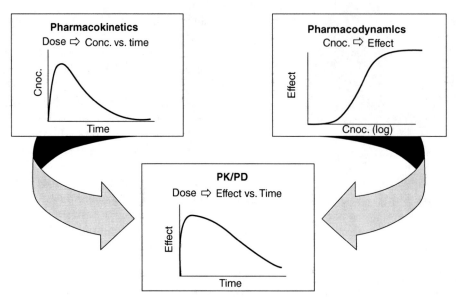

FIGURE 8.8. Pharmacokinetic/pharmacodynamic (PK/PD) modeling as a combination of the classic pharmacological disciplines of pharmacokinetics and pharmacodynamics.

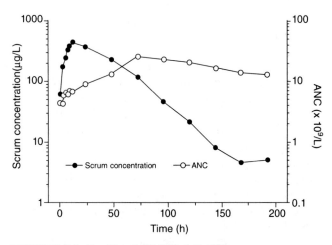

FIGURE 8.9. Profile of PEG30-rhG-CSF mean serum concentration and ANCs after a single subcutaneous injection (200 µg/kg).

polyethylene glycol (PEG) can alter the PK and PD properties of the protein to significantly increase the time the modified protein remains effective in the circulation. A study was done to examine the PK and PD of PEG30-rhG-CSF in comparison with PEG20-rhG-CSF to assess the relationship between serum concentrations of PEG30-rhG-CSF and granulopoietic effects of the drug as well as to define the relationship between neutrophil counts, clearance, and serum concentration of PEG30-rhG-CSF measured while absolute neutrophil counts (ANCs) were monitored after a single subcutaneous injection. As a result, the concentration–

effect–time curve was obtained to show the relevant relationship between PK and PD. Serum concentration and ANC following a single 200 µg/kg injection showed the characteristics of the concentration–time–effect relationship of PEG30-rhG-CSF (Figure 8.9). The results suggested that PEG30-rhG-CSF serum clearance is directly related to the number of neutrophils. The PK/PD results in our study of PEG30-rhG-CSF suggests that the drug exerts its effect by binding to cell surface receptors on hematopoietic progenitor cells to stimulate the proliferation, differentiation, and maturation of neutrophil precursors. Due to the time required for this process to occur, PD effects should occur after the PK effects. The PK of PEG30-rhG-CSF therefore depends on neutrophil recovery and may be characterized as a self-regulating mechanism (Cai et al., 2008).

8.7 STUDIES ON THE PK OF TCMs

8.7.1 The Challenge in PK Research of TCMs

TCM has an ancient history and a unique system, including theory, methodology, prescription formulation, and drugs. It has been widely used clinically for over thousands of years in China. Therefore, TCM played a very important role in health care in China. The early studies on the metabolism and PK of TCM were initially reported in 1960s in China. Since 1980, the research scope and quality of drug metabolism and PK of TCM has made considerable progress.

There are differences between TCM and chemical drugs. For TCM, (1) recipe-derived components *in vivo* are possibly to be detected; (2) the number of components is relatively restricted; (3) they could represent the therapeutic effect of the parent recipe; (4) the concentration and PK could be affected by the combination of herbs in recipe; (5) effects of new bioactive components related with those of their parent recipe; and (6) the PK can be affected by the pathological state in TCM treatment significantly. All the ideas of disease and recipe PK exhibited the characteristics of combining TCM theory and PK.

The PK of TCM, including quantitative descriptions of the kinetic changes of ADME/E of TCM, as well as the theories and technologies involved, is a collection of pharmacology, chemistry of TCM, analytical chemistry, and mathematics. Study of TCM requires mutual collaboration and efforts of interdisciplinary experts. The challenge in the PK field of TCM mainly comes from three ways: (1) the relationship between fundamental theory and traditional medicine; (2) the difficulty in research methods, characterization of minute amount of research objects, and controllability of quality; and (3) the application of advanced technology requires a relatively higher standard in the PK research, with the challenge coming from other countries (Liu, 2005a,b). Our analysis suggests that three aspects of drug metabolism and PK studies can be utilized in the modernization of TCM: (1) the oral absorption of TCM needs to be clarified; (2) activity of different components as well as herb–herb and herb–drug interactions can be determined; and (3) PK studies, together with well-developed analytical techniques, can provide a potential approach for the standardization of TCM *in vivo* (Liu, 2005b).

X. J. Wang and his team at Heilongjiang University of Traditional Chinese Medicine, Harbin, finished the study on the "Effective components in serum and PK after oral administration of Liuwei Dihuang (LWDH) pill, a traditional Chinese medicine formulation." Based on the theory of serum pharmacochemistry of Chinese medicine, the compatibility principle of the LWDH pill used for treating the syndrome of kidney deficiency were preliminarily elucidated not only by analyzing the components in serum after oral administration of the LWDH pill but also by evaluating the pharmacological effects and the PK of the detected components *in vivo*. Among 8 of 11 compounds detected after oral administration, 5-hydroxymethyl-furoic acid (5-HMFA) was identified as either the original form of 5-HMFA existing in *Radix rehmanniae* as a main composition crude drug or the metabolite of 5-hydroxymethyl-2-furfural (5-HMF) existing in both *Fructus corni* and *Rhizoma alismatis*, another two with the main composition herbal drugs of the LWDH pill. Furthermore, the kidney toxicity and related effects of these compounds were evalu-

ated, using both the rat kidney deficiency model and aging model combined with blood stasis. These bioactivities were proven to be consistent with the therapeutic effects of the LWDH pill. It was found that 5-HMFA could dramatically increase content of superoxide dismutase (SOD), decrease whole blood viscosity, plasma viscosity, hematocrit, spleen index, and contents of MDA FIB and ICAM-1. As a result, validated antiaging and improving hemorrheology bioactivities of 5-HMFA could be well understood, whereas 5-HMFA is one of the main effective components of LWDH Pill *in vivo*. In addition, PK of 5-HMFA after oral administration of LWDH was also investigated. It was suggested that not only the absorption but also the distribution of 5-HMFA were rapid ($t_{1/2k\alpha} = 0.1$ h, $t_{1/2\alpha} = 2.62$ h); however, the elimination was very slow ($t_{1/2\beta} = 32.66$ h). These properties might be partially attributed to its multiple sources of 5-HMFA *in vivo*, deriving from both the small amount of prototype 5-HMFA in *R. rehmanniae* and the metabolites of 5-HMF in *F. corni* and *R. alismatis*. Thus, LWDH could be considered as a natural sustained releaser of 5-HMFA, and this might contribute to the long-lasting therapeutic effect of this formula; such PK properties again proved bioactive superiority of the LWDH pill (Wang et al., 2007).

BBR, abundant in the root of rhizome and stem bark of many plants, is one of the most popular natural products in eastern Asian countries. As a new class of antibacterial drug, BBR has also been found to possess many other pharmacological activities. In a recent study, Liu et al. (2009) explored the roles of metabolites in the disposition of BBR in rats after intravenous administration. The results showed that the major circulating metabolites of BBR were oxidative metabolites M1 (via demethylation), M2 (via demethylenation) (Fig 8.10), and their corresponding glucuronides, with M2-glucuronide approximately 24-fold higher than that of M1-glucuronide. Incubations with rat liver microsomes were conducted to examine the formation kinetics of two oxidative metabolites—M1 and M2—as well as the depletion kinetics, leading to the formation of glucuronide conjugates. Efforts were also made to examine the roles of key CYP and UDP-glucuronosyltansferase (UGT) isoforms responsible for BBR metabolism using known chemical inhibitors and/or substrates. *In vitro*, the formation of M1 and M2 were comparable and multiple CYP enzymes were involved. In contrast, the glucuronidation of M2 was much faster than that of M1. Inhibition studies using well-characterized UGT substrates suggested that both M1 and M2 could be glucuronidated by UGT1A1 and UGT2B1, while M2 glucuronidation was catalyzed by UGT1A1. In summary, oxidative demethylenation and the subsequent glucuronidation were the main metabolic pathways of BBR in rats.

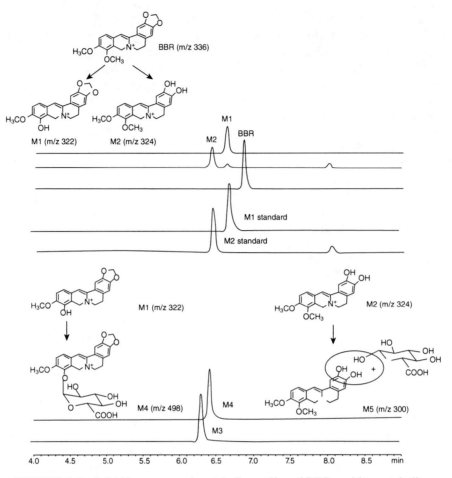

FIGURE 8.10. LC-MS spectra and metabolic profiles of BBR and its metabolites.

8.7.2 New Concept on PK Markers

Identification of components of an herbal product with favorable drug-like properties will extend our knowledge in terms of pharmacological efficacy and safety. An herbal component can be defined as drug-like, if desired features can be obtained, for example, required potency, a wide safety range, and appropriate PK properties, and abundant in the herbal product. A deficiency of these properties mentioned limits the application of the components of the herbal product. The author proposed a concept of PK markers on "The challenge of drug metabolism and PK research in traditional Chinese medicines (TCM)" (Second APR-ISSX Meeting, Shanghai, China, May 11–14, 2008). It was suggested that the PK markers should be active compounds presented in TCM complex preparation or their metabolites, and that the markers could be determined in biological sample after administration of the preparation, and related with pharmacological activities in therapy with TCM (Xiao et al., 2008).

For a drug, the pharmacological effect is attained when the drug or its active metabolite reaches and sustains an adequate concentration at an appropriate site of action; this hypothesis should be applied to the herbal product as well. Both the dosage levels and fates of active components in the body govern their target site concentrations after administration of an herbal product. The relevant PK properties include the ability of an herbal chemical to be absorbed from the site of administration and to pass through multiple biological barriers to reach the site of action, sufficient metabolic stability to achieve therapeutically meaningful systemic and target site concentrations, and appropriate metabolic liability to be eliminated effectively by the excretory process. PK markers may be used to show systemic exposure to the herbal product in animals and/or humans. For a multiherb product, identification of PK markers derived from each component herb is important for evaluating the combination rationality and for investigating possible synergistic interactions between the component herbs. Such studies are also relevant for designing rational dosage regimens, evaluating potential herb–drug or herb–herb interactions, and developing new formulations. C. Li et al. studied the PK properties of the main active components from Danshen (*Radix salviae miltiorrhizae*) and Sanqi.

In the study on the PK of Danshen, PK properties of putatively active phenolic acids were examined to identify suitable compounds that could indicate systemic exposure to oral cardiotonic pills. Researchers found that plasma tanshinol (TSL) was a suitable PK marker and urinary TSL was a surrogate PK marker for oral cardiotonic pills. The other Danshen phenolic acids were unsuitable due to the unfavored PK properties. In addition, sublingual administration of the pills neither changed the absorption rate and bioavailability of TSL nor improved the poor bioavailability of the other tested Danshen compounds in terms of oral administration. Three points are noteworthy: (1) in addition to PK properties, the dose level, dosage form, and route of administration may affect the suitability of an herbal component measured in plasma or urine as a PK marker for the herbal product; (2) PK markers from the other component herbs also need to be identified, and a combination of PK markers could provide a more complete characterization of the systemic exposure to cardiotonic pills; and (3) a substantial fraction (60%) of intravenous injection was eliminated in intact form by renal excretion, the biotransformation of this Danshen phenolic compound remains to be revealed (Lu et al., 2008).

Li et al. suggested that the PK properties of active components, including a significant dose-dependent systemic exposure and an appropriate elimination half-life, can be considered as PK markers for a specific herbal product. When the active component is unknown or a suitable assay is not feasible, the main chemical constituents or the metabolites detected in plasma or urine may be evaluated as substituted PK markers. PK markers may be used to show systemic exposure to the herbal product in animals and/or humans. For a multiherb product, identification of PK markers derived from each component herb is important for evaluating the combination rationality and for investigating possible synergistic interactions between the component herbs. Such studies are also relevant for designing rational dosage regimens, evaluating potential herb–drug or herb–herb interactions, and developing new formulations.

Panax notoginseng (Sanqi) is a cardiovascular herb containing ginsenosides that are believed to be responsible for the therapeutic effects of Sanqi. To gain understanding of the comparative systemic exposure and PK properties of various ginsenosides from Sanqi and of the key factors that affect their absorption and disposition and to investigate the influence of colonic deglycosyl-ation on systemic exposure to the ginsenosides, researchers carried out many studies using multiple approaches *in vivo*, *in vitro*, and *in silico* to determine ginsenoside exposure, absorption, and disposition.

The most abundant fecal compounds in rats were GRg1 (*Cum.Ae*, 39.1 mol/kg), C-K (20.9 mol/kg), and GF1 (7.1 mol/kg). The high level of these compounds in feces resulted from the colonic microflora stripping the sugar moieties from the ginsenosides in the extract was observed, supported by their absence (C-K) or presence at a very low level (GF1) in Sanqi extract. The contribution of colonic deglycosylation of major Ppt-type ginsenosides to the fecal GRg1 was confirmed by two lines of evidence: (1) fecal recovery of GRg1 after oral administration of pure GRg1 was significantly lower than that of after oral administration of Sanqi extract (at the same GRg1 dose) in the same rats; and (2) oral administration of pure 20 g Rf, GRe, and NGR1 generated fecal GRg1. Meanwhile, poor intestinal absorption and rapid biliary excretion accounted for the relatively low plasma levels of GRg1, C-K, and GF1. Although both of the aglycones (Ppd and Ppt) were detected in fecal samples after oral administration of Sanqi extract, their fecal *Cum.Ae* values were quite low (0.4 and 0.2 mol/kg, respectively). Ppd and Ppt were not detected in plasma, bile, or urine, suggesting that they were two minor metabolites. Due to the metabolism of intravenously administered GRb1, the monooxidized metabolite, the deglycosylated metabolite of GRd were detected in rat bile and urine. These metabolic reactions were catalyzed by enzymes expressed in rat organs such as liver.

The ginsenosides Ra3, Rb1, Rd, Re, Rg1, and notogin-senoside R1 were the major saponins present in the herbal extract. Systemic exposure to ginsenosides Ra3, Rb1, and Rd after oral administration of the extract was significantly greater than that of other compounds. Considerable colonic deglycosylation of the ginsenosides occurred, but the plasma levels of deglycosylated metabolites were low in rats. Poor membrane permeability and active biliary excretion are the two primary factors limiting systemic exposure to most ginsenosides and their deglycosylated metabolites. In contrast with other ginsenosides, biliary excretion of ginsenosides Ra3 and Rb1 was passive. Meanwhile, the active biliary excretion of ginsenoside Rd was significantly slower than that of other saponins. Slow biliary excretion, inefficient metabolism, and slow renal excretion led to long circulating and thus relatively high exposure levels of these three ginsenosides. For these reasons, plasma ginsenosides Ra3, Rb1, and Rd were identified as PK markers for systemic exposure to Sanqi extract in rat. The investigation of the absorption and disposition of ginsenosides from an herb was systematic, the information gained from which is important for building connections between Sanqi administration and its medicinal effects (Liu et al., 2009).

Components in an herbal medicine are often structurally related and may be divided into more than one class. Analysis of quantitative structure–PK relationships (QSPKRs) of herbal chemicals may provide insight into the role of molecular properties and/or functional group presentation in the PK of compounds and may help understand the PK properties within the compound series and predict which homologues have favorable PK properties compatible with the pharmacological activities. The key role that structure-based differentiation can play in the absorption and disposition of herbal constituents highlights the fact that QSPKR analysis is a vital component of the multicomponent PK study. In the study carried out by Liu et al. (2009), QSPKR analysis helped identify key factors affecting the systemic exposure to different ginsenosides and their deglycosylated products after administration of *Radix notoginseng* extract.

8.7.3 Identification of Nontarget Components from Herbal Preparations

The World Health Organization estimated that 65–80% of the world population used herbal medicines as the primary therapy. However, it has been widely accepted that the identification of components of herbal medicines is of great significance to their quality control and the disclosure of the secret underlying their effectiveness. Accordingly, qualitative and quantitative determination of components existing in herbal medicines has now become a very heated issue. Therefore, the rapid and reliable identification of chemical components contained in herbal preparations remains as a great challenge, despite recent progress in various analytical technologies.

G. J. Wang et al. proposed a novel and generally applicable approach for identifying nontarget components from herbal preparations, based on the usage of liquid chromatography ion trap time-of-flight mass spectrometry (LC/MS-IT-TOF). A simple program was originally developed for searching the common diagnostic ions from all experimentally generated ions. The components sharing the exact same ions (mass error < 5 mDa) were classified into the same family. All families were then connected into a coherent network through the bridging components that are present in two or more families. With the benefit from such a network, it is feasible to sequentially characterize the structures of all diagnostic ions once a single component has been *de novo* identified. The structures of the diagnostic ions could then be used as a priori information for selecting the exact candidates containing the substructures of the

corresponding diagnostic ions from the primary database hits. This strategy enables a nearly sevenfold narrowing of the database hits, and thus substantially enhances the analytical efficiency. With the use of such an approach, 43 out of 53 components incorporated into the network have been successfully identified from the test herbal preparation. For the others that cannot be identified, a complementary approach to screen by sequential loss of specific chemical groups proposed from the accurate mass differences between fragments was established to narrow down the database hits.

The novel method for determination and identification has been successfully applied to the global identification of nontarget components in Mai-Luo-Ning (MLN) injection, a well-known herbal prescription widely used in China for over 30 years to treat cerebral thrombosis, vascular occlusion of angeitides, and deep vein thrombosis of the lower limbs. All of the 87 peaks detected have been successfully identified by combining the use of both of approaches mentioned above, apart from failure to distinguish some isomers. It is a completely different case in this study for nontarget identification since little was known about the diagnostic ions for nontarget compounds. A novel strategy for determining the diagnostic ions and classifications of the complete nontarget components is reported. An approach of database querying by chemical formula combined with fragment comparison is proved to be very useful for the identification of components, which would be useful for the identification of complicated nontarget components from various complex mixtures such as herbal preparations (Hao et al., 2008).

Scutellarin is widely used in treatment of various cardiovascular diseases. Limited data are available regarding its metabolism and PK in humans. Zhong et al. developed a method to identify major metabolites of scutellarin in human urine and plasma and to determine simultaneously the parent drug and its major metabolites in human plasma for PK studies. Four metabolites were detected in urine samples by liquid chromatography coupled with electrospray multistage MS, but only one of them was found in plasma. Its structure was confirmed as scutellarin 6-*O*-β-D-glucuronide by MS, nuclear magnetic resonance (NMR) and ultraviolet (UV) absorbance spectra. The plasma concentrations of scutellarin and the major metabolite were simultaneously determined using LC-MS. It was indicated that scutellarin could be absorbed into the intestine after hydrolysis to its aglycone by bacterial enzymes, followed by reconjugation in the intestinal cell and/or liver with glucuronic acid catalyzed by the Phase II enzyme, which showed regioselectivity and species difference, suggesting that regioselectivity of glucuronoconjugation for scutellarin may be of importance for pharmacological activity and that plasma concentration of isoscutellarin can be used as a biomarker of scutellarin intake (Chen et al., 2006).

8.8 PK AND BIOAVAILABILITY OF NANOMATERIALS

8.8.1 Research and Development of Nanopharmaceuticals

Nanotechnologies, in particular nanomedicine, including nanopharmaceuticals, are changing significantly at present. Nanoparticles have been already used in many products (mainly in cosmetics), but other areas such as pharmaceutics and general medicine have not. The key issues of druggablity of nanopharmaceuticals are drug targeting and safety. The drug delivery system is developed to make the life-saving drug available at the target where required the most. However, these systems fail to work efficiently under most circumstances because the particles of the drug molecule are too large for the cells to absorb or have the potential to cause tissue damage. Due to their extremely small size, nanoparticles can be easily taken by cells. In addition, they are completely soluble and harmless to the tissues. Nanoscience and nanotechnology and their applications in China have been developed rapidly and yet are still at their early stage (Liu, 2009).

In recent years, an increasing number of nanomaterials and nanotechnology applications have been used as novel imaging, diagnostic, and therapeutic agents in the treatment of cancer and other diseases. The safe and effective development and use of these new nanotechnologies requires collaboration among diverse groups of international experts including basic scientists, toxicologists, clinical researchers, policymakers, and others. The goal of the first Joint China-U.S. Symposium on Nanobiotechnology and Nanomedicine on October 21–23, 2008, in Beijing, China, was to share experience and exchange information with respect to the development of nanotechnology for medical use including prevention, detection, and treatment of disease, and ensuring the safety of the general public and of personnel working in the field. The symposium also provided a platform for exploring the development of new research collaborations among top Chinese and American scientists (Wang et al., 2009).

8.8.2 Biopharmaceutics and Therapeutic Potential of Engineered Nanomaterials

Engineered nanomaterials are at the leading edge of the rapidly developing nanosciences, and are an important class of new materials with specific physicochemical properties that are different from bulk materials with the same compositions. Progress has been made to address the course of ADME of engineered nanoparticles *in vitro* and *in vivo*. The distinctive methodology used for studying the biopharmaceutics of nanoparticles has been recently updated (Sun et al., 2007; Cheng et al., 2009a and 2009b).

On the basis of understanding of nanopharmaceutical knowledge and our laboratory conditions, we proposed the following evaluation approach of research projects and contents including biodistribution, metabolic fate, persistence of nondegradable systems, specific therapeutic issues, and immunogenicity, which are shown in Table 8.2 (Liu, 2009).

8.8.3 Biodistribution and Biodegradation

Drug targeting distribution of nanopharmaceuticals is important for selecting PD and PK characteristics. Cytostatics are bonded to the iron oxides. This form of administration, also known as sluicing, is particularly suitable for cytostatics since the intention is not to achieve a high concentration of the cytostatic at the site of action (e.g., tumor) but to reduce the harmful effect on other tissues. The toxicity of currently available anticancer drugs and the inefficiency of chemotherapeutic treatments, especially for advanced stages of the disease, have limited the optimization of clinical drug combinations and effective chemotherapeutic protocols. Nanomedicine, however, allows the release of drugs by biodegradation and self-regulation of nanomaterials *in vitro* and *in vivo*. Nanotechnologies are characterized by effective drug encapsulation, controllable

TABLE 8.2. The Biomedical Evaluation of Nanopharmaceuticals

Evaluation Terms	Evaluation Contents
Biodistribution	Whole organism, cellular level
Metabolism and pharmacokinetics	Absorption, distribution, metabolism, and excretion
Immunogenicity, immunopharmacology	IgG/IgM production, cytokine induction
Persistence of nondegradable systems	Possibility of lysosomal storage disease
Biocompatibility	Biological environment and toxicology and adverse effect to patients
Specific therapeutic and toxicological issues	Therapeutic index of nanomedicines and its delivery systems in drug delivery relates to toxicity of drug payload

IgG: immunoglobulin G; IgM: immunoglobulin M.

self-assembly, specificity, and biocompatibility as a result of their own material properties. Nanotechnology has the potential to overcome current chemotherapeutic barriers in cancer treatment because of the unique nanoscale size and the distinctive bioeffects of nanomaterials.

Inspired by the discovery of dimeric motor proteins capable of undergoing transportation in living cells, significant efforts have been expended recently to the fabrication of track-walking nanomotors possessing two foot-like components that each can bind or detach from an array of anchorage groups on the track in response to local events of reagent consumption. The key problem in fabricating bipedal nanomotors is how the motor as a whole can gain the synergic capacity of directional track-walking, given the fact that each pedal component alone is often incapable of any directional drift. Implemented bipedal motors to date solve this thermodynamically intricate problem by an intuitive strategy that requires a heteropedal motor, multiple anchorage species for the track, and multiple reagent species for motor operation. Realistic molecular mechanism calculations on molecular-scale models were performed to identify a detailed molecular mechanism by which motor-level directionality form a homopedal motor along a minimally heterogeneous track. Optimally, the operation may be reduced to a random supply of a single species of reagents to allow the motor's self-functioning. The mechanism suggests that a distinct class of fabrication targets of drastically reduced system requirements. Intriguingly, a defective form of the mechanism falls into the realm of the well-known Brownian motor mechanism, yet distinct features emerge from normal working of the mechanism (Wang, 2007).

8.8.4 Doxorubicin Polyethylene Glycol-Phosphatidylethnolamine (PEG-PE) Nanoparticles

Solid tumors account for more than 85% of cancer mortality. To obtain nutrients for growth and metastasis, cancer cells must grow around existing vessels or stimulate formation of new blood vessels. These new vessels are abnormal in structure and characterized by leakage, tortuousness, dilation, and a haphazard pattern of interconnection. Tumor structure and blood flow hinder the treatment of solid tumors. To reach cancer cells in optimal quantity, a therapeutic agent must pass through an imperfect blood vasculature to the tumor, cross vessel walls into the interstitium, and penetrate multiple layers of solid tumor cells. Recent studies have demonstrated that poor penetration and limited distribution of doxorubicin in solid tumors are the main causes of its inadequacy as a chemotherapeutic agent.

PEG-PE micelles containing doxorubicin are an important contribution to nanomedicine development (which is called "nanoparticles carrying chemotherapy drug deeper into solid tumors"). Doxorubicin encapsulated into PEG-PE micelles increased the accumulation and penetration into tumors in terms of the number of cells and the intracellular levels. This phenomenon can be attributed to the efficient internalization of the drug-containing micelles by the endocytotic cell uptake mechanism and enhanced permeability and retention of tumors with leaky vasculature. High intracellular retention is especially important because doxorubicin must be internalized into tumor cells to achieve the effective treatment for tumors. The doxorubicin-containing PEG-PE micelles had greatly increased antitumor activity in both subcutaneous and lung metastatic Lewis lung cancer (LLC) tumor models compared with free doxorubicin, and are also less toxic. This drug packaging technology may provide a new strategy for design of cancer treatment (Tang et al., 2007). This research was given high remark by the *Journal of the National Cancer Institute* (Dreher and Chilkoti, 2007).

Investigation on the PK properties of micelle-encapsulated doxorubicin (M-Dox), liposome-encapsulated doxorubicin (L-Dox), and general doxorubicin (G-Dox) was carried out by Wei et al. Although, the PK characteristics of both M-Dox and G-Dox were very similar, while L-Dox was significantly different, in terms of Vd, CL, and AUC, the accumulation of M-Dox in the heart, spleen, kidney, lung, muscle, and skin decreased dramatically after three intravenous injections. The observation suggested that elimination of doxorubicin could be enhanced by nanomicelles and that side effects of doxorubicin may be reduced (Xu et al., 2008; Wei et al, 2008).

8.8.5 Micelle-Encapsulated Alprostadil (M-Alp)

Alprostadil (Alp) is an endogenous substance that is able to dilate vessels and inhibit platelet aggregation, and therefore be potentially active for the treatment of chronic arterial obstruction and microcirculation disturbance. Due to its weak stability in biological matrixes, new delivery systems such as nanoparticle micelle need to be developed. Li et al. studied the PK and tissue distribution of M-Alp and free alprostadil (F-Alp) by intravenous injection at a dosage of 200 µg/kg in rats. Though the PK results showed that both of the Alps were eliminated quickly with a half-life time of 4.39 and 4.76 min for M-Alp and F-Alp, respectively, the tissue distribution results indicated that the novel M-Alp tended to accumulate into tissues such as the heart and liver. This nanocarrier assembling technology may provide a new strategy for delivery of Alp for the treatment of chronic arterial obstruction and microcirculation disturbance (Li et al., 2008).

8.8.6 Paclitaxel Magnetoliposomes

Paclitaxel, belonging to the taxane class of anticancer agents, is probably the most important chemotherapeutic agent against cancer over the past decades. In clinical trials, paclitaxel has been successfully used for the treatment of ovarian, breast, lung, and head/neck cancers, and of acquired immunodeficiency syndrome (AIDS) related Kaposi's sarcoma. To improve the water solubility of paclitaxel and explore the best type of paclitaxel liposomes, lyophilized paclitaxel loaded with negatively charged and magnetic liposomes were evaluated for the preferred presentation to the tumor under neoplastic condition, and enhanced drug level in tumor and plasma, as well as decreased drug uptake in heart, liver, and spleen.

Zhang et al. (2005) reported the biological evaluation *in vitro* of lyophilized negatively charged paclitaxel magnetic liposomes as a potential carrier for breast carcinoma via parenteral administration. Remarkable differences including the shorter half-life of paclitaxel in magnetoliposomes and higher concentration in tumor cell, as well as lower in heart, were observed by comparison of PK properties with other paclitaxel dosage forms. Also, lyophilized paclitaxel magnetic liposomes presented higher potency in the treatment of breast cancer than other formulations via subcutaneous and intravenous administration. The study demonstrated that paclitaxel magnetoliposomes can effectively be delivered to tumors and exert a significant anticancer activity with fewer side effects in the xenograft model.

In addition, Cui et al. successfully prepared anti-hypoxia inducible factor 1a (HIF-1a) antibody-conjugated nanomicelles (NMs) filled with paclitaxel, and confirmed that the prepared nanodrugs can specifically target and selectively kill MGC-803 cancer cells, and dramatically decrease the toxic side effects of paclitaxel. Because of intensive expression of hypoxia inducible factor 1a (HIF-1a) in almost all cancer cells, and pluronic P123 polymer being capable of inhibiting MDR, this unique nanopharmaceutical, anti-HIF-1a antibody-conjugated NMs filled with paclitaxel, exhibit an attractive technological prospect and have great potential in tumor molecular imaging and targeting therapy in the near future (Song et al., 2010).

REFERENCES

Cai YM, Chen ZM, Jiang L, Li M, Liu CX (2008) Pharmacokinetics and pharmacodynamics of subcutaneous single doses of pegylated human G-CSF mutant (PEG30-rhG-CSF) in beagle dogs. *Chin J Clin Oncol* 5:326–332.

Chen XY, Cui L, Duan XT, Ma B, Zhong DF (2006) Pharmacokinetics and metabolism of the flavonoid scutellarin in humans after a single oral administration. *Drug Metab Dispos* 34:1345–1352.

Cheng TF, Sun YD, Si DY, Liu CX (2009a) Attention on research of pharmacology and toxicology of nanomedicines. *Asian J Pharmacodyn Pharmacokinet* 9(1):25–27.

Cheng TF, Si DY, Liu CX (2009b) Rapid and sensitive LC-MS method for pharmacokinetic study of vinorelbine in rats. *Biomed Chromatogr* 23:909–911.

Dreher MR, Chilkoti A (2007) Toward a systems engineering approach to cancer drug delivery. *J Natl Cancer Inst* 99:983–985.

Franke RM, Gardner ER, Sparreboom A (2010) Pharmacogenetics of drug transporters. *Curr Pharm Des* 16(2):220–230.

Hao HP, Cui N, Wang GJ, Xiang BM, Liang Y, Xu XY, Zhang H, Yang J, Zheng CN, Wu L, Gong P, Wang W (2008) Global detection and identification of nontarget components from herbal preparations by liquid chromatography hybrid ion trap time-of-flight mass spectrometry and a strategy. *Anal Chem* 80(21):8187–8194.

Hong Kong Medical Publisher (2008) 2007 Ten top news on pharmacological science in China. *Asian J Pharmacodyn Pharmacokinet* 8(1):L7–14.

Hop CE, Cole MJ, Davidson RE, Duignan DB, Federico J, Janiszewski JS, Jenkins K, Krueger S, Lebowitz R, Liston TE, Mitchell W, Snyder M, Steyn SJ, Soglia JR, Taylor C, Troutman MD, Umland J, West M, Whalen KM, Zelesky V, Zhao SX (2008) High throughput ADME screening: Practical considerations, impact on the portfolio and enabler of *in silico* ADME models. *Curr Drug Metab* 9(9):847–853.

Jin J, Wang FP, Wei H, Liu GT (2005) Reversal of multidrug resistance of cancer through inhibition of P-glycoprotein by 5-bromotetrandrine. *Cancer Chemother Pharmacol* 55(2):179–188.

Li C, Du FF, Chen Y, Xu X, Zheng J, Xu F, Zhu DY (2004) A sensitive method for the determination of the novel chloliesterase inhibitor ZT-1 and its active metabo;ite huperzine A in rat blood suing liquid chromatography/tandem mass spectrometry. *Rapid Commun Mass Spectrom* 18:651–656.

Li QS, Cheng TF, Huang YR, Si DY, Liu CX (2008) Comparative pharmacokinetics and tissue distribution between free and micelle-encapsulated alprostadil. *Drug Metab Rev* 40(S2):80–81.

Liu CX (2005a) The understanding drug metabolism and pharmacokinetics of traditional Chinese medicines. *Asian J Drug Metab Pharmacokinet* 5(1):25–38.

Liu CX (2005b) Difficulty and hot-points on pharmacokinetics of traditional Chinese medicines. In *Science and Technology at the Frontier in China*, Vol. 8, Xu KD, ed., pp. 137–160. High Education Press, Beijing.

Liu CX (2008) Challenges in drug metabolism and pharmacokinetics for research on traditional Chinese medicines. *Drug Metab Rev* 40(S2):19.

Liu CX (2009) Research and development of nanopharmaceuticals in China. *Nano Biomed Eng* 1(1):1–15.

Liu CX (2010) Translational research in discovery, research and development of new drugs. *Chongqing Med* 39(1): 1–3.

Liu CX, Wei GL, Li QS (2001a) Current status of guidelines for bioeuivalenece studies in difference countries. *Asian J Drug Metab Pharmacokinet* 1(3):207–215.

Liu CX, Wei GL, Li QS (2001b) Methodology study of validation for bioanalysis in studies on pharmacokinetics and bioavailability. *Asian J Drug Metab Pharmacokinet* 1(4):279–286.

Liu HF, Yang JL, Du FF, Gao XM, Ma XT, Huang YH, Xu F, Niu W, Wang FQ, Mao Y, Sun Y, Lu T, Liu CX, Zhang BL, Li C (2009) Absorption and disposition of ginsenosides after oral administration of panax notoginseng extract to rats. *Drug Metab Dispos* 37:1–9.

Liu HY, Xu X, Yang ZH, Deng YX, Liu XD, Xie L (2006) Impaired function and expression of P-glycoprotein in blood–brain barrier of streptozotocin-induced diabetic rats. *Brain Res* 1123:245–252.

Liu HY, Yang HW, Wang DL, Liu YC, Liu XD, Li Y, Xie L, Wang GJ (2009) Insulin regulates P-glycoprotein in rat brain microvessel endothelial cells via an insulin receptor-mediated PKC/NF-κB pathway but not a PI3K/Akt pathway. *Eur J Pharmacol* 602:277–282.

Liu L, Deng Y, Yu S, Lu S, Xie L, Liu X (2008) Berberine attenuates intestinal disaccharidases in streptozotocin-induced diabetic rats. *Pharmazie* 63:384–388.

Liu XD, Yang ZH, Yang JS, Yang HW (2007) Increased P-glycoprotein expression and decreased phenobarbital distribution in the brain of pentylenetetrazole-kindled rats. *Neuropharmacology* 53:657–663.

Liu XD, Yang ZH, Yang HW (2008) Repetitive/temporal hypoxia increased P-glycoprotein expression in cultured rat brain microvascular endothelial cells *in vitro*. *Neurosci Lett* 432:184–187.

Liu YT, Hao HP, Xie HG, Lv H, Liu CX, Wang GJ (2009) Oxidative demethylenation and subsequent glucuronidation are the major metabolic pathways of berberine in rats. *J Pharm Sci* 98(11):4391–4401.

Lu T, Wang JL, Gao XM, Chen P, Du FF, Sun Y, Wang FQ, Xu F, Shang HC, Huang YH, Wang Y, Wan RZ, Liu CX, Zhang BL, Li C (2008) Can be used as plasma and urinary tanshinol from salvia miltiorrhiza (Danshen) pharmacokinetic markers for cardiotonic pills, a cardiovascular herbal medicine. *Drug Metab Dispos* 36:1578–1586.

Meibohm B (2006) The role of pharmacokinetics and pharmacodynamics in the development of biotech drugs. In *Pharmacokinetics and Pharmacodynamics of Biotech Drugs*, Meibohm B, ed., pp. 5–12. Wiley-VCH verlagGmbH Co. KGaA, Weinheim, Germany.

Meng LH, Ding J (2007) Salvicine, a novel topoisomerase II inhibitor, exerts its potent anticancer activity by ROS generation. *Acta Pharmacol Sin* 28(9):1460–1465.

Meng LH, Liu CX, Ding J (2007) A touch on some preclinical experiences with salvicine. *Asian J Pharmacodyn Pharmacokinet* 7(4):253–258.

Niu Z, Chen F, Sun J, Liu X, Wang Y, Chen D, He Z (2008) High-performance liquid chromatography for the determination of 3-n-butylphthalide in rat plasma by tandem quadrupole mass spectrometry: Application to a pharmacokinetic study. *J Chromatogr B Analyt Technol Biomed Life Sci* 870(1):135–139.

Selick HE, Beresford AP, Tarbit MH (2002) The emerging importance of predictive ADME simulation in drug discovery. *Drug Discov Today* 7(2):109–116.

Song H, He R, Wang K, Ruan J, Bao CC, Li N, Ji JJ, Cui DX (2010) Anti-HIF-1a antibody-conjugated pluronic tri-block copolymers encapsulated with paclitaxel for tumor targeting therapy. *Biomaterials* 39(8):2302–2312.

Sun YD, Chen ZM, Wei H, Liu CX (2007) Nanotechnology challenge: Safety of nanomaterials and nanomedicines. *Asian J Pharmacodyn Pharmacokinet* 7(1):17–31.

Tang N, Du G, Wang N, Liu C, Hang H, Liang W (2007) Improving penetration in tumors with nano-assemblies of phospholipids and doxorubicin. *J Natl Cancer Inst* 99:1004–1015.

Wang FP, Wang L, Yang JS, Nomura M, Kenichi Miyamoto K (2005) Reversal of P-glycoprotein-dependent resistance to vinblastine by newly synthesized bisbenzylisoquinoline alkaloids in mouse leukemia P388 cells. *Biol Pharm Bull* 28(10):1979–1982.

Wang PC, Blumenthal RP, Zhao YL, Schneider JA, Miller N, Grodzinski P, Gottesman MM, Tinkle S, Wang K, Wang C, Liang XJ (2009) Building scientific progress without borders: Nanobiology and nanomedicine in China and the U.S. *Cancer Res* 69(13):5294–5295.

Wang XJ, Sun WJ, Zhang N, Sun H, Geng F, Piao CY (2007) Isolation and identification of constituents absorbed into blood after oral administration of Liuwei Dihuang pill. *Chin J Nat Med* 5(4):277–280.

Wang ZS (2007) Synergic mechanism and fabrication target for bipedal nanomotors. *Proc Natl Acad Sci USA* 104(46): 17921–17926.

Wei GL, Xiao SH, Lu R, Liu CX (2006a) Simultaneous determination of ZT-1 and its metabolite huperzine A in plasma by high-performance liquid chromatography with ultraviolet detection. *J Chromatogr B* 830:120–125.

Wei GL, Xiao SH, Lu R, Liu CX (2006b) Pharmacokinetics of ZT-1 in experimental animals. *Drug Metab Rev* 37(S2):329.

Wei GL, Xiao SH, Si DY, Liu CX (2008) Comparative pharmacokinetics of micelle- or liposomal encapsulated doxorubicin formulations to healthy rats and tumor-bearing mice. *Drug Metab Rev* 40(S2):81–82.

Xia JH (2001) Validation of analytical methods for pharmacokinetics, bioavailability and bioequivalence studies. *Asian J Drug Metab Pharmacokinet* 1(2):95–100.

Xiao SH, Wei GL, Liu CX, Wang FP (2004) Pharmacokinetics of bromotetrandrin (W198) in rats and beagle dogs. *Acta Pharmacol Sin* 39(4):301–304.

Xiao SH, Wei GL, Liu CX, Wang FP (2005) Tissue distribution and excretion of bromotetrandrine in rats. *Acta Pharmacol Sin* 40(5):453–456.

Xiao XF, Qiao XL, Hou WB, Shan Q, Zhang XX, Liu CX (2008) Studies on pharmacokinetics of pharmacokinetic-markers in Huanglianjiedu decoction to cerebral ischemia reperfusion model mice. *Asian J Pharmacodyn Pharmacokinet* 8(4):287–298.

Xu YY, Xiao SH, Wei GL, Liu CX, Si DY (2008) Nanoparticle of doxorubicin eliminate the accumulation in tissues of tumor-bearing mice. *Drug Metab Rev* 40(S2):193.

Zhang JQ, Zhang ZR, Yang H, Tan QY, Qin SR, Qiu XL (2005) Lyophilized paclitaxel magneto-liposomes as a potential drug delivery system for breast carcinoma via parenteral administration: *In vitro* and *in vivo* studies. *Pharm Res* 22(4):573–583.

Zhang JW, Ge GB, Liu Y, Wang LM, Liu XB, Zhang YY, Li W, He YQ, Wang ZT, Sun J, Xiao HB, Yang L (2008) Taxane's substituents at C3′ affect its regioselective metabolism: Different *in vitro* metabolism of cephalomannine and paclitaxel. *Drug Metab Dispos* 36:418–426.

Zhang L, Zhang YD, Zhao P, Huang SM (2009a) Predicting drug-drug interactions: An FDA perspective. *AAPS J* 11(2):300–306.

Zhang XY, Yang J, Yin XF, Liu XD, Wang GJ (2009b) Hepatobiliary transport of glutathione and its role in cholestasis. *Yao Xue Xue Bao* 44(4):327–332.

Zhao C, He Z, Cui S, Zhang R (2003) Determination of 3-n-butylphthalide in rabbit plasma by HPLC with fluorescence detection and its application in pharmacokinetic study. *Biomed Chromatogr* 17(6):391–395.

Zhou HH (2003) *Pharmacogenetics. From Molecular to Clinical*. People's Military Medical Press, Beijing.

Zhou HH (2004) Pharmacogentics: Genotype-directed personalized drug therapy. In *Science and Technology at the Frontier in China*, Vol. 7, Zhou HH, ed., pp. 299–335. High Education Press, Beijing.

PART B

ADME SYSTEMS AND METHODS

9

TECHNICAL CHALLENGES AND RECENT ADVANCES OF IMPLEMENTING COMPREHENSIVE ADMET TOOLS IN DRUG DISCOVERY

JIANLING WANG AND LESLIE BELL

9.1 INTRODUCTION

Discovering and developing new drugs have become extremely challenging and are as risky as ever. Medicinal chemists have been considerate of the chemical space of new chemical entities (NCEs) not only from a ligand affinity and drug efficacy standpoint but also from an absorption, distribution, metabolism, elimination, and toxicity (ADMET) and druggability outlook. Relevant and comprehensive ADMET profiles have seldom been brought into critical evaluation for potential risk and decision-making processes so early in the discovery phase of drug candidates. ADMET has become one of the three indispensable legs of the drug discovery tripod, along with biology (for therapeutic target finding) and chemistry (for candidate optimization). As a result, a panel of *in silico*, *in vitro*, and *in vivo* tools is being developed and integrated into current drug discovery flowcharts from hit finding to lead identification and optimization, all the way to the nomination of development candidates. This enables project teams to identify the most optimal drug candidate(s) for an unmet medical need while, in parallel, mitigating the possible ADMET risks (Figure 9.1). A suite of comprehensive *in vitro* ADMET assays, as summarized in Table 9.1, is then crucial to assess the ADMET properties and druggability in parallel with preclinical efficacy evaluations. This profiling offers a broad look at the ADMET properties of NCEs and raises flags for those with potential ADMET risks, which can be addressed by

medicinal chemists via additional structure–activity relationships (SARs).

ADMET profiling plays a critical role along with biology and medicinal chemistry. Therefore, it is extremely important that reliable, reproducible, and validated high-quality assays be developed and deployed in support of drug discovery and optimization. However, many technical challenges are present for *in vitro* ADMET profiling in early drug discovery. This chapter discusses various ADMET assays for support of discovery and the technical challenges involved in developing these assays, and the merits of implementing a tier-based profiling portfolio to meet diverse needs.

9.2 "A" IS THE FIRST PHYSIOLOGICAL BARRIER THAT A DRUG FACES

9.2.1 Solubility and Dissolution

Solubility is a key property affecting both *in vitro* and *in vivo* ADMET properties. Multiple solubility expressions (e.g., thermodynamic solubility, equilibrium solubility, kinetic or apparent solubility, and intrinsic solubility) are available but quite distinctive in terms of the data collection and their potential applications. In general, unless otherwise specified, solubility refers to a thermodynamic property (or equilibrium solubility). Thermodynamic solubility, also referred to as "saturation shake-flask" method, reflects not only the molecular interactions of NCEs in the solution phase but also

ADME-Enabling Technologies in Drug Design and Development, First Edition. Edited by Donglu Zhang and Sekhar Surapaneni.
© 2012 John Wiley & Sons, Inc. Published 2012 by John Wiley & Sons, Inc.

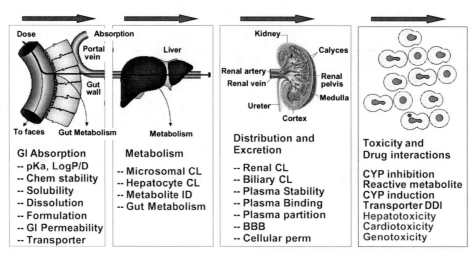

FIGURE 9.1. New ADMET profiling paradigm in drug discovery and development in support of an integrated strategy for the assessment of ADMET and druggability of drug candidates in parallel to their efficacy. The new strategy will require a comprehensive suite of predictive *in silico* and *in vitro* ADMET profiling tools in the early discovery phase.

the solid packing insights (polymorph character) (Yalkowsky, 1999; Avdeef, 2003; Wang et al., 2007; Zhou et al., 2007). This method, as the "gold standard" for solubility determination by industry and the U.S. Food and Drug Administration (FDA), is widely utilized to project the *in vivo* gastrointestinal (GI) absorption and bioavailability of NCEs together with *in vitro* permeability and metabolic clearance data. This approach, however, is time consuming and labor intensive, and thereby applicable for compounds in the late discovery and development phases, where compounds are well characterized for crystalline forms and a sufficient amount of material is available. Thermodynamic solubility is frequently employed to assess the *in vivo* exposure of test compounds as well as to understand the interplay among ADMET parameters in physiology. In contrast, kinetic or apparent solubility is measured for characterizing and qualifying *in vitro* biological and ADMET profiling assays (Wang et al., 2007). Intrinsic solubility is applied for assessing solubility of NCEs as neutral species and predicting their impact on permeability, which is believed to correlate better with the availability of neutral molecules in solution.

9.2.1.1 Thermodynamic or Equilibrium Solubility
The "shake-flask" thermodynamic solubility assay typically starts with dry powders of the test compound, with a shaking at room temperature for 24 h. This is generally acceptable as most of discovery NCEs are prepared in amorphous format and typically present relatively fast dissolution curve to reach >90% of its maximal solubility within 24 h (Zhou et al., 2007). The obtained test solution or solution/precipitation mixture is then separated either by a filtration or centrifugation procedure. A calibration curve (in 0.1–500 μM) is derived (using high-performance liquid chromatography (HPLC)/

ultraviolet (UV) at multiple wavelengths such as 210, 254, 280, and 360 nm) from standard solution of the compound in an organic solvent such as dimethyl sulfoxide (DMSO), acetonitrile, or methanol (100 μM). The equilibrium solubility of the test compound is determined by quantifying the concentration of the filtrate (via the filtration procedure) or supernatant (via the centrifugation protocol) of the test sample against the calibration curve. Lately industry is shifting toward utility of liquid chromatography-mass spectrometry (LC-MS) for quantification, which is beneficial for low soluble compounds and also for those lacking an appropriate UV chromophore.

9.2.1.2 High-Throughput (HT) Kinetic Solubility
In contrast to thermodynamic solubility measured under the equilibrium condition between drug solid and saturated solution, kinetic solubility typically offers a snapshot of drug's dissolution at a specific time point (e.g., with 15–30 min incubation) approaching their equilibrium. The typical turbidity-based HT kinetic solubility, initiated by Lipinski et al. (1997), involves nephelometric titration using either optical (Thermo [Thermo Scientific, Hudson, NH, USA]) or laser nephelometry (BMG [BMG LABTECH, Inc., Cary, NC, USA]) in 96-well microplates (384-well plates for laser nephelometry) to monitor the concentration where NCEs start precipitating out of solution (see Figure 9.2). While HT kinetic solubility provides a fast and cost-effective surrogate solubility filter for *in vitro* ADMET assays, its application to project the solubility impact on formulation and *in vivo* exposure is significantly limited by its poor correlation to the conventional solubility data derived under thermodynamically equilibrated condition (Zhou et al., 2007).

TABLE 9.1. Integrated Tools for *In Silico*, *In Vitro*, and *In Vivo* ADMET Profiling in Drug Discovery

Property	In Silico (Virtual Screening)	In Vitro (HT Screening)	In Vitro (Low Throughput, Gold Standard)	In Vivo
Solubility and physicochemical				
Equilibrium solubility	ACD[a], ADMET Predictor[TMc], Absolv[d], Cerius[i], ADME[e]	Miniaturized shake-flask + LC-UV/LC-MS using reconstituted samples	Shake-flask + LC-UV/LC-MS from solid, or intrinsic solubility	N/A
pK_a	ACD[a], MoKa[g], Marrin[b], ADME Boxes[d]	SGA profiler, CE, RP-HPLC	GLpKa and T3 (dual-phase potentiometric titration)	N/A
LogP/D	Biobyte, ACD[a], Cerius[i]	eLogP (HPLC), MEEKC-CE, artificial membrane, IAM	Shake-flask, GLpKa, and T3 (dual-phase potentiometric titration)	N/A
Absorption				
GI permeability	Cerius[i] ADME[e], ADMET Predictor[TMc] Absolv[d]	PAMPA, 3–7-day Caco-2 cell line	21-day Caco-2 with mechanistic study	Fractional absorption
GI transporters		21-day Caco-2 cell line Pgp-transfected MDCK	Substrate and inhibition studies using 21-day Caco-2 cell or Pgp-, BCRP-, or MRP2-transfected cell lines	Transporter gene knockout model
Distribution				
Plasma stability			*In vitro* plasma stability	
Protein binding	Simcyp[h], GastroPlus[c]	RED	Ultracentrifugation	
Blood/plasma partition	N/A	Plasma depletion	*Ex vivo* whole blood incubation with LC-MS of RBC and plasma fractions	C_{blood}/C_{plasma}
Metabolism				
Hepatic metabolic CL	MetaSite[f], Volsurf[f]/SIMCA[i] - driven local models	Microsomal/S9/cytosol/ cryopreserved hepatoctye CL	Fresh hepatocyte CL, liver slice	AUC/C_{max}, F(%), portal vein infusion
Elimination				
Renal clearance	cLogP/D[a,i] (low-resolution filter), Volsurf[f]/SIMCA[i] - driven local models	Transporter vesicles or transfected cell lines, kidney microsomal CL (metabolism)	(No standard)	f_e, renal vein catheterization
Biliary clearance	N/A	Transporter vesicles or transfected cell lines	SCH biliary excretion model	Bile duct-cannulated models

(Continued)

TABLE 9.1. (*Continued*)

Property	In Silico (Virtual Screening)	In Vitro (HT Screening)	In Vitro (Low Throughput, Gold Standard)	In Vivo
Liability/toxicity				
CYP inhibition	Cerius[i], ADME[e]	HLM + LC-MS, % inhibition at 1 concentration, rCYP + fluorogenic, IC_{50}	HLM + LC-MS, IC_{50} or K_i	AUC/C_{max} increase; humanized mouse model
Time-dependent CYP Inhibition	Simcyp[h]	IC_{50}-shift or k_{obs} at single [I] using HLM or rCYP + LC- or RapidFire[j]-MS	Characterization of K_i and k_{inact} using HLM or rCYP + LC-MS	AUC/C_{max} increase
CYP induction	N/A	PXR and CAR TR-FRET competitive binding or reporter gene assays	Liver slices, primary human hepatocytes (mRNA and enzymatic)	AUC/C_{max} decrease
Reactive metabolites	MetaSite[f]	GSH (or other nucleophilic) trapping assay	Covalent binding	Human toxicity

[a] ACD (http://www.acdlabs.com/).
[b] ChemAxon (http://www.chemaxon.com/).
[c] SimulationsPlus Inc. (http://www.simulations-plus.com/).
[d] Pharma Algorithms (http://www.pharma-algorithms.com/).
[e] Accelrys (http://www.accelrys.com/).
[f] Molecular Discovery (http://www.moldiscovery.com/index.php).
[g] MoKa, Molecular Discovery (http://www.moldiscovery.com/).
[h] Simcyp Ltd. (http://www.simcyp.com/).
[i] Umetrics AB (http://www.umetrics.com/).
[j] Biocius LifeSciences (http://www.biocius.com/).
[k] Agilent Technologies (http://www.chem.agilent.com/en-US/Pages/Homepage.aspx).
AUC, area under the curve; C_{max}, maximum concentration; eLogP, LogP determined by HPLC; SGA, special gradient analyzer; RP-HPLC, reversed-phase high-performance liquid chromatography.

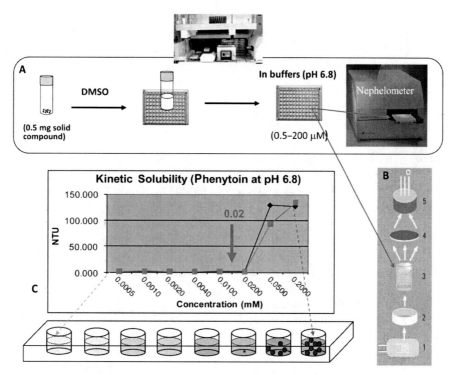

FIGURE 9.2. Schematic presentation of the nephelometric kinetic solubility assay. In an automated workstation, small aliquots of the DMSO stock solution are dispensed in the Cl-, particle- and bubble-free assay buffer in a 96-well plate (A) and the formation of particles is monitored by nephelometry (B). Each compound is tested at eight to nine discrete concentrations in the 0.5–200 μM range and solubility is determined as the last concentration before two consecutive readings higher than the background (C).

9.2.1.3 HT Equilibrium Solubility

For HT equilibrium solubility measurement (Figure 9.3), samples reconstituted (on a GeneVac [Genevac, Inc., Gardiner, NY, USA] HT-4X evaporator) from the DMSO stock solution of test compounds (e.g., 10 mM) in 96-well microplates, instead of dry powder, are used to start the assay, thereby evading the burden to dispatch solid powder and drastically reducing sample consumption. The reconstituted sample plates are filled with designated buffers and sealed by aluminum foil before loading onto shakers for agitated incubation (24–72 h). After equilibrium, a centrifugation procedure is run and the supernatant liquid is transferred into an analytical plate for analysis by HPLC or other HT mass spectrometry (MS) analysis (Brown et al., 2010). The HT equilibrium solubility data collected using the microtiter plate format show reasonable predictivity to conventional thermodynamic solubility using the shake-flask approach using a set of Novartis (Novartis Institutes for Biomedical Research, Cambridge, MA, USA) discovery NCEs (Figure 9.3).

9.2.1.4 Intrinsic Solubility and Solubility pH Profile

Intrinsic solubility (S_O) refers to the equilibrium

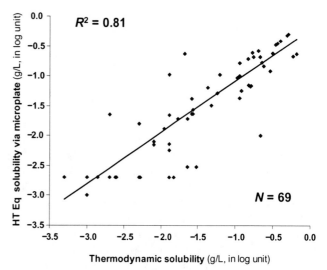

FIGURE 9.3. The HT equilibrium solubility data collected using the microtiter plate format show reasonable predictivity to conventional thermodynamic solubility using the shake-flask approach. The data were from a set of 69 Novartis discovery NCEs, where the regression reflects the exclusion of two outliers (circled) and the inclusion of "qualified data" from 27 NCEs reaching the limit of detection (LOD) in either of the two measurements. (Reprinted from Wang, 2009; with permission.)

solubility of NCEs in the neutral form. Conventionally, intrinsic solubility may be estimated by identifying the lowest value from the multiple equilibrium solubility data collected under a broad pH range. Nowadays, intrinsic solubility for ionizable NCEs may be determined using the potentiometric titration approach named pSOL introduced by pION Inc. (Woburn, MA) and Sirius Analytical Inc. (Beverly, MA), respectively (Avdeef 2003; Box et al., 2009).

The pSOL protocol, requiring ionization constant (pK_a) data, is initiated by a partition coefficient (LogP)-based solubility estimate (Eq. 9.1) of solubility (Ran et al., 2001):

$$LogS_O = 1.17 - 1.38 \, LogP. \quad (9.1)$$

The titration preferably starts from the low soluble to the high soluble species of the test compound (i.e., from acidic to alkaline pH for an acid) in order to avoid supersaturation. Typically, the compound solid is first dissolved using strong acid (for a base) or strong base (for an acid) and then quickly reprecipitated in the titration solution, allowing for equilibrium to establish before the data collection starts. The intrinsic solubility and the solubility pH profile are calculated from the difference between the titration curves in the presence (causing precipitation) and absence (calculated using pK_a) of an excess of the test compound (Wang et al., 2007).

9.2.1.5 Solubility and Dissolution in Biorelevant Media

It should be noted that thermodynamic or equilibrium solubility in aqueous buffers offers simplified determinants to flag solubility risks to oral absorption of NCEs. Practically, the composition of the intestine is far more complex than aqueous buffers. Therefore, the latest industry trend is to project solubility impact on the *in vivo* performance subsequent to an oral dose using simulated intestinal fluid (SIF), simulated gastric fluid (SGF), fasted-state simulated intestinal fluid (FaSSIF), and fed-state simulated intestinal fluid (FeSSIF) (Galia et al., 1998). FaSSIF and FeSSIF contain a mixture of bile salts and surfactants (such as sodium taurocholate) mimicking the physiological media in human stomach and intestine and promoting wetting process in GI tract (Galia et al., 1998). Therefore, those biorelevant media are crucial in identifying the dissolution limiting factors to absorption and ultimately in establishing *in vitro–in vivo* correlation (IVIVC). Indeed, many NCEs with poor equilibrium solubility derived from either shake-flask or potentiometric approaches in aqueous media exhibited drastically improved solubility in FaSSIF, most importantly, toward a better agreement with *in vivo* exposure derived in rat

pharmacokinetic (PK) studies (Wang, 2009). In addition, comparison of solubility in FaSSIF and FeSSIF can also help forecast the food effects *in vivo*. Determination of solubility in the biorelevant media is essential for the accurate assignment of Biopharmaceutical Classification System (BCS) Class II (and also Class I) NCEs where solubility or dissolution is rate-limiting to absorption and also for lead optimization for establishing structure–property relationship (SPR) and IVIVC in early discovery.

From a practical perspective, the solubility data in FaSSIF or FeSSIF can be readily measured by utilizing the HT equilibrium protocol (9.2.1.3) on the same automated workstation where FaSSIF or FeSSIF media will replace the aqueous buffers, and a specially designed analytical method may be employed to deal with the additives in both bioequivalent media.

9.2.2 GI Permeability and Transporters

The absorption of oral drugs is a complex phenomenon involving passive and active mechanisms across the GI mucosa (Artursson and Tavelin, 2003; Hämäläinen and Frostell-Karlsson, 2004). The former includes a paracellular pathway through the tight junction between cells (for small and polar molecular) and/or a transcellular mechanism. For the latter, drug permeation is either promoted by specific transporters (e.g., peptide transporter 1 [PEPT1], organic anion-transporting polypeptide [OATP]) using "carrier-mediated" or uptake mechanisms. In contrast, drug absorption could be limited by efflux transporters such as P-glycoprotein (Pgp)/multidrug resistance 1 (MDR1), breast cancer resistance protein (BCRP), or multidrug resistance-associated protein (MRP), which extrude drugs that partition into enterocytes back into lumen of the GI tract. Some NCEs that are substrates of above efflux transporters may experience difficulties in penetrating the intestinal membrane to reach systemic circulation and could potentially become the victims of drug–transporter interaction (Giacomini et al., 2010).

The existence of potential multiple transport mechanisms in drug absorption presents a challenge for accurate assessment of permeability and its impact on oral absorption. In addition, permeability of NCEs depends not only on the chemical structure or intrinsic properties such as charges (pK_a) and LogP but also on their interaction with biological cells and issues. While Egan et al. (2000) reported an *in silico* absorption model using polar surface area (PSA) and calculated LogP (cLogP), various *in vitro* models, including parallel artificial membrane permeability assay (PAMPA) and cell-based systems such as human colon adenocarcinoma cells (Caco-2) and Madin–Darby canine kidney (MDCK) cells become more popular in assessing GI permeability and the contributions of active transporters in the permeation process (Hämäläinen and Frostell-Karlsson, 2004; Balimane et al., 2006) in the drug discovery phase. Additional models include tissue-based Ussing chamber, intestinal single-pass perfusion, and *in vivo* animal absorption studies (Wang and Skolnik, 2010).

9.2.2.1 PAMPA

As indicated by the name, an artificial membrane permeability assay utilizes a chemical membrane immobilized on a 96-well filter plate with sample analysis by a UV or LC-MS (Kansy et al., 1998). Distinct PAMPA models were established by Kansy et al. (1998), Avdeef et al. (2001), Faller (Wohnsland and Faller, 2001), Sugano et al. (2001), Zhu et al. (2003), and Di et al. (2003) to mimic passive diffusion of NCEs across the GI tract. The major differences among various assays include the composition and preparation of artificial membranes, the type of substrate filter material, the pH variation and incubation time of permeability assay, and the pH gradient between the donor and acceptor chambers in the assay (known as a sink in the acceptor), as well as the method for quantification. For instance, by utilizing a hydrophilic polyvinylidene fluoride (PVDF) substrate, Zhu's work was able to drastically lessen the incubation time from 15 h to 2 h (Zhu et al., 2003). Faller and coworkers successfully established an artificial model using hexadecane without involving cumbersome phospholipids (Wohnsland and Faller, 2001). Avdeef introduced the "double-sink" model that simulates the concentration and pH gradient across the GI membrane (Avdeef et al., 2001). Quantification using LC-MS (Liu et al., 2003; Wang and Faller, 2007) or HPLC (Liu et al., 2003) appreciably improved the sensitivity and robustness of PAMPA by extending the limit of detection for NCEs with low solubility. It also prevents interference originating from impurity with high solubility and/or strong UV chromophore. Overall, PAMPA, with its excellent robustness and reasonable predictivity toward NCEs using passive diffusion mechanisms, ultimately offers a fast and relatively cost-effective method to estimate permeability in early drug discovery.

9.2.2.2 Caco-2

For years, the Caco-2 (from human colon adenocarcinoma) cell model has served as the gold standard for *in vitro* permeability assessment in industry (Artursson and Tavelin, 2003). This is due not only to its morphological (e.g., tight junction and brush-border) and functional similarities (e.g., multiple transport mechanisms) to human intestinal enterocytes but also to the ability to investigate the interplay among different transport systems and differentiate the relative contributions from passive and active transport mecha-

nisms to the overall permeability across the human GI tract. Earlier gene expression work using microchip technology revealed that more than 1000 genes present in human duodenum were also expressed in Caco-2 cell lines (Sun et al., 2002). Recently, Hilgendorf et al. (2007) quantitatively compared the messenger ribonucleic acid (mRNA) expression of 36 drug transporters in human jejunum, colon, liver, and kidney with most commonly utilized *in vitro* models including Caco-2. An excellent agreement was observed for human jejunum and Caco-2 cells, where most of the transporters identified in human jejunum were expressed in Caco-2 cell line (Hilgendorf et al., 2007). Furthermore, the relative gene expression levels in human tissue from jejunum also agreed considerably with those in Caco-2 (e.g., $R^2 = 0.85$ derived from a "rank correlation analysis"). It should be noted that for solute-linked carrier (SLC) or carrier-mediated transporters, relative gene expression levels in Caco-2 cell line match those in jejunum. For adenosine triphosphate (ATP)-binding cassette (ABC) or efflux transporters, the gene expression levels for Pgp and BCRP were lower in Caco-2 cell line than in human jejunum despite comparable results found for multidrug resistance-related proteins (MRP2 and MRP3). The unique enterocytes-like nature, robust cell culturing, and flexibility for molecular engineering warrant Caco-2 as a tool for mechanistic studies. These include transport pathway, determination of transporter substrates, inhibitors and inducers, and interplay between transporters and metabolic enzymes and potential drug–drug interactions (DDIs) using Pgp-, BCRP-, MRP2-, cytochrome P450 (CYP) 3A4-, and pregnane X receptor (PXR)-induced or transfected models.

Several recent manuscripts reviewed the advantages and limitations of Caco-2 models as well as its latest development, especially in the industrial perspective (Wang et al., 2007; Press and Grandi, 2008; Wang and Skolnik, 2010). Typically, Caco-2 cell models with 21-day culturing are employed due to full expression of transporters. Although Caco-2 models using accelerated cell culturing procedures (e.g., 3–7 days) are used to a limited extent, they are not suitable for studying active transport of NCEs due to inadequate expression of transporters (Liang et al., 2000; Alsenz and Haenel, 2003; Lakeram et al., 2007). The short-term Caco-2 culturing system exhibited a reasonable agreement with 21-day Caco-2 model for "apical-to-basolateral" (A-B) apparent permeability (P_{app}) but not efflux ratio (ER) (data not shown), supporting its feasibility for permeability assessment instead of transporter characterization.

Caco-2 has been successfully validated and widely implemented in 24- or 96-well plate formats to assess the permeability and drug interactions associated with

GI-related transporters (Alsenz and Haenel, 2003; Ungell and Karlsson, 2003; Marino et al., 2005; Skolnik et al., 2010). It should be noted that as a cellular model, the outcomes may vary significantly depending on the conditions used such as cell culturing (e.g., the source, passage number and density of Caco-2 cells, the length of culturing, the type of substrate filter plates, and the composition of feeding media), assay conditions (e.g., the length and temperature for incubation, initial loading concentration at donor chambers, usage of "sink" condition, presence of bovine serum albumin [BSA], the pH and composition of incubation media), and bioanalysis (e.g., the sensitivity of assay readouts—LC-UV vs. LC-MS vs. liquid chromatography-tandem quadrupole mass spectrometry [LC-MS/MS]—specific liquid chromatography [LC] method and MS ionization method used, single vs. pooling). The obtained permeability rank or calculated fraction absorbed can thus be compared between laboratories. Similarly to PAMPA, precaution should be taken when dealing with low-solubility compounds frequently occurred in early drug discovery to choose agents inert to the permeability and transport process of the Caco-2 model. Transepithelial electrical resistance (TEER) and impermeable control standards (e.g., Lucifer yellow and mannitol) are routinely utilized to monitor the integrity and tight junction of Caco-2 monolayer. Introduction of BSA to the basolateral compartment is useful to mimic the *in vivo* sink condition and to help minimize the nonspecific binding (NSB) of NCEs to the cells and labware but the effect appeared to vary greatly depending on the mechanism by which NCEs are transported across the monolayer (Saha and Kou, 2002; Neuhoff et al., 2006). In the Caco-2 model, permeability is typically measured from "apical" to "basolateral" compartments (absorptive permeability, $P_{app}(A-B)$) and in the reverse direction (secretory permeability, $P_{app}(B-A)$). It should not be surprising that the "uncalibrated" permeability coefficients such as $P_{app}(A-B)$ and $P_{app}(B-A)$ may vary significantly between laboratories. The current practice in industry is to establish a correlation between "uncalibrated" $P_{app}(A-B)$ results derived within a specific lab and corresponding human fraction absorbed data (Wang, 2009).

Both paracellular and transcellular routes contribute to passive permeability, with the dominant route depending on molecular properties as well as cell monolayer characteristics. Therefore, it is useful to identify the primary permeation pathway as a paracellular flag that indicates potentially incomplete absorption of drugs *in vivo*. The transport pathway is also valuable for medicinal chemists to build SAR and execute pathway-specific optimization. While it is widely anticipated that low molecular weight (MW), hydrophilic compounds can cope with the tight-junction regulated pathway, several exceptions were observed when classifying paracellular mechanism using MW and cLogP (Skolnik et al., 2010). Skolnik et al. reported an approach to differentiate the paracellular compounds by using lipophilicity (<1), alongside permeability $LogP_{app}(A-B)$ (<−5.5 cm/s) measured in the Caco-2 assay.

"Bidirectional" approach is typically utilized to assess the transport mechanism in the Caco-2 model by quantifying the ratio between $P_{app}(B-A)$ and $P_{app}(A-B)$ (Artursson and Tavelin, 2003; Wang and Skolnik, 2010). The NCEs that are substrates for efflux transporters pose a major concern for drug absorption and deposition as efflux may significantly limit molecules from absorption into the enterocytes and GI membrane and eventually retard the exposure. ER, expressed as $P_{app}(B-A)/P_{app}(A-B)$, has been utilized to identify the NCEs with potential efflux issues. While the classification boundary may vary by lab, NCEs with ER >> 1 are characteristic of potential efflux substrates and those with ER close to unity are dominated by passive mechanism. Once oral absorption of a drug candidate or scaffold is limited by efflux-dependent GI permeability, Caco-2 mechanistic study may help establish SPR, allowing medicinal chemists to dial out the efflux issue by structural optimization (Hochman et al., 2002; Troutman and Thakker, 2003; Wang, 2009). Experimentally, one can identify the transporters (e.g., Pgp, MRP2, and BCRP) responsible for NCE efflux and also the potential enhancement in oral absorption when primary transporters are inhibited (Wang, 2009). In addition, many drugs that were identified as efflux substrates by *in vitro* data are in fact completely absorbed *in vivo* (Lennernäs, 2007). In this case, ER and $P_{app}(A-B)$ can be assessed at elevated NCE concentrations to investigate the potential of saturated transporters at high dose and eventually build IVIVC for highly soluble or formulated NCEs (Hochman et al., 2002; Wang, 2009).

Apical to basolateral apparent permeability data are compared with experimental human fraction absorbed (FA_{exp}) data reported in literature (Skolnik et al., 2010) by fitting a nonlinear regression according to the Boltzmann model (Eq. 9.2):

$$Y = \frac{[\text{Minimum} + (\text{Maximum} - \text{Minimum})]}{[1 + (\exp((P_{50} - X)/\text{Slope}))]}, \quad (9.2)$$

where Y is FA_{exp} and X is $LogP_{app}(A-B)$. Minimum and maximum values are constrained to 1 and 100 FA_{exp}, respectively. The P_{50} is the $LogP_{app}(A-B)$ value at 50% absorption in humans. Both the P_{50} and slope are derived from the model fit and subsequently used to determine the calculated human fraction absorbed (FA_{calc}) of test compounds using $LogP_{app}(A-B)$, as shown in the previous report (Skolnik et al., 2010). The FA_{calc} is more intuitive than a $LogP_{app}(A-B)$ value for understanding potential oral intestinal absorption, and thus Caco-2 permeability ranking for NCEs is binned in Novartis as follows: FA_{calc} 0–35% low, 35–75% medium, and 75–100% high. As $LogP_{app}(A-B)$ derived from different labs may vary substantially, FA_{calc} offers additional benefit for assessing the predictivity of Caco-2 permeability assays across labs as well as comparing the Caco-2 permeability data with those derived from PAMPA using passive mechanisms (9.2.2.1).

9.2.2.3 MDCK-MDR1

The MDCK model has also been utilized for permeability evaluation of NCEs using passive diffusion mechanisms. As originating from dog kidney, the general concern limiting its application lies in its expression levels and substrate specificity of various transporters that may be different from the *in vivo* situation in humans (Irvine et al., 1999; Balimane et al., 2000). Recent development using MDR1-transfected MDCK-MDR1 model (Bohets et al., 2001) allows for estimating the contribution of efflux transporters with reduced cell culturing cycle time (3–5 days). By regulating the level of Pgp expression, the sensitivity of NCEs to Pgp in the MDCK assay may be amplified, although the relevance of this to GI physiology has yet to be established.

9.2.2.4 Integrated Strategy to Assess GI Permeability in Drug Discovery

The analysis on Novartis drug discovery NCEs reveals that the majority (60–70%) of discovery candidates transport across GI membrane via passive mechanisms. PAMPA has proven to be a robust, flexible, and cost-effective HT permeability screening tool despite the lack of transporter function. Caco-2, albeit requiring cell culturing procedure that can be handled by full automation, offers a reliable and informative approach for transporter screening and also for subsequent in-depth mechanistic studies. Given the advantages and limitations of each approach, the latest consensus appears to favor a strategy that combines all three *in vitro* approaches with *in silico* models to ensure high-quality assessment of permeability in early discovery (Wang and Skolnik, 2010). First, a number of *in silico* parameters (cLogP, PSA, and MW) are useful to project the probability for an NCE to permeate the GI membrane, ideally applied at the triaging and selection of HT screening hits, and for designing a new scaffold in early discovery. The absorption model developed by Egan et al. (2000) demonstrates to be a fast and decent *in silico* predictive tool. For instance, analysis of Novartis discovery NCEs using the absorption model reveals that ~94% of NCEs in the absorption model "good" zone

are categorized as either "high" (same rank) or "medium" (with one rank difference) in PAMPA using the LC-MS approach, leading to decent predictivity. In contrast, NCEs in "poor" or "borderline" zones of *in silico* prediction show mixed results in PAMPA, assuming the necessity of *in vitro* assessment. Here PAMPA serves as an HT *in vitro* permeability tool used for lead selection and optimization, especially for those NCEs showing high risk by *in silico* prediction and for scaffolds using passive diffusion mechanisms. Using Novartis models, the PAMPA data exhibiting a "low" permeable rank agree well with those in the same Caco-2 rank. The Caco-2 model should be applied to challenging scaffolds involving active transport mechanisms or with higher MW (e.g., >600). The former (potential substrates or inhibitors for efflux transporters), most likely, will exhibit higher permeability in PAMPA and may have a poor IVIVC. Caco-2 mechanistic studies are valuable to identify the major transporters such as Pgp, MRP2, and BCRP and to appraise the impact of shutting down active transporters via either inhibitory (Varma et al., 2003) or saturation mechanisms (Bourdet and Thakker, 2006). Specific transporter-transfected cell lines may be ideal to tackle the impact of an individual transporter on drug's PK/PD (relationship of pharmacokinetics and pharmacodynamics) performance subsequent to Caco-2 transporter assays.

9.3 "M" IS FREQUENTLY CONSIDERED PRIOR TO DISTRIBUTION DUE TO THE "FIRST-PASS" EFFECT

9.3.1 Hepatic Metabolism

The ability to achieve good oral drug exposure is, by and large, heralded as an essential criterion for successful selection of a drug candidate for most disease areas. Once the compound is in solution and permeates the gut wall, an orally absorbed drug begins its pursuit of systemic distribution with an initial travel through the hepatic portal vein, feeding drug directly into the liver, which is central to key metabolic and elimination processes. In this capacity, "first-pass" metabolism/elimination represents a critical barrier differentiating oral drug exposure from centrally administered (intravenous) drug exposure, and the bioavailability term, $F(\%)$, can be significantly impacted by the magnitude of "first-pass" clearance. Noticeably, the events of "first-pass" clearance continue to be important for both centrally (intravenous) and orally administered drugs, as each subsequent pass of drug circulated through the liver provides an opportunity for metabolic and excretory elimination mechanisms to further extract drug from the central (blood/plasma) compartment.

In vitro models that can be used to reliably and efficiently characterize a drug's potential for liver extraction stand to improve the overall pace of drug discovery. In identifying compounds with minimal clearance liabilities, the most promising candidates may be selected for more time- and resource-consuming *in vivo* studies. Furthermore, *in vitro* metabolic stability systems offer one of the earliest and best opportunities to investigate the potential for metabolic liabilities in higher species including human (Chaturvedi et al., 2001; Masimirembwa et al., 2001; Kariv et al., 2002; Masimirembwa et al., 2003). This section offers some points to consider when selecting technologies on which to build an absorption, distribution, metabolism, and elimination (ADME) profiling strategy around metabolic liability.

9.3.2 CYPs and Drug Metabolism

The CYPs represent a family of broad-selectivity monooxygenases that have been implicated, to some extent, in the oxidative metabolism of ~75% of drug substances and are estimated to provide the major clearance route for about half of all marketed drugs (Wienkers and Heath, 2005; Guengerich, 2005a). CYP isoforms have been identified in every species and in virtually every animal tissue, thus far, examined (Nelson et al., 1993). Both species- and tissue-specific distributions have been characterized. With broad, overlapping substrate recognition across isoforms (within a species) and within subfamilies (across species), CYPs are ensured a sustained importance in drug discovery efforts. In addition to CYPs, there are many other relevant enzyme systems that contribute to drug metabolism (e.g., soluble oxidases, conjugative enzymes, esterases). The differential localization of these various enzyme systems is of key importance in the selection of an appropriate *in vitro* metabolism model. Table 9.2 provides an overview of enzyme enrichment among liver tissue fractions. Some *in vitro* models (e.g., tissue slices, fresh or cyropreserved hepatocytes) support a broad spectrum of drug metabolizing enzyme (DME) activities. These systems may prove useful when significant non-CYP metabolism is involved or when clearance pathways are largely undefined for the drug of interest. Many *in vitro* models employing hepatic subcellular fractions (e.g., cytosol and microsomes), however, are suitable for monitoring only one or two major enzyme systems at a time. Each model comes with unique advantages and limitations, so the intent for utilization to meet the particular discovery needs must be weighed to select the most appropriate technology.

TABLE 9.2. Summary of Distribution of Relevant DMEs and Their Activities in Various Liver Fractions

Enzyme	Activity	Preferred Target	Enriched Subcellular Fraction	Special Needs
CYPs	Hydroxylation, dealkylation	-C, -N	Microsomes	NADPH
UGTs	Glucuronide conjugation	-O, -N, -S; phenol/carboxylic acid > alcohol/amine	Microsomes	UDPGA + alamethacin pretreatment
FMOs	Oxidation	2° and 3° amines or thiols	Microsomes	NADPH; inactivated at 40°C, optimum activity at pH 9
Alcohol/aldehyde dehydrogenases and soluble oxidases	Dehydrogenation, oxidation	-OH, -CHO	Cytosol	No cofactors needed
Sulfotransferases (ST)	Sulfate conjugation	-OH, -NH$_2$	Cytosol	PAPS
Glutathione *S*-transferases (GST)	Glutathione conjugation	Epoxides, arene oxides, nitro groups, hydroxylamines	Cytosol, microsomes	Glutathione
N-Acetyltransferases (NATs)	Acetylation	-OH, -NH$_2$	Cytosol	Acetyl coenzyme A
Methyl transferase	-OH, -NH$_2$, -SH, heterocyclic N	Catecholamines, phenols, amines	Cytosol	SAM
Esterases, amidases	Hydrolysis	Ester/amide bonds	Plasma/whole blood, cytosol, Microsomes	No cofactors needed
Monoamine oxidases (MAOs)	Oxidative deamination	Hydrophilic amines	Mitochondria	No cofactors needed

NADPH, β-nicotinamide adenine dinucleotide phosphate; PAPS, 3′-phosphoadenosine-5′-phosphosulfate; SAM, *S*-adenosyl methionine; UDPGA, uridine 5′-diphosphoglucuronic acid.

9.3.2.1 Liver Microsomes (LM) A number of *in vitro* models are available to characterize the contributions of CYPs to drug clearance, but the favored model in the pharmaceutical industry, especially for routine *in vitro* assessment of drug metabolism, is LM, the subcellular fraction of liver enriched for CYP content.

High-quality LM preparations are abundantly available from commercial suppliers at relatively low cost. Preparations from several preclinical species, as well as humans, can be readily purchased and incubations require only buffer, cofactor (β-nicotinamide adenine dinucleotide phosphate [NADPH]), and compound for conducting stability assays. In addition to ease of use, the scalability of *in vitro* results for estimation of *in vivo* clearance and the ability to simultaneously monitor multiple major DMEs represent the major advantages of LM over other systems, such as individual recombinant CYP (rCYP) preparations (Obach et al., 1997; Yan and Caldwell, 2003).

In addition to the drug metabolizing CYPs, the LM fraction contains UDP-glucuronosyltransferases (UGTs) and flavin-containing monooxygenases (FMOs), altogether representing the primary clearance routes for ~60% of marketed drugs (Wienkers and Heath, 2005). FMOs, like CYPs, are integral membrane proteins that utilize NADPH. However, the overall contribution to drug metabolism for this family of enzymes is generally considered to be minor, with a few noted exceptions (Bhamre et al., 1995; Chung and Cha, 1997). Assay approaches that take advantage of differential pH and temperature sensitivities to distinguish between CYP and FMO activities in microsomes have been well characterized (Grothusen et al., 1996). UGTs are a family of membrane-associated proteins, natively localized to the lumen of the endoplasmic reticulum. UGTs carry a prominent role in drug metabolism by catalyzing the conjugative modification of drugs (or their metabolites) with glucuronide to facilitate excretion. UGT activity, commonly referred to as "Phase II metabolism," is easily separated from CYP activity, which is known as "Phase I metabolism," in microsomal preparations, as the former requires the cofactor uridine 5′-diphosphoglucuronic acid (UDPGA) rather than NADPH. It is possible, when needed, to coactivate CYP and UGT pathways through dual supplementation with both NADPH and UDPGA (Yan and Caldwell, 2003). In early drug discovery, it is often desirable to quickly identify the NCEs with the highest metabolic liability, in line with *in vivo* clearance data. In this paradigm, liver microsomal fractions offer a fast, practical, and cost-effective screening option for first-pass clearance assessments, considering the collective impact of most CYP and UGT enzymes. In light of the prominent role of hepatic CYPs in drug metabolism, this model has the

greatest potential to yield useful correlations with *in vivo* clearance.

9.3.2.2 Hepatocyte Suspensions Apart from CYP and UGT activities, little other metabolic information can be gained from the simple LM model. Alternatively, isolated hepatocytes retain functionality for a number of DME systems without the need for cofactor supplementation. However, hepatocytes offer unique practical challenges relative to liver tissue fractions, including higher interlot variability and greater expense. Nevertheless, hepatocytes become an appealing alternative to microsomes, for example, when *in vitro* CYP- and/or UGT-mediated microsomal clearance fails to account for high clearance *in vivo*. With the whole hepatocyte model, if handled properly, the vast majority of the enzymes outlined in Table 9.2 will be present in native environments with endogenous levels of cofactors and accessory proteins, a distinct advantage when the predominant clearance pathway is unknown. Even when a predominant clearance pathway has been clearly identified, there may still be something to gain from whole cell hepatocyte studies. For example, while UGT activity can, technically, be reconstituted from microsomal preparations, this is achieved under somewhat contrived conditions, demanding membrane permeation through the use of pore-forming agents (Fisher et al., 2000; Lin and Wong, 2002). It has been suggested that the native environment provided by the intact hepatocyte allows for more appropriate evaluation of UGTs and their contributions to hepatic clearance.

9.3.2.3 Rate of Metabolic Clearance (Loss-of-Parent Approach) Whether used with LM or whole hepatocyte preparations, the accepted practice in drug discovery is to characterize rates of drug metabolism by monitoring the loss of parent drug, rather than formation of metabolite(s), from the incubation system. With appropriate assay design, the half-life ($t_{1/2}$) for clearance of parent drug from the incubation medium should reflect the intrinsic rate of metabolic clearance (CL_{int}) for the drug. CL_{int} values (often expressed as μL of incubation medium cleared/min/mg microsomal protein) may be readily scaled to generate estimates of *in vivo* hepatic clearance (CL_h) for a species (Obach, 1999). A few considerations of the biological assay design will help plan for a robust standard biological protocol that should be applicable across multiple species (e.g., human, rat, mouse, dog, and/or monkey) and should be fully automatable, if needed.

1. *Test Article Concentration:* For best estimates of CL_{int}, the rate of clearance should be evaluated at a test concentration that is below its K_m to avoid any saturation of enzymatic processes and reflect

first-order elimination. Since the apparent K_m for metabolism of NCEs is usually not known, a very low initial NCE concentration (approximately 1 μM) is recommended. This is a reasonably sound starting concentration since even prototypical and selective CYP probe substrates, including midazolam, felodipine, bufuralol, and diclofenac, are metabolized by HLM with low-micromolar K_m values (*approximately* 1–5 μM). There are few examples of submicromolar K_m kinetics reported for the major drug metabolizing CYP isoforms. A low-micromolar starting concentration, however, calls for a sensitive mode of detection, which can be achieved by LC-MS/MS techniques.

2. *Timing:* The *in vitro* metabolic clearance rate, whether derived from LM or hepatocyte suspensions, is generated from remaining parent drug concentrations collected over multiple time points from a reaction including appropriate cofactor(s) (e.g., NADPH and/or UDPGA). For the most accurate and reproducible assessments of compounds with medium-to-short half-lives ($t_{1/2} < 30$ min), it is best to use at least four time points, including $T = 0$, with the earliest time point acquired by ~5 min (e.g., 0, 5, 15, and 30 min). Therefore, automation that supports rapid turnaround to collecting the earliest time point is advantageous toward characterizing high-clearance compounds.

3. *Bioanalysis:* Due to the very low NCE concentrations often employed in metabolic stability studies (~1 μM), LC-MS/MS is currently the standard analytical approach for HT, sensitive quantification of parent drugs. The persistent disadvantage of this technology is that bioanalytical time must be invested up front to tune each separate test compound before the actual reaction samples can be analyzed. However, introduction of network interfacing software (e.g., Microsoft's [Microsoft, Redmond, WA, USA] Access® or Oracle Corp.'s [Redwood Shores, CA, USA] Oracle® databases) to recognize redundancies in compound's tune files generated on networked instruments allow tune files to be pulled from a shared database rather than forfeiting instrument time in replicating a tune analysis. These files might represent new batches of a previously tested compound or compounds that were previously analyzed in a different ADME study with an MS readout. Recent advances in high-resolution exact mass detection using time-of-flight (e.g., Exactive™, Thermo Scientific, San Jose, CA, USA), in contrast to triple quadrupole, mass analysis may eventually eliminate the need for up-front tuning altogether (Chan

et al., 2009; Lu et al., 2010). RapidFire MS technology (RapidFire™, Biocius LifeSciences, Woburn, MA) is another progressing innovation that may yet negate the bioanalytical burden for these kinds of assays by dispensing with the time dedicated to chromatographic separation of analytes preceding mass analysis (Brown et al., 2010). LDTD ion source technology likewise eliminates the LC time investment by introducing an infrared laser to thermally desorb dried analytes from a stainless steel sample well (Wu et al., 2007). There is keen potential for successes in rapid-fire and LDTD technologies to significantly reduce the analytical burden and improve the turnaround times for these and other ADME profiling assays.

9.4 "D" IS CRITICAL FOR CORRECTLY INTERPRETING PK DATA

9.4.1 Blood/Plasma Impact on Drug Distribution

Protein binding can reduce the effective concentration ("free ligand") of a drug that is available to elicit pharmacological or biological activity. If the drug is highly bound (>99%) to proteins, then the actual inhibitory potency may be up two orders of magnitude greater than an "apparent" K_i or IC_{50} might suggest. It is often unknown whether the *in vitro* protein binding influences are reflective of the *in vivo* scenario. The *in vivo* scenario is much more complex and there are a number of ways in which drug distribution within and between tissues can impact the manifestation of ADME events. In this respect, the effort to understand drug equilibrium between blood and the individual organ tissues can contribute to reconciling disconnects between *in vitro* and *in vivo* observations.

Blood has the distinction of serving as the body's "mass transit," carrying drugs from the point of entry (site of injection or absorption) to widespread points of organ and tissue perfusion. Once drugs have access to the bloodstream, they generally rapidly distribute into interstitial and intracellular fluids throughout the body. Cardiac output, regional blood flow, and individual tissue volumes, as well as the physicochemical properties of drugs can influence the rate and extent to which drugs distribute extravascularly. In addition to lipid solubility and ionization properties that can influence distribution, protein binding seems to be a critical determinant of blood-to-tissue partitioning. The "unbound" fraction of drugs (f_u) is presumably free to equilibrate with extravascular fluids and drive target, as well as off-target, responses. Protein binding becomes particularly influential for distribution and clearance when there are significant differences in a drug's binding affinity for elements in the bloodstream relative to the macromolecules of individual tissues (Tillement et al., 1984; Pacifici and Viani, 1992; Schmidt et al., 2009). Unfortunately, it has proven exceedingly challenging to directly, experimentally quantify extravascular drug distribution. By comparison, blood is a simpler matrix that can be separated into its individual components without extensive homogenization or invasive solvent extraction. Plasma protein binding (PPB) is, therefore, the most common point of reference in beginning to develop drug distribution hypotheses. Plasma, itself, is the liquid component of blood that remains when whole blood is devoid of blood cells. Since handling and extraction of whole blood, as well as bioanalysis of whole cell extracts, can be difficult, PK sampling from the plasma compartment is frequently deemed the simplest avenue for collecting representative *in vivo* drug exposure data and for modeling parameters such as drug clearance (CL) and volume of distribution (V_d). Observed plasma concentrations further serve as a starting point for building early PK/PD relationships. Most often, it is assumed that the concentration of a drug that is detected in plasma, at any given point in time, is equal to the concentration of the drug in whole blood. A major caveat with plasma sampling, alone, is that the resulting "exposure" profiles confirm only that a given fraction of the administered dose was measurable in plasma over the time course of the study. It provides no insight into the further distribution of drug outside of the plasma. As discussed above, without further testing, one can only speculate on the distribution and local concentrations of the remaining fraction of administered dose.

Nevertheless, the information gained from plasma sampling is indispensible for assessing the *in vivo* ADMET for NCEs. To establish confidence in the validity of information gained from plasma, it is important to consider that plasma is far from an inert vehicle just for collecting drugs. Even *ex vivo*, plasma presents constitutively active enzymatic proteins, bulk serum proteins, lipids, and cell waste products (Skeaff et al., 2006). As such, it can become important to evaluate the effect that plasma, itself, is capable of exerting on the drug. Furthermore, subsequent to the fact that plasma is only a representative fraction of whole blood, there are times when it is worth examining whether there is asymmetry in the partition of drugs between plasma and whole blood cells. Several approaches relevant to characterizing the dynamics of drugs in blood and plasma are discussed in this section.

9.4.2 Plasma Stability

Plasma hydrolysis is a liability in drug development and has important implications. Drugs with poor plasma sta-

bility often present high clearance rates, short half-lives, and insufficient exposure *in vivo*, resulting in failure to demonstrate efficacy. Additionally, very rapid plasma hydrolysis compromises the quality of the PK profiles that can be obtained for a drug, when *ex vivo* degradation in the plasma becomes a variable for bioanalysis of PK samples.

There are ways though that plasma DMEs can be exploited to benefit drug discovery. Prodrug strategies, to overcome solubility- and GI permeability-limited exposure, for example, have harnessed the activities of plasma enzymes to rapidly "release" active drug from an adequately soluble or permeable precursor molecule (Barthel et al., 2009). This strategy can also be applied to regulate the rate at which a rapidly absorbed prodrug is "activated" by selectively targeting slower plasma hydrolysis pathways (e.g., extended release) (Song et al., 2002; Mishra et al., 2008).

Most plasma instability is a result of enzymatic hydrolysis attributed, largely, to plasma esterases, but can also be due to other circulating enzymes including amidase, lipase, phosphatase, and peptidase (Parkinson, 2001; Di et al., 2005; Kerns and Di, 2008). Although esters pose the greatest structural risk for hydrolysis, other chemical linkages are susceptible to plasma degradation, such as amide, carbonate, carbamate, lactam, lactone, and sulfonamide (D'Souza and Topp, 2004; Kerns and Di, 2008).

Overall, the *in vitro* plasma-mediated rate of drug hydrolysis is quite robust and is relatively insensitive to the concentration of drugs (linearity expected up to ~20 μM), drug solvent (DMSO tolerated up to 2.5%, v/v), and buffer dilution (up to 1:1 plasma) (Di et al., 2005). Although plasma has a very high catalytic capacity and is difficult to saturate, a test compound concentration, again, of ~1 μM is a good starting point for *in vitro* measurements. This concentration represents physiologically relevant *in vivo* plasma concentrations and is generally low enough to avoid potential solubility issues, yet high enough to robustly support "loss-of-parent" bioanalysis. Usually, *in vitro* plasma stability is used to survey drugs for very rapid decomposition in an attempt to diagnose poor *in vivo* exposure. In this case, time points comparable with microsomal clearance (<1, 5, 10, 20 min) should yield reliable hydrolysis half-lives ($t_{1/2}$) to support comparison of drug series or species differences. The *in vitro* assay design is relatively simple. Even with sustained incubation at 37°C, plasma enzyme activity remains stable for almost 24 h. Most plasma enzymes, including esterases, do not require additional cofactors for activation, so hydrolysis is initiated simply with addition of drugs into the plasma sample. Hydrolysis can usually be terminated with an excess (two- to threefold) of organic solvent, with centrifugation introduced to remove precipitated protein prior to bio-analysis. The major bottleneck is similar to that of microsomal metabolic stability assessments, bioanalytical optimization, and throughput.

9.4.3 PPB

As already mentioned, the extent to which drugs bind to plasma proteins can have significant impact on *in vivo* PK and pharmacodynamics (PD) (Schmidt et al., 2009). Rarely, clinically relevant drug interactions may even emerge as a direct consequence of PPB displacement *in vivo* (Sellers, 1979; Rolan, 1994). However, *in vivo* protein binding displacement to precipitate a DDI seems to require somewhat of a "perfect storm" scenario, in that the susceptible victim drugs are characterized with a very low therapeutic index, low clearance, and a small V_d (Rolan, 1994). Still, the extent to which protein binding influences drug distribution remains a strong interest in drug discovery.

Most small drug molecules exhibit some degree of reversible protein binding throughout the body such that the binding equilibrium between free, unbound drug and protein-bound drug can be expected to follow the law of mass action. In human plasma, human serum albumin (HSA) and alpha-1-acid glycoprotein (AAG) provide the majority of nonspecific, reversible binding sites for circulating drugs (Israili and Dayton, 2001; Bertucci and Domenici, 2002), although other proteins, such as globulins and lipoproteins, are also present in plasma. The broad observation is that acidic drugs tend to occupy binding sites on HSA through electrostatic and hydrophobic bonds, while basic drugs tend to partition with AAG (Tillement et al., 1984). Because levels of AAG are considerably lower than those of serum albumin, the mechanism of plasma binding interaction can have implications with regard to linearity and saturation of binding *in vitro*.

In recent years, the introduction of the rapid equilibrium device (RED) has greatly enhanced our capability for assessing protein binding (Pacifici and Viani, 1992; Waters et al., 2008). This technology employs a multi-well format that incorporates individual dialysis cells where drug-spiked plasma (added inside the individual dialysis cell) is immersed in a protein-free dialysis buffer, such as phosphate-buffered saline.

During incubation, free drug redistributes across the dialysis membrane into the dialysis buffer. Within 2–4 h, aliquots can be removed directly from the plasma-containing insert or the dialysis buffer chamber and extracted with organic solvent containing MS internal standard. With this approach, the fraction of drug unbound in plasma (f_{up}) equals the concentration of drug measured in buffer (which mirrors the free con-

centration, C_{free}, in the plasma chamber) relative to the concentration of drug measured in plasma (where $C_{total} = C_{free} + C_{bound}$):

$$f_{up} = \frac{[\text{Drug}]_{buffer}}{[\text{Drug}]_{plasma}} = \frac{C_{free}}{C_{total}}. \qquad (9.3)$$

The combined recovery of a drug from both the buffer and plasma compartments relative to the total, initial concentration of drug in plasma (sampled at the beginning of the incubation) can be a useful measure of drug stability in plasma.

This RED technology is highly compatible with automation and recent improvements in the apparatus include introducing up to 96-sample cells per plate to maximize a single plate capacity. To further increase the throughput, one option is to pool test compounds in the incubations. This should be well tolerated as long as care is taken to avoid saturating protein binding with additive drug concentrations (Wan and Rehngren, 2006).

Alternate methods, such as ultrafiltration or ultracentrifugation, are also used in discovery support. Ultrafiltration utilizes high-MW cutoff (e.g., 10 kDa) filters under centrifugal pressure to separate free from protein-bound drug. This approach has proven amenable to HT configuration; however, nonspecific drug binding (to membrane filters and multiwell plate polymers) and potential for centrifugal "edge effects" across multiwell assay plates are noted concerns (Zhang and Musson, 2006). Ultracentrifugation, on the other hand, circumvents nonspecific binding issues, but comparatively large plasma volumes are required and throughput capacity is generally low relative to other approaches.

9.4.4 Blood/Plasma Partitioning

In vivo exposures and clearance rates for drugs that are highly sequestered by whole blood cells might not be adequately captured with plasma sampling alone, and a blood/plasma partitioning study may be useful for interpreting data generated from plasma. Furthermore, an observation of nonlinear drug distribution may be explained, at least in part, by saturation of drug uptake and/or binding in different tissues, including red blood cells (RBCs) (van den Bongard et al., 2003).

Plasma itself accounts for ~45% of total animal blood volume, with RBCs almost entirely fulfilling the remaining fraction of blood volume (Wilkinson, 2001). Infrequently, leukocytes, platelets, granulocytes, and lymphocytes may be targets for nonspecific drug binding or uptake. A few physicochemical properties have been associated with a drug's tendency for RBC partitioning (Fagerholm, 2007). Lipid solubility, degree of ionization, molecular size, and ability to form hydrogen bonds have all been identified as relevant factors. Highly lipophilic, largely un-ionized drugs favor RBC uptake, while larger molecules with high potential for hydrogen bonding are poorer candidates for RBC partitioning, indicative of a passive permeability-driven partitioning process.

In theory, the evaluation of blood/plasma partitioning is quite simple. Plasma and cellular constituents from drug-spiked blood can be readily separated by centrifugation. Certainly, bioanalytical determination of drug concentrations in diverse biological matrices is routinely practiced. Still, there remain two primary challenges for this type of assay. The first is the requirement of bioanalytical standards prepared separately for plasma, RBC fractions, and whole blood. This necessity is driven by potential variability in the extraction efficiency of drug from the different matrices, as well as variable matrix effects on MS ionization during bioanalysis. The second is the particular challenge of handling the RBC fraction. These fractions are viscous and difficult to accurately pipet. Even following successful pipetting, these samples require time-consuming cycles of freeze-thaw to achieve cell lysis and lysates must then undergo extensive cleanup prior to bioanalysis. The unique advantage of this conventional approach is that it permits analysis of blood partitioning *ex vivo*, enabling studies to follow a drug that has already been dosed *in vivo*. However, it is a fairly labor-intensive investment for routine drug discovery work. A robust, alternative approach that circumvents the need to handle whole blood or RBC blood fractions has been described (Yu et al., 2005). Rather than monitoring appearance of drug in RBC fractions, this approach is based on monitoring drug disappearance from the plasma fraction. The assumption is that the drug encounters no stability issues either in the blood cell or plasma fraction.

Similar to the conventional method, a known volume of fresh, heparinized (or ethylenedinitrilotetraacetic acid [EDTA]-treated) whole blood is spiked with a test compound. In parallel, an identical volume of plasma freshly isolated from the same whole blood is spiked with the same amount of drug so that the total drug concentration in plasma mimics the total drug concentration in whole blood. Although either sample could serve as the reference for total blood concentration (C_{total}), the plasma sample is a much more convenient matrix as a point of reference. Following an incubation of whole blood with drugs to equilibrium, the whole blood is separated by centrifugation into RBC and plasma fractions and the concentration of drug in the plasma fraction (C_{PL}) alone is used to infer the concentration of drug in the RBC fraction (C_{RBC}). The partition coefficient (K) of the drug in RBCs is equal to the ratio of drug concentration in each fraction:

$$K_{RBC/PL} = \frac{C_{RBC}}{C_{PL}} = \frac{C_{total} - C_{PL}}{C_{PL}}. \qquad (9.4)$$

When $K_{RBC/PL} \sim 1$, this is an indication that the test drug distributes equally between the blood cell fraction and plasma and reaffirms that plasma is, in fact, a fair, representative matrix from which to derive PK profiles. When $K_{RBC/PL} \gg 1$, this may be evidence that the test drug favors uptake by, or association with, the RBC fraction (Sun et al., 1987; Yu et al., 2005).

Less commonly, when $K_{RBC/PL} < 0.5$, this may indicate that the test drug exhibits high plasma retention and is, at least partially, excluded from distributing into RBCs (Yu et al., 2005). A bias toward plasma distribution might be an indication of very tight PPB or may reflect very poor cellular permeability. In general though, partitioning favoring plasma retention may not significantly impact interpretation of PK profiles derived from plasma analysis. Since plasma makes up more than half of total blood volume, plasma concentrations will still reflect total blood concentration within twofold, despite the plasma bias. Nonetheless, this result may have implications regarding the ability of drugs to redistribute outside of plasma to reach target or drug clearance tissues, and additional studies to understand the impact of the observation on PK/PD may be warranted.

Overall, the less-intensive plasma-depletion approach offers a robust alternative to the conventional RBC sampling protocol for assessment of blood-to-plasma drug partitioning. This method is better suited to meet the needs of drug discovery and still yields comparable results with the direct RBC sampling method (Yu et al., 2005). With some optimization, this approach should be amenable to automation. While the *in vitro* approach benefits from a simplified, indirect matrix sampling, efforts to assess blood/plasma distribution could still potentially benefit from future innovation in fractionation and/or extraction technologies that might support clean and direct matrix analysis of drug distribution *ex vivo*.

9.5 "E": THE ELIMINATION OF DRUGS SHOULD NOT BE IGNORED

Most gaps in clearance IVIVC derived from liver metabolism models can be explained, in part, by the fact that nearly one-third of drug clearance can be attributed to renal and biliary excretion and esterase-mediated hydrolysis. Furthermore, CYPs and other hepatic DMEs, including UGTs, have been shown to play a significant role in drug metabolism in extrahepatic organs, including lung, brain, intestine, and kidney. Due to the prevailing likelihood and expectation that hepatic CYP metabolism will dominate drug clearance, these additional clearance mechanisms are often underrepresented in most discovery ADME portfolios.

In reality, metabolism is not necessarily a prerequisite to hepatic extraction of drug for elimination. Rapid hepatic uptake and extensive hepatobiliary excretion can contribute to quite high clearance rates, reducing systemic exposure. Hepatobiliary recycling can be associated with atypical *in vivo* time–exposure profiles, exhibiting multiple exposure peaks and troughs, confounding half-life and clearance estimates and challenging models of dosing intervals. The role of drug transporters in the processes of hepatobiliary excretion highlights a clearance mechanism that is susceptible to both DDI and interspecies differences. Depending on the extent of hepatobiliary excretion, one or more of these outcomes may be expected and the ability to predict these outcomes has the potential to improve optimization of clearance, offer insight into interspecies differences, support human dose projections, and flag drug interaction liabilities. As such, a robust ADME strategy for evaluating drug clearance should, ideally, incorporate diverse technologies to enable exploration of CYP-independent clearance hypotheses.

The most comprehensive *in vitro* model that is currently available is derived from culturing primary human or rat hepatocytes in a gelled-collagen sandwich-cultured hepatocyte (SCH) configuration (Dunn et al., 1989; LeCluyse et al., 1994; Liu et al., 1999a,b,c). These cultures develop several key functional characteristics of liver including establishment of normal cell polarity (apical and basolateral membranes), development of extensive canalicular networks, and expression of multiple functional transporter proteins, including apical uptake transporters, OATP and sodium-taurocholate cotransporting polypeptide (NTCP), and basolateral efflux transporters, MRP2, MDR1 (Pgp), and bile salt export pump (BSEP).

The polarized expression of native transporters in the SCH model creates an ideal model for studying hepatobiliary transport. After 3–4 days in culture, these hepatocytes form an integrated monolayer with compartmentalized bile pockets that can be visualized with electron microscopy. The bile pockets (mimicking bile canaliculi) are sealed by tight junctions formed between the hepatocytes. The formation and maintenance of these tight junctions is dependent on Ca^{2+} and, as such, bile pockets are disrupted when Ca^{2+} is removed from the incubation medium (Liu et al., 1999b).

The utility of this SCH biliary excretion model takes advantage of the sensitivity of the tight junctions forming the bile pockets to the presence of Ca^{2+}. When intact (Ca^{2+}-supplemented) sandwich cultures are challenged with a drug, the drug is first taken up into the hepato-

cytes. For drugs with low passive permeability, drug uptake across the sinusoidal membrane may be the rate-limiting step (Treiber et al., 2004). If biliary efflux is involved in hepatic elimination, drugs that permeate the sinusoidal membrane will be further effluxed across the canalicular membrane where it remains sequestered within the microscopic bile pockets.

The benefits of implementing an *in vitro* model for biliary clearance are currently still in competition with the high cost of running this assay and the relatively low screening capacity (currently only commercially available in 24-well format), limiting its broad application in early discovery. Nonetheless this should not prevent the SCH model from being a powerful tool for drug discovery, especially when tackling NCEs or chemical scaffolds where the *in vitro* metabolic data (e.g., from microsomal or hepatocyte clearance assays) significantly underpredict *in vivo* clearance.

9.6 METABOLISM- OR TRANSPORTER-RELATED SAFETY CONCERNS

It is common clinical scenario for two or more drugs to be administered in concert in order to effectively interrupt a challenging therapeutic target (e.g., oncology or infectious diseases) or to manage the multitude of symptoms that might be associated with a particular disease area (e.g., cardiovascular or diabetes). Most often to the detriment of the patient, coadministered drugs frequently compete for the biological processes that regulate their uptake, distribution, and elimination. For example, a drug that is normally effluxed during intestinal absorption might be characterized with a low fraction absorbed when dosed alone. However, coadministration of a drug that competes for intestinal efflux can precipitate a boost in the fraction absorbed for one, or both, of the drugs, leading to a higher systemic exposure than is normally seen. This potential for DDIs similarly extends to competition for drug clearance pathways, such as metabolic enzymes and biliary transporters, having a profound effect on systemic exposure of a drug. Occasionally, DDIs at the gene transcription level can occur and these interactions typically manifest as a decrease, rather than increase, in drug exposure following coadministration of another drug. These unexpected changes in exposure, whether increased or decreased, share a common outcome in that they generally present a challenge to the therapeutic window for the drug. DDIs that increase drug exposure(s) can exceed drug safety margins, inducing "off-target" toxicities in the patient. Although DDIs that decrease drug exposure(s) are generally not viewed as severely as exposure-increasing DDIs, these effects can, nevertheless, undermine drug efficacies, triggering failures in clinical efficacy studies or, worse, failure of life-saving treatments to succeed in critical-need patients, sometimes even with the expense of increasing risk for toxic metabolites.

The addition of reliable *in vitro* models for assessing drug interaction potential can impart a critical advantage in a discovery program for lead compound selection and ongoing drug development. In some disease areas, such as oncology, management of biological targets with novel therapeutics continues to show only marginal gains in drug efficacy. The real clinical advantage that drives new drug approval may come down to improvements in safety and tolerability profiles. Safety liabilities also bring along significant time investments and costly demands for clinical DDI studies throughout development. Such investments are ideally avoided. If they cannot be altogether avoided, early identification of DDI liability can at least help direct development planning for detailed risk assessment. The following four sections will explore the safety concerns with respect to reversible and time-dependent CYP inhibitions and CYP induction as well as reactive metabolites.

9.7   REVERSIBLE CYP INHIBITION

The prominent role of CYP isoforms in the metabolism of diverse drugs and the exceptionally broad substrate recognition for metabolism by these enzymes mean that they are frequently the focus for potential DDIs. Often, as a consequence of one drug directly inhibiting the metabolism of the other, one or both of the drugs will experience an increase in plasma exposure, relative to individual dosing, which precipitates a clinically relevant adverse reaction (Patsalos and Perucca, 2003; Obach et al., 2006). It has become a common strategy for pharmaceutical companies to attempt to minimize late-stage attrition, potential market restraints, and serious clinical complications by assessing undesirable drug interaction risks early in drug discovery (Obach et al., 2006).

9.7.1   *In vitro* CYP Inhibition

Many discovery drugs will be scrutinized for their potential to become perpetrators of CYP-mediated drug interactions. This is achieved by evaluating the effect of the NCE on the clearance of one or more "CYP-sensitive" probe substrates. CYP-sensitive substrates for *in vivo* DDI studies are characterized by a dominant, single CYP-mediated clearance pathway that has been shown to be effectively blocked by codosing with a prototypical selective CYP inhibitor in humans. The strong correlation observed between *in vitro* CYP inhibition of the clearance of such CYP-sensitive

TABLE 9.3. Validated Isoform-Selective CYP Probe Reactions

Isoform	Substrate	Probe Reaction
CYP3A4/5	Midazolam	1′-hydroxylation
	Felodipine	Dehydrogenation
	Testosterone	6β-hydroxylation
	Dextromethorphan	N-demethylation
CYP2D6	Bufuralol	1′-hydroxylation
	Dextromethorphan	O-demethylation
CYP2C9	Diclofenac	4′-hydroxylation
	Tolbutamide	Hydroxylation
CYP1A2	Phenacetin	O-deethylase
	Caffeine	N3-demethylation
CYP2C19	*S*-Mephenytoin	4′-hydroxylation
CYP2C8	Amodiaquine	N-deethylase
	Paclitaxel	6α-hydroxylation

substrates (K_i or IC_{50}) and the potential for an NCE to precipitate a clinically relevant DDI is compelling. For any drug reaching an *in vivo* level of exposure that equals or exceeds its *in vitro* K_i for CYP isoform inhibition, conditions are ripe for eliciting an increase in exposure for any coadministered drug principally eliminated through that isoform.

9.7.2   Human Liver Microsomes (HLM) + Prototypical Probe Substrates with Quantification by LC-MS

The recognized "gold standard" approach to generating CYP inhibition data utilizes HLM preparations and prototypical drug substrates as selective probes of individual CYP isoforms (Bjornsson et al., 2003). Table 9.3 summarizes some potential choice reactions to probe isoforms of interest. Potential inhibitors added to the optimized assay design are monitored for a dose-dependent decrease in the metabolism of a chosen probe substrate to a specific metabolite. The HLM approach has been highly regarded for its retention of multiple metabolic pathways, characteristic of *in vivo* metabolism, and use of compatible *in vivo* probe substrates. As a result, this approach is endorsed by the FDA for the evaluation of new drugs.

1. *Probe Substrate Selection:* Six major human CYP isoforms are most often implicated in the metabolism of drugs (CYP1A2, 2C8, 2C9, 2C19, 2D6, and 3A4). Among these, CYPs 3A4, 2D6, and 2C9 are associated with the vast majority of clinically relevant DDIs and generally warrant priority allocation of resources in an early DDI profiling paradigm. The CYP selective reactions for these three major isoforms tend to yield robust analyte signals with modest amounts of microsomal

protein (<0.2 mg/mL) and relatively short incubations (<15 min). For risk assessment, IC_{50} values obtained from *in vitro* assays using HLM coupled with selective CYP marker substrates serve as reasonable predictors of an inhibitor's potency (K_i) and the use of *in vitro* CYP inhibition data for the prediction of *in vivo* DDIs has been demonstrated (Tucker et al., 2001; Bjornsson et al., 2003; Obach et al., 2006).

The dependence on LC-MS/MS to monitor reaction progression historically implied labor-intensive bioanalysis with long analytical run times. Recent innovation in bioanalytical technologies has helped mitigate this bottleneck to accommodate the HT demands of early drug discovery. As an alternative to the MS readout, fluorogenic CYP substrates have been employed by the industry, in combination with rCYP isoforms, to gain rapid, HT CYP inhibition estimates. However, most fluorogenic substrates are not CYP selective and are therefore not advantageous for LM-based IC_{50} determinations.

2. *Liquid Handling:* One of the persistent challenges in ADME is in dealing with poorly soluble NCEs. ADME assays that rely on a dose response can require relatively high test compound concentrations (>10 μM) to elicit the desired effect. At Novartis, all test compounds that arrive for HT ADME profiling are supplied as 10 mM DMSO stock solutions. For the CYP inhibition assays, standardized, robotic dilution procedures are in place for generating the sample dilutions and generating assay-ready plates for IC_{50} determinations. With the limitations of robotic liquid transfer through pipeting, the resulting DMSO concentration in final biological incubations may be as high as 0.5% (*v/v*), which is potentially problematic given that individual CYPs can exhibit differing sensitivities to the presence of solvent (Chauret et al., 1998; Busby et al., 1999). Although this lends some amount of background CYP inhibition, DMSO (up to ~0.5% *v/v*) can be tolerated in CYP inhibition assays with minimal effect seen on IC_{50}s providing that the solvent concentration remains low and is maintained across the entire dilution series for a test compound, including the reference (100% activity) incubation lacking any test compound. Coincidentally, the presence of this small amount of DMSO cosolvent certainly helps maintain most of discovery NCEs dissolved in the reaction media.

Even when automated, the process of serial dilution can still be cumbersome, contributing hours to the overall assay procedure time. Acoustic dispensing, as performed by Labcyte's Echo 550 (Labcyte Inc., Sunnyvale, CA), is a relatively recent technology that can accurately deliver submicroliter volumes of DMSO solutions, as little as ~2.5 nL delivered directly into the individual reaction wells of a 96-, 384-, or 1536-well assay plate, without the need for prior dilution steps. This technique is compatible over solutions of broad viscosity and liquid vapor pressure ranges. An internal evaluation showed that assay plates prepared using acoustic dispense versus traditional liquid dilution methods yielded comparable CYP IC_{50}s ($R^2 = 0.8$) using both marketed and proprietary compounds, indicating that acoustic liquid dispensing is reasonable as a time-saving investment for integration into ADMET workflow.

3. *Automating the Assay:* Using a standard format to assess the inhibition potential against the major CYP isoforms (3A4, 2D6, and 2C9), NCEs are measured at seven concentrations up to 50 μM plus a negative (0 μM) control. The IC_{50} values of the NCEs reflect a decrease in overall MS signal of a major metabolite from the probe substrate caused by the inhibition of the enzymatic activity by the NCE. This assay design, applicable to any isoform selective probe reaction, has yielded robust signals and reproducible results with only singlet dilutions of NCEs. All liquid handling, temperature control, and reaction timing sequences can be automated. In Novartis, we have successfully automated the entire run sequence using the Tecan Freedom EVO200 automation platform (Durham, NC) with 96-channel multichannel arm (MCA) head, equipped with on-deck thermally controlled platform shakers, an integrated Liconic STX-44 shaking incubator (Woburn, MA), a below-deck Hettich refrigerated centrifuge (Tuttlingen, Germany), a Liconic LPX220 storage carousel, and a Velocity11 PlateLoc Plate Sealer (Menlo Park, CA). This automation configuration is designed to perform every step of the incubation protocol, from serial dilution of test compounds and preparation of HLM suspensions to the centrifugation of quenched reactions and sealing of the final bioanalytical assay plate. A successful, "walk-away" automated configuration ensures that the biological assay burden for the CYP inhibition assays is low and the lack of manual intervention minimizes the potential for user error.

4. *Bioanalysis:* In recent years, advances in LC-MS/MS, including introduction of the RapidFire technology, have enabled increased throughput,

expanding the opportunity for implementation of this high-quality, HT approach to support the earliest phases of compound assessment (Kim et al., 2005; Brown et al., 2010). The recent introduction of RapidFire MS has practically eliminated the analytical bottleneck for these assays by reducing the time investment by almost 90%. Single-analyte RapidFire analyses can be decreased from ~2 min/injection to <10 s/injection without compromising data quality (Brown et al., 2010). The success of this application for CYP DDI bioanalysis has been due, in part, to both the availability of heavy-labeled analytes for use as MS internal standards and the limited number of analytes being monitored in each assay (e.g., 1′hydroxymidazolam, 4′-hydroxydiclofenac or 1′-hydroxybufuralol).

In order to conserve bioanalytical resources without impacting assay biology, postincubation analyte pooling has proven very successful in reducing the analytical burden. In contrast to pooled probe substrates in biological incubations, postincubation analyte pooling circumvents the potential hazard for pooled substrates to complicate reaction biochemistry. The combination of postincubation analyte pooling and RapidFire bioanalysis (for CYPs 2D6 and 2C9) is valid for further optimization of resource management (Brown et al., 2010). A good correlation has also been demonstrated comparing conventional LC-MS/MS with laser diode thermal desorption (LDTD)-MS bioanalysis, supporting another viable option in easing the bioanalytical burden for this assay approach (Wu et al., 2007).

9.7.3 Implementation Strategy

For research paradigms in which bioanalytical resources are critically limiting, a modified HLM + LC-MS approach (e.g., single inhibitor concentration rather than full IC_{50}) can be used to establish a cascaded DDI risk assessment for drug discovery. An initial, high "% inhibition" readout can be presumed to flag most of the NCEs with high DDI risk, while an HLM + LC-MS-derived IC_{50} may be allocated to more clearly resolve risk for compounds passing this filter. This strategy may be reasonable for supporting early discovery depending on the organization's follow-up strategy. Since >85% of all Novartis NCEs exhibit IC_{50}s against the major CYP isoforms that are >1 μM, a significant fraction of NCEs would require follow-up attention, and as such, the integration of a single-concentration prescreen does not represent an efficient strategy. With an overly aggressive threshold for filtering out potent inhibitors from a single-concentration prescreen, project teams run the risk

of deprioritizing high-quality compounds as "false-positives." From this point of view, there is some merit in confirming "positive" hits from the initial screen with a full IC_{50}. In this scenario, a "% inhibition" prefilter with a follow-up to confirm IC_{50} for the seemingly most potent ~15% of NCEs may be a more manageable scenario.

As mentioned, one of the major challenges in characterizing CYP inhibition is the poor solubility of most NCEs. Because most CYP inhibition occurs with mid- to high-micromolar IC_{50}s, the escalating dose poses potential for compound to crash out of the incubation reaction. This can lead to "false-negative" readouts for compounds with a relatively high K_i and poor solubility under the assay conditions. A custom kinetic solubility assay designed to filter out poorly soluble compounds, as applied prior to Caco-2 cellular permeability assay (see Section 9.2.1.2), can be an effective way to mitigate such false-negatives, but this can also prevent many potent inhibitors from being characterized. Perhaps a more appropriate strategy would be to retrospectively cite the kinetic solubility observations to help flag potential false-negatives.

9.8 MECHANISM-BASED (TIME-DEPENDENT) CYP INHIBITION

In simplest terms, the inhibition of CYP-dependent metabolism can be broadly addressed as being either reversible competitive (as discussed in Section 9.7) or essentially irreversible (also known as mechanism-based inhibition). Truly irreversible inhibition most often occurs when a compound metabolism proceeds through a reactive intermediate, or a reactive metabolite, that is permanently retained in the CYP active site as a covalent modification, thus inactivating the enzyme (Silverman, 1988; Kent et al., 2001). This inactivation is characterized by a time dependency such that prolonged incubation of inhibitor with the active enzyme (CYP cofactor is required) produces greater inactivation. Quasi-irreversible inhibition develops similarly through active site modification, but is not characterized by covalent binding. Rather, it is associated with metabolite inhibitory complexes having very low dissociation rates that render this inhibition "essentially irreversible." This mechanism, nevertheless, also displays cofactor-mediated time dependency. In practice, there is no clinical distinction between irreversible and quasi-irreversible mechanism-based inactivation when assessing DDI risk. The combined phenomena of quasi-irreversible and irreversible inhibition (also termed "time-dependent inhibition," or TDI), however, do represent a safety liability that is functionally distinct from simple reversible inhibition (Hollenberg, 2002).

Compared with reversible inhibition, the impact of TDI has the potential to be more challenging to manage clinically. Once enzymes are irreversibly inactivated, restoration of enzyme activity is rate-limited by *de novo* protein synthesis (Lim et al., 2005). Therefore, this type of CYP inhibition poses a severe safety threat to drug development and has even been responsible for drug withdrawal from the market, illustrated by the calcium channel blocker mibefradil. Mibefradil was introduced into medical practice in 1997 but was withdrawn only a year later as a consequence of potent irreversible inactivation of CYP3A, which triggered serious DDIs with other substrates of CYP3A (Backman et al., 1999; Zhou, 2008).

The binding event that terminally modifies the CYP active site may involve a reaction intermediate or product becoming "trapped" in stable coordination with the catalytic heme or may involve local adduction of heme nitrogens and/or active site residues by a released reactive product (Fontana et al., 2005). Usually, the intensity of resources involved in establishing the mechanism of inactivation is not warranted. During early drug discovery, the focus is rather on characterizing a series or scaffold for TDI potential so that it can be dialed out early. The following sections describe the *in vitro* approaches that have been useful toward identifying and characterizing this risk.

9.8.1   Characteristics of CYP3A TDI

CYP3A, dominant in the liver as the most abundant CYP isoform, has shown the highest involvement in drug metabolism reactions and has been associated with more clinically relevant TDI events than any other CYP isoform (Zhou et al., 2005; Johnson, 2008). Experimentally, we have observed that time-dependent CYP3A inhibition is detected *in vitro* at a frequency along the discovery pipeline that is comparable with, if not exceeding, reversible CYP3A inhibition. Therefore, implementing mechanisms to identify CYP3A time-dependent inactivation during drug discovery is particularly prudent.

Since both irreversible and quasi-irreversible inactivation arises from the events of catalytic turnover, all TDI will have some similar features. As such, three key hallmarks of CYP TDI can be exploited as a basis for screening and/or characterizing the potential for drugs to be time-dependent CYP inhibitors. First, inactivation is associated with a time dependency characterized by a rate constant (k_{inact}). Second, there is a concentration dependency, such that the apparent rate of inactivation, k_{obs} (or the amount of inhibition observed within a fixed time interval), increases with increasing inhibitor concentration. Because CYP isoforms require NADPH for catalytic activity, NADPH dependency is the third criteria for establishing CYP TDI. Consequently, *in*

vitro systems lacking NADPH will not support catalytic inactivation.

9.8.2   *In vitro* Screening for CYP3A TDI

There are at least two viable approaches to screen for CYP TDI potential (Grimm et al., 2009). Both screening approaches rely on coupling an initial "preincubation" of the enzyme in the presence of inhibitor plus NADPH (to permit inactivation to occur) with a "marker" substrate incubation to assess the fraction of CYP activity remaining. One screening method evaluates the rate of NADPH-dependent activity loss observed using one or two inhibitor concentrations (which we will refer to as the "inactivation rate [k_{obs}]" approach). The other is to compare the NADPH and concentration dependency of the inhibitor for enzyme inactivation following a single inactivation interval (commonly referred to as the "IC$_{50}$-shift" approach).

9.8.3   Inactivation Rate (k_{obs})

The rate of CYP inactivation observed using one or two concentrations of test compound can serve as a very effective and rapid prescreen for TDI. In this approach, samples from an initial mixture of HLM, test compound, and NADPH are removed over several time points (e.g., 0, 10, 20, 30 min) and further incubated for 2–3 min with a high concentration of marker CYP substrate (e.g., midazolam for CYP3A) to drive CYP turnover at a rate approaching V_{max} (Figure 9.4). Under these marker substrate conditions, the amount of marker substrate converted to product (e.g., 1'-hydroxymidazolam) is directly proportional to the amount of active CYP isoform remaining in the HLM fraction following the preincubation. A log-linear plot of the preincubation time versus the % CYP activity remaining yields a line with the slope = $-k_{obs}$. Theoretically, any measurable k_{obs} may be indicative of TDI potential. However, the threshold k_{obs} that is preselected as a filter for flagging a test compound as "positive" for TDI should be defined by the confidence limits around substrate-independent background rates of CYP inactivation under the same incubation conditions (Grimm et al., 2009).

One advantage of this screening approach is that the relationship between the observed k_{obs} and the TDI inactivation parameters k_{inact} and K_i is clearly defined:

$$k_{obs} = \frac{k_{inact}*[I]}{K_I + [I]} \qquad (9.5)$$

where k_{obs} is expressed as rate of inactivation per minute and [I] equals the incubation concentration (μM) of inhibitor. It is straightforward to identify the sensitivity limits of the assay with regard to the choice of test compound concentration. When [I] >> K_i, any compounds

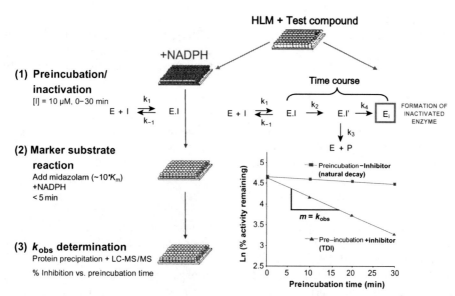

FIGURE 9.4. Schematic overview of the k_{obs} assay for screening TDI, where [I] is the concentration of test compounds (or potential CYP inhibitors), E refers to CYP proteins, E-I and E-I* are the reaction intermediates, E_i is the inactivated CYP, and P is the reaction product.

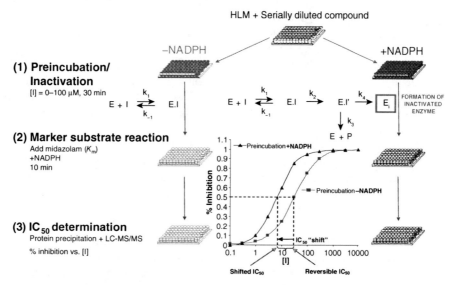

FIGURE 9.5. Schematic overview of the IC_{50}-shift assay for screening TDI.

for which $k_{inact} > k_{obs}$ can be expected to manifest TDI in this screen. As K_i increases, however, k_{inact} must also increase in order for the observed rate of inactivation to exceed the predetermined k_{obs} threshold for identifying TDI-positive compounds. It is best practice to include test compounds at more than one concentration (e.g., 10 μM, 50 μM) to work around the concentration sensitivity of this model.

9.8.4 IC_{50}-Shift

As an alternative to the k_{obs} approach, the extent to which a preincubation of HLM, test compound, and NADPH may "shift" the apparent inhibition potential (IC_{50}) of the test compound can signal TDI. In this approach, samples from an initial mixture of HLM, test compound (over multiple concentrations), and NADPH (or buffer for TDI negative control) are removed following an initial incubation period (e.g., 30 min) and further subjected to the standard conditions for measuring reversible CYP inhibition with a marker CYP substrate (see Section 2.5.1). Figure 9.5 provides an overview of the IC_{50}-shift principle. The relationship between k_{inact}, K_i, and fold-IC_{50}-shift is less intuitive than the k_{obs} readout. However, the exploitation of broad concentration dependency together with a substantial preincubation period (e.g., 30 min) has potential to lend greater sensitivity in identifying TDI.

9.8.5 Implementation Strategy

The TDI screens described here as abbreviated versions of a full k_{inact} and K_i determination can compliment early safety liability assessments performed during drug discovery. Any approach, however, should be carefully validated for performance and the interpretation of readout should be developed in line with a corporate strategy for addressing TDI. According to current FDA guidances, clinical DDI studies are expected to be performed for any drug candidate that shows time-dependent CYP inactivation kinetics *in vitro*, regardless of the inactivation potency. Rapid TDI screening can be a useful mechanism for surveying chemical scaffolds/ chemotypes for TDI liability very early in the drug discovery pathway. This effort is intended to avoid, laborious, and often inadequate, SPR to "dial out" TDI postlead nomination. It should be noted that there has been mixed success in directly correlating either a fold-shift in IC_{50} or k_{obs} with the individual inactivation kinetic parameters of the test compound (k_{inact} and K_i). Because time-dependent inactivation of CYP3A tends to cluster with k_{inact} ~0.01–0.05/min, and K_i in the mid to low micromolar range, the majority of TDI positive discovery compounds have revealed fold-shifts of ~2- to 5-fold and k_{obs} rates that cluster in the range of 0.01–0.05/min. It is not advised to use fold-shifts or k_{obs} within these ranges to rank-order a series of TDI-positive compounds. However, more potent inactivators with very low K_i and/or very high k_{inact} do tend to present more dramatic shifts and higher k_{obs}. Ultimately, *in vitro–in vivo* predictions and risk assessment for clinical TDI depend on multiple *in vivo* factors (including the rate of enzyme resynthesis, the free concentration of inhibitor *in vivo*, and the fraction of victim drug cleared through the particular CYP isoform) in relation to the kinetics for CYP inactivation (k_{inact} and K_i) (Venkatakrishnan and Obach, 2007).

9.9 CYP INDUCTION

As previously alluded, some potential also exists for DDIs to result in systemic underexposure. In contrast to drug interactions stemming from inhibition of metabolic clearance, such as reversible and time-dependent CYP inhibition, interactions that precipitate decreases in drug exposure can be attributed to modification of CYP protein expression. In addition to increasing drug clearance, these changes, typically up-regulation of CYP protein expression, can have other detrimental out-comes, such as increased generation of reactive or toxic metabolites. Changes in protein levels most often occur when a drug effectively activates the relevant nuclear transcription pathway that drives the protein's expression. The CYP protein family is exquisitely evolved to respond to the biological challenge of xenobiotics. Not only are the xenobiotic metabolizing isoforms enriched in the prominent defense organs (e.g., gut, liver) to minimize initial exposure to xenobiotics, but many of the same chemical structures recognized as CYP substrates have also proven capable of triggering CYP transcriptional up-regulation.

The breadth of human drug metabolizing CYP isoforms (as well as several other families of DMEs and drug transporters) appears mostly to be controlled by three major nuclear receptors: the PXR, the constitutive androstane receptor (CAR), and the aryl hydrocarbon receptor (AhR) (Fuhr, 2000; Jones et al., 2000; LeCluyse, 2001; Moore et al., 2002; Willson and Kliewer, 2002; Luo et al., 2004; Okey et al., 2005; Martinez-Jimenez et al., 2007). Functionally, a drug that is recognized by the primary nuclear receptor induces dimerization of the nuclear receptor with a binding partner(s), initiating translocation of the complex to the nucleus. The translocated complex binds to and activates the genomic responsive element that, in turn, induces gene transcription and, finally, protein synthesis.

Traditionally, it was common in CYP induction studies to characterize expression levels of all of the major CYP (or even other DME family) isoforms, individually, following a period of induction. However, with clearer insight into how groups of DMEs are regulated, in parallel, through the actions of PXR, CAR, and AhR (Olinga et al., 2008), a commonly accepted approach has evolved to focus on individual, well-characterized pathways under regulation by each of the nuclear receptors as the assay readout. From this information, the likely changes in expression of other similarly controlled enzymes and/or transporters may be inferred. Table 9.4

TABLE 9.4. Major CYP Nuclear Hormone Receptors and Associated Inducible DMEs

Major Nuclear Hormone Receptors	Associated Inducible DMEs or Drug Transporters
PXR (NR1/2) (pregnane X receptor)	CYP3A, CYP2B, CYP2Cs, OATP2, UGT1A, MRP2, MDR1, GST-A2, BSEP, SULT2A
CAR (NR1/3) (constitutive androstane receptor)	CYP2B, CYP2Cs, CYP3A, CYP1A2, UGT1A, MRP2, OATP2
AhR (aryl hydrocarbon receptor)	CYP1A1/2, CYP1B1, GST, UGT, ALDH, NQO1

summarizes the groups of DMEs and drug transporters regulated by each of these three major nuclear receptors.

To evaluate the potential a drug carries to induce CYP expression, the "gold standard" *in vitro* approach is to incubate freshly isolated hepatocytes with the drug inducer for up to 3 days and then measure increases in enzyme activity (with a marker substrate reaction), protein expression (e.g., Western blotting), or mRNA expression (e.g., polymerase chain reaction [PCR], Northern blotting, microarrays) (Hewitt et al., 2007; Lake et al., 2009). However, this approach is labor intensive and time consuming, requiring several days to achieve a maximum effect. Since drug binding to one of the three major nuclear receptors (PXR, CAR, AhR) usually represents the key initial event in triggering CYP induction, an alternative technology exploiting the critical initial event of receptor binding has been introduced for assessing CYP activation potential with fast, indirect readouts.

The potential for CYP inducers to participate in the initial step including receptor recognition and binding is readily explored using competitive receptor binding technologies designed to assess drug receptor binding affinities. A time-resolved fluorescence resonance energy transfer (TR-FRET) approach has developed as a sensitive indirect screening readout for PXR and CAR competitive receptor binding (Shukla et al., 2009). This technology relies on a change in fluorescent emission to reflect the proximity of two individual and distinct fluorophores. When two suitable fluorophores are close enough (e.g., receptor–ligand binding), the excitation energy of the first fluorophore (the donor) is transferred to the second fluorophore (the acceptor), which generates a dominant fluorescence acceptor emission (minimal contribution from the donor). As the two donors separate, the ratio of their relative fluorescent intensities shifts. By tagging a nuclear receptor molecule and a known receptor ligand with compatible fluorophores, any competitor-induced ligand displacement can be measured as a change in FRET signal. This approach is tractable in multiwell formats and is automatable for HT. In principle, this method is comparable with radioligand binding (RLB) displacement, but the need for postincubation sample workup and the hazard of using radioactivity are circumvented.

While this simple, competitive binding-displacement type of assay can quantitatively capture receptor–ligand affinities (e.g., IC_{50}), the major limitation in this design is that there is no qualitative readout to describe the functional implications of the receptor–ligand interaction. Ligand binding to the PXR, CAR, or AhR nuclear receptors may result in other functional outcomes apart from activation or nonspecific binding. Subsequent to

ligand binding, nuclear receptors undergo conformational changes, resulting in a cascade of events including dissociation of repressor proteins, association of coactivator proteins, and assembly of transcriptional factors for activation of target genes. Several coactivators are known to be recruited by these receptors, and agonist, antagonist, and inverse agonist ligand interactions have all been reported (Lempiäinen et al., 2005). The FRET technology described above can be further adapted to deliver a pseudofunctional assay readout. In reconfiguring the fluorescent tag from the receptor ligand to the coactivator peptide, one can monitor recruitment of the coactivator peptide to the receptor in response to ligand binding (antagonist), disruption of any constitutive association of nuclear receptor with a fluorescein-labeled coactivator peptide (inverse agonists) or the enhancement of an interaction between the receptor and coactivator peptide (agonists). The coregulator peptide assay can be performed with either PXR or CAR.

9.10   REACTIVE METABOLITES

While most drugs undergo biotransformation or metabolism prior to elimination, their respective metabolites can be quite different. For instance, the majority of them are stable as excreted, but some are considered "active" or "reactive" toxic species. These pose the threat in off-target pharmacology and trigger a number of severe drug reactions including mechanism-based CYP inhibition (see Section 9.8), hepatotoxicity, genotoxicity, carcinogenicity, and immune-mediated toxicity, causing a number of drug withdrawals (Antoine et al., 2008). Thus, there is a general need to assess the potential risk of NCEs with respect to their metabolite-mediated toxicities in early drug discovery in order to advance only those with the least toxicity probability, as suggested by the FDA.

9.10.1   Qualitative *in vitro* Assays

As there is currently no efficient probe to reliably predict metabolite-mediated toxicities in human (Uetrecht, 2003; Caldwell and Yan, 2006; Kumar et al., 2008), most of the current tactics to evaluate such risks are to assess the reactive metabolites formed, based on the belief that prevention of the formation of reactive intermediates that bind to proteins will reduce potential risk for idiosyncratic drug reactions (IDRs) (Guengerich, 2005b; Baillie, 2008; Kumar et al., 2008). The glutathione (GSH) trapping assay captures the reactive intermediates via GSH, a major scavenger of reactive metabolites (Caldwell and Yan, 2006; Baillie, 2008). Albeit fast and HT screening amenable in drug discovery, the GSH trapping assay, however, typically delivers qualitative information that is difficult to quantitatively evaluate against numerous NCEs (Kumar et al., 2008). In addition, not all metabolites are readily seized by GSH owing to the appreciably divergent nature of the metabolites such as "hard" versus "soft" metabolites (Yan et al., 2007) or suboptimal stability of the GSH conjugates formed with reactive metabolites (Uetrecht, 2003). This may be compensated by switching to modified trapping agents (Yan et al., 2007) or utilizing additional trapping agents over GSH, such as mercaptoethanol (thiol nucleophiles), cyanide ion (an effective trap for iminium ions) (Gorrod and Aislaitner, 1994), and methoxylamine (a trapping agent for reactive carbonyl species) to catch all reactive metabolites (Kumar et al., 2008).

9.10.1.1 Semiquantitative in vitro Assays

For semiquantitative identification of reactive metabolites, alternative methodologies with improved sensitivity are available using either fluorescent traps (dansyl-GSH) (Gan et al., 2005) or radiolabeled trapping agents such as ^{3}H-GSH, ^{35}S-GSH, ^{35}S-mercaptoethanol, and ^{14}C-cyanide (Mulder and Le, 1988; Meneses-Lorente et al., 2006; Masubuchi et al., 2007). Utilizing 10 commercial drugs with reported metabolite-mediated hepatotoxicity, the ^{35}S-GSH conjugate formation by Masubuchi et al. (2007) has led to excellent correlations ($R^2 = 0.93$ in both rat and human LM) with covalent binding studies using radioactively labeled NCEs.

9.10.2   Quantitative in vitro Assay

The covalent binding assay, referred to as "gold standard" in vitro assay to assess reactive metabolites (Kumar et al., 2008), is widely accepted in late discovery or early development. It quantitatively monitors the extent and rate of covalent binding of reactive metabolites to rat and human liver microsomal proteins in vitro (or to hepatocytes when needed) and in vivo (rat liver and plasma) using radiolabeled NCEs (Hop et al., 2006; Kumar et al., 2008).

It should be noted that the trapping and covalent binding assays are designed only to foresee the propensity of the drug candidates to undergo metabolic activation, not necessarily predict toxicity (Baillie, 2008; Obach et al., 2008). For instance, these methods can give false-negatives if the in vitro system used does not contain all the enzymes responsible for the formation of the major reactive metabolites in vivo. In addition, some reactive metabolites (e.g., free radicals) could lead to IDRs without any significant covalent binding. Conversely, they more frequently result in false-positives

since many reactive metabolites efficiently scavenged by trapping agents may not necessarily cause toxicity in vivo (Uetrecht, 2003; Baillie, 2008). As a result, mixed findings have been reported thus far. Gan et al. (2009) confirmed a general trend between the extent of thiol adduct formation and the potential for drug-induced toxicity (DIT) when in vitro thiol trapping of reactive metabolites was applied to 50 marketed drugs (10 DIT positive and 40 DIT negative). In contrast, Obach et al. (2008) revealed that prediction to metabolite toxicities using the covalent binding approach may not be very straightforward, given the variability in daily dose, the fraction of metabolism undergoing bioactivation, and competing metabolic pathways that scavenge reactive metabolites. Other recent work also indicated that the covalent binding amount itself may not be sufficient enough to discriminate positive from negative compounds for idiosyncratic drug-induced liver injury risk in human (Usui et al., 2009). Rather, the collective consideration of covalent binding amount and daily dose can help estimate the risk of idiosyncratic drug-induced liver injury in human. Meanwhile, a separate study reported that most of the "problematic" drugs, including "withdrawn" and "warning" drugs, exhibited higher HLM in vitro covalent binding yields than the "safe" drugs (Takakusa et al., 2008). Interestingly, the problematic drugs missed by the in vitro HLM covalent binding assay were identified by a rat in vivo covalent binding study, confirming the collaborative roles of covalent binding and tissue distribution in the risk assessment of metabolite-related toxicity (Takakusa et al., 2008).

9.11   CONCLUSION AND OUTLOOK

Undoubtedly, the ADMET-enabling technologies, albeit emerging, have been drastically advanced in the past decade, which has greatly leveraged our capability to evaluate discovery candidates and shifted our drug assessment paradigm. Indeed numerous tier-based approaches have materialized for divergent ADMET areas in line with regulatory guidance and strategy. It should be noted that the effectiveness and ultimate impact of the in vitro ADMET tools rely greatly on their quality and predictivity to the gold standards and PK/PD models. Whereas there is still room to continue improving our innovative technologies in automation, miniaturization, and bioanalysis, it is imperative to enhance our predictive power by establishing in silico–in vitro–in vivo correlation by taking advantage of existing massive ADMET data from discovery NCEs with divergent chemical space. Furthermore, greater dedication may be committed in the integrated utilization of comprehensive ADMET data for risk assessment and boost our understanding of large PK/PD picture by monitoring the interplay of multidimensional ADMET data Wang and Collis, 2011). Ultimately, all above effort in the development of innovative ADMET-enabling technologies needs to be translated into improved quality (e.g., safer), productivity (e.g., more new drugs for unmet medical needs), and efficiency (reduced attrition rate and development time) in the discovery and development of new drug candidates.

ACKNOWLEDGMENTS

The authors would like to thank Linhong Yang and Drs. Guoyu Pan and Liping Zhou for providing assay protocols and for the exciting discussions.

REFERENCES

Alsenz J, Haenel E (2003) Development of a 7-day, 96-well Caco-2 permeability assay with high-throughput direct UV compound analysis. *Pharm Res* 20(12):1961–1969.

Antoine DJ, Williams DP, Park BK (2008) Understanding the role of reactive metabolites in drug-induced hepatotoxicity: State of the science. *Expert Opin Drug Metab Toxicol* 4(11):1415–1427.

Artursson P, Tavelin S (2003) Caco-2 and emerging alternatives for prediction of intestinal drug transport: A general overview. In *Drug Bioavailability*, van de Waterbeemd H, Lennernas H, Artursson P, eds., pp. 72–89. Wiley-VCH, Verlag GmbH & Co. KGaA, Weinheim, Germany.

Avdeef A (2003) *Absorption and Drug Development, Solubility, Permeability and Charge State*. Wiley-Interscience, Hoboken, NJ.

Avdeef A, Strafford M, Block E, Balogh MP, Chambliss W, Khan I (2001) Drug absorption *in vitro* model: Filter-immobilized artificial membranes. 2. Studies of the permeability properties of lactones in *Piper methysticum* Forst. *Eur J Pharm Sci* 14:271–280.

Backman JT, Wang JS, Wen X, Kivistö KT, Neuvonen PJ (1999) Mibefradil but not isradipine substantially elevates the plasma concentrations of the CYP3A4 substrate triazolam. *Clin Pharmacol Ther* 66:401–407.

Baillie TA (2008) Metabolism and toxicity of drugs. Two decades of progress in industrial drug metabolism. *Chem Res Toxicol* 21(1):129–137.

Balimane PV, Chong S, Morrison RA (2000) Current methodologies used for evaluation of intestinal permeability and absorption. *J Pharmacol Toxicol* 44:301–312.

Balimane PV, Han YH, Chong S (2006) Current industrial practices of assessing permeability and P-glycoprotein interaction. *AAPS J* 8(1):E1–E13.

Barthel BL, Zhang Z, Rudnicki DL, Coldren CD, Polinkovsky M, Sun H, Koch GG, Chan DCF, Koch TH (2009) Preclini-cal efficacy of a carboxylesterase 2-activated prodrug of doxazolidine. *J Med Chem* 52:7678–7688.

Bertucci C, Domenici E (2002) Reversible and covalent binding of drugs to human serum albumin: Methodological approaches and physiological relevance. *Curr Med Chem* 9:1463–1482.

Bhamre S, Bhagwat SV, Shankar SK, Boyd MR, Ravindranath V (1995) Flavin-containing monooxygenase mediated metabolism of psychoactive drugs by human brain microsomes. *Brain Res* 672:276–280.

Bjornsson TD, Callaghan JT, Einolf HJ, Fischer V, Gan L, Grimm S, Kao J, King SP, Miwa G, Ni L, Kumar G, McLeod J, Obach RS, Roberts S, Roe A, Shah A, Snikeris F, Sullivan JT, Tweedie D, Vega JM, Walsh J, Wrighton SA (2003) The conduct of *in vitro* and *in vivo* drug-drug interaction studies: A Pharmaceutical Research and Manufacturers of America (PhRMA) perspective. *Drug Metab Dispos* 31:815–832.

Bohets H, Annaert P, Mannens G, Beijsterveldt VL, Anciaux K, Verboven P, Meuldermans W, Lavrijsen K (2001) Strategies for absorption screening in drug discovery and development. *Curr Top Med Chem* 1:367–383.

Bourdet DL, Thakker DR (2006) Saturable absorptive transport of the hydrophilic organic cation ranitidine in Caco-2 cells: Role of pH-dependent organic cation uptake system and p-glycoprotein. *Pharm Res* 23(6):1165–1177.

Box K, Comer JE, Gravestock T, Stuart M (2009) New ideas about the solubility of drugs. *Chem Biodivers* 6:1767–1788.

Brown A, Bickford S, Hatsis P, Amin J, Bell L, Harriman S (2010) High-throughput analysis of *in vitro* cytochrome p450 inhibition samples using mass spectrometry coupled with an integrated liquid chromatography/autosampler system. *Rapid Commun Mass Spectrom* 24:1207–1210.

Busby WF, Ackermann JM, Crespi CL (1999) Effect of methanol, ethanol, dimethyl sulfoxide and acetonitrile on *in vitro* activities of cDNA-expresssed human cytochromes P-450. *Drug Metab Dispos* 27:246–249.

Caldwell GW, Yan Z (2006) Screening for reactive intermediates and toxicity assessment in drug discovery. *Curr Opin Drug Discov Devel* 9(1):47–60.

Chan EC, New LS, Yap CW, Goh LT (2009) Pharmaceutical metabolite profiling using quadrupole/ion mobility spectrometry/time-of-flight mass spectrometry. *Rapid Commun Mass Spectrom* 23:384–394.

Chaturvedi PR, Decker CJ, Odinecs A (2001) Prediction of pharmacokinetic properties using experimental approaches during early drug discovery. *Curr Opin Chem Biol* 5:452–463.

Chauret N, Gauthier A, Nicoll-Griffith DA (1998) Effect of common organic solvents on *in vitro* cytochrome P450-mediated metabolic activities in human liver microsomes. *Drug Metab Dispos* 26:1–4.

Chung WG, Cha YN (1997) Oxidation of caffeine to theobromine and theophylline is catalyzed primarily by flavin-containing monooxygenase in liver microsomes. *Biochem Biophys Res Commun* 235:685–688.

D'Souza AJM, Topp EM (2004) Release from polymeric prodrugs: Linkages and their degradation. *J Pharm Sci* 93:1962–1979.

Di L, Kerns EH, Fan K, Mcconnell OJ, Carter GT (2003) High throughput artificial membrane permeability assay for blood-brain barrier. *Eur J Med Chem* 38:223–232.

Di L, Kerns EH, Hong Y, Chen H (2005) Development and application of high throughput plasma stability assay for drug discovery. *Int J Pharm* 297:110–119.

Dunn JC, Yarmush ML, Koebe HG, Tompkins RG (1989) Hepatocyte function and extracellular matrix geometry: Long-term culture in a sandwich configuration. *FASEB J* 3:174–177.

Egan WJ, Merz KM, Baldwin JJ (2000) Prediction of drug absorption using multivariate statistics. *J Med Chem* 43: 3867–3877.

Fagerholm U (2007) Prediction of human pharmacokinetics—Evaluation of methods for prediction of volume of distribution. *J Pharm Pharmacol* 59:1181–1190.

Fisher MB, Campanale K, Ackermann BL, VandenBranden M, Wrighton SA (2000) In vitro glucuronidation using human liver microsomes and the pore-forming peptide alamethicin. *Drug Metab Dispos* 28:560–566.

Fontana E, Dansette PM, Poli SM (2005) Cytochrome p450 enzymes mechanism based inhibitors: Common substructures and reactivity. *Curr Drug Metab* 6:413–454.

Fuhr U (2000) Induction of drug metabolizing enzymes: Pharmacokinetic and toxicological consequences in humans. *Clin Pharmacokinet* 38:493–504.

Galia E, Nicolaides E, Horter D, Löbenberg R, Reppas C, Dressman JB (1998) Evaluation of various dissolution media for predicting in vivo performance of class I and II drugs. *Pharm Res* 15:698–705.

Gan J, Harper TW, Hsueh MM, Qu Q, Humphreys WG (2005) Dansyl glutathione as a trapping agent for the quantitative estimation and identification of reactive metabolites. *Chem Res Toxicol* 18(5):896–903.

Gan J, Ruan Q, He B, Zhu M, Shyu WC, Humphreys WG (2009) In vitro screening of 50 highly prescribed drugs for thiol adduct formation—Comparison of potential for drug-induced toxicity and extent of adduct formation. *Chem Res Toxicol* 22(4):690–698.

Giacomini KM, Huang SM, Tweedie DJ, Benet LZ, Brouwer KLR, Chu X, Dahlin A, Evers R, Fischer V, Hillgren KM, Hoffmaster KA, Ishikawa T, Keppler D, Kim RB, Lee CA, Niemi M, Polli JW, Sugiyama Y, Swaan PW, Ware JA, Wright SH, Wah Yee S, Zamek-Gliszczynski MJ, Zhang L (2010) Membrane transporters in drug development. *Nat Rev Drug Discov* 9(3):215–236.

Gorrod JW, Aislaitner G (1994) The metabolism of alicyclic amines to reactive iminium ion intermediates. *Eur J Drug Metab Pharmacokinet* 19:209–217.

Grimm SW, Einolf HJ, Hall SD, He K, Lim HK, Ling KH, Lu C, Nomeir AA, Seibert E, Skordos KW, Tonn GR, Van Horn R, Wang RW, Wong YN, Yang TJ, Obach RS (2009) The conduct of in vitro studies to address time-dependent inhibition of drug-metabolizing enzymes: A perspective of the pharmaceutical research and manufacturers of America. *Drug Metab Dispos* 37:1355–1370.

Grothusen A, Hardt J, Bräutigam L, Lang D, Böcker R (1996) A convenient method to discriminate between cytochrome P450 enzymes and flavin-containing monooxygenases in human liver microsomes. *Arch Toxicol* 71:64–71.

Guengerich FP (2005a) Human cytochrome P450 enzymes. In *Cytochrome P450: Structure, Mechanism, and Biochemistry*, 3rd ed., Ortiz de Montellano PR, ed. Kluwer Academic/Planum Publishers, New York.

Guengerich FP (2005b) Principles of covalent binding of reactive metabolites and examples of activation of bis-electrophiles by conjugation. *Arch Biochem Biophys* 433(2):369–378.

Hämäläinen MD, Frostell-Karlsson A (2004) Predicting the intestinal absorption potential of hits and leads. *Drug Discov Today Technol* 1:397–406.

Hewitt NJ, Lecluyse EL, Ferguson SS (2007) Induction of hepatic cytochrome P450 enzymes: Methods, mechanisms, recommendations, and in vitro-in vivo correlations. *Xenobiotica* 37:1196–1224.

Hilgendorf C, Ahlin G, Seithel A, Artursson P, Ungell AL, Karlsson J (2007) Expression of thirty-six drug transporter genes in human intestine, liver, kidney, and organotypic cell lines. *Drug Metab Dispos* 35(8):1333–1340.

Hochman JH, Yamazaki M, Ohe T, Lin JH (2002) Evaluation of drug interactions with P-glycoprotein in drug discovery: In vitro assessment of the potential for drug-drug interactions with P-glycoprotein. *Curr Drug Metab* 3(3):257–273.

Hollenberg PF (2002) Characteristics and common properties of inhibitors, inducers, and activators of CYP enzymes. *Drug Metab Rev* 34:17–35.

Hop CECA, Kalgutkar AS, Soglia JR (2006) Importance of early assessment of bioactivation in drug discovery. *Annu Rep Med Chem* 41:369–381.

Irvine JD, Takahashi L, Lockhart K, Cheong J, Tolan JW, Selick HE, Grove JR (1999) MDCK (Madin-Darby canine kidney) cells: A tool for membrane permeability screening. *J Pharm Sci* 88:28–33.

Israili ZH, Dayton PG (2001) Human alpha-1-glycoprotein and its interactions with drugs. *Drug Metab Rev* 33:161–235.

Johnson WW (2008) Cytochrome P450 inactivation by pharmaceuticals and phytochemicals: Therapeutic relevance. *Drug Metab Rev* 40:101–147.

Jones SA, Moore LB, Shenk JL, Wisely GB, Hamilton GA, McKee DD, Tomkinson NC, LeCluyse EL, Lambert MH, Willson TM, Kliewer SA, Moore JT (2000) The pregnane X receptor: A promiscuous xenobiotic receptor that has diverged during evolution. *Mol Endocrinol* 14:27–39.

Kansy M, Senner F, Gubernator K (1998) Physicochemical high throughput screening: Parallel artificial membrane permeation assay in the description of passive absorption processes. *J Med Chem* 41:1007–1010.

Kariv I, Rourick RA, Kassel DB, Chung TD (2002) Improvement of "hit-to-lead" optimization by integration of in vitro

HTS experimental models for early determination of pharmacokinetic properties. *Comb Chem High Throughput Screen* 5:459–472.

Kent UM, Juschyshyn MI, Hollenberg PF (2001) Mechanism-based inactivators as probes of cytochrome P450 structure and function. *Curr Drug Metab* 2:215–243.

Kerns EH, Di L (2008) *Drug-like Properties: Concepts, Structure, Design and Methods: From ADME to Toxicity Optimization,* 1st Ed. Academic Press, New York.

Kim MJ, Kim H, Cha IJ, Park JS, Shon JH, Liu KH, Shin JG (2005) High-throughput screening of inhibitory potential of nine cytochrome P450 enzymes *in vitro* using liquid chromatography/tandem mass spectrometry. *Rapid Commun Mass Spectrom* 19:2651–2658.

Kumar S, Kassahun K, Tschirret-Guth RA, Mitra K, Baillie TA (2008) Minimizing metabolic activation during pharmaceutical lead optimization: Progress, knowledge gaps and future directions. *Curr Opin Drug Discov Devel* 11(1):43–52.

Lake BG, Price RJ, Giddings AM, Walters DG (2009) *In vitro* assays for induction of drug metabolism. *Methods Mol Biol* 481:47–58.

Lakeram M, Lockley DJ, Sanders DJ, Pendlington R, Forbes B (2007) Paraben transport and metabolism in the biomimetic artificial membrane permeability assay (BAMPA) and 3-day and 21-day Caco-2 cell systems. *J Biomol Screen* 12(1):84–91.

LeCluyse EL (2001) Human hepatocyte culture systems for the *in vitro* evaluation of cytochrome P450 expression and regulation. *Eur J Pharm Sci* 13:343–368.

LeCluyse EL, Audus KL, Hochman JH (1994) Formation of extensive canalicular networks by rat hepatocytes cultured in collagen-sandwich configuration. *Am J Physiol* 266:C1764–C1774.

Lempiäinen H, Molnár F, Macias Gonzalez M, Peräkylä M, Carlberg C (2005) Antagonist- and inverse agonist-driven interactions of the vitamin D receptor and the constitutive androstane receptor with corepressor protein. *Mol Endocrinol* 19:2258–2272.

Lennernäs H (2007) Modeling gastrointestinal drug absorption requires more *in vivo* biopharmaceutical data: Experience from *in vivo* dissolution and permeability studies in humans. *Curr Drug Metab* 8:645–657.

Liang E, Chessic K, Yazdanian M (2000) Evaluation of an accelerated Caco-2 cell permeability model. *J Pharm Sci* 89(3):336–345.

Lim HK, Duczak N Jr., Brougham L, Elliot M, Patel K, Chan K (2005) Automated screening with confirmation of mechanism-based inactivation of CYP3A4, CYP2C9, Cyp2C19, CYP2D6, and CYP1A2 in pooled human liver microsomes. *Drug Metab Dispos* 33:1211–1219.

Lin JH, Wong BK (2002) Complexities of glucuronidation affecting *in vitro-in vivo* extrapolation. *Curr Drug Metab* 3:623–646.

Lipinski CA, Lombardo F, Dominy BW, Feeney PJ (1997) Experimental and computational approaches to estimate solubility and permeability in drug discovery and development settings. *Adv Drug Deliv Rev* 23:3–25.

Liu H, Sabus C, Carter GT, Du C, Avdeef A, Tischler M (2003) *In vitro* permeability of poorly aqueous soluble compounds using different solubilizers in the PAMPA assay with liquid chromatography/mass spectrometry detection. *Pharm Res* 20(11):1820–1826.

Liu X, Chism JP, LeCluyse EL, Brouwer KR, Brouwer KLR (1999a) Correlation of biliary excretion in sandwich-cultured rat hepatocytes and *in vivo* in rats. *Drug Metab Dispos* 27:637–644.

Liu X, LeCluyse EL, Brouwer KR, Gan LL, Lemasters JJ, Stieger B, Meier PJ, Brouwer KLR (1999b) Biliary excretion in primary rat hepatocytes cultured in a collagen-sandwich configuration. *Am J Physiol Gastrointest Liver Physiol* 277:12–21.

Liu X, LeCluyse EL, Brouwer KR, Lightfood RM, Lee JI, Brouwer KLR (1999c) Use of Ca2+ Modulation to evaluate biliary excretion in sandwich-cultured rat hepatocytes. *J Pharmacol Exp Ther* 289:1592–1599.

Lu W, Clasquin MF, Melamud E, Amador-Noguez D, Caudy AA, Rabinowitz JD (2010) Metabolomic analysis via reversed-phase ion-pairing liquid chromatography coupled to a stand alone orbitrap mass spectrometer. *Anal Chem* 82:3212–3221.

Luo G, Guenthner T, Gan LS, Humphreys WG (2004) CYP3A4 induction by xenobiotics: Biochemistry, experimental methods and impact on drug discovery and development. *Curr Drug Metab* 5:483–505.

Marino AM, Yarde M, Patel H, Chong S, Balimane PV (2005) Validation of the 96-well Caco-2 cell culture model for high throughput permeability assessment of discovery compounds. *Int J Pharm* 297(1–2):235–241.

Martinez-Jimenez CP, Jover R, Donato MT, Castell JV, Gomez-Lechon MJ (2007) Transcriptional regulation and expression of CYP3A4 in hepatocytes. *Curr Drug Metab* 8:185–194.

Masimirembwa CM, Thompson R, Andersson TB (2001) *In vitro* high throughput screening of compounds for favorable metabolic properties in drug discovery. *Comb Chem High Throughput Screen* 4:245–263.

Masimirembwa CM, Bredberg U, Andersson TB (2003) Metabolic stability for drug discovery and development: Pharmacokinetic and biochemical challenges. *Clin Pharmacokinet* 42:515–528.

Masubuchi N, Makino C, Murayama N (2007) Prediction of *in vivo* potential for metabolic activation of drugs into chemically reactive intermediate: Correlation of *in vitro* and *in vivo* generation of reactive intermediates and *in vitro* glutathione conjugate formation in rats and humans. *Chem Res Toxicol* 20(3):455–464.

Meneses-Lorente G, Sakatis MZ, Schulz-Utermoehl T, De Nardi C, Watt AP (2006) A quantitative high-throughput trapping assay as a measurement of potential for bioactivation. *Anal Biochem* 351(2):266–272.

Mishra A, Veerasamy R, Jain PK, Dixit VK, Agrawal RK (2008) Synthesis, characterization and pharmacological evaluation of amide prodrugs of flurbiprofen. *J Braz Chem Soc* 19:89–100.

Moore LB, Maglich JM, McKee DD, Wisely B, Willson TM, Kliewer SA, Lambert MH, Moore JT (2002) Pregnane X receptor (PXR), constitutive androstane receptor (CAR), and benzoate X receptor (BXR) define three pharmacologically distinct classes of nuclear receptors. *Mol Endocrinol* 16:977–986.

Mulder GJ, Le CT (1988) A rapid, simple *in vitro* screening test, using [3H]glutathione and L-[35S]cysteine as trapping agents, to detect reactive intermediates of xenobiotics. *Toxicol In Vitro* 2(3):225–230.

Nelson DR, Kamataki T, Waxman DJ, Guengerich FP, Estabrook RW, Feyereisen R, Gonzalez FJ, Coon MJ, Gunsalus IC, Gotoh O, Okuda K, Nebert DW (1993) The P450 superfamily: Update on new sequences, gene mapping, accession numbers, early trivial names of enzymes, and nomenclature. *DNA Cell Biol* 12:1–51.

Neuhoff S, Artursson P, Zamora I, Ungell AL (2006) Impact of extracellular protein binding on passive and active drug transport across Caco-2 cells. *Pharm Res* 23(2):350–359.

Obach R, Walsky R, Venkatakrishnan K, Gaman E, Houston J, Tremaine L (2006) The utility of *in vitro* cytochrome P450 inhibition data in the prediction of drug-drug interactions. *J Pharmacol Exp Ther* 316:336–348.

Obach RS (1999) Prediction of human clearance of twenty-nine drugs from hepatic microsomal intrinsic clearance data: An examination of *in vitro* half-life approach and nonspecific binding to microsomes. *Drug Metab Dispos* 27:1350–1359.

Obach RS, Baxter JG, Liston TE, Silber BM, Jones BC, MacIntyre F, Rance DJ, Wastall P (1997) The prediction of human pharmacokinetic parameters from preclinical and *in vitro* metabolism data. *J Pharmacol Exp Ther* 283:46–58.

Obach RS, Kalgutkar AS, Soglia JR, Zhao SX (2008) Can *in vitro* metabolism-dependent covalent binding data in liver microsomes distinguish hepatotoxic from nonhepatotoxic drugs? An analysis of 18 drugs with consideration of intrinsic clearance and daily dose. *Chem Res Toxicol* 21(9):1814–1822.

Okey AB, Boutros PC, Harper PA (2005) Polymorphisms of human nuclear receptors that control expression of drug-metabolizing enzymes. *Pharmacogenet Genomics* 15:371–379.

Olinga P, Elferink MGL, Draaisma AL, Merema MT, Castell JV, Perez G, Groothius GM (2008) Coordinated induction of drug transporters and phase I and II metabolism in human liver slices. *Eur J Pharm Sci* 33:380–389.

Pacifici G, Viani A (1992) Methods of determining plasma and tissue binding of drugs. *Clin Pharmacol* 23:449–468.

Parkinson A (2001) Biotransformation of Xenobiotics. In *Casarett and Doull's Toxicology the Basic Science of Poisons*, 6th ed., Klaassen CD, ed. McGraw-Hill, New York.

Patsalos PN, Perucca E (2003) Clinically important drug interactions in epilepsy: Interactions between antiepileptic drugs and other drugs. *Lancet Neurol* 2:473–481.

Press B, Grandi DD (2008) Permeability for intestinal absorption: Caco-2 assay and related issues. *Curr Drug Metab* 9:893–900.

Ran Y, Jain N, Yalkowsky SH (2001) Prediction of aqueous solubility of organic compounds by the general solubility equation (GSE). *J Chem Inf Comput Sci* 41(5):1208–1217.

Rolan PE (1994) Plasma protein binding displacement interactions-why are they still regarded as clinically important? *Br J Clin Pharmacol* 37:125–128.

Saha P, Kou JH (2002) Effect of bovine serum albumin on drug permeability estimation across Caco-2 monolayers. *Eur J Pharm Biopharm* 54(3):319–324.

Schmidt S, Gonzalez D, Derendorf H (2009) Significance of protein binding in pharmacokinetics and pharmacodynamics. *J Pharm Sci* 99:1107–1122.

Sellers EM (1979) Plasma protein displacement interactions are rarely of clinical significance. *Pharmacology* 18:225–227.

Shukla SJ, Nguyen DT, Macarthur R, Simeonov A, Frazee WJ, Hallis TM, Marks BD, Singh U, Eliason HC, Printen J, Austin CP, Inglese J, Auld DS (2009) Identification of pregnane X receptor ligands using time-resolved fluorescence resonance energy transfer and quantitative high-throughput screening. *Assay Drug Dev Technol* 7:143–169.

Silverman RB (1988) Mechanism-based enzyme inactivation. In *Chemistry and Enzymology*, Vol. 1. CRC Press Inc., Boca Raton, FL.

Skeaff CM, Hodson L, McKenzie JE (2006) Dietary-induced changes in fatty acid composition of human plasma, platelet, and erythrocyte lipids follow a similar time course. *J Nutr* 136:565–569.

Skolnik S, Lin X, Wang J, Chen XH, He T, Zhang B (2010) Towards prediction of *in vivo* exposure using a 96-well Caco-2 permeability assay. *J Pharm Sci* 99(7):3246–3265.

Song H, Griesgraber GW, Wagner CR, Zimmerman C (2002) Pharmacokinetics of amino acid phosphoramidate monoesters of zidovudine in rats. *Antimicrob Agents Chemother* 46:1357–1363.

Sugano K, Hamada H, Machida M, Ushio H, Saitoh K, Terada K (2001) Optimized conditions of bio-mimetic artificial membrane permeability assay. *Int J Pharm* 228:181–188.

Sun D, Lennernas H, Welage LS, Barnett JL, Landowski CP, Foster D, Fleisher D, Lee KD, Amidon GL (2002) Comparison of human duodenum and Caco-2 gene expression profiles for 12,000 gene sequences tags and correlation with permeability of 26 drugs. *Pharm Res* 19:1400–1416.

Sun JX, Embil K, Chow DS, Lee CC (1987) High-performance liquid chromatographic analysis, plasma protein binding and red blood cell partitioning of phenprobamate. *Biopharm Drug Dispos* 8:341–351.

Takakusa H, Masumoto H, Yukinaga H, Makino C, Nakayama S, Okazaki O, Sudo K. (2008) Covalent binding and tissue distribution/retention assessment of drugs associated with idiosyncratic drug toxicity. *Drug Metab Dispos* 36(9):1770–1779.

Tillement JP, Houin G, Zini R, Urien S, Albengres E, Barre J, Lecomte M, D'Athis P, Sebille B (1984) The binding of drugs to blood plasma macromolecules: Recent advances and therapeutic significance. *Adv Drug Res* 13:59–94.

Treiber A, Schneiter R, Delahaye S, Clozel M (2004) Inhibition of organic anion transporting polypeptide-mediated hepatic uptake is the major determinant in the pharmaco-kinetic interaction between bosentan and cyclosporin A in the rat. *J Pharmacol Exp Ther* 308:1121–1129.

Troutman MD, Thakker DR (2003) Efflux ratio cannot assess P-glycoprotein-mediated attenuation of absorptive transport: Asymmetric effect of P-glycoprotein on absorptive and secretory transport across Caco-2 cell monolayers. *Pharm Res* 20:1200–1209.

Tucker GT, Houston JB, Huang SM (2001) Optimizing drug development: Strategies to assess drug metabolism/transporter interaction potential—Toward a consensus. *Pharm Res* 18:1071–1080.

Uetrecht J (2003) Screening for the potential of a drug candidate to cause idiosyncratic drug reactions. *Drug Discov Today* 8(18):832–837.

Ungell AL, Karlsson J (2003) Cell culture in drug discovery: An industrial perspective. In *Drug Bioavailability*, van de Waterbeemd H, Lennernas H, Artursson P, eds., pp. 90–131. Wiley-VCH, Verlag GmbH & Co. KGaA, Weinheim, Germany.

Usui T, Mise M, Hashizume T, Yabuki M, Komuro S (2009) Evaluation of the potential for drug-induced liver injury based on *in vitro* covalent binding to human liver proteins. *Drug Metab Dispos* 37(12):23823–22392.

Van den Bongard HJGD, Pluim D, van Waardenburg RCAM, Ravic M, Beijnen JH, Schellens JHM (2003) *In vitro* pharmacokinetic study of the novel anticancer agent E7070: Red blood cell and plasma protein binding in human blood. *Anticancer Drugs* 14:405–410.

Varma MVS, Ashokraj Y, Chinmoy SD, Panchagnula R (2003) P-glycoprotein inhibitors and their screening: A perspective from bioavailability enhancement. *Pharmacol Res* 48(4):347–359.

Venkatakrishnan K, Obach RS (2007) Drug-drug interactions via mechanism-based cytochrome P450 inactivation: Points to consider for risk assessment from *in vitro* data and clinical pharmacologic evaluation. *Curr Drug Metab* 8:449–462.

Wan H, Rehngren M (2006) High-throughput screening of protein binding by equilibrium dialysis combined with liquid chromatography and mass spectrometry. *J Chromatogr A* 1102:125–134.

Wang J (2009) Comprehensive assessment of ADMET risks in drug discovery. *Curr Pharm Des* 15(19):2195–2219.

Wang J, Collis A (2011) Maximizing the outcome of early ADMET models: Strategies to win the drug-hunting battles? *Expert Opin Drug Metab Toxicol* 7(4):381–386.

Wang J, Faller B (2007) Progress in bioanalytics and automation robotics for ADME screening. In *Comprehensive Medicinal Chemistry*, Vol. 5, 2nd ed., ADME-Tox Approaches, Testa B, van de Waterbeemd H, eds., pp. 341–356. Elsevier Ltd, Oxford.

Wang J, Skolnik S (2010) Mitigating permeability-mediated risks in drug discovery. *Expert Opin Drug Metab Toxicol* 6(2):171–187.

Wang J, Urban L, Bojanic D (2007) Maximising use of *in vitro* ADMET tools to predict *in vivo* bioavailability and safety. *Expert Opin Drug Metab Toxicol* 3(5):641–665.

Waters NJ, Jones R, Williams G, Sohal B (2008) Validation of a rapid equilibrium dialysis approach for the measurement of plasma protein binding. *J Pharm Sci* 97:4586–4595.

Wienkers LC, Heath TG (2005) Predicting *in vivo* drug interactions from *in vitro* drug discovery data. *Nat Rev Drug Discov* 4:825–833.

Wilkinson GR (2001) The dynamics of drug absorption, distribution, and elimination. In *Goodman and Gilman's: The Pharmacological Basis of Therapeutics*, 10th ed., Hardman JG and Limbird LE, eds. McGraw-Hill, New York.

Willson TM, Kliewer SA (2002) PXR, CAR and drug metabolism. *Nat Rev Drug Discov* 1:259–266.

Wohnsland F, Faller B (2001) High-throughput permeability pH profile and high-throughput alkane/water LogP with artificial membranes. *J Med Chem* 44:923–930.

Wu J, Hughes CS, Picard P, Letarte S, Gaudreault M, Lévesque JF, Nicoll-Griffith DA, Bateman KP (2007) High-throughput cytochrome P450 inhibition assays using laser diode thermal desorption-atmospheric pressure chemical ionization-tandem mass spectrometry. *Anal Chem* 79:4657–4665.

Yalkowsky SH (1999) *Solubility and Solubilization in Aqueous Media*. ACS-Oxford Univ. Press, New York.

Yan Z, Caldwell GW (2003) Metabolic assessment in liver microsomes by co-activating cytochrome P450s and UDP-glycosyltransferases. *Eur J Drug Metab Pharmacokinet* 28:223–232.

Yan Z, Maher N, Torres R, Huebert N (2007) Use of a trapping agent for simultaneous capturing and high-throughput screening of both "soft" and "hard" reactive metabolites. *Anal Chem* 79(11):4206–4214.

Yu S, Li S, Yang H, Lee F, Wu JT, Qian MG (2005) A novel liquid chromatography/tandem mass spectrometry based depletion method for measuring red blood cell partitioning of pharmaceutical compounds in drug discovery. *Rapid Commun Mass Spectrom* 19:250–254.

Zhang J, Musson DG (2006) Investigation of high-throughput ultrafiltration for the determination of an unbound compound in human plasma using liquid chromatography and tandem mass spectrometry with electrospray ionization. *J Chromatogr B Analyt Technol Biomed Life Sci* 843:47–56.

Zhou L, Yang L, Tilton S, Wang J (2007) Development of high throughput equilibrium solubility assay using miniaturized shake-flask method in early drug discovery. *J Pharm Sci* 98(11):3052–3071.

Zhou S, Yung Chan S, Cher Goh B, Chan E, Duan W, Huang M, McLeod HL (2005) Mechanism-based inhibition of cytochrome P450 3A4 by therapeutic drugs. *Clin Pharmacokinet* 44:279–304.

Zhou SF (2008) Potential strategies for minimizing mechanism-based inhibition of cytochrome P450 3A4. *Curr Pharm Des* 14:990–1000.

Zhu C, Jiang L, Chen TM, Hwang KK (2003) A comparative study of artificial membrane permeability assay for high throughput profiling of drug absorption potential. *Eur J Med Chem* 37(5):399–407.

10

PERMEABILITY AND TRANSPORTER MODELS IN DRUG DISCOVERY AND DEVELOPMENT

Praveen V. Balimane, Yong-Hae Han, and Saeho Chong

10.1 INTRODUCTION

The path to successful drug discovery and development in a pharmaceutical company is a very long and tortuous one, fraught with uncertainty and risks, and demanding massive resources and cost. A recent report has pegged the final price tag of bringing a molecule from lab to market at higher than a billion dollars, with an estimated research time running into multiple years [1]. Despite the considerable investment in terms of finance and resources, the number of drug approvals per year have held steady or decreased for the last few years. The success rate of progressing from initial clinical testing to final approval has remained disappointingly low, with less than 10% of the compounds entering Phase I clinical testing eventually reaching the patients [2]. Efforts to increase the success rate of drug discovery and development have led to an industry-wide adoption of the parallel matrix approach, where pharmacological properties (i.e., potency and efficacy) are screened in parallel with absorption, distribution, metabolism, excretion, and toxicity (ADMET) profiling (i.e., developability) to maximize the ability to select superior drug candidates with the best chances of making it to market. The availability of highly accurate, low-cost, and high-throughput screening (HTS) techniques that can provide fast and reliable data on the developability characteristics of drug candidates is crucial for the new strategy to succeed.

Screening discovery compounds for their biopharmaceutical properties (e.g., solubility, intestinal permeability, cytochrome P450 [CYP] inhibition, metabolic stability, and more recently, drug–drug interaction [DDI] potential involving drug transporters) has become a critical step that can make or break the fortune of a company. When defining a preferred absorption, metabolism, distribution, and excretion (ADME) space, fast and reliable determination of the permeability/ absorption properties and drug–transporter interaction potentials of drug candidates is quickly becoming the key characterization study performed during lead selection and lead optimization.

Currently, a variety of experimental models is available for evaluating the intestinal permeability and transporter interaction potential of drug candidates [3, 4]. The most popular models for assessing permeability/ absorption include *in vitro* methods (an artificial lipid membrane such as is used in a parallel artificial membrane permeability assay [PAMPA], cell-based systems such as Caco-2 cells, Mardin–Darby canine kidney [MDCK] cells, etc., the tissue-based Ussing chamber); *in situ* methods (intestinal single-pass perfusion); and *in vivo* methods (whole-animal absorption studies). Models to investigate transporter interaction potential (also referred to as transporter phenotyping) include *in vitro* methods (intact cells such as hepatocytes, transient and stable transfected cells, insect vector (sf9) vesicles,

ADME-Enabling Technologies in Drug Design and Development, First Edition. Edited by Donglu Zhang and Sekhar Surapaneni.
© 2012 John Wiley & Sons, Inc. Published 2012 by John Wiley & Sons, Inc.

Xenopus oocytes); *ex vivo* methods (organ slices and perfusions); and *in vivo* methods (transgenic knockout animal models, etc.).

10.2 PERMEABILITY MODELS

Despite tremendous innovations in drug delivery methods in the last few decades, the oral route still remains as the most preferred route of administration for most new chemical entities (NCEs). The oral route is preferred by virtue of its convenience, low cost, and high patient compliance compared with alternative routes. However, compounds intended for oral administration must have adequate aqueous solubility and intestinal permeability in order to achieve therapeutic concentrations. With the explosive growth in the field of genomics and combinatorial chemistry, coupled with technological innovations in the last few years, synthesizing a large number of potential drug candidates is no longer a bottleneck in the drug discovery process. Instead, the task of screening compounds simultaneously for biological activity and biopharmaceutical properties (e.g., solubility, permeability/absorption, stability) has become the major challenge. This has provided a great impetus within the pharmaceutical industry to implement appropriate screening models that are high capacity, cost-effective, and highly predictive of *in vivo* permeability and absorption. Typically, a combination of models is used synergistically in assessing intestinal permeability. A tiered approach is the most popular design, which involves high-throughput (but less predictive) models for primary screening followed by low-throughput (but more predictive) models for secondary screening and mechanistic studies. PAMPA and cell culture-based models offer the right balance between predictability and throughput, and currently enjoy wide popularity throughout the pharmaceutical industry.

10.2.1 PAMPA

The PAMPA model was first introduced in 1998 [5], and since then numerous reports have been published illustrating the general applicability of this model as a high-throughput permeability screening tool. The model consists of a hydrophobic filter material coated with a mixture of lecithin/phospholipids dissolved in an inert organic solvent such as dodecane, creating an artificial lipid membrane barrier that mimics the intestinal epithelium. The rate of permeation across the membrane barrier was shown to correlate well with the extent of drug absorption in humans. The use of 96-well microtiter plates coupled with rapid analysis using a spectrophotometric plate reader makes this system a very attractive model for screening a large number of compounds and libraries. PAMPA is much less labor-intensive than cell culture methods, but it appears to show similar predictability. One of the main limitations of this model is that PAMPA underestimates the absorption of compounds that are actively absorbed via drug transporters. Despite this limitation, PAMPA serves as an invaluable primary permeability screen during the early drug discovery process because of its high throughput capability. Lately there have been significant improvements to the model to accommodate its broader use at the discovery stage (stable PAMPA plates for long-term usage, pH adaptability, specialized plates for estimating permeability into various target organs—brain, liver, etc.) [6].

10.2.2 Cell Models (Caco-2 Cells)

Varieties of cell monolayer models that mimic *in vivo* intestinal epithelium in humans have been developed and currently enjoy widespread popularity. Unlike enterocytes, human immortalized (tumor) cells grow rapidly into confluent monolayers followed by a spontaneous differentiation, providing an ideal system for transport studies. A few of these cell models that are most commonly used are Caco-2, MDCK, LLC-PK1, and HT-29.

The Caco-2 cell model has been the most popular and most extensively characterized cell-based model in examining the permeability of drugs in both the pharmaceutical industries and academia. Caco-2 cells, a human colon adenocarcinoma, undergo spontaneous enterocytic differentiation in culture and become polarized cells with well-established tight junctions, resembling intestinal epithelium in humans. It has also been demonstrated that the permeability of drugs across Caco-2 cell monolayers correlated very well with the extent of oral absorption in humans. In the last 10–15 years, Caco-2 cells have been widely used as an *in vitro* tool for evaluating the permeability of discovery compounds and for conducting in-depth mechanistic studies [3, 7] (Figure 10.1).

10.2.3 P-glycoprotein (Pgp) Models

Adequate permeability is required not only for oral absorption but also for sufficient drug distribution to pharmacological target organs (e.g., tumor, liver). In addition to simple passive diffusion across lipid bilayers, numerous transporters appear to play a critical role in selective accumulation and distribution of drugs into target organs. Pgp is one of the most extensively studied transporters that have been unequivocally known to impact the ADMET characteristics of drug molecules.

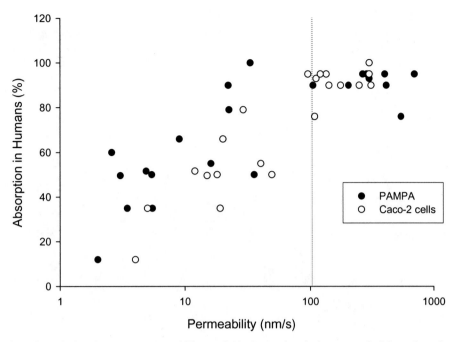

FIGURE 10.1. Correlation between permeability and % absorption in humans of ~25 marketed compounds.

It is a ubiquitous transporter, which is present on the apical surface of the enterocytes, canalicular membrane of hepatocytes, and the apical surface of kidney, placenta, and endothelial cells of brain membrane. Because of its strategic location, it is widely recognized that Pgp is a major determinant in disposition of a wide array of drugs in humans. The oral bioavailability of fexofenadine increased significantly when erythromycin or ketoconazole (a well-known inhibitor of Pgp) was coadministered in humans, suggesting Pgp as a permeability barrier at the absorption site. Similarly, Pgp at the blood–brain barrier limits the entry of drugs into the brain. The biliary elimination of vincristine decreased significantly in the presence of verapamil (a known Pgp substrate/inhibitor). Therefore, the early screening of drug candidates for their potential to interact with Pgp (either as a substrate or inhibitor) is becoming necessary and critical. There are various *in vitro* and *in vivo* models used for assessing Pgp interaction [8]. *In vitro* assays such as ATPase activity [9], rhodamine-123 uptake [10], calcein AM uptake [11], cell-based bidirectional transport, and radioligand binding along with *in vivo* models such as transgenic (knockout mice) animals [12] are often used to assess the involvement of Pgp. The cell-based bidirectional permeability assay is the most popular method for identification of Pgp substrate in drug discovery labs [3, 8]. This cell model provides the right balance of adequate throughput and functional utility (Figure 10.2).

10.3 TRANSPORTER MODELS

Lately it has been realized that in addition to CYPs (which control the metabolism of compounds and thus its disposition), transporter proteins expressed in several key absorption and elimination organs (liver, kidney, intestine, brain, etc.) can also play a leading role in dictating the disposition of drugs. There is a plethora of these transporter proteins that are broadly classified into the ATP-binding cassette (ABC) transporter family (involved in efflux transport) and the solute carrier (SLC) transporter family (involved in influx transport). The ones that have demonstrated unequivocal clinical significance are the ones that have strategic expression at inlet and outlet locations in key organs and generally have broad substrate specificity. Some prominent ones in each family are as follows: ABC family = Pgp, breast cancer resistance protein (BCRP), and multi-drug resistance protein 2 (MRP2); SLC family = organic anion-transporting polypeptides (OATPs), organic cation transporter (OCT), and organic anion transporter (OAT) [13]. Study of transporters is becoming exceedingly important because they have the potential to influence the ADME properties of drugs, they can significantly impact the safety and toxicity profiles of drugs (via accumulation in target organs), they can lead to clinical DDIs [14], and they also help explain clinical variability due to polymorphisms [15]. Additionally, there are increasing market and regulatory pressures that are also

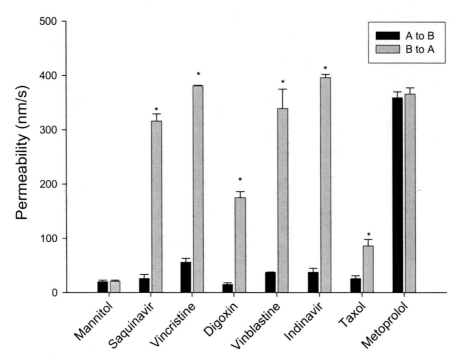

FIGURE 10.2. Bidirectional efflux (B to A/A to B) ratio of classical Pgp substrates in Caco-2 cells (mannitol and metoprolol are used as negative controls). * = significantly different.

TABLE 10.1. Major Drug Transporters in Key Target Organs

	Uptake/SLC Type	Efflux/ABC Type
Liver	OATP1B1/1B3/2B1, OCT1, OAT2	MRP2, Pgp, BCRP, BSEP
Kidney	OAT1/3, OCT2/3, OCTN1/2, MATE1/2, PEPT1/2	MRP2/4, Pgp
Intestine	OATPs, PEPT1	MRP1/2/3, Pgp, BCRP
Brain	OAT3, OCT2, OATP1A2	Pgp, MRPs, BCRP

encouraging companies to integrate transporter interaction models earlier during compound profiling paradigms [16]. Most companies are beginning to apply coordinated transporter interaction screening strategies beginning in the discovery stage and carrying on up to the late development stage to help anticipate interactions, design clinical studies, provide guidance on labeling, and help manage the interactions in the clinic. There are a plethora of *in vitro* and *in vivo* models for transporter interaction studies that are discussed extensively in several review articles [17–19]. Listed below are the most popular methodologies that are utilized in discovery and development for transporter interactions (Table 10.1).

10.3.1 Intact Cells

Using fresh hepatocytes to model hepatic transport processes in humans and preclinical species is fast being recognized as a reliable first-line assessment method [20]. Since fresh hepatocytes are equipped with all hepatic transporters as well as Phase I and Phase II enzymes, the hepatocyte system provides an appropriate model for evaluating transporter- and enzyme-mediated hepatic clearance [21]. Generally, the uptake into the hepatocytes is measured in suspension by a centrifugation method using oil-layered tubes to separate hepatocytes from the incubation medium. Alternatively, the assay can be run in semi-high throughput fashion with plated hepatocytes (in 24- and 96-well plates). Hepatocytes are also sometimes used in primary culture mode after growing for several days. However, there are scientific limitations inherent to this configuration, such as the decrease in functionality of uptake transporters (e.g., OATPs) as well as Phase I and Phase II enzyme activities during culture. Recently, cryopreserved hepatocytes have become popular in a drug discovery environment, and several laboratories have shown that transporters and CYP activities can be maintained in rat, dog, cynomolgus monkey, and human cryopreserved hepatocytes [21]. Due to the obvious advantages of cryopreserved hepatocytes (availability, resource consistency, etc.), the use of cryopreserved hepatocytes is likely to increase in the future.

10.3.2 Transfected Cells

Wild-type cell lines such as Caco-2, MDCK, and LLC-PK1 are popular tools for mechanistic transporter studies but suffer from a major drawback in that they express significant levels of several different transporters (Papt1, Pgp, BCRP, MRPs, etc.) that make delineation of the role of individual transporters almost impossible [18]. To overcome this issue, transfected cell lines have been developed that are essentially engineered cells overexpressing a single transporter of interest and thus maximizing the role of a single transporter in a particular study. The molecular-level process involves incorporation of the transporter cDNA into the naive mock cell line leading to the enhanced protein expression of the single transporter in the cells. Once DNAs are stably transfected in cell lines, the cell lines maintain transporter activity over subsequent passages. The most widely utilized cell lines for stable transfection are COS-7, CHO, HEK-293, and MDCK cells since they express low levels of endogenous transporter proteins. Many stable cell lines for the uptake and efflux transporters are successfully established, and their functional characterizations are well documented in the literature. Although development of stably transfected cells is a time-consuming and labor-intensive process, the stably transfected cells generally offer advantages in cost, data variation, and analytical flexibility, as compared with other transporter models (Figure 10.3). Therefore, stably transfected cells are very attractive transporter phenotyping tools.

10.3.3 *Xenopus* Oocyte

Drug transporters can be transiently expressed in immature eggs (oocytes) from the South African clawed frog, *Xenopus laevis*, by injecting the cRNA of transporters into the cytoplasm of the cells [22]. This leads to the overexpression of the single transporter at the oocyte plasma membrane layer, thus providing a suitable model for transporter studies. To evaluate transporter-mediated uptake into oocytes, water-injected oocytes are used as a reference, and the uptake amounts are compared between water- and transporter cRNA-injected oocytes. Many models for major uptake transporters (OATPs, OATs, OCTs, etc.) in *Xenopus* oocytes are commercially available from various vendors. However, extreme care should be taken while performing such studies because oocytes are very fragile and can lead to higher variability than other models.

10.3.4 Membrane Vesicles

Membrane vesicles prepared from a variety of intact tissues have been used for several decades to assess the transporter mechanism in major tissues such as liver, kidney, and intestinal tissue [23]. The blood side of cell membranes (liver-sinusoidal, kidney, and intestine-basolateral) and lumenal side of cell membranes (liver-bile canalicular, the intestine, and kidney-brush border) can be prepared separately by a series of centrifugations and/or precipitations with metal ions (e.g., Ca2+). Membrane vesicles prepared from intact tissues express multiple transporters, thus the observed transport can be attributed to the mixed roles of different transporters. Therefore, the use of membrane vesicles prepared from the cells overexpressing single transporter cells provide us with unequivocal information supporting our understanding of transport mechanisms. Various cells such as Sf9, HEK-293, MDCK, and LLC-PK1 cells are commonly used as host cells to prepare vesicles that are transfected with ABC transporters (MRPs, Pgp, BCRP, BSEP, etc.) for mechanistic studies [18]. Since membrane vesicles do not contain metabolic enzymes, the

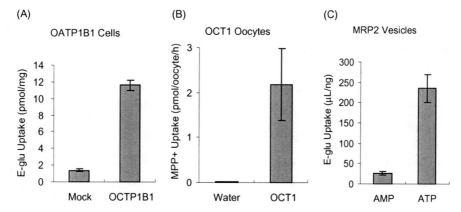

FIGURE 10.3. (A) Uptake of estradiol-17β-D-glucuronide (E-glu, 1 μM) into stably transfected OATP1B1 or mock/HEK-293 cells. (B) Uptake of 1-methyl-4-phenylpyridinium (MPP+, 1 μM) into *Xenopus* oocytes injected with OCT1 cRNA or water. (C) Uptake of estradiol-17β-D-glucuronide (E-glu, 1 μM) into MRP-expressing membrane vesicles in the presence of 5 mM AMP or ATP.

TABLE 10.2. Comparison of Popular Transporter Models

	Cell Lines	Membrane Vesicles	*Xenopus* Oocytes
Availability	Limited	Commercial	Commercial
Cost	Inexpensive	Expensive	Expensive
Labeled compound requirement	Preferred	Yes	Preferred
Throughput	Low	High	Low
Nonspecific binding	Negligible	Significant	Negligible
Quality of data	High	Medium	Low

model presents a significant advantage over other models (cell based or *in vivo*) when dealing with metabolically labile compounds. Recently, assay methods with membrane vesicles have been greatly improved, and a study can be performed in a high-throughput setting by using 96-well plates and cell harvesting devices [18]. However, the nonspecific binding to membrane filters can often be a significant technical issue, especially for lipophilic compounds (Table 10.2).

10.3.5 Transgenic Animal Models

Although many *in vitro* transporter assay tools have been developed so far, the prediction often lacks an *in vitro–in vivo* bridge. Transgenic animal models are genetically engineered animals that have undergone a targeted gene mutation (gene knockout) leading to a complete absence of a selected transporter protein. Parallel studies in a wild type (that expresses all transporters) along with knockout (where one targeted transporter has been deleted) provides an attractive *in vivo* model to tease out the role of transporter proteins. Knockout mice models of various transporters have been established to date, and their usefulness has been demonstrated for Mdr1a, Mdr1b, Mdr1a/1b, Mrp1, Mrp2, Mrp4, Bcrp, Bsep, and so on [24–26].. Knockout mice for SLC transporters have also been established and characterized functionally: Oct1, Oct2, Oct3, Octn2, Oat1, Oat3, Pept1, Pept2, and so on [24, 27, 28].

10.4 INTEGRATED PERMEABILITY–TRANSPORTER SCREENING STRATEGY

Drug discovery and development is an endeavor that is both resource heavy as well as extremely time sensitive. In various stages of the discovery cycle, compounds are selected via profiling their properties through several *in vitro* and *in vivo* models. For the discovery process to be efficient and eventually financially rewarding for the companies, the screening models need to be extremely high throughput and inexpensive at the early stage (where one deals with thousands of compounds) and more mechanistic and predictive at the later stages (where one deals with only a few leads). The process of selecting the right compounds to progress through the discovery stage to development and finally to navigate to market hinges on utilizing the "right models" at the "right time" to understand the benefit/liability ratio of a compound and making a decision for the next step.

The "ideal strategy" for compound selection would involve the use of a combination of *in vitro* models that are high throughput (but less predictive) with *in vivo* models that are low throughput (but more predictive) to effectively evaluate the intestinal permeability and transport characteristics of a large number of drug candidates during the lead selection and lead optimization processes. PAMPA and Caco-2 cells are the most frequently utilized *in vitro* models to assess intestinal permeability. The popularity of these models stems from their potential for high throughput, cost-effectiveness, and adequate predictability of absorption potential in humans (Table 10.2). However, several caveats associated with these models, such as poor predictability for transporter-mediated and paracellularly absorbed compounds, significant nonspecific binding to cells/devices leading to poor recovery, and variability associated with experimental factors, need to be considered carefully to realize their full potential. Transporter interaction studies with Pgp, the most well-studied and pharmaceutically relevant transporter, forms the next layer of profiling to ensure that Pgp transporter-mediated DDIs would not be a clinical issue. Assays such as ATPase, inhibition assays, and the bidirectional assay in cell models form the mainstay for Pgp studies. The investigative studies for interaction with other transporters are often performed based on critical information generated by other functional areas (chemistry, biology, preclincal in drug metabolism and pharmacokinetics [DMPK], etc.). OATPs and efflux transporters are studied if the liver is the target organ (efficacy or toxicity) or if a compound (or its metabolites) is eliminated via biliary excretion. OATs and OCTs are studied if the kidney is the target organ, or drug elimination is primarily via urinary excretion. Efflux transporters are studied if the target is the central nervous system or tumors. Both influx and efflux intestinal transporters are studied to investigate nonlinear PK as well as to make other preclinical PK observations of interest.

A typical transporter interaction study involves identification of specific drug transporter(s) involved in the disposition of test compounds using one or more of the

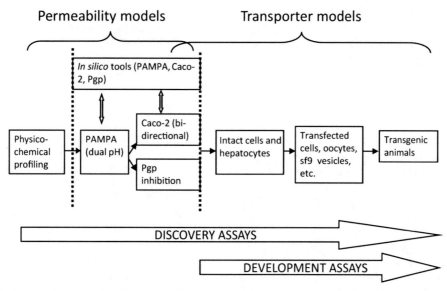

FIGURE 10.4. Typical screening paradigm incorporating permeability and transporter assays in various stages of discovery and development cycle.

models discussed earlier in this chapter. The objective of transporter interaction studies is to provide unequivocal evidence that the test compound interacts with specific transporter(s) so that we may anticipate such an interaction at the clinical level. Transporter phenotyping studies are performed at the preclinical stage, and the results are often incorporated in designing appropriate clinical studies to determine the extent of interaction in humans. These specialized clinical DDI studies play a critical role in devising an effective strategy in managing transporter-mediated DDIs in the clinic and thus making the approved drugs safer (Figure 10.4).

REFERENCES

1. FDA (2004) Challenges and opportunities on the critical path to new medical products. In *FDA Report*. Rockville, MD: Food and Drug Administration.

2. Kola I, Landis J (2004) Can the pharmaceutical industry reduce attrition rates? *Nat Rev Drug Discov* 3(8): 711–715.

3. Balimane PV, Han YH, Chong S (2006) Current industrial practices of assessing permeability and P-glycoprotein interaction. *AAPS J* 8(1):E1–13.

4. Kerns EH, et al. (2004) Combined application of parallel artificial membrane permeability assay and Caco-2 permeability assays in drug discovery. *J Pharm Sci* 93(6): 1440–1453.

5. Kansy M, Senner F, Gubernator K (1998) Physicochemical high throughput screening: Parallel artificial membrane permeation assay in the description of passive absorption processes. *J Med Chem* 41(7):1007–1010.

6. Faller B (2008) Artificial membrane assays to assess permeability. *Curr Drug Metab* 9(9):886–892.

7. Artursson P, Karlsson J (1991) Correlation between oral drug absorption in humans and apparent drug permeability coefficients in human intestinal epithelial (Caco-2) cells. *Biochem Biophys Res Commun* 175(3):880–885.

8. Polli JW, et al. (2001) Rational use of *in vitro* P-glycoprotein assays in drug discovery. *J Pharmacol Exp Ther* 299(2): 620–628.

9. Hrycyna CA, et al. (1998) Mechanism of action of human P-glycoprotein ATPase activity. Photochemical cleavage during a catalytic transition state using orthovanadate reveals cross-talk between the two ATP sites. *J Biol Chem* 273(27):16631–16634.

10. Feller N, et al. (1995) Functional detection of MDR1/P170 and MRP/P190-mediated multidrug resistance in tumour cells by flow cytometry. *Br J Cancer* 72(3):543–549.

11. Hollo Z, et al. (1994) Calcein accumulation as a fluorometric functional assay of the multidrug transporter. *Biochim Biophys Acta* 1191(2):384–388.

12. Wijnholds J, et al. (1997) Increased sensitivity to anticancer drugs and decreased inflammatory response in mice lacking the multidrug resistance-associated protein. *Nat Med* 3(11):1275–1279.

13. Shitara Y, Horie T, Sugiyama Y (2006) Transporters as a determinant of drug clearance and tissue distribution. *Eur J Pharm Sci* 27(5):425–446.

14. Lin JH (2007) Transporter-mediated drug interactions: Clinical implications and *in vitro* assessment. *Expert Opin Drug Metab Toxicol* 3(1):81–92.

15. Ieiri I, Higuchi S, Sugiyama Y (2009) Genetic polymorphisms of uptake (OATP1B1, 1B3) and efflux (MRP2, BCRP) transporters: Implications for inter-individual differences in the pharmacokinetics and pharmacodynamics

of statins and other clinically relevant drugs. *Expert Opin Drug Metab Toxicol* 5(7):703–729.

16. Giacomini KM, et al. (2010) Membrane transporters in drug development. *Nat Rev Drug Discov* 9(3):215–236.

17. Oswald S, et al. (2007) Transporter-mediated uptake into cellular compartments. *Xenobiotica* 37(10–11):1171–1195.

18. Sahi J (2005) Use of *in vitro* transporter assays to understand hepatic and renal disposition of new drug candidates. *Expert Opin Drug Metab Toxicol* 1(3):409–427.

19. Kitamura S, Maeda K, Sugiyama Y (2008) Recent progresses in the experimental methods and evaluation strategies of transporter functions for the prediction of the pharmacokinetics in humans. *Naunyn Schmiedebergs Arch Pharmacol* 377(4–6):617–628.

20. Hewitt NJ, et al. (2007) Primary hepatocytes: Current understanding of the regulation of metabolic enzymes and transporter proteins, and pharmaceutical practice for the use of hepatocytes in metabolism, enzyme induction, transporter, clearance, and hepatotoxicity studies. *Drug Metab Rev* 39(1):159–234.

21. Houle R, et al. (2003) Retention of transporter activities in cryopreserved, isolated rat hepatocytes. *Drug Metab Dispos* 31(4):447–451.

22. Pritchard JB, Miller DS (2005) Expression systems for cloned xenobiotic transporters. *Toxicol Appl Pharmacol* 204(3):256–262.

23. Ruetz S, et al. (1987) Isolation and characterization of the putative canalicular bile salt transport system of rat liver. *J Biol Chem* 262(23):11324–11330.

24. Glaeser H, Fromm MF (2008) Animal models and intestinal drug transport. *Expert Opin Drug Metab Toxicol* 4(4):347–361.

25. Vlaming ML, Lagas JS, Schinkel AH (2009) Physiological and pharmacological roles of ABCG2 (BCRP): Recent findings in Abcg2 knockout mice. *Adv Drug Deliv Rev* 61(1):14–25.

26. Xia CQ, Milton MN, Gan LS (2007) Evaluation of drug-transporter interactions using *in vitro* and *in vivo* models. *Curr Drug Metab* 8(4):341–363.

27. Ahn SY, et al. (2009) Interaction of organic cations with organic anion transporters. *J Biol Chem* 284(45):31422–31430.

28. Ocheltree SM, et al. (2005) Role and relevance of peptide transporter 2 (PEPT2) in the kidney and choroid plexus: In vivo studies with glycylsarcosine in wild-type and PEPT2 knockout mice. *J Pharmacol Exp Ther* 315(1):240–247.

11

METHODS FOR ASSESSING BLOOD–BRAIN BARRIER PENETRATION IN DRUG DISCOVERY

Li Di and Edward H. Kerns

11.1 INTRODUCTION

The blood–brain barrier (BBB) is the membrane surrounding the microblood capillary vessels in the brain. It plays a critical role in maintaining the homeostasis in the brain, where a fine balance is critical for all species. The BBB is made of endothelial cells with very tight junctions. There are about 400 mi of the blood capillaries in the brain with strong P-glycoprotein (Pgp) efflux activity at the apical membrane of the BBB cells. Pgp plays a very important role in preventing toxic compounds from entering into the brain. The characteristics of the BBB, such as tight intercellular junctions, absence of fenestrations, and Pgp efflux transport, make it difficult for drug molecules to enter the brain and interact with central nervous system (CNS) targets. It has been reported that only 2% of drug discovery compounds can cross the BBB and potentially become successful CNS agents [1]. The BBB is one of the major challenges for CNS therapy, leading to a low success rate in the clinic [2].

Besides the BBB, there is a blood cerebrospinal fluid barrier (BCSFB) interface with the brain. It is in a separate compartment. Delivery of drug molecules to the brain via cerebrospinal fluid (CSF) at the BCSFB is difficult because the flow of interstitial fluid (ISF) in the brain to the CSF is very fast. The surface area of BCSFB is also 5000 times smaller than the BBB. Taken together, the BBB is the most important delivery route and barrier for drugs targeting CNS diseases [3, 4].

There are many mechanisms for brain penetration (Figure 11.1) [5, 6]. Most drugs enter the brain by transcellular passive diffusion through the lipid membrane into the brain. There is very limited paracellular transport because of the tight junctions between the endothelial cells of the BBB. Compounds that are substrates for uptake transporters (e.g., LAT1, PEPT1) can enter the brain at a higher rate than passive diffusion alone. Efflux transporters, such as Pgp, oppose entry of drug molecules into the brain and can reduce compound concentration in the brain through efflux mechanisms. Plasma protein binding and brain tissue binding can affect distribution of drugs into the brain. Metabolism in the liver and brain can also contribute to the amount of drug present in the brain.

The multiple mechanisms and bioprocesses of the brain make it challenging to develop a simple assay to address the complicated system. There are many ways to evaluate brain penetration. There are currently two concepts evolving in the pharmaceutical industry and academia to describe brain penetration. They are rate and extend [5, 7–9]. Rate is a measure of the initial slope of brain penetration, which is important for drugs that require rapid onset, such as anesthesia. Extent measures the amount of drug in the brain at steady state, which is critical for indications requiring sustained drug effects during chronic dosing. Rate is not the same as extent, but they also have common features related to molecular properties of compounds, such as lipophilicity, hydrogen

ADME-Enabling Technologies in Drug Design and Development, First Edition. Edited by Donglu Zhang and Sekhar Surapaneni.
© 2012 John Wiley & Sons, Inc. Published 2012 by John Wiley & Sons, Inc.

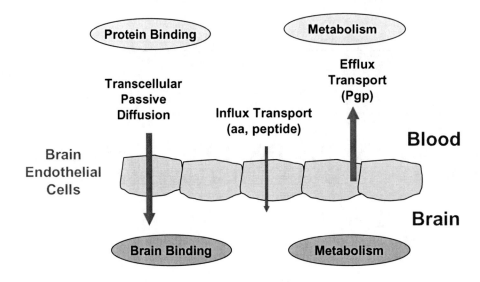

FIGURE 11.1. Mechanisms for brain penetration (modified from References 5, 6).

bonding, and polar surface area (PSA) [5]. Ultimately, it is the free drug concentration at the site of action that exerts the pharmacological activity [10, 11].

Many methods for assessing BBB penetration have been developed and are well documented [12–15]. It has been proposed that three variables are most relevant to predict CNS drug action: rate, extent, and brain tissue binding [10]. This chapter will cover the methods commonly used in drug discovery in assessing BBB penetration.

11.2   COMMON METHODS FOR ASSESSING BBB PENETRATION

Many methods have been developed to measure and predict the brain penetration potential of drug candidates [12–15] (Table 11.1). Because of the multiple mechanisms of brain penetration for small molecules [6], the methods are designed to address specific aspects of BBB penetration, such as BBB membrane permeability, distribution of drugs between brain and blood, free drug concentration in the brain, effects of transporters on the BBB, and others. Which BBB method should be used for a particular drug discovery program is really dependent on what questions that project team is trying to answer. There is really no one-method-fits-all approach for the BBB. Methods are specific to particular mechanisms that are most relevant to a project. Here we discuss the common methods applied in drug discovery to address specific questions of project teams with regard to BBB penetration.

TABLE 11.1. Commonly Used BBB Methods in Drug Discovery

Interest	Methods
Free drug concentration in brain	Brain PK & brain tissue binding
	CSF concentration
BBB permeability	*In situ* brain perfusion
	PAMPA-BBB
	LogD
Pgp efflux	MDR1-MDCK
	Caco-2

11.3   METHODS FOR DETERMINATION OF FREE DRUG CONCENTRATION IN THE BRAIN

If the question of a project is how to predict whether a compound will be efficacious *in vivo* based on *in vitro* activity data, it is important to determine the free drug concentration in the brain [5, 9, 16]. Determination of free drug concentration in the biophase (brain in this case) is critical to predicting *in vivo* efficacy, developing pharmacokinetic (PK)/pharmacodynamic (PD) relationships and selecting dose and dosing frequency [10]. Total drug concentration in the brain is usually not relevant to *in vivo* activity [11, 16]. Drug molecules bound nonspecifically to lipids and proteins in the brain are nonproductive and not available to interact with the therapeutic target to produce efficacy.

The gold standard method for measuring free drug concentration in the brain is *in vivo* microdialysis [17–20]. However, the assay is very low throughput, labor intensive, expensive, and technically challenging for lipophilic compounds due to nonspecific binding of test compounds to the dialysis probe membranes/equipment, which leads to low recovery and unreliable data. As a result, *in vivo* microdialysis has not been routinely applied in early drug discovery to determine free drug concentration in the brain, where higher throughput and rapid turnaround time is needed to support large number of compounds and programs. It is more commonly used for late-stage drug discovery and development, when a limited number of compounds are studied in depth. Microdialysis is also widely used to determine neurotransmitters, such as dopamine, glutamate, and γ-aminobutyric acid (GABA) [21].

Free drug concentration in the brain is a complicated variable affected by many different processes *in vivo*, such as absorption, metabolism, and distribution. Therefore, optimization of free drug concentration involves multiple parameters. For orally administered drugs, the strategy is to optimize solubility and permeability, to optimize fraction absorbed, and metabolic stability to optimized blood concentrations. Optimization of fraction unbound, through structural modification, is counterproductive and should be avoided in drug discovery [11].

11.3.1 *In vivo* Brain PK in Combination with *in vitro* Brain Homogenate Binding Studies

In vivo brain PK experiments are typically used to determine total brain exposure of drug candidates (bound plus unbound). Brain PK studies are usually performed by dosing animals of selected species (typically rat or mouse) with test compounds at a certain dose via a desirable route (e.g., intravenous, intraperitoneally, or orally) [12]. At different time points after dosing, blood samples and whole brains are taken. The compounds are extracted from plasma and brain tissue homogenates and analyzed with light chromatography-mass spectrometry (LC-MS). PK data, such as area under the curve (AUC) and C_{max} in plasma and brain, and brain exposure/plasma exposure (B/P) ratio, can be derived from the study using statistical software. The exposure (e.g., C_{max}) obtained from brain PK studies is the total drug concentration. To obtain the free drug concentration in the brain, the fraction unbound in the brain ($f_{u, brain}$) is determined by using in the following calculation:

$$C_{max,u} = C_{max} * f_{u,brain}$$

Fraction unbound in the brain is often determined *in vitro* using equilibrium dialysis with brain homogenates in drug discovery [22–27]. Typically, brain homogenates with test compounds and buffers are added into the

donor and acceptor wells, respectively, for dialysis. Once equilibrium is reached (typically about 5 h), the compound is extracted from the matrix material with organic solvent and analyzed using LC-MS. The fraction unbound is the concentration in the acceptor divided by the concentration in the donor and corrected by the dilution factor. Fraction unbound, measured using brain homogenates, gives good correlation to *in vivo* data [25]. It is a rapid, high-throughput, and low-cost approach to obtaining free drug concentration in the brain when combined with brain PK studies. Reports suggest that there is little difference in brain binding among species [9]; therefore, one can estimate free drug concentration using fraction unbound from a single species early in drug discovery. Brain tissue binding studies using brain slice have also been reported and a high-throughput method has been developed [28–30]. It has been suggested that data from brain slices is slightly more reliable in predicting free drug concentration *in vivo* than using brain homogenates [30], owing to unmasking of cellular material during homogenization. The protocol using brain slices to study binding is still quite involved and laborious [29]. Brain tissue binding study with brain homogenate is the most common and cost-effective approach in drug discovery to measure fraction unbound. Cassette dosing further improves the throughput [27].

The free drug concentration in the brain for an orally administered drug is independent of brain tissue binding [7, 11, 31]. Optimization of fraction unbound is nonproductive [11].

11.3.2 Use of CSF Drug Concentration as a Surrogate for Free Drug Concentration in the Brain

CSF concentration is frequently used as a surrogate measure for *in vivo* free drug concentration in the brain [32–34]. CSF drug concentration has been shown to have good prediction of brain free drug concentration and is better than using unbound plasma concentration [24, 35]. Human and rat have similar rank ordering based on CSF information, which makes rat CSF concentration a viable tool in drug discovery. Any difference between rat and human CSF data is presumably owing to human data that was obtained under human disease states [10, 34].

CSF contains very little protein and the sampling is much more amenable in drug discovery compared with *in vivo* microdialysis. Catheters are inserted into the cisterna magna or lumber intrathecal space for serial CSF sampling, and drug concentrations in CSF are determined using LC-MS. For large animals, studies can be repeated with the same catheterized animals to minimize variations from individual animals. CSF concentration is currently the only source of information on brain

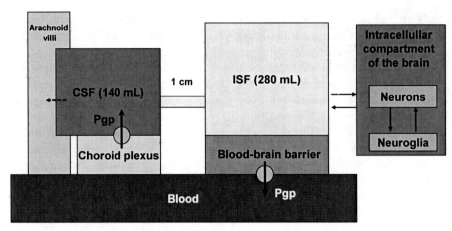

FIGURE 11.2. Direction of efflux for Pgp on BBB and BCSFB [3, 4, 12].

exposure in human. Timing of the sampling is important since the concentration–time profiles may be different between CSF and brain. Cassette dosing can be applied to improve throughput [34]. When compounds have good membrane permeability, using CSF as a surrogate for ISF concentration works well because equilibrium is rapidly established between CSF and ISF in the brain. However, in some cases, CSF concentration can either overpredict or underpredict free drug concentration in the brain. CSF tends to underestimate free drug concentration for compounds with very high brain penetration. CSF overestimates ISF concentration for compounds with very low brain penetration [10, 34]. For compounds that are substrates of efflux transporters, such as Pgp substrates, the CSF drug concentrations tend to be higher than free drug concentration in the brain. This is because Pgp pumps from brain side to the blood side on the BBB, but pumps from the blood side to the CSF on the BCSFB (see Figure 11.2). For this reason, Pgp substrates can accumulate at the CSF and lead to overestimation of free drug concentration in the brain. On the other hand, for compounds that are substrates for uptake transporters, the free drug concentration will be higher in the brain than in the CSF. Despite the complexity of the brain physiology, CSF can be used a surrogate to predict free drug concentration in the brain when compounds are not substrates of transporters.

11.4 METHODS FOR BBB PERMEABILITY

If one is interested in the ability of a compound to cross the BBB, permeability methods should be used. Permeability is a rate of how fast a compound crosses the BBB and it is important for drugs requiring a rapid onset, such as anesthesia. BBB permeability can also affect the free drug concentration in the brain.

Low-permeability compounds have a low rate of crossing the BBB, and thus equilibrium is usually not fully established between the blood and the brain within common experimental conditions. In this case, the free drug concentration is usually higher in the blood than in the brain.

Highly permeable compounds have a high rate of crossing the BBB. Equilibrium is fully established in a short amount of time and the free drug concentration is the same on the blood and the brain side of the BBB. Substrates for uptake or efflux transporters have a proportionately higher or lower free drug concentration on the brain side, respectively, than if BBB permeability was only via passive diffusion.

BBB permeability is a pure measure of a compound's characteristics and is not confounded by other mechanisms, such as plasma protein or brain tissue binding and metabolism. Characteristics such as B/P, which in the past was commonly used in CNS research, result from multiple mechanisms. Since passive BBB permeability is a single variable parameter of a compound, medicinal chemists can develop structural–BBB permeability relationships and optimize BBB penetration potential.

11.4.1 *In situ* Brain Perfusion Assay

The *in situ* brain perfusion assay is the most common *in situ*/*in vivo* method for studying BBB permeability in whole animals [36, 37]. Wild-type and knockout mice can both be used in order to evaluate the roles of transporters in BBB transport. The assay is performed by infusing a test compound with a reference in a perfusate to the external carotid artery at 5–20 mL/min to keep the arterial pressure at 80–120 mm Hg. The perfusion time is very short, typically around 30 s and no more than 2 min [15, 38]. At the end of the experiment, animals are sacrificed and the brain is taken out quickly, homogenized, extracted with organic solvents, and analyzed with LC-MS. Because the perfusion is done rapidly, there is negligible metabolism or tissue binding. The assay is designed to measure permeability of compound across BBB without the impact of other mechanisms (plasma protein or tissue binding, metabolism, etc.). Some of the studies use high concentrations of test compounds (e.g., 50 μM), owing to the low detection limit caused by the short perfusion time. The high concentrations of test compounds often saturate transporter (e.g., Pgp) activities at the BBB and lead to higher BBB permeability. When transporters are saturated, this method essentially measures the passive diffusion component of BBB permeability. The *in situ* brain perfusion assay is not widely used in the pharmaceutical industry because it does not provide drug concentration in the brain in an actual dosing setting. BBB permeability is often determined using *in vitro* and *in silico* methods in the industry.

11.4.2 High-throughput PAMPA-BBB

Since most drugs enter the brain through passive diffusion, the passive permeability of the BBB is a very important component for drug design. PAMPA-BBB has been developed to evaluate the passive BBB permeability in high throughput [6]. PAMPA-BBB data has been demonstrated to have good correlation to *in situ* brain perfusion data and is a valuable tool for screening BBB permeability in early drug discovery [6, 39]. PAMPA-BBB uses an artificial membrane made of polar brain lipids in dodecane. The low fluidity nature of PAMPA-BBB is very similar to the lipid properties of brain endothelial cells. The assay is performed by dissolving test compounds in buffer, which is added to the donor wells. The acceptor has a porous filter bottom, which is coated with brain lipid in dodecane and filled with buffer. The acceptor plate is placed on the donor plate to form a "sandwich." Test compounds diffuse from the donor well through the lipid membrane and into the acceptor well. After overnight incubation, the compound concentrations in the donor and acceptor wells are measured and the BBB permeability is determined [6].

Many cell-based models, using cells of cerebral or noncerebral origin, have been developed to mimic the BBB and predict BBB permeability [40–43]. The major limitations of cell-based methods are as follows: (1) they typically have unstable transporter expression and do not give reliable prediction of transporter activity, usually over- or underexpressing transporters of the BBB; (2) certain cell-based models do not have tight-enough junctions, making them much leakier than the BBB; and (3) cells with noncerebral nature usually have very different lipid composition than the BBB, which

gives very different selectivity [39]. As a result, the prediction of BBB permeability using cell-based models is usually unsatisfactory. In addition, cell-based assays are typically much more expensive and laborious than PAMPA-BBB. They involve cell cultures in transwell filter plates. Cell-based assays are best used as models for transporters of the BBB, rather than as BBB permeability models.

11.4.3 Lipophilicity (LogD₇.₄)

Lipophilicity ($LogD_{7.4}$) is one of the dominant factors for BBB permeability [44]. LogD has been shown to have good correlations with *in situ* brain perfusion data [39]. Calculated LogD can be used early in the drug discovery process to guide structural modification in order to enhance BBB permeability. It is important to note that one should use $LogD_{7.4}$ rather than LogP since it is more physiologically relevant. LogP has been shown to have poor correlation with *in situ* brain perfusion data for BBB permeability [38].

11.5 METHODS FOR PGP EFFLUX TRANSPORT

If the question is whether a compound is a substrate for Pgp and how to "dial out" the efflux properties, a Pgp efflux transporter assay should be applied. Even though several efflux transporters have been identified on the BBB, such as, Pgp, breast cancer resistance protein (BCRP), multi-drug resistance protein 4 (MRP4), and multi-drug resistance protein 5 (MRP5) [45], Pgp by far is the most prevalent transporter preventing drugs from entering the brain. Most commercial CNS drugs do not have Pgp efflux activity [46, 47]. Therefore, screening of Pgp efflux is important for CNS drug therapy.

Many different approaches are available to assay for Pgp substrates [48]. The most commonly used methods are cell-based monolayer bidirectional transport assays, such as MDR1-MDCK, LLC-PK1, and Caco-2. Cells are grown on filter membranes to confluence. Pgp transporters are expressed on the apical membrane. The transport assay is performed bidirectionally, from apical to basal and from basal to apical directions. The permeability ratio of b-a over a-b is the efflux ratio. For certain cells that express multiple efflux transporters (e.g., Caco-2), a Pgp inhibitor is added to confirm if a compound is a Pgp substrate. Most organizations prefer using MDR1-MDCK than Caco-2 due to much shorter culture time (3 days vs. 21 days) and much stronger signal and sensitivity. Even though the exact expression level of Pgp in the cell lines has no relevance to BBB, they are great tools to diagnose Pgp issues and guide structural modification to overcome Pgp efflux and enhance brain penetration.

11.6 CONCLUSIONS

BBB penetration is still a rapidly growing field with new concepts and technologies continuously evolving. New tools are becoming available to help drug discovery scientists develop better strategies to enhance the success of CNS drugs. Commonly used BBB methods in the industry include determination of free drug concentration in the brain using brain PK study and brain tissue binding; BBB permeability with PAMPA-BBB or LogD; and determination Pgp efflux activity using MDR1-MDCK or Caco-2. New delivery technologies and methods for the BBB will continue to be developed to meet the challenges of CNS diseases.

REFERENCES

1. Pardridge WM (2001) Crossing the blood-brain barrier: Are we getting it right? *Drug Discov Today* 6:1–2.

2. Kola I, Landis J (2004) Opinion: Can the pharmaceutical industry reduce attrition rates? *Nat Rev Drug Discov* 3: 711–716.

3. Audus KL. In *Designing Drugs to Minimize or Maximize Exposure to the Brain.* Princeton, NJ: Residential School on Medicinal Chemistry, Drew University, 2002.

4. Pardridge WM (1995) Transport of small molecules through the blood-brain barrier: Biology and methodology. *Adv Drug Deliv Rev* 15:5–36.

5. Di L, Kerns EH, Carter GT (2008) Strategies to assess blood-brain barrier penetration. *Expert Opin Drug Discov* 3:677–687.

6. Di L, Kerns EH, Fan K, McConnell OJ, Carter GT (2003) High throughput artificial membrane permeability assay for blood-brain barrier. *Eur J Med Chem* 38:223–232.

7. Liu X, Chen C (2005) Strategies to optimize brain penetration in drug discovery. *Curr Opin Drug Discov Devel* 8:505–512.

8. Hammarlund-Udenaes M, Friden M, Syvanen S, Gupta A (2008) On the rate and extent of drug delivery to the brain. *Pharm Res* 25(8):1737–1750.

9. Read KD, Braggio S (2010) Assessing brain free fraction in early drug discovery. *Expert Opin Drug Metab Toxicol* 6:337–344.

10. Hammarlund-Udenaes M (2010) Active-site concentrations of chemicals—Are they a better predictor of effect than plasma/organ/tissue concentrations? *Basic Clin Pharmacol Toxicol* 106:215–220.

11. Smith DA, Di L, Kerns EH (2010) The effect of plasma protein binding on *in vivo* efficacy: Misconceptions in drug discovery. *Nat Rev Drug Discov* 9(12):929–939.

12. Kerns EH, Di L *Drug-like Properties: Concepts, Structure Design and Methods: From ADME to Toxicity Optimization.* London: Elsevier, 2008.

13. Gumbleton M, Audus KL (2001) Progress and limitations in the use of *in vitro* cell cultures to serve as a permeability screen for the blood-brain barrier. *J Pharm Sci* 90: 1681–1698.

14. Pardridge WM, Triguero D, Yang J, Cancilla PA (1990) Comparison of *in vitro* and *in vivo* models of drug transcytosis through the blood-brain barrier. *J Pharmacol Exp Ther* 253:884–891.

15. Hammarlund-Udenaes M, Bredberg U, Friden M (2009) Methodologies to assess brain drug delivery in lead optimization. *Curr Top Med Chem* 9:148–162.

16. Liu X, Vilenski O, Kwan J, Apparsundaram S, Weikert R (2009) Unbound brain concentration determines receptor occupancy: A correlation of drug concentration and brain serotonin and dopamine reuptake transporter occupancy for eighteen compounds in rats. *Drug Metab Dispos* 37: 1548–1556.

17. de Lange ECM, Danhof M, de Boer AG, Breimer DD (1997) Methodological considerations of intracerebral microdialysis in pharmacokinetic studies on drug transport across the blood-brain barrier. *Brain Res Rev* 25:27–49.

18. Hammarlund-Udenaes M, Paalzow LK, de Lange ECM (1997) Drug equilibration across the blood-brain barrier—Pharmacokinetic considerations based on the microdialysis method. *Pharm Res* 14:128–134.

19. Elmquist WF, Sawchuk RJ (1997) Application of microdialysis in pharmacokinetic studies. *Pharm Res* 14:267–288.

20. Hammarlund-Udenaes M (2000) The use of microdialysis in CNS drug delivery studies. Pharmacokinetic perspectives and results with analgesics and antiepileptics. *Adv Drug Deliv Rev* 45:283–294.

21. Watson CJ, Venton BJ, Kennedy RT (2006) *In vivo* measurements of neurotransmitters by microdialysis sampling. *Anal Chem* 78:1391–1399.

22. Mano Y, Higuchi S, Kamimura H (2002) Investigation of the high partition of YM992, a novel antidepressant, in rat brain—*In* vitro and *in vivo* evidence for the high binding in brain and the high permeability at the BBB. *Biopharm Drug Dispos* 23:351–360.

23. Kalvass JC, Maurer TS (2002) Influence of nonspecific brain and plasma binding on CNS exposure: Implications for rational drug discovery. *Biopharm Drug Dispos* 23:327–338.

24. Maurer TS, DeBartolo DB, Tess DA, Scott DO (2005) Relationship between exposure and nonspecific binding of thirty-three central nervous system drugs in mice. *Drug Metab Dispos* 33:175–181.

25. Liu X, et al. (2009) Unbound drug concentration in brain homogenate and cerebral spinal fluid at steady state as a surrogate for unbound concentration in brain interstitial fluid. *Drug Metab Dispos* 37:787–793.

26. Summerfield SG, et al. (2006) Improving the *in vitro* prediction of *in vivo* central nervous system penetration:

Integrating permeability, P-glycoprotein efflux, and free fractions in blood and brain. *J Pharmacol Exp Ther* 316: 1282–1290.

27. Wan H, Rehngren M, Giordanetto F, Bergstroem F, Tunek A (2007) High-throughput screening of drug-brain tissue binding and *in silico* prediction for assessment of central nervous system drug delivery. *J Med Chem* 50:4606–4615.

28. Becker S, Liu X (2006) Evaluation of the utility of brain slice methods to study brain penetration. *Drug Metab Dispos* 34:855–861.

29. Friden M, et al. (2009) Development of a high-throughput brain slice method for studying drug distribution in the central nervous system. *Drug Metab Dispos* 37:1226–1233.

30. Friden M, Gupta A, Antonsson M, Bredberg U, Hammarlund-Udenaes M (2007) *In vitro* methods for estimating unbound drug concentrations in the brain interstitial and intracellular fluids. *Drug Metab Dispos* 35:1711–1719.

31. Benet LZ, Hoener B-A (2002) Changes in plasma protein binding have little clinical relevance. *Clin Pharmacol Ther* 71:115–121.

32. Shen DD, Artru AA, Adkison KK (2004) Principles and applicability of CSF sampling for the assessment of CNS drug delivery and pharmacodynamics. *Adv Drug Deliv Rev* 56:1825–1857.

33. Lin JH (2008) CSF as a surrogate for assessing CNS exposure: An industrial perspective. *Curr Drug Metab* 9:46–59.

34. Friden M, et al. (2009) Structure-brain exposure relationships in rat and human using a novel data set of unbound drug concentrations in brain interstitial and cerebrospinal fluids. *J Med Chem* 52:6233–6243.

35. Liu X, et al. (2006) Evaluation of cerebrospinal fluid concentration and plasma free concentration as a surrogate measurement for brain free concentration. *Drug Metab Dispos* 34:1443–1447.

36. Takasato Y, Rapoport SI, Smith QR (1984) An *in situ* brain perfusion technique to study cerebrovascular transport in the rat. *Am J Physiol* 247:H484–H493.

37. Smith QR, David AD (2003) *In situ* brain perfusion technique. *Methods Mol Med* 89:209–218.

38. Summerfield SG, et al. (2007) Central nervous system drug disposition: The relationship between *in situ* brain permeability and brain free fraction. *J Pharmacol Exp Ther* 322:205–213.

39. Di L, Kerns EH, Bezar IF, Petusky SL, Huang Y (2009) Comparison of blood-brain barrier permeability assays: *In situ* brain perfusion, MDR1-MDCKII and PAMPA-BBB. *J Pharm Sci* 98:1980–1991.

40. Garberg P, et al. (2005) *In vitro* models for the blood-brain barrier. *Toxicol In Vitro* 19:299–334.

41. Cecchelli R, et al. (2007) Modelling of the blood-brain barrier in drug discovery and development. *Nat Rev Drug Discov* 6:650–661.

42. Hansen DK, Scott DO, Otis KW, Lunte SM (2002) Comparison of *in vitro* BBMEC permeability and *in vivo* CNS uptake by microdialysis sampling. *J Pharm Biomed Anal* 27:945–958.

43. Bachmeier CJ, Trickler WJ, Miller DW (2006) Comparison of drug efflux transport kinetics in various blood-brain barrier models. *Drug Metab Dispos* 34:998–1003.

44. Liu X, Tu M, Kelly RS, Chen C, Smith BJ (2004) Development of a computational approach to predict blood-brain barrier permeability. *Drug Metab Dispos* 32:132–139.

45. Giacomini KM, et al. (2010) Membrane transporters in drug development. *Nat Rev Drug Discov* 9:215–236.

46. Doran A, et al. (2005) The impact of P-glycoprotein on the disposition of drugs targeted for indications of the central nervous system: Evaluation using the MDR1A/1B knockout mouse model. *Drug Metab Dispos* 33:165–174.

47. Liu X, Chen C, Smith BJ (2008) Progress in brain penetration evaluation in drug discovery and development. *J Pharmacol Exp Ther* 325(2):349–356.

48. Polli JW, et al. (2001) Rational use of *in vitro* P-glycoprotein assays in drug discovery. *J Pharmacol Exp Ther* 299: 620–628.

12

TECHNIQUES FOR DETERMINING PROTEIN BINDING IN DRUG DISCOVERY AND DEVELOPMENT

Tom Lloyd

12.1 INTRODUCTION

Plasma protein binding plays an important role in the whole-body disposition of drugs. Pharmacokinetic (PK) properties such as hepatic metabolism rate, renal excretion rate, membrane transport rate, and distribution volume are functions of the ratio of free fraction. As such, drug protein binding data can be useful in allometric scaling of PK parameters from animal studies to human subjects. From a pharmacodynamic viewpoint, knowing the ratio of free fraction of drugs in plasma is important because unbound drugs can easily reach the target organ, whereas bound drugs pass the blood capillary wall with great difficulty. This principle can often be extended to other compartments that are in equilibrium with plasma as long as there is no active transport. Therefore, precise information on the free drug fraction is essential for drug development and for determining the safety of a drug in clinical trials [1–12].

Intersubject variations in plasma protein concentrations have been documented [13–16]. However, the importance of these differences is minimized by the predominance of two protein components in plasma. Albumin and α_1-acid glycoprotein (AGP) comprise 60% of the total protein in plasma and account for the majority of drug binding [17]. Still, cases where specific drugs have exhibited varied inter- and intraindividual subject binding characteristics include instances of high-affinity binding site saturation, disease-induced variations, genetically determined modifications of proteins,

diurnal variation in concentration of transport proteins, metabolite protein binding effects, and even the presence of exogenous contaminants from plasticizers and smoking. There may also be stereoselective differences as a consequence of chiral discriminative properties of the binding sites of human serum albumin and AGP. These can account for significant differences between species [18].

Most drugs bind to proteins in a reversible manner by means of weak chemical bonds such as ionic, van der Waals, hydrogen, and hydrophobic bonds with the hydroxyl, carboxyl, or other reversible sites available in the amino acids that constitute the protein chain [19]. Albumin makes up roughly half of the total plasma proteins (normal human albumin range is 34–54 mg/mL) with a molecular weight ranging from 65,000 to 69,000 Da. Acidic drugs are known to bind tightly to human serum albumin. AGP has a concentration in plasma of between 0.4 and 1 mg/mL and a molecular weight of roughly 40,000 Da. AGP primarily binds to basic and neutral drugs [20].

Literature protein binding values for selected drugs in human plasma are presented in Table 12.1 [21]. These are typically calculated as

$$PPB\% = (C_{total} - C_{free}/C_{total}) \times 100,$$

where PPB% is the percentage of binding to plasma proteins, C_{free} is the concentration of the free drug (measured at equilibrium in buffer correcting for any

ADME-Enabling Technologies in Drug Design and Development, First Edition. Edited by Donglu Zhang and Sekhar Surapaneni.

TABLE 12.1. Human Protein Binding Measurements for Selected Drugs

Drug	Bound in Plasma (%)
Ibuprofen	>99
Warfarin	99
Verapamil	90 ± 2
Propranolol	93.3 ± 1.2
Digoxin	25 ± 5
Caffeine	36 ± 7
Furosemide	98.8 ± 0.2

Selected from *Goodman and Gilman's The Pharmacological Basis of Therapeutics* [21].

dilutions or protein-free plasma), and C_{total} is the initial total concentration of drug in plasma. Conversely, the free fraction ($f_{unbound}$) can be expressed as

$$f_{unbound} = 1 - (PPB\%/100).$$

The fraction unbound often relates directly to the pharmacological activity of the drug. It is determined by the drug's affinity for the protein, the concentration of the binding protein, and the concentration of the drug relative to the binding protein. Since the binding of drugs to plasma proteins is rather nonselective, many drugs with similar physicochemical characteristics compete with each other and with endogenous substances for these binding sites. This can lead to a type of drug interaction based on one drug displacing another, particularly significant for highly bound drugs. Examples include tolbutamide and warfarin, where the free concentration can be significantly elevated in the presence of nonsteroidal anti-inflammatory agents. Warfarin activity can also be elevated leading to an increased risk of bleeding in the presence of sulfonamides such as furosemide. Risk of adverse effect is greatest if the displaced drug has a limited volume of distribution, if the competition extends to the drug bound in tissues, if elimination of the drug is also reduced, or if the displacing drug is administered in high dosage by rapid intravenous injection [21].

12.2 OVERVIEW

There are many different techniques that have been used to assess the free fraction of drugs in the presence of proteins. The most common are equilibrium dialysis, ultrafiltration, and ultracentrifugation [18, 22, 23]. Such *in vitro* techniques can then be verified *in vivo* by microdialysis to answer specific questions. Other chromatography and capillary electrophoresis techniques have

their place as tools for screening or rapid rank ordering of compounds in terms of their protein binding. They can also be useful for understanding binding interactions with specific components, but are generally not used to represent the overall protein environment. Also, due to challenges in sensitivity, they do not always allow for measurements at the therapeutically relevant concentration. Spectroscopic tools can also play an important role addressing specific binding issues.

Other parameters to control that have been shown to alter the *in vitro* fractional binding of a ligand to plasma components regardless of the technique being applied include assay temperature, drug concentration, ligand stability, buffer source (pH, concentration, and composition), plasma pH, and freshness of the plasma source [24–26]. Whenever possible, incubation with 10% CO_2 should be used for the assay of plasma protein binding (especially for basic drugs that may become more ionized with a rising pH) since it produces pH values close to physiological conditions (pH 7.35–7.42) while leaving plasma composition unchanged [27, 28]. Alternatively, a 10-fold dilution of plasma with isotonic buffer could be used, as long as the concentration of drug remains at least 10 times lower than that of the proteins.

Pooling sample matrix to reduce the number of calibration curves and protect against analyte adsorption during processing is commonly applied. A free fraction sample is added to a well already loaded with an equivalent volume of control plasma. Similarly, plasma sample aliquots are diluted with an equivalent volume of buffer. The standards also are prepared in a 50% plasma matrix so that all the generated unknown samples match up with a single mixed matrix curve [29–31]. The same approach has been applied in determinations of drug protein binding in microsomes [29, 32] and homogenized brain tissue [29].

An extension of this approach has been applied in cases where substantial compound loss has been incurred due to nonspecific binding during ultrafiltration and the volume of filtrate is not known up front. In this case, the collection tubes are weighed twice to measure the volume of filtrate. A control plasma sample is prepared in parallel. Then the sample and control retentates are crossed over with the respective filtrates. The result is a sample representing the drug in the filtrate and another representing the drug in the retentate. By comparing the total drug recovery of the filtrate and retentate samples with the starting plasma concentration, one determines the degree of nonspecific binding. The volume of filtrate produced is calculated from the weights of the collection vials measured both pre- and postfiltrate generation assuming a density of 1.0. The concentration of the drug in the filtrate is then determined by correcting for dilution of the filtrate with control retentate back to the total sample volume applied initially [33].

As the application area moves upstream into the drug discovery stage, pooling analytes in protein binding experiments has been shown to increase throughput [31]. One example involved ultrafiltration in a 96-well format, pooling four different compounds and assessing their binding at 10 μM in the presence of each other. Because of the much greater protein concentrations relative to analyte, this technique was validated showing strong agreement with individual analyte controls as well as the literature values for these analytes [34]. Similarly, pooling of as many as 25 compounds in homogenized brain tissue drug protein binding studies has been reported [35].

Analysis can be accomplished by liquid scintillation counter for radiolabeled compounds or by light chromatography-tandem mass spectrometry (liquid chromatography [LC]-MS/MS). In the case of radiolabeled compounds with either tritium or carbon isotope labeling, one needs to ensure the purity and stability of the label. This is especially important with tritium, which while easier to synthesize, may not be as stable to exchange. Liquid chromatography (LC) analysis with radiochemical detection may be required to assess compound stability for radiolabeled analytes. Although requiring additional extraction and separation steps, LC-MS/MS affords more selective detection.

The most common techniques, equilibrium dialysis, ultrafiltration, and ultracentrifugation have all been reformatted in recent years to reduce matrix volumes consumed and facilitate the use of automation. Many, if not all, of the process steps can incorporate liquid handler technology and even robotic handling to increase throughput and reliability. Such steps may include diluting stocks, spiking analytes into the matrix, loading matrix and possibly buffer into the desired format, sampling free fraction and matrix, diluting or pooling samples after collection, and finally, microplate extraction processing in the case of LC-MS/MS analysis. With increased capacity to execute such studies comes the ability to assess binding at different concentrations, across different species, and in different matrices (e.g., microsomes and different tissue homogenates).

For ultrafiltration and equilibrium dialysis devices, one must monitor for the possibility of protein leakage, which would compromise the validity of the experiments. In the case of ultracentrifugation, one must similarly monitor the protein content of the free fraction sampled after centrifugation.

The Bradford assay offers an easily applied, spectrophotometric method for the determination of unknown protein concentration. The Bradford protein assay is based on the observation that the absorbance maximum for an acidic solution of Coomassie Brilliant Blue shifts from 465 nm to 595 nm when binding occurs [36]. This assay is sensitive in measuring protein concentration from 0.2 to 1.5 mg/mL (linear range of the assay). An immunoglobulin G (IgG) protein standard solution (e.g., bovine gamma globulin, lyophilized IgG, Bio-Rad, Hercules, CA) is used to generate a standard curve for protein determination. When mixed with the protein assay dye reagent, absorbance measurements can be made using a spectrophotometer at 595 nm. Unknown protein concentration is calculated by plotting a standard curve , calculating the best fit for that curve using a least squares linear regression and reading the protein content from the standard curve. Such a technique can be used to assess the actual protein concentration of ostensibly protein free fractions from various techniques.

12.3   EQUILIBRIUM DIALYSIS

Equilibrium dialysis is the most frequently used method and is often considered the gold standard for determining drug protein binding. In such an experiment, two compartments are separated by a semipermeable membrane typically with an 8000–12,000-Da molecular weight cutoff. The membrane allows the analyte being studied to pass but retains the protein content on one side (Figure 12.1). A time course experiment predetermines the amount of time required to reach equilibrium in the system. This can vary by compound and is typically carried out at 37°C. Reported equilibration times range from 2 to 24 h depending on the setup. The volume-to-surface area ratio on the dialysis membrane governs the time required for the analyte to reach equilibrium between the bound fraction, protein sample side, and the free fraction, dialysate buffer side. Vertical alignment of the dialysis membrane reduces the potential for trapped air pockets. Such air pockets would reduce the surface-to-volume ratio and slow the time to reach equilibrium. The capability to shake the apparatus can speed the time to reach equilibrium. The individual wells or chambers are covered to prevent evaporation and pH change. Measurements of the compound stability in matrix at 37°C for the duration of the incubation typically accompany such experiments. High-throughput formats with 24-, 48- and 96-well designs accessible by automated liquid handlers have been introduced in recent years to increase the capacity to perform such determinations [17, 29, 30, 37].

Most devices for equilibrium dialysis are constructed out of Teflon™ (DuPont, Wilmington, DE, USA). Although this offers an improvement over plastics used in disposable ultrafiltration devices with respect to nonspecific binding, it does not eliminate the concern.

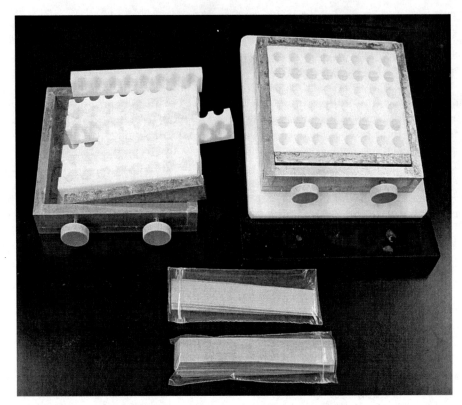

FIGURE 12.1. Wyeth (formerly Collegeville, PA, USA) custom equilibrium dialysis assembly with 500-μL chambers.

Nonspecific binding can also take place on the membrane itself. There is particular concern over nonspecific binding for lipophilic compounds. Measuring concentrations from both sides of the membrane and utilizing the ratio will further reduce the effect of nonspecific binding. The magnitude of the nonspecific binding can be inferred as a part of total recovery measurements comparing matrix stability concentration with the sum of the recoveries from either side of the membrane after dialysis. Nonspecific binding can also be evaluated by spiking the compound into dialysate buffer alone, adding it to both sides of the dialysis apparatus and measuring the concentration recovered.

Frequently, although not in all cases [30], the membranes require a series of soaking pretreatments to remove preservatives and prepare it for use. An example procedure would be soaking the membrane in water for 20 min, then in 30% ethanol for 15 min, rinsing with water three times and then soaking in isotonic sodium phosphate buffer until the time of use [37]. Once wetted, the membrane must not be allowed to dry out.

Such devices have the potential to incur volume shifts from the dialysate buffer to the plasma side due to differences in osmotic pressure. This shifting results in plasma protein dilution and has been shown to reduce the *in vitro* fractional binding of ligand to plasma proteins [37–41]. Extra osmotic pressure may need to be corrected for in a case where a Donnan effect is observed. In equilibrium dialysis, with a nondiffusible, charged protein on one side of the membrane, Donnan equilibrium leads to a pH difference across the membrane. With an ionizable drug, the concentration dissolved in water will be different on the two sides of the membrane to an extent dependent on the pH difference and the pK_a of the drug. This must be allowed for in calculating the concentration bound to protein [42]. Poor aqueous solubility of an analyte can also pose problems for this technique.

12.4 ULTRACENTRIFUGATION

A simpler approach involves ultracentrifugation where the plasma sample is separated into three distinct layers by a centrifugal field at very high speeds (typically 40,000–100,000 rpm generating several hundred thousand *g*). The centrifuge is operated under a vacuum to reduce friction spinning at such high speeds. This also allows for maintaining a constant temperature. At the very surface after ultracentrifugation, one finds chylomicrons and very low-density lipoproteins [43]. Just beneath this surface layer is the protein-free layer with the protein-containing phase found at the bottom. Any concerns over nonspecific binding are reduced as there is no membrane-separated compartment. The assumptions of this technique are that there is no protein in the zone being sampled (which can be measured as well as qualitatively characterized) and that there is no alteration of the binding equilibrium during ultracentrifugation (this has been typically demonstrated indirectly by comparing values with other techniques). There have been a few reports of a protein-contaminated fraction that was used for the determination of free drug concentration so this fraction should be tested [44, 45]. There are reports in the literature of physical phenomena such as sedimentation, back diffusion, and viscosity leading to differences between free drug concentrations determined by ultracentrifugation compared with equilibrium dialysis [42, 46]. In one report, the error due to sedimentation of the drug can be as large as 10% for drugs at 300 Da and up to 40% for high-mass drugs such as suramin (1297 Da), so comparisons across techniques would be indicated [47].

In addition to the need to have such a specialized centrifuge, disadvantages of this technique included lengthy centrifugation times of 4–6 h, consumption of larger volumes of sample (e.g., 2 mL), and smaller capacity rotors (e.g., 20–40 tubes) [48]. Recent developments have scaled down these requirements with faster speeds requiring 1.5–3.5 h of centrifugation, smaller sample volumes (e.g., 200 μL), and enlarged rotor capacity up to 72 tubes (e.g., available from Beckman Coulter Inc., Fullerton, CA) [49, 50]. Even in these scaled-down examples, studies to assess the protein content found in the free fraction samples suggested a negligible effect for compounds ≤99% protein bound. Samples must be gathered carefully and in a timely manner as jostling of the tubes or diffusion over time will undo the protein concentration gradient. This step is well suited for a robotic liquid handler that can quickly and reproducibly sample at exactly the same depth each time. This is particularly true in the scaled-down version of this technique where a sample of 20–35 μL from a reinforced polycarbonate wall ultracentrifugation tube containing a total of 200 μL is required.

12.5 ULTRAFILTRATION

Ultrafiltration also involves the use of a molecular weight cutoff membrane. Protein samples are typically incubated with shaking at 37°C for 30–60 min to allow for equilibration of the drug protein binding. Matrix stability samples are usually collected before and after this incubation. The equilibrated protein sample is then applied to a filtration device, and pressure (typically centrifugal force or positive pressure) is applied to force the solution through the membrane.

One parameter to pay attention to in comparing various devices and formats for ultrafiltration is the molecular weight cutoff of the filter being used and the amount of protein it allows to pass. One finds cutoff ranges in the literature from 10,000 to 30,000 Da being employed. In cases where one is measuring very highly bound compounds, obviously using a membrane that allows 1% or perhaps even 5% protein breakthrough is going to bias measurements of free fraction. Conversely, using a finer filtration membrane will reduce the filtrate yield in terms of the volume as well as require a longer centrifugation step. The volume of filtrate harvested will also vary according to the g-force being applied across an array of tubes or microplate well positions as the radius varies.

As an example, Amicon centrifree micropartition devices with a filter membrane of 30,000 molecular weight cutoff from Millipore (Billerica, MA) were used in one instance. To achieve equilibrium between the drug and plasma proteins, the spiked protein samples were incubated at 4°C for 20 min prior to ultrafiltration. Samples of 1 mL volume were transferred to Amicon centrifree micropartition devices and centrifuged at 3000g (4°C) for 15 min. Approximately 200 μL of the ultrafiltrates were then collected [4].

The primary disadvantage of the ultrafiltration method is nonspecific binding of the drug to the filtration apparatus. Another is concentration of the plasma proteins during centrifugation.

Online, automated versions of these techniques include a system for continuous ultrafiltration, affording calculation of the protein binding over a wide range of different drug/protein ratios in one single experiment [51]. Examples of protein binding determinations by online equilibrium dialysis have also been described [52, 53]. Solid-phase microextraction, which uses an extraction phase that dissolves or adsorbs the drug of interest and rejects proteins, is another technique recently reported that measures the protein binding at equilibrium. Advantages of this approach include requiring a relatively small sample size, short analysis time, and the ability to directly study complex samples such as whole blood [27].

All three of these most commonly used in vitro techniques yield protein binding values that closely agree for the majority of drugs assessed with each methodology having associated shortcomings as noted. Of course, the best way to corroborate these determinations is to measure in vivo directly within the system being studied. The only tool currently available that can intercede and explicitly provide information on the extracellular free fraction is microdialysis. However, this too would require some application specific validation as there are numerous cases where the absolute concentration values differ between *in vitro* and *in vivo* free fraction determinations. These can be due to microdialysis experimental challenges that may vary according to drug class and tissue region [54].

12.6 MICRODIALYSIS

Microdialysis involves the implantation of a small probe into a specific region of a tissue- or fluid-filled space [55]. A variety of probe designs have been used, including linear, U-shaped, or concentric geometries. Semipermeable membrane materials used in probe construction range from low to high molecular weight cutoff. During microdialysis, a physiologically compatible perfusion fluid is delivered through the probe at a low and constant flow rate (typically between 0.1 and 5 µL/min). This sampling technique is volume neutral in most applications so there are no volume limitations even when working with small animals. In static mode, sample volume is limited to 0.1–0.2 µL/min. Experimental conditions that affect probe recovery include the perfusion rate, temperature, probe membrane composition and surface area, nature of the dialyzed tissue, physicochemical properties of the analyte, and other factors that influence molecular diffusion characteristics [56].

As microdialysis probes are usually perfused with aqueous solutions, the technique is conceptually limited to the study of water-soluble drugs. In order to enable the measurement of lipophilic compounds routinely, *in vitro* experiments have demonstrated the usefulness of lipid emulsion as perfusate instead of aqueous solution. Microdialysate samples consist primarily of relatively small hydrophilic analytes in highly ionic aqueous samples. Typical sample volume is a few microliters or less with analyte concentrations in the picomolar or nanomolar range. However, microdialysis samples are generally protein free requiring no sample preparation prior to high-performance liquid chromatography (HPLC) analysis. Since microdialysis is labor intensive and requires specialized skills, it is not suitable for high-throughput screening of large numbers of compounds. Rather it is used to address specific questions and confirm *in vitro* screening models [54].

12.7 SPECTROSCOPY

Other approaches are available to study drug–protein interactions in greater detail such as various spectroscopic techniques including ultraviolet (UV)-visible, fluorescence, infrared, nuclear magnetic resonance,

optical rotary dispersion, and circular dichroism. Such measurements can be performed in solution and monitor changes of the electronic and spectroscopic energy levels caused by the binding event with the ligand or protein. Such methods provide a better understanding of the binding mechanism and insight into three-dimensional protein structure [57]. They do not characterize binding with multiple components however, and such methods often lack sufficient sensitivity to allow study in the therapeutic range.

Surface plasmon resonance has allowed the interaction of drugs with albumin and AAG to be characterized both thermodynamically and kinetically [58]. Such a system monitors changes in the refractive index that occur as molecule complexes form or break during the binding reaction at the sensor surface anchoring one of the interaction partners [59]. In this case, a solution of the drug would be passed over the surface protein. However, the immobilization linkage to the surface may alter the binding activity of some proteins. Still, especially for protein–protein interactions (which cannot be measured by conventional methods such as equilibrium dialysis, ultracentrifugation, and ultrafiltration that rely on gross differences in molecular size), surface plasmon resonance, which senses changes proportional to the mass of the bound material on the sensor chip, may be a sound alternative [57].

For early discovery screening purposes seeking to rank-order compounds on the basis of drug protein binding, other *in vitro* and chromatographic tools have been applied. One such *in vitro* technique is the parallel artificial membrane permeability assay (PAMPA) recently adapted to measure serum protein binding constants [60]. Either a hexadecane or 1-octanol membrane is used to separate two compartments allowing free drug to pass while measuring rates with and without human serum albumin in the donor chamber. The assay is in a 96-well format, requires no equilibration period, self-corrects for nonspecific adsorption, and has no potential for volume shift as the chemical membranes are not water permeable. Another approach is the erythrocyte partition method, which consists of measuring the partitioning of the drug between plasma and blood cells and the partitioning between buffer and blood cells. The ratio of the two partition coefficients yields the free fraction [7]. A recent *in vitro* microplate format involves human serum albumin immobilized on Transil™ beads (Nimbus Biotechnology, Leipzig, Germany) [61]. The process involves a 2-min incubation in the presence of the beads followed by a centrifugation or filtration step to separate out the beads. The resulting supernatant can then be read by a plate reader. Alternatively, one can instead purchase attached egg yolk phosphatidylcholine to these silica beads, which will serve as a lipid membrane. The solid-supported lipid membrane can then be used as a substitute for the blood cells in the aforementioned erythrocyte partition method [62].

12.8 CHROMATOGRAPHIC METHODS

In a similar approach, affinity chromatography uses immobilized biopolymer (enzymes, receptors, ion channels, or antibodies) protein stationary phases. The stability and constant binding behavior of such columns provide a tool for studying interactions between small ligands and biomacromolecules. However, nonphysiological experimental conditions (pH adjustment, presence of organic modifiers) may alter the conformation and natural binding behavior of the attached protein (similar to the attachment in surface plasmon resonance). This approach may provide information on the relative affinity of ligand binding and on the area(s) where the interaction takes place [18]. Again most of these efforts have involved human serum albumin [63, 64], with limited success studying binding to AGP [65] and at least one instance of evaluating binding to high-density lipoproteins [66].

High-performance size exclusion chromatography is based on packing that has sorbent only inside the pores. Equilibrated analytes in plasma are injected onto the column. Proteins are too large to enter these pores and interact with the sorbent, thereby passing unretained through the column. Free drug, however, is able to diffuse into the pores and interact with the sorbent resulting in later elution. Zonal elution involves a small-plug injection where the retention time is used to obtain the association constants. In frontal analysis, a large-plug injection is made and quantification is based on plateau height. Drug and protein are mixed together and injected into either the mixture of drug and protein (direct separation method), into one of the interaction partners while the other species is dissolved in the eluent (Hummel–Dreyer method) [67], or into buffer while both components are dissolved in the mobile phase (vacancy peak method) [68].

Capillary electrophoresis is also used to study drug protein binding, providing the possibility to evaluate interactions in free solution [69]. Capillary electrophoretic methods have been applied successfully for both 1:1 and 1:*n* combination ratios between receptor and ligand, but they are restricted to a single protein at a time and do not allow precise control of the temperature [27]. This technique offers fast separations and analysis while consuming small amounts of sample and reagent. However, there is the potential for protein adsorption on the capillary walls and the detection

limits often do not allow for measurements at therapeutic levels. Typically, these applications are performed with UV detection, which allows for operating at physiological conditions (buffer pH and ionic strength). One can turn to mass spectrometry (MS) detection for greater sensitivity; however, this greatly restricts the buffers that can be used [57].

12.9 SUMMARY DISCUSSION

Although such abbreviated *in vitro* methods and chromatography techniques can be applied in early screening for the purposes of ranking the binding affinity and quantitative structure–activity relationship modeling, when the protein binding data is intended for *in vivo* scaling, a single compound measurement using equilibrium dialysis is strongly recommended [12]. This is particularly true for highly bound compounds where small absolute differences in percent unbound become amplified. Although the main drug binding proteins are albumin and AGP, plasma contains many other proteins. There is a high probability that many small molecules will exhibit some levels of binding. Therefore, to determine the extent of plasma protein binding, the molecule should be tested directly in a protein binding assay using plasma or serum [27].

Having compared over the past decade eight different formats and variations of the three most widely accepted techniques (equilibrium dialysis, ultrafiltration, and ultracentrifugation) for determining drug protein binding, despite all the caveats in their application noted previously, we found considerable agreement for most compounds studied. At therapeutic concentrations in plasma and microsomes, a combination of equilibrium dialysis and ultracentrifugation techniques matched with liquid scintillation counting or LC-MS/MS detection offered a powerful combination of tools. For the definitive determinations to be used in early drug development for allometric PK scaling, custom equilibrium dialysis devices similar to those described by Banker et al. [37], but having a larger 500 μL volume on either side of the membrane (Figure 12.1) offered sufficient working volume to be able to accurately measure free fraction even for potent, high binding compounds. These devices were reusable and proved very reliable and cost-effective. A complementary technique was ultracentrifugation particularly useful for the outlying instances where nonspecific binding rendered equilibrium dialysis data questionable.

As demands for *in vitro* data on drug protein binding in plasma and microsomes expanded upstream to the compound optimization stage, a scaled-down version of the ultracentrifugation format (Figure 12.2) [29, 49,

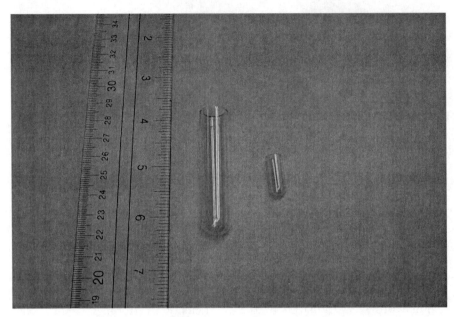

FIGURE 12.2. Miniaturization of the ultracentrifugation format.

FIGURE 12.3. Automated liquid handling for the Thermo Scientific Pierce RED Device for Rapid Equilibrium Dialysis (Rockford, IL, USA).

50] met the data quality and throughput requirements feeding into the human PK modeling and candidate selection process.

At an even earlier drug discovery stage supporting central nervous system therapeutic area research efforts, the equilibrium dialysis device shown in Figure 12.1 was applied to assess free fractions in plasma and homogenized brain tissue collected from rodent PK studies. In some instances, a commercial version, deemed not sufficiently reliable for definitive drug development stage experiments due to occasional membrane leakage observations (Figure 12.3), was also applied. These values, combined with the drug concentrations determined in plasma and brain, provided a much more complete comparison of the drug candidates being considered. As the free drug principle suggests, only the collision of free drug with the target is likely to contribute to binding [11]. Traditional brain-to-plasma concentrations alone can be misleading, especially in cases where the compounds are P-glycoprotein efflux substrates (can be assessed based on Caco-2 cell permeability asymmetry) [70]. Drug delivery to the brain can be

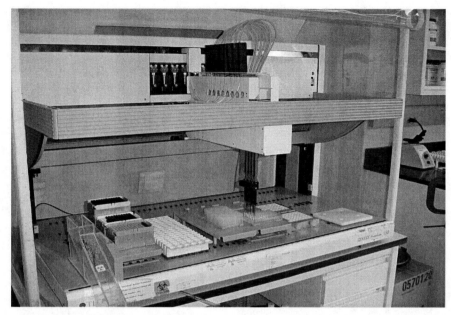

FIGURE 12.4. Automated liquid handling for ultracentrifugation or equilibrium dialysis with plasma, microsomes, and homogenized brain tissue using the Tecan Freedom (Tecan, Durham, NC, USA).

comprehensively described by the concentration ratio of unbound drug in brain to blood, permeability clearance into the brain, and intrabrain distribution [71].

All of these formats can be run with fairly high throughput utilizing automation to homogenize tissue samples in parallel with an isotonic solution (e.g., Autogizer™, Tomtec®, Hamden, CT) and to automate liquid handling steps (Figures 12.3 and 12.4). The effect of diluting tissue concentration as part of the homogenization and dialysis process was studied and found to be consistent up to at least a 20-fold tissue dilution [35].

There are allometric PK scaling software tools such as Simcyp™ (Simcyp Ltd., Sheffield, UK) that apply not only plasma protein binding but also values for microsomal and hepatocyte free fraction into the models predicting human PK and pharmacodynamics from animal and *in vitro* data. These predictions include potential drug–drug interactions. Microsomal binding (specific, nonspecific, or a combination of both) may have a major effect on estimation of inhibitory potency of P450 inhibitors [32].

The extent to which a compound is bound to plasma proteins is a critical component to predicting how a potential drug will interact with its intended target *in vivo* and with the clearance mechanisms of the organism [6]. The literature suggests that plasma protein binding controls the free drug concentrations in plasma and in those compartments in steady-state equilibrium with plasma. As such, it is a key factor in the translation of *in vitro* biochemical activity into *in vivo* pharmacological activity [11].

ACKNOWLEDGMENT

This chapter is dedicated to the memory of Joe McDevitt, who was the lead in this application area at Wyeth over the past decade and a guiding force.

REFERENCES

1. Kwong TC (1985) Free drug measurements: Methodology and clinical significance. *Clin Chim Acta* 151:193.

2. Levy RH, Moreland TA (1984) Rational for monitoring free drug level. *Clin Pharmacokinet* 9(Suppl 1):1.

3. Svensson CK, Woodruff MN, Baxter JG, Lalka D (1986) Free drug concentration monitoring in clinical practice: Rationale and current status. *Clin Pharmacokinet* 11:450.

4. Tang Y, Zhu H, Zhang Y, Huang C (2006) Determination of human plasma protein binding of baicalin by ultrafiltration and high-performance liquid chromatography. *Biomed Chromatogr* 20:1116–1119.

5. Greenblatt DJ, Sellers EM, Koch-Weser J (1982) Importance of protein binding for the interpretation of serum and plasma drug concentrations. *J Clin Pharmacol* 22:259–263.

6. Herve F, Urien S, Albengres E, Duche J (1994) Drug binding in plasma. *Clin Pharmacokinet* 25(1):44–58.

7. Schuhmacher J, Buhner K, Witt-Laido A (2000) Determination of the free fraction and relative free fraction of drugs strongly bound to plasma proteins. *J Pharm Sci* 89:1008–1021.

8. Swift CG, Ewin JM, Clark P, Stevenson IH (1985) Responsiveness to oral diazepam in the elderly: Relationship to

total and free plasma concentrations. *Br J Clin Pharmacol* 20(2):111–118.

9. Wilkinson GR (1983) Plasma and tissue binding considerations in drug disposition. *Drug Metab Rev* 14:427–465.

10. Tozer TN (1981) Concepts basic to pharmacokinetics. *Pharmacol Ther* 12:109–131.

11. Trainor GL (2007) The importance of plasma protein binding in drug discovery. *Expert Opin Drug Discov* 2(1): 51–64.

12. Wan H, Holmen AG (2007) High throughput screening of physiochemical properties and *in vitro* ADME profiling in drug discovery. *Comb Chem High Throughput Screen* 50:4606–4615.

13. Jusko WJ, Gretch M (1976) Plasma and tissue protein binding of drugs in pharmacokinetics. *Drug Metab Rev* 5(1):42–139.

14. Wallace WM, Verbeek RK (1987) Plasma protein binding of drugs in the elderly. *Clin Pharmacokinet* 12:40–72.

15. Zini R, Riant P, Barre J, Tillement JP (1990) Disease-induced variations in plasma protein levels. *Clin Pharmacokinet* 19(2):147–159.

16. Routledge PA, Barchowsky A, Bjornsson A, Kitchell BB, Shand DG (1980) Lidocaine plasma protein binding. *Clin Pharmacol Ther* 26(3):336–347.

17. Kariv I, Cao H, Oldenburg KR (2001) Development of a high throughput equilibrium dialysis method. *J Pharm Sci* 90:580–587.

18. Oravcova J, Bohs B, Lindner W (1996) Drug-protein binding studies new trends in analytical and experimental methodology. *J Chromatogr B* 677:1–28.

19. Singh SS, Mehta J (2006) Measurement of drug-protein binding by immobilized human serum albumin-HPLC and comparison with ultrafiltration. *J Chromatogr B* 834:108–116.

20. Naranjo CA, Sellers EM *Drug-Protein Binding.* New York, NY: Praeger, 1986:233–251.

21. Gilman AG, Goodman LS, Rall TW, Murad F, eds. *Goodman and Gilman's The Pharmacological Basis of Therapeutics*, 7th ed., New York, NY: MacMillan, 1985:12, 54, 889, 1664–1735.

22. Pacifici GM, Viani A (1992) Methods of determining plasma and tissue binding of drugs. *Clin Pharmacokinet* 23:449–468.

23. Sebille B (1990) Methods of drug protein binding determinations. *Fundam Clin Pharmacol* 4(Suppl 2):151–161.

24. Kristensen CB, Gram LF (1982) Equilibrium dialysis for determination of protein binding of imipramine-evaluation of a method. *Acta Pharmacol Toxicol (Copenh)* 50:129–135.

25. Dayton PG, Stiller RL, Cook DR, Perel JM (1983) The binding of ketamine to plasma proteins: Emphasis on human plasma. *Eur J Clin Pharmacol* 24:825–830.

26. Ponganis KV, Stanski DR (1985) Factors affecting the measurement of lidocaine protein binding by equilibrium dialysis in human serum. *J Pharm Sci* 74(1):57–60.

27. Musteata FM, Pawliszyn J, Qian MG, Wu JT, Miwa GT (2006) Determination of drug plasma protein binding by solid phase microextraction. *J Pharm Sci* 95(8):1712–1722.

28. Fura A, Harper TW, Zhang H, Fung L, Shyu WC (2003) Shift in pH of biological fluids during storage and processing: Effect on bioanalysis. *J Pharm Biomed Anal* 32: 513–522.

29. Lloyd TL, Ghupur A, Chun A, Sears S, Liu Z (2009) Alternative Automated Methods for Free Fraction Determination in Tissue and Plasma. Applied Pharmaceutical Analysis Conference on September 16, 2009, in Boston, MA.

30. Waters NJ, Jones R, Williams G, Sohal B (2007) Validation of a rapid equilibrium dialysis approach for the measurement of plasma protein binding. *J Pharm Sci* 97(10): 4586–4595.

31. Allen MC, Shah TS, Day WW (1998) Rapid determination of oral pharmacokinetics and plasma free fraction using cocktail approaches: Methods and applications. *Pharm Res* 15(1):93–97.

32. Tran TH, Von Moltke LL, Venkatakrishnan K, Granda BW, Gibbs MA, Obach RS, Harmatz JS, Greenblatt DJ (2002) Microsomal protein concentration modifies the apparent inhibitory potency of CYP3A inhibitors. *Drug Metab Dispos* 30(12):1441–1445.

33. Taylor S, Harker A (2005) Modification of the ultrafiltration technique to overcome solubility and non-specific binding challenges associated with the measurement of plasma protein binding of corticosteroids. *J Pharm Biomed Anal* 41:299–303.

34. Fung EN, Chen YH, Lau YY (2003) Semi-automatic high-throughput determination of plasma protein binding using a 96-well plate filtrate assembly and fast liquid chromatography-tandem mass spectrometry. *J Chromatogr B* 795:187–194.

35. Wan H, Rehngren M, Giordanetto F, Bergstrom F, Tunek A (2007) High-throughput screening of drug-brain tissue binding and *in silico* prediction for assessment of central nervous system drug delivery. *J Med Chem* 50: 4606–4615.

36. Bradford M (1976) A rapid and sensitive method for the quantitation of microgram quantities of protein utilizing the principle of protein-dye binding. *Anal Biochem* 72: 248–254.

37. Banker MJ, Clark TH, Williams JA (2002) Development and validation of a 96-well equilibrium dialysis apparatus for measuring plasma protein binding. *J Pharm Sci* 92: 967–974.

38. Huang J (1983) Errors in estimating the unbound fraction of drugs due to the volume shift in equilibrium dialysis. *J Pharm Sci* 72:1358–1359.

39. Lima J, MacKichan JJ, Libertin N, Sabino J (1983) Influence of volume shifts on drug binding during equilibrium dialysis: Correction and attenuation. *J Pharmacokinet Biopharm* 11:483–498.

40. Boudinot FD, Jusko WJ (1984) Fluid shifts and other factors affecting plasma protein binding of prednisolone by equilibrium dialysis. *J Pharm Sci* 73:774–780.

41. Khor SP, Wu HJ, Boxenbaum H (1985) Simultaneous correction for volume shifts and leakage in equilibrium dialysis protein binding experiments. *Int J Pharm* 23:109–113.

42. Mapleson WW (1987) Computation of the effect of Donnan equilibrium on pH in equilibrium dialysis. *J Pharmacol Methods* 17:231–242.

43. Fisher WR, Hammond MG, Warmke GL (1972) Measurements of the molecular weight variability of plasma low density lipoproteins among normals and subjects with hyper-b-lipoproteinemia. *Biochemistry* 11:519–525.

44. Barre J, Chamouard JM, Houin G, Tillement JP (1985) Equilibrium dialysis, ultrafiltration and ultracentrifugation compared for determining the plasma-protein-binding characteristics of valproic acid. *Clin Chem* 31:60–64.

45. Bombardt PA, Brewer JE, Johnson MG (1994) Protein binding of tirilazad (U-74006) in human, Sprague-Dawley rat, beagle dog and cynomolgus monkey serum. *J Pharmacol Exp Ther* 269:145–150.

46. Zini R Methods in drug protein binding analysis. In *Human Pharmacology. The Basis of Clinical Pharmacology*, Kuemmerle H, Shibuya T, Tillement JP eds. Oxford, UK: Elsevier Science, 1991:235–282.

47. Kurz H, Trunk H, Weitz B (1977) Evaluation of methods to determine protein-binding of drugs. Equilibrium dialysis, ultrafiltration, ultracentrifugation, gel filtration. *Arzneimittelforschung* 27:1373–1380.

48. Legg B, Rowland M (1987) Cyclosporin: Measurement of fraction unbound in plasma. *J Pharm Pharmacol* 39:599–603.

49. Nakai D, Kumamoto K, Sakikawa C, Kosaka T, Tokui T (2003) Evaluation of the protein binding ratio of drugs by a micro-scale ultracentrifugation method. *J Pharm Sci* 93(4):847–854.

50. Adams W, Ghupur A, Lloyd TL (2009) Cross-validation of a reduced volume, semi-automated ultracentrifugation process for assessing protein binding in plasma and microsomes to improve efficiency and capacity. Eastern Analytical Symposium, Somerset, NJ, November 18, 2009.

51. Heinze A, Holzgrabe U (2006) Determination of the extent of protein binding of antibiotics by means of an automated continuous ultrafiltration method. *Int J Pharm* 311:108–112.

52. Johansen K, Krogh M, Andresen AT, Christophersen AS, Lehne G, Rasmussen KE (2005) Automated analysis of free and total concentrations of three antiepileptic drugs in plasma with on-line dialysis and high-performance liquid chromatography. *J Chromatogr B* 669(2):281–288.

53. Mandla R, Line PD, Midtvedt K, Bergan S (2003) Automated determination of free mycophenolic acid and its glucuronide in plasma from renal allograft recipients. *Ther Drug Monit* 25(3):407–414.

54. Chaurasia CS, et al. (2007) AAPS-FDA workshop white paper: Microdialysis principles, application, and regulatory perspectives report from the joint AAPS-FDA workshop, November 4–5, 2005, Nashville, TN. *AAPS J* 9(1):48–59.

55. Ungerstedt U (1991) Microdialysis—principles and applications for studies in animals and man. *J Intern Med* 230:365–373.

56. Smith AD, Justice JB (1994) The effect of inhibition of synthesis, release, metabolism and uptake on the microdialysis extraction fraction of dopamine. *J Neurosci* 54:75–82.

57. Vuignier K, Schappler J, Veuthey JL, Carrupt PA, Martel S (2010) Drug-protein binding: A critical review of analytical tools. *Anal Bioanal Chem* 398:53–66.

58. Cimitan S, Lindgren MT, Bertucci C, Danielson UH (2005) Early absorption and distribution analysis of antitumor and anti-AIDS drugs: Lipid membrane and plasma protein interactions. *J Med Chem* 48(10):3536–3546.

59. Rich RL, Myszka DG (2004) Why you should be using more SPR biosensor technology. *Drug Discov Today Technol* 1:301–308.

60. Lazaro E, Lowe PJ, Briand X, Faller B (2008) New approach to measure protein binding based on a parallel artificial membrane assay and human serum albumin. *J Med Chem* 51:2009–2017.

61. Vogel HG *Drug Discovery and Evaluation: Safety and Pharmacokinetic Assays.* New York, NY: Springer, 2006.

62. Schuhmacher J, Kohlsdorfer C, Buhner K, Brandenburger T, Kruk R (2004) High-throughput determination of the free fraction of drugs strongly bound to plasma proteins. *J Pharm Sci* 93:816–830.

63. Domenici E, Bertucci C, Salvadori P, Felix G, Cahagne I, Motellier S, Wainer IW (1990) Synthesis and chromatographic properties of an HPLC chiral stationary phase based upon human serum-albumin. *Chromatographia* 29:170–176.

64. Bertucci C, Bartolini M, Gotti R, Andrisano V (2003) Drug affinity to immobilized target bio-polymers by high-performance liquid chromatography and capillary electrophoresis. *J Chromatogr B* 797:111–129.

65. Xuan H, Hage DS (2005) Immobilization of alpha(1)-acid glycoprotein for chromatographic studies of drug-protein binding. *Anal Biochem* 346:300–310.

66. Chen S, Sobansky MR, Hage DS (2010) Analysis of drug interactions with high-density lipoprotein by high-performance affinity chromatography. *Anal Biochem* 397:107–114.

67. Soltes L (2004) The Hummel-Dreyer method: Impact in pharmacology. *Biomed Chromatogr* 18:259–271.

68. Hage DS, Tweed SA (1997) Recent advances in chromatographic and electrophoretic methods for the study of drug-protein interactions. *J Chromatogr B* 699:499–525.

69. Heegaard NHH, Kennedy RT (1999) Identification, quantitation and characterization of biomolecules by capillary electrophoretic analysis of binding interactions. *Electrophoresis* 20:3122–3133.

70. Kalvass JC, Maurer TS (2002) Influence of nonspecific brain and plasma binding on CNS exposure: Implications for rational drug discovery. *Biopharm Drug Dispos* 23:327–338.

71. Hammarlund-Udenaes M, Friden M, Syvanen S, Gupta A (2007) On the rate and extent of drug delivery to the brain. *Pharm Res* 25(8):1737–1750.

13

REACTION PHENOTYPING

Chun Li and Nataraj Kalyanaraman

13.1 INTRODUCTION

Most of drugs absorbed by the body will be metabolized to some degree, and a large number of enzymes exist to metabolize them. Drug metabolism is the major clearance mechanism listed for approximately three-quarters of the top 200 prescribed drugs in the United States in 2002, and of the drugs cleared via metabolism, about three-quarters are metabolized by members of the cytochrome P450 (CYP) superfamily (Williams et al., 2004; Wienkers and Heath, 2005). Understanding how investigated drug is cleared and which drug metabolizing enzymes are responsible for its drug clearance allow one to predict the risks of drug–drug interaction (DDI) with coadministered drugs and the consequences of interindividual variability in metabolism and pharmacokinetics. Reaction phenotyping or isozyme mapping are studies designed to identify and characterize the enzyme(s) responsible for the metabolism of new chemical entities (NCEs).

Several of the CYP enzymes involved in drug metabolism are known to exhibit polymorphic expression in human populations (Nakamura et al., 1985), with most notably CYP2D6, CYP2C9, CYP2C19, and CYP2A6 contributing to variable drug exposures (Rodrigues and Rushmore, 2002). These four CYPs may be responsible for between 35% and 40% of the metabolic clearance of drugs (Guengerich, 2000). If the *in vivo* clearance of a drug is largely mediated by a polymorphic CYP, poor metabolizers (PMs) will produce elevated plasma area under the curve (AUC) and/or an increased half-life, and can adversely affecting the safety and efficacy of

drugs, while extensive metabolizers (EMs) will have higher clearance and lower exposure of the drug and possibly resulting in potential lack of efficacy. Large dosing adjustments may be necessary to achieve the safe and effective use of the drug when a genetic polymorphism affects an important metabolic route of elimination (Sanderson et al., 2005). In addition, significant DDIs in which one agent alters the exposure of a concomitantly administered drug through effects on CYP-mediated metabolism are well documented (Lin and Lu, 1998; Tanaka, 1998; Dresser et al., 2000). DDIs can lead to severe side effects and have resulted in early termination of development, refusal of approval, severe prescribing restrictions, and withdrawal of drugs from the market. The magnitude of drug interaction is dependent on the concentration and potency of the inhibitor and the number of elimination pathways associated with the victim drug (Tucker et al., 2001; Houston and Galetin, 2003; Ito et al., 2005). If drug clearance is largely dependent on a single enzyme, inhibition or induction of that enzyme will have a profound effect on the drug exposure in comparison with a drug that has several metabolic routes of elimination. Given the limited number of P450 enzymes that contribute to the clearance of marketed small-molecule drugs, P450s represent the primary source of clinically observed drug interactions in patients. Reaction phenotyping results help to define the DDI potential of drug candidate as a victim.

Consequently, most pharmaceutical companies now screen their NCEs *in vitro* for induction and inhibition of drug metabolizing enzymes and transporters. NCEs are also screened for metabolic stability, and studies are

ADME-Enabling Technologies in Drug Design and Development, First Edition. Edited by Donglu Zhang and Sekhar Surapaneni.
© 2012 John Wiley & Sons, Inc. Published 2012 by John Wiley & Sons, Inc.

conducted to identify the enzyme(s) responsible for metabolic turnover. Knowledge that a particular drug is not a substrate for certain metabolic pathways is helpful. For example, if it is learned early in drug development that a molecule is not a substrate for CYP3A4 or that this pathway represents only a minor contribution to overall metabolism, then concern is lessened or eliminated for possible inhibition of CYP3A4 metabolism by drugs such as ketoconazole and erythromycin or possible induction of metabolism by drugs such as rifampin and anticonvulsants, and clinical interactions studies with these agents may not be necessary. The information obtained in definitive reaction phenotyping can be used for the planning of clinical studies and in support of claims in the product label.

In this chapter, *in vitro* phenotyping approaches and strategies for drug metabolizing enzymes at both drug discovery and development stages are reviewed. Major emphasis is placed on P450 reaction phenotyping, but roles of other important non-P450 drug metabolizing enzymes (flavin-containing monooxygenases [FMOs], monoamine oxidases [MAOs], aldehyde oxidase [AO]) and conjugation enzymes such as uridine 5'-diphospho (UDP)-glucuronosyltransferases (UGTs) are also discussed. And more recently, the role of hepatic uptake and efflux transporters and their interplay with drug metabolizing enzymes have become increasingly important in drug elimination (Giacomini et al., 2010). This represents a whole new area of research and is beyond the discussion of this chapter.

13.2   INITIAL CONSIDERATIONS

The main purpose of reaction phenotyping is to determine the relative contributions of individual drug metabolizing enzymes to the overall clearance of a drug candidate, and therefore to assess its victim potential for DDI. Before conducting detailed reaction phenotyping experiments, it is important to first define the predominant clearance mechanisms for the drug candidate and the most appropriate *in vitro* system in which to study these clearance mechanisms. There is no rationale for performing P450 reaction phenotyping for a drug candidate if P450-mediated reactions only contribute a minor role in its overall clearance (<30%), as indicated in the Pharmaceutical Research and Manufacturers of America (PhRMA) paper (Bjornsson et al., 2003). A thorough understanding of the overall metabolic profile of the drug candidate is also essential and an obvious prerequisite to conducting drug metabolizing enzyme phenotyping. Furthermore, the appropriate substrate concentration and enzymatic reaction conditions need to be carefully assessed, and the enzyme kinetics needs

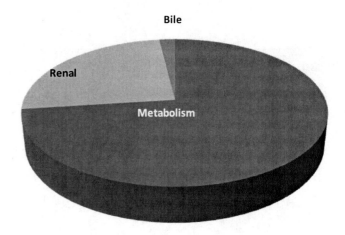

FIGURE 13.1. Routes of elimination of the top 200 most prescribed drugs in 2002. Metabolism represents the listed clearance mechanism for ~73% of the top 200 drugs. Of the drugs cleared via metabolism, about three-quarters are metabolized by members of the cytochrome P450 (CYP) superfamily. (Adapted from Williams et al., 2004 and Wienkers and Heath, 2005).

to be properly determined to ensure success of the reaction phenotyping studies.

13.2.1   Clearance Mechanism

As illustrated in Figure 13.1, of many drugs on the market and many drug candidates in drug development, metabolism, especially P450-mediated reactions, represents the major route of elimination (Williams et al., 2004; Wienkers and Heath, 2005). Data obtained after administration of radiolabeled investigational compound to humans provides the most definitive information on the routes of drug clearance. In many cases, however, the human radiolabel study is conducted at late drug development stage and often after the initiation of reaction phenotyping experiments, and therefore, major human clearance mechanisms may need to be predicted.

An important aspect in predicting clearance mechanisms and conducting appropriate reaction phenotyping experiments is the examination of the structure and physicochemical properties of the compounds being studied (Williams et al., 2003). Physicochemical characteristics such as lipophilicity can affect metabolism and influence drug clearance (Lewis et al., 2004; Varma et al., 2009). Drug candidate with a $LogD_{7.4}$ greater than 1 will likely needed to be converted into a more polar entity in order to facilitate excretion in urine or bile. Compounds of high hydrophilicity will more likely be cleared unchanged. A fundamental understanding of different types of metabolic reactions and possible enzymes involved in these reactions are critical. The knowledge of CYP and other drug metabolizing enzymes, and the reactions they catalyze, has been growing over the years and well established (Guengerich, 2007; Brown et al., 2008). In addition, predictive softwares such as Metasite (Molecular Discovery Ltd., Pinner, UK) (Cruciani et al., 2005) and Meteor (Lhasa Ltd., Leeds, UK) (Testa et al., 2004) are now available, which allow some preliminary prediction of the potential metabolic soft spots and type of biotransformation of drug candidates. Furthermore, information on the rate and extent of *in vitro* metabolism in humans and other animal species and clearance mechanism determined in animal models can help predict the clearance mechanism in humans. Careful consideration of all these information will help to make a successful prediction of whether the compound is likely to be extensively metabolized in humans.

13.2.2 Selecting the Appropriate *in vitro* System

If human *in vivo* absorption, distribution, metabolism, and excretion (ADME) data is available, and the major initial routes of drug metabolism are known, the data will indicate the most appropriate *in vitro* system for the study of reaction phenotyping. In many cases, however, the human radiolabel ADME study is unavailable at the time of reaction phenotyping. The U.S. Food and Drug Administration (FDA) and PhRMA guidance recommended that the metabolism of the drug candidate be examined in as "complete" an *in vitro* system as is possible (Bjornsson et al., 2003). Hepatocytes may be the best *in vitro* system to use for providing the overall assessment of metabolic profile of the drug candidate. Hepatocytes are an integrated metabolic system, closer to *in vivo* environment, and contain many enzymes and enzyme cofactors not present in microsomal fraction, and therefore offer advantages over liver microsomes in overall *in vitro* metabolic profiling of drug candidate. The contribution of P450 enzymes ($f_{m,cyp}$) versus other cytosolic or conjugative enzyme systems can be assessed in hepatocytes using a pan nonspecific P450 inhibitor such as 1-aminobenzotriazole (ABT), which is thought to inactivate P450 enzymes nonselectively by covalent modification of the heme prosthetic group following bioactivation (Ortiz de Montellano and Mathews, 1981). If the rate of metabolism of drug candidate is inhibited significantly after preincubation of hepatocytes with 1 mM ABT, CYP likely plays a prominent role in the metabolism of drug candidate (Dalmadi et al., 2003; Williams et al., 2003). However, quantitatively distinguishing P450 from non-P450-mediated metabolism based on inhibition of ABT needs to be cautioned. Recent reports showed that ABT displayed differential inactivation of CYP isozymes, whereas P450 2A6 and 3A4 activity was

essentially eliminated upon 30-min pretreatment with ABT, and the other human P450s were less affected, especially CYP2C9, with roughly 60% activity remaining after pretreatment (Linder et al., 2009).

Once the overall *in vitro* metabolic profile has been examined and major routes of metabolism identified, mechanistic reaction phenotyping to identify the key enzymes responsible for metabolism of selected drug candidate can be planned. In hepatocytes, cell membrane can present a barrier for both substrate and selective inhibitors to entry into the cell, making it difficult to study reaction phenotyping. Subcellular fractions, in particular, human liver microsomes (HLM), may be the optimal *in vitro* system for this purpose. HLM are a rich source of heme-containing metabolic enzymes, including CYP enzymes, FMOs, UGTs, glutathione *S*-transferases (GSTs), esterases, microsomal epoxide hydrolase, and possible contamination of MAOs. Hepatic cytosol contains a good source of molybdenum cofactor enzymes, sulfotransferases (SULTs) (Pacifici, 2004), and AO (Lake et al., 2002). Mitochondria contain MAOs and aldehyde dehydrogenases. These subcellular fractions can be readily used to evaluate the possible involvement of different drug metabolizing enzymes in the metabolism of drug candidate.

13.2.3 Substrate Concentration

The selection of substrate concentration used in the *in vitro* incubation is a key consideration in reaction phenotyping experimental design. It is common to have multiple enzymes involved in the same metabolic reaction, and if the *in vitro* reaction phenotyping experiment is carried out using pharmacologically irrelevant substrate concentrations, major isozyme(s) identified *in vitro* may not reflect the same enzyme responsible for its metabolism *in vivo*. A good example is the 5-hydroxylation of lansoprazole, which is mainly catalyzed by two kinetically distinct CYP enzymes, CYP3A4 and CYP2C19 (Pearce et al., 1996). When HLM were incubated with high lapoprazole concentrations (100 µM), the reaction appeared to be catalyzed by CYP3A4, a low-affinity, high-capacity enzyme. However, at the pharmacologically relevant substrate concentration of 1 µM, lansoprazole 5-hydroxylation is primarily catalyzed by CYP2C19, as it is *in vivo*. PhRMA has recommended that *in vitro* CYP reaction phenotyping experiments should be conducted with pharmacologically relevant drug concentrations. And in cases where clinical data are not available, and what constitutes a pharmacologically relevant concentration is not known, then substrate concentration should be $<K_m$, with K_m representing the Michaelis–Menten constant associated with the enzyme activity of the greatest intrinsic clearance if more than one enzyme

is involved. In the absence of knowledge of K_m, reaction phenotyping experiments are typically conducted at a substrate concentration of 1 µM. In most cases, it is below K_m, because CYP and other drug metabolizing enzymes tend to catalyze reactions with relatively low affinity. This 1 µM low substrate concentration permits reaction phenotyping studies to be conducted under the initial rate conditions and allows the identification of the high-affinity enzyme responsible for metabolism of drug candidate.

13.2.4   Effect of Incubation Time and Protein Concentration

Conditions for the linear production of metabolite with respect to microsomal concentration and incubation time must be established before conducting detailed reaction phenotyping experiments. The incubation should be conducted under the initial rate conditions, where the concentration of parent drug candidate was not significantly depleted (>15%). With the depletion of >15% of substrate, reaction kinetics should be analyzed by a modified form of the Michaelis–Menten equation that accounts for the reduced substrate concentration with time (Segel, 1972). Preliminary experiments should be conducted using a fixed substrate concentration and protein concentration for multiple time periods (e.g., 0, 5, 10, 15, 20, 30, 45, and 60 min). In general, reactions catalyzed by P450 enzymes do not maintain linearity past 45 min of incubation time. In contrast, glucuronidation reactions catalyzed by UGTs can be linear for up to 4 h.

Protein concentrations in the *in vitro* incubations will also impact the rate of metabolism. Typically, a microsomal protein concentration range of 0.1–1 mg/mL is used for incubation. At concentrations greater than 2 mg/mL, nonspecific binding to drug candidate may become an issue. The high nonspecific binding of substrate (drug candidate) to microsomal protein can produce altered kinetic estimates, with increased K_m at increasing protein concentrations (Kalvass et al., 2001). This will result in a lower estimate of intrinsic clearance (V_{max}/K_m), and an underestimate of *in vivo* drug clearance. This phenomenon of substrate binding to microsomal protein may also account for the discrepancies measured by different labs and in different *in vitro* systems. Advances in measurement and prediction of protein binding in microsomes can potentially fill the gap (Obach, 1997; Wring et al., 2002).

13.2.5   Determination of Kinetic Constant K_m and V_{max}

Reaction phenotyping entails an initial assessment of Michaelis–Menten kinetics in HLM or recombinant enzymes. The goal of reaction phenotyping study is to derive an estimate of *in vivo* intrinsic clearance by a given enzymatic pathway from *in vitro* data (Houston, 1994). The successful *in vitro*-to-*in vivo* extrapolation relies on the accuracy of the calculated *in vitro* intrinsic clearance (V_{max}/K_m) of a given enzymatic pathway (Houston, 1994; Obach, 1999, 2001). Experiments are typically conducted under incubation conditions that minimize substrate depletion (<15%), and that are linear with respect to microsomal protein concentration and incubation time. Initial rates of metabolite formation are measured over a wide range of substrate concentration [S] if solubility permits. Experiments described in the previous section for evaluation of the effect of incubation time and protein concentration on the formation of metabolites provides the preliminary basis for selecting an appropriate range of substrate concentrations and conditions to determine K_m and V_{max} of a particular enzymatic reaction. The substrate concentrations typically range from one-tenth of K_m to 10 times K_m. After measurement of reaction rates for diverse range of substrate concentrations, enzyme kinetic parameters K_m and V_{max} can be obtained by fitting the data directly to the Micahaelis–Menten equation using nonlinear regression (Segel, 1972; Tracy and Hummel, 2004). When kinetic parameters are determined with individual human liver microsomal samples, V_{max} values can vary enormously from one sample to the other because the rate of reaction depends on concentration of both substrate and enzyme, and levels of enzymes in human liver samples can vary significantly. K_m values, on the other hand, should remain relatively constant, and is independent of the specific content of the enzyme. K_m is an intrinsic property of an enzyme, and is a measure of the affinity of an enzyme for a given substrate. K_m represents the substrate concentration at which half of the enzyme molecules are bound with substrate.

It is also often useful to perform a linear transformation of the original kinetic data, by either Lineweaver–Burk (1/V against 1/[S] plot) method (Lineweaver and Burk, 1934) or Eadie–Hofstee (V against V/[S] plot) method (Hofstee, 1952). The Eadie–Hofstee plot is very sensitive to deviation of simple Michaelis–Menten single enzyme kinetics. A curved Eadie–Hofstee plot would suggest more than one enzyme contributes to the metabolic reaction, while a straight-line plot suggests a single enzyme reaction with K_m determined from the slope of the line, or possible two enzymes with similar K_m values. The Eadie–Hofstee graphic analysis is routinely used as a diagnostic tool to detect multiple enzyme kinetics.

13.2.6   Development of Analytical Methods

Before conducting detailed reaction phenotyping experiments, sensitive and robust analytical methods should be developed to measure the rate of metabolite formation or possibly the disappearance of drug candidate. Two analytical approaches are often employed, liquid chromatography (LC) with radiometric detection for radiolabeled substrate, and liquid chromatography-tandem mass spectrometry (LC-MS/MS) using synthetic metabolite standards. When reaction phenotyping is conducted at drug development stage, radiolabeled drug candidate is typically available, and high-performance liquid chromatography (HPLC) coupled with online radio flow detection (RFD) is the primary radiochromatographic technique used (Morovjan et al., 2002; Zhu et al., 2005a). The major disadvantage of HPLC-RFD is its relatively poor sensitivity due to short residence times (5–15 s) of the radioactive peaks in the radio flow detector cell (Zhu et al., 2005b). The limit of quantitation of RFD ranges from 750 to 1500 disintegrations per minute (dpm), which will limit its use in cases where the experiments are conducted at low substrate concentrations, or low turnover metabolites, or with a drug at a low specific radioactivity. Newer techniques have been introduced in recent years including the use of HPLC coupled with off-line microplate scintillation counting (MSC) and stop-flow liquid radiochromatographic detection technique, called accurate radioisotope counting or ARC. In HPLC-MSC analysis (Boernsen et al., 2000; Kiffe et al., 2003; Wallace et al., 2004; Zhu et al., 2005b; Bruin et al., 2006), the HPLC effluent is collected into 96-well microplates (Luma plates in which yttrium silicate scintillators are deposited at the bottom of each well, or Scintiplates embedded with solid scintillators), and then evaporated using a speed vacuum system. The radioactivity of the residue in the plates is determined by counting up to 12 wells at a time with a microplate scintillation counter using either TopCount (Perkin Elmer, Waltham, MA, USA) or MicroBeta (Perkin Elmer, Waltham, MA, USA) counter (Nedderman et al., 2004; Zhu et al., 2005b). HPLC-MSC provides excellent sensitivity, but has relatively low throughput. Stop-flow HPLC-RFD (Nassar et al., 2003, 2004) technique has the flexibility to be operated in different modes (by-fraction, by-level, nonstop), the technique greatly extended the capacity of RFD by improving its sensitivity. It can be used in an automated fashion and has been used successfully in enzyme kinetics studies of muraglitazar glucuronidation (Zhao et al., 2008).

In the cases where synthetic metabolites are available, measurement of the rate of metabolite formation and subsequent reaction phenotyping are facilitated by the use of tandem mass spectrometry and multiple reaction monitoring (MRM) approach (Williams et al., 2003). Similarly, if reaction phenotyping needs to be conducted during drug discovery stage where radiolabeled drug candidate and synthetic drug metabolites are not

available, a substrate depletion approach (Obach and Reed-Hagen, 2002; Williams et al., 2003) is often used with LC-MS/MS by MRM monitoring of the investigational drug candidate. LC-MS/MS is a very powerful, sensitive, and selective technique, and allows detection of drug candidates and metabolites at subnanomolar concentrations. In the absence of synthetic metabolite standard, however, the use of LC/MS/MS for quantitation of metabolites needs to be cautioned, because the relative rates of metabolite formation may not be estimated based on its peak response due to different ionization efficiencies from the drug candidate. Sometimes, combination of detection methods from UV, radiometric, and LC-MS/MS can be very useful in metabolite quantitation and reaction phenotyping (Li et al., 2009).

13.3 CYP REACTION PHENOTYPING

Drug metabolism and biotransformation is the major route of elimination for many drugs, and oxidative metabolism by P450 enzymes is the most common metabolic pathway (Rendic, 2002; Wienkers and Heath, 2005). The P450 enzymes are membrane proteins expressed in the endoplasmic reticulum of mammalian cells, with highest expression in the liver, and also in extrahepatic tissues such as intestine, lung, and kidney (Krishna and Klotz, 1994). In humans, different CYP genes have been identified and categorized into 18 families (Nelson, 1999). However, enzymes from CYP1, CYP2, and CYP3 families are known to be expressed sufficiently in human liver, and carry out most xenobiotics biotransformations. Based on the importance played in the metabolism of drugs, CYP1A2, 2C9, 2C19, 2D6, and 3A4 are the isozymes of major importance, together they constitute approximately 50% of total hepatic P450 protein, but are involved in the majority of P450-catalyzed drug biotransformation reactions (Smith et al., 1998; Bjornsson et al., 2003; Wienkers and Heath, 2005). CYP2C8, 2B6, and 3A5 are P450 enzymes of emerging importance, and have recently received more attention in the scientific literature due to involvement in the metabolism of specific drugs or classes of drugs (Walsky et al., 2005, 2006; Bjornsson et al., 2003). In the case of CYP3A5, it may be involved more in the metabolism of CYP3A substrates *in vivo* than previously reported. CYP1A1, 1B1, 2A6, 2E1, 4A11, and so on do not have a major role in drug metabolism; they have been shown to be involved in the metabolism of few, if any, therapeutic agents.

Over recent years, a variety of *in vitro* reagents and tools have been developed, and CYP reaction phenotyping approaches to determine which specific CYP isozymes are involved in the metabolism of a given compound have been established and standardized and widely reviewed (Rodrigues, 1999; Madan 2002; Lu et al., 2003; Williams et al., 2003, 2005; Zhang et al., 2007b; Harper and Brassil, 2008; McGinnity et al., 2008). The FDA and the PhRMA also published guidelines for the pharmaceutical industry to conduct these experiments (Bjornsson et al., 2003). P450 reaction phenotyping is performed using a combination of three basic approaches, and will be discussed in the subsequent sections. The objective of *in vitro* CYP reaction phenotyping is not to simply identify all the P450 enzymes able to oxidize a particular substrate but rather to characterize the enzyme(s) that are associated with key metabolic pathways that control or influence the clearance of the drug.

13.3.1 Specific Chemical Inhibitors

CYP reaction phenotyping is used to determine fraction of drug metabolized by specific P450 enzymes and is possible in part because of the accumulation of compounds identified as potent and selective inhibitors of individual P450 enzymes (Newton et al., 1995; Mancy et al., 1996; von Moltke et al., 1998; Suzuki et al., 2002; Cai et al., 2004; Stresser et al., 2004; Walsky et al., 2005). A list of FDA-recommended and commonly used selective chemical inhibitors for P450s is shown in Table 13.1 (Tucker et al., 2001; Bjornsson et al., 2003; Zhang et al., 2007b). Once appropriate substrate concentration is chosen and enzyme kinetics has been determined in HLM, chemical inhibition studies with CYP isoform

selective inhibitors are relatively straightforward. However, one needs to keep in mind several factors before conducting inhibition studies. First, several inhibitors listed such as furafylline, methoxalen, and troleandomycin are mechanism-based inactivators, and would require a preincubation of the inhibitor with NADPH and HLM for 15 min or more prior to the addition of drug candidate to evaluate the residue P450 enzyme activity. Second, appropriate controls should be included in each chemical inhibition experiment, and the volume of organic solvent such as DMSO, methanol, and acetonitrile in the incubation should be kept to a minimum (≤0.5%, v/v) because of their inhibitory effects (Chauret et al., 1998; Hickman et al., 1998). Finally, inhibitor concentrations should be chosen to ensure maximum inhibitory effect yet maintaining selective inhibition of individual P450 enzymes.

The selectivity of chemical inhibition is often dependent on inhibitor concentration (Ono et al., 1996). For example, ketoconazole is a potent and selective inhibitor of CYP3A4 ($K_i < 20$ nM) at 1 μM, but at higher concentrations, it also inhibits several other CYP enzymes such as CYP1A1, CYP2B6, CYP2C8, and CYP2C9 (with K_i values in the micromolar range) (Newton et al., 1995). In most cases, the specificity of chemical inhibitors is limited to a particular concentration range. When conducting CYP phenotyping experiments using chemical inhibitors, it is highly desirable to evaluate the effect of an inhibitor on a particular reaction pathway with a wide range of inhibitor con-

TABLE 13.1. Examples of the Specific Chemical Inhibitors of Various CYP Isoforms

P450 Isoform	Specific Chemical Inhibitor	K_i (μM)	[I] Used in CYP Reaction Phenotyping
CYP1A2	Furafylline	0.6–0.73	10–30[a]
	α-Napththoflavone	0.01	1
CYP2A6	Tranylcypromine	0.02–0.2	1–10
	Methoxalen	0.01–0.2	1[a]
CYP2B6	N, N′, N′-triethylenethiophosphoramide (thioTEPA)	4.8	50
CYP2C8	Montelukast	0.0092–0.15	0.1
	Quercetin	1.1	10–30
CYP2C9	Sulfaphenazole	0.3	10
CYP2C19	Benzylnivanol	0.25	1
CYP2D6	Quinidine	0.027–0.4	<5
CYP2E1	Clomethiazole	12	50
	Diethyldithiocarbamate	9.8–34	50
CYP3A	Ketoconazole	0.037–0.18	1
	Troleandomycin	17	25–100[a]

K_i is the inhibition constant of inhibitor, determined from human liver microsomes or appropriate recombinant CYPs. [I] is the concentration of inhibitor (μM) required to ensure that [I]/K_i ≥ 10 (when [S] ≤ K_m), and that the corresponding CYP form is selectively inhibited (% inhibition ≥ 83%) in human liver microsomes.
[a] Mechanism-based inhibitor and will require a 15–30-min preincubation in the presence of NADPH before adding substrate.
Source: References from FDA drug interaction website; Bourrie et al., 1996; Rodrigues and Roberts, 1997; Rae et al., 2002; Suzuki et al., 2002; Walsky et al., 2005.

centrations. This concentration range can shift with concentration of microsomal protein used due to non-specific binding of investigated drug candidate, as illustrated by the decreased inhibitory effect of montelukast on CYP2C8 with increasing microsomal protein concentrations (Walsky et al., 2005). In general, when substrate concentration [S] ≤ K_m, relatively high concentrations of inhibitor with $[I]/K_i \geq 10$ are used to ensure inhibition is extensive (≥80%). For some inhibitors such as ketoconazole, sulaphenazole, and quinidine, $[I]/K_i$ ratios up to 30 can be used, which will enable inhibition of ≥93% while remaining relatively selective (Bourrie et al., 1996; Rodrigues, 1999). Alternatively, a concentration range of inhibitor can be used to generate an IC_{50} (Newton et al., 1995). When the involvement of multiple CYPs is suspected, multiple inhibitors can be coincubated, assuming the inhibitory effects are independent and additive (Rodrigues et al., 1996).

The effect of CYP chemical inhibitors on multiple CYP reaction pathways of the investigated drug candidate can be evaluated at the same time in HLM when radiolabeled drug candidate is available. As shown in Figure 13.2, the effect of several CYP isoform selective chemical inhibitors on metabolism of a highly selective angiogenesis inhibitor motesanib was evaluated (Li et al., 2009). The involvement of several CYP isozymes in the metabolism of multiple oxidative pathways of motesanib, including N-oxidation (M3) and indoline ring oxidation (M4 and M5), can be readily identified.

13.3.2   Inhibitory CYP Antibodies

Potent, specific, and inhibitory antibodies against various human P450 isoforms represent one of the most valuable tools for P450 identification. Specific monoclonal antibodies against most, if not all, of human P450 isoforms relevant to drug metabolism are now commercially available (Gelboin et al., 1999; Mei et al., 1999; Stresser and Kupfer, 1999; Krausz et al., 2001; Shou and Lu, 2009). These specific CYP antibodies have expanded the diversity of inhibitors available to determine the involvement of different P450 pathways to a drug's clearance. CYP inhibition by antibodies is noncompetitive in nature and is therefore independent of the substrate concentration.

The conduct of experiments to determine the effect of a P450 antibody on a particular metabolic reaction in HLM is relatively simple, and shares many characteristics of chemical inhibition. Inhibition of enzyme activity is typically conducted using a pooled human liver microsomal sample, and a fixed relevant concentration of drug candidate. To conduct a good antibody inhibition

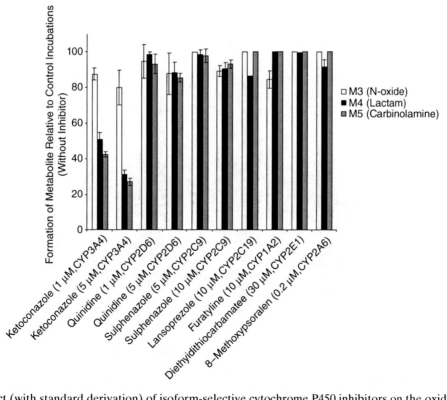

FIGURE 13.2. Effect (with standard derivation) of isoform-selective cytochrome P450 inhibitors on the oxidative metabolism of motesanib by human liver microsomes.

study, it is highly desirable to examine the effect of a particular antibody in a concentration-dependent manner on the metabolism of the drug of interest (Wang and Lu, 1997). It should be noted that proper control incubations be also conducted to measure the ability of the antibody to inhibit an established marker P450 activity and the effect of control antibody on the metabolism of drug candidate of interest. An example of the effect of CYP antibodies on the oxidative metabolism of highly selective angiogenesis inhibitor motesanib (Li et al., 2009) is shown in Figure 13.3.

A primary limitation of anti-P450 antibodies for reaction phenotyping has been the cross-reactivity with related P450s, especially in the case of polyclonal or antipeptide antibodies. Monoclonal antibodies to CYP2C8, 2C9, and 2C19 are potent and specific (Krausz et al., 2001), while antibodies to CYP3A isoforms are less specific and exhibit cross-reactivity between CYP3A4 and 3A5 isoforms (Mei et al., 1999). Another limitation is that many CYP antibodies do not completely inhibit the activity of microsomal CYP enzyme or the corresponding recombinant CYP enzyme. Similarly, chemical CYP inhibitors can also rarely achieve complete inhibition of a specific CYP isozyme when used at well-established concentrations that ensure adequate selectivity and maximal potency, and typically

results in inhibition of up to 80%. This phenomenon can confound the interpretation and add some uncertainties in the quantitative assessment of contribution from each specific CYP isozyme. The remaining residue CYP activity reduces the predictive nature of *in vitro* extrapolation techniques and increases the uncertainty regarding the potential activity from other drug metabolizing enzymes. Furthermore, remaining activity can bring into question the P450 activity from less abundant hepatic P450s such as 2B6 and 2C8, which are showing increased metabolic activity toward numerous commercial drugs (Walsky et al., 2005, 2006). An approach using the combination of commercial antibody and chemical inhibitor against P450 3A has been reported (Rock et al., 2008), and showed a superior inhibition profile of CYP3A4 compared with the use of antibody and chemical inhibitors individually.

13.3.3 Recombinant CYP Enzymes

Another very useful approach to CYP reaction phenotyping involves the use of purified or recombinant (cDNA-expressed) human CYP enzymes. Numerous human CYP enzymes have been cloned and heterologously expressed individually in various cell types (Remmel and Burchell, 1993; Guengerich et al., 1997;

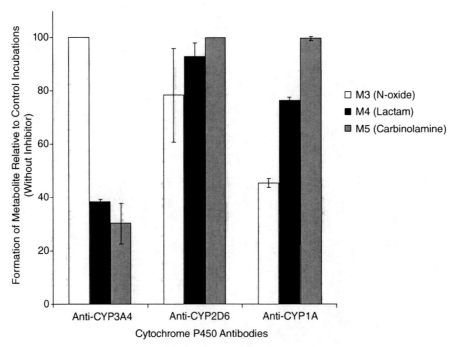

FIGURE 13.3. Effect (with standard derivation) of isoform-selective cytochrome P450 antibodies (10 μg/0.1 mg HLM) on the oxidative metabolism of motesanib by HLM (Li et al., 2009). In positive control experiments using anti-CYP antibodies (10 μg/0.1 mg of HLM) against known substrate reactions, CYP3A4 antibody inhibited 1′-OH midazolam formation from midazolam (5 μM) by 41.3%; CYP2D6 antibody inhibited 1′-OH bufuralol formation from bufuralol (15 μM) by 76.2%; and CYP1A antibody inhibited acetaminophen formation from phenacetin (50 μM) by 90.5%.

Friedberg et al., 1999). Microsomes from these cells, which contain a single human CYP enzyme with NADPH-CYP reductase with or without cytochrome b_5, are commercially available (e.g., Gentest Corp., Panvera Corp., and Oxford Biomedical Research Inc.). One of the attractive features of recombinant enzymes is that metabolism studies with drug candidate can be greatly simplified and the involvement of particular CYPs can be quickly assessed. In P450 reaction phenotyping, the investigational drug candidate is incubated with a battery of expressed enzymes at a well-defined substrate concentration. Valuable information on which CYP enzymes can and which ones cannot metabolize the drug candidate can be readily obtained (Mankowski, 1999). An example is shown in Figure 13.4, in which the metabolism of motesanib, a highly selective angiogenesis inhibitor, was evaluated in a panel of recombinant CYP enzymes (Li et al., 2009), and several recombinant CYP enzymes (CYP3A, CYP2D6, CYP1A1, CYP2B1, and CYP2B6) exhibited activity in catalyzing the formation of three oxidative metabolites of motesanib.

However, the major disadvantage of recombinant CYP enzymes is that they differ in their catalytic competency, and they are not expressed in cells at concentra-

tions that reflect their levels in native HLM. Furthermore, the recombinant CYP enzymes are usually expressed with much higher levels of NADPH-CYP reductase than those present in HLM. In addition, cytochrome b_5 coexpression can affect the kinetics of drug metabolism by certain CYP enzymes (Yamazaki et al., 1999). Therefore, simple evaluation of metabolism by recombinant CYP enzyme does not necessarily provide the extent to which that particular CYP contributes to the metabolism of the drug candidate. For those P450 recombinant enzymes for which activity is observed above a recombinant control, further examination of the enzyme kinetics to measure intrinsic clearance is warranted for any quantitative assessment of a particular CYP enzyme to the overall CYP-mediated reaction.

In order to develop a continuum for studies with recombinant systems to HLM and ultimately to *in vivo* metabolic clearance, two basic approaches have been proposed and used increasing over the years. One approach is the use of the "relative activity factor" (RAF) for quantitative scaling of single recombinant enzyme data to multienzyme systems such as microsomes or hepatocytes (Crespi and Miller, 1999; Roy et al., 1999; Stormer et al., 2000; Venkatakrishnan et al.,

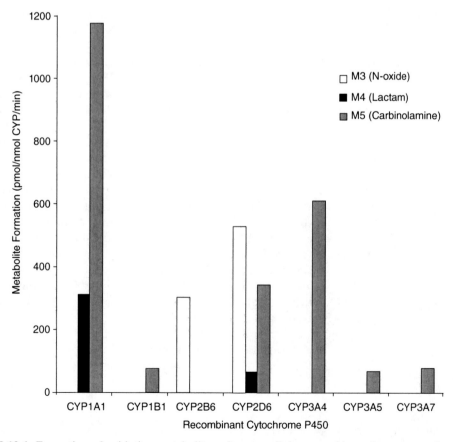

FIGURE 13.4. Formation of oxidative metabolites of motesanib by recombinant human cytochrome P450.

2000, 2001). In this approach, a standard marker CYP probe substrate is used to normalize an isoform-specific reaction activity in recombinant enzyme to that in HLM. RAFs are usually obtained by dividing the maximal activity (V_{max}) in HLM by the V_{max} of the same reaction catalyzed by the recombinant CYP enzyme. Thus, difference in substrate turnover by the recombinant CYP versus HLM can be factored. And the recombinant enzyme data for the investigation drug candidate can be converted to a microsomal equivalent using this established RAF. One should be well aware of the assumptions used in the method and use caution to interpret data generated. Appropriate care should be used in the choice of probe substrates used for the determination of RAFs, especially for enzymes that demonstrate substrate-dependent kinetics (e.g., CYP3A4). And RAFs should be generated in the same laboratory for both recombinant CYPs and HLM, and use the same lots of enzyme sources as used for the determination of enzyme kinetics of the investigated drug candidate.

Alternately, another quantitative approach scales the expressed enzyme activity using the relative abundance of specific P450 enzymes in liver microsomes determined by immunoquantitation (Rodrigues, 1999). The normalization process is relatively straightforward; the rate of reaction as measured by the metabolite formation in a recombinant CYP is multiplied by the mean specific content of the corresponding CYP in native HLM. The normalized rates for each CYP (which is expressed as pmol/min/mg microsomal protein) are summed to obtain the total normalized rate, and the normalized rate for each CYP is expressed as a percentage of the total normalized rate (Rodrigues, 1999). The normalized data can be related to chemical or antibody inhibition data and correlation regression analysis obtained with HLM, and be used as integrated assessment of reaction phenotyping. A modified scaling approach incorporating a scaling factor (intersystem extrapolation factor [ISEF]) to account for the potential differences in intrinsic activity per unit enzyme between recombinant enzyme and HLM has also been reported (Proctor et al., 2004). A comparison of the relative abundance approach, RAF based on V_{max}, and intrinsic clearance CL_{int} has also been published utilizing the substrate depletion approach, where the most accurate prediction method was found to be RAF based on CL_{int} (Emoto et al., 2006).

13.3.4 Correlation Analysis for CYP Reaction Phenotyping

Correlation analysis involves measuring the reaction velocity of the biotransformation pathway of interest in several individual human liver microsomal samples and correlating the reaction rates with the activity of the

individual CYP enzymes in the same bank of microsomal samples (Bjornsson et al., 2003; Lu et al., 2003). Typically, more than 10 human liver microsomal samples with well-characterized and large sample-to-sample variation in CYP activities are used. And for statistical purposes, it is important to select a bank of HLM in which the CYP enzyme activities do not correlate highly with each other. The study should be conducted under initial rate conditions and use pharmacologically relevant concentrations of drug candidate (Kassahun et al., 2001). Measured velocities of a biotransformation reaction for the investigated drug candidate are subjected to correlation analysis with P450 isoform-specific marker activities by simple regression analysis (r^2 coefficient of determination). If one P450 is responsible for the formation of a metabolite, the rates of metabolite formation will have a linear relationship with the catalytic activities of that particular P450 across the bank of human microsomal samples.

Correlation analysis provides valuable information on the extent to which the metabolism of a drug candidate will vary from one subject to another, and thereby gives an initial estimate of potential pharmacokinetic variability in the clinic. Correlation analysis works very well when a single enzyme is involved in a particular biotransformation of drug candidate. When two or more CYP enzymes contribute significantly to metabolite formation at pharmacologically relevant concentrations, correlation analysis may lack the statistical power to establish the identity of each enzyme. Multivariate regression analysis (Clarke, 1998) can be used in this case with some success as demonstrated for 1-hydroxylation of bufuralol (Sharer and Wrighton, 1996). Selective chemical inhibitors or inhibitory antibodies can also be used in the correlation analysis experiment (Yumibe et al., 1996; Johnson et al., 2000; Tang et al., 2000) to help delineate the relative contributions of CYP enzymes when more than one CYP is involved in the metabolism. The method allows for the residue microsomal catalytic activity to represent the contribution of the second P450 when the activity from the major contributing CYP is inhibited. While some investigators view correlation analysis as a less reliable method for CYP reaction phenotyping, and suggest only using it for confirmation study from inhibition and recombinant enzyme studies (Tucker et al., 2001; Lu et al., 2003), correlation analysis still represents a valuable tool in characterizing the metabolism of drug candidate by multiple P450s, and provides clinically relevant information.

13.3.5 CYP Reaction Phenotyping in Drug Discovery versus Development

Definitive CYP reaction phenotyping is often carried out in drug development stage. Typically, the study is conducted using multiple approaches, namely, antibody inhibition, chemical inhibition, metabolism of drugs by recombinant P450 enzymes, and correlation analysis (Zhang et al., 2007a). Measuring the formation of individual metabolites is often used, and it represents a scientifically superior approach in CYP reaction phenotyping as compared with the substrate depletion method. Individual reactions can be monitored, and conditions can be used such that linear initial reaction velocities can be measured. This approach, however, requires that either an authentic standard of the metabolite is available for construction of calibration curves or that radiolabeled material is available for radiometric HPLC quantitation of metabolites.

With the availability of commercial recombinant CYPs and potent and selective CYP inhibitors or antibodies, many companies are now conducting CYP reaction phenotyping studies early in drug discovery (Williams et al., 2003; Emoto et al., 2006). Experiments are typically conducted using pooled HLM and recombinant CYP isoforms with drug candidate (typically at low concentration of 1 μM), using the substrate depletion method. The effect of selective chemical inhibitors and/or inhibitory antibodies (sometimes in combination) can also be evaluated. These studies are typically done to evaluate the role of the five major CYPs (CYP1A2, 2C9, 2C19, 2D6, and 3A4) in the microsomal metabolism of the drug candidate. Information regarding as to whether enzymes with known genetic polymorphisms such as CYP2D6 and 2C19 play a major role in the oxidative metabolism of a lead drug candidate is useful. The objective of these studies at drug discovery stage is focused to select the best candidates for further development and to develop NCEs with no or minimal DDI liabilities and less dependency on polymorphic CYPs for clearance, and thereby reduce the risk of potential clinical DDIs or significant interpatient differences in pharmacokinetics in drug development. However, this method is only suitable for relatively high-turnover compounds. For metabolically stable compounds with less than 20% of depletion over the incubation period, it will not be practically possible to determine the relative contributions of each CYPs. Furthermore, one need to use caution and not overinterpret this data generated in drug discovery because the clearance mechanism of drug candidate in human is unknown, and contributions from other routes of elimination and the role of non-P450-mediated clearance may be significant.

13.4   NON-P450 REACTION PHENOTYPING

P450 enzymes play a dominant role in the metabolism of small-molecule drugs on the market and in develop-

ment (Williams et al., 2004). Other non-P450 drug metabolizing enzymes such as FMOs, MAOs, AO, esterases, and amidases can also contribute to the overall clearance of drug candidate (Beedham, 1997, Benedetti et al., 2006). Reaction phenotyping approaches for elucidating the contributions of non-P450 enzymes are still developing or remain to be developed. Similar to CYP reaction phenotyping, identification of contributions of non-P450 enzymes requires an understanding of the particular enzyme substrate specificity, mechanism of reaction, tissue expression and subcellular location, selective inhibitors, experimental conditions, and so on. As the underlying biochemistry of these enzymes and their impact on drug clearance *in vivo* becomes better understood, and more *in vitro* tools such as selective inhibitors and inhibitory antibodies become available, the principles applied for P450 reaction phenotyping could be applied to these enzymes as well.

13.4.1   FMOs

FMOs are NADPH-dependent oxidative enzymes, which much like the CYPs are present in the liver microsomal subcellular fraction. These enzymes catalyze oxidation of nitrogen-, sulfur-, selenium-, and phosphorus-containing compounds (Cashman, 2000). There are five FMO isozymes currently identified in humans: FMO1, FMO2, FMO3, FMO4, and FMO5 (Cashman, 2000). FMO1 is a major FMO isozyme in the fetal liver; its expression in the adult liver relative to the other FMOs is low. However, there is significant extrahepatic expression of FMO1 in the adult kidney and intestine (Yeung et al., 2000; Koukouritaki et al., 2002). The expression of FMO2 is also low in the liver, but it appears to be the major FMO isozyme in the adult lung (Dolphin et al., 1998; Krueger and Williams, 2005). FMO4 expression in the liver is the lowest of the known FMOs. FMO3 and FMO5 are considered to be the most highly expressed FMO isozymes in the adult liver, with FMO3 expression being greater than FMO5 (Lomri et al., 1992; Krueger and Williams, 2005). Due to its high hepatic expression and broad substrate specificity, FMO3 is generally considered the most important FMO isozyme for drug metabolism, and most published accounts relating to FMO reactions typically pertain to FMO3.

The experimental considerations involving FMO reactions have been expertly reviewed previously (Krueger and Williams, 2005; Cashman, 2008). The salient features will be discussed briefly here. FMOs are active under reaction conditions optimal for P450s (i.e., pH 7.4), but it has been suggested that the FMO activity can be enhanced significantly at pH 9–10 (Cashman, 2005). Buffer selection and incubation temperature for optimal FMO activity do not appear to vary significantly from those discussed for P450s. Because FMO is thermally labile in the absence of NADPH, an ideal incubation protocol to maximize FMO activity is to preincubate the enzyme source in the presence of NADPH and initiate the reaction by addition of substrate. It is essential to use well-characterized microsomes for these studies; well-characterized HLM are available from a variety of commercial sources. In cases where the preparation and handling of the microsomes is questionable, experiments to verify FMO activity are recommended.

FMOs are expressed in the same hepatic subcellular fraction as the CYPs, and use the same cofactor, NADPH. Hence both enzyme families can potentially contribute to the oxidative metabolism of drug candidates in HLM incubations supplemented with NADPH. Therefore, in certain cases, it is important to distinguish the contributions from FMOs versus P450s in order to have a better estimate of the fraction metabolized by a certain CYP isoform ($f_{m,cyp}$). Differentiation of FMO and P450 activity can be achieved by either selectively inactivating the P450 or the FMO enzymes (Grothusen et al., 1996). A pan P450 inhibitor such as ABT or detergent can be used to inactivate P450 activity. FMOs are considerably less susceptible to the degradation in the presence of detergents than P450s (Rettie et al., 1994). Thermal degradation and chemical inhibition can be used for inactivate FMO activity. FMOs are susceptible to thermal degradation, especially in the absence of NADPH. Approximately 60–80% of FMO3 activity is lost when the enzyme source is heated at 55°C for about a minute. The majority (~85%) of P450 activity, however, remains intact following the heat treatment. No FMO isozyme-specific chemical inhibitors have been identified. Methimazole, a FMO substrate, is often used *in vitro* as a competitive inhibitor at saturating concentration of ~200 μM (Attar et al., 2003; Wynalda et al., 2003). The results from a methimazole inhibition experiment should, however, be interpreted with caution, as FMO-mediated oxidation of methimazole has been shown to result in a reactive metabolite that inhibits P450 activity (Kedderis and Rickert, 1985). Recombinant FMOs are now commercially available, and as in the case of P450s and UGTs, kinetic studies with recombinant protein can be used for identification of FMO contribution to drug clearance.

Multiple polymorphisms in FMOs have been identified and have been reviewed in detail (Krueger and Williams, 2005; Motika et al., 2007). FMO3 polymorphisms are most studied in this regard. A large number of FMO3 variants result in a decrease in FMO3 activity, which often manifests itself in a condition called trimethylaminuria (TMAU) (Motika et al., 2007; Motika et al., 2009). FMO3 is the primary FMO isozyme involved with the oxidation of trimethylamine (TMA),

an endogenous compound formed as a product of choline metabolism. A deficiency in FMO3 results in a decreased clearance of TMA, due to which TMA is excreted in large amounts in breath, sweat, and urine. The severity of TMAU is an indicator of decrease in FMO3 activity. The incidence of this condition is, however, considered relatively rare (Mitchell and Smith, 2001). No inducers for FMOs have been reported, and given the enzyme mechanism, strong inhibition of FMOs, especially mechanism-based inhibition, is unlikely (Cashman, 2008). These aspects, coupled with the fact that there are currently very few drugs where FMO-mediated metabolism is the major clearance mechanism (Williams et al., 2004), suggest a low risk for FMO-related DDIs.

13.4.2 MAOs

MAOs oxidize amines, particularly biogenic amines involved as neurotransmitters such as dopamine and serotonin (Cashman, 1997). Primary, secondary, and tertiary amines with adjacent carbon atoms that have two abstractable hydrogen atoms are likely substrates for MAOs (Beedham, 2002). MAOs are favoprotein oxidases located in the mitochondria, and liver microsomes prepared from frozen liver tissue are often contaminated with mitochondrial membranes and therefore can contain MAO activity. Two forms of MAO are known, A and B, which differs in substrate selectivity and inhibitor sensitivity (Edmondson et al., 2004). Purified recombinant human MAO-A and MAO-B are now available (Edmondson et al., 2004). Relatively selective inhibitors have also been reported (Fowler et al., 1981); parglycine is a broad MAO inhibitor, clorgyline is a selective irreversible inhibitor of MAO-A, and L-deprenyl and selegiline are selective irreversible inhibitors of MAO-B. Identification of MAO involved in oxidative deamination reactions can, therefore, be readily assessed using these tools (Kalgutkar et al., 2001; Salva et al., 2003; Erickson et al., 2007).

Oxidative deamination of xenobiotics catalyzed by MAOs does not require NADPH as cofactor. The aldehyde metabolite formed can be further metabolized to alcohols either by aldehyde reductases or alcohol dehydrogenases, or to acids by aldehyde dehydrogenases or AOs (Strolin-Benedetti and Dostert, 1994; Beedham, 1997).

13.4.3 AO

AO is a molybdozyme present in the cytosolic fraction of the liver and other tissues of several mammalian species (Beedham, 2002; Garattini et al., 2008). Despite its name, AO not only catalyzes the oxidation of aldehydes to carboxylic acid but is also involved in nucleophilic oxidation of N-heterocycles. AO-mediated oxidation reactions result in metabolites different from CYP-mediated electrophilic oxidation (Garattini et al., 2008). AO is also involved in reductive reactions, for example, nitrosoaromatic to hydroxylamines, and isoxazoles to keto alcohols (Obach et al., 2004a). In contrast to CYPs, AO does not require NADPH as cofactor. In the absence of added cofactor, oxidative reactions in the cytosol are likely catalyzed by AO (and/or xanthine oxidase). AO plays an important role in the metabolism of a variety of drugs, including the antipsychotic agent ziprasidone (Beedham et al., 2003) and the antiepileptic agent zonisamide (Sugihara et al., 1996).

There is currently a single AO isozyme identified in humans, AOX1. AO expression varies profoundly between species, with highest AO activities reported in monkey and human, while lower activity is reported in rat. No AO expression has been detected in dog (Austin et al., 2001; Garattini et al., 2008). AO is considered a relatively minor enzyme involved in drug metabolism (Williams et al., 2004). However, there are compelling reasons to discuss AO reaction phenotyping in some detail. Due to the fact that AO is expressed in cytosol, AO metabolism is likely to go undetected in preliminary liver microsome clearance experiments. Moreover, even in metabolite identification experiments in systems such as human hepatocytes and S9 where AO activity is present, AO metabolism does not result in a diagnostic product like that for other cytosolic enzymes such as N-acetyltransferases (NATs), and so AO involvement may go unnoticed. Furthermore, due to a marked decrease in AO metabolism in commonly used preclinical species, rat and dog, AO metabolism can be potentially underestimated, which can lead to underprediction of human clearance predictions and/or overestimation of CYP-mediated contributions.

Identification of AO involvement in the metabolism of drug candidate can be facilitated using selective inhibitors such as menadione, chlorpromazine, hydralazine, and raloxifene in the liver cytosolic incubations (Lake et al., 2002; Obach et al., 2004a). Raloxifene is an especially potent inhibitor of AO, with a K_i of 0.87–51 nM, which has also been shown to be selective for AO over xanthine oxidase (Obach, 2004b).

There is currently no evidence of clinical drug interactions due to induction or inhibition of AO. In an *in vitro* context, many compounds appear to show strong AO inhibition. A study testing the effects of 239 marketed compounds and drugs on AO showed that 36 of those compounds inhibited AO activity by at least 80% (Obach et al., 2004a). There are currently no reports of functional polymorphisms involving AO (Beedham et al., 2003).

13.5 UGT CONJUGATION REACTION PHENOTYPING

The UGTs are a class of membrane-bound enzymes that catalyze the transfer of sugars (glucuronic acid, glucose, and xylose) to a variety of molecules in a reaction that is described as glucuronidation. The UGT substrates are typically small, hydrophobic molecules (either endogenous or exogenous), which are also often referred to as the aglycone. The sugar moiety (i.e., the glycone), which is typically glucuronic acid, can be conjugated to aliphatic and aromatic alcohols, thiols, primary, secondary, tertiary, and aromatic amino groups as well as acidic carbon atoms (Kiang et al., 2005). Conjugation with glucuronic acid results in an increase in water solubility and therefore excretion. This reaction follows an SN_2 mechanism involving a backside attack of the aglycone by the anomeric carbon of the glucuronic acid resulting in the formation of a β-glucuronide. Glucuronidation of parent drug is a listed clearance mechanism for 1 in 10 of the top 200 prescribed drugs in 2002 (Williams et al., 2004). The discussion of UGT phenotyping is limited to direct glucuronidation reactions as primary clearance pathway of parent drug, not secondary pathways of clearance of oxidative metabolites, which are catalyzed mostly by CYP enzymes.

In humans, the known UGT isozymes are split into two gene families, UGT1 and UGT2, based on amino acid sequence (Mackenzie et al., 1997 and 2005). Members of these two families are further split into three subfamilies, UGT1A, UGT2A, and UGT2B. The majority of the UGT isozymes involved in the metabolism of xenobiotics reside in the UGT1A and UGT2B subfamilies, and these isozymes will be the focus of the discussion from this point forward. The UGT1A subfamily consists of nine functional proteins: UGT1A1 and UGT1A3-UGT1A10. The UGT2B subfamily consists of seven functional proteins: UGT2B4, UGT2B7, UGT2B10, UGT2B11, UGT2B15, UGT2B17, and UGT2B28. The liver is the major site of glucuronidation. All of the isozymes from subfamilies UGT1A and UGT2B with the exception of UGT1A7, UGT1A8, and UGT1A10 are expressed in the liver (Strassburg et al., 1997; Tukey and Strassburg, 2000; Fisher et al., 2001). The intestine and kidney also show UGT expression. All UGT1A and UGT2B isozymes with the exception of UGT1A9, UGT2B11, UGT2B17, and UGT2B28 are expressed in the intestine (Strassburg et al., 1997; Cheng et al., 1998; Fisher et al., 2001). The UGTs expressed in the kidney include UGT1A9 and UGT2B7 (Fisher et al., 2001).

Based on current evidence, the UGT isoforms important in xenobiotic metabolism are UGT1A1, UGT1A3, UGT1A4, UGT1A6, UGT1A9, UGT2B7, and UGT2B15

(Minors et al., 2006). Of particular importance are UGT1A1 and UGT2B7. UGT1A1 is the only known human UGT to glucuronidate bilirubin (Burchell et al., 1995), a critical physiological metabolic reaction. UGT2B7 is considered a very important enzyme for the glucuronidation of many drugs (Williams et al., 2004).

13.5.1 Initial Considerations in UGT Reaction Phenotyping

There are two specific clinical considerations around UGT involvement in metabolism of drug candidates. The first is UGT1A1-catalyzed metabolism; UGT1A1 plays an important role in the elimination of bilirubin. Bilirubin is primarily eliminated by a UGT1A1-catalyzed glucuronidation followed by MRP2-mediated biliary efflux of the glucuronide (Kamisako et al., 2000). The inhibition of bilirubin glucuronidation by atazanavir, indinavir, and erlotinib has been hypothesized to play a role in the hyperbilirubinemia caused by these drugs (Zhang et al., 2005; Liu et al., 2010). The extent of hyperbilirubinemia observed with indinavir was, in some cases, severe enough to warrant discontinuation of therapy (Zucker et al., 2001). UGT1A1 is also responsible for the metabolism of several other endogenous and exogenous substrates, including 15% drugs that have glucuronidation as a clearance mechanism of the top 200 drugs in the United States in 2002 (Williams et al., 2004). The inhibition of UGT1A1 is particularly important for substrates with narrow therapeutic index, such as irinotecan and etoposide (Kawato et al., 1991; Wen et al., 2007). Therefore, drug candidates that are UGT1A1 substrates and/or inhibitors pose a potential risk of disrupting bilirubin homeostasis, and detailed reaction phenotyping studies can help define this risk.

The second consideration before UGT reaction phenotyping is UGT polymorphisms. There are polymorphisms reported with virtually every known UGT isozyme. These polymorphisms and their functional consequences have been reviewed elsewhere (Guillemette, 2003; Nagar and Remmel, 2006; Court, 2010). Some of the reported polymorphisms result in alterations in enzyme activity *in vitro*; however, the most profound clinically observed effects involve UGT1A1 polymorphisms. The most common UGT1A1 variant is UGT1A1*28, a variant with a promoter polymorphism that results in lower expression of the enzyme. Individuals with homozygous UGT1A1*28 tend to show mild, asymptomatic hyperbilirubinemia, a condition referred to as Gilbert's syndrome. The bilirubin clearance (an indicator of UGT1A1 activity) is decreased by approximately 40% (Tukey and Strassburg, 2000). It is estimated that the effects associated with Gilbert's syndrome are found in 6–12% of the population (Guillemette, 2003).

The UGT1A1*28 genotype has also been shown to be associated with decreased *in vivo* clearance of UGT1A1 substrates such as SN-38, the active metabolite of irinotecan (Ramchandani et al., 2007). Genotyped HLM with the UGT1A1*28*28 variant are now available commercially, and kinetic studies can be performed with these microsomes to assess the potential changes to metabolic clearance of new drug candidates.

13.5.2 Experimental Approaches for UGT Reaction Phenotyping

Three basic approaches can be used for UGT reaction phenotyping (Bauman et al., 2005), that is, the use of chemical inhibitors, recombinant enzymes, and correlation analysis using a phenotyped human liver bank, similar to those used in CYP reaction phenotyping. However, tools are not as complete and methods are not as refined as compared with CYP reaction phenotyping due to the relative lack of potent and specific chemical inhibitors, and a complete lack of commercially available inhibitory antibodies for UGTs.

13.5.2.1 Optimizing Conditions for Glucuronidation Incubations
Before the three phenotyping approaches are discussed in detail, the factors affecting UGT enzyme activity *in vitro* must be noted. UGT catalyzed reaction kinetics are considerably influenced by incubation conditions, much more so than reactions catalyzed by other drug metabolizing enzymes. In addition to the general experimental considerations as discussed for the CYP enzymes, there are some additional conditions to be optimized for glucuronidation incubations. These conditions often vary from substrate to substrate and so they should be considered on a case-by-case basis.

Unlike the CYP class of enzymes, the majority of the active UGT enzyme is expressed in the luminal side of the endoplasmic reticulum, which is believed to impede the access of the cofactor UDPGA to the enzyme (Shepherd et al., 1989; Yokota et al., 1992). Therefore, in order to measure UGT activity in microsomes, mechanisms to disrupt the endoplasmic reticulum (ER) membrane are required to facilitate access of cofactor. Chief among these are the use of mechanical methods such as sonication or chemical agents such as detergents or the pore-forming peptide alamethicin (Fisher et al., 2000). Detergents have been shown to affect enzyme activity in addition to increasing membrane permeability (Fulceri et al., 1994), and therefore alamethicin has fast become the method of choice for "activating"

UGT-catalyzed reactions (Boase and Miners, 2002). Alamethicin concentrations used are typically normalized to milligram of protein in the reaction and should be optimized for each individual reaction. The reaction velocity tends to increase proportionally with increasing concentration of alamethicin up to a concentration of about 25–50 µg/mg protein, after which minimal gains in reaction velocity are obtained (Soars et al., 2003). UDPGA, the cofactor for glucuronidation, is typically included in the reaction at saturating concentrations (2–5 mM). A source of magnesium, typically magnesium chloride (1–5 mM), appears to improve enzyme activity (Boase and Miners, 2002; Soars et al., 2003). Saccharolactone, an inhibitor of the enzyme β-glucuronidase, can be included in the *in vitro* reaction, although there are reports that the addition may have limited benefit (Boase and Miners, 2002).

Most commonly, glucuronidation incubations are conducted in phosphate buffer (0.1 M, pH 7.4). However, UGT activity can depend largely on incubation pH, buffer type, and ionic strength (Boase and Miners, 2002; Soars et al., 2003; Engtrakul et al., 2005). For substrates containing carboxylic acid, the acyl glucuronide metabolite formed is chemically unstable at pH values >7, and enzyme incubations can be conducted at pH 6.8 to minimize product hydrolysis (Miners et al., 1997). In a recent publication, glucuronidation enzyme kinetic parameters for acidic, basic, and neutral compounds at varying pH values were measured, and in all three cases the reaction rates were highest at the pH where the substrate was likely to be unionized and hence most permeable (Chang et al., 2009). In another example, UGT2B7-catalyzed 3′-azido-3′-deoxythymidine (AZT) glucuronidation showed roughly a fourfold difference among six buffer systems tested, and glucuronidation rates were greater in carbonate-based buffer systems when compared with phosphate or tris-based buffer systems (Engtrakul et al., 2005).

Finally, the addition of albumin to *in vitro* incubations should be considered, especially for glucuronidation reactions involving UGT1A9 and UGT2B7. It is postulated that long-chain unsaturated fatty acids present in both HLM and recombinant enzyme preparations are potent competitive inhibitors of these isozymes (Tsoutsikos et al., 2004). The addition of 2% bovine serum albumin (BSA) or 2% fatty acid free human serum albumin (HSA-FAF) has been shown to increase UGT2B7-catalyzed AZT glucuronidation CL_{int} by approximately 10-fold due to a decrease in K_m of the reaction without affecting V_{max} (Rowland et al., 2007). The observed effect is hypothesized to result from the fact that albumin binds these fatty acids, thereby resulting in a reduced free fraction of fatty acids that are available to inhibit the glucuronidation reaction.

13.5.2.2 Use of Recombinant UGT Enzymes

To date, at least 18 human UGTs have been identified, and 12 recombinantly expressed UGTs are commercially available. The use of recombinant enzymes plays an important role in UGT reaction phenotyping. After evaluation of glucuronidation kinetics of investigated drug candidate in HLM, glucuronidation reactions can be monitored and evaluated in a panel of recombinant UGTs to identify which UGTs are capable of catalyzing the reaction. Of the specific recombinant UGTs that demonstrated glucuronidation activities, the extent or significance of the involvement can be further evaluated by examining the kinetics of glucuronidation in recombinant UGTs. It should be noted that glucuronidation of many drugs by HLM and recombinant enzymes can exhibit "atypical" or non-Michaelis–Menten kinetics. Proper model fitting of experimental data is essential for calculation of kinetic data (Miners et al., 2010).

However, the absence of knowledge on the relative expression levels of UGTs in HLM or recombinant UGTs makes it difficult to quantitatively extrapolate the relative contributions of individual UGTs to the glucuronidation reaction in HLM. Data from recombinant UGTs should be integrated with information from selective chemical inhibition and possibly correlation analysis (Bauman et al., 2005).

13.5.3 Use of Chemical Inhibitors for UGTs

For phenotyping experiments in HLM or hepatocytes, UGT isozyme-specific inhibitors are required. The availability of UGT isozyme-specific inhibitors is currently limited, but the literature around this area is growing. On the other hand, a fair number of probe substrates are now available for the different UGT isozymes (see Table 13.2), and in some cases these compounds can be used as competitive inhibitors at saturating concentrations. There are currently potent and relatively selective inhibitors characterized for four UGT isozymes UGT1A1, 1A4, 1A9, and 2B7. Erlotinib has been shown to inhibit UGT1A1, with a K_i of 0.64 µM (Liu et al., 2010). The selectivity of the UGT1A1 inhibition at low concentrations was not determined, but at 100 µM concentration erlotinib inhibited UGT1A1-catalyzed 4-MU glucuronidation by ~90%, and was at least fivefold more selective for UGT1A1 than any other isozyme. Favonoids such as hesamethoxy-flavon and tangeretin (Williams et al., 2002, 2004) have also been identified as potent inhibitors of UGT1A1, but they also inhibit UGT1A6, 1A9, and possibly other UGT1A enzymes. Hecogenin is well characterized as a selective UGT1A4 inhibitor (Uchaipichat et al., 2006a). At 10 µM concentration, hecogenin inhibited >80% of UGT1A4 activity and was at least fourfold more selective for UGT1A4 than any other isozyme.

Niflumic acid has been shown as a selective inhibitor at UGT1A9 at low concentrations (~2.5 μM), but at higher concentrations it also inhibits UGT1A1 (Gaganis et al., 2007; Miners et al., 2010). Fluconazole (Uchaipichat et al., 2006b) has been proposed as an inhibitor of UGT2B7, although a recent report suggests that it may be acting as an alternate substrate for the enzyme (Bourcier et al., 2010). At concentrations of 1–2.5 mM, fluconazole inhibited UGT2B7-catalyzed AZT glucuronidation by 50–75%, and the inhibition was approximately twofold more selective for UGT2B7 than any other isozyme.

13.5.4 Correlation Analysis for UGT Reaction Phenotyping

Correlation analysis can be performed for UGT enzyme activities in individual HLM, and approach is similar to that described for CYP. This approach has been used successfully in some instances (Kaji and Kume, 2005; Katoh et al., 2007; Chen et al., 2009). For example, in an HLM bank of 22 livers, Katoh et al. found that translast glucuronidation correlated most significantly with bilirubin and estradiol glucuronidation, which are considered to be primarily catalyzed by UGT1A1. However, commercially available specific probe substrates have only been identified for certain human UGTs, as shown in Table 13.2. Furthermore, correlation analyses with most of these probe substrates are limited by the fact that there is often only three- to fivefold difference in the glucuronidation rates among the individual human liver samples.

13.6 REACTION PHENOTYPING FOR OTHER CONJUGATION REACTIONS

In addition to glucuronidation catalyzed by UGTs, other conjugation reactions catalyzed by SULTs, GSTs, and NATs) can also contribute to clearance of some drugs, but to a much lesser extent (Williams et al., 2004). Sulfonation of chemicals involves the conjugation of the substrate with a sulfonyl (SO_3^-) group. The cosubstrate 3′-phosphoadenosine-5′-phosphosulfate (PAPS) acts as the sulfonyl donor and the reaction is catalyzed by a SULT enzyme. Conjugation can occur at –C-OH, –N-OH, and –NH side chains to yield O-sulfates and N-sulfates. SULTs are cytosolic enzymes expressed mainly in the liver and small intestine, but also in other organs. To date, 11 SULT isoforms have been discovered in humans, and recombinant SULT enzymes are commercially available (Schneider and Glatt, 2004). Some SULT inhibitors have been reported (Pacifici and Coughtrie, 2005).

GSTs catalyze the reaction of endogenous nucleophile glutathione with electrophiles (e.g., epoxides, quinines, quinonemethides) to form glutathione conjugates. Reactive electrophilic metabolites of drugs, often formed by the CYP enzymes, are common substrates for the GSTs. There are six major classes of the human cytosolic GST enzymes (Mannervik et al., 2005). Methods for SULT, GST, and NAT reaction phenotyping are not well developed, and their contributions to overall clearance of most drug candidates are typically not significant.

TABLE 13.2. Examples of Selective Substrates and Inhibitors of Major UGTs Isozymes Involved in Drug Metabolism

UGT Isozyme	Substrate	Inhibitors	Inhibitor Concentration Used for UGT Phenotyping
UGT1A1	Bilirubin β-Estradiol Etoposide	Erlotinib Atazanavir	100 μM
UGT1A3	NorUDCA 23-glucuronidation R-Lorazepam		
UGT1A4	Trifluoperazine Imipramine	Hecogenin	10 μM
UGT1A6	Serotonin Deferiprone	Serotonin	
UGT1A9	Propofol	Niflumic acid	2.5 μM
UGT2B7	Zidovudine (AZT) Morphine Epirubicin	Fluconazole	2500 μM
UGT2B15	S-Oxazepam S-Lorazepam Esculetin		

Sources: Bosma et al., 1994; Innocenti et al., 2001; Nakajima et al., 2002; Court, 2005; Court et al., 2003; Krishnaswamy et al., 2003; Zhang et al., 2005; Uchaipichat et al., 2006a; Gaganis et al., 2007; Benoit-Biancamano et al., 2009; Miners et al., 2010; Trottier et al., 2010.

13.7 INTEGRATION OF REACTION PHENOTYPING AND PREDICTION OF DDI

As already discussed, the goal of metabolic reaction phenotyping is to determine the relative contribution of a given drug metabolizing enzyme to the overall metabolism of a drug candidate, that is, $f_{m,enz}$ ($f_{m,cyp}$ or $f_{m,UGT}$). Reaction phenotyping allows an assessment of the victim potential of a drug candidate. Victim potential can be quantified on the basis of fractional metabolism, as shown in the following equations:

$$\frac{\text{AUC}_{po,I}}{\text{AUC}_{po,ctr}} = \left(\frac{f_{ab}'}{f_{ab}}\right) \cdot \left(\frac{f_g'}{f_g}\right) \cdot \left(\frac{f_h'}{f_h}\right) \cdot \left(\frac{\text{CL}}{\text{CL}'}\right) \quad (13.1)$$

$$\frac{\text{AUC}_{po,I}}{\text{AUC}_{po,ctr}} \approx \frac{1}{\dfrac{f_m \cdot f_{m,enz}}{1 + \dfrac{[I]}{K_i}} + (1 - f_m \cdot f_{m,enz})}, \quad (13.2)$$

where $\text{AUC}_{po,I}$ and $\text{AUC}_{po,ctl}$ are plasma AUC values in the presence or absence of an enzyme inhibitor, respectively. In addition, $f_{ab}', f_g', f_h', \text{CL}',$ and $f_{ab}, f_g, f_h, \text{CL}$ represent the fraction of drug absorbed, fraction of drug escaped gut metabolism, fraction of drug escaped hepatic extraction, and intrinsic clearance in the presence of inhibitor and in the absence of inhibitor, respectively. Assuming the inhibitor does not affect the absorption or gut metabolism of a victim drug, the magnitude of drug interaction by reversible inhibition as measured by the fold AUC increase of the victim drug can be estimated by Equation 13.2 (Ito et al., 1998, 2005; Rodrigues et al., 2001; Brown et al., 2005; Obach et al., 2006). The magnitude of interaction depends on the inhibitor concentration [I] and its inhibitory constant K_i, but it also largely depends on the product of f_m and $f_{m,enz}$, which defines the victim potential of the affected drug. f_m and $f_{m,enz}$ represent fraction of dose eliminated by metabolism, and fraction catalyzed by the affected drug metabolizing enzyme. The product $f_m*f_{m,enz}$ will also be the major factor governing the impact of a polymorphism on total clearance of the drug candidate in investigation. As shown in Figure 13.5, victim drugs with higher $f_m*f_{m,cyp}$ will be more sensitive to inhibition of that CYP-mediated pathway, and have greater magnitude of exposure increase in the presence of a perpetrator.

Overall contribution of metabolism to the total drug clearance in human (f_m) can be obtained from human ADME studies, and sometimes can be estimated based on various *in vitro* approaches and *in vivo* ADME studies of preclinical species. Fractional contribution from a specific metabolizing enzyme ($f_{m,enz}$) can be determined by reaction phenotyping approaches as dis-

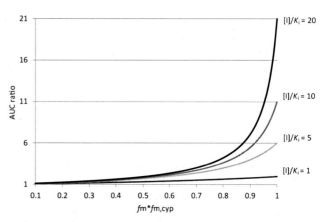

FIGURE 13.5. The effect of fractional metabolism by a particular CYP, $f_m*f_{m,cyp}$, on the magnitude of drug interaction (as measured by fold increase in AUC) to a victim drug with increasing inhibition by a CYP inhibitor.

cussed in above sections. Because discrepancies between different *in vitro* approaches may arise, a combination of two to three approaches is always recommended. The *in vitro* data set should be aligned or integrated whenever possible with appropriate clinical information from human radiolabel study and drug interaction studies with probe drugs (Zhang et al., 2007a). With the integrated information, the fraction of the dose cleared via all CYPs (f_m), and the contribution of each CYP to total CYP-dependent metabolism ($f_{m,cyp}$) can be estimated (Zhang et al., 2007a). The magnitude of a drug interaction, and the impact of a CYP polymorphism on the pharmacokinetic profile of the victim drug, which are governed by the product of f_m and $f_{m,cyp}$ (i.e., $f_m \times f_{m,cyp}$), can then be more accurately assessed. Majority of reported clinically significant DDIs are CYP mediated, which is consistent with the fact that CYP enzymes represent majority of clearance mechanisms for small-molecule drug candidates (Williams et al., 2004; Wienkers and Heath, 2005). For drugs cleared primarily by glucuronidation, the likelihood of significant pharmacokinetic drug-drug interaction is low due to low affinity of drugs for UGTs and multiple enzyme families contributing to the metabolism. In clinic, the exposure increases of UGT substrates are rarely greater than 2 in the presence of UGT inhibitors (Williams et al., 2004).

13.8 CONCLUSION

Reaction phenotyping is an important activity during drug development process, and the data is often used in support of claims in product label, and for planning of clinical studies such as drug interaction studies or assessment of special patient populations. Ideally, drug candidate should have a balanced clearance mechanism to minimize the potential as a victim of high magnitude of DDI or significant interindividual variability in pharmacokinetics due to polymorphism. Conducting reaction phenotyping experiments in drug discovery and development to assess the relative contributions of various drug metabolizing enzymes to the overall metabolism helps to define these potential risks.

P450 enzymes play a prominent role in the clearance of many drugs and drug candidates, and experimental approaches are well defined for CYP reaction phenotyping. Other drug metabolizing enzymes such as FMOs, MAOs, and UGTs can also be important when the drug candidate is metabolized primarily by non-P450 enzymes. Tools and methods for non-P450 reaction phenotyping are not as well established; studies are typically conducted to answer specific questions regarding the potential involvement of these enzymes in the metabolism of drug candidate, often not quantitative. Several *in vitro* approaches are available for reaction phenotyping, and it is extremely important to use integrated and combined approaches for a complete overall assessment. The *in vitro* data should also be integrated with clinical data from human ADME study to allow a rational prediction of DDI potential, and to help plan the potential follow-up clinical DDI studies.

REFERENCES

Attar M, Dong D, Ling KH, Tang-Liu DD (2003) Cytochrome P450 2C8 and flavin-containing monooxygenases are involved in the metabolism of tazarotenic acid in humans. *Drug Metab Dispos* 31:476–481.

Austin NE, Baldwin SJ, Cutler L, Deeks N, Kelly PJ, Nash M, Shardlow CE, Stemp G, Thewlis K, Ayrton A, Jeffrey P (2001) Pharmacokinetics of the novel, high-affinity and selective dopamine D3 receptor antagonist SB-277011 in rat, dog and monkey: *In vitro/in vivo* correlation and the role of aldehyde oxidase. *Xenobiotica* 31:677–686.

Bauman JN, Goosen TC, Tugnait M, Peterkin V, Hurst SI, Menning LC, Milad M, Court MH, Williams JA (2005) UDP-glucuronosyltransferase 2b7 is the major enzyme responsible for gemcabene glucuronidation in human liver microsomes. *Drug Metab Dispos* 33:1349–1354.

Beedham C (1997) The role of non-P450 enzymes in drug oxidation. *Pharm World Sci* 19:255–263.

Beedham C (2002) Molybdenum hydrolyases. In *Enzyme Systems that Metabolise Drugs and Other Xenobiotics*, Ioannides C, ed., pp. 147–187. Wiley, Hoboken, NJ.

Beedham C, Miceli JJ, Obach RS (2003) Ziprasidone metabolism, aldehyde oxidase, and clinical implications. *J Clin Psychopharmacol* 23:229–232.

Benedetti MS, Whomsley R, Baltes E (2006) Involvement of enzymes other than CYPs in the oxidative metabolism of xenobiotics. *Expert Opin Drug Metab Toxicol* 2:895–921.

Benoit-Biancamano MO, Connelly J, Villeneuve L, Caron P, Guillemette C (2009) Deferiprone glucuronidation by human tissues and recombinant UDP glucuronosyltransferase 1A6: An *in vitro* investigation of genetic and splice variants. *Drug Metab Dispos* 37:322–329.

Bjornsson TD, Callaghan JT, Einolf HJ, Fischer V, Gan L, Grimm S, Kao J, King SP, Miwa G, Ni L, Kumar G, McLeod J, Obach RS, Roberts S, Roe A, Shah A, Snikeris F, Sullivan JT, Tweedie D, Vega JM, Walsh J, Wrighton SA (2003) The conduct of *in vitro* and *in vivo* drug-drug interaction studies: A Pharmaceutical Research and Manufacturers of America (PhRMA) perspective. *Drug Metab Dispos* 31:815–832.

Boase S, Miners JO (2002) *In vitro-in vivo* correlations for drugs eliminated by glucuronidation: Investigations with the model substrate zidovudine. *Br J Clin Pharmacol* 54:493–503.

Boernsen KO, Floeckher JM, Bruin GJ (2000) Use of a microplate scintillation counter as a radioactivity detector for miniaturized separation techniques in drug metabolism. *Anal Chem* 72:3956–3959.

Bosma PJ, Seppen J, Goldhoorn B, Bakker C, Oude Elferink RP, Chowdhury JR, Chowdhury NR, Jansen PL (1994) Bilirubin UDP-glucuronosyltransferase 1 is the only relevant bilirubin glucuronidating isoform in man. *J Biol Chem* 269:17960–17964.

Bourcier K, Irvine N, Maximilien J, Hyland R, Kempshall S, Jones RM, Jones B (2010) Investigation into UDP-glucuronosyltransferase enzyme kinetics of imidazole and triazole containing antifungal drugs in human liver microsomes and recombinant UGT enzymes. *Drug Metab Dispos* 38:923–929.

Bourrie M, Meunier V, Berger Y, Fabre G (1996) Cytochrome P450 isoform inhibitors as a tool for the investigation of metabolic reactions catalyzed by human liver microsomes. *J Pharmacol Exp Ther* 277:321–332.

Brown CM, Reisfeld B, Mayeno AN (2008) Cytochrome P450: A structure-based summary of biotransformation using representative substrates. *Drug Metab Rev* 40:1–100.

Brown HS, Ito K, Galetin A, Houston JB (2005) Prediction of *in vivo* drug–drug interactions from *in vitro* data: Impact of incorporating parallel pathways of drug elimination and inhibitor absorption rate constant. *Br J Clin Pharmacol* 60:508–518.

Bruin GJ, Waldmeier F, Boernsen KO, Pfaar U, Gross G, Zollinger M (2006) A microplate solid scintillation counter as a radioactivity detector for high performance liquid chromatography in drug metabolism: Validation and applications. *J Chromatogr A* 1133:184–194.

Burchell B, Brierley C, Rance D (1995) Specificity of human UDP-glucuronosyltransferases and xenobiotic glucuronidation. *Life Sci* 57:1819–1831.

Cai X, Wang RW, Edom RW, Evans DC, Shou M, Rodrigues AD, Liu W, Dean DC, Baillie TA (2004) Validation of (-)-N-3-benzyl-phenobarbital as a selective inhibitor of CYP2C19 in human liver microsomes. *Drug Metab Dispos* 32:584–586.

Cashman JR (1997) Monoamine oxidase and flavin-containing monooxygenases. In *Comprehensive Toxicology: Biotrans-*

formation, Vol. 3, Guengerich FP, ed., pp. 69–96. Elsevier Science, Oxford.

Cashman JR (2000) Human flavin-containing monooxygenase: Substrate specificity and role in drug metabolism. *Curr Drug Metab* 1:181–191.

Cashman JR (2005) Some distinctions between flavin-containing and cytochrome P450 monooxygenases. *Biochem Biophys Res Commun* 338:599–604.

Cashman JR (2008) Role of flavin-containing monooxygenase in drug development. *Expert Opin Drug Metab Toxicol* 4:1507–1521.

Chang JH, Yoo P, Lee T, Klopf W, Takao D (2009) The role of pH in the glucuronidation of raloxifene, mycophenolic acid and ezetimibe. *Mol Pharm* 6:1216–1227.

Chauret N, Gauthier A, Nicoll-Griffith DA (1998) Effect of common organic solvents on *in vitro* cytochrome P450-mediated metabolic activities in human liver microsomes. *Drug Metab Dispos* 26:1–4.

Chen M, LeDuc B, Kerr S, Howe D, Williams DA (2009) Identification of human UGT2B7 as the major isoform involved in the O-glucuronidation of chloramphenicol. *Drug Metab Dispos* 38:368–375.

Cheng Z, Radominska-Pandya A, Tephly TR (1998) Cloning and expression of human UDP-glucuronosyltransferase (UGT) 1A8. *Arch Biochem Biophys* 356:301–305.

Clarke SE (1998) *In vitro* assessment of human cytochrome P450. *Xenobiotica* 28:1167–1202.

Court MH (2005) Isoform-selective probe substrates for *in vitro* studies of human UDP-glucuronosyltransferases. *Methods Enzymol* 400:104–116.

Court MH (2010) Interindividual variability in hepatic drug glucuronidation: Studies into the role of age, sex, enzyme inducers, and genetic polymorphism using the human liver bank as a model system. *Drug Metab Rev* 42:202–217.

Court MH, Krishnaswamy S, Hao Q, Duan SX, Patten CJ, Von Moltke LL, Greenblatt DJ (2003) Evaluation of 3′-azido-3′-deoxythymidine, morphine and codeine as probe substrates for UDP-glucuronosyltransferase 2B7 (UGT2B7) in human liver microsomes: Specificity and influence of the UGT2B7*2 polymorphism. *Drug Metab Dispos* 31:1125–1133.

Crespi CL, Miller VP (1999) The use of heterologously expressed drug metabolizing enzymes—State of the art and prospects for the future. *Pharmacol Ther* 84:121–131.

Cruciani G, Carosati E, De Boeck B, Ethirajulu K, Mackie C, Howe T, Vianello R (2005) MetaSite: Understanding metabolism in human cytochromes from the perspective of the chemist. *J Med Chem* 48:6970–6979.

Dalmadi B, Leibinger J, Szeberényi S, Borbás T, Farkas S, Szombathelyi Z, Tihanyi K (2003) Identification of metabolic pathways involved in the biotransformation of tolperisone by human microsomal enzymes. *Drug Metab Dispos* 31:631–636.

Dolphin CT, Beckett DJ, Janmohamed A, Cullingford TE, Smith RL, Shephard EA, Phillips IR (1998) The flavin-containing monooxygenase 2 gene (FMO2) of humans, but not of other primates, encodes a truncated, nonfunctional protein. *J Biol Chem* 273:30599–30607.

Dresser GK, Spence JD, Bailey DG (2000) Pharmacokinetic pharmacodynamic consequences and clinical relevance of cytochrome P450 3A4 inhibition. *Clin Pharm* 38:41–57.

Edmondson DE, Mattevi A, Binda C, Li M, Hubalek F (2004) Structure and mechanism of monoamine oxidase. *Curr Med Chem* 11:1983–1993.

Emoto C, Murase S, Iwasaki K (2006) Approach to the prediction of the contribution of major cytochrome P450 enzymes to drug metabolism in the early drug-discovery stage. *Xenobiotica* 36:671–683.

Engtrakul JJ, Foti RS, Strelevitz TJ, Fisher MB (2005) Altered AZT (3′-azido-3′-deoxythymidine) glucuronidation kinetics in liver microsomes as an explanation for underprediction of *in vivo* clearance: Comparison to hepatocytes and effect of incubation environment. *Drug Metab Dispos* 33:1621–1627.

Erickson DA, Hollfelder S, Tenge J, Gohdes M, Burkhardt JJ, Krieter PA (2007) *In vitro* metabolism of the analgesic bicifadine in the mouse, rat, monkey, and human. *Drug Metab Dispos* 35:2232–2241.

Fisher MB, Campanale K, Ackermann BL, VandenBranden M, Wrighton SA (2000) *In vitro* glucuronidation using human liver microsomes and the pore-forming peptide alamethicin. *Drug Metab Dispos* 28:560–566.

Fisher MB, Paine MF, Strelevitz TJ, Wrighton SA (2001) The role of hepatic and extrahepatic UDP-glucuronosyltransferases in human drug metabolism. *Drug Metab Rev* 33:273–297.

Fowler CJ, Oreland L, Callingham BA (1981) The acetylenic monoamine oxidase inhibitors, clorgyline, deprenyl, pargyline and J-508: Their properties and applications. *J Pharm Pharmacol* 33:341–347.

Friedberg T, Pritchard MP, Bandera M, Hanlon SP, Yao D, McLaughlin LA, Ding S, Burchell B, Wolf CR (1999) Merits and limitations of recombinant models for the study of human P450-mediated drug metabolism and toxicity: An inter-laboratory comparison. *Drug Metab Rev* 31:523–544.

Fulceri R, Banhegyi G, Gamberucci A, Giunti R, Mandl J, Benedetti A (1994) Evidence for the intraluminal positioning of p-nitrophenol UDP-glucuronosyltransferase activity in rat liver microsomal vesicles. *Arch Biochem Biophys* 309:43–46.

Gaganis P, Miners JO, Knights KM (2007) Glucuronidation of fenamates: Kinetic studies using human kidney cortical microsomes and recombinant UDP-glucuronosyltransferase (UGT) 1A9 and 2B7. *Biochem Pharmacol* 73:1683–1691.

Garattini E, Fratelli M, Terao M (2008) Mammalian aldehyde oxidases: Genetics, evolution and biochemistry. *Cell Mol Life Sci* 65:1019–1048.

Gelboin HV, Krausz KW, Gonzalez FJ, Yang TJ (1999) Inhibitory monoclonal antibodies to human cytochrome P450 enzymes: A new avenue for drug discovery. *Trends Pharmacol Sci* 20:432–438.

Giacomini KM, Huang SM, Tweedie DJ, et al. (2010) Membrane transporters in drug development. *Nat Rev Drug Discov* 9:215–236.

Grothusen A, Hardt J, Bräutigam L, Lang D, Böcker R (1996) A convenient method to discriminate between cytochrome

P450 enzymes and flavin-containing monooxygenases in human liver microsomes. *Arch Toxicol* 71:64–71.

Guengerich FP (2000) Pharmacogenomics of cytochrome P450 and other enzymes involved in biotransformation of xenobiotics. *Drug Dev Res* 49:4–16.

Guengerich FP (2007) Oxidative, reductive, and hydrolytic metabolism of drugs. In *Drug Metabolism in Drug Design and Development*, Zhang D, Zhu M, Humphreys WG, eds., pp. 15–35. Wiley-Interscience, Hoboken, NJ.

Guengerich FP, Parikh A, Johnson EF, Richardson TH, von Wachenfeldt C, Cosme J, Jung F, Strassburg CP, Manns MP, Tukey RH, Pritchard M, Fournel-Gigleux S, Burchell B (1997) Heterologous expression of human drug-metabolizing enzymes. *Drug Metab Dispos* 25:1234–1241.

Guillemette C (2003) Pharmacogenomics of human UDP-glucuronosyltransferase enzymes. *Pharmacogenomics J* 3:136–158.

Harper TW, Brassil PJ (2008) Reaction phenotyping: Current industry efforts to identify enzymes responsible for metabolizing drug candidates. *AAPS J* 10(1):200–207.

Hickman D, Wang J, Wang Y, Unadkat JD (1998) Evaluation of the selectivity of *in vitro* probes and suitability of organic solvents for the measurement of human cytochrome P450 monooxygenase activities. *Drug Metab Dispos* 26:207–215.

Hofstee BHJ (1952) On the evaluation of the constants V_m and K_m in enzyme reactions. *Science* 116:329–331.

Houston JB (1994) Utility of *in vitro* drug metabolism data in predicting *in vivo* metabolic clearance. *Biochem Pharmacol* 47:1469–1479.

Houston JB, Galetin A (2003) Progress towards prediction of human pharmacokinetic parameters from *in vitro* technologies. *Drug Metab Rev* 35:393–415.

Innocenti F, Iyer L, Ramírez J, Green MD, Ratain MJ (2001) Epirubicin glucuronidation is catalyzed by human UDP-glucuronosyltransferase 2B7. *Drug Metab Dispos* 29:686–692.

Ito K, Iwatsubo T, Kanamitsu S, Ueda K, Suzuki U (1998) Prediction of pharmacokinetic alterations caused by drug–drug interactions: Metabolic interaction in the liver. *Pharmacol Rev* 50:387–412.

Ito K, Hallifax D, Obach RS (2005) Impact of parallel pathways of drug elimination and multiple cytochrome P450 involvement on drug–drug interactions: CYP2D6 paradigm. *Drug Metab Dispos* 33:837–844.

Johnson JA, Herring VL, Wolfe MS, Relling MV (2000) CYP1A2 and CYP2D6 4-hydroxylate propranolol and both reactions exhibit racial differences. *J Pharmacol* 293:453–459.

Kaji H, Kume T (2005) Characterization of afloqualone N-glucuronidation: Species differences and identification of human UDP-glucuronosyltransferase isoform(s). *Drug Metab Dispos* 33:60–67.

Kalgutkar AS, Dalvie DK, Castagnoli N Jr., Taylor TJ (2001) Interactions of nitrogen-containing xenobiotics with monoamine oxidase (MAO) isozymes A and B: SAR studies on MAO substrates and inhibitors. *Chem Res Toxicol* 14:1139–1162.

Kalvass JC, Tess DA, Giragossian C, Linhares MC, Maurer TS (2001) Influence of microsomal concentration on apparent intrinsic clearance: Implications for scaling *in vitro* data. *Drug Metab Dispos* 29:1332–1336.

Kamisako T, Kobayashi Y, Takeuchi K, Ishihara T, Higuchi K, Tanaka Y, Gabazza EC, Adachi Y (2000) Recent advances in bilirubin metabolism research: The molecular mechanism of hepatocyte bilirubin transport and its clinical relevance. *J Gastroenterol* 35:659–664.

Kassahun K, McIntosh IS, Shou M, Walsh DJ, Rodeheffer C, Slaughter DE, Geer LA, Halpin RA, Agrawal N, Rodrigues AD (2001) Role of human liver cytochrome P4503A in the metabolism of etoricoxib, a novel cyclooxygenase-2 selective inhibitor. *Drug Metab Dispos* 29:813–820.

Katoh M, Matsui T, Yokoi T (2007) Glucuronidation of antiallergic drug, Tranilast: Identification of human UDP-glucuronosyltransferase isoforms and effect of its phase I metabolite. *Drug Metab Dispos* 35:583–589.

Kawato Y, Aonuma M, Hirota Y, Kuga H, Sato K (1991) Intracellular roles of SN-38, a metabolite of the camptothecin derivative CPT-11, in the antitumor effect of CPT-11. *Cancer Res* 51:4187–4191.

Kedderis GL, Rickert DE (1985) Loss of rat liver microsomal cytochrome P-450 during methimazole metabolism. Role of flavin-containing monooxygenase. *Drug Metab Dispos* 13:58–61.

Kiang TK, Ensom MH, Chang TK (2005) UDP-glucuronosyltransferases and clinical drug-drug interactions. *Pharmacol Ther* 106:97–132.

Kiffe M, Jehle A, Ruembeli R (2003) Combination of high-performance liquid chromatography and microplate scintillation counting for crop and animal metabolism studies: A comparison with classic on-line and thin-layer chromatography radioactivity detection. *Anal Chem* 75:723–730.

Koukouritaki SB, Simpson P, Yeung CK, Rettie AE, Hines RN (2002) Human hepatic flavin-containing monooxygenases 1 (FMO1) and 3 (FMO3) developmental expression. *Pediatr Res* 51:236–243.

Krausz KW, Goldfarb I, Buters JTM, Yang TJ, Gonzalez FJ, Gelboin HV (2001) Monoclonal antibodies specific and inhibitory to human cytochromes P450 2C8, 2C9 and 2C19. *Drug Metab Dispos* 29:1410–1423.

Krishna DR, Klotz U (1994) Extrahepatic metabolism of drugs in humans. *Clin Pharmacokinet* 26:144–160.

Krishnaswamy S, Duan SX, von Moltke LL, Greenblatt DJ, Court MH (2003) Validation of serotonin (5-hydroxytryptamine) as an *in vitro* substrate probe for human UDP-glucuronosyltransferase (UGT) 1A6. *Drug Metab Dispos* 31:133–139.

Krueger SK, Williams DE (2005) Mammalian flavin-containing monooxygenases: Structure/function, genetic polymorphisms and role in drug metabolism. *Pharmacol Ther* 106:357–387.

Lake BG, Ball SE, Kao J, Renwick AB, Price RJ, Scatina JA (2002) Metabolism of zaleplon by human liver: Evidence for involvement of aldehyde oxidase. *Xenobiotica* 32:835–847.

Lewis DFV, Jacobs MN, Dickin M (2004) Compound lipophilicity for substrate binding to human P450s in drug metabolism. *Drug Discovery Today* 9(12):530–537.

Li C, Kuchimanchi M, Hickman D, Poppe L, Hayashi M, Zhou Y, Subramanian R, Kumar G, Surapaneni S (2009) In vitro metabolism of the novel, highly selective oral angiogenesis inhibitor motesanib diphosphate in preclinical species and in humans. *Drug Metab Dispos* 37:1378–1394.

Lin JH, Lu AYH (1998) Inhibition and induction of cytochrome P450 and the clinical implications. *Clin Pharmacokinet* 35:361–390.

Linder C, Renaud N, Hutzler J (2009) Is 1-aminobenzotriazole an appropriate in vitro tool as a nonspecific cytochrome P450 inactivator? *Drug Metab Dispos* 37:10–13.

Lineweaver H, Burk D (1934) The determination of enzyme dissociation constants. *J Am Chem Soc* 56:658–666.

Liu Y, Ramirez J, House L, Ratain MJ (2010) Comparison of the drug-drug interactions potential of erlotinib and gefitinib via inhibition of UDP-glucuronosyltransferases. *Drug Metab Dispos* 38:32–39.

Lomri N, Gu Q, Cashman JR (1992) Molecular cloning of the flavin-containing monooxygenase (form II) cDNA from adult human liver. *Proc Natl Acad Sci U S A* 89:1685–1689.

Lu AY, Wang RW, Lin JH (2003) Cytochrome P450 in vitro reaction phenotyping: A re-evaluation of approaches used for P450 isoform identification. *Drug Metab Dispos* 31:345–350.

Mackenzie PI, Owens IS, Burchell B, Bock KW, Bairoch A, Belanger A, Fournel-Gigleux S, Green M, Hum DW, Iyanagi T, Lancet D, Louisot P, Magdalou J, Chowdhury JR, Ritter JK, Schachter H, Tephly TR, Tipton KF, Nebert DW (1997) The UDP glycosyltransferase gene superfamily: Recommended nomenclature update based on evolutionary divergence. *Pharmacogenetics* 7:255–269.

Mackenzie PI, Bock KW, Burchell B, Guillemette C, Ikushiro S, Iyanagi T, Miners JO, Owens IS, Nebert DW (2005) Nomenclature update for the mammalian UDP glycosyltransferase (UGT) gene superfamily. *Pharmacogenet Genomics* 15:677–685.

Madan AJ (2002) In vitro approaches for studying the inhibition of drug-metabolizing enzymes and identifying the drug metabolizing enzymes responsible for the metabolism of drugs. In *Drug-Drug Interactions*, Rodrigues AD, ed., pp. 217–294. Informa Healthcare, New York.

Mancy A, Dijols S, Poli S, Guengerich P, Mansuy D (1996) Interaction of sulfaphenazole derivatives with human liver cytochromes P450 2C: Molecular origin of the specific inhibitory effects of sulfaphenazole on CYP 2C9 and consequences for the substrate binding site topology of CYP 2C9. *Biochemistry* 35:16205–16212.

Mankowski DC (1999) The role of CYP2C19 in the metabolism of (+/−) bufuralol, the prototypic substrate of CYP2D6. *Drug Metab Dispos* 27:1024–1028.

Mannervik B, Board PG, Hayes JD, Listowsky I, Pearson WR (2005) Nomenclature for mammalian soluble glutathione transferases. *Methods Enzymol* 401:1–8.

McGinnity DF, Waters NJ, Tucker J, Riley RJ (2008) Integrated in vitro analysis for the in vivo prediction of cytochrome P450 mediated drug-drug interactions. *Drug Metab Dispos* 36:1126–1134.

Mei Q, Tang C, Assang C, Lin Y, Slaughter D, Rodrigues D, Baillie TA, Rushmore TH, Shou M (1999) Role of a potent inhibitory monoclonal antibody to cytochrome P450 in assessment of human drug metabolism. *J Pharmacol Exp Ther* 291:749–759.

Miners JO, Valente L, Lillywhite KJ, Mackenzie PI, Burchell B, Baguley BC, Kestell P (1997) Preclinical prediction of factors influencing the elimination of 5,6-dimethylxanthenone-4-acetic acid, a new anticancer drug. *Cancer Res* 57:284–289.

Miners JO, Mackenzie PI, Knights KM (2010) The prediction of drug-glucuronidation parameters in humans: UDP-glucuronosyltransferase enzyme-selective substrate and inhibitor probes for reaction phenotyping and in vitro-in vivo extrapolation of drug clearance and drug-drug interaction potential. *Drug Metab Rev* 42:196–208.

Minors J, Knights K, Houston J, Mackenzie P (2006) In vitro–in vivo correlation for drugs and other compounds eliminated by glucuronidation in humans: Pitfalls and promises. *Biochemical Pharmacology* 71:1531–1539.

Mitchell SC, Smith RL (2001) Trimethylaminuria: The fish malodor syndrome. *Drug Metab Dispos* 29:517–521.

von Moltke LL, Greenblatt DJ, Schmider J, Wright CE, Harmatz JS, Shader RI (1998) In vitro approaches to predicting drug interactions in vivo. *Biochem Pharmacol* 55:113–122.

Morovjan G, Dalmadi-Kiss B, Klebovich I, Mincsovics E (2002) Metabolite analysis, isolation, and purity assessment using various liquid chromatographic techniques combined with radioactivity detection. *J Chromatogr Sci* 40:603–608.

Motika MS, Zhang J, Cashman JR (2007) Flavin-containing monooxygenase 3 and human disease. *Expert Opin Drug Metab Toxicol* 3:831–845.

Motika MS, Zhang J, Zheng X, Riedler K, Cashman JR (2009) Novel variants of the human flavin-containing monooxygenase 3 (FMO3) gene associated with trimethylaminuria. *Mol Genet Metab* 97:128–135.

Nagar S, Remmel RP (2006) Uridine diphosphoglucuronosyltransferase pharmacogenetics and cancer. *Oncogene* 25:1659–1672.

Nakajima M, Tanaka E, Kobayashi T, Ohashi N, Kume T, Yokoi T (2002) Imipramine N-glucuronidation in human liver microsomes: Biphasic kinetics and characterization of UDP-glucuronosyltransferase isoforms. *Drug Metab Dispos* 30:636–642.

Nakamura K, Goto F, Ray WA, McAllister CB, Jacqz E, Wilkinson GR, Branch RA (1985) Interethnic differences in genetic polymorphism of debrisoquin and mephenytoin hydroxylation between Japanese and Caucasian populations. *Clin Pharmacol Ther* 38:402–408.

Nassar AEF, Bjorge SM, Lee DY (2003) On-line liquid chromatography-accurate radioisotope counting coupled with a radioactivity detector and mass spectrometer for metabolite identification in drug discovery and development. *Anal Chem* 75:785–790.

Nassar AE, Parmentier Y, Martinet M, Lee DY (2004) Liquid chromatography-accurate radioisotope counting and microplate scintillation counter technologies in drug metabolism studies. *J Chromatogr Sci* 42:348–353.

Nedderman ANR, Savage ME, White KL, Walker DK (2004) The use of 96-well scintiplates to facilitate definitive metabolism studies for drug candidates. *J Pharm Biomed Anal* 34:607–617.

Nelson DR (1999) Cytochrome P450 and the individuality of species. *Arch Biochem Biophys* 369:1–10.

Newton DJ, Wang RW, Lu AY (1995) Cytochrome P450 inhibitors. Evaluation of specificities in the *in vitro* metabolism of therapeutic agents by human liver microsomes. *Drug Metab Dispos* 23:154–158.

Obach RS (1997) Nonspecific binding to microsomes: Impact on scale-up of *in vitro* intrinsic clearance to hepatic clearance as assessed through examination of warfarin, imipramine, and propranolol. *Drug Metab Dispos* 25:1359–1369.

Obach RS (1999) Prediction of human clearance of twenty-nine drugs from hepatic microsomal intrinsic clearance data: An examination of *in vitro* half-life approach and nonspecific binding to microsomes. *Drug Metab Dispos* 27:1350–1359.

Obach RS (2001) The prediction of human clearance from hepatic microsomal metabolism data. *Curr Opin Drug Disc Dev* 4:36–44.

Obach RS (2004b) Potent inhibition of human liver aldehyde oxidase by raloxifene. *Drug Metab Dispos* 32:89–97.

Obach RS, Reed-Hagen AE (2002) Measurement of Michaelis constants for cytochrome P450-mediated biotransformation reactions using a substrate depletion approach. *Drug Metab Dispos* 30:831–837.

Obach RS, Huynh P, Allen MC, Beedham C (2004a) Human liver aldehyde oxidase: Inhibition by 239 drugs. *J Clin Pharmacol* 44:7–19.

Obach RS, Walsky RL, Venkatakrishnan K, Gaman EA, Houston JB, Tremaine LM (2006) The utility of *in vitro* cytochrome P450 inhibition data in the prediction of drug–drug interactions. *J Pharmacol Exp Ther* 316(1):336–348.

Ono S, Hatanaka T, Hotta H, Satoh T, Gonzalez FJ, Tsutsui M (1996) Specificity of substrate and inhibitor probes for cytochrome P450s: Evaluation of *in vitro* metabolism using cDNA-expressed human P450s and human liver microsomes. *Xenobiotica* 26:681–693.

Ortiz de Montellano PR, Mathews JM (1981) Autocatalytic alkylation of the cytochrome P-450 prosthetic haem group by 1-aminobenzotriazole. Isolation of an NN-bridged benzyneprotoporphyrin IX adduct. *Biochem J* 195:761–764.

Pacifici G (2004) Inhibition of human liver and duodenum sulfotransferases by drugs and dietary chemicals: A review of the literature. *Int J Clin Pharmacol Ther* 42:488–495.

Pacifici GM, Coughtrie MWH, eds. (2005) *Human Cytosolic Sulfotransferases*. CRC Press, Boca Raton, FL.

Pearce RE, Rodrigues AD, Goldstein JA, Parkinson A (1996) Identification of the human P450 enzymes involved in lasoprazole metabolism. *J Pharmacol Exp Ther* 277:805–816.

Proctor NJ, Tucker GT, Rostami-Hodjegan A (2004) Predicting drug clearance from recombinantly expressed CYPs: Inter-system extrapolation factors. *Xenobiotica* 34:151–178.

Rae JM, Soukhova NV, Flockhart DA, Desta Z (2002) Triethylenethiophosphoramide is a specific inhibitor of cytochrome P450 2B6: Implications for cyclophosphamide metabolism. *Drug Metab Dispos* 30:525–530.

Ramchandani RP, Wang Y, Booth BP, Ibrahim A, Johnson JR, Rahman A, Mehta M, Innocenti F, Ratain MJ, Gobburu JV (2007) The role of SN-38 exposure. UGT1A1*28 polymorphism, and baseline bilirubin in predicting severe irinotecan toxicity. *J Clin Pharmacol* 47:78–86.

Remmel RP, Burchell B (1993) Validation and use of cloned, expressed human drug-metabolizing enzymes in heterologous cells for analysis of drug metabolism and drug–drug interactions. *Biochem Pharmacol* 46:559–566.

Rendic S (2002) Summary of information on human CYP enzymes: Human P450 metabolism data. *Drug Metab Rev* 34:83–448.

Rettie AE, Lawton MP, Sadeque AJ, Meier GP, Philpot RM (1994) Prochiral sulfoxidation as a probe for multiple forms of the microsomal flavin-containing monooxygenase: Studies with rabbit FMO1, FMO2, FMO3, and FMO5 expressed in *Escherichia coli*. *Arch Biochem Biophys* 311:369–377.

Rock D, Foti RS, Pearson JT (2008) The combination of chemical and antibody inhibitors for superior P450 3A inhibition in reaction phenotyping studies. *Drug Metab Dispos* 32:105–112.

Rodrigues AD (1999) Integrated cytochrome P450 reaction phenotyping: Attempting to bridge the gap between cDNA-expressed cytochromes P450 and native human liver microsomes. *Biochem Pharmacol* 57:465–480.

Rodrigues AD, Roberts EM (1997) The *in vitro* interaction of dexmedetomidine with human liver microsomal cytochrome P4502D6 (CYP2D6). *Drug Metab Dispos* 25:651–655.

Rodrigues AD, Rushmore TH (2002) Cytochrome P450 pharmacogenetics in drug development: *In vitro* studies and clinical consequences. *Curr Drug Metab* 3:289–309.

Rodrigues AD, Winchell GA, Dobrinska MR (2001) Use of *in vitro* drug metabolism data to evaluate metabolic drug-drug interactions in man: The need for quantitative databases. *J Clin Pharmacol* 41:368–373.

Rodrigues AD, Kukulka MJ, Roberts EM, Ouellet D, Rodgers TR (1996) [O-methyl 14C]Naproxen O-demethylase activity in human liver microsomes: Evidence for the involvement of cytochrome P4501A2 and P4502C9/10. *Drug Metab Dispos* 24:126–136.

Rowland A, Gaganis P, Elliot DJ, Mackenzie PI, Knights KM, Miners JO (2007) Binding of inhibitory fatty acids is responsible for the enhancement of UDP-glucuronosyltransferase 2B7 activity by albumin: Implications for *in vitro-in vivo* extrapolation. *J Pharmacol Exp Ther* 321:137–147.

Roy P, Yu LJ, Crespi CL, Waxman DJ (1999) Development of a substrate-activity based approach to identify the major human liver P450 catalysts of cyclophosphamide and ifosfamide activation based cDNA-expressed activities and liver microsomal P450 profiles. *Drug Metab Dispos* 25:651–655.

Salva M, Jansat JM, Martinez-Tobed A, Palacios J (2003) Identification of the human liver enzymes involved in the metabolism of the antimigraine agent almotriptan. *Drug Metab Dispos* 31:404–411.

Sanderson S, Emery J, Higgins J (2005) CYP2C9 gene variants, drug dose, and bleeding risk in warfarin-treated patients: A HuGEnet systematic review and meta-analysis. *Genet Med* 7:97–104.

Schneider H, Glatt H (2004) Sulpho-conjugation of ethanol in humans *in vivo* and by individual sulphotransferase forms *in vitro*. *Biochem J* 383:543–549.

Segel IH (1972) *Enzyme Kinetics*. John Wiley and Sons, New York.

Sharer J, Wrighton S (1996) Identification of the human hepatic cytochrome P450 involved in the *in vitro* oxidation of antipyrine. *Drug Metab Dispos* 24:487–494.

Shepherd SR, Baird SJ, Hallinan T, Burchell B (1989) An investigation of the transverse topology of bilirubin UDP-glucuronosyltransferase in rat hepatic endoplasmic reticulum. *Biochem J* 259:617–620.

Shou M, Lu AY (2009) Antibodies as a probe in cytochrome P450 research. *Drug Metab Dispos* 37:925–931.

Smith DA, Abel SM, Hyland R, Jones BC (1998) Human cytochrome P450s: Selectivity and measurement *in vivo*. *Xenobiotica* 28:1095–1128.

Soars MG, Ring BJ, Wrighton SA (2003) The effect of incubation conditions on the enzyme kinetics of UDP-glucuronosyltransferases. *Drug Metab Dispos* 31:762–767.

Stormer E, von Moltke LL, Greenblatt DJ (2000) Scaling drug biotransformation data from cDNA-expressed cytochrome P-450 to human liver: A comparison of relative activity factors and human liver abundance in studies of mirtazapine metabolism. *J Pharmacol Exp Ther* 295:793–801.

Strassburg CP, Oldhafer K, Manns MP, Tukey RH (1997) Differential expression of the UGT1A locus in human liver, biliary, and gastric tissue: Identification of UGT1A7 and UGT1A10 transcripts in extrahepatic tissue. *Mol Pharmacol* 52:212–220.

Stresser DM, Kupfer D (1999) Monospecific antipeptide antibody to cytochrome P-450 2B6. *Drug Metab Dispos* 27:517–525.

Stresser DM, Broudy MI, Ho T, et al. (2004) Highly selective inhibition of human CYP3Aa *in vitro* by azamulin and evidence that inhibition is irreversible. *Drug Metab Dispos* 32:105–112.

Strolin-Benedetti M, Dostert P (1994) Contribution of amine oxidases to the metabolism of xenobiotics. *Drug Metab Rev* 26:507–535.

Sugihara K, Kitamura S, Tatsumi K (1996) Involvement of mammalian liver cytosols and aldehyde oxidase in reductive metabolism of zonisamide. *Drug Metab Dispos* 24:199–202.

Suzuki H, Kneller B, Haining RL, Trager WF, Rettie AE (2002) -N-3-Benzylnirvanol and (-)-N-3-benzyl-phenobarbital: New potent and selective *in vitro* inhibitors of CYP2C19. *Drug Metab Dispos* 30:235–239.

Tanaka E (1998) Clinically important pharmacokinetic drug–drug interactions: Role of cytochrome P450 enzymes. *J Clin Pharm Ther* 23:403–416.

Tang C, Shou M, Mei Q, Rushmore TH, Rodrigues AD (2000) Major role of human liver microsomal cytochrome P450 2C9 (CYP2C9) in the oxidative metabolism of celecoxib, a novel cyclooxygenase-II inhibitor. *J Pharmacol* 293:453–459.

Testa B, Balmat A, Long A (2004) Predicting drug metabolism: Concepts and challenges. *Pure Appl Chem* 76(5):907–914.

Tracy T, Hummel MA (2004) Modeling kinetic data from *in vitro* drug metabolism enzyme experiments. *Drug Metab Rev* 40:1–100.

Trottier J, El Husseini D, Perreault M, Pâquet S, Caron P, Bourassa S, Verreault M, Inaba TT, Poirier GG, Bélanger A, Guillemette C, Trauner M, Barbier O (2010) The human UGT1A3 enzyme conjugates norursodeoxycholic acid into a C23-ester glucuronide in the liver. *J Biol Chem* 285:1113–1121.

Tsoutsikos P, Miners JO, Stapleton A, Thomas A, Sallustio BC, Knights KM (2004) Evidence that unsaturated fatty acids are potent inhibitors of renal UDP-glucuronosyltransferases (UGT): Kinetic studies using human kidney cortical microsomes and recombinant UGT1A9 and UGT2B7. *Biochem Pharmacol* 67:191–199.

Tucker GT, Houston JB, Huang S-M (2001) Optimising drug development: Strategies to assess drug metabolism/transporter interaction potential—Toward a consensus. *Clin Pharmacol Ther* 70:103–114.

Tukey RH, Strassburg CP (2000) Human UDP-glucuronosyltransferases: Metabolism, expression, and disease. *Annu Rev Pharmacol Toxicol* 40:581–616.

Uchaipichat V, Mackenzie PI, Elliot DJ, Miners JO (2006a) Selectivity of substrate (trifluoperazine) and inhibitor (amitriptyline, androsterone, canrenoic acid, hecogenin, phenylbutazone, quinidine, quinine, and sulfinpyrazone) "probes" for human udp-glucuronosyltransferases. *Drug Metab Dispos* 34:449–456.

Uchaipichat V, Winner LK, Mackenzie PI, Elliot DJ, Williams JA, Miners JO (2006b) Quantitative prediction of *in vivo* inhibitory interactions involving glucuronidated drugs from *in vitro* data: The effect of fluconazole on zidovudine glucuronidation. *Br J Clin Pharmacol* 61:427–439.

Varma MVS, Feng B, Obach RS, Troutman MD, Chupka J, Miller HR, El-Kattan A (2009) Physicochemical determinants of human renal clearance. *J Med Chem* 52:4844–4852.

Venkatakrishnan K, von Moltke LL, Court MH, Harmatz JS, Crespi CL, Greenblatt DJ (2000) Comparison between cytochrome P450 (CYP) content and relative activity approaches to scaling from cDNA-expressed CYPs to human liver microsomes: Ratios of accessory proteins as sources of discrepancies between the approaches. *Drug Metab Dispos* 28:1493–1504.

Venkatakrishnan K, von Moltke LL, Greenblatt DJ (2001) Application of the relative activity factor approach in scaling from heterologously expressed cytochromes P450 to human liver microsomes: Studies on amitriptyline as a model substrate. *J Pharmacol Exp Ther* 297:326–337.

Wallace D, Hildeshiem A, Pinto LA (2004) Comparison of bench top microplate beta counters with the traditional gamma counting method for measurement of chromium-51 release in cytotoxic assays. *Clin Diagn Lab Immunol* 11:255–260.

Walsky RL, Obach RS, Gaman EA, Gleeson JR, Proctor WR (2005) Selective inhibition of human cytochrome P450 2C8 by montelukast. *Drug Metab Dispos* 33:413–418.

Walsky RL, Astuccio AV, Obach RS (2006) Evaluation of 227 drugs for *in vitro* inhibition of cytochrome P450 2B6. *J Clin Pharmacol* 46:1426–1438.

Wang RW, Lu AYH (1997) Inhibitory anti-peptide antibody against human CYP3A4. *Drug Metab Dispos* 25:762–767.

Wen Z, Tallman MN, Ali SY, Smith PC (2007) UDP-glucuronosyltransferase 1A1 is the principal enzyme responsible for etoposide glucuronidation in human liver and intestinal microsomes: Structural characterization of phenolic and alcoholic glucuronides of etoposide and estimation of enzyme kinetics. *Drug Metab Dispos* 35:371–380.

Wienkers LC, Heath TG (2005) Predicting *in vivo* drug interactions from *in vitro* drug discovery data. *Nat Rev Drug Discov* 4:825–833.

Williams J, Hurst S, Bauman J, Jones B, Hyland R, Gibbs J, Obach R, Ball S (2003) Reaction phenotyping in drug discovery: Moving forward with confidence? *Curr Drug Metab* 4:527–534.

Williams J, Hyland R, Jones B, Smith D, Hurst S, Goosen T, Peterkin V, Koup J, Ball S (2004) Drug-drug interactions for UDP-glucuronosyltransferase substrates: A pharmacokinetic explanation for typically observed low exposure (AUCi/AUC) ratios. *Drug Metab Dispos* 32:1201–1208.

Williams J, Bauman JN, Cai H, Conlon K, Hansel S, Hurst S, Sadagopan N, Tugnait M, Zhang L, Sahi J (2005) *In vitro* ADME phenotyping in drug discovery: Current challenges and future solutions. *Curr Opin Drug Discov Devel* 8:78–88.

Williams JA, Ring BJ, Cantrell VE, Campanale K, Jones DR, Hall SD, Wrighton SA (2002) Differential modulation of UDP-glucuronosyltransferase 1A1 (UGT1A1)-catalyzed estradiol-3-glucuronidation by the addition of UGT1A1 substrates and other compounds to human liver microsomes. *Drug Metab Dispos* 30:1266–1273.

Wring SA, Silver IS, Serabjit-Singh CJ (2002) Automated quantitative and qualitative analysis of metabolic stability: A process for compound selection during drug discovery. *Methods Enzymol* 357:285–296.

Wynalda MA, Hutzler JM, Koets MD, Podoll T, Wienkers LC (2003) *In vitro* metabolism of clindamycin in human liver and intestinal microsomes. *Drug Metab Dispos* 31:878–887.

Yamazaki H, Nakajima M, Nakamura M, Asahi S, Shimada N, Gillam EMJ, Guengerich FP, Shimada T, Yokoi T (1999) Enhancement of cytochrome P-450 3A4 catalytic activities by cytochrome b5 in bacterial membranes. *Drug Metab Dispos* 27:999–1004.

Yeung CK, Lang DH, Thummel KE, Rettie AE (2000) Immunoquantitation of FMO1 in human liver, kidney, and intestine. *Drug Metab Dispos* 28:1107–1111.

Yokota H, Yuasa A and Sato R (1992) Topological disposition of UDP-glucuronyltransferase in rat liver microsomes. *J Biochem* 112:192-196.

Yumibe N, Huie K, Chen KJ, Snow M, Clement RP, Cayen MN (1996) Identification of human liver cytochrome P450 enzymes that metabolize the nonsedating antihistamine loratadine: Formation of descarboethoxyloratadine by CYP3A4 and CYP2D6. *Biochem Pharmacol* 51:165–172.

Zhang D, Chando TJ, Everett DW, Patten CJ, Dehal SS, Humphreys WG (2005) *In vitro* inhibition of UDP-glucuronosyltransferases by atazanavir and other HIV protease inhibitors and the relationship of this property to *in vivo* bilirubin glucuronidation. *Drug Metab Dispos* 33:1729–1739.

Zhang D, Wang L, Chandrasena G, Ma L, Zhu M, Zhang H, Davis CD, Humphreys WG (2007a) Involvement of multiple cytochrome P450 and UDP-glucuronosyltransferase enzymes in the *in vitro* metabolism of muraglitazar. *Drug Metab Dispos* 35:139–149.

Zhang H, Davis CD, Sinz MW, Rodrigues AD (2007b) Cytochrome P450 reaction phenotyping: An industrial perspective. *Expert Opin Drug Metab Toxicol* 3:667–687.

Zhao W, Wang L, Zhang D, Zhu M (2008) Rapid and sensitive determination of enzyme kinetics of drug metabolism using HPLC coupled with a stop-flow radioactivity flow detector. *Drug Metab Lett* 2:41–46.

Zhu M, Zhao W, Jimenez H, Zhang D, Yeola S, Dai R, Vachharajani N, Mitroka J (2005a) Cytochrome P450 3A-mediated metabolism of buspirone in human liver microsomes. *Drug Metab Dispos* 33:1500–1507.

Zhu M, Zhao W, Vazquez N, Mitroka JG (2005b) Analysis of low level of radioactive metabolites in biological fluids using high-performance liquid chromatography with microplate scintillation counting: Method validation and application. *J Pharm Biomed Anal* 39:233–245.

Zucker S, Qin X, Rouster S, Yu F, Green R, Keshavan P, Feinberg J, Sherman KE (2001) Mechanism of indinavir-induced hyperbilirubinemia. *Proc Natl Acad Sci U S A* 98:12671–12676.

14

FAST AND RELIABLE CYP INHIBITION ASSAYS

Ming Yao, Hong Cai, and Mingshe Zhu

14.1 INTRODUCTION

Cytochrome P450 (CYP) enzymes are a superfamily of hemethiolate enzymes that catalyze oxidative, reductive, and possibly hydrolytic metabolism of drugs. As a result, they generate a wide variety of metabolites that are usually more polar than the parent drugs and are eliminated quickly from humans and animals. Metabolic clearance of drugs is considered a detoxification process, although some metabolites are toxic or pharmacologically active. Among the great number of CYP enzymes that have been discovered, CYP1A2, CYP2A6, CYP2B6, CYP2C9, CYP2C19, CYP2D6, and CYP3A4/5 are responsible for most metabolic clearance of marketed drugs in humans. Inhibition of a CYP enzyme that is responsible for a major clearance pathway (>25% of the total clearance) of a drug (victim) by a coadministered drug that is a chemical inhibitor (perpetrator) can lead to significant increases in exposure to the victim drug. CYP inhibition is the most common mechanism underlying drug–drug interactions (DDIs) (Nettleton and Einolf, 2011), which has caused several drugs to be withdrawn from the market or placed with warnings on their labels.

Traditionally, the pharmaceutical industry conducted clinical DDI studies only in the later clinical trials mainly for drug registration purposes. In the 1990s, several regulatory guidelines on *in vitro* and *in vivo* DDI studies were issued by regulatory authorities in Europe and the United States. Consequently, workshops and symposiums on conducting DDI studies were organized by pharmaceutical industrial organizations and regulatory agencies (Bjornsson et al., 2003). Rec-

ommendations from the guidance documents and the workshops; better understanding of CYP enzymology and enzyme kinetics; the availability of analytical technology and CYP reagents, including expressed human CYP enzymes and probe substrates; and selective chemical inhibitors of CYP enzymes motivated and enabled pharmaceutical companies to assess CYP inhibitory liability in lead optimization (Obach et al., 2006). *In vitro* CYP inhibition assays developed for the purpose, including those using fluorogenic substrates and recombined human CYP enzymes (Figure 14.1 and Table 14.1), are characterized by high throughput capability and low operation cost. Results from the assays allow for the termination of new chemical entities that are strong inhibitors of major CYP enzymes for progression. In addition, structure–activity relationships can be developed for a large set of CYP inhibition data, which can facilitate the design of alternate agents that have no or reduced CYP inhibitory potency. In the later stages of drug discovery, CYP inhibition assays that employ human liver microsomes (HLM) and CYP probe substrates are often performed (Figure 14.1 and Table 14.1). The data generated from these assays are mainly used to support selection of clinical candidates, regulatory submissions for investigational new drug (IND) applications, and the design of clinical DDI studies (Obach et al., 2005).

In this chapter, we briefly describe various *in vitro* CYP inhibition assays developed in the past 20 years. Their advantages, limitations, and current applications are discussed. The main focus of this chapter is the methodology for *in vitro* assessment of CYP inhibitory potency, which is currently employed in the

ADME-Enabling Technologies in Drug Design and Development, First Edition. Edited by Donglu Zhang and Sekhar Surapaneni.
© 2012 John Wiley & Sons, Inc. Published 2012 by John Wiley & Sons, Inc.

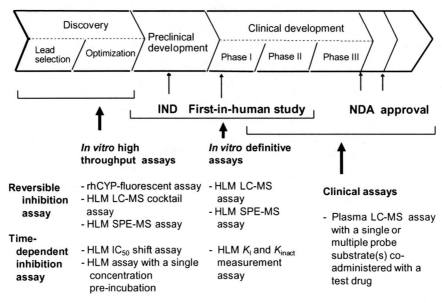

FIGURE 14.1. Assessment of CYP inhibition in drug discovery and development.

TABLE 14.1. Comparison of *In Vitro* CYP Inhibition Assays

Assay	Advantages	Limitations	Current Applications
Fluorescent	High throughput	Not suited for HLM	Rank ordering of a large number of compounds in the same chemotypes
	Low cost	rhCYP data showed false-positive and not suited for regulatory filing	Used when HLM availability or cost is an issue
	Easy to operate	Not applicable to fluorescent compounds	
Radiometric	Fast	Requires specifically designed radiolabeled probe substrate	Hardly used
	Suited for HLM	Requires specific facilities and personnel to handle radioactive materials	
Luminescence	High throughput	Require specifically designed probe substrates	Hardly used
	Easy to operate	Not available for most CYP enzymes	
	Suited for HLM		
LC-MS with a single substrate	Suited for HLM	Requires an LC-MS instrument and operator	Widely used at late discovery and development stages
	Applicable to various compounds	Medium throughput	
	Suited for time-dependent inhibition analysis	Costs relatively more than a cocktail assay	
	Suited for regulatory filing		
LC-MS with multiple substrates (cocktail assay)	High throughput	Requires an LC-MS instrument and operator	Used at the early discovery stage for fast and low-cost screening
	Relatively low cost	May not be suited for regulatory filing	
	Suited for HLM	Not suited for time-dependent inhibition analysis	
MS without LC	High throughput	Requires an expensive MS instrument	Widely applied from early discovery to development
	Suited for HLM	Costs relatively more than a cocktail assay	
	Suited for regulatory filing		
	Suited for time-dependent inhibition analysis		

pharmaceutical industry. Descriptions of detailed method development strategies, reagents used, experimental protocols, and the analytical technologies of these assays are included. The assays are also compared on the basis of their speed, reliability, and utility. Data interpretation of *in vitro* CYP inhibition experiments and prediction of clinical DDI potentials based on *in vitro* data, which have been extensively described in the literature (Houston and Galetin, 2008; Zhou and Zhou, 2009; Templeton et al., 2010; VandenBrink and Isoherranen, 2010), are not covered in this chapter.

14.2 CYP INHIBITION ASSAYS IN DRUG DISCOVERY AND DEVELOPMENT

To address the needs for evaluation of CYP inhibitory potency for lead compounds and clinical candidates at different stages of drug discovery and development, a variety of *in vitro* CYP inhibition assays have been developed (Stresser et al., 2002; Smith et al., 2007; Fowler and Zhang, 2008; Lahoz et al., 2008). Working principles, advantages, limitations, and current applications of common *in vitro* CYP inhibition assays are summarized in Table 14.1. Microtiter plate assays are carried out using a nonspecific fluorescent substrate and a single recombinant human CYP (rhCYP) enzyme. After 10–30-min incubations, results are quickly recorded using a plate reader. The rhCYP-fluorescent assay is a fast, easy operation that does not use mass spectrometry (MS), and it has been widely used for high-throughput screening for a large number of compounds in lead selection and optimization (Moody et al., 1999; Bapiro et al., 2001; Stresser et al., 2002; Cohen et al., 2003; Donato et al., 2004; Di et al., 2007). Fluorescent assays are especially applicable to the determination of structure, CYP inhibitory activity relationships, and rank order of compounds that have the same chemotypes, knowledge of which helps medicinal chemists to design compounds with better CYP inhibition profiles. However, since most probe substrates employed in fluorescent assays are not CYP enzyme-specific, these assays cannot be used with HLM. It has been reported that the rhCYP-fluorescent assay generated more false-positive results than HLM-based assays in comparison with CYP inhibition data observed in clinical studies (Bell et al., 2008). In addition, rhCYP-fluorescent assays are not suited for compounds that have strong fluorescent responses (Fowler and Zhang, 2008). As a result, HLM high-throughput assays, including the HLM cocktail assay and HLM MS assays, have replaced rhCYP-fluorescent assays for lead optimization in most pharmaceutical companies (Figure 14.1 and Table 14.1).

CYP inhibition assays using selective probe substrates and HLM followed by liquid chromatography-mass spectrometry (LC-MS) analysis of metabolites of the probe substrates have been developed and validated for the definitive evaluation of drug candidates for their inhibitory effects in late discovery and drug development (Jenkins et al., 2004; Walsky and Obach, 2004; Yao et al., 2007). Results from HLM-LC-MS assays provide some key drug metabolism and pharmacokinetic (DMPK) information affecting the selection of clinical candidates, the design of clinical DDI studies, and regulatory filings for new drugs. Fully validated CYP inhibition assays following "in the spirit of good laboratory practice (GLP)" have been recommended by representatives of regulatory agencies, academia, and industry (Bjornsson et al., 2003). Walsky and Obach (2004) reported GLP-like CYP inhibition assays for 10 human CYP enzymes. In the experiment, a probe substrate of CYP enzyme was incubated with HLM in the presence of various concentrations of a test compound. Formation rates of a probe substrate metabolite were measured by a validated liquid chromatography-tandem mass spectrometry (LC-MS/MS) method that used a stable isotope analog of the metabolite as an internal standard (IS). We have reported the development and full validation of an HLM-LC-MS assay for the assessment of the inhibitory effects of drug candidates on five CYP enzymes (CYP1A2, CYP2C9, CYP2C19, CYP2D6, and CYP3A) (Yao et al., 2007). The selective substrates used, the related metabolic reactions, and the internal standards used in these assays are shown in Table 14.2. The incubations were run in a fully automated fashion (96-well format) and under kinetically rigorous conditions. Metabolites of probe substrates were analyzed by LC–MS/MS. The HLM-LC-MS assay provides two main advantages: (1) treatment of incubation plates to overcome nonspecific binding issues and detailed methods; and (2) the use of a newly designed filtration plate to accelerate sample processing and reduce assay variability.

Although HLM-LC-MS assays provide high-quality, reliable results, medium throughput and high operation cost have greatly limited routine applications of these assays in lead optimization, in which a large number of active compounds are subjected to screening (Figure 14.1 and Table 14.1). To improve throughput and reduce cost, several HLM CYP inhibition assays, such as radiometric (Rodrigues et al., 1996; Zhang and Thomas, 1996; Riley and Howbrook, 1997; Draper et al., 1998; Grand et al., 2002; Di Marco et al., 2007), luminescence (Garcia et al., 2008), and cocktail assays (Bu et al., 2001a,b; Dierks et al., 2001; Zhang et al., 2002; Testino and Patonay, 2003; Yin et al., 2004; Turpeinen et al., 2005; He et al., 2007; Lin et al., 2007; O'Donnell et al., 2007;

TABLE 14.2. FDA-Preferred and Acceptable Chemical Substrates for *in vitro* Experiments[a]

CYP	Substrate Preferred	K_m (µM)	Substrate Acceptable	K_m (µM)
1A2	Phenacetin O-deethylation	1.7–152	7-Ethoxyresorufin-O-deethylation	0.18–0.21
			Theophylline-N-demethylation	280–1230
			Caffeine-3-N-demethylation	220–1565
			Tacrine 1-hydroxylation	2.8, 16
2A6	Coumarin-7-hydroxylation	0.30–2.3		
	Nicotine C-oxidation	13–162		
2B6	Efavirenz hydroxylase	17–23	Propofol hydroxylation	3.7–94
	Bupropion-hydroxylation	67–168	S-mephenytoin N-demethylation	1910
2C8	Taxol 6-hydroxylation	5.4–19	Amodiaquine N-deethylation	2.4,
			Rosiglitazone *para*-hydroxylation	4.3–7.7
2C9	Tolbutamide methylhydroxylation	67–838	Flurbiprofen 4′-hydroxylation	6–42
	S-warfarin 7-hydroxylation	1.5–4.5	Phenytoin-4-hydroxylation	11.5–117
	Diclofenac 4′-hydroxylation	3.4–52		
2C19	S-mephenytoin 4′-hydroxylation	13–35	Omeprazole 5-hydroxylation	17–26
			Fluoxetine O-dealkylation	3.7–104
2D6	(±)-Bufuralol 1′-hydroxylation	9–15	Debrisoquine 4-hydroxylation	5.6
	Dextromethorphan O-demethylation	0.44–8.5		
2E1	Chlorzoxazone 6-hydroxylation	39–157	*p*-Nitrophenol 3-hydroxylation	3.3
			Lauric acid 11-hydroxylation	130
			Aniline 4-hydroxylation	6.3–24
3A4/5[b]	Midazolam 1-hydroxylation	1–14	Erythromycin N-demethylation	33–88
	Testosterone 6β-hydroxylation	52–94	Dextromethorphan N-demethylation	133–710
			Triazolam 4-hydroxylation	234
			Terfenadine C-hydroxylation	15
			Nifedipine oxidation	5.1–47

[a] Note that this is not an exhaustive list. For an updated list, see the following link: http://www.fda.gov/drugs/developmentapprovalprocess/developmentresources/druginteractionslabeling/ucm093664.htm (FDA, 2006; EMA, 2010).

[b] Recommend use of two structurally unrelated CYP3A4/5 substrates for evaluation of *in vitro* CYP3A inhibition. If the drug inhibits at least one CYP3A substrate *in vitro*, then *in vivo* evaluation is warranted.

Tolonen et al., 2007; Rainville et al., 2008; Youdim et al., 2008; Zientek et al., 2008), have been developed (Table 14.1). CYP substrate cocktail assays employ a mixture of probe substrates in HLM incubation and allow for simultaneous assessment of the reversible inhibitory potency of several CYP forms by a test compound. Cocktail assays are especially useful for a small or medium drug discovery organization. Due to the lack of suitable substrates for all major CYP enzymes, radiometric and luminescence assays are not routinely used in pharmaceutical research.

As an alternative, high-throughput assays consisting of online solid-phase extraction (SPE) and MS without an LC column (HLM SPE-MS, Table 14.1) are widely applied to analysis of metabolites of probe substrates formed in HLM incubation with individual substrates (Lim et al., 2010). Recently, a device called "RapidFire" was introduced that combines sampling in 96- or 384-well plates, online SPE, and quantification by a mass spectrometer that reached analytical speed at 5 s per injection (Holt et al., 2009). The HLM SPE-MS assays greatly enhanced throughput of metabolite quantifica-

tion and produce the same quality of result as HLM-LC-MS inhibition assays. Therefore, HLM SPE-MS assays can be employed for both lead optimization in drug discovery and regulatory filing in drug development. A major limitation of the assay is its inability to analyze metabolites that have one or multiple isomer(s), such as metabolites of testosterone, a CYP3A4 probe substrate. HLM SPE-MS assays are well suited for large discovery organizations.

In addition to the assessment of reversible CYP inhibitory potency, time-dependent inhibition (TDI) of CYP enzymes by a drug candidate is determined at different stages of drug discovery and development (Figure 14.1) (Grimm et al., 2009). The HLM LC-MS assay with a single probe substrate has been widely adapted for the determination of TDI, such as single concentration preincubation assays, IC_{50} shift assays, inhibition curve shift, and K_I and K_{inact} measurement experiments (Table 14.1 and Figure 14.1). The HLM SPE-MS assay is also useful for the determination of IC_{50} shift analysis (Lim et al., 2010). The HLM cocktail assay may not suit for TDI even though its throughput is excellent (Table 14.1).

14.3 HLM REVERSIBLE CYP INHIBITION ASSAY USING INDIVIDUAL SUBSTRATES

Many drug metabolism labs have established and validated HLM CYP reversible inhibition (RI) assays (Ayrton et al., 1998; Chu et al., 2000; Peng et al., 2003; Walsky and Obach, 2004; Yin et al., 2004). In Sections 14.3.1–14.3.5, detailed incubation conditions, experimental protocols and automated procedures, LC-MS analysis, and data calculation of HLM LC-MS assays routinely conducted in our lab are described (Yao et al., 2007).

14.3.1 Choice of Substrate and Specific Inhibitors

Preferred isoform-specific probe substrates and inhibitors are recommended by U.S. and European regulatory agencies (FDA, 2006; EMA, 2010) and are wildly used in the pharmaceutical industry (Table 14.2). CYP probe substrates, substrate metabolites, and CYP chemical inhibitors as positive controls employed in HLM reversible CYP inhibition assays conducted in our lab are listed in Table 14.3.

14.3.2 Optimization of Incubation Conditions

To ensure that high-quality data are generated from HLM CYP RI experiments, three key incubation parameters—microsomal protein concentration, incubation time, and probe substrate concentration—should be optimized. With an optimized incubation time, protein concentration in incubation should be in linear ranges with respect to metabolism reaction rates. In addition, protein concentrations should be kept close to 0.15 mg/mL or lower if possible to minimize the effects of nonspecific protein binding. Optimal substrate concentrations are slightly lower or close to the Michaelis–Menten constant (K_m) value determined under optimized incubation conditions. Furthermore, the consumption of substrate should be less than 20% after incubation is completed. In the K_m determination experiment, eight substrate concentrations were often used in our studies, and the substrate concentrations span a range from $1/3K_m$ to $3K_m$. K_m values were determined by nonlinear regression of enzyme activity versus substrate concentration. Substrate saturation curves and inhibition data were analyzed using the enzyme kinetics module of GraFit version 5.0 (Erithacus Software Ltd., Horley Surrey, U.K.). Metabolite formation kinetic data of eight CYP inhibition assays determined in our lab are presented in Table 14.4 and are consistent with those reported in the literature for those eight CYP enzymes. (Table 14.3). The formation of major metabolites was linear with incubation times up to 20 min for the CYP1A2, CYP2D6, CYP2C9, and CYP3A4 (testoster-one) assays, up to 10 min for the CYP3A4 (midazolam) assay, and up to 50 min for the CYP2C19 assay. The formation of major metabolites was linear with protein concentrations from 0.1 to 0.3 mg/mL (CYP1A2, CYP2D6, CYP2C9, and CYP3A4 assays), and from 0.1 to 0.45 mg/mL (CYP2C19 assay).

14.3.3 Incubation Procedures

14.3.3.1 Preparation of Substrates, Positive Controls, Test Compounds, Standards, and Quality Control (QC) of Samples

1. Working solutions of HLM were prepared by diluting pooled HLM (20 mg/mL, purchased from BD Biosciences, Sparks, MD, USA) with a 100 mM phosphate buffer (pH 7.4) to form solutions of 0.11 to 0.28 mg/mL (solutions were referred to as HLM-1A).

2. Stock solutions of the metabolites were prepared in acetonitrile/water and then further diluted with HLM-1A to obtain the highest concentration standard and QC samples (Table 14.3).

3. A second HLM working solution (referred to as HLM-2) was prepared by diluting a probe substrate with HLM-1A to a concentration close to its K_m value.

4. Positive control (inhibitor) or test compound stock solutions were prepared in dimethyl sulfoxide (DMSO), and then 2.5 µL of the stock solutions was dissolved in HLM-2 at the highest concentration used. All working solutions were stored on ice before being transferred and diluted.

5. Serial dilutions were performed by a TECAN liquid handler (Tecan Group Ltd., Durham, NC, USA) for all samples. Seven concentrations for the standard and four concentrations for QC samples were prepared for calibration and quality control. Eight concentrations were prepared for positive inhibitors and test compounds.

14.3.3.2 Incubation Procedure for IC₅₀ Determination

1. *Incubation Condition.* A phosphate buffer (100 mM KH_2PO_4, pH 7.4) containing 1 mM ethylene diamine tetraacetic acid (EDTA) was prepared from 400 mM mono- and dibasic potassium phosphate stock solutions that were prepared fresh every 6 months and stored at 4°C. Frozen stock solutions of liver microsomes (BD Biosciences) were used once after thawing. Nicotinamide adenine dinucleotide phosphate (NADPH) stock solutions (10 mM) in phosphate buffer were made fresh daily.

TABLE 14.3. Preparation of Stock and Working Solution of Substrate, Standard, QC, Positive Control, Test Compound, and Internal Standard

	Substrate	Standard	QC	MRM Transition	Positive Control	Internal Standard
CYP1A2 assay	Phenacetin	Acetaminophen	Acetaminophen	152.1 > 109.9	α-Naphthoflavone	4OH-Butyranilide
Final concentration (μM)	45	5	4		1	
CYP2A6 assay	Coumarin	7OH-Coumarin	7OH-Coumarin	163 > 107	Tranylcypromine	6,7-Dihydroxycoumarin
Final concentration (μM)	0.65	2	1.5		5	
CYP2B6 assay	Bupropion	OH-Bupropion	OH-Bupropion	256 > 238	Orphenadrine	Trazodone
Final concentration (μM)	100	0.1	0.08		2000	
CYP2C8 assay	Taxol	6α-Hydroxytaxol	6α-Hydroxytaxol	914.5 > 541.3	Montelukast	Deacetyltaxol-C
Final concentration (μM)	5	0.1	0.08		5	
CYP 2C9 assay	Diclofenac	4OH-Diclofenac	4OH-Diclofenac	312.10 > 265.8	Sulfaphenazole	Flufenamic acid
Final concentration (μM)	10.00	10.00	7.00		20	
CYP2C19 assay	(S)-Mephenytoin	4OH-Mephenytion	4OH-Mephenytion	235.25 > 149.99	N-3-Benzylnirvanol	Phenytoin
Final concentration (μM)	55.00	2.5	2.0		20	
CYP2D6 assay	Dextromethorphan	Dextrorphan	Dextrorphan	258.2 > 157.1	Quinidine	Propranolol
Final concentration (μM)	10	10	7		10	
CYP3A4 assay	Midazolam	1OH-Midazolam	1OH-Midazolam	342 > 324	Ketoconazole	α-OH-Triazolam
Final concentration (μM)	5	1.25	1.0		5	
CYP3A4 assay	Testosterone	6OH-testosterone	6OH-Testosterone	305.3 > 269.2	Ketoconazole	6OH-Progesterone
Final concentration (μM)	75	36	24		5	

TABLE 14.4. Summary of Enzyme Kinetic Parameters (Mean ± SE) for Five Human CYP Activities in Pooled Human Liver Microsomes

| Enzyme | Assay | Incubation Condition | | K_m/V_{max} Determination | |
		Time (min)	Protein Concentration (mg/mL)	V_{max} (pmol/mg/min)	K_m (μM)
CYP1A2	Phenacetin O-deethylase	10	0.15	722 ± 65	45.0 ± 3.8
CYP2A6	Coumarin 7-hydroxylation	5	0.05	1005 ± 28	0.662 ± 0.11
CYP2B6	Bupropion hydroxylation	5	0.05	306 ± 14	125 ± 9
CYP2C8	Taxol 6α-hydroxylation	5	0.05	344 ± 36.5	4.82 ± 0.34
CYP2C9	Diclofenac 4′-hydroxylase	7	0.15	5300 ± 190	9.8 ± 0.5
CYP2C19	(S)-Mephenytoin 4′-hydroxylase	40	0.25	56 ± 2.3	55.6 ± 2.8
CYP2D6	Dextromethorphan O-demethylase	7	0.15	493 ± 38	10.9 ± 2.2
CYP3A4	Midazolam 1′-hydroxylase	5	0.1	1756 ± 274	4.13 ± 0.3
CYP3A4	Testosterone 6β-hydroxylase	10	0.15	5147 ± 296	83.3 ± 3.3

2. *Stock solution preparation.* Analytes (i.e., metabolites) were prepared in solvent and stored at −20°C or 4°C. Internal standards were dissolved in acetonitrile and further diluted with acetonitrile or 3% formic acid and acetonitrile (7:3, v/v) to prepare working solutions (Table 14.3).

3. *Sample Preparation and Incubation.* The assay was designed to run six compounds (five test compounds and one positive control) for each CYP enzyme. A known inhibitor for each CYP enzyme was run alongside the test compounds. Eight concentrations of each inhibitor run in triplicate were used to calculate the IC_{50} value. Two plates were used to determine IC_{50} values for five test compounds. One plate contained a standard, two test compounds, and one positive control, and the other plate contained QC samples and three additional test compounds. Figure 14.2A details the plate layout for the process of sample preparation, incubation, and filtration.

4. *Preparing the Dilution Plate.* The highest concentration of standard, positive control, and two test compounds were manually prepared and spiked into the last well (H) of each column (1, 3, 5, and 7) of a 2-mL 96-well preparation plate (I) for serial dilution by a TECAN (Figure 14.2A). To the serial dilution samples in the preparation plate, 1–1.33 μL of DMSO was added to maintain the same amount of organic solvent, 0.16% (v/v). Then blank HLM-1A was transferred to column 1 for standard dilution, and HLM-2 was transferred to columns 3, 5, and 7 except the last well of each column. The test compounds were diluted serially to form eight concentrations and mixed well by the TECAN before transferring.

5. *TECAN Incubation.* After serial dilution by the TECAN, 180 μL of mixtures located in columns

1, 3, 5, and 7 were transferred to an incubation plate in triplicate. After preincubation at 37°C in a 96-well temperature-controlled heater block for 5 min, 20 μL of NADPH (10 mM in 100 mM phosphate buffer) was added to each well of the reaction plate to give a final volume of 200 μL and initiate the reaction. The plates were maintained at 37°C for the time period defined (Table 14.3).

6. *Sample Filtration.* To prepare the filter plates, 240 μL of acetonitrile containing an internal standard was transferred into a filter plate (or 100 μL of 30% acetonitrile in 1% formic acid for the CYP1A2 assay). After incubation, 120 μL of the reaction mixtures (or 150 μL for the CYP1A2 assay) from the wells containing positive control and test compound were transferred in the filter plate to stop the reaction. An aliquot (108 μL) from the wells containing standard samples was then transferred to the filter plate along with an additional 12 μL of NADPH. Also, 1 μL of the five test compounds (5 mM) was added to the blank at positions A-1 to A-3 and B-1 to B-2 in the filtration plate, respectively, as a control to monitor the interference of the test compound with the analysis of the corresponding metabolite of each substrate. The filter plate containing terminated incubation mixtures was then stacked on a 2-mL 96-well receiver plate that was preloaded with 360 μL of 0.1% formic acid in water, vortexed for 30 s, and all mixtures were passed through a 0.45-μm hydrophobic or hydrophilic (CYP1A2 assay only) polytetrafluoroethylene (PTFE) membrane by centrifugation for 5 min at 2000 g into the receiving plate. Finally, the receiver plate was vortexed and sealed with a polypropylene film, and 10–25 μL of sample was injected into the LC-MS/MS for quantitation.

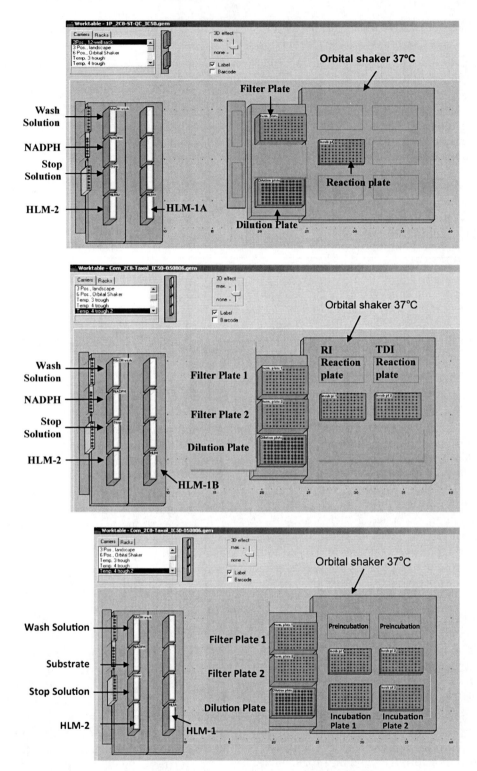

FIGURE 14.2. (A) TECAN Genesis platform layout for reversible inhibition. (B) TECAN Genesis platform layout for simultaneous reversible and time-dependent inhibition. (C) TECAN Genesis platform layout for K_I and K_{inact} determination.

The second set of preparation plate and reaction plate (plate II) was generated in the same fashion as preparation plate I except that the standard, the positive control, and two test compounds were replaced by QC samples and three other test compounds (Table 14.3). If not using a TECAN, all samples can be diluted in the dilution plate manually, then transferred to reaction plate using a multichannel pipette.

14.3.4   LC-MS/MS Analysis

LC-MS (4000 Qtrap mass spectrometer, AB Sceix, Framingham, MA, USA) was employed for metabolite quantification. Multiple reaction monitoring (MRM) in a positive electrospray was employed to monitor a pre-defined metabolite of each substrate and an internal standard, with a dwell time set to 150 ms for each transition. Heated nebulizer parameters were set as follows (arbitrary units): IS = 5000, curtain gas 25; and temperature 350°C. The flow rates of gas 1 and gas 2 were set at 45. The mass transition and collision energy for each metabolite and internal standard can be found in Table 14.3. Data were collected and processed using Sciex Analyst 1.4.1 data collection and integration software.

14.3.5   Data Calculation

14.3.5.1   Data Calculation Tools   Peak area ratios of the analyte to internal standard were calculated within the Analyst software. A calibration curve was generated by quadratic regression, weighted by 1/x. The equation of this curve was then used to calculate the concentrations in all samples. Between- and within-assay rela-

tive standard deviation (RSD) for QC samples were calculated within EXCEL (Microsoft Office 2007, Microsoft, Redmond, WA, USA) using a one-way analysis of variance (ANOVA). K_m and V_{max} values were estimated using XLfit™ (ID Business Solutions Inc., Guildford, UK). When inhibition of CYP enzymes was observed, IC_{50} values and standard errors were calculated using the transformed Michaelis–Menten equation (4-parameter logistic) for competitive inhibition with XLfit and displayed on an appropriate plot.

14.3.5.2   Chromatography and Specificity   Since some probe substrates, such as testosterone, have multiple isomeric metabolites, attempts were made to achieve maximal chromatographic resolution in order to minimize interferences. For example, multiple isomers of hydroxytestosterone (hydroxylation at the 2-, 6-, 15-, and 16-positions) are generated when testosterone is used as a substrate in HLM. Under the optimized high-performance liquid chromatography (HPLC) conditions described, four isomers were completely separated by a SB C18 Zorbax column (2.1 × 150 mm, 5 μm) within a 6-min run time (Figure 14.3A). Also, we confirmed that 6β-hydroxytestosterone and 2-hydroxytestosterone accounted for 82% and 13% of the total metabolites, respectively (Figure 14.3B), when testosterone was incubated in HLM under incubation conditions employed in our lab.

14.3.5.3   Determination of IC_{50} Values   An IC_{50} value of a known specific CYP inhibitor was measured under optimized incubations (Table 14.5), and was

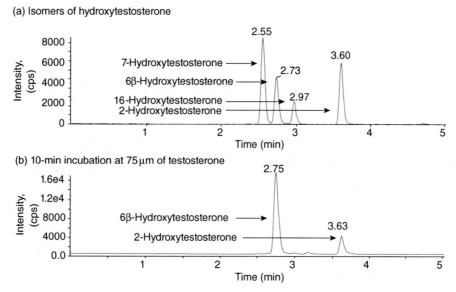

FIGURE 14.3. MRM chromatograms from the analysis of monohydroxyl metabolites of testosterone. (A) Separation of four standards of monohydroxyl testosterone. (B) An incubation of testosterone with HLM in the presence of NADPH.

TABLE 14.5. Summary of IC$_{50}$ Values of Inhibitors for Eight Human Cytochrome P450 Enzymes in Pooled Human Liver Microsomes for Reversible and Time-Dependent Inhibition

Enzyme (Substrate)	Reversible Inhibition		Time-Dependent Inhibition	
	Inhibitor (Range)	Mean^a ± SD	Inhibitor (Range)	Mean^a ± SD
CYP1A2 (phenacetin)	α–Naphthoflavone (0–1 μM)	0.0141 ± 0.0016	Furafylline	0.34 ± 0.13
CYP2A6 (coumarin)	Tranylcypromine (0–5 μM)	0.0741 ± 0.0126	N/D^b	N/D^b
CYP2B6 (bupropion)	Orphenadrine (0–2000 μM)	507.7	Phencyclidine (0–120 μM)	2.86 ± 0.28
CYP2C8 (taxol)	Montelukast (0–5 μM)	0.039 ± 0.0104	Phenelzine (0–2500 μM)	50.1 ± 18.1
CYP2C9 (diclofenac)	Sulfaphenazole (0–20 μM)	0.478 ± 0.085	Tienilic acid (0–12.5 μM)	0.27 ± 0.011
CYP2C19 ((S)-mephenytoin)	(+)-N-3-Benzylnirvanol (0–20 μM)	0.395 ± 0.079	Ticlopidine (0–45 μM)	1.07 ± 0.026
CYP2D6 (dextromethorphan)	Quinidine (0–10 μM)	0.076 ± 0.022	Paroxetine (0–1.25 μM)	0.085 ± 0.007
CYP3A4 (midazolam)	Ketoconazole (0–5 μM)	0.0323 ± 0.0015	Troleandomycin (0–5 μM)	0.67 ± 0.096
CYP3A4 (testosterone)	Ketoconazole (0–5 μM)	0.0477 ± 0.007	Troleandomycin (0–5 μM)	0.41 ± 0.08

^{a} $n = 5$.
^{b} Not determined (N/D).

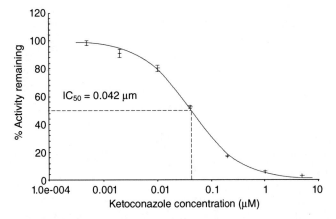

FIGURE 14.4. Inhibition of formation of 6β-hydroxytetostorne by ketoconazole.

consistent with values reported in the literature. A typical CYP RI curve for inhibition of 6β-hydroxytetostorne by ketoconazole is exhibited in Figure 14.4. To ensure the assay quality, the acceptance criterion was set for the resulting IC$_{50}$ value of each positive control inhibitor within 0.5- to 2-fold of a mean IC$_{50}$ value established during the assay validation. The HLM LC-MS single substrate assay is a golden CYP inhibition assay that can be applied for the determination of IC$_{50}$ and K_I values of reversible CYP inhibition as well as determination of IC$_{50}$ shifts and K_I/K_{inact} values for CYP TDI. Results generated from these assays are mainly used for support in the selection of clinical candidates, regulatory registrations, and design of clinical DDI studies.

14.4   HLM RI ASSAY USING MULTIPLE SUBSTRATES (COCKTAIL ASSAYS)

HLM cocktail CYP inhibition assays have been often applied to fast screening for RI of major human CYP

enzymes in lead optimization (Figure 14.1 and Table 14.1). In this section, key components in the development and performance of cocktail assays are discussed.

14.4.1   Choice of Substrate and Specific Inhibitors

One of the key components in the success of a cocktail assay is the selection of probe substrates. Table 14.6 is a summary of probe substrates commonly employed in HLM CYP reversible cocktail assays. Typically, the preferred probe substrates for HLM reversible CYP inhibition assays (Table 14.2) are applicable to cocktail assays. However, some probe substrates of CYP isoforms or their metabolites are inhibitors of other CYP isoforms. In those cases, use of these probe substrates in a cocktail assay should be avoided. For example, amodiaquine is a commonly used CYP2C8 substrate in single substrate inhibition assays (Table 14.2). It inhibits CYP2D6 activity. To avoid the impact of amodiaquine on CYP2D6 inhibition, a strategy using two separate cocktail assays was developed, in which inhibition of CYP1A2 (phenacitin), CYP2B6 (bupropion), CYP2C8 (amodiaquine), and CYP2C19 (omperazole) were evaluated in one cocktail assay, and inhibition of CYP3A4/5 (testosterone), CYP2C9 (omperazole), and CYP2D6 (dextromethorphan) were examined in another cocktail assay (Table 14.6) (Dixit et al., 2007). If one particular substrate has a very slow turnover rate under conditions of a cocktail assay, or if a marker metabolite of a probe substrate is not sensitive to electrospray LC-MS detection, use of this particular probe substrate is an issue. For example, S-mephenytoin is a low-turnover substrate of CYP2C19, and its 4-hydroxymephenytoin metabolite is not ionized well in positive ion electrospray. These both make the detection of 4-hyddroxymephenytoin challenging. As an alternative, CPY2C19 substrate omeperazole can be effectively used for the evaluation of the inhibition of CYP2C19 in cocktail assays (Table 14.6).

14.4.2 Optimization of Incubations

Similar to HLM reversible CYP inhibition assays with a single probe substrate, key incubation conditions in a cocktail assay should be optimized, including substrate concentration, incubation time, HLM protein concentration, and depletion of substrates after incubation. Detailed procedures of the optimization have been described in Section 14.3.2. However, since multiple substrates are involved in a cocktail assay, it would be very difficult to make all aspects of the incubation conditions as good as those used in a single substrate assay. Results from a cocktail assay are acceptable as long as IC$_{50}$ values of positive controls determined by a cocktail assay are similar to those generated from HLM assays with a single substrate. As indicated in Figure 14.1, CYP inhibition data of a lead compound determined from a cocktail assay are further confirmed or refined in later discovery or early development by using the HLM assay with a single substrate.

14.4.3 Incubation Procedures

In our lab, an HLM cocktail assay was developed for fast screening of RI of major CYP enzymes. The assay employed five probe substrates, including phenacetin for CYP1A2, midazolam for CYP3A4, diclofenac for CYP2C9, dextromethorphan for CYP2D6, and S-mephenytoin for CYP2C19. Substrate metabolites, including acetaminophen (CYP1A2), 1-hydroxymidazolam (CYP3A4), 4-hydroxydiclofenac (CYP2C9), dextrorphan (CYP2D6), and 4-hydroxymephenytoin (CYP2C19), were monitored using full MS scan analysis on an LC-LTQ-Orbitrap. An example of analyzing the inhibitory effects of ticlopidine on CYP enzymes using the cocktail assay is described below.

14.4.3.1 Preparation of Incubation Mixtures

1. A phosphate buffer (100 mM KH$_2$PO$_4$, pH 7.4) containing 1 mM MgCl2 was prepared from 1 M mono- and dibasic potassium phosphate stock solutions that were purchased from Sigma-Aldrich (SIgma-Aldrich, St. Louis, MO, USA) and stored at 4°C. Frozen pooled HLM (BD Biosciences) was used only once after thawing. NADPH stock solutions (10 mM) in phosphate buffer were made fresh daily.

2. Working solutions of HLM were prepared by diluting pooled HLM (20 mg/mL, purchased from BD Biosciences) with 100 mM of phosphate buffer (pH 7.4) to form solutions with HLM concentration at 0.11 mg/mL.

3. Five substrate probes phenacetin (60 mM), midazolam (20 mM), diclofenac (30 mM), dextromethorphan (10 mM), and S-mephenytoin (40 mM) in a mixture of acetonitrile:water (1:1) or methanol were spiked into a previously made HLM solution (0.11 mg/mL) to final concentrations of 45, 2, 5, 5, and 40 μm for phenacetin, midazolam, diclofenac, dextromethorphan, and S-mephenytoin, respectively, in the incubation mixture.

4. The test compound, ticlopidine, was diluted in serial with a mixture of acetonitrile and water (1:1) to final incubation concentrations of 0, 0.001, 0.004, 0.02, 0.08, 0.4, 2, and 10 μM in the incubation mixtures. The final DMSO concentration in the incubations was 0.1%.

14.4.3.2 Incubation Procedure for IC$_{50}$ Determination

1. An aliquot of ticlopidine (5 μL) at eight concentrations from 0 to 10 μM was manually spotted into 1.5-mL incubation tubes in triplicate in a 48-well temperature-controlled heater block.

2. Incubation mixtures (445 μL) containing five probe substrates at final concentrations of 45 μM (phenacetin), 2 μM (midazolam), 5 μM (diclofenac), 5 μM (dextromethorphan), and 40 μM (S-mephenytoin) were transferred to the incubation tubes in triplicate by a TECAN. After preincubation at 37°C in a 48-well temperature-controlled heater block for 5 min, 50 μL of NADPH (10 mM in 100 mM phosphate buffer) was added to each well of the reaction plate (at a final volume of 500 μL) to initiate the reaction. The incubations were maintained at 37°C for 10 min.

3. To prepare the filter plates, 50 μL of a mixture of water, acetonitrile, and formic acid (47:50:3) containing an internal standard (propranolol) (1750 nM) was transferred into a filter plate (Strata impact protein precipitation filter plate, Phenomenex, Torrance, CA, USA) stacked on a 96-well receiver plate. After 10 min of incubation, 300 μL of the reaction mixtures from the incubation tubes were transferred into the filter plate and mixed with the quench solution to stop the reaction. The quenched reaction mixtures were passed through a 0.45-μm filter plate by centrifugation (10 min at 3000 g) into a receiver plate. Finally, the receiver plate was vortexed and placed on LC-MS for the metabolite quantification.

14.4.3.3 Preparation of Standard and QC Solutions

Stock solutions of the metabolites (1 mM) were prepared in a mixture of acetonitrile and water. The standard and QC solutions were prepared and further

TABLE 14.6. A Summary of Some *in vitro* Cocktail Assays and Their Incubation Conditions Using HLM or Combined Recombinant P450s Published in the Literature

CYP	(Dierks et al., 2001)	(Bu et al., 2001a)	(Testino and Patonay, 2003)	(Weaver et al., 2003)[a]
1A2	Ethoxyresorufin		Phenacetin	Phenacetin
2A6	Coumarin	Coumarin		
2B6				
2C19	(S)-Mephenytoin		Omeprazole	(S)-Mephenytoin
2C8	Paclitaxel			
2C9	Diclofenac	Tolbutamide	Tolbutamide	Diclofenac
2D6	Bufuralol	Dextromethorphan	Bufuralol	Bufuralol
2E1		Chlorzoxazone		
3A4	Midazolam	Midazolam	Midazolam	Midazolam
3A4/5				
HLM (mg/mL)	0.5	0.25	0.5	1A2/2C9/2C19/2D6/3A4 (15/5/2.5/5/5)*
Incubation time (min)	20	20	20	10
Analytical time (min)	4	2.5	5.5	2.5

[a] Recombinant P450 enzyme cocktail assay. *Units: pmol/ml.

[b] Bupropion showed inhibition of the formation of S-mephenytoin 4-hydroxylation and dextromethorphan O-demethylation. Two groups of substrates were incubated separately, and then the samples were pooled for LC-MS/MS analysis. (A), Group A; (B), Group B.

[c] Recombinant P450 enzyme cocktails with above substrates were also tested.

[d] Two cocktail assays were used to measure CYP activities. When combined, there was significant inhibition of CYP2D6 by 2C8 substrate amodiaquine. No postincubation extraction was required. This assay was used for hepatocyte induction determination. (A), Group A; (B), Group B.

diluted with an HLM working solution to a serial of standard solutions (0.69–10,000 nM) and QC samples (3.75–7000 nM), respectively. The standard solutions and QC samples were prepared in the same HLM matrix as the incubation mixtures. After 10 min of incubation at 37°C, 270 µL of the standard and QC samples were transferred into a filter plate preloaded with 50 µL of a mixture of water, acetonitrile, and formic acid (47:50:3) containing the internal standard (propranolol). Finally, 30 µL of an NADPH solution (10 mM) was added to the standard solutions, and QC samples. The mixtures were passed through a 0.45-µm filter by centrifugation into the receiver plate in the same way as the reaction mixtures were processed.

14.4.4 LC-MS/MS Analysis

Metabolites from five probe substrates and an internal standard in the cocktail assay incubations described above were quantitatively determined using ultrahigh-pressure liquid chromatography (UPLC) coupled with an LTQ-Orbitrap mass spectrometer (Thermo Scientific, Franklin, MA, USA) with a UPLC column (Acquity UPLC HSS T3 2.1 × 100 mm, 1.8-µM particle size). The mobile phase A was water (0.1% formic acid), and the mobile phase B was acetonitrile (0.1% formic acid). The flow rate was 600 µL/min, with a gradient from 5% B to 50% B in 6.4 min. The total run time was 10 min.

The high-resolution mass spectrometer LTQ-Orbitrap was run in the positive electrospray mode. Full-scan analysis was performed in the profile mode from 130 to 350 amu and with resolution at 15,000 (full width at half maximum [FWHM]). After optimization, the capillary temperature was set at 375°C with sheath gas flow at 40 and auxiliary gas at 20. The electrospray ionization (ESI) voltage was set at 5 kV. The capillary voltage was 20 V with the tube lens set at 59 V. Figure 14.5 illustrates extracted ion chromatograms (±5 ppm window) of the five probe metabolites and propranolol (internal standard). Data were collected and processed using the Xcalibur data collection and integration software (Thermo Scientific, Franklin, MA, USA).

14.4.5 Data Calculation

Data calculation tools used in the determination of IC$_{50}$ values are the same as those employed in data calculation for an HLM LC-MS assay with a single probe substrate (see Section 14.3.4). However, unlike HLM single substrate assays, results from cocktail assays mainly support discovery programs, including termination of lead compounds for progression if they have strong *in vitro* CYP inhibition potency as well as determination of the structure–activity relationship for the design of new compounds that have no or reduced CYP inhibition potential.

Continuing Table 14.6

(Kim et al., 2005)[b]	(Tolonen et al., 2007)	(Di et al., 2007)[c]	(Tolonen et al., 2007)[d]	(Youdim et al., 2008; Zientek et al., 2008)
Phenacetin (A)	Melatonin		Phenacetin (A)	Tacrine
Coumarin (A)	Coumarin			
Bupropion (B)	Bupropion		Bupropion (A)	
(S)-Mephenytoin (A)	Omeprazole		Omeprazole (A)	(S)-Mephenytoin
Paclitaxel (A)	Amodiaquine		Amodiaquine (A)	
Tolbutamide (B)	Tolbutamide	Diclofenac	Tolbutamide (B)	Diclofenac
Dextromethorphan (A)	Dextromethorphan	Bufuralol	Dextromethorphan (B)	Dextromethorphan
Chlorzoxazone (B)	Chlorzoxazone			
Midazolam (A)	Omeprazole, midazolam, testosterone	Midazolam		Midazolam
			Testosterone (B)	
0.25	0.5	0.1/0.5	0.5	0.1
15	30	20	30	8, 10
6.5	8			1, 2, or 4

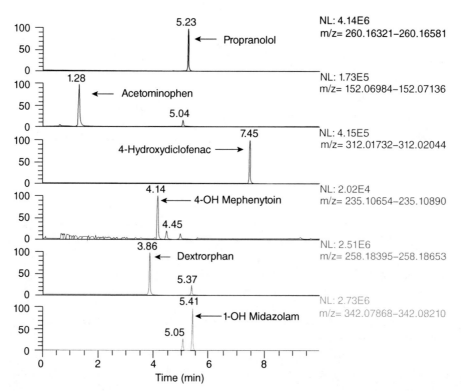

FIGURE 14.5 HR-MS analysis of metabolite concentrations of multiple probe substrates in HLM. Extracted MS chromatograms of metabolites of a mixture of substrates (phenacetin, diclofenac, (S)-mephenytoin, dextromethorphan, and midazolam) and an internal standard, propranolol, were obtained with a ±5 ppm mass accuracy window. HLM incubation was carried out with the five probe substrates after 10 min. The reaction mixture (200 μL) was quenched with a 20 μL mixture of water, acetonitrile, and formic acid (92:5:3) containing 1000 nM of propranolol. Full-scan-MS LC-MS data were acquired using an LC-LTQ-Oribtrap as described in Section 14.4.4.

14.5   TIME-DEPENDENT CYP INHIBITION ASSAY

As illustrated in Figure 14.1, TDI assays are often conducted from discovery in support of lead optimization and clinical candidate selection, and in development prior to clinical DDI studies (Fowler and Zhang, 2008; Grime et al., 2009; Grimm et al., 2009). In this section, an HLM IC_{50} shift assay and an HLM K_I and K_{inact} measurement experiment performed in our lab are discussed (Chang et al., 2010).

14.5.1 IC_{50} Shift Assay

The assay was designed to evaluate one test compound per incubation. Two plates were used to determine IC_{50} values for both reversible and time-dependent inhibition simultaneously. Results allow for the calculation of IC_{50} shifts. Both the RI plate and the TDI plate contained the same test compound, a positive control for RI, and a positive control for TDI. A procedure to determine CYP2C8 IC_{50} shift is discussed below.

14.5.1.1   *Preparation of Stock Solutions and All Related Reagents*

1. Stock solution
   - 100 mM phosphate buffer (pH 7.4), containing 1 mM EDTA
   - 50 mM Taxol (Bristol-Myers Squibb, New York, NY, USA) in DMSO:MeOH (1:1)
   - 100 µM 6α-hydroxytaxol in DMSO
   - 50 µM 10-deacetyltaxol-C (internal standard)
   - 3 mM montelukast (positive control, reversible) in 50% acetonitrile (ACN)
   - 1 M phenelzine (positive control, metabolism) in DMSO
   - 20 mg/mL HLM (BD Biosciences, Sparks, MD, USA)
   - NADPH 10 mM, 10 mL
   - 30 mM testing compound in DMSO
2. Reagent preparation
   - Prepared 90 mL methanol containing 100 nM 10-deacetyltaxol-C in stop container (internal standard) pipet 180 µL of 50 µM 10-Ddeacetyltaxol-C into 90 mL methanol
   - Prepared HLM at 0.0625 mg/mL (final: 0.05 mg/mL)—pipet 281.3 µL of 20 mg/mL HLM into 90 mL 100 mM phosphate buffer

14.5.1.2   *Incubation Procedure for IC_{50} Determination of RI and TDI*

1. The highest concentrations of the calibration standard, QC sample, test compound, and posi-

tive controls were manually transferred into the last well (H) of each column (1, 2, 4, 6, and 8) in a 2-mL 96-well dilution plate (Table 14.4 and Figure 14.2B).

2. Appropriate volume of DMSO was added to each well containing the test compound and positive controls to maintain the same percentage of DMSO in all wells. The final DMSO concentration was around 0.15% (v/v).

3. Paclitaxel solution for simultaneous analysis of RI and TDI was transferred (50 µM, 600 µL) into each well of column 12.

A TECAN performed the following procedures:

4. Preparation of standard curve, QC samples, and solutions of montelukast and phenelzine in an HLM-1B solution by following the procedure described in Section 14.3.3.

5. Aliquots of the mixtures (160 µL) located in columns 4, 6, and 8 were transferred to two incubation plates in triplicate for RI and TDI.

6. NADPH solution (10 mM, 20 µL) was added to the wells in the TDI reaction plate containing positive controls and the test compound to initiate the preincubation reaction (37°C for 30 min).

7. While the reaction was ongoing, 360 µL of methanol containing the internal standard was transferred to the filter plates for RI and TDI.

8. Paclitaxel (50 µM, 20 µL) and NADPH (10 mM, 20 µL) were added to the wells containing the positive control and test compound in the reaction plate for RI analysis to initiate the reaction at 37°C.

9. After incubation for 5 min, the reaction mixture (120 µL) containing the positive control and test compounds was transferred onto the filter plate to stop the reversible reaction.

10. Aliquots of calibration standards (108 µL) were transferred to the filter plate for RI analysis, and aliquots of QC samples (108 µL) were transferred to the filter plate for TDI analysis (Table 14.3).

11. NADPH (10 mM, 12 µL) was added to each well containing the calibration standards and QC samples.

12. After a 30-min preincubation in the reaction plate for TDI analysis, paclitaxel (50 µM, 20 µL) was added to the wells containing the positive controls and test compound in the reaction plate for TDI to initiate the reaction at 37°C.

13. After incubation for 5 min, an aliquot of the reaction mixture (120 µL) containing positive con-

trols and test compound was transferred onto the filter plate for TDI analysis to stop the reaction.

14. The filter plates for RI and TDI analysis were stacked on two 2-mL 96-well receiver plates. After vortexing for 30 s, these mixtures were loaded on filter plates and passed through a 0.45-μm hydrophobic PTFE membrane by centrifugation for 5 min at a speed of 2000 g. Filtrates were collected into the receiving plates.

15. The receiving plates were vortexed again, and then sealed with a polypropylene film. The solution in the receiving plates (5–10 μL) was injected onto LC-MS/MS for metabolite quantification.

14.5.1.3 Data Calculation

Calculation tools used were the same as those described in Section 14.3.5. The IC$_{50}$ curves for RI and TDI from this experiment were constructed in Figure 14.5. The IC$_{50}$ and IC$_{50}$ shift values of montelukast and phenelzine (shown below) were calculated from the plots (Figure 14.6). Montelukast is a very potent reversible CYP2C8 inhibitor, but it did act as a significant TDI inhibitor of CYP2C8 so that the IC$_{50}$ shift was less than 1. On the other hand, phenelzine is a time-dependent inhibitor of CYP2C8, but did not show RI inhibitor of CYP2C8. Therefore, its IC50 shift was 3.2. IC$_{50}$ values for inhibitors of other CYP enzymes that are used as positive controls of TDI in our lab are summarized in Table 14.5.

	IC$_{50}$ (μM)		
	Reversible (1)	Time-Dependent (2)	Fold Changes
Inhibitor	No Preincubation	30-Min Preincubation	(1)/92)
Montelukast	0.054 ± 0.0061	0.072 ± 0.0071	0.8
Phenelzine	162 ± 46.8	50.1 ± 18.1	3.2

14.5.2 K_I and K_{inact} Measurements

K_I and K_{inact} measurements have been utilized at the development state to provide more defined TDI data to support clinical DDI studies (Figure 14.1). These experiments are typically conducted by using the preferred probe substrates and LC-MS methods for HLM LC-MS assays with a single substrate (Table 14.1) except multiple time point, concentrations of the test compound and the probe substrate are used. In the following sections, the procedure and data calculation of an assay for determining CYP3A K_I and K_{inact} values are discussed, in which midazolam was used as a probe substrate of CYP3A4, and domperidone was examined as a time-dependent inhibitor of CYP3A4. A layout of a TECAN

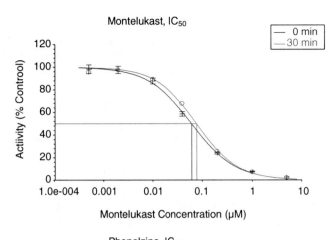

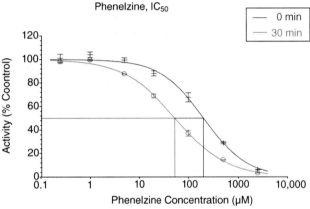

FIGURE 14.6. (A) Inhibition of formation of 6α-hydroxypaclitaxle by montelukast. (B) Inhibition of formation of 6α-hydroxypaclitaxel by phenelzine.

experiment for automated incubation is displayed in Figure 14.2C.

14.5.2.1 Reagent Preparation

1. Prepared 60 mL of acetonitrile containing 1 μM of 1-hydroxymidazolam-D4 in stop container (internal standard).

2. Prepared 30 mL of HLM-1 of 1.25 mg/mL in 100 mM phosphate buffer (final: 1 mg/mL preincubation, 0.1 mg/mL activity).

3. Prepared standard and QC in 0.111 mg/mL HLM-2 for (final: 0.1 mg/mL).

4. Prepared 12 mL of 10 mM NADPH (final 1 mM) in 100 mM phosphate buffer.

5. Prepared 40 mL of 31.25 μM midazolam in 100 mM phosphate buffer (final: 25 μM).

6. Prepared 100 mL of 0.1% formic acid by dissolving 100 μL in 100 mL of water.

14.5.2.2 Sampler Preparation

1. Mixed 2.78 μL of 1 mM 1-OH-midazolam and 2497 μL of HLM-2 for standard and place the

resulting solution in H-1 position of the dilution plate. After serial dilution with HLM-2 by the TECAN, the final concentrations for the standard curve are 0, 2, 5, 25, 100, 250, 500, and 1000 nM.

2. Mixed 2.67 µL of 1 mM 1-OH-midazolam and 2997 µL of HLM-2 for QC dilution and placed 1.5 mL of resulting solution in G-2 and H-2 positions of the dilution plate, respectively. After serial dilution with HLM-2 by the TECAN, the final concentrations of QC samples are 10, 150, 400, and 800 nM.

3. Added 1.13 µL DMSO to column 4 and 6 wells from A to G to keep the same amount of the solvent.

4. Prepared 4.8 mL of 125 µM of domperidone in HLM-1 and placed 2 mL of the solution in H-4 and H-6 position of the dilution plate. After serial dilution by the TECAN, the final concentrations of domperidone were 1.56, 3.13, 6.25, 12.5, 25, 50, and 100 µM, respectively.

5. The NADPH solution was transferred (10 mM, 800 µL) to each well of column 12.

14.5.2.3   Reaction Procedure

1. Transferred 80 µL of the prepared inhibitor (domperidone) solution from columns 4 and 6 (positions A–H) of the dilution plate to columns 4–12 of the preincubation plates 1 and 2.

2. Added 20 µL of the NADPH solution to columns 4–12 of the preincubation plates 1 and 2 (positions A–H).

3. Transferred 160 µL of substrate and 20 µL of the NADPH solution to the activity reaction plates 1 and 2.

4. After 0, 10, 15, 20, 25, and 30 min of preincubation, 20 µL of the preincubated mixture was transferred to an activity plate 1 or 2 to start the metabolism reaction for 3 min.

5. Transferred 120 µL of the reaction solution from activity reaction plates 1 and 2 to 2-mL hydrophobic filter plates 1 and 2 (column 4–12) to stop the reaction.

14.5.2.4   Standard Curve and QC Preparation

1. Transferred 108 µL of the standard solutions from column 1 of the dilution plate to columns 1–3 of filter plate 1.

2. Transferred 108 µL of QC from column 2 of the dilution plate to columns 1–3 of filter plate 2.

3. Added 12 µL of the NADPH solution to each well of the standard and QC samples.

14.5.2.5   Filtration Procedure   Stacked the filter plate with a 2-mL 96-well injection plate containing 360 µL of 0.1% formic acid solution. After vortexing for 30 s, these two plates were centrifuged at 2250 g for 5 min. Covered and vortexed the injection plate before injecting 5 µL of filtered incubation solution into LC-MS/MS.

14.5.3   Data Calculation

IC50 values were calculated using XLfit from the percentage activity versus inhibitor concentration plots. The observed rates of CYP3A inactivation (k_{obs}) were calculated from the initial slopes of the linear regression lines of semilogarithmic plots (remaining CYP3A activity versus preincubation time). The inhibitor concentration that supports half the maximal rate of inactivation (K_I) and the maximal rate of enzyme inactivation (K_{inact}) values were calculated in GraphPad Prizm (GraphPad Software, Inc., La Jolla, CA, USA) by nonlinear regression based on the following equation:

$$k_{obs} = K_{inact} \times S/(K_I + S),$$

where S is the initial domperidone concentration. Pharmacokinetic parameters were calculated in Kinetica™ (Thermo Fisher Scientific, Philadelphia, PA) using noncompartmental analysis.

The detail kinetics of CYP3A4 inactivation by domperidone were measured and described by a maximal rate (K_{inact}) of 0.037/min (Figure 14.7A) and an apparent K_I of 12 µM (Figure 14.7B). When CYP3A4 inhibition by domperidone was determined via preincubation with NADPH-fortified HLM in the absence of the CYP3A4 substrate for 30 min, the IC$_{50}$ was 3.2 µM. In comparison, the IC$_{50}$ value of CYP3A4 by domperidone without preincubation was 10.1 µM. These results indicate that domperidone is a time-dependent CYP3A4 inhibitor.

14.6   SUMMARY AND FUTURE DIRECTIONS

In the past 20 years, several types of *in vitro* CYP inhibition assays have been developed in support of fast, reliable, and cost-effective evaluation of human CYP inhibitory potential of pharmaceutical compounds in the pharmaceutical industry (Table 14.1) (Obach et al., 2005; Lahoz et al., 2008). During this period of time, regulatory requirements for the selection and performance of *in vitro* CYP inhibition assays have been fully defined, and all reagents to be employed in CYP inhibition assays, including probe substrates and selective

(A)

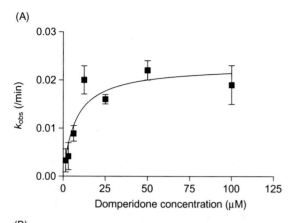

(B)

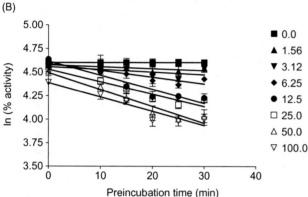

FIGURE 14.7. Time- and domperidone concentration-dependent inactivation of CYP3A-catalyzed midazolam 1'-hydroxylation activity in HLM. (A) Plots of ln (% activity) versus time at different domperidone concentrations in HLM; the slope of each linear regression line is $-k_{obs}$. (B) Plots of k_{obs} versus domperidone concentrations at preincubation, K_{inact}, and K_I were estimated by nonlinear regression.

chemical inhibitors of individual CYP enzymes, are available. Furthermore, some *in vitro* CYP inhibition data generated from various assays have been validated and verified via clinical DDI studies. As a result, analytical strategies and methodologies for the assessment of *in vitro* CYP inhibition from early discovery to preclinical development are maturing and becoming more unified (Walsky and Obach, 2004; Fowler and Zhang, 2008; Grimm et al., 2009). The rhCYP-fluorescence assay was the first one widely applied to high-throughput analysis of CYP inhibition in lead optimization. Currently, the assays are mainly employed for rank order of discovery compounds with the same chemotypes that often process CYP inhibitory ability. HLM assays with a specially designed luminescence or radioactive probe substrate are capable of providing reliable CYP inhibitory data in a high throughput manner. However, due to the lack of probe substrates for all of the major CYP enzymes and other limitations, these assays are no longer employed in most drug metabolism labs in the pharmaceutical industry.

The HLM assay, using commonly accepted probe substrates and MS, have become the methods of choice for *in vitro* evaluation of potency of both reversible and time-dependent CYP inhibition in discovery and development. These assays include the HLM LC-MS assay with a single probe substrate, HLM cocktail assays, and HLM SPE-MS assays with a single substrate (Table 14.1). Currently, two-tiered approaches are often applied in the industry. For small or medium pharmaceutical research organizations, HLM cocktail assays can be mainly placed in lead optimization that require medium throughput. These assays are able to handle 20–30 compounds per week per LC-MS instrument. In late discovery and development where the number of compounds for analysis is reduced greatly, while the assay reliability becomes very important, the HLM LC-MS assay with individual probe substrates can play a dominant role. The same assay can be applied to the evaluation of TDI. For a large pharmaceutical company or contract research organization that deals with more than a hundred compounds per week per mass spectrometer, HLM SPE-MS assays, which are often performed in a specific high-throughput screening lab, would be ideal. Data generated from the assays usually are of high quality and are eligible to support design of clinical DDI studies and regulatory registrations. Assessments of TDI including the measurement of K_I and K_{inact} in development can be carried out using HLM LC-MS assays with individual substrates in a drug metabolism lab (Figure 14.1).

Although *in vitro* CYP inhibition experiments are standardized across different drug metabolism labs, continued technical improvements of HLM/mass spectrometer-based CYP inhibition assays are expected. First, cost-effective, high-performance online SPE technology will be available for a variety of MS platforms. As a result, HLM SPE-MS assays will replace HLM LC-MS assays to increase analytical throughput in some labs. Furthermore, nanochip MS technology is getting practical, which would significantly reduce volumes of incubation and samples for analysis. Consequently, the improvement would greatly reduce the use of HLM and increase productivity of CYP inhibition assays. Finally, there is an increased need for the use of a single LC-MS platform in all drug metabolism studies, including drug metabolite identification and quantitative analysis in enzyme kinetic experiments among small research groups. Two types of mass spectrometer can serve the purpose. The triple quadrupole-linear ion trap instrument is routinely employed for quantification of drugs and metabolites *in vitro* and *in vivo* by using its triple quadrupole scanning functions. Recently, it has been widely applied to metabolite profiling and identification by using its unique information-dependent product ion spectral acquisitions (Yao et al., 2008; Yao et al., 2009). Another MS platform that has shown great potential in both qualitative and quantitative analysis is the high-resolution mass spectrometry (HR-MS). HR-MS has become an instrument of choice in drug metabolite identification (Zhang et al., 2009). Its capability in quantification has recently received a lot of attention. HR-MS has been applied in quantitative analysis in support of discovery pharmacokinetics and metabolic stability experiments (Nagele and Fandino, 2007; Scigelova and Makarov, 2009). Recently, it has been successfully applied to sensitive analysis of metabolites in an HLM cocktail assay (Cai et al., 2011).

REFERENCES

Ayrton J, Plumb R, Leavens WJ, Mallett D, Dickins M, Dear GJ (1998) Application of a generic fast gradient liquid chromatography tandem mass spectrometry method for the analysis of cytochrome P450 probe substrates. *Rapid Commun Mass Spectrom* 12:217–224.

Bapiro TE, Egnell AC, Hasler JA, Masimirembwa CM (2001) Application of higher throughput screening (HTS) inhibition assays to evaluate the interaction of antiparasitic drugs with cytochrome P450s. *Drug Metab Dispos* 29:30–35.

Bell L, Bickford S, Nguyen PH, Wang J, He T, Zhang B, Friche Y, Zimmerlin A, Urban L, Bojanic D (2008) Evaluation of fluorescence- and mass spectrometry-based CYP inhibition assays for use in drug discovery. *J Biomol Screen* 13:343–353.

Bjornsson TD, Callaghan JT, Einolf HJ, Fischer V, Gan L, Grimm S, Kao J, King SP, Miwa G, Ni L, Kumar G, McLeod J, Obach RS, Roberts S, Roe A, Shah A, Snikeris F, Sullivan JT, Tweedie D, Vega JM, Walsh J, Wrighton SA (2003) The conduct of *in vitro* and *in vivo* drug-drug interaction studies: A Pharmaceutical Research and Manufacturers of America (PhRMA) perspective. *Drug Metab Dispos* 31:815–832.

Bu HZ, Knuth K, Magis L, TEITELBAUM P (2001a) High-throughput cytochrome P450 (CYP) inhibition screening via cassette probe-dosing strategy: III. Validation of a direct injection/on-line guard cartridge extraction-tandem mass spectrometry method for CYP2C19 inhibition evaluation. *J Pharm Biomed Anal* 25:437–442.

Bu HZ, Magis L, Knuth K, Teitelbaum P (2001b) High-throughput cytochrome P450 (CYP) inhibition screening via a cassette probe-dosing strategy. VI. Simultaneous evaluation of inhibition potential of drugs on human hepatic isozymes CYP2A6, 3A4, 2C9, 2D6 and 2E1. *Rapid Commun Mass Spectrom* 15:741–748.

Cai H, Yao M, Joseph J, Zhu M (2011) Application of high-resolution mass spectrometry (HR-MS) to sensitive metabolite quantification for CYP inhibition cocktail assay. 59th ASMS Conference on Mass Spectrometry and Allied Topics. Denver, CO,, June 3–9, 2011.

Chang SY, Fancher RM, Zhang H, Gan J (2010) Mechanism-based inhibition of human cytochrome P4503A4 by domperidone. *Xenobiotica* 40:138–145.

Chu I, Favreau L, Soares T, Lin C, Nomeir AA (2000) Validation of higher-throughput high-performance liquid chromatography/atmospheric pressure chemical ionization tandem mass spectrometry assays to conduct cytochrome P450s CYP2D6 and CYP3A4 enzyme inhibition studies in human liver microsomes. *Rapid Commun Mass Spectrom* 14:207–214.

Cohen LH, Remley MJ, Raunig D, Vaz AD (2003) *In vitro* drug interactions of cytochrome p450: An evaluation of fluorogenic to conventional substrates. *Drug Metab Dispos* 31:1005–1015.

Di L, Kerns EH, Li SQ, Carter GT (2007) Comparison of cytochrome P450 inhibition assays for drug discovery using human liver microsomes with LC-MS, rhCYP450 isozymes with fluorescence, and double cocktail with LC-MS. *Int J Pharm* 335:1–11.

Dierks EA, Stams KR, Lim HK, Cornelius G, Zhang H, Ball SE (2001) A method for the simultaneous evaluation of the activities of seven major human drug-metabolizing cytochrome P450s using an *in vitro* cocktail of probe substrates and fast gradient liquid chromatography tandem mass spectrometry. *Drug Metab Dispos* 29:23–29.

Di Marco A, Cellucci A, Chaudhary A, Fonsi M, Laufer R (2007) High-throughput radiometric CYP2C19 inhibition assay using tritiated (S)-mephenytoin. *Drug Metab Dispos* 35:1737–1743.

Dixit V, Hariparsad N, Desai P, Unadkat JD (2007) *In vitro* LC-MS cocktail assays to simultaneously determine human cytochrome P450 activities. *Biopharm Drug Dispos* 28:257–262.

Donato MT, Jimenez N, Castell JV, Gomez-Lechon MJ (2004) Fluorescence-based assays for screening nine cytochrome P450 (P450) activities in intact cells expressing individual human P450 enzymes. *Drug Metab Dispos* 32:699–706.

Draper AJ, Madan A, Latham J, Parkinson A (1998) Development of a non-high pressure liquid chromatography assay to determine [14C]chlorzoxazone 6-hydroxylase (CYP2E1) activity in human liver microsomes. *Drug Metab Dispos* 26:305–312.

EMA (2010) Draft guideline on the investigtion of drug ineractions.

FDA (2006) Guidance for industry. Drug interaction studies—Study design.data analysis, and implicatons for dosing and labeling.

Fowler S, Zhang H (2008) *In vitro* evaluation of reversible and irreversible cytochrome P450 inhibition: Current status on methodologies and their utility for predicting drug-drug interactions. *AAPS J* 10:410–424.

Garcia MC, Ma D, Dicioccio AT, Cali J (2008) The use of a high-throughput luminescent method to assess CYP3A enzyme induction in cultured rat hepatocytes. *In Vitro Cell Dev Biol Anim* 44:129–134.

Grand F, Kilinc I, Sarkis A, Guitton J (2002) Application of isotopic ratio mass spectrometry for the *in vitro* determination of demethylation activity in human liver microsomes using N-methyl-13C-labeled substrates. *Anal Biochem* 306:181–187.

Grime KH, Bird J, Ferguson D, Riley RJ (2009) Mechanism-based inhibition of cytochrome P450 enzymes: An evaluation of early decision making *in vitro* approaches and drug-drug interaction prediction methods. *Eur J Pharm Sci* 36:175–191.

Grimm SW, Einolf HJ, Hall SD, He K, Lim HK, Ling KH, Lu C, Nomeir AA, Seibert E, Skordos KW, Tonn GR, Van Horn R, Wang RW, Wong YN, Yang TJ, Obach RS (2009) The conduct of *in vitro* studies to address time-dependent inhibition of drug-metabolizing enzymes: A perspective of the pharmaceutical research and manufacturers of America. *Drug Metab Dispos* 37:1355–1370.

He F, Bi HC, Xie ZY, Zuo Z, Li JK, Li X, Zhao LZ, Chen X, Huang M (2007) Rapid determination of six metabolites from multiple cytochrome P450 probe substrates in human liver microsome by liquid chromatography/mass spectrometry: Application to high-throughput inhibition screening of terpenoids. *Rapid Commun Mass Spectrom* 21:635–643.

Holt TG, Choi BK, Geoghagen NS, Jensen KK, Luo Q, Lamarr WA, Makara GM, Malkowitz L, Ozbal CC, Xiong Y, Dufresne C, Luo MJ (2009) Label-free high-throughput screening via mass spectrometry: A single cystathionine quantitative method for multiple applications. *Assay Drug Dev Technol* 7:495–506.

Houston JB, Galetin A (2008) Methods for predicting *in vivo* pharmacokinetics using data from *in vitro* assays. *Curr Drug Metab* 9:940–951.

Jenkins KM, Angeles R, Quintos MT, Xu R, Kassel DB, Rourick RA (2004) Automated high throughput ADME assays for metabolic stability and cytochrome P450 inhibition profiling of combinatorial libraries. *J Pharm Biomed Anal* 34:989–1004.

Kim MJ, Kim H, Cha IJ, Park JS, Shon JH, Liu KH, Shin JG (2005) High-throughput screening of inhibitory potential of nine cytochrome P450 enzymes *in vitro* using liquid chromatography/tandem mass spectrometry. *Rapid Commun Mass Spectrom* 19:2651–2658.

Lahoz A, Donato MT, Castell JV, Gomez-Lechon MJ (2008) Strategies to *in vitro* assessment of major human CYP enzyme activities by using liquid chromatography tandem mass spectrometry. *Curr Drug Metab* 9:12–19.

Lim KB, Ozbal CC, Kassel DB (2010) Development of a high-throughput online solid-phase extraction/tandem mass spectrometry method for cytochrome P450 inhibition screening. *J Biomol Screen* 15:447–452.

Lin T, Pan K, Mordenti J, Pan L (2007) *In vitro* assessment of cytochrome P450 inhibition: Strategies for increasing LC/MS-based assay throughput using a one-point IC(50) method and multiplexing high-performance liquid chromatography. *J Pharm Sci* 96:2485–2493.

Moody GC, Griffin SJ, Mather AN, McGinnity DF, Riley RJ (1999) Fully automated analysis of activities catalysed by the major human liver cytochrome P450 (CYP) enzymes: Assessment of human CYP inhibition potential. *Xenobiotica* 29:53–75.

Nagele E, Fandino AS (2007) Simultaneous determination of metabolic stability and identification of buspirone metabolites using multiple column fast liquid chromatography time-of-flight mass spectrometry. *J Chromatogr A* 1156:196–200.

Nettleton DO, Einolf HJ (2011) Assessment of cytochrome p450 enzyme inhibition and inactivation in drug discovery and development. *Curr Top Med Chem* 11:382–403.

Obach RS, Walsky RL, Venkatakrishnan K, Houston JB, Tremaine LM (2005) *In vitro* cytochrome P450 inhibition data and the prediction of drug-drug interactions: Qualitative relationships, quantitative predictions, and the rank-order approach. *Clin Pharmacol Ther* 78:582–592.

Obach RS, Walsky RL, Venkatakrishnan K, Gaman EA, Houston JB, Tremaine LM (2006) The utility of *in vitro* cytochrome P450 inhibition data in the prediction of drug-drug interactions. *J Pharmacol Exp Ther* 316:336–348.

O'Donnell CJ, Grime K, Courtney P, Slee D, Riley RJ (2007) The development of a cocktail CYP2B6, CYP2C8, and CYP3A5 inhibition assay and a preliminary assessment of utility in a drug discovery setting. *Drug Metab Dispos* 35:381–385.

Peng SX, Barbone AG, Ritchie DM (2003) High-throughput cytochrome p450 inhibition assays by ultrafast gradient liquid chromatography with tandem mass spectrometry using monolithic columns. *Rapid Commun Mass Spectrom* 17:509–518.

Rainville PD, Wheaton JP, Alden PG, Plumb RS (2008) Sub one minute inhibition assays for the major cytochrome P450 enzymes utilizing ultra-performance liquid chromatography/tandem mass spectrometry. *Rapid Commun Mass Spectrom* 22:1345–1350.

Riley RJ, Howbrook D (1997) *In vitro* analysis of the activity of the major human hepatic CYP enzyme (CYP3A4) using [N-methyl-14C]-erythromycin. *J Pharmacol Toxicol Methods* 38:189–193.

Rodrigues AD, Kukulka MJ, Roberts EM, Ouellet D, Rodgers TR (1996) [O-methyl 14C]naproxen O-demethylase activity in human liver microsomes: Evidence for the involvement of cytochrome P4501A2 and P4502C9/10. *Drug Metab Dispos* 24:126–136.

Scigelova M, Makarov A (2009) Advances in bioanalytical LC-MS using the Orbitrap mass analyzer. *Bioanalysis* 1:741–754.

Smith D, Sadagopan N, Zientek M, Reddy A, Cohen L (2007) Analytical approaches to determine cytochrome P450 inhibitory potential of new chemical entities in drug discovery. *J Chromatogr B Analyt Technol Biomed Life Sci* 850:455–463.

Stresser DM, Turner SD, Blanchard AP, Miller VP, Crespi CL (2002) Cytochrome P450 fluorometric substrates:

Identification of isoform-selective probes for rat CYP2D2 and human CYP3A4. *Drug Metab Dispos* 30:845–852.

Templeton I, Peng CC, Thummel KE, Davis C, Kunze KL, Isoherranen N (2010) Accurate prediction of dose-dependent CYP3A4 inhibition by itraconazole and its metabolites from *in vitro* inhibition data. *Clin Pharmacol Ther* 88:499–505.

Testino SA Jr., Patonay G (2003) High-throughput inhibition screening of major human cytochrome P450 enzymes using an *in vitro* cocktail and liquid chromatography-tandem mass spectrometry. *J Pharm Biomed Anal* 30:1459–1467.

Tolonen A, Petsalo A, Turpeinen M, Uusitalo J, Pelkonen O (2007) *In vitro* interaction cocktail assay for nine major cytochrome P450 enzymes with 13 probe reactions and a single LC/MSMS run: Analytical validation and testing with monoclonal anti-CYP antibodies. *J Mass Spectrom* 42:960–966.

Turpeinen M, Uusitalo J, Jalonen J, Pelkonen O (2005) Multiple P450 substrates in a single run: Rapid and comprehensive *in vitro* interaction assay. *Eur J Pharm Sci* 24:123–132.

Vandenbrink BM, Isoherranen N (2010) The role of metabolites in predicting drug-drug interactions: Focus on irreversible cytochrome P450 inhibition. *Curr Opin Drug Discov Devel* 13:66–77.

Walsky RL, Obach RS (2004) Validated assays for human cytochrome P450 activities. *Drug Metab Dispos* 32:647–660.

Weaver R, Graham KS, Beattie IG, Riley RJ (2003) Cytochrome P450 inhibition using recombinant proteins and mass spectrometry/multiple reaction monitoring technology in a cassette incubation. *Drug Metab Dispos* 31:955–966.

Yao M, Zhu M, Sinz MW, Zhang H, Humphreys WG, Rodrigues AD, Dai R (2007) Development and full validation of six inhibition assays for five major cytochrome P450 enzymes in human liver microsomes using an automated 96-well microplate incubation format and LC-MS/MS analysis. *J Pharm Biomed Anal* 44:211–223.

Yao M, Ma L, Humphreys WG, Zhu M (2008) Rapid screening and characterization of drug metabolites using a multiple ion monitoring-dependent MS/MS acquisition method on a hybrid triple quadrupole-linear ion trap mass spectrometer. *J Mass Spectrom* 43:1364–1375.

Yao M, Ma L, Duchoslav E, Zhu M (2009) Rapid screening and characterization of drug metabolites using multiple ion monitoring dependent product ion scan and postacquisition data mining on a hybrid triple quadrupole-linear ion trap mass spectrometer. *Rapid Commun Mass Spectrom* 23:1683–1693.

Yin OQ, Lam SS, Lo CM, Chow MS (2004) Rapid determination of five probe drugs and their metabolites in human plasma and urine by liquid chromatography/tandem mass spectrometry: Application to cytochrome P450 phenotyping studies. *Rapid Commun Mass Spectrom* 18:2921–2933.

Youdim KA, Lyons R, Payne L, Jones BC, Saunders K (2008) An automated, high-throughput, 384 well cytochrome P450 cocktail IC50 assay using a rapid resolution LC-MS/MS end-point. *J Pharm Biomed Anal* 48:92–99.

Zhang H, Zhang D, Ray K, Zhu M (2009) Mass defect filter technique and its applications to drug metabolite identification by high-resolution mass spectrometry. *J Mass Spectrom* 44:999–1016.

Zhang T, Zhu Y, Gunaratna C (2002) Rapid and quantitative determination of metabolites from multiple cytochrome P450 probe substrates by gradient liquid chromatography-electrospray ionization-ion trap mass spectrometry. *J Chromatogr B Analyt Technol Biomed Life Sci* 780:371–379.

Zhang XJ, Thomas PE (1996) Erythromycin as a specific substrate for cytochrome P4503A isozymes and identification of a high-affinity erythromycin N-demethylase in adult female rats. *Drug Metab Dispos* 24:23–27.

Zhou ZW, Zhou SF (2009) Application of mechanism-based CYP inhibition for predicting drug-drug interactions. *Expert Opin Drug Metab Toxicol* 5:579–605.

Zientek M, Miller H, Smith D, Dunklee MB, Heinle L, Thurston A, Lee C, Hyland R, Fahmi O, Burdette D (2008) Development of an *in vitro* drug-drug interaction assay to simultaneously monitor five cytochrome P450 isoforms and performance assessment using drug library compounds. *J Pharmacol Toxicol Methods* 58:206–214.

15

TOOLS AND STRATEGIES FOR THE ASSESSMENT OF ENZYME INDUCTION IN DRUG DISCOVERY AND DEVELOPMENT

ADRIAN J. FRETLAND, ANSHUL GUPTA, PEIJUAN ZHU, AND CATHERINE L. BOOTH-GENTHE

15.1 INTRODUCTION

Induction of drug metabolism is a process whereby the activity of enzymes responsible for drug metabolism is increased relative to their basal states within an individual. Most often this is a result of *de novo* synthesis of new enzymes, but in rare instances it is a result of the stabilization of preexisting enzymes (Tompkins and Wallace, 2007). All Phase I, Phase II, and Phase III (transporter) proteins have been shown to be inducible (Lin, 2006, 2007); however, the Phase I enzymes, cytochrome P450s (P450s), are the most inducible when considering fold change from baseline (Martin et al., 2008). The P450 isoforms that are the most inducible in magnitude are CYP3A4, CYP2B6, and CYP1A2 (Lin, 2006). These P450 isoforms are responsible for the oxidative metabolism of approximately 85% of the marketed drugs (Lynch and Price, 2007). Thus, induction of drug metabolism is a major concern in the development of new drugs.

Importantly, regulatory agencies in both North America and Europe have issued guidance for the assessment of induction prior to the filing of a new drug application (NDA). The concerns related to induction of drug metabolism are due not only to potential drug–drug interactions (DDIs) through loss of efficacy of coadministered drugs but also to the loss of efficacy of the administered drug through the induction of its own clearance pathways, a process commonly referred to as autoinduction. Unlike the other common DDI form,

P450 inhibition, induction DDIs rarely are linked to induced toxicities of coadministered drugs. However, one could hypothesize that increased metabolic activity could lead to increased production of circulating metabolic products with an increase in the likelihood for metabolite-associated toxicity and/or DDI.

The molecular events leading to induction have been characterized over the past decades (reviewed in Whitlock, 1999; Wilson and Kliewer, 2002). Assays to identify and screen for induction have evolved with molecular biology and have enabled the early identification and ability to screen for this potential liability (Figure 15.1). Additionally, a greater understanding of the molecular events and the development of *in vitro : in vivo* extrapolations have allowed for early assessments of the risk for clinical DDIs. This chapter describes the most used *in vitro* technologies used in the modern drug discovery and development industry, and includes a brief discussion of risk assessment strategies.

15.2 UNDERSTANDING INDUCTION AT THE GENE REGULATION LEVEL

The three principle receptors responsible for the induction of drug metabolism include the pregnane X receptor (PXR), the constitutive androstane receptor (CAR), and the aryl hydrocarbon receptor (AhR), which regulate the induction of CYP3A4, CYP2B6, and CYP1A2, respectively. Other receptors are also known

ADME-Enabling Technologies in Drug Design and Development, First Edition. Edited by Donglu Zhang and Sekhar Surapaneni.
© 2012 John Wiley & Sons, Inc. Published 2012 by John Wiley & Sons, Inc.

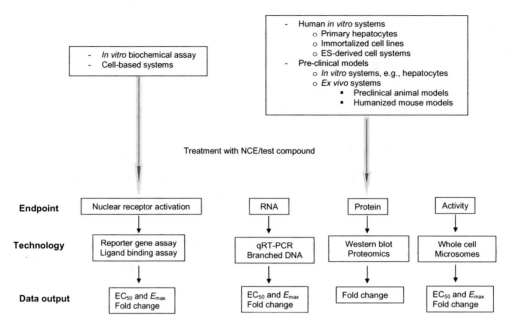

FIGURE 15.1. Schematic of the *in vitro* techniques to assess induction-mediated clinical DDI of NCEs. qRT-PCR, quantitative reverse transcriptase-polymerase chain reaction.

to participate in the regulation of CYP induction but play a modulatory role in nature or are limited to very selective cases. Several published reviews explain the regulation of P450s by receptors in depth (Hewitt et al., 2007; Pavek and Dvorak, 2008). As PXR, CAR, and AhR are the most prominent regulators of CYP induction, the majority of the published literature has focused on the molecular mechanisms by which induction occurs.

The principle mechanism by which induction of drug metabolizing enzymes occurs is via ligand:receptor interactions resulting in an increase in mRNA expression. In principle, measuring the activation of the receptors, either directly via classical ligand binding techniques or indirectly using reporter gene assays, is the simplest and most straightforward approach to develop structure–activity relationships for early drug discovery programs. As understanding of the responsible receptors has increased, the ability to predict ligand binding *in silico* has progressed and can provide a method for a rapid *in silico* screen with optimized models. Used in combination, they provide an invaluable toolbox for the characterization and screening of small-molecule drug discovery compounds for CYP induction.

15.3 *IN SILICO* APPROACHES

15.3.1 Model-Based Drug Design

With the increase in computational power, computational biology has advanced, and the use of *in silico*

approaches has gained popularity in all aspects of drug discovery. Along with computational advances, protein crystallization techniques have allowed for several crystal structures to be solved for the prominent drug metabolism receptor PXR with a variety of ligands (Watkins et al., 2001; Chrencik et al., 2005; Xue et al., 2007) as well as CAR (Xu et al., 2004). As yet, no crystal structure has been solved or reported for AhR. Structural determinants of PXR ligand binding have been difficult to characterize as the ligand binding pocket is described as large and flexible, contributing to its promiscuity. In fact, the structural diversity of the compounds with which crystals have been solved is quite large and includes the macrolide antibiotic rifampin, herbal medicine hyperforin, and the endobiotic 17β-estradiol. As a result, model-based drug design approaches using crystal structures have been limited in early drug discovery. In addition, molecular docking approaches do not differentiate compounds that bind to PXR and may not be efficient activators of transcription, thus complicating the interpretation and design of new molecules.

15.3.2 Computational Models

An alternate approach to model-based drug design using protein crystallization is molecular modeling. The majority of the work with molecular modeling to date has focused on PXR (Ekins and Erickson, 2002; Schuster and Langer, 2005; Ekins et al., 2007; Gao et al., 2007;

Lemaire et al., 2007; Khandelwal et al., 2008; Yasuda et al., 2008; Ekins et al., 2009) with only limited studies on CAR (Windshügel et al., 2005). The goal of molecular modeling is to generate a set of molecular descriptors that describe ligand binding. Using these descriptors, medicinal chemists can rapidly modify new drug candidates for reduced ligand binding affinity, which can be tested rapidly. The ultimate goal would be a "virtual" screen for PXR ligand binding, which would be advantageous due to its near instantaneous turnaround and small resource investment.

The generation of molecular models requires robust input data to generate a reliable model. The use of either crystal structural data or *in vitro* data is needed to develop, test, and validate a molecular model. As the ligand binding pocket of PXR is promiscuous, models using crystal structure data are limited, most likely due to difficulties in interpretation and uncertainty of the input data (Ekins and Erickson, 2002; Schuster and Langer, 2005). The use of *in vitro* data provides for a more robust data set with which to develop a model, and this approach is the most popular approach used for PXR reported in the literature (Ekins et al., 2007; Gao et al., 2007; Lemaire et al., 2007; Khandelwal et al., 2008; Yasuda et al., 2008; Ekins et al., 2009). Even with robust *in vitro* data and advanced computational methods, the predictability of published models has been limited (Khandelwal et al., 2008; Ekins et al., 2009), and the general applicability to chemical scaffolds outside of the test sets has been difficult. The application of such molecular models requires generation of unique training sets for each chemical scaffold in question using multiple modeling approaches to build the most robust model. These limitations may limit the feasibility and applicability of such molecular approaches broadly. At this time, the use of molecular models with rapid *in vitro* screening methods provides a rapid optimization tool for medicinal chemistry.

15.4 *IN VITRO* APPROACHES

15.4.1 Ligand Binding Assays

As the modulation of drug metabolism induction is primarily due to receptor-mediated activities, the evaluation of ligand binding using classical biochemical techniques is a valid procedure. Several ligand binding assays for PXR have been developed to aid in the screening of molecules in drug discovery programs (Jones et al., 2000; Moore et al., 2000; Zhu et al., 2004; Shukla et al., 2009). In addition to PXR, ligand binding assays have been developed for CAR (Moore et al., 2000) and AhR (Gasiewicz and Neal, 1982; Bradfield and Poland, 1988), but their use in routine screening

applications has been limited due to their limited role in drug metabolism. However, with the potential interest of AhR agonists as carcinogens, the ligand binding assay for AhR may be of particular interest in the field of toxicology. Due to its importance in drug metabolism, this discussion will be limited to PXR, but in principle the assays discussed could be applied with modifications to the other receptors.

A commonly used receptor binding method is the scintillation proximity assay (SPA). This method has been developed to examine the receptor binding characteristics of PXR ligands (Jones et al., 2000; Moore et al., 2000; Zhu et al., 2004). The principle of the SPA assay utilizes a recombinant expressed ligand binding domain of PXR linked covalently to a bead containing scintillant. A high-affinity radiolabeled ligand, commonly [^{3}H]-SR12813, is incubated with the SPA beads and allowed to reach equilibrium. Upon binding, the proximity of the radioligand activates the scintillant, and the resultant binding is measured as light using scintillation counting. The nonradiolabeled compound of interest and radiolabeled compound are then coincubated, and the degree of receptor binding is quantified by scintillation counting. The inhibition of binding of the radiolabeled compound is an indication of receptor binding, and a quantitative relationship can be derived. Depending on the amount of characterization needed, a single concentration or multiple concentrations of test compound may be used.

The SPA assay has several advantages over conventional radioligand displacement assays in that it does not require washing or filtration (Alouani, 2000); thus, it is amenable to high-throughput applications, from 96- to 384-well, depending on the throughput required and density of concentrations needed. As with all assays, proper setup and validation is required, including incubation times and radiolabeled ligand concentration. Proper interpretation and use of SPA data is critical in assessing risk of induction. The SPA assay only measures ligand binding, and does not measure transactivation mediated by PXR. This requires care in interpreting high-affinity ligands identified in the SPA assay. Zhu et al. (2004) correlated affinity as measured in a high-throughput SPA assay with transactivation in a reporter gene assay. The majority of compounds showed a good correlation between affinity and transactivation; however, there was a subset of compounds composed of high-affinity ligands as measured by the SPA assay that did not transactivate in the reporter gene assay. Awareness of this potential for molecules that bind PXR but do not result in transcription is important for interpretation and may require follow-up in other *in vitro* assays, such as a reporter gene assay or a suitable hepatocyte model.

The time-resolved fluorescence resonance energy transfer (TR-FRET) assay provides an alternative method to the SPA assay to measure ligand binding activity that is not dependent on the use of radioactive materials. The TR-FRET assay has been developed to study the binding characteristics of several nuclear receptors, including CAR and FXR (Zhou et al., 1998; Moore et al., 2000). A high-throughput TR-FRET assay was developed for PXR based in the 1536-well format (Shukla et al., 2009). The principle of the TR-FRET assay is similar to that of the SPA assay. Both assays use recombinant expressed ligand binding domains and are also competition-based assays where the test compound competes for the ligand binding domain of PXR. The difference lies in the method of detection. The SPA assay uses a radiolabel ligand and scintillant-based beads, and the TR-FRET assay uses fluorescently labeled probes. In the published high-throughput method, a terbium (Tb^{3+})-labeled glutathione S-transferase (GST) antibody is incubated with a GST-tagged PXR ligand binding domain and a fluorescein-labeled PXR ligand. The Tb^{3+}-labeled antibody acts as the fluorescent donor when excited at a specific wavelength. Upon excitation, the energy is transferred to the acceptor molecule (fluorescein-labeled PXR ligand), and light is emitted at a specific wavelength. When coincubated with the test compound, the labeled PXR ligand can be displaced, the transfer of fluorescent energy transfer is disrupted, and the Tb^{3+} does not transfer its energy to an adjacent labeled PXR molecule that emits light at a different wavelength. The monitoring of both processes allows for more robust data analysis as well as less interference from nonspecific effects. The high-throughput nature of the TR-FRET assay allows for the screening of numerous compounds and the generation of robust kinetic data for the characterization of molecules. As with the SPA assay, the potential for false-positives (tight receptor binding with little or no transactivation) requires follow-up in secondary assays to confirm the relative risk.

15.4.2 Reporter Gene Assays

The reporter gene assay has been an invaluable tool for probing the molecular mechanisms of transcription. The assay can be credited with defining the nature of core elements required for transcription of mRNA for an innumerable number of genes. Additionally, it has allowed researchers to define the mechanism by which enhancer elements in DNA interact with receptors that are required for the inducible nature of many genes, including several common drug metabolizing enzymes, for example, *CYP3A4*, *CYP2B6*, and *CYP1A2* (Garrison et al., 1996; Moore et al., 2000; Dickins, 2004). The

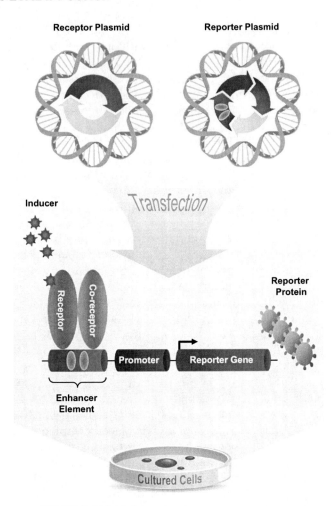

FIGURE 15.2. Schematic of a reporter gene assay.

basis of a reporter gene assay is transfection of an engineered plasmid containing receptor response elements and a core promoter coupled to a protein that is easily assayable (Figure 15.2). Oftentimes, the receptor and/or coreceptor of interest must also be transfected as many common immortalized cell lines have lost their endogenous expression of the receptor of interest (Wang and Negishi, 2003; Naspinski et al., 2008). Common reporters include luciferase, secreted alkaline phosphatase (SEAP), and β-lactamase. Reporter gene assays have proved invaluable in the basic science of gene regulation as well as the screening of molecules for drug induction liabilities.

There are several considerations and options in the design of reporter gene assays. The first option is which reporter to use. The use of luciferase genes from several species of firefly has become quite popular over the past decade. Luciferase catalyzes the hydrolysis of luciferin to oxyluciferin and light. Emitted light is quantified using commonly available luminometers. Genetic engineering of native firefly luciferase genes has resulted in several variants of commonly available luciferase genes with more optimal properties for use in reporter gene assays (Fan and Wood, 2007). The enzymatic reaction catalyzed by luciferase is quite efficient, thus providing a very sensitive method for quantification. This sensitivity has allowed the development of cell-based reporter gene assays based on luciferase in a 384-well format (Herbst et al., 2009). As the luciferase protein is commonly expressed in the intracellular compartment, assays utilizing luciferase require a cell lysis step before analysis. Systems using secretable luciferase are being explored (Huang et al., 2009).

Another reporter gene utilized is SEAP. The SEAP protein is derived from the human placental alkaline phosphatase (AP) protein, which is expressed natively on the cell surface. In its recombinant form it is truncated, and this form is secreted into the cell media in reporter gene assays. Unlike many mammalian enzymes, SEAP is thermostable up to 65°C. As the reporter is secreted into the media, assays utilizing SEAP do not require a cell lysis step, thus preserving the cell monolayer for other assays, for example, a cytotoxicity or a multiplexed reporter assay. A complicating factor in the use of SEAP is high levels of non-SEAP-related AP activity; however, as SEAP is stable at high temperatures, nonspecific AP is eliminated by heating samples.

Additional reporters have been described in the literature and are commonly used, such as reporters based on β-lactamase and chloramphenicol acetyltransferase (CAT). CAT is an older technology based on the use of radiolabeled chloramphenicol. With the advent of sensitive and robust nonradiolabel methods, the use of CAT has been largely replaced in routine screening applications. The use of β-lactamase relies on FRET technology and the use of cell-permeable fluorescently labeled probes.

A reporter construct has two essential constituents, an enhancer element and a promoter element, within the regulatory region, which can vary widely. The promoter element can be derived from a gene of interest, *CYP3A4*, *CYP1A2*, *CYP2B6*, and so on, or from a general promoter, commonly thymidine kinase (TK) or cytomegalovirus (CMV). Opinions vary on the advantages and disadvantages of each, but for drug metabolism induction studies, the core promoter from a gene of interest is the most commonly used (Raucy et al., 2002; Zhu et al., 2007; Cui et al., 2008). The important element is the enhancer element, which is the element to which the receptor of interest will bind. The critical elements within the *CYP3A4*, *CYP2B6*, and *CYP1A2* genes have been identified and characterized (Fujisawa-Sehara et al., 1987; Fujisawa-Sehara et al., 1988; Honka-koski et al., 1998; Goodwin et al., 1999; Sueyoshi et al., 1999). Interestingly, the reporter constructs from

CYP3A4 and *CYP2B6* are interchangeable within reporter gene assays (Faucette et al., 2006; Faucette et al., 2007), underscoring the promiscuity and cross-regulation of PXR and CAR in the induction of drug metabolism.

An additional method to consider for the design and implementation of a cell-based reporter gene assay is transfection, which can be transient or stable. A reporter gene assay requires the introduction of the components (reporter plasmid, nuclear receptor, control reporter plasmid) of the assay into a cell. This can be performed on an assay-by-assay basis (transient transfection) or through the production of a stable cell line that incorporates the different components of the reporter gene assay into the host cell's genome through antibiotic selection coexpressed on the plasmid of interest. In general, because of differences in transfection efficiency between assays, the transient method is subject to more assay-to-assay variability. Some groups have reported the use of batch transfection with subsequent cryopreservation as a method to reduce variability (Zhu et al., 2007; Herbst et al., 2009). The production of stable cell lines is limited by the number of antibiotic selection markers that can be incorporated within a single cell line, often just two. With that, the inclusion of an additional control reporter construct is not often feasible, thus requiring additional steps to correct for cell number or cytotoxicity within an experiment.

Proper interpretation of reporter gene assay data is critical for follow-up activities/assays and placing the data in the proper context. With ligand binding assays, the binding of compound to receptor does not necessarily translate to activation of transcription or induction, and with reporter gene assays, the increase in reporter gene activity, transactivation, does not always translate directly into increases in mRNA or protein in hepatocytes and/or *in vivo* (Luo et al., 2002). The reasons for this may include time-dependent inactivation of enzyme activity, poor exposure of the test compound due to rapid metabolism, or other confounding factors. This often requires follow-up in an immortalized hepatocyte cell line or primary hepatocytes. Conversely, if a compound does not increase reporter gene activity, inclusion of one or more cytotoxicity markers is important to insure the negative result is truly a result of the compound's transactivation properties and not of toxicity. Normalization of reporter gene activity to a positive control is often used as a metric to determine potential risk.

Reporter gene assays based on PXR have been developed and are well characterized. The use of PXR reporter gene assays is common in early drug discovery programs. As PXR is not expressed in most if not all transformed cell lines, reporter assays require the transfection of PXR. Several high-throughput methods based on both transient and stable transfection protocols have been developed based on not only human PXR but also other preclinical species (Raucy et al., 2002; Vignati et al., 2004; Trubetskoy et al., 2005). As there are large differences in ligand selectivity between species, use of the proper species in addressing PXR activation is critical. Assays based on higher-density plate formats (384- or 1536-well) have provided a convenient high-throughput assay to screen molecules for PXR transactivation. Additionally, the correlation of PXR activity in reporter gene assays with enzymatic activity and mRNA in primary human hepatocyte cultures is reasonable (Luo et al., 2002), thus validating its use as an early screen for induction liabilities.

The CAR presents an interesting challenge in the design and interpretation of a reporter gene assay and data. Since CAR is not expressed to an appreciable extent in transformed cell lines, transfection of exogenous CAR is a necessity for a reporter gene assay (Wang and Negishi, 2003). The complication arises in that CAR is constitutively active upon transfection into transformed cell lines (Masahiko and Honkakoski, 2000). This high basal level of reporter protein activity lowers the dynamic range for which activators of transcription can be identified. A splice variant, termed CAR3, has been indentified that does not possess constitutive activity. This variant retains ligand-activated reporter gene activity in cell-based assays and may prove valuable for the screening of CAR ligands in drug discovery programs (Auerbach et al., 2005; Chen et al., 2010). It should be noted that the CAR3 variant does not retain the CAR activator function associated with induction-like activities associated with phenobarbital (PB), but it has been shown to transactivate CAR3 in a reporter gene assay, suggesting a complex regulatory mechanism for PB-like inducers (Chen et al., 2010). A recent method utilizing a fluorescently labeled CAR adenoviral vector in primary human hepatocytes may provide an alternative method to identify CAR ligands and activators in a reporter gene-like assay (Li et al., 2009). If coupled with high-throughput microscopy techniques, it may provide a rapid and efficient method to screen for CAR activity. Like PXR, CAR does exhibit species-specific ligands, and this fact should be considered in addressing questions related to induction observed in preclinical species during the drug development process.

Reporter gene assays for AhR have been developed and validated (Garrison et al., 1996; Murk et al., 1996; Yueh et al., 2005). Unlike PXR and CAR, many transformed cell lines express sufficient AhR for reporter gene assays. This requires selection of the proper species cell line. Unlike PXR and CAR, there does not appear to be a large species difference in AhR. Any species

difference seems to be related to affinity, and not ligand selectivity (Garrison et al., 1996; Ramadoss et al., 2005; Nohara et al., 2006).

15.5 *IN VITRO* HEPATOCYTE AND HEPATOCYTE-LIKE MODELS

15.5.1 Hepatocyte Cell-Based Assays

Beyond the use of cell-based reporter gene assays, several *in vitro* hepatocyte models have been characterized for their ability to accurately assess new chemical entities (NCEs) for their induction liabilities. Primary human hepatocytes are the most well-characterized system to assess induction, and to date they remain the gold standard and most accepted method by the regulatory agencies for this purpose. Several reports have demonstrated the usefulness of such cells to assess the induction potential of NCEs (Kostrubsky et al., 1999; LeCluyse et al., 2000; Hewitt, et al. 2007). Although fresh human hepatocytes have the advantage of being closest to liver cells in terms of having the necessary volume of molecular elements crucial for induction, they suffer from certain disadvantages such as limited and unpredictable availability, interindividual variability, uncertainty in quality, inability to proliferate, and a rapid loss of expression of several metabolizing enzymes in culture conditions. Hence, the establishment of a more reliable and readily available *in vitro* model system would be valuable for routine drug discovery activities.

Over the past several years, the use of cryopreserved primary hepatocytes has become popular. The advantage of cryopreserved hepatocytes is that they overcome the limited availability and uncertainty associated with fresh hepatocytes, as well as allow researchers to create banks of well-characterized hepatocytes that can be easily stored. Several studies have compared the response to prototypic inducers in fresh versus cryopreserved hepatocytes and have shown little difference with respect to inducibility both at the mRNA and enzyme activity level of P450s (Hengstler et al., 2000; Garcia et al., 2003). Some differences in the appropriate expression and targeting of some drug transporters have been observed (Bi et al., 2006). The absolute effect of these differences on the inducibility of P450s is unknown but may lead to an under- or overestimation of risk, depending on the individual characteristics of test compounds. Emphasizing the general acceptability of cryopreserved hepatocytes is the suitability of their use in regulatory filings. Despite the apparent advantages of cryopreserved hepatocytes, their costs may be prohibitive for routine screening applications; thus, the use of alternative models that are hepatocyte-like would be beneficial.

15.5.2 Hepatocyte-Like Cell-Based Assays

In addition to primary human hepatocytes, several established cell lines have been shown to have potential for assessing induction potential. The draft Guidance for Industry issued by the U.S. Food and Drug Administration (http://www.fda.gov/ohrms/dockets/ac/06/briefing/2006-4248B1-03-FDA-Topic-2-guidance.pdf) allows for P450 induction data generated in immortalized cells provided that it is demonstrated that positive controls have been used. Although this draft guidance gives flexibility to the sponsor to utilize immortalized cell lines based on the positive control data for two major P450 enzymes, it does not take into consideration induction of other important P450 enzymes.

The nontumorigenic immortalized cell line Fa2N-4 has been evaluated extensively for its potential use as an *in vitro* model system to assess enzyme induction. The induction of various important genes such as CYP1A1/2, CYP3A4, CYP2C9, UGT1A, and MDR1 as measured by mRNA expression levels and P450 activity has been demonstrated (Mills et al., 2004). Ripp et al. (2006) demonstrated that Fa2N-4 cells can be used successfully to predict CYP3A4 induction-mediated clinical DDIs by calculating the relative induction score (RIS) factoring in the *in vivo* free plasma concentrations. In this study data generated in Fa2N-4 cells using 24 compounds, 18 compounds known to be inducers and 6 compounds known to be negative for induction in primary human hepatocytes had considerably less variability (21–37% depending on the parameter) than the primary human hepatocytes (200% as reported in hepatocytes obtained from 62 individual donors by Madan et al. (2003). This decrease in variability increases confidence in the data and facilitates interpretation from experiment to experiment. Although these advantages of reproducibility of the data and availability of the cell line make the Fa2N-4 cells a very attractive system, a recent report suggests that these cells have significant differences in the levels of expression of major drug metabolizing enzymes, nuclear receptors, and transporters (Hariparsad et al., 2008). The levels were as much lower as 50-fold for important receptors such as CAR. Hence, induction of enzymes (such as CYP2B6 and CYP3A4), which involve CAR as the crucial nuclear receptor, will be underestimated from such cells. This study demonstrates the limitations of use of Fa2N-4 cells and points to the fact that the data generated in such systems should be interpreted with caution. Several other hepatic cell lines such as BC2 (Gómez-Lechón et al., 2001) have also been used with limited success.

Another immortalized hepatoma cell line, HepaRG, shows superiority over the Fa2N-4 cells (McGinnity et al., 2009) and the HepG2 cells (Hart et al., 2010; Jennen et al., 2010) as HepaRG cells expresses molecular receptors (specifically CAR) and P450 enzyme expression levels closer to primary human hepatocytes. Anthérieu et al. (2010) reported that HepaRG cells not only accurately respond to prototypical inducers of CYP1A1, CYP2B6, and CYP3A4 (mRNA and activity endpoints) but also express basal expression levels of several Phase II enzymes, transporters, and nuclear receptors similar to primary human hepatocytes. Based on these recent findings, HepaRG cells seem to be a much more promising cell line compared with the other hepatic cell lines used so far for assessing induction *in vitro*.

More recently, human embryonic stem cells (hESC) have been used to generate hepatocyte-like cells. The hESCs are immortal, pluripotent, and possess the inherent capacity to proliferate, thus they could be considered a potential powerful "biotool" for industrial applications. Several groups have reported successful generation of a homogenous population of hepatocyte-like cells from stem cells (Rambhatla et al., 2003; Hay et al., 2010; Kim et al., 2011). Using hESC may have certain advantages over using traditional immortalized cell lines as they have demonstrated phenotype, expression, activity levels, and inducibility of various metabolic enzymes comparable at least in part to primary human hepatocytes (Ruhnke et al., 2005; Ek et al., 2007). Additionally, most reports show similarities in morphology, and liver markers such as albumin production and urea synthesis; however, to date only a few have been able to demonstrate metabolic capacity equivalent to the primary human hepatocytes (Ek et al., 2007; Duan et al., 2010). Although stem cell technology shows promise for use in enzyme induction studies, it may be considered as still in the early stages of development and characterization. Thus, improvement and further characterization of this technology is imperative before it can be used as substitute for primary human hepatocytes.

15.6 EXPERIMENTAL TECHNIQUES FOR THE ASSESSMENT OF INDUCTION IN CELL-BASED ASSAYS

Induction can be assessed at three different levels: mRNA, protein, and enzymatic activity. Assessment of each has its advantages and disadvantages and can be more appropriate depending on the stage of drug development and/or the issue to be addressed. However, it should be noted that regulatory agencies use the induction of enzyme activity as the baseline with which to assess the risk of clinical induction. Here we describe various methods to assess the three markers of enzyme induction.

15.6.1   mRNA Quantification

The quantification of mRNA has progressed substantially in the past decade, and newer quantification methods are invaluable and critical in the assessment of induction in drug discovery programs of both *in vitro* and *in vivo* samples. The quantification has progressed significantly from low-throughput methods that relied on gel electrophoresis and the hybridization of radiolabeled probes to high-throughput multiplexed methods based on polymerase chain reaction (PCR) or other higher-throughput hybridization/amplification techniques, for example, branched DNA (bDNA) technology. Depending on the throughput required, selection of the appropriate method is important.

The use of singleplex methods has been well described and is commonly used in all aspects of drug discovery, particularly in the study of induction. Currently, two techniques are the predominant methods used for the quantification of mRNA in drug induction studies. The first is bDNA technology. This technology is a hybridization-based method that is analogous to an enzyme-linked immunoassay (ELISA) assay (Urdea et al., 1991; Collins et al., 1997), and a general schematic of the assay format is shown in Figure 15.3. Cell lysates are applied to a plate that has been coated with oligonucleotides with a specific sequence not related to the target mRNA, the capture probe. Two additional oligonucleotide probe sets are added, one that hybridizes with the capture probe and mRNA of interest (the capture extender probe), and another that hybridizes with the mRNA of interest and has a sequence that can hybridize with a label probe (the bDNA). Both probe sets provide specificity within the assay and allow for robust washing of the plate. The label probe, a probe that binds the bDNA, is added. The label probe is coupled with an enzyme that allows for quantification of the mRNA, commonly AP. Branch DNA technology is exquisitely sensitive, and does not require an amplification step; that is, mRNA can be quantified directly from cell lysates or other matrices. It is commonly used in clinical applications for the detection of HIV, and the reagents for quantification of the mRNAs for the common drug metabolizing enzymes in humans as well as preclinical species are available through commercial vendors. A disadvantage of bDNA lies in that it requires specialized reagents and know-how to develop novel probe sets for mRNAs that may not be commercially available.

The other commonly used singleplex technique to quantify mRNA is based on quantitative reverse transcriptase-polymerase chain reaction (qRT-PCR)

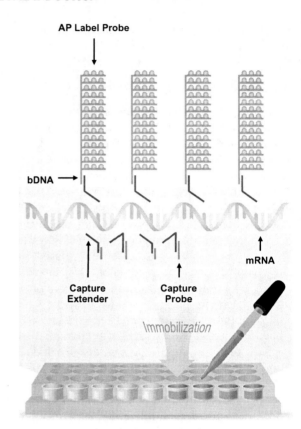

FIGURE 15.3. Schematic of mRNA quantitation.

using FRET technology. RT-PCR requires the isolation of RNA using any number of commonly available protocols or commercial kits. After isolation, RNA is reverse-transcribed to DNA using the enzyme reverse transcriptase. After reverse transcription, a PCR reaction is conducted utilizing three probes; two are the amplification probes used in all PCR reactions that provide amplification and specificity. The third is the quantification probe, which has linked to it a fluorescent reporter and a quencher probe. In its intact state, the fluorescence from the fluorescent probe is quenched due to its proximity to the quencher probe, and only background fluorescence is detected. If the probe binds to DNA, the exonuclease activity of the PCR enzyme, Taq polymerase, cleaves the reporter from the oligonucleotide quantification probe, thus allowing fluorescent light to be emitted and quantified. Due to the amplifying nature of the method, assays employing quantitative qRT-PCR are exquisitely sensitive. Additionally, the design and optimization of new probe sets is well described in the literature, and the specific oligonucleotide reagents are easily synthesized by any number of commercial vendors. Assays utilizing RT-PCR do require the lysis and isolation of RNA from samples.

Recent development of multiplex analyses methods for the quantification of mRNA have further increased throughput and decreased turnaround time. The ability to multiplex RT-PCR methods has been described in the literature (Vansant et al., 2006; Northwood et al., 2010), and the process utilizes slightly different techniques. Additionally, a multiplexed method based on bDNA has also been developed and described in the literature (Flagella et al., 2006). Multiplexed mRNA analyses add technical complexity, and may be difficult to develop without sufficient technical expertise. Several commercial vendors have developed optimized kits that provide easy access to multiplex mRNA quantification (http://www.htgenomics.com; http://www.beckman.com).

Available equipment and costs are often the deciding factor in the final choice of analysis technique. Independent of analysis technique chosen, proper optimization, experimental technique, and experimental design are critical for the delivery of robust experimental data. The availability of numerous commercial vendors provides a good source of experimental reagents and design considerations.

15.6.2 Protein Quantification

An alternative to the measurement mRNA for the assessment of induction is the quantitation of the amount of protein after drug treatment. Immunoanalytical techniques are by far the most popular techniques for protein quantitation. They utilize antibodies or antibody-related reagents to selectively determine the amount of target proteins/antigens. The selectivity of an immunoanalytical method is highly dependent on the selectivity and quality of the antibody used. Many immunoanalytical methods can be adapted to higher-throughput applications, thus further expanding their application in drug discovery and clinical testing. Immunoanalytical methods have been applied in the investigation of enzyme induction. The most popular techniques include Western blot, enzyme immunoassay/enzyme-linked immunoassay (EIA/ELISA), multiplexed immunoassay with beads, and quantitative proteomics.

Western blot was first introduced with radio detection by George Stark in 1979 (Renart et al., 1979). Nowadays, most Western blots are performed with chemiluminescence detection (Kricka, 2003). A general procedure of Western blot starts with gel electrophoresis of a protein sample followed by transfer to a membrane (typically nitrocellulose or polyvinylidene fluoride [PVDF]). The membrane is then probed for the antigen of interest with a primary antibody, and subsequently, a secondary antibody is applied to recognize the primary antibody. Most often, the secondary antibody is linked to a reporter enzyme such as horseradish peroxidase that allows detection of the antibody–antigen complex using chemiluminescence. The luminescence can be captured by a photographic film or other electronic-based luminescence detectors. To correct for potential differences in the amount of sample loaded, a reference protein such as β-actin or glyceraldehyde-3-phosphate dehydrogenase (GAPDH) is often used to normalize the signal intensity of the target protein. Western blot is one of the most used immunoanalytical techniques in biomedical research due to its relatively low cost. However, owing to the labor intensive nature of the Western blot assay, it is low throughput, and can only provide semiquantitative and relative measurements of protein levels. Many commercial Western blot reagents are available with antibodies against specific CYP450 isozymes or drug transporters (Table 15.1). Several studies have used Western blot analysis in combination with probe substrates and RT-PCR to confirm the increase in expression levels of particular CYP450 isozymes or drug transporters (Fardel et al., 1996; Ramaiah et al., 2001; Nieth and Lage, 2005; Barnes et al., 2007; Skupinska et al., 2007; Deng et al., 2008; Coburger et al., 2009).

An alternative method to measure protein levels is ELISA. The assay was first described in 1971 (Engvall and Perlmann, 1971). In the same year, Van Weemen and Schuurs (1971) reported their work on EIA, a similar method to quantify human gonadotropin concentrations in urine using horseradish peroxidase as the reporter. A general procedure for an EIA/ELISA assay is as follows: A sample containing an unknown amount of antigen (or target protein) is fixed onto a microtiter plate either nonspecifically (through surface binding) or specifically (via capture by another antibody specific to the same antigen). The detection antibody is added and binds to the antigen. The detection antibody is most often linked to a reporter enzyme such as AP or horseradish peroxidase. In the final step, a substrate is added, which is acted upon by the reporter enzyme to generate a quantifiable signal, for example, fluorescence or luminescence. If standard proteins/antigens are available, a calibration curve can be prepared along with the unknown samples and the concentrations of the proteins/antigens in the unknown samples can be extrapolated from the calibration curve.

EIA/ELISA is usually highly sensitive and can be easily adapted to high-throughput formats. It can provide accurate protein quantitation with a relatively small amount of sample with rapid turnaround time. In drug metabolism studies, EIA/ELISA has been applied to confirm drug-induced changes in expression levels of drug metabolizing enzymes such as P450 isozymes, GSTs, and drug transporters (Rodrigues et al., 1989; Huang and Gibson, 1991; Makowska et al., 1992; Geng and Strobel, 1998; Kyokawa et al., 2001; Devi and Devaraj, 2006).

TABLE 15.1. Commercial Sources of Antibodies for Enzyme/Transporter Induction Studies

Target Protein	Antibody Commercial Sources	Application
CYP1A1	Abcam, Cambridge, UK	ELISA, Western
	LifeSpan BioSciences, Seattle, WA, USA	ELISA, Western
	Thermo Scientific Pierce Antibodies, Rockford, IL, USA	Western
CYP1A2	Abcam	ELISA, Western
	LifeSpan BioSciences	ELISA, Western
	Thermo Scientific Pierce Antibodies	Western
	Millipore, Billerica, MA, USA	Western
	Sigma, St. Louis, MO, USA	Western
CYP2B6	Abcam	ELISA, Western
	LifeSpan BioSciences	Western
CYP2C8	Abcam	ELISA, Western
	LifeSpan BioSciences	ELISA, Western
	Thermo Scientific Pierce Antibodies	Western
CYP2C9	Abcam	ELISA, Western
	LifeSpan BioSciences	ELISA, Western
	Thermo Scientific Pierce Antibodies	Western
	Novus Biologicals, Littleton, CO, USA	Western
CYP2C19	Abcam	ELISA, Western
	LifeSpan BioSciences	ELISA, Western
	Thermo Scientific Pierce Antibodies	Western
CYP2D6	Abcam	ELISA, Western
	LifeSpan BioSciences	ELISA, Western
	Thermo Scientific Pierce Antibodies	Western
CYP3A4	Abcam	ELISA, Western
	LifeSpan BioSciences	ELISA, Western
	Thermo Scientific Pierce Antibodies	Western
CYP3A5	Abcam	ELISA, Western
	LifeSpan BioSciences	ELISA, Western
	Sigma	ELISA, Western
	Thermo Scientific Pierce Antibodies	ELISA
P-glycoprotein/MDR1, MDR3	Abcam	Western
	Sigma	Western
	Novus Biologicals	ELISA, Western
	Santa Cruz Biotechnology, Santa Cruz, CA, USA	ELISA, Western
OATP1A2, OATP1B1, OATP1C1, OATP1B3, OATP2B1, OATP4A1, OATP6A1	Santa Cruz Biotechnology	ELISA, Western
MRP1, MRP2, MRP3, MRP4, MRP5, MRP6, MRP7	Santa Cruz Biotechnology	ELISA, Western
BCRP/ABCG2	Santa Cruz Biotechnology	ELISA, Western
	Abcam	Western
	Protein Tech, Chicago, IL, USA	ELISA, Western
OCT1, OCT2, OCT3, OCT6	Santa Cruz Biotechnology	ELISA, Western
	Abcam	ELISA, Western
	GenWay Biotech, San Diego, CA, USA	ELISA, Western
OAT1, OAT2, OAT3	Abbiotec	ELISA, Western
	Santa Cruz Biotechnology	ELISA, Western
	Antibodies-Online Inc., Atlanta, GA, USA	ELISA, Western
PEPT1, PEPT2	Santa Cruz Biotechnology	ELISA, Western
	Abbiotec, San Diego, CA, USA	ELISA, Western
CNT1, CNT2, CNT3	Santa Cruz Biotechnology	ELISA, Western
ENT1, ENT2, ENT3, ENT4	Santa Cruz Biotechnology	ELISA, Western

A multiplexed immunoassay enables simultaneous quantitation of many proteins/antigens in one analysis. Luminex technology has introduced a technique that allows for the analysis of several proteins simultaneously utilizing a multicolor bead-based technology (Vignali, 2000; Joos et al., 2002). In this assay, beads of the same color are loaded with a primary antibody against a specific protein/antigen. After the protein/antigen binds to the beads, a secondary antibody that recognizes the same protein/antigen is applied. The secondary antibody is attached to a fluorescent reporter dye label that enables the quantitation of target protein/antigen in the sample. The multiplex nature of the assay is achieved through the use of beads of different colors (loaded with antibodies against different protein/antigen) mixed and incubated with sample to achieve simultaneous assessment of several target proteins. A multicolor flow cytometer is then used to sort the beads based on their color in one channel, and then detect the fluorescent reporter dye in the alternate channel. The Luminex technology has been successfully applied to cytokine assays and infectious diseases (Prabhakar et al., 2002; Rouquette et al., 2003). However, its application in drug metabolism enzyme induction research has been limited as reagents for this application are scarce. Considering the larger number of enzymes/isozymes that are involved in drug metabolism, for example, P450s, Phase II enzymes, and transporters, a successful multiplex immunoassay would have the possibility to offer great value in the research of drug-induced enzyme induction.

A challenge of enzyme induction research is to quantify the enzyme of interest either through absolute protein concentration or relative fold changes between different experimental conditions. The accuracy and sensitivity of immunoassays (Western blot or ELISA) are heavily dependent on the quality and specificity of the antibodies used (Hage, 1993). Antibodies not only are expensive to make and/or purchase but also may lack sufficient specificity to distinguish close isoforms or to differentiate posttranslationally modified versions of the enzyme target. Recently, mass spectrometers have progressed to the point where their sensitivity and reliability allow for their increasing use in protein quantitation. Newer generation accurate-mass instruments can easily distinguish different isoforms or different posttranslationally modified versions of a given protein/peptide by the detection of protonated ions of proteins/peptides by mass-to-charge ratio (m/z). Furthermore, due to the fast scan rate, mass spectrometers can simultaneously monitor ion signals for tens, sometimes hundreds of proteins when optimized conditions exist. The evolution of the technology has led to the emergence of the quantitative proteomics field.

Mass spectrometry (MS) can be used for both absolute quantitation and relative quantitation of proteins and peptides. The absolute quantitation approach usually quantifies proteins by quantifying their tryptic peptides using multiple reaction monitoring (MRM) or selected ion monitoring (SIM) as introduced by Gygi in 2003 (Hage, 1993; Gerber et al., 2003; Kirkpatrick et al., 2005). The generic workflow of this method involves use of a stable-isotope-labeled (usually ^{2}H, ^{13}C, or ^{15}N) peptide or analog peptide as an internal standard. Typically, the sample containing the proteins of interest is first digested to produce tryptic peptide fragments. Stable-isotope-labeled versions of the same fragment peptides are added as internal standards. Samples for a calibration curve can be prepared by mixing different concentrations of synthesized unlabeled peptides with the internal standards. The samples are then analyzed using a reverse-phase high-performance liquid chromatography (HPLC)-MS system or a matrix-assisted laser desorption ionization–time-of-flight (MALDI-TOF) system. The peptides are detected in SIM mode in single-quad and TOF mass spectrometers, or as MRM transitions in triple-quadrupole mass spectrometers. The endogenous and internal standard peptides coelute in liquid chromatography (LC) and behave similarly in mass spectrometers with the exception of the mass difference. Thus, based on the amount of internal standard added to each sample, one can measure the concentration of the corresponding endogenous peptides.

The design of MRM/SIM methods for peptide quantitation can be a challenging task, especially for complex matrices such as plasma or whole cell lysates. The peptides and MRM/SIM methods must be unique to the target proteins. To aid in the design and optimization of proteomic experiments, several software packages facilitate the method design by searching for unique peptide sequences and predicting peptides that should be consistently observed for a given protein (Mallick et al., 2007; Kohl et al., 2008). In addition, software packages exist that predict MRM transitions (Kohl et al., 2008; Martin et al., 2008; Mead et al., 2009; Prakash et al., 2009; Sherwood et al., 2009). These software packages have greatly reduced the time and effort required to design and optimize novel proteomic methods.

The absolute protein quantitation approach has been successfully applied in enzyme induction research. For example, in 2004 Galeva used MALDI-TOF to quantify P450 isozymes in various rat and rabbit liver microsomes (Galeva et al., 2003). In this study, the highly homologous P450s such as CYP2B1 and CYP2B2 were easily differentiated. In addition, it was discovered that CYP2A10, which had been previously reported only in rabbit olfactory and respiratory nasal mucosa, was present in PB-induced rabbit liver microsomes. In 2005,

Alterman developed a similar MALDI-TOF assay to quantify P450 isozymes based on their unique isozyme-specific tryptic peptides for both in-gel digestion and in-solution digestion samples (Alterman et al., 2005). Using this approach, simultaneous quantification of three human P450 isozymes (CYP1A2, CYP2E1, and CYP2C19) was achieved. In 2008, Kamiie et al. built a quantitative atlas for membrane transporter proteins in the blood–brain barrier, liver, and kidney of mice using an LC-MRM method (Kamiie et al., 2008). Recently, a detailed comparison between immunoassays and a nano-LC-MS assay was conducted for quantitation of CYP2D6.1 and CYP2D6.2 allelic isozymes (Yu et al., 2009). It was found that the results of immunoassay varied significantly using different allelic forms from different sources, but nano-LC-MS assay provided accurate quantitative results for both isozymes. These studies demonstrate the utility of MS-based quantitative proteomics and its potential as a tool to investigate enzyme induction.

Relative quantitation usually relies on the incorporation of a set of differential isotopic tags that are chemically identical but different in masses to a set of samples that are to be compared. After the labeling, samples with different tags are mixed and analyzed by MS. The relative abundance of the ion signals from the differentially labeled peptides indicates the relative amount of the peptides (or the corresponding proteins) in the samples. This type of quantitation was successfully applied in a P450 induction study using ^{18}O labels introduced by tryptic digestion in ^{18}O-labeled water (Lane et al., 2007). The use of isotope-coded affinity tags (ICAT) is an alternate method for relative quantitation (Gygi et al., 1999). The ICAT label includes an iodoacetamide moiety that reacts with thiol and a biotin moiety that allows avidin affinity purification. Later a cleavable version of ICAT (cICAT) with an acid-cleavable biotin moiety was developed (Hansen et al., 2003; Li et al., 2004). The ICAT labels can be made with multiple deuterium or ^{13}C to generate the differential masses. ICAT technology has been a popular choice for relative quantitation of proteins/peptides that possess reactive cysteine residues. In 2001, Han et al. successfully used this technique to profile microsomal proteins in the differentially induced human myeloid leukemia (HL-60) cells (Han et al., 2001). Because ICAT-labeled peptides/proteins can be easily purified and concentrated, it can be used to quantify low-concentration proteins in highly complex samples. A disadvantage of ICAT quantitation is that it can only compare two samples because only two versions of the label are available. To bypass this limitation, isobaric tags for relative and absolute quantification (iTRAQ) are now available to compare up to eight samples in a single analysis (Ross et al., 2004). The

iTRAQ label is an isobaric tagging reagent containing a reporter group with variable mass, a balance group, and an amino-reactive group. Although iTRAQ technology has been used in general proteomics profiling and biomarker research (DeSouza et al., 2007; Ruppen et al., 2010), its use in enzyme induction studies has not been reported.

15.6.3   Assessment of Enzyme Activity

Despite the recent advancement and application of novel technologies to assess enzyme induction at earlier points in the induction process, assessment of the increases in enzyme activity remains the gold standard and a requirement for all regulatory submissions. MS technology has progressed to allow the measurement of enzyme activity from a small sample size, that is, a 96- or 384-well hepatocyte culture, as well as cocktail analysis (Lahoz et al., 2008; Feidt et al., 2010). Most published methods utilize drug-like substrates, for example, midazolam and bupropion, or well-established *in vitro* probes such as testosterone; however, the use of highly selective fluorescent probes may provide certain advantages over standard analytical procedures, specifically speed of analysis and cost savings related to LC-MS equipment (Donato et al., 2004). To date, the regulatory agencies have not provided specific guidance on the requirements for substrates used in any assay to assess induction *in vitro* but only on the recommendations for positive control inducers. Recent advances in metabonomics research has identified endogenous biomarkers of CAR and PXR pathways and may provide an integrated approach for screening induction along with cytotoxicity in the future (Croixmarie et al., 2010).

15.7   MODELING AND SIMULATION AND ASSESSMENT OF RISK

When induction issues are identified, it is important to assess what the potential risk for induction is. Currently, the regulatory requirements for assessment of risk relate an increase in enzyme activity to a positive control, termed percent positive control, in primary human hepatocytes. A threshold has been set at 40% of positive control. Anything below this threshold can be considered a low likelihood for induction with no clinical study follow-up required. However, above this level, a clinical DDI study is highly recommended. From a regulatory perspective, the guidance is clear; however, an important question relates to what is the true risk of induction taking into consideration more factors such as dose, pharmacokinetics, and potency of induction. This has been addressed using several models including

TABLE 15.2. Assessment and Prediction of Induction Clinical Drug Interactions from *In Vitro* Systems

Sophistication	Method	Data Required	Output	Reference
	F2	Fold change in cell system Calibration set Estimated C_{max}	Simple metric	Kanebratt and Andersson (2008)
	RIS	EC_{50}/E_{max} Calibration set Estimated C_{max} Free fraction	Simple metric	Ripp et al. (2006)
	PBPK	EC_{50}/E_{max} Modeling software, (e.g., Simcyp) PBPK model	Pharmacokinetic prediction	Almond et al. (2009); Fahmi et al. (2009)

PBPK, physiologically based pharmacokinetics.

reporter gene assays, immortalized hepatocytes, and primary hepatocytes (Ripp et al., 2006; Kanebratt and Andersson, 2008; Shou et al., 2008; Almond et al., 2009; Chu et al., 2009; Fahmi et al., 2009; Fahmi et al., 2010; Fahmi and Ripp, 2010). These models span the range from static assessments using induction potency, C_{max} and fraction unbound to more advanced dynamic models using modeling and simulation software such as Simcyp or GastroPlus. A simplistic summary of the most used models is provided in Table 15.2. Inclusion of such analyses is an important part of an integrated drug interaction assessment for induction and to aid in the design of clinical drug interaction studies.

15.8 ANALYSIS OF INDUCTION IN PRECLINICAL SPECIES

Since humans are the focus of the majority of drug discovery and development activities, they have been the focal point of the discussion to this point. Still, it is important to note that induction can occur in all preclinical species used to assess pharmacokinetics and toxicology. From a strict drug development perspective, induction in these species offers little relevance to the human situation. However, it is important to note that induction in preclinical toxicology species may complicate the assessment of safety margin due to ambiguity in the pharmacokinetics over time. If reagents are available, that is, antibodies, primers, and probes, then the *in vitro* tools described above can also be used to assess the induction potential in these species as well. One could consider assessing the induction potential in a

proactive manner in the preclinical toxicology species of record as a way to address any potential loss of exposure before the study and consider dose adjustments and/or prioritizing one species over another. Additionally, livers can be collected for an *ex vivo* analysis of mRNA, protein levels, and enzyme activity using the technologies described above. Analysis of enzyme induction in preclinical species can be integrated into an overall strategy for the preclinical development of novel drugs.

15.9 ADDITIONAL CONSIDERATIONS

Although *in vitro* tools have been demonstrated to accurately predict clinical DDIs, one of the major caveats of using such tools is assessing the effect of metabolites. If the metabolites are potent inducers and have significant exposure *in vivo*, the induction data from *in vitro* studies on the parent compound may underpredict the induction liability, especially when levels of metabolites *in vitro* do not correspond to the levels *in vivo*. Hence, such result data require cautious interpretation. Another complication in *in vitro* studies is that if the compound of interest is a time-dependent inhibitor, such a property may confound the results generated in the experiment. LS180, a human colon adenocarcinoma cell line, was used to demonstrate such an effect with HIV protease inhibitors (Gupta et al., 2008), which are known potent time-dependent inhibitors of CYP3A4. *In vivo* studies in preclinical animal species will help in such complex cases if there are no interspecies differences in the mechanism of induction. In fact, the cynomolgus monkey has been demonstrated to be a predictive animal model for PXR mediated human CYP3A4 induction (Kim et al., 2010). Literature on data from various animal models is needed as scientists focus on moving away from induction in drug development.

15.10 CONCLUSION

As with all aspects of modern drug discovery and development, our understanding and knowledge of induction and induction assay techniques have evolved over time. The technologies used to assess induction preclinically have progressed, and today the modern drug metabolism scientist has many tools to address the issue early in the drug discovery process. This is important, as once a molecule enters the clinical development phase, drug metabolism/DDI liabilities associated with it cannot be fixed. Therefore, it is important to have full knowledge of the characteristics of the molecule as it progresses. If possible, early screening strategies can limit induction liabilities. However, important advancements in the understanding of *in vitro*:*in vivo* correlations may allow assessment of actual clinical induction earlier and may allow the progression of molecules with low clinical risk of induction. Outlining a clear strategy to address induction throughout the drug discovery and development process is an important responsibility of drug metabolism scientists. Hopefully, this chapter provided an overview of all tools available to aid the drug metabolism scientist in defining the strategy that works best.

REFERENCES

Almond LM, Yang J, Jamei M, Tucker GT, Rostami-Hodjegan A (2009) Towards a quantitative framework for the prediction of DDIs arising from cytochrome P450 induction. *Curr Drug Metab* 10:420–432.

Alouani S (2000) Scintillation proximity binding assay. *Methods Mol Biol* 138:135–141.

Alterman MA, Kornilayev B, Duzhak T, Yakovlev D (2005) Quantitative analysis of cytochrome p450 isozymes by means of unique isozyme-specific tryptic peptides: A proteomic approach. *Drug Metab Dispos* 33:1399–1407.

Anthérieu S, Chesné C, Li R, Camus S, Lahoz A, Picazo L, Turpeinen M, Tolonen A, Uusitalo J, Guguen-Guillouzo C, Guillouzo A (2010) Stable expression, activity, and inducibility of cytochromes P450 in differentiated HepaRG cells. *Drug Metab Dispos* 38:516–525.

Auerbach SS, Stoner MA, Su S, Omiecinski CJ (2005) Retinoid X receptor-alpha-dependent transactivation by a naturally occurring structural variant of human constitutive androstane receptor (NR1I3). *Mol Pharm* 68:1239–1253.

Barnes SN, Aleksunes LM, Augustine L, Scheffer GL, Goedken MJ, Jakowski AB, Pruimboom-Brees IM, Cherrington NJ, Manautou JE (2007) Induction of hepatobiliary efflux transporters in acetaminophen-induced acute liver failure cases. *Drug Metab Dispos* 35:1963–1969.

Bi Y, Kazolias D, Duignan DB (2006) Use of cryopreserved human hepatocytes in sandwich culture to measure hepatobiliary transport. *Drug Metab Dispos* 34:1658–1665.

Bradfield CA, Poland A (1988) A competitive binding assay for 2,3,7,8-tetrachlorodibenzo-*p*-dioxin and related ligands of the Ah receptor. *Mol Pharmacol* 34:682–688.

Chen T, Tompkins LM, Li L, Li H, Kim G, Zheng Y, Wang H (2010) A single amino acid controls the functional switch of human constitutive androstane receptor (CAR) 1 to the xenobiotic-sensitive splicing variant CAR3. *J Pharmacol Exp Ther* 332:106–115.

Chrencik JE, Orans J, Moore LB, Xue Y, Peng L, Collins JL, Wisely GB, Lambert MH, Kliewer SA, Redinbo MR (2005) Structural disorder in the complex of human pregnane X receptor and the macrolide antibiotic rifampicin. *Mol Endocrinol* 19:1125–1134.

Chu V, Einolf HJ, Evers R, Kumar G, Moore D, Ripp S, Silva J, Sinha V, Sinz M, Skerjanec A (2009) *In vitro* and *in vivo* induction of cytochrome P450: A survey of the current practices and recommendations: A pharmaceutical research and manufacturers of America perspective. *Drug Metab Dispos* 37:1339–1354.

Coburger C, Lage H, Molnar J, Hilgeroth A (2009) Impact of novel MDR modulators on human cancer cells: Reversal activities and induction studies. *Pharm Res* 26:182–188.

Collins ML, Irvine B, Tyner D, Fine E, Zayati C, Chang C, Horn T, Ahle D, Detmer J, Shen L-P, Kolberg J, Bushnell S, Urdea MS, Ho DD (1997) A branched DNA signal amplification assay for quantification of nuclei acid targets below 100 molecules/ml. *Nucleic Acids Res* 25:2979–2984.

Croixmarie V, Umbdenstock T, Cloarec O, Moreau A, Pascussi J-M, Parmentier Y, Boursier-Neyret C, Walther B (2010) Metabonomic studies on human hepatocyte in primary culture. In *Hepatocytes, Methods in Molecular Biology*, Maurel P, ed., pp. 355–374. Springer Science, Berlin, Germany.

Cui X, Thomas A, Gerlach V, White RE, Morrison RA, Cheng KC (2008) Application and interpretation of hPXR screening data: Validation of reporter signal requirements for prediction of clinically relevant CYP3A4 inducers. *Biochem Pharmacol* 2008(76):680–689.

Deng Y, Bi HC, Zhao LZ, He F, Liu YQ, Yu JJ, Ou ZM, Ding L, Chen X, Huang ZY, Huang M, Zhou SF (2008) Induction of cytochrome P450s by terpene trilactones and flavonoids of the *Ginkgo biloba* extract EGb 761 in rats. *Xenobiotica* 38:465–481.

DeSouza LV, Grigull J, Ghanny S, Dub V, Romaschin AD, Colgan TJ, Siu KWM (2007) Endometrial carcinoma biomarker discovery and verification using differentially tagged clinical samples with multidimensional liquid chromatography and tandem mass spectrometry. *Mol Cell Proteomics* 6:1170–1182.

Devi A, Devaraj H (2006) Induction and expression of GST-Pi foci in the liver of cyclophosphamide-administered rats. *Toxicology* 217:120–128.

Dickins M (2004) Induction of cytochrome P450. *Curr Top Med Chem* 4:1745–1766.

Donato MT, Jiménez N, Castell JV, Gómez-Lechón MJ (2004) Fluorescence-based assays for screening nine cytochrome P450 (P450) activities in intact cells expressing individual human P450 enzymes. *Drug Metab Dispos* 32:699–706.

Duan Y, Ma X, Zou W, Wang C, Bahbahan IS, Ahuja TP, Tolstikov V, Zern MA (2010) Differentiation and characterization of metabolically functioning hepatocytes from human embryonic stem cells. *Stem Cells* 28:674–686.

Ek M, Söderdahl T, Küppers-Munther B, Edsbagge J, Andersson TB, Björquist P, Cotgreave I, Jernström B, Ingelman-Sundberg M, Johansson I (2007) Expression of drug metabolizing enzymes in hepatocyte-like cells derived from human embryonic stem cells. *Biochem Pharmacol* 74:496–503.

Ekins S, Erickson JA (2002) A pharmacophore for human pregnane X receptor ligands. *Drug Metab Dispos* 30:96–99.

Ekins S, Chang C, Mani S, Krasowski MD, Reschly EJ, Iyer M, Kholodovych V, Ai N, Welsh WJ, Sinz M, Swaan PW, Patel R, Bachmann K (2007) Human pregnane X receptor antagonists and agonists define molecular requirements for different binding sites. *Mol Pharmacol* 72:592–603.

Ekins S, Kortagere S, Iyer M, Reschly EJ, Lill MA, Redinbo MR, Krasowski MD (2009) Challenges predicting ligand-receptor interactions of promiscuous proteins: The nuclear receptor PXR. *PLoS Comput Biol* 5:1–12.

Engvall E, Perlmann P (1971) Enzyme-linked immunosorbent assay (ELISA) quantitative assay of immunoglobulin G. *Immunochemistry* 8:871–874.

Fahmi OA, Ripp SL (2010) Evaluation of models for predicting drug-drug interactions due to induction. *Expert Opin Drug Metab Toxicol* 6:1399–1416.

Fahmi OA, Hurst S, Plowchalk D, Cook J, Guo F, Youdim K, Dickins M, Phipps A, Darekar A, Hyland R, Obach RS (2009) Comparison of different algorithms for predicting clinical drug-drug interactions, based on the use of CYP3A4 *in vitro* data: Predictions of compounds as precipitants of interaction. *Drug Metab Dispos* 37:1658–1666.

Fahmi OA, Kish M, Boldt S, Obach RS (2010) Cytochrome P450 3A4 mRNA is a more reliable marker than CYP3A4 activity for detecting pregnane X receptor-activated induction of drug-metabolizing enzymes. *Drug Metab Dispos* 38:1605–1611.

Fan F, Wood KV (2007) Bioluminescent assays for high-throughput screening. *Assay Drug Dev Technol* 5:127–136.

Fardel O, Lecureur V, Corlu A, Guillouzo A (1996) P-glycoprotein induction in rat liver epithelial cells in response to acute 3-methylcholanthrene treatment. *Biochem Pharmacol* 51:1427–1436.

Faucette SR, Sueyoshi T, Smith CM, Negishi M, Lecluyse EL, Wang H (2006) Differential regulation of hepatic CYP2B6 and CYP3A4 genes by constitutive androstane receptor but not pregnane X receptor. *J Pharmacol Exp Ther* 317:1200–1209.

Faucette SR, Zhang TC, Moore R, Sueyoshi T, Omiecinski CJ, LeCluyse EL, Negishi M, Wang H (2007) Relative activation of human pregnane X receptor versus constitutive androstane receptor defines distinct classes of CYP2B6 and CYP3A4 inducers. *J Pharmacol Exp Ther* 320:72–80.

Feidt DM, Klein K, Hofman U, Riedmaier S, Knobeloch D, Thasler WE, Weiss TS, Schwab M, Zanger UM (2010) Profiling induction of cytochrome P450 enzyme activity by statins using a new liquid chromatography-tandem mass spectrometry cocktail assay in human hepatocytes. *Drug Metab Dispos* 38:1589–1597.

Flagella M, Bui S, Zheng Z, Nguyen CT, Zhang A, Pastor L, Ma Y, Yang W, Crawford KL, McMaster GK, Witney F, Luo Y (2006) A multiplex branched DNA assay for parallel quantitative gene expression profiling. *Anal Biochem* 352:50–60.

Fujisawa-Sehara A, Sogawa K, Yamane M, Fujii-Kuriyama Y (1987) Characterization of xenobiotic responsive elements upstream from the drug-metabolizing cytochrome P-450c gene: A similarity to glucocorticoid regulatory elements. *Nucleic Acids Res* 15:4179–4191.

Fujisawa-Sehara A, Yamane M, Fujii-Kuriyama Y (1988) A DNA-binding factor specific for xenobiotic responsive elements of P-450c gene exists as a cryptic form in cytoplasm: Its possible translocation to nucleus. *Proc Natl Acad Sci U S A* 85:5859–5863.

Galeva N, Yakovlev D, Koen Y, Duzhak T, Alterman M (2003) Direct identification of cytochrome P450 isozymes by matrix-assisted laser desorption/ionization time of flight-based proteomic approach. *Drug Metab Dispos* 31:351–355.

Gao Y-D, Olson SH, Balkovec JM, Zhu Y, Royo I, Yabut J, Evers R, Tan EY, Tang W, Hartley DP, Mosley RT (2007) Attenuating pregnane X receptor (PXR) activation: A molecular modeling approach. *Xenobiotica* 37:124–138.

Garcia M, Rager J, Wang Q, Strab R, Hidalgo IJ, Owen A, Li J (2003) Cryopreserved human hepatocytes as alternative *in vitro* model for cytochrome p450 induction studies. *In Vitro Cell Dev Biol Anim* 39:283–287.

Garrison PM, Tullis K, Aarts JM, Brouwer A, Giesy JP, Denison MS (1996) Species-specific recombinant cell lines as bioassay systems for the detection of 2,3,7,8-tetrachlorodibenzo-p-dioxin-like chemicals. *Fundam Appl Toxicol* 30:194–203.

Gasiewicz TA, Neal RA (1982) The examination and quantitation of tissue cytosolic receptors for 2,3,7,8-tetrachlorodibenzo-p-dioxin using hydroxylapatite. *Anal Biochem* 124:1–11.

Geng J, Strobel HW (1998) Expression, induction and regulation of the cytochrome P450 monooxygenase system in the rat glioma C6 cell line. *Brain Res* 784:276–283.

Gerber SA, Rush J, Stemman O, Kirschner MW, Gygi SP (2003) Absolute quantification of proteins and phosphoproteins from cell lysates by tandem MS. *Proc Natl Acad Sci U S A* 100:6940–6945.

Gómez-Lechón MJ, Donato T, Jover R, Rodriguez C, Ponsoda X, Glaise D, Castell JV, Guguen-Guillouzo C (2001) Expression and induction of a large set of drug-metabolizing enzymes by the highly differentiated human hepatoma cell line BC2. *Eur J Biochem* 268:1448–1459.

Goodwin B, Hodgson E, Liddle C (1999) The orphan human pregnane X receptor mediates the transcriptional activation of CYP3A4 by rifampicin through a distal enhancer module. *Mol Pharmacol* 56:1329–1339.

Gupta A, Mugundu GM, Desai PB, Thummel KE, Unadkat JD (2008) Intestinal human colon adenocarcinoma cell line LS180 is an excellent model to study pregnane X receptor, but not constitutive androstane receptor, mediated CYP3A4 and multidrug resistance transporter 1 induction: Studies with anti-human immunodeficiency virus protease inhibitors. *Drug Metab Dispos* 2008(36):1172–1180.

Gygi SP, Rist B, Gerber SA, Turecek F, Gelb MH, Aebersold R (1999) Quantitative analysis of complex protein mixtures using isotope-coded affinity tags. *Nat Biotechnol* 17:994–999.

Hage D (1993) Immunoassays. *Anal Chem* 65:420–424.

Han DK, Eng J, Zhou H, Aebersold R (2001) Quantitative profiling of differentiation-induced microsomal proteins using isotope-coded affinity tags and mass spectrometry. *Nat Biotechnol* 19:946–951.

Hansen KC, Schmitt-Ulms G, Chalkley RJ, Hirsch J, Baldwin MA, Burlingame AL (2003) Mass spectrometric analysis of protein mixtures at low levels using cleavable 13C-isotope-coded affinity tag and multidimensional chromatography. *Mol Cell Proteomics* 2:299–314.

Hariparsad N, Carr BA, Evers R, Chu X (2008) Comparison of immortalized Fa2N-4 cells and human hepatocytes as *in vitro* models for cytochrome P450 induction. *Drug Metab Dispos* 36:1046–1055.

Hart SN, Li Y, Nakamoto K, Subileau EA, Steen D, Zhong XB (2010) A comparison of whole genome gene expression profiles of HepaRG cells and HepG2 cells to primary human hepatocytes and human liver tissues. *Drug Metab Dispos* 38:988–994.

Hay DC, Pernagallo S, Diaz-Mochon JJ, Medine CN, Greenhough S, Hannoun Z, Schrader J, Black JR, Fletcher J, Dalgetty D, Thompson AI, Newsome PN, Forbes SJ, Ross JA, Bradley M, Iredale JP (2010) Unbiased screening of polymer libraries to define novel substrates for functional hepatocytes with inducible drug metabolism. *Stem Cell Res* 2010 Dec 10 [Epub ahead of print] PMID: 21277274.

Hengstler JG, Utesch D, Steinberg P, Platt KL, Diener B, Ringel M, Swales N, Fischer T, Biefang K, Gerl M, Böttger T, Oesch F (2000) Cryopreserved primary hepatocytes as a constantly available *in vitro* model for the evaluation of human and animal drug metabolism and enzyme induction. *Drug Metab Rev* 32:81–118.

Herbst J, Anthony M, Stewart J, Connors D, Chen T, Banks M, Petrillo EW, Agler M (2009) Multiplexing a high-throughput liability assay to leverage efficiencies. *Assay Drug Dev Technol* 7:294–303.

Hewitt NJ, Lecluyse EL, Ferguson SS (2007) Induction of hepatic cytochrome P450 enzymes: Methods, mechanisms, recommendations, and in vitro-in vivo correlations. *Xenobiotica* 37:1196–1224.

Honkakoski P, Zelko I, Sueyoshi T, Negishi M (1998) The nuclear orphan receptor CAR-retinoid X receptor heterodimer activates the phenobarbital-responsive enhancer module of the CYP2B gene. *Mol Cell Biol* 18:5652–5658.

Huang PC, Chen CY, Yang FY, Au LC (2009) A multisampling reporter system for monitoring microRNA activity in the same population of cells. *J Biomed Biotechnol* 2009:104716.

Huang S, Gibson GG (1991) Differential induction of cytochromes P450 and cytochrome P450-dependent arachidonic acid metabolism by 3,4,5,3′,4′-pentachlorobiphenyl in the rat and the guinea pig. *Toxicol Appl Pharmacol* 108:86–95.

Jennen DG, Magkoufopoulou C, Ketelslegers HB, van Herwijnen MH, Kleinjans JC, van Delft JH (2010) Comparison of HepG2 and HepaRG by whole-genome gene expression analysis for the purpose of chemical hazard identification. *Toxicol Sci* 115:66–79.

Jones SA, Moore LB, Shenk JL, Wisely GB, Hamilton GA, McKee DD, Tomkinson NCO, LeCluyse EL, Lambert MH, Willson TM, Kliewer SA, Moore JT (2000) The pregnane X receptor: A promiscuous xenobiotic receptor that has diverged during evolution. *Mol Endocrinol* 14:27–39.

Joos TO, Stoll D, Templin MF (2002) Miniaturised multiplexed immunoassays. *Curr Opin Chem Biol* 6:76–80.

Kamiie J, Ohtsuki S, Iwase R, Ohmine K, Katsukura Y, Yanai K, Sekine Y, Uchida Y, Ito S, Terasaki T (2008) Quantitative atlas of membrane transporter proteins: Development and application of a highly sensitive simultaneous LC/MS/MS method combined with novel *in-silico* peptide selection criteria. *Pharm Res* 25:1469–1483.

Kanebratt KP, Andersson TB (2008) HepaRG cells as an *in vitro* model for evaluation of cytochrome P450 induction in humans. *Drug Metab Dispos* 36:137–145.

Khandelwal A, Krasowski MD, Reschly EJ, Sinz MW, Swaan PW, Ekins S (2008) Machine learning methods and docking for predicting human pregnane x receptor activation. *Chem Res Toxicol* 21:1457–1467.

Kim N, Kim H, Jung I, Kim Y, Kim D, Han YM (2011) Expression profiles of miRNAs in human embryonic stem cells during hepatocyte differentiation. *Hepatol Res* 41:170–183.

Kim S, Dinchuk JE, Anthony MN, Orcutt T, Zoeckler ME, Sauer MB, Mosure KW, Vuppugalla R, Grace JE Jr, Simmermacher J, Dulac HA, Pizzano J, Sinz M (2010) Evaluation of cynomolgus monkey pregnane X receptor, primary hepatocyte, and *in vivo* pharmacokinetic changes in predicting human CYP3A4 induction. *Drug Metab Dispos* 38:16–24.

Kirkpatrick DS, Gerber SA, Gygi SP (2005) The absolute quantification strategy: A general procedure for the quantification of proteins and post-translational modifications. *Methods* 35:265–273.

Kohl M, Gorden R, Eisenacher M, Schnabel A, Meyer HE, Marcus K, Stephan C (2008) Automated calculation of unique peptide sequences for unambiguous identification of highly homologous proteins by mass spectrometry. *J Proteomics Bioinform* 1:6–10.

Kostrubsky VE, Ramachandran V, Venkataramanan R, Dorko K, Esplen JE, Zhang S, Sinclair JF, Wrighton SA, Strom SC (1999) The use of human hepatocyte cultures to study the induction of cytochrome P-450. *Drug Metab Dispos* 27: 887–894.

Kricka LJ (2003) Clinical applications of chemiluminescence. *Analytica Chimica Acta* 500:279–286.

Kyokawa Y, Nishibe Y, Wakabayashi M, Harauchi T, Maruyama T, Baba T, Ohno K (2001) Induction of intestinal cytochrome P450 (CYP3A) by rifampicin in beagle dogs. *Chem Biol Interact* 134:291–305.

Lahoz A, Donato MT, Picazo L, Castell JV, Gómez-Lechón MJ (2008) Assessment of cytochrome P450 induction in human hepatocytes using the cocktail strategy plus liquid chromatography tandem mass spectrometry. *Drug Metab Lett* 2:205–209.

Lane CS, Wang Y, Betts R, Griffiths WJ, Patterson LH (2007) Comparative cytochrome P450 proteomics in the livers of immunodeficient mice using 18O stable isotope labeling. *Mol Cell Proteomics* 6:953–962.

LeCluyse E, Madan A, Hamilton G, Carroll K, DeHaan R, Parkinson A (2000) Expression and regulation of cytochrome P450 enzymes in primary cultures of human hepatocytes. *J Biochem Mol Toxicol* 14:177–178.

Lemaire G, Benod C, Nahoum V, Pillon A, Boussioux A-M, Guichou J-F, Subra G, Pascussi J-M, Bourguet W, Chavanieu A, Balaguer P (2007) Discovery of a highly active ligand of human pregnane x receptor: A case study from pharmacophore modeling and virtual screening to "*in vivo*" biological activity. *Mol Pharmacol* 72:572–581.

Li C, Hong Y, Tan YX, Zhou H, Ai JH, Li SJ, Zhang L, Xia QC, Wu JR, Wang HY, Zeng R (2004) Accurate qualitative and quantitative proteomic analysis of clinical hepatocellular carcinoma using laser capture microdissection coupled with isotope-coded affinity tag and two-dimensional liquid chromatography mass spectrometry. *Mol Cell Proteomics* 3:399–409.

Li H, Chen T, Cottrell J, Wang H (2009) Nuclear translocation of adenoviral-enhanced yellow fluorescent protein-tagged-human constitutive androstane receptor (hCAR): A novel tool for screening hCAR activators in human primary hepatocytes. *Drug Metab Dispos* 37:1098–1106.

Li J, Steen H, Gygi SP (2003) Protein profiling with cleavable isotope-coded affinity tag (cICAT) reagents. *Mol Cell Proteomics* 2:1198–1204.

Lin JH (2006) CYP induction-mediated drug interactions: *In vitro* assessment and clinical implications. *Pharm Res* 23:1089–1116.

Lin JH (2007) Transporter-mediated drug interactions: Clinical implications and *in vitro* assessment. *Expert Opin Drug Metab Toxicol* 3:81–92.

Luo G, Cunningham M, Kim S, Burn T, Lin J, Sinz M, Hamilton G, Rizzo C, Jolley S, Gilbert D, Downey A, Mudra D, Graham R, Carroll K, Xie J, Madan A, Parkinson A, Christ D, Selling B, LeCluyse E, Gan LS (2002) CYP3A4 induction by drugs: Correlation between a pregnane X receptor reporter gene assay and CYP3A4 expression in human hepatocytes. *Drug Metab Dispos* 30:795–804.

Lynch T, Price A (2007) The effect of cytochrome P450 metabolism on drug response, interactions, and adverse events. *Am Fam Physician* 76:391–396.

Madan A, Graham RA, Carroll KM, Mudra DR, Burton LA, Krueger LA, Downey AD, Czerwinski M, Forster J, Ribadeneira MD, Gan LS, LeCluyse EL, Zech K, Robertson P Jr., Koch P, Antonian L, Wagner G, Yu L, Parkinson A (2003) Effects of prototypical microsomal enzyme inducers on cytochrome P450 expression in cultured human hepatocytes. *Drug Metab Dispos* 31:421–431.

Makowska JM, Gibson GG, Bonner FW (1992) Species differences in ciprofibrate induction of hepatic cytochrome P450 4A1 and peroxisome proliferation. *J Biochem Toxicol* 7:183–191.

Mallick P, Schirle M, Chen SS, Flory MR, Lee H, Martin D, Ranish J, Raught B, Schmitt R, Werner T, Kuster B, Aebersold R (2007) Computational prediction of proteotypic peptides for quantitative proteomics. *Nat Biotechnol* 25:125–131.

Martin DB, Holzman T, May D, Peterson A, Eastham A, Eng J, McIntosh M (2008) MRMer, an interactive open source and cross-platform system for data extraction and visualization of multiple reaction monitoring experiments. *Mol Cell Proteomics* 7:2270–2278.

Masahiko N, Honkakoski P (2000) Induction of drug metabolism by nuclear receptor CAR: Molecular mechanisms and implications for drug research. *Eur J Pharm Sci* 11: 259–264.

McGinnity DF, Zhang G, Kenny JR, Hamilton GA, Otmani S, Stams KR, Haney S, Brassil P, Stresser DM, Riley RJ (2009) Evaluation of multiple *in vitro* systems for assessment of CYP3A4 induction in drug discovery: Human hepatocytes, pregnane X receptor reporter gene, and Fa2N-4 and HepaRG cells. *Drug Metab Dispos* 37:1259–1268.

Mead JA, Bianco L, Ottone V, Barton C, Kay RG, Lilley KS, Bond NJ, Bessant C (2009) MRMaid, the web-based tool for designing multiple reaction monitoring (MRM) transitions. *Mol Cell Proteomics* 8:696–705.

Mills JB, Rose KA, Sadagopan N, Sahi J, de Morais SM (2004) Induction of drug metabolism enzymes and MDR1 using a novel human hepatocyte cell line. *J Pharmacol Exp Ther* 309:303–309.

Moore LB, Parks DJ, Jones SA, Bledsoe RK, Consler TG, Stimmel JB, Goodwin B, Liddle C, Blanchard SG, Willson TM, Collins JL, Kliewer SA (2000) Orphan nuclear receptors constitutive androstane receptor and pregnane X receptor share xenobiotic and steroid ligands. *J Biol Chem* 275:15122–15127.

Murk AJ, Legler J, Denison MS, Giesy JP, van de Guchte C, Brouwer A (1996) Chemical-activated luciferase gene expression (CALUX): A novel *in vitro* bioassay for Ah receptor active compounds in sediments and pore water. *Fundam Appl Toxicol* 33:149–160.

Naspinski C, Gu X, Zhou GD, Mertens-Talcott SU, Donnelly KC, Tian Y (2008) Pregnane X receptor protects HepG2 cells from BaP-induced DNA damage. *Toxicol Sci* 104:67–73.

Nieth C, Lage H (2005) Induction of the ABC-transporters Mdr1/P-gp (Abcb1), mrp1 (Abcc1), and bcrp (Abcg2) during establishment of multidrug resistance following exposure to mitoxantrone. *J Chemother* 17:215–223.

Nohara K, Ao K, Miyamoto Y, Ito T, Suzuki T, Toyoshiba H, Tohyama C (2006) Comparison of the 2,3,7,8-tetrachloro-dibenzo-p-dioxin (TCDD)-induced CYP1A1 gene expression profile in lymphocytes from mice, rats, and humans: Most potent induction in humans. *Toxicology* 225: 204–213.

Northwood EL, Elliott F, Forman D, Barrett JH, Wilkie MJ, Carey FA, Steele RJ, Wolf R, Bishop T, Smith G (2010) Polymorphisms in xenobiotic metabolizing enzymes and diet influence colorectal adenoma risk. *Pharmacogenet Genomics* 20:315–326.

Pavek P and Dvorak Z (2008) Xenobiotic-induced transcriptional regulation of xenobiotic metabolizing enzymes of the cytochrome P450 superfamily in human extrahepatic tissues. *Curr Drug Metab* 9:129–143.

Prabhakar U, Eirikis E, Davis HM (2002) Simultaneous quantification of proinflammatory cytokines in human plasma using the LabMAP assay. *J Immunol Methods* 260:207–218.

Prakash A, Tomazela DM, Frewen B, Maclean B, Merrihew G, Peterman S, Maccoss MJ (2009) Expediting the development of targeted SRM assays: Using data from shotgun proteomics to automate method development. *J Proteome Res* 8:2733–2739.

Ramadoss P, Marcus C, Perdew GH (2005) Role of the aryl hydrocarbon receptor in drug metabolism. *Expert Opin Drug Metab Toxicol* 1:9–21.

Ramaiah SK, Apte U, Mehendale HM (2001) Cytochrome P4502E1 induction increases thioacetamide liver injury in diet-restricted rats. *Drug Metab Dispos* 29:1088–1095.

Rambhatla L, Chiu CP, Kundu P, Peng Y, Carpenter MK (2003) Generation of hepatocyte-like cells from human embryonic stem cells. *Cell Transplant* 12:1–11.

Raucy J, Warfe L, Yueh M-F, Allen SW (2002) A cell-based reporter gene assay for determining induction of CYP3A4 in a high-volume system. *J Pharmacol Exp Ther* 303:412–423.

Renart J, Reiser J, Stark GR (1979) Transfer of proteins from gels to diazobenzyloxymethyl-paper and detection with antisera: A method for studying antibody specificity and antigen structure. *Proc Natl Acad Sci U S A* 76:3116–3120.

Ripp SL, Mills JB, Fahmi OA, Trevena KA, Liras JL, Maurer TS, de Morais SM (2006) Use of immortalized human hepatocytes to predict the magnitude of clinical drug-drug interactions caused by CYP3A4 induction. *Drug Metab Dispos* 34:1742–1748.

Rodrigues AD, Ayrton AD, Williams EJ, Lewis DF, Walker R, Ioannides C (1989) Preferential induction of the rat hepatic P450 I proteins by the food carcinogen 2-amino-3-methyl-imidazo[4,5-f]quinoline. *Eur J Biochem* 181:627–631.

Ross PL, Huang YN, Marchese JN, Williamson B, Parker K, Hattan S, Khainovski N, Pillai S, Dey S, Daniels S, Purkayastha S, Juhasz P, Martin S, Bartlet-Jones M, He F, Jacobson A, Pappin DJ (2004) Multiplexed protein quantitation in *Saccharomyces cerevisiae* using amine-reactive isobaric tagging reagents. *Mol Cell Proteomics* 3:1154–1169.

Rouquette AM, Desgruelles C, Laroche P (2003) Evaluation of the new multiplexed immunoassay, FIDIS, for simultaneous quantitative determination of antinuclear antibodies and comparison with conventional methods. *Am J Clin Pathol* 120:676–681.

Ruhnke M, Ungefroren H, Nussler A, Martin F, Brulport M, Schormann W, Hengstler JG, Klapper W, Ulrichs K, Hutchinson JA, Soria B, Parwaresch RM, Heeckt P, Kremer B, Fändrich F (2005) Differentiation of *in vitro*-modified human peripheral blood monocytes into hepatocyte-like and pancreatic islet-like cells. *Gastroenterology* 128:1774–1786.

Ruppen I, Grau L, Orenes-Piero E, Ashman K, Gil M, Algaba F, Bellmunt J, Sínchez-Carbayo M (2010) Differential protein expression profiling by iTRAQ-2DLC-MS/MS of human bladder cancer EJ138 cells transfected with the metastasis suppressor KiSS-1 gene. *Mol Cell Proteomics* 9:2276–2291.

Schuster S, Langer T (2005) The identification of ligand features essential for PXR activation by pharmacophore modeling. *J Chem Inf Model* 45:431–439.

Sherwood CA, Eastham A, Lee LW, Peterson A, Eng JK, Shteynberg D, Mendoza L, Deutsch EW, Risler J, Tasman N, Aebersold R, Lam H, Martin DB (2009) MaRiMba: A software application for spectral library-based MRM transition list assembly. *J Proteome Res* 8:4396–4405.

Shou M, Hayashi M, Pan Y, Xu Y, Morrissey K, Xu L, Skiles GL (2008) Modeling, prediction, and *in vitro in vivo* correlation of CYP3A4 induction. *Drug Metab Dispos* 36:2355–2370.

Shukla SJ, Nguyen D-T, MacArthur R, Simeonov A, Frazee WJ, Hallis TM, Marks BD, Singh U, Eliason HC, Printen J, Austin CP, Inglese J, Auld DS (2009) Identification of pregnane X receptor ligands using time-resolved fluorescence resonance energy transfer and quantitative high-throughput screening. *Assay Drug Dev Technol* 7:143–159.

Skupinska K, Misiewicz I, Kasprzycka-Guttman T (2007) A comparison of the concentration-effect relationships of PAHs on CYP1A induction in HepG2 and Mcf7 cells. *Arch Toxicol* 81:183–200.

Sueyoshi T, Kawamoto T, Zelko I, Honkakoski P, Negishi M (1999) The repressed nuclear receptor CAR responds to phenobarbital in activating the human CYP2B6 gene. *J Biol Chem* 274:6043–6046.

Thompkins LM and Wallace AD (2007) Mechanisms of cytochrome P450 induction. *J Biochem Mol Toxicol* 21:176–181.

Trubetskoy O, Marks B, Zielinski T, Yueh M-F, Raucy J (2005) A simultaneous assessment of CYP3A4 metabolism and induction in the DPX-2 cell line. *AAPS J* 7:E6–E13.

Urdea MS, Horn T, Fultz TJ, Anderson M, Running JA, Hamren S, Ahle D, Chang CA (1991) Branch DNA amplification multimers for the sensitive, direct detection of human hepatitis viruses. *Nucleic Acids Symp Ser* 24:197–200.

Vansant G, Pezzoli P, Saiz R, Birch A, Duffy C, Ferre F, Monforte J (2006) Gene expression analysis of troglitazone reveals its impact on multiple pathways in cell culture: A case for *in vitro* platforms combined with gene expression analysis for early (idiosyncratic) toxicity screening. *Int J Toxicol* 25:85–94.

Van Weemen BK, Schuurs AHWM (1971) Immunoassay using antigen—Enzyme conjugates. *FEBS Lett* 15:232–236.

Vignali DA (2000) Multiplexed particle-based flow cytometric assays. *J Immunol Methods* 243:243–255.

Vignati LA, Bogni A, Grossi P, Monshouwer M (2004) A human and mouse pregnane X receptor reporter gene assay in combination with cytotoxicity measurements as a tool to evaluate species-specific CYP3A induction. *Toxicology* 199:23–33.

Wang H, Negishi M (2003) Transcriptional regulation of cytochrome P450 2B gene by nuclear receptors. *Curr Drug Metab* 4:515–525.

Watkins RE, Wisely GB, Moore LB, Collins JL, Lambert MH, Williams SP, Willson TM, Kliewer SA, Redinbo MR (2001) The human nuclear xenobiotic receptor PXR; Structural determinants of directed promiscuity. *Science* 292: 2329–2333.

Whitlock JP Jr (1999) Induction of cytochrome P4501A1. *Annu Rev Pharmacol Toxicol* 39:103–125.

Wilson TM and Kliewer SA (2002) PXR, CAR and drug metabolism. *Nat Rev Drug Discov* 1:259–266.

Windshügel B, Jyrkkärinne J, Poso A, Honkakoski P, Sippl W (2005) Molecular dynamics simulation of the human CAR ligand-binding domain: Deciphering the molecular basis for constitutive activity. *J Mol Model* 11:69–79.

Xu RX, Lambert MH, Wisely BB, Warren EN, Weinert EE, Waitt GM, Williams JD, Collins JL, Moore LB, Wilson TM, Moore JT (2004) A structural basis for the constitutive activity in the human CAR/RXRα hetrodimer. *Mol Cell* 16:919–928.

Xue Y, Moore LB, Orans J, Peng L, Bencharit S, Kliewer SA, Redinbo MR (2007) Crystal structure of the pregnane X receptor-estradiol complex provides insights into endobiotic recognition. *Mol Endocrinol* 21:1028–1038.

Yasuda K, Ranade A, Venkataramanan R, Strom S, Chupka J, Ekins S, Schuetz E, Bachmann K (2008) A comprehensive *in vitro* and *in silico* analysis of antibiotics that activate pregnane x receptor and induce CYP3A4 in liver and intestine. *Drug Metab Dispos* 36:1689–1697.

Yu AM, Qu J, Felmlee MA, Cao J, Jiang XL (2009) Quantitation of human cytochrome P450 2D6 protein with immunoblot and mass spectrometry analysis. *Drug Metab Dispos* 37:170–177.

Yueh MF, Kawahara M, Raucy J (2005) Cell-based high-throughput bioassays to assess induction and inhibition of CYP1A enzymes. *Toxicol In Vitro* 19:275–287.

Zhou G, Cummings R, Li Y, Mitra S, Wilkinson HA, Elbrecht A, Hermes JD, Schaeffer JM, Smith RG, Moller DE (1998) Nuclear receptors have distinct affinities for coactivators: Characterization by fluorescence resonance energy transfer. *Mol Endocrinol* 12:1594–1604.

Zhu Z, Kim S, Chen T, Lin J-H, Bell A, Bryson J, Dubaquie Y, Yan N, Yanchunas J, Xie D, Stoffel R, Sinz M, Dickinson K (2004) Correlation of high-throughput pregnane X receptor (PXR) transactivation and binding assays. *J Biomol Screen* 9:533–540.

Zhu Z, Puglisi J, Connors D, Stewart J, Herbst J, Marino A, Sinz M, O'Connell J, Banks M, Dickinson K, Cacace A (2007) Use of cryopreserved transiently transfected cells in high-throughput pregnane X receptor transactivation assay. *J Biomol Screen* 12:248–254.

16

ANIMAL MODELS FOR STUDYING DRUG METABOLIZING ENZYMES AND TRANSPORTERS

Kevin L. Salyers and Yang Xu

16.1 INTRODUCTION

Drug metabolizing enzymes (DMEs) and transporters exhibit a fairly high degree of structural and functional similarity across species. Data from laboratory animal species, such as rodents, canines, and nonhuman primates (NHP), have played a key role in our understanding of drug absorption, distribution, metabolism, and excretion (ADME) processes. We continue to utilize animal models to understand the mechanisms of drug disposition, predict metabolism and toxicity, drug–drug interactions (DDIs), and determine pharmacokinetics-pharmacodynamics (PK-PD) relationships. The direct extrapolation of animal data to humans is often limited by species differences in expression level, substrate specificity, activity, regulation, and genetic polymorphism of DMEs and transporters. Our challenge is to develop and select suitable models to address complex questions *in vivo*. Advances in molecular genetics and genetic engineering techniques make it possible to examine human DMEs and transporters in intact experimental animals, with the potential to overcome certain species differences. This chapter provides an overview to the applications of animal models in studying various important aspects of DMEs and transporters.

16.2 ANIMAL MODELS OF DMEs

16.2.1 Section Objectives

Most xenobiotics that enter the body are subjected to biotransformation that facilitates their elimination. *In vitro* studies using human recombinant enzymes, liver microsomes, human hepatocytes, and liver slices are important methods to assess and predict human drug metabolism. However, they alone cannot predict with confidence how the variable interplay of ADME processes will modulate the pharmacological activity and toxicity of xenobiotics *in vivo*. Studies in intact animals have additive value for predicting such modulation.

While some success has been seen in the *in vitro–in vivo* extrapolation (IVIVE) of human DDIs, these models cannot fully replicate the complex interplay of pharmacokinetic effects of a drug in the body, and sometimes fail to predict DDIs for a variety of reasons. For example, the true inhibitor concentration at the effector site cannot be predicted based on plasma concentrations and protein binding alone, especially if the inhibitor accumulates in the liver because of active transport (Yao and Levy, 2002).

Marked species differences exist in the expression and regulation of DMEs. The reader is referred to several excellent reviews for the basic differences in DMEs between humans and the most commonly employed preclinical animal models (Guengerich, 1997; Martignoni et al., 2006; Turpeinen et al., 2007; DeKeyser and Shou, 2011). It has been a challenge to obtain reliable animal models to accurately reflect human drug metabolism. Humanized mouse models were developed in an effort to create more reliable *in vivo* systems to study and predict human responses to xenobiotics. A summary of major applications and limitations of commonly used animal models is listed in Table 16.1. In this chapter, due to the limitation of space, we will mainly

ADME-Enabling Technologies in Drug Design and Development, First Edition. Edited by Donglu Zhang and Sekhar Surapaneni.
© 2012 John Wiley & Sons, Inc. Published 2012 by John Wiley & Sons, Inc.

TABLE 16.1. Animal Models Commonly Used to Study Human DMEs

Animal Model	Major Applications	Potential Limitations
Native animal model (wild type)	• Study first-pass metabolism • Study highly conserved CYPs in drug metabolism, toxicity, and physiological homeostasis • Predict clinical DDIs of conserved CYPs using prototypical inducers or inhibitors • Study regulation of DMEs • Generate metabolites *in vivo*	• Marked species differences exist in the expression, substrate specificity, catalytic activity, and regulation of DMEs
Knockout mouse model	• Study a highly conserved DME or nuclear receptor in drug metabolism, toxicity, regulation, and physiological homeostasis • Assess drug toxicity, efficacy, and DDIs • Study the role of DMEs in human diseases	• Species differences in metabolic profiles and susceptibility to chemical toxicity • Confounding influence of endogenous mouse gene • Compensatory changes in gene expression and functional redundancy among similar genes
Humanized mouse model	• Study and predict the DMEs of interest in drug metabolism, toxicity, DDIs, and physiological homeostasis • Study regulation of DMEs • Predict tissue-specific (e.g., intestine vs. liver) first-pass metabolism in humans using tissue-selective transgenic model • Study genetic polymorphism of human DME genes	• Confounding influence of endogenous mouse gene—can be partially overcome by making humanized mice on target gene knockout background • Compensatory changes in gene expression • Site and levels of transgene expression: what is relevant to humans is often not clear • Species differences of other DMEs
Chimeric mouse model with humanized liver	• Study hepatic first-pass metabolism • Study drug metabolism, toxicity, DDIs, and physiological homeostasis • Study regulation of DMEs • Study human genetic polymorphisms of DMEs	• Limited capability in modeling oral drug pharmacokinetics in humans • Confounding influence of endogenous DMEs in other organs (e.g., intestine, kidney) • Cannot propagate animal models • Limited availability and high variability of donor hepatocytes
Disease animal model	• Study the effects and mechanisms of diseases on expression of DMEs and drug pharmacokinetics • Help optimize the dose and schedule	• Relevance of the disease model to human diseases • Species differences in physiology and pathology

Part of the contents compiled from Muruganandan and Sinal (2008).

discuss the application of animal models in the study of the role of key DMEs, that is, cytochrome P450s (CYPs or P450s), in oral bioavailability, metabolism, toxicity, DDIs, and human diseases. CYP3A4 is undoubtedly one of the most important players in drug metabolism in humans, and is implicated in numerous metabolic DDIs. Thus, CYP3A-related animal models and study examples on small-molecule oral drugs will be emphasized. For a comprehensive review of the applications of animal models in studying DMEs, the reader is referred to several recent detailed reviews (Marathe and Rodrigues, 2006; Cheung and Gonzalez, 2008; Muruganandan and Sinal, 2008; Wang and Xie, 2009).

16.2.2 *In vivo* Models to Study the Roles of DMEs in Determining Oral Bioavailability

Most drugs are taken orally, and oral route continues to be the preferred option in drug discovery and develop-

ment. For those intended to act systemically, a significant fraction of the dose can be lost during its absorption and first-pass through a sequence of organs prior to entering the systemic circulation. F_{oral} can be viewed as the product of the fractions of the dose that escape first-pass metabolism by two major metabolizing organs (liver and intestine):

$$F_{oral} = F_a \times F_I \times F_H, \tag{16.1}$$

where F_a is the fraction of an oral dose absorbed intact through the apical membrane of the enterocyte, and F_I and F_H are the fractions of the absorbed dose that escape metabolism by the intestine (enterocytes) and liver, respectively. Figure 16.1 describes the major factors that can influence oral bioavailability of small-molecule drugs. Low and variable oral bioavailability of some drugs has been attributed to high first-pass extraction by the intestine and liver. Efficacious drugs with

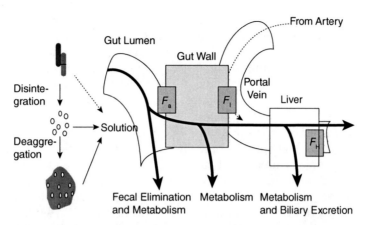

FIGURE 16.1. Factors that influence oral bioavailability (modified from Rowland and Tozer, 1995). F_a is the fraction of an oral dose absorbed intact through the apical membrane of the enterocyte, and F_I and F_H are the fractions of the absorbed dose that escape metabolism by the intestine (enterocytes) and liver, respectively.

low first-pass elimination and relatively high F_{oral} are very attractive for clinical development.

If the low systemic exposure of a compound is attributed to low oral bioavailability, it is important to distinguish between poor oral absorption and high first-pass metabolism and whether the latter is by liver and/or intestine. Models such as cell monolayers, *in situ* rat intestinal perfusion, rat everted gut sac, and Ussing chamber techniques have been used to estimate the absorption potential of the compound and will not be discussed herein (for review, see Bohets et al., 2001).

A commonly used *in vivo* model to determine hepatic and intestinal extraction is to measure systemic plasma concentration after intravenous and oral (or intraduodenal) administration to intact animals. Hepatic extraction (E_H) can be estimated indirectly from the observed systemic clearance (CL_s) by assuming that extrahepatic metabolism is negligible after intravenous administration (Eq. 16.2), where $Q_{H,plasma}$ is liver plasma flow:

$$E_H = CL_s / Q_{H,plasma}. \qquad (16.2)$$

Intestinal extraction (E_I) can be indirectly estimated by assuming a complete absorption ($F_{abs} = 1$) and that only the gastrointestinal epithelium and the liver are involved in the first-pass effect (Eq. 16.3):

$$E_I = 1 - F_I = 1 - F_{oral}/F_H = 1 - F_{oral}/(1 - E_H). \quad (16.3)$$

However, this indirect approach would not distinguish between intestinal metabolism (most commonly mediated by CYP3A) and P-glycoprotein (Pgp)-mediated efflux.

Portal vein (pv)-cannulated animals is another *in vivo* model system that measures directly hepatic and intestinal first-pass metabolism. Placement of a catheter

into the portal vein provides a way to accurately quantify drug absorption from the gut lumen into portal venous circulation and distinguish the intestinal extraction of drugs from that of the liver. In this model, hepatic extraction (E_H) is determined following oral administration by collecting blood samples *simultaneously* from catheters placed in the portal vein (presystemic) and in the jugular or other veins (systemic). E_H is calculated by dividing the systemic area under the curve (AUC) (AUC_s) by the presystemic AUC in portal vein (Eq. 16.4), assuming that liver blood flow is not disturbed by catheterization:

$$E_H = 1 - AUC_s/AUC_{pv}. \qquad (16.4)$$

The intestinal first-pass extraction (E_I) can be estimated from the AUC data of both parent and metabolite(s) using Equation 16.5 (Paine et al., 1996). E_I represents the fraction of the absorbed parent dose subjected to intestinal mucosal metabolism during first pass. For simplicity, the Equation 16.5 was developed by assuming that only one major metabolite X (metab X) was formed *in vivo*. Q_{pv} refers to total plasma flow in portal vein:

$$
\begin{aligned}
E_I &= \text{Amount metabolite X formed/} \\
&\quad (\text{Amount metabolite X formed} + \\
&\quad \text{Amount parent drug available}) \\
&= \frac{Q_{pv} \cdot \left(AUC_{pv}^{metab\,X} - AUC_s^{metab\,X} \right)}{\begin{array}{l} Q_{pv} \cdot \left(AUC_{pv}^{metab\,X} - AUC_s^{metab\,X} \right) + \\ Q_{pv} \cdot \left(AUC_{pv}^{parent} - AUC_s^{parent} \right) \end{array}}
\end{aligned}
\qquad (16.5)
$$

In addition, infusion of drug via a portal vein cannula can simulate presentation of drug to the liver by controlled-release devices, and multiple doses may be used to determine enzyme kinetics (e.g., enzyme saturation) *in vivo*.

The above two approaches were successfully applied to determine the role of intestinal versus hepatic CYP3A-mediated metabolism of cyclosporine (Kolars et al., 1991) or midazolam (Paine et al., 1996) in humans. Patients undergoing liver transplantation were given intraduodenal doses during their anhepatic phase. Under these conditions, the estimated intestinal first-pass extraction ratios in patients rival the hepatic extraction measured in healthy subjects (Wu et al., 1995; Paine et al., 1996), which was measured from the intravenous data and calculated using Equation 16.2 (Wu et al., 1995; Thummel et al., 1996).

A third *in vivo* model to examine incomplete bioavailability issues involves bile duct cannulation of animals and collection of bile, urine, and feces following dosing. The amount of parent and metabolites in the collected matrices can be quantified and the major route of elimination be inferred. It is important to ensure that circulating metabolites formed in humans are covered in selected toxicology species at appropriate levels. Radiolabeled ADME studies are commonly used to determine the distribution, metabolic profile, and elimination of the drug candidate and facilitate quantitative measurement of parent drug and metabolites in both animals and humans. ADME studies in humans are intended to better define the elimination characteristics of a drug, particularly in the target therapeutic population where disease may have an effect. Bile can be collected and measured too, though this is not done routinely in humans (see Chapter 23 for details).

The models above can be used in combination with selective modulators of absorption or metabolism to answer mechanistic questions regarding permeability and/or metabolism. For example, ketoconazole (a potent and selective CYP3A inhibitor) (Marathe and Rodrigues, 2006) or 1-aminobenzotriazole (ABT; a nonselective inhibitor of most CYP enzymes) (Balani et al., 2002) was commonly coadministered to modulate metabolism *in vivo*. ABT has favorable physicochemical, toxicity, and pharmacokinetic properties, making it an ideal tool for pharmacology and toxicology studies. Both ketoconazole and ABT inhibit intestinal and hepatic CYP enzymes after oral administration (Balani et al., 2002; Marathe and Rodrigues, 2006). It is expected that the metabolic clearance will decrease and the *in vivo* pharmacological activity will increase in the presence of the inhibitor. If not, then factors other than P450-mediated metabolism are inferred to be affecting the systemic exposure and the elimination of the drug.

ABT can distinguish absorption from metabolism processes in pharmacokinetic screens in rats and has

been suggested as an alternative to *in vivo* cannulation studies due to the ease of preparing and interpreting pharmacokinetic data (Caldwell et al., 2005). In a recent study, various dosing routes of ABT (i.e., oral vs. systemic administration) were shown as an effective method to elucidate gut and liver contributions to first-pass and systemic clearance (Strelevitz et al., 2006).

The rat provides the most flexibility and has been successfully used in the models described above for oral absorption studies. Use of portal vein catheterization to assess the intestinal and/or hepatic extraction of drugs has been explored in species other than rats (Hoffman et al., 1995), including dogs (Tam-Zaman et al., 2004), rabbits (Kunta et al., 2004), and monkeys (Ward et al., 2001).

Rhesus monkey CYP3A64 has 93% homology in its protein sequence to human CYP3A4 and 83% homology to CYP3A5 (Carr et al., 2006). Recent results indicate that cynomolgus monkeys express CYPs functionally similar to human CYPs that are important in drug metabolism (Iwasaki and Uno, 2009). Thus, monkeys have been tested as an *in vivo* model to estimate the absorption and first-pass hepatic or intestinal extraction of xenobiotics, with the combined use of ketoconazole (Ward et al., 2001, 2004; Ogasawara et al., 2007). Ketoconazole was characterized as a prototypic dual Pgp/CYP3A inhibitor and this further demonstrated the use of the monkey to investigate the role of Pgp/CYP3A in limiting the oral bioavailability of new drug candidates (Ward et al., 2004). The strategy of employing selective modulators in animal studies was also used to predict human DDIs, which will be discussed in more detail in a later section.

Caution must be exercised when extrapolating these results into humans using these native models, especially in a quantitative way. There are significant species differences in the expression, substrate specificity, and catalytic activity of DMEs present in various metabolizing tissues (Guengerich, 1997). For example, through intravenous and oral administration of midazolam, Kotegawa et al. (2002) concluded that the low oral bioavailability of midazolam in the rat is the result of hepatic rather than gastrointestinal first-pass metabolism, contrary to humans (Paine et al., 1996). Similarly, midazolam is only extracted in the liver, but not the intestine of female rats treated by dexamethasone, a CYP3A inducer (Kanazu et al., 2005). It is worth noting that the liver microsomes prepared from such female rats had metabolic properties similar to those of humans (Kanazu et al., 2004b).

In an elegant study of intestine- and liver-selective transgenic expression of human CYP3A4 in a complete mouse Cyp3a gene knockout background, it was reported that first-pass metabolism of the anticancer drug docetaxel by the gut wall, and not the liver, is likely to be a major cause of its low oral bioavailability in humans (van Herwaarden et al., 2007). In this case, all eight mouse Cyp3a genes were disrupted and replaced with a human CYP3A4 cassette expressed specifically in the intestine or the liver. The genetic models used in this study are a powerful new tool for preclinical predictions of hepatic versus intestinal first-pass metabolism mediated by human CYP3A4 for drug candidates.

There has been growing interest in defining the role of conjugation enzymes, such as UDP-glucuronosyltransferase (UGT), sulfotransferase, and glutathione S-transferase (GST), in *in vivo* metabolism. The dietary monoterpene alcohol borneol is a broad-spectrum inhibitor of UGT activity (Watkins and Klaassen, 1983). In a study of propofol in mice, borneol was found to prolong the anesthetic effect by fourfold following a 100 mg/kg intraperitoneal dose of propofol in mice. Borneol pretreatment (100 mg/kg, intraperitoneal) was associated with increased propofol exposure, presumably by inhibiting propofol glucuronidation (Lin et al., 2006). However, isoform-selective and potent inhibitors for these conjugation enzymes are not yet identified for use in modulation of metabolism. In this regard, genetically modified animal models of DMEs provide an alternative tool to investigate drug metabolism issues. For example, the Gunn rat model lacks expression of UGT1A, as the result of recessive genes. This animal model showed much slower excretion of glucuronides and was used frequently to determine the consequence of UGT1A deficiency *in vivo* (Watanabe et al., 2000). However, a recent study showed compensatory up-regulation of intestinal UGT2Bs and hepatic anion efflux transporters in Gunn rats that has complicated the usage of this model to study the role of UGT1A in drug disposition (Wang et al., 2009). Therefore, a good understanding of these models and their applications require further characterization and validation.

16.2.3 *In vivo* Models to Predict Human Drug Metabolism and Toxicity

Drug toxicity is one of the most common barriers in drug development, and P450 and the conjugation enzymes can modulate the severity of toxicity. Animal models do not, in general, provide ADME information that can be directly extrapolated to humans due to interspecies variations in DMEs (Guengerich, 1997). However, drug metabolism, efficacy, and safety studies in humans will all benefit from the selection of an animal model that best approximates human metabolism.

Efforts have been made to identify the best animal model for human metabolic activities and for toxicity studies. For instance, Bogaards et al. (2000) demonstrated the value and relevance of mouse CYP1A2- and CYP2E1-dependent metabolism to the corresponding metabolism in humans. These results have permitted studies related to these two CYPs with high relevance to humans, particularly with respect to xenobiotic and drug toxicity (Cheung and Gonzalez, 2008).

A recent *in vitro* study revealed that the profile of hepatic CYP activities that best resemble the human was seen in mouse, followed by monkey, minipig, and dog liver microsomes, with rats displaying the most divergence (Turpeinen et al., 2007). The authors suggested that if hepatic xenobiotic metabolizing characteristics were to be the sole reason for the selection of animal species for toxicity studies, then the rat might not be the most appropriate model to mimic human CYP activity patterns (Turpeinen et al., 2007).

It may be argued that there is no need to seek the metabolically "closest-to-man" animal model for all safety investigations of a new drug. Rather, it might be appropriate that when it is known which human CYPs are involved in the metabolism of a compound, the available comparative data can be used for selecting the most suitable species for *in vivo* experiments (Bogaards et al., 2000; Turpeinen et al., 2007).

In vivo animal models have provided mechanistic understanding of drug toxicity. For instance, the *in vivo* determination of glutathione adducts, as well as qualitative and quantitative determination of covalently bound proteins in animal models, either by a radiolabeled compound or immunological methods can predict toxicological liability of xenobiotics. For instance, efavirenz at high doses showed renal toxicity in rats that was not observed in monkeys or humans (Mutlib et al., 1999). The nephrotoxicity was attributed to the formation of a glutathione adduct that was further metabolized to a mixture of cysteinyl-glycine adducts and cysteine adducts. The cysteinyl-glycine adduct was detected in significant quantities only in rat and guinea pig urine and was not detected in other species (Mutlib et al., 1999). Rats dosed with labeled efavirenz and coadministration of acivicin (a gamma-glutamyltranspeptidase inhibitor) confirmed that species-specific formation of glutathione-derived conjugate(s) from efavirenz was involved in producing nephrotoxicity in rats (Mutlib et al., 2000). It was concluded that nephrotoxicity was not a safety concern in humans.

ABT has been used frequently to diagnose whether metabolites were the culprit for drug toxicity. For example, acrylamide-induced dominant lethal effects in spermatids of male mice are inhibited by ABT, supporting a hypothesis that the metabolite of acrylamide is the ultimate clastogen (Adler et al., 2000).

Marked interspecies differences in CYPs limit the value of many CYP knockout mouse models for predicting human drug metabolism. However, knockout mouse models for conserved Cyp1a2 and Cyp2e1 have proven useful in the context of studying drug metabolism and safety in human subjects. The absence of Cyp1a2 in mice exhibited no overt phenotype, yet it increased the paralysis time in a zoxazolamine paralysis test compared with wild-type mice. These results demonstrated clearly a role for Cyp1a2 in the metabolism of this drug *in vivo* (Liang et al., 1996). This CYP-null mouse model has since been used to characterize the role of human CYP1A2 in the metabolism of various carcinogens and environment contaminants (for review, see Muruganandan and Sinal, 2008). Given that CYP2E1 is expressed in the renal proximal tubules, this isoform has been implicated in the nephrotoxicity of some drugs. For instance, Cyp2e1-null mouse model exhibited marked protection against the nephrotoxic effects of cisplatin, a chemotherapeutic agent that undergoes extensive renal elimination, thus indicating an important role of CYP2E1 in cisplatin-induced renal injury in humans (Liu and Baliga, 2003). Cyp2e1-null mice have also been used to investigate the role of this isoform in the metabolism of chemical carcinogens and other toxic xenobiotics (for review, see Muruganandan and Sinal, 2008).

Caution still needs to be taken when extrapolating data generated from CYP-null mouse models to humans, even in the case of CYPs that exhibit a comparatively high degree of interspecies conservation. As an example, Zhang et al. (2002) used constitutive androstane receptor (CAR) knockout mice and transgenic mice expressing human CAR to reveal a role of Cyp2e1, Cyp1a2, and Cyp3a in acetaminophen-induced toxicity in mice. However, in humans, only CYP2E1 is considered to be the predominant isoform for the bioactivation of acetaminophen, and the most relevant to acetaminophen-induced hepatotoxicity (Nelson et al., 2003).

Given the overall limited value of wild-type and CYP knockout mouse models in advancing the understanding of drug metabolism in humans, transgenic mice expressing human metabolic enzymes are invaluable experimental models that can replicate to various degrees many aspects of human drug metabolism (Cheung and Gonzalez, 2008). Many transgenic mice that express human DMEs have become generated in recent years, including human CYP2D6, CYP3A4, CYP2E1, CYP1A1, CYP1A2, CYP1B1, CYP reductase, glutathione transferases, and so on (Gonzalez and Yu, 2006). Mice expressing human CYP2D6 efficiently metabolized debrisoquine, a probe substrate of CYP2D6 activity in humans, reflective of human extensive metabolizers of debrisoquine, while wild-type mice were similar to human poor metabolizers of debrisoquine

(Corchero et al., 2001). Another example, CYP3A4 transgenic mice showed a higher rate of midazolam clearance after oral administration than wild-type mice (Granvil et al., 2003). It is surprising that, in this CYP3A4 transgenic model, expression was found in the small intestine but not in the liver, the major site of CYP3A4 expression in humans. This could limit the value of the model. A more recent CYP3A4 transgenic mouse model, with human CYP3A4 expressed in the liver and small intestine of both immature and adult females, should facilitate in the *in vivo* analysis of the metabolism of oral drugs (Cheung et al., 2006).

A significant limitation of most of transgenic models expressing human enzymes is the presence and expression of homologous mouse genes. One way to mitigate this confounding effect is to introduce the human transgene into a knockout mouse model where the relevant endogenous genes have been disrupted. In mouse lines that express human CYP3A4 on a mouse Cyp3a cluster knockout background (van Herwaarden et al., 2007), CYP3A4 expression in the intestine decreased the absorption of docetaxel into the bloodstream, while hepatic expression increased the systemic docetaxel clearance. This model can help determine the relative contribution of CYP3A4 to metabolic clearance of drug candidates and endogenous molecules, without the confounding effects of endogenous mouse Cyp3a enzymes. Background mice, that is, those that lack all murine Cyp3a genes, displayed higher exposure levels, exhibited severely impaired detoxification capacity and increased sensitivity to docetaxel, compared to the wild type (van Herwaarden et al., 2007). The data suggested a primary role for CYP3A in xenobiotic detoxification. One caveat is that compensatory changes in gene expression can occur, for example, up-regulation of CYP2C enzymes in the Cyp3a knockout mouse model discussed above (van Waterschoot et al., 2008), which can still complicate the interpretation of the data when using these models.

A relatively new approach to generating humanized mice does not involve genetic manipulation, but relies instead on transplantation of human hepatocytes into mouse liver. This creates a chimeric mouse model with a humanized liver called "chimeric mice" or "hepatocyte-humanized mice" (Tateno et al., 2004). The end result is to replace essentially most of the endogenous hepatic drug metabolism capabilities with the human equivalents. In the liver of the chimeric mouse, human Phase I and Phase II enzymes are expressed and have a similar drug metabolizing capacity as the donor (Katoh and Yokoi, 2007). Furthermore, human-specific metabolites can be detected in the serum, suggesting that chimeric mice might be used as a human ADME model for both *in vitro* and *in vivo* studies (Katoh and Yokoi, 2007). Chimeric mice with humanized liver are expected to have advantages over other methods in predicting human drug metabolism and toxicity. One caveat is that the chimeric mouse model only recapitulates the contribution of the liver to drug metabolism, and not those from other organs including intestine. Metabolism by the catalytic activity of endogenous mouse CYPs in the intestine may confound the prediction. Thus, the chimeric mice may not be a useful model for the pharmacokinetics of oral drugs if the intestinal metabolism is significant and different from humans.

16.2.4 *In vivo* Models to Study the Regulation of DMEs

DMEs exhibit considerable interindividual variation in expression, due in part to developmental, hormonal, and genetic regulation, and to factors such as age, gender, and disease state. Nuclear receptors are the key regulators of the expression of the majority of DMEs (Wang and Xie, 2009). Transcriptional regulation of DMEs by either endobiotics or xenobiotics acting through receptors, such as pregnane X receptor (PXR), CAR, and aryl hydrocarbon receptor (AHR), can also contribute significantly to this variability. Enzyme up-regulation by nuclear receptors and their ligands is the basis for induction-associated DDIs and an important consideration in clinical practice and drug discovery. Elucidation of the sources of interindividual differences of constitutive intestinal and hepatic expression of DMEs is critical to understand and accurately predict interpatient variability in drug response. However, the cause of such variability is not completely understood despite a substantial amount of effort. In this section, we will discuss the animal models used to study both the constitutive and inductive regulation of DMEs. The impact of disease state on the expression of DMEs will be covered briefly.

Striking species differences exist in the response to xenobiotics, notably between rodents and humans. For instance, rifampin induces the expression of DMEs in humans but not in rats. Consequently, typical rodent models may not be useful in predicting the human response. Our knowledge in understanding the mechanisms of the regulation of DMEs has, however, been greatly enhanced by gene-modified animal models described previously. These *in vivo* models are dynamic, with drug ADME processes occurring concurrently along with the inclusion of other *in vivo* factors.

Humanized mouse models expressing human genes of P450 isoforms and nuclear receptors, or a combination of both, have been developed and evaluated to study the regulation of DMEs (Xie et al., 2000; Gonzalez and Yu, 2006; Ma et al., 2008; Wang and Xie, 2009).

Nuclear receptor-modified mice were among the first *in vivo* transgenic models used to assess the regulation of DMEs. A large body of *in vivo* and *in vitro* evidence

has been accumulated to link PXR to the regulation of CYP3A gene expression. "Loss-of-function" gene knockout and "gain-of-function" transgenic mouse studies, plus several lines of *in vitro* evidence, have provided convincing genetic and pharmacological evidence to support the role of PXR in the up-regulation of CYP3A gene expression in response to xenobiotics. Targeted disruption of the mouse PXR locus abolished the CYP3A xenobiotic response to prototypic inducers such as pregnenolone-16a-carbonitrile (PCN) or dexamethasone in these animals (Xie et al., 2000; Staudinger et al., 2001). In contrast, hepatic expression of an activated form of human PXR (hPXR) in transgenic mice resulted in sustained induction of CYP3A enzymes and enhanced protection against chemicals, such as the sedatives tribromoethanol and zoxazolamine that are metabolized by CYP3A (Xie et al., 2000). Recently, Ma and coworkers reported rifaximin as a gut-specific hPXR activator and inducer of Cyp3a11 in PXR-humanized mice due to its poor absorption and consequent high concentration of the compound in the intestine (Ma et al., 2007). This study revealed an important aspect of transcriptional regulation of intestinal CYPs via PXR and suggested that the pharmacokinetic profile of a ligand/drug may determine tissue-specific effect on transactivation of CYPs.

In addition to nuclear receptor-modified mice, transgenic mice expressing human metabolic enzymes also showed utility in understanding the regulation of DMEs. For example, the CYP2D6-humanized mice was used to determine the mechanism of regulation of the human gene (Corchero et al., 2001) by introducing the CYP2D6 transgene into liver-specific hepatocyte nuclear factor 4α (HNF4α)-null mice (Hayhurst et al., 2001). Mice that lack expression of HNF4α in the liver have a decreased expression of the CYP2D6 transgene, indicating a role for this factor in regulation of the liver-specific expression of CYP2D6 (Corchero et al., 2001). However, CYP2D6 expression was not completely lost in the absence of HNF4α; the results indicated that other liver-enriched factors may also have a role in its expression.

These are just a few examples to illustrate the power of gene-modified animal models to address complicated questions in the regulation of DMEs. The reader is referred to a previous review for a discussion on the value and limitations of mouse models for studying drug metabolism in humans (Muruganandan and Sinal, 2008).

Understanding the causes of intestinal and hepatic CYP3A4 variation is very important to fully optimize individual drug therapy, especially for those drugs with a narrow therapeutic window. A transgenic mouse model was created with intestinal and hepatic expression of the entire human CYP3A4 gene, including known 5′-regulatory elements (Cheung et al., 2006). Although it is not in an organ-selective way and not on a mouse Cyp3a-deficient background, the model can permit investigation of the sources of interindividual variability in intestinal and hepatic CYP3A4 expression. In this model, continuous infusion of recombinant growth hormone in transgenic male mice increased hepatic CYP3A4 mRNA and protein to normal female levels. This model could help elucidate the role of growth hormone in determining the sex-dependent expression of CYP3A4 in human liver (Cheung et al., 2006). Transgenic mice that contain the regulatory promoter for human CYP1A2, CYP3A4, and CYP27B1 have also been generated to investigate the regulation of the CYP gene in mice (Gonzalez and Yu, 2006).

The active form of vitamin D, 1alpha,25-dihydroxyvitamin D_3 $(1,25(OH)_2D_3)$ and vitamin D receptor (VDR) induced expression of CYP3A4 in intestinal cells *in vitro* (Thummel et al., 2001). To provide *in vivo* evidence of $1,25(OH)_2D_3$ in regulation of CYP3A, studies were carried out in both native (Xu et al., 2006) and PXR knockout rodents (Makishima et al., 2002). Rodent orthologs Cyp3a11 and CYP3A23 are up-regulated by $1,25(OH)_2D_3$ in the intestine *in vivo* (Makishima et al., 2002; Xu et al., 2006). In addition, it was demonstrated that $1,25(OH)_2D_3$ strongly induces CYP3A23 only in rat intestine, but not in liver, which corresponds to a 366-fold higher mRNA expression of VDR in rat intestine in comparison with the liver (Xu et al., 2006). Consistently, a 77-fold higher expression of VDR mRNA was reported in the human jejunal mucosa compared with the liver (Xu et al., 2006). Because of the conservation of the VDR binding element between rat CYP3A23 and human CYP3A4, these data imply that the human liver will be less sensitive than the intestine to the transcriptional effects of $1,25(OH)_2D_3$ and that this regulatory pathway may contribute to interindividual variability in constitutive intestinal CYP3A4 expression. This example illustrated the power of combining both *in vitro* and *in vivo* data to bridging the animal findings into humans. As shown in this example, despite known species differences in nuclear receptors and enzymes, results from native animal models can still shed some insights on the regulation in humans and help formulate the hypothesis for further testing in humans.

The expression of DMEs is also regulated by disease and pregnancy state.

In various animal models and clinical reports, inflammation or infection has been associated with decreased hepatic expression and/or activities of P450s (Renton, 2004; Aitken et al., 2006; Morgan et al., 2008). For example, CYP3A4 activity, measured by the erythromycin breath test, was reduced 20–60% following surgical stress in 16 patients (Haas et al., 2003). Chronic diseases are often associated with inflammation. Cancer patients have shown elevated inflammatory mediators associated with reduced CYP3A4 metabolism and increased interindividual variability and drug toxicity have been reported (Kacevska et al., 2008; Morgan et al., 2008). A humanized CYP3A4 transgenic mouse model was used to delineate tumor-mediated changes in the transcriptional regulation of CYP3A4 (Robertson et al., 2008). Tumor-mediated down-regulation in the hepatic transcription of the human CYP3A gene provided a mechanistic explanation of clinical observations. Animal model studies suggest that this regulation may be mediated through IL-6 originating from the tumor itself (Morgan et al., 2008). However, caution should be taken when interpreting pharmacokinetic, pharmacodynamic, and safety data obtained in inflammatory models such as collagen-induced arthritis and comparing with pharmacokinetics in normal animals due to complicated changes in DMEs (Aitken et al., 2006; Morgan et al., 2008).

Changes (mainly decreases) in CYP-associated hepatic drug clearance have also been documented in animals and patients with chronic or end-stage renal disease (ESRD) (Leblond et al., 2001; Nolin et al., 2008). Studies employing experimental models of ESRD helped understand the underlying mechanism(s) of altered DME (and drug transporter) function in kidney disease. The current prevailing explanation is that cumulated uremic toxins (e.g., urea, parathyroid hormone, indoxyl sulfate, and cytokines) are the culprit for the altered pathways (Leblond et al., 2001; Nolin et al., 2008).

Clinical data on the effect of disease on CYPs and other enzymes remain scarce, and preclinical studies can be helpful in understanding how human drug disposition is regulated in pregnancy or various disease states. Recently, both mouse and NHP were shown to reproduce the phenomenon of increased oral clearance for HIV protease inhibitors in pregnant women (Mathias et al., 2006; Zhang et al., 2009). These models should be further investigated for studying the mechanisms of DME (and transporter) regulation during pregnancy.

16.2.5 *In vivo* Models to Predict Induction-Based DDIs in Humans

There are two major concerns associated with enzyme induction: a reduction in the therapeutic efficacy of comedications and an induction in reactive metabolite-induced toxicity. Because of species differences in drug disposition, *in vivo* animal studies alone do not predict human DDIs. However, the information obtained from validated *in vivo* animal models does prove useful when used in conjunction with IVIVE methods. In discovery, drug candidates can be rank-ordered based on the magnitude of the *in vivo* interaction and progressed depending on the assessment of the risk of potential DDIs. Animal models are also used for the evaluation of drug–herbal, and drug–food interactions; for example, warfarin clearance was shown to be increased in rats by Chinese herbs (Mu et al., 2006).

Native animal models are generally not amenable to assessing the induction potential in humans due to differences in the ligand binding domain of most nuclear receptors that mediate the transcriptional gene activation of numerous DMEs. For instance, the time-dependent exposure decreases of a compound in nonclinical multiple dose toxicology studies are generally due to autoinduction of DMEs. However, because of interspecies differences in CYP induction, this observation may not be readily extrapolated to humans.

Recently, rhesus and cynomolgus monkeys were used to evaluate *in vivo* drug interactions. CYP1A, 2B, and 3A in the cynomolgus monkey respond to known human inducers (Bullock et al., 1995). In rhesus and cynomolgus monkeys, the pharmacokinetics of midazolam was significantly altered with rifampin coadministration, resulting in reduced systemic exposure and hepatic bioavailability (Prueksaritanont et al., 2006; Kim et al., 2010), comparable with those in humans (Backman et al., 1996). These data suggest that monkeys have the potential to predict human CYP3A-mediated drug interactions; nevertheless, more validation is needed to refine the induction models that can be used to assess potential human DDIs due to induction at therapeutically relevant concentrations.

Humanized transgenic animal models have received greater attention for enzyme induction studies since they represent dynamic *in vivo* situation for the human gene(s) of interest. For example, it has been shown in the case of the humanized PXR mouse that rifampin activated the hPXR *in vivo* and initiated transcription of mouse Cyp3a11 (Xie et al., 2000). At a similar dose of rifampin in the wild-type mice, no induction of Cyp3a11 was found because rifampin is not an effective ligand for mouse PXR. hPXR "humanized" mice offers a unique screening tool to evaluate DDIs *in vivo* and represents an important step in the development of safer human drugs. In a recent study, a double transgenic (Tg) mouse expressing hPXR and CYP3A4 (Tg-CYP3A4/hPXR) was generated (Ma et al., 2008). In Tg-CYP3A4/hPXR mice, the hPXR-CYP3A4-mediated rifampin-protease inhibitor interactions were recapitulated: The metabolic stability of three protease inhibitors decreased 52–99% in liver microsomes prepared from Tg-CYP3A4/hPXR mice pretreated with rifampin. In this model, *in vivo* rifampin pretreatment produced an approximately 80% decrease in the area under the serum amprenavir concentration–time curve (Ma et al., 2008). The Tg-CYP3A4/hPXR mouse model could

serve as another useful tool to study CYP3A4 function and induction *in vivo*.

Humanized mouse models expressing one or more of these nuclear receptors and DMEs are useful for assessing the impact of drug candidates on enzyme expression (Wang and Xie, 2009). However, quantitative prediction of human induction with the humanized models remains to be a challenge, largely due to the differences in physiology between mouse and human.

In chimeric mice (i.e., hepatocyte-humanized mice), CYP3A4 and CYP1A1/2 mRNA, protein content, and catalytic activity were induced by known human inducers *in vivo* and in hepatocytes isolated from humanized chimeric mice (Katoh et al., 2005). The chimeric mouse could predict the *in vivo* induction of liver CYPs in humans. As mentioned before, due to the absence of a humanized intestine, suitability of the chimeric mouse model for accurate prediction of human induction for oral drugs and xenobiotics needs to be further examined.

16.2.6 *In vivo* Models to Predict Inhibition-Based DDIs in Humans

Mice and rats have been used to predict inhibition-based DDI in humans. Several groups have conducted animal studies using midazolam or triazolam as a CYP3A substrate and ketoconazole as a CYP3A inhibitor to assess the potential applicability of an *in vivo* model for study of the interactions related to CYP3A (for review, see Marathe and Rodrigues, 2006). It has been noted that most CYP3A substrates are also substrates of Pgp, an efflux transporter. Since ketoconazole can effectively inhibit both CYP3A and Pgp, the interaction between a CYP3A substrate and ketoconazole can be partially attributed to the effect on Pgp, depending on the dose and concentration of ketoconazole. Data suggest that neither midazolam nor triazolam is a substrate of mouse Pgp (Von Moltke et al., 2004; Marathe and Rodrigues, 2006). It is worth noting that triazolam is a better choice as a CYP3A probe in the mouse compared with midazolam. Formation of 1'-OH midazolam has a major CYP2C component in addition to CYP3A in mice, and ketoconazole does not inhibit 1'-OH midazolam formation in mouse liver microsomes (Perloff et al., 2000).

Kotegawa et al. (2002) studied the interactions of midazolam and ketoconazole in the male Sprague Dawley (SD) rat. Following pretreatment with ketoconazole, rats were administered midazolam orally (Kotegawa et al., 2002). Ketoconazole increased the AUC of oral midazolam sixfold. There is a more pronounced 16-fold increase in midazolam AUC in human subjects because of both gut and liver significant first-pass effects

(Tsunoda et al., 1999). Other rat models developed include dexamethasone-pretreated female rats to study the interaction between midazolam and ketoconazole (Kanazu et al., 2004b). This animal model is thought to be more representative of human CYP3A4-mediated metabolism owing to very low involvement of other CYPs (e.g., CYP 2C11). Recently, Mandlekar et al. (2007) showed that the rat is a useful animal model to rank-order compounds in the lead optimization stage based on the magnitude of increase in AUC of midazolam in the presence versus absence of ketoconazole. Tang et al. (2008) illustrated in a case study that it is possible to provide semiquantitative estimates of clinical DDIs resulting from time-dependent inhibition of CYP3A in a rat model.

Several CYP gene knockout and humanized mouse lines have been established to study DDIs. Coadministration of the antibiotic ciprofloxacin in humans causes a significant elevation of serum concentrations of the cytokine antagonist pentoxifylline (PTX) (Peterson et al., 2004). Experiments using the CYP1A2-specific inhibitor furafylline in mice showed that PTX is a substrate for CYP1A2 and that ciprofloxacin inhibition of this isoform was primarily responsible for the elevation of PTX levels. Complementary experiments with Cyp1a2-null mice revealed increased serum levels of PTX compared with wild-type mice and confirmed the major role of CYP1A2 in PTX metabolism (Peterson et al., 2004). In another example, Granvil et al. (2003) were able to show that pretreatment with ketoconazole increased C_{max} and AUC of orally administered midazolam more significantly in humanized Tg-CYP3A4 mice than in wild-type mice.

The chimeric mouse model could also be a useful tool for assessing the drug interactions via enzyme inhibition. For instance, after pretreatment with quinidine, a specific inhibitor of human CYP2D6, AUC of a CYP2D6 metabolite, 4′-hydroxydebrisoquine, was significantly decreased in the chimeric mice but not in the control mice (Katoh and Yokoi, 2007).

Monkeys have a high sequence similarity in CYP3A to that in humans (Carr et al., 2006). They have been used to study CYP3A enzyme inhibition and *in vivo* DDIs between midazolam and CYP3A inhibitors such as ketoconazole, erythromycin and diltiazem, as well as several preclinical development candidates (Kanazu et al., 2004a; Prueksaritanont et al., 2006; Zhang et al., 2007). It was observed that the pharmacokinetic parameters of midazolam were changed by treatment with CYP3A inhibitors only when midazolam was given orally but not when given intravenously (Kanazu et al., 2004a; Prueksaritanont et al., 2006). Thus, it appears that CYP3A inhibitors modified the first-pass metabolism in the liver and/or intestine, but not the systemic metabo-

lism. It was concluded that monkeys given midazolam orally can predict CYP3A inhibition DDIs in humans (Kanazu et al., 2004a; Prueksaritanont et al., 2006).

16.2.7 *In vivo* Models to Study the Function of DMEs in Physiological Homeostasis and Human Diseases

DMEs not only play a major role in the clearance of xenobiotics but also have diverse biological and clinical roles, including the metabolism and synthesis of endogenous hydrophobic lipids such as cholesterol, bile acids, steroid hormones, and fatty acids (for review, see Nebert and Russell, 2002).

Studies in different animal models have shown that several nuclear receptors and DMEs are involved in physiological homeostasis and in treatment and prevention of human diseases, such as cholestasis and hyperbilirubinemia (for review, see Wang and Xie, 2009). For example, activation of PXR in transgenic mice is sufficient to prevent lithocholic acid-induced cholestatic liver damage (Xie et al., 2001). In contrast, mice with a combined loss of PXR and CAR had a heightened lithocholic acid sensitivity (Uppal et al., 2005). Furthermore, the cholestatic preventive effect of PXR was reasoned to be due to the activation of CYP3A, SULT2A, a bile acid detoxification hydroxysteroid sulfotransferase, and several other PXR target genes (Wang and Xie, 2009).

Many roles in cell growth and differentiation have been attributed to $1,25(OH)_2D_3$, including a central role in calcium homeostasis and skeletal metabolism. To investigate the *in vivo* functions of $1,25(OH)_2D_3$ and the molecular basis of its actions, a mouse model deficient in 1alpha(OH)ase, which synthesizes $1,25(OH)_2D_3$ from its precursor, was developed (Panda et al., 2001). The abnormalities developed in these mice were similar to those described in humans with the genetic disorder vitamin D-dependent rickets type I. Female mutant mice were infertile and exhibited uterine hypoplasia and absent corpora lutea; furthermore, immune dysfunction was also seen in these mice (Panda et al., 2001). These findings established a critical role for the 1alpha(OH)ase enzyme in mineral and skeletal homeostasis as well as in female reproduction and point to an important role for 1alpha(OH)ase in regulating immune function.

Data from DME-null mice reveal that loss of these enzymes, in general, did not result in an adverse effect on either mammalian development or physiological homeostasis. However, subtle phenotypes were noted in a few enzyme-null mice. For example, CYP1B1 deficiency is associated with primary congenital glaucoma in humans and CYP1B1 knockout mice have ocular drainage structure abnormalities that resemble those in patients (WuDunn, 2002). Another example, mice with targeted disruption of soluble epoxide hydrolase (sEH) revealed a role in blood pressure regulation (Sinal et al., 2000), and also identified sEH as a novel target for therapeutic intervention in hypertension. Animal models thus provide useful insight into the mechanisms and consequences of disease-related mutations in DMEs in humans.

Because CYPs have the ability to metabolize many endogenous substances, the CYP2D6-humanized mouse model has been further applied to the search for potential endogenous substrates for CYP2D6. It was revealed that CYP2D6 in this model catalyzed the O-demethylation of a number of endogenous, psychotropic methoxyin-dolethylamines derived from serotonin (Yu et al., 2004). Polymorphic CYP2D6 may therefore exert an influence on mood and behavior. Thus, the CYP2D6-humanized mouse is another useful tool to further investigate the physiological and pathological significance of CYP2D6 endogenous substrates. The utility is not, however, always clear. For example, data from CYP3A4-humanized and wild-type mice suggested that CYP3A4 may play an important role in the homeostasis of estradiol (Yu et al., 2004). Yet, a knockout mouse model with eight mouse Cyp3a genes disrupted and loss of all mouse Cyp3a enzymes had no apparent impact on the disposition of the steroidal substrates testosterone and estradiol, nor did it have an effect on the viability and fertility of these animals (van Herwaarden et al., 2007). The cause of the discrepancy in the role of CYP3A4 in the homeostasis of estradiol between these two mouse models is not clear. One potential limitation of the latter model was revealed in microarray analysis of total hepatic gene expression, with significant increases and decreases in the mRNA levels for a number of genes observed in the knockout animals that may provide some compensatory changes not seen in the humanized model (van Herwaarden et al., 2007).

16.2.8 Summary

Animal models replicate a more dynamic *in vivo* situation to study and predict human responses to xenobiotics. Humanized mouse models provide an approach to overcome species differences in drug metabolism. Recent advances in transgenic technology enable creation of "humanized rats" providing novel *in vivo* models for studying DMEs. Animal models, in combination with metabolism modulators, offer broad applications in the evaluation and prediction of human drug and carcinogen metabolism, pharmacokinetics, pharmacodynamics, drug toxicities, and DDIs, and also provide a sound foundation for clinical trials. In addition, these models also help us to understand how the expression of DMEs is regulated by either exogenous or endogenous factors. There are accepted limitations with each model, as described and summarized in Table 16.1. Choosing the right model and combination of models and *in vitro* methods can enhance our understanding of the role of DMEs in drug candidates and ultimately improve drug safety and efficacy.

16.3 ANIMAL MODELS OF DRUG TRANSPORTERS

16.3.1 Section Objectives

Transporters are proteins that transport endogenous compounds (i.e., electrolytes, bile acids, lipids, sugars, hormones, and amino acids) and xenobiotics (i.e., drugs and environmental toxins) across biological membrane and maintain cellular and extracellular balance as well as serve to detoxify xenobiotics in cells. They are generally divided into two large families of proteins: adenosine triphosphate (ATP) binding cassette (ABC) transporter superfamily, and the solute carrier (SLC) family. ABC efflux transporters are directly coupled to ATPase activity and use the energy from the hydrolysis of ATP to extrude substrates across the cell membrane. The SLC transporter proteins are divided into two classes, active and facilitative transport. The active SLC transporters use energy provided by an ion exchanger (difference in pH), or they are indirectly coupled to Na^+/K^+ ATPase that causes a change in membrane potential that drives the transporter. Examples of active SLC transporters include SLC22A6 (OAT1), SLC01B1 (OATP1B1), SLC22A1 (OCT1), and SLC15A1 (PEPT1). The SLC facilitative transporters are not coupled to an energy source and passively facilitate the diffusion of substrates across a membrane down a concentration gradient.

The ABC transporters are present in the body where they can influence the exposure of drugs. Examples of ABC transporters include ABCB1 (MDR1), ABCC2 (MRP2), and ABCG2 (BCRP). In the intestinal epithelium, they can reduce the uptake of orally administered drugs; in liver canaliculi, they contribute to drug elimination by biliary excretion, and in the placenta, testes, and blood–brain barrier (BBB), they can help reduce drug penetration. Efflux and uptake transporters interact to mediate the concentration and translocation of endogenous substrates as well as xenobiotics across cellular membranes. In the following section, *in vivo* models used to characterize these uptake and efflux transporters in drug absorption, distribution, and elimination are reviewed. For further information on transporters and *in vivo* models, the reader is referred to several recent reviews (Shitara et al., 2006; Takano et al., 2006; You and Morris, 2007; Xia et al., 2008; Salyers, 2009).

16.3.2 *In vivo* Models to Characterize Transporters in Drug Absorption

Patient compliance and manufacturing cost make peroral the most popular route of drug administration. While transporters can limit uptake of xenobiotics in the intestines, they have been shown to limit oral bioavailability of drugs. Problems associated with low oral bioavailability drugs include: high inter- and intrasubject variability, large dose requirement, and increased development cost. Absorption of drugs can occur by either passive or active transport. Passive absorption is directly affected by a compound's physiochemical properties, in particular molecular weight, lipophilicity, polar surface area, and hydrogen bond potential. Active transport occurs mainly by transporters that can affect drug absorption in the small intestine, and elimination by the liver and kidney. The inhibition or lack of transporter function can significantly increase or decrease exposure of drugs to tissues and result in either reduced efficacy or increased toxicity. Drug candidates are therefore screened for transporter involvement or potential for DDIs early in the discovery phase of drug development. The role of transporters in oral absorption can be evaluated by using transporter knockout or knockin, deficient strains of animals, or by using chemical inhibitors of transporters in preclinical species and humans. A list of the animal models, chemical inhibitors, and substrates cited in the following absorption studies are summarized in Table 16.2.

Pgp is encoded by the multidrug resistance (*MDR1*) gene and is partially responsible for the multidrug resistance observed with chemotherapeutic agents. It is a drug efflux transporter localized in the apical membrane of intestinal epithelium and many other tissues. The role of Pgp in drug absorption has been documented well using *Mdr1* knockout mice (Schinkel et al., 1996, 1997). An early example of this model includes Sparreboom et al. (1997); they used *Mdr1a* knockout and wild-type mice to determine the effect of intestinal Pgp on the absorption of paclitaxel. Following oral administration, the systemic AUC of paclitaxel was sixfold higher in Pgp-deficient mice compared with control animals. However, oral bioavailability only increased about threefold, suggesting that a decrease in systemic clearance was also involved in the change of AUC (Sparreboom et al., 1997). Based on these data, the authors concluded that intestinal Pgp limits the oral bioavailability of paclitaxel and may serve to protect against oral exposure of various drugs and other xenobiotics. The *Mdr1a* knockout mouse model is very useful for studying intestinal Pgp activity because the *Mdr1a* gene is predominately expressed (mRNA) in mouse intestine, although increased mRNA expression of *Mdr1b* was detected in kidney and liver of *Mdr1a* knockout mice (Schinkel et al., 1994). Further investigation determined that the expression of breast cancer resistance protein (BCRP) gene was not up-regulated in *Mdr1a* knockout mouse (Jonker et al., 2000). As a consequence of the up-regulation of *Mdr1b* gene, distribution and elimination studies in the *Mdr1a* knockout model should be evaluated accordingly.

The oral bioavailability of fexofenadine in rodents is very poor. To investigate if Pgp was responsible for the low oral bioavailability, Tahara et al. (2005) examined the oral exposure of fexofenadine in *Mdr1a/b* dual knockout and wild-type mice. Following oral administration, the absence of Pgp caused a sixfold increase in fexofenadine AUC compared with wild

TABLE 16.2. Examples of Animal Models and Chemical Inhibitors Used to Study Transporters Involved in Absorption

Target		Substrate	Animal Model and/or Inhibitors	Reference
Absorption	Pgp	Paclitaxel	*Mdr1a* KO and WT mice	Sparreboom et al. (1997)
	Pgp	Fexofenadine	*Mdr1a/b* and WT mice	Tahara et al. (2005)
	Pgp, BCRP	Topotecan	*Mdr1a/b, Bcrp1* KO and WT mice; gefitinib	Leggas et al. (2006)
	Pgp	Amitriptyline	Rabbit: quinidine	Abaut et al. (2007)
	Pgp	Erythromycin	Mouse, rat, dog, NHP: elacridar	Ward and Azzarano (2004)
	MRP2	Furosemide, probenecid, methotrexate	EHBR and SD control rat	Chen et al. (2003a)
	MRP2	PhIP (food toxin)	TR- and Wistar control rat	Dietrich et al. (2001)
	MRP2, BCRP	4-Methyl-umbelliferone, E3040	EHBR and SD control rat, *Bcrp1* KO and WT mice	Adachi et al. (2005)

KO, knockout; WT, wild type.

type. However, following intravenous administration, similar systemic clearance was observed between the two mouse models (Tahara et al., 2005). The results suggest that Pgp was responsible for the increase in intestinal absorption and not a reduction in hepatic clearance (Tahara et al., 2005).

Early studies with coadministration of topotecan, a BCRP substrate, with elacridar (GF120918), a BCRP/Pgp inhibitor, significantly increased the oral bioavailability of topotecan in animal model studies (Jonker et al., 2000). To investigate a potential DDI with Pgp, the pharmacokinetics of topotecan was evaluated with and without pretreatment of gefitinib in *Mdr1a/b* and *Bcrp1* knockouts and corresponding wild-type mice (Leggas et al., 2006). The systemic clearance of topotecan was decreased, and apparent oral bioavailability was significantly increased, in both *Mdr1a/b* and *Bcrp1* knockout mice compared to their corresponding WT mice (Leggas et al., 2006). The results clearly suggest that at clinically relevant concentrations, gefitinib inhibits both Pgp and Bcrp intestinal transporters, thereby allowing a significant increase in topotecan absorption (Leggas et al., 2006). Based on these results, the authors suggest that the clinical use of gefitinib may interfere with the transport of cytotoxic agents when used in combination, and the potential for DDIs should be evaluated (Leggas et al., 2006).

To evaluate the impact of Pgp on the low oral exposure of amitriptyline (a tricyclic antidepressant), Abaut et al. (2007) pretreated rabbits with quinidine, (a Pgp inhibitor) and demonstrated a threefold increase in oral bioavailability. The data clearly suggests the involvement of Pgp in the absorption of amitriptyline in the rabbit; however, involvement of other transporter(s) could not be ruled out.

Recently, Ward and Azzarano (2004) characterized elacridar (GF120918) following oral administration in commonly used preclinical species (mouse, rat, dog, and NHP). The pharmacokinetic and exposure parameters were obtained for various doses and formulations of elacridar, in all four preclinical species. Portal and systemic concentrations were reported in rat, dog, and NHPs, along with hepatic extraction and theoretical maximum absorption, were calculated for two oral doses of elacridar. The authors also evaluated the efficacy of elacridar in the NHP. They selected erythromycin as a known substrate for Pgp and CYP3A4 (Wacher et al., 1995), and propranolol was included as a negative control. Elacridar (3 mg/kg) caused a 10-fold increase in erythromycin concentrations in hepatic portal and systemic exposure and led to a significant increase in absorption from 6% to 42%, while hepatic extraction was unchanged. In contrast, the same dose of elacridar (3 mg/kg) had no significant effect on absorption or

hepatic extraction of propranolol in the NHP (Ward and Azzarano, 2004). The maximum concentrations of inhibitor was less than 1 μM in portal and systemic circulation (Ward and Azzarano, 2004), therefore elacridar should not inhibit CYP3A4 ($IC_{50} = 11$ μM) at concentrations examined in this *in vivo* model. The data suggest that erythromycin exposure following oral administration in preclinical species may be dependent on Pgp/BCRP activity, not Pgp/CYP3A as has been suggested (Ward and Azzarano, 2004).

The role of multidrug resistance-associated protein 2 (MRP2) in oral absorption is less well studied, despite two mutant strains of rats being available, a Wistar (TR⁻) and Sprague Dawley eisai hyperbilirubinemic (EHBR). The major challenge has been that substrates of MRP2 are often glucuronide, sulfate, and glutathione conjugates, which are usually charged and hydrophilic with limited membrane permeability. In general, these conjugated substrates are poorly absorbed following oral administration. Similar to Pgp and CYP3A, the coexpression of conjugating enzymes, UGT and GST, along with MRP2 efflux transporter in the enterocytes, suggest a coordinated host defense mechanism against drugs and other xenobiotics.

The role of MRP2 in the oral bioavailability of a food-derived heterocyclic amine, 2-amino-1-methyl-6-phenylimidazole[4,5-b]pyridine (PhIP) was determined in MRP2-deficient rats, TR⁻ and Wistar control rats (Dietrich et al., 2001). Following oral administration of PhIP, the systemic exposure was about twofold higher in TR⁻ rats compared with controls. A similar twofold increase was observed in portal vein concentrations following intraduodenal administration. These results confirm that the higher systemic exposure of PhIP was due to increased absorption, and not altered metabolism in the TR⁻ rat (Dietrich et al., 2001).

The *in situ* intestinal perfusion model is another option to examine the role of MRP2 and BCRP in the extrusion of sulfate and glucuronide conjugates into the intestinal lumen (Adachi et al., 2005). The authors used 4-methylumbelliferone (4MU) and E3040 as substrates in the *in situ* intestinal perfusion model in EHBR rat, *Bcrp1* knockout mice, and corresponding wild-type animals. Both substrates are metabolized to their corresponding glucuronide and/or sulfate conjugates by enzymes in the enterocytes. In the *in situ* intestinal perfusion model, there was no significant difference in absorption of 4MU and E3040 between control and EHBR rats, but a significant decrease was observed in the efflux rate of glucuronide and sulfate conjugates in *Bcrp1* knockout mice compared with wild type (Adachi et al., 2005). The authors concluded that Bcrp1 has an important role in the extrusion of conjugates produced in the enterocytes, with a synergistic role between the

TABLE 16.3. Examples of Animal Models and Chemical Inhibitors Used to Study Transporters Involved in CNS Penetration

Target		Substrate	Animal Model and/or Inhibitors	Reference
CNS penetration	Pgp	Digoxin	*Mdr1a/b* KO and WT mice	Schinkel et al. (1994, 1997)
	Pgp	Indinavir, nelfinir, saquinar	*Mdr1a* KO and WT mice	Kim et al. (1998)
	Pgp	6 antihistamines	*Mdr1a/b* KO and WT mice	Chen et al. (2003b)
	Pgp	Fexofenadine	*Mdr1a/b* KO and WT mice	Tahara et al. (2005)
	Pgp	Oseltamivir	*Mdr1a/b* KO and WT mice	Morimoto et al. (2008)
	MRP4, OAT3	Oseltamivir and metabolite	*Mrp4, Oat3* KO and WT mice	Ose et al. (2009)
	Pgp	HIV protease inhibitors	Mice: zosuquidar	Choo et al. (2000)
	Pgp	Nelfinavir	Rat: elacridar	Savolainen et al. (2002)
	Pgp	Nelfinavir	NHP: zosuquidar	Kaddoumi et al. (2007)
	Pgp, BCRP	Imatinib	*Mdr1a/b, Bcrp1* KO and WT mice; elacridar, pantoprazole	Breedveld et al. (2005)
	Pgp, BCRP	Imatinib	*Mdr1a/b, Bcrp1, Mdr1a/b-Bcrp* KO and WT mice; elacridar, pantoprazole	Oostendorp et al. (2009)
	Pgp	Verapamil	NHP: valspodar	Lee et al. (2006)

KO, knockout; WT, wild type.

conjugating enzymes and the efflux transporter in elimination of xenobiotics (Adachi et al., 2005).

From the examples reviewed above, efflux transporters may play an important role in limiting oral bioavailability. However, poor oral exposure may also be due to hepatic extraction, intestinal metabolism, and/or intrinsic properties of the molecule. Physical chemical properties of the molecule can yield low solubility and/or poor membrane permeability, which lead to limited passive transport. The interplay of passive and active transport of many drugs has recently been reviewed (Sugano et al., 2010).

16.3.3 *In vivo* Models Used to Study Transporters in Brain Penetration

Adequate exposure in the brain is essential for the treatment of central nervous system (CNS) diseases, while minimal CNS exposure is optimal for compounds intended for peripheral disorders. The pharmaceutical industry quickly recognized that an efficient strategy for enhancing CNS penetration of potential therapeutics was to increase their lipophilicity. However, this strategy only worked for small molecules that were dependent on passive diffusion to cross the BBB. The ability of a drug to cross the BBB is related to the drug's molecular weight, lipid solubility, degree of ionization, protein and tissue binding, and its affinity for specific active transporters. Generally, increased lipid solubility enhances the rate of penetration of drugs into the CNS, whereas ionization greatly diminishes it. An effective strategy to increase BBB permeability for the development of CNS drugs was to confirm that potential drug candidates were not substrates for one of the major efflux trans-

porters in the BBB. Many active transporters have been identified in the BBB and include several members of the ABC (Pgp, BCRP, and MRP2) and many from the SLC family (amino acid, organic cation, and organic anion transporters). A list of the animal models, chemical inhibitors, and substrates used in the following CNS penetration studies are summarized in Table 16.3.

Cordon-Cardo et al. (1989) used mouse monoclonal antibodies (mAb) to demonstrate that Pgp was highly expressed in endothelial cells of the BBB, and postulated a physiological role for Pgp in regulating CNS entry of some small lipophilic molecules. The multidrug-resistant (*Mdr*) gene in mice is expressed in a tissue-specific manner; both *Mdr1a* and *Mdr1b* are expressed in the liver and kidney, while the brain predominantly expresses the *Mdr1a* gene (Croop et al., 1989).

In a series of studies, Schinkel et al. (1994, 1995, 1997) demonstrated the use of the multidrug-resistant knockout mouse model. The group demonstrated significantly higher CNS levels of ^{3}H-digoxin (17-fold) and ^{3}H-cyclosporine (55-fold) following intravenous administration in *Mdr1a* knockout mice compared with wild type (Schinkel et al., 1995). As discussed previously, the mouse has two genes that code for Pgp, and the authors produced a mouse with both genes deleted or the dual knockout model (*Mdr1a/b*). This model was used in subsequent studies to demonstrate a further increase in CNS levels of ^{3}H-digoxin (27-fold) could be obtained in the dual knockout compared with the 17-fold increase in the single knockout mouse model (Schinkel et al., 1997). These studies clearly identified the transgenic mouse model as an important tool for studying the role of Pgp in the brain uptake of drugs. Since that time, numerous studies have used the transgenic *Mdr1a/b* knockout mouse model to determine the role of Pgp in the CNS distribution of drug candidates. An example would be the HIV protease inhibitors, many of which display poor CNS distribution. Following intravenous administration of indinavir, nelfinavir, and saquinavir, brain concentrations of the three HIV protease inhibitors were elevated 7–36-fold in *Mdr1a* knockout mice relative to wild-type mice (Kim et al., 1998). These data clearly demonstrate that Pgp limits the brain penetration of these agents and the authors suggested the simultaneous use of Pgp inhibitors to enhance the brain penetration of CNS therapeutics (Kim et al., 1998).

Antihistamines are another interesting example. Chen et al. (2003a) demonstrated that sedating H_1-receptor antagonists, such as hydroxyzine, diphenhydramine, and triprolidine, are not Pgp substrates and their distribution into the CNS is comparable in *Mdr1a/b* knockout mice and wild type. Whereas, the nonsedating H_1-antagonists cetirizine, loratadine, and desloratadine are Pgp substrates and the brain-to-plasma AUC was clearly enhanced in *Mdr1a/b* knockout mice compared with wild type. Another laboratory used the same Pgp-deficient mouse model to demonstrate that fexofenadine, another nonsedating antihistamine, was also a Pgp substrate and the transporter likely limits its CNS distribution (Tahara et al., 2005).

Oseltamivir is an ester-type prodrug of Ro-640802 and is used for the treatment and prophylaxis of infectious diseases caused by the influenza virus. Recently, CNS side effects and abnormal behavior in teenagers (associated with the drug) has motivated further CNS penetration studies of the drug and its active metabolite. Following a single oral dose of oseltamivir to *Mdr1a/b* knockout and wild-type mice, a dose-dependent increase of oseltamivir CNS levels was observed in *Mdr1a/b* knockout and to a lesser extent in wild-type mice (Morimoto et al., 2008). The results suggest that oseltamivir, but not its active metabolite (Ro-640802), is a substrate of Pgp and that other transporters may be involved in its CNS penetration (Morimoto et al., 2008).

To determine if MRP4 (ABCC4) and organic anion transporter (OAT3; SLC22A8) transporters were involved in the brain distribution of the active metabolite of oseltamivir, Ose et al. (2009) used two new mouse knockout models, *Mrp4* and *Oat3*. To determine if these transporters are involved in the efflux of Ro-640802 from the brain, the authors directly injected the active metabolite into the cerebral cortex and measured the amount remaining in the brain. The elimination of Ro-640802 from the brain was delayed in both *Mrp4*- and *Oat3*-deficient mice compared with wild type (Ose et al., 2009). To determine if these two transporters limit brain penetration, the active metabolite was administered as a subcutaneous infusion (Alzet pump from DURECT

Corporation, Cupertino, CA), and brain and plasma levels were measured. There was no significant difference in brain levels between the *Oat3*-deficient mice compared with wild type; but Ro-640802 brain levels were fourfold higher in *Mrp4* knockout mice compared with wild type (Ose et al., 2009). The data suggest that Oat3 is not involved with the efflux of Ro-640802 from the luminal membrane of the BBB, whereas Mrp4 clearly limits the brain penetration by extruding it from the luminal membrane back into the blood.

Transporter inhibitors or the "chemical knockout" model can be used to determine the role of transporters in the BBB in both preclinical and clinical studies. The potential benefit of reversing Pgp-mediated multidrug resistance has led to intensive efforts to develop potent and selective Pgp inhibitors. Early studies by Kim et al. (1998) demonstrated that the HIV protease inhibitors, indinavir, nelfinavir, and saquinavir, were Pgp substrates (as discussed above). Choo et al. (2000) demonstrated a significant increase in CNS and testes concentrations of HIV protease inhibitors following pretreatment with Pgp inhibitors, zosuquidar (LY-335979) or valspodar (PSC833). At the highest doses of the Pgp inhibitors (50 mg/kg), CNS and testes concentrations of nelfinavir were similar to those observed in Mdr1/a knockout animals (Choo et al., 2000). Following the work by Choo and colleagues, Savolainen et al. (2002) demonstrated that the potent, but less selective Pgp inhibitor elacridar (GF120918) could significantly increase brain concentrations of nelfinavir in the rat, while plasma concentrations remained unchanged. In another study with the protease inhibitor, monkeys were pretreated with zosuquidar (intravenous, 3 mg/kg) or vehicle, followed by intravenous administration of nelfinavir (6 mg/kg), and brain and plasma concentrations of nelfinavir were measured (90 min postdose). The results suggest that zosuquidar, a specific Pgp inhibitor, caused a significant increase (146-fold) in the distribution of nelfinavir in the brain of monkey but did not significantly change plasma concentrations of the drug (Kaddoumi et al., 2007). This striking increase in brain concentrations of nelfinavir in monkeys far exceeded that observed with the same inhibitor in rodents (Choo et al., 2000). Further studies are needed to understand this remarkable species difference.

The combination of knockout animals and inhibitors can be used together to better determine the role of transporters. Imatinib is a potent and selective tyrosine kinase inhibitor with potential to treat primary tumors of the CNS, but has inadequate CNS penetration due to efflux transporters. Breedveld et al. (2005) used *Bcrp1* and *Mdr1a/1b* knockout mice to identify transporters involved in the brain penetration of imatinib and two inhibitors, pantoprazole or elacridar, to modulate its

CNS penetration. The brain penetration of imatinib can be enhanced by coadministration of Pgp or BCRP inhibitors, and the combination of both inhibitors led to a greater CNS penetration (Breedveld et al., 2005). More recently, the same laboratory used both of these inhibitors and/or single knockout mice or the dual knockout mice (*Mdr1a/1b/Bcrp1*) to further refine the role of each transporter in the brain penetration of imatinib. Results indicate that both efflux transporters are involved, but Pgp has a greater role in CNS penetration and systemic clearance of imatinib (Oostendorp et al., 2009).

Recent progress in positron emission tomography (PET) imaging techniques have made it possible to evaluate the BBB transporters in preclinical species, as well as human studies. Hendrikse et al. (1999) first described the use of PET imaging to study Pgp-mediated efflux of ^{11}C-verapamil in rodents. The authors showed a significant increase in ^{11}C-verapamil in the CNS with cyclosporine-A pretreatment.

More recently, Sasongko et al. (2005) used healthy volunteers who received ^{11}C-verapamil (<0.12 µg/kg) with and without cyclosporine-A pretreatment (2.5 mg/kg/h; Food and Drug Administration [FDA] approved for clinical use as Pgp inhibitor). They observed about a twofold increase in the brain uptake of ^{11}C-verapamil with cyclosporine-A pretreatment. Lee et al. (2006) studied the function of Pgp at the BBB in rhesus monkeys by comparing the brain uptake of ^{11}C-verapamil in the presence and absence of valspodar (PSC833), a Pgp inhibitor. They observed a four- to fivefold increase in the brain uptake of ^{11}C-verapamil with valspodar (Lee et al., 2006). These findings support use of PET imaging as a useful and noninvasive technique to determine the role of transporters in the BBB.

From the examples reviewed above, brain penetration by various transporters affects brain distribution more significantly compared with oral absorption. Generally, oral bioavailability ratio of knockout : wild type range between 1 and 3, whereas brain : plasma ratio of knockout : wild type can range from 5 to 30. For example, nelfinavir has a bioavailability ratio of 3.7 and a brain : plasma ratio of 31 (Kim et al., 1998, Sugano et al., 2010). Other examples include verapamil and cyclosporine; both of these drugs have bioavailability ratios close to 1, whereas brain : plasma ratios range from 8 to 30 (Lee et al., 2006, Sugano et al., 2010). Being a substrate for an efflux transporter in the intestine can decrease oral absorption, and is generally not a desirable characteristic of a drug. Whereas, being a substrate for an efflux transporter in the brain can add value, for example, the nonsedating antihistamines (Tahara et al., 2005). However, adequate exposure for treatment of CNS diseases is necessary and being a substrate for an efflux transporter, like Pgp, is not favorable.

16.3.4   *In vivo* Models to Assess Hepatic and Renal Transporters

Elimination of drugs from the liver or biliary excretion starts with the uptake across the sinusoidal membrane into the hepatocytes, and metabolism and/or diffusion to the canalicular membrane where transporters can efflux the parent drug or its metabolite into the bile. Therefore, potential uptake transporters, Phase I and/or Phase II metabolism, and efflux transporters all need to be considered when evaluating hepatobiliary excretion of a drug. A list of the animal models, chemical inhibitors, and substrates used in the following hepatic and renal studies are summarized in Table 16.4.

MRP2 (ABCC2) has an important role in the elimination of endogenous substrates, as well as drugs and other xenobiotics, primarily as their conjugates (glucuronide, sulfate, or glutathione). To evaluate the role of MRP2 in mediating biliary excretion of furosemide, probenecid, and methotrexate, Chen et al. (2003b) used

TABLE 16.4. Example of Animal Models and Chemical Inhibitors Used to Study Hepatic and Renal Transporters

Target		Substrate	Animal Model and/or Inhibitors	Reference
Hepatic	MRP2	Furosemide, probenecid, methotrexate	EHBR and SD control rat	Chen et al. (2003a)
	MRP2, BCRP	Pitavastin	*Bcrp1* KO and WT mice; EHBR and SD rat	Hirano et al. (2005)
	BCRP	Nitrofurantoin	*Bcrp1* KO and WT mice	Merino et al. (2005)
	OATP1B1	Methotrexate	*OATP1B1* KO and WT mice	van de Steeg et al. (2009)
Renal	BCRP	4-Methylumbelliferone sulfate, E3040	*Bcrp1* KO and WT mice	Mizuno et al. (2004)
	OCT1	Metformin	*Oct1* KO and WT mice; cimetidine	Wang et al. (2002)
	Pgp, BCRP	Imatinib	*Mdr1a/b, Bcrp1, Mdr1a/b-Bcrp1* KO and WT mice; elacridar, pantoprazole	Oostendorp et al. (2009)

KO, knockout; WT, wild type.

bile duct-cannulated Mrp2-mutant EHBR rat and its control (SD). All three substrates reached steady-state plasma concentration within 2 h in both strains. Biliary clearance of furosemide was similar in both strains of rats, but the biliary clearance of probenecid and methotrexate was decreased about 40-fold in Mrp2-deficient rats compared with control animals (Chen et al., 2003b). The results suggest that Mrp2 is responsible for much of the biliary clearance of probenecid and methotrexate, but not furosemide (Chen et al., 2003b).

To estimate the contribution of BCRP1 and MRP2 efflux transporters in the biliary clearance of pitavastatin, Hirano et al. (2005) used *Bcrp1* knockout mice, EHBR rats, and their corresponding control animals. The authors demonstrated that the Bcrp1 transporter was responsible for the majority of biliary excretion of pitavastatin in mice, and Mrp2 played a minor role in the hepatobiliary excretion in the rat (Hirano et al., 2005).

Nitrofurantoin is a widely used antibiotic to treat urinary tract infections and has been prescribed to lactating mothers. Early research demonstrated that nitrofurantoin was actively transported into human and rat milk; concentrations of nitrofurantoin in milk were many times higher than those in serum (Gerk et al., 2001). *In vitro* studies by Merino et al. (2005) demonstrated that nitrofurantoin was transported by murine and human BCRP, but not by MDR1 or MRP2 efflux transporters. For *in vivo* confirmation, the same authors administered nitrofurantoin orally to *Bcrp1* knockout and wild-type mice, the systemic AUC was increased fourfold in *Bcrp1* knockout compared with wild-type mice; and twofold higher following intravenous administration (Merino et al., 2005). To determine the role of this efflux transporter in the hepatobiliary excretion, nitrofurantoin (intravenous) was administered to *Bcrp1* knockout and wild-type mice with cannulated gall bladder, with bile and plasma collected over time. Biliary excretion of nitrofurantoin in *Bcrp1* knockout mice was significantly decreased (98% in the first hour) compared with wild-type mice (Merino et al., 2005). Further studies with lactating *Bcrp1* knockout and wild-type mice, demonstrated that milk concentrations of nitrofurantoin following intravenous administration were about 15-fold lower in *Bcrp1*-deficient animals compared with wild type. Overall, the data suggests that *Bcrp1* has a significant role in the hepatobiliary excretion and milk secretion of nitrofurantoin in mice (Merino et al., 2005).

Human organic anion-transporting polypeptide 1B1 (OATP1B1/SLC01B1), a sodium-independent bile-acid transporter, is expressed on the sinusoidal (basolateral) membrane of human hepatocytes and is responsible for the uptake of a variety of endogenous and xenobiotics from the portal vein into hepatocytes (Abe et al., 1999).

Further *in vitro* studies by the same author and colleagues (Abe et al., 2001) demonstrated that methotrexate was a substrate for human OATP1B1. Recently, van De Steeg et al. (2009) generated a transgenic mouse with expression of human OATP1B1 only in the liver, not expressed in kidney or small intestine. Using this new model, the authors showed the plasma AUC of methotrexate was decreased (1.5-fold) and liver : plasma ratios were at least twofold increased at 15, 30, and 60 min following intravenous administration in mice expressing human OATP1B1 compared with wild type (van De Steeg et al., 2009). The new liver-specific OATP1B1 transgenic mouse model appears to be an important *in vivo* tool to determine the role of human OATP1B1 in the distribution and pharmacokinetics of endogenous compounds and potential drug candidates. Because direct human orthologs of OATPs may not exist in mice, Oatp genes may need to be deleted and relevant human SLCO genes introduced to more fully mimic human hepatic drug uptake capabilities.

The kidney plays an important role in the distribution, metabolism, and/or excretion of xenobiotics, including many different drugs and/or their metabolites. The mechanisms that contribute to the renal excretion of drugs and other xenobiotics are closely related to glomerular filtration, secretion, and reabsorption. Urinary excretion starts with uptake from blood across the basolateral membrane into the epithelial cell, and metabolism and/or diffusion to the luminal brush-border membrane where transporters can efflux the parent drug or its metabolite into the urine. Active transporters are mainly involved in tubular secretion (apical efflux) and reabsorption (apical uptake) of drugs; and both processes need to be considered when evaluating urinary excretion of a drug. For example, Mrp2, Pgp, and Bcrp are localized on the luminal brush-border membrane of proximal tubules and can function to efflux endogenous compounds as well as drugs and/or their metabolites into the urine (for review, see Xia et al., 2008). To evaluate the role of Bcrp1 in the urinary excretion of 6-hydroxy-5,7-dimethyl-2-methylamino-4-(3-pyridylmethyl) benzothiazole (E3040) and 4-methylumbelliferone sulfate (4MUS), Mizuno et al. (2004) used Bcrp knockout and wild-type mice. Following intravenous administration, the urinary clearance of E3040S (sulfate conjugate of E3040) was 2.4-fold lower in Bcrp knockout compared with wild-type mice (Mizuno et al., 2004). Data also suggest that the renal clearance of unbound E3040S was greater than glomerular filtration rate (GFR); therefore, tubular secretion accounts for the majority of urinary excretion. In contrast, the absence of *Bcrp1* transporter did not affect the pharmacokinetic profile or the urinary excretion of the metabolite 4MUS (Mizuno et al., 2004).

Metformin is widely used in the treatment of type II diabetes mellitus and appears to work by increasing peripheral sensitivity to insulin and by reducing glucose production in the liver. In humans, coadministration of cimetidine led to a significant increase in systemic exposure of metformin (Somogyi et al., 1987). Considering both compounds are positively charged at physiological pH, Wang et al. (2002) proposed that both compounds share a specific cationic transporter. To determine the role of organic cation transporter (*Oct1*) in the disposition and elimination of metformin, Jonker et al. (2001) used *Oct1*-transfected CHO cells, *Oct1* knockout mice and cimetidine (substrate of *Oct1*). The in vitro studies demonstrated that metformin uptake was much higher in Oct1-transfected cells compared with vector-transfected cells; these results suggest that metformin is a substrate of rat Oct1 (Wang et al., 2002). Following intravenous administration, the distribution of metformin to the liver was about 30-fold higher in wild type compared with *Oct1* knockout mice (Wang et al., 2002). However, plasma concentration–time profiles and renal elimination of metformin were similar between Oct1-deficient and wild-type mice. In the rat, coadministration of cimetidine inhibited the urinary excretion and led to much higher systemic exposure of metformin (Wang et al., 2002). Taken together, the human and rat data suggest that a cation-specific mechanism may be involved in the urinary excretion of metformin (Wang et al., 2002).

A comprehensive ^{14}C-ADME study can identify the major elimination pathways of a drug, but an ADME study in transporter deficient animals, or pretreatment with a chemical inhibitor, can further clarify the role of drug transporters for a particular elimination pathway. To study the role of Pgp and BCRP in the ADME of imatinib, Oostendorp et al. (2009) used *Mdr1a/b*, *Bcrp1* knockout, and *Mdr1a/b/Bcrp1* dual knockout mouse models, with and without transporter inhibitors. All mice were pretreated with vehicle, elacridar (GF120918) or pantoprazole, and administered an intravenous or oral dose of ^{14}C-imatinib and plasma and excreta collected (Oostendorp et al., 2009). The AUC_{iv} for imatinib was 1.6-fold higher in the dual knockouts than that in wild type. Pretreatment with elacridar or pantoprazole led to a 2.0- and 3.5-fold increase in AUC_{iv} in wild-type mice (Oostendorp et al., 2009). Interestingly, pretreatment with either inhibitor, elacridar or pantoprazole, significantly increased exposure of imatinib in both wild-type and Pgp/BCRP-deficient mice. The data suggest that other transporters, or metabolic enzyme systems, are involved in the DDI between imatinib and the inhibitors (Oostendorp et al., 2009). Following cannulation of the gall bladder of wild-type and *Mdr1a/b/Bcrp1* dual knockout mice, the biliary excretion of ^{14}C-imatinib in *Mdr1a/b/Bcrp1* was significantly lower than wild type, and pretreatment of wild-type mice with elacridar also reduced biliary excretion similar to *Mdr1a/b/Bcrp1*-deficient mice. Results from the biliary excretion studies confirm that Pgp and Bcrp are the major transporters involved in the hepatobiliary excretion of imatinib (Oostendorp et al., 2009).

16.3.5 Summary

In conclusion, transporters are essential to normal cellular physiology. Besides their endogenous substrates, many drugs and other xenobiotics are substrates and/or modulators of these transporters. Once thought to be an adaptive response of cancer cells to chemotherapeutics, now we better understand their role in oral bioavailability, hepatobiliary, and renal excretion of drugs. Furthermore, transporters can contribute to pharmacological refuge sites such as the brain, testes, and fetus, where they protect these tissues from potential therapeutic drug penetration. For these reasons, candidate drugs are routinely evaluated for possible interactions with influx and efflux transporters. A well-designed set of studies in one or more of the animal models discussed can help identify transporter(s) involved. Once identified, appropriate in vitro screens can focus on one or more transporters to rapidly screen potential drug candidates. While the selection of an in vitro assay or animal model will depend on the questions being asked, it is highly recommended that a combination of in vitro and in vivo methods be used for an accurate assessment of transporter activity.

To predict transporter-mediated DDI, the transporter(s) involved need to be fully characterized. With the current transporter tools, many examples of DDIs have been identified in animal models, as well as clinical studies. However, accurate predictions of DDIs or a more robust in vitro–in vivo correlation (IVIVC) will require further development of our transporters tools. New and commercially available animal models such as tissue-selective knockouts, new transporter knockouts and/or human gene knockins, multiple transporter knockouts, and chimeric or humanized animal models will be necessary to facilitate a better understanding of transporters in the disposition of drugs.

16.4 CONCLUSIONS AND THE PATH FORWARD

The goal of this chapter has been to provide an overview of the preclinical animal models used in understanding the role of enzymes and transporters in optimizing pharmacokinetic parameters. Many of the DMEs and transporters have been cloned and their activity assessed individually in *in vitro* assays. Although many advances in *in vitro* methods have occurred, animal studies are still essential to understand the pharmacokinetics of drugs. A combination of *in vitro* assays and animal models is necessary to investigate both the function and regulation of DMEs and transporters in an integrated biological system.

The cellular colocalization of transporters and enzymes suggest a coordinated host defense mechanism against drugs and other xenobiotics. Besides CYP3A and Pgp transport–metabolism interplay, additional efflux transporters, such as BCRP or MRP2, and uptake transporters (e.g., OATPs), can interact with other CYPs and Phase II enzymes such as UGT or GST. Examples of the coexpression of conjugating enzymes (UGT and GST) along with the MRP2 efflux transporter have been reviewed in this chapter. The potential for DDIs to alter therapeutic drug levels increases with polypharmacy practices. DDIs can cause enough of a change in tissue concentrations of drugs, especially those with a narrow therapeutic window, to cause serious toxic effects. Most DDIs reported occur with DMEs, though transporter-mediated DDIs are better recognized today; both types of DDIs have been discussed in this review.

The development of transgenic animals with one or more transporter or enzyme genes deleted (knockouts) have been instrumental to our current understanding of specific isoforms of transporters and enzymes in the disposition of drug candidates. The role of enzymes and transporters in the disposition of drug candidates has been evaluated by using genetically modified animals or by using chemical inhibitors in both animals and humans. Animal models and inhibitors discussed in this review are summarized in tables for easily referencing. The recently developed humanized mouse models may overcome some of the species differences caused by the intrinsic genes. However, new and commercially available animal models such as tissue-selective knockouts, new transporter and enzyme knockouts, multiple-gene knockouts, and chimeric or humanized animal models will be necessary to facilitate a better understanding of enzymes and transporters in the disposition of drugs.

ACKNOWLEDGMENTS

The authors would like to thank Drs. Magang Shou, Carl Davis, and Sekhar Surapaneni for their critical review of the manuscript.

REFERENCES

Abaut A-Y, Chevanne F, Corre PL (2007) Oral bioavailability and intestinal secretion of amitriptyline: Role of P-glycoprotein? *International Journal of Pharmaceutics* 330:121–128.

Abe T, Kakyo M, Tokui T, Nakagomi R, Nishio T, Nakai D, Nomura H, Unno M, Suzuki M, Naitoh T, Matsuno S, Yawo H (1999) Identification of a novel gene family encoding human liver-specific organic anion transporter LST-1. *The Journal of Biological chemistry* 274:17159–17163.

Abe T, Unno M, Onogawa T, Tokui T, Kondo TN, Nakagomi R, Adachi H, Fujiwara K, Okabe M, Suzuki M, Numoki K, Sato E, Seki M, Date F, Ono K, Kondo Y, Shiba K, Suzuki M, Ohtani H, Shimosegawa T, Linuma K, Nagura H, Ito S, Matsun S (2001) LST-2, a human liver-specific organic anion transporter, determines methotrexate sensitivity in gastrointestinal cancers. *Gastroenterology* 120:1689–1699.

Adachi Y, Suzuki H, Schinkel AH, Sugiyama Y (2005) Role of breast cancer resistance protein (Bcrp1/Abcg2) in the extrusion of glucuronide and sulfate conjugates from enterocytes to intestinal lumen. *Molecular Pharmacology* 67:923–928.

Adler ID, Baumgartner A, Gonda H, Friedman MA, Skerhut M (2000) 1-Aminobenzotriazole inhibits acrylamide-induced dominant lethal effects in spermatids of male mice. *Mutagenesis* 15:133–136.

Aitken AE, Richardson TA, Morgan ET (2006) Regulation of drug-metabolizing enzymes and transporters in inflammation. *Annual Review of Pharmacology and Toxicology* 46:123–149.

Backman JT, Olkkola KT, Neuvonen PJ (1996) Rifampin drastically reduces plasma concentrations and effects of oral midazolam. *Clinical Pharmacology and Therapeutics* 59:7–13.

Balani SK, Zhu T, Yang TJ, Liu Z, He B, Lee FW (2002) Effective dosing regimen of 1-aminobenzotriazole for inhibition of antipyrine clearance in rats, dogs, and monkeys. *Drug Metabolism and Disposition* 30:1059–1062.

Bogaards JJ, Bertrand M, Jackson P, Oudshoorn MJ, Weaver RJ, van Bladeren PJ, Walther B (2000) Determining the best animal model for human cytochrome P450 activities: A comparison of mouse, rat, rabbit, dog, micropig, monkey and man. *Xenobiotica* 30:1131–1152.

Bohets H, Annaert P, Mannens G, Van BL, Anciaux K, Verboven P, Meuldermans W, Lavrijsen K (2001) Strategies for absorption screening in drug discovery and development. *Current Topics in Medicinal Chemistry* 1:367–383.

Breedveld P, Pluim D, Cipriani G, Weilinga P, van Tellingen O, Schinkel AH, Schellens HM (2005) The effect of Bcrp (Abcg2) on the *in vivo* pharmacokinetics and brain penetration of imatinib mesylate (Gleevec): Implications for the use of breast cancer resistance protein and P-glycoprotein inhibitors to enable the brain penetration of imatinib in patients. *Cancer Research* 65:2577–2582.

Bullock P, Pearce R, Draper A, Podval J, Bracken W, Veltman J, Thomas P, Parkinson A (1995) Induction of liver microsomal cytochrome P450 in cynomolgus monkeys. *Drug Metabolism and Disposition* 23:736–748.

Caldwell GW, Ritchie DM, Masucci JA, Hageman W, Cotto C, Hall J, Hasting B, Jones W (2005) The use of the suicide

CYP450 inhibitor ABT for distinguishing absorption and metabolism processes in *in-vivo* pharmacokinetic screens. *European Journal of Drug Metabolism and Pharmacokinetics* 30:75–83.

Carr B, Norcross R, Fang Y, Lu P, Rodrigues AD, Shou M, Rushmore T, Booth-Genthe C (2006) Characterization of the rhesus monkey CYP3A64 enzyme: Species comparisons of CYP3A substrate specificity and kinetics using baculovirus-expressed recombinant enzymes. *Drug Metabolism and Disposition* 34:1703–1712.

Chen C, Hanson E, Watson JW, Lee JS (2003a) P-glycoprotein limits the brain penetration of nonsedating but not sedating H1-antagonist. *Drug Metabolism and Disposition* 31:312–318.

Chen C, Scott D, Hanson E, Franco J, Berryman E, Volberg M, Liu X (2003b) Impact of Mrp2 on the biliary excretion and intestinal absorption of furosemide, probenecid, and methotrexate using Eisai hyperbilirubinemic rats. *Pharmaceutical Research* 20:31–37.

Cheung C, Gonzalez FJ (2008) Humanized mouse lines and their application for prediction of human drug metabolism and toxicological risk assessment. *The Journal of Pharmacology and Experimental Therapeutics* 327:288–299.

Cheung C, Yu AM, Chen CS, Krausz KW, Byrd LG, Feigenbaum L, Edwards RJ, Waxman DJ, Gonzalez FJ (2006) Growth hormone determines sexual dimorphism of hepatic cytochrome P450 3A4 expression in transgenic mice. *The Journal of Pharmacology and Experimental Therapeutics* 316:1328–1334.

Choo EF, Leake B, Wandel C, Imamura H, Wood AJ, Wilkinson GR, Kim RB (2000) Pharmacological inhibition of P-glycoprotein transport enhances the distribution of HIV-1 protease inhibitors into the brain and testes. *Drug Metabolism and Disposition* 28:655–660.

Corchero J, Granvil CP, Akiyama TE, Hayhurst GP, Pimprale S, Feigenbaum L, Idle JR, Gonzalez FJ (2001) The CYP2D6 humanized mouse: Effect of the human CYP2D6 transgene and HNF4alpha on the disposition of debrisoquine in the mouse. *Molecular Pharmacology* 60:1260–1267. [Erratum appears in Molecular Pharmacology 2002;61(1):248]

Cordon-Cardo C, O'Brien JP, Casals D, Rittman-Grauer L, Biedler JL, Melamed MR, Bertino JR (1989) Multidrug-resistance gene (P-glycoprotein) is expressed by endothelial cells at blood-brain barrier sites. *Proceedings of the National Academy of Sciences of the United States of America* 86:695–698.

Croop JM, Raymond M, Haber D, Devault A, Arceci RJ, Gros P, Housman DE (1989) The three mouse multidrug resistance (mdr1) genes are expressed in a tissue specific manner in normal mouse tissue. *Molecular and Cellular Biology* 9:1346–1350.

DeKeyser JG, Shou M (2011) Species differences of drug metabolizing enzymes. In *Encyclopedia of Drug Metabolism*, Vol. 1. Sinz M, Rodrigues D, eds. John Wiley & Sons, Hoboken, NJ. (In press).

Dietrich CG, Rudi de Waart D, Ottenhoff R, Scoots IG, Oude Elerink RP (2001) Increased bioavailability of the food-derived carcinogen 2-amino-1-methy-6-phenylimidazo[4,5-b]pyridine in MRP2-deficient rats. *Molecular Pharmacology* 59:974–980.

Gerk PM, Oo CY, Paxton EW, Moscow JA, McNamara PJ (2001) Interactions between cimetidine, nitrofurantoin and probenecid active transport into rat milk. *The Journal of Pharmacology and Experimental Therapeutics* 296:175–180.

Gonzalez FJ, Yu AM (2006) Cytochrome P450 and xenobiotic receptor humanized mice. *Annual Review of Pharmacology and Toxicology* 46:41–64.

Granvil CP, Yu AM, Elizondo G, Akiyama TE, Cheung C, Feigenbaum L, Krausz KW, Gonzalez FJ (2003) Expression of the human CYP3A4 gene in the small intestine of transgenic mice: *In vitro* metabolism and pharmacokinetics of midazolam. *Drug Metabolism and Disposition* 31:548–558.

Guengerich FP (1997) Comparisons of catalytic selectivity of cytochrome P450 subfamily enzymes from different species. *Chemico-Biological Interactions* 106:161–182.

Haas CE, Kaufman DC, Jones CE, Burstein AH, Reiss W (2003) Cytochrome P450 3A4 activity after surgical stress. *Critical Care Medicine* 31:1338–1346.

Hayhurst GP, Lee YH, Lambert G, Ward JM, Gonzalez FJ (2001) Hepatocyte nuclear factor 4alpha (nuclear receptor 2A1) is essential for maintenance of hepatic gene expression and lipid homeostasis. *Molecular and Cellular Biology* 21:1393–1403.

Hendrikse NH, de Vries EG, Eriks-Fluks L, van der Graaf WT, Hospers GA, Willemsen AT, Vaalburg W, Franssen EJ (1999) A new *in vivo* method to study P-glycoprotein transport in tumors and the blood-brain barrier. *Cancer Research* 59:2411–2416.

van Herwaarden AE, Wagenaar E, van der Kruijssen CM, van Waterschoot RA, Smit JW, Song JY, van der Valk MA, van Tellingen O, van der Hoorn JW, Rosing H, Beijnen JH, Schinkel AH (2007) Knockout of cytochrome P450 3A yields new mouse models for understanding xenobiotic metabolism. *The Journal of Clinical Investigation* 117:3583–3592.

Hirano M, Maeda K, Matsushima S, Nozaki Y, Kusuhara H, Sugiyama Y (2005) Involvement of BCRP (ABCG2) in the biliary excretion of pitavastatin. *Molecular Pharmacology* 68:800–807.

Hoffman DJ, Seifert T, Borre A, Nellans HN (1995) Method to estimate the rate and extent of intestinal absorption in conscious rats using an absorption probe and portal blood sampling. *Pharmaceutical Research* 12:889–894.

Iwasaki K, Uno Y (2009) Cynomolgus monkey CYPs: A comparison with human CYPs. *Xenobiotica* 39:578–581.

Jonker JW, Smit JW, Brinkhuis RF, Maliepaard M, Bijen JH, Schellens JHM, Schinkel AH (2000) Role of breast cancer resistance protein in the bioavailability and fetal penetration of topotecan. *Journal of the National Cancer Institute* 92:1651–1656.

Jonker JW, Wagenaar E, Mol CA, Buitelaar M, Koepsell H, Smit JW, Schinkel AH (2001) Reduced hepatic uptake and intestinal excretion of organic cations in mice with a targeted disruption of the organic cation transporter 1 (Oct1 [Slc22a1] gene. *Molecular and Cellular Biology* 21:5471–5477.

Kacevska M, Robertson GR, Clarke SJ, Liddle C (2008) Inflammation and CYP3A4-mediated drug metabolism in advanced cancer: Impact and implications for chemotherapeutic drug dosing. *Expert Opinion On Drug Metabolism and Toxicology* 4:137–149.

Kaddoumi A, Choi S, Kinman L, Whittington D, Tsai C-C, Ho RJ, Anderson BD, Unadkat JD (2007) Inhibition of P-gp activity at the primate blood-brain barrier increases the distribution of nelfinavir into the brain but not into the CSF. *Drug Metabolism and Disposition* 35:1459–1462.

Kanazu T, Yamaguchi Y, Okamura N, Baba T, Koike M (2004a) Model for the drug-drug interaction responsible for CYP3A enzyme inhibition. I: Evaluation of cynomolgus monkeys as surrogates for humans. *Xenobiotica* 34:391–402.

Kanazu T, Yamaguchi Y, Okamura N, Baba T, Koike M (2004b) Model for the drug-drug interaction responsible for CYP3A enzyme inhibition. II: Establishment and evaluation of dexamethasone-pretreated female rats. *Xenobiotica* 34:403–413.

Kanazu T, Okamura N, Yamaguchi Y, Baba T, Koike M (2005) Assessment of the hepatic and intestinal first-pass metabolism of midazolam in a CYP3A drug-drug interaction model rats. *Xenobiotica* 35:305–317.

Katoh M, Yokoi T (2007) Application of chimeric mice with humanized liver for predictive ADME. *Drug Metabolism Reviews* 39:145–157.

Katoh M, Matsui T, Nakajima M, Tateno C, Soeno Y, Horie T, Iwasaki K, Yoshizato K, Yokoi T (2005) *In vivo* induction of human cytochrome P450 enzymes expressed in chimeric mice with humanized liver. *Drug Metabolism and Disposition* 33:754–763.

Kim RB, Fromm MF, Wandel C, Leake B, Wood AJ, Roden DM (1998) The drug transporter P-glycoprotein limits oral absorption and brain entry of HIV-1 protease inhibitors. *The Journal of Clinical Investigation* 101:289–294.

Kim S, Dinchuk JE, Anthony MN, Orcutt T, Zoeckler ME, Sauer MB, Mosure KW, Vuppugalla R, Grace JE Jr., Simmermacher J, Dulac HA, Pizzano J, Sinz M (2010) Evaluation of cynomolgus monkey pregnane X receptor, primary hepatocyte, and *in vivo* pharmacokinetic changes in predicting human CYP3A4 induction. *Drug Metabolism and Disposition* 38:16–24.

Kolars JC, Awni WM, Merion RM, Watkins PB (1991) First-pass metabolism of cyclosporin by the gut. *Lancet* 338:1488–1490.

Kotegawa T, Laurijssens BE, Von Moltke LL, Cotreau MM, Perloff MD, Venkatakrishnan K, Warrington JS, Granda BW, Harmatz JS, Greenblatt DJ (2002) *In vitro*, pharmacokinetic, and pharmacodynamic interactions of ketoconazole and midazolam in the rat. *The Journal of Pharmacology and Experimental Therapeutics* 302:1228–1237.

Kunta JR, Lee SH, Perry BA, Lee YH, Sinko PJ (2004) Differentiation of gut and hepatic first-pass loss of verapamil in intestinal and vascular access-ported (IVAP) rabbits. *Drug Metabolism and Disposition* 32:1293–1298.

Leblond F, Guevin C, Demers C, Pellerin I, Gascon-Barre M, Pichette V (2001) Downregulation of hepatic cytochrome P450 in chronic renal failure. *Journal of the American Society of Nephrology* 12:326–332.

Lee YJ, Maeda J, Kusuhara H, Okauchi T, Inaji M, Nagai Y, Obayashi S, Nakao R, Suzuki K, Sugiyama Y, Suhara T (2006) *In vivo* evaluation of P-glycoprotein function at the blood-brain barrier in nonhuman primates using ^{11}C-verapamil. *The Journal of Pharmacology and Experimental Therapeutics* 316:647–653.

Leggas M, Panetta JC, Zhuang Y, Schuetz JD, Johnston B, Bai F, Sorrentino B, Zhou S, Houghton PJ, Stewart CF (2006) Gefitinib modulates the function of multiple ATP-binding cassette transporters *in vivo*. *Cancer Research* 66:4802–4807.

Liang HC, Li H, McKinnon RA, Duffy JJ, Potter SS, Puga A, Nebert DW (1996) Cyp1a2(-/-) null mutant mice develop normally but show deficient drug metabolism. *Proceedings of the National Academy of Sciences of the United States of America* 93:1671–1676.

Lin LA, Shangari N, Chan TS, Remirez D, O'Brien PJ (2006) Herbal monoterpene alcohols inhibit propofol metabolism and prolong anesthesia time. *Life Sciences* 79:21–29.

Liu H, Baliga R (2003) Cytochrome P450 2E1 null mice provide novel protection against cisplatin-induced nephrotoxicity and apoptosis. *Kidney International* 63:1687–1696.

Ma X, Shah YM, Guo GL, Wang T, Krausz KW, Idle JR, Gonzalez FJ (2007) Rifaximin is a gut-specific human pregnane X receptor activator. *The Journal of Pharmacology and Experimental Therapeutics* 322:391–398.

Ma X, Cheung C, Krausz KW, Shah YM, Wang T, Idle JR, Gonzalez FJ (2008) A double transgenic mouse model expressing human pregnane X receptor and cytochrome P450 3A4. *Drug Metabolism and Disposition* 36:2506–2512.

Makishima M, Lu TT, Xie W, Whitfield GK, Domoto H, Evans RM, Haussler MR, Mangelsdorf DJ (2002) Vitamin D receptor as an intestinal bile acid sensor. *Science* 296:1313–1316.

Mandlekar SV, Rose AV, Cornelius G, Sleczka B, Caporuscio C, Wang J, Marathe PH (2007) Development of an *in vivo* rat screen model to predict pharmacokinetic interactions of CYP3A4 substrates. *Xenobiotica* 37:923–942.

Marathe PH, Rodrigues AD (2006) *In vivo* animal models for investigating potential CYP3A- and Pgp-mediated drug-drug interactions. *Current Drug Metabolism* 7:687–704.

Martignoni M, Groothuis GM, de Kanter R (2006) Species differences between mouse, rat, dog, monkey and human CYP-mediated drug metabolism, inhibition and induction. *Expert Opinion On Drug Metabolism and Toxicology* 2:875–894.

Mathias AA, Maggio-Price L, Lai Y, Gupta A, Unadkat JD (2006) Changes in pharmacokinetics of anti-HIV protease inhibitors during pregnancy: The role of CYP3A and

P-glycoprotein. *The Journal of Pharmacology and Experimental Therapeutics* 316:1202–1209.

Merino G, Jonker JW, Wagenaar E, van Herwaarden AE, Schinkel AH (2005) The breast cancer resistance protein (BCRP/ABCG2) affects pharmacokinetics, hepatobiliary excretion, and milk secretion of the antibiotic nitrofurantoin. *Molecular Pharmacology* 67:1758–1764.

Mizuno N, Suzuki M, Kusuhara H, Suzuki H, Takeuchi K, Niwa T, Jonker JW, Sugiyama Y (2004) Impaired renal excretion of 6-hydroxy-5,7-dimethyl-2-methylamino-4-(3-pyridylmethyl) benzothiazole (E3040) sulfate in breast cancer resistance protein (BCRP1/ABCG2) knockout mice. *Drug Metabolism and Disposition* 32:898–901.

Morgan ET, Goralski KB, Piquette-Miller M, Renton KW, Robertson GR, Chaluvadi MR, Charles KA, Clarke SJ, Kacevska M, Liddle C, Richardson TA, Sharma R, Sinal CJ (2008) Regulation of drug-metabolizing enzymes and transporters in infection, inflammation, and cancer. *Drug Metabolism and Disposition* 36:205–216.

Morimoto K, Nakakariya M, Shirasaka Y, Kakinuma C, Fujita T, Tamai I, Ogihara T (2008) Oseltamer (Tamiflu) efflux transport at the blood-brain barrier via P-glycoprotein. *Drug Metabolism and Disposition* 36:6–9.

Mu Y, Zhang J, Zhang S, Zhou HH, Toma D, Ren S, Huang L, Yaramus M, Baum A, Venkataramanan R, Xie W (2006) Traditional Chinese medicines Wu Wei Zi (*Schisandra chinensis* Baill) and Gan Cao (*Glycyrrhiza uralensis* Fisch) activate pregnane X receptor and increase warfarin clearance in rats. *The Journal of Pharmacology and Experimental Therapeutics* 316:1369–1377.

Muruganandan S, Sinal CJ (2008) Mice as clinically relevant models for the study of cytochrome P450-dependent metabolism. *Clinical Pharmacology and Therapeutics* 83:818–828.

Mutlib AE, Chen H, Nemeth GA, Markwalder JA, Seitz SP, Gan LS, Christ DD (1999) Identification and characterization of efavirenz metabolites by liquid chromatography/mass spectrometry and high field NMR: Species differences in the metabolism of efavirenz. *Drug Metabolism and Disposition* 27:1319–1333.

Mutlib AE, Gerson RJ, Meunier PC, Haley PJ, Chen H, Gan LS, Davies MH, Gemzik B, Christ DD, Krahn DF, Markwalder JA, Seitz SP, Robertson RT, Miwa GT (2000) The species-dependent metabolism of efavirenz produces a nephrotoxic glutathione conjugate in rats. *Toxicology and Applied Pharmacology* 169:102–113.

Nebert DW, Russell DW (2002) Clinical importance of the cytochromes P450. *Lancet* 360:1155–1162.

Nelson SD, Slattery JT, Thummel KE, Watkins PB (2003) Car unlikely to significantly modulate acetaminophen hepatotoxicity in most humans. *Hepatology* 38:254–257.

Nolin TD, Naud J, Leblond FA, Pichette V (2008) Emerging evidence of the impact of kidney disease on drug metabolism and transport. *Clinical Pharmacology and Therapeutics* 83:898–903.

Ogasawara A, Kume T, Kazama E (2007) Effect of oral ketoconazole on intestinal first-pass effect of midazolam and fexofenadine in cynomolgus monkeys. *Drug Metabolism and Disposition* 35:410–418.

Oostendorp RL, Buckle T, Beijnen JH, Tellingen OV, Schellens JH (2009) The effect of P-gp (Mdr1a/b) BCRP (Bcrp1) and P-gp/BCRP inhibitors on the *in vivo* absorption, distribution, metabolism and excretion of imatinib. *Investigational New Drugs* 27:31–40.

Ose A, Ito M, Kusuhara H, Yamatsugu K, Kanai M, Shibasaki M, Hosokawa M, Schuetz JD, Sugiyama Y (2009) Limited brain distribution of [3R,4R,5S]-4-acetamido-amino-3-(1-ethylpropoxy)-1-cyclohexene-1-carboxylate phosphate (Ro 64-0802), a pharmacologically active form of oseltamivir, by active efflux across the blood-brain barrier mediated by organic anion transporter 3 (Oat3/Slc22a8) and mulidrug resistance-associated protein 4 (Mrp4/Abcc4). *Drug Metabolism and Disposition* 37:315–321.

Paine MF, Shen DD, Kunze KL, Perkins JD, Marsh CL, McVicar JP, Barr DM, Gillies BS, Thummel KE (1996) First-pass metabolism of midazolam by the human intestine. *Clinical Pharmacology and Therapeutics* 60:14–24.

Panda DK, Miao D, Tremblay ML, Sirois J, Farookhi R, Hendy GN, Goltzman D (2001) Targeted ablation of the 25-hydroxyvitamin D 1alpha -hydroxylase enzyme: Evidence for skeletal, reproductive, and immune dysfunction. *Proceedings of the National Academy of Sciences of the United States of America* 98:7498–7503.

Perloff MD, Von Moltke LL, Court MH, Kotegawa T, Shader RI, Greenblatt DJ (2000) Midazolam and triazolam biotransformation in mouse and human liver microsomes: Relative contribution of CYP3A and CYP2C isoforms. *The Journal of Pharmacology and Experimental Therapeutics* 292:618–628.

Peterson TC, Peterson MR, Wornell PA, Blanchard MG, Gonzalez FJ (2004) Role of CYP1A2 and CYP2E1 in the pentoxifylline ciprofloxacin drug interaction. *Biochemical Pharmacology* 68:395–402.

Prueksaritanont T, Kuo Y, Tang C, Li C, Qiu Y, Lu B, Strong-Basalyga K, Richards K, Carr B, Lin JH (2006) *In vitro* and *in vivo* CYP3A64 induction and inhibition studies in rhesus monkeys: A preclinical approach for CYP3A-mediated drug interaction studies. *Drug Metabolism and Disposition* 34:1546–1555.

Renton KW (2004) Cytochrome P450 regulation and drug biotransformation during inflammation and infection. *Current Drug Metabolism* 5:235–243.

Robertson GR, Liddle C, Clarke SJ (2008) Inflammation and altered drug clearance in cancer: Transcriptional repression of a human CYP3A4 transgene in tumor-bearing mice. *Clinical Pharmacology and Therapeutics* 83:894–897.

Rowland M, Tozer TN (1995) *Clinical Pharamcokinetics: Concepts and Application.* Lippincott Williams & Wilkins, Philadelphia.

Salyers KL (2009) Preclinical pharmacokinetic models for drug discovery and development. In *Handbook of Drug Metabolism*, 2nd ed., Pearson PG, Wienkers LC, eds., pp. 659–673. Informa Healthcare, New York.

Sasongko L, Link JM, Muzi M, Mankoff DA, Yang X, Collier AC, Shoner SC, Unadkat JD (2005) Imaging P-glycoprotein transport activity at the human blood-brain barrier with positron emission tomography. *Clinical Pharmacology and Therapeutics* 77:503–514.

Savolainen J, Edwards JE, Morgan ME, McNamara PJ, Anderson BD (2002) Effects of a P-glycoprotein inhibitor on brain and plasma concentrations of anti-human immunodeficiency virus drugs administered in combination in rats. *Drug Metabolism and Disposition* 30:479–482.

Schinkel AH, Smit JJ, van Tellingen O, Beijnen JH, Wagenaar E, van Deemter L, Mol CA, van der Valk MA, Robanus-Maandag EC, te Riele HP, Berns AJ, Borst P (1994) Disruption of the mouse mdr1a P-glycoprotein gene leads to a deficiency in the blood-brain barrier and to increased sensitivity to drugs. *Cell* 77:491–502.

Schinkel AH, Wagenaar E, van Deemter L, Mol CA, Borst P (1995) Absence of the mdr1a P-glycoprotein in mice affects tissue distribution and pharmacokinetics of dexamethasone, digoxin, and cyclosporine A. *The Journal of Clinical Investigation* 96:1698–1705.

Schinkel AH, Wagenaar E, van Deemter L (1996) P-glycoprotein in the blood-brain barrier of mice influences the brain penetration and pharmacological activity of many drugs. *The Journal of Clinical Investigation* 97:2517–2524.

Schinkel AH, Mayer U, Wagenaar E (1997) Normal and altered pharmacokinetics in mice lacking Mdr1-type (drug transporting) P-glycoproteins. *Proceedings of the National Academy of Sciences of the United States of America* 94:4028–4033.

Shitara Y, Horie T, Sugiyama Y (2006) Transporters as a determinant of drug clearance and tissue distribution. *European Journal of Pharmaceutical Sciences* 27:425–446.

Sinal CJ, Miyata M, Tohkin M, Nagata K, Bend JR, Gonzalez FJ (2000) Targeted disruption of soluble epoxide hydrolase reveals a role in blood pressure regulation. *The Journal of Biological Chemistry* 275:40504–40510.

Somogyi A, Stockley C, Keal J, Rolan P, Bochner F (1987) Reduction of metformin renal tubular secretion by cimetidine in man. *British Journal of Clinical Pharmacology* 23:545–551.

Sparreboom A, Van Asperen J, Mayer U, Schinkel AH, Smit JW, Meijer DK, Borst P, Nooijen WJ, Beijnen JH, Van Tellingen O (1997) Limited oral bioavailability and active epithelial excretion of paclitaxel (Taxol) caused by P-glycoprotein in the intestine. *Pharmacology* 94: 2031–2035.

Staudinger JL, Goodwin B, Jones SA, Hawkins-Brown D, MacKenzie KI, LaTour A, Liu Y, Klaassen CD, Brown KK, Reinhard J, Willson TM, Koller BH, Kliewer SA (2001) The nuclear receptor PXR is a lithocholic acid sensor that protects against liver toxicity. *Proceedings of the National Academy of Sciences of the United States of America* 98:3369–3374.

van de Steeg E, van Der Kruijssen CM, Wagenarr E, Burggraaff JE, Mesman E, Kenworthy KE, Schinkel AH (2009) Methotreate pharmacokinetics in transgenic mice with liver-specific expression of human organic anion-transporting polypeptide 1B1 (SLCo1B1). *Drug Metabolism and Disposition* 37:277–281.

Strelevitz TJ, Foti RS, Fisher MB (2006) *In vivo* use of the P450 inactivator 1-aminobenzotriazole in the rat: Varied dosing route to elucidate gut and liver contributions to first-pass and systemic clearance. *Journal of Pharmaceutical Sciences* 95:1334–1341.

Sugano K, Kansy M, Artursson P, Avdeef A, Bendels S, Di L, Ecker GF, Faller B, Fischer H, Gerebtoff G, Lennernaes H, Senner F (2010) Coexistence of passive and carrier-mediated processes in drug transport. *Nature reviews. Drug discovery* 9:597–614.

Tahara H, Kusuhara H, Fuse E, Sugiyama Y (2005) P-glycoprotein plays a major role in the efflux of fexofenadine in the small intestine and blood-brain barrier, but only a limited role in its biliary excretion. *Drug Metabolism and Disposition* 33:963–968.

Takano M, Yumoto R, Murakami T (2006) Expression and function of efflux drug transporters in the intestine. *Pharmacology and Therapeutics* 109:137–161.

Tam-Zaman N, Tam YK, Tawfik S, Wiltshire H (2004) Factors responsible for the variability of saquinavir absorption: Studies using an instrumented dog model. *Pharmaceutical Research* 21:436–442.

Tang W, Stearns RA, Wang RW, Miller RR, Chen Q, Ngui J, Bakshi RK, Nargund RP, Dean DC, Baillie TA (2008) Assessing and minimizing time-dependent inhibition of cytochrome P450 3A in drug discovery: A case study with melanocortin-4 receptor agonists. *Xenobiotica* 38:1437–1451.

Tateno C, Yoshizane Y, Saito N, Kataoka M, Utoh R, Yamasaki C, Tachibana A, Soeno Y, Asahina K, Hino H, Asahara T, Yokoi T, Furukawa T, Yoshizato K (2004) Near completely humanized liver in mice shows human-type metabolic responses to drugs. *The American Journal of Pathology* 165:901–912.

Thummel KE, O'Shea D, Paine MF, Shen DD, Kunze KL, Perkins JD, Wilkinson GR (1996) Oral first-pass elimination of midazolam involves both gastrointestinal and hepatic CYP3A-mediated metabolism. *Clinical Pharmacology and Therapeutics* 59:491–502.

Thummel KE, Brimer C, Yasuda K, Thottassery J, Senn T, Lin Y, Ishizuka H, Kharasch E, Schuetz J, Schuetz E (2001) Transcriptional control of intestinal cytochrome P-4503A by 1alpha,25-dihydroxy vitamin D3. *Molecular Pharmacology* 60:1399–1406.

Tsunoda SM, Velez RL, Von Moltke LL, Greenblatt DJ (1999) Differentiation of intestinal and hepatic cytochrome P450 3A activity with use of midazolam as an *in vivo* probe: Effect of ketoconazole. *Clinical Pharmacology and Therapeutics* 66:461–471.

Turpeinen M, Ghiciuc C, Opritoui M, Tursas L, Pelkonen O, Pasanen M (2007) Predictive value of animal models for human cytochrome P450 (CYP)-mediated metabolism: A comparative study *in vitro*. *Xenobiotica* 37:1367–1377.

Uppal H, Toma D, Saini SP, Ren S, Jones TJ, Xie W (2005) Combined loss of orphan receptors PXR and CAR heightens sensitivity to toxic bile acids in mice. *Hepatology* 41:168–176.

Von Moltke LL, Granda BW, Grassi JM, Perloff MD, Vishnuvardhan D, Greenblatt DJ (2004) Interaction of triazolam and ketoconazole in P-glycoprotein-deficient mice. *Drug Metabolism and Disposition* 32:800–804.

Wacher VJ, Wu CY, Benet LZ (1995) Overlapping substrate specificities and tissues distribution of cytochrome P450 3A and P-glycoprotein: Implications for drug delivery and activity in cancer chemotherapy. *Molecular Carcinogenesis* 13:129–134.

Wang D, Jonker JW, Kato Y, Kusuhara H, Schinkel AH, Sugiyama Y (2002) Involvement of organic cation transporter 1 in hepatic and intestinal distribution of metformin. *The Journal of Pharmacology and Experimental Therapeutics* 302:510–515.

Wang H, Xie W (2009) Nuclear receptor-mediated gene regulation in drug metabolism. In *Drug Metabolism Handbook: Concepts and Applications*, Nassar AF, Hollenberg PF, Scatina J, eds., pp. 449–478. John Wiley & Sons, Hoboken, NJ.

Wang SW, Kulkarni KH, Tang L, Wang JR, Yin T, Daidoji T, Yokota H, Hu M (2009) Disposition of flavonoids via enteric recycling: UDP-glucuronosyltransferase (UGT) 1As deficiency in Gunn rats is compensated by increases in UGT2Bs activities. *The Journal of Pharmacology and Experimental Therapeutics* 329:1023–1031.

Ward KW, Azzarano LM (2004) Preclinical pharmacokinetic properties of the P-glycoprotein inhibitor GF 120918A (HCl salt of GF 120918, 9, 10-dihydro-5-methoxy9-oxo-N-[4-[2-(1,2,3,4-tetrahydro-6,7-dimethoxy-2-isoquinolinyl) ethyl] phenyl]-4-acridine-carboxamide) in the mouse, rat, dog, and monkey. *The Journal of Pharmacology and Experimental Therapeutics* 310:703–709.

Ward KW, Proksch JW, Levy MA, Smith BR (2001) Development of an *in vivo* preclinical screen model to estimate absorption and bioavailability of xenobiotics. *Drug Metabolism and Disposition* 29:82–88.

Ward KW, Stelman GJ, Morgan JA, Zeigler KS, Azzarano LM, Kehler JR, Surdy-Freed JE, Proksch JW, Smith BR (2004) Development of an *in vivo* preclinical screen model to estimate absorption and first-pass hepatic extraction of xenobiotics. II. Use of ketoconazole to identify P-glycoprotein/CYP3A-limited bioavailability in the monkey. *Drug Metabolism and Disposition* 32:172–177.

Watanabe T, Furukawa T, Sharyo S, Ohashi Y, Yasuda M, Takaoka M, Manabe S (2000) Effect of troglitazone on the liver of a Gunn rat model of genetic enzyme polymorphism. *The Journal of Toxicological Sciences* 25:423–431.

van Waterschoot RA, van Herwaarden AE, Lagas JS, Sparidans RW, Wagenaar E, van der Kruijssen CM, Goldstein JA, Zeldin DC, Beijnen JH, Schinkel AH (2008) Midazolam metabolism in cytochrome P450 3A knockout mice can be attributed to up-regulated CYP2C enzymes. *Molecular Pharmacology* 73:1029–1036.

Watkins JB, Klaassen CD (1983) Chemically-induced alteration of UPD-glucuronic acid concentration in rat liver. *Drug Metabolism and Disposition* 11:37–40.

Wu CY, Benet LZ, Hebert MF, Gupta SK, Rowland M, Gomez DY, Wacher VJ (1995) Differentiation of absorption and first-pass gut and hepatic metabolism in humans: Studies with cyclosporine. *Clinical Pharmacology and Therapeutics* 58:492–497.

WuDunn D (2002) Genetic basis of glaucoma. *Current Opinion in Ophthalmology* 13:55–60.

Xia CQ, Yang JJ, Balani S (2008) Drug transporters in drug disposition, drug interactions, and drug resistance. In *Drug Metabolism in Drug Design and Development*, Zhang D, Zhu M, Humphreys WG, eds., pp. 137–202. John Wiley & Sons, Hoboken, NJ.

Xie W, Barwick JL, Downes M, Blumberg B, Simon CM, Nelson MC, Neuschwander-Tetri BA, Brunt EM, Guzelian PS, Evans RM (2000) Humanized xenobiotic response in mice expressing nuclear receptor SXR. *Nature* 406: 435–439.

Xie W, Radominska-Pandya A, Shi Y, Simon CM, Nelson MC, Ong ES, Waxman DJ, Evans RM (2001) An essential role for nuclear receptors SXR/PXR in detoxification of cholestatic bile acids. *Proceedings of the National Academy of Sciences of the United States of America* 98:3375–3380.

Xu Y, Iwanaga K, Zhou C, Cheesman MJ, Farin F, Thummel KE (2006) Selective induction of intestinal CYP3A23 by 1alpha,25-dihydroxyvitamin D3 in rats. *Biochemical Pharmacology* 72:385–392.

Yao C, Levy RH (2002) Inhibition-based metabolic drug-drug interactions: Predictions from *in vitro* data. *Journal of Pharmaceutical Sciences* 91:1923–1935.

You G, Morris ME (2007) Overview of drug transporter families. In *Drug Transporters: Molecular Characterization and Role in Drug Disposition*. You G, Morris ME, eds., pp. 1–10. John Wiley & Sons, Hoboken, NJ.

Yu AM, Idle JR, Gonzalez FJ (2004) Polymorphic cytochrome P450 2D6: Humanized mouse model and endogenous substrates. *Drug Metabolism Reviews* 36:243–277.

Zhang H, Zhang D, Li W, Yao M, D'Arienzo C, Li YX, Ewing WR, Gu Z, Zhu Y, Murugesan N, Shyu WC, Humphreys WG (2007) Reduction of site-specific CYP3A-mediated metabolism for dual angiotensin and endothelin receptor antagonists in various *in vitro* systems and in cynomolgus monkeys. *Drug Metabolism and Disposition* 35:795–805.

Zhang H, Wu X, Chung F, Naraharisetti SB, Whittington D, Mirfazaelian A, Unadkat JD (2009) As in humans, pregnancy increases the clearance of the protease inhibitor nelfinavir in the nonhuman primate *Macaca nemestrina*. *The Journal of Pharmacology and Experimental Therapeutics* 329:1016–1022.

Zhang J, Huang W, Chua SS, Wei P, Moore DD (2002) Modulation of acetaminophen-induced hepatotoxicity by the xenobiotic receptor CAR. *Science* 298:422–424.

17

MILK EXCRETION AND PLACENTAL TRANSFER STUDIES

MATTHEW HOFFMANN AND ADAM SHILLING

17.1 INTRODUCTION

Nonclinical placental transfer and lacteal excretion studies are regularly performed as part of the development of drugs that may be administered to women of child-bearing potential. The purpose of these studies is to determine the potential exposure and risk to the fetus and to the breastfeeding infant. These studies are typically performed in conjunction with nonclinical reproductive and developmental toxicity studies so that any toxicity observed in the fetus or the neonate can be linked to drug exposure. However, absence of fetal or neonate toxicity does not eliminate the need to perform these studies. Conversely, if toxicity is observed in the reproductive and/or developmental toxicity studies but drug exposure in the fetus or neonate cannot be established, it may be difficult to explain the mechanism by which the toxicity arises.

There is minimal regulatory guidance regarding the study design for nonclinical lacteal excretion and placental transfer studies. International Conference on Harmonization (ICH) guidance S3A addresses the assessment of exposure in toxicity studies and specifies that exposure should be determined in dams, embryos, fetuses, or newborns, and further states that "secretion in milk may be assessed to define its role in the exposure to newborns" (ICH Guidance for Industry, 1995). This guidance allows for additional studies to be performed as necessary to study embryo/fetal transfer and secretion in milk. Two Food and Drug Administration (FDA) guidance documents comment on the potential exposure of neonates via breast milk, but provide minimal information regarding how or when to assess this potential exposure using nonclinical studies (FDA Guidance for Industry, 2011: FDA Guidance for Industry, 2005). This chapter will discuss the most common methods for obtaining these data, the pros and cons of each approach, some drug characteristics that may affect placental transfer and/or milk excretion, and what should be considered when planning a study, and will provide some sample data from nonclinical and clinical studies.

17.2 COMPOUND CHARACTERISTICS THAT AFFECT PLACENTAL TRANSFER AND LACTEAL EXCRETION

The placenta acts as a link between maternal and fetal circulation, allowing nutrients and oxygen to be transferred to the fetus while often providing some protection to the fetus from potentially harmful xenobiotics in maternal blood. To accomplish these functions, the placenta is a highly perfused, complex, and dynamic structure that changes throughout pregnancy, and also exhibits significant interspecies differences (detailed description of placenta structure is reviewed by Enders and Blankenship, 1999). Similarly, breast milk is a link between the mother and newborn, supplying nourishment to the infant, but also providing a potential route of exposure for xenobiotics present in maternal blood. The lactating mammary gland consists of a branching network of ducts formed by epithelial cells ending in extensive alveolar clusters with milk secreted by alveolar epithelial cells into the lumen by contraction of the

ADME-Enabling Technologies in Drug Design and Development, First Edition. Edited by Donglu Zhang and Sekhar Surapaneni.
© 2012 John Wiley & Sons, Inc. Published 2012 by John Wiley & Sons, Inc.

surrounding myoepithelial cells (review by McManaman and Neville, 2003). As with drug distribution into any tissue, there are multiple processes affecting both drug transfer through the placenta into the fetus and drug secretion by the mammary gland into milk, and ultimately to the breastfeeding infant. The major factors involved in determining the extent of placental transfer and lacteal excretion overlap considerably and are described below and depicted in Figure 17.1.

17.2.1   Passive Diffusion

It is thought that nearly all drugs cross the placenta and are excreted in milk to some degree, with passive diffu-

sion being the most common mechanism by which this occurs. Therefore, the physicochemical properties of the drug, such as lipophilicity, molecular weight, ionization, and plasma protein binding, can affect the extent of placental transfer and milk excretion (Table 17.1) (Reynolds and Knott, 1989; Atkinson and Begg, 1990; Pacifici and Nottoli, 1995; Audus, 1999; Agatonovic-Kustrin et al., 2002; Ito and Lee, 2003; Berlin and Briggs, 2005). Using these properties, a good estimate of the milk-to-plasma ratio can be made, but for placental transfer, these parameters only allow for a prediction of whether a drug will readily cross the placenta. For a more quantitative estimate of placental transfer, additional parameters are needed. Assuming diffusion across

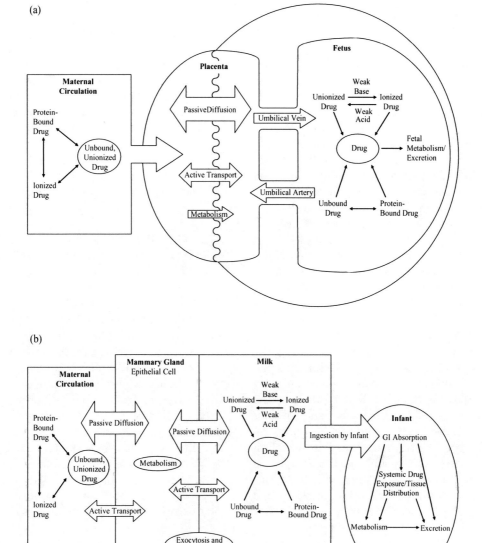

FIGURE 17.1. Schematic representation of (A) placental transfer and (B) lacteal excretion showing the variables that can influence exposure to the fetus or infant. (Placental transfer figure adapted from Syme et al., 2004.)

TABLE 17.1. Physicochemical Properties of Drugs That Affect Placental Transfer and Excretion into Milk

	Effect on Placental Transfer	Lacteal Secretion
Lipophilicity	Increased lipophilicity increases transfer	Increased lipophilicity increases transfer
Molecular weight	Molecular weight >500 Da leads to incomplete transfer; >1000 Da leads to very poor transfer	High transfer at <300–500 Da; minimal transfer >800 Da
Ionization	Low pK_a (<4) for weak acids and high pK_a (>6) for weak bases usually limit transfer	Weak bases, particularly with pK_a >8 have enhanced transfer due to ion trapping (more ionized in milk); weak acids transfer poorly (more ionized in plasma)
Maternal plasma protein binding	Not a good predictor of transfer	High protein binding limits transfer

the placenta is not slower than clearance, the predicted fetal/maternal circulating drug concentration ratio (F/M_{pred}) resulting from passive diffusion can be calculated using the following equation (Garland, 1998):

$$F/M_{pred} = ((f_{pm} \times f_{pm,un\text{-}ionized})/(f_{pf} \times f_{pf,un\text{-}ionized})) \times (CL_{MF}/(CL_{FM} + CL_F)), \quad (17.1)$$

where f_{pm} and f_{pf} are unbound fractions in maternal and fetal plasma, $f_{pm,un\text{-}ionized}$ and $f_{pf,un\text{-}ionized}$ are un-ionized fractions in maternal and fetal plasma, CL_{MF} is clearance from maternal to fetal circulation, CL_{FM} is clearance from fetal to maternal circulation, and CL_F is clearance from the fetal circulation. These clearance parameters can be difficult to obtain and are often unavailable, so the frequency with which this calculation can be used is minimal, but the equation does show how each parameter can affect placental transfer. It is important to note that pH of both maternal and fetal plasma play a role in determining placental transfer. The un-ionized fractions are calculated using the pH of fetal and maternal plasma (obtained from literature values or measured experimentally) by the following equations from Fleishaker and McNamara (1988):

$$f_{un\text{-}ionized}(\text{weak base}) = 1/(1 + 10^{(pK_a\text{-}pH)}) \quad \text{or}$$
$$f_{un\text{-}ionized}(\text{weak acid}) = 1/(1 + 10^{(pH\text{-}pK_a)}) \quad (17.2)$$

Since fetal plasma is slightly more acidic than maternal plasma (approximately 0.1 pH units), weak bases such as chloroquine (CQ), morphine, and local anesthetics may be more ionized in fetal circulation than in maternal circulation. This can lead to more extensive placental transfer of these drugs to the fetus because the un-ionized form of the drug equilibrates across the placenta with a certain percentage of the drug becoming ionized in fetal blood, thus allowing additional un-ionized drug to equilibrate across the placenta. Under normal conditions, this ion trapping is minimal due to the small difference between fetal and maternal plasma

pH. However, fetal plasma pH decreases when the fetus is stressed, as during labor and delivery, which has been linked to increases in fetal lidocaine concentrations (Syme et al., 2004).

An equation similar to Equation 17.1 can be used to predict the milk-to-plasma ratio (M/P_{pred}) (Fleishaker et al., 1987):

$$M/P_{pred} = (f_p * f_{p,un\text{-}ionized})/((f_m * f_{m,un\text{-}ionized}) \times S/M), \quad (17.3)$$

where f_m and f_p are unbound fractions in milk and maternal plasma, $f_{m,un\text{-}ionized}$ and $f_{p,un\text{-}ionized}$ are un-ionized fractions in milk and plasma, and S/M is the skim-to-whole milk concentration ratio. Milk is generally slightly acidic (6.6–7.2) compared with maternal plasma (7.4) so like placental transfer, the ionization characteristics of a drug can play a major role in its secretion into milk. Therefore, acidic compounds (penicillins, nonsteroidal anti-inflammatory drugs [NSAIDs]) are less likely to preferentially partition into milk (more ionized in plasma), while the concentrations of basic compounds (e.g., beta-blockers) are more likely to be higher in milk compared with plasma due to ion trapping. Since protein binding and fat content affect drug partitioning into milk, it is also important to understand the differences in milk composition between species. The physicochemical properties of a compound combined with differences between human milk and plasma with regard to lipid and protein content can result in differences between plasma and milk concentrations.

17.2.2 Drug Transporters

Transfer across the placental and secretion into milk can be affected by whether or not the drug is a substrate for one or more of the drug transporters expressed in the placenta or mammary tissue (Table 17.2) (for reviews, see Ito and Alcorn, 2003; Leazer and Klaassen, 2003; Syme et al., 2004; Ganapathy and Prasad, 2005; Prouillac and Lecoeur, 2010). Most drug transporters identified in

TABLE 17.2. Drug Transporters Expressed in Human Placenta and Mammary Tissue

Transporter	Placental Orientation	Mammary
P-glycoprotein (Pgp)	Maternal-facing	Present at low levels
Breast cancer resistance protein (BCRP)	Maternal-facing	Present
Multidrug resistance proteins (MRPs)	Maternal-facing (MRP1, 2, 3, 8)	Present (MRP1, 2, 5)
Carnitine transporter and organic cation transporter	Maternal-facing (OCTN1, 2)	Present (OCTN1, OCTN2)
Serotonin transporter (SERT)	Maternal-facing	Present
Norepinephrine transporter (NET)	Maternal-facing	Present
Concentrated nucleoside transporters (CNTs)	—[a]	Present (CNT1, CNT3)
Organic cation transporters (OCTs)	Fetal-facing (OCT3)	Present (OCT1, OCT3)
Reduced folate transporter (RFT-1)	Fetal-facing	—
Monocarboxylate transporters (MCTs)	Orientation unknown	—
Organic anion transporters	Fetal-facing (OAT4)	—
Organic anion transporting polypeptides (OATPs)	Maternal-facing (OATP-A, C, D, E) Fetal-facing (OATP-B)	Present (OATP-A, B, D, E)

[a] Indicates transporter has not been identified in this tissue.

human placenta are located on the brush-border membrane (maternal-facing) with the function of minimizing exposure of xenobiotics to the fetus. Two such transporters are P-glycoprotein (Pgp) and breast cancer resistance protein (BCRP), both of which transport a wide variety of drugs. Therefore, drugs that are strong substrates for these transporters, such as many antivirals and chemotherapeutic agents, generally do not cross the placenta to a significant extent (Young et al., 2003). Conversely, drugs that are substrates for basal membrane (fetal-facing) transporters may cross the placenta to a greater extent than predicted based on physico-chemical properties. Methotrexate, a substrate for the fetal-facing reduced folate transporter, is a commonly used drug that is a good example of such transport; in which fetal exposure is higher than corresponding maternal exposure (Ganapathy and Prasad, 2005). Pinocytosis and phagocytosis can also occur at the placenta, but the rate at which they occur is considered too slow to significantly affect the transfer of drugs.

For partitioning of drugs into milk and ingestion by the suckling neonate, predictions provide a reasonable estimate of the lacteal transfer for most drugs since passive diffusion is the primary mechanism of partitioning from plasma to milk across the mammary epithelium. The main reason for inaccurate predictions is when active transport mechanisms are involved. Fewer drug transporters are known to be expressed in mammary tissue as compared with the placenta, and their involvement generally results in the accumulation of drug in the milk leading to higher-than-predicted milk/plasma ratios. Examples include cimetidine, nitrofurantoin, and probenecid, for which the observed accumulation into the milk of lactating rats was 5- to 100-fold higher than predicted. These drugs may be substrates for OCT1 and OCT3, two organic cation transporters shown to be present in the mammary gland at relatively

high amounts (Gerk et al., 2001). Accumulation of cimetidine and nitrofurantoin may also be mediated by BCRP, present in mammary tissue and responsible for the lacteal excretion of topotecan and acyclovir (Jonker et al., 2005; Merino et al., 2005). Although probenecid is also a substrate for organic anion transporters (OATs), these transporters are not present in mammary tissue.

While the above focuses on the potential for small molecules to cross the placenta or be secreted into milk, it does not address the ever increasing area of biopharmaceuticals. Several monoclonal antibodies have been approved for human use and numerous others are currently being developed. Most of these antibodies are derived from human immunoglobulin G (IgG). Since native maternal IgG is actively transported across the placenta (Pitcher-Wilmott et al., 1980) and secreted into milk in humans (McClelland et al., 1978), it is reasonable to anticipate that the same will be true for the majority of these monoclonal IgG antibodies. In fact, the monoclonal IgG antibodies herceptin (trastuzumab) and rituxan (rituximab) cross the placenta in monkeys and are detectable in milk of lactating monkeys. There is minimal information regarding the placental transfer or milk secretion of other biopharmaceuticals (e.g., proteins), but the typically high molecular weight (>1000 Da) of these drugs would generally preclude extensive passive diffusion or efficient transport by drug transporters such as Pgp or BCRP. Since these products have significant structural diversity, a universal set of characteristics is yet to be identified for predicting placental transfer or milk secretion of these compounds.

17.2.3 Metabolism

Metabolism of drugs by the placenta and mammary gland may also affect transfer into the fetus or newborn through placental and/or lacteal transfer. The mRNA and/or protein from multiple CYP isozymes (1A1, 1B1, 2C, 2E1, 3A, and 4B1), various UGTs (2B4, 2B7, 2B10, 2B11, 2B15, and 2B17), sulfotransferases, epoxide hydrolases, N-acetyltransferases (mammary gland only), and glutathione transferases have been detected in the placenta (Syme et al., 2004) and mammary gland (Williams and Phillips, 2000). In general, the amount of these enzymes present in placenta and mammary tissue are low as compared with the maternal and fetal liver and likely make only a minor contribution to the amount of drug that crosses the placenta or is secreted in milk.

17.3 STUDY DESIGN

17.3.1 Placental Transfer Studies

17.3.1.1 Overview
The general study design for determining the extent to which a compound crosses the placenta is to administer the test compound to pregnant animals, then measure drug concentrations in maternal blood, fetal blood, whole fetuses, amniotic fluid, and/or placental tissue at various time points postdose. Samples are usually collected during organogenesis and/or post-organogenesis, near full term (Table 17.3). Samples collected during organogenesis can often help explain any teratogenecity observed during the reproductive/developmental toxicology studies, while those collected postorganogenesis generally have an increased likelihood of showing placental transfer since placental blood flow and surface area increase, while placental thickness decreases, as pregnancy progresses (Kaufmann and Scheffen, 1998). Additionally, for studies performed in rodents, samples collected postorganogenesis are near full term and maximize the sample volume that can be obtained as fetus size increases. This study design specifically addresses placental transfer and does not address possible exposure of the embryo to drug prior to the placenta being formed. The specific design of the study is dependent on the purpose and scope of the study and the available analytical methods. Additionally, it is important to understand the potential regulatory implications of these studies and of lacteal excretion

studies, as the results may impact product labeling (Table 17.4).

Placental transfer studies should be conducted in one of the species used in the Segment II and Segment III developmental toxicity studies using the same route of administration, so the results from the two studies can be linked and appropriate conclusions made. If developmental toxicity occurs in one species, it is more useful to perform the placental transfer study in the species in which toxicity was observed. The dose(s) used should be similar to those in the developmental toxicity studies, preferably at levels that were not maternally toxic, but resulted in developmental toxicity, if applicable. For these placental transfer studies, rat and rabbit are the most commonly used species, but others, such as mouse or monkey, may also be used. While animal models of placental transfer are widely used to predict potential human exposure, there may be instances where in vitro or clinical studies may be more appropriate as some interspecies differences in placental transfer of drugs have been reported (Syme et al., 2004).

17.3.1.2 In vitro Models
Some in vitro systems have been used to predict human placental transfer, with the best characterized and physiologically most relevant being the placenta perfusion method, first described by Panigel et al. (1967). In this method, a single intact cotyledon from a human placenta collected following a full-term pregnancy is perfused with buffer or plasma in both the maternal and fetal circulation. Drug, as well as a freely diffusible marker such as antipyrine, is added to either the maternal or fetal perfusate and concentrations in both perfusates are measured at various times to determine the extent of placental transfer. System integrity must be monitored throughout the experiment. Studies can be performed using a closed (recirculating) or open (single-pass or nonrecirculating) system. The closed system is useful for measuring the initial, most rapid rate of drug transfer, as well as any metabolite formation. The open system is used to measure steady-state conditions following equilibration, which can then be used to determine clearance parameters. The perfusion method allows for placental transfer of a compound to be measured in vitro using intact human

TABLE 17.3. Average Length of Gestation in Days and Time of Organogenesis (Including Typical Gestation Day on Which Single-Dose Placental Transfer Studies Are Performed, Where Applicable)

	Rat	Rabbit	Mouse	Dog[a]	Monkey[b]	Human
Length of gestation	22	33	19	63	164	267
Organogenesis	6–15 (12)	6–18 (13)	6–15 (12)	14–30	20–45	21–56

[a] Rarely used for developmental toxicity testing.
[b] Cynomolgus monkey.

TABLE 17.4. Examples of Product Label Wording for Compounds That Cross the Placenta and/or Are Excreted into Milk

Drug/class	Product Label
Propranolol/beta-blocker	**Distribution:** Propranolol crosses the blood–brain barrier and the placenta, and is distributed into breast milk. **Nursing Mothers:** Propranolol is excreted in human milk. Caution should be exercised when administered to a nursing woman.
Morphine/opiate analgesic	**Labor and Delivery:** Intravenous morphine readily passes into the fetal circulation and may result in respiratory depression in the neonate. **Nursing Mothers:** Morphine is excreted in maternal milk in amounts that may cause sedation of a nursing infant. Use in nursing mothers should be individualized based on the clinical situation.
Acyclovir/antiviral	**Nursing Mothers:** Acyclovir concentrations have been documented in breast milk in two women following oral administration of acyclovir and ranged from 0.6 to 4.1 times corresponding plasma levels. These concentrations would potentially expose the nursing infant to a dose of acyclovir up to 0.3 mg/kg/day. Acyclovir should be administered to a nursing mother with caution and only when indicated.
Metformin/anti-hyperglycemic	**Teratogenic Effects: Pregnancy Category B:** Metformin was not teratogenic in rats and rabbits at doses up to 600 mg/kg/day. Determination of fetal concentrations demonstrated a partial placental barrier to metformin. **Nursing Mothers:** Studies in lactating rats show that metformin is excreted into milk and reaches levels comparable with those in plasma. Similar studies have not been conducted in nursing mothers. Because the potential for hypoglycemia in nursing infants may exist, a decision should be made whether to discontinue nursing or discontinue the drug, taking into account the importance of the drug to the mother.
Fluoxetine/antidepressant	**Labor and Delivery:** The effect of fluoxetine on labor and delivery in humans is unknown. However, because fluoxetine crosses the placenta and because of the possibility that fluoxetine may have adverse effects on the newborn, fluoxetine should be used during labor and delivery only if the potential benefit justifies the potential risk to the fetus. **Nursing Mothers:** Because fluoxetine is excreted in human milk, nursing while on fluoxetine is not recommended. In one breast milk sample, the concentration of fluoxetine plus norfluoxetine was 70.4 ng/mL. The concentration in the mother's plasma was 295.0 ng/mL. No adverse effects on the infant were reported. In another case, an infant nursed by a mother on fluoxetine developed crying, sleep disturbance, vomiting, and watery stools. The infant's plasma drug levels were 340 ng/mL of fluoxetine and 208 ng/mL of norfluoxetine on the second day of feeding.
Gefitinib/anticancer	**Teratogenic Effects: Pregnancy Category D**: A single-dose study in rats showed that gefitinib crosses the placenta after an oral dose of 5 mg/kg (30 mg/m², about one-fifth the recommended human dose on a mg/m² basis). When pregnant rats were treated with 5 mg/kg from the beginning of organogenesis to the end of weaning gave birth, there was a reduction in the number of offspring born alive. In rabbits, a dose of 20 mg/kg/day (240 mg/m², about twice the recommended dose in humans on a mg/m² basis) caused reduced fetal weight. There are no adequate and well-controlled studies in pregnant women using gefitinib. If gefitinib is used during pregnancy or if the patient becomes pregnant while receiving this drug, she should be apprised of the potential hazard to the fetus or potential risk for loss of the pregnancy. **Nursing mothers:** Following oral administration of carbon-14-labeled gefitinib to rats 14 days postpartum, concentrations of radioactivity in milk were higher than in blood. Levels of gefitinib and its metabolites were 11- to 19-fold higher in milk than in blood after oral exposure of lactating rats to a dose of 5 mg/kg. Because many drugs are excreted in human milk and because of the potential for serious adverse reactions in nursing infants, women should be advised against breastfeeding while receiving gefitinib.

placental tissue. Overall, this approach has been predictive of whether or not a compound will cross the placenta, but it cannot predict what will happen after the compound enters the fetal circulation. For example, if a compound is rapidly metabolized by the fetus, drug concentrations in fetal blood may be negligible even if the compound easily crosses the placenta. Conversely, if the drug accumulates in the fetus due to a mechanism other than active transport, such as the high fetal plasma protein binding of valproate, the method may not be predictive. For these reasons, the placental perfusion method has not been accepted as an alternative to performing *in vivo* assessments of placental transfer, but can be useful for carrying out mechanistic studies such transporter interaction studies.

The additional *in vitro* models for placental transfer consist of tissue fractions isolated from human placenta, plasma membrane vesicles prepared from human placenta, and cultured human placenta cells (see Rama Sastry, 1999, for review). Each of these models can be useful for measuring certain aspects of placental transfer but are not considered adequate to consistently predict *in vivo* transfer of drugs across the placenta.

17.3.1.3 In vivo *Models*
17.3.1.3.1 Animal Studies From an analytical perspective, placental transfer studies can be performed using either radiolabeled or unlabeled test material. There are many factors to consider prior to determining which approach is most appropriate for a given compound, first of which is the availability of the radiolabeled material, which may be difficult and/or expensive to synthesize and formulate. Additionally, the extent to which the compound is metabolized and the presence of any pharmacologically active metabolites may influence how the study is designed and if radiolabeled material should be used.

If radiolabeled material is not being used, it is common practice to include a measurement of placental transfer in the Segment II reproductive toxicity studies, where fetal blood is collected from some litters postorganogenesis, at the time of maternal sacrifice, so that drug concentrations can be determined. This approach eliminates the need to perform a separate placental transfer study, allows for fetal exposure to be determined following repeat dosing to the mother, and provides a direct comparison of fetal exposure to any possible toxicity observed in the study. However, this study design measures fetal blood exposures typically at one or two time points following the final dose, and usually measures only parent compound and occasionally any metabolites for which a validated bioanalytical assay is available. Therefore, compounds that cross the placenta and are rapidly excreted or metabolized by the

fetus may not be detected using this approach. Additionally, these are drug safety studies typically performed under good laboratory practices (GLP) conditions, which may limit any exploratory investigations measuring drug and/or metabolite concentrations in blood or tissue. This study design is best suited for compounds where placental transfer is expected, either based on embryo-fetal developmental toxicity data or data from a similar class of compounds, and the purpose is to confirm the presence of drug in fetal circulation.

A typical design for this type of study in monkeys would be daily administration of test article to animals throughout gestation, with the final dose occurring at approximately gestation day (GD) 100. Maternal toxicokinetic samples are collected from study animals prior to GD 100 to determine the full maternal exposure profile. All study animals are sacrificed at 4 h following the final dose with maternal and fetal (from the umbilical cord) blood samples collected at this time. Drug concentrations in maternal and fetal blood, serum, or plasma are determined using a liquid chromatography-mass spectrometry (LC-MS) assay and maternal-to-fetal drug concentration ratios determined. Since fetal blood is collected only at a single time point using this design, it is important to have an understanding of the test article pharmacokinetics to maximize the likelihood of observing drug in the fetal circulation. Additionally, this study design only allows for fetal sample collection postorganogenesis; if fetal exposure must be measured during organogenesis, an additional set of animals should be added.

Placental transfer studies using radiolabeled test compound (^{14}C or ^3H) are usually performed with a single dose administered to the mother, followed by collection of maternal blood, amniotic fluid, placentas, and/or fetuses at multiple time points postdose. The benefits of this study design are that multiple tissues can be collected, all drug-related material can be measured by determining total radioactivity concentrations, and since it is a stand-alone study, it can be designed to address any compound-specific concerns (i.e., transfer across the placenta at different gestation days or drug concentrations in a particular fetal tissue). If necessary, high-performance liquid chromatography (HPLC) with radioflow detection can be used to measure concentrations of parent compound as well as individual metabolites in the various samples collected. The disadvantages to using radiolabeled material are that it requires an additional study using more animals and is usually performed following a single dose rather than repeat dosing. This study design is best suited for compounds where placental transfer is unknown, for highly metabolized compounds, and/or when the purpose of the study is more exploratory in nature.

A typical design for this type of study includes a single administration of radiolabeled material (200 μCi/kg for ^{14}C material, 1 mCi/kg for ^{3}H material) to pregnant rats on GD 12 (during organogenesis) and GD 18 (postorganogenesis, near full term). At various time points up to 24 h postdose, animals are sacrificed and maternal blood, embryos/fetuses, amniotic fluid, and/or placentas are collected so total radioactivity concentrations can be determined, with embryos/fetuses being pooled by dam prior to processing. If fetal blood is required, it can be collected by decapitation of the fetus. Radioactivity in plasma and amniotic fluid can be determined directly by liquid scintillation counting. For placentas and embryos/fetuses, the tissues must be homogenized and oxidized prior to liquid scintillation counting. Additional analysis using HPLC with radioflow detection and LC-MS can be performed if it is necessary to measure parent drug and/or metabolites rather than total radioactivity concentrations. If exposure to a particular fetal tissue is a concern, that tissue can be dissected and analyzed or quantitative whole-body autoradiography can be performed on the pregnant animals with the fetuses still *in utero*. Data from these studies are usually expressed as drug or radioactivity concentrations in each tissue collected, and are also expressed as a ratio relative to maternal blood concentrations. Pharmacokinetic parameters such as AUC, C_{max}, and T_{max} may also be calculated. Due to potential species differences in placental transfer, the data generated are generally treated as qualitative rather than quantitative.

17.3.1.3.2 Human Studies

The placental transfer of drugs can be evaluated in humans, with the most popular approach being collection of maternal blood (from a peripheral vein) and fetal blood (from the umbilical cord) at the time of placental expulsion. This allows for a ratio of the drug concentrations in maternal and fetal circulation to be determined. Fetal elimination and metabolism can also be estimated if drug concentrations are measured in both umbilical venous and arterial blood. Additional measurements of placental transfer in humans can be obtained by collecting amniotic fluid and/or a sample placental tissue. The limitations of this study design are that samples can be collected only at a single time point and the timing of the collection can not be predetermined as it is dependent on when delivery and placental expulsion occur.

Collection of umbilical blood while the fetus is *in utero* can also be performed allowing for a more thorough investigation of placental transfer. Using this approach, multiple samples can be collected at any time between midgestation through full term, with collection times being based on drug pharmacokinetics rather than being restricted to the time of delivery. However, this approach is used rather sparingly as it is a rather invasive procedure.

17.3.1.4 Sample Data

A thorough nonclinical study to evaluate the placental transfer of fluoxetine in sheep was performed by Kim et al. (2004). Using catheterized sheep, it was possible to administer fluoxetine to either the ewe or the fetus and collect multiple maternal and fetal blood samples. Maternal and fetal plasma concentrations of fluoxetine, norfluoxetine, their individual enantiomers, and their glucuronide conjugates were then determined, as was *in vitro* and *ex vivo* protein binding of fluoxetine and norfluoxetine in maternal and fetal plasma. This approach allowed for the evaluation of stereoselective placental transfer of parent compound and a pharmacologically active metabolite; placental transfer of glucuronide conjugates; and fetal metabolism and clearance of parent compound.

Following administration to the ewe, there was rapid placental transfer of fluoxetine and norfluoxetine, with maternal-to-fetal plasma AUC ratios of approximately 0.6 for both compounds. These data are consistent with the physicochemical properties of fluoxetine (lipophilic with a log D of 2.08 at pH 7.4 and weakly basic with a pK_a of 8.7) and the higher protein binding in maternal plasma (2.2–5.7% free fraction) as compared with fetal plasma (5.8–10.5%). Stereoselective differences in exposure were observed, with the R/S enantiomer ratios in both maternal and fetal plasma approximately 1.7; this suggests that both enantiomers crossed the placenta to a similar extent. Using data generated following administration of fluoxetine to the fetus, there was minimal fetal clearance of fluoxetine and no fetal glucuronidation or metabolism to norfluoxetine. By generating such extensive data, it was possible to investigate how plasma protein binding, *in vitro* and *in vivo* metabolism, and stereoselectivity could affect placental transfer in sheep, and assess the potential impact of these parameters on placental transfer in humans.

A clinical placental transfer study was performed by Gerdin et al. (1990) to investigate the transfer of morphine and its glucuronide metabolite in human. Morphine was administered intravenously to pregnant women and maternal and fetal blood samples were collected. Morphine rapidly crossed the placenta, with a fetal-to-maternal plasma ratio of 0.96 at 5 min after dosing, and the ratio remained close to 1 in most of the subsequent samples. These data are not surprising since morphine readily crosses most biological membranes. The concentration of the glucuronide metabolite in fetal blood was less than 0.002 the concentration in maternal circulation at 12 min; and the ratio generally remained below 0.6, indicating significantly less placental transfer for the metabolite and/or minimal fetal metabolism. These data clearly demonstrated placental transfer of morphine, as well as fetal exposure, and are reflected in the product label (Table 17.4).

17.3.1.5 Summary Using only the physicochemical properties of a compound, it is difficult to accurately predict the extent of placental transfer. Therefore, it is standard practice to perform an *in vivo* assessment of placental transfer, typically with one of the animal species used in the Segment II drug safety studies. When performed correctly, these studies can help explain results from embryo-fetal developmental toxicity testing and are routinely used to support drug registration without the need to perform additional work. If desired, further mechanistic investigations (e.g., transporter interaction) can be performed using *in vitro* or *in vivo* systems, or clinical studies can be performed to directly measure placental transfer in humans.

17.3.2 Lacteal Excretion Studies

17.3.2.1 Overview Lactation studies have been performed in a wide range of mammalian species such as rats, mice, rabbits, dogs, monkeys, goats, and cows, as well as in the clinical setting. The most common species used for absorption, distribution, metabolism, and excretion (ADME) purposes are rats due to their extensive use in toxicological and metabolic evaluations during drug development. Rats have short gestation periods along with relatively short maturation periods allowing for reproductive toxicity studies to be conducted within a reasonable time period for drug safety evaluation. Rabbits have also been used extensively given their routine selection in reproductive toxicological studies and their larger size compared with rats that allows for the collection of larger sample volumes.

There are several factors to consider when choosing a study design for a lactation study (species, time points, dose, etc.). It is important to understand the differences in milk composition between species, which can influence drug partitioning and subsequent concentration in

milk. Mouse, rat, and rabbit milk contain higher relative amounts of protein and lipids compared with humans, while human milk has a higher percentage of carbohydrates (Table 17.5). Since the lipophilicity and protein binding properties of drugs can play a major role in their propensity to be secreted into milk, these species differences should be accounted for. In addition, the composition of milk can change over the course of lactation. In humans, high protein levels are observed in colostrum (secreted shortly after birth) until lactation day (LD) 5 after which protein concentrations decrease and fat content increases until stabilizing by LD 15. Even during a single feeding, the milk composition in humans has been shown to change, with fat content and pH typically increasing throughout a feeding. In rats, milk protein levels remain fairly stable during lactation, although timing of sample collection can be an important factor since rats eat primarily at night, which can impact milk composition. The effect of lactation on drug metabolism, pharmacokinetics, and clearance of the test compounds should also be considered given the differential gene expression and hepatic and intestinal protein levels in pregnant and nonpregnant rodents and human (reviews by Ito and Lee, 2003; McNamara and Abbassi, 2004).

17.3.2.2 In vitro Models It is often necessary and beneficial to estimate the potential for drugs to partition into breast milk. Given the caveats described in Section 17.2.1, the predicted milk-to-blood (serum or plasma) concentration ratio (M/P_{pred}) can be calculated for many drugs based on physicochemical properties, as well as data derived from relatively simple *in vitro* techniques. Equilibrium dialysis as described by Notarianni et al. (1995) involves using baby formula, which has similar composition to human milk (protein, fat, and pH). Studies can also be performed in formula with the addition of bovine serum albumin (BSA) and/or triglycerides. The most common methods for predicting the milk-to-plasma ratios in animals and humans involve variations of the diffusional methods reviewed by Fleishaker (2003), which utilize plasma protein binding, pH measurements in plasma and milk, unbound

TABLE 17.5. Composition of Milk from Common Mammalian Species Used in Drug Development (% or g/100 mL or g/100 g)

	Rat	Rabbit	Mouse	Dog	Monkey	Cow	Human
Fat (lipids)	10.9–17.5	10.5–18.3	13.1	8.0–13.7	4.6–5.4	3.6–3.7	3.2–4.6
Protein	8.0–12.5	10.4–15.5	9.0	5.2–10.2	1.5–2.5	3.4	1.0–1.8
Casein	8.5–12.0	6.2	7.0	6.6–7.5	1.1	2.8	0.2–0.4
Carbohydrates (lactose)	1.2–3.7	1.8–2.1	3.0	2.6–4.5	7.8–8.1	4.6–4.8	6.9–7.1
pH	6.6–6.7	7.36	nd	nd	7.0–7.2	6.4–6.8	7.0–7.1

nd, no data.

fractions in skim milk and plasma and the skim-to-whole milk ratio. One disadvantage is the need to collect milk from animals or humans, although it does not require dosing of drug (and unnecessary exposure) or extensive sampling regimens.

Certain parameters are needed for estimating milk-to-plasma ratios, such as the pH of blood and milk, protein concentration in blood and milk, protein binding of drug in blood and milk, milk fat content, and skim-to-whole milk drug concentration ratio. Total milk and plasma protein concentrations can be determined by commercially available kits. It is preferable to measure pH experimentally, particularly for human samples (e.g., anaerobically using a clinical blood analyzer), but estimations based from literature values can be used. The % fat in each sample may be determined as creamatocrit, which is well correlated to lipid concentration. In a common method described by Lucas et al. (1978), milk samples are placed in nonheparinized hematocrit tubes (i.e., 75×1.5 mm) and centrifuged at $12{,}000\,g$ for 15 min. The tubes are immediately removed from the centrifuge and placed vertically. The % of cream layer is measured by comparing the length of the cream layer in the column compared with the whole milk layer and represents the % creamatocrit. Since primarily only unbound drug is secreted into milk protein, data on the unbound fraction of drug in plasma and milk may be needed. For protein binding determinations, the milk sample should be centrifuged for 10–20 min at a relatively low speed (e.g., $5000\,g$) to separate the fat from the skim milk layer (bottom) layer. Equilibrium dialysis can be conducted in a similar manner as would be done for compound binding in plasma protein, except that the phosphate buffer pH should match that of the milk determined as described above.

In cases where the involvement of active transporting mechanisms are suspected (i.e., drug has higher milk/plasma ratios than predicted), *in vitro* methods utilizing mammary epithelial cell lines that synthesize casein in response to prolactin can be employed to investigate whether a drug is transported from plasma to milk by active transport. To assess potential drug transport across the human mammary gland, cultured trypsin-resistant human mammary epithelial cell (HMEC) monolayers are utilized, which also produce β-casein, form a tight monolayer system useful for assessing passive and carrier-mediated transport, and express hOCT1 and hOCT3, two drug transporters found in human mammary tissue (Kimura et al., 2006). In this technique, normal HMECs are grown in human mammary epithelial growth medium supplemented with insulin, epidermal growth factor, hydrocortisone, amphotericin B, gentamicin, and bovine pituitary extract. The epidermal growth factor is then replaced with prolactin to differentiate the cells

after which the cells are grown on collagen-coated inserts with a pore size of 3 μm using a transwell system, seeded at half-density in an incubator at 37°C with a 5% CO_2 atmosphere. When the cells are 80% confluent (approximately 7 days), 0.025% trypsin/0.1% EDTA is added and detached cells are removed by centrifugation. Attached cells are allowed to grow back to 80% confluence and the trypsin treatment is performed a total of three times. The cells (inserts) are then transferred to plates containing secretion medium for 7 more days at which point approximately 50% of the cells are responsive to prolactin and the transepithelial resistance is high (800–1500 ohm*cm^2). Incubations are performed with test drug placed on the apical or basolateral side of the well with gentle shaking at 37°C. Samples are taken at various time points from the side of the well that did not have drug to assess the amount of drug that has crossed from the donor side to the receiver side. It is recommended to replace the volume removed with control media. As with other systems involving cell monolayers, mannitol can serve as a typical control for integrity of the monolayer accounting for transcellular permeation. It is also recommended to measure the concentrations of drug in the donor and receiver sides to account for any metabolism of the test compound as well as binding to the culture plate or to cells.

17.3.2.3 In vivo *Models*
17.3.2.3.1 Animal Studies In rodents, a typical evaluation for assessment of lacteal transfer involves (1) measurement of drug concentrations in milk of the lactating dams, and (2) comparison of drug exposure in both the dam and pups. The first arm is important since the concentration of drug in the milk, expressed as C_{max} and AUC_{0-t}, has more clinical relevance than milk/plasma ratio. The second arm assesses the drug bioavailability to the infant. This is critical given that just because a drug is in milk at high levels, it does not necessary mean exposure to the infant will be clinically significant, since infant exposure is dependent on absorption, metabolism, distribution, and clearance of the drug in the infant after ingestion.

To determine the drug concentrations in milk and evaluate the profile of drug-derived material, pregnant females are allowed to deliver their litters naturally with the day of birth being designated as postnatal day (PD) or LD 0. After a few days, the litters can be culled to a designated number of pups per litter (normally 6–10). Preferences can be made to select for one gender of offspring, if deemed necessary. Alternatively, dams with designated litter sizes (and gender) can be acquired commercially. Dams are housed with pups until approximately LD 10–14, at which time the dams are administered radiolabeled (^{14}C) drug by the designated route of administration and placed back in a cage without the pups. A typical dose will contain 100–400 µCi/kg (~25–150 µCi/rat). Pups are removed from the cage 1–4 h prior to milk collection as longer periods of separation can affect the milk composition and possibly impact partitioning of the test drug. The dams are then given a single intraperitoneal dose of oxytocin (1–5 IU/kg) approximately 30–60 min prior to milk collection, which allows for a maximal amount of milk to be collected for analyses (e.g., 1 mL/rat over a 10-min collection period). If mice are utilized, the milk volumes will be 5- to 10-fold lower. The dams are anesthetized (e.g., ketamine, xylazine, acepromazine, or a mixture of these) and milk collected at selected time points either manually or using a milking machine. Normally each rat is milked only once with as much collected as possible to allow for assessment of drug and metabolite levels. After milk collection, blood can be collected for drug and metabolite analysis to determine the milk-to-plasma ratio for the analyte(s). The milk can be analyzed directly for radioactivity by liquid scintillation counting. For determination of individual analytes, milk samples normally require extraction (e.g., methanol/acetonitrile) prior to analyses by LC-MS. It is important to warm milk samples to 37°C to liquefy fat in order to have a homogenous sample.

To determine and compare drug exposure in the dam and pups, pregnant females are allowed to deliver their litters. At approximately LD 10, the dams are dosed with radiolabeled drug (typically 100–400 µCi/kg) by the designated route of administration. After dosing, the dams are placed back into the cage with the pups and allowed free access to food and water, while the pups are allowed to suckle normally. At designated time points after dosing, a certain number of pups per dam are sacrificed for sample collection. Pups can be bled terminally to determine circulating concentrations. The carcasses can be utilized to quantitate individual tissue concentrations in the pups by tissue extraction and radioactivity determination or by quantitative whole-body autoradiography. The entire pup (or pool of pups) can also be homogenized for determination of radioactivity and metabolite profiling. Blood samples can be collected from each dam at designated time points. If larger sample volumes are needed, blood can be collected terminally from each dam. At the final time point, dams can be sacrificed for assessment of maternal plasma and tissue concentrations.

Assessment of lacteal transfer can also be performed as part of a Segment III perinatal-postnatal reproductive and developmental toxicology study. In this study design, the F_0 generation female rats are continually dosed with drug starting on GD 6 and spanning through lactation (often to approximately postpartum day 20).

Therefore, the offspring may be exposed to drug *in utero* and postnatally through lactation. Although the primary objective of this study design is to assess the safety of drug during embryonic and fetal development, parturition, and lactation, quantitation of drug levels in the milk and subsequent transfer to pups can provide context for any toxicological finding and help establish safety margins. The disadvantages of such a study design are the complexity of incorporating additional endpoints to this type of toxicological evaluation. Furthermore, pharmacokinetic measurements are unlikely to be in the same animals undergoing toxicological evaluation as additional satellite groups are needed for collection of blood from dams and pups. These studies are performed with unlabeled drug and extensive metabolite profiling and quantitation of individual metabolites is typically not performed. Therefore, this study design is not normally recommended for highly metabolized drugs or those with multiple metabolites, particularly when metabolites retain some degree of pharmacological activity, unless reference standards and an adequate bioanalytical method are available for quantitation. The design is similar to radiolabeled studies described above with blood collected from females at several time points on a single day after the first week of lactation (i.e., postnatal day [PD] 8) to determine the maternal plasma AUC_{0-24}. On a day during the second week of lactation (i.e., PD 13), the pups are removed from the dam's cage, the dams are given a dose of oxytocin, and milk is collected (under anesthesia). The pups can be sampled for blood and/or tissues to determine drug and metabolite concentrations. After milking, blood can also be collected from the dams. Since this is normally performed as a terminal bleed, other sampling can be done as well (tissue harvesting, sectioning, etc.).

Using rabbits in this study design allows for larger sample collections. The offspring can be separated the day before a kinetic study is performed. Using a catheter in the ear vein, several (e.g., 10) blood samples of 1 mL each can be collected from a single rabbit over a 24-h period. Milk can be collected at the same times, manually into a centrifuge tube connected to the vacuum system. At each time point, the selected gland should be emptied completely. In addition to precautions taken when considering drug stability, milk samples should be stored at ≤−20°C until analyses and not refrozen after thawing (i.e., no more than one freeze/thaw cycle) as the integrity of milk, such as lipid content can be affected, confounding data obtained from compromised samples. A previous study indicated that creamatocrit values decreased significantly in samples after 45 min at room temperature and after 30 min at 37°C as well as for samples subjected to two freeze/thaw cycles (Silprasert et al., 1986). For longer-term storage (>1 week), samples should be stored at $\leq-70°C$ as lipolysis may occur at higher temperatures (Lavine and Clark, 1987).

In cases where drug transporter involvement is suspected due to a significantly higher milk/plasma ratio than predicted and/or demonstration of active transport in the HMEC model described above, *in vivo* studies can be performed to assess the contribution of active transport processes in lactation. This design can also be used to assess the contribution of other mechanisms that may affect the lacteal transfer of test drug and/or its metabolites (e.g., enzyme inhibition, induction). Regardless, the study can be carried out as a typical milk excretion study in a similar manner as described above. A separate group of lactating dams is administered test drug and an interacting drug (pretreated, given at the same, etc.). The concentrations of test compound, radioactivity, metabolite, and so on in milk and plasma in the presence and absence of interacting drug are compared.

17.3.2.3.2 Human Studies In many cases, pregnant and breastfeeding women on medications are better served to remain on treatment, while still being advised to breastfeed their infants. Additionally, the breastfeeding infant's only source of nutrition at a time during rapid growth and development may be breast milk. Therefore, establishing the drug pharmacokinetics in milk and quantitating the exposure to the infant can be important in assessing the risk-benefit of breastfeeding while on medication(s). The most common method is to determine the drug concentrations in milk and plasma of the mother, estimate the daily exposure to the infant and when possible, measure circulating levels of the drug in the breastfed infant. It is preferable to perform these determinations in subjects at steady state; however, for drugs that do not accumulate with chronic dosing or are used acutely, a single-dose study may be sufficient. Maternal venous blood samples (5–10 mL) are collected from a cannula at various time points between 0 and 24 h, depending on the plasma/blood pharmacokinetics of the drug. During the same intervals, both breasts are emptied via a manual or electric breast pump. A larger aliquot (15 mL) is taken for assessment of drug levels and a smaller aliquot (1 mL) is taken using a blood-gas syringe for pH measurement, which can be performed using a blood gas analyzer. The % fat (creamatocrit) in each sample should be determined. To quantitate the transfer of drug to the infant, establish the infant exposure in relation to maternal exposure, and estimate drug clearance in the infant, the remaining milk can be bottle-fed to the infant, and venous blood and/or urine sample(s) collected from the infant. Milk/plasma ratios are calculated based on the AUCs in plasma and milk of the mother. Assuming an oral bio-

availability in the infant of 100%, the cumulative drug excretion over 24 h can be divided by the infant's body weight to give a dose in milligrams per kilogram (mg/kg) (Kristensen et al., 1998). Alternatively, assuming an average daily milk intake of 0.15 L/kg/day (Bennett, 1996), the infant exposure can be expressed as a percentage of the maternal weight normalized dose. The dose in human milk can be calculated as the maternal plasma concentration at steady state multiplied by the milk-to-plasma ratio and the infant milk intake (0.15 L/kg/day or value obtained experimentally). One proposed threshold for concern is when the weight-adjusted drug transfer exceeds 10% (Bennett, 1996). When the infant's drug clearance can be calculated or estimated, Ito and Lee (2003) has proposed the use of "exposure index" to link the milk-to-plasma ratio, milk intake, and infant's drug clearance expressed as a percentage equal to as follows: $100 \times M/P \times$ Milk intake (150 mL/kg/day = 0.1 mL/kg/min)/Infant drug clearance (mL/min/kg). Since this type of study is usually performed during the latter stages of drug development, this information can be put into context with existing clinical data to weigh the risks to infant and mother through discontinuing treatment during breastfeeding or restricting breastfeeding during drug administration.

17.3.2.4 Sample Data Excretion of dasatinib into milk was investigated following oral administration of [^{14}C]dasatinib to lactating rats approximately 8 or 9 days postpartum, followed by collection of milk and plasma samples (He et al., 2008). Radioactivity was extensively secreted into milk, with milk-to-plasma ratios at the individual time points ranging from 2.4 to 37.2, and an AUC ratio of nearly 25. Using HPLC with radioactivity flow detection, the majority of the radioactivity was characterized as parent compound. Based on the physicochemical properties of dasatinib (free fraction of 97% in rat plasma; basic compound with a pK_a of 6.8; lipophilic with a log D of 3.1 at pH 7), a milk–to-plasma ratio of approximately 2 was predicted. A subsequent study demonstrated that dasatinib was a substrate for BCRP-mediated transport (Chen et al., 2009), and this likely accounted for the higher-than-expected secretion into milk. Similar data have been reported for other BCRP substrates such as apixaban (Wang et al., 2011).

Secretion of the antimalaria drug CQ, and its pharmacologically active metabolite desethylchloroquine (DECQ), into human breast milk was investigated following three oral doses of CQ (750 mg of CQ phosphate, 465 mg free base) to women on days 1–3 following delivery (Law et al., 2008). CQ is a weak base (pK_a of 10.2) that is not particularly lipophilic (log D of 0.96 at pH 7.4). Concentrations of CQ and DECQ were measured in milk (hind- and fore-milk) and infant exposures estimated based on average milk intake (0.15 L/kg/day). While the median creamatocrit was higher in hind-milk as compared with fore-milk, only relatively small differences in CQ and DECQ concentrations between hind- and fore-milk were observed. Median CQ and DECQ milk concentrations over the 17-day sampling time were 226 and 97 µg/mL, respectively, resulting in a relative infant dose of 2.3% for CQ (34 µg/kg/day) and 1.0% for DECQ (15 µg/kg/day). These predicted doses to the infant were well below the recommended therapeutic pediatric (25 mg/kg over 48 h) and infant (up to 50 mg/kg/day) doses. Therefore, it was concluded that breast-feeding following CQ administration to the mother should have little potential risk for the newborn, but is also unlikely to provide the infant any protection against malaria.

17.3.2.5 Summary Based on the physicochemical properties of a compound and data generated using *in vitro* techniques, it is possible to estimate the milk-to-plasma ratio if passive diffusion is the primary mechanism of secretion. However, since other factors such as drug transporters may be involved in secretion into milk and the acceptance that *in vivo* milk excretion studies in laboratory animals are relatively inexpensive and straightforward, an *in vivo* assessment of milk excretion is typically performed to support drug registration. Occasionally, a clinical study may be warranted to directly measure milk excretion in humans and exposure to the infant; this is more common for drug classes that are typically administered to women postpartum, such as antidepressants, antipsychotics, and antihypertensives, to aid in the risk-benefit assessment of drug treatment to determine whether it is necessary to discontinue drug therapy while breastfeeding.

17.4 CONCLUSIONS

This chapter details the key issues and considerations to properly plan and perform placental transfer and milk excretion studies. It is important to incorporate the available ADME and toxicology information into the study design and to understand the ultimate goal of the study. While *in vitro* studies may be useful for mechanistic studies, *in vivo* animal studies are currently the most accepted approach for assessing placental transfer and milk excretion, and are often required by regulatory agencies as part of the marketing application for many drugs. When performed correctly, these studies can be extremely useful in rationalizing any developmental toxicity observed in the Segment II and/or Segment III toxicity studies and for predicting potential exposure

and risk to the human fetus or breastfeeding neonate. Clinical studies assessing the placental transfer of drugs are usually not a requirement, but can be performed when needed. Clinical milk excretion studies are more common, and are performed to assess potential drug exposure to the breastfeeding infant and to better estimate risk of concomitant drug therapy to the infant.

REFERENCES

Agatonovic-Kustrin S, Ling LH, Tham SY, Alany RG (2002) Molecular descriptors that influence the amount of drugs transfer into human breast milk. *J Pharm Biomed Anal* 29:103–119.

Atkinson HC, Begg EJ (1990) Prediction of drug distribution into human milk from physicochemical characteristics. *Clin Pharmacokinet* 18(2):151–167.

Audus KL (1999) Controlling drug delivery across the placenta. *Eur J Pharm Sci* 8:161–165.

Bennett PN (1996) Use of the monographs on drugs. In *Drugs and Human Lactation*, Bennett PN, ed., pp. 67–74. Elsevier, Amsterdam.

Berlin CM, Briggs GG (2005) Drugs and chemicals in human milk. *Semin Fetal Neonatal Med* 10:149–159.

Chen Y, Agarwal S, Shaik NM, Chen C, Yang Z, Elmquist WF (2009) P-glycoprotein and breast cancer resistance protein influence brain distribution of dasatinib. *J Pharmacol Exp Ther* 330:956–963.

Enders AC, Blankenship TN (1999) Comparative placental structure. *Adv Drug Deliv Rev* 38:3–15.

FDA Guidance for Industry (2005) Clinical lactation studies—Study design, data analysis, and recommendations for labeling.

FDA Guidance for Industry (2011) Reproductive and developmental toxicities integrating results to assess concerns.

Fleishaker JC (2003) Models and methods for predicting drug transfer into human milk. *Adv Drug Deliv Rev* 55:643–652.

Fleishaker JC, McNamara PJ (1988) *In vivo* evaluation in the lactating rabbit of a model for xenobiotic distribution into breast milk. *J Pharmacol Exp Ther* 244(3):919–924.

Fleishaker JC, Desai N, McNamara PJ (1987) Factors affecting the milk-to-plasma drug concentration ratio in lactating women: Physical interactions with protein and fat. *J Pharm Sci* 76:189–193.

Ganapathy V, Prasad PD (2005) Role of transporters in placental transfer of drugs. *Toxicol Appl Pharmacol* 2(Suppl.):381–387.

Garland M (1998) Pharmacology of drug transfer across the placenta. *Obstet Gynecol Clin North Am* 25:21–42.

Gerdin E, Rane A, Lindberg B (1990) Transplacental transfer of morphine in man. *J Perinat Med* 18:305–312.

Gerk PM, Oo CY, Paxton EW, Moscow JA, McNamara PJ (2001) Interactions between cimetidine, nitrofurantoin, and probencid active transport into rat milk. *J Pharmacol Exp Ther* 296:175–180.

He K, Lago MW, Iyer RA, Shyu W-C, Humphreys WG, Christopher LJ (2008) Lacteal secretion, fetal and maternal tissue distribution of dasatinib in rats. *Drug Metab Dispos* 36:2564–2570.

ICH Guidance for Industry (1995). Toxicokinetics: The assessment of systemic exposure in toxicity studies.

Ito S, Alcorn J (2003) Xenobiotic transporter expression and function in the human mammary gland. *Adv Drug Deliv Rev* 55:653–665.

Ito S, Lee A (2003) Drug excretion into breast milk—Overview. *Adv Drug Deliv Rev* 55:617–627.

Jonker JW, Merino G, Musters S, van Herwaarden AE, Bolscher E, Wagenaar E, Mesman E, Dale TC, Schinkel AH (2005) The breast cancer resistance protein BCRP (ABCG2) concentrates drugs and carcinogenic xenotoxins into milk. *Nat Med* 11:127–129.

Kaufmann P, Scheffen I (1998) Placental development. In *Fetal and Neonate Physiology*, Polin RA, Fox WW, eds., pp. 59–70. WB Saunders, Philadelphia.

Kim J, Riggs KW, Rurak DW (2004) Stereoselective pharmacokinetics of fluoxetine and norfluoxetine enantiomers in pregnant sheep. *Drug Metab Dispos* 32:212–221.

Kimura S, Morimoto K, Okamoto H, Ueda H, Kobayashi D, Kobayashi J, Morimoto Y (2006) Development of a human mammary epithelial cell culture model for evaluation of drug transfer into milk. *Arch Pharm Res* 29(5): 424–429.

Kristensen JH, Ilett KF, Dusci LJ, Hackett LP, Yapp P, Wojnar-Horton RE, Roberts MJ, Paech M (1998) Distribution and excretion of sertraline and N-desmethylsertraline in human milk. *Br J Clin Pharmacol* 45:453–457.

Lavine M, Clark RM (1987) Changing patterns of free fatty acids in breast milk during storage. *J Pediatr Gastroenterol Nutr* 6:769–774.

Law I, Ilett KF, Hackett LP, Page-Sharp M, Baiwag F, Gomorrai S, Mueller I, Karunajeewa HA, Davis TME (2008) Transfer of chloroquine and desethylchloroquine across the placenta and into milk in Melanesian mothers. *Br J Clin Pharmacol* 65:674–679.

Leazer TM, Klaassen CD (2003) The presence of xenobiotic transporters in rat placenta. *Drug Metab Dispos* 31:153–167.

Lucas A, Gibbs JAH, Lyster RLJ, Baum JD (1978) Creamatocrit: Simple clinical technique for estimating fat concentration and energy value of human milk. *Br Med J* 1:1018–1020.

McClelland DBL, McGrath J, Samson RR (1978) Antimicrobial factors in human milk: Studies of concentration and transfer to the infant during early stages of lactation. *Acta Paediatr Scand* 67(Suppl. 271):1–20.

McManaman JL, Neville MC (2003) Mammary physiology and milk secretion. *Adv Drug Deliv Rev* 55:629–641.

McNamara PJ, Abbassi M (2004) Neonatal exposure to drugs in breast milk. *Pharm Res* 21:555–566.

Merino G, Jonker JW, Wagenaar E, van Herwaarden AE, Schinkel AH (2005) The breast cancer resistance protein (BCRP/ABCG2) affects pharmacokinetics, hepatobiliary excretion, and milk secretion of the antibiotic nitrofurantoin. *Mol Pharmacol* 67:1758–1764.

Notarianni LJ, Belk D, Aird SA, Bennett PN (1995) An *in vitro* technique for the rapid determination of drug entry into breast milk. *Br J Clin Pharmacol* 40:333–337.

Pacifici GM, Nottoli R (1995) Placental transfer of drugs administered to the mother. *Clin Pharmacokinet* 28:235–269.

Panigel M, Pascaud M, Brun JL (1967) Une nouvelle technique de perfusion de l'espace intervilleux dans le placenta humain isole. *Pathol Biol* 15:821.

Pitcher-Wilmott RW, Hindocha P, Woods CBS (1980) The placental transfer of IgG subclasses in human pregnancy. *Clin Exp Immunol* 41:404–408.

Prouillac C, Lecoeur S (2010) The role of the placenta in fetal exposure to xenobiotics: Importance of membrane transporters and human models for transfer studies. *Drug Metab Dispos* 38:1623–1635.

Rama Sastry BV (1999) Techniques to study human placental transport. *Adv Drug Deliv Rev* 38:17–39.

Reynolds F, Knott C (1989) Pharmacokinetics in pregnancy and placental drug transfer. *Oxf Rev Reprod Biol* 11:389–449.

Silprasert A, DeJsarai W, Keawvichit R, Amatayakul K (1986) Effect of storage on the creamatocrit and total energy content in human milk. *Hum Nutr Clin Nutr* 40C:31–36.

Syme MR, Paxton JW, Keelan JA (2004) Drug transfer and metabolism by the human placenta. *Clin Pharmacokinet* 43:487–514.

Wang L, He K, Maxwell B, Grossman SJ, Tremaine LM, Humphreys WG, Zhang D (2011) Tissue distribution and elimination of [^{14}C]apixaban in rats. *Drug Metab Dispos* 39:256–264.

Williams JA, Phillips DH (2000) Mammary expression of xenobiotic metabolizing enzymes and their potential role in breast cancer. *Cancer Res* 60:4667–4677.

Young AM, Allen CE, Audus KL (2003) Efflux transporters of the human placenta. *Adv Drug Deliv Rev* 55:125–133.

18

HUMAN BILE COLLECTION FOR ADME STUDIES

Suresh K. Balani, Lisa J. Christopher, and Donglu Zhang

18.1 INTRODUCTION

Biliary excretion studies provide invaluable insight into the disposition of xenobiotics. These studies determine the extent of the excretory route of a compound and its metabolites in bile and, together with urinary data, more accurately establish the total metabolism of the compound (Zhang and Comezoglu, 2007). Results from biliary excretion studies help define the need for the conduct of clinical studies in subjects with hepatic insufficiency and also aid in understanding clearance pathways, enterohepatic recirculation, hepatobiliary toxicity, and the magnitude of drug–drug interactions (DDIs). While bile collection is routine in animal absorption, distribution, metabolism, and excretion (ADME) studies, it is rarely collected in human studies. The primary reason for not collecting bile in human ADME studies has been the limited availability of patients with T-tube bile drainage or those undergoing exploration of the biliary tree for the diagnosis of cholecystitis or cholelithiasis (gallstones). However, in recent years, a less invasive, nonsurgical procedure has become available for the collection of bile from healthy volunteers that can be applied to the conduct of routine human ADME studies. The availability of this procedure may facilitate collection of human bile data on a more routine basis. This chapter describes the methodologies used to collect bile and their utility, as well as their current limitations.

18.2 PHYSIOLOGY

Bile is a viscous, slightly alkaline, greenish-yellow liquid produced by hepatocytes in the liver. It contains bile salts, phospholipids, cholesterol, proteins, amino acids, nucleotides, vitamins, bile pigments, water, and various organic anions and inorganic substances. The overall composition of bile varies between species and also within each species based on nutritional status; it can be altered under certain disease states, such as cholelithiasis, Crohn's disease, cirrhosis, and malignancies of the hepatobiliary system. Bile is released from the gallbladder into the intestine in response to food consumption. It acts as an emulsifying agent, thereby facilitating digestion and the absorption of fats and fat-soluble vitamins (Klassen and Watkins, 1984). Another important function of bile is to serve as a carrier for removal of some drugs and toxins and their metabolic products, so that they may be eliminated from the body via excretion into the intestine.

In humans, approximately 1 L of bile per day is produced by the liver. The bile flows from hepatocytes, through bile canaliculi, eventually draining into right and left hepatic ducts, which join to form the common hepatic duct (Figure 18.1). Bile, drained into the common hepatic duct, is delivered via cystic duct to the gallbladder, a 7–10-cm-long sac under the liver, where it is stored and concentrated up to 90% over time. When food enters the duodenum, the hormone cholecystokinin is

ADME-Enabling Technologies in Drug Design and Development, First Edition. Edited by Donglu Zhang and Sekhar Surapaneni.
© 2012 John Wiley & Sons, Inc. Published 2012 by John Wiley & Sons, Inc.

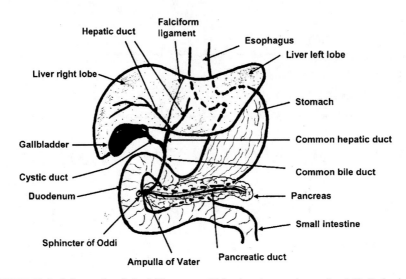

FIGURE 18.1. Schematic of the biliary tree. This sketch was drawn by S.K. Balani, 2011.

released from the mucosal epithelium of duodenum, stimulating gallbladder contraction and ejection of bile as well as secretion of pancreatic juice. The stored bile flows out of the gallbladder into the common bile duct, which joins with the pancreatic duct, and enters the duodenum at the ampulla of Vater. The flow of fluids from the common bile duct into the intestine is regulated by the sphincter of Oddi.

The gallbladder is not necessary for human survival; humans can live without serious effects following its removal. In these individuals, bile flow into the intestine is not efficiently regulated. Bile is produced at a slow rate and secreted directly into the intestine, rather than being concentrated in the gallbladder and released in large quantities in response to food. The gallbladder is absent in some animals, such as rats, pigeons, and horses.

18.3 UTILITY OF THE BILIARY DATA

Biliary excretion data can be utilized in conjunction with urinary and fecal excretion data to achieve a more thorough understanding of the ADME properties of an investigational drug. The knowledge of the extent of human biliary excretion and the amount and types of metabolites present in bile aid in the planning for future clinical studies. For example, if a compound is cleared largely via hepatic pathways, that is, metabolism and/or excretion into bile, clinical studies in subjects with hepatic insufficiency are recommended in the clinic. This information may also facilitate the design of appropriate DDI studies by providing information about which enzyme systems are important to study and the potential/anticipated magnitude of those drug interactions.

While most metabolites can be detected intact in bile, thus providing an accurate accounting of their presence and abundance, the metabolite profiles of fecal samples can be misleading. Fecal profiles must be evaluated carefully due to the potential for nonsystemic metabolism of drug-related components by the gut microflora (Steinmetz and Balko, 1973; Drasar and Hill, 1974). For example, compounds that undergo direct glucuronidation in the liver and secretion in bile would mostly be detected as their corresponding aglycones, and not as glucuronides in the fecal samples, as a result of hydrolase activity of the gut microbes (Parker et al., 1980). Similarly, N-oxides would likely undergo reduction in the gut (Mitchell et al., 1997; Jaworski et al., 1991; Christopher et al., 2008), and would be misinterpreted as unchanged parent compound in the feces. For compounds administered orally, this would confound the interpretation of how much metabolism had occurred, since parent compound detected in feces could be the result of unabsorbed drug or systemic metabolites excreted in bile being converted back to the parent compound via metabolism by gut microflora. Thus, relying on fecal data alone could lead to an underestimation of the extent of metabolism of the compound or the importance of a particular metabolic pathway. Both the extent of human metabolism (i.e., the overall fraction of the drug metabolized *in vivo*) and the fraction metabolized via a particular pathway are required in order to quantitatively understand the extent of different enzymes involved in the clearance of the compound (Lu et al., 2007, 2009, 2010), which in turn helps to more accurately predict the DDI potential for those enzymes.

Bile collection can also be used to assess the extent of metabolism through conjugation or reactive metabolite(s) formation. For example, both muraglitazar and peliglitazar primarily formed acyl glucuronide metabolites in hepatocyte incubations; however, only very low levels of these metabolites were detected in urine and feces of humans following oral administration. Bile collection in humans indicated that 20–40% of dose was excreted as acyl glucuronide metabolites in the bile (Wang et al., 2006, 2011). These data suggested that the hydrolysis of glucuronide metabolites by the gut microflora would have prevented accurate evaluation of the extent of glucuronidation *in vivo*.

Biliary excretion studies can also help determine whether a compound may be subject to enterohepatic recirculation (Wang et al., 2006). If a significant portion of the dose is cleared as intact drug or its direct conjugates in the bile, a drug is a candidate for enterophepatic circulation. In addition, subjects undergoing bile collection may show reduced systemic exposure as compared with subjects without bile collection due to interruption of the enterohepatic recirculation.

For intravenously administered drugs, bile collection can provide indirect evidence for the involvement of a secretory route of elimination from gut mucosa by means of efflux transporters (Zhang et al., 2009). If a portion of the administered dose is detected in the feces after intravenous administration of a drug to bile duct-cannulated animals, direct intestinal secretion is most likely responsible. There is currently no technology that can be used for ^{3}H- or ^{14}C-labeled compounds to assess the intestinal secretion of small molecules in humans, as bile can only be collected for a maximum of a few hours. In this case, data from bile duct-cannulated animals were required to gain insight into the human situation.

18.4 BILE COLLECTION TECHNIQUES

Studies in bile duct-cannulated animals are routinely included as part of a drug development program. In these studies, the bile duct is surgically cannulated, resulting in complete recovery of the total biliary output. In addition, bile can be collected for prolonged periods with supplementation of bile acids (Balani et al., 1999) (over the study duration, which is typically a few days) after drug administration to gain a more complete assessment of biliary excretion and metabolism. Bile collected via duct cannulation is free of gastric juices and other intestinal secretions.

The procedures available for the collection of bile in humans include both invasive and noninvasive techniques. Invasive methods generally involve patients undergoing surgical procedures to treat or diagnose underlying disease. The availability of such patients for involvement in exploratory drug studies is extremely rare. Noninvasive methods may utilize patients or healthy volunteers, making these techniques more amenable to drug disposition studies. With the existing methodologies, it is not currently possible to achieve prolonged bile collection in a human study; at a maximum, bile can only be collected for a few hours.

18.4.1 Invasive Methods

Invasive techniques for bile collection involve ultrasonically guided percutaneous, sub-, or transhepatic fine-needle puncture of the gallbladder (Hussaini et al., 1995; Wee et al., 1995), hepaticojejunostomy (Brookman et al., 1997), or T-tube percutaneous biliary drainage in patients after cholecystectomy (Rollins and Klaassen, 1979; Lorenz et al., 1984; Jüngst et al., 2001; Ghibellini et al., 2006a and 2006b). Availability of such patients is scarce for exploratory drug studies. However, when available, these studies provide extremely valuable information on the disposition of drugs *in vivo*. A major shortcoming of these techniques is that a complete collection of bile is not possible; the bile can only be collected for a spot check or a short interval. Of the techniques described above, T-tube drainage of the common bile duct provides more quantitative collection, albeit for a short period of time. For the cephalosporin antibiotics cefamandole, cefazolin, and cephalothin, used in the treatment of biliary tract disease, the total T-tube bile was collected for 6 h postintravenous administration (Ratzan et al., 1978). The study revealed that cefamandole, which showed the highest biliary concentrations, would be the drug of choice to treat postoperative biliary tree infections.

Due to the surgical nature of these procedures, the possibility of infection, bile leakage, hemorrhage, and right upper quadrant pain exist (Nahrwold, 1986; Hosking et al., 1992; Williams et al., 1994; Hussaini et al., 1995; Tudyka et al., 1995). Also, these patients generally are on multiple coadministered drugs for treatment of their condition or to facilitate the surgical procedures, which may sometimes preclude an accurate assessment of the metabolism of the investigational drug.

18.4.2 Noninvasive Methods

18.4.2.1 Oro-Gastroduodenal Aspiration

18.4.2.1.1 *Case Study 1* A nonsurgical procedure was developed for partial collection of bile from the duodenum of normal healthy volunteers (Balani et al., 1997). This procedure involved the use of a modified naso/orogastric (NG) tube for suction of the fluids from two separate locations—one from the stomach and the other next to ampulla of Vater, the location where bile enters the duodenum. For collection of bile and duodenal fluids from the latter location, two polyethylene (PE) tubes were inserted through the NG tube, exiting at the distal end. The length of the PE tubes extending from the gastric end of the NG tube is adjusted to facilitate placement of the end of the PE tubes close to the ampulla of Vater. Additionally, a fenestrated metal weight is attached to the end of the PE tubes to fine-tune the positioning by X-ray fluoroscopy. The gastric juices are separately aspirated from the stomach. The intubation is effected by the volunteer, advancing at a convenient pace, and the tube is finally secured by tape on the cheek, to avoid slippage or migration due to peristaltic action of the NG. The timing for installation of the nasogastric tube is typically scheduled prior to the observed T_{max}, which has been established from previous pharmacokinetic studies. Following continuous collection for 4–6 h, the subjects are infused with cholecystokinin carboxyl-terminal octapeptide (20 ng/kg, 5 min) 2 h before the end of collection to stimulate gallbladder contraction and collection of the maximum amount of concentrated bile. Simultaneous collection of plasma and other excreta is generally performed.

This procedure clearly has some benefits as it can be done in healthy volunteers, with little risk to the subjects. In addition, unlike the surgical methods, this method of bile collection does not require any concomitant medications, thus the data generated would not be impacted by DDIs. There are, however, obvious limitations to this technique, as bile can only be collected for a maximum of few hours. The procedure is limited by the subjects' tolerance of the device and the fact that they remain fasted over the course of bile collection. It should be noted that the collected fluid is not pure bile but rather a mixture of bile, duodenal secretions, and residual gastric juices. In addition, the collection of fluids may not be comprehensive, as some of the fluids may continue through the digestive tract, particularly if the placement of the tube is suboptimal. Finally, if the test compound is administered via the oral route, there is some potential to collect unabsorbed drug from the intestine if bile collection is initiated too early.

The above procedure was applied to the study of [^{14}C]montelukast (Singulair [Merck & Co., Rahway, NJ, USA]), an antagonist of the cyteinyl leukotriene receptor, to understand its disposition in humans thoroughly (Balani et al., 1997). Montelukast showed an oral bioavailability of about 60% (Cheng et al., 1996), and is cleared minimally via the renal route. Thus, at least 60% of dose was expected to be cleared through bile and/or gut secretion. Characterization of the biliary pathway of elimination thus was considered important to understand drug disposition. Since continuous collection of bile for long periods of time was not possible, for reasons discussed above, an innovative protocol was used to

segment the collection periods consecutively in subsets of volunteers to provide a longer-term picture of the biliary excretion. Specifically, in one group of subjects, bile was collected for 2–8 h postdose to capture the time around the plasma T_{max}, which is approximately 4 h. This collection provided an early picture of the biliary metabolic profile. In a second group of subjects, bile was collected from 8–12 h postdose, providing a later metabolic profile. The former collection time allowed for the transit of the drug solution past the duodenum (Wilson and Washington, 1989), thus reducing any possibility of capturing unabsorbed dose from the duodenum. The initial time frame for bile collection may be well suited for compounds that are excreted rapidly in bile, exemplified by short plasma half-life. However, compounds with long half-lives may require longer collection times, which would pose tolerance issues with the volunteers. In such situations, or when a compound shows temporal variations in pharmacokinetics, it may be useful to alter the timing of bile collection. With montelukast, it was shown that a maximum of about 20% of the orally dosed radioactivity was recovered over the two collection periods, which, considering the bioavailability (60%) of the compound, represented good recovery. The drug was extensively metabolized, with only a small amount remaining as intact compound in the bile samples. Several hydroxylated, some diastereomeric, and acyl glucuronide metabolites were detected in bile. The metabolic profiles of bile were similar across both collection periods, thus metabolic profiles generated from these samples may represent the bulk of the dose. Thus, definitive human metabolite identification and quantitation was achieved by this method, without the confounding metabolic contributions from gastrointestinal microflora.

18.4.2.1.2 Case Study 2 A similar human bile collection procedure was used to study disposition of [^{14}C] muraglitazar (Pargluva [Bristol-Myers Squibb, New York, NY, USA]), a novel dual alpha/gamma peroxisome proliferator activated receptor (PPAR) activator (Wang et al., 2006). A major modification, compared with the earlier described method, involved use of a commercially available Miller-Abbott Type Intestinal Tube (Model AN-22, Anderson Products Inc., Haw River, NC) for bile collection. This device comprised a weighted bilumenal PE nasogastric tube with an inflatable latex balloon tip at the distal end (Figure 18.2). One side of the tubing is perforated at the gastric end to permit aspiration of fluids; the other side terminates in the balloon and is utilized for injection of air to inflate the balloon. After intubation, and placement of the tube in the duodenum below the ampulla of Vater, the balloon is inflated to block the flow of intestinal fluids, permit-

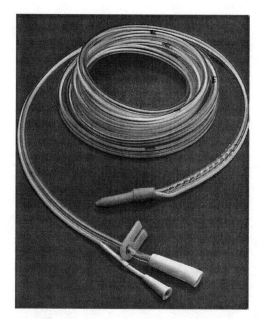

FIGURE 18.2. Preweighted Miller–Abbott-type intestinal tube for bile collection. This device was used for collection of bile in human ADME studies with muraglitazar, apixaban, and peliglitazar, as described in the text. (Model AN22 Intestinal Tube, Anderson Products Inc., Haw River, NC.)

ting a more complete collection of the biliary secretions. As this apparatus does not have a separate port to suction gastric fluids, the fluid collected represents a mixture of gastric, biliary, and intestinal secretions.

Nonclinical data for muraglitazar in animals revealed that direct glucuronidation was the major pathway for elimination of the drug, with the majority of the administered dose excreted in the feces (Zhang et al., 2009). In order to better understand the contribution of direct glucuronidation to the metabolism of muraglitazar in humans, the metabolism and disposition of [^{14}C] muraglitazar was investigated in healthy male subjects with and without bile collection after a single 20 mg oral dose (Wang et al., 2006). For the subjects undergoing bile collection, bile was collected from 3–8 h postdose. In addition, an intravenous dose of Kinevac® (Bracco Diagnostics, Princeton NJ, USA) (sincalide, a synthetically-prepared C-terminal octapeptide of cholecystokinin) was administered approximately 7 h postdose to stimulate gallbladder contraction and maximize bile secretion.

For subjects that had only feces and urine collected, 91% of the administered radioactivity was recovered in the feces. For the bile collection group, 40% of the administered radioactivity was recovered in the bile collected from 3–8 h postdose and 51% was recovered in the feces. The glucuronide of muraglitazar (15% of

dose) and the glucuronides of its oxidative metabolites (together 16% of dose) accounted for approximately 80% of the biliary radioactivity. In contrast, fecal samples only contained muraglitazar and its oxidative metabolites, suggesting hydrolysis of biliary glucuronides by gut/colonic microflora (Figure 18.3). Thus, bile collection enabled the identification of glucuronidation as a major pathway for the metabolism of muraglitazar; with conventional urine and fecal collection, the importance of this pathway would not have been recognized.

18.4.2.1.3 Case Study 3 The bile collection technique used for muraglitazar was also used to study the disposition of [^{14}C]apixaban, an orally bioavailable, highly selective, and direct acting/reversible factor Xa inhibitor (Raghavan et al., 2009). Nonclinical data showed that following oral administration of [^{14}C]apixaban, the majority of the radioactive dose (>73%) was excreted in the feces in intact mice, rats, and dogs; in bile duct-cannulated rats, <3% of the dose was excreted in bile. Intravenous administration of apixaban to bile duct-cannulated (BDC) rats indicated that approximately 22% of the dose was recovered in feces as parent drug, suggesting significant direct intestinal secretion of apixaban (Zhang et al., 2009).

In the human study, the disposition of [^{14}C]apixaban was studied after oral administration to healthy male subjects. In addition to the traditional urine and fecal collections, bile was collected from four subjects in the 3–8-h interval after dosing. The T_{max} for both drug and total radioactivity was similar for both dose groups and ranged from 0.5 to 2 h. The respective recovery of the administered radioactivity from subjects in the nonbile group and the bile collection groups were 56.0% and 46.7% in feces, and 24.5% and 28.8% in urine; the biliary excretion of radioactivity was <3%. Across the two groups, the parent drug represented approximately half of the recovered radioactivity in the combined excreta. The metabolic pathways for apixaban identified across all matrices included O-demethylation, hydroxylation, and sulfation of hydroxylated O-demethyl apixaban. Taken together, the preclinical and human ADME data suggest that there were multiple major pathways for the elimination of apixaban, including oxidative metabolism, renal excretion, and direct intestinal secretion, each representing a significant fraction of the dose. Consistent with the limited biliary secretion of drug-related radioactivity observed in bile duct-cannulated rats, biliary secretion of apixaban was also a minor pathway for elimination of the drug in humans. Only a small portion of the dose (<3%) was recovered in bile in the 5-h collection period, indicating a limited role of biliary excretion in the overall elimination of apixaban.

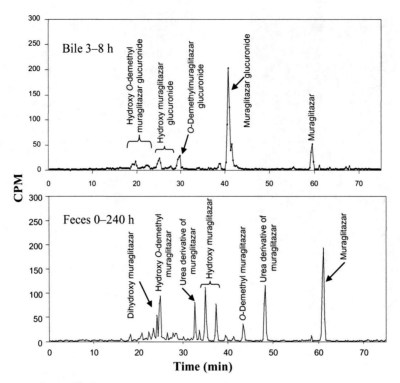

FIGURE 18.3. Radioactivity profile of [^{14}C] muraglitazar in human bile (3–8 h postdose collection) and feces (0–240 h) collection following oral administration. CPM, counts per minute.

Together with the nonclinical information on fecal excretion after intravenous administration in BDC rats and the known role of apixaban as a P-glycoprotein (Pgp) substrate (Zhang et al., unpublished data), the overall data suggest that direct intestinal secretion may also play an important role in the elimination of apixaban in humans.

These case studies demonstrate that relative to ADME studies where only urine and feces were collected, the biliary collection provides a more accurate assessment of hepatic involvement in the biliary secretion of the parent and its metabolites. In addition, although direct intestinal secretion cannot be directly assessed with the current methodology, when used in conjunction with data from animal ADME studies, data from the human bile study may lead to a better understanding of the role of intestinal secretion in drug elimination in humans.

18.4.2.2 Entero-Test®: String Collection Method

Recently, a novel technique for sampling bile has been applied to a study of drug metabolism in dogs (Guiney et al., 2010). The technique involves the use of the Entero-Test® (HDC Corp., Mountainview, CA, USA), a commercially available device that is utilized in the clinical setting to sample upper gastrointestinal fluid for diag-

nostic purposes. The device consists of a weighted gelatin capsule containing 90 cm of highly adsorbent nylon string (pediatric version; an adult version contains 140 cm of string). At a time appropriate to investigate metabolism for the route of drug administration, the capsule is swallowed and one end of the string is taped to the corner of the mouth. The gelatin capsule dissolves in the stomach and the string which is weighted at the distal end passes into the duodenum. Following an appropriate sampling period, approximately 3–4 h, the string, stained with adsorbed bile and gastrointestinal fluid, is withdrawn through the mouth. The weight is released during removal of the string, and is eliminated in the stool. The string is then extracted and metabolic profiles examined. Using this procedure, the authors were able to confirm the presence of known metabolites of simvastatin after oral dosing and simvastatin hydroxyl acid after intravenous dosing.

The application of the Entero-Test technique for obtaining human biliary information on drug candidates is currently under investigation. Although this technique does not provide quantitative information on the percentage of dose eliminated via the bile, it may provide a qualitative "snapshot" of the levels of metabolites in bile. Due to its simplicity, the technique could be easily implemented and, in conjunction with more comprehensive biliary studies in animals, the Entero-Test technique may be used to gain a better understanding of human biliary metabolism.

18.5 FUTURE SCOPE

Noninvasive methods for *in vivo* bile collection in human drug disposition studies will need to be further refined. Surgical approaches will continue to provide one alternative as long as such patients are available and willing to participate in studies with exploratory drugs. However, such data needs to be evaluated with caution due to coadministration of multiple other drugs and physiological changes resulting from the surgery. For diagnostic purposes, the most impressive existing technique for collection of pure bile appears to be video-endoscopic guided nasobiliary drainage (NBD). This involves insertion of a very thin tube into the common bile duct via the ampulla of Vater in the second portion of the duodenum during upper gastrointestinal endoscopy. This technique has demonstrated feasibility as it has been successfully used for the treatment of acute cholangitis (Itoi et al., 2008; Singh et al., 2009). NBD can be performed for extended periods of time, and bile collection over 3–4 days is not uncommon (Stapelbroek et al., 2006). The NBD technique therefore has the potential to be extended for use in drug disposition studies and may provide a more quantitative assessment of biliary secretion. An alternative approach that may have a role in the future would be to affix a nasogastroduodenal tube around the opening of the ampulla of Vater via suction. Like NBD, this method would also result in the collection of bile that is free from gastric secretions. In theory, this approach has the potential to provide an even more quantitative capture of bile. However, the potential complications of this procedure (e.g., cholangitis, pancreatitis) are currently unknown. The use of NBD to collect bile from the common bile duct is therefore encouraged.

The major limitation of all existing and proposed techniques for human bile collection is the inability to obtain quantitative bile collection, as no technique permits continuous and comprehensive collection of bile over a long period of time. The data obtained from these experiments can simply be used as a "snapshot" of the biliary elimination process or can be extrapolated cautiously to give more quantitative results. Thus, the implementation of bile collection techniques similar to the ones described in this chapter in human ADME studies will help investigators gain a better understanding of drug disposition. It is our hope that less invasive techniques for bile collection will be developed, making the conduct of biliary secretion studies more routine.

ACKNOWLEDGMENT

The authors would like to show appreciation of Dr. Asit Parikh of Clinical Research at Millennium: The Takeda Oncology Company for his invaluable discussions and guidance on biliary tree diseases and nasobiliary drainage.

REFERENCES

Balani SK, Xu X, Pratha V, Koss MA, Amin RD, Dufresne C, Miller RR, Arison BH, Doss GA, Chiba M, Freeman A, Holland SD, Schwartz JI, Lasseter K, Gertz B, Isenberg JI, Rogers JD, Lin JH, Baillie TA (1997) Metabolic profiles of montelukast sodium (Singulair), a potent leukotriene$_1$ receptor antagonist, in human plasma and bile. *Drug Metab Dispos* 25:1282–1287.

Balani SK, Kauffman LR, DeLuna FA, Lin JH (1999) Nonlinear pharmacokinetics of efavirenz (DMP-266), a Potent HIV-1 reverse transcriptase inhibitor, in rats and monkeys. *Drug Metab Dispos* 27:41–45.

Brookman LJ, Rolan PE, Benjamin IS, Palmer KR, Wyld PJ, Lloyd P, Flesch G, Waldmeier F, Sioufi A, Mullins F (1997) Pharmacokinetics of valsartan in patients with liver disease. *Clin Pharmacol Ther* 62:272–278.

Cheng H, Leff JA, Amin R, Gertz BJ, DeSmet M, Noonan N, Rogers JD, Malbecq W, Meisner D, Somers G (1996) Pharmacokinetics, bioavailability, and safety of montelukast sodium (MK-0476) in healthy males and females. *Pharm Res* 13:445–448.

Christopher LJ, Cui D, Li W, Barros A Jr., Arora VK, Zhang H, Wang L, Zhang D, Manning JA, He K, Fletcher AM, Ogan M, Lago M, Bonacorsi SJ, Humphreys WG, Iyer RA (2008) Biotransformation of [14C]dasatinib: *In vitro* studies in rat, monkey an human and disposition after administration to rats and monkeys. *Drug Metab Dispos* 36:1341–1356.

Drasar BS, Hill MJ (1974) *Human Intestinal Flora.* Academic Press, New York.

Ghibellini G, Leslie EM, Brouwer KL (2006a) Methods to evaluate biliary excretion of drugs in humans: An updated review. *Mol Pharm* 3:198–211.

Ghibellini G, Vasist LS, Hill TE, Heizer WD, Kowalsky RJ, Brouwer KLR (2006b) Determination of the biliary excretion of piperacillin in humans using a novel method. *Br J Clin Pharmacol* 62:304–308.

Guiney WJ, Beaumont C, Thomas SR (2010) Use of the Entero-Test®, a novel approach for the non-invasive capture of biliary metabolites in dogs. *Drug Metab Dispos* 38:851–856.

Hosking SW, Hacking CN, Herbetko J, Dewbury KC (1992) Safety and feasibility of percutaneous ultrasound guided puncture of the gallbladder for crystal analysis. *J Gastroenterol Hepatol* 7:379–381.

Hussaini SH, Kennedy C, Pereira SP, Wass JA, Dowling RH (1995) Ultrasound-guided percutaneous fine needle puncture of the gall bladder for studies of bile composition. *Br J Radiol* 68:271–276.

Itoi T, Kawai T, Sofuni A, Itokawa F, Tsuchiya T, Kurihara T, Kusano C, Saito Y, Gotoda T (2008) Efficacy and safety of 1-step transnasal endoscopic nasobiliary drainage for the treatment of actute cholangitis in patients with previous endoscopic sphincterotomy (with videos). *Gastrointest Endosc* 68:84–90.

Jaworski TJ, Hawes EM, Hubbard JW, McKay G, Midha KK (1991) The metabolites of chlorpromazine N-oxide in rate bile. *Xenobiotica* 21:1451–1459.

Jüngst D, Niemeyer A, Müller I, Zündt B, Meyer G, Wilhelmi M, Pozo RD (2001) Mucin and phospholipids determine viscosity of gallbladder bile in patients with gallstones. *World J Gastroenterol* 7:203–207.

Klassen CD, Watkins JB III (1984) Mechanisms of bile formation, hepatic uptake, and biliary excretion. *Pharmacol Rev* 36:1–67.

Lorenz D, Lucker PW, Mennicke WH, Wetzelsberger N (1984) Pharmacokinetic studies with silymarin in human serum and bile. *Methods Find Exp Clin Pharmacol* 6:655–661.

Lu C, Miwa GT, Prakash SR, Gan L-S, Balani SK (2007) A novel model for the prediction of drug-drug interactions in humans based on *in vitro* cytochrome P450 phenotypic data. *Drug Metab Dispos* 35:79–85.

Lu C, Balani SK, Qian MG, Prakash SR, Ducray PS, von Moltke LL (2010) Quantitative prediction and clinical observation of a CYP3A inhibitor-based drug-drug interactions with MLN3897, a potent C-C chemokine receptor-1 antagonist. *J Pharmacol Exp Ther* 332:562–568.

Mitchell SC, Zhang AQ, Noblet JM, Gillespie S, Jones N, Smith RL (1997) Metabolic disposition of [14C]-trimethylamine N-oxide in rat: Variation with dose and route of administration. *Xenobiotica* 27:1187–1197.

Nahrwold DL (1986) The biliary system. In *The Textbook of Surgery*, Sabiston DC, ed., pp. 1128–1169. W. B. Saunders, Philadelphia.

Parker RJ, Hirom PC, Millburn P (1980) Enterohepatic recycling of phenolphthalein, morphine, lysergic acid diethylamide (LSD) and diphenylacetic acid in the rat. Hydroylsis of glucuronic acid conjugates in the gut lumen. *Xenobiotica* 10:689–703.

Raghavan N, Frost C, Yu Z, He K, Zhang H, Humphreys W, Pinto D, Chen S, Bonacorsi S, Wong P, Zhang D (2009) Apixaban metabolism and pharmacokinetics after oral administration to humans. *Drug Metab Dispos* 37:74–81.

Ratzan KR, Baker HB, Lauredo I (1978) Excretion of cefamandole, cefazolin and cephalothin into T-tube bile. *Antimicrob Agents Chemother* 13:985–987.

Rollins DE, Klaassen CD (1979) Biliary excretion of drugs in man. *Clin Pharmacokinet* 4:368–379.

Singh V, Bhalla A, Sharma N, Dheerendra PC, Agarwal R, Mahi SK (2009) Nasobiliary drainage in acute cholestatic hepatitis with pruritus. *Dig Liver Dis* 41:442–445.

Stapelbroek JM, van Erpecum KJ, Klomp LWJ, Venneman NG, Schwartz TP, van Berge Henegouwen GP, Devlin J, van Nieuwkerk CMJ, Knisely AS, Houwen RHJ (2006) Nasobiliary drainage induces long-lasting remission in benign recurrent intrahepatic cholestatis. *Hepatology* 43:51–53.

Steinmetz PR, Balko C (1973) Therapeutic implications of the intestinal microflora. *N Engl J Med* 289:623–628.

Tudyka T, Kratzer W, Kuhn K, Janowitz P, Wechsler JG, Adler G (1995) Diagnostic value of fine-needle puncture of gall bladder: Side effects, safety, and prognostic value. *Hepatology* 21:1303–1307.

Wang L, Zhang D, Swaminathan A, Xue Y, Cheng PT, Wu S, Mosqueda-Garcia R, Aurang C, Everett DW, Humphreys WG (2006) Glucuronidation as a major clearance pathway of muraglitazar in humans: Different metabolic profiles in subjects with and without bile collection. *Drug Metab Dispos* 34:427–439.

Wang L, Munsick C, Chen S, Bonacorsi S, Cheng PT, Humphreys WG, and Zhang D (2011) Metabolism and disposition of 14C-labeled Peliglitazar in humans. *Drug Metab Dispos* 39:228–238.

Wee A, Nilsson B, Wang TL, Siew PY (1995) Tuberculous pseudotumor causing biliary obstruction: Report of a case with diagnosis by fine needle aspiration biopsy and bile cytology. *Acta Cytol* 39:559–562.

Williams JA, Treacy PJ, Sidey P, Worthley CS, Townsend NC, Russell EA (1994) Primary duct closure versus T-tube drainage following exploration of the common bile duct. *Aust N Z J Surg* 64:823–826.

Wilson CG, Washington N (1989) *Physiological Pharmaceutics. Biological Barriers to Drug Absorption.* John Wiley & Sons, New York.

Zhang D, Comezoglu N (2007) Design and ADME study. Metabolite profiling and identification. In *Drug Metabolism in Drug Design and Development: Principle and Applications*, Zhang D, Zhu M, Humphreys WG, eds., pp. 573–604. Wiley, Hoboken, NJ.

Zhang D, Wang L, Raghavan N, Zhang H, Li W, Cheng PT, Yao M, Zhang L, Zhu M, Bonacorsi S, Mitroka J, Hariharan N, Hosagrahara V, Chandrasena G, Shyu W, Humphreys GW (2007) Comparative metabolism of radiolabeled muraglitazar in animals and humans by quantitative and qualitative metabolite profiling. *Drug Metab Dispos* 35:150–167.

Zhang D, He K, Raghavan N, Wang L, Mitroka J, Maxwell BD, Knabb RM, Frost C, Schuster A, Hao F, Gu Z, Humphreys W, Grossman SJ (2009) Comparative metabolism of C-14 apixaban in mice, rats, rabbits, dogs, and humans. *Drug Metab Dispos* 37:1738–1748.

PART C

ANALYTICAL TECHNOLOGIES

19

CURRENT TECHNOLOGY AND LIMITATION OF LC-MS

Cornelis E.C.A. Hop

19.1 INTRODUCTION

Mass spectrometers have been around for about 100 years and have progressed from being an academic scientific curiosity to being mainstream instruments used on a daily basis by thousands of companies worldwide. In the 1940s, mass spectrometry was used to separate uranium isotopes, but the first significant industrial application was in the petroleum industry starting in the 1950s. High-resolution data obtained with magnetic sector mass spectrometers were used to characterize the constituents in petroleum distillates. This did not provide complete and quantitative characterization, but it did provide molecular formulas of the various petroleum constituents. In the 1970s, gas chromatography-mass spectrometry (GC-MS) instruments became more ubiquitous and they were used to identify volatile compounds. Some nonvolatile compounds (such as monosaccharides) could be identified as well following chemical derivatization. It was of some use in the pharmaceutical industry, but GC-MS focused mainly on applications in the beverage, flavor, and fragrance industries and in detection of environmental contaminants. The first breakthrough came with the introduction of fast atom bombardment (FAB; also called liquid secondary ionization mass spectrometry) by Prof. Barber in the early 1980s. FAB allowed ionization of nonvolatile compounds such as oligopeptides and small-molecule drugs. However, FAB generates abundant matrix ions that can interfere with detection of the analyte(s) of interest and does not allow online chromatographic separation. (The moving belt interface was introduced to allow chromatographic separation, but its

capabilities were limited to say the least.) Both factors limited its practical use. The breakthrough that revolutionized the field and greatly increased the use of mass spectrometry as a mainstream bioanalytical tool was the introduction of atmospheric pressure ionization—both atmospheric pressure chemical ionization (APCI) and electrospray ionization (ESI)—in the mid-1980s. Indeed, Prof. Fenn received the Nobel Prize in Chemistry in 2002 for "the development of methods for identification and structure analysis of biological macromolecules" (The Nobel Prize in Chemistry 2002, 2002) using ESI mass spectrometry. The ability of ESI to ionize polypeptides and proteins as intact entities enabled the proteomics field and contributed to advances in biotechnology. In addition, the routine liquid chromatography-mass spectrometry (LC-MS) capabilities of APCI and ESI revolutionized the small-molecule pharmaceutical industry. It was greatly beneficial for characterization of reaction mixtures generated by synthetic chemists. Moreover, it allowed accurate and specific quantitation of nonvolatile drugs in biological matrices and identification of metabolites. The latter greatly enhanced the role and importance of determining the absorption, distribution, metabolism, and excretion (ADME) characteristics of compounds in drug discovery and development.

In the 1980s, inappropriate ADME characteristics were a leading cause of attrition in drug development (about 40% of drugs failed due to drug metabolism and pharmacokinetic reasons) and this has been reduced to less than 10% in the 1990s (Kola and Landis, 2004). Although it is impossible to identify a single causative reason for this reduction, it is quite likely that the

ADME-Enabling Technologies in Drug Design and Development, First Edition. Edited by Donglu Zhang and Sekhar Surapaneni.
© 2012 John Wiley & Sons, Inc. Published 2012 by John Wiley & Sons, Inc.

detection capabilities provided by LC-MS contributed to this decline because it allowed earlier assessment and, ultimately, optimization of ADME properties in drug discovery.

APCI and ESI have been interfaced successfully with single and triple quadrupole instruments, three-dimensional and linear ion traps, time-of-flight instruments, Fourier transform instruments, orbitraps, and hybrids thereof. In particular, the ease of use of quadrupole mass spectrometers contributed to the viral spread of its use in the pharmaceutical industry. Right now, the benefits of these instruments far outweigh the still relatively high cost.

In this chapter, a brief overview of the advantages/disadvantages and applications of various types of mass spectrometers used in the pharmaceutical industry is presented, with an emphasis on applications in the area of ADME. Subsequent chapters will offer a more detailed picture about specific applications. The overall process of sample analysis by LC-MS can be separated in three unique but interdependent steps: sample preparation, chromatographic separation, and mass spectrometric analysis. Each step will be described in detail.

19.2 SAMPLE PREPARATION

Due to the increased sensitivity and selectivity of the mass spectrometer as the bioanalytical detector, sample preparation for plasma samples is frequently limited to precipitation of abundant endogenous proteins such as albumin, and the same applies for samples from *in vitro* studies. However, if interference or variable ion suppression/enhancement occurs, a more robust sample cleanup method such as liquid-liquid extraction (LLE) or solid-phase extraction (SPE) may be required (Henion et al., 1998). LLE or SPE may also be beneficial if a very low limit of detection is required as may be the case for low-dose studies in humans. It is possible after any extraction method to transfer the desired liquid phase, dry it down, and reconstitute it in a smaller volume of solvent to increase the concentration of the analyte in the sample and thereby lower the limit of detection and quantitation. Sample preparation for urine is usually limited to sample dilution and centrifugation. Sample preparation for feces is more complex and usually involves homogenization and LLE. For preclinical studies, tissue collection to determine the local concentration at the site of action (efficacy or toxicity) has become popular and necessitates tissue homogenization using mechanical homogenization, digestion, or extraction.

For quantitative bioanalysis, it is possible to monitor and optimize the recovery of the analyte of interest.

However, this is not possible for metabolite identification. The samples may contain a range of metabolites that can vary greatly in their lipophilicity, and moreover, authentic standards are usually not available. Thus, sample preparation for biotransformation studies is frequently limited to protein precipitation to prevent changing the relative abundance of parent compound and metabolites due to differences in recovery. If radiolabeled material is available, it is possible to quantify the extraction recovery.

A very different approach of sample preparation is dried blood spot analysis. This technique has been extensively used for detection of metabolic disorders in newborns, but recently it has been combined with LC-MS/MS for quantitative bioanalysis (Spooner et al., 2009). A small amount of blood (<100 μL) is deposited on absorbent paper and allowed to dry. Next, a small hole is punched out of the paper and the analyte is extracted with a solvent for subsequent LC-MS/MS analysis. The advantages of this technique is that the small amount of blood required facilitates serial bleeding in mice, eliminates the need for a parallel pharmacokinetics (PK) group in toxicology studies, and enables pediatric studies. Another advantage is that storage and shipment of the paper containing the dried blood spots do not require refrigeration.

19.3 CHROMATOGRAPHY SEPARATION

Although gas chromatography and capillary electrophoresis are occasionally used, liquid chromatography is currently by far the most common separation technique interfaced with mass spectrometers. Traditionally, isocratic and gradient chromatographic conditions have been used for elution, but gradients usually have superior chromatographic resolving power. Fortunately, technology did not stand still in this area either. First, significant progress has been made with column technology. A more diverse range of stationary phases is available. Particles smaller than 2 μm are available and that allows more efficient separation as illustrated by the Van Deemter equation:

$$H = A\mu^{0.33} + B/\mu + C\mu,$$

where H is the plate height associated with the column; A, B, and C are constants that are dependent on the physical characteristics of the column packing material; and μ is the linear velocity of the mobile phase. A is the contribution of eddy diffusion, B is the contribution due to axial diffusion, and C is the broadening due to mass transfer. The net effect of a reduction in particle size is that the optimum plate height is lower (i.e., better chromatographic resolution) and it occurs at a higher flow rate. The disadvantage of a higher flow rate is increased column back pressure, but this can be overcome with pumps able to withstand higher pressure. The ultimate advantage is increased chromatographic efficiency and shorter cycle times and the latter will increase the capacity of the LC-MS instrument (Mazzeo et al., 2005; Plumb et al., 2008). This is frequently referred to as ultrahigh-pressure liquid chromatography (UHPLC).

Other column advances are (1) the availability of monolithic silica rod columns, which reduce the back-pressure and operating them at higher flow rates will reduce the cycle time (Hsieh et al., 2002), and (2) particles with a solid, fused core surrounded by porous silica, which offers increased chromatographic efficiency without the need for higher-pressure pumps (Cunliffe et al., 2009). Turbulent flow chromatography also offers specific advantages (Ayrton et al., 1997). The high flow rate combined with the larger particle size allows small-molecule analytes to be trapped on the column, but prevents retention of larger molecules, such as proteins. Thus, samples can be injected without prior protein precipitation and this approach has been successfully applied to analysis of plasma samples and samples from *in vitro* studies. Turbulent flow chromatography technique is frequently accompanied by column switching and back-flushing the analyte of interest (and other retained small molecules) to an analytical column for chromatographic separation. Finally, systems with multiple, parallel liquid chromatography (LC) columns have become popular (Wu, 2001; King et al., 2002). A column switching device is inserted that allows the operator to send the column effluent to the mass spectrometer

exclusively around the time the compound of interest elutes. The principle is illustrated in Figure 19.1. It greatly increases the duty cycle of the instrument and it allows a significant increase in sample throughput per mass spectrometer. The latter is particularly important for quantitative bioanalysis where the data acquisition time may be the bottleneck. Although there is a continuous desire to increase throughput and reduce the cycle time, adequate chromatographic separation is still required to prevent coelution with labile metabolites that could interfere with detection of the parent compound (e.g., acyl glucuronides that could fragment in the ion source to the parent compound) (Jemal and Xia, 1999).

For metabolite identification, the bottleneck is usually data interpretation and therefore the impetus to switch to short cycle times is less pronounced. Moreover, short gradients may come at the expense of chromatographic resolution, which could be problematic for identification of isobaric metabolites (e.g., if multiple hydroxylated metabolites are formed). Nevertheless, UHPLC has become popular for metabolite identification as well, because it maintains chromatographic separation but with a shorter cycle time. The latter is especially important for metabolite identification studies in drug discovery to keep pace with the desire by synthetic chemists to get information about metabolic liabilities to address in the next generation of compounds. Since the mass spectrometric output for metabolite identification studies is not quantitative (*vide infra*), splitting the column effluent and directing a fraction thereof to a UV or diode array detector is imperative. If the compound is radiolabeled, a radio flow detector can be used in parallel with mass spectrometric detection.

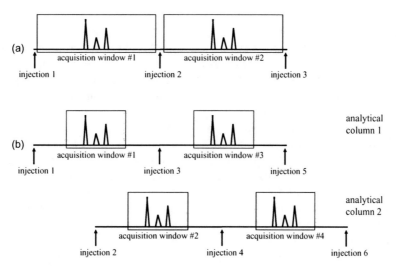

FIGURE 19.1. Schematic representation of conventional, serial analysis (a) and staggered parallel analysis (b) using two analytical columns and column switching.

19.4 MASS SPECTROMETRIC ANALYSIS

Mass spectrometers are the detectors of choice for most quantitative and qualitative studies for small-molecule drugs. Large-molecule drugs such as proteins, antibodies, and RNAi can be detected by mass spectrometers as well, but most quantitative studies still involve techniques such as enzyme-linked immunosorbent assay (ELISA). The first step is ionization of the sample. In the past, most ionization was by electron impact. However, that technique is not compatible with liquid chromatography. In the mid-1980s, ESI and APCI were introduced and they revolutionized LC-MS analysis. The next step is mass analysis and several types of mass spectrometers are available for that purpose.

19.5 IONIZATION

As mentioned above, ESI and APCI are the mainstream ionization techniques for analysis. In case ESI is used, the column effluent elutes from a narrow capillary carrying a high positive or negative voltage. This results in a Taylor cone and the formation of small droplets with an excess positive or negative charge. A countercurrent of heated drying gas will result in evaporation of solvent from the droplets and the droplets will become enriched in protonated ([M + H]$^+$ ions) or deprotonated ([M – H]$^-$ ions) analyte ions. The subsequent steps continue to be the topic of debate. One possibility is that the droplets fragment due to coulombic repulsion and this process proceeds until a single ionized analyte molecule is left. Alternatively, ionized analyte molecules are expelled from the droplets due to coulombic repulsion associated with the excess charge on the droplets. In the case of APCI, the column effluent is very rapidly evaporated and a discharge from a high-voltage needle positioned nearby results in a "cloud" of reagent ions formed by the solvent. The charged solvent ions can transfer a positive or negative charge to the analyte of interest via proton transfer. Both ESI and APCI occur under atmospheric pressure conditions and the ions are subsequently transferred via a potential gradient to the vacuum of the mass spectrometer. Whether to use APCI or ESI for quantitative bioanalysis has been discussed and debated for a long time. It is argued that the high temperature used by APCI may sometimes lead to thermal degradation of the analyte (e.g., acyl glucuronides). Others argue that the ion suppression effects are more pronounced with ESI. The literature shows that both have been used successfully for quantitation of small-molecule drugs.

Both ESI and APCI allow ionization of a wide range of nonvolatile compounds. Ionization is considered mild and in the positive ion mode [M + H]$^+$ ions are generated without much fragmentation occurring upon ionization, which contrasts electron impact ionization used for GC-MS. Ionization can be enhanced by lowering the pH of the mobile phase with formic or acetic acid. Sometimes, the proton affinity of the analyte is insufficient to detect [M + H]$^+$ ions, but even in those cases, characteristic [M + Na]$^+$ or [M + K]$^+$ ions are frequently detected. Direct interference from isobaric ions resulting in lack of selectivity is possible, but this can be addressed by chromatographic resolution (*vide supra*) or acquiring MS/MS data (*vide infra*). However, indirect interference may occur as well by reducing or enhancing the ionization efficiency of the analyte of interest. The impact of coeluting components on the ionization efficiency of the analyte of interest can be visualized by injecting a blank extracted matrix sample while infusing and monitoring the analyte of interest (Bonfiglio et al., 1999). Frequently, highly polar compounds eluting close to the chromatographic solvent front will reduce the ionization efficiency of the coeluting analyte of interest. Most analysts address this by increasing the chromatographic retention of the analyte of interest. Alternatively, a more selective sample extraction method can be employed. That said, plasma protein precipitation is still by far the most ubiquitous (and simplest) sample cleanup method. Finally, an internal standard can compensate to some extent for variable matrix effects. Nevertheless, the matrix effect must be explored and good addressed in detail for bioanalysis supporting good laboratory practice (GLP) toxicology and clinical studies.

It is also possible to generate ions in the negative ion mode. However, many pharmaceutical compounds contain one or more basic nitrogen atoms and therefore these compounds do not ionize well in the negative ion mode. However, many nonsteroidal anti-inflammatory drugs and peroxisome proliferator-activated receptor (PPAR) agonists contain carboxylic acid functionalities or mimetics thereof, which allow efficient ionization in the negative ion mode. A distinct, but not universal, advantage of the negative ionization mode is that background signals due to the matrix tend to be less abundant, which increases the signal-to-noise ratio for the analyte of interest and its metabolites. All examples presented in this chapter refer to the positive ionization mode unless otherwise mentioned.

Another inherent feature of APCI and ESI is the need for a standard curve to quantify an analyte. The ionization efficiency varies greatly from analyte to analyte and is also affected by the mobile phase, and consequently, observing two signals of equal abundance does not mean that both analytes are equally abundant. This implies that it is not possible to quantitate metabolites or impurities without authentic standards. This effect can be overcome to some extent—albeit not completely—by reducing the flow rate to 100 nL/min or less (Hop et al., 2005; Hop, 2006). The simplistic interpretation of the latter phenomenon is that at much lower flow rates the droplets are smaller, but still highly charged. Thus, in the positive ion mode, there is an increased amount of protons available to ionize analytes negating differences in ionization efficiency due to different proton affinities among analytes. However, liquid chromatographic equipment that can generate reliable gradients at such low flow rates is not routinely available.

Finally, it is possible to fragment ions in the ion source itself (i.e., prior to mass analysis) by changing the potential gradient in the ion source. However, all ions generated in the ion source will fragment, and consequently, it is hard to unambiguously establish precursor ion–fragment ion relations in the mass spectrum, which can complicate interpretation of the data.

Matrix-assisted laser desorption/ionization (MALDI) can be used as well, but the lack of a direct interface with a chromatographic separation method limits its use to analysis of relatively clean samples from *in vitro* studies (Gobey et al., 2005). Figure 19.2 shows data obtained by taking samples of a metabolic stability assay and adding a MALDI matrix for ionization. MS/MS data need to be acquired to prevent interference from metabolites and/or the more abundant matrix ions. The speed with which samples can be analyzed is its main advantage; it is possible to analyze a 96-well plate within minutes.

Other ionization techniques such as atmospheric pressure photoionization (APPI), desorption electrospray ionization (DESI), and direct analysis in real time (DART), are available as well, but have not been adopted broadly in the pharmaceutical industry.

19.6 MS MODE VERSUS MS/MS OR MS^n MODE

If data are acquired in the MS mode, only a single stage of mass separation is applied. The selectivity of the measurement is determined by the chromatographic separation and the single stage of mass analysis. It is possible that endogenous components having the same mass-to-charge (m/z) ratio as the analyte of interest coelute. This can be addressed by changing the chromatography, but this could result in rather long cycle times. Thus, it is more common to acquire data in the MS/MS mode. The first mass analyzer is used to select and transmit the m/z ratio of the ions of interest. Next the mass-selected ions are fragmented by collision with an inert gas (usually nitrogen, helium, or argon). The second stage of mass analysis is used to separate all fragment ions according to their m/z ratio. Either a full-scan MS/MS spectrum can be obtained or a specific fragment ion can be monitored. The latter is called selected reaction monitoring (SRM), and if multiple transitions are monitored, it is called multiple reaction monitoring (MRM). Thus, MS/MS provides an additional degree of selectivity and this allows shorter chromatographic cycle times. Gradients lasting 30 min were quite common with UV detection. However, LC-MS/MS analysis usually involves isocratic

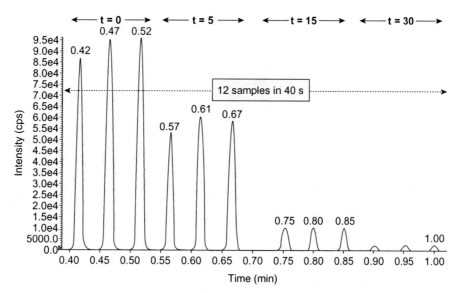

FIGURE 19.2. Profile of the metabolic stability of an analyte (in triplicate) obtained by MALDI using a triple quadrupole mass spectrometer operating in the single reaction monitoring MS/MS mode.

or gradient separation lasting 1–4 min per sample. With ion trap instruments, it is possible to consecutively isolate and fragment ions, thereby allowing MS^n data to be generated.

19.7 MASS SPECTROMETERS: SINGLE AND TRIPLE QUADRUPOLE MASS SPECTROMETERS

Up until the late 1970s, the most ubiquitous mass spectrometers were magnetic sector instruments. The greatest disadvantage of those instruments was that they required a very high skill level of the operator. Tremendous progress has been made in the field of mass spectrometry and the types of mass spectrometers currently in routine use are relatively easy to operate. The most common instruments are single and triple quadrupole mass spectrometers. Single quadrupoles combined with a liquid chromatography system are frequently used in synthetic chemistry labs to monitor reactions and the purity of the product. It is also possible to use the mass spectrometer to isolate fractions in a complex mixture. The response of the mass spectrometer is used to trigger a fraction collector ensuring isolation of the desired components in the mixture. These advances have greatly enhanced the productivity of synthetic chemists and the purity of the reaction products.

Virtually all quantitation of drugs in biological matrices is performed with triple quadrupole instruments in the MS/MS mode. After chromatographic elution, the analyte of interest is ionized, mass-selected with the first quadrupole, and fragmented in the collision cell. One of the fragment ions is transmitted by the third quadrupole and detected with an electron multiplier detection system, which invariably is connected to a sophisticated data acquisition system. This is called SRM. If multiple analytes are monitored (including the internal standard), it is referred to as MRM. Thus, there are three levels that influence and enhance the selectivity of the bioanalytical method: (1) sample extraction, (2) chromatographic separation, and (3) MS/MS detection. This has made it possible to generate selective methods with a detection limit of 1–10 ng/mL (usually sufficient for routine drug discovery studies) within 10 min or less. However, developing quantitative methods for GLP toxicology or clinical studies that will be submitted to the Food and Drug Administration (FDA) (Vishwanathan et al., 2007; Savoie et al., 2009) are still quite time-consuming, in particular if they require a low limit of quantitation (sub ng/mL).

Triple quadrupole instruments are also heavily used for metabolite identification. Regular product ion MS/MS mass spectra (see Fig. 19.3) are quite useful to obtain structural information for metabolites. The m/z values of the most abundant fragment ions can be related to distinct features of the analyte. Table 19.1 contains information about the most common biotransformation pathways and each is associated with a specific change in the m/z value of the $[M + H]^+$ ions. For example, formation of a glucuronide metabolite increases the m/z value by 176 units. Detailed analysis of the product ion MS/MS spectra of the parent compound and its metabolite, allows the operator to assign the structural modification to a particular moiety in the drug candidate. Aside from specific biotransformation pathways, such as N-dealyklation, it is frequently impossible to identify

Triple Quadrupole Scan Modes

	ionization	fragmentation	detection
production spectrum	ABC^+	$A^+ + BC$	A^+
		$AB + C^+$	C^+
	ABD^+	$A^+ + BD$	A^+
		$AB + D^+$	D^+
precursor ion spectrum	ABC^+	$A^+ + BC$	A^+
		$AB + C^+$	
	ABD^+	$A^+ + BD$	A^+
		$AB + D^+$	
constant neutral loss spectrum	ABC^+	$A^+ + BC$	
		$AB + C^+$	C^+
	ABD^+	$A^+ + BD$	
		$AB + D^+$	D^+

FIGURE 19.3. Principles of product ion, precursor ion, and constant neutral scan modes feasible with a triple quadrupole mass spectrometer.

TABLE 19.1. Common Phase I and II Biotransformations and the Corresponding Change in Mass of the Parent Compound

Mass Change (Da)	Type of Biotransformation
−28	Deethylation ($-C_2H_4$)
−14	Demethylation ($-CH_2$)
−2	Two-electron oxidation ($-H_2$)
+2	Two-electron reduction ($+H_2$)
+14	Addition of oxygen + two-electron oxidation ($+O -H_2$)
+16	Addition of oxygen ($+O$)
+18	Hydration ($+H_2O$)
+30	Addition of two oxygen atoms + two-electron oxidation ($+2O -H_2$)
+32	Addition of two oxygen atoms ($+2O$)
+80	Sulfate conjugation
+107	Taurine conjugation
+176	Glucuronide conjugation
+305 or 307	Glutathione adduct

the exact structure of the metabolite. Isolation and nuclear magnetic resonance (NMR) analysis may be required to obtain the exact structure.

If it is known that the parent compound loses a characteristic neutral moiety upon collisional activation, the first and third quadrupoles can be scanned with the mass of the neutral moiety as the constant offset between the first and third quadrupole to facilitate identification of metabolites closely related to the parent compound. This is termed a constant neutral loss MS/MS spectrum and the principle is illustrated in Figure 19.3. Two common constant neutral loss scans for metabolites are the loss of 176 Da to detect glucuronides and the loss of 129 Da for detection of glutathione adducts. Of course, if the parent compound undergoes significant metabolic modification or if the mass of the neutral moiety that is monitored in the constant neutral loss MS/MS mode is modified by a metabolic process, the neutral loss MS/MS spectrum will fail to detect these metabolites. Finally, precursor ion MS/MS spectra provide value as well. In this scan mode, a common fragment ion—common to the parent compound and its metabolites—is monitored (see Fig. 19.3). However, if biotransformations affect the part of the drug candidate that gives rise to the common fragment ion, the metabolites will not be detected using this scan mode. However, the latter disadvantage is more than made up by the increase in selectivity obtained in both constant neutral loss and precursor ion modes compared with the full-scan MS mode. This is illustrated in Figure 19.4, which shows the total ion current chromatograms for a dog hepatocyte incubation obtained in the MS mode and the

m/z 42 precursor ion MS/MS mode. Metabolites are much easier to detect (i.e., with increased signal-to-noise ratio) in the precursor ion MS/MS mode. However, metabolites that would involve modification of the thiazolidinedione moiety would be missed.

There are two distinct disadvantages to quadrupole mass spectrometers. First, they usually operate at unit mass resolution, and therefore, it is not possible to measure the mass of ions within 5–10 ppm, which would allow assignment of the molecular formula. Second, although the sensitivity in the SRM mode is very impressive, the sensitivity in the full-scan mode (either MS or product ion, precursor ion, or constant neutral loss MS/MS) is limited due to the relatively slow scan speed. (The mass spectrometer monitors ions that are of no interest to the analyst for a significant fraction of the scan time.) The latter disadvantage is partly overcome in the latest generation of hybrid quadrupole-ion trap triple quadrupole mass spectrometers that allow collection of ions in the second quadrupole/collision cell prior to mass separation in the third quadrupole. A greater abundance of fragment ions will increase the signal-to-noise ratio and overall quality of the MS/MS mass spectra.

Advantages of triple quadrupole mass spectrometer:

- Easy to operate
- Very sensitive for quantitative bioanalysis using the SRM or MRM mode
- Several useful scan modes for metabolite identification such as constant neutral loss and precursor ion scans

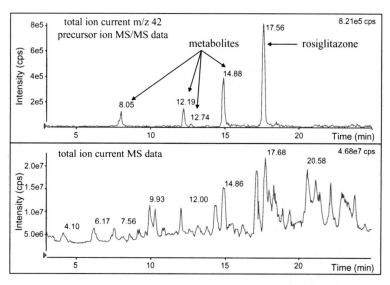

FIGURE 19.4. Negative ion mode total ion current traces for the m/z precursor ion MS/MS data (top) and MS data (bottom) from dog hepatocytes incubated with rosiglitazone.

Disadvantages of triple quadrupole mass spectrometers:

- Unit mass resolution
- Limited full-scan sensitivity

19.8 MASS SPECTROMETERS: THREE-DIMENSIONAL AND LINEAR ION TRAPS

The first three-dimensional ion traps were introduced in the mid-1990s. Although quite powerful for metabolite identification, their use for drug quantitation is limited. Compared with the SRM mode on a triple quadrupole mass spectrometer, the detection limit is usually not as good because of the lower duty cycle. Moreover, the linear range is limited because a high abundance of ions in the ion trap will inevitably give rise to space charge effects and a distinctly nonlinear standard curve. It is possible to compensate for space charge effects with automatic gain control, which reduces the amount of time the ions can enter the ion trap, but that may impact the reproducibility. The space charge effect is reduced with linear ion traps, which have a larger volume to store ions.

Although an ion trap does not allow constant neutral loss and precursor ion scan modes, a key advantage for metabolite identification is the possibility to acquire MS^n data. Sometimes, the nature of fragment ions in a product ion MS/MS spectrum may be in doubt. One way to address this is by increasing the collision energy and causing secondary fragmentation of the primary fragment ions. However, interpretation of the spectrum may be problematic because all primary fragment ions will dissociate further, thereby obfuscating the relationship between specific primary and secondary fragment ions. With ion traps, it is possible to isolate specific fragment ions and then fragment those ions further—MS^3 and beyond. This will provide an additional degree of structural characterization, although the mass accuracy is limited to single mass units. Several examples have been presented in the literature using MS^3, MS^4, and beyond to identify the specific moiety of the drug candidate that was metabolically modified.

Advantages of ion trap mass spectrometers:

- Easy to operate
- Compact
- Relatively cheap
- MS^n capabilities for metabolite identification

Disadvantages of ion trap mass spectrometers:

- Less suitable for drug quantitation because of reduced sensitivity (SRM is not feasible)

- Constant neutral loss and precursor ion mode scanning not feasible
- Limited full-scan sensitivity
- Unit mass resolution

19.9 MASS SPECTROMETERS: TIME-OF-FLIGHT MASS SPECTROMETERS

The increased duty cycle is a unique advantage of time-of-flight mass spectrometers. A single, full-scan mass spectrum can be obtained in the submillisecond time range, although many spectra need to be summed to obtain a spectrum with an adequate signal-to-noise ratio. Frequently, the time-of-flight section is preceded by a quadrupole and a collision cell to allow acquisition of MS/MS spectra. Product ion MS/MS data can be acquired, but these quadrupole time-of-flight hybrid mass spectrometers do not allow constant neutral loss and precursor ion scan modes. A second advantage of a time-of-flight mass analyzer is that it allows accurate mass determination (usually within 5–10 ppm), but it requires internal calibration to get the best mass accuracy. Accurate mass measurements are based on the exact mass of the various atoms present in drug molecules (see Table 19.2). Accurate mass data can be obtained for the $[M + H]^+$ ions of the parent compound and its metabolites. Availability of the accurate mass of ions allows the user to reduce the number of possible molecular formulas for metabolites considerably. Accurate mass data can also be obtained for fragment ions in MS/MS mass spectra, and this greatly enhances the ability to interpret MS/MS mass spectra and assign more specific structures to metabolites. The high resolution and accurate mass capability also enables mass defect filtering for metabolite identification (Zhang

TABLE 19.2. Exact Mass and Abundance of Specific Isotopes of Atoms Commonly Found in Drug Molecules

Atom	Exact Mass (Da)	Isotope Abundance (%)
^1H	1.0078	99.985
^{12}C	12.0000	98.9
^{13}C	13.0034	1.1
^{14}N	14.0031	99.6
^{16}O	15.9949	99.8
^{19}F	18.9984	100
^{32}S	31.9721	95.0
^{33}S	32.9715	0.8
^{34}S	33.9679	4.2
^{35}Cl	34.9689	75.5
^{37}Cl	36.9659	24.5

et al., 2007, 2009). Each drug candidate has a specific fractional mass; for example, the exact mass of the $[M + H]^+$ ions of nefazodone is 470.2323 and the corresponding fractional mass is 0.32323. Most Phase I metabolites will have a fractional mass close to that of the parent compound. Thus, the full-scan high-resolution MS data can be filtered for a mass window close to that of the fractional mass. In the case of nefazodone, a filter of 0.1923–0.2723 has been successfully used to identify several Phase I metabolites (Zhang et al., 2009). Note that for detection of Phase II conjugates a different mass defect filter is required because of the addition of multiple C, H, N, O, and S atoms. Therefore, it is preferred to set up different mass defect filters for detection of different types of metabolites. This technique is particularly useful for detection of unanticipated metabolites that can easily escape detection if nonradiolabeled material is used.

Time-of-flight and quadrupole time-of-flight mass spectrometers are rarely used for quantitative bioanalysis because it is still hard to compete with the ease of use of triple quadrupole mass spectrometers and their sensitivity in the SRM mode. One attractive feature of using a time-of-flight mass spectrometer to quantify drugs in biological matrices is that it is inherently done in the full-scan mode. Thus, it is possible at any moment in time to go back to the raw data and interrogate the data for the presence of specific metabolites. (The latter is not possible when data are acquired with a triple quadrupole mass spectrometer in the SRM mode.) The enhanced mass resolution compensates to some extent for the significant drop in selectivity in the MS mode. Nevertheless, the dynamic range and accuracy and precision do not meet the high standards set by triple quadrupole mass spectrometers for quantitation.

Advantages of quadrupole time-of-flight mass spectrometers:

- Improved duty cycle, which increases the sensitivity in full-scan MS and MS/MS mode
- Enhanced mass resolution
- Accurate mass determination (within 5–10 ppm)

Disadvantages of quadrupole time-of-flight mass spectrometers:

- Less suitable for drug quantitation because of reduced sensitivity (SRM is not feasible)
- Constant neutral loss and precursor ion mode scanning not feasible
- Harder to operate
- More expensive

19.10   MASS SPECTROMETERS: FOURIER TRANSFORM AND ORBITRAP MASS SPECTROMETERS

Fourier transform mass spectrometers have been available since the late 1960s, but they were mainly used in academia to study ion chemistry. Although their accurate mass determination capability is very impressive, they are quite expensive, hard to operate, and having to fill the superconducting magnet regularly with liquid helium and nitrogen is inconvenient. The more recently introduced orbitrap seems to offer all the advantages of Fourier transform instruments without the disadvantages. Currently, the most common configuration of the orbitrap is with a linear ion trap on the front end and this type of instrument has taken the metabolite identification field by storm. The instrument is relatively easy to use and does not require internal calibration; 5 ppm mass accuracy can be achieved on a routine basis with external calibration. The latter greatly facilitates acquisition of routine accurate mass data. In contrast to triple quadrupole mass spectrometers, constant neutral loss and precursor ion scan are not possible, but this deficit is overcome by the accurate mass capability and the ability to acquire MS^n spectra.

For the same reasons as described above for time-of-flight mass spectrometers, linear ion trap orbitrap hybrid mass spectrometers are rarely used for quantitative bioanalysis.

Advantages of Fourier transform and orbitrap mass spectrometers:

- Sensitivity in full-scan MS, MS/MS, and MS^n mode
- Enhanced mass resolution
- Accurate mass determination (within 5 ppm)
- MS^n capabilities for metabolite identification

Disadvantages of Fourier transform and orbitrap mass spectrometers:

- Less suitable for drug quantitation because of reduced sensitivity (SRM is not feasible)
- Constant neutral loss and precursor ion mode scanning not feasible
- Harder to operate
- More expensive

19.11   ROLE OF LC-MS IN QUANTITATIVE *IN VITRO* ADME STUDIES

Sensitive and selective quantitation of drugs and/or their metabolites by LC-MS/MS has greatly enhanced the ability of drug metabolism and pharmacokinetics (DMPK) scientists to influence drug discovery and development. An integral component of drug discovery is early incorporation of *in vitro* ADME endpoints such as the following:

- Permeability in Caco-2 or Madin-Darby canine kidney (MDCK) cells or parallel artificial membrane permeability assay (PAMPA)
- Efflux in MDCK or LLCPK cells overexpressing transporters such as P-glycoprotein and BCRP
- Metabolic stability in microsomes, S9, hepatocytes, or recombinant cytochrome P450 enzymes
- Competitive cytochrome P450 inhibition
- Mechanism-based or time-dependent cytochrome P450 inhibition
- Cytochrome P450 induction
- Plasma protein binding or drug binding in other matrices such as microsomes or brain
- Blood-to-plasma partitioning

Each of these *in vitro* ADME assays should be incorporated in the overall screening cascade for discovery projects, although the order may vary from project to project. The desire is to incorporate appropriate ADME endpoints as early as possible and to have it weighted as important as endpoints such as potency and selectivity. Traditionally, potency and selectivity still dominate decision making, but on occasion it may be easier to optimize potency and selectivity than ADME properties. Ideally, *in vitro* ADME data are acquired in the very early stages of drug discovery when multiple chemical series are still under consideration. In that case, the ADME data can be taken into consideration when deciding which series to pursue further in lead optimization. An attractive alternative approach in early drug discovery is the use of *in silico* models because these models can be applied prior to synthesis. Most companies have a range of commercial and in-house models available. Many of these *in silico* models are based on *in vitro* ADME data generated by higher-throughput LC-MS/MS assays (Hop et al., 2008).

Most pharmaceutical companies have introduced various routine, higher-throughput *in vitro* ADME assays to aid the drug discovery process with a capacity of 200 compounds or more per week (Hop et al., 2008). The sample preparation for these *in vitro* samples is usually limited to protein precipitation without a concentration step. Using column switching and/or UHPLC, the cycle times are very short and it is possible to obtain data for as much as four samples in 1 min. To increase the throughput even further, it is possible to pool samples with different analytes and use their MRM transitions to selectively monitor the various analytes of interest. Frequently, these assays are heavily automated

and use 96- or 384-well format (Janiszewski et al., 2008). Although these data can and should be used to influence design in drug discovery, they are not rigorous enough for regulatory submissions and more customized assays are required for that purpose.

Caco-2 or MDCK cells or PAMPA provide a measurement of the degree of permeability of the drug candidates, with higher permeability being indicative of a higher likelihood of good absorption. However, the *in vivo* situation is more complex and additional factors affect the percentage of the dose reaching the systemic circulation (e.g., the occurrence of intestinal metabolism). The liver is the most prominent organ for clearance and excretion of drugs. Thus, cellular fractions such as microsomes and S9 or intact hepatocytes are used to determine the metabolic stability of the drug candidates. Many projects suffer from poor metabolic stability of the leads, and therefore this assay is usually the first *in vitro* ADME assay in the screening cascade. Since a large percentage of drugs on the market are metabolized by the ubiquitous cytochrome P450 enzymes, it is important to determine if the drug candidates are inhibitors of these enzymes. This can be determined by looking at the inhibition of metabolism of probe substrates that are exclusively metabolized by specific cytochrome P450 (CYP) enzymes. Examples are provided in Table 19.3. To speed up analysis, it is possible to pool samples of incubations with different probe substrates and use the power of MRM on a triple quadrupole to provide sufficient selectivity for the assay. It is also valuable to monitor for mechanism-based inhibition of CYP enzymes by the drug candidates. This can be caused by tight noncovalent binding in the active site of the cytochrome P450 enzyme or due to covalent modification of the heme group or the apoprotein of the CYPs. This can be addressed by incubating microsomes with the drug candidates prior to adding the probe substrates. The preincubation may or may not give rise to a marked decrease in activity of the cytochrome P450 enzyme, and hence, reduced turnover of the probe substrate specific for that enzyme. Again, pooling of samples is possible.

Since most efficacy is determined by the free drug level at the site of action (Smith et al., 2010), it is important to determine the extent of binding of drug candidates to plasma proteins (albumin and α-acid glycoprotein) and/or tissue homogenates. Although ultrafiltration and ultracentrifugation are still used, the most frequently used technique is dialysis. A 96-well block is used with buffer on one side and plasma or tissue homogenate on the either side. After incubation, both sides are sampled. This assay usually requires a more sensitive assay because binding is quite extensive for higher molecular weight and/or lipophilic drug candidates (>95% bound). For proper pharmacokinetics/pharmacodynamics modeling, it is important to know

TABLE 19.3. Probe Substrates and LC-MS/MS Detection Method Used for Inhibition of Specific Cytochrome P450 Enzymes

Enzyme	Reaction	Ionization Mode	MRM Transition
CYP1A2	Phenacetin O-deethylation	+	$152 \rightarrow 110$
CYP2B6	Bupropion hydroxylation	+	$256 \rightarrow 238, 139$
CYP2C8	Paclitaxel 6α-hydroxylation	+	$870 \rightarrow 286, 105$
CYP2C9	S-warfarin 7'-hydroxylation	+	$325 \rightarrow 179$
CYP2C9	S-warfarin 7'-hydroxylation	−	$323 \rightarrow 177$
CYP2C9	Tolbutamide 4'-hydroxylation	+	$287 \rightarrow 171, 89$
CYP2C9	Tolbutamide 4'-hydroxylation	−	$285 \rightarrow 186$
CYP2C19	S-mephenytoin 4'-hydroxylation	+	$235 \rightarrow 150$
CYP2C19	S-mephenytoin 4'-hydroxylation	−	$233 \rightarrow 190$
CYP2D6	Dextrometorphan O-demethylation	+	$258 \rightarrow 199, 157$
CYP2E1	Chlorzoxazone 6-hydroxylation	+	$186 \rightarrow 130$
CYP2E1	Chlorzoxazone 6-hydroxylation	−	$184 \rightarrow 120$
CYP3A4/5	Midazolam 1'-hydroxylation	+	$342 \rightarrow 324, 297, 203$
CYP3A4/5	Testosterone 6β-hydroxylation	+	$305 \rightarrow 269$

the free tissue concentration at the site of action, and this can be obtained by multiplying the total concentration and the free fraction in the relevant tissue.

Studying drug transport has also become more common (Giacomini et al., 2010). Many drugs are to some extent substrates of uptake or efflux transporters. The most extensively studied transporter is P-glycoprotein and it can (1) limit intestinal absorption, (2) enhance biliary excretion, and (3) limit brain and/or tumor penetration. For central nervous system (CNS) targets where considerable levels in the brain are necessary, it is of vital importance to incorporate the P-glycoprotein assay early in the screening cascade to weed out compounds that would be efficiently effluxed from the brain. Cell lines for other transporters such as breast cancer resistance proteins (BCRPs), organic anion transporters (OATs), organic cation transporters (OCTs), organic anion transporting polypeptides (OATPs) and multidrug resistance-associated proteins (MRPs) are available as well.

19.12   QUANTITATIVE *IN VIVO* ADME STUDIES

LC-MS/MS is used extensively for bioanalysis of drugs and/or their metabolites in matrices such as plasma, blood, urine, and feces from *in vivo* studies. As described above, sample preparation is usually limited to protein precipitation, but may be more elaborate for GLP toxicology and clinical studies. In addition, this methodology can be used to determine drug levels in specific compartments of pharmacological interest such as the brain, tumor, or liver. Sample preparation for tissues involves thorough tissue homogenization, precipitation of endogenous components, and centrifugation. Triple quadrupole mass spectrometers are used to monitor the

MRM transitions of the parent compound as well as the internal standard, and if desired, metabolites can be monitored as well. An isotopically labeled analog of the analyte is preferred as the internal standard to eliminate differences in ion suppression/enhancement between the analyte and the internal standard. However, an isotopically labeled analog is usually not available in drug discovery, and therefore, a structural analog is used as internal standard.

For drug discovery studies, throughput is a critical factor and several approaches have been proposed to increase throughput for pharmacokinetic screening of drug candidates:

Sample Pooling: Samples from animals dosed with different compounds are pooled and all MRM transitions are monitored (e.g., cassette-accelerated rapid rat screen; Korfmacher et al., 2001).

Cassette or "N-in-One" Dosing: A mixture of drugs is administered to the same animals and all MRM transitions are monitored (Berman et al., 1997; Olah et al., 1997). It is possible that one of the components in the mixture inhibits the metabolism of another component in the mixture (White and Manitpisitkul, 2001). To reduce this risk, it is advisable to lower the dose. Frequently, a known reference compound is included in the cassette.

Multiplexed Analysis: Instruments are available that have multiple sprayers that allow introduction of the column effluent coming from different LC columns (Yang et al., 2001). By rotating between the different sprayers, it is possible to monitor the mobile phase coming from each column in isolation. One batch of samples can be analyzed using multiple sprayers or multiple batches of samples can be analyzed simultaneously each using a unique sprayer.

Staggered Parallel Analysis: A column switching device is used to allow the mobile phase to enter the ion source exclusively when the analytes of interest elute (Wu, 2001; King et al., 2002). Before and after this time window, the effluent from another LC column is monitored. This principle is illustrated in Figure 19.1.

The potential disadvantage of sample pooling and cassette dosing is that complex mixtures—containing multiple drug candidates and many metabolites—are analyzed. Thus, the risk of interference is more pronounced. The risk can be reduced to some extent by avoiding parent compounds with the same molecular weight or the same molecular weight as a potential metabolite of another compound (e.g., +16 Da or +32 Da metabolites).

19.13   METABOLITE IDENTIFICATION

LC-MS/MS is a very powerful tool for metabolite identification. Metabolites generated from *in vitro* incubations or *in vivo* studies can be detected with relative ease. The metabolism information can be conveyed to chemists to address specific metabolic liabilities in the next generation of compounds. It can also be used to facilitate assessment of the exposure to metabolites in humans versus those generated in preclinical studies (usually rat and dog) as recommended by the Metabolites in Safety Testing (MIST) guidance (US Food and Drug Administration Guidance for Industry, 2008; ICH Harmonized Tripartite Guideline, 2009).

Interpretation of the MS/MS data from product ion, neutral loss, or precursor ion scans allows the user to identify the specific site or, more frequently, the moiety of the molecule that was metabolically transformed. Sometimes, the exact site of metabolism may be unambiguous (e.g., for simple N-dealkylations), but for common pathways, such as oxidation, it is frequently required to isolate the metabolite and obtain NMR data to identify the exact structure. For example, it was not possible by LC-MS/MS to identify the exact sites of hydroxylation of a PPARγ agonist containing a cyclohexyl substituent (Agrawal et al., 2005). Subsequent NMR studies identified the predominant metabolic pathways as axial hydroxylation of the para and meta positions mediated by CYP2C19 and equatorial hydroxylation of the para and meta positions mediated by CYP2C8 (see Fig. 19.5). Additional tools available to facilitate metabolite identification are as follows:

- Accurate mass measurements (*vide supra*)
- Isotope patterns of [M + H]⁺ ions (metabolite identification is greatly facilitated by the presence of

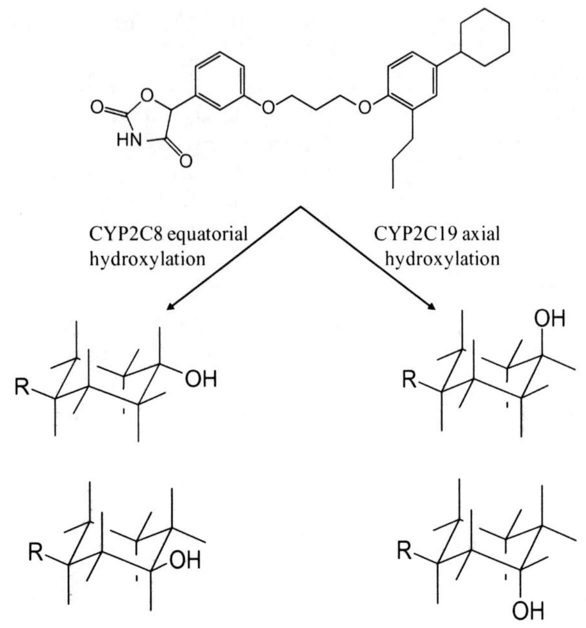

FIGURE 19.5. Stereospecific and regiospecific metabolism of the cyclohexyl group of a PPARγ agonist. The exact locations of hydroxylation required isolation of the metabolites and analysis by NMR.

atoms with a unique isotope pattern such as chlorine and bromine; see Table 19.2)
- Presence of a radiolabel (usually ³H or ¹⁴C)
- Chemical derivatization
- Hydrogen–deuterium exchange (Liu and Hop, 2005)
- Studies with close-in analogs

Another area of metabolite identification that is receiving more and more attention is the detection of adducts that are indicative of bioactivation (Evans et al., 2004; Hop et al., 2006). Many drugs are bioactivated and there is some (mostly circumstantial) evidence that reactive intermediates may give rise to (idiosyncratic) toxicity. Although reactive intermediates cannot be detected, it is possible to trap them with agents such as glutathione (abundant in the liver and an effective nucleophile by virtue of its cysteine sulfhydryl group), potassium cyanide (a "hard" nucleophile), and methoxylamine (to detect aldehydes). LC-MS/MS assays have been implemented to facilitate detection of these adducts. For example, glutathione adducts can be detected by a 129-Da neutral loss scan in the positive ion mode or a m/z 272 precursor ion scan in the negative ion mode (Dieckhaus et al., 2005). Recently, Gan et al. (2009) showed that there is a relationship between the estimated total human *in vivo* thiol adduct burden (i.e., normalized for the daily dose) and drug-induced toxicity. In the latter study, dansyl glutathione was used to facilitate quantitation of the adducts.

Finally, it is worth emphasizing that the data obtained in metabolite identification studies are qualitative only. The ionization efficiencies can vary widely across metabolites, in particular if a basic center has been eliminated from the drug candidate by biotransformation. Splitting the column effluent and acquiring UV or, if the compound is radiolabeled, radioactivity data will provide more reliable quantitative data. It is possible to use nanospray to reduce differences in ionization efficiency (Hop et al., 2005; Hop, 2006), but this technique is not routinely available.

Although great advances have been made in the instruments available for metabolite identification and some software tools are available for interpretation of the spectra, it generally requires involvement from a highly skilled scientist and can be time-consuming.

19.14 TISSUE IMAGING BY MS

To inform teams about tissue distribution of drugs, animals can be dissected and the tissue concentration can be determined in whole organs by routine LC-MS/MS assays as described above. However, it does not tell you how the drug is distributed in that tissue. It is possible to obtain that information by a quantitative whole-body autoradiography (QWBA) study. However, a QWBA study requires radiolabeled material that may not be available in drug discovery and it does not distinguish levels of the parent drug versus those of metabolites. Prof. Caprioli pioneered MALDI imaging and this technique has been successfully used to detect drugs and their metabolites in tissues (Khatib-Shahidi et al., 2006). First, a matrix solution is sprayed on the tissue or whole-body slice. After evaporation of the solvent, this may be repeated multiple times to get good coverage of the matrix. The matrix solution extracts the analytes out of the tissue and the analytes are subsequently ionized by MALDI. To prevent interference by the rather abundant matrix ions, MS/MS detection is preferred. By scanning the laser across the tissue or whole-body slice, it is possible to obtain information about the distribution of the parent drug and its metabolites at a spatial resolution of about 50 μm. Figure 19.6 shows the MALDI image of a compound dosed in rats at 150 mg/kg. It is clear that the compound accumulates in the skin (possibly by binding to melanin), which could help explain the clinically observed rash. The physicochemical properties of the compound, in particular the topological polar surface area, explain why it does not cross the blood–brain barrier.

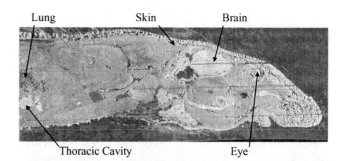

FIGURE 19.6. MALDI image displaying tissue distribution of a compound administered orally to a rat at 150 mg/kg (bottom), and the corresponding optical image (top). The compound is present in the skin and lung, but is excluded from the brain. (Data courtesy of S. Khatib-Shahidi.)

The disadvantages of MALDI imaging include limited sensitivity and the time it takes to acquire the data (24 h for a whole-body image is not uncommon). If metabolites are monitored and detection in the MS/MS mode is used, the time it takes to acquire the data multiplies. The latter can be addressed by acquiring data in high-resolution MS mode and extracting the images for the parent drug and its metabolites postacquisition (Cornett et al., 2008).

19.15 CONCLUSIONS AND FUTURE DIRECTIONS

It is realistic to state that the atmospheric pressure ionization techniques and the ability to interface them with LC equipment have revolutionized ADME sciences. The speed and ease with which quantitative and qualitative *in vitro* and *in vivo* ADME studies can be performed allow early incorporation of ADME feedback in drug design and it prevents late-stage failures due to poor pharmacokinetic and/or drug metabolism properties. Triple quadrupole mass spectrometers are the preferred instruments for quantitative analysis and more and more laboratories are switching to orbitrap mass spectrometers for metabolite identification. In both fields, the push for shorter cycle times will intensify to enhance efficiencies. MALDI imaging could become more ubiquitous provided the sensitivity, spatial resolution, and acquisition speed can be improved further.

One area that has not been embraced by vendors of mass spectrometers is miniaturization. This could take bioanalysis directly into the animal facility or the clinic to allow real-time analysis. It would also facilitate integration with miniaturized, integrated ADME devices, such as multicompartmental micro cell culture chips where cells are embedded in three-dimensional hydrogels in separate chambers representing different organs (such as the liver, bone marrow, and tumor tissue), connected by narrow channels that mimic the blood flow (Sung and Shuler, 2009; Sung et al., 2010). The latter devices will enhance our understanding of tissue distribution and the dynamic interplay between these tissues, and hence, increase the predictivity of physiologically based pharmacokinetic, pharmacokinetic-pharmacodynamic, and pharmacokinetic-toxicodynamic modeling.

REFERENCES

Agrawal AK, Hop CECA, Pang J, Silva Elipe MV, Desai RC, Leung KH, Franklin RM. (2005) *In vitro* metabolism of a new oxazolidinedione hypoglycemic agent utilizing liver microsomes and recombinant human cytochrome P450 enzymes. *J Pharm Biomed Anal* 37:351–358.

Ayrton J, Dear GJ, Leavens WJ, Mallett DN, Plumb RS (1997) The use of turbulent flow chromatography/mass spectrometry for the rapid direct analysis of a novel pharmaceutical compound in plasma. *Rapid Commun Mass Spectrom* 11:1953–1958.

Berman J, Halm K, Adison K, Shaffer J (1997) Simultaneous pharmacokinetic screening of a mixture of compounds in the dog using API LC/MS/MS analysis for increased throughput. *J Med Chem* 40:827–829.

Bonfiglio R, King RC, Olah TV, Merkle K (1999) The effects of sample preparation methods on the variability of the electrospray ionization response for model drug compounds. *Rapid Commun Mass Spectrom* 13:1175–1185.

Cornett DS, Frappier SL, Caprioli RM (2008) MALDI-FTICR imaging mass spectrometry of drugs and metabolites in tissue. *Anal Chem* 80:5648–5653.

Cunliffe JM, Noren CF, Hayes RN, Clement RP, Shen JX (2009) A high-throughput LC-MS/MS method for quantitation of posaconazole in human plasma: Implementing fused core silica liquid chromatography. *J Pharm Biomed Anal* 50:46–52.

Dieckhaus CM, Fernandez-Metzler CL, King R, Krilikowski PH, Baillie TA (2005) Negative ion tandem mass spectrometry for the detection of glutathione conjugates. *Chem Res Toxicol* 18:630–638.

Evans DC, Watt AP, Nicoll-Griffith DA, Baillie TA (2004) Drug-protein adducts: An industry perspective on minimizing the potential for drug bioactivation in drug discovery and development. *Chem Res Toxicol* 17:3–16.

Gan J, Ran Q, He B, Zhu M, Shyu WC, Humphreys WG (2009) *In vitro* screening of 50 highly prescribed drugs for thiol adduct formation—Comparison of potential for drug-induced toxicity and extent of adduct formation. *Chem Res Toxicol* 22:690–698.

Giacomini KM, Huang S-M, et al. (2010) Membrane transporters in drug development. *Nat Rev Drug Discov* 9:215–236.

Gobey J, Cole M, Janiszewski J, Covey T, Chau T, Kovarik P, Corr J (2005) Characterization and performance of MALDI on a triple quadrupole mass spectrometer for analysis and quantification of small molecules. *Anal Chem* 77:5643–5654.

Henion J, Brewer E, Rule G (1998) Sample preparation for LC/MS/MS: Analyzing biological and environmental samples. *Anal Chem* 70:650A–656A; and references therein.

Hop CECA (2006) Use of nano-electrospray for metabolite identification and quantitative absorption, distribution, metabolism and excretion studies. *Curr Drug Metab* 7:557–563.

Hop CECA, Chen Y, Yu LJ (2005) Uniformity of ionization response of structurally diverse analytes using a chip-based nanoelectrospray ionization source. *Rapid Commun Mass Spectrom* 19:3139–3142.

Hop CECA, Kalgutkar AS, Soglia JR (2006) Importance of early assessment of bioactivation in drug discovery. In *Annual Reports in Medicinal Chemistry*, Vol. 41, Doherty AM, ed., pp. 369–381. Elsevier, Amsterdam, The Netherlands.

Hop CECA, Cole MJ, Davidson RE, Duignan DB, Federico J, Janiszewski JS, Jenkins K, Krueger S, Lebowitz R, Liston TE, Mitchell W, Snyder M, Steyn SJ, Soglia JR, Troutman MD, Umland J, West M, Whalen KM, Zelesky V, Zhao SX (2008) High throughput ADME screening: Practical considerations, impact on the portfolio and enabler of *in silico* ADME models. *Curr Drug Metab* 9:847–853.

Hsieh Y, Wang G, Wang Y, Chackalamannil S, Brisson J-M, Ng K, Korfmacher WA (2002) Simultaneous determination of a drug candidate and its metabolite in rat plasma samples using ultrafast monolithic column high-performance liquid chromatography/tandem mass spectrometry. *Rapid Commun Mass Spectrom* 16:944–950.

ICH Harmonized Tripartite Guideline. (2009) *Guidance on nonclinical safety studies for the conduct of human clinical trials and marketing authorization for pharmaceuticals M3(R2) June 2009.* http://www.ich.org/fileadmin/Public_Web_Site/ICH_Products/Guidelines/Multidisciplinary/M3_R2/Step4/M3_R2__Guideline.pdf.

Janiszewski JS, Liston TE, Cole MJ (2008) Perspectives on bioanalytical mass spectrometry and automation in drug discovery. *Curr Drug Metab* 9:986–994.

Jemal M, Xia Y-Q (1999) The need for adequate chromatographic separation in the quantitative determination of drugs in biological samples by high performance liquid chromatography with tandem mass spectrometry. *Rapid Commun Mass Spectrom* 13:97–106.

Khatib-Shahidi S, Andersson M, Herman JL, Gillespie TA, Caprioli RM (2006) Direct molecular analysis of whole-body animal tissue sections by imaging MALDI mass spectrometry. *Anal Chem* 78:6448–6456.

King RC, Miller-Stein C, Magiera DJ, Brann J (2002) Description and validation of a staggered parallel high performance liquid chromatography system for good laboratory practice level quantitative analysis by liquid chromatography/tandem mass spectrometry. *Rapid Commun Mass Spectrom* 16:43–52.

Kola I, Landis J (2004) Can the pharmaceutical industry reduce attrition rates? *Nat Rev Drug Discov* 3: 711–715.

Korfmacher WA, Cox KA, Ng KJ, Veals J, Hsieh Y, Wainhaus S, Broske L, Prelusky D, Nomeir A, White RE (2001) Cassette-accelerated rapid rat screen: A systematic procedure for the dosing and liquid chromatography/atmospheric pressure ionization tandem mass spectrometric analysis of new chemical entities as part of new drug discovery. *Rapid Commun Mass Spectrom* 15:335–340.

Liu DQ, Hop CECA (2005) Strategies for characterization of drug metabolites using liquid chromatography-tandem mass spectrometry in conjunction with chemical derivatization and on-line H/D exchange approaches. *J Pharm Biomed Anal* 37:1–18.

Mazzeo JR, Neue UD, Kele M, Plumb RS (2005) Advancing LC performance with smaller particles and higher pressure. *Anal Chem* 77:460A–467A.

Olah TV, McLoughlin DA, Gilbert JD (1997) The simultaneous determination of mixtures of drug candidates by liquid chromatography/atmospheric pressure chemical ionization mass spectrometry as an *in vivo* drug screening procedure. *Rapid Commun Mass Spectrom* 11:17–23.

Plumb RS, Potts WB III, Rainville PD, Alden PG, Shave DH, Baynham G, Mazzeo JR (2008) Addressing the analytical throughput challenges in ADME screening using ultra-performance liquid chromatography/tandem mass spectrometry methodologies. *Rapid Commun Mass Spectrom* 22:2139–2152.

Savoie N, Booth BP, Bradley T, Garofolo F, Hughes NC, Hussain S, King SP, Lindsay M, Lowes S, Ormsbey E, Phull R, Rocci JML Jr, Vallano PT, Viau A, Zhu Z (2009) The 2nd calibration and validation workshop on recent issues in good laboratory practice bioanalysis. *Bioanalysis* 1:19–30.

Smith DA, Di L, Kerns EH (2010) The effect of plasma protein binding on *in vivo* efficacy: Misconceptions in drug discovery. *Nat Rev Drug Discov* 9:929–939.

Spooner N, Lad R, Barfield M (2009) Dried blood spots as a sample collection technique for the determination of pharmacokinetics in clinical studies: Considerations for the validation of a quantitative bioanalytical method. *Anal Chem* 81:1557–1563.

Sung JH, Shuler ML (2009) A micro cell culture analog (μCCA) with 3-D hydrogel culture of multiple cell lines to assess metabolism-dependent cytotoxicity of anti-cancer drugs. *Lab Chip* 9:1385–1394.

Sung JH, Kam C, Shuler ML (2010) A microfluidic device for a pharmacokinetic-pharmacodynamic (PK-PD) model on a chip. *Lab Chip* 10:446–455.

The Nobel Prize in Chemistry 2002. (2002) *Nobelprize.org Apr 5, 2011.* http://nobelprize.org/nobel_prizes/chemistry/laureates/2002.

US Food and Drug Administration Guidance for Industry. (2008) *Safety testing of drug metabolites. February 2008.* http://www.fda.gov/downloads/Drugs/GuidanceCompliance RegulatoryInformation/Guidances/ucm079266.pdf.

Vishwanathan CT, Bansal S, Booth B, DeStefano AJ, Rose MJ, Sailstad J, Shah VP, Skelly JP, Swann PG, Weiner R (2007) Quantitative bioanalytical methods validation and implementation: Best practices for chromatographic and ligand binding assays. *Pharm Res* 24:1962–1973.

White RE, Manitpisitkul P (2001) Pharmacokinetic theory of cassette dosing in drug discovery screening. *Drug Metab Dispos* 29:957–966.

Wu J-T (2001) The development of a staggered parallel separation liquid chromatography/tandem mass spectrometry system with on-line extraction for high-throughput screening of drug candidates in biological fluids. *Rapid Commun Mass Spectrom* 15:73–81.

Yang L, Mann TD, Little D, Wu N, Clement RP, Rudewicz P (2001) Evaluation of a four-channel multiplexed electrospray triple quadrupole mass spectrometer for the simultaneous validation of LC/MS/MS methods in four different preclinical matrices. *Anal Chem* 73:1740–1747.

Zhang D, Cheng PT, Zhang H (2007) Mass defect filtering on high resolution LC/MS data as a methodology for detecting metabolites with unpredictable structures: Identification of oxazole-ring opened metabolites of muraglitazar. *Drug Metab Lett* 1:287–292.

Zhang H, Zhang D, Ray K, Zhu M (2009) Mass defect filter technique and its application to drug metabolite identification by high resolution mass spectrometry. *J Mass Spectrom* 44:999–1016.

20

APPLICATION OF ACCURATE MASS SPECTROMETRY FOR METABOLITE IDENTIFICATION

Zhoupeng Zhang and Kaushik Mitra

20.1 INTRODUCTION

Clinical efficacy of a drug is generally dependent on its plasma concentration. One of the factors that determine the exposure of drugs is metabolism, a complex process where the parent molecule is converted into new structural entities called metabolites. Identification of metabolites has evolved to be a key component of drug discovery studies with multifaceted applications: (1) to understand how the compound is cleared from the body (early intervention would allow chemical modifications in lead compound classes to optimize pharmacokinetic properties); (2) to identify structural liabilities that may lead to reactive intermediates, via metabolism, that are suspects for adverse effects; (3) to discover the beneficial pharmacodynamic and adverse off-target activities of metabolites, which may contribute, respectively, to the overall efficacy or the toxicity of the drug; (4) to assure that the safety assessment studies are conducted in those preclinical species that have similar metabolism profiles as predicted for humans, thereby ensuring adequate "metabolite coverage"; and (5) to improve confidence in human pharmacokinetic prediction (Benfenati et al., 1991; Takamura et al., 2006; Zhang and Gan, 2007; Zhang et al., 2009).

Technically, metabolite identification involves three steps: separation, detection, and structural analysis. High-performance liquid chromatography is one of the commonly utilized approaches to separate metabolites from biological matrices, and mass spectrometry is the preferred choice for detection and structural analysis of

metabolites (Zhang et al., 2009). Mass spectrometers that identified metabolites only by measuring their nominal masses were used for metabolite identification in past years; however, measurement of nominal masses suffered from the drawback of generating false positive results as well as from limitations in applying postacquisition software to facilitate the structural assignments of molecular fragments. In recent years, high-resolution/ accurate mass spectrometers have been utilized more often in metabolite identification studies to overcome the disadvantages associated with nominal mass spectrometers. Also, accurate mass measurements can be now coupled with sophisticated software to facilitate the detection of metabolites, thus increasing efficiency and throughput. This review is intended to summarize major utilities of high-resolution/accurate mass spectrometers in metabolite identification.

20.2 HIGH-RESOLUTION/ACCURATE MASS SPECTROMETERS

Mass spectrometers are instruments that can measure the masses and relative concentrations of molecular ions generated in a system. High resolution refers to the ability of the instrument to measure the small differences deriving from different elemental components that have the same nominal mass. Accurate mass measurement allows the instrument to assign a precise relative molecular mass to a signal detected in the system. High-resolution/accurate mass spectrometers

ADME-Enabling Technologies in Drug Design and Development, First Edition. Edited by Donglu Zhang and Sekhar Surapaneni.

have the ability to accurately measure ions in a high-resolution mode. The major platforms of high-resolution/accurate mass spectrometers are briefly discussed below.

20.2.1 Linear Trap Quadrupole-Orbitrap (LTQ-Orbitrap) Mass Spectrometer

The LTQ-Orbitrap mass spectrometer was introduced by Thermo Scientific (San Jose, CA, USA) several years ago. It is capable of high resolution (R = 60,000) full mass spectrometry (MS) scan as well as sequential MS/MS (tandem mass spectrometry [MS^n]) analysis of selected ions. The function of the data-dependent MS^n is to obtain full scan, as well as MS^n spectra of unknown metabolites in a single run. The accuracy of molecular ions and their fragments is generally <2 ppm. However, the LTQ/Orbitrap is not able to perform true neutral loss (NL), product ion (PI), and multiple reaction monitoring (MRM) scans that are employed in the detection of uncommon or unpredicted metabolites with traditional triple quadrupole or triple quadrupole-linear ion trap instruments.

In 2009, the benchtop high-resolution/accurate mass spectrometer, Exactive, was introduced by Thermo Scientific. The Exactive is capable of performing a full MS scan as well as a higher-energy collisional dissociation (HCD) scan with high resolution (R = 100,000) and a mass accuracy of <3 ppm. The Exactive's HCD scan obtains all ion fragmentations and is similar to the MS^e scan with the Q-tof (quadrupole time-of-flight mass spectrometer) instrument (Waters Corp. Milford, MA, USA). In addition, the Exactive is able to perform scans in alternating positive and negative modes (polarity switch) during the same LC-MS run. Because the Exactive has no ion trap in the front, it is not capable of performing MS^n scans.

20.2.2 Q-tof and Triple Time-of-Flight (TOF)

The TOF mass spectrometer was first discovered about half a century ago (Stephens, 1946). In a TOF mass spectrometer, ions of different mass-to-charge ratios are accelerated by an electric field and travel into a field-free drift tube. The time for an ion to subsequently reach a detector depends on the mass-to-charge ratio of the particle. Based on this time and other known experimental parameters, the mass-to-charge ratio of the ion (accurate mass) is obtained (Ma and Chowdhury, 2007). The hybrid quadrupole time-of-flight accurate mass spectrometer (Q-tof) from Waters Corp. has an additional quadrupole in front of a TOF analyzer, enabling the collision-induced dissociation (CID) of molecular ions. It is capable of performing a high-resolution (R = 20,000) full-scan MS scan as well as analysis of MS/MS scan of selected ions, precursor ion scan, and neutral

loss scan. The mass accuracy of <5 ppm is achieved by using internal continuous lockspray calibration. By coupling with ultra-performance liquid chromatography (UPLC), the Q-tof is also capable of performing "all-in-one" analyses with full scan and the MS^e functions, obtaining the information needed for structural assignment of metabolites from a single LC-MS run.

In May 2010, AB Sciex (Foster City, CA, USA) introduced a TripleTOF 5600 system, which is also a hybrid quadrupole time-of-flight accurate mass spectrometer. It can reach a resolution of >40,000 with a mass accuracy of <2 ppm without continuous internal calibration. This system also can perform neutral loss scanning, precursor scanning, and multiple mass defect filter-triggered information-dependent acquisition. The linear dynamic range is about four orders of magnitude for quantitative accuracy. The TripleTOF 5600 system is also capable of performing "all-in-one" analyses (MS/MS^{all}) with full scan and the MS^e functions.

20.2.3 Hybrid Ion Trap Time-of-Flight Mass Spectrometer (IT-tof)

The hybrid IT-tof, recently introduced by Shimadzu (Columbia, MD, USA), has a frontal ion trap to focus ions before ejection into the TOF for accurate mass measurement. Similar to the LTQ-Orbitrap, the IT-tof is capable of high-resolution (R = 10,000) full MS scanning as well as sequential MS/MS (MS^n) analysis of selected ions. The function of the data-dependent scanning obtains a full scan as well as MS^n spectra of unknown metabolites in a single run. The accuracy of measurement of molecular ions and their fragments is generally <5 ppm. The IT-tof can also perform neutral loss survey scan (MS^3) if the specified neutral loss is observed in the MS^2 spectrum. In addition, its high speed ion polarity-switching feature enables obtaining a pair of positive and negative ion MS spectra at a fast rate (2.5 times per second).

20.3 POSTACQUISITION DATA PROCESSING

Postacquisition data processing is an approach to utilize software in the detection of metabolites. In the era of nominal mass spectrometers where the accuracy of the masses of the detected ions was low, postacquisition processing of data files did not add significant value to the final analysis of the raw data files. However, the high mass accuracy of ions detected by the current high-resolution/accurate mass spectrometers makes postacquisition data processing practicable and reliable. In recent years, several postacquisition data processing techniques, including mass defect filtering (MDF)

(Zhang et al., 2003; Zhang, 2006; Zhu et al., 2006), background subtraction (Zhang and Yang, 2008; Zhang et al., 2008), neutral loss filtering (NLF) (Lim et al., 2008; Ruan et al., 2008), product ion filtering (PIF) (Ruan et al., 2008), and isotope pattern filtering (IPF) (Lim et al., 2008), have been developed for mining metabolites whose spectra are already recorded in accurate mass full-scan MS and MS/MS databases. MDF and background subtraction techniques are two of the most commonly utilized postacquisition data processing approaches in metabolism studies today.

20.3.1 MDF

The exact mass of a molecule is the sum of the calculated masses of individual isotopic elements that compose a single molecule. The mass defect of a molecule is defined as the difference of the exact mass from its integer. For example, water (H_2O) and oxygen (O_2) molecules have exact masses of 18.0106 and 31.9898 with mass defects of +0.0106 Da and −0.0102 Da, respectively. Common Phase I and Phase II metabolites generally result in mass defect shifts of <50 mDa from the parent molecules (Zhang et al., 2003; Zhu et al., 2006). For example, monohydroxylation of the parent introduces a mass defect of −5.1 mDa, methylation of +15.7 mDa, glucuronidation of 32.1 mDa. Thus, when the mass defect filter window of ±50 mDa is applied to the acquired accurate mass data, the molecular ions whose decimal portions are not within the predefined MDF window are excluded, and the ions whose decimal portions lie in the window are retained. This DMF approach therefore removes the majority of interference ions from the raw data files. The metabolites remaining in the resulting filtered data files are relatively easily identified. In cases of large changes in molecular formula through metabolism (such as breaking of the parent compound into two metabolites by hydrolysis, or a glutathione [GSH] conjugation), the mass defects may be >50 mDa. However, the majority of such biotransformations are predictable, and the

mass defect changes can be built into the processing software (Zhang et al., 2009).

In the example below, the MDF technique was applied to analyze a bile sample of rats dosed with an experimental drug. The total ion chromatogram (TIC) of the raw LC-MS data was not able to provide the needed information for the metabolite profiling due to the presence of a significant amount of endogenous matrix components from the *in vivo* sample (Figure 20.1A). Following the application of the MDF window of ±50 mDa to the original data file, several drug-related peaks were observed (Figure 20.1B). Particularly noteworthy is the metabolite ion of m/z 757.2470, at a retention time of 23.9 min, embedded among the compound-independent endogenous matrix components in the original mass spectrum (Figure 20.2A,B). Thus, MDF provides a powerful semiautomated technique to identify metabolites, both expected ones such as oxidations and unexpected/novel ones, from matrices that may contain large amounts of hindering components.

20.3.2 Background Subtraction Software

Background subtraction is a postacquisition data processing approach by which molecular ions detected in the background of the sample matrix can be subtracted from the analyte. Subtraction of the background of the sample matrix components can significantly clean up the analyte sample, enabling the ease of detection of metabolites of interest. Use of this software was exemplified by the detection of GSH adducts of clozapine, diclofenac, imipramine, and tacrine in human liver microsomal incubations supplemented with GSH (Zhang and Yang, 2008). First, a control scan time window of ±1.0 min around an analyte scan was defined to check the ions from background or matrix. Such a checking algorithm was looped throughout all the analyte scans in a data set. Any ion in the analyte scan was subtracted by identifying the maximum intensity of the ions within a specified tolerance window (e.g., ±5 ppm) toward that ion in the control scans. This

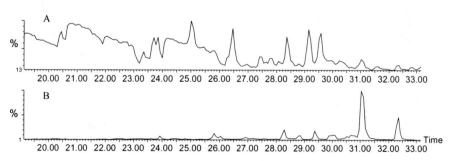

FIGURE 20.1. Total ion chromatograms (TIC) of a test compound in rat bile: (A) TIC of raw data; (B) TIC of the processed data (MDF of ±50 mDa).

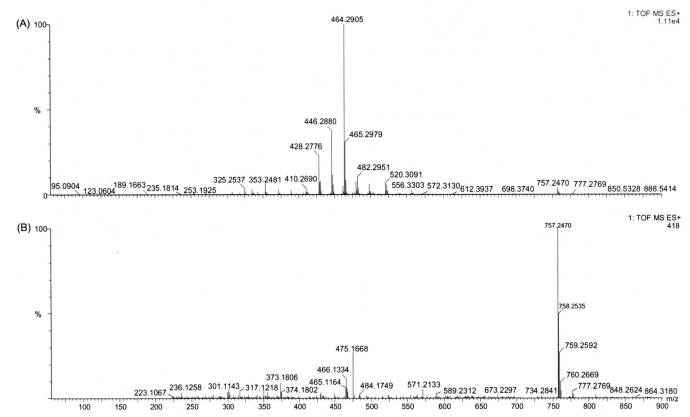

FIGURE 20.2. Mass spectra of a metabolite at retention time of 23.92 min: (A) mass spectra by combining scans over the peak and subtracting scans before and after the peak of raw data; (B) mass spectra by combining scans over the peak and subtracting scans before and after the peak of the processed data (MDF of ±50 mDa).

intensity was then multiplied with a user-specified scaling factor and was directly subtracted from that of the analyte ion in the analyte scan. Various GSH adducts were easily detected. Many of them were doubly charged species that would be difficult to be detected using nominal mass spectrometers. This background subtraction approach was proved to be very useful in the detection of troglitazone metabolites in rat plasma, bile, and urine samples (Zhang et al., 2008). A noise reduction algorithm (NoRA), an add-on function to the background subtraction approach that incorporates removing ion signals that are inconsistent across adjacent scans, was recently reported for effective detection of *in vivo* metabolites (Zhu et al., 2009).

20.4 UTILITIES OF HIGH-RESOLUTION/ ACCURATE MASS SPECTROMETRY (HRMS) IN METABOLITE IDENTIFICATION

In the past decades, nominal mass spectrometers such as triple quadrupoles, ion traps, and hybrid triple quadrupole-linear ion traps were the commonly utilized

instrumentations for metabolite identification. Generally, multiple injections of a given sample are required to obtain efficient mass spectra for structural elucidation. The major drawback for nominal mass spectrometers is lack of high accuracy of detected ions, resulting in a significant numbers of false positive results. This complicates the structural assignment of metabolites as well as reduces efficiency in the metabolite identification processes, limiting the usage of mass spectrometers for fast screening of large numbers of compounds.

In the past few years, high-resolution/accurate mass spectrometry has been gradually utilized in all aspects of traditional metabolite identification areas (Zhang et al., 2009), thanks to advances in mass spectrometry technologies and postacquisition processing software. The major uses of high-resolution/accurate mass spectrometry in metabolite identification are reviewed below.

20.4.1 Fast Metabolite Identification of Metabolically Unstable Compounds

In vitro metabolic stability studies in microsomes or hepatocytes, often having the capacity to support a large number of compounds on a weekly basis, are among the first screening studies done on discovery compounds. This type of study is generally followed by experiments to identify metabolic soft spots in unstable compounds in order to design follow-up molecules devoid of such liabilities. Thus, there is a need for conducting metabolite identification studies at a high throughput.

Recent technical advances in the hardware and software of mass spectrometers allow such a high throughput. The hybrid LTQ-Orbitrap mass spectrometer has high trapping capacity, an MS^n scan function, and produces accurate mass measurement. Its fast data-dependent acquisition of accurate MS^n spectra on an LC timescale enables the increased throughput of metabolite identification (Peterman et al., 2006). General full-scan MS and MS^n data sets of *in vitro* samples are acquired with a generic data-dependent method (Ruan et al., 2008). The data are processed using multiple post-acquisition data mining techniques, including extracted-ion chromatography (EIC), MDF, PIF, and NLF techniques. The high-resolution EIC approach searches the acquired full-scan MS data set with a list of accurate m/z values calculated from common metabolism reactions with an assigned mass accuracy (e.g., 5 ppm). The MDF process, which searches for metabolites based on the similarity of the mass defects of the metabolites to those of the test compounds and their core substructures, is used to find common metabolites as well as those uncommon metabolites that are not detected by the EIC processing. The high-resolution PIF and NLF processes selectively detect metabolites with predefined product ions and neutral losses of fragments from parent molecules, assuming that fragmentation patterns of metabolites are similar to those of the test compounds (Ruan et al., 2008). After the metabolites are detected by any of the above data processing techniques, the MS^n spectra of the metabolites can be retrieved from the originally acquired data set for the purpose of structure assignment. This integrated approach using LTQ-Orbitrap generally requires only a single LC-MS/MS run of *in vitro* samples to obtain enough full MS and MS^n spectra for structural elucidation, enabling fast metabolite identification of large numbers of metabolically unstable compounds.

It was shown that the use of Q-tof coupled with UPLC significantly increases the throughput, sensitivity, and chromatographic resolution in metabolite identification of samples from *in vitro* and *in vivo* samples. In a reported study with *in vitro* incubations of compounds in liver microsomes, about a fivefold increase in the signal-to-noise (S/N) ratio in UPLC over high-performance liquid chromatography (HPLC) was observed in comparison analyses, mainly due to a combination of reduced peak width (and consequent increase in concentration of the analyte at the detector) and a reduction in ion suppression (due to the efficient separation of metabolites from biological matrix components). "All-in-one" analyses using UPLC/Q-Tof with two collision energies in two scan functions for metabolite identification have been reported (Wrona et al., 2005; Bateman et al., 2007). In this approach, CID spectra are acquired using two different collision energies. Molecular ions of interest were detected in the full-scan MS spectra obtained at low collision energy, and the fragment ion information was obtained at high-energy collision. Due to the lack of specific precursor ion selection, the MS^e spectra (where "E" represents collision energy) at high energy are really "pseudo" MS/MS because of the full-scan MS function. Thus, the MS^e spectra obtained in this approach may have ions not related to the parent molecule or its metabolites. Use of UPLC with significantly increased resolving power and sensitivity substantially ameliorates the possibility of coelution of metabolites/impurities in analysis (Castro-Perez et al., 2005; Mortishire-Smith et al., 2005). By applying MDF using mass range and biotransformation selective filters in combination with the "all-in-one" UPLC/Q-tof approach, the endogenous (matrix) interference in biological samples was greatly minimized (Mortishire-Smith et al., 2005). This postacquisition processing greatly improved the selectivity for identifying metabolites.

Compound A (Figure 20.3) was metabolically unstable in rat liver microsomes in the presence of NADPH, and only about 9% of the parent compound remained at 45 min (Figure 20.4). The accurate mass spectrometry analysis (Q-tof) coupled with UPLC revealed that nine oxidative metabolites were formed in rat liver microsomal incubations (Figure 20.5). A monooxygenated

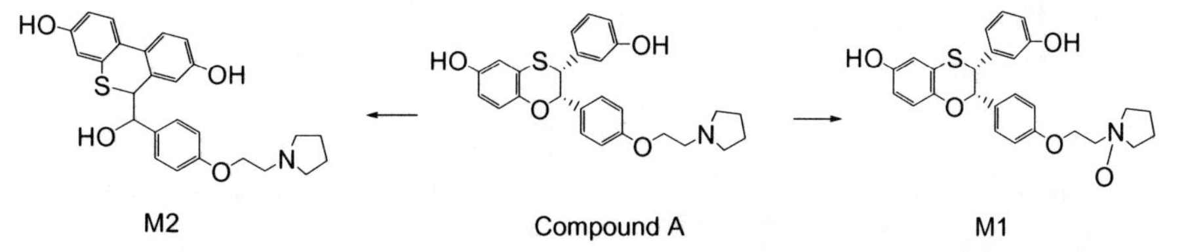

FIGURE 20.3. Structures of compound A and its metabolites M1 and M2.

metabolite (M1), which eluted behind the parent molecule, was identified as a major metabolite in both HRMS (Figure 20.5) and UV (Figure 20.6) analysis. The ion of m/z 310.1856 in the MS^e spectrum of M1 (Figure 20.7B) may have resulted from the loss of an oxygen atom (−15.9949) from the ion of m/z 326.1792. Thus, by comparing the MS^e spectrum of parent compound A

(Figure 20.7A) with that of M1 (Figure 20.7B), it can be speculated that the structure of this major metabolite M1 is likely the *N*-oxide derivative (Figure 20.3). Thus, the metabolic soft spot of compound A in rat liver microsomes was quickly identified as the nitrogen atom on the pyrrolidine ring at the discovery screening stage to guide chemists for further modification of this lead compound.

20.4.2 Identification of Unusual Metabolites

Detection and characterization of unusual metabolites is the main challenge faced by drug metabolism scientists. Traditional approaches using nominal mass spectrometers to detect unusual metabolites include scanning of the precursor ions of a given unmodified portion of the parent molecule, scanning for the neutral loss of a known portion of a parent molecule, or manual comparison of extra ions in analyte samples with the control samples. These approaches are very time-consuming, and it is very common to have false positive or false negative results.

High-resolution/accurate mass spectrometry together with postacquisition processing software is a useful approach to identifying unusual metabolites. The improved sensitivity with high-resolution/accurate mass spectrometers often significantly minimizes false results in the detection of unusual metabolites. Also, due to the high accuracy of mass measurement, the m/z values of the detected molecular ions and their MS/MS fragments can provide direct information regarding the structural

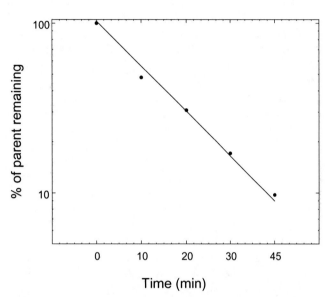

FIGURE 20.4. Metabolic stability of compound A (1 μM) in rat liver microsomes (0.5 mg protein/mL) in the presence of NADPH.

Expected Metabolites:

Metabolites Name	Formula	m/z Found	mDa	Time	Area Abs	Area %
−H2	C28H25NO4S	448.1592	1.0	5.49	2.20	0.10 (0.04)
−H2	C28H25NO4S	448.1566	−1.6	7.01	26.50	1.24 (0.51)
M2 ⟶ Parent	C28H27NO4S	450.1744	0.5	5.63	44.70	2.10 (0.86)
Parent	C28H27NO4S	450.1737	−0.2	7.13	533.00	25.03 (10.29)
+O	C28H27NO5S	466.1700	1.2	5.98	64.20	3.02 (1.24)
+O	C28H27NO5S	466.1700	1.2	6.44	98.10	4.61 (1.89)
M1 ⟶ +O	C28H27NO5S	466.1692	0.4	7.54	1098.90	51.61 (21.21)
+H2O	C28H29NO5S	468.1830	−1.4	4.80	2.10	0.10 (0.04)
+O2	C28H27NO6S	482.1647	1.0	6.75	96.80	4.55 (1.87)
+O2	C28H27NO6S	482.1638	0.1	6.89	162.70	7.64 (3.14)

Total metabolites: 10

Combined Metabolite Peaks (All Found and Unexpected Peaks)

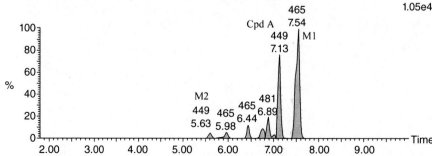

FIGURE 20.5. Metabolynx report of metabolite profile of compound A in incubation with rat liver microsomes.

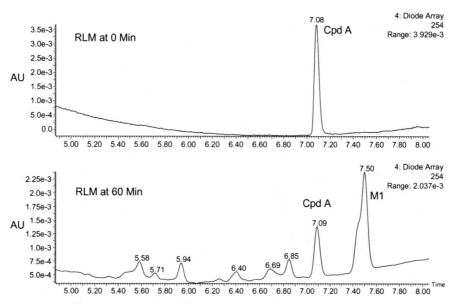

FIGURE 20.6. The UV spectrum of the sample from incubation of compound A with rat liver microsomes (RLM) at 254 nm.

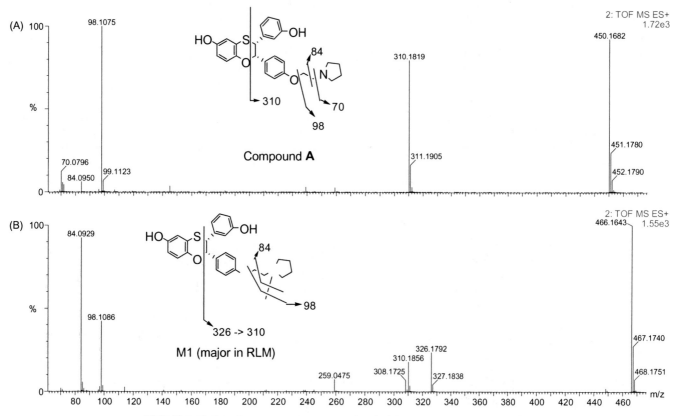

FIGURE 20.7. MS^e spectra of compound A (A) and its metabolite M1 (B).

information of the unusual metabolites, and software-assisted elementary analysis of the detected ions may be used in the characterization of the molecular formula of unusual metabolites. Furthermore, the use of postacquisition processing functions, including MDF, search of a library of combinations of different predicted Phase I and Phase II metabolites with or without various dealkylations (Mortishire-Smith et al., 2009) of the parent molecules, comparison of background ions from control samples, and isotopic pattern searching, and so on make the detection of unusual metabolites more practicable.

In the previous example of Q-tof analysis of samples from incubations of compound A in rat liver microsomes, an additional "parent" molecular ion at 5.63 min was detected in full-scan MS analysis (Figure 20.5). This additional "parent" molecular ion has the same accurate mass (450.1744) as the parent molecule (450.1737), but a different retention time (5.63 min) from the parent molecule (7.13 min). Also, this additional "parent" molecular ion was not seen in the control sample (0 min incubation). Further analysis of MS^e spectra indicates that the fragments of m/z 70 and 98 were present from both molecular ions, indicating the intact form of the pyrrolidine ring functional group in the additional "parent" molecule. However, the diagnostic fragment of m/z 310 of the parent molecule, which resulted from the cleavage of the -C-S and -C-O bonds of the 2,3-dihydro-1,4-benzoxathiin ring, was not shown in the additional "parent" molecule. Data above suggest that this additional "parent" molecular ion is a drug-related metabolite that may be derived from an unusual biotransformation mechanism (ring rearrangement). The fragments also suggest that the rearrangement might have occurred at or around the 2,3-dihydro-1,4-benzoxathiin ring. Indeed, based on the nuclear mag-

netic resonance (NMR) analysis as well as comparison with a synthetic standard, this additional "parent" molecule is an extended hydroquinone metabolite (M2; Figure 20.3). The formation of hydroquinone M2 is catalyzed by cytochrome P450 3A4 through an initial two-electron oxidation process to afford the quinonium cation intermediate I (Figure 20.8). The quinonium cation I is then hydrolyzed to the *para* quinone intermediate II. The *para* carbon atom of the phenolic ring of II undergoes intramolecular nucleophilic addition to the quinone carbonyl, followed by dehydration of the product III to give the extended quinone intermediate IV. The intermediate IV then is reduced to afford M2 (Zhang et al., 2005). Indeed, the high-resolution/accurate mass spectrometry approach can significantly improve the detection of unusual metabolites from *in vitro* and *in vivo* biological samples.

When unusual metabolism reactions are discovered in analyst labs or published, the new reactions may be immediately added into the library of known reactions for postacquisition processing of accurate mass data files. For example, it was reported that 4-(methylnitrosamino)-l-(3-pyridyl)-l-butanone (NNK) reacted with NADP$^+$ in rat liver microsomes to form the NNK-ADP$^+$ adduct (Figure 20.9) (Peterson et al., 1994). Mechanistically, NADP$^+$, one of the components of the NADPH regenerating system, is hydrolyzed by NAD glycohydrolases to form ADP-ribose. The nitrogen of the NNK molecule is then added onto the carbon of the ADP-ribose to form the corresponding NNK-ADP$^+$ adduct (Figure 20.9). It is expected that other nitrogen-containing compounds may undergo the same mechanism to form the corresponding ADP adducts. Thus, the formula of C15H23N5O16P3 (molecular mass of 622.0353), which represents the accurate mass of the

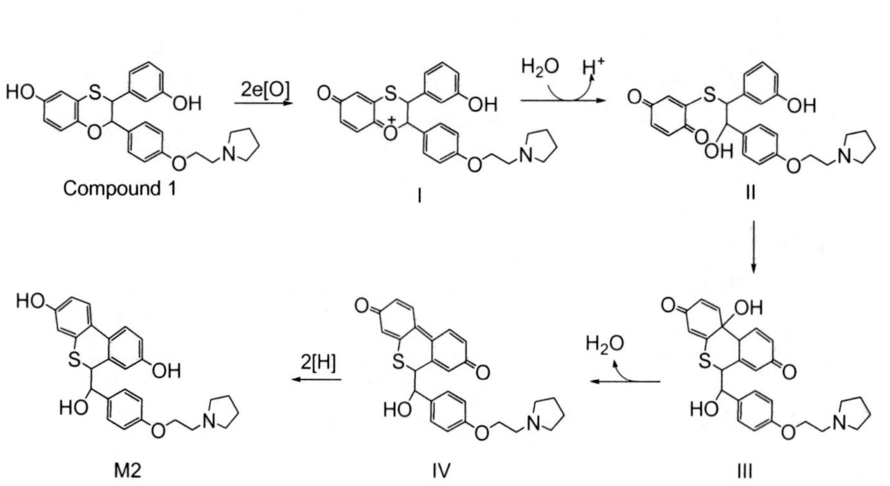

FIGURE 20.8. Proposed mechanism for the formation of metabolite M2 from incubation of compound A with rat liver microsomes.

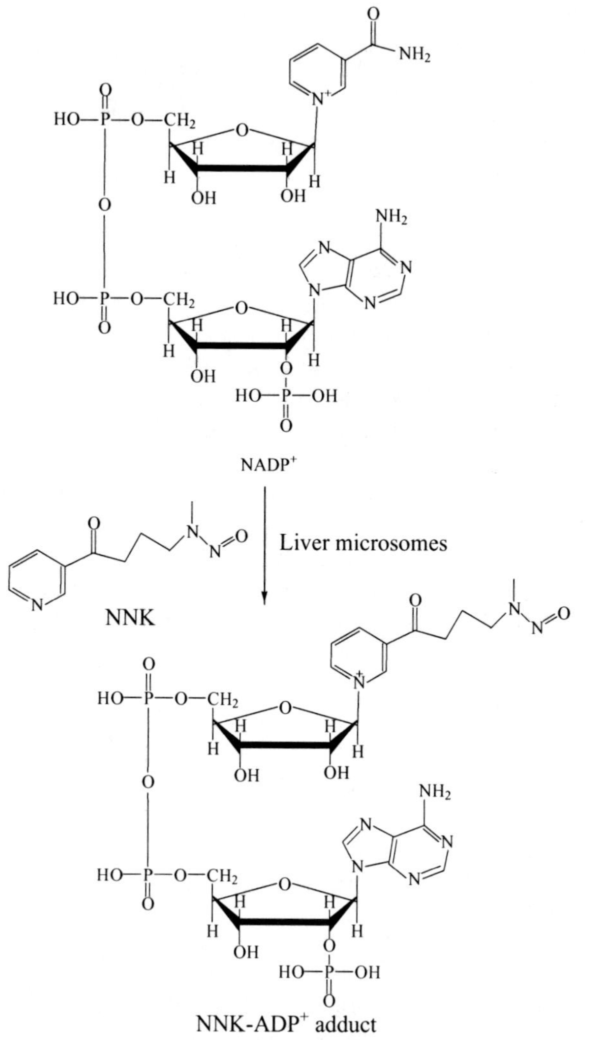

FIGURE 20.9. Formation of the NNK-ADP⁺ adduct from NADP⁺ and NNK.

ADP-ribose portion, can be added into the expected list of metabolites for future detection of this kind of unusual metabolite (Figure 20.10).

20.4.3 Identification of Trapped Adducts of Reactive Metabolites

In most cases, metabolism of a drug is considered a detoxification process. However, in some cases, clinically observed adverse effects might be associated, partly or wholly, with "toxic" metabolic processes. Even though there is not much direct correlation reported, it is speculated that bioactivation of drugs in some cases might be related to some clinically observed adverse effects. Bioactivation is a process where xenobiotics are metabolized to proximate and ultimate reactive intermediates (Kalgutkar et al., 2002,2005). For example, acetaminophen is bioactivated by cytochrome P450 enzymes to reactive metabolites, which might be responsible for the observed acute liver damage in animals and humans (James et al., 2003). In order to develop safer drugs, lack of bioactivation might be considered as one of the property parameters for selecting compounds in drug discovery for further development.

In general, reactive intermediates formed in bioactivation processes are electrophiles. They are not stable in biological systems and react readily with electrophilic functional groups in proteins to form drug-protein adducts. Thus, reactive intermediates themselves are rarely detected using routine analytical techniques. The most common approach is to use natural or synthetic nucleophilic "trapping" agents to react with reactive intermediates to form corresponding adducts. GSH, N-acetylcysteine (NAc), and potassium cyanide are commonly utilized agents for *in vitro* trapping studies. Their corresponding adducts can be detected and

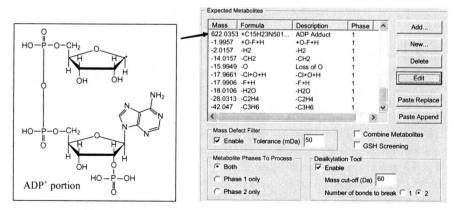

FIGURE 20.10. Addition of the ADP⁺ portion to the "expected" metabolite list in the Metabolynx method setup for further fast screening of the ADP adducts.

characterized by LC-MS/MS and NMR analysis. Based on the structures of these "trapped" adducts, the structures of reactive intermediates can be proposed. Subsequently, structural modification of the substructures liable for bioactivation can be pursued to minimize the formation of reactive intermediates.

LC-MS/MS analysis is the most commonly used method for detecting and characterizing the trapped adducts (Zhang and Gan, 2007). Due to the fact that both GSH and NAc molecules of the corresponding adducts lose m/z 129 Da by CID of the adducts, it is common to use constant neutral loss scanning (m/z 129 Da or other related neutral losses) of triple quadrupole mass spectrometers to analyze biological samples. In the case of the cyanide trapping studies, the constant neutral loss scanning of m/z 27 Da is used to detect the cyanide adducts. Subsequent MS/MS analysis by CID of the detected MH⁺ species of adducts in a second LC-MS injection is required for characterizing the structures of the adducts. The other approach is to use ion trap mass spectrometers to screen potential adducts in MS and data-dependent MS^n modes. The m/z values commonly used in screening are the molecular weights of the trapping agents (GSH, NAc, or cyanide) plus MH⁺ (molecular ion of the test compound), or plus molecular weights of modified test compounds (MH⁺+14, MH⁺+16, MH⁺+18, MH⁺+32, any other part of the parent molecule). Any molecular ions of potential adducts detected in the full MS scan are further characterized in the MS^n analysis. The major disadvantages of using nominal mass spectrometers (triple quadrupole mass spectrometers or ion trap mass spectrometers) include that they are time-consuming and yield false positives. The use of high-resolution/accurate mass spectrometry has significantly minimized the above disadvantages.

It has been reported that high resolution/accurate mass spectrometer LTQ-Orbitrap has been used to analyze samples from in vitro incubations of acetaminophen, diclofenac, carbamazepine, clozapine, p-cresol, 4-ethylphenol, and 3-methylindole with GSH in human liver microsomes (Zhu et al., 2007). Accurate mass full-scan MS acquisition, and data-dependent MS^2 and MS^3 acquisition were used, and the acquired data files were processed using a mass defect filter software with a window of ±40 mDa around the mass defect of a GSH adduct template (MH⁺ of a drug + GSH – 2H) over a mass range of ±50 Da around the mass of the filter template. All previously reported GSH adducts of acetaminophen, diclofenac, carbamazepine, clozapine, p-cresol, 4-ethylphenol, and 3-methylindole were detected. In addition, several previously unreported GSH adducts were also detected using the above approach. The results indicated that accurate mass spectrometry coupled with the mass defect filter software is more

sensitive and selective in detecting GSH adducts with minimized false positives and false negatives than the neutral loss approach using a nominal triple quadrupole mass spectrometer (TSQ from Thermo Scientific).

Using a high-resolution/accurate mass spectrometer Q-tof coupled with UPLC is another powerful approach for fast screening of the trapping adducts of reactive metabolites from in vitro and in vivo samples, and its utility is illustrated in the following example. When compound A (Figure 20.3) (50 µM) was incubated with pooled human liver microsomes (1 mg protein/mL) in the presence of NAc (5 mM) and NADPH (1.2 mM) for 0 min (control) and 60 min (analyte), the incubations were quenched with two volumes of acetonitrile. The supernatants were evaporated to dryness, and the residues were dissolved in acetonitrile:water:formic acid (95:5:0.05) for UPLC/Q-tof analysis. Samples (10 µL) were loaded onto an Acquity UPLC HSS T3 column (2.1 × 50 mm, 1.8 µm [Waters Corp., Milford, MA, USA]). The flow rate was set at 0.5 mL/min. The mobile phase consisted of solvent A (water with 0.1% formic acid) and solvent B (acetonitrile with 0.1% formic acid). The UPLC runs were programmed by a linear increase from 5% to 40% of solvent B during a 10-min period. The mass spectra were recorded in full scan and a subsequent full scan at a ramped low to high collision energy (15–45%). The acquired data files were processed using Metabolynx software (Waters Corp.) with a mass defect (MDF) window of ±50 mDa around the accurate mass of the parent molecule. The Metabolynx software compared the detected ions from analyte samples against a library of ions predicted from known metabolism reactions. The detected metabolites were grouped as "expected metabolites." Metabolynx also compared the detected ions in analytes against the ions in the corresponding retention time regions in control samples. Any ions in analytes which had more than predefined threefold ions in controls were considered as "unexpected" metabolites. Three NAc adducts were detected as "expected metabolites" in incubations of compound A, with human liver microsomes in the presence of NAc (Figure 20.11). The first NAc adduct had a molecular ion of m/z 611.1923 at 4.60 min, suggesting that this NAc molecule was added to the parent molecule or the metabolite M2, which had the same molecule weight of the parent molecule. The other two NAc adducts had molecular ions of m/z 627.1841 at 5.44 min and 627.1836 at 5.70 min, suggesting that a NAc molecule and an oxygen atom were added to the parent molecule or the metabolite M2, which had the same molecule weight of the parent molecule. The structures of these NAc adducts might be further assigned based on the MS^e spectra, MS/MS spectra, or NMR analysis. This UPLC/Q-tof approach provides a fast screening

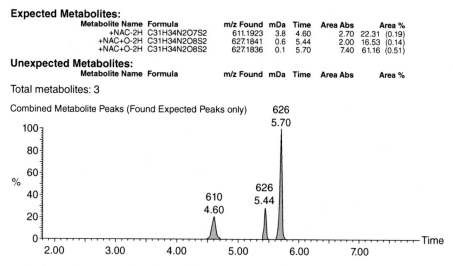

Expected Metabolites:

Metabolite Name	Formula	m/z Found	mDa	Time	Area Abs	Area %
+NAC-2H	C31H34N2O7S2	611.1923	3.8	4.60	2.70	22.31 (0.19)
+NAC+O-2H	C31H34N2O8S2	627.1841	0.6	5.44	2.00	16.53 (0.14)
+NAC+O-2H	C31H34N2O8S2	627.1836	0.1	5.70	7.40	61.16 (0.51)

Unexpected Metabolites:

Metabolite Name	Formula	m/z Found	mDa	Time	Area Abs	Area %

Total metabolites: 3

FIGURE 20.11. Metabolynx report for the detection of NAc adducts of compound A in incubation with human liver microsomes in the presence of NAc.

method to detect the trapped adducts of reactive intermediates with minimized false positive and false negative results compared with the traditional nominal mass spectrometry-HPLC approach (Zhang et al., 2005).

20.4.4 Analysis of Major Circulating Metabolites of Clinical Samples of Unlabeled Compounds

The exposure of a parent drug as well as its major circulating metabolites in humans is closely related to the safety profile of the drug. In the case where the exposure of the major circulating metabolites identified in human plasma exceeds that in animals, there might be a need to perform additional safety studies of those metabolites in animals prior to proceeding further with clinical studies of a developmental drug candidate. This may significantly delay the timeline for commercializing a drug. Guidance documents from regulatory agencies encourage the identification of human circulating metabolites as early as possible during the drug development process. Until a few years ago, the identification of major circulating metabolites of a developmental drug candidate in human plasma was performed as part of a radiolabeled human absorption, disposition, metabolism, and elimination (ADME) study at a significantly later stage (e.g., Phase IIb). The availability of sophisticated mass spectrometry techniques has now allowed the detection of circulating metabolites to be conducted early on in Phase I studies.

Even though nominal mass spectrometers using precursor ion and neutral loss scanning can be used to detect the circulating metabolites of unlabeled drug candidates, false negative results may be obtained when the fragmentation patters of metabolites are significantly different from that of the parent, or unusual metabolites are formed during metabolism processes. Application of high-resolution/accurate mass spectrometry overcomes the above limitations and provides a useful tool to effectively detect circulating metabolites in human plasma.

In a recent example, time-proportionally pooled human plasma samples were analyzed using a UPLC/Q-tof accurate mass spectrometer system, and the acquired data files were processed using the MDF tool in the Metabolynx software package (Tiller et al., 2008). The filter for metabolites derived from Phase I metabolism (mainly oxidative metabolism) covered a mass range of +50 to −100 Da relative to the mass of the test drug, with an MDF range of −50 to +20 mDa around the exact mass of the parent. The filter for metabolites derived from Phase II metabolism covered a mass range of +50 to +250 Da relative to the mass of the test drug, with an MDF range of −50 to +20 mDa around the exact mass of the parent after addition of 30 mDa to account for the positive mass defect associated with the inclusion of a glucuronide moiety. Several Phase I (demethylation) and Phase II (glucuronidation of a parent and a demethylated derivative) metabolites were effectively detected. Direct comparison of the relative amount (MS responses relative to the parent molecule) of circulating metabolites in human plasma to that of the time-proportionally pooled animal plasma samples from safety assessment studies may provide an estimate of the coverage of the human metabolites in animals. If needed, external calibration of MS responses of parent and metabolites may be performed using *in vitro* and *in vivo* samples of radiolabeled drugs to obtain the radioactivity/MS response factor. This approach can be utilized to analyze selected plasma samples from first-in-human clinical studies to provide preliminary information for the major circulating metabolite of developmental drug candidates, enabling the early finding of potential safety issues associated with the metabolites.

20.4.5   Applications in Metabolomics

Metabolomics is a systematic study of the metabolic responses of living systems to physiological or pathophysiological stimuli. These stimuli could be caused by diseases, environmental chemicals, or drugs, and so on. Thus, changes in endogenous metabolite profiles may be closely related to the presence of disease, specific lesions in the organs, or drug treatments, and could be used as biomarkers to guide preclinical or clinical diagnosis, and efficacy and safety studies.

NMR spectroscopy and mass spectrometry are the two most commonly used techniques to study the structures of endogenous metabolites. The advantages of NMR include high reproducibility, simplicity of sample preparation, and recovery of the analytical samples for further analysis. However, the major disadvantage is the relatively low sensitivity and selectivity because any hydrogen or carbon atoms of the endogenous components may interfere with the detection of the same kind of atoms of metabolites. This makes it relatively difficult to characterize unknown metabolites. On the other hand, the mass spectrometry technique overcomes the above limitations associated with NMR. Even though nominal mass spectrometry has relatively high analytical sensitivity, it is also a challenge for identifying unknown metabolites where much more "fingerprint" information is needed for structural assignments.

High-resolution/accurate mass spectrometry makes metabolomics studies more practicable. Due to the nature of its high mass accuracy, high-resolution/accurate mass spectrometry provide much more "fingerprint" information for identifying unknown metabolites (Takahashi et al., 2008; Draper et al., 2009) by searching the available metabolomics database, analyzing elemental compositions, and so on. It was reported that samples extracted from human urine, blood, and hair were analyzed using a micrOTOF (Bruker Daltonics, Billerica, MA, USA) accurate mass spectrometer combined with either HPLC or capillary electrophoresis separations (Liotta et al., 2010). The chromatographic/electrophoretic peaks pertaining to possible unknown compounds/ metabolites were processed/searched using background subtraction and peak deconvolution/mass spectral purification tools. The accurate mass value (molecular weight) of interest was searched (±10 ppm tolerance)

against a compound database containing +50,000 compounds of toxicological interest and their Phase I and Phase II metabolites. The chemical formulae of the retrieved candidate (hit) compounds were listed, and mass errors were calculated to exclude structural isomers, salts, and duplicates, and so on. The accurate mass of the unknown was searched against a predefined list of accurate mass lists derived from the 23 common biotransformations. Based on the match of the accuracy of the molecular ions and their biotransformation pathways with corresponding functional groups, the unknown compounds/metabolites can be finally identified. This approach has significantly reduced the numbers of false positive and false negative results in metabolomics analysis.

20.5   CONCLUSION

Traditionally, metabolite identification studies using conventional nominal mass spectrometers were plagued by low turnaround time, primarily as a result of significant manual searching of mass ions of metabolites obtained in full scans as well as the manual analysis of mass fragments obtained in MS/MS scans. High-resolution/accurate mass spectrometers make it possible to use various software to automatically search metabolites as well as to assign mass fragments (due to the high accuracy of the MS and the MS/MS data). Thus, the efficiency of metabolite identification processes can be significantly improved (Baillie, 2008). While there is still a need for manual assignment of structures from the identified mass fragments, a lot of progress has been made in designing software that will compile the information from mass fragments, and to a broad approximation be able to assign potential structures of metabolites.

In addition to its general usage in qualitative analysis of metabolites in biological systems, high-resolution/ accurate mass spectrometry can be used for quantitative purposes. Recently, it was reported that high-resolution/ accurate mass spectrometers were used to perform simultaneous quantitative and qualitative analyses of drugs and metabolites (Bateman, et al., 2009; Zhang et al., 2009). In a single LC-MS run, both quantitative and qualitative data from human microsomal incubation and rat plasma samples from pharmacokinetic studies were obtained, which makes simultaneous quantitative–qualitative workflow a possibility. This application is expected to be used widely with improvement in the dynamic ranges and scan speeds of mass analyzers, as well as the separation efficiency of liquid chromatography by UPLC.

In summary, high-resolution/accurate mass spectrometry has wide application in metabolism studies, including rapid identification of metabolic liabilities, easy detection of unusual metabolites, characterization of trapped adducts of reactive intermediates, analysis of clinical samples, and in metabolomics. It is expected that high-resolution/accurate mass spectrometry will be a major platform of analytical technology in metabolite identification in the coming years.

REFERENCES

Baillie TA (2008) Metabolism and toxicity of drugs. Two decades of progress in industrial drug metabolism. *Chem Res Toxicol* 21(1):129–137.

Bateman KP, Castro-Perez J, Wrona M, Shockcor JP, Yu K, Oballa R, Nicoll-Griffith DA (2007) MS^E with mass defect filtering for *in vitro* and *in vivo* metabolite identification. *Rapid Commun Mass Spectrom* 21:1485–1496.

Bateman KP, Kellmann M, Muenster H, Papp R, Taylor L (2009) Quantitative-qualitative data acquisition using a benchtop Orbitrap mass spectrometer. *J Am Soc Mass Spectrom* 20:1441–1450.

Benfenati E, Fanelli R, Bosone E, Biffi C, Caponi R, Cianetti M, Farina P (1991) Mass spectrometric identification of urinary and plasma metabolites of 6-(6'-carboxyhexyl)-7-n-hexyl-1,3-diazaspiro-[4-4]-nonan-2,4-dione, a new cytoprotective agent. *Drug Metab Dispos* 19:913–916.

Castro-Perez J, Plumb R, Granger JH, Beattie I, Joncour K, Wright A (2005) Increasing throughput and information content for *in vitro* drug metabolism experiments using ultra-performance liquid chromatography coupled to a quadrupole time-of-flight mass spectrometer. *Rapid Commun Mass Spectrom* 19:843–848.

Draper J, Enot DP, Parker D, Beckmann M, Snowdon S, Lin W, Zubair H (2009) Metabolite signal identification in accurate mass metabolomics data with MZedDB, an interactive m/z annotation tool utilising predicted ionisation behaviour "rules". *BMC Bioinformatics* 10:227–243.

James LP, Mayeux PR, Hinson JA (2003) Acetaminophen-induced hepatotoxicity. *Drug Metab Dispos* 31:1499–1506.

Kalgutkar AS, Dalvie DK, O'Donnell JP, Taylor TJ, Sahakian DC (2002) On the diversity of oxidative bioactivation reactions on nitrogen-containing xenobiotics. *Curr Drug Metab* 3:379–424.

Kalgutkar AS, Gardner I, Obach RS, Shaffer CL, Callegari E, Henne KR, Mutlib AE, Dalvie DK, Lee JS, Nakai Y, O'Donnell JP, Boer J, Harriman SP (2005) A comprehensive listing of bioactivation pathways of organic functional groups. *Curr Drug Metab* 6:161–225.

Lim HK, Chen J, Cook K, Sensenhauser C, Silva J, Evans DC (2008) A generic method to detect electrophilic intermediates using isotopic pattern triggered data-dependent high-resolution accurate mass spectrometry. *Rapid Commun Mass Spectrom* 22:1295–1311.

Liotta E, Gottardo R, Bertaso A, Polettini A (2010) Screening for pharmaco-toxicologically relevant compounds in biosamples using high-resolution mass spectrometry: A "metabolomic" approach to the discrimination between isomers. *J Mass Spectrom* 45(3):261–271.

Ma S, Chowdhury S (2007) Application of liquid chromatography/mass spectrometry for metabolite identification. In *Drug Metabolism in Drug Design and Development: Basic Concepts and Practice*, Zhang D, Zhu M, Humphreys WG, eds., pp. 319–367. John Wiley & Sons, Hoboken, NJ.

Mortishire-Smith RJ, O'Connor D, Castro-Perez JM, Kirby J (2005) Accelerated throughput metabolic route screening in early drug discovery using high-resolution liquid chromatography/quadrupole time-of-flight mass spectrometry and automated data analysis. *Rapid Commun Mass Spectrom* 19:2659–2670.

Mortishire-Smith RJ, Castro-Perez JM, Yu K, Shockcor JP, Goshawk J, Hartshorn MJ, Hill A (2009) Generic dealkylation: A tool for increasing the hit-rate of metabolite rationalization, and automatic customization of mass defect filters. *Rapid Commun Mass Spectrom* 23:939–948.

Peterman SM, Duczak N Jr, Kalgutkar AS, Lame ME, Soglia JR (2006) Application of a linear ion trap/Orbitrap mass spectrometer in metabolite characterization studies: Examination of the human liver microsomal metabolism of the non-tricyclic anti-depressant nefazodone using data-dependent accurate mass measurements. *J Am Soc Mass Spectrom* 17:363–375.

Peterson LA, Ng DK, Stearns RA, Hecht SS (1994) Formation of NADP(H) analogs of tobacco-specific nitrosamines in rat liver and pancreatic microsomes. *Chem Res Toxicol* 7:599–608.

Ruan Q, Peterman S, Szewc MA, Ma L, Cui D, Humphreys WG, Zhu M (2008) An integrated method for metabolite detection and identification using a linear ion trap/Orbitrap mass spectrometer and multiple data processing techniques: Application to indinavir metabolite detection. *J Mass Spectrom* 43:251–261.

Stephens W (1946) Pulsed mass spectrometer with time dispersion. *Bull Am Phys Soc* 21:22.

Takahashi H, Kai K, Shinbo Y, Tanaka K, Ohta D, Oshima T, Altaf-Ul-Amin M, Kurokawa K, Ogasawara N, Kanaya S (2008) Metabolomics approach for determining growth-specific metabolites based on Fourier transform ion cyclotron resonance mass spectrometry. *Anal Bioanal Chem* 391(8):2769–2782.

Takamura F, Tanaka A, Takasugi H, Taniguchi K, Nishio M, Seki J, Hattori K (2006) Metabolism investigation leading to novel drug design 2: Orally active prostacyclin mimetics. Part 5. *Bioorg Med Chem Lett* 16:4475–4478.

Tiller PR, Yu S, Bateman KP, Castro-Perez J, McIntosh IS, Kuo Y, Baillie TA (2008) Fractional mass filtering as a means to assess circulating metabolites in early human clinical studies. *Rapid Commun Mass Spectrom* 22:3510–3516.

Wrona M, Mauriala T, Bateman KP, Mortishire-Smith RJ, O'Connor D (2005) "All-in-one" analysis for metabolite identification using liquid chromatography/hybrid quadrupole time-of-flight mass spectrometry with collision energy switching. *Rapid Commun Mass Spectrom* 19:2597–2602.

Zhang H (2006) Detection and characterization of metabolites in biological matrices using mass defect filtering of liquid chromatography/high resolution mass spectrometry data. *Drug Metab Dispos* 34:1722–1733.

Zhang H, Yang Y (2008) An algorithm for thorough background subtraction from high-resolution LC/MS data: Application for detection of glutathione-trapped reactive metabolites. *J Mass Spectrom* 43:1181–1190.

Zhang H, Zhang D, Ray K (2003) A software filter to remove interference ions from drug metabolites in accurate mass liquid chromatography/mass spectrometric analyses. *J Mass Spectrom* 38(10):1110–1112.

Zhang H, Ma L, He K, Zhu M (2008) An algorithm for thorough background subtraction from high-resolution LC/MS data: Application to the detection of troglitazone metabolites in rat plasma, bile, and urine. *J Mass Spectrom* 43:1191–1200.

Zhang NR, Yu S, Tiller P, Yeh S, Mahan E, Emary WB (2009) Quantitation of small molecules using high-resolution accurate mass spectrometers—A different approach for analysis of biological samples. *Rapid Commun Mass Spectrom* 23(7):1085–1094.

Zhang Z, Gan J (2007) Protocols for assessment of *in vitro* and *in vivo* bioactivation potential of drug candidates. In *Drug Metabolism in Drug Design and Development: Basic Concepts and Practice*, Zhang D, Zhu M, Humphreys WG, eds., pp. 447–476. John Wiley & Sons, Hoboken, NJ.

Zhang Z, Chen Q, Li Y, et al. (2005) *In vitro* bioactivation of dihydrobenzoxathiin selective estrogen receptor modulators by cytochrome P450 3A4 in human liver microsomes: Formation of reactive iminium and quinone type metabolites. *Chem Res Toxicol* 18:675–685.

Zhang Z, Zhu M, Tang W (2009) Metabolite identification and profiling in drug design: Current practice and future directions. *Curr Pharm Des* 15:2220–2235.

Zhu M, Ma L, Zhang D, Ray K, Zhao W, Humphreys WG (2006) Detection and characterization of metabolites in biological matrices using mass defect filtering of liquid chromatography/high resolution mass spectrometry data. *Drug Metab Dispos* 34:1722–1733.

Zhu M, Ma L, Zhang H, Humphreys WG (2007) Detection and structural characterization of glutathione-trapped reactive metabolites using liquid chromatography-high-resolution mass spectrometry and mass defect filtering. *Anal Chem* 79:8333–8341.

Zhu P, Ding W, Tong W, Ghosal A, Alton K, Chowdhury S (2009) A retention-time-shift-tolerant background subtraction and noise reduction algorithm (BgS-NoRA) for extraction of drug metabolites in liquid chromatography/mass spectrometry data from biological matrices. *Rapid Commun Mass Spectrom* 23:1563–1572.

21

APPLICATIONS OF ACCELERATOR MASS SPECTROMETRY (AMS)

Xiaomin Wang, Voon Ong, and Mark Seymour

21.1 INTRODUCTION

A relative newcomer to the arsenal of bioanalytical techniques in the pharmaceutical industry is accelerator mass spectrometry (AMS) (Turteltaub and Vogel, 2000; Prakash et al., 2007; Lappin and Stevens, 2008; Tuniz and Norton, 2008). AMS was originally developed for radiocarbon dating in the late 1970s by Nelson (*Science*, 1977, 198, 508) and Bennett (*Science*, 1977, 198, 509). The methods were able to detect natural ^{14}C at part-per-trillion levels by using tandem Van de Graaff accelerators. Nearly 20 years later, applications involving AMS in biomedical research began appearing in scientific literature. The studies included determination of DNA adducts by ^{14}C isotope ratios (Felton, 1991; Turteltaub, 1990), bone resorption by using ^{41}Ca isotope ratios (Elmore, 1990), and Al bound to blood proteins by ^{26}Al tracer were demonstrated (Day, 1991).

AMS is used in biomedical analysis by exploiting its ^{14}C detection sensitivity. The technique requires a ^{14}C-labeled analyte, which is used as a tracer. Due to its high sensitivity, AMS requires only a small amount of radioactivity, typically 100–250 nCi compared with 50–100 µCi for a conventional human absorption, disposition, metabolism, and elimination (ADME) study, to deliver detection limits that can be in the femtogram to attogram levels (10^{-15} to 10^{-18} g), depending on the specific activity of the ^{14}C-labeled compound administered (Lappin and Garner, 2003). The safety concerns inherent in the use of radioisotopes are thus significantly reduced, and the regulatory hurdles associated with dosing radiolabeled compounds to humans are effectively removed.

A number of distinct applications for AMS in the biomedical field have emerged. Much of the early work focused on the microdosing approach, whereby a very low, subtherapeutic dose of compound is administered. This is attractive as it requires much less investment in chemistry, manufacturing, and controls (CMC) and toxicology resources prior to the conduct of a human study. The U.S. Food and Drug Administration (FDA) defined microdosing in a guidance document titled "Exploratory IND Studies" that was finalized on January 2006. This concept has created a new clinical phase of drug development, dubbed Phase 0, which precedes the traditional, IND-driven, Phase 1 testing. A microdose is defined as "less than 1/100th of the dose of a test substance calculated (based on animal data) to yield a pharmacologic effect of the test substance with a maximum dose of <100 micrograms."

AMS is also increasingly finding utility in Phase 1, firstly as a means to obtain early human data to address the issue of metabolites in safety testing and secondly as an elegant and cost-effective way to obtain fundamental human pharmacokinetic data (including volume of distribution, clearance, and absolute bioavailability) as early as possible during clinical development (Lappin, 2006).

21.2 BIOANALYTICAL METHODOLOGY

21.2.1 Sample Preparation

Prior to AMS analysis, the carbon in the sample must be isolated. For example, for some instruments, each sample is converted to graphite in a two-stage process. First, the sample is oxidized with copper (II) oxide at 900°C to produce carbon dioxide gas, which is cryogenically transferred then reduced to elemental carbon using a mixture of titanium hydride and zinc with cobalt as catalyst at 500°C. The graphite/cobalt mixture produced is then pressed into aluminum cathodes, which are loaded into the ion source of the AMS instrument (see Figure 21.1).

Many samples (e.g., plasma, whole blood, feces, and tissues) contain sufficient carbon to be graphitized without addition. However, for samples that do not contain sufficient endogenous carbon to produce the 2 mg of graphite required for AMS analysis, carrier carbon must be added prior to graphitization. Compounds commonly used as carrier carbon include liquid paraffin, tributyrin, and salt sodium benzoate. All are obtained from petrochemical sources and are therefore heavily ^{14}C-depleted, virtually all of the ^{14}C they originally contained having decayed over millions of years. Hence, the addition of carrier carbon does not contribute additional background ^{14}C to the sample.

21.2.2 AMS Instrumentation

Figure 21.1 shows an example of a Pelletron (National Electrostatics Corp., Middleton, WI, USA) tandem accelerator mass spectrometer used for biomedical analysis, located at Xceleron (York, U.K.). In the ion source, the graphitized samples are bombarded by a beam of Cs$^+$ ions at 3–10 keV to generate negative ions, including C$^-$. The ions are accelerated to approximately 80 keV from the ion source by a combination of an extractor (~20 kV) and a preacceleration tube (60 kV). At these energies, ^{12}C$^-$, ^{13}C$^-$, and ^{14}C$^-$ can be resolved. Interfering ions with similar masses (e.g., ^{16}O$^-$) will also be present at this stage. However, the isobar ^{14}N$^-$ is unstable and therefore cannot interfere with the ^{14}C measurement.

Upon leaving the preacceleration tube, the ions enter a spherical electrostatic analyzer at a 45-degree angle. Although the ion beam that exits the analyzer still contains ions with various masses, the analyzer allows only ions with a predetermined kinetic energy to pass through the analyzer chamber. From here, the ions travel through an injection magnet at a 90-degree angle, which provides sequential injection for the various carbon ions (i.e., ^{14}C$^-$, ^{13}C$^-$, and ^{12}C$^-$) into the Pelletron accelerator tube. The duty cycle of the sequential injection is about 99% for the target isotope, that is, ^{14}C, consisting of 150 μs for ^{12}C$^-$, 600 μs for ^{13}C$^-$, and 100 ms for ^{14}C$^-$.

In the Pelletron accelerator, negative carbon ions and molecular isobars of ^{14}C$^-$ (e.g., ^{12}CH$_2^-$, ^{13}CH$^-$) accelerate toward the positive terminal (~5 MV), and collide with argon gas in a low-pressure cell, where they undergo molecular dissociation and are stripped of electrons to give positively charged ions ($^{12,13\ \&\ 14}$C$^{+1\ to\ +6}$). C^{4+} ions are selected for measurement as they are generally the most abundant at this high energy. These ions are then directed toward the high-energy analyzing magnet.

Immediately past the analyzing magnet, the ^{12}C^{4+} and ^{13}C^{4+} ions are detected by offset Faraday cups, whereas the ^{14}C^{4+} ions are detected by a gas ionization detector. In general, interfering non-^{14}C^{4+} ions (e.g., ^{7}Li$_2^{4+}$) are

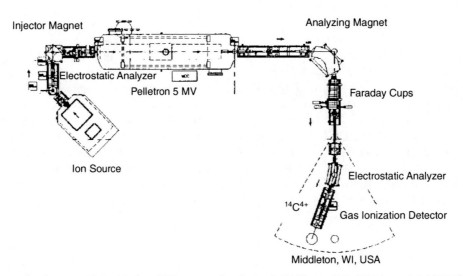

FIGURE 21.1. Schematic diagram of the National Electrostatics Corp. (Middleton, WI, USA) model 15SDH-2 5 MV Pelletron tandem accelerator mass spectrometer (AMS).

prevented from reaching the gas ionization detector by a combination of electrostatic and magnetic analyzers as well as charge state separation (in the Pelletron tube).

21.2.3 AMS Analysis

21.2.3.1 AMS Standards AMS measures the ratio of $^{14}C:^{13}C:^{12}C$ in a sample. A range of standards with defined carbon isotope ratios are available, such as the Australian National University (ANU) sugar, which is a crop of sugar harvested in the 1950s and having an elevated ^{14}C content due to atomic bomb tests carried out in the Southern hemisphere at around that time. Graphite prepared from ANU sugar (or another certified standard) is used to calibrate the AMS instrument. Further ANU standards are graphitized alongside unknown samples as process controls to confirm that nothing in the sample preparation procedure has perturbed the $^{12}C:^{13}C:^{14}C$ ratio.

Since AMS is based on isotope ratio measurement, AMS requires the presence of at least two isotopes for a given analyte (e.g., $^{14}C:^{12}C$). In addition, because most biological samples contain background ^{14}C, standards will need to be introduced for AMS calibration, process control, and background corrections.

21.2.3.2 High-Performance Liquid Chromatography (HPLC)-AMS Analysis

Any sample that contains carbon and that can be introduced into the small glass sample tubes used may be graphitized. However, AMS measures the overall isotope ratio in the graphite sample. Thus, analysis of intact samples produces total ^{14}C-content data, without discriminating between individual ^{14}C-labeled species. This is analogous to total radioactivity data from a conventional ADME study, obtained using liquid scintillation counting, which represent the sum of all compound-related material, including any metabolites. In order to quantify specific analytes, it is necessary to isolate the analyte prior to graphitization. Typically, this is achieved through sample extraction and/or chromatographic separation.

Currently, no online liquid chromatography (LC)-AMS interface is commercially available. Therefore, sample analysis is carried out in a stepwise fashion, with the LC eluant being collected as fractions and the fraction(s) corresponding to the target analyte being graphitized as described above. Samples are fortified with the nonradiolabeled analyte prior to fractionation, and the retention time of the added standard is used to identify fraction(s) for analysis.

As described above, the AMS instrument is independently calibrated using certified standards. Therefore, it is not necessary to construct a calibration curve in the same way as is required, for example, for an LC-MS

assay. Nevertheless, AMS analysis of HPLC fractions only yields information on the amount of ^{14}C in that fraction: Any procedural losses that occur prior to graphitization (i.e., during extraction or chromatographic separation) must be accounted for when calculating levels in the original sample. Some practitioners argue that, as long as the overall recovery of ^{14}C is kept close to 100%, extraction efficiency can be ignored. However, not only is it commonly difficult to achieve quantitative recovery but, since the neither the composition of the sample nor the recovery of individual components is known, high overall recovery does not necessarily equate to high recovery of the target analyte.

One method used to account for procedural losses involves spiking a series of plasma standard samples with a range of concentrations of ^{14}C-labeled analyte. A constant amount of non-radiolabeled analyte is subsequently added, as internal standard (IS), prior to sample extraction (e.g., using protein precipitation or liquid-liquid extraction). The concentration of IS used is high enough that it can be quantified using a conventional UV detector during the fractionation procedure. A recovery curve is then constructed by plotting the known concentration of ^{14}C-labeled analyte spiked into the sample on the *x*-axis against the ratio of the AMS result for the HPLC fraction and the IS UV response on the *y*-axis. Since AMS provides a linear response (Fifield, 1999), the slope of the recovery curve provides the instrument response for the determination of the ^{14}C concentration in the plasma sample according to the following equation (Lappin, 2004):

$$K = \frac{\left[\dfrac{R_A}{mU} + C\right]\Phi}{L_{mass}},$$

where K is the amount of analyte in the original sample,

R_A is the isotope ratio measured in the HPLC fraction,

Φ is the amount of carbon added as carbon carrier,

L_{mass} is the mass specific radioactivity of the analyte,

m is the slope of the recovery curve,

U is the UV response for the IS, and

C is the intercept of the recovery curve.

With the increasing adoption of LC-AMS in the bioanalytical sphere, and the inclusion of data generated using this technique in dossiers submitted to regulatory authorities in support of marketing applications for new pharmaceuticals, practitioners have begun to turn their attention to analytical method validation. As with any quantitative method, it is clear that parameters such as accuracy, precision, and limits of quantification must be defined and assured (Wang et al., 2009, Deck et al, 2010 and Lappin et al., 2011). However, there are differences between LC-AMS and other quantitative methods, such as LC-mass spectrometry/mass spectrometry (MS/MS), that require a distinct approach to that commonly used for conventional techniques. Thus, for AMS the sample is destroyed prior to analysis so that it is not possible to discriminate between analyte and other ^{14}C-labeled components (i.e., metabolites) based on mass. Instead, selectivity comes exclusively from chromatographic separation. On the other hand, because the ultimate analyte is always carbon, AMS detection is not sensitive to the effects of matrix interferences (i.e., ion suppression) in the way conventional mass spectrometry is. Currently, LC-AMS is an off-line technique, which introduces additional points where controls are necessary. However, the most fundamental difference between LC-AMS and LC-MS/MS is that AMS gives an absolute response, that is, a defined isotope ratio, whereas the signal from MS/MS is relative and the response factor varies between analytes and from run to run. Therefore, it is not necessary to prepare a separate calibration line with each analytical run for LC-AMS. Practitioners are currently engaged in a number of international initiatives, for example, through the Global Bioanalysis Consortium (http://www.globalbioanalysisconsortium. org) and the European Bioanalytical Forum (http://www. europeanbioanalysisforum.eu), aimed at developing recommendations for assay validation specific for LC-AMS. It is anticipated that these will eventually form the basis of regulatory guidelines.

21.3 AMS APPLICATIONS IN MASS BALANCE/METABOLITE PROFILING

The first report on the use of AMS for a human mass balance and metabolite profiling study was published in 2002 for a novel farnesyl transferase inhibitor, R115777 (Garner et al., 2002). The study was conducted using 34 nCi of [^{14}C]R115777 mixed with 50 mg of nonlabeled R115777 and administered as a suspension in 20 mL of water to four male subjects. This radioactive dose was estimated to result in a radiation dose of less than 1 μSv (0.1 millirem), which was exempt from regulatory approval in the United Kingdom where the study was conducted. Whole blood, urine, and fecal samples were collected over 7 days to obtain mass balance, plasma pharmacokinetic, and metabolite profiling information. It is noteworthy that, although AMS was capable of analyzing radioactivity levels far below the detection limits of liquid scintillation counting, in this study maximal plasma total ^{14}C concentrations (i.e., 1.6–2.9 dpm/mL, analyzing

60 μL of sample) were only five times above background levels. All contemporary biological samples contain background ^{14}C. The amount of background ^{14}C present in a sample depends on its carbon content; that is, the higher the total carbon, the higher the level of background ^{14}C. Thus, reducing the amount of carbon in a sample reduces background ^{14}C. For example, in the case of plasma, protein precipitation reduces background by about 25-fold, thereby enhancing the signal-to-noise ratios.

Metabolite profiling was performed by employing timed fraction collection of the effluent from LC analysis and subsequent AMS analysis of the collected fractions as a way to resolve the various components making up the total radioactivity. Authentic reference standards of potential metabolites were available for assignment of ^{14}C peaks by LC co-chromatography. AMS radiochromatograms from this off-line approach compared favorably with profiles subsequently obtained using LC-MS/MS with one exception. There was a discrepancy in the profile of the fecal extract as LC-MS/MS analysis had suggested that the loss of a methylimidazole moiety was a minor metabolic pathway, whereas AMS had indicated otherwise. The authors proposed that the reason for this observation may be due to instability of the metabolite in methanolic fecal extracts, although no attempt was made to resolve this discrepancy. A clear limitation of AMS technology for metabolite profiling is the lack of a liquid chromatography interface, so fractions need to be collected and analyzed off-line. Therefore, AMS analysis, while very sensitive, is labor-intensive, and hence, profiling using LC-AMS costs more than using conventional scintillation counting-based methods. This cost must be weighed against the benefit of obtaining human metabolism data much earlier than is otherwise possible.

More recently, a similar strategy employing AMS was instrumental in determining the metabolite and excretion profiles of ixabepilone (Beumer et al., 2007; Cömezoğlu et al., 2009), a semisynthetic analog of epothilone B being developed for use in the treatment of cancer, following administration to cancer patients. The authors reported that a conventional radiolabel human ADME study could not be carried out due to the radiolytic instability of [^{14}C]ixabepilone at a typical specific activity (generally 1–10 mCi/mg) required for conventional detection but that it was sufficiently stable at low specific activity (1–2 nCi/mg), which was optimal for AMS detection. As with the previous example, total radioactivity in collected excreta samples were measured by AMS to quantify the excretion profile while pooled plasma, urine, and feces extracts were further subjected to LC with fraction collection for metabolite profiling by off-line AMS (to determine the number and relative amounts of metabolites) followed by traditional LC-MS/MS analysis to identify ^{14}C-labeled components.

Despite the advantage of not needing extensive regulatory approval for radiation safety, there has not been widespread adoption of AMS for the conduct of human mass balance or metabolic profiling studies since the first publication in 2002, as evidenced from the dearth of reports in the scientific literature. However, this will likely change in the coming years as the regulatory landscape has shifted toward earlier quantitative metabolite profiling driven primarily by the introduction of newer guidance documents by the FDA and European Medicines Agency (EMAE). One in particular, entitled "Safety Testing of Drug Metabolites" (or MIST guidance, as it has been referred to), was finalized in 2008, and proposed that human oxidative metabolites, the exposure of which exceeds 10% of the parent drug exposure at steady state, are of concern and that additional toxicological testing may be required on metabolites that display relatively higher exposure in humans than in preclinical animal species. Details of this guidance have been treated in a separate chapter of this book, but the implication is that there is a need for metabolite profiling beyond qualitative analysis or structural elucidation. Hence more quantitative data are needed of the different metabolites to ensure that there is adequate coverage of each metabolite (in terms of plasma exposures) by the selected animal toxicology species. Moreover, the more recently issued EMEA M3 guidelines ("Non-Clinical Safety Studies for the Conduct of Human Clinical Trials and Marketing Authorization for Pharmaceuticals," EMEA M3 (R2), 2009) include a threshold for safety concerns for metabolites based on their exposure relative to total compound-related material, that is, parent plus all metabolites, known and unknown. At the present time, definitive quantitative information on the latter can only be derived using tracer techniques and AMS provides a means of obtaining this information at a very early stage during development.

Another advancement that would be expected to increase the use of AMS for metabolite profiling would be the successful online coupling of AMS with liquid chromatography. The authors are aware of a number of research groups working in this area, and a recent publication (Prakash et al., 2007) describes exactly that outcome. However, without commercialization, it remains to be seen if this platform will enjoy extensive success.

21.4 AMS APPLICATIONS IN PHARMACOKINETICS

While the widespread use of AMS for human mass balance and drug metabolism research is uncertain, its remarkable sensitivity for quantitative analysis as applied to pharmacokinetics and oral bioavailability assessment is not (Lappin et al., 2006, Vuong et al., 2007 and Zhou et al., 2009).

As discussed above, many of the early AMS-enabled studies involved administration of a microdose. A fairly complete human ADME study of zidovudine, a reverse transcriptase inhibitor with antiviral activity, using AMS was carried out by Vuong and coworkers in 2007 (Vuong et al., 2007). Following administration of a microdose to a healthy volunteer, the authors determined the pharmacokinetics, mass balance, and metabolite exposure by AMS. The authors found that zidovudine pharmacokinetic parameters from the microdose compared favorably with published values for therapeutic doses. Similar to the preceding study, the authors also used AMS to measure concentrations of $[^{14}C]$zidovudine in peripheral blood mononuclear cells (PBMCs). The uptake of zidovudine into PBMCs was measured using PBMC harvested from 10 mL of whole blood. The authors commented that, although specific analysis of the forms of the label within the cell was not quantified, it was not beyond the reach of AMS sensitivity. In a later study, intracellular concentrations of zidovudine triphosphate following a microdose (100 µg) or a therapeutic dose (300 mg) of zidovudine were measured in PBMCs using LC-AMS and LC-MS/MS. The results obtained using the two techniques were comparable; however, LC-AMS was calculated to be at least 30,000 times more sensitive (Chen et al., 2010). Indeed, AMS has been used in biomedical research to study the kinetics of DNA adduct formation by anticancer drugs (Turteltaub et al., 1990; Hah et al., 2006; Hah et al., 2007).

Since it was first proposed for use in ADME studies, there has been the perception that pharmacokinetic data from a microdose may not be predictive of a therapeutic dose due to the potential for nonlinear absorption and/or metabolism. However, while no comprehensive comparative database between microdose and therapeutic dose pharmacokinetics exists to date, there have been several comparisons made and an acceptable degree of agreement has been found for a majority of drugs investigated (Lappin et al., 2006; Lappin and Garner, 2008). Lappin and Garner reviewed all available published studies through 2008 that compared the pharmacokinetics of microdose with therapeutic dose and reported that 15 of the 18 drugs (or 83%) showed linear pharmacokinetics (within a factor of 2) when extrapolated. It was, however, not immediately clear what factors were responsible for nonlinearity. More recently, the results from the study of an additional seven compounds funded by the European Union, called "The European Microdosing AMS Partnership Programme" (EUMAPP), involving a pan-European collaboration between industry and academia, became available. The study found that microdose pharmacokinetics compared well (within a factor of 2) with those reported in literature, bringing the current level of agreement between microdose and therapeutic dose close to 90%. Particularly good agreement was noted for intravenous administrations. More importantly, where oral microdose data was not predictive, it was possible to explain the discrepancy based on the known metabolic or chemical properties of the drug. These studies represent the start of a database that appears to support the utility of microdosing for human pharmacokinetics and perhaps will aid proponents of this approach in the ongoing debate that is occurring internally within many pharmaceutical companies (Karara et al., 2010).

One strategy employed to evaluate the potential for nonlinear pharmacokinetics was to examine pharmacokinetics in an animal model prior to the conduct a human microdose study, which would cost more to perform. In 2004, Sandhu and coworkers (Sandhu et al., 2004) compared the pharmacokinetics of a compound in development (7-deaza-2′-C-methyl-adenosine; compound A) at pharmacological versus microdose in dogs. AMS was used to quantitate samples from microdose administration, whereas LC-MS/MS was employed for samples from pharmacological dose administration. The data indicated that the pharmacokinetic properties of compound A in dogs following dosing at 0.02 mg/kg were similar to those at 1 mg/kg, indicating linear kinetics across the 50-fold dose range. The authors also commented that only the exceptional sensitivity of AMS was capable of exposing the multiphasic pharmacokinetic profile of compound A, even after a microdose, that had been previously not been revealed by LC-MS/MS.

As a "poor man's" option, the same strategy could be applied with quantitation by conventional LC-MS/MS systems using the lowest possible doses that would be considered a microdose. With the advent of highly sensitive triple quadrupole mass spectrometers operating under selected reaction monitoring mode, indeed this strategy has been utilized successfully (Balani et al., 2006; Ni et al., 2008). Balani and coworkers commented that since AMS is still expensive to be used routinely, they had conducted an oral microdosing assessment in rats using quantitation by LC-MS/MS. The authors showed that pharmacokinetics of model drugs, fluconazole and tolbutamide, can be readily characterized by LC-MS/MS at a microdose of 1 µg/kg in rats. The authors further validated the pharmacokinetic linearity that was expected between the microdose and >1000-fold higher doses in rats for these two compounds prior to applying the same strategy for a proprietary compound, MLNX, under development and being considered for human microdosing. The authors found that MLNX pharmaco-

kinetics in rats were not linear and suggested that MLNX pharmacokinetics in humans would likely not be linear as well, although Lappin and Garner had noted in a separate published review (Lappin and Garner, 2008) that only the highest dose tested fell away from linearity and questioned whether this would be reflected in a human study at therapeutic doses.

While it is clear that the potential for nonlinear pharmacokinetics will continue to be an issue for the widespread adoption of microdosing, AMS technology also has utility in Phase 1 and in recent years there has been significant growth in this area. It is likely that this growth will continue, as drug developers become increasingly aware of the potential to obtain fundamental human pharmacokinetic data, which can inform critical "go/no-go" decisions, at the earliest possible juncture.

AMS has found a niche in the conduct of absolute bioavailability studies. A "piggyback" approach is taken, whereby an extravascular dose of the nonradiolabeled test compound is administered at a pharmacologically relevant dose, followed by an intravenous dose of ^{14}C-labeled compound administered at approximately the T_{max} of the extravascular dose. A common approach is to include a microtracer dose as part of an early clinical study (e.g., a single or multiple ascending dose study) within a flexible clinical protocol. Although the intravenous microtracer dose is administered at or below the microdosing threshold (typically around 10 µg), it is distributed and cleared after dilution into the extravascular dose and therefore pharmacokinetic linearity is not an issue (Sarapa et al., 2005; Stevens et al., 2009). Because the intravenous dose administered is so low, no additional safety data over and above those required to administer the extravascular dose are required.

This piggyback approach of dosing can be applied to distinguish between the contribution of formulation performance, absorption, first-pass metabolism, and systemic clearance for drugs with low and highly variable oral bioavailability. For example, it was utilized to learn that the moderate bioavailability of nelfinavir, an antiviral drug, was due to significant first-pass metabolism rather than poor absorption, limiting the potential of formulation improvement to decrease pill burden (Sarapa et al., 2005). The authors showed that AMS allowed several thousandfold dose reduction of [^{14}C] nelfinavir relative to that required for liquid scintillation counting and reduces the need to formulate high intravenous doses, which may require substantial efforts in the case of drugs with low aqueous solubility. More recently, the absolute bioavailability of an investigational antiviral drug (RDEA806) was determined from an intravenous ^{14}C-microtracer dose administered with a 200 mg oral dose. Having the absolute bioavailability of the oral dose in tablet form early during clinical development allowed further development of dosage forms with the knowledge of what constitutes maximum bioavailability improvement. Interestingly, in the same study, the intravenous pharmacokinetics of RDEA806 coadministered with the oral dose were found to be equivalent to those of an intravenous ^{14}C-microdose (without oral dose), suggesting "dose linearity" from 80 μg intravenous to 200 mg oral/intravenous combination (Stevens et al., 2009).

Boddy et al. (2007) employed the piggyback approach to investigate the pharmacokinetic profile of imatinib under steady-state conditions in patients with chronic myeloid myeloma (CML). In addition, the pharmacokinetic/pharmacodynamic relationship of imatinib was examined in this small pilot study of six patients. Plasma and peripheral blood lymphocyte samples were obtained after administration of a microdose of [^{14}C]imatinib in addition to a normal therapeutic dose. The authors reported that imatinib was detectable in three of six patients, but there was no correlation between the peripheral blood lymphocyte concentrations and corresponding plasma concentrations. The authors concluded that although there was no clear association between imatinib pharmacokinetics and clinical response in this small patient group, the study did highlight the application of AMS in clinical pharmacology.

21.5 CONCLUSION

Even though most of the studies reported in this chapter were published within the last 5 years, the applications of AMS methodologies in drug metabolism and pharmacokinetics are established. Although much of the work using this technique reported to date has focused on the use of microdosing (Phase 0), AMS is also a powerful tool in Phase 1 and beyond. In particular, its utility in obtaining early human intravenous pharmacokinetic and metabolism data is likely to be the basis of significant growth in the uptake of the technology over the next period. Recent advances in quantitative LC-AMS methods are fueling this growth, and the potential development of a commercially available direct LC-AMS interface would be a major step forward. Although AMS does not provide any direct structural information on the metabolites, the extent of metabolism can be assessed by comparing total ^{14}C levels with those of parent and/or known metabolites, and quantitative data on the number and relative amounts of metabolites can be obtained from metabolite profiles. Armed with this information, conventional techniques, such as LC-MS and nuclear magnetic resonance (NMR), can be strategically employed to elucidate structures.

In summary, AMS is a state-of-the-art analytical technique that plays an important role in drug metabolism and pharmacokinetics research. The continued development of these technologies, as part of an integrated approach with complementary techniques, will provide valuable information to assist in the development of more effective drugs in the future.

REFERENCES

Balani SK, Nagaraja NV, Qian MG, Costa AO, Daniels JS, Yang H, Shimoga PR, Wu JT, Gan LS, Lee FW, Miwa GT (2006) Evaluation of microdosing to assess pharmacokinetic linearity in rats using liquid chromatography-tandem mass spectrometry. *Drug Metab Dispos* 34(3):384–388.

Beumer JH, Garner RC, Cohen MB, Galbraith S, Duncan GF, Griffin T and Beijnen JH, Schellens JHM (2007) Human mass balance study of the novel anticancer agent ixabepilone using accelerator mass spectrometer. *Invest New Drugs* 25:327–334.

Boddy AV, Sludden J, Griffin MJ, Garner C, Kendrick J, Mistry P, Dutreix C, Newell DR, O'Brien SG (2007) Pharmacokinetic investigation of imatinib using accelerator mass spectrometry in patients with chronic myeloid leukemia. *Clin Cancer Res* 13(14):4164–4169.

Chen J, Garner C, Lee L, Seymour M, Fuchs E, Hubbard W, Parsons T, Pakes G, Fletcher CV, Flexner C (2010) Accelerator mass spectrometry measurement of intracellular concentrations of active drug metabolites in human target cells *in vivo*. *Clin Pharmacol Ther* 88:796–800.

Cömezoğlu SN, Ly VT, Zhang D, Humphreys WG, Bonacorsi SJ, Everett DW, Cohen MB, Gan J, Beumer JH, Beijnen JH, Schellens HM, Lappin G (2009) Biotransformation profiling of [(14)C]ixabepilone in human plasma, urine and feces samples using accelerator mass spectrometry (AMS). *Drug Metab Pharmacokinet* 24(6):511–522.

Day JP, Barker J, Evans LJ, Perks J, PJ, Lilley JS, Drumm PV, Newton GW (1991) Aluminum absorption studied by 26Al tracer. *Lancet* 337(8753):1345.

Deck BD, Ognibene T, Vogel JS (2010) Analytical validation of accelerator mass spectrometry for pharmaceutical development. *Bioanalysis* 2(3):469–485.

Elmore D, et al., (1990) Calcium-41 as a long-term biological tracer for bone resorption. *Nucl Instr Meth* B52:536.

Felton JS, Turteltaub KW, Gledhill BL, Vogel JS, Buonarati MH, Davis JC (1991) DNA dosimetry following carcinogen exposure using accelerator mass spectrometry and 32P-postlabeling. *Prog Clin Biol Res* 372:243–253.

Fifield LK (1999) Accelerator mass spectrometry and its applications. *Rep Prog Phys* 62:1223–1274.

Garner RC, Goris I, Laenen AA, Vanhoutte E, Meuldermans W, Gregory S, Garner JV, Leong D, Whattam M, Calam A, Snel CA (2002) Evaluation of accelerator mass spectrometry in a human mass balance and pharmacokinetic study-experience with 14C-labeled (R)-6-[amino(4-chlorophenyl)

(1-methyl-1H-imidazol-5-yl)methyl]-4-(3-chlorophenyl)-1-methyl-2(1H)-quinolinone (R115777), a farnesyl transferase inhibitor. *Drug Metab Dispos* 30(7):823–830.

Hah SS, Stivers KM, de Vere White RW, Henderson PT (2006) Kinetics of carboplatin-DNA binding in genomic DNA and bladder cancer cells as determined by accelerator mass spectrometry. *Chem Res Toxicol* 19:622–626.

Hah SS, Mundt JM, Ubick EA, Turteltaub KW, Gregg JP, Henderson PT (2007) A sample preparation protocol for quantification of radiolabeled nucleoside incorporation into DNA by accelerator mass spectrometry. *Nucl Instrum Methods Phys Res B* 259:763–766.

Karara AH, Edeki T, McLeod J, Tonelli AP, Wagner JA (2010) PhRMA survey on the conduct of first-in-human clinical trials under exploratory investigational new drug applications. *J Clin Pharmacol* 50(4):380–391.

Lappin G (2004) Current perspectives of 14C-isotope measurement in biomedical accelerator mass spectrometry. *Anal Bioanal Chem* 378(2):356–364.

Lappin G (2006) The use of isotopes in determination of absolute bioavailability of drugs in human. *Expert Opin Drug Metab Toxicol* 2:419–427.

Lappin G, Garner RC (2003) Ultra-sensitive detection of radiolabeled drugs and their metabolites using accelerator mass spectrometry. In *Handbook of Analytical Separations*, Wilson I, ed., pp. 331–349. Elsevier, Amsterdam.

Lappin G, Garner RC (2008) The utility of microdosing over the past 5 years. *Expert Opin Drug Metab Toxicol* 4(12):1499–1506.

Lappin G, Stevens L (2008) Biomedical accelerator mass spectrometry: Recent applications in metabolism and pharmacokinetics. *Expert Opin Drug Metab Toxicol* 4:1021–1033.

Lappin G, Kuhnz W, Jochemsen R, Kneer J, Chaudhary A, Oosterhuis B, Drijfhout WJ, Rowland M, Garner RC (2006) Use of microdosing to predict pharmacokinetics at the therapeutic dose: Experience with 5 drugs. *Clin Pharmacol Ther* 80:203–215.

Lappin G, Seymour M, Young G, Higton D, Hill HM (2011) AMS method validation for quantification in pharmacokinetic studies with concomitant extravascular and intravenous administration. *Bioanalysis* 3(4):393–405.

Ni J, Ouyang H, Aiello M, Seto C, Borbridge L, Sakuma T, Ellis R, Welty D, Acheampong A (2008) Microdosing assessment to evaluate pharmacokinetics and drug metabolism in rats using liquid chromatography-tandem mass spectrometry. *Pharm Res* 25(7):1572–1582.

Prakash C, Shaffer CL, Tremaine LM, Liberman RG, Skipper PL, Flarakos J, Tannenbaum SR (2007) Application of liquid chromatography-accelerator mass spectrometry (LC-AMS) to evaluate the metabolic profiles of a drug candidate in human urine and plasma. *Drug Metab Lett* 1(3):226–231.

Sandhu P, Vogel JS, Rose MJ, Ubick EA, Brunner JE, Wallace MA, Adelsberger JK, Baker MP, Henderson PT, Pearson PG, Baillie TA (2004) Evaluation of microdosing strategies for studies in preclinical drug development: Demonstration of linear pharmacokinetics in dogs of a nucleoside analog over a 50-fold dose range. *Drug Metab Dispos* 32(11):1254–1259.

Sarapa N, Hsyu PH, Lappin G, Garner RC (2005) The application of accelerator mass spectrometry to absolute bioavailability studies in humans: Simultaneous administration of an intravenous microdose of 14C-nelfinavir mesylate solution and oral nelfinavir to healthy volunteers. *J Clin Pharmacol* 45(10):1198–1205.

Stevens LA, Evans P, Shen Z, Yeh LT, Ong V, Densel M, Rowlings C, Dueker SR, Lohstroh PN, Tu D, Giacomo J, Quart B (2009) Microdose and microtracer intravenous pharmacokinetics of RDEA806 in healthy subjects. *110th Annual Meeting of the American Society for Clinical Pharmacology and Therapeutics, March 18-21, 2009.*

Tuniz C, Norton G (2008) Accelerator mass spectrometry: New trends and applications. *Nucl Instrum Methods Phys Res B* 266:1837–1845.

Turteltaub KW, Vogel JS (2000) Applications of accelerator mass spectrometry for pharmaceutical research. *Curr Pharm Des* 6:991–1007.

Turteltaub KW, Felton JS, Gledhill BL, Vogel JS, Southon JR, Caffee MW, Finkel RC, Nelson DE, Proctor ID, Davis JC (1990) Accelerator mass spectrometry in biomedical dosimetry: Relationship between low-level exposure and covalent binding of heterocyclic amine carcinogens to DNA. *Proc Natl Acad Sci USA* Jul; 87(14):5288–5292.

Vuong LT, Ruckle JL, Blood AB, Reid MJ, Wasnich RD, Synal H-A, Dueker SR (2007) Use of accelerator mass spectrometry to measure the pharmacokinetics and peripheral blood mononuclear cell concentrations of zidovudine. *J Pharm Sci* 97:2833–2843.

Wang X, Seymour M, Bryan P, Kumar G (2009) Development and Validation of HPLC-AMS method for quantification of [14C]-CC-11050 in human plasma. *AAPS J.* 11(S2): 3833.

Zhou XJ, Garner RC, Nicholson S, Kissling CJ, Mayers D (2009) Microdose pharmacokinetics of IDX899 and IDX989, candidate HIV-1 non-nucleoside reverse transcriptase inhibitors, following oral and intravenous administration in healthy male subjects. *J Clin Pharmacol* 49(12):1408–1416.

22

RADIOACTIVITY PROFILING

Wing Wah Lam, Jose Silva, and Heng-Keang Lim

22.1 INTRODUCTION

Compounds radiolabeled with ^{14}C or ^{3}H are routinely used in *in vitro* and *in vivo* drug metabolism studies in the course of drug development (Chang et al., 1998; Egnash and Ramanathan, 2002; Zhang et al., 2007). One of the advantages of using radioisotopes is the ability to determine accurately the percent composition of metabolites in a particular sample matrix without the need for a reference metabolite standard. The amount of radioactivity present in *in vitro* samples and *in vivo* excretory samples usually is high, and the amount of samples is normally not a limiting factor. Thus, detecting the presence of a metabolite using conventional methods usually does not pose any difficulties. However, detecting low abundant circulating metabolites can be challenging, especially when the metabolites are present in low levels of radioactivity. This is most critical when pharmaceutical industries are moving toward developing more potent drugs as a means to reduce drug attrition due to toxicity by reducing the body burden of reactive metabolites from lower doses. Furthermore, a definitive human absorption excretion metabolism (AEM) study conducted in Europe tended to use a lower radioactive dose (~50 µCi) compared with that used in United States (Penner et al., 2009). Thus, radioprofiling metabolites in plasma samples from such human studies in order to understand definitively the metabolism and disposition of a drug candidate prior to new drug application (NDA) filing poses tremendous analytical challenges. Earlier quantitative information on circulatory human metabolites at steady state is necessary to ensure coverage of human metabolites in safety assessment in

preclinical species, which is driven largely by the metabolites in safety testing (MIST) guidance from the FDA (2008) and ICH M3 guidance (2009) from EMEA. These regulatory requirements prompted a revision of the strategies used for drug metabolism studies from the preclinical to the clinical phases to include capture of both qualitative and quantitative information on metabolites from first-in-human (FIH) studies for earlier assessment of exposure coverage of human metabolites in preclinical species. This has resulted in the exploration of various methods to quantitate metabolites without reference standards, which has included calibration of mass spectrometer (MS) response with radiolabeled analogs (Yu et al., 2007) and near equimolar MS response via low-flow nanospray (Valaskovic et al., 2006). None of these techniques provides quantitative information related to the mass or body burden of metabolites as stipulated by the ICH M3 guidance. One potential solution that has drawn considerable attention from the pharmaceutical industry is to obtain earlier quantitative information on metabolites using a ^{14}C-microdosing study with accelerator mass spectrometry (AMS) detection (Garner et al., 2000). However, the incorporation of ^{14}C-microdosing study with AMS detection in the development plans for all compounds is unrealistic and prohibitively expensive. Hence, it is desirable to have highly sensitive instruments and techniques that can circumvent the profiling challenges and afford accurate quantitation results for individual metabolites. Because of these needs, various analytical techniques for the detection of trace amounts of radioactivity have been developed and are discussed in this chapter.

ADME-Enabling Technologies in Drug Design and Development, First Edition. Edited by Donglu Zhang and Sekhar Surapaneni.
© 2012 John Wiley & Sons, Inc. Published 2012 by John Wiley & Sons, Inc.

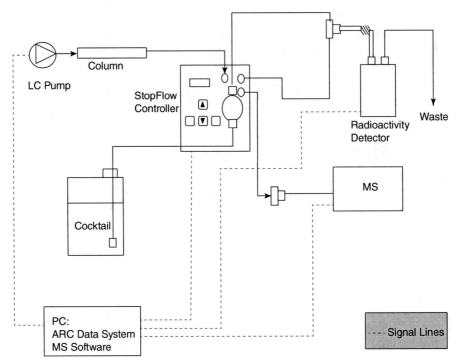

FIGURE 22.1. Schematic of LC-ARC (StopFlow) setup.

22.2   RADIOACTIVITY DETECTION METHODS

Typically, two approaches are currently used to detect radioactivity from a high performance liquid chromatography (HPLC) sample analysis, online or off-line. Online detection offers real-time detection, and thus an analyst has the advantage of visualizing the results during an analysis. As shown in Figure 22.1, the postcolumn eluate and scintillation cocktail were mixed at the "T" position right before radioactivity detection occurs.

On the other hand, off-line detection is set up with a fraction collector. For example, postcolumn HPLC eluates are collected into 96- or 384-well Lumaplates or Scintiplates. The duration for each fraction is usually collected in seconds. The shorter the time collected for each fraction, the higher is its resolution. If necessary, a repeat injection can be collected into the same wells for detection of low levels of radioactivity in the sample. The eluates in the plate are dried down prior to counting using a TopCount instrument (PerkinElmer, Inc., Downers Grove, IL, USA). Current TopCount instruments can have up to 12 channels for simultaneous detection, but radioactivity counting is still the bottleneck and severely impacts its throughput. The counting results can be exported, plotted, and presented as a radiochromatogram using ProFSA (PerkinElmer, Inc., Waltham, MA, USA), ARC (AIM Research Co., Hockessin, DE, USA), or other software applications.

Regardless of whether an online with real-time detection or off-line fraction collection method is used,

the instrument can be used as a standalone unit or coupled to a mass spectrometer. Thus, with the MS coupled technique, it becomes liquid chromatography-radioactivity detection-mass spectrometry (LC-RAD-MS); and both radioprofiling of samples as well as mass spectral data can be acquired simultaneously through a postcolumn split valve. Obviously, the advantage is that in a single run, both radioprofiling and mass spectral data of metabolites are acquired. This method is appealing and is the preferred method for metabolism studies. It not only is more efficient in analyzing samples, but also minimizes the discrepancies between separate runs, especially when the metabolic profiles are complex. However, when radioactivity in the sample is limited, stand-alone radioactivity detection may be preferred. With this approach, the entire radioactivity from the sample injected will be used for radioprofiling. Thus, stronger radioactive peaks can be obtained. If the amount of radioactivity is below the limit of detection (LOD) of the online radioactivity detection, repeated fraction collection for off-line counting using a Top-Count instrument may provide a solution.

22.2.1   Conventional Technologies

22.2.1.1   Conventional   Online   Radiochemical   Detection   A conventional online HPLC radioactivity detector operates in the continuous flow detection mode. This detector can be used as a stand-alone for radioactivity detection or coupled with a mass spec-

trometer as a hyphenated LC-RAD-MS instrument. The postcolumn LC eluate is split using a splitter to deliver anywhere from 50% to 80% of the flow to the radioactivity detector, and the remaining goes to the mass spectrometer for metabolite identification. This conventional detector is relatively less sensitive in detecting radiolabeled metabolites, as illustrated by the reported limit of detection in the range of 250–500 dpm (Zhu et al., 2007). The sensitivity of such detectors can be improved by using a combination of larger flow cells together with an optimized ratio of the LC to scintillation flow rate. The longer path length associated with a larger detection flow cell results in a longer residence time for detection of a radioactive sample, which leads to improved signal to noise ratio and translates into an increased detection sensitivity. However, the increased sensitivity obtained with a larger flow cell is accompanied by peak broadening and thus led to decreased resolution. This conventional radioactivity detection technique is still widely applicable for *in vitro* and *in vivo* (excreta) sample analyses wherein the amount of radioactivity is not a limiting factor (Egnash and Ramanathan, 2002).

22.2.2   Recent Technologies

22.2.2.1   Online Detection
22.2.2.1.1   Liquid Chromatography-Accurate Radioisotope Counting (StopFlow)-Mass Spectrometry (LC-ARC [StopFlow]-MS) Technique   The StopFlow method of detection was developed by the AIM Research Co. (Hockessin, DE) and represents a dramatic improvement in sensitivity of detection as compared with conventional radioactivity detection. Based on the equation below, the improvement in sensitivity of detection is accomplished by a dramatic increase in residence time by stopping flow without the undesired peak broadening from diffusion of chromatographic peak:

$$T = \frac{V}{L \times (1+R)},$$

where T is the counting time of an analyte in the flow cell (min), V is the flow cell volume (μL), L is the LC flow rate (μL/min), and R is the scintillant/LC ratio.

This online technique with marked improvement in sensitivity for detecting radioactive peaks is ideally suited for profiling samples with low level of radioactivity, especially in plasma. The sensitivity in detecting metabolites with low levels of radioactivity increases dramatically (Lee, 2003a,b; Nassar et al., 2003). Both ^{3}H and ^{14}C samples have been used to demonstrate the utility of this technique (Gaddamidi et al., 2004; Colizza et al., 2007; Lam et al., 2007, 2008; Zhu et al., 2007). A comparison of radioprofiles of rat plasma containing ^{3}H-labeled mefenamic acid using conventional beta-

RAM technique and LC-ARC (StopFlow) illustrates the advantages of using this newer LC-ARC StopFlow technique. An equal amount of radioactivity, approximately 7600 dpm, in rat plasma was analyzed in each case. The upper and lower panes in Figure 22.2 were the radiochromatograms acquired using the conventional beta-radioactivity monitor (RAM) and StopFlow techniques, respectively. The signal-to-noise ratio obtained for the strongest peak is no more than 3 to 1. In fact, some artifacts were present at around 47 min in the chromatogram (Figure 22.2). Without the better quality radiochromatogram acquired from StopFlow, it would have been difficult to distinguish whether the artifact peaks were real. The radiochromatogram acquired using the StopFlow method clearly showed not only the major metabolite peaks but also some minor metabolites with lower level of radioactivity (~130 dpm; Lam et al., 2007).

The schematic setup for StopFlow technique is shown in Figure 22.1. The StopFlow Controller in the setup has two major functions. One of the main functions is to maintain the back-pressure of the column during the stop-flow data collection period. This is crucial for a successful stop-flow analysis, since it will maintain the peak shape and resolution of the analytes present in the sample by preservation of chromatography fidelity via minimizing peak broadening from diffusion of the chromatographic peak. The second function is to deliver an appropriate amount of cocktail optimized for peak detection in the flow cell. The StopFlow technique can be set up in three different modes: the by-level, by-fraction, and non-StopFlow modes. When the by-level mode is used, the LC flow will continue to run until a certain threshold of radioactivity is detected in the flow cell. The ARC instrument will stop both the LC and the scintillation cocktail flow and allow the radiochemical detector to count the radioactivity in a predefined length of time, typically, 15–60 s per fraction. The length of each fraction being counted is also predefined (in seconds). The shorter the length of time each fraction is counted, the better is the peak resolution and the longer is the chromatographic run time. The counting will continue until the radioactivity in the flow cell falls below the threshold. The LC and the cocktail flow will continue in a nonstop fashion until the instrument detects radioactivity above the threshold again, and the ARC instrument stops both the LC and cocktail flows and resumes counting as before. Another way of setting up the stop-flow analysis is called the by-fraction mode. The setup simply allows counting each predefined region in a stop-flow fashion regardless of the amount of radioactivity present during the run.

The count zone, which is the predefined region within the entire run time, can be defined in the method prior to analysis for both by-level and by-fraction modes. The

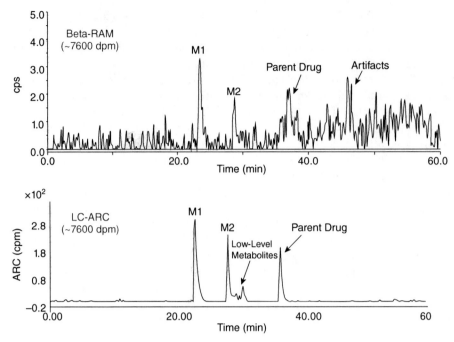

FIGURE 22.2. Comparison of radiochromatograms of rat plasma samples (8 h) acquired using conventional beta-RAM (upper panel), and LC-ARC StopFlow (lower panel) techniques following oral administration of ^{3}H-mefenamic acid (150 mg/kg) (Lam et al., 2007).

defined count zone is the region to be counted during an analysis, and non-StopFlow will apply if it is outside the count zone. For the nonstop-flow setup, it can be visualized as equivalent to the conventional way of detecting radioactivity using LC-RAD-MS analysis. Zhu et al. (2007) present a good illustration and comparison of all three modes of radioactivity detection.

As in any technique, there are various advantages and disadvantages when applying the StopFlow method. A clear advantage of using the StopFlow technique is the dramatic gain in sensitivity in detecting low levels of radioactive metabolites from a good-quality radiochromatogram, which is hard to obtain through conventional online detection. One of the major disadvantages of this technique is the lengthy time required to analyze each sample. It is not uncommon to take several hours to complete just one LC analysis. The actual length of the analysis is dependent on the chromatographic run time, StopFlow mode used for the analysis, and count zone chosen, as well as the count time defined in the method setup. The other major disadvantage is that when MS is used during the StopFlow analysis (LC-RAD-MS), the mass spectral file becomes so large that it is practically impossible to open the file. Thus, there is a need to separate LC-RAD profiling and MS analysis, which can significantly impact the throughput of analysis. To circumvent these problems, AIM Research Company has developed an alternative sensitive technique, called DynamicFlow™, which will be discussed below.

22.2.2.1.2 Liquid Chromatography-Accurate Radioisotope Counting (DynamicFlow)-Mass Spectrometry (LC-DynamicFlow-MS) Technique This is a relatively new technique, which was developed to overcome the two major limitations of the StopFlow technology. As discussed above, the two major drawbacks of the StopFlow technique are the long analytical times needed to complete each run and the difficulty in simultaneous acquisition of mass spectral data. The DynamicFlow technique was developed to overcome these problems, but at the same time it retains or surpasses the sensitivity of the StopFlow method. With this development, the run time was similar to that of conventional LC-RAD analysis, and the advantage of simultaneous acquisition of mass spectral data was added. This technique was illustrated using human urine sample and is shown in Figure 22.3 (Chen et al., 2008). In addition to normal-flow LC-DynamicFlow-MS, the compatibility of this technique with ultra performance liquid chromatography (UPLC) was recently evaluated as UPLC-DynamicFlow-MS, which is based on peak heights, and provided more than a twofold gain in sensitivity as indicated by a much narrower peak (peak width at half height of UPLC [7.2 s] vs. normal-flow [16.2 s]) with an improvement of chromatographic resolution when coupled with the UPLC (Figure 22.4). Similar results have also been reported from coupling UPLC with a conventional radioactivity detector capable of variable scintillation flow but with a limit of detection

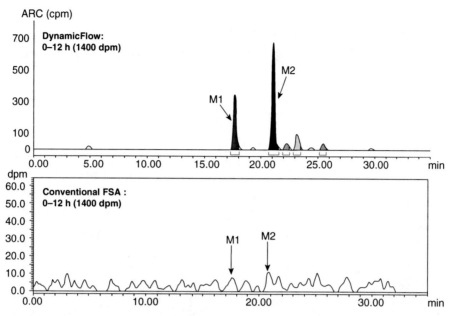

FIGURE 22.3. Comparison of analysis of a reconstituted pooled 0–12 h human urine sample extract, containing 1400 dpm total radioactivity of a[14]C-compound A and its metabolites, using both dynamic- and convention-flow radioactivity detection (Chen et al., 2008). Chromatographic separation using a Thermo AquaStar column (Thermo Fisher Scientific, Inc., Bellefonte, PA, USA) (150 × 2.1 mm ID, 3 μm), thermoregulated at 50°C, and eluted with a nonlinear gradient of 2.5 mM ammonium acetate and acetonitrile at 400 μL/min. The eluant was split 1:1 postcolumn for simultaneous acquisition of radioactivity and MS data.

[signal/noise ratio, S/N = 3] of 200 dpm on-column (Cuyckens et al., 2008).

To further explore the utilities of DynamicFlow technology, different aspects of the application were investigated and included a lower flow rate, a smaller column internal diameter (ID), and a smaller radioactivity detection flow cell. The utility has been illustrated using an *in vitro* incubate of rat liver microsomes of ^{14}C-diclofenac and in the presence of nicotinamide adenine dinucleotide phosphate, reduced (NADPH) and uridine 5′-diphosphoglucuronic acid/uridine 5′-diphospho-N-acetylglucosamine (UDPGA/UDPAG). In this study, the ^{14}C-diclofenac and its metabolites were chromatographically separated on a Betasil Phenyl/Hexyl column (Thermo Fisher Scientific, Inc., Bellefonte, PA, USA) (150 × 1.0 mm ID, 3 μm) using a flow rate of 100 μL/min, and the radioactivity was detected with a 60-μL flow cell (Figure 22.5). Three radioactive components corresponding to diclofenac metabolites were detected with impressive signal-to-noise ratio for each component following injection of a total radioactivity of 144 dpm on column. In a similar study, the limit of detection was found to be 15 dpm (Lam, 2009). This gain in sensitivity observed with a smaller ID column, a lower flow rate, and a smaller cell provided insight into a future direction for optimizing the sensitivity of DynamicFlow technology. A better appreciation of the gain in sensitivity for the DynamicFlow technology relative to conventional detection can be obtained by understanding the theory behind radioactivity detection with DynamicFlow technology.

Radioactivity decay follows the Poisson distribution, which essentially promises higher sensitivity with longer counting time (Currie, 1968; Zhu et al., 2007):

$$Ld = \frac{\left(2.71 + 4.65\sqrt{B \times \frac{E}{100} \times T}\right)}{T \times \frac{E}{100}},$$

where Ld is the limit of detection, T is the counting time (min), E is the counting efficiency of the detector (%), and B is the background of radioactivity detector (dpm).

The longer the counting time for each radioactive sample, the higher the detection sensitivity will be. This principle is applicable to both online and off-line radioactivity detection. However, there is a practical limit to the counting time that can be spent on each analyte peak, which therefore imposes a limit on the detection sensitivity of radioactive peaks using conventional online radioactivity detectors.

However, DynamicFlow radioactivity detection technology employs a novel technique (Chen et al., 2008; Lee, 2009) that varies either the scintillation cocktail flow rate, the LC flow rate, or both during the analytical run as a means to significantly increase the radioactivity detection sensitivity while maintaining the run time the same or close to conventional LC run times.

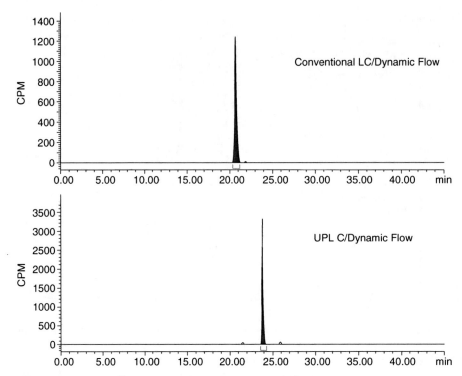

FIGURE 22.4. Radiochromatograms of ¹⁴C-compound C (7236 dpm) acquired using conventional LC and UPLC (Chen et al., 2008) methods and dynamic-flow radioactivity detection. Chromatographic separation was achieved using a Thermo AquaStar column (150 × 2.1 mm ID, 3 µm) for the conventional LC and Waters Acquity BEH UPLC column (150 × 2.1 mm ID, 1.7 µm ID; Waters Corporation, Milford, MA, USA), thermoregulated at 50°C, and eluted with a nonlinear gradient of 2.5 mM ammonium acetate and acetonitrile at 400 µL/min. The eluant was split 1:1 postcolumn for simultaneous acquisition of radioactivity and MS data.

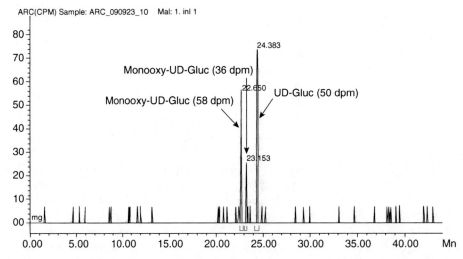

FIGURE 22.5. Radiochromatogram of rat liver microsomes (RLM) incubate of ¹⁴C-diclofenac (10 µM) in the presence of NADPH and UDPGA/UDPAG (144 dpm). Chromatographic separation using a Thermo Betasil Phenyl/Hexyl column (150 × 1.0 mm ID, 3 µm), thermoregulated at 50°C, and eluted with a nonlinear gradient of 2.5 mM ammonium acetate and acetonitrile at 100 µL/min; the radioactivity was detected with a 60 µL flow cell. The eluant was split 1:1 postcolumn for simultaneous acquisition of radioactivity and MS data.

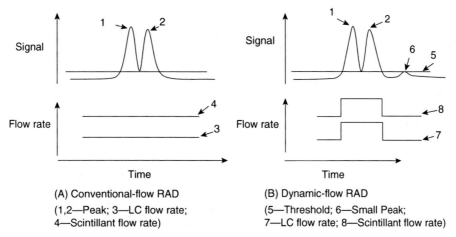

(A) Conventional-flow RAD
(1,2—Peak; 3—LC flow rate;
4—Scintillant flow rate)

(B) Dynamic-flow RAD
(5—Threshold; 6—Small Peak;
7—LC flow rate; 8—Scintillant flow rate)

FIGURE 22.6. Conventional-flow radioactivity detection (RAD) (A) and DynamicFlow RAD (B) (Lam et al., 2009).

The principle of DynamicFlow technology is illustrated in Figure 22.6. Figure 22.6A illustrates the traditional radio-HPLC method, where the scintillation cocktail flow rate remains normally unchanged irrespective of signal level. The ratio of flow rate of scintillation cocktail to LC eluate is fixed to a certain value to ensure certain resolution of radioactive peaks and sensitivity. Under this scenario, the counting time for all portions of eluate is relatively unchanged during an analytical run. Typically, conventional methods do not afford enough sensitivity for detection of low-level radioactive peaks, with the limit of detection lying between 250 and 500 dpm (Zhu et al., 2007).

Operation under DynamicFlow mode is illustrated in Figure 22.6B, whereby the scintillation cocktail flow rate and/or LC flow rate can be slowed down in a certain retention time region where the signal is weak to increase the counting time for weak signals, leading to increased sensitivity in detection of the weak signal. This also has the added benefit that low radioactive peaks are not diluted out by the scintillant, and thereby, the technique is capable of achieving a higher limit of detection. Conversely, the scintillation cocktail and/or LC flow rate can be increased to reduce residence time, leading to higher resolution. Thus, the amount of residence time needed for detection is dynamically adjusted according to the concentration of the analyte present in the sample. Therefore, DynamicFlow methods can successfully increase sensitivity and maintain resolutions in the same run but without changing the run time.

22.2.2.2 Off-Line Detection
22.2.2.2.1 TopCount and Microbeta Detection As mentioned earlier, off-line radioactivity detection is an alternative to generate semiquantitation of metabolites (Boernsen, 2000; Boernsen et al., 2000). The entire post column LC eluant can be collected directly into 96- or 384-well plates. Alternatively, an LC system can be coupled with a fraction collector as well as a mass spectrometer (LC-Fraction Collector-MS) to collect fractions for counting of radioactivity and MS data simultaneously. For the latter setup, a split valve is used to direct the desirable percentage of eluant into the fraction collector and mass spectrometer. Thus, with this setup, both radioactivity profiling of metabolites and MS data can be acquired in a single run.

Typically, two types of plates are used for this purpose, and they are the Lumaplates and Scintiplate. A layer of solid yttrium silicate is embedded at the bottom of the Lumaplates, and the plate is counted using a TopCount instrument (Floeckher, 1991a,b). Similarly, a layer of solid scintillant is incorporated at the bottom of the Scintiplate, and the plate is counted using a Microbeta instrument (PerkinElmer, Inc., Downers Grove, IL, USA) (Yin et al., 2000). These plates are commonly dried using a Savant SpeedVac system (Thermo Electron Corporation, Holbrook, NY, USA) prior to counting. The advantage of centrifugal drying using a SpeedVac system is the ability to concentrate the majority of the radioactivity from the sample to the bottom of the plate. This ensures maximum contact of the radioactive sample with the coated solid scintillant during counting. This step will minimize the loss of radioactivity sticking to the wall of the plate, which has no scintillant and therefore will not be counted. Thus, a maximum amount of radioactivity will interact with the solid scintillant at the bottom of the plate when centrifugal drying is used. The advantage of using Lumaplates is their sensitivity over Scintiplates (Nedderman et al., 2004; Lam et al., 2007). The limit of detection for Lumaplates was reported to be 4.2 dpm by Kiffe et al. (2007). However, the radioactivity associated with the solid scintillant in a Lumaplate cannot be conveniently recovered for additional interrogation by MS for structure elucidation. The advantage of using Scintiplates is that the sample can be recovered for subsequent MS analysis. Recently, another type of 96- or 384- well plate called Cytostar-T plate (PerkinElmer, Inc., Downers Grove, IL, USA) was evaluated by Kiffe et al. (2007) for off-line radioactivity counting. The parameters for counting these plates can be optimized to gain a significant lower background, and this results in higher sensitivity, but it was approximately threefold less sensitive than Luma-plate (Figure 22.7). The advantage of using a Cytostar-T plate is that it also allows recovery of metabolites from the wells for additional MS work to identify their structures.

Both DynamicFlow and TopCount techniques are current methods of choice for detection of low-level radioactivity. The limits of detection are 4.2 dpm (Top-Count) as reported by Kiffe et al. (2007) and 15 dpm (DynamicFlow) by Lam et al. (2009).

22.3   AMS

AMS is a highly sensitive technique that is capable of detecting ultralow levels of radioactivity in samples at parts-per-trillion to quadrillion. This is because AMS measurement is based on single atom counting of radio-nuclides with very high signal-to-noise ratio capability due to low background interference from low natural abundance of radionuclide together with high discrimination of interference from atomic and molecular isobars. This is best appreciated by an understanding of the operation of the AMS, which is briefly described below since it has been reviewed elsewhere by Dueker et al. (2010), and is shown in a schematic illustrating a lower-energy, smaller-footprint, user-friendly AMS instrument (Figure 22.8) In principle, the detection of ^{14}C by an AMS study involves the combustion of individual HPLC fractions collected, followed by reduction of the CO_2 generated to elemental carbon to form graphite pellets for analysis by cesium sputtering. This leads to formation of the negatively charged ions of ^{12}C, ^{13}C, which are separated based on mass by a low-field magnet to remove interference by atomic isobars. A tandem electrostatic accelerator is then used to accelerate negative carbon ions for high-energy collision with an inert gas or a foil to convert the negative ions to positively charged species to eliminate interference by molecular isobars. These positive ions are then

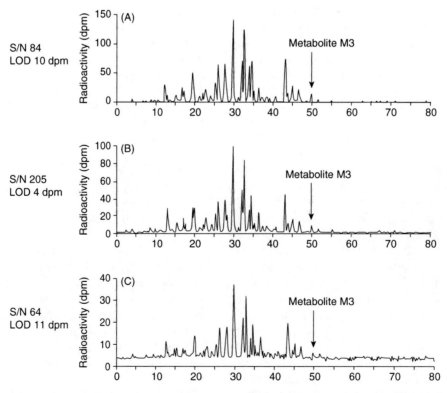

FIGURE 22.7. Comparison of metabolite patterns resulting from a urine sample (injection each 2000 dpm) by (A) liquid scintillation counting (counting time/fraction: 10 min); (B) Luma plate scintillation counting (counting time/well: 3 × 20 min); (C) Cytostar-T microplate scintillation counting (counting time/well: 3 × 20 min). (Reprinted with permission from Kiffe et al., 2007.)

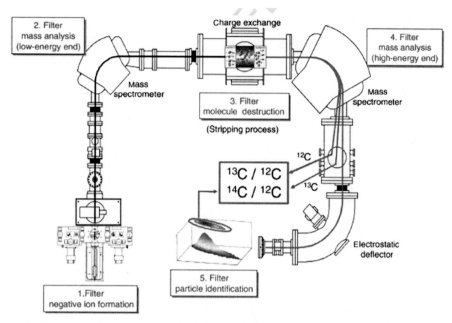

FIGURE 22.8. Schematic describing the operating principle of a lower-energy, smaller-footprint, and user-friendly accelerator mass spectrometer (Reprinted with permission from Dueker et al., 2010.)

separated into individual ions by an electrostatic analyzer based on energy and charge for subsequent identification and counting by a detector for specificity of response. The ^{14}C is normalized to stable carbon isotopes to reduce variability from ion emission and transmission.

Typically, this level of sensitivity of AMS is several orders of magnitude higher than the limit of detection of even the most sensitive decay-counting detector, DynamicFlow. This is because only a small fraction of the decay period is measured due to the long half-life of radionuclides such as ^{3}H or ^{14}C, which are typically used in drug metabolism pharmacokinetics (DMPK) studies. Although this technique was initially developed for radiocarbon dating of samples from archeological and earth sciences (Bennett et al., 1977; Nelson et al., 1977), the inherently high sensitivity of AMS has recently been taken advantage of for earlier investigation of pharmacokinetics and mass balance studies in humans via microdosing using ^{14}C tracer (Litherland, 1980; Elmore and Phillips, 1987; Elmore et al., 1990; Cupid and Garner, 1998; Gilman et al., 1998; Barker and Garner, 1999; Garner et al., 2000, 2002; Young et al., 2001; Lappin and Garner, 2003; Sarapa, 2003; Sarapa et al., 2005). The concept of microdosing was first described in a position paper by the European Medicines Agency (2004) on nonclinical safety studies in support of clinical trials with a single microdose, which typically involved dosing at ≤1% of the pharmacological dose or 100 μg, whichever is lower, allowing earlier investigation of

drug candidates in humans without extensive toxicity testing in animals. This is equivalent to administration of low nCi (≤100 nCi) per subject.

This application of the AMS technique has been nicely evaluated by Garner et al. (2002) in a human mass balance and pharmacokinetic study. In this study, ^{14}C-labeled (*R*)-6-[amino(4-chlorophenyl)(1-methyl-1*H*-imidazol-5-yl)methyl]-4-(3-chlorophenyl)-1-methyl-2(1*H*)-quinolinone ([^{14}C]R115777) was dosed to healthy human subjects. Each subject received a single oral dose of 50 mg of ^{14}C-R115777, which contained only 34.35 nCi/subject. Plasma, urine, and feces samples were initially separated by HPLC, and fractions were collected, followed by individual fraction analysis by AMS. Results from analysis of these samples afforded not only the mass balance and plasma concentration data but also radioactivity profiles from different matrices (Figure 22.9). In general, there was reasonably good agreement of pharmacokinetics (PK) data generated from microdose and therapeutic dose studies of seven drugs (European Microdosing AMS Partnership Programme, 2008. In addition, microdosing permitted opportunity for earlier assessment of exposure coverage of human metabolites in preclinical species following the guidances on safety testing of metabolites from both FDA and ICH M3, taking into consideration the potential disconnect between metabolite profiles at microdose compared with at steady state.

The low sample throughput from manual graphitization of samples, which resulted in high cost per sample

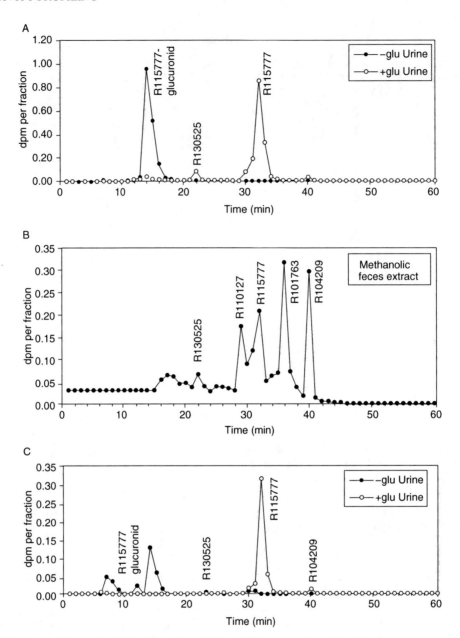

FIGURE 22.9. HPLC metabolite profiles determined by AMS analysis of mobile phase fractions of a combined pool of 0–24-h urine samples before and after β-glucuronidase treatment (A), a combined pool of methanolic extracts of a selection of feces samples (B), and a combined pool of 3-h plasma samples before and after β-glucuronidase treatment (C). (Reprinted with permission from Garner et al., 2002.)

together with initial high cost, large footprint, and the need for a nuclear physicist to operate the AMS instrument, has prevented wide adoption of AMS in the laboratories of pharmaceutical industry. There have been many advances in AMS, which include the recent commercialization of a user-friendly, compact, lower-cost, higher-throughput, and lower-energy accelerator mass spectrometer (BioMICADAS, Vitalea Science, Davis, CA, USA) equipped with a cesium sputtering ion source that can analyze both graphite and CO_2 samples

(Schulze-Konig et al., 2010). Furthermore, Tannenbaum's group at the Biological Engineering Accelerator Mass Spectrometry (BEAMS) Laboratory at Massachusetts Institute of Technology (MIT) developed interfaces for direct coupling with gas chromatography (GC) via modification of the interface for stable isotope MS or with HPLC via a new laser-induced combustion interface (Skipper et al., 2002). The traditional AMS graphitization step was eliminated using the laser-induced combustion interface, which permitted direct conversion of liquid samples into gaseous carbon dioxide in the front-end for the AMS instrument (Lieberman, 2004). This technique has been successfully applied to quantify low amounts of ^{14}C in a wide variety of samples, such as plasma and urine (Prakash 2007).

22.4   INTRACAVITY OPTOGALVANIC SPECTROSCOPY

A new emerging laser-based intracavity optogalvanic spectroscopy has been reported by Murnick et al. (2008, 2010) to be capable of quantitating $^{14}CO_2$ at the zeptomole level from oxidation of a $\geq 10 \, \mu g$ sample. The high sensitivity of detection (10^{15} or better), a >2-order of magnitude of linear dynamic range (5×10^{-15} to >1.5×10^{-12} in $^{14}C/^{12}C$ ratios), low microgram (μg) sample requirement, together with capability of online analysis by coupling with GC or LC make this technique an attractive alternative to AMS in analysis of trace radioactive samples. Unfortunately, this analytical technique is not yet commercially available.

22.5   SUMMARY

Highly sensitive techniques to quantify circulating metabolites following administration of radiolabeled compounds are a necessity for understanding the safety and efficacy of novel drug candidates. The techniques discussed above are currently available for radioactivity profiling, and each has its strengths and weaknesses. Further developments resulting in improvement of these techniques are much needed. For DynamicFlow detection, continued development to increase sensitivity is desirable. One of the considerations is to further reduce the background noise during analysis to give a superior signal to noise ratio. Another aspect is to improve both hardware and software so that it becomes more rugged and affords more reproducible results, especially when metabolites with little radioactivity are analyzed.

The attractiveness of off-line counting using Top-Count lies in the high limit of detection due to a very low background noise of ~2 cpm. The availability of the 384-well Lumaplates makes it convenient to collect small fractions (e.g., 5-s fractions) to ensure closer alignment of the reconstructed radiochromatogram with that from online detection. The typical count time is 5–10 min per 6 or 12 channel detections, and it usually takes several hours to complete counting each plate. The overall process from sample injection to plotting out a radiochromatogram is still lengthy, and it is desirable to shorten the overall time needed to increase throughput.

Another concern is the potential loss of volatile metabolite(s) during sample drying under reduced pressure. In addition, the lack of user-friendly software for processing data generated by TopCount requires the use of a third-party software for data processing such as plotting out radiochromatograms for subsequent analysis.

AMS is an ultrasensitive instrument for detection of low levels of radioactivity, and the recent trend toward commercialization of smaller-footprint, lower-energy, and user-friendly AMS should increase accessibility to this technology. However, it is desirable to develop smaller and less expensive AMS instruments to make the technology even more affordable for implementation in laboratories across the pharmaceutical industry (Ognibene et al., 2002). Unfortunately, the cost of analysis per sample is high due to the low throughput resulting from the labor-intensive manual process of AMS analysis. The recent development of a laser-induced combustion interface as an alternative to graphitization permitted rapid online LC-AMS analysis (Lieberman, 2004). This LC-AMS interface has the capability to increase sample throughput and to generate high-resolution chromatographic profiles (Prakash et al., 2007).

In addition to improving these detection technologies, a combination of using smaller-ID UPLC columns and low LC flow rate should also be further researched for optimal sensitivity detection. Because of the availability of these sensitive detection techniques, applications can be widened to studies requiring low concentrations of drugs, such as reaction phenotyping and enzyme kinetics (Zhao et al., 2004; Zhu et al., 2007).

ACKNOWLEDGMENTS

The authors would like to thank Dr. Dian Lee and Aaron Young for their helpful discussion of LC-ARC (both StopFlow and DynamicFlow).

REFERENCES

Barker J, Garner RC (1999) Biomedical applications of accelerator mass spectrometry-isotope measurements at the level of the atom. *Rapid Commun Mass Spectrom* 13: 285–293.

Bennett CL, Beukens RP, Clover MR, Gove HE, Liebert RB, Litherland AE, Purser KH, Sondheim WE (1977) Radiocarbon dating using electrostatic accelerators: Negative ions provide the key. *Science* 198:508–510.

Boernsen KO (2000) Using the TopCount microplate scintillation and luminescence counter and deep-well Lumaplate microplates in combination with micro-separation techniques for metabolic studies. Application Note AN004-TC, Packard Instrument Co.

Boernsen KO, Floeckher JM, Bruin GJM (2000) Use of a microplate scintillation counter as a radioactivity detector for miniaturized separation techniques in drug metabolism. *Anal Chem* 72:3956–3959.

Chang M, Sood VK, Wilson GJ, Kloosterman DA, Sanders PE, Schuette MR, Judy RW, Voorman RL, Maio SM, Slatter JG (1998) Absorption, distribution, metabolism, and excretion of atevirdine in the rat. *Drug Metab Dispos* 26(10): 1008–1018.

Chen J, Lam W, Xu F, Silva J, Lee D, Lim HK (2008) Application of UPLC/dynamic-flow RAD/MS for metabolite identification and profiling. The 56th American Society for Mass Spectrometry (ASMS). Denver, CO, June 1–5.

Colizza K, Awad M, Kamel A (2007) Metabolism, pharmacokinetics, and excretion of the substance P receptor antagonist CP-122,721 in humans: structural characterization of the novel major circulating metabolite 5-Trifluoromethoxy salicylic acid by high-performance liquid chromatography-tandem mass spectrometry and NMR spectroscopy. *Drug Metab Dispos* 35(6):884–897.

Cupid BC, Garner RC (1998) Accelerator mass spectrometry—A new tool for drug metabolism studies. In *Drug metabolism: Towards the Next Millennium*, Gooderham NJ, ed., pp. 175–187. Ion Press, Amsterdam.

Currie LA (1968) Limits for qualitative detection and quantitative determination. *Anal Chem* 40:568–593.

Cuyckens F, Koppen V, Kembuegler R, Leclercq L (2008) Improved liquid chromatography-online radioactivity detection for metabolite profiling. *J Chromatogr A* 1209: 128–135.

Dueker SR, Vuong LT, Lohstroh PN, Giacomo JA, Vogel JS (2010) Quantifying exploratory low dose compounds in humans with AMS. *Adv Drug Deliv Rev* 63(7):518–531.

Egnash LA, Ramanathan R (2002) Comparison of heterogeneous and homogeneous radioactivity flow detectors for simultaneous profiling and LC-MS/MS characterization of metabolites. *J Pharm Biomed Anal* 27:271–284.

Elmore D, Phillips FM (1987) Accelerator mass spectrometry for measurement of long-lived radioisotopes. *Science* 236:543–550.

Elmore D, Bhattacharyya MH, Sacco-Gibson N, Peterson DP (1990) Calcium-41 as a long-term biological tracer for bone resorption. *Nucl Instrum Methods Phys Res B* 52:531–535.

European Medicines Agency (2004) In Position Paper on Non-Clinical Safety Studies to Support Clinical Trials with a Single Microdose CPMP/SWP/2599/02/Rev1, http://www.ema.europa.eu/docs/en_GB/document_library/Scientific_guideline/2009/09/WC500002720.pdf.

European Microdosing AMS Partnership Programme (EUMAPP) (2008) Outcomes from EUMAPP—A Study Comparing In Vitro, In Silico, Microdose and Pharmacological Dose Pharmacokinetics, http://www.eumapp.com.

Floeckher J (1991a) Solid scintillation counting. Application Note TCA-002, Packard Instrument Co.

Floeckher J (1991b) Theory of TopCount operation. Application Note TCA-003, Packard Instrument Co.

Gaddamidi V, Scott MT, Swain SR, Brown AM, Young G, Hashinger BM, Lee D, Bookhart W (2004) Sensitive on-line detection of radioisotopes using LC-ARC applications in agrochemical research. *Drug Metab Rev* 36(Suppl. 1):256.

Garner RC, Barker J, Flavell C, Garner JV, Whattam M, Young GC, Cussans N, Jezequel S, Leong D (2000) A validation study comparing accelerator MS and liquid scintillation counting for analysis of ^{14}C-labelled drugs in plasma, urine and faeces. *J Pharm Biomed Anal* 24:197–209.

Garner RC, Goris I, Laenen AAE, Vanhoutte E, Meuldermans W, Gregory S, Garner JV, Leong D, Whattam M, Calam A, Snel CAW (2002) Evaluation of accelerator mass spectrometry in a human mass balance and pharmacokinetic study—Experience with ^{14}C-labeled (R)-6-[amino(4-chlorophenyl)(1-methyl-1H-imidazol-5-yl) methyl]-4-(3-chlorophenyl)-1-methyl-2(1H)-quinolinone (R115777), a farnesyl transferase inhibitor. *Drug Metab Dispos* 30:823–830.

Gilman SD, Gee SJ, Hammock BD, Vogel JS, Haack K, Buchholz BA, Freeman SPH, Wester RC, Hui X, Maibach HI (1998) Analytical performance of accelerator mass spectrometry and liquid scintillation counting for detection of ^{14}C-labeled atrazine metabolites in human urine. *Anal Chem* 70:3463–3469.

ICH M3 Guidance (2009) Guidance on nonclinical safety studies for the conduct of human clinical trials and marketing authorization for pharmaceuticals.

Kiffe M, Nufer R, Trunzer M, Graf D (2007) Cytostar-T plates—A valid alternative for microplate scintillation counting of low radioactivity in combination with high-performance liquid chromatography in drug metabolism studies? *J Chromatogr A* 1157:65–72.

Lam W, Loi CM, Atherton J, Stolle W, Easter J, Mutlib A (2007) Application of in-line liquid chromatography-accurate radioisotope counting-mass spectrometry (LC-ARC-MS) to evaluate metabolic profile of [^{3}H]-mefenamic acid in rat plasma. *Drug Metab Lett* 1:179–188.

Lam W, Lim HK, Silva J, Young A, Lee DY (2009) Increase in sensitivity of on-line radioactivity detection with microbore LC-MS coupled with DynamicFlow ARC system. The 17th North American International Society for the Study of Xenobiotics (ISSX), Baltimore, MD, October 18–22.

Lam WW, Loi CM, Nedderman A, Walker D (2008) Applications of high-sensitivity mass spectrometry and radioactivity detection techniques in drug metabolism studies. In *Mass Spectrometry in Drug Metabolism and Pharmacokinetics*, 1st ed., Ramanathan R, ed., pp. 253–273. John Wiley & Sons, New York.

Lappin G, Garner RC (2003) Big physics, small doses. The use of AMS and PET in human microdosing of development drugs. *Nat Rev Drug Discov* 2:233–240.

Lee D (2003a) Methods and apparatus for detection of radioactivity in liquid samples. U.S. Patent 6,546,786.

Lee D (2009) Dynamic flow liquid chromatography. U.S. Patent 7,624,626.

Lee DY (2003b) Advances in radio-HPLC. 12th International Society for the Studies of Xenobiotics, Rhode Island, 2003.

Liberman RG, Tannenbaum SR, Hughey BJ, Shefer RE, Klinkowstein RE, Prakash C, Harriman SP, Skipper PL (2004) An interface for direct analysis of 14C in non-volatile samples by accelerator mass spectrometry. *Anal Chem* 76:328–334.

Litherland AE (1980) Ultrasensitive mass spectrometry with accelerators. *Annu Rev Nucl Part Sci* 30:437–473.

Murnick DE, Dogru O, Ilkmen E (2008) Intracavity optogalvanic spectroscopy, a new ultra-sensitive analytical technique for ^{14}C analysis. *Anal Chem* 80(13):4820–4824.

Murnick DE, Dogru O, Ilkmen E (2010) ^{14}C analysis via intracavity optogalvanic spectroscopy. *Nucl Instrum Methods Phys Res B* 268:708–711.

Nassar AEF, Bjorge SM, Lee DY (2003) On-line liquid chromatography—Accurate radioisotope counting coupled with a radioactivity detector and mass spectrometer for metabolite identification in drug discovery and development. *Anal Chem* 75:785–790.

Nedderman ANR, Savage ME, White KL, Walker DK (2004) The use of 96-well Scintiplates to facilitate definitive metabolism studies for drug candidates. *J Pharm Biomed Anal* 34:607–617.

Nelson DE, Korteling RG, Scott WR (1977) Carbon-14: Direct detection at natural concentrations. *Science* 198:507–508.

Ognibene TJ, Bench G, Brown TA, Peaslee GF, Vogel JS (2002) A new accelerator mass spectrometry system for 14C-quantification of biochemical samples. *Int J Mass Spectrom* 218:255–264.

Penner N, Klunk LJ, Prakash C (2009) Human radiolabeled mass balance studies: Objectives, utilities and limitations. *Biopharm Drug Dispos* 30:185–203.

Prakash C, Shaffer CL, Tremaine LM, Liberman RG, Skipper PL, Flarakos J, Tannenbaum SR (2007) Application of liquid chromatography-accelerator mass spectrometry (LC-AMS) to evaluate the metabolic profiles of a drug candidate in human urine and plasma. *Drug Metab Lett* 1:226–231.

Sarapa N (2003) Early human microdosing to reduce attrition in clinical drug development. *Am Pharm Outsourcing* 4:42–47.

Sarapa N, Hsyu PH, Lappin G, Garner RC (2005) The application of accelerator mass spectrometry to absolute bioavailability studies in humans: simultaneous administration of an intravenous microdose of ^{14}C-nelfinavir mesylate solution and oral nelfinavir to healthy volunteers. *J Clin Pharmacol* 45:1198–1205.

Schulze-Konig T, Dueker SR, Giacomo J, Suter M, Vogel JS, Synal H-A (2010) BioMICADAS: Compact next generation AMS system for pharmaceutical science. *Nucl Instrum Methods Phys Res B* 268:891–894.

Skipper PL, Hughey BJ, Liberman RG, Choi MH, Wishnok JS, Klinkowstein RE, Shefer RE, Tannenbaum SR, Harriman SP, Prakash C (2002) Interfaces for direct analysis of isotopes by accelerator mass spectrometry. Abstracts of Papers, 224th ACS National Meeting, Boston, MA, August 18–22, 2002.

U.S. Food and Drug Administration (2008) *Guidance for Industry: Safety Testing of Drug Metabolites*, FDA, Rockville, MD. http://www.fda.gov/downloads/Drugs/GuidanceCompliance RegulatoryInformation/Guidances/ucm079266.pdf.2008.

Valaskovic GA, Utley L, Lee MS, Wu J-T (2006) Ultra-low flow nanospray for the normalization of conventional liquid chromatography/mass spectrometry through equimolar response: Standard-free quantitative estimation of metabolite levels in drug discovery. *Rapid Commun Mass Spectrom* 20:1087–1096.

Yin H, Greenberg GE, Fischer V (2000) Application of Wallac Microbeta radioactivity counter and Wallac Scintiplate in metabolite profiling and identification studies. Application Note, PerkinElmer Life Sciences.

Young G, Ellis W, Ayrton J, Hussey E, Adamkiewicz B (2001) Accelerator mass spectrometry (AMS): Recent experience of its use in a clinical study and the potential future of the technique. *Xenobiotica* 31:619–632.

Yu CP, Chen CL, Gorycki FL, Neiss TG (2007) A rapid method for quantitatively estimating metabolites in human plasma in the absence of synthetic standards using a combination of liquid chromatography/mass spectrometry and radiometric detection. *Rapid Commun Mass Spectrom* 21:497–502.

Zhang D, Wang L, Raghavan N, Zhang H, Li W, Cheng P, Yao M, Zhang L, Zhu M, Bonacorsi S, Yeola S, Mitroka J, Hariharan N, Hosagrahara V, Chandrasena G, Shyu WC, Humphreys WG (2007) Comparative metabolism of radiolabeled muraglitazar in animals and humans by quantitative and qualitative metabolite. *Drug Metab Dispos* 35:150–167.

Zhao W, Wang L, Zhang D, Zhu M (2004) Rapid and sensitive determination of enzyme kinetics of drug metabolism using HPLC coupled with a stop-flow radioactivity flow detector. *Drug Metab Rev* 36(Suppl. 1):257.

Zhu M, Zhao W, Humphreys WG (2007) Applications of liquid radiochromatography techniques in drug metabolism studies. In *Drug Metabolism in Drug Design and Development*, 1st ed., Zhang D, Zhu M, Humphreys WG, eds., pp. 289–317. John Wiley & Sons, New York.

23

A ROBUST METHODOLOGY FOR RAPID STRUCTURE DETERMINATION OF MICROGRAM-LEVEL DRUG METABOLITES BY NMR SPECTROSCOPY

KIM A. JOHNSON, STELLA HUANG, AND YUE-ZHONG SHU

A robust method for *in vitro* metabolite generation and facile sample preparation on analytical high-performance liquid chromatography (HPLC) was established for rapid structure determination of microgram-level drug metabolites by using high-field nuclear magnetic resonance (NMR) equipped with a cryoprobe. A single 2–5-mL incubation of drug candidate (10–30 μM) in microsomes, hepatocytes, or recombinant drug metabolizing enzymes, typically cytochrome P450s and uridine 5′-diphospho (UDP)-glucuronosyltransferases, was used for metabolite formation. Following precipitation of proteins and solvent removal, metabolite mixtures were chromatographed with 5–10 injections onto an HPLC-mass spectrometry (MS) system. Metabolites were collected into a 96-well plate, dried, and reconstituted in deuterated NMR solvents. NMR spectra of isolated metabolites were acquired on a 500 MHz spectrometer equipped with a 5-mm cryogenic probe. The methodology has been successfully employed as an extension of HPLC-tandem mass spectrometry (MS/MS)-based metabolite identification and applied frequently to 0.5–10 μg quantities of metabolite. Most structure determinations were achieved rapidly by ^{1}H-NMR with satisfactory signal-to-noise ratio, whereas some required 2-dimensional (2D) NMR data analysis. This report describes the method development and metabolite structure determination using the model compound trazodone. In addition to trazodone, a large number of examples from our laboratories have proven that the microgram-level NMR method avoids time-consuming preparative-scale metabolite generation and purification and circumvents technical complications associated with online liquid chromatography (LC)-NMR. Most importantly, the turnaround time of metabolite structure determination for metabolically unstable compounds using the present methodology is more in sync with the cycle time during which medicinal chemists modify the identified metabolic soft spot while performing other iterative lead optimization activities, demonstrating a real impact to the drug discovery process.

23.1 INTRODUCTION

A major aim in drug discovery is the identification of safe and efficacious medicines that can be administered according to convenient once- or twice-daily dosing regimens. Since metabolic processes influence many other parameters that are relevant in this regard, including bioavailability, systemic clearance, and toxicology, issues relating to drug metabolism are important in the selection of viable drug candidates. As metabolism-related liabilities, such as metabolic soft spots and reactive and potentially toxic metabolites, continue to be a major cause of attrition for drug candidates, biotransformation studies are becoming increasingly important in guiding the refinement of a lead series during drug discovery and in characterizing lead candidates prior to clinical

ADME-Enabling Technologies in Drug Design and Development, First Edition. Edited by Donglu Zhang and Sekhar Surapaneni.
© 2012 John Wiley & Sons, Inc. Published 2012 by John Wiley & Sons, Inc.

evaluation (Shu et al., 2008). The fast-paced drug discovery process requires biotransformation scientists to identify drug metabolites efficiently in sync with the speed at which medicinal chemists synthesize drug candidates during iterative cycles of lead-optimizing activities (Watt et al., 2003; Ma et al., 2006). HPLC coupled with MS/MS is widely used as the front-line analytical tool for metabolite identification, and the fragmentation of molecules of interest in tandem MS by collision-induced dissociation (CID) generally enables the assignment of Markush-type structures, which may be useful in some instances to assist with drug design effort of medicinal chemistry. However, the product ion spectra are often not informative enough for localizing the site of metabolic change, particularly for unknown and unpredicted drug metabolites. In addition, LC-MS has inherent limitations in elucidating regio- and stereochemistry.

As a tool for structure elucidation, NMR spectroscopy offers details on atom connectivities within a molecule, and is particularly useful when the exact site of metabolism is of interest for structural modification, but cannot be readily assigned by MS. The application of NMR for metabolite structure determination was a subject of comprehensive review in a recent monograph, where relevant hardware, key parameters and methodology, and select metabolite examples were illustrated (Huang et al., 2007). However, within the community of biotransformation, especially in the drug discovery setting, NMR plays a much smaller role than does MS. This is mostly attributed to two reasons. First, while tandem MS allows the identification of metabolites present in complex *in vitro* incubations and *in vivo* biofluids, the sample for NMR has to be reasonably pure and largely free of endogenous contaminants to allow reliable and unambiguous assignment, typically requiring time-consuming metabolite scale-up and purification that ultimately restrict the frequent use of NMR. Second, despite significant advances in hardware, such as the cryogenic probe with three- to fivefold lower thermal noise by cooling the receiver electronics to near-liquid helium temperature, resulting in dramatic increases in sensitivity and resolution, NMR still is a relatively insensitive technique (orders of magnitude less sensitive than that of MS), requiring a larger sample size (≥ 0.5 µg) to observe a simple 1D NMR spectrum, which may not even be achievable with a typical online LC-NMR method. Although the expectation set for online LC-NMR since its emergence more than a decade ago had been high, to practicing biotransformation scientists, this technique has not lived up to its promise to simplify the structure determination of drug metabolites primarily due to limitations in chromatographic resolution and sample size going to the NMR flow probe cell (Chen et al., 2007).

In order to overcome the common obstacles and impracticality of metabolite structure determination by NMR, we developed an improved analytical HPLC-based method for sample preparation, followed by structure determination by an NMR spectrometer equipped with a cryoprobe. Since the process is based on well-established analytical scale chromatographic and NMR technologies, it allowed us to optimize each technique independently, provide sufficient NMR information on 0.5–10 µg of metabolite samples, and assign their structures on a regular basis, without having to perform preparative HPLC separations. Herein, this chapter is intended to describe the methodology for rapidly purifying microgram quantities of and biologically relevant metabolites for structure determinations using trazodone as a model compound.

23.2 METHODS

23.2.1 Liver Microsome Incubations of Trazodone

Incubations (5 mL total volume) were performed in a shaking water bath at 37°C. Trazodone (30 µM final concentration, 0.15 µmol) was preincubated for 5 min in Tris-HCl buffer (50 mM, pH 7.4) containing microsomes (1 mg/mL protein content). Nicotinamide adenine dinucleotide phosphate (NADPH) (3 mM) and glutathione (GSH) (5 mM) were added to initiate reactions. Reactions were stopped at 90 min by addition of cold acetonitrile (3X volumes). Chemical stability of trazodone was assessed in incubations in which liver microsomes were not added. After centrifugation at 1500 g for 10 min, the extent of metabolite formation was monitored by LC-MS analysis, while the reminder of the supernatants were evaporated overnight on a SpeedVac (ThermoSavant, Holbrook, NY, USA) and reconstituted in 0.8 mL of the Tris buffer for metabolite purification.

23.2.2 HPLC and Metabolite Purification

HPLC system consisted of Thermo Accela quaternary pump, autosampler, and photodiode array detector (ThermoFisher Scientific, San Jose, CA). Chromatography was carried out on a Sunfire C-18 column (Waters Corp., Milford, MA, USA) (3 mm × 150 mm, 3 µm) with a binary mixture of 10 mM ammonium acetate adjusted to pH 5.0 with formic acid and containing 5% acetonitrile (solvent A) and acetonitrile (solvent B). The mobile phase initially consisted of A/B (90:10), it was then linearly programmed to A/B (10:90) over 25 min, held for 5 min, and then programmed to A/B (90:10) over 1 min. The column was equilibrated at A/B (90:10) for 5 min before making the next injection. A flow rate of 0.4 mL/min was used for all analyses. For metabolite purifications, fractions from 12–22 min during the LC run were collected into a deep-well microtiter plate containing 0.7 mL glass inserts at 0.125 min per well. Multiple injections (60 μL each) and chromatographies (up to 10) were repeated under the same conditions in an automated fashion, and eluant from the column was collected into the same plate. The solvent in collected metabolite wells was evaporated overnight on a Centrivap prior to dissolution in deuterated solvent for NMR.

23.2.3 HPLC-MS/MS

HPLC-MS/MS was conducted on a Thermo Orbitrap Discovery (ThermoFisher Scientific, San Jose, CA, USA) using electrospray ionization. The eluant from the HPLC column was routed into the flow cell of a Thermo Surveyor photodiode array detector (ThermoFisher Scientific, San Jose, CA, USA), then into the ion source of the mass spectrometer. The delay in response between the two detectors was about 0.2 min. The electrospray interface was operated at 3 kV, sheath pressure at 80 psi, capillary temperature at 275°C, and the mass spectrometer was operated in the positive mode. High-resolution full-scan MS measurements were obtained on the Orbitrap at 30,000 resolution (at 500 Da) with daily external calibrations using a mixture of ultramark, L-methionyl-arginyl-phenylalanyl-alanine acetate (MRFA), and caffeine as specified by the vendor. MS^n studies were performed using the linear ion trap (LTQ) with an isolation width of 3, normalized collision energy of 30 V, an activation Q of 0.25, and an activation time of 30 ms. MS^n fragments were passed into the Orbitrap for high-resolution mass measurements.

23.2.4 NMR

Each metabolite sample, collected and dried in the 0.7 mL glass insert from HPLC runs, was dissolved in 50 μL DMSO-d_6 and transferred to a 1.7-mm NMR tube. The use of a 1.7-mm tube for metabolite analysis was driven by the reduction of interference noise from NMR solvent and residual water, although the mass sensitivity gain from changing a 3-mm to a 1.7-mm tube in a 5-mm probe was insignificant since the radio frequency (RF) coil was further away from the sample (Martin and Hadden, 2000; de Swiet, 2005). All NMR data were acquired on a Bruker (Bruker Daltonics Inc., Billerica, MA, USA) 500 MHz spectrometer equipped with a 5-mm triple resonance cryoprobe (TCI) cryoprobe. ^{1}H-NMR experiments were performed by using the wet pulse sequence for water suppression (Ogg et al., 1994). Typical NMR parameters include 256 numbers of scan; 1 Hz line broadening; 1-s delay time; and 1.6-s acquisition time. Metabolite quantity was calculated by comparing the integral of ^{13}C satellite peaks of deuterated solvent with the integrals of proton signal from metabo-

lites (Dalisay and Molinski, 2009). Rotating-frame Overhauser effect spectroscopy (ROESY) spectrum of ~26 nmol of Met4 dissolved in 50 μL of dimethyl sulfoxide (DMSO)-d_6 was acquired as a 2048 × 128 matrix, with 64 scans per increment. A 2D ROESY with continue wave (CW) spin lock for mixing pulse sequence were used for this experiment (Bax and Davis, 1985). The data was processed to 2048 × 2048 points by zero-filling the second frequency domain. Metabolite quantity was calculated by comparing the integral of proton signals of metabolites with the integral of ^{13}C satellite peaks of DMSO-d_6, namely the quantitation by solvent ^{13}C-satellites (QSCS) method described by Dalisay and Molinski (2009). Briefly, a quantification curve was established by using 2.0, 1.0, 0.5, 0.2, and 0.1 mM of trazodone as an external standard. Using the equation $A_{CH}/A_{sat} = c \times m$, where A_{CH}/A_{sat} is the integral ratio of an aromatic proton peak of trazodone and the ^{13}C satellite peaks of DMSO-d_6, m is number of moles of trazodone, the slope (c) was determined as 0.34 nmol^{-1} ($R = 0.989,27$) and subsequently used for the calculation of metabolite quantity. The quantity of metabolites, m, was obtained from the same equation where A_{CH}/A_{sat} was the integral ratio of the aromatic proton peak of the metabolite and the ^{13}C satellite peaks of DMSO-d_6. To assist in the interpretation of NMR spectra, the ACD ^{1}H-NMR Predictor (Advanced Chemistry Development, Toronto, ON) was used.

23.3 TRAZODONE AND ITS METABOLISM

Trazodone is a second-generation antidepressant and anxiolytic. It inhibits the reuptake of serotonin, but possesses a lower affinity for the serotonin transporter (SERT) than selective serotonin reuptake inhibitors (SSRIs), such as fluoxetine. Trazodone's anxiolytic and antidepressant effects are likely due to its antagonism at the 5-HT$_{2A}$ and 5-HT$_{2C}$ receptors. It is clinically used for the treatment of major depression, often in conjunction with fluoxetine or other SSRIs, or to control sleep disturbance symptoms when using SSRIs and norepinephrine reuptake inhibitors (SNRIs) because of its sedating side effects (Haria et al., 1994).

Trazodone is extensively metabolized in humans via hydroxylation, N-dealkylation, and N-oxidation pathways (Baiocchi et al., 1974; Yamato et al., 1974; Jauch et al., 1976; Fujiwara et al., 1989). Early biotransformation studies of trazodone led to the identification of several major metabolites in the human and animal excreta, including a triazolopyridinone dihydrodiol metabolite (Met5), a chlorophenyl hydroxyl metabolite (Met4) and its glucuronide, whereas in human plasma, 1-(3'-chlorophenyl)piperazine or *m*-CPP (Met3), was a major circulating metabolite. Several recent investigations focused on the metabolic activation of trazodone to electrophilic quinine-imine and epoxide intermediates at the chlorophenyl and triazolopyridinone rings, respectively (Kalgutkar et al., 2005; Wen et al., 2008). A number of GSH conjugates were identified from NADPH-supplemented microsomes in the presence of GSH. However, except Met3-5, most metabolite structures were assigned tentatively based on the combination of LC-MS/MS spectra and proposed bioactivation reaction mechanism. Without NMR evidence or direct comparison with synthesized metabolite standards, the regiochemistry of these metabolites remained unclear. The pattern of multisite metabolism and formation of both oxidative and conjugative metabolites make trazodone a good model compound to apply our methodology to microgram-scale metabolites for rapid structure determination using NMR.

23.4 TRAZODONE METABOLITE GENERATION AND NMR SAMPLE PREPARATION

The generation of oxidative metabolites and GSH conjugates of trazodone in NADPH-supplemented microsomes in the presence of GSH was achieved by a single 5-mL incubation using 30 μM of trazodone. The incubation conditions, postincubation work-up, and sample preparation by HPLC closely resembled a typical biotransformation experiment. The HPLC-MS/MS system used mild or low concentration of mobile phase modifier (10 mM ammonium acetate). Other HPLC conditions for purification of metabolite were also very similar to typical metabolite identification analysis except that multiple (5–10) injections and separations of metabolites were made overnight in an automated fashion to reduce column overloading. Fractions from HPLC runs were collected repeatedly into a deep-well microtiter plate containing 0.7-mL glass inserts for drying in Centrivap while heat and high-intensity light were avoided. Glass inserts were used for metabolite collection and transfer to a small volume (50 μL) of deuterated solvents for NMR since they had significant advantage in minimizing sample contamination from phthalates, which were frequently encountered in commercially available microtiter plates and caused interference in NMR spectra. If metabolites or conjugates are anticipated to be unstable due to acidic or basic HPLC mobile phase modifier during sample work-up, addition of buffer solution, such as 1 M ammonium acetate (pH 6.5, 100 μL) into the glass inserts prior to fraction collection was found to greatly reduce the sample decomposition. The established HPLC purification process reliably yielded 0.5–10 μg of metabolites (Dalisay and Molinski, 2009) of trazodone and other

drug candidates in our laboratories, which were sufficient for most 1D ^{1}H-NMR and some 2D NMR with satisfactory signal-to-noise ratio. Most structure determinations were achieved by interpreting diagnostic 1D ^{1}H-NMR spectral information, such as chemical shifts, ^{1}H-^{1}H coupling patterns, and integrations, that is associated with the site of metabolism and complementary to the LC-MS/MS data. For proper spectral comparison among metabolites while avoiding data inconsistency caused by extraneous pH, solvent, and other HPLC-related effects, the NMR spectra of metabolites were directly compared with the spectrum of trazodone, which was collected from the same HPLC experiment.

Seven fractions containing a total of eight metabolites and glutathione conjugates of trazodone in human liver microsomes were obtained from the HPLC runs, and were subjected to NMR data acquisition using a 5-mm cryoprobe (1.7-mm tube). Both LC-MS/MS and NMR spectra indicated that a pair of metabolites, Met1a/Met1b, were not fully separable on HPLC and collected into a single fraction.

23.5 METABOLITE CHARACTERIZATION

The accurate mass spectrum (MS), MS/MS, and NMR spectral data characteristic to the metabolites are shown in Table 23.1, and the expanded ^{1}H-NMR spectra of aromatic region in Figure 23.1. For NMR signal assignment purposes, the numbering of structural elements is intended for illustration only and does not comply with International Union of Pure and Applied Chemistry (IUPAC) rules.

The fraction (1.8 μg/0.007 μmol) containing Met1a and Met1b showed that the 1D ^{1}H-NMR spectrum originated almost completely from Met1a, which exhibited a protonated molecular ion (MH$^+$) at m/z 262.1661 and an empirical molecular formula of $C_{13}H_{19}N_5O$ from accurate MS, indicating the loss of chlorophenyl moiety due to metabolism. The loss of the moiety was also evident in the ^{1}H-NMR (Figure 23.1) where only four aromatic proton resonances of the triazolopyridinone moiety, H5 (7.20 ppm), H6 (7.2 ppm), H7 (6.6 ppm), and H8 (7.85 ppm), were observed. Interestingly, Met1a has never been reported as a trazodone metabolite prior to the present study.

Met1b was a minor metabolite in the fraction. Its accurate MS/MS showed a molecular ion (MH$^+$) at m/z 693.2203 and empirical molecular formula of $C_{29}H_{38}O_8N_8ClS$, indicative of a GSH conjugate in conjunction with hydroxylation. In addition to fragment ions at m/z 564 (−129 Da, -glutamyl) and m/z 420 (−273 Da, -glutathionyl-lacking sulfur) diagnostic for an

TABLE 23.1. Characterization of Trazodone Metabolites and Glutathione Conjugates in Human Liver Microsomes Based on Integrated Data Analysis of HPLC-MS/MS and NMR

Met No.	HPLC Rt (min)[d]	Observed MH+ (m/z)	Mass Error Δppm[a]	Empirical Formula	Estimated Quantity (μmol)[b]	Key MS/MS Fragment Ions (m/z)	Key NMR Information	Combined Data Interpretation on Metabolite Structure	Regiochemistry Determination by NMR
Met1a	11.7	262.1661	0.5	$C_{13}H_{19}N_5O$	0.007[c]	197, 176, 165, 148	H5, H6, H7, and H8 resonances of the triazolopyridinone moiety	N-dealkylation of chlorophenyl ring	Yes
Met1b	11.7	693.2203	1.9	$C_{29}H_{38}O_8N_8ClS$		675, 564, 420, 388, 176		GSH conjugate on chlorophenyl	No
Met2	16.5	695.2358	2.1	$C_{29}H_{40}O_8N_8ClS$	0.009	677, 659, 566, 422, 237	H5, H6, H7, and H8 on partially reduced triazolopyridinone moiety from 1D NMR and ^{1}H-^{1}H COSY	7-OH, 8-GSH conjugate on partially reduced triazolopyridinone	Proposed
Met3	17.2	197.0850	5.0	$C_{10}H_{14}ClN_2$	0.048	154	Intact chlorophenyl ring, but no proton signals due to triazolopyridinone	1-(3'-chlorophenyl) piperazine or m-CPP	Yes
Met4	18.2	388.1531	0.9	$C_{19}H_{22}O_2N_5Cl$	0.026	350, 251, 176, 148, 133	H18, H21, H22 on chlorophenyl ring, H18/H15 and H22/ H15 NOEs in ROESY	para-Hydroxyl (20-OH) on chlorophenyl	Yes
Met5	20.2	406.1634	1.6	$C_{19}H_{24}O_3N_5Cl$	0.003	386, 235, 210, 192, 182, 164	H5, H6, H7 and H8 assigned to partially reduced triazolopyridinone from 1D NMR and ^{1}H-^{1}H COSY	7-OH, 8-OH dihydrodiol on triazolopyridinone	Yes
Met6	21.3	406.1637	0.7	$C_{19}H_{24}O_3N_5Cl$	0.0002	386, 210, 192, 182	^{1}H-NMR not interpretable	Suspected diastereomer of Met5	No
Met7	21.6	388.1530	1.2	$C_{19}H_{22}O_2N_5Cl$	0.013	253, 192, 176, 148, 133	^{1}H-NMR similar to Met 4, but H18, H20, and H21 on chlorophenyl ring in a different ABX spin system from that of Met4	22-OH on chlorophenyl	Proposed

[a] The mass error Δppm = 1×10^6 (Observed mass of MH+ —Theoretical mass of MH+)/Theoretical mass of MH+.
[b] Dalisay and Molinski (2009).
[c] Combined weight of predominant Met1a and minor Met1b.
[d] Under LC-MS conditions, trazodone had an HPLC retention time (Rt) of 22.5 min.

357

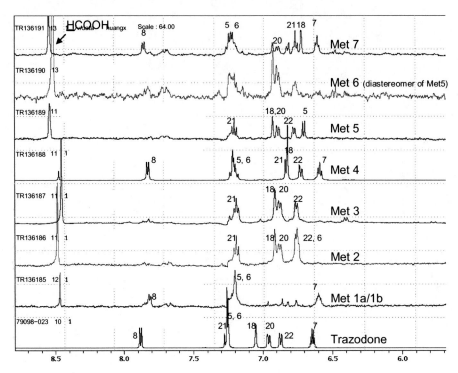

FIGURE 23.1. Expanded ^{1}H-NMR spectra of aromatic region of trazodone metabolites in human liver microsomes.

aromatic (sp^2) carbon-linked GSH conjugate, the product ion at m/z 176 corresponded to an intact *N*-propyl-triazolopyridinone piece, suggesting that hydroxylation and subsequent GSH conjugation had occurred on the 3-chlorophenyl ring system. Met1b appeared to be identical to the "conjugate 6" reported by Kalgutkar et al. (2005). However, the regiochemistry of GSH conjugation and hydroxylation remained undefined due to an insufficient quantity of sample.

Met2 proved to be another GSH conjugate of trazodone as indicated by its molecular ion (MH$^+$) at m/z 695.2358 and empirical molecular formula of $C_{29}H_{30}O_8N_8ClS$. Its MS/MS product ion spectra, including the diagnostic fragment ion at m/z 237 representing an intact 3-chlorophenyl-*N*-propylpiperazine moiety, led to the assignment of Met2 as a "dihydrodiol"-type GSH conjugate at the triazolopyridinone motif, which seemed to be similar to the "conjugate 3" reported by Kalgutkar et al. (2005) but without regiochemistry characterization. The ^{1}H-NMR spectrum of Met2 exhibited resonances of two vinyl protons H5 (6.75 ppm, d) and H6 (5.59 ppm, m) that significantly shifted upfield compared with those of trazodone (Figures 23.1 and 23.2), whereas those of H7 and H8 protons also shifted upfield due to loss of the double bond and were masked by the intense water and solvent signals in the 1D ^{1}H-NMR spectrum. A subsequent H-H correlation spectroscopy (COSY) spectrum revealed H7 (4.9 ppm) and H8

(5.38 ppm) (Figure 23.2), and further extended the H5/H6 spin system to H6/H7 and H7/H8. Using the ACD software for the prediction of ^{1}H chemical shifts based on NMR data of similar substructures, the observed chemical shifts of H5, H6, H7, and H8 were more consistent with the predicted values of a 7-hydroxyl, 8-glutathionyl substitution rather than a 7-glutathionyl, 8-hydroxyl substitution (Table 23.2). Met2 is thus proposed to have 7-hydroxyl and 8-glutathionyl regiochemistry on the partially reduced triazolopyridinone moiety.

The major metabolite Met3 (9.4 μg/0.048 μmol) demonstrated a molecular ion (MH$^+$) at m/z 197.0850 and empirical molecular formula of $C_{10}H_{14}ClN_2$, identical with those reported for 1-(3′-chlorophenyl)piperazine or *m*-CPP, an *N*-dealkylation product present in human plasma. The assignment of Met3 to m-CPP was also supported by its ^{1}H-NMR spectrum where only proton signals from an intact chlorophenyl ring were observed in the aromatic region (Figure 23.1).

Met4 was among the most abundant metabolite (9.9 μg/0.026 μmol) obtained from the human liver microsomal incubation of trazodone. Its molecular ion (MH$^+$, m/z 388.1531), empirical molecular formula ($C_{19}H_{22}O_2N_5Cl$), and MS/MS data (Table 23.1) were consistent with a monohydroxylated metabolite on the chlorophenyl ring. The ^{1}H-NMR spectrum of Met4 exhibited a number of key features; a recognizable spin system that has a coupling pattern among three remain-

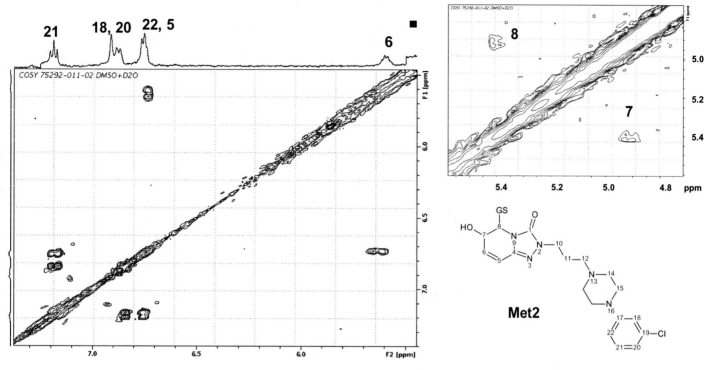

FIGURE 23.2. Diagnostic region of the ¹H, ¹H-COSY of Met2.

TABLE 23.2. Observed and Calculated ¹H-NMR Chemical Shifts of Triazolopyridinone Region of Met 2

Atom No.	Observed Met2	Calculated by ACD[a] 7-OH, 8-S-G	Calculated by ACD[a] 7-S-G, 8-OH
5	6.75, d	6.32	6.32
6	5.59, m	5.72	5.72
7	4.90	4.83	4.01
8	5.38	5.11	4.45

[a] ACD ¹H-NMR Predictor (Advanced Chemistry Development, Toronto, ON) was used.

ing protons on the chlorophenyl ring of nearly first-order (AMX), and marked upfield shifts of H21 (6.83 ppm) and H18 (6.81 ppm) from corresponding H21 (7.20 ppm) and H18 (7.02 ppm) in trazodone (Figure 23.1), which suggested that Met4 is either a 20-hydroxyl or a 22-hydroxyl metabolite. The concrete regiochemistry evidence was obtained by 2D-ROESY (rotating-frame Overhauser enhancement spectroscopy) spectrum, which clearly indicated through space interactions (nuclear Overhauser effects [NOEs]) between H18 and H15, and H22 and H15 (Figure 23.3), leading to the final assignment of 20-hydroxyl or *para*-hydroxy trazodone

structure for Met4, which was previously detected in human urine (Baiocchi et al., 1974; Jauch et al., 1976), rat excreta (Yamato et al., 1974), and liver microsomal incubations (Kalgutkar et al., 2005).

Both Met5 and Met6 showed the same molecular ion (MH⁺, m/z 406) and empirical molecular formula ($C_{19}H_{24}O_3N_5Cl$), suggesting two dihydrodiol metabolites. The observed product ions at m/z 210, 192, 182, 164 (Table 23.1) further suggested the location of dihydrodiol at the triazolopyridinone ring (Kalgutkar et al., 2005). However, only Met5 exhibited interpretable ¹H-NMR spectrum (Figure 23.1). Four triazolopyridinone proton signals of Met5, H5 (6.60 ppm), H6 (5.51 ppm), H7 (4.15 ppm), and H8 (4.38 ppm), were found to belong to one spin system from the ¹H, ¹H-COSY spectrum (data not shown), and were also significantly shifted upfield compared with corresponding protons in trazodone. Furthermore, the lack of the characteristic trazodone H8 signal near 7.88 ppm in Met5 implied cytochrome P450-mediated oxidation at C7 and C8 positions. The combined evidence strongly supported a 7-, 8-dihydrodiol rather than a 5-, 6-dihydrodiol for Met5, although the relative stereochemistry of C7 versus C8 could not be determined in this study. The assigned regiochemistry is also consistent with that reported for a major trazodone metabolite in human urine (Baiocchi et al., 1974; Jauch et al., 1976; Fujiwara et al., 1989).

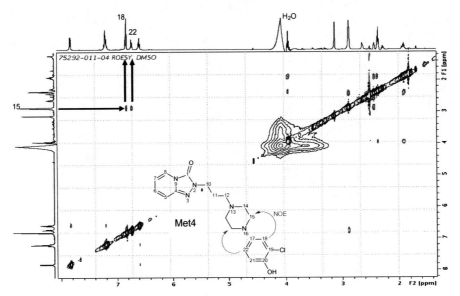

FIGURE 23.3. Diagnostic region of ROESY of Met4.

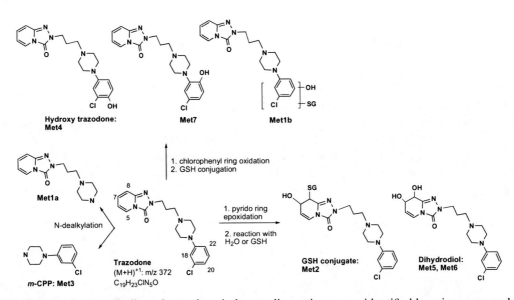

FIGURE 23.4. *In vitro* metabolites of trazodone in human liver microsomes identified by microgram-scale NMR.

Met7 showed identical molecular ion (MH+, m/z 388.1531), empirical molecular formula ($C_{19}H_{22}O_2N_5Cl$), and product ion spectra (Table 23.1) to Met4. Met7 is thus a monohydroxylated metabolite on the chlorophenyl ring and a regioisomer of Met4. The ^{1}H-NMR spectrum of Met7 was similar to that of Met 4, except that the ABX spin system of the chlorophenyl protons in Met7, comprising H18 (6.68 ppm), H20 (6.79 ppm), and H21 (6.73 ppm), shifted noticeably from those of Met4. Consequently, Met7 was proposed as the 22-hydroxyl metabolite of trazodone (Figure 23.1).

In summary and as shown in Figure 23.4, the structure and regiochemistry of four metabolites, Met1a, Met3, Met4, and Met5, were unambiguously characterized using ^{1}H-NMR, ^{1}H, ^{1}H-COSY, and ROESY techniques on 1–10 μg of sample. Regiochemistry assignments for another two metabolites (Met2 and Met7) were proposed based on observed chemical shifts and similarities to other metabolites. Two minor metabolites, Met1b and Met6, were not structurally defined beyond the gross structure assigned from MS/MS data primarily due to insufficient quantity.

Among the reported trazodone metabolites, only the regiochemistry of Met4 and Met5 was previously characterized (Baiocchi et al., 1974; Jauch et al., 1976; Fujiwara et al., 1989; Kalgutkar et al., 2005). Thus, the present study provided the first regiochemical evidence for Met2 and Met7 from NMR. In addition, the structure of Met1a has not been published previously and is thus established here for the first time.

It is important to point out that all the NMR data acquisitions and structure elucidations were achieved on microgram-scale metabolites originated from a single 5-mL microsomal incubation of trazodone (30 μM), and that completion time for the entire workflow was within a few consecutive days while sample stability was closely monitored.

23.6 COMPARISON WITH FLOW PROBE AND LC-NMR METHODS

As a comparison, we also evaluated microflow probe NMR (10 μL active volume, Protasis Probe) with a sample isolated from HPLC. It required typically a larger sample size (5 μg or more) for ^{1}H-NMR spectrum due to dilution effect from the push solvent and a less sensitive receiver operating at the room temperature. An overnight ROESY data acquisition on a 5-μg trazodone sample failed to provide interpretable data set. The shortcoming due to the dilution effect of the microflow probe may be improved with segmented flow technology (Kautz et al., 2005) when it becomes more accessible. Similar to LC-NMR, the microflow method suffered from greater difficulty in sample recovery.

Experience suggests that the most significant challenge in online LC-NMR, either in a stop-flow or loop-storage mode, remains the design of an appropriate chromatographic system to separate interfering substances while capturing the eluting peak of interest into the flow probe cell. Optimally, the peak volume should be matched to the flow cell volume, which often requires extremely fine control of system timings. In addition, the typical flow cell volume (~120 μL) is at least two times larger than that of a 1.7-mm NMR tube, leading to reduced sensitivity due to sample dilution. Another disadvantage of stop-flow mode occurs if multiple metabolites are analyzed. The metabolite peaks still remaining on the HPLC column may deteriorate during the NMR data acquisition time. LC-solid-phase extraction (SPE)-NMR addresses some limitations of LC-NMR since SPE provides a simpler approach to collect and concentrate individual HPLC peaks prior to washing off the analyte directly into a conventional NMR tube (Godejohann et al., 2004; Sandvoss et al., 2005).

However, not all metabolites, particularly GSH conjugates and glucuronides, can be readily trapped onto an SPE cartridge.

23.7 METABOLITE QUANTIFICATION BY NMR

We estimated metabolite quantities using the recently published QSQC method by comparing the integral of proton signal of metabolites to the integral of ^{13}C satellite peaks of DMSO-d$_6$ (Dalisay and Molinski, 2009). It should be pointed out that although this method provides useful quantity information for most metabolites, there are complications when applied to microgram level of metabolites. Unlike neat chemicals that can be accurately weighed and serially diluted in a nonaqueous NMR solvent (Protasis, Marlboro, MA, USA), the metabolites were prepared from reversed-phase HPLC fractions. The wetness of samples, particularly those early eluted metabolite peaks, is often significant and detrimental to quality ^{1}H-NMR spectra. Many NMR factors, such as signal to noise (S/N) ratio, resolution, and sharpness of the proton signal used for calculation, degree of water suppression, duration of the relaxation delay, and imperfect peak shape due to potential intermediate exchange, can inevitably introduce error for metabolite quantification. Nevertheless, in the drug discovery environment, the estimated metabolite quantity using this method offers valuable information in the rank-order of metabolites without synthetic standards.

23.8 CONCLUSION

In the present study, we successfully developed a robust methodology for rapid structure determination of microgram-scale drug metabolites using cryoprobe NMR spectroscopy. Interpretable 1H-NMR spectra of metabolites can be obtained in a time- and cost-effective manner starting from routine scale incubation (2–5 mL) with a parent compound (10–30 μM) in microsomes, hepatocytes, or recombinant drug metabolizing enzymes, typically cytochrome P450s and UDP-glucuronosyltransferases. The utilization of a high-field NMR equipped with a 5.0-mm cryoprobe (1.7-mm tube) extended the limit of detection and resulted in adequate 1H-NMR with ~1 μg or more sample after isolation by analytical HPLC. 2D NMR experiments may be needed for structure elucidation of some metabolites and sample requirements would increase to ~5–10 μg depending on the complexity of the structure (e.g., conjugates), sharpness of NMR signals, and the need of unmasking solvent interference. Since our off-line methodology is based on well-established chromatographic and NMR technologies, it allowed us to optimize each technique independently, and made it practical to obtain sufficient NMR information fast and on a regular basis without having to develop preparative or NMR-compatible HPLC separations. A large number of examples, including that of trazodone, from our laboratories have proven that the microgram-level NMR method avoids time-consuming preparative-scale metabolite generation and purification and circumvents technical complications associated with online LC-NMR. Most importantly, the turnaround time of the metabolite structure determination workflow becomes in sync with the cycle time during which medicinal chemists modify the identified metabolic soft spot while performing other iterative lead optimization activities, demonstrating a real impact to the drug discovery process.

REFERENCES

Baiocchi L, Frigerio A, Giannangeli M, Palazzo G (1974) Basic metabolites of trazodone in humans. *Arzneimittelforschung* 24:1699–1706.

Bax AD, Davis DG (1985) Practical aspects of two-dimensional transverse NOE spectroscopy. *J Magn Reson* 63:207–213.

Chen Y, Monshouwer M, Fitch WL (2007) Analytical tools and approaches for metabolite identification in early drug discovery. *Pharm Res* 24:248–257.

Dalisay DS, Molinski TF (2009) NMR quantitation of natural products at the nanomole scale. *J Nat Prod* 72:739–744.

Fujiwara S, Noumi K, Kawashima T, Awata N (1989) Absorption, metabolism and excretion of trazodone hydrochloride in humans. *Yakuri to Chiryo* 17:1365–1382.

Godejohann M, Tseng LH, Braumann U, Fuchser J, Spraul M (2004) Characterization of a paracetamol metabolite using on-line LC-SPE-NMR-MS and a cryogenic NMR probe. *J Chromatogr A* 1058:191–196.

Haria M, Fitton A, McTavish D (1994) Trazodone. A review of its pharmacology, therapeutic use in depression and therapeutic potential in other disorders. *Drugs Aging* 4:331–355.

Huang X, Powers R, Tymiak AE, Espina R, Roongta V (2007) Introduction to NMR and its application in metabolite structure determination. In *Drug Metabolism in Drug Design and Development*, Zhang D, Zhu M, Humphreys WG, eds., pp. 369–409. John Wiley & Sons, New York.

Jauch R, Kopitar Z, Prox A, Zimmer A (1976) Pharmacokinetics and metabolism of trazodone in man. *Arzneimittelforschung* 26:2084–2089.

Kalgutkar AS, Henne KR, Lame ME, Vaz AD, Collin C, Soglia JR, Zhao SX, Hop CE (2005) Metabolic activation of the nontricyclic antidepressant trazodone to electrophilic quinone-imine and epoxide intermediates in human liver microsomes and recombinant P4503A4. *Chem Biol Interact* 155:10–20.

Kautz RA, Goetzinger WK, Karger BL (2005) High-throughput microcoil NMR of compound libraries using zero-dispersion segmented flow analysis. *J Comb Chem* 7:14–20.

Ma S, Chowdhury SK, Alton KB (2006) Application of mass spectrometry for metabolite identification. *Curr Drug Metab* 7:503–523.

Martin GE, Hadden CE (2000) Long-range (1)H-(15)N heteronuclear shift correlation at natural abundance. *J Nat Prod* 63:543–585.

Ogg RJ, Kingsley PB, Taylor JS (1994) WET, a T1- and B1-insensitive water-suppression method for *in vivo* localized 1H NMR spectroscopy. *J Magn Reson B* 104:1–10.

Sandvoss M, Bardsley B, Beck TL, Lee-Smith E, North SE, Moore PJ, Edwards AJ, Smith RJ (2005) HPLC-SPE-NMR in pharmaceutical development: Capabilities and applications. *Magn Reson Chem* 43:762–770.

Shu YZ, Johnson BM, Yang TJ (2008) Role of biotransformation studies in minimizing metabolism-related liabilities in drug discovery. *AAPS J* 10:178–192.

de Swiet TM (2005) Optimal electric fields for different sample shapes in high resolution NMR spectroscopy. *J Magn Reson* 174:331–334.

Watt AP, Mortishire-Smith RJ, Gerhard U, Thomas SR (2003) Metabolite identification in drug discovery. *Curr Opin Drug Discov Devel* 6:57–65.

Wen B, Ma L, Rodrigues AD, Zhu M (2008) Detection of novel reactive metabolites of trazodone: Evidence for CYP2D6-mediated bioactivation of m-chlorophenylpiperazine. *Drug Metab Dispos* 36:841–850.

Yamato C, Takahashi T, Fujita T (1974) Studies on metabolism of trazodone. I. Metabolic fate of (14C)trazodone hydrochloride in rats. *Xenobiotica* 4:313–326.

24

SUPERCRITICAL FLUID CHROMATOGRAPHY

JUN DAI, YINGRU ZHANG, DAVID B. WANG-IVERSON, AND ADRIENNE A. TYMIAK

24.1 INTRODUCTION

A substance becomes a supercritical fluid at conditions above its critical temperature (T_C) and critical pressure (P_C). In this regime, there is no difference between gas and liquid since both phases have the same properties (e.g., density, viscosity, diffusion coefficient, etc.). The unique physicochemical properties of supercritical fluids have drawn the attention of separation scientists with the promise of new mobile phases and other chromatographic improvements. In comparison with liquid chromatography (LC), supercritical fluid chromatography (SFC) has the potential for higher separation efficiency and faster speed due to the high diffusivity and low back-pressure of the mobile phase. When compared to gas chromatography (GC), SFC offers greater versatility for separating a wide variety of solutes at lower temperatures. Some of the theoretical advantages of SFC have been demonstrated in a variety of important applications, but the full potential of this technology is yet to be realized.

24.2 BACKGROUND

Utilizing a supercritical fluid as the mobile phase for chromatography was first suggested by Lovelock in 1958 (Gere, 1983). Four years later, Klesper et al. (1962) demonstrated the pioneering application of SFC for separating nickel porphyrins using chlorofluoromethanes as the mobile phase. Subsequent early research of SFC was conducted by Sie et al. (Sie et al., 1966; Sie and Rijnders, 1967) and Giddings et al. (1968).

During the 1970s, the development of SFC was relatively slow, in part due to the success and rapid advances in high performance liquid chromatography (HPLC) instrumentation and applications (Sanagi and Smith, 1988). Interest in SFC was renewed in the 1980s with the use of capillary columns as a primary application in the petrochemical industry. In these "GC-like" early uses, SFC presented its advantages over GC by extending separations to a wider array of compounds, especially those that are nonvolatile or thermally unstable. However, the development of SFC as a general separation technique for a broad range of compounds has not progressed as rapidly as expected, despite the significant interest generated by the promise of increased efficiency and speed. This is detailed in the comprehensive and informative review by Smith (1999). In the early 1990s, the technology was nearly abandoned as being inefficient and without practical applications in the pharmaceutical industry (Harris, 2002; Berger, 2007).

A major obstacle in the development of SFC as a chromatography technique has been the lack of satisfactory commercial instrumentation with reliable reproducibility and high sensitivity. The early SFC instruments delivered poor chromatographic performance in comparison to other chromatographic techniques (LC, GC, etc.), thereby reducing interest in the technology. With few researchers actively involved in testing the limits of the technology, gaps remained in understanding the physicochemical properties of supercritical fluids and how they behaved during chromatography of diverse solutes. These deficiencies in the theory behind SFC presented additional challenges for developing routine chromatographic separations.

ADME-Enabling Technologies in Drug Design and Development, First Edition. Edited by Donglu Zhang and Sekhar Surapaneni.
© 2012 John Wiley & Sons, Inc. Published 2012 by John Wiley & Sons, Inc.

The renaissance of SFC was marked by the development of packed column SFC and its supporting instrumentation (Taylor, 2009). The first packed column SFC instrument was introduced by Hewlett-Packard in 1982. In 1995, Terry Berger bought the SFC section from Hewlett-Packard and funded Berger Instruments. In the late 1990s, Berger Instruments had most of the global market for SFC instruments (Mukhopadhyay, 2008). Owing much to the systematic research on packed column SFC conducted by Terry Berger (1995), SFC became recognized as a separation technique that is suitable for a wide variety of compounds. The Chromatographic Society briefly summarized the contribution of Terry Berger to modern SFC when he was awarded the 2004 Martin Gold Medal (The Chromatographic Society, 2004). Specifically, Berger and coworkers studied the density and solvent strength effects (Berger and Deye, 1990; Deye et al., 1990), introduced the use of additives and systematically studied their effects on peak shape and retention (Berger and Deye, 1991), applied SFC separation to broad classes of compounds including small drug-like molecules (Berger, 1997), and demonstrated the feasibility of long columns with large pressure drops (Berger and Wilson, 1993). Berger and his team also developed the separator technology to allow quantitative recovery of solutes without the need for cyclone separators or issues with aerosol generation.

Today, three decades after the introduction of the first commercial packed column SFC instrument, great progress has been made in optimizing the performance of the instrumentation in a number of areas such as back-pressure regulation, constant flow rate control, methods for modifier addition, in-line sample injection, and automation. Practical knowledge of columns chemistries, organic modifiers, and solvent additives has been acquired along with a better understanding of fundamental mechanisms of retention. SFC has become an accessible alternative separation technique that is complementary to CG and LC with its own significant and unique advantages. The success of SFC is evidenced by its common use in the pharmaceutical industry for preparative chiral resolution of drug discovery lead compounds, where the speed of separation for a wide variety of chiral molecules has been demonstrated and highly valued. Examples of SFC applications have expanded to include both analytical and preparative separations for drug discovery and drug development processes. With the commercial availability of a wide variety of chiral and achiral stationary phases, coupling of SFC with other separation techniques and different detectors has enabled a range of pharmaceutical, biological, and metabolic sample types to be separated.

24.3   SFC INSTRUMENTATION AND GENERAL CONSIDERATIONS

The block diagram in Figure 24.1 illustrates the basic design of a typical SFC system. Commercial SFC

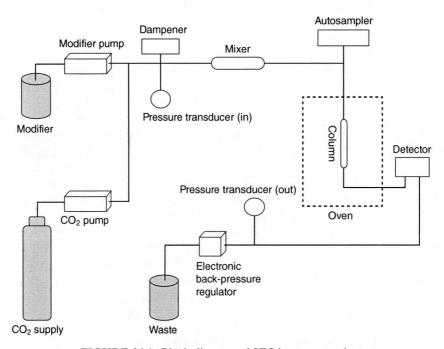

FIGURE 24.1. Block diagram of SFC instrumentation.

instruments rely on many of the same components that comprise traditional HPLC systems. However, there are some important differences between these two instrument types. One of the key requirements for an SFC instrument is the back-pressure regulator (BPR), which is used to control the outlet pressure and prevent expansion of the eluant into a gas within the system.

The pumps are operated as flow sources only and the electronic BPR controls the system pressure independent of the flow rate. To deliver liquid carbon dioxide and minimize the compressibility, the pump head for CO_2 needs to be chilled (Sanagi and Smith, 1988). Pump control algorithms are used to empirically optimize the calculated nominal compressibility compensation (Berger, 1997). The mobile phase is then pumped as a liquid and the pressurized fluid is preheated above or near the critical temperature in the oven before entering the column. For binary fluids, a secondary pump is required to deliver the mobile phase modifier.

Most modern SFC instruments use reciprocating piston pumps specifically designed for compressible fluids and an electromechanical BPR downstream of the column for independent control of pressure. These pumps enable the accurate mixing of compressible fluids and liquids, which is essential for reliable chromatography. In addition, an injection system capable of introducing the sample into a high-pressure environment without disturbing system pressure is required. The injector must provide accurate and reproducible injection of a sample under high pressure without any sample carryover. Since mobile phase density is critical for operating SFC, temperature and pressure controls are very important for the reproducibility and sensitivity of SFC.

24.3.1 Detectors Used in SFC

Almost all conventional LC and GC detectors have been adapted to SFC. However, mostly due to the compressibility of the mobile phase, baselines in SFC under comparable conditions typically exhibit higher noise levels than those observed in LC for most standard detectors. In addition to mechanical noise from the BPR, the overall noise in SFC could include contributions from thermal, electronic, and chemical sources.

Traditionally, HPLC is preferred over SFC in that it has high sensitivity of detection with a wide dynamic range. Lower sensitivity has been a major hurdle for the application of SFC for drug impurity profiling where detection of very low levels of impurities is required. However, significant progress has been made in reducing noise sources and increasing the overall signal-to-noise ratio in SFC. For detectors that are located before the pressure regulator, such as ultraviolet (UV) and circular dichroism detectors, the detection flow cell must

be capable of withstanding elevated pressures. In the case of detectors located after or not in line with the pressure regulator, such as mass spectrometry (MS) and evaporative light-scattering detectors (ELSD), mobile phase evaporation occurs through natural depressurization of CO_2.

UV is the most commonly used detector in SFC due to its good sensitivity, large dynamic range, as well as low cost. However, since the system remains pressurized until after the detector, a high-pressure UV detector cell is required. Improvements in UV detection for SFC have included increasing the light path and ensuring thermal equilibrium in order to increase signal-to-noise ratios (Berger, 2007).

MS is another widely used detector for SFC. Supercritical carbon dioxide is complete transparency to a mass spectrometer. The interfacing of packed column SFC with MS was initially reported in 1985 by Crowther and Henion for the rapid analysis of multicomponent mixtures of nonpolar and polar compounds (Crowther and Henion, 1985). In the past decade, the advancement of instrumentation in both SFC and MS has greatly increased the reliability and robustness of the SFC-MS technique. Useful information for a variety of SFC-MS interfaces has been established. Consequently, it has been widely used for chemical and bioanlytical analysis and purification of diverse types of molecules from small drugs to polymers, polypeptides, and proteins in different sample matrices (Combs et al., 1997; Bolaños et al., 2004; Li and Hsieh, 2008). Atmospheric pressure ionization (API) sources, both electrospray and atmospheric pressure chemical ionization (APCI), have been used in SFC-MS. APCI is the most popular ionization in SFC due to its high-flow compatibility. Detection sensitivity also favors APCI because the interface can be used without splitting (Pinkston et al., 2006). The advantages and drawbacks of various SFC-API interfacing approaches were compared, including direct effluent introduction with no active BPR in high-speed bioanalytical applications, pre-BPR-split interface, and an interface which provides total-flow-introduction with a mechanical BPR (Pinkston, 2005). Because there is no photoionization of CO_2 by the krypton lamp, lower noise can be obtained when using an atmospheric pressure photoionization (APPI) interface in SFC.

Flame ionization detection (FID) has been widely used for open-tubular SFC analysis. In addition to allowing pressure or density programming, FID provides universal, sensitive detection of carbon compounds with a uniform response factor. However, FID is limited to pure CO_2 or mobile phase mixtures with only a very small amount of modifier since nearly all organic modifiers produce significant FID signals (Sanagi and Smith, 1988).

Various selective detectors have been reported to provide very sensitive detection in SFC. Strode and Taylor (1996) studied the optimization of an electron-capture detector for SFC. Despite the fact that CO_2 and some modifiers contribute to electron capture, sensitive detection was possible. In the presence of 5% methanol as a modifier, low limit of detection (LOD) in the picogram range was achieved. Chemiluminescent nitrogen detector and sulfur-selective chemiluminescence detection for SFC with methanol as modifier (Shi et al., 1997a,b) have also been reported. Response with high sensitivity, selectivity, and wide dynamic range was obtained for nitrogen- and sulfur-containing compounds.

Dressman et al. (1996) reported electrochemical detection in SFC with neat and methanol-modified CO_2 mobile phases. The electrochemical detector provided low-nanogram detection limits and responded linearly over two decades of injected quantities. However, the detector was compatible with only 1% (v/v) methanol.

Amperometric detection at a naked platinum microelectrode was reported for packed capillary column SFC using CO_2 modified with 1% acetonitrile (Wallenborg et al., 1997). Detection based on oxidation or reduction was found to be compatible with the use of methanol as a sole modifier, with a detection limit in the picogram range.

Brunelli et al. (2007) reported the use of charged aerosol detection (CAD) for drug analysis in SFC. Mobile phase flow compensation before the BPR and temperature control was used to reduce the noise and increase reproducibility. Relatively uniform response was found for nonvolatile analytes with the use of methanol in the mobile phase. The average LOD for this configuration was reported to be 4.5 ng.

Takahashi et al. (2008) reported a quantitative comparison of CAD and ELSD in SFC using certified reference material of polyethylene glycol (PEG) and an equimass mixture of uniform PEG oligomers. CAD was able to detect 10-fold diluted solutions of uniform oligomers compared to ELSD. Recently, the same group reported a condensation nucleation light-scattering detector (CNLSD) used for SFC (Takahashi et al., 2009). In comparison with ELSD, the CNLSD was able to detect a 10-fold less concentrated solution of uniform oligomers. In combination with SFC's capacity to completely separate every oligomer with different polymerization indexes, CNLSD could precisely determine the polymer mass distribution of PEG 1000 without calibration.

Mah and Thurbide (2008) reported a novel method of interfacing the acoustic flame detector (AFD) with SFC as an alternative universal detector in separations requiring an organic cosolvent in the mobile phase. By applying resistive heating directly to the burner region between the restrictor outlet and the acoustic flame, they were able to eliminate infrequent severe noise, baseline drifting, and peak deformations. For different levels of methanol-modified CO_2, the interface was able to reduce detector noise to a common minimal range near 10–25 Hz when an appropriate temperature was achieved.

24.3.2 Mobile Phases Used in SFC

Different supercritical/superheated fluids including water (Greibrokk and Andersen, 2003), nitrous oxide (Wright et al., 1985), ammonia (Lauer et al., 1983), and haloalkanes (Ong et al., 1990) were reported in the literature as chromatography mobile phases. However, the most commonly used supercritical fluid for SFC is carbon dioxide due to its nearly ideal combination of properties. As shown in the phase diagram in Figure 24.2, carbon dioxide has a near ambient T_C at 31°C and a relatively low P_C at 73 bar. Compared with the other supercritical fluids mentioned above, carbon dioxide is also chemically inert, nontoxic, noncorrosive, nonflammable, and nonexplosive. It can be readily available at low cost as an industrial by-product. In addition, carbon dioxide does not generate interfering responses in most common detectors including UV, FID, infrared, MS, and ELSD. All these excellent properties make carbon dioxide the predominant supercritical fluid for SFC.

However, as a solvent, CO_2 is relatively nonpolar. At a density of 0.25 g/cm³, CO_2 has a similar solvent strength to perfluorinated alkanes; and at a density of 0.98 g/cm³, it is slightly more polar than hexane (Gere, 1983). Although the density of a supercritical fluid can be changed with pressure and temperature, which in turn

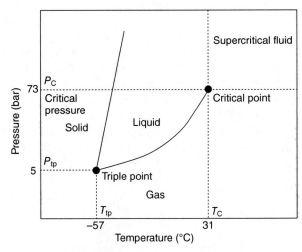

FIGURE 24.2. Phase diagram of CO_2.

alters its solvation power, pure CO_2 is unable to elute very polar compounds. This limitation of CO_2 is often compensated by adding modifiers to increase the polarity and alter the solvent strength. Berger and Deye (1990) showed that the retention of polar solutes changed dramatically with polar modifiers, and that a smaller effect on retention was observed with the changes in the density of binary fluids.

Thanks to the miscibility of CO_2 with most organic solvents, almost all organic modifiers used in LC can also be used in SFC. A wide variety of solvents, such as methanol, ethanol, isopropanol, acetonitrile, methyl *tert*-butyl ether, dichloromethane, and dimethylsulfoxide, have been reported as modifiers in SFC. Similar to LC, polar ionic additives (i.e., acid, base, and salt) have been used to improve peak shape for the separation of acids and amines. Addition of a small amount of water has also been reported (France et al., 1991; Pyo, 2000; Ashraf-Khorassani and Taylor, 2010).

Addition of a modifier to CO_2 alters the critical temperature and pressure. Consequently, under the commonly used SFC conditions, most binary mobile phases will be in a near (sub- or super-) critical state. Although the effects of chromatographic variables on separation may vary depending on whether the mobile phase is under a sub- or supercritical condition, a distinction between sub- and supercritical fluid condition is less important in practice. Furthermore, the change of the physicochemical properties at near critical condition is continuous, and the diffusion coefficient in modified CO_2 is higher than that in a liquid (Sassiat et al., 1987; Berger, 1997). Therefore, improvements in speed and efficiency are still commonly achieved even when working in the subcritical region. Phase separation of modified carbon dioxide eluants is rare under most operating conditions in packed column SFC. SFC is used to refer to both sub- and supercritical chromatography in this chapter.

24.3.3   Stationary Phases Used in SFC

Both open-tubular and packed columns have been used in SFC. Open-tubular SFC is more similar to GC and it can be employed with all the detectors used for GC. Open-tubular columns are typically long narrow capillary tubes with a coating film of stationary phase on the wall. Open-tubular SFC is a powerful analytical tool for low volatility and/or thermally unstable analytes that are not suitable for analysis by GC, as well as for some analytes that cannot be sufficiently separated by LC.

Packed columns, albeit with lower efficiency, provide higher sample capacity and better reproducibility compared with open-tubular columns. Consequently, packed columns have become the most commonly used station-ary phases in today's SFC. Almost all the packed columns that are used in LC can be used in SFC. Silica-based stationary phases with various functional moieties are commonly used in achiral SFC. Popular phases include 2-ethylpyridine, cyano, diol, and amino columns. Derivatized polysaccharide chiral stationary phases are the most popular for chiral SFC separations.

In SFC, retention behavior of different compounds depends largely on the stationary phase. Column classification based on quantitative structure–retention relationships with different solutes can be used to guide the column selection in SFC (Lesellier and West, 2007; West and Lesellier, 2008a; Lesellier, 2009).

24.3.4   Comparison of SFC with Other Chromatographic Techniques

Density, diffusion coefficient, and viscosity are important chromatographic properties of a mobile phase. Table 24.1 compares these physical properties among gas, supercritical fluid, and liquid (Bartle, 1988). The relationship of mobile phase density and diffusion coefficient of solute is illustrated in Figure 24.3 (Schoenmakers, 1988). As seen in Table 24.1 and Figure 24.3, a supercritical fluid acts as an intermediate between gas and liquid in that it has gas-like properties at low pressure/high temperature and liquid-like properties at high pressure/low temperature. These intrinsic properties of supercritical fluids give SFC unique advantages with separation power that is complementary to both GC and LC.

In SFC, the diffusion coefficient and viscosity are functions of density. Due to the high compressibility of the supercritical fluid, both pressure and temperature can have a greater effect in SFC than in LC, which makes them useful tools for tuning selectivity in SFC. A supercritical fluid has a much smaller viscosity compared with that of a liquid. The resulting reduced column pressure drop allows operation at higher flow rates and the use of longer columns in SFC. A supercritical fluid has a much higher diffusion rate than a liquid. The optimum flow rate of SFC is about three times faster than that of LC (Gere, 1983). With the faster diffusion rate and lower viscosity of its mobile phase, SFC can provide faster separations with better efficiency than those in LC.

SFC mobile phases have a higher solvating power than those used in GC. With its higher density mobile phase and reduced operating temperature, SFC is also suitable for analytes with high molecular weight and low volatility, which are often difficult for GC. Thermally unstable compounds that are not suitable for GC can also be separated by SFC.

On the other hand, a supercritical fluid is more viscous than a gas and has lower solvating power than

TABLE 24.1. Order of Magnitude of Physical Properties for Gaseous, Supercritical, and Liquid States

Properties	Density (g/cm^3)	Viscosity(g/cm/s)	Diffusion(cm^2/s)
Gas	10^{-3}	10^{-5}–10^{-4}	10^{-2}–1
Supercritical fluid	10^{-1}–1	10^{-4}–10^{-3}	10^{-4}–10^{-5}
Liquid	1	10^{-3}–10^{-2}	10^{-6}–10^{-5}

Data obtained from reference (Bartle, 1988).

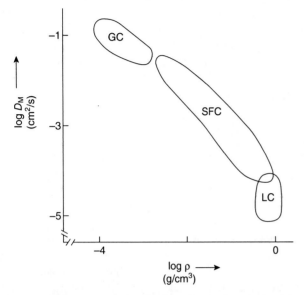

FIGURE 24.3. Schematic diagram illustrating the relationship between the mobile phase density and the diffusion coefficients of the solutes. Areas indicate typical gases (GC), liquids (LC), and supercritical fluids (SFC). Reproduced with permission of the Royal Society of Chemistry (Schoenmakers, 1988).

a liquid. Therefore, the speed and efficiency of SFC cannot compete with GC, and the solvating capacity is lower than that in LC. In addition, the low polarity of carbon dioxide leads to low solubility of certain analytes, and often makes the separation of very polar compound more difficult comparing with traditional reversed-phase liquid chromatography (RPLC). Addition of organic modifiers and/or ionic additives to supercritical fluid CO_2 provides a trade-off between solvating capacity and separating speed. Meanwhile, the modifiers also increase the viscosity of the mobile phase.

24.3.5 Selectivity in SFC

SFC is commonly considered to be a normal-phase technique for most applications, that is, using a polar stationary phase and a relatively nonpolar mobile phase. However, nonpolar stationary phases such as the very popular octadecylsilane (ODS) columns have also been used in SFC with the reversed-phase mechanism (Lesellier, 2009). Therefore, SFC offers a wider domain of retention mechanisms by combining modified CO_2 with columns of different polarities. A series of studies have reported column classification and selectivity comparisons for achiral SFC (West and Lesellier, 2008a,b).

Due to the fluid compressibility and the varied amount of mobile phase adsorbed onto the stationary phase in SFC, selectivity adjustment in SFC is more versatile, and the interpretation is more complicated when compared with LC. Often, very different selectivity can be obtained in SFC.

Slonecker et al. (1996) quantitatively studied the informational orthogonality of two-dimensional (2D) chromatographic separations. They used the retention behavior of up to 46 solutes to compare the 2D range-scaled retention time plots and information entropy calculations for different techniques, including SFC, RPLC, gas–liquid chromatography, and micellar electrokinetic capillary chromatography. Over 100 combinations of technique/stationary phase pairs were used to simulate the 2D chromatography. With the solutes tested, six of the seven system combinations that demonstrated zero informational similarity (i.e., informational orthogonality) utilized SFC.

This characteristic of orthogonal selectivity can be used to couple SFC with other separation techniques to obtain effective multidimensional chromatography with the additional resolving power needed for complicated sample systems (François and Sandra, 2009; Adam et al., 2010).

In addition, column coupling in SFC can be easier to apply for selectivity tuning due to the reduced viscosity and lower column pressure drops compared with those in LC (Berger and Wilson, 1993; Gaudin et al., 2000). Different types of columns can be coupled to provide additional selectivity, or multiple columns of the same type can be connected in series to increase column length and thereby increase resolution. Coupling of achiral and chiral columns was reported to address the limited achiral selectivity of chiral stationary phases (Phinney et al., 1998). Serial connection of five 25-cm-long cyanopropyl columns generated ~100,000 plates and enabled the separation of analytes that coeluted on a single column (Brunelli et al., 2008). Due to the concomitant increase in pressure, retention on coupled columns is shorter than the sum of the retention times on each individual column (Lesellier, 2009).

The use of organic modifiers in CO_2 can affect selectivity, efficiency, and eluotropic strengths depending on the operating conditions used for SFC (Cantrell et al., 1996; Blackwell et al., 1997). The addition of organic modifier in SFC alters the solvent strength of the mobile phase and generally improves the solubility of analytes in the mobile phase. Meanwhile, the added modifier can change the specific interactions between solutes and the stationary phases. Furthermore, modifiers can also modify the stationary phase by adsorbing onto its surface. Studies have shown that the proportion of modifier adsorbed onto the stationary phase depends on the mobile phase composition (Strubinger et al., 1991; Lesellier, 2009). Therefore, modifiers in SFC do not only compete for active sites on the stationary phase but also influence analyte retention by changing the phase ratio.

Various additives are used in SFC to improve peak shape and adjust selectivity for polar compounds (Ashraf-Khorassani et al., 1988; Steuer et al., 1990; Blackwell et al., 1997; Lavison and Thiébaut, 2003; Zheng et al., 2005a,b). Most additives used are highly polar ionic compounds including acid, base, and water. Strong bases, such as diethylamine and isopropylamine, have been used for the separation of amines, and strong acids, including trifluoroacetic acid and formic acid, are used for elution of acidic compounds. Salts, such as ammonium acetate, were reported to improve peak shape for samples containing both acidic and basic solutes (Anton et al., 1991; Pinkston et al., 2004; Cazenave-Gassiot et al., 2009). Additives can affect solutes, mobile phase, and stationary phase. First, additives increase the polarity and solvating capacity of carbon dioxide to facilitate elution of polar solutes. Second, additives can modify and deactivate the stationary phase. The amount of additives adsorbed increases with increasing stationary phase polarity and also depends on the concentration of the modifier (Berger, 1997). Furthermore, additives can interact with the solute by ion pairing or suppressing ionization (Steuer et al., 1990; Berger and Deye, 1991). These effects can be beneficial for the separation of polar ionic solutes.

Due to the compressibility of the supercritical fluid, both pressure and temperature are important parameters in SFC: they can change the fluid properties such as density, viscosity, and diffusivity. Such changes will be smaller for binary fluids due to their higher intrinsic densities and lower compressibility compared with single-component fluids.

Change in the flow rate can lead to an increase in the pressure and also result in a slight change in separation (Lesellier, 2009). At a fixed mass flow rate, retention decreases with increasing back-pressure. Considerable loss in column efficiency was observed when lower back-pressure was used. This effect is dependent on the percentage of organic modifier (Rajendran et al., 2005). For CO_2-based binary fluids, studies have shown that lower back-pressures produce lower plate counts and the effect is more pronounced with a lower percentage of modifier (Rajendran et al., 2008).

The temperature effect on retention in SFC is not as straightforward as that for pressure. For pure CO_2, the effect of temperature on retention and efficiency decreases with increasing density (Ibáñez et al., 1994). Weckwerth and Carr (1998) studied the retention process in SFC using linear solvation energy relationships. Dispersion interactions and cavity formation processes were found to dominate retention in an open tubular SFC with pure CO_2 as the mobile phase. A temperature increase at constant pressure decreases dispersion interactions with both the stationary phase and the mobile phase due to a decrease in carbon dioxide density. Retention can vary depending on the relative contributions of these interactions with temperature. With different amounts of organic modifier, the competition effect of mobile phase interactions and solvated stationary phase alters the retention behavior by impacting the enthalpy of transfer of the solute between the fluid mobile phase and the solvated stationary phase (Yonker et al., 1987). The temperature effect is solute specific and is the result of multiple mechanisms arising from the experimental conditions, including the mobile phase and the stationary phase (Lou et al., 1997). Therefore, temperature can be a very useful selectivity tool to optimize SFC separations.

Due to the compressibility of the mobile phase and the adsorption of the mobile phase on the stationary phase, the use of gradient methods (e.g., gradient in temperature, composition, pressure) in SFC can provide even greater leverage in altering retention behavior. On the other hand, a comprehensive modeling under gradient conditions has not been practical due to the changes in many chromatographic properties.

Compared with LC, SFC holds the advantage of fast reequilibration when changing mobile phase and temperature in selectivity tuning. It has also been found that SFC has a minimal memory effect from additives.

24.4 SFC IN DRUG DISCOVERY AND DEVELOPMENT

Starting from high throughput screening, the search for an effective drug against a disease can encounter the broadest spectrum of chemotypes. For chromatographic analysis of such a broad spectrum of compounds, HPLC is currently the dominant separation tool due to its versatility and robustness. In recent years, new developments in instrumentation and column technologies have focused on overcoming the slow diffusion rate and high viscosity of HPLC. Examples include the use of monolithic columns at high flow rate, the use of small particles at very high pressure (i.e., so-called ultraperformance liquid chromatography), and the use of thermally stable columns at high temperature. In contrast to the need to increase speed in LC, SFC holds distinct advantages with its inherent fast diffusion rate and low viscosity mobile phase. Compared with HPLC, SFC separations for complicated biological samples can be much faster than typical LC method. For this reason, SFC is increasingly prevalent for applications requiring high throughput.

Selectivity is another aspect for considering SFC. Consequently, SFC has been increasingly explored as an alternative or complementary technique to HPLC in the areas of drug discovery and drug development. Although packed column SFC has been most widely applied for chiral resolution in the pharmaceutical industry, it is also suitable for analysis of achiral small drug molecules. SFC often offers orthogonal selectivity to that of RPLC, and sometimes provides differential selectivity from normal-phase liquid chromatography (NPLC). In addition, SFC is more versatile than NPLC in that the selectivity adjustment is more flexible and its mobile phase has better compatibility with MS.

SFC is also considered to be a green technology. The CO_2 used as the bulk mobile phase is reclaimed as a by-product from other industrial processes. The CO_2 replaces hexanes or heptanes typically used in NPLC. Since it evaporates after SFC separation, only a small amount of modifying organic solvent remains and therefore less hazardous waste handling is required. SFC is categorized as green chemistry for the exact reason.

In drug discovery and development, while packed column SFC has been mostly focused on purification and purity determinations (mainly for stereoisomer resolution), SFC applications are emerging for a wide variety of drug molecules and their metabolites in different sample matrices (Abbott et al., 2008). Major advantages of using SFC as a tool to support pharmaceutical research include the following: improved throughput and efficiency for existing LC analyses; unique selectivity that is different from LC; highly efficient screening for difficult or demanding method development; fast analysis with high resolution and high efficiency to improve throughput for bioanalysis; and effective preparative chromatography for chiral and achiral drug molecules at reduced cost.

24.4.1 SFC Applications for Pharmaceuticals and Biomolecules

Separations of pharmaceutical and biological samples on both open-tubular and packed column SFC have been reported. Early separations were focused on open-tubular SFC with GC-like detectors (mainly FID). In the past decade, the use of packed columns with mobile phases containing modifiers has greatly increased the applications of SFC to different type of samples by overcoming the limitations associated with open-tubular SFC, particularly for analysis of polar compounds with low solubility in pure carbon dioxide. Many of the reported SFC separations show distinct advantages compared to LC and GC.

24.4.1.1 SFC Applications for Metabolites in Natural Products
Capillary SFC separation of alpha- and beta-carotene and lycopene in carrot and tomato was demonstrated when HPLC separation was reported to be insufficient (Schmitz et al., 1989). All *trans* beta- and alpha-carotene were separated from their respective *cis*-isomers with capillary SFC using 1% ethanol as modifier. All beta-carotene *cis*-isomers were separated with an SB-cyanopropyl-25-polymethylsiloxane column, while alpha-carotene isomers were separated with two SB-cyanopropyl-50-polymethyl-siloxane columns.

A useful SFC method for the analysis of structurally close polyprenol analogs of rubber plant metabolites was reported (Bamba et al., 2001). High-resolution separation of polyprenols from *Eucommia ulmoides* leaves was achieved by SFC-UV with an Inertsil ODS-3 column (GL Sciences, Torrance, CA, USA). In this case, a twofold increase in resolution of octadecaprenol and nonadecaprenol was obtained in SFC compared to LC. Complicated geometric isomers and polymers with molecular weights greater than 7000 Da were separated.

Simultaneous SFC analysis of carotenes, vitamin E, sterols, and squalene metabolites in palm oil was achieved in 20 min by using a silica column and 4% ethanol as the modifier (Choo et al., 2005). The analysis time was reduced by half compared to LC and GC. Meanwhile, information about the individual vitamin E isomers was obtained. Benefiting from the lower operation temperature of SFC, thermal decomposition of the metabolites was negligible compared to GC and sample preparation was simpler since derivatization was not required. The quantification results were comparable with those obtained in LC and GC.

Natural plant metabolites furocoumarins are used in many cosmetic and fragrance products. Desmortreux et al. (2009) reported packed column SFC separation of furocoumarins of essential oils. The separation was optimized using different columns and operating conditions.

Satisfactory isocratic separation of 16 furocoumarins was achieved in 10 min and separation was further improved by a two-step gradient method within 11 min, which was a significant improvement compared to the six-step gradient LC method in 55 min.

Recently, Bamba et al. (2008) reported the comprehensive analysis of lipids by SFC-MS including phospholipids, neutral lipids, glycolipids, and sphingolipids. Good separation of each lipid class was observed on a cyanopropyl silica-based column, while an ODS column separated lipids with different unsaturation of the fatty acid side chains as well as isomers. The reported SFC method showed advantages over GC and LC for complex lipids samples with different polarity components by using simple, mild conditions. The same group also reported the successful separation of seven carotenoids including structural isomers (beta-carotene, lycopene, lutein, zeaxanthin, antheraxanthin, neoxanthin, and violaxanthin) by SFC-MS using a monolithic ODS column (Matsubara et al., 2009). Serial connection of three 10-cm-long ODS columns was possible due to the low back-pressure of SFC and resulted in improved metabolite profiling and identification for an actual biological sample.

24.4.1.2 SFC Applications for Active Pharmaceuticals and Metabolites

Pinkston et al. (2006) compared LC-MS and SFC-MS for screening of over 2000 pharmaceutically relevant compounds including nonpolar aliphatics, aromatics, carotenoids, amine hydrohalides, quaternary ammonium salts, multicarboxylate salts, sulfonates, sulfates, sulfamic acid salts, phosphates, phosphonates, multiphosphonate salts, polyhydroxy compounds, and nitro compounds. Although SFC is commonly regarded to be more suitable for nonpolar or relative low polarity compounds, the authors reported that the percentage of samples eluted and detected by SFC-MS (75%) were comparable with those by LC-MS (79.4%). SFC-MS was able to identify 3.7% of the compounds that were not detected by HPLC-MS, while 8.1% of the compounds were observed using HPLC-MS, but not SFC-MS. Compounds that contain a phosphate, a phosphonate, or a bisphosphonate were the only classes that appeared to be consistently detected in LC-MS, but not in SFC-MS under the studied conditions. They also found that the APCI source required less cleaning during the SFC-MS separations than it did during LC-MS. They concluded that SFC-MS method was at least as durable, reliable, and user-friendly as the LC-MS method.

SFC has been used for the analysis of pharmaceutical formulations. The direct assay of research compounds in 100% aqueous formulations by chiral SFC was reported by Mukherjee et al. (Mukherjee and Cook, 2006; Mukherjee, 2007). These feasibility studies indicated that the SFC approach has the potential to significantly reduce the typical sample processing time prior to analysis. The results showed that SFC could assay aqueous formulations for both acidic and basic pharmaceutical compounds with a high degree of selectivity, accuracy, precision, robustness, sensitivity, and linearity over a wide range of concentrations.

SFC has been used for bioanalysis with different biological matrices. Simmons et al. (1995) investigated the separation of nonsteroidal anti-inflammatory drug phenylbutazone and its major metabolite oxyphenbutazone in human serum using SFC-UV. No interference from endogenous components was observed. Detection limits of the SFC assay were 0.1 µg/mL for phenylbutazone and 1.0 µg/mL for oxyphenbutazone. The SFC quantitative results were found to be comparable with that of HPLC. SFC/UV was used for a pharmacological study of racemic camazepam and its rat liver microsomal metabolites (Wang et al., 1995). Considerable improvement of analysis time, selectivity, and efficiency was obtained with SFC. With the enhanced separation, SFC offered metabolite profiling information that cannot be obtained by LC. SFC-UV was used to quantitatively determine phenylglyoxylic and mandelic acids (metabolites of styrene) in urine samples (Simon and Nicot, 1996). The developed SFC method was used to evaluate the toxic effect of styrene and to monitor occupational exposures. The SFC-UV assay was able to quantitatively monitor styrene metabolites down to the level of one-twentieth of the lowest biological exposure index values.

24.4.1.3 SFC Applications for Biomolecules

Both packed column and open-tubular SFC has been reported for the separation of cholesterol and cholesteryl esters in human serum (Nomura et al., 1993; Kim et al., 1994). Akira Nomura et al. (1993) reported the determination of cholesterol and cholesteryl esters in human serum on ODS-silica gel column with pure CO_2 as mobile phase with simultaneous FID and UV detection. Acceptable quantitative analysis with an LOD of 4–6 pg were reported for the simultaneous analysis of cholesterol and its esters in biological fluids using capillary SFC with a FID detection without thermal degradation and derivatization (Kim et al., 1994).

SFC was evaluated for the development of a rapid method for estrogen metabolites associated with breast cancer. Xu et al. (2006) reported the separation of 15 estrogen metabolites using SFC with tandem mass spectrometry (MS/MS). A linear gradient with methanol/CO_2 at 2 mL/min was run on columns with 2.1 mm internal diameter and 150 mm in length. Separation and quantification of all 15 estrogens was achieved in less than 10 min by coupling of a cyanopropyl silica with a diol column. A comparable detection limit to that of RPLC (70 min) was obtained but with a much faster analysis time. The baseline peak width in SFC was 20–30 s compared to ~1.8 min in HPLC. The study demonstrated the feasibility of SFC as a broad, fast, sensitive analysis method for resolving and quantifying closely related estrogens that differ by functional groups (proton, methoxy, hydroxy, or keto) or position of substitution.

Polypeptides are traditionally analyzed by LC-MS, but SFC-MS is emerging as a useful alternative with its potential for fast analysis and short method development times. Blackwell and Stringham (1999) found that elution of polypeptides in SFC depended on the acidity of the mobile phase additive. According to their study, a strong acid was needed for elution and improved peak shape. Bolaños et al. (2004) reported the analytical and preparative separation of gramicidin D. They also reported the fast SFC separation of peptides from 2 to 20 amino acids in length, as well as full-length rabbit and bovine cytochrome C with SFC-MS flow-injection analysis. Later, Zheng et al. (2006) evaluated the feasibility of SFC-MS for polypeptides up to 40 mers. They showed that relatively large peptides, containing a variety of acidic and basic residues, can be eluted in SFC. Trifluoroacetic acid was used as an additive in a CO_2/methanol mobile phase to suppress deprotonation of peptide carboxylic acid groups and to protonate peptide amino groups. SFC was also found to be very promising to significantly reduce hydrogen/deuterium back exchange in solution-phase MS analysis of peptides (Emmett et al., 2006). SFC was able to recover information lost in HPLC for the short exchange period due to its reduced hydrogen/deuterium back-exchange, which resulted from a combination of the nonexchanging CO_2 mobile phase, fast flow rates, and short retention times.

Recently, Ashraf-Khorassani and Taylor (2010) reported SFC separation of four water-soluble nucleobases (thymine, uracil, adenine, and cytosine) with alcohol- and additive-modified CO_2 as mobile phase. The separations were studied on diol, cyanopropyl, and 2-ethyl pyridine columns with methanol, ethanol, 1-propanol, and 1-butanol examined as modifiers. Formic acid, ammonium acetate, and water were investigated as additives. The use of water as the additive with an alcohol modifier was found to be effective in terms of adjusting selectivity and improving peak shape. Separations with water were comparable to those with ammonium acetate and both gave separations superior to those of formic acid.

24.4.2 SFC Chiral Separations

It has been well documented that the enantiomers of a drug can display different properties in terms of activity and/or toxicity. Each enantiomer can have a unique therapeutic or adverse effects profile in that the isomers can possess different pharmacological activities and/or be metabolized at a different rate or via a different pathway.

To improve the safety and efficacy of pharmaceutical products, drug chirality and chiral purity has drawn attention from both the pharmaceutical industry and regulatory agencies. During toxicological and pharmacological tests, all drugs containing chiral centers must have chiral assays in order to fully interpret study results. For compounds containing a chiral center, enantiomeric separation is needed for monitoring enantioselective synthesis, separating racemates to obtain pure enantiomers for drug testing, determining the chiral purity of enantiomers, and monitoring chiral stability in formulations and in biological fluids.

Chiral resolution has been one of the most difficult tasks in separation sciences. Due to the unpredictable nature of chiral selectivity, chiral method development is largely a trial-and-error process that involves screening of various chiral selectors under different conditions. Optimizing the separation can require a significant amount of time for each set of screening experiments and for column equilibration between studies. The application of packed column SFC to enantiomeric separations was first reported by Mourier et al. in 1985. In the past decade, remarkable progress in SFC separation for chiral molecules has been made in the pharmaceutical industry. In many cases, including the separation of enantiomers, diastereomers, and atropisomers (Qian-Cutrone et al., 2008), SFC is clearly superior to LC.

The process of developing SFC chiral analytical methods is similar to that employed for LC, but SFC offers more flexibility in parameter selection (Lynam and Blackwell, 1997). Similar to LC, most chiral separations in SFC use a chiral selector in the stationary phase. A wide variety of chiral stationary phases including polysaccharides, macrocyclic antibiotics, Pirkle type, and polymeric columns have been used in SFC. Most SFC chiral applications have been based on derivatized polycaccharide chiral stationary phases (Mangelings and Heyden, 2008), which seem to have the broadest range of enantioselectivity among the commercially available chiral stationary phases. Alcohols, especially methanol, are the most commonly used modifier in chiral SFC. As an alternative, 2,2,2-trifluoroethanol has been reported as modifier for alcohol-sensitive chiral compounds (Byrne et al., 2008). Various additives have been used in chiral SFC (Blackwell, 1999). Some of the additives, such as ethanesulfonic acid and cyclic amines, provide unique selectivities (Ye et al., 2004; Stringham, 2005). An increase in column temperature typically decreases the enantiomeric selectivity by decreasing the effective difference in enantiomer binding enthalpy, often resulting in coelution of enantiomers. A more dramatic increase in temperature can lead to "entropically driven" chiral separation with the possibility of elution order reversal (Stringham and Blackwell, 1996).

Compared with conventional HPLC chiral separation, SFC offers the following advantages: fast separations at higher flow rate, comparable or unique selectivity, relatively straightforward and rapid method development, fast reequilibration, less of a memory effect when using additives (Ye et al., 2004), and reduced usage of organic solvents. For example, SFC was used to evaluate chiral recognition capabilities of three macrocyclic glycopeptide chiral selectors with a set of 111 chiral compounds including heterocycles, analgesics, β-blockers, sulfoxides, N-protected amino acids, and native amino acids (Liu et al., 2002). All the SFC separations were completed in less than 15 min and 70% were eluted in less than 4 min. Sometimes SFC can offer unique selectivity that cannot be obtained in HPLC (Bargmann-Leyder et al., 1995a, 1995b). The fast reequilibration time in SFC allows quick selectivity adjustment, which is beneficial for efficient chiral method screening.

Currently, the major applications of analytical SFC separation for chiral molecules in pharmaceutical, biological, and metabolic analysis include the following: automated fast SFC chiral screening systems to provide efficient chiral method development; column coupling or column switching to gain resolution for difficult chiral separations without incurring high pressure or mobile phase incompatibility; and SFC high-throughput chiral bioanalytical assays to support pharmacokinetic studies.

An effective screening system is the key for chiral method development. Welch et al. (2007) reported a chiral SFC tandem column screening system. With modification of a commercial analytical SFC instrument, two banks of five different columns were individually controlled with two six-position high-pressure column selection valves. The resulting instrument allowed chiral screening of 10 individual columns and 25 unique tandem column arrangements. Zeng et al. (2007) reported a parallel SFC-MS screening system for high-throughput enantioselective optimization and separation. Eight chiral columns were grouped into two sets for the four-column parallel mode with multiple mobile phases. A column control valve was used for converting from the four-column screening mode to a one-column optimization mode. Di et al. (2004) reported an automated screening system using a robotic system and chiral SFC for monitoring enantioselectivity of enzymatic transformations. A rapid method at a runtime of 1.5 min on a Chiralcel OD (Daicel Chemical Industries, Tokyo, Japan) column with 2% MeOH/CO_2 was used. The fast, high-throughput chiral SFC analysis provided information for optimal reaction conditions in the study of enzymatic transformations.

Enantiomers of different drugs, for example, antiulcer (Toribio et al., 2005) and anti-inflammatory drugs (Yang et al., 2005), have been successfully separated by Chiral SFC with better separation and shorter analysis time compared to HPLC. Chiral SFC was also used to explore biotransformation mechanisms and drug metabolite profiling. To understand the metabolic biotransformation of nobiletin, SFC was used to separate two of its primarily regioisomer metabolites 3'-demethylnobiletin and 4'-demethylnobiletin (Li et al., 2006; Wang et al., 2006). In the cases of both chiral and achiral columns, SFC offered superior separation of the two nobiletin metabolites as compared to HPLC. The optimized method was used to profile the nobiletin metabolites in mouse urine and quantitatively determined that 4'-demethylnobiletin was the major metabolite. Pereillo et al. (2002) reported the separation of clopidogrel-related metabolites by chiral SFC. Useful metabolites profiling information was obtained by SFC separation of four enantiomeric pairs of acrylonitrile-derivatized clopidgrel-related metabolites.

As a result of the low pressure drop in SFC, column coupling can be used for difficult chiral separations. Baseline separation of a mixture of chiral pharmaceutical compounds with four stereoisomers was achieved by the SFC tandem column method using Chiralpak AD-H (Daicel Chemical Industries, Tokyo, Japan) and Chiralcel OD-H (Barnhart et al., 2005). Baseline separation could not be obtained with either individual column. Coupling of achiral/chiral columns was reported to address the limited achiral selectivity of chiral stationary phases. An amino or cyano column was coupled in series with a Chiralcel OD column to adjust the selectivity (Phinney et al., 1998). Separations of both achiral and chiral components were demonstrated for β-blockers, benzodiazepines, as well as for a multidrug medication. Alexander and Staab (2006) reported the coupling of Chiralcel OD-H with an achiral silica-based column to enable the simultaneous separation of four steroisomers from cinnamonitrile and hydrocinnamonitrile.

Due to the fast diffusion rate and reduced viscosity of SFC mobile phases, column switching can be very effective to enhance the resolution of difficult chiral separations. We reported the novel combination of "simulated moving columns" (SMC) with chiral SFC (Zhang et al., 2007). SMC uses two or three short chiral columns connected in series, and enables unresolved enantiomers to separate repeatedly and exclusively through each of the columns until sufficient resolution is attained. The technique is significantly enhanced through the use of SFC. The number of SMC cycles in SFC can be dramatically increased with significantly less band broadening compared to HPLC. The greatly enhanced efficiency with increasing column virtual length was useful for overcoming difficult chiral separations.

In a comparison study, SFC was investigated with polar organic solvent chromatography and NPLC for chiral separations on polysaccharide-based stationary phases (Matthijs et al., 2006). Complementary separation results were obtained for three techniques. SFC provided fast, sometime unique, chiral separations compared with LC. However, as a complementary technique to LC, SFC is not always the best choice. Anton et al. (1994) reported several cases where apparent interaction of basic solutes with CO_2 led to poor chiral resolution in SFC.

24.4.3 SFC Applications for High-Throughput Analysis

High-throughput analysis and quantification are important parts of drug discovery and development. With its distinct advantages in analysis speed, SFC, especially combined with mass spectrometers, has been increasingly recognized as a very effective tool for high-throughput analysis.

Hsieh et al. (2005) studied the effects of the nebulizer temperature, eluant flow rates, and compositions on the chromatographic performance and ionization efficiency of analytes in positive ion mode under packed column SFC conditions. A higher percentage of modifier caused loss of the relative APCI response. The use of basic additives was found to have a strong impact on the ionization efficiency using the APCI source. *N,N*-dimethylethylamine resulted in a large decrease in APCI response, but this was not observed for isopropylamine. The chiral SFC/APCI-MS/MS approach at ~2 min/sample was applied for simultaneous determination of two pairs of racemic drugs for *in vitro* samples at low nanogram per milliliter concentrations.

Bolaños et al. (2003) reported the use of SFC/APCI-MS with time-of-flight (TOF) analyzer for high-throughput analysis using six model compounds. Dramatic improvement using TOF-MS high-speed sampling rates was obtained in comparison with acquisition by a quadrupole-MS system. The authors concluded that the union of SFC with TOF-MS provided the maximum density of chemical information per unit time.

Ventura et al. (1999a) demonstrated a threefold reduction in analysis time of SFC/APCI-MS compared with that of LC/APCI-MS for high-throughput multicomponent combinatorial library mixtures. In addition, separation of library components was found to be improved in SFC compared to LC, albeit with modestly reduced sensitivity of SFC/APCI-MS relative to LC/

APCI-MS. An auxiliary solvent to the APCI source was used to improve the mass spectral sensitivity (Ventura et al., 1999b).

Hoke et al. (2000) compared SFC-MS/MS and LC-MS/MS for bioanalytical determination of R- and S-ketoprofen in human plasma. Both SFC and LC chromatographic methods were used to generate validation data and study sample data from patients dosed with either orally or topically administered ketoprofen. Comparable analytical attributes such as specificity, linearity, sensitivity, accuracy, precision, and ruggedness were obtained for both techniques. Approximately threefold less analysis time was needed for SFC compared to LC, which resulted in significant time savings for a large batch of pharmacokinetic samples. Later, they demonstrated high-throughput bioanalytical quantitation using SFC-MS/MS for pharmacokinetics applications using dextromethorphan in human plasma as a model compound (Hoke et al., 2001). Fast, accurate, and quantitative analysis was achieved at a speed of ~6 s/sample using a 2×10 mm cyano column with isocratic 35/65 CO_2/methanol at a flow rate of 7.5 mL/min.

The antimetabolite antineoplastic agent cytarabine (ara-C) is used to treat leukemia. Hsieh et al. (2007) reported the separation of ara-C from endogenous compounds in mouse plasma by SFC/APCI-MS/MS. The SFC-MS/MS method facilitated the analysis of both polar and nonpolar compounds in the biological fluid. Determination of ara-C at nanograms per milliliter concentrations in mouse plasma was achieved on bare silica stationary phase using an isocratic CO_2/methanol mobile phase with ammonium acetate as additive at ~2.5 min/sample. The mouse plasma levels of ara-C obtained by the SFC method were found to be consistent with those determined by LC. With equivalent accuracy, SFC provided reliable and higher throughput data in support of an *in vivo* study.

Chen et al. (2006) reported the enantioselective determination of betablockers (R,S)-propranolol in mouse blood by SFC/APCI-MS/MS with a Chiralcel OD-H column. The proposed chiral SFC/MS achieved ~3 min/sample for the determination of (R, S)-propranolol at a low nanogram per milliliter level and was applied to support a pharmacokinetic study. Coe et al. (2006) reported fast bioanalytical SFC/APCI-MS/MS for separation and quantification of R/S-warfarin in human plasma. Compared to an LC method, the SFC method developed on Chiralpak AD column was at least two times faster. In addition, a twofold increase in resolution of the enantiomers was achieved with improved sensitivity. The validated SFC-MS/MS method was applied to bioanalysis in a drug interaction clinical study of approximately 460 samples within 2 days.

24.4.4 Preparative Separations

In drug discovery, it is important to provide stereoiso-merically pure chiral compounds for pharmacological testing. Several approaches are being used to meet the increasing demands of producing enantiomerically pure synthetic intermediates and products, such as asymmetric synthesis, chiral separation or purification, and kinetic resolution. With effective chiral preparative chromatography separation, it could be more time efficient to synthesize racemic mixtures and then chromatographically separate the isomers than to develop enantiomeric syntheses of the drug candidates of interest.

Today, SFC has become one of the favorite techniques for preparative separations in the pharmaceutical industry due to its high speed and resolution, high productivity, and low operational costs.

The implementation of supercritical fluid CO_2 as part of the mobile phase has distinct advantages over 100% liquid mobile phase. The cost is significantly lowered due to the reduced consumption of organic solvents. The dry down process of the isolated fraction is simplified and the evaporation time is reduced because of the lower solvent volume. At the same time, waste handling and disposal costs are reduced.

The preparation of large amounts (from milligrams to kilograms) of pure material has been enabled by stacked-mode injection and preparative mode injection systems in SFC. Stacked injection increases productivity by allowing multiple injections of the same sample before the end of a run, thereby decreasing the overall time for purification. By installing the sample loop in the modifier stream before the modifier/CO_2 mixing point, the preparative mode injection enables injecting a large volume/amount of sample with minimal impact of sample solvent on the column equilibrium. Implementation of these technologies has led to the development of successful preparative instruments, which offer the best performance under isocratic conditions for chiral separations, and are also effective for sequential gradient runs. Meantime, overoptimum flow rate can be used in preparative SFC to increase productivity by speeding up the separation without high back pressure. An example is the isocratic chiral separation reported by Kraml et al. (2005). The separation of over 20 g of carbobenzyloxy-derivatized chiral compound was accomplished in 21 min by stacked injections using an injection interval of 160 s on a 5 cm × 25 cm Whelk-O1 (R, R) column (Regis Technologies, Morton Grove, IL, USA). The mobile phase of 43/57 isopropanol/CO_2 was run at 350 mL/min with an average sample injection of 4.35 mL (670 mg/mL). The reported sample recovery was 98.1% and the enantiomeric excesses for the first and second enantiomers were 99.9% and 99.4%, respectively.

One of the main problems encountered in the development of preparative SFC instrumentation, as required for high-throughput purification of diverse compounds, is the technology available to collect fractions. Fraction collection at nearly atmospheric pressure has been used to decrease the complexity of instrument plumbing and improve operational safety. Makeup flow at the fraction collector and use of a nonpolar cosolvent (such as hexane) modifier in combination with a polar modifier have been used as strategies to increase recoveries for samples that only require a small percentage of polar modifier.

Continued efforts to improve options have been focused on SFC mass-directed open-bed fraction collection preparative systems. Wang et al. (2001) reported an automated mass-directed packed column SFC semi-preparative system. A partially closed foil design was used for the fraction collection. This platform provided an average sample recovery of 77%. Zhang et al. (2006) reported the development of a mass-directed SFC purification system with accurate peak detection and reliable fraction collection. Issues on software compatibility, the interface between the preparative SFC and the mass spectrometer, and fraction collection were addressed. Fraction collection at atmospheric pressure with >85% recovery was demonstrated at flow rates up to 30 mL/min. The SFC-MS purification system was used in support of high-throughput library purification and proved a valuable tool in complementing an RPLC-MS-based technology platform. Recently, Waters Corp. (Milford, MA, USA) launched the Prep SFC 100 mass-directed purification system. A gas/liquid separator and makeup flow are used to overcome the challenge of aerosol formation at high flow rates that is caused by the depressurization of CO_2. A comparison study of mass-directed purification of LC-MS at 50 mL/min and SFC-MS at 100 g/min showed similar performance in terms of success rate and purity, with 27% overall success for SFC and 25% for RPLC (Mich et al., 2010). Different separations were obtained indicating "orthogonality" in selectivity for SFC and RPLC. The batch process time was reduced 20% by using SFC in place of RPLC due to its shorter postpurification dry down time.

24.5 FUTURE PERSPECTIVE

The above examples illustrate a variety of successful applications of SFC in pharmaceutical, biological, and clinical studies. Its clear utility as a complementary separation tool with distinct advantages over LC and GC suggests that more SFC applications can be expected for an even wider spectrum of compounds in different sample matrices.

Benefits offered by SFC, including short analysis times, high efficiency, cost saving, and environment friendliness, have made it a highly valued and widely used technology in the pharmaceutical industry, especially for drug discovery. Use of more sophisticated detectors and hyphenated SFC techniques (i.e., multidimensional separation) are expected to be particularly useful to fully realize the advantages of this technology platform. Further improvement in the instrumentation, particularly the analytical instruments for reproducibility, sensitivity, automation, and robustness, will allow more versatile and accurate analyses in clinical and biological applications, including regulated analyses. More extensive test sets of molecules as well as systematic validation of developed methods will be needed to widen the application of SFC in drug development. Advancement in column technologies with development of more suitable achiral SFC columns will also be required.

In addition, basic research on the technique by scientists outside of applied industries will be necessary and welcome to advance our understanding of SFC. Research studies from academia and other nonprofit organizations, which currently are notably missing (Chester and Pinkston, 2004; Mukhopadhyay, 2008), will be needed to clarify the fundamental principles behind the many apparent advantages of SFC and to provide a theoretical basis for prediction of SFC separations.

REFERENCES

Abbott E, Veenstra TD, Issaq HJ (2008) Clinical and pharmaceutical applications of packed-column supercritical fluid chromatography. *Journal of Separation Science* 31:1223–1230.

Adam F, Thiébaut D, Bertoncini F, Courtiade M, Hennion M-C (2010) Supercritical fluid chromatography hyphenated with twin comprehensive two-dimensional gas chromatography for ultimate analysis of middle distillates. *Journal of Chromatography A* 1217:1386–1394.

Alexander AJ, Staab A (2006) Use of achiral/chiral SFC/MS for the profiling of isomeric cinnamonitrile/hydrocinnamonitrile products in chiral drug synthesis. *Analytical Chemistry* 78:3835–3838.

Anton K, Bach M, Geiser A (1991) Supercritical fluid chromatography in the routine stability control of antipruritic preparations. *Journal of Chromatography A* 553:71–79.

Anton K, Eppinger J, Frederiksen L, Francotte E, Berger TA, Wilson WH (1994) Chiral separations by packed-column super- and subcritical fluid chromatography. *Journal of Chromatography A* 666:395–401.

Ashraf-Khorassani M, Fessahaie MG, Taylor LT, Berger TA, Deye JF (1988) Rapid and efficient separation of PTH-amino acids employing supercritical CO2 and an ion pairing agent. *Journal of High Resolution Chromatography* 11:352–353.

Ashraf-Khorassani M, Taylor LT (2010) Subcritical fluid chromatography of water soluble nucleobases on various polar stationary phases facilitated with alcohol-modified CO2 and water as the polar additive. *Journal of Separation Science* 33:1–10.

Bamba T, Fukasaki E, Kajiyama S, Ute K, Kitayama T, Kobayashi A (2001) The occurrence of geometric polyprenol isomers in the rubber-producing plant, *Eucommia ulmoides* Oliver. *Lipids* 36:727–732.

Bamba T, Shimonishi N, Matsubara A, Hirata K, Nakazawa Y, Kobayashi A, Fukusaki E (2008) High throughput and exhaustive analysis of diverse lipids by using supercritical fluid chromatography-mass spectrometry for metabolomics. *Journal of Bioscience and Bioengineering* 105:460–469.

Bargmann-Leyder N, Sella C, Bauer D, Tambute A, Caude M (1995a) Supercritical fluid chromatographic separation of .beta.-blockers on chyrosine-A: Investigation of the chiral recognition mechanism using molecular modeling. *Analytical Chemistry* 67:952–958.

Bargmann-Leyder N, Tambuté A, Caude M (1995b) A comparison of LC and SFC for cellulose- and amylose-derived chiral stationary phases. *Chirality* 7:311–325.

Barnhart WW, Gahm KH, Thomas S, Notari S, Semin D, Cheetham J (2005) Supercritical fluid chromatography tandem-column method development in pharmaceutical sciences for a mixture of four stereoisomers. *Journal of Separation Science* 28:619–626.

Bartle KD (1988) Theory and principles of supercritcal fluid chromatography. In *Supercritical Fluid Chromatography*, Smith RM, ed. Royal Society of Chemistry, Cambridge, UK.

Berger D (2007) Supercritical fluid chromatography (SFC): A review of technical developments. *LCGC Europe* 20: 164–165.

Berger TA, ed. (1995) *Packed Column SFC*. Royal Society of Chemistry, Letchworth, UK.

Berger TA (1997) Separation of polar solutes by packed column supercritical fluid chromatography. *Journal of Chromatography A* 785:3–33.

Berger TA, Deye JF (1990) Composition and density effects using methanol/carbon dioxide in packed column supercritical fluid chromatography. *Analytical Chemistry* 62:1181–1185.

Berger TA, Deye JF (1991) Role of additives in packed column supercritical fluid chromatography: Suppression of solute ionization. *Journal of Chromatography A* 547:377–392.

Berger TA, Wilson WH (1993) Packed column supercritical fluid chromatography with 220,000 plates. *Analytical Chemistry* 65:1451–1455.

Blackwell JA (1999) Effect of acidic mobile phase additives on chiral selectivity for phenylalanine analogs using subcritical fluid chromatography. *Chirality* 11:91–97.

Blackwell JA, Stringham RW (1999) Effect of mobile phase components on the separation of polypeptides using carbon dioxide-based mobile phases. *Journal of High Resolution Chromatography* 22:74–78.

Blackwell JA, Stringham RW, Weckwerth JD (1997) Effect of mobile phase additives in packed-column subcritical and supercritical fluid chromatography. *Analytical Chemistry* 69:409–415.

Bolaños B, Greig M, Ventura M, Farrell W, Aurigemma CM, Li H, Quenzer TL, Tivel K, Bylund JMR, Tran P, Pham C, Phillipson D (2004) SFC/MS in drug discovery at Pfizer, La Jolla. *International Journal of Mass Spectrometry* 238:85–97.

Bolaños BJ, Ventura MC, Greig MJ (2003) Preserving the chromatographic integrity of high-speed supercritical fluid chromatography separations using time-of-flight mass spectrometry. *Journal of Combinatorial Chemistry* 5:451–455.

Brunelli C, Górecki T, Zhao Y, Sandra P (2007) Corona-charged aerosol detection in supercritical fluid chromatography for pharmaceutical analysis. *Analytical Chemistry* 79:2472–2482.

Brunelli C, Zhao Y, Brown M-H, Sandra P (2008) Development of a supercritical fluid chromatography high-resolution separation method suitable for pharmaceuticals using cyanopropyl silica. *Journal of Chromatography A* 1185:263–272.

Byrne N, Hayes-Larson E, Liao W-W, Kraml CM (2008) Analysis and purification of alcohol-sensitive chiral compounds using 2,2,2-trifluoroethanol as a modifier in supercritical fluid chromatography. *Journal of Chromatography B* 875:237–242.

Cantrell GO, Stringham RW, Blackwell JA, Weckwerth JD, Carr PW (1996) Effect of various modifiers on selectivity in packed-column subcritical and supercritical fluid chromatography. *Analytical Chemistry* 68:3645–3650.

Cazenave-Gassiot A, Boughtflower R, Caldwell J, Hitzel L, Holyoak C, Lane S, Oakley P, Pullen F, Richardson S, Langley GJ (2009) Effect of increasing concentration of ammonium acetate as an additive in supercritical fluid chromatography using CO2-methanol mobile phase. *Journal of Chromatography A* 1216:6441–6450.

Chen J, Hsieh Y, Cook J, Morrison R, Korfmacher WA (2006) Supercritical fluid chromatography-tandem mass spectrometry for the enantioselective determination of propranolol and pindolol in mouse blood by serial sampling. *Analytical Chemistry* 78:1212–1217.

Chester TL, Pinkston JD (2004) Supercritical fluid and unified chromatography. *Analytical Chemistry* 76:4606–4613.

Choo YM, Ng MH, Ma AN, Chuah CH, Hashim MA (2005) Application of supercritical fluid chromatography in the quantitative analysis of minor components (carotenes, vitamin E, sterols, and squalene) from palm oil. *Lipids* 40:429–432.

Coe RA, Rathe JO, Lee JW (2006) Supercritical fluid chromatography-tandem mass spectrometry for fast bioanalysis of R/S-warfarin in human plasma. *Journal of Pharmaceutical and Biomedical Analysis* 42:573–580.

Combs MT, Ashraf-Khorassani M, Taylor LT (1997) Packed column supercritical fluid chromatography-mass spectroscopy: A review. *Journal of Chromatography A* 785:85–100.

Crowther JB, Henion JD (1985) Supercritical fluid chromatography of polar drugs using small-particle packed columns with mass spectrometric detection. *Analytical Chemistry* 57:2711–2716.

Desmortreux C, Rothaupt M, West C, Lesellier E (2009) Improved separation of furocoumarins of essential oils by supercritical fluid chromatography. *Journal of Chromatography A* 1216:7088–7095.

Deye JF, Berger TA, Anderson AG (1990) Nile Red as a solvatochromic dye for measuring solvent strength in normal liquids and mixtures of normal liquids with supercritical and near critical fluids. *Analytical Chemistry* 62:615–622.

Di L, McConnell OJ, Kerns EH, Sutherland AG (2004) Rapid, automated screening method for enzymatic transformations using a robotic system and supercritical fluid chromatography. *Journal of Chromatography B* 809:231–235.

Dressman SF, Simeone AM, Michael AC (1996) Supercritical fluid chromatography with electrochemical detection of phenols and polyaromatic hydrocarbons. *Analytical Chemistry* 68:3121–3127.

Emmett MR, Kazazic S, Marshall AG, Chen W, Shi SDH, Bolaños B, Greig MJ (2006) Supercritical fluid chromatography reduction of hydrogen/deuterium back exchange in solution-phase hydrogen/deuterium exchange with mass spectrometric analysis. *Analytical Chemistry* 78: 7058–7060.

France JE, Snyder JM, King JW (1991) Packed-microbe supercritical fluid chromatography with flame ionization detection of abused vegetable oils. *Journal of Chromatography A* 540:271–278.

François I, Sandra P (2009) Comprehensive supercritical fluid chromatography x reversed phase liquid chromatography for the analysis of the fatty acids in fish oil. *Journal of Chromatography A* 1216:4005–4012.

Gaudin K, Lesellier E, Chaminade P, Ferrier D, Baillet A, Tchapla A (2000) Retention behaviour of ceramides in sub-critical fluid chromatography in comparison with non-aqueous reversed-phase liquid chromatography. *Journal of Chromatography A* 883:211–222.

Gere D (1983) Supercritical fluid chromatography. *Science* 222:253–259.

Giddings JC, Myers MN, McLaren L, Keller RA (1968) High pressure gas chromatography of nonvolatile species. *Science* 162:67–73.

Greibrokk T, Andersen T (2003) High-temperature liquid chromatography. *Journal of Chromatography A* 1000:743–755.

Harris CM (2002) Product review: The SFC comeback. *Analytical Chemistry* 74:87 A–91 A.

Hoke SH, Pinkston JD, Bailey RE, Tanguay SL, Eichhold TH (2000) Comparison of packed-column supercritical fluid chromatography-tandem mass spectrometry with liquid chromatography-tandem mass spectrometry for bioanalytical determination of (R)- and (S)-ketoprofen in human plasma following automated 96-well solid-phase extraction. *Analytical Chemistry* 72:4235–4241.

Hoke SH, Tomlinson JA, Bolden RD, Morand KL, Pinkston JD, Wehmeyer KR (2001) Increasing bioanalytical throughput using pcSFC-MS/MS:10 minutes per 96-well plate. *Analytical Chemistry* 73:3083–3088.

Hsieh Y, Favreau L, Cheng K-C, Chen J (2005) Chiral supercritical fluid chromatography/tandem mass spectrometry for the simultaneous determination of pindolol and propranolol in metabolic stability samples. *Rapid Communications in Mass Spectrometry* 19:3037–3041.

Hsieh Y, Li F, Duncan CJG (2007) Supercritical fluid chromatography and high-performance liquid chromatography/tandem mass spectrometric methods for the determination of cytarabine in mouse plasma. *Analytical Chemistry* 79:3856–3861.

Ibáñez E, Tabera J, Herraiz M, Reglero G (1994) Effect of temperature and density on the performance of micropacked columns in supercritical fluid chromatography. *Journal of Chromatography A* 667:249–255.

Kim D, Lee K, Heo G (1994) Analysis of cholesterol and cholesteryl esters in human serum using capillary supercritical fluid chromatography. *Journal of Chromatography B: Biomedical Sciences and Applications* 655:1–8.

Klesper E, Corwin AH, Turner DA (1962) High pressure gas chromatography above critical temperature. *The Journal of Organic Chemistry* 27:700.

Kraml CM, Zhou D, Byrne N, McConnell O (2005) Enhanced chromatographic resolution of amine enantiomers as carbobenzyloxy derivatives in high-performance liquid chromatography and supercritical fluid chromatography. *Journal of Chromatography A* 1100:108–115.

Lauer HH, McManigill D, Board RD (1983) Mobile-phase transport properties of liquefied gases in near critical and supercritical fluid chromatography. *Analytical Chemistry* 55:1370–1375.

Lavison G, Thiébaut D (2003) Evaluation of a ristocetin bonded stationary phase for subcritical fluid chromatography of enantiomers. *Chirality* 15:630–636.

Lesellier E (2009) Retention mechanisms in super/subcritical fluid chromatography on packed columns. *Journal of Chromatography A* 1216:1881–1890.

Lesellier E, West C (2007) Description and comparison of chromatographic tests and chemometric methods for packed column classification. *Journal of Chromatography A* 1158:329–360.

Li F, Hsieh Y (2008) Supercritical fluid chromatography-mass spectrometry for chemical analysis. *Journal of Separation Science* 31:1231–1237.

Li S, Wang Z, Sang S, Huang M-T, Ho C-T (2006) Identification of nobiletin metabolites in mouse urine. *Molecular Nutrition & Food Research* 50:291–299.

Liu Y, Berthod A, Mitchell CR, Xiao TL, Zhang B, Armstrong DW (2002) Super/subcritical fluid chromatography chiral separations with macrocyclic glycopeptide stationary phases. *Journal of Chromatography A* 978:185–204.

Lou X, Janssen H-G, Cramers CA (1997) Temperature and pressure effects on solubility in supercritical carbon dioxide and retention in supercritical fluid chromatography. *Journal of Chromatography A* 785:57–64.

Lynam KG, Blackwell JA (1997) Optimization of chiral resolution using packed columns with carbon dioxide-based mobile phases. *Chirality* 9:672–677.

Mah C, Thurbide KB (2008) An improved interface for universal acoustic flame detection in modified supercritical fluid chromatography. *Journal of Separation Science* 31:1314–1321.

Mangelings D, Heyden YV (2008) Chiral separations in sub- and supercritical fluid chromatography. *Journal of Separation Science* 31:1252–1273.

Matsubara A, Bamba T, Ishida H, Fukusaki E, Hirata K (2009) Highly sensitive and accurate profiling of carotenoids by supercritical fluid chromatography coupled with mass spectrometry. *Journal of Separation Science* 32:1459–1464.

Matthijs N, Maftouh M, Heyden YV (2006) Chiral separation strategy in polar organic solvent chromatography and performance comparison with normal-phase liquid and supercritical-fluid chromatography. *Journal of Separation Science* 29:1353–1362.

Mich A, Matthes B, Chen R, Buehler S (2010) Using mass-directed preparative SFC and RPLC for library compound purification. In *Application Note*, http://chromatographyonline.findanalytichem.com/lcgc/data/articlestandard//lcgc/082010/657505/article.pdf.

Mourier PA, Eliot E, Caude MH, Rosset RH, Tambute AG (1985) Supercritical and subcritical fluid chromatography on a chiral stationary phase for the resolution of phosphine oxide enantiomers. *Analytical Chemistry* 57:2819–2823.

Mukherjee PS (2007) Validation of direct assay of an aqueous formulation of a drug compound AZY by chiral supercritical fluid chromatography (SFC). *Journal of Pharmaceutical and Biomedical Analysis* 43:464–470.

Mukherjee PS, Cook SE (2006) A feasibility study on direct assay of an aqueous formulation by chiral supercritical fluid chromatography (SFC). *Journal of Pharmaceutical and Biomedical Analysis* 41:1287–1292.

Mukhopadhyay R (2008) SFC: Embraced by industry but spurned by academia. *Analytical Chemistry* 80:3091–3094.

Nomura A, Yamada J, Takatsu A, Horimoto Y, Yarita T (1993) Supercritical fluid chromatographic determination of cholesterol and cholesteryl esters in serum on ODS-silica gel column. *Analytical Chemistry* 65:1994–1997.

Ong CP, Lee HK, Li SFY (1990) Chlorodifluoromethane as the mobile phase in supercritical fluid chromatography of selected phenols. *Analytical Chemistry* 62:1389–1391.

Pereillo J-M, Maftouh M, Andrieu A, Uzabiaga M-F, Fedeli O, Savi P, Pascal M, Herbert J-M, Maffrand J-P, Picard C (2002) Structure and stereochemistry of the active metabolite of clopidogrel. *Drug Metabolism and Disposition* 30:1288–1295.

Phinney KW, Sander LC, Wise SA (1998) Coupled achiral/chiral column techniques in subcritical fluid chromatography for the separation of chiral and nonchiral compounds. *Analytical Chemistry* 70:2331–2335.

Pinkston JD (2005) Advantages and drawbacks of popular supercritical fluid chromatography interfacing approaches—A user's perspective. *European Journal of Mass Spectrometry* 11:189–197.

Pinkston JD, Stanton DT, Wen D (2004) Elution and preliminary structure-retention modeling of polar and ionic substances in supercritical fluid chromatography using volatile ammonium salts as mobile phase additives. *Journal of Separation Science* 27:115–123.

Pinkston JD, Wen D, Morand KL, Tirey DA, Stanton DT (2006) Comparison of LC/MS and SFC/MS for screening of a large and diverse library of pharmaceutically relevant compounds. *Analytical Chemistry* 78:7467–7472.

Pyo D (2000) Separation of vitamins by supercritical fluid chromatography with water-modified carbon dioxide as the mobile phase. *Journal of Biochemical and Biophysical Methods* 43:113–123.

Qian-Cutrone J, Dasgupta B, Kozlowski ES, Dalterio R, Wang-Iverson D, Vrudhula VM (2008) Separation of maxi-K channel opening 3-substitued-4-arylquinolinone atropisomers by enantioselective supercritical fluid chromatography. *Journal of Pharmaceutical and Biomedical Analysis* 48:1120–1126.

Rajendran A, Kräuchi O, Mazzotti M, Morbidelli M (2005) Effect of pressure drop on solute retention and column efficiency in supercritical fluid chromatography. *Journal of Chromatography A* 1092:149–160.

Rajendran A, Gilkison TS, Mazzotti M (2008) Effect of pressure drop on solute retention and column efficiency in supercritical fluid chromatography. Part 2: Modified carbon dioxide as mobile phase. *Journal of Separation Science* 31:1279–1289.

Sanagi MM, Smith RM (1988) The emergence and instrumenation of SFC. In *Supercritical Fluid Chromatography*, Smith RM, ed. Royal Society of Chemistry, Cambridge, UK.

Sassiat PR, Mourier P, Caude MH, Rosset RH (1987) Measurement of diffusion coefficients in supercritical carbon dioxide and correlation with the equation of Wilke and Chang. *Analytical Chemistry* 59:1164–1170.

Schmitz HH, Artz WE, Poor CL, Dietz JM, Erdman JW Jr. (1989) High-performance liquid chromatography and capillary supercritical-fluid chromatography separation of vegetable carotenoids and carotenoid isomers. *Journal of Chromatography A* 479:261–268.

Schoenmakers PJ (1988) Open columns or packed columns for supercritical fluid chromatography-a comparison. In *Supercritical Fluid Chromatography*, Smith RM, ed. Royal Society of Chemistry, Cambridge, UK.

Shi H, Taylor LT, Fujinari EM (1997a) Chemiluminescence nitrogen detection for packed-column supercritical fluid chromatography with methanol modified carbon dioxide. *Journal of Chromatography A* 757:183–191.

Shi H, Taylor LT, Fujinari EM, Yan X (1997b) Sulfur-selective chemiluminescence detection with packed column supercritical fluid chromatography. *Journal of Chromatography A* 779:307–313.

Sie ST, Beersum WV, Rijnders GWA (1966) High-pressure gas chromatography and chromatography with supercritical fluids. I. The effect of pressure on partition coefficients in gas-liquid chromatography with carbon dioxide as a carrier gas. *Separation Science and Technology* 1:459–490.

Sie ST, Rijnders GWA (1967) High-pressure gas chromatography and chromatography with supercritical fluids. IV. fluid-solid chromatography. *Separation Science and Technology* 2:755–777.

Simmons BR, Jagota NK, Stewart JT (1995) A supercritical fluid chromatographic method using packed columns for phenylbutazone and oxyphenbutazone in serum, and for phenylbutazone in a dosage form. *Journal of Pharmaceutical and Biomedical Analysis* 13:59–64.

Simon P, Nicot T (1996) Capillary electrophoresis and supercritical chromatography, complementary and alternative techniques for the determination of urinary metabolites of styrene. *Journal of Chromatography B: Biomedical Sciences and Applications* 679:103–112.

Slonecker PJ, Li X, Ridgway TH, Dorsey JG (1996) Informational orthogonality of two-dimensional chromatographic separations. *Analytical Chemistry* 68:682–689.

Smith RM (1999) Supercritical fluids in separation science—The dreams, the reality and the future. *Journal of Chromatography A* 856:83–115.

Steuer W, Baumann J, Erni F (1990) Separation of ionic drug substances by supercritical fluid chromatography. *Journal of Chromatography A* 500:469–479.

Stringham RW (2005) Chiral separation of amines in subcritical fluid chromatography using polysaccharide stationary phases and acidic additives. *Journal of Chromatography A* 1070:163–170.

Stringham RW, Blackwell JA (1996) "Entropically driven" chiral separations in supercritical fluid chromatography. confirmation of isoelution temperature and reversal of elution order. *Analytical Chemistry* 68:2179–2185.

Strode JTB, Taylor LT (1996) Optimization of electron-capture detector when using packed-column supercritical fluid chromatography with modified carbon dioxide. *Journal of Chromatography A* 723:361–369.

Strubinger JR, Song H, Parcher JF (1991) High-pressure phase distribution isotherms for supercritical fluid chromatographic systems. 2. Binary isotherms of carbon dioxide and methanol. *Analytical Chemistry* 63:104–108.

Takahashi K, Kinugasa S, Senda M, Kimizuka K, Fukushima K, Matsumoto T, Shibata Y, Christensen J (2008) Quantitative comparison of a corona-charged aerosol detector and an evaporative light-scattering detector for the analysis of a synthetic polymer by supercritical fluid chromatography. *Journal of Chromatography A* 1193:151–155.

Takahashi K, Kinugasa S, Yoshihara R, Nakanishi A, Mosing RK, Takahashi R (2009) Evaluation of a condensation nucleation light scattering detector for the analysis of synthetic polymer by supercritical fluid chromatography. *Journal of Chromatography A* 1216:9008–9013.

Taylor LT (2009) Supercritical fluid chromatography for the 21st century. *The Journal of Supercritical Fluids* 47:566–573.

The Chromatographic Society (2004) Dr Terry Berger—2004 Martin Gold Medal. In *The Chromatographic Society*, http://www.chromsoc.com/drterryberger.aspx.

Toribio L, del Nozal MJ, Bernal JL, Alonso C, Jiménez JJ (2005) Comparative study of the enantioselective separation of several antiulcer drugs by high-performance liquid chromatography and supercritical fluid chromatography. *Journal of Chromatography A* 1091:118–123.

Ventura MC, Farrell WP, Aurigemma CM, Greig MJ (1999a) Packed column supercritical fluid chromatography/mass spectrometry for high-throughput analysis. *Analytical Chemistry* 71:2410–2416.

Ventura MC, Farrell WP, Aurigemma CM, Greig MJ (1999b) Packed column supercritical fluid chromatography/mass spectrometry for high-throughput analysis. Part 2. *Analytical Chemistry* 71:4223–4231.

Wallenborg SR, Markides KE, Nyholm L (1997) Development of an amperometric detector for packed capillary column supercritical fluid chromatography. *Analytical Chemistry* 69:439–445.

Wang MZ, Klee MS, Yang SK (1995) Achiral and chiral analysis of camazepam and metabolites by packed-column supercritical fluid chromatography. *Journal of Chromatography B: Biomedical Sciences and Applications* 665:139–146.

Wang T, Barber M, Hardt I, Kassel DB (2001) Mass-directed fractionation and isolation of pharmaceutical compounds by packed-column supercritical fluid chromatography/mass spectrometry. *Rapid Communications in Mass Spectrometry* 15:2067–2075.

Wang Z, Li S, Jonca M, Lambros T, Ferguson S, Goodnow R, Ho C-T (2006) Comparison of supercritical fluid chromatography and liquid chromatography for the separation of urinary metabolites of nobiletin with chiral and non-chiral stationary phases. *Biomedical Chromatography* 20:1206–1215.

Weckwerth JD, Carr PW (1998) Study of interactions in supercritical fluids and supercritical fluid chromatography by solvatochromic linear solvation energy relationships. *Analytical Chemistry* 70:1404–1411.

Welch CJ, Biba M, Gouker JR, Kath G, Augustine P, Hosek P (2007) Solving multicomponent chiral separation challenges using a new SFC tandem column screening tool. *Chirality* 19:184–189.

West C, Lesellier E (2008a) Orthogonal screening system of columns for supercritical fluid chromatography. *Journal of Chromatography A* 1203:105–113.

West C, Lesellier E (2008b) A unified classification of stationary phases for packed column supercritical fluid chromatography. *Journal of Chromatography A* 1191:21–39.

Wright BW, Kalinoski HT, Smith RD (1985) Investigation of retention and selectivity effects using various mobile phases in capillary supercritical fluid chromatography. *Analytical Chemistry* 57:2823–2829.

Xu X, Roman JM, Veenstra TD, Van Anda J, Ziegler RG, Issaq HJ (2006) Analysis of fifteen estrogen metabolites using packed column supercritical fluid chromatography-mass spectrometry. *Analytical Chemistry* 78:1553–1558.

Yang Y, Su B, Yan Q, Ren Q (2005) Separation of naproxen enantiomers by supercritical/subcritical fluid chromatography. *Journal of Pharmaceutical and Biomedical Analysis* 39:815–818.

Ye YK, Lynam KG, Stringham RW (2004) Effect of amine mobile phase additives on chiral subcritical fluid chromatography using polysaccharide stationary phases. *Journal of Chromatography A* 1041:211–217.

Yonker CR, McMinn DG, Wright BW, Smith RD (1987) Effect of temperature and modifier concentration on retention in supercritical fluid chromatography. *Journal of Chromatography A* 396:19–29.

Zeng L, Xu R, Laskar DB, Kassel DB (2007) Parallel supercritical fluid chromatography/mass spectrometry system for high-throughput enantioselective optimization and separation. *Journal of Chromatography A* 1169:193–204.

Zhang X, Towle MH, Felice CE, Flament JH, Goetzinger WK (2006) Development of a mass-directed preparative supercritical fluid chromatography purification system. *Journal of Combinatorial Chemistry* 8:705–714.

Zhang Y, Dai J, Wang-Iverson DB, Tymiak AA (2007) Simulated moving columns technique for enantioselective supercritical fluid chromatography. *Chirality* 19:683–692.

Zheng J, Glass T, Taylor LT, Pinkston JD (2005a) Study of the elution mechanism of sodium aryl sulfonates on bare silica and a cyano bonded phase with methanol-modified carbon dioxide containing an ionic additive. *Journal of Chromatography A* 1090:155–164.

Zheng J, Pinkston JD, Zoutendam PH, Taylor LT (2006) Feasibility of supercritical fluid chromatography/mass spectrometry of polypeptides with up to 40-Mers. *Analytical Chemistry* 78:1535–1545.

Zheng J, Taylor LT, Pinkston JD, Mangels ML (2005b) Effect of ionic additives on the elution of sodium aryl sulfonates in supercritical fluid chromatography. *Journal of Chromatography A* 1082:220–229.

25

CHROMATOGRAPHIC SEPARATION METHODS

Wenying Jian, Richard W. Edom, Zhongping (John) Lin, and Naidong Weng

25.1 INTRODUCTION

25.1.1 A Historical Perspective

Not too long ago, liquid chromatographic (LC) techniques coupled with ultraviolet (UV), fluorescence, or electrochemical detection were still the default methods for separation and detection of drug candidates and their metabolites in biological samples. However, these detection methods were usually characterized by lower sample throughput, and, in general, they had limited sensitivity and poor specificity. Extensive chromatographic separation of analytes from interfering matrix components was needed. Fluorescence and electrochemical detection are also dependent on the chemical/physical properties of the analytes, and derivitization may be necessary for analytes without conducive properties. In comparison, mass spectrometry (MS) provides exceptional sensitivity and selectivity, as well as the capability to determine the molecular weights and fragmentation patterns, which can assist in the structure elucidation of analytes. Gas chromatography (GC)-MS has been a widely used technique for separating and quantifying trace components in complex samples. However, the analytes must be volatile and have a large degree of thermal stability to enable separation by GC and detection by MS. A time-consuming and tedious process of chemical derivatization of nonvolatile compounds is often required before analysis by GC-MS. Even though GC-MS is now seldom used for metabolite profiling and quantitation, it remains a useful tool for some special applications, such as separation of volatile positional isomers, due to the highly

efficient chromatographic resolution offered by GC. The application of atmospheric pressure ionization (API) techniques has provided a breakthrough for the combination of LC techniques with MS (LC-MS), and has brought LC-MS to the forefront of analytical techniques. LC-MS analysis has been routinely used in absorption, distribution, metabolism, and excretion (ADME) labs in the pharmaceutical industry since the mid-1990s for the qualitative and quantitative analysis of pharmaceutical compounds in biological matrices. For metabolite identification studies, LC-MS has also been coupled with in-line radioactivity flow detectors or UV detectors for simultaneous acquisition of MS data and radio- or UV-chromatograms, which enables identification of components within the specific retention time regions with relevant MS data for structure elucidation.

25.1.2 The Need for Separation in ADME Studies

For metabolite identification work, good chromatographic separation is essential. Metabolites can be structurally diverse and they are often present in relatively low concentrations on top of a large excess of endogenous components. To complicate matters further, several metabolites might have the same molecular weight (isobaric), such as those generated by multiple sites of hydroxylation. These positional isomers could coelute chromatographically, making the task of identifying the position of the biotransformation challenging. Separating these metabolites chromatographically would result in cleaner MS/MS spectra, which in turn would simplify the structure elucidation process.

ADME-Enabling Technologies in Drug Design and Development, First Edition. Edited by Donglu Zhang and Sekhar Surapaneni.
© 2012 John Wiley & Sons, Inc. Published 2012 by John Wiley & Sons, Inc.

For quantitative analysis of compounds in biological fluids (i.e., bioanalysis), LC separation followed by tandem MS (MS/MS) analysis using multiple reaction monitoring (MRM) is commonly employed. Specificity afforded by MS/MS often minimizes the need for extensive chromatographic separation and decreases the time required to develop rigorous LC conditions. However, there are a number of sources of interference that can potentially originate from *in vivo* biological samples, and if not adequately separated from target compounds, may compromise performance of the methods: (1) Endogenous components in the biological matrices such as salts, proteins, fatty acids, and phospholipids may cause suppression or enhancement of ionization in the LC-MS interface, artificially changing the signal intensity; (2) Drug-derived components such as metabolites and prodrugs may compete with the parent drug for ionization and cause suppression. Conjugate metabolites (e.g., acyl glucuronide, sulfate conjugate), N-oxides, and prodrugs may also undergo in-source conversion to generate ions that are identical to that of the parent drug and therefore interfere with quantitation (Matuszewski et al., 1998; Jemal and Xia, 1999); (3) Dosing vehicles (e.g., polyethylene glycol, cyclodextrin, Tween) used in formulations for animal studies have been shown to cause MS signal suppression and introduce bias in the concentration results (Tong et al., 2002; Shou and Naidong, 2003). Therefore, adequate chromatographic resolution between the target analytes and potentially interfering components is still of high priority in method development for quantitative bioanalysis assays in order to ensure accurate and precise results.

It is commonly believed that matrix from *in vitro* ADME screening assays (e.g., P450 inhibition/induction, permeability, protein binding) is relatively clean, and therefore, short LC run times without optimal chromatographic separation are often used. While this may be acceptable in some cases, adequate separation is still needed for accurate measurements in most applications. For example, buffers used in protein binding assays may cause ion suppression and affect quantitation if not separated from the analytes (Chu and Nomeir, 2006).

Analysis of chiral compounds is another case where adequate separation needs to be emphasized. According to the Food and Drug Administration (FDA) guideline issued in 1992 for the development of new stereoisomeric drugs, in order to evaluate the pharmacokinetics of a single enantiomer, or a mixture of enantiomers, manufacturers should develop quantitative assays for individual enantiomers in *in vivo* samples early in drug development (FDA, 1992). Stereoisomers cannot be differentiated by MS in most cases due to their identical molecular weights. Therefore, chromatographic separation must be achieved for the individual isomers. This will allow the assessment of potential interconversion, and the activity, toxicity, and ADME properties of the individual isomers.

25.1.3   Challenges for Current Chromatographic Techniques in Support of ADME Studies

It is very challenging to establish reliable analysis methods for polar compounds because they show little or no retention on traditional reversed-phase columns. In addition, highly aqueous mobile phases that are used in an attempt to retain polar compounds on reversed-phase columns are not conducive for achieving high sensitivity on LC-MS/MS systems, and could also lead to LC column phase collapse (dewetting of the stationary phase). Matrix effects could also be more serious for polar compounds because they are prone to coeluting with interfering components early in the run time. All the above-mentioned factors could result in compromised performance of analytical methods for polar compounds. The recent FDA guidance for "Safety Testing of Drug Metabolites" (MIST) that requires measurement of disproportionate drug metabolites (i.e., metabolites that present only in humans, or present at significantly higher plasma concentrations in humans than in the animals used in nonclinical studies) highlights the need for development of reliable analytical methods for metabolites (FDA, 2008). The need to separate and quantify metabolites, which are generally more polar than the parent drugs, makes the chromatographic issues for polar compounds more prominent.

The ever-growing number of new chemical entities that enter drug discovery and development present the challenge of improving analytical throughput. Miniaturized columns have been widely adopted for increasing chromatographic speed. For quantitative LC-MS/MS analytical methods, short, narrow-bore (20–100 mm × ~2.0 mm) columns with small particle sizes (3–5 µm), relatively high flow rates (e.g., >0.5 mL/min), and fast gradient elution have been routinely used to achieve analytical run times in single-digit minute time. A practical challenge for development of new analytical methodology for higher throughput is to retain or improve the separation efficiency while increasing the speed of analysis.

Sample preparation is a key step in LC-MS/MS methodology and can often be a bottleneck in development and application of quantitative assays. Extraction procedures for the analytes of interest from biological matrices have evolved from manual processing of individual tubes to automated operation in 96-well plate format. Online extraction procedures can further reduce the time for sample pretreatment and simplify sample processing. Further increasing the throughput for sample preparation, boosting the effectiveness of removing matrix inferences, improving the recovery of target analytes, as well as following the trends of low sample volume, ease of use, automation, and an environment-friendly approach still remains a challenge for LC-MS/MS analysis of pharmaceutical compounds in complex biological fluids.

25.2 LC SEPARATION TECHNIQUES

LC is the predominant separation technique used in current ADME studies for qualitative and quantitative analysis of pharmaceutical compounds. LC separates compounds based on their chemical/physical characteristics (e.g., hydrophobicity, polarity, charge, size), and their interactions with stationary phase and mobile phase.

25.2.1 Basic Practical Principles of LC Separation Relevant to ADME Studies

The basic relationship between separation efficiency and speed in LC can be depicted by the Van Deemter equation: $H = A + B/\mu + C\mu$ (Figure 25.1). In this equation, H is the theoretical peak plate height, which represents the distance for the analytes to transfer between the mobile phase and stationary phase and to reequilibrate. With smaller plate height, higher separation efficiency can be achieved with a given column length. The term μ is the linear velocity of the mobile phase (cm/s), which is proportional to the flow rate with a given column dimension (cross section). The A-term represents the contribution to plate height from eddy diffusion, which results from turbulent flow as the analyte moves between the mobile phase and the stationary phase. A smaller A-term means less peak diffusion and better resolution. The A-term is independent of linear velocity and therefore has no contribution to the shape

of the Van Deemter curve. The B-term represents the longitudinal diffusion of the analyte. Contribution of the B-term is negligible under regular or high-flow LC due to the fact that longitudinal diffusion in a liquid medium is minimal under these operational conditions. The C-term represents contributions from resistance to mass transfer in the stationary and mobile phases. The C-term increases linearly with mobile phase velocity and its contribution to the peak height is therefore considerable. A smaller C-term will lead to a fairly flat ascending portion of the Van Deemter curve at higher mobile phase linear velocity, which means that the separation can be carried out at higher speed without sacrificing separation efficiency.

In practice, column efficiency can be measured by plate number (N), a reciprocal of plate height (H), which is often calculated based on retention time and the peak width. Efficiency of a chromatographic system can also be expressed by peak capacity (n_c), which is defined as the maximum number of peaks that can fit side by side between the first (or unretained peak) and the peak of interest, with unit resolution (Francois et al., 2009). Chromatographic systems with higher efficiency provide sharper peaks than those with lower column efficiency, under a given flow rate. Theoretically, peak capacity of a two-dimension separation is equal to the product of the peak capacities of the individual dimensions, if they are 100% independent of each other (orthogonal). Therefore, separation efficiency can be drastically improved by combining two modes of separations that are based on orthogonal mechanisms.

Selectivity (α) refers to the capacity of a chromatographic system to retain certain types of analytes to a significantly greater extent than others. It does not take into account the width of the peaks. Therefore, sufficient selectivity does not guarantee good efficiency as two well-separated peaks may have very broad peak shape, and vice versa, as two sharp peaks may not be fully resolved.

Providing acceptable column efficiency and selectivity has been achieved, one has to pay attention to retention factor (k') to ensure that analytes are sufficiently retained on the column. Retention factor k' is calculated by $(t'-t_0)/t_0$, where t' is the retention time of the peak of interest and t_0 is the retention time of an unretained compound. A retention capacity of 3–5 is generally considered sufficient for the purpose of quantitative analysis using LC-MS because the mass spectrometer can differentiate coeluting components by their unique molecular weight and fragmentation. In qualitative metabolite identification work, higher retention factors are usually needed in order to obtain selective information for the unknown peaks in the chromatogram. Retention capacity can be improved by

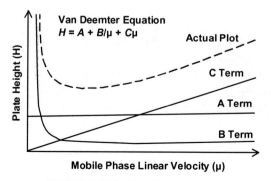

FIGURE 25.1. Van Deemter plot.

changing stationary phase, mobile phase composition, and column temperature.

Besides selection of the stationary phase and mobile phase for the specific analytical requirements of the analyte of interest, several LC conditions can be optimized to achieve a best balance between speed, specificity, sensitivity, and simplicity. These conditions include column temperature, gradient system, column dimension, flow rate, and injection solvent: (1) Column temperature: Elevated temperature can reduce the viscosity of the mobile phases as well as improve mass transfer. As a result, flow rate and column reequilibrium can be strongly accelerated without loss in efficiency as a favorable consequence of increased mass transfer. Therefore, high-temperature LC can be used in combination with high flow rate for fast separations. However, the stability of the stationary phase and the on-column stability of the analytes are of major concern when high temperature is used. (2) Isocratic versus gradient elution: Isocratic (constant composition) mobile phase facilitates fast runs by avoiding column conditioning/reequilibration steps. It also decreases the negative effects associated with rapid changes of the mobile phase composition such as system carryover or precipitation of analytes. However, as the peak width is proportional to the retention time in isocratic systems, the late-eluting peaks will appear broad and flat. In addition, strongly retained components may elute late and interfere with subsequent injections, and they may also accumulate on the column and cause column deterioration. In comparison, gradient (increasing elution strength either linearly or stepwise) mobile phase reduces the retention of the late-eluting components and improves their peak shapes. It can also help to wash the column after injection and to prevent retaining of the late-eluting compounds on the column. Peak capacity under gradient elution is higher than isocratic as the peak width is reduced for the late eluting analytes. (3) Column dimension: Narrow-bore (~2.0 mm in diameter) columns with column lengths of 20–100 mm are most frequently used in quantitative LC-MS/MS analysis in ADME studies. The small column diameter tends to increase ionization efficiency for MS detection by providing a low volumetric flow rate and sample preconcentration. The short column length can provide fast run time with acceptable selectivity when MRM is used for detection. In qualitative work such as metabolite identification, longer (150–250 mm) columns are often used to provide improved separation efficiency. Larger diameter (4.6 mm) is used when the column effluent needs to be split between the mass spectrometer and UV or radioactivity detectors. (4) Flow rate: In general, it is often advantageous to use higher flow rates than the recommended ones based on column dimensions for the sake of time-saving and better peak shape. For example, for a 2.1-mm-diameter column with a recommended flow rate of 0.2 mL/min, it is practical to use a flow rate of 0.4–0.6 mL/min, which provides fast analysis and narrow peaks. The minor loss in separation efficiency is still acceptable under this condition. Column back-pressure and compatibility with the MS interface should be considered with high flow rates. Post–column splitting of the effluent can be applied to avoid overflowing the ion source. (5) Injection solvent: Injection solvent plays a significant role in chromatographic efficiency, in particular with isocratic elution. The elution strength of the injection solvent should be slightly weaker than that of the isocratic mobile phase to facilitate on-column focusing of the analyte band. Otherwise, peak distortion and fronting can be observed (Naidong et al., 2001). Approaches for achieving this compatibility include dry down and reconstitution with a weaker injection solvent, dilution with a weaker elution solvent, and the use of a different chromatographic mode. The advantage of dry down and reconstitution is the potential for analyte enrichment in a lower volume. However, one must pay special attention to the stability of the analyte as well as the potential artifacts (Section 25.3.1). Dilution with a weaker elution solvent (e.g., water or buffer for a reversed-phase LC-MS) might be counterintuitive since it does seem the analyte is diluted. One would appreciate the fact that after 1:1 dilution with a weaker elution solvent (e.g., 0.1 mL acetonitrile precipitant diluted with 0.1 mL ammonium acetate buffer), the maximal allowed injection volume would be increased by fourfold. In other words, one would be able to double the absolute amount of the analyte injected onto the column after the 1:1 dilution (Cheng et al., 1999).

A common issue in the application of LC-MS/MS analysis in support of quantitative studies is system carryover, which has an impact on the accuracy and precision of quantitation, especially at lower concentrations (Hughes et al., 2007). Carryover is caused by residual analyte from a previous sample that is adsorbed on, or trapped to the active surfaces of the autoinjector system, solvent lines, extraction columns (e.g., online extraction), or the analytical column. Carryover can be assessed by injecting one or more matrix blank samples after a high concentration sample or standard, and is often expressed as a percentage of the lower limit of quantitation (LLOQ). It is dependent on the dynamic range of the assay and the issue is often exacerbated when a large dynamic range is used. Carryover can often be minimized by an appropriate choice of needle wash solutions, injector needles, and wash methods. If carryover is persistent, certain procedures should be provided in the analytical method to handle known carryover, such as injecting matrix blanks after certain samples and avoiding randomization of the samples. In addition to the LC-MS system, if an automatic liquid handler with fixed tips is used for sample handling or preparation, a similar assessment for carryover should also be employed.

25.2.2 Major Modes of LC Frequently Used for ADME Studies

25.2.2.1 Reversed-Phase Liquid Chromatography (RPLC)
RPLC is the most frequently used LC method in the pharmaceutical industry for separation of small-molecule drug candidates and their metabolites in ADME studies. RPLC has a nonpolar stationary phase and an aqueous, moderately polar mobile phase, and the analytes are retained by their hydrophobic interaction with the stationary phase. The most commonly used stationary phase is silica treated with organosilane $(R(CH_3)_2Si\cdot)$, where R represents different bonded phases (e.g., C18, phenyl) to afford specific selectivity characterization. Due to steric hindrance, the underivatized silica often still retains a rather large number of unreacted silanol groups, which could mediate secondary interactions with analytes. Endcapping processes that use a very small organosilane reagent that can reach the surface and react with the remaining silanols are employed to further reduce the number of free silanol groups. Another type of RPLC stationary phase consists of a polymer support, such as polystyrene divinylbenzene (PSDVB). The polymeric materials allow better column stability under a wider range of pH and higher temperatures than silica-based columns, but the column efficiency is usually significantly lower than C18 columns. Commonly used mobile phase for RPLC is any miscible combination of water (as a weaker solvent) and various organic liquids such as acetonitrile and methanol (as stronger solvents). Buffers or additives such as ammonium acetate or formic acid are often added to the mobile phase to modify the ionic strength and pH, and they may have a significant impact on the retention, selectivity, and sensitivity of ionizable analytes. Only volatile buffers or additives (e.g., ammonium acetate, formic acid) can be used for LC-MS application because nonvolatile ones (e.g., dipotassium phosphate, sodium hydroxide) are not compatible with the MS ion source.

In RPLC, compounds are retained based on their hydrophobicity. Therefore, it is very challenging to retain polar compounds such as oligonucleotides and peptides under traditional RPLC conditions. Ion pairing chromatography has been explored as an effective alternative. The ion pairing reagents added into the mobile phase can improve chromatographic retention by formation of neutral ion pairs with analytes. Commonly used volatile ion pairing reagents for LC-MS analysis include triethylamine (TEA), dibutylamine, and per-fluorinated carboxylic acids with n-alkyl chains such as trifluoroacetic acid (TFA), heptafluorobutanoic acid (HFBA), and nonafluoropentanoic acid (NFPA) (Gao et al., 2006; Lin et al., 2007). Due to the formation of ion pairs between the analyte and the ion pairing reagent, ionization efficiency can be impaired and low sensitivity might be observed in LC-MS analysis. Post–column infusion of propionic acid and isopropanol has been shown to alleviate ion suppression caused by TFA (Apffel et al., 1995). Alternatively, acetic acid (0.5%) or propionic acid (1%) directly added to the TFA-containing mobile phase can also effectively reduce ion suppression (Shou and Naidong, 2005).

Porous graphitic carbon (PGC) has been utilized in LC-MS analysis for polar drugs because it has been shown to provide better retention for polar compounds than traditional RPLC (Knox et al., 1986). The stationary phase of PGC is composed of large and flat sheets of hexagonally arranged carbon atoms. The mobile phase usually consists of water, methanol, or acetonitrile but it requires higher organic content than silica-based RPLC for elution of the analytes, which results in better chromatographic retention and improved ionization efficiency in mass spectrometers (Hsieh et al., 2007b). The retention mechanism on PGC involves the dispersive interaction of analyte–mobile phase and analyte–graphite surface, and dipolar and ionic interaction of the polar analyte with the polarizable graphite. PGC has been shown to produce better resolution for diastereomers than silica-based columns (Xia et al., 2006).

25.2.2.2 Hydrophilic Interaction Chromatography (HILIC)
HILIC has been gaining interest and utilization since this concept was first introduced by Alpert in 1990 (Alpert, 1990). It has become a favored choice for separation of drugs and/or metabolites in support of ADME studies due to its superior ability to retain polar compounds (Naidong, 2003; Xu et al., 2009; Jian et al., 2010). In HILIC, the analyte interacts with a hydrophilic stationary phase and is eluted with a relatively hydrophobic binary mobile phase. Bare silica is the most commonly used stationary phase. In addition, bonded phase columns such as cyano, diol, amino, and zwitterions have also become commercially available for HILIC use. HILIC typically employs water-miscible polar organic solvents such as acetonitrile or methanol, and water is usually the stronger eluting solvent. The gradient elution typically starts with 5–10% aqueous in mobile phase and increases to 50–60% (even higher in extreme cases) for elution of the analytes. In HILIC, polar compounds are more highly retained than non-polar compounds, and the elution order is usually the reverse of that on RPLC columns. An initial water content of higher than

5% is crucial for the retention mechanism, which involves the partitioning of analytes between the bulk eluant and a water-rich layer partially immobilized on the stationary phase. A common practice for HILIC is to use at least 10 mM buffer concentration, or 0.2% additive concentration, in the mobile phase to achieve the best peak shape and retention reproducibility. The theory is that the buffer or additive could facilitate the retaining of the water-rich layer, which is the basis for hydrophilic retention (Grumbach et al., 2004). In addition, ion exchange, hydrogen bond formation, dipole–dipole interactions, and other interactions could also play roles in the retention mechanism, depending on the particular conditions employed (Hemstrom and Irgum, 2006).

HILIC is complimentary to RPLC and provides good retention and unique selectivity for polar compounds. The highly volatile organic mobile phase used in HILIC can provide increased sensitivity with electrospray ionization MS. Figure 25.2 demonstrates a comparison of LC-MS analysis of deoxyguanosine (dG) and 8-hydroxy-2′-deoxyguanosine (8-OHdG) under HILIC and RPLC conditions. Being highly polar compounds, both analytes showed poor retention under RPLC conditions and eluted at the solvent front. In comparison, HILIC conditions provided excellent retention and separation of the two analytes. The signal intensity under HILIC was 50–70 times higher than RPLC.

Organic solvents used for sample extraction, such as acetonitrile for protein precipitation (PPT) or solid-phase extraction (SPE) elution, and methyl t-butyl ether, diethyl ether, or ethyl acetate for liquid-liquid extraction (LLE) are stronger elution solvents than the mobile phase in reversed-phase mode, but they are weaker than the mobile phase in the HILIC mode.

Therefore, the organic extracts from PPT, SPE, and SPE can often be directly injected onto a HILIC column, without the problem of mismatching with the mobile phase and causing chromatographic peak shape distortion (Naidong et al., 2002, 2004; Li et al., 2004; Song and Naidong, 2006; Xue et al., 2006). Direct injection not only streamlines the process and facilitates higher throughput, but also, and probably more importantly, minimizes potential artifacts (Section 25.3.1) or contamination.

25.2.2.3 Normal-Phase Liquid Chromatography (NPLC)

NPLC employs a polar stationary phase and a nonpolar, nonaqueous mobile phase. Typically, bare silica, amino-bonded, or cyano-bonded silica is used as stationary phases. The retention mechanism on NPLC involves adsorption of the analyte to the stationary surface by hydrophilic interactions (e.g., hydrogen bonding) and retention increases as the polarity of the analytes increase. The use of more polar solvent (e.g., methanol) in the mobile phase will decrease retention, whereas more hydrophobic solvent (e.g., hexane) tends to increase retention times. NPLC is more difficult to operate because any trace levels of moisture in the mobile phase can significantly impact the chromatography and must be strictly controlled. Nevertheless, NPLC can effectively retain and separate analytes that are readily soluble in nonpolar solvents, such as fatty acids and hormones. The interaction strength on NPLC relies not only on the functional groups in the analyte molecule, but also on steric factors, which can be exploited to resolve structural isomers. Therefore, an important field of application for NPLC in the pharmaceutical industry is chiral separations, which will be discussed in more detail in Section 25.2.3.

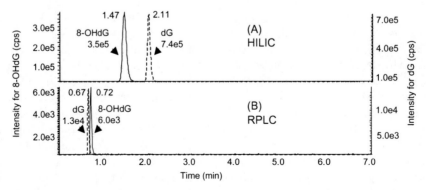

FIGURE 25.2. Total ion chromatograms of a 10 µL injection of a dG/8-OHdG (100/100 ng/mL) neat solution under (A) HILIC conditions on a Luna silica (2) column (50 × 2.0 mm, 3 µm, Phenomenex) and RPLC on a Luna C18 (2) (50 × 2.0 mm, 3 µm, Phenomenex). Mobile phase A was 0.1% (v:v) formic acid in water and mobile phase B was 0.1% (v:v) formic acid in acetonitrile. The flow rate was 0.3 mL/min. The gradient for (A) was 90% B to 65% B in 5 min, and for (B) was 10%B to 65%B in 5 min.

25.2.2.4 Other Modes of Chromatography Ion exchange chromatography (IEC) and size exclusion chromatography (SEC) are two important modes of separation for large molecules, but they have not been extensively used in ADME studies for small-molecule drug candidates. With the recent emergence of biomolecules such as peptides, proteins, or oligonucleotides as therapeutic agents, they have the potential to be the methods of choice for chromatographic separation. IEC retains and separates ions or polar compounds based on their charges. The stationary phase is typically a resin or gel matrix consisting of agarose or cellulose beads with covalently bonded, charged functional groups. The analytes (anions or cations) are retained on the stationary phase by attraction to functional groups with opposite charge, and can be eluted by increasing the concentration of a similarly charged species that will displace the analyte ions from the stationary phase. IEC is most often used for separation of biomolecules including proteins, peptides, amino acids, and nucleotides. It is rarely directly interfaced with mass spectrometers because the mobile phase used to elute analytes normally contains high concentrations of nonvolatile buffers/salts and is not compatible with the ion source. Instead, due to its orthogonality to RPLC, it is often adopted as the first dimension of separation to couple with RPLC in two-dimensional liquid chromatography (2D-LC) systems for comprehensive analysis of complex components. SEC is a chromatographic method in which molecules in solution are separated as a result of their permeation into the matrix of a stationary phase such as agarose gel or polyacrylamide. Large molecules may be excluded from some or all of the porous matrices of the packing by their physical size. Therefore, they may elute from the column before smaller molecules, which have the opportunity to enter the pores. SEC is often used for analysis and purification of synthetic and natural polymers, such as proteins, polysaccharides, or nucleotides.

25.2.3 Chiral LC

In the past, chiral separation methods usually involved derivatization of analytes with chiral reagents to generate diastereoisomers, which can be resolved on achiral columns. The enantiomeric purity of these reagents became very critical for the correct assessment of chiral compounds. With the availability of commercial chiral stationary phases (CSPs), direct chiral separation using LC has grown significantly as a simple and practical approach for resolving mixtures of enantiomers. Resolution on chiral columns is achieved by different interaction behavior of the two enantiomers (analytes) with the single enantiomer (chiral selector) immobilized on the CSP. The forces involved in the interaction are polar/ ionic interactions, π-π interactions, hydrophobic effects and hydrogen bonding. These forces are very weak and are subject to many variables such as additives, pH, temperature, and composition of the mobile phases, all of which require careful optimization in order to achieve maximal selectivity.

The chiral selector used as CSP for LC can be classified into two types (Okamoto and Ikai, 2008). The first type consists of optically active small molecules that are immobilized on silica gel or organic polymer gel. A variety of small molecules have been used as the chiral selector, such as macrocyclic antibiotics, cyclodextrin, and crown ether. The second type consists of optically active polymers, which are further divided into synthetic polymers (such as polyamide) and natural polymers (such as polysaccharide and protein). Polymer-based CSPs are usually prepared by coating the polymers in silica gel in order to improve the resolution efficiency and mechanical strength. Polysaccharide-based derivatives are currently the most popular chiral selectors for pharmaceutical applications due to their versatility, durability, and loading capacity. According to the literature, more than 80% of enantio-separations of pharmaceutical molecules are successful with this type of column (e.g., Chiralpak column from Daicel Chemical Industries, Osaka, Japan) (Morin, 2009). Chiral separations using polysaccharide columns are typically performed using normal phase mode because the interactions involved in enantiomeric resolution on these CSPs are stronger under normal phase conditions. One of the challenges of interfacing NPLC with MS is the possible explosion hazard when a high flow of flammable solvent, such as hexane, is introduced into the heated ionization source. Three approaches have been explored to overcome this problem. The first is to minimize air penetration into the ion source environment via exclusive use of inert gases such as nitrogen for the nebulizing and desolvation processes. The second is to reduce the source temperature (e.g., to 250°C). The disadvantage of this approach is incomplete desolvation of the column effluent, leading to poor sensitivity in LC-MS/MS analysis. The third approach is to incorporate a large aqueous postcolumn makeup flow to reduce the hexane concentration in the mobile phase prior to the ion source. The major drawbacks of this approach are dilution of the analytes and potential loss of separation resolution due to extracolumn band broadening effects. From this perspective, RPLC is simpler to operate than NPLC, and the mobile phase is more compatible with the mass spectrometric detection. Most macrocyclic CSPs work well under reversed-phase conditions. In addition, polysaccharide-based and protein-based reversed-phase chiral columns are also commercially available and have been successfully utilized for quantitative bioanalysis of stereoisomeric pharmaceutical compounds (Chen et al., 2005).

25.3 SAMPLE PREPARATION TECHNIQUES

It is critical to develop sample preparation procedures that effectively clean up the sample and concentrate the analytes of interest in order to ensure acceptable selectivity, resolution, and sensitivity for LC-MS analysis. There has been a great contrast between fast chromatographic analysis and conventional sample preparation, which is often the most labor-intensive and time-consuming step in an analytical method. The introduction of automated off-line and online sample preparation and advances in sample preparation techniques have greatly improved the efficiency of sample preparation.

25.3.1 Off-Line Sample Preparation

Direct injection ("dilute and shoot") of samples is simple and efficient and is most suitable for matrices that do not contain a large amount of cellular components and protein. For example, urine and bile can be directly injected onto LC columns following centrifugation and/or filtration, if concentration of the drug-related material is not needed. For more proteineous biological samples, such as microsomal incubation samples or plasma samples, a quick and generic preparation method is PPT with an organic solvent, with or without the addition of acid, followed by centrifugation and/or filtration. Both "dilute and shoot" and PPT are easy to automate in 96-well format. Even though they are simple and straightforward, these types of sample preparation methods may not sufficiently remove endogenous compounds such as salts, phospholipids, and fatty acids, and could result in significant matrix interference. Therefore, sample dilution and injecting a small volume of PPT samples are recommended. On the other hand, they are the extraction procedures least subject to variable recovery of drug-related components and are often the method of choice for metabolite profiling studies. However, one should always be aware not to lose drug or metabolites as a result of coprecipitation with the proteins or as a result of limited solubility.

SPE provides better efficiency than PPT for selective removal of matrix components and for concentrating the analytes of interest from complex biological samples. SPE sorbents are commonly based on chemically bonded silica, cross-linked polymers, or graphitized carbon, with bonded silica-based materials dominating the field. Particles of the SPE phase can be packed in a cartridge, plate, or column. Currently, SPE in 96-well plate format is the most popular for LC-MS analysis in support of various ADME studies due to its compatibility for automation. In general, SPE mechanisms include reversed-phase, normal phase, HILIC, ion exchange, and mixed mode. There are four steps involved in the use of SPE sorbents: (1) conditioning of the sorbent bed; (2) loading of the sample onto the sorbent allowing selective retention of the analytes; (3) washing the sorbent to remove the undesired matrix components; and (4) eluting the retained compounds using a strong solvent. Most often, the strong solvent of the eluant (except for that from HILIC SPE) is not compatible for direct injection into RPLC and requires evaporation and reconstitution into a weaker solvent. SPE often gives better sample cleanup than PPT but may not be as cost-effective due to labor and material costs.

LLE is a mass transfer procedure where an aqueous sample is mixed with an immiscible solvent (e.g., ethyl acetate, methyl *tert*-butyl ether, hexane) that exhibits preferential solubility/selectivity toward the analytes of interest in the sample. One of the key determinants in achieving high recovery in the LLE process is charge neutralization of the analyte by pH manipulation and subsequent extraction by the organic solvent. Following phase separation, often by centrifugation, the organic layer containing the extracted analytes is usually evaporated and the samples are reconstituted into an appropriate solvent for injection onto the RPLC system. LLE gives excellent sample cleanup and has been shown to remove phospholipids more effectively than PPT and SPE. LLE can be conducted in an automated manner in 96-well plate format. In this case, it is critical to choose plates and covers that are proven to prevent well-to-well contamination during the mixing process. LLE is more cost-effective than SPE, but it may result in generation of larger amounts of organic waste. An alternative to traditional LLE is supported LLE (SLE), in which the extraction interface occurs between the sample absorbed onto an inert solid support and a water-immiscible solvent passing through the support. SLE is formatted in 96-well plates and therefore is easy to automate. It has also been shown to have better extraction efficiency and is less prone to contamination than regular LLE.

In sample preparation, caution is necessary to avoid degradation of analytes and/or artifactual generation of metabolites. For example, modifications can be observed for certain chemical structures following PPT with acetonitrile, evaporation, and reconstitution as shown in following table (Leclercq et al., 2009). In addition, analytes can also be lost during dry down due to evaporation or adsorption to the container.

Structural Group	Conversion
Carbamate	Carbamate migration to an alcohol on a β-carbon (Mannens et al., 2007)
Basic nitrogen	*N*-oxidation
Ester	Hydrolysis
Acyl glucuronide	Hydrolysis, acyl migration
Primary amine	N-glycosidation
N-oxide	Reduction

25.3.2 Online Sample Preparation

Direct online extraction has an advantage over off-line extraction in that it affords minimized sample preparation, reduces loss of the analyte during extraction, and decreases the chance of artifacts, contamination, and dilution during the extraction. A typical configuration for online extraction is depicted in Figure 25.3, which consists of an autosampler, two sets of binary pumps, an extraction column, an analytical column, and a divert valve. In loading/extraction stage, a plasma sample is injected to the extraction column with a 100% aqueous mobile phase at a high flow rate. During this period, the analytes of interest are retained on the head of the extraction column while the undesired matrix components such as macromolecules and salts are removed and diverted to waste. In the elution stage, the valve is switched so that the extraction column is in-line with the analytical column and the mass spectrometer. The retained analytes are eluted to the analytical column for separation and subsequent detection by the mass spectrometer, using an organic/aqueous mobile phase at a regular flow rate. The system can be configured to have elution either in forward-flush mode, in which direction of flow for the extraction column in the elution stage is the same as that in the loading stage, or opposite, that is, back-flush mode. In forward-flush mode, the analytes may exhibit chromatographic behavior when they pass through the extraction column. Strongly retained components may also accumulate in the extraction column and cause deterioration of performance. Therefore, many applications adopted the back-flush mode for better peak shape and longer lifetime for the extraction column. In addition, special attention often needs to be paid to reduce carryover, which tends to be a prominent issue for online extraction procedures.

Currently available supporting phases for online extraction include turbulent flow chromatography (TFC) columns, SPE cartridges or columns, and restricted access materials (RAM): (1) TFC can be achieved with narrow-bore columns (typically 1 × 50 mm) packed with large-diameter particles (typically 30–50 μm). They are normally operated at a high flow rate (4–6 mL/min),

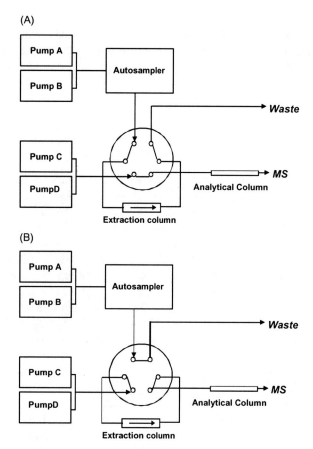

FIGURE 25.3. A typical online extraction configuration: (A) loading/extraction state; (B) elution state.

under which conditions the small molecules are rapidly bound to the stationary phase while proteins are being removed by the flow. TFC columns marketed by Cohesive Technologies (Thermo Fisher, Waltham, MA, USA) are widely used for this purpose. (2) A variety of online SPE extraction cartridges based on reversed-phase mechanism, ion exchange mechanism, or mixed-mode mechanism are commercially available. The automated on-line SPE system (Symbiosis™) from Spark Holland (Emmen, The Netherlands) features disposable, single-use cartridges contained in a 96-well cassette and is less prone to problems such as clogging and carryover. Monolithic phases have also been explored as extraction support for high flow rate online extraction due to their high permeability, and they have demonstrated satisfying results for analysis of pharmaceutical compounds in plasma (Zang et al., 2005; Xu et al., 2006). (3) RAM excludes the retention of macromolecules either by creating a physical diffusion barrier (based on particle pore size) or with a chemical diffusion barrier, consisting of a bonded polymer/protein network on the surface of the particles. Low-molecular-weight compounds are retained via hydrophobic interaction to the interior of the phase. Application of RAM for direct analysis of pharmaceutical compounds in biological fluids has been demonstrated on commercialized supporting materials such as those externally modified with alkyl diol silica (C4, C8, or C18) or with proteins (alpha-1 acid glycoprotein) (Mullett, 2007).

25.3.3 Dried Blood Spots (DBS)

The collection of whole blood samples on paper, known as DBS, dates back to the early 1960s in newborn screening for inherited metabolic disorders. As a less invasive sampling method requiring a smaller blood volume (typically less than 100 µL), DBS offers simpler sample collection and storage, easier transfer, and reduced risk of infection from various pathogens. The small sample volume allows for several bleeds to be taken from the same animal and decreases the number of the animals required in preclinical ADME studies. The typical DBS sample preparation includes punching one or more disks from DBS cards/paper into assay tubes or wells of a 96-well plate, adding a certain amount of extraction solvent (methanol, acetonitrile, or a mixture of water/organic) containing the internal standard(s), extracting with gentle shaking/vortexing and centrifugation, and then direct injection onto the LC-MS/MS system for analysis. To date, DBS-LC-MS/MS is emerging as an important method for quantitative analysis of small molecules (Spooner et al., 2009; Li and Tse, 2010) and as a means for metabolite identification as well (Mauriala et al., 2005; Thomas et al., 2010).

Special attention needs to be paid on stability of drugs and metabolites during the method development of DBS assays. Several applications have demonstrated the advantages of DBS over other approaches for preserving some unstable analytes from degradation, or delaying the degradation process (Alfazil and Anderson, 2008; Garcia Boy et al., 2008). On the other hand, some compounds (e.g., ester- or amide-containing compounds) are prone to very quick enzymatic degradation. The degradation could take place during the drying process of DBS sampling or even before the blood samples are spotted onto the cards/paper (Garcia Boy et al., 2008). Enzyme inhibitors cannot be conveniently mixed during blood sample collection prior to being spotted onto the cards/paper. Therefore, the stability of drugs and metabolites in whole blood without any pretreatment for a minimum 2–3 h to cover the blood collection and drying process must be assessed during method development before a DBS assay could be implemented. If instability is observed, DBS sampling should not be used (Spooner et al., 2009; Li and Tse, 2010).

25.4 HIGH-SPEED LC-MS ANALYSIS

The large number of samples generated in various ADME studies demands the development of high-speed separation technologies to enable higher-throughput sample analysis. Current trends in high-speed LC separations involve ultrahigh-pressure liquid chromatography (UHPLC), monolith technology, fused-core columns, and high-speed HILIC.

25.4.1 UHPLC

UHPLC has recently become a widespread analytical technique and is unequivocally leading the fast chromatography approach in many laboratories that focus on fast separation and high-sensitivity bioanalytical assays (Mazzeo et al., 2005). The key advantages of UHPLC are the increased speed of analysis, higher separation efficiency, higher sensitivity, and much lower solvent consumption as compared with other analytical approaches. This is all enabled by sub-2-µm-particle analytical columns and specially designed instruments. The Van Deemter curves of sub-2-µm-particle-packed columns show lower plate height at higher linear velocities, and over a wider range of linear velocities, compared with conventional particle columns. This is because small particles have shorter diffusion path lengths, allowing analyte molecules to travel in and out of the particle faster with less resistance (smaller C-term) and spend less time inside the particle where peak diffusion can occur (smaller A-term). Therefore, small-particle columns provide increased separation efficiency and the ability to work at increased linear velocity without a loss of efficiency, leading to both better resolution and higher speed. In addition, MS detection can be enhanced by using small-particle columns as reduced peak width and increased peak concentration provide better sensitivity. However, the benefit of utilizing small-particle materials is accompanied by increased column backpressure, which is inversely proportional to the square of particle size. In the past decade, UHPLC systems have been developed to cope with the high backpressure resulting from small-particle columns. In 2004, Waters (Milford, MA, USA) commercialized the ACQUITY™ system, which is able to handle pressures up to ~1000 bar. Other manufacturers followed this approach, such as the JASCO (Easton, MD, USA) Xtreme LC system (maximum pressure ~1000 bar), the Agilent (Santa Clara, CA, USA) 1200 Series Rapid Resolution LC system (maximum pressure ~600 bar), and the Shimadzu (Columbia, MD, USA) Prominence system (maximum pressure ~300 bar). Reversed-phase sub-2-µm-particle columns have become available from almost all major column manufacturers. The recent introduction of UHPLC columns for HILIC mode offers prospects for similar improvements in the separation of polar molecules such as glycans (Grumbach et al., 2008; Ahn et al., 2010).

To date, approximately 700 papers have been published on the subject of UHPLC and sub-2-μm particles. Major applications of UHPLC include pharmaceutical analysis (drug discovery and development, quality control of drug products, bioequivalence studies, therapeutic drug monitoring, etc.), proteomics, metabolomics, environmental analysis, food analysis, plant analysis, and chiral separations. In bioanalytical applications, UHPLC has demonstrated that separations 2–20 times faster can be achieved while maintaining or improving resolution and sensitivity. The optimal mobile phase flow rate for UHPLC-MS experiments should be comprised between 300 and 600 μL/min for 2.1-mm-diameter columns, while a higher flow rate could generate a loss in sensitivity and slightly reduced separation efficiency. A typical UHPLC analysis of a single compound in a biological matrix using an internal standard requires 1–2 min. This high rate of analyses is extremely important for high-throughput ADME laboratories.

Use of UHPLC for enhanced chromatographic separation may be a method of choice for the reduction of matrix effects by resolving the analyte peak from the ionization suppression components such as phospholipids. Nevertheless, one must be aware of the potential risk of using solely the chromatographic resolution power of UHPLC. The sharp analyte peak is sometimes impressively separated within seconds from the equally sharp matrix peaks, which are typically not monitored during routine sample analysis. Upon repeated injections of biological samples, even with sample precleanup via extraction, the column chemistry may slightly change and efficiency may become lower. The previously resolved matrix peaks (ionization suppression components) may then partially coelute with the analyte peak, resulting in significant matrix effects, which oftentimes could be even more significant than a conventional column due to the sharper matrix peaks (i.e., higher concentration of ionization suppression components). A good balance of sample preparation and chromatography (such as orthogonal approaches) should be used to minimize matrix effects.

25.4.2 Monolithic Columns

One way to overcome some of the challenges associated with fast LC-MS analysis, such as high column back-pressure, is the use of monolithic columns. Compared with particle-packed columns, monolithic columns consist of one piece of a continuous, porous material that is sealed against the wall of a tube. There are two types of monolithic supporting materials: cross-linked organic polymers and monolithic silica (Wu et al., 2008a). Organic polymer monoliths are typically used for separation of peptides and proteins, or isolation and purification of biopolymers. In comparison, monolithic silica columns are more commonly used in separation of low-molecular-weight compounds and thus have found application in pharmaceutical industry. Currently, the only two commercial monolithic silica columns are Chromolith™ from Merck KGaA (Darmstadt, Germany) and Onyx™ from Phenomenex (Torrance, CA). Three types of surface chemistry are available: silica, C8, and C18.

The major chromatographic features of monolithic silica columns are attributed to the high porosity and the large through-pore size-to-skeleton size ratio, giving rise to high permeability and a small diffusion path length, which in turn result in lower back-pressure and lower plate height compared with a particle-packed column. The Van Deemter curves of monolithic columns are flatter than particle-packed columns, which means the columns can be operated at higher flow rates without losing resolution power (Unger et al., 2008). Meanwhile, monolithic silica columns usually have a column back-pressure that is three to five times lower than particle-packed columns operated under same conditions as shown in Figure 25.4 (Shou et al., 2002). Therefore, monolithic columns in LC can be utilized to conduct high-speed separations with fast mass transfer kinetics and low column back-pressure. The typical flow rate on a 4.6 × 50-mm-diameter monolithic silica column is 4.0 mL/min and it can go up to 9 mL/min, which is much higher than the optimal operating flow rate for the MS ion source (typically <1.0 mL/min). Therefore, a post-column splitter is required in order to reduce the flow entering the mass spectrometer.

Monolithic silica columns have been used in ADME studies for high-throughput quantitative and qualitative LC-MS/MS analysis of drugs and their metabolites. In one example, methylphenidate and its metabolite ritalinic acid were baseline separated in 15 s with a flow rate of 3.5 min/mL. Overall, 768 protein-precipitated rat plasma samples (eight 96-well pates) were analyzed for quantitation of methylphenidate and ritalinic acid within 3 h and 45 min (Barbarin et al., 2003). It is evident that monolithic columns can significantly improve the throughput of ADME sample analysis without the problems associated with high back-pressure, which are usually encountered for sub-2 μm-particle columns. In addition, monolithic silica stationary phase requires a shorter regeneration time and has quicker equilibration in a gradient elution, thus further increasing sample throughput. The large through-pore (2 μm macropore) has less potential for clogging, making it ideal for "dilute

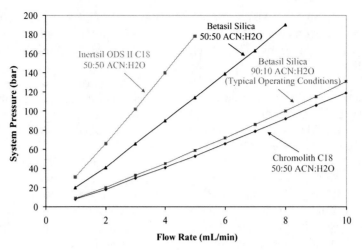

FIGURE 25.4. Back-pressure comparison between a Betasil silica column (Thermo Fisher, Waltham, MA, USA), an Inertsil ODS II C18 column (GL Sciences, Torrance, CA, USA), and a Chromolith C18 column. All columns were new, of a 50 × 4.6 mm dimension. Betasil and Inertsil columns are of 5 μm particle size (Shou et al., 2002).

and shoot" applications as well as PPT samples from complex matrices such as plasma and urine. This results in less time needed for sample preparation. In comparison with conventional particle-packed columns, which are usually contained in a stainless steel shell with a frit on each end to prevent bleeding of column material, monolithic silica columns are constructed in polyether ether ketone (PEEK) with no frits. This can eliminate some potential issues such as carryover associated with stainless steel, and column clogging related to accumulation of materials on the inlet frit. In the case when higher separation efficiency is needed, such as in metabolite profiling studies, monolithic silica columns can be linked in series to give a higher plate number than particle-packed columns, while producing back-pressure well below the limit of regular LC systems.

The application of monolithic silica columns in routine analytical work is to some extent limited by high mobile phase consumption, and a limited selection of commercially available column chemistry and dimensions. In addition, the stationary phase surface area of monolithic silica columns is lower than conventional particle-packed columns, resulting in lower loading capacity. Finally, the lack of pH stability of monolithic silica columns prevents their use for separation of basic compounds (e.g., tricyclic antidepressants) at high pH (Atia et al., 2009).

25.4.3 Fused-Core Silica Columns

Recently, partially porous or fused-core materials have been recognized as an alternative to sub-2-μm particles for ultrafast separations. The fused-core particle consists of a thin layer of porous shell fused to a solid particle. Currently, there are three commercially available brands

of fused-core columns: the HALO™ from Advanced Materials Technology (Wilmington, DE), Ascentis Express™ from Supelco (St. Louis, MO), and the newly marketed Kinetex™ Core-shell from Phenomenex. The HALO and Acentis Express columns have a particle size of 2.7 μm, which is composed of a 1.7-μm solid core and a 0.5-μm porous shell. The Kinetex comes with two types of particles: 2.6 μm consisting of a 1.9-μm solid core and a 0.35-μm porous shell, and 1.7 μm consisting of a 1.25-μm solid core and a 0.23-μm porous shell. The surface chemistry of current fused-core columns includes C18, C8, reversed phase-amide, and HILIC, providing selectivity for a broad range of compounds.

Compared with totally porous particles, the fused-core particles have a much shorter diffusion path for a solute into and out of the stationary phase, and therefore, peak broadening is minimized at higher mobile phase flow rates. Fused-core particles have a narrower particle size distribution than totally porous particles, which, in combination with the high density of the material, facilitate packing of the columns with a higher plate number, leading to high separation efficiency. Because of their larger particle size, the back-pressure generated from a fused-core column is usually approximately half of that from sub-2-μm column with the same dimensions and operating conditions (Cunliffe and Maloney, 2007). The above characteristics allow reduction in analysis time, with a subsequent increase in sample throughput compared with conventional LC columns, without sacrificing resolution and without the need to change the LC system to an ultrahigh-pressure system.

Fused-core columns are claimed to have better ruggedness than sub-2-μm columns. The narrow particle size distribution of the fused-core particles permits the use of frits with 2-μm pores (except for the Kinetex

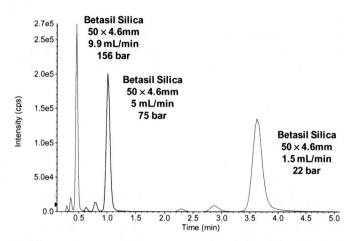

FIGURE 25.5. Total ion chromatogram of a 15 µL injection of a morphine/M6G/M3G (50/100/1000 ng/mL) neat solution under different mobile phase flow rates. Mobile phase A was 0.01% TFA in water and mobile phase B was 0.01% TFA in acetonitrile. The isocratic mobile phase was 89% B (Shou et al., 2002).

1.7-µm-particle columns), the same as those used on most columns packed with 5-µm particles. In comparison, sub-2-µm particles require frits with a pore size of 0.5 µm or less that are prone to fouling, leading to shortened column lifetime. Another feature of fused-core particles that contribute to their ruggedness is that they are denser than totally porous particles, and therefore, they form more highly stable beds in the packed column.

A number of studies have been published to demonstrate the use of fused-core columns in ADME studies for separation of drugs and their metabolites. A typical study was described by Hseih et al. for determination of rimonabant in mouse plasma using an LC-MS/MS assay (Hsieh et al., 2007a). The HALO C18 column (2.1 × 50 mm, 2.7 µm) was operated in gradient mode with a flow rate of 1.2 mL/min. The column effluent was directly connected to the atmospheric pressure chemical ionization (APCI) source for MS detection. The method required only 1.5 min per sample, giving throughput comparable with that achieved using sub-2-µm columns on a UHPLC system. In another example, Song et al. demonstrated that by using a HALO C18 column (2.1 × 30 mm, 2.7 µm), a 34% reduction in assay time and a two- to threefold increase in efficiency (N) was achieved compared to a conventional C18 column (5 µm) for quantitation of imipramine and desipramine in rat plasma. The observed back-pressure from a flow rate of 1.0 mL/min did not exceed ~220 bar, making it compatible with standard LC equipment (Song et al., 2009).

25.4.4 Fast Separation Using HILIC

Due to low viscosity of the mobile phase and high permeability of the column, HILIC shows much lower back-pressure than RPLC operated under similar conditions including column dimensions, particle size, and flow rate, as demonstrated in Figure 25.4. This enables the use of high flow rates under HILIC conditions, which provides a technique for ultrafast separation of polar compounds complementary to other reversed-phase ultrafast separation techniques. As demonstrated in an example of the quantitative analysis of morphine, morphine-6-glucuronide (M6G), and morphine-3-glucuronide (M3G) on a 50 × 4.6 mm Betasil (Thermo Fisher, Waltham, MA, USA) silica column (5 µm), by increasing the mobile phase flow rate from 1.5 to 9.9 mL/min, the analysis time was reduced from about 4.5 min to about 0.6 min, while the separation was not significantly affected (Figure 25.5). The system back-pressure increased from 22 bar at 1.5 mL/min to 156 bar at 9.9 mL/min, which can be handled readily by regular LC systems (Shou et al., 2002). The final method for quantitation of morphine, M6G, and M3G extracted from human plasma utilizing a 50 × 3.0-mm Betasil silica column operated at 4 mL/min gave a back-pressure of 122 bar and a total injection-to-injection cycle time of 48 s. For optimal MS sensitivity, the column effluent was split and about 500 µL/min of the flow was introduced into the electrospray ionization (ESI) source.

Reducing particle size in HILIC separations has similar improvements in efficiency as seen in RPLC. The Van Deemter curve under HILIC conditions demonstrated that the separation efficiency of 1.7-µm bridged ethyl hybrid (BEH) silica is well maintained with increased linear velocity, while that of 3-µm silica is compromised when the linear velocity is above 2.0 mm/s (Grumbach et al., 2008). Due to the lower back-pressure, sub-2-µm silica can be operated with regular LC pressures at flow rates comparable with those used in UHPLC.

The feasibility of using sub-2-μm unmodified silica stationary phase under HILIC conditions for drug quantitation has been demonstrated by Hseih et al. for quantitation of everolimus in mouse plasma (Hsieh et al., 2009). Acceptable resolution for the analyte and its internal standard was obtained in less than 1 min on a 1.7-μm BEH silica column with a flow rate of 1.0 mL/min operated on a conventional LC system. The assay performance of the sub-2-μm HILIC method, in terms of sample throughput, resolution, sensitivity, and accuracy was comparable with that of reversed-phase UHPLC, and the analytical results obtained by the two approaches were in good agreement. This study demonstrated that HILIC using a sub-2-μm-particle stationary phase has comparable chromatographic efficiency as reversed phase UHPLC but demands neither expensive ultrahigh-pressure instrumentation nor new laboratory protocols. Therefore, it is feasible to utilize the sub-2-μm HILIC approach as an alternative to reversed-phase UHPLC for ultrafast separation of polar drugs and their metabolites.

Fused-core particles are available in HILIC mode for fast separation of polar compounds. One study has shown that fused-core HILIC generated twofold lower back-pressure, equivalent mass transfer resistance, and 30% less efficiency than sub-2-μm HILIC columns (Chauve et al., 2010). As the back-pressure is always relatively low in HILIC, these features are not very attractive when compared with sub-2-μm HILIC. However, separation efficiency can be improved by coupling various fused-core HILIC columns in series to reach the highest possible efficiencies under back-pressure compatible with conventional instrumentation (McCalley, 2008).

Another strategy for rapid separation of polar compounds is the use of HILIC monolithic support. In RPLC, monolithic columns are attractive because of their high permeability, leading to low back-pressure and fast mass transfer. However, as the back-pressure obtained in the HILIC mode is already very low, and as the performance of monoliths is generally not better than that of columns packed with 3.5-μm particles, there is less interest in using monoliths in HILIC mode (Chauve et al., 2010).

25.5 ORTHOGONAL SEPARATION

Orthogonal separation refers to separation modes based on different mechanisms, with marked changes in relative retention of the analytes so that peaks which are unresolved in one mode will likely be separated in the second dimension. The benefit of using orthogonal separation is the reduction of component overlap, which can potentially alleviate interference originating from complex biological samples, improve the selectivity and sensitivity, as well as increase the number of detected compounds in unknown component analysis.

25.5.1 Orthogonal Sample Preparation and Chromatography

In order for a sample preparation procedure to achieve the best cleanup effect, it ideally should be orthogonal to the chromatography separation. However, some of the conventional combinations of sample preparation procedure and separation mode lack the desired orthogonality. For example, reversed-phase SPE extraction columns poorly retain polar analytes, which are also poorly retained on reversed-phase analytical columns. Therefore, the impact of sample cleanup is minimal because the analytes may coelute with matrix interferences leading to potential ion suppression issues on LC-MS systems. In contrast, a combination of sample extraction techniques with chromatography techniques that are based on orthogonal mechanisms (e.g., reversed-phase SPE with HILIC, or ion exchange SPE with RPLC) can potentially provide better sample cleanup and improved performance of the assay. A recent study exploring different combinations of stationary phases, including SEC-RPLC, strong cation exchange (SCX)-RPLC, RPLC-RPLC, and HILIC-RPLC, has demonstrated that the highest degree of orthogonality was achieved by the HILIC-RPLC combination (Gilar et al., 2005). Therefore, HILIC is a good candidate to be used in combination with reversed-phase-based sample preparation either off-line or online to obtain better assay performance for polar compounds.

There have been many applications of off-line coupling of HILIC with different modes of sample extraction, such as reversed-phase SPE, mixed-mode ion exchange SPE, or LLE, for quantitation of polar drugs and metabolites in complex biological fluids. Besides improved assay performance, an additional advantage of using HILIC is that the effluent from the extraction steps can often be directly injected onto the column because it is a weaker eluant for HILIC. Elimination of the evaporation and reconstitution steps not only improves assay throughput but also avoids potential artifacts and contamination.

HILIC can also be combined with different modes of sample preparation in an online extraction setting. For example, weak cation exchange (WCX) extraction columns have been used in conjunction with HILIC in an online extraction system for quantitation of basic compounds such as metanephrine in plasma (de Jong et al., 2007). WCX cartridges contain a mixed-mode weak carbonyl exchange moiety, which retains strong bases at neutral pH, permitting the cartridge to be washed thoroughly with both water and 100% acetonitrile without loss of the analyte of interest. The sorbent is then neutralized by acid, and the analyte is eluted with the high-organic elution solvent, which is a weak solvent for HILIC. The buffer in the solvent is favorable for maintaining good peak shape and retention in HILIC conditions, and the pH is well tolerated by the silica column. Therefore, WCX extraction columns and HILIC are a good match for online extraction setups, and coupling these two modes of retention maximizes the efficiency of sample cleanup. For online coupling of reversed-phase-based extraction columns with HILIC analytical columns, attention needs to be paid to optimize the elution conditions of the extraction column in order to achieve the best retention of the analytes on the HILIC analytical column. Highly aqueous mobile phases used for conventional online extraction conditions may cause peak distortion and "breakthrough" on HILIC analytical columns because water is a stronger solvent on HILIC. A gradient elution starting from high organic content (e.g., 90% acetonitrile in water) can provide better peak shape and retention (Deng et al., 2005).

25.5.2 2D-LC

2D-LC by coupling orthogonal separation mechanisms is often used for analysis of multicomponent mixtures of extreme complexity, such as metabolomics or proteomics components. 2D-LC could be either off-line or online. In off-line 2D-LC mode, the effluent from the first-dimension column is collected manually or via a fraction collector, concentrated (if necessary) and reinjected on the secondary column. Either a part or the entire components from the first dimension can be analyzed on the second dimension. Online 2D-LC can be divided into heart-cutting or comprehensive LC, both of these two approaches involve connecting the two dimensions via an appropriate interface. In heart-cutting 2D-LC, only the relevant part of the effluent, containing the target compounds, is directed to the second dimension, whereas in comprehensive LC, the entire effluent from the first dimension is subjected to analysis on the second dimension. Online 2D-LC often involves complicated valve switching configurations and complex instrument and software settings.

The orthogonal separation mechanism, together with the water-miscible mobile phases, makes HILIC a favorable candidate for online coupling with RPLC for 2D-LC analysis. Special consideration for mobile phase compatibility is required when interfacing the two modes. The retention on RPLC is provided by a low organic content in the mobile phase, while HILIC requires the opposite. Therefore, the relatively weak eluting mobile phases for RPLC are rather strong for HILIC and vice versa, making concentration of the analytes on the column head in the second dimension difficult. One way to resolve this issue is post–column addition of solvent to modify the polarity of the effluent from the first column (Louw et al., 2008). Another approach is to add a "transitional" SPE column after the first-dimension column to trap and concentrate the analytes before introducing them to the second-dimension column (Mihailova et al., 2008).

It should be noted that 2D-LC is rarely necessary in qualitative or quantitative analysis of drugs and their metabolites in support of ADME studies. However, it might provide a practical solution in (but not limited to) the following situations: (1) when multiple analytes with differing polarity (e.g., analyte and its metabolites) need to be identified and/or quantified in a single run; (2) when coeluting components cannot be separated on single-dimensional chromatography. For example, for separation of mixtures of enantiomers, first dimension may be used to separate diastereoisomers, and correctly timed transfer of the peaks to an enantioselective second dimension will resolve the enantiomers; (3) when extensive separation of the analyte of interest from matrix interference cannot be achieved using conventional sample preparation procedures; and (4) when complementary confirmation needs to be conducted to evaluate the separation effects from single-mode chromatography. For example, a single peak from RPLC can be transferred to HILIC for orthogonal separation to inspect if there are any components that have been coeluting in that single peak under RPLC conditions.

25.6 CONCLUSIONS AND PERSPECTIVES

LC-MS has been, and will most likely continue to be, the most favored choice of analytical methodology for ADME studies. Tremendous progress has been made in recent years to improve the separation efficiency and analysis speed of LC-MS techniques for qualitative and quantitative analysis of drugs and metabolites. Emerging new column technologies and instrumentation will continue to reshape the ways of conducting analysis in support of ADME studies, and thus will impact new drug discovery and development.

Introduction of novel chromatographic techniques such as HILIC has brought better separation, increased sensitivity, and higher throughput for analysis of polar compounds and metabolites. Use of HILIC has been also recently extended to relatively nonpolar compounds, biomarkers, proteins and peptides. With the emergence of large-molecule medicines, there is a growing need for exploring the potential of HILIC or other novel separation techniques as tools for oligonu-cleotide, peptide, and protein analysis. Novel column technologies have brought a new variety of columns for improved separation efficiency. For example, mixed-mode reversed-phase columns that have embedded ion pair groups in the hydrophobic stationary phase can provide the capability for retention of polar, ionic com-pounds by ion exchange interactions and hydrophobic compounds by reversed-phase mechanisms. Mixed-mode HILIC columns featuring packing with an alkyl long chain and a hydrophilic polar terminus have dem-onstrated potential for separating a wide range of both polar and nonpolar compounds, in either HILIC or RPLC mode. Orthogonal separation employing chro-matographic modes based on diverse mechanisms can significantly improve the capacity to analyze extremely complex components and will greatly facilitate meta-bolomics and proteomics analysis for identification of novel biomarkers in support of drug discovery and development.

Utilization of small particles (sub-2 μm) as stationary phase has significantly improved separation speed and efficiency. Monolithic silica columns, fused-core particle columns, and HILIC provide alternatives for ultrahigh-speed LC and do not require ultrahigh-pressure instrumentation. New materials such as nano-particles (<100 nm in diameter) have the potential to further reduce analysis time and improve separation efficiency (Zhang et al., 2006). Multiplexing or parallel LC-MS/MS that allows the introduction of effluent from multiple simultaneously operated LC systems to a mass spectrometer is an alternative approach for high throughput without sacrificing chromatographic integrity (Korfmacher et al., 1999; Hsieh and Korfm-acher, 2006).

One preeminent problem remaining for LC-MS analysis of biological samples is how to eliminate the detrimental effects of matrix interference. New sample preparation products for selectively binding of phos-pholipids such as the HybridSPE™ from Supelco, Captiva™ ND[lipids] from Varian (Palo Alto, CA, USA), or Ludox AS-40 reagent (Sigma, St. Louis, MO, USA) can potentially attenuate matrix effects and improve the performance of assays (Wu et al., 2008b). New SPE materials such as molecular imprinted polymers (MIPs) that are synthesized to contain binding sites for a single target analyte can potentially provide superior selectiv-ity and cleanup (Haginaka, 2009). Immunoaffinity columns contain antibodies immobilized on supporting materials and have shown the capability of extracting proteins as well as small molecules with superior selec-tivity. Ionization suppression in the electrospray source is inversely proportional to the LC flow rate, and has been shown to be practically absent at flow rate below

20 nL/min in a model study (Schmidt et al., 2003). Due to its ability to significantly reduce ion suppression, nanoflow LC (typically <100 nL/min) coupled with ele-trospray ("nanospray") has been successfully used for metabolite identification studies for detection of metab-olites of low abundance (Prakash et al., 2007). Nano-spray has also been demonstrated to give significantly less variation in ionization efficiency than regular-flow LC-MS for a range of compounds and therefore has the potential to be utilized for quantitation of drugs and/or metabolites when authentic standards are not available (Hop et al., 2005). Nanospray-MS is also attractive from environmental perspective due its low consumption of mobile phases.

Miniaturization for lower sample volume and smaller analytical devices is the trend in current LC-MS analy-sis. Microsampling techniques, such as DBS analysis and microdialysis sampling, have aroused great interest in the bioanalysis community due to their small sample volume requirements, convenience for sample collec-tion and storage, and potential cost savings. Microscale sample preparation techniques, such as microextraction by packed sorbent (MEPS), allow SPE to be performed with a very small sample volume (<10 μL) and sig-nificantly reduce the extraction solvent volume (Altun et al., 2004). Chip-based infusion apparatuses have become commercially available to combine nanoscale chromatographic separation and delivery of analytes into the mass spectrometer in a small chip, thereby enabling sample analysis from submicroliter sample volumes (Yin and Killeen, 2007).

REFERENCES

Ahn J, Bones J, Yu YQ, Rudd PM, Gilar M (2010) Separation of 2-aminobenzamide labeled glycans using hydrophilic interaction chromatography columns packed with 1.7 microm sorbent. *J Chromatogr B Analyt Technol Biomed Life Sci* 878:403–408.

Alfazil AA, Anderson RA (2008) Stability of benzodiazepines and cocaine in blood spots stored on filter paper. *J Anal Toxicol* 32:511–515.

Alpert AJ (1990) Hydrophilic-interaction chromatography for the separation of peptides, nucleic acids and other polar compounds. *J Chromatogr* 499:177–196.

Altun Z, Abdel-Rehim M, Blomberg LG (2004) New trends in sample preparation: On-line microextraction in packed syringe (MEPS) for LC and GC applications. Part III: Determination and validation of local anaesthetics in human plasma samples using a cation-exchange sorbent, and MEPS-LC-MS-MS. *J Chromatogr B Analyt Technol Biomed Life Sci* 813:129–135.

Apffel A, Fischer S, Goldberg G, Goodley PC, Kuhlmann FE (1995) Enhanced sensitivity for peptide mapping with electrospray liquid chromatography-mass spectrometry in the presence of signal suppression due to trifluoroacetic acid-containing mobile phases. *J Chromatogr A* 712: 177–190.

Atia NN, York P, Clark BJ (2009) Comparison between monolithic and particle-packed platinum C18 columns in HPLC determination of acidic and basic test mixtures. *J Sep Sci* 32:2732–2736.

Barbarin N, Mawhinney DB, Black R, Henion J (2003) High-throughput selected reaction monitoring liquid chromatography-mass spectrometry determination of methylphenidate and its major metabolite, ritalinic acid, in rat plasma employing monolithic columns. *J Chromatogr B Analyt Technol Biomed Life Sci* 783:73–83.

Chauve B, Guillarme D, Cléon P, Veuthey JL (2010) Evaluation of various HILIC materials for the fast separation of polar compounds. *J Sep Sci* 33:752–764.

Chen J, Korfmacher WA, Hsieh Y (2005) Chiral liquid chromatography-tandem mass spectrometric methods for stereoisomeric pharmaceutical determinations. *J Chromatogr B Analyt Technol Biomed Life Sci* 820:1–8.

Cheng YF, Neue UD, Woods LL (1999) Novel high-performance liquid chromatographic and solid-phase extraction methods for quantitating methadone and its metabolite in spiked human urine. *J Chromatogr B Biomed Sci Appl* 729:19–31.

Chu I, Nomeir AA (2006) Utility of mass spectrometry for in-vitro ADME assays. *Curr Drug Metab* 7:467–477.

Cunliffe JM, Maloney TD (2007) Fused-core particle technology as an alternative to sub-2-microm particles to achieve high separation efficiency with low backpressure. *J Sep Sci* 30:3104–3109.

de Jong WH, Graham KS, van der Molen JC, Links TP, Morris MR, Ross HA, de Vries EG, Kema IP (2007) Plasma free metanephrine measurement using automated online solid-phase extraction HPLC tandem mass spectrometry. *Clin Chem* 53:1684–1693.

Deng Y, Zhang H, Wu JT, Olah TV (2005) Tandem mass spectrometry with online high-flow reversed-phase extraction and normal-phase chromatography on silica columns with aqueous-organic mobile phase for quantitation of polar compounds in biological fluids. *Rapid Commun Mass Spectrom* 19:2929–2934.

FDA (1992) Development of new stereoisomeric drugs.

FDA (2008) Guidance for industry-safety testing of drug metabolites.

Francois I, Sandra K, Sandra P (2009) Comprehensive liquid chromatography: Fundamental aspects and practical considerations–a review. *Anal Chim Acta* 641:14–31.

Gao S, Bhoopathy S, Zhang ZP, Wright DS, Jenkins R, Karnes HT (2006) Evaluation of volatile ion-pair reagents for the liquid chromatography-mass spectrometry analysis of polar compounds and its application to the determination of methadone in human plasma. *J Pharm Biomed Anal* 40:679–688.

Garcia Boy R, Henseler J, Mattern R, Skopp G (2008) Determination of morphine and 6-acetylmorphine in blood with use of dried blood spots. *Ther Drug Monit* 30:733–739.

Gilar M, Olivova P, Daly AE, Gebler JC (2005) Orthogonality of separation in two-dimensional liquid chromatography. *Anal Chem* 77:6426–6434.

Grumbach ES, Wagrowski-Diehl DM, Mazzeo JR, Alden B, Iraneta PC (2004) Hydrophilic interaction chromatography using silica columns for the retention polar analytes and enhanced ESI-MS sensitivity. *LCGC North Am* 22:1010–1023.

Grumbach ES, Diehl DM, Neue UD (2008) The application of novel 1.7 microm ethylene bridged hybrid particles for hydrophilic interaction chromatography. *J Sep Sci* 31:1511–1518.

Haginaka J (2009) Molecularly imprinted polymers as affinity-based separation media for sample preparation. *J Sep Sci* 32:1548–1565.

Hemstrom P, Irgum K (2006) Hydrophilic interaction chromatography. *J Sep Sci* 29:1784–1821.

Hop CE, Chen Y, Yu LJ (2005) Uniformity of ionization response of structurally diverse analytes using a chip-based nanoelectrospray ionization source. *Rapid Commun Mass Spectrom* 19:3139–3142.

Hsieh Y, Korfmacher WA (2006) Increasing speed and throughput when using HPLC-MS/MS systems for drug metabolism and pharmacokinetic screening. *Curr Drug Metab* 7:479–489.

Hsieh Y, Duncan CJ, Brisson JM (2007a) Fused-core silica column high-performance liquid chromatography/tandem mass spectrometric determination of rimonabant in mouse plasma. *Anal Chem* 79:5668–5673.

Hsieh Y, Duncan CJ, Brisson JM (2007b) Porous graphitic carbon chromatography/tandem mass spectrometric determination of cytarabine in mouse plasma. *Rapid Commun Mass Spectrom* 21:629–634.

Hsieh Y, Galviz G, Long BJ (2009) Ultra-performance hydrophilic interaction liquid chromatography/tandem mass spectrometry for the determination of everolimus in mouse plasma. *Rapid Commun Mass Spectrom* 23:1461–1466.

Hughes NC, Wong EY, Fan J, Bajaj N (2007) Determination of carryover and contamination for mass spectrometry-based chromatographic assays. *AAPS J* 9:E353–E360.

Jemal M, Xia YQ (1999) The need for adequate chromatographic separation in the quantitative determination of drugs in biological samples by high performance liquid chromatography with tandem mass spectrometry. *Rapid Commun Mass Spectrom* 13:97–106.

Jian W, Edom RW, Xu Y, Weng N (2010) Recent advances in application of hydrophilic interaction chromatography for quantitative bioanalysis. *J Sep Sci* 33:681–697.

Knox JH, Kaur B, Millward GR (1986) Structure and performance of porous graphitic carbon in liquid chromatography. *J Chromatogr A* 352:3–25.

Korfmacher WA, Veals J, Dunn-Meynell K, Zhang X, Tucker G, Cox KA, Lin CC (1999) Demonstration of the capabilities of a parallel high performance liquid chromatography tandem mass spectrometry system for use in the analysis of drug discovery plasma samples. *Rapid Commun Mass Spectrom* 13:1991–1998.

Leclercq L, Cuyckens F, Mannens GS, de Vries R, Timmerman P, Evans DC (2009) Which human metabolites have we MIST? Retrospective analysis, practical aspects, and perspectives for metabolite identification and quantification in pharmaceutical development. *Chem Res Toxicol* 22:280–293.

Li AC, Junga H, Shou WZ, Bryant MS, Jiang XY, Naidong W (2004) Direct injection of solid-phase extraction eluents onto silica columns for the analysis of polar compounds isoniazid and cetirizine in plasma using hydrophilic interaction chromatography with tandem mass spectrometry. *Rapid Commun Mass Spectrom* 18:2343–2350.

Li W, Tse FL (2010) Dried blood spot sampling in combination with LC-MS/MS for quantitative analysis of small molecules. *Biomed Chromatogr* 24:49–65.

Lin ZJ, Li W, Dai G (2007) Application of LC-MS for quantitative analysis and metabolite identification of therapeutic oligonucleotides. *J Pharm Biomed Anal* 44:330–341.

Louw S, Pereira AS, Lynen F, Hanna-Brown M, Sandra P (2008) Serial coupling of reversed-phase and hydrophilic interaction liquid chromatography to broaden the elution window for the analysis of pharmaceutical compounds. *J Chromatogr A* 1208:90–94.

Mannens GS, Hendrickx J, Janssen CG, Chien S, Van Hoof B, Verhaeghe T, Kao M, Kelley MF, Goris I, Bockx M, Verreet B, Bialer M, Meuldermans W (2007) The absorption, metabolism, and excretion of the novel neuromodulator RWJ-333369 (1,2-ethanediol, [1-2-chlorophenyl]-, 2-carbamate, [S]-) in humans. *Drug Metab Dispos* 35:554–565.

Matuszewski BK, Chavez-Eng CM, Constanzer ML (1998) Development of high-performance liquid chromatography-tandem mass spectrometric methods for the determination of a new oxytocin receptor antagonist (L-368,899) extracted from human plasma and urine: A case of lack of specificity due to the presence of metabolites. *J Chromatogr B* 716:195–208.

Mauriala T, Chauret N, Oballa R, Nicoll-Griffith DA, Bateman KP (2005) A strategy for identification of drug metabolites from dried blood spots using triple-quadrupole/linear ion trap hybrid mass spectrometry. *Rapid Commun Mass Spectrom* 19:1984–1992.

Mazzeo JR, Neue UD, Kele M, Plumb RS (2005) Advancing LC performance with smaller particles and higher pressure. *Anal Chem* 77:460 A–467 A.

McCalley DV (2008) Evaluation of the properties of a superficially porous silica stationary phase in hydrophilic interaction chromatography. *J Chromatogr A* 1193:85–91.

Mihailova A, Malerod H, Wilson SR, Karaszewski B, Hauser R, Lundanes E, Greibrokk T (2008) Improving the resolution of neuropeptides in rat brain with on-line HILIC-RP compared to on-line SCX-RP. *J Sep Sci* 31:459–467.

Morin P (2009) Separation of chiral pharmaceutical drugs by chromatographic and electrophoretic techniques. *Ann Pharm Fr* 67:241–250.

Mullett WM (2007) Determination of drugs in biological fluids by direct injection of samples for liquid-chromatographic analysis. *J Biochem Biophys Methods* 70:263–273.

Naidong W (2003) Bioanalytical liquid chromatography tandem mass spectrometry methods on underivatized silica column with aqueous/organic mobile phase. *J Chromatogr B Analyt Technol Biomed Life Sci* 796:209–224.

Naidong W, Chen YL, Shou W, Jiang X (2001) Importance of injection solution composition for LC-MS-MS methods. *J Pharm Biomed Anal* 26:753–767.

Naidong W, Shou WZ, Addison T, Maleki S, Jiang X (2002) Liquid chromatography/tandem mass spectrometric bioanalysis using normal-phase columns with aqueous/organic mobile phases—A novel approach of eliminating evaporation and reconstitution steps in 96-well SPE. *Rapid Commun Mass Spectrom* 16:1965–1975.

Naidong W, Zhou W, Song Q, Zhou S (2004) Direct injection of 96-well organic extracts onto a hydrophilic interaction chromatography/tandem mass spectrometry system using a silica stationary phase and an aqueous/organic mobile phase. *Rapid Commun Mass Spectrom* 18:2963–2968.

Okamoto Y, Ikai T (2008) Chiral HPLC for efficient resolution of enantiomers. *Chem Soc Rev* 37:2593–2608.

Prakash C, Shaffer CL, Nedderman A (2007) Analytical strategies for identifying drug metabolites. *Mass Spectrom Rev* 26:340–369.

Schmidt A, Karas M, Dulcks T (2003) Effect of different solution flow rates on analyte ion signals in nano-ESI MS, or: When does ESI turn into nano-ESI? *J Am Soc Mass Spectrom* 14:492–500.

Shou WZ, Naidong W (2003) Post-column infusion study of the "dosing vehicle effect" in the liquid chromatography/tandem mass spectrometric analysis of discovery pharmacokinetic samples. *Rapid Commun Mass Spectrom* 17:589–597.

Shou WZ, Naidong W (2005) Simple means to alleviate sensitivity loss by trifluoroacetic acid (TFA) mobile phases in the hydrophilic interaction chromatography-electrospray tandem mass spectrometric (HILIC-ESI/MS/MS) bioanalysis of basic compounds. *J Chromatogr B Analyt Technol Biomed Life Sci* 825:186–192.

Shou WZ, Chen YL, Eerkes A, Tang YQ, Magis L, Jiang X, Naidong W (2002) Ultrafast liquid chromatography/tandem mass spectrometry bioanalysis of polar analytes using packed silica columns. *Rapid Commun Mass Spectrom* 16:1613–1621.

Song Q, Naidong W (2006) Analysis of omeprazole and 5-OH omeprazole in human plasma using hydrophilic interaction chromatography with tandem mass spectrometry (HILIC-MS/MS)—Eliminating evaporation and reconstitution steps in 96-well liquid/liquid extraction. *J Chromatogr B Analyt Technol Biomed Life Sci* 830:135–142.

Song W, Pabbisetty D, Groeber EA, Steenwyk RC, Fast DM (2009) Comparison of fused-core and conventional particle size columns by LC-MS/MS and UV: Application to pharmacokinetic study. *J Pharm Biomed Anal* 50:491–500.

Spooner N, Lad R, Barfield M (2009) Dried blood spots as a sample collection technique for the determination of pharmacokinetics in clinical studies: Considerations for the validation of a quantitative bioanalytical method. *Anal Chem* 81:1557–1563.

Thomas A, Déglon J, Steimer T, Mangin P, Daali Y, Staub C (2010) On-line desorption of dried blood spots coupled to hydrophilic interaction/reversed-phase LC/MS/MS system for the simultaneous analysis of drugs and their polar metabolites. *J Sep Sci* 33:873–879.

Tong XS, Wang J, Zeng S, Pivnichny JV, Griffin PR, Shen X, Donnelly M, Cakerich K, Nunes C, Fenyk-Melody J (2002) Effect of signal interference from dosing excipients on pharmacokinetic screening of drug candidates by liquid chromatography/mass spectrometry. *Anal Chem* 74:6305–6313.

Unger KK, Skudas R, Schulte MM (2008) Particle packed columns and monolithic columns in high-performance liquid chromatography-comparison and critical appraisal. *J Chromatogr A* 1184:393–415.

Wu R, Hu L, Wang F, Ye M, Zou H (2008a) Recent development of monolithic stationary phases with emphasis on microscale chromatographic separation. *J Chromatogr A* 1184:369–392.

Wu ST, Schoener D, Jemal M (2008b) Plasma phospholipids implicated in the matrix effect observed in liquid chromatography/tandem mass spectrometry bioanalysis: Evaluation of the use of colloidal silica in combination with divalent or trivalent cations for the selective removal of phospholipids from plasma. *Rapid Commun Mass Spectrom* 22:2873–2881.

Xia YQ, Jemal M, Zheng N, Shen X (2006) Utility of porous graphitic carbon stationary phase in quantitative liquid chromatography/tandem mass spectrometry bioanalysis: Quantitation of diastereomers in plasma. *Rapid Commun Mass Spectrom* 20:1831–1837.

Xu RN, Fan L, Kim GE, El-Shourbagy TA (2006) A monolithic-phase based on-line extraction approach for determination of pharmaceutical components in human plasma by HPLC-MS/MS and a comparison with liquid-liquid extraction. *J Pharm Biomed Anal* 40:728–736.

Xu RN, Reiser R, El-Shourbagy TA (2009) Bioanalytical hydrophilic interaction chromatography: Recent challenges, solutions and applications. *Bioanalysis* 1:239–253.

Xue YJ, Liu J, Unger S (2006) A 96-well single-pot liquid-liquid extraction, hydrophilic interaction liquid chromatography-mass spectrometry method for the determination of muraglitazar in human plasma. *J Pharm Biomed Anal* 41:979–988.

Yin H, Killeen K (2007) The fundamental aspects and applications of Agilent HPLC-Chip. *J Sep Sci* 30:1427–1434.

Zang X, Luo R, Song N, Chen TK, Bozigian H (2005) A novel on-line solid-phase extraction approach integrated with a monolithic column and tandem mass spectrometry for direct plasma analysis of multiple drugs and metabolites. *Rapid Commun Mass Spectrom* 19:3259–3268.

Zhang Z, Wang Z, Liao Y, Liu H (2006) Applications of nanomaterials in liquid chromatography: Opportunities for separation with high efficiency and selectivity. *J Sep Sci* 29:1872–1878.

26

MASS SPECTROMETRIC IMAGING FOR DRUG DISTRIBUTION IN TISSUES

Daniel P. Magparangalan, Timothy J. Garrett, Dieter M. Drexler, and Richard A. Yost

26.1 INTRODUCTION

The drug discovery and development process explores novel drug candidates to address unmet medical needs and to improve current treatments. However, the discovery of a new lead is only the first step. After initial *in silico* and *in vitro* predictive absorption, distribution, metabolism, and excretion (ADME) screens to assess potential compound liabilities, drug efficacy and potency must be established at the site of the pharmacological target. The developability and safety of a drug candidate is assessed with detailed *in vivo* absorption, distribution, metabolism, excretion, and toxicology (ADMET) studies spanning from preclinical trials to postlaunch monitoring. One key aspect of ADMET studies involves determining the disposition of a drug within the body's tissues.

26.1.1 Imaging Techniques for ADMET Studies

Imaging techniques are employed for determining the path a drug takes within the body and its distribution within tissue. Autoradiography, being the most prevalent, traces a radiolabeled drug as it is absorbed into the blood system, distributed to the target organ, and excreted from the body (Solon et al., 2010); however, drawbacks of autoradiography exist. As this technique relies upon indirect analysis, that is, detection of the radiolabel or chemical tag rather than the drug, potential metabolism of the dosed active pharmaceutical ingredient (API) must be considered. Metabolic modi-

fication of the API could cleave the radiolabel from the drug, thus losing any means of detecting the metabolite; but even if the radiolabel is retained, the detection method cannot distinguish between the API and metabolite. Other labeling techniques, such as fluorescence microscopy, also have this problem, as well as an additional problem: modification to the drug with a chemical tag could change the uptake of the drug (or its metabolism) within the body. For example, a large fluorescence tag attached to a drug could hinder passage of an antipsychotic drug as it moves through the blood–brain barrier. A further disadvantage of these approaches is the requirement that the drug be synthesized with the radiolabel or fluorescent label (at potentially great cost). Other imaging techniques have been used to follow drug distribution (e.g., positron emission tomography [PET] [Schiffer et al., 2007] and magnetic resonance imaging [MRI] [Sosnovik and Weissleder, 2007]); however, these techniques suffer from low sensitivity, low specificity, limited molecular information, and/or insufficient spatial resolution (Stoeckli and Farmer, 2004).

26.1.2 Mass Spectrometric Imaging (MSI) Background

A complementary technique to established imaging techniques is MSI (Caprioli et al., 1997). MSI is an imaging technique that employs a microscopically focused ionization source to desorb and ionize molecules at a spatially defined point on the tissue; the ions

ADME-Enabling Technologies in Drug Design and Development, First Edition. Edited by Donglu Zhang and Sekhar Surapaneni.
© 2012 John Wiley & Sons, Inc. Published 2012 by John Wiley & Sons, Inc.

are then transferred to the mass analyzer and a mass spectrum is generated for that point (similar to a pixel). The tissue sample is rastered underneath the microprobe and the mass spectrum is collected at each defined interval. Images are generated by selecting an ion of interest and creating a position-specific, two-dimensional ion map, typically with a color gradient used to indicate the third dimension (ion signal intensity).

The most significant advantage of MSI over other imaging techniques is molecular specificity without the need for radiolabels or other tags, especially when performed in conjunction with tandem mass spectrometry or high-resolution accurate mass analysis. Incorrect identification of a cleaved florescence tag or radiolabel is greatly reduced, and metabolites are readily differentiated from the parent drug. In addition to direct analysis of the drug molecule, observation of other physiological effects due to drug administration is possible. For example, Atkinson et al. (2007) imaged an antitumor prodrug and its active metabolite in a dosed solid tumor section and verified the containment of the cytotoxic metabolite by comparing the localization of endogenous adenosine triphosphate (ATP) within the tissue.

The ability to probe discrete sections of tissue for drugs, metabolites, and endogenous compounds is a powerful investigative tool for ADMET studies. Compound-specific images generated from MSI offer insights into drug distribution and drug interaction within the body with little sample preparation and with high molecular specificity. However, MSI does suffer from a few drawbacks. First, quantitative MSI is limited; at this point, MSI is primarily a qualitative technique. Second, MSI is limited to *in situ* studies, thus complicating real-time studies of a drug's fate in the body that could be avoided if *in vivo* studies were performed. Despite these two limitations, MSI offers a complementary analytical technique to established methods to ascertain drug efficacy and safety.

A typical workflow for MSI employing ionization by matrix-assisted laser desorption/ionization (MALDI) is illustrated in Figure 26.1. Tissue samples are sectioned at a specific thickness, mounted onto a sample stage, after which a MALDI matrix is applied. In the source region, a microprobe interrogates a series of spatially defined spots on the tissue, where the spot is most often defined by the diameter of the focused laser beam

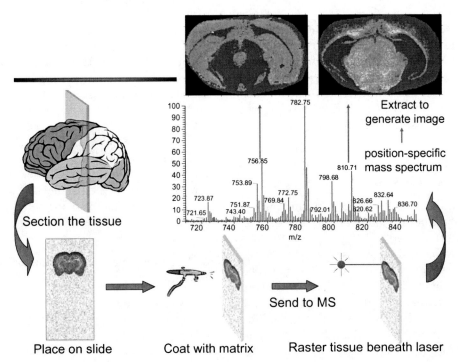

FIGURE 26.1. Process flow of an MSI experiment by MALDI-MS. Fresh frozen tissue samples are sectioned on a cryostat and then mounted onto a microscope slide (or MALDI target plate). The tissue is coated with an appropriate MALDI matrix using an airbrush or other device that promotes an even crystal layer, with crystals ideally smaller than the spot size of the laser. After insertion into the instrument, the sample is rastered beneath the laser spot as position specific mass spectra are collected. Images are generated by extracting a specific m/z and plotting the ion's intensity versus its position.

(typically 50–200 μm). The mass spectrum is then collected for each spot and saved for future data processing, resulting in mass-to-charge (m/z) specific ion maps illustrating the distribution of the analyte within the tissue (Garrett et al., 2007).

26.2 MSI INSTRUMENTATION

Two primary components are necessary for MSI: (1) a microprobe (a microscopically focused ionization source to generate ions at a spatially defined point in the tissue section); and (2) a mass analyzer. In the following sections, various microprobe ion sources and mass analyzers typically associated with MSI are presented. In addition, sample preparation and postacquisition data processing techniques are discussed.

26.2.1 Microprobe Ionization Sources

Three aspects that must be considered when selecting an ionization source are ionization efficiency, spatial resolution, and sampling rate. In general, it is important to optimize ionization efficiency to generate the greatest number of ions from a given spot without causing in-source fragmentation. The size of the microprobe spot defines the spatial resolution, sensitivity, and overall image quality of the MS image. Finally, the sampling rate will affect the overall analysis time for each tissue section.

Optimization of each parameter is important; however, the interconnectivity of the parameters needs to be considered. For example, decreasing the spot size to increase spatial resolution will result in less analyte ablated, thus increasing the detection limit for the instrument. Furthermore, a decrease in step size results in an increase in total analysis time (e.g., given identical scan areas, a 50% decrease of step size would result in a four times increase in analysis time). The three most common sources for MSI are MALDI, secondary ion mass spectrometry (SIMS), and desorption electrospray ionization (DESI). Although SIMS offers spatial resolution superior to the other two sources, most of the SIMS-MSI work focuses on the analysis of endogenous compounds within tissue. For this reason, SIMS will not be discussed in this chapter (although more information can be found in a review paper by Solon et al., 2010). The following sections will focus on the two ionization sources (MALDI and DESI) most commonly found for drug analysis by MSI as well as emerging ionization sources.

26.2.1.1 MALDI The most commonly used ionization source for MSI (especially for small-molecule

analysis) is MALDI. MALDI is a soft ionization technique that employs a laser to desorb predominantly singly charged ions from a solid. The key to MALDI is the application of a matrix to the sample. Prior to the introduction of MALDI (Karas and Hillenkamp, 1988), laser desorption of intact analytes from tissue was difficult due to the hard ionization process that caused a high degree of fragmentation because of the energies needed for desorption of the molecules into the gas phase. Karas and Hillenkamp found that addition of a large excess of an organic matrix that absorbs at the laser wavelength greatly increased ionization efficiency and the likelihood of forming intact molecular ion species. With this discovery, the analysis of high-molecular-weight biomolecules (e.g., peptides and proteins) became the most common application.

The MALDI matrix typically employed is a small organic acid that cocrystallizes with the analyte and absorbs light at the wavelength of the laser. As the majority of the energy is absorbed by the organic matrix, sample fragmentation is minimized. MALDI matrices typically used are 2,5-dihydroxybenzoic acid (DHB), α-cyano-4-hydroxycinnamic acid (CHCA), and sinapinic acid (SA). The selection of a specific MALDI matrix is based on the tissue type and analyte of interest. Further discussion of MALDI matrix selection can be found in the tissue preparation section (Section 26.3.3.2).

During the laser ablation of the matrix surface, charged analytes (i.e., ions) are generated and then transferred to the mass spectrometer. The singly charged ions produced by MALDI are preferred as the mass spectrum is cleaner and easier to interpret (as opposed to an electrospray ionization (ESI) source that can generate a variety of multiply charged species). This is important for tissue analysis as there is an abundance of endogenous molecules that can interfere with trace drug analysis.

MALDI experiments typically use a pulsed UV laser, for example, N_2 laser with a 337-nm wavelength, as the ionization source. The laser energy and the number of laser shots required for a MALDI experiment are optimized with respect to MALDI matrix selection and tissue type for each experiment. For example, DHB typically requires more laser energy when compared with a "hotter" matrix, such as CHCA or SA.

Although the repetition rate of the laser does not play a direct role in the image quality, higher repetition rates increase sample throughput, with optimal rates employed near the scan rate of the mass analyzer. This is important when performing experiments on larger tissue sections, such as MSI over whole-body tissue sections. For example, depending on the rastering pattern and repetition rate of the laser, whole-body tissue experiments can take upward of 5 days (~115,000 pixels over the entire tissue section at a lateral resolution of

250 μm with an acquisition rate of ~4 s/pixel; Khatib-Shahidi et al., 2006).

26.2.1.2 *DESI*

An alternative method to MALDI is DESI (Takats et al., 2004; Cooks et al., 2006), an atmospheric pressure surface sampling technique that, unlike MALDI, does not require a matrix. Charged droplets are emitted from the electrospray source and strike the tissue surface at an angle, thus desorbing compounds as ions that are then transferred into the mass spectrometer. The tissue surface is moved beneath the DESI emitter in a horizontal manner (along the *x*-axis); a lane scan consists of a single scan from one edge to the other of a tissue section. At the completion of a lane scan, the tissue can either be returned to its original *x* position and then moved a fixed distance along the *y*-axis (lane spacing) before performing another lane scan (unidirectional scanning) or the tissue can be moved a fixed distance along the *y*-axis and then scanned in a direction opposite to the previous lane scan (raster scanning) (Kertesz and Van Berkel, 2008).

The primary advantages of the DESI technique are that there is little sample preparation required (no MALDI matrix application) and experiments are performed at atmospheric pressure. A further advantage is that specimen size is not limited, so subdivision of whole-body tissue sections is unnecessary for a DESI-MS experiment. In contrast, a MALDI-MSI experiment may require the division of a whole-body section into pieces that will fit onto the MALDI target plate. This may require computer-stitching programs that can degrade image quality.

A disadvantage of DESI is that many parameters must be optimized prior to a DESI experiment (multiple spray angles and sample distance from both the ESI emitter and to the mass spectrometer; Takats et al., 2005). Also, the sampling rate must be optimized to balance length of analysis time versus signal response. Finally, the image quality is much lower than MALDI-MSI experiments due to the poor spatial resolution (250 μm; Ifa et al., 2007). In addition, inaccurate localization of desorbed ions is a possibility due to the DESI process. For example, during a lane scan performed from left to right on a surface, it was noted that analytes were washed toward the right (Kertesz and Van Berkel, 2008). The washing affect is due to the impact of the DESI plume onto the surface as well as analyte–surface interaction, which can result in an MSI representation that is shifted in the direction of the DESI spray (Pasilis et al., 2007). Consequently, a raster pattern can cause a zigzag effect on the MSI image (Kertesz and Van Berkel, 2008). Although unidirectional lane scans could eliminate the zigzag effect, analyte delocalization in the direction of the DESI spray is still an issue. Despite these challenges, DESI has potential for increased use within the MSI field.

26.2.1.3 *Emerging Ionization Sources*

An advantage of atmospheric pressure ionization sources for surface sample analysis is the ability to sample a surface without moving it into the vacuum region, which is usually small to limit the volume to be evacuated for a standard MALDI target plate, typically ~8 × ~13 cm. Additionally, atmospheric pressure analysis permits sampling tissue in its native condition. Another alternative atmospheric pressure source amendable to MSI experiments is a liquid microjunction surface-sampling probe (LMJ-SSP; Van Berkel et al., 2008; Kertesz and Van Berkel, 2010). This probe consists of an inner and outer sampling tube that forms a microjunction at the surface of the tissue section as solvent is delivered to the surface and then drawn away toward the mass spectrometer. Thus, the effective resolution of this probe is the diameter of the sampling tube (~500 μm). This technique has been shown to sample various organs in whole-body tissue sections; however, no MS images have been published using this technique. A disadvantage of this technique is that the sampling solvent will bias against analytes not compatible with the solvent. As a consequence, incorrect assumptions may be drawn if ion maps of two analytes are compared.

Another alternative microprobe source is nano-structure-initiator mass spectrometry (NIMS; Northen et al., 2007). Although it is not an atmospheric pressure technique, this technique has a similar advantage as SIMS in that no matrix application step is necessary. But unlike SIMS, NIMS is a soft ionization technique; thus, analytes are less likely to undergo fragmentation during the ionization process. And with a similar lateral spatial resolution (150 μm), NIMS has the potential to generate image maps comparable with MALDI-MSI. This technique, however, is still relatively new; thus, only one application of NIMS to drugs and metabolites has appeared (Patti et al., 2010).

26.2.2   Mass Analyzers

Several mass analyzers are in use for MSI experiments. The most common mass analyzer for MSI found in the literature is the time-of-flight (ToF) mass analyzer (Reyzer et al., 2003). However, other mass analyzers offer their own advantages that should not be discounted. For experiments targeting lower–molecular-weight species (i.e., most ADMET studies), tandem MS and/or high-resolution mass spectrometry (MS) is almost a necessity. This is due to lack of cleanup separation steps prior to mass analysis, such as extraction and liquid chromatography (LC) prior to MS, to remove interfering compounds. Therefore, linear ion trap (LIT; Garrett et al., 2007), Fourier transform-ion cyclotron resonance (FT-ICR; Cornett et al., 2008), Orbitrap (Landgraf et al., 2009), QqToF (Reyzer et al., 2003), and ToF/ToF (Stoeckli et al., 2007) instruments are effective when performing MSI experiments for ADMET studies of tissue.

26.2.2.1 ToF Mass Analyzers
ToF mass analyzers offer reasonably high mass accuracy (below 20 ppm), mass resolution (>10,000), and a large m/z window (up to m/z 10,000 in reflectron mode and up to m/z 100,000 in linear mode). Fast "scan times" of the ToF allow for increased sample throughput when paired with a high-repetition laser. In addition, the pulsed nature of the MALDI laser is well suited for use with a ToF, as a complete mass spectrum can be recorded for each laser pulse. However, these advantages are balanced by an increased number of laser shots necessary to obtain an adequate signal-to-noise ratio due to the spot-to-spot variability noted in MALDI experiments (typically 80 to over 700 laser shots per spot in MS mode, to 400 to 8000 laser shots in MS/MS mode; Reyzer et al., 2003; Bouslimani et al., 2010).

A second drawback of the ToF mass analyzer is the limited structural elucidation capabilities and limited ability to distinguish between analyte and interfering species at the same nominal m/z (isobaric species). Most MSI experiments employ a single ToF analyzer, and thus provide only molecular weight information (no structural information). At most, a ToF instrument can provide two stages of MS by coupling a second ToF analyzer to form a ToF/ToF (Stoeckli et al., 2007) or adding a quadrupole mass filter to form a QqToF (Reyzer et al., 2003; Khatib-Shahidi et al., 2006; Atkinson et al., 2007; Chen et al., 2008; Trim et al., 2008; Li et al., 2009). QqToF and ToF/ToF instruments offer better analyte specificity than stand-alone ToFs because they provide collision-induced dissociation (CID) for monitoring product ions in MS/MS. However in some cases, MS^3 (or even further stages) may be required to elucidate a compound's structure or to distinguish between isobaric species (Garrett et al., 2007). In this case, an LIT is necessary.

26.2.2.2 LIT Mass Analyzers
Although it lacks the wide mass range of a ToF mass analyzer, an ion trap (either a three-dimensional [3D] quadrupole ion trap or an LIT) offers advanced structural elucidation capability and analyte specificity by providing two or more stages of tandem mass spectrometry (MS/MS and MS^n) capabilities. Some of the earliest MALDI-MSI experiments were performed on a 3D quadrupole ion trap, studying the levels of paclitaxil spiked into rat liver

sections (Troendle et al., 1999). Selecting an ion of interest and inducing fragmentation enables the identification of an analyte or differentiation from a potentially interfering peak (Garrett et al., 2007). Positive identification is especially important in MALDI analysis as there are typically many interfering MALDI matrix-related peaks in the mass range of a potential drug compound. This overlap makes distinguishing analyte from matrix difficult or impossible, particularly at trace levels common in ADMET studies. An additional advantage is that intermediate pressure MALDI-LIT (~70 mTorr) requires fewer laser shots per pixel (~10 laser shots or less) (Garrett et al., 2007). This both decreases analysis time and reduces the amount of matrix ablated from the tissue surface, which in turn increases the number of times that a tissue sample may be analyzed.

Ion traps are not without drawbacks. Most commercial systems have an upper m/z limit of 4000 (although some instruments have been modified to scan up to m/z 5500; Magparangalan et al., 2010). This upper m/z cutoff could preclude detection of peptide/protein-based therapeutic drugs or detection of proteins affected by the administration of a drug. In this case, on-tissue enzymatic digestion is an option (Groseclose et al., 2007). However, most drug molecules have molecular weights of less than 1000 Da, and thus are well within the m/z range of the LIT instrument.

In addition to the m/z limitation, space charging of ions within the trap can cause mass shifts in the mass spectrum. Although using an LIT as opposed to a 3D trap largely alleviates the space-charging issues (Schwartz et al., 2002), highly abundant background compounds (such as those found during tissue imaging experiments) may impact analyte identification. Despite these challenges, ion traps are a remarkably useful tool for mass spectrometric images because of the ease with which they perform MS/MS and MS^n.

26.2.2.3 Triple Quadrupole (QqQ) and Other Hybrid Quadrupole Mass Analyzers
Although QqQ mass analyzers are widely utilized for pharmaceutical drug analysis, QqQ instruments are not typically coupled to a MALDI source. An early laser microprobe system employed laser desorption on a QqQ system (Perchalski, 1985). The strength of a QqQ instrument is the use of MS/MS to perform selected reaction monitoring (SRM) and/or multiple reaction monitoring (MRM) experiments. In particular, MRM scans on a QqQ instrument allow for high selectivity for detection of a drug compound with 100% duty cycle. Coupling a high-repetition rate laser (1 kHz) to a QqQ instrument increases sample throughput (Gobey et al., 2005; Sleno and Volmer, 2005).

Whereas there are few MALDI-QqQ instruments, hybrid quadrupole instruments have been employed for MSI experiments, for example, a QqLIT (Kertesz et al., 2008; Hopfgartner et al., 2009). Here, an LIT replaces Q_3 to serve as either a mass filter or an ion trap. This is key because tandem mass spectrometry is almost a necessity for drug analysis due to the abundance of interfering ions with low molecular weights in tissue. QqLIT instruments offer precursor ion scans that can be helpful for identifying the parent drug or any structural analogs.

26.2.2.4 High-Resolution Mass Analyzers

Another way to positively identify an analyte is via accurate mass analysis with a high-resolution mass analyzer (mass resolution >100,000). Two options are to use an Orbitrap (Landgraf et al., 2009; Strupat et al., 2009) or an FT-ICR (Cornett et al., 2008) instrument. Both mass analyzers offer superior mass accuracy (<2 ppm) and mass resolution, when compared with an ion trap or ToF. The higher mass accuracy and mass resolution allows the user to distinguish between analyte and background peaks at the same nominal mass (Watson and Sparkman, 2007). This is of particular importance with ADMET studies since drug compounds often have molecular ions in the same m/z range as MALDI matrix ions, as well as endogenous compounds (e.g., lipids).

One drawback to using an instrument with high mass accuracy is the low sample throughput for the instrument. Sample throughput for MSI experiments can be defined as the number of pixels (i.e., spatially defined data points) per unit time. ToF instruments typically provide spatially defined mass spectra at a rate of 30–50 pixels/min (Cornett et al., 2008), with ion traps having a similar throughput (60–70 pixels/min) (Garrett et al., 2007). In contrast, 4 pixels/min is the norm for an FT-ICR experiment. Thus, for a profiling experiment, it could take 4–8 h, or even more at higher spatial resolution, to obtain FT-ICR mass spectrometric data over an entire tissue section containing 1000–1700 pixels (Cornett et al., 2008). An Orbitrap experiment requires much less time than an FT-ICR (25 pixels/min, especially if the scanned mass range is narrowed); however, the sample throughput is still not comparable with either a ToF or an LIT (Landgraf et al., 2009).

26.3 MSI WORKFLOW

MSI requires significantly less sample preparation than other imaging techniques such as autoradiography or fluorescence techniques; however, optimizing sample preparation and instrument parameters is key to generating quality mass spectrometric images. Several topics such as sample handling and sectioning, matrix selection, and MSI analysis modes will be discussed in the following sections.

26.3.1 Postdissection Tissue/Organ Preparation and Storage

When surgically removing tissue and whole organs from the host animal, care must be taken to preserve the original shape of the tissue, lest spatial information is lost. After excision of the tissue, the tissue may be loosely wrapped in aluminum foil prior to immersion in liquid nitrogen for 30–60 s. Quick submersion is not suggested, as tissue may undergo tissue cracking and degradation. In addition, direct submersion is discouraged to prevent sample adhesion to the walls of the Dewar. Finally, freshly dissected tissue should not be immediately placed in small plastic tubing, otherwise the tissue may mold to conform to the shape of the tube when frozen. Whole tissues and organs can be stored at −80°C for at least a year with little sample degradation (Schwartz et al., 2003).

26.3.2 Tissue Sectioning and Mounting

Although histological sectioning is typically performed on formalin-fixed, paraffin-embedded tissue, fresh-frozen sectioning of the tissue is the preferred method for MSI experiments. Use of paraffin or optimal cutting temperature (OCT) polymer for tissue mounting to the sample stage can cause ion suppression during MSI data collection. To avoid contamination of the tissue sections from the embedding medium, one can deposit several drops of deionized water to the sample stage and then position the tissue onto the drops, thereby freeze-mounting the tissue to the sample stage (Landgraf et al., 2009). Fresh-frozen sectioning of tissue thus avoids additional sample preparation steps.

Tissue sectioning is performed in a cryostat with a microtome. The cryostat is held at a temperature ranging from −20 to −30°C, depending on the tissue type. Tearing of the tissue usually occurs at warmer temperatures; thus, lowering the cryostat temperature should improve tissue section quality. In addition, tissue section thickness can affect the amount of analyte extracted from the tissue (and in turn analyte response) during MALDI experiments because a molecule must travel a longer distance with thicker tissue sections (Crossman et al., 2006).

Tissue sections are transferred to the sample plate (or microscope glass slide) by picking up the tissue with a set of forceps or using an artistic brush (Schwartz et al., 2003). For ToF experiments, a conductive surface must be used to avoid inaccuracies in mass measurement due to surface charging. Other mass analyzers do not have this limitation and can use either conductive or nonconductive surfaces; thus, conventional microscope slides are an option as well as stainless steel MALDI sample plates (Garrett et al., 2007). Tools used for the transfer and mounting of the tissue should be kept within the cryostat to avoid warming of the tissue.

After the tissue section has been transferred to the sample plate, the tissue is gently heated (e.g., placing a finger against the back of the microscope slide), then placed back into the cryostat to thaw-mount the tissue to the sample plate surface. After thaw-mounting, tissue sections should be stored at −80°C until analysis.

26.3.3 Tissue Section Preparation, MALDI Matrix Selection, and Deposition

Tissue sample preparation is the last step prior to insertion into the MSI instrument. This section will focus on the selection and application of the MALDI matrix to the tissue section. Although a single general purpose protocol will often generate adequate MSI results, ideally each MSI experiment should be individually optimized to tune the experiment for varying drug molecule properties and tissue types.

26.3.3.1 *Tissue Preparation* Prior to matrix application, the tissue may undergo washing followed by drying in a desiccator to remove excess water. The washing step can be performed to remove salts endogenous to the tissue section that can potentially inhibit MALDI matrix crystal formation. In addition, washing the tissue can remove some of the endogenous compounds (e.g., lipids) and improve the signal-to-noise ratio. For example, a tissue section can be rinsed (or immersed) in 80–100% ethanol for 30 s (Schwartz et al., 2003). However, care must be taken to limit the immersion, as the drug compound may migrate or, worse, be washed away. It is often useful to save the wash solutions to determine if any analyte has been extracted during the wash process. As a final step prior to matrix application, the tissue section must be placed in a desiccator or under vacuum to remove any excess water (typically from 30 min to an hour).

26.3.3.2 *MALDI Matrix Selection* Two properties are necessary for performing optimal MALDI experiments: (1) the matrix should absorb at the wavelength of the laser to limit in-source fragmentation; and (2) the matrix should cocrystallize with the target analyte so that the analyte is desorbed into the gas phase upon ablation of the matrix. UV-absorbing compounds that are commonly used with MALDI, and in turn with MSI, are DHB, SA, and CHCA (although the use of ionic

matrices have been reported to have the advantage of continuous tissue coverage; Lemaire et al., 2006; Meriaux et al., 2010). Selection of a specific MALDI matrix is often based on the analyte of interest. SA is typically used for proteins and large peptides, CHCA is typically used for smaller molecules, and DHB is used as a general-purpose matrix. However, a drawback to using these matrices is the abundance of low m/z matrix and cluster ions. Alternative higher–molecular-weight matrices such as porphyrins (Ayorinde et al., 1999) also absorb in the UV region but produce less interference in the low m/z region (Cohen and Gusev, 2002).

26.3.3.3 *MALDI Matrix Application Methods* Crystal size and coating uniformity play a larger role for MSI applications than standard MALDI experiments. Spatial resolution is not limited by just the raster step size across the tissue and the diameter of the laser spot; rather, the size of individual MALDI matrix crystals limits the effective pixel dimension of an image. Thus, DHB, with its elongated crystals, potentially provides a poorer spatial resolution than CHCA or SA, which have small spherical crystals. In addition, coating uniformity will affect the variability of the ion signal across the tissue.

MALDI matrix coating uniformity and crystal size can be optimized by the choice of matrix and the MALDI matrix application technique. Although dried-droplet matrix application requires the least amount of equipment (a pipettor), it is by far the most irreproducible and time-consuming of the matrix deposition techniques. Common matrix application techniques for MSI found in the literature include pneumatic spraying (Garrett et al., 2007), inkjet printing (Baluya et al., 2007), acoustic matrix deposition (Aerni et al., 2006), sublimation (Hankin et al., 2007), and solvent-free matrix dry-coating (Puolitaival et al., 2008). Of these methods, pneumatic spraying is by far the most common technique. Pneumatic spraying can be performed with a thin layer chromatography (TLC) sprayer, a Meinhard nebulizer, or even an artistic airbrush, as each can produce even, homogenous layers of small matrix crystals across the entire tissue section (Kaletas et al., 2009). However, care must be taken to obtain an even coating of matrix across the entire tissue section. Commercial devices can further increase matrix crystal reproducibility by automating the process.

26.3.4 Spatial Resolution: Relationship between Laser Spot Size and Raster Step Size

Depending on the size of the tissue sample, different methodologies can be used for MSI. Typically, the raster step size in an imaging experiment is set equivalent to the spot size of the laser. The spatial resolution is then limited by the spot size of the laser (assuming that the MALDI matrix crystals are not larger than the spot size). To increase the spatial resolution, the raster and the laser spot size can both be decreased, or oversampling may be performed. In this case, the raster step size is reduced to be smaller than diameter of the laser spot. This has the disadvantages, however, of increasing the analysis time and providing different spatial resolution in the x and y directions. However, the increase in analysis time can be mitigated by using a more directed approach of interrogating a smaller region of the tissue.

In some cases, the raster step size is increased to be greater than the diameter of the laser spot. Whole-body tissue sections commonly utilize such an approach due to the large area that must be examined. For example, Khatib-Shahidi et al. (2006) used a raster step size of 500 μm for analyzing whole-body rat tissue sections. The trade-off in increasing the raster step size is a loss of spatial resolution.

26.4 APPLICATIONS OF MSI FOR *IN SITU* ADMET TISSUE STUDIES

Traditional ADMET studies of drug compounds employ both imaging studies by an indirect measurement technique (e.g., fluorescence measurement of a tagged drug) and trace analysis of the drug and metabolites (e.g., LC-MS of tissue extracts). Together, these methods can offer a comprehensive picture of a drug's distribution within the body. However, individually, these techniques can offer misleading information due to poor specificity (analyte and/or spatial). Although MSI does not offer a "one stop" solution to ADMET studies, MSI does offer additional confirmation to support the results from other parallel experiments. The following section will describe several examples that demonstrate the utility of MSI for ADMET studies.

26.4.1 Determination of Drug Distribution and Site of Action

After determining drug efficacy, verification that a drug is delivered to its intended location is often the next step for ADMET studies. Drug administration is futile if the drug is irreversibly retained, metabolized, or excreted prior to reaching its intended target. For example, neurological drug compounds often target sections of the brain. However, the drug will be ineffective if it cannot cross the blood–brain barrier.

MSI is an ideal analytical technique to determine a drug's disposition within the body, as MSI can directly distinguish between a drug and its metabolite. In contrast, techniques that rely on radiolabels or fluorescent labels may mistakenly track a labeled metabolite (that may or may not have any pharmacological activity)

rather than the drug itself. This point is illustrated by Li et al. (2009). They examined the distribution of the antihistamine drug astemizole and its major metabolite, desmethylastemizole, in rat brain sections to determine whether the drug or its metabolite is causing central nervous system (CNS)-related side effects. Experiments were performed in MS/MS mode on a MALDI-QqToF to further differentiate drug and metabolites from endogenous compounds. The researchers determined that the metabolite localized around the ventricle regions of the brain, whereas the drug molecule was evenly distributed throughout the brain regions. From this result and additional LC-MS data, the researchers concluded that the drug, not the metabolite, was likely the cause of the CNS side effects in rats.

Another example of determining the site of action was described by Atkinson et al. (2007). Here they examined both a prodrug (AQ4N) and its active metabolite (AQ4) by employing a MALDI-QqToF in MS mode. AQ4N is a bioreductive prodrug that is administered in conjunction with other antitumor drugs for cancer treatments (AQ4 has been shown to sensitize tumors to existing cancer treatments; McKeown et al., 1995, 1996; Friery et al., 2000; Gallagher et al., 2001). As a bioreductive drug, AQ4N is an inactive prodrug that is converted into a potent cytotoxin in the presence of high levels of specific reductases or in regions deprived of an adequate oxygen supply (such as those found in solid tumors, which exhibit regions of hypoxia due to the high, unchecked growth within the tumor that outpaces its blood supply). This specificity for conversion of the prodrug into the active metabolite is ideal, as it will limit side effects in cancer treatments.

MSI is particularly well suited to verifying the transformation of AQ4N to AQ4 in hypoxic regions of solid tumors. For their experiments, Atkinson et al. (2007) injected AQ4N directly into solid tumors growing within tumor-bearing mice. Following excision of the tumor, sectioning, and matrix application, MSI was performed on the tumor sections. MS images illustrated areas of AQ4N and AQ4 (Figure 26.2). Figure 26.2a is an image of m/z 184, the phosphatidylcholine head group, to illustrate the position of the tumor section on the sample sheet (right side of image). Figure 26.2b illustrates how an overlay MSI can be used to differentiate between regions of prodrug (predominantly in the top half of the image) and its metabolite (predominantly in the bottom half of the image) and the areas of overlap (white). Researchers expected the metabolite to be present only in hypoxic regions of the tumor (as opposed to well-oxygenated regions). The researchers concluded from the marked delineation between the regions and the limited amount of overlap in the image that AQ4N was confined to the hypoxic regions of the solid tumor tissue. To verify this conclusion, the overlapping images of

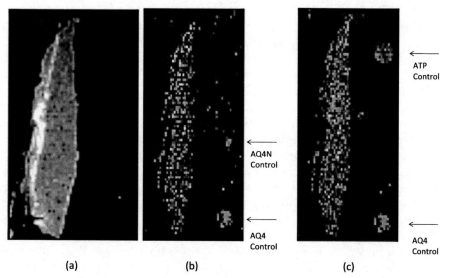

FIGURE 26.2. MALDI-MS images illustrating the distribution of the bioreductive prodrug AQ4N, its active metabolite (AQ4), and ATP within solid tumor tissue. Standards were deposited to the right of the tissue sections in panels (b) and (c). The position of the tumor section is outlined by imaging m/z 184 (phosphatidyl-choline head group) (a). The locations of AQ4N ([M + Na]$^+$ m/z 467; top half of the image) and AQ4 ([M + H]$^+$ m/z 413; bottom half of the image) (b) were determined and illustrated in the MS image. Control spots of AQ4N and AQ4 were applied to the right of the tissue section. White indicates areas of overlap between AQ4N and AQ4. AQ4N should be metabolized to AQ4 under hypoxic conditions, whereas the prodrug should remain intact in the well-oxygenated regions of the tumor. ATP ([M-H$_3$PO$_3$]$^-$ m/z 409; top half of the image) and AQ4 ([M + H]$^+$ m/z 413; bottom half of the image) (c) is imaged to illustrate differentiation between hypoxic and well-oxygenated regions within the tumor tissue. Depleted ATP levels are expected in hypoxic regions of tumor tissue. White regions indicate areas of overlap between ATP and AQ4. (Adapted with permission from Atkinson et al., 2007.)

ATP and AQ4 were generated (Figure 26.2c). ATP was imaged because regions of hypoxia exhibit low concentrations of ATP (Kribben et al., 2003). Due to the distinct regions of ATP and AQ4 and lack of areas indicating overlap (white), the researchers concluded that the conversion of AQ4 was limited to the hypoxic solid tumor regions of the tissue section. Thus, with MSI, the prodrug AQ4N and its active metabolite AQ4 were readily distinguished and imaged, and the results were verified by examining the localization of an endogenous molecule, ATP, in solid tumor sections.

These conclusions could not have been drawn if autoradiography alone were used, as it could not distinguish between the prodrug (AQ4N) and the active metabolite (AQ4). Furthermore, ATP would have been invisible by that technique. MSI, however, is not limited to examining small tissue sections. Analysis of whole-body tissue sections by MSI offers similar advantages, as will be discussed in the following section.

26.4.2 Analysis of Whole-Body Tissue Sections Utilizing MSI

Whole-body autoradiography (WBA) is a quantitative imaging technique in which radiolabeled drugs (i.e., drugs tagged with beta emitters such as ^{3}H, ^{14}C, or ^{125}I)

are administered to the target animal (Solon et al., 2010). Postdose, the animals are sacrificed, snap frozen, and then sectioned on a large cryostat to generate single sections encompassing the animal from head to tail. After sectioning, the sections are then placed adjacent to a phosphor-imaging plate or X-ray film along with a set of calibration standards, with exposure times ranging from a couple of days to weeks (depending on the concentration of the labeled drug; longer exposures allow for greater analyte sensitivity). Whereas WBA offers excellent quantitation, sensitivity, and resolution, WBA suffers from limited analyte specificity, as WBA detects all compounds with the radiolabel (e.g., the labeled drug and any metabolites that also contain the label). LC-MS is typically used in conjunction with WBA to examine the drug and its metabolites; however, spatial resolution is lost as LC-MS requires the homogenization and extraction of the drug/metabolite from the tissue section. Laser capture microdissection (LCM) of thin tissue sections can target specific sections of tissue; however, the spatial resolution is limited by the size of section excised (Emmert-Buck et al., 1996; Drexler et al., 2007). MSI has the potential to combine the analyte specificity of LC-MS and the spatial localization of WBA for the detection of drugs and their metabolites in whole-body thin tissue sections (Rohner et al., 2005; Khatib-Shahidi et al., 2006; Stoeckli et al., 2007; Chen et al., 2008; Kertesz et al., 2008; Trim et al., 2008).

An early example of using MSI for whole-body analysis was detailed by Khatib-Shahidi et al. (2006). Whole-body sections were examined using a ToF/ToF instrument in MS/MS mode for the detection of olanzapine (OLZ) in orally dosed rats. Note that an entire whole-body tissue section will rarely fit within the vacuum chamber of a MALDI ionization source; in this study, whole-body tissue sections were quartered to fit onto the MALDI target plate, individually coated with tissue, and then analyzed. During the data work-up stage, the images are then stitched together using either the instrument's imaging software or a commercial image-processing program (e.g., Adobe Photoshop). An optical image of the whole-body tissue section is illustrated in Figure 26.3a. The authors reported the detection of OLZ (Figure 26.3b) and its major metabolites, *N*-desmethyl OLZ (Figure 26.3c) and 2-hydroxymethyl OLZ (Figure 26.3d), within the tissue section. Specifically, the authors noted the presence of the parent drug throughout the tissue section, whereas the two metabolites were absent from the brain and CNS. From this information, the authors inferred that the metabolites do not reach the CNS.

Another example of MSI applied to whole-body tissue sections was presented by Chen et al. (2008). They used a QqToF in MS/MS mode to determine the bioavailability of terfenadine in rats. Rats were administered terfenadine and subsequently sacrificed 1 or 4 h postdose. After tissue sectioning and matrix application, the researchers found that the terfenadine was localized in the stomach and intestine of the rat, whereas the major metabolite, fexofenadine, was found in the liver, intestine, and stomach. The authors indicated that the 4-h postdose MS images exhibited terfenadine in the small intestines and its metabolite in the small and large intestines. From the lack of terfenadine in the rest of the whole-body section, the authors concluded that metabolism of the drug in the liver and (to a lesser extent) the small intestine prevented general distribution of the drug throughout the body. In summary, the ability to differentiate drug and metabolite to determine the cause for low bioavailability is a powerful tool for ADMET studies.

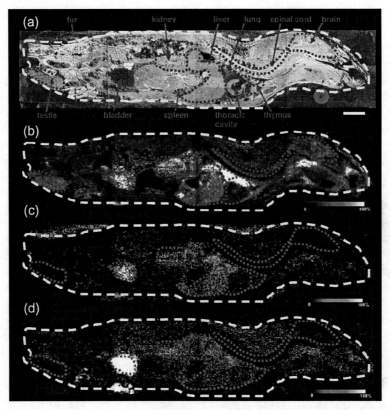

FIGURE 26.3. Whole-body tissue experiments by MALDI-MS/MS. Panel (a) is the optical image of the whole-body rat tissue section from a rat sacrificed 2 h postadministration of OLZ. The MS/MS images of OLZ (m/z 313 → 256) (b), *N*-desmethyl OLZ (m/z 299 → 256) (c), and 2-hydroxymethyl OLZ (m/z 329 → 272) (d) are presented. The scale bar in panel (a) represents 1 cm. (Adapted with permission from Khatib-Shahidi et al., 2006.)

Despite the analyte specificity and relatively short analysis times, MSI is more likely to be used as a complementary technique to WBA. This is due to the better sensitivity and quantitation capabilities that WBA offers over MSI. Although there have been some studies examining the quantitative capabilities of MSI (Stoeckli et al., 2007; Reich et al., 2010), MSI remains primarily a qualitative technique.

26.4.3 Increasing Analyte Specificity for Mass Spectrometric Images

A literature search for MSI indicates that the majority of papers use a single stage of mass analysis (e.g., a ToF mass analyzer). A ToF mass analyzer is useful for peptide and protein applications; however, analysis of drugs and metabolites is more readily performed with a tandem mass analysis system, as illustrated in all but one of the previous example applications. Tissue is a complex mixture of lipids, peptides, and other biological compounds. Addition of MALDI matrix molecules complicates the MS analysis even further. To determine drug distributions within tissue, MSI techniques that use enhanced analyte selectivity to aid in detection of drugs from tissue are critical. These are discussed below.

One method to enhance analyte selectivity is to use multiple stages of mass spectrometry (MS^n) to achieve the necessary specificity, as has been demonstrated with an LIT (Garrett et al., 2007). An example specific to ADMET studies of this was demonstrated by Drexler et al. (2007). The researchers used an LIT to identify crystalline features found in the spleen of rats undergoing a high-dosage toxicological study of a proprietary drug candidate. These rats were fed a high-dose regimen of a prodrug over the course of 2 weeks. An MS/MS experiment examining the fragmentation of the parent drug and imaging a product ion over the entire spleen surface displayed response in both the regions containing the birefigent crystals and the regions free of the crystals (Figure 26.4). The researchers concluded that the crystals were due to the free parent drug and this hypothesis was supported by performing LCM on both the microcrystalline and microcrystalline-free sections and then comparing each sample's MS^3 spectrum with that of a standard.

MSI-MS^n experiments require prior knowledge of the analyte. In addition, only one precursor ion per experiment can be examined (unless alternating scans or multinotch isolation is employed; Reich et al., 2008, 2009). An alternative to this strategy is to utilize high-resolution MS. High-resolution MS allows the user to differentiate multiple ions that may fall at the same nominal mass. For example, an FT-ICR was used to distinguish 2-hydroxymethyl OLZ (m/z 329.069) from three other interferants at nominal mass m/z 329 in kidney tissue (Figure 26.5) (Cornett et al., 2008). If this experiment were performed by MS/MS, multiple experiments would be required to generate images of both 2-hydroxymethyl OLZ and the parent drug. One drawback of using a high-resolution instrument for MSI is the lengthy experiment duration. Data collection on an FT-ICR can take upward of 10 times as long when

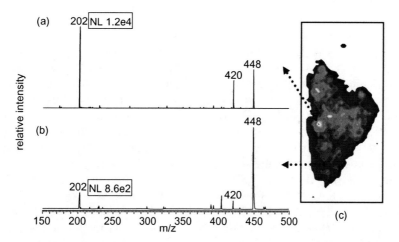

FIGURE 26.4. MS^2 image and spectra of a spleen tissue section, harvested from a rat given a high dose of a proprietary prodrug over the course of 2 weeks. The MS^2 mass spectra of areas where the birefigent crystals are present (a) and absent (b) suggest that the same molecule is localized at both areas (however at differing concentrations). The image of ion m/z 202 (c), extracted from the MS^2 experiment monitoring the fragmentation of the [M + H]$^+$ ion of the active drug (m/z 448) demonstrates high intensities in the area where the birefigent crystals are localized. (Adapted with permission from Drexler et al., 2007.)

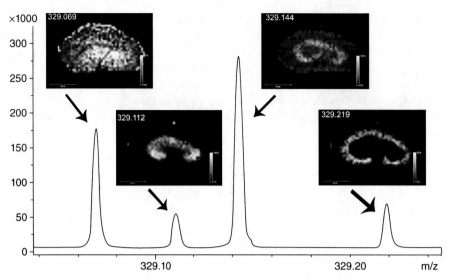

FIGURE 26.5. MS images of kidney tissue derived from an FT-ICR MSI experiment. Four different images were generated from the ions with the nominal m/z 329. The image of m/z 329.069 indicates the localization of the metabolite 2-hydroxymethyl OLZ. The other three isobaric species are unrelated background ions. (Adapted with permission from Cornett et al., 2008.)

compared with a ToF-MSI experiment. However, the information gathered with high-resolution MS affords a greater pool of information for data mining.

Another method to improve analyte specificity is to add an additional stage of analysis, such as a separation step. Addition of LC or gas chromatography (GC) to the instrument design for MSI is incompatible with MS imaging; however, other alternatives are emerging. One method is to incorporate ion mobility separation. Ion mobility spectrometry (IMS) is a gas-phase technique that separates ions based on their collisional cross section (Jackson et al., 2007). IMS coupled with a MALDI-MS instrument affords additional analyte specificity by using drift times to help differentiate between drug ions and isobaric background ions. Trim et al. (2008) demonstrated the use of MALDI-IMS-MS for the detection of vinblastine in whole-body rat tissue sections. Their experiments involved rats dosed with the drug and then sacrificed 1 h postdose. An IMS-QqToF instrument in MS and MS/MS mode was used to detect the vinblastine from the tissue. Figure 26.6 illustrates the potential of IMS-MSI experiments. Specifically, there is a decrease in signal intensity for the IMS-MS and IMS-MS/MS images in the areas indicated by the white arrow in Figures 26.6a and 26.6b, respectively. By extracting ions with drift times that match vinblastine and then imaging the drug by MS or MS/MS, response from isobaric background ions are removed from the image and a more accurate image arises in the renal portion of the whole body section. More importantly, the MSI images are similar to the WBA images (Figure 26.6c),

further establishing MSI as a complementary technique to established imaging methods.

Interferences from endogenous compounds and MALDI matrix ions complicate most MSI experiments. Increasing analyte specificity from MS/MS, MS^n, and high-resolution MS can decrease the complexity of the data analysis step. Thus, the use of instruments with multiple stages of analysis or high resolution is vital for MSI experiments focusing on the distribution of drugs within thin tissue sections.

26.4.4 DESI Applications for MSI

MALDI is the most common ionization source for MSI due to its familiarity and widespread use in the general MS community. However, MALDI is not without its limitations. Sample preparation can prove to be a challenge for novice MSI users, as the selection and application of MALDI matrix for a given tissue and analyte requires much thought and experimentation. In addition, the MALDI matrix can hinder analysis due to isobaric matrix clusters/analyte-matrix clusters that can obscure an analyte peak or lead to an incorrect analyte assignment.

As the matrix poses the greatest disadvantage in a MALDI analysis, methods that do not require a matrix would seem to be ideal. For example, an infrared (IR)-MALDI laser can use water native in the tissue as its matrix. Some studies have shown this technique to be useful (Nemes et al., 2010), but no studies yet have been performed for MSI of pharmaceuticals in tissue.

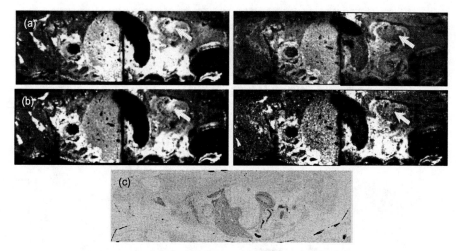

FIGURE 26.6. Contrasting MS images from MALDI-IMS-MS and MALDI-IMS-MS/MS experiments. Reduced response of the m/z 811 ion in the renal pelvis region of the IMS-MS image (a, right) (indicated by the white arrow) versus the conventional MS image (a, left) suggests that IMS reduces the amount of isobaric species unrelated to vinblastine. The MS/MS images (b) of m/z 751 from m/z 811 further separates out interfering ions. The WBA image (c) of the ^{3}H-vinblastin compound is presented to verify drug distribution in the MALDI-IMS-MS images. (Adapted with permission from Trim et al., 2008.)

Another approach to matrix-less MSI involves the use of DESI, as DESI-MSI does not utilize a matrix and requires little sample preparation. However, optimizing the sampling parameters is more difficult as the optimal spray angles and distances from the spray nozzle to the mass spectrometer must be determined. In addition, the spatial resolution of DESI-MSI is not on par with MALDI-MSI or SIMS imaging experiments. Despite these challenges, DESI-MSI has the potential to serve as a viable tool for ADMET studies. Two studies in particular demonstrate the feasibility of DESI for drug studies (Kertesz et al., 2008; Wiseman et al., 2008).

Wiseman et al. (2008) described the detection of clozapine and its major metabolites from rat organs using a DESI source that was coupled to an LIT mass spectrometer. Clozapine was administered to rats; after sacrifice and tissue sectioning, the DESI source was used to scan the tissue with a lateral resolution of approximately 250 μm. Using DESI-MS/MS, clozapine could be detected in the brain, lung, kidney, and testis; in addition, clozapine and one of its metabolites (N-desmethylclozapine) were located in the lung tissue. Separate LC-MS studies involving tissue homogenization and analyte extraction confirmed the presence of clozapine and its metabolite in tissue. Unfortunately, the other expected metabolite (clozapine-N-oxide) was not detected by either DESI-MS or MS/MS images.

Another example of applying DESI to whole-body tissue sections was presented by Kertesz et al. (2008). Here, DESI images using a hybrid QqLIT mass spectrometer were compared with WBA for the analysis of

propranolol in whole-body tissues. In addition to eliminating the MALDI matrix application, DESI experiments had a higher acquisition rate than a comparable MALDI-MSI experiment on a QqLIT. For example, the MALDI-QqLIT instrument described in Section 26.4.1 (Li et al., 2009) acquired spectra at a rate of one spectrum every 4 s at a lateral resolution of 100–150 μm (spot size of the laser). This is in contrast to their DESI experiment that collected 50 spectra every second at a lateral resolution of 140 μm (lane scan rate of 7 mm/s) (Figure 26.7). This, however, is scanning at the highest scan rate available. The researchers noted a decrease in the analyte response when compared with slower scan rates. In addition, the DESI images were of lower quality as the width of the lane was 200 μm as opposed to the MALDI laser spot size.

26.5   CONCLUSIONS

Comprehensive and rigorous ADMET studies are necessary for development of any potential drug. A series of experiments must be performed to confirm the efficacy and safety of a drug candidate. For example, autoradiography is an established imaging technique for visualizing drug distribution within the body, and LC-MS of tissue homogenates permit targeted drug analysis in addition to quantitative results. While these methods and tools for ADMET studies are well established, new technology is being developed to help shorten and improve drug development.

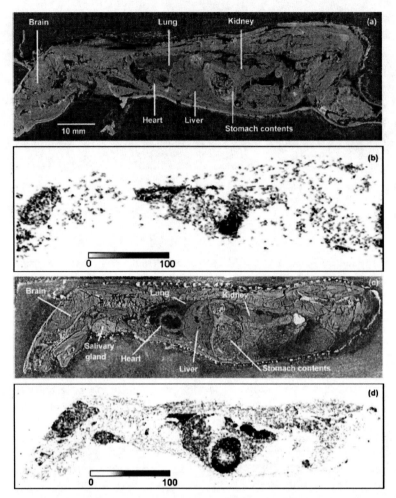

FIGURE 26.7. Images depicting the localization of propranolol and ^{3}H-propranolol utilizing DESI-MS and WBA, respectively. Optical image (a) of a rat dosed with propranolol (7.5 mg/kg, sacrificed 20 min postdose). Localization of the m/z 260→116 fragment ion of propranolol (b) was demonstrated using MSI-DESI-MS/MS. The tissue section was moved beneath the DESI emitter at a rate of 7 mm/s (total analysis time of 79 min). Lower intensity of the propranolol was noted when compared with slower lane scans. Optical image (c) of a different whole-body rat section originating from a rat dosed with ^{3}H-propranolol (7.5 mg/kg, sacrificed 20 min postdose). The localization of propranolol in various organs and regions of the rat in the WBA image (d) compares favorably well with the MSI image (b) of propranolol. (Adapted with permission from Kertesz et al., 2008.)

Although still in the developmental stage, MSI is emerging as a complementary technique that combines the advantages of both autoradiography and LC-MS into a single imaging experiment. With a single *in situ* imaging experiment that offers selectivity to monitor drug compounds and metabolites and even affected endogenous compounds, within a shorter period and with less costly sample preparation, MSI can reduce the amount of time required for ADMET studies. However, MSI is not without its limitations. MSI is limited to *in situ* samples, and is less quantitative than autoradiography and LC/MS. Despite these limitations, developments in MSI are pushing the technique toward becoming a complementary analytical tool for ADMET studies.

REFERENCES

Aerni H-R, Cornett DS, Caprioli RM (2006) Automated acoustic matrix deposition for MALDI sample preparation. *Anal Chem* 78:827–834.

Atkinson SJ, Loadman PM, Sutton C, Patterson LH, Clench MR (2007) Examination of the distribution of the bioreductive drug AQ4N and its active metabolite AQ4 in solid tumours by imaging matrix-assisted laser desorption/ionisation mass spectrometry. *Rapid Commun Mass Spectrom* 21:1271–1276.

Ayorinde FO, Hambright P, Porter TN, Keith QL Jr. (1999) Use of meso-Tetrakis(pentafluorophenyl)porphyrin as a matrix for low molecular weight alkylphenol ethoxylates in laser desorption/ionization time-of-flight mass spectrometry. *Rapid Commun Mass Spectrom* 13:2474–2479.

Baluya DL, Garrett TJ, Yost RA (2007) Automated MALDI matrix deposition method with inkjet printing for imaging mass spectrometry. *Anal Chem* 79:6862–6867.

Bouslimani A, Bec N, Glueckmann M, Hirtz C, Larroque C (2010) Matrix-assisted laser desorption/ionization imaging mass spectrometry of oxaliplatin derivatives in heated intraoperative chemotherapy (HIPEC)-like treated rat kidney. *Rapid Commun Mass Spectrom* 24:415–421.

Caprioli RM, Farmer TB, Gile J (1997) Molecular imaging of biological samples: Localization of peptides and proteins using MALDI-TOF MS. *Anal Chem* 69:4751–4760.

Chen J, Hsieh Y, Knemeyer I, Crossman L, Korfmacher WA (2008) Visualization of first-pass drug metabolism of terfenadine by MALDI-imaging mass spectrometry. *Drug Metab Lett* 2:1–4.

Cohen LH, Gusev AI (2002) Small molecule analysis by MALDI mass spectrometry. *Anal Bioanal Chem* 373:571–586.

Cooks RG, Ouyang Z, Takats Z, Wiseman JM (2006) Ambient mass spectrometry. *Science* 311:1566–1570.

Cornett DS, Frappier SL, Caprioli RM (2008) MALDI-FTICR imaging mass spectrometry of drugs and metabolites in tissue. *Anal Chem* 80:5648–5653.

Crossman L, McHugh NA, Hsieh Y, Korfmacher WA, Chen J (2006) Investigation of the profiling depth in matrix-assisted laser desorption/ionization imaging mass spectrometry. *Rapid Commun Mass Spectrom* 20:284–290.

Drexler DM, Garrett TJ, Cantone JL, Diters RW, Mitroka JG, Prieto Conaway MC, Adams SP, Yost RA, Sanders M (2007) Utility of imaging mass spectrometry (IMS) by matrix-assisted laser desorption ionization (MALDI) on an ion trap mass spectrometer in the analysis of drugs and metabolites in biological tissues. *J Pharmacol Toxicol Methods* 55:279–288.

Emmert-Buck MR, Bonner RF, Smith PD, Chuaqui RF, Zhang Z, Goldstein SR, Weiss RA, Liotta LA (1996) Laser capture microdissection. *Science* 274:998–1001.

Friery OP, Gallagher R, Murray MM, Hughes CM, Galligan ES, McIntyre IA, Patterson LH, Hirst DG, McKeown SR (2000) Enhancement of the antitumor effect of cyclophosphamide by the bioreductive drugs AQ4N and tirapazamine. *Br J Cancer* 82:1469–1473.

Gallagher R, Hughes CM, Murray MM, Friery OP, Patterson LH, Hirst DG, McKeown SR (2001) The chemopotentiation of cisplatin by the novel bioreductive drug AQ4N. *Br J Cancer* 85:625–629.

Garrett TJ, Prieto-Conaway MC, Kovtoun V, Bui H, Izgarian N, Stafford G, Yost RA (2007) Imaging of small molecules in tissue sections with a new intermediate-pressure MALDI linear ion trap mass spectrometer. *Int J Mass Spectrom* 260:166–176.

Gobey J, Cole M, Janiszewski J, Covey T, Chau T, Kovarik P, Corr J (2005) Characterization and performance of MALDI on a triple quadrupole mass spectrometer for analysis and quantification of small molecules. *Anal Chem* 77:5643–5654.

Groseclose MR, Andersson M, Hardesty WM, Caprioli RM (2007) Identification of proteins directly from tissue: *In situ* tryptic digestions coupled with imaging mass spectrometry. *J Mass Spectrom* 42:254–262.

Hankin JA, Barkley RM, Murphy RC (2007) Sublimation as a method of matrix application for mass spectrometric imaging. *J Am Soc Mass Spectrom* 18:1646–1652.

Hopfgartner G, Varesio E, Stoeckli M (2009) Matrix-assisted laser desorption/ionization mass spectrometric imaging of complete rat sections using a triple quadrupole linear ion trap. *Rapid Commun Mass Spectrom* 23:733–736.

Ifa DR, Wiseman JM, Song Q, Cooks RG (2007) Development of capabilities for imaging mass spectrometry under ambient conditions with desorption electrospray ionization (DESI). *Int J Mass Spectrom* 259:8–15.

Jackson SN, Ugarov M, Egan T, Post JD, Langlais D, Schultz JA, Woods AS (2007) MALDI-ion mobility-TOFMS imaging of lipids in rat brain tissue. *J Mass Spectrom* 42:1093–1098.

Kaletas BK, van der Wiel IM, Stauber J, Dekker LJ, Guzel C, Kros JM, Luider TM, Heeren RMA (2009) Sample preparation issues for tissue imaging by imaging MS. *Proteomics* 9:2622–2633.

Karas M, Hillenkamp F (1988) Laser desorption ionization of proteins with molecular masses exceeding 10,000 daltons. *Anal Chem* 60:2299–2301.

Kertesz V, Van Berkel GJ (2008) Scanning and surface alignment considerations in chemical imaging with desorption electrospray mass spectrometry. *Anal Chem* 80:1027–1032.

Kertesz V, Van Berkel GJ (2010) Fully automated liquid extraction-based surface sampling and ionization using a chip-based robotic nanoelectrospray platform. *J Mass Spectrom* 45:252–260.

Kertesz V, Van Berkel GJ, Vavrek M, Koeplinger KA, Schneider BB, Covey TR (2008) Comparison of drug distribution images from whole-body thin tissue sections obtained using desorption electrospray ionization tandem mass spectrometry and autoradiography. *Anal Chem* 80:5168–5177.

Khatib-Shahidi S, Andersson M, Herman JL, Gillespie TA, Caprioli RM (2006) Direct molecular analysis of whole-body animal tissue sections by imaging MALDI mass spectrometry. *Anal Chem* 78:6448–6456.

Kribben A, Feldkamp T, Horbelt M, Lange B, Pietruck F, Herget-Rosenthal S, Heemann U, Philipp T (2003) ATP protects, by way of receptor-mediated mechanisms, against hypoxia-induced injury in renal proximal tubules. *J Lab Clin Med* 141:67–73.

Landgraf RR, Conaway MCP, Garrett TJ, Stacpoole PW, Yost RA (2009) Imaging of lipids in spinal cord using intermediate pressure matrix-assisted laser desorption-linear ion trap/orbitrap MS. *Anal Chem* 81:8488–8495.

Lemaire R, Tabet JC, Ducoroy P, Hendra JB, Salzet M, Fournier I (2006) Solid ionic matrixes for direct tissue analysis and MALDI imaging. *Anal Chem* 78:809–819.

Li F, Hsieh Y, Kang L, Sondey C, Lachowicz J, Korfmacher WA (2009) MALDI-tandem mass spectrometry imaging of astemizole and its primary metabolite in rat brain sections. *Bioanalysis* 1:299–307.

Magparangalan DP, Garrett TJ, Drexler DM, Yost RA (2010) Analysis of large peptides by MALDI using a linear quadrupole ion trap with mass range extension. *Anal Chem* 82:930–934.

McKeown SR, Hejmadi MV, McIntyre IA, McAleer JJA, Patterson LH (1995) AQ4N: An alkylaminoanthraquinone N-oxide showing bioreductive potential and positive interaction with radiation *in vivo*. *Br J Cancer* 72:76–81.

McKeown SR, Friery OP, McIntyre IA, Hejmadi MV, Patterson LH, Hirst DG (1996) Evidence for a therapeutic gain when AQ4N or tirapazamine is combined with radiation. *Br J Cancer Suppl* 74:S39–S42.

Meriaux C, Franck J, Wisztorski M, Salzet M, Fournier I (2010) Liquid ionic matrixes for MALDI mass spectrometry imaging of lipids. *J Proteomics* 73:1204–1218.

Nemes P, Woods AS, Vertes A (2010) Simultaneous imaging of small metabolites and lipids in rat brain tissues at atmospheric pressure by laser ablation electrospray ionization mass spectrometry. *Anal Chem* 82:982–988.

Northen TR, Yanes O, Northen MT, Marrinucci D, Uritboonthai W, Apon J, Golledge SL, Nordstrom A, Siuzdak G (2007) Clathrate nanostructures for mass spectrometry. *Nature* 449:1033–1036.

Pasilis SP, Kertesz V, Van Berkel GJ (2007) Surface scanning analysis of planar arrays of analytes with desorption electrospray ionization-mass spectrometry. *Anal Chem* 79: 5956–5962.

Patti GJ, Woo H-K, Yanes O, Shriver L, Thomas D, Uritboonthai W, Apon JV, Steenwyk R, Manchester M, Siuzdak G (2010) Detection of carbohydrates and steroids by cation-enhanced nanostructure-initiator mass spectrometry (NIMS) for biofluid analysis and tissue imaging. *Anal Chem* 82:121–128.

Perchalski R (1985) Characteristics and application of a laser ionization/evaporation source for tandem mass spectrometry. Chemistry. Gainesville, University of Florida. PhD: 195.

Puolitaival SM, Burnum KE, Cornett DS, Caprioli RM (2008) Solvent-free matrix dry-coating for MALDI imaging of phospholipids. *J Am Soc Mass Spectrom* 19:882–886.

Reich RF, Cudzilo K, Yost RA (2008) Quantitative imaging of cocaine and its metabolites in postmortem brain tissue by intermediate-pressure MALDI/linear ion trap tandem mass spectrometry. 56th ASMS Conference on Mass Spectrometry and Allied Topics, Denver, CO.

Reich RF, Cromwell KN, Yost RA (2009) MALDI-MSn quantitation by selective isolation of analyte and internal standard ions using a Multi-Notch SWIFT waveform. 57th ASMS Conference on Mass Spectrometry and Allied Topics, Philadelphia, PA.

Reich RF, Cudzilo K, Levisky JA, Yost RA (2010) Quantitative MALDI-MSn analysis of cocaine in the autopsied brain of a human cocaine user employing a wide isolation window and internal standards. *J Am Soc Mass Spectrom* 21:564–571.

Reyzer ML, Hsieh Y, Ng K, Korfmacher WA, Caprioli RM (2003) Direct analysis of drug candidates in tissue by matrix-assisted laser desorption/ionization mass spectrometry. *J Mass Spectrom* 38:1081–1092.

Rohner TC, Staab D, Stoeckli M (2005) MALDI mass spectrometric imaging of biological tissue sections. *Mech Ageing Dev* 126:177–185.

Schiffer WK, Liebling CNB, Patel V, Dewey SL (2007) Targeting the treatment of drug abuse with molecular imaging. *Nucl Med Biol* 34:833–847.

Schwartz JC, Senko MW, Syka JEP (2002) A two-dimensional quadrupole ion trap mass spectrometer. *J Am Soc Mass Spectrom* 13:659–669.

Schwartz SA, Reyzer ML, Caprioli RM (2003) Direct tissue analysis using matrix-assisted laser desorption/ionization mass spectrometry: Practical aspects of sample preparation. *J Mass Spectrom* 38:699–708.

Sleno L, Volmer DA (2005) Some fundamental and technical aspects of the quantitative analysis of pharmaceutical drugs by matrix-assisted laser desorption/ionization mass spectrometry. *Rapid Commun Mass Spectrom* 19: 1928–1936.

Solon EG, Schweitzer A, Stoeckli M, Prideaux B (2010) Autoradiography, MALDI-MS, and SIMS-MS imaging in pharmaceutical discovery and development. *AAPS J* 12:11–26.

Sosnovik DE, Weissleder R (2007) Emerging concepts in molecular MRI. *Curr Opin Biotechnol* 18:4–10.

Stoeckli M, Farmer TB (2004) MALDI-MS imaging in biomedical research. In *Biomedical Applications of Proteomics*, Sanchez J-C, Corthals GL, Hochstrasser DF, eds., pp. 373–388. Wiley-VCH, Weinheim.

Stoeckli M, Staab D, Schweitzer A (2007) Compound and metabolite distribution measured by MALDI mass spectrometric imaging in whole-body tissue sections. *Int J Mass Spectrom* 260:195–202.

Strupat K, Kovtoun V, Bui H, Viner R, Stafford G, Horning S (2009) MALDI produced ions inspected with a linear ion trap-orbitrap hybrid mass analyzer. *J Am Soc Mass Spectrom* 20:1451–1463.

Takats Z, Wiseman JM, Gologan B, Cooks RG (2004) Mass spectrometry sampling under ambient conditions with desorption electrospray ionization. *Science* 306:471–473.

Takats Z, Wiseman JM, Cooks RG (2005) Ambient mass spectrometry using desorption electrospray ionization (DESI): Instrumentation, mechanisms and applications in forensics, chemistry, and biology. *J Mass Spectrom* 40:1261–1275.

Trim PJ, Henson CM, Avery JL, McEwen A, Snel MF, Claude E, Marshall PS, West A, Princivalle AP, Clench MR (2008) Matrix-assisted laser desorption/ionization-ion mobility separation-mass spectrometry imaging of vinblastine in whole body tissue sections. *Anal Chem* 80:8628–8634.

Troendle FJ, Reddick CD, Yost RA (1999) Detection of pharmaceutical compounds in tissue by matrix-assisted laser desorption/ionization and laser desorption/chemical ionization tandem mass spectrometry with a quadrupole ion trap. *J Am Soc Mass Spectrom* 10:1315–1321.

Van Berkel GJ, Kertesz V, Koeplinger KA, Vavrek M, Kong A-NT (2008) Liquid microjunction surface sampling probe electrospray mass spectrometry for detection of drugs and metabolites in thin tissue sections. *J Mass Spectrom* 43:500–508.

Watson JT, Sparkman OD (2007) *Introduction to Mass Spectrometry: Instrumentation, Applications, and Strategies for Data Interpretation*. Wiley, Hoboken, NJ.

Wiseman JM, Ifa DR, Zhu Y, Kissinger CB, Manicke NE, Kissinger PT, Cooks RG (2008) Desorption electrospray ionization mass spectrometry: Imaging drugs and metabolites in tissues. *Proc Natl Acad Sci USA* 105: 18120–18125.

27

APPLICATIONS OF QUANTITATIVE WHOLE-BODY AUTORADIOGRAPHY (QWBA) IN DRUG DISCOVERY AND DEVELOPMENT

LIFEI WANG, HAIZHENG HONG, AND DONGLU ZHANG

27.1 INTRODUCTION

Radiolabeled drugs have been widely used in drug discovery and development studies since radioactivity can easily be detected and quantified using liquid scintillation techniques, quantitative whole-body autoradiography (QWBA), and other methods (Shaffer et al., 2006; Christopher et al., 2008; Wang et al., 2006, 2010a,b; Zhang et al., 2009). Tissue distribution studies with radiolabeled drugs in animals are important in drug discovery and development and provide distribution and pharmacokinetic information of the test drug or its metabolites in animal tissues or organs (Igari et al., 1982; Xiang et al., 2004; He et al., 2008). Before starting a human absorption, distribution, metabolism, and excretion (ADME) study, a tissue distribution study in rats with the radiolabeled drug is used to project human organ exposures and estimate the safety after dosing the radiolabeled drug. In addition, full tissue distribution data in male, female, and pregnant rats are required for regulatory filing of a new drug application.

QWBA is a well-established imaging technique using a radiotracer to determine the whole-body tissue distribution of radioactivity or the localization of radiolabeled drugs in specific tissues of animals. It is widely used in drug discovery and development to provide detailed and comprehensive information on the distribution and localization of labeled drugs in various tissues (Solon et al., 2002a; Potchoiba and Nocerini,

2004; Skotland et al., 2006; Yu et al., 2007). QWBA has several advantages in tissue distribution study: (1) QWBA provides a visible whole-body picture of the distribution of radioactivity in animals; (2) QWBA can accurately determine the radioactivity concentration in most major organs or tissues including the fetus in pregnant animals; and (3) QWBA can detect the potential sites of drug accumulation and provide additional information on disposition and localization within very small tissues that could be the potential toxicological or pharmacological sites of action.

27.2 EQUIPMENT AND MATERIALS

Several types of equipment and materials are needed for QWBA studies. A cryomicrotome (e.g., Leica CM3600 Cryomacrocut, Nussloch, Germany) is used for sectioning frozen animal carcasses and drying whole-body sections. Phosphor imaging plates (e.g., Fuji Biomedical, Stamford, CT) are used for capturing the whole-body autoradiogram image. Radioactive standards, for example, ^{14}C-glucose or ^{3}H-glucose, are used to build a standard curve for calculation of the radioactivity concentration. An imaging system (e.g., Fuji FLA imaging system, Fuji Biomedical) is used for whole-body image acquisition. Image analysis software (e.g., MCID image analysis software 7.0, InterFocus Imaging Ltd., Cambridge, U.K.) is used for quantification of

ADME-Enabling Technologies in Drug Design and Development, First Edition. Edited by Donglu Zhang and Sekhar Surapaneni.
© 2012 John Wiley & Sons, Inc. Published 2012 by John Wiley & Sons, Inc.

radioactivity concentrations in tissues of a whole-body autoradiogram.

27.3 STUDY DESIGNS

27.3.1 Choice of Radiolabel

The choice of radioisotope, the position of a radiolabel in a drug compound, the radiochemical purity, and the specific activity are important parameters in designing the QWBA studies. ^{14}C and ^{3}H are the most often used isotopes for QWBA studies since both ^{14}C-labeled and ^{3}H-labeled drugs have a long radioactive half-life and can provide good resolution for whole-body imaging. ^{14}C-labeled drugs require a shorter exposure time than ^{3}H-labeled drugs, while ^{3}H-labeled drugs have high specific activity and multiple labeled sites, and thus achieve better imaging resolution than other isotopes if a longer exposure time is allowed. Other isotopes, such as ^{35}S, ^{32}P, and ^{125}I, which have short radioactive half-lives and high specific activities, are also suitable for QWBA (Nair et al., 1992; Osaka et al., 1996; Solon and Kraus, 2002b; Riccobene et al., 2003; Coro et al., 2005). Radiolabeled drugs with high specific activity and purity are needed for QWBA. The radiolabeled drugs usually should have a radiochemical purity of ≥98%, but in certain cases, ≥95% is acceptable. The purity of a radiolabeled drug should be checked before preparation of the dosing solution. The stability of the radiolabeled drug under conditions of administration should also be checked predose and postdose.

27.3.2 Choice of Animals

Many animal species can be used for QWBA studies. Small rodents (rat or mouse) are the most frequently used animals for QWBA studies, but rabbits, small dogs, and monkeys (~5 kg) are also used under specific indications. The key for animal selection is that the animal size must fit into the microtome stage used for whole-body section. The selection of animal should also depend on the objectives of the QWBA study and should match the species used in the toxicological studies. Male animals are normally used for QWBA studies; female animals are also used to compare any difference in the distribution of radioactivity between males and females.

Before dose administration, animals usually complete at least 14 days of quarantine or conditioning. In general, a certified canine or primate diet such as LabChow™ or LabDiet™ (Purina Mills, Inc., St. Louis, MO, USA) is fed to dogs and monkeys, and a certified rodent diet such as LabDiet (Purina Mills Inc.) is given to rats and mice *ad libitum* daily. Tap water is provided to animals *ad libitum*. Mice are fasted for 4 h, and rats, dogs, and monkeys are usually fasted overnight before dose administration.

27.3.3 Dose Selection, Formulation, and Administration

The dose level of drug used for animals in QWBA studies should be toxic to the animals and should provide sufficient circulating radioactivity. For a high-dose drug, a well-tolerated mid-dose should be chosen. The drug vehicle used in the toxicity studies is preferred for the QWBA studies. Dose formulation is usually prepared on the day of dosing. If the drug is stable, a dose solution may be prepared the day before dosing and stored at ≤−20°C. The target dose level (mg/kg), volume (mL/kg), and drug concentration (mg/mL) should be defined in the study protocol.

The radioactivity concentration of a drug (μCi/mg) in formulation should be verified by analyzing triplicate aliquots (~100 μL; diluted if necessary) from the top, middle, and bottom of the dose solution by a liquid scintillation counter (LSC) at predose and postdose. The specifications (including identification, manufacturer, physical description, lot number, specific activity for radiolabeled drugs, expiration date, and storage temperature) of the components of the dose solution should be documented in the study file. Radioactive dose levels should be selected based on the isotope type and dose route to be used. For ^{14}C, ^{35}S, or ^{32}P, which has a relatively higher β-energy, the radioactive dose level range is 50–100 μCi/kg; for ^{3}H, which has a lower β-energy, the radioactive dose level range is 1–2 mCi/kg. The volume of the dose formulation for each animal should be calculated based on the animal body weights. If a syringe is used for dosing, the actual amount of administered dose to each animal is determined by weighing the dosing syringe before and after dose administration.

The dosing route of the drug in animals is usually the same as the route proposed for clinical use. Oral and intravenous administrations are commonly used. Other dosing routes include intraperitoneal, intramuscular, intranasal, and inhalation. Single doses are normally used for QWBA studies, but repeated doses are also used in special cases. One animal is usually used for one time point.

27.4 QWBA EXPERIMENTAL PROCEDURES

27.4.1 Embedding

After dosing animals with radiolabeled drugs, each animal is euthanized immediately at a specified time point. The carcasses of the animals are frozen in a hexane/dry-ice bath at approximately –70°C for at least 15–20 min to reduce the possibility of drug diffusion in tissues and organs. Then each frozen carcass is embedded in a chilled 2% carboxymethylcellulose matrix and mounted on a microtome stage (Leica CM3600 Cryomacrocut) maintained in a hexane/dry-ice bath at approximately –70°C for about 1–2 h.

27.4.2 Whole-Body Sectioning

The frozen block inside which the animal is embedded is mounted in a large cryomicrotome (such as Leica CM3600 Cryomicrotome). The frozen animal block is then trimmed until the tissues and organs of interest appear. When an appropriate level with desired tissues and organs is identified, a transparent collection tape (Scotch Tape No. 8210, 3M Ltd., St. Paul, MN) is adhered to the surface of the animal block. The whole-body section (about 40 μm thick) is taken in a sagittal plane and captured on an adhesive tape After all whole-body sections, which cover all major tissues and organs, are collected, sections are allowed to freeze dry at –20°C in a cryomicrotome for at least 48 h.

27.4.3 Whole-Body Imaging

After drying, whole-body sections are trimmed and a set of whole-body sections for each animal is mounted on a piece of cardboard and covered with a thin plastic wrap. Then the whole-body sections are exposed along with ^{14}C or ^{3}H-glucose as calibration standards (American Radiolabeled Chemicals, St. Louis, MO) to a ^{14}C-sensitive phosphor imaging plate (BAS-MS 2040, Fuji Biomedical) or a ^{3}H-sensitive phosphor imaging plate (BAS TR2040s, Fuji Biomedical). The whole-body sections are exposed on the imaging plate in a way similar to that for X-ray film. The imaging plates with whole-body sections are placed in light-tight exposure cassettes, and then the cassettes are kept in a lead chamber at room temperature to reduce the radiation background noise generated during the exposure. The exposure time is determined based on the type of radioisotope and radioactivity in whole-body sections. For ^{14}C-drugs, the exposure time usually is from a couple of hours to 1 day, and for ^{3}H-drugs, the exposure time could be up to 3–5 weeks.

The imaging plate is a flexible image sensor plate. Numerous small crystals of photostimulable phosphor of barium fluorobromide containing a trace amount of bivalent europium as a luminescence center are uniformly coated with a 150–300-μm-thick polyester support film. After exposure, the imaging plate is scanned using an image scanner with a He-Ne laser beam, which produces red light at the wavelength of 633 mm as the excitation light. A bluish purple (400 nm) photostimulated luminescence (PSL) is then emitted by laser excitation, directed to the photomultiplier tube (PMT), and converted to analog electric signals in chronological order. These signals are subsequently converted to digital signals and the whole-body imaging can be viewed on a computer screen (Shigematsu et al., 1999).

27.4.4 Quantification of Radioactivity Concentration

The concentrations of radioactivity in tissues and organs of whole-body images are quantified using an image analysis software, such as the MCID image analysis software 7.0 (Imaging Research, Inc., St. Catherines, ON). In this software, the PSL values (PSL/mm^2) of the ^{14}C or ^{3}H-calibration standards are determined and then plotted against the concentrations of the standards to construct a calibration curve, which is then used to calculate the concentrations of radioactivity in tissues or organs in whole-body images. Under normal conditions of whole-body autoradiography, the quantification of radioactivity concentration can be achieved at 1–4 nCi/g of wet tissue.

27.5 APPLICATIONS OF QWBA

QWBA has many applications in drug discovery and development. In drug development, QWBA data is used to support regulatory submissions. The QWBA data are used to determine drug tissue distribution and pharmacokinetics, and to predict human radiation dosimetry that might occur during human ADME studies. QWBA is also used in drug discovery to support the selection of new drug candidates and identify, evaluate, and address toxicology issues. In addition, QWBA is a good technique to study placental and brain penetration of drugs in rodents. With its high imaging resolution, the drug distribution in small fetal tissue can also be identified and quantified. Moreover, QWBA is used to answer very specific questions related to a toxic or pharmacologic target or to address ADME issues during drug discovery.

27.5.1 Case Study 1: Drug Delivery to Pharmacology Targets

Tumor uptake of novel folate receptor-targeted epothilone in CD2F1 mice after systemic administration of [^{3}H]drug A (Gan et al., 2009).

27.5.1.1 Study Design [^{3}H]drug A had a specific activity of 20.5 Ci/mmol and a radiochemical purity of >99%. Six to eight-week-old female mice (CD2F1 strain)

TABLE 27.1. Radioactivity in Major Mouse Tissues and Tumors after Administration of [^{3}H]Drug A to Tumor-Bearing Mice

	Radioactivity in Tissue (µg eq./g tissue) (mean ± SD)			
	30 min	2 h	24 h	48 h
M109 (FR-)	1.73 ± 0.59	1.17 ± 0.40	1.20 ± 0.67	0.53 ± 0.13
98M109 (FR+)	5.97 ± 1.44	NC	4.11 ± 2.24	2.29 ± 1.07
Brain	0.11 ± 0.03	LLQ	LLQ	LLQ
Blood	4.08 ± 0.48	0.35 ± 0.08	LLQ	LLQ
Bone marrow	2.27 ± 0.64	1.73 ± 0.21	0.93 ± 0.16	0.40 ± 0.08
Heart	5.63 ± 0.61	2.40 ± 0.11	0.91 ± 0.16	0.59 ± 0.05
Liver	8.45 ± 1.39	4.08 ± 0.53	1.41 ± 0.24	1.07 ± 0.21
Lung	9.28 ± 1.15	2.35 ± 0.48	0.80 ± 0.13	0.40 ± 0.08
Kidney	11.33 ± 4.48	6.03 ± 3.01	1.49 ± 0.27	1.41 ± 0.37
Muscle	0.80 ± 0.35	0.29 ± 0.05	0.35 ± 0.08	0.24 ± 0.05
Skin	4.13 ± 0.99	0.80 ± 0.08	0.40 ± 0.08	0.43 ± 0.05
Salivary gland	3.12 ± 0.48	3.09 ± 0.64	1.87 ± 0.27	1.41 ± 0.19
Spinal cord	0.19 ± 0.03	0.11 ± 0.03	LLQ	LLQ
Intestine/contents	52.61 ± 21.65	45.76 ± 32.27	37.01 ± 26.53	4.69 ± 2.69

LLQ, lowest limit of quantification (0.08 µg-eq./g tissue); NC, not collected.

were purchased from Harlan Laboratories Inc. (Indianapolis, IN), and fed *ad libitum* with normal or folate-deficient rodent chow (Harlan Teklad, Madison, WI) for the duration of the experiment. Since normal rodent chow contains a high concentration of folate acid (6 mg/kg chow), folate receptor overexpressed mice used in these studies were maintained on the folate-free diet for 2 weeks before tumor implantation to achieve serum folate concentrations close to the range of normal human serum. For tumor cell inoculation, 1×10^6 98M109 cells in 100 µL culture medium were injected into the subcutaneous area of the mice's left flank and 1×10^6 M109 cells in 100 µL medium were injected into the subcutaneous area of the mice's right flank. After inoculation, tumors were allowed to grow for 14 days in the mice.

Female mice bearing both a 98M109 tumor (FR+, folate receptor-positive tumor) and a M109 tumor (FR−, folate receptor-negative tumor) for 2 weeks were used for this study. A [^{3}H]drug A dosing solution was freshly prepared in Delbecco's phosphate saline buffer, and the concentration of drug and radioactivity was 1 mg/mL and 2.5 mCi/mL, respectively. Four mice each received a single intravenous dose of [^{3}H]drug A at 4 mg/kg by a tail vein. After dosing, one mouse per time point was euthanatized at 0.5, 2, 24, and 48 h postdose and was frozen in a dry ice-hexane bath at −70°C for 10 min. The tissue distributions of radioactivity in the animals at various times were determined by QWBA.

27.5.1.2 Results
After administration of single intravenous doses of [^{3}H]drug A (4 mg/kg) to the test mice, radioactivity in the mouse tissues and tumors was measured using QWBA. The distribution of radioactivity in mouse tissues and tumors is listed in Table 27.1 and shown in Figure 27.1. In all mice, the highest radioactivity appeared in the gastrointestinal (GI) tract, followed by kidney, lung liver, heart, blood, salivary gland, skin, muscle, and other tissues (Figure 27.1). The highest concentrations of radioactivity in most tissues were reached at 30 min postdose. At 2 h postadministration, radioactivity was eliminated quickly in kidney, lung, liver, heart, and other tissue. And at 48 h after dosing, most administered radioactivity was eliminated from mouse livers.

For tumor sites, at 30 min after administration, a moderate level of radioactivity was detected in 98M109 tumors (FR+) (6 µg eq./g) (Table 27.1 and Figure 27.1). At 24 and 48 h after dosing, radioactivity was 4.1 and 2.3 µg eq./g, respectively, in 98M109 tumors, which declined slightly from the 30-min time point. However, the levels of radioactivity in M109 tumors (FR−) were significantly lower than in 98M109 tumors, and the values were 1.7, 1.2, 1.2, and 0.5 eq./g for 0.5, 2, 24, and 48 h, respectively.

In summary, after an intravenous administration of [^{3}H]drug A to tumor-bearing mice, the tissue distribution of [^{3}H]drug A was extensive. All results indicated the preferential distribution of [^{3}H]drug A from blood into the folate receptor-positive tumor, 98M109.

27.5.2 Case Study 2: Tissue Distribution and Metabolite Profiling

Determination of the radioactivity distribution and metabolite profiles of buspirone in rat tissues using QWBA and LC-MS (Zhu et al., 2003).

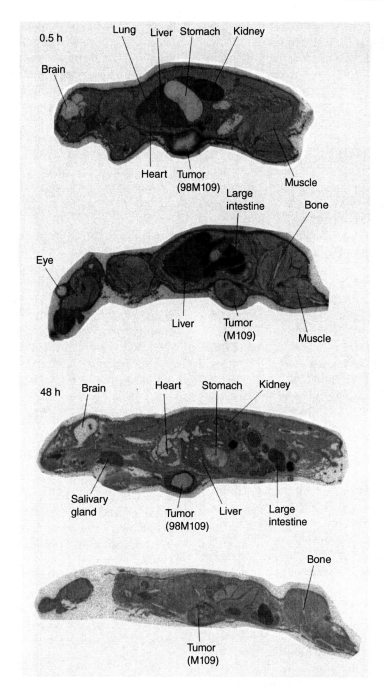

FIGURE 27.1. Representative whole-body autoradiograms of tumor-bearing mice at 0.5 and 48 h after dosing with [^{3}H]drug A.

27.5.2.1 Study Design [^{14}C]Buspirone had a specific activity of 27 μCi/mg and a radiochemical purity of >97.5%. Two bile duct-cannulated male Sprague Dawley rats (body weight: 250–280 g) were used for this study. Animals were administered [^{14}C]buspirone at 20 mg/kg (100 μCi/kg), in a vehicle of 6 mM hydrochloric acid in water at 2 mg/mL. Rats were euthanized by CO_2 inhalation at 1 and 6 h after dosing. Blood was collected into tubes containing K_2EDTA via right jugular venipunc-

ture, and plasma was prepared by centrifugation at 1300 g for 10 min. Each rat carcass was frozen and embedded. Whole-body sections were collected at 40 μm for whole-body autoradiography.

27.5.2.2 Brain Tissue Sampling Whole-body sections at 200-μm thickness were collected on adhesive tape using a Leica CM3600 Cryomicrotome at approximately −20°C. Major organs and tissues (brain, heart, kidney, liver, and lung) were represented in these sections. The brain tissue was scissored out from two or three sections at −20°C and collected into a scintillation vial. The brain sample (about 200 mg) was homogenized in a glass homogenizer with 1 mL of distilled water and the homogenized sample was extracted twice with 1 mL water/acetonitrile (50/50, v/v) and then centrifuged at 3000 rpm for 10 min at 5°C in a Heraeus Biofuge Fresco refrigerated centrifuge (Heraeus Sepatech GmbH, Osterode, Germany). The supernatants were combined and transferred to a glass tube, dried under a stream of nitrogen gas, and then reconstituted with water to approximate 200 μL. Aliquots of concentrated samples were transferred to autosampler vials and injected into the HPLC-MS for metabolite profiling and identification.

27.5.2.3 Metabolite Profiling and Identification

Metabolite profile for plasma and brain samples was performed on a Shimadzu Class VP System (Shimadzu, Columbia, MD) equipped with two pumps (model LC-10AD), an autoinjector (model SIL-10AD), and a diode array detector (model SPD-MA10A) at a flow rate of 1 mL/min. A Zorbax RX-C8 column (Phenomenex, Torrance, CA, USA) (4.6 × 250 mm) and a linear stepwise gradient with solvent A (0.01% trifluoroacetic acid in water) and solvent B (acetonitrile) were used. Solvent B in the gradient started at 8% and then changed as follows: 8% (8 min), 40% (30 min), 90% (35 min), and 8% (40 min). HPLC eluant was collected into Deepwell LumaPlate™-96-well plates coated with solid scintillators (PerkinElmer Life and Analytical Sciences, Waltam, MA, USA) at a rate of 0.25 min per well with a Gilson Model FC 204 fraction collector (Gilson, Middleton, WI), and then dried using a speed vacuum system (SpeedVac AES-2010; Savant Instruments, Holbrook, NY). Radioactivity of the residues in the 96-wells was determined with a TopCount microplate scintillation counter (PerkinElmer Life and Analytical Sciences, Waltham, MA, USA) with a counting time of 10 min/well. Metabolite identification for plasma and brain samples was carried out for metabolite identification by using a capillary LC coupled to an LTQ mass spectrometer. The LTQ mass spectrometer (Thermo Fisher Scientific, San Jose, CA) was operated in positive electrospray ionization mode. The heated capillary temperature was maintained at 300°C; the sheath gas and auxiliary gas flow rates were set at 60 and 15 units, respectively. The ion spray voltage, capillary voltage, and tube lens offset voltage were adjusted to 2, 9, and 85 V, respectively. The normalized collision energy was 30% during MS/MS acquisition, and helium was used as the collision gas.

27.5.2.4 Results

The tissue distribution of drug-related radioactivity is shown in Figure 27.2. At 1 h after dosing, radioactivity distributed extensively in rat tissues. The highest radioactivity was detected in the GI tract and liver, and the radioactivity in the brain, heart, lung, and kidney were relatively low. At 6 h after dosing, radioactivity was still present in the GI tract and liver in high concentrations, and radioactivity in brain, heart, lung, kidney, and other tissues was still quantifiable.

[^{14}C]Buspirone and metabolites in rat plasma and brain samples were profiled at 1 and 6 h (Figure 27.3). The metabolic profiles of plasma were qualitatively similar between 1 and 6 h samples, and metabolites, including 1-pyrimidinylpiperazine (1-PP), 6′-hydroxybuspirone (6′-OH-Bu), and 5,6′-dihydoxybuspirone (5,6′-di-OH-Bu), were the major components accounting for more than 85% of plasma radioactivity. Buspirone was a minor component in plasma samples. The metabolic profiles of brain samples were qualitatively similar between 1 and 6 h. Metabolites 1-PP and 6′-OH-Bu were the major components in brain samples, with buspirone as a minor component. In summary, multiple metabolites were presented in rat blood, but only 1-PP, 6′-OH-Bu, and buspirone can penetrate into rat brain and appeared to be active components.

27.5.3 Case Study 3: Tissue Distribution and Protein Covalent Binding

The covalent binding of [^{14}C]rofecoxib to the arterial elastin (Oitate et al., 2006).

27.5.3.1 Study Design

[^{14}C]Rofecoxib was orally administered at 2 mg/kg as a solution in polyethylene glycol 400. The dosing solutions were prepared shortly before dosing at the same specific radioactivity within the same study (range: 68–370 μCi/kg body weight). Rats were euthanized at 48 h after 3-day repeated administrations of [^{14}C]rofecoxib (2 mg/kg) and at 48 h and 10 days after 7-day repeated administrations of [^{14}C]rofecoxib (2 mg/kg). Then the rats were frozen and whole-body sections were made and imaged.

27.5.3.2 Results

QWBA results showed that at 48 h after 3- or 7-day repeated administrations of [^{14}C]rofecoxib, radioactivity appeared to have accumulated in the aorta and interspinal ligaments in a dosing frequency-dependent manner. In addition, the radioactivity in these tissues was still to be clearly observed 10 days after the 7-day repeated administrations (Figure 27.4), when the radioactivity was almost completely cleared from the rest of the body. In summary, whole-body autoradioluminography and quantitative determination of the tissue concentrations showed that considerable radioactivity is retained and accumulated in the thoracic aorta of rats after oral administration of [^{14}C]rofecoxib, These findings suggested that [^{14}C]rofecoxib

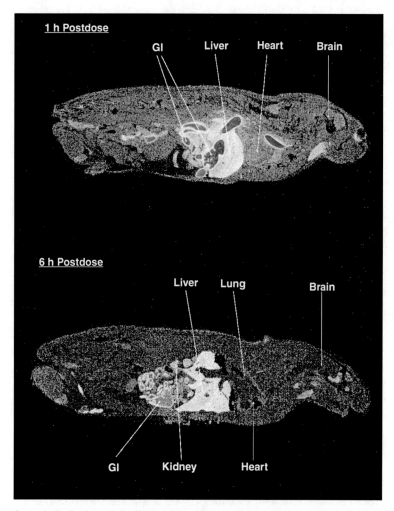

FIGURE 27.2. Representative whole-body autoradiography of a rat after a single oral administration of [^{14}C]buspirone at 20 mg/kg.

and/or its metabolite(s) are covalently bound to elastin in the arteries.

27.5.4 Case Study 4: Rat Tissue Distribution and Human Dosimetry Calculation

27.5.4.1 Human Dosimetry Prediction Method A
Quantitative tissue distribution of drug-related material and tissue distribution of [^{14}C]razaxaban to support a human ADME study. Eleven adult male Sprague Dawley rats were used for this study. Each rat received a single oral dose of [^{14}C]razaxaban at 5 mg/kg (100 µCi/ kg). One rat per time point was euthanatized at 0.5, 1, 3, 6, 10, 12, 18, 24, 30, 48, and 96 h postdose. Whole-body sections were collected to include major tissues and organs. After drying, whole-body sections were exposed along with ^{14}C-calibration standards to a ^{14}C-sensitive

phosphor imaging plate (BAS-MS 2040). After exposure, the imaging plates were scanned and concentrations of radioactivity in tissues and organs were quantified. After single oral doses of [^{14}C]razaxaban to rats, the distribution of radioactivity in rat tissues from 0.5 to 48 h was determined using QWBA and the results are listed in Table 27.2.

Pharmacokinetic and radiation dosimetry parameters (half-life [$t_{1/2}$], maximum concentration C_{max} [µg eq./g], and dose exposure [D]) in tissues were calculated based on the tissue concentration data. The tissue concentration data in the unit of µCi/g-tissue were converted to the unit of µg-equivalents of [^{14}C]-drug/g tissue using Microsoft Excel. The pharmacokinetic and radiation dosimetry parameters were calculated using WinNonLin™ Professional Edition© 2003 v. 4.2 (Pharsight Corp., Mountain View, CA). Effective dose (in mrem) for individual tissues and the whole-body were calculated based on a

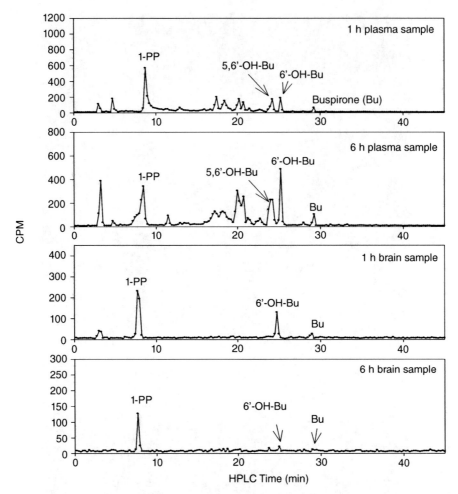

FIGURE 27.3. Metabolic profiles in plasma and brain samples of rats after a single oral administration of [^{14}C]buspirone at 20 mg/kg.

proposed human dose of 100 μCi of radiocarbon given to a 70-kg human.

The predicted radioactive exposure to human tissues following a 100 μCi dose of the radioisotope-labeled test compound was estimated using the following equation (Dain et al., 1994; Solon and Lee, 2002c):

$$D = (73.8)(E_\beta)(C_{max\,(human)})(t_{1/2}) \times 1000,$$

where D = radioactive exposure (mrem); E_β = average β particle energy for ^{14}C (0.156 MeV was used for this determination to provide a conservative estimate of radioactive exposure); $C_{max\,(human)}$ = Maximum radioactivity concentration (μCi/g) in humans, determined by extrapolation as follows: $C_{max\,(human)} = C_{max\,(rat)}$ (μg eq./g) × Human dose (μCi/g)/Rat dose (μg/g). The rat dose used was the mean actual dose administered to all rats in the test (5.03 μg/g). For the planned human study, the radioactive dose for an adult (70-kg male subject) is approxi-

mately 100 μCi (approximately 0.00143 μCi/g for a 70-kg male); $t_{1/2}$ = biological effective half-life (days) of ^{14}C-labeled compound in the rat.

The human dosimetry results are showed in Table 27.3. Based on the QWBA study in Sprague Dawley rats, the whole-body radiation dose in men was estimated to be 0.51942 mrem. This value is well below the exposure limit of 3 rem after a single dose for human isotope studies. In Sprague Dawley rats, the stomach, small intestine, liver, large intestine, and pancreas were the five matrices expected to be exposed to the highest dose of radiation. In humans, these organs are estimated to be exposed to 0.0306, 0.286, 0.0637, 0.0471, and 0.0112 mrem, respectively, at the expected target oral dose of 100 μCi of [^{14}C]razaxaban. All values were well below the exposure limit of 3–5 rem limits specified by the Code of Federal Regulations (21 CFR: part 361.1, 2009) of the United States. Based on these predictions, it is safe to administer 100 μCi of radioactivity to human subjects.

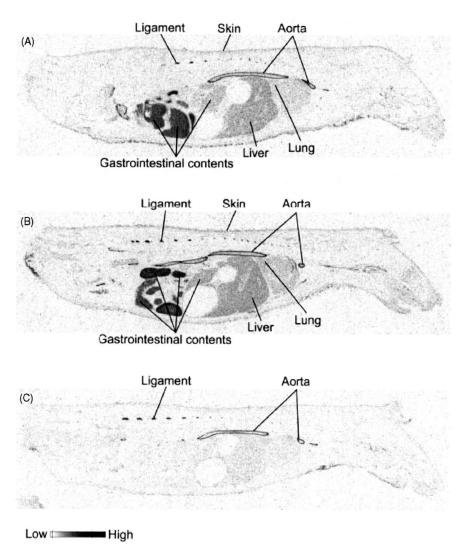

FIGURE 27.4. Whole-body autoradiograms after repeated oral administrations of [^{14}C]rofecoxib to rats at a dose of 2 mg/kg. (A) 48 h after 3-day repeated administrations; (B) 48 h after 7-day repeated administrations; (C) 10 days after 7-day repeated administrations. (Reprinted with permission from *Drug Metabolism and Disposition*.)

27.5.4.2 Human Dosimetry Prediction Method B

When no clinical pharmacokinetic data are available for a compound, human dosimetry can be obtained from prediction method A. When clinical data are available and elimination pathways in preclinical or clinical species are known, prediction method B can be used to obtain potentially more accurate dosing data. In the following section, we will demonstrate a human dosimetry prediction using method B.

The tissue distribution of radioactivity was examined by whole-body autoradiography following a single oral administration of [^{14}C]drug B at 5 mg/kg to male Long Evans rats (one animal/time point, sacrificed at 0.5, 1, 2, 4, 8, 12, 24, 48, 72, 96, and 168 h postdose). The drug

concentrations in rat tissues (µg equivalents of [^{14}C] drug/g tissue) determined by QWBA were calculated using Microsoft Excel. The area under the curve (AUC$_{inf}$ (µg eq.·h/g)), C_{max}, T_{max}, and terminal half-life ($t_{1/2}$) of the drug in each tissue or organ were determined by Kinetica™ 4.4 (Thermo Fisher Scientific Inc., Waltham, MA).

Rat tissue distribution results were used to calculate pharmacokinetic and radiation dosimetry parameters according to recommendations of the Internal Commission on Radiological Protection (ICRP publication 103).

The effective dose (mSv) for individual human organs and the total effective dose for the whole body were calculated based on a dosimetry estimation using an 80 µCi dose in human subjects of 70-kg body weight:

TABLE 27.2. Radioactivity (µg eq. of razaxaban/g tissue) in Tissues of Male Sprague Dawley Rats after a Single PO Dose of [^{14}C]Razaxaban at 5 mg/kg

Tissue	µg eq. of [^{14}C]Razaxaban/g									
	0.5 h	1 h	3 h	6 h	10 h	12 h	18 h	24 h	30 h	48 h
Adrenal gland	0.045	0.038	0.057	NS	NS	NS	NS	NS	NS	NS
Blood (cardiac)	NS	NS	NS	NS	NS	NS	NS	NS	NS	NS
Bone	NS	NS	NS	NS	NS	NS	NS	0.040	0.033	0.034
Bone marrow	NS	NS	NS	NS	NS	NS	NS	BLQ	BLQ	BLQ
Cecum	NS		0.074	0.159	0.259	0.186	0.065	0.0648	0.051	NS
Cecum contents	NS	NS	NS	NS	NS	NS	NS	NS	NS	NS
(Cerebellum)	NS	NS	NS	NS	NS	NS	NS	NS	NS	NS
(Cerebrum)		NS	NS	NS	NS	NS	NS	NS	NS	NS
Diaphragm	NS	NS	NS	NS	NS	NS	NS	NS	NS	NS
Epididymis	NS	NS	NS	NS	NS	NS	NS	NS	NS	NS
Esophagus contents	2.73	9.54	2.30	2.62	0.131	0.107	NS	0.116	NS	NS
Esophagus	0.052	0.084	0.176	0.176	0.095	0.112	NS	0.045	NS	NS
Eye	NS	NS	NS	NS	NS	NS	NS	NS	NS	NS
Fat (abdominal)	NS	NS	NS	NS	NS	NS	NS	NS	NS	NS
Fat (brown)	NS	NS	NS	NS	NS	NS	NS	NS	NS	NS
Harderian gland	NS	NS	NS	NS	NS	NS	NS	NS	NS	NS
Kidney	0.055	0.065	0.040	NS	NS	NS	NS	NS	NS	NS
Large intestinal contents	BLQ	BLQ	BLQ	NS	73.3	145	39.3	7.12	4.32	0.055
Large intestine	NS	NS	NS	NS	0.116	0.229	0.157	0.202	0.085	BLQ
Liver	0.091	0.168	0.281	0.119	0.317	0.108	0.051	0.028	0.043	NS
Lung	NS	NS	NS	NS	NS	NS	NS	NS	NS	NS
Muscle	NS	NS	NS	NS	NS	NS	NS	NS	NS	NS
Myocardium	NS	NS	NS	NS	NS	NS	NS	NS	NS	NS
Pancreas	0.027	0.041	0.033	NS	NS	NS	NS	NS	NS	NS
Pituitary gland	NS	NS	NS	NS	NS	NS	NS	NS	NS	NS
Prostate	NS	NS	NS	NS	NS	NS	NS	NS	NS	NS
Renal cortex	0.050	0.060	0.037	NS	NS	NS	NS	NS	NS	NS
Renal medulla	0.059	0.074	0.041	NS	NS	NS	NS	NS	NS	NS
Salivary gland	NS	NS	NS	NS	NS	NS	NS	NS	NS	NS
Seminal vesicle	NS	NS	NS	NS	NS	NS	NS	NS	NS	NS
Skin	NS	NS	NS	NS	NS	NS	NS	NS	NS	NS
Small intestinal contents	10.9	29.7	164	50.1	12	0.445	0.06	0.541	0.053	NS
Small intestine	0.135	0.299	0.248	0.303	0.134	0.051	BLQ	0.035	BLQ	NS
Spinal cord	NS	NS	NS	NS	NS	NS	NS	NS	NS	NS
Spleen	0.031	0.030	0.032	NS	NS	NS	NS	NS	NS	NS
Stomach	0.152	0.122	0.150	0.069	0.071	BLQ	NS	NS	NS	NS
Stomach contents	107	86.4	69.3	14.3	0.281	0.249	BLQ	0.464	BLQ	NS
Testis	NS	BLQ	NS	NS	NS	NS	NS	NS	NS	NS
Thymus	NS	NS	NS	NS	NS	NS	NS	NS	NS	NS
Thyroid	NS	NS	NS	NS	NS	NS	NS	NS	NS	NS
Urinary bladder	NS	BLQ	BLQ	BLQ	NS	NS	NS	NS	NS	NS

BLQ, below the limit of quantitation (<0.0257 µg eq. [^{14}C]razaxaban/g); NS, not sampled (sample shape not discernible from background).

1. Human dose: 50 mg (80 µCi or 2.96 MBq) of [^{14}C] drug B will be administered as a single oral dose to a human subject. The specific activity is 1.6 µCi/mg or 0.0592 MBq/mg.

   The clinical data has shown that the AUC$_{0-inf}$ after a 50-mg oral dose of drug B is 5668 ng·h/mL in humans. It is assumed that the plasma AUC of total radioactivity of [^{14}C]drug B would not differ significantly from the plasma AUC of the parent drug at the selected dose. Therefore, AUC$_{(0-t)}$ (^{14}C) = 5668 ng eq.·h/mL.

2. Excretion of ^{14}C via renal and fecal tract: The excretion via feces is 80% and via kidney/urine, it is 20% in rats.

TABLE 27.3. Predicted Radiation Dose (mrem) in Male Humans Given 100 μCi of [^{14}C]Razaxaban

Matrix	C_{max} in Rat (μg equiv/g)	C_{max} in Human (μCi/g)	$t_{1/2}$ in Rat (h)	$t_{1/2}$ in Rat (days)	Predicted ^{14}C Exposure in Human (mrem)
Adrenal gland	0.057	0.0000162	4.70	0.196	0.0115
Blood	NA	NA	NA	NA	NA
Bone	0.040	0.0000114	4.70	0.196	0.00804
Bone marrow	NA	NA	NA	NA	NA
Cecum	0.648	0.00018	1.64	0.0682	0.0454
Cerebellum	NA	NA	NA	NA	NA
Cerebrum	NA	NA	NA	NA	NA
Epididymis	NA	NA	NA	NA	NA
Esophagus	NA	NA	NA	NA	NA
Eye	NA	NA	NA	NA	NA
Fat (abdominal)	NA	NA	NA	NA	NA
Fat (brown)	NA	NA	NA	NA	NA
Kidney	0.065	0.0000185	2.86	0.119	0.00794
Large intestine	0.229	0.0000650	4.80	0.200	0.0471
Liver	0.317	0.0000900	4.70	0.196	0.0637
Lung	NA	NA	NA	NA	NA
Muscle	NA	NA	NA	NA	NA
Myocardium	NA	NA	NA	NA	NA
Pancreas	0.041	0.0000116	6.39	0.266	0.0112
Plasma (LSC)	0.0075	0.00000213	4.70	0.196	0.00151
Prostate	NA	NA	NA	NA	NA
Salivary gland	NA	NA	NA	NA	NA
Skin	NA	NA	NA	NA	NA
Small intestine	0.303	0.0000861	22.1	0.921	0.286
Spinal cord	NA	NA	NA	NA	NA
Spleen	0.032	0.00000909	4.70	0.196	0.00643
Stomach	0.152	0.0000432	4.70	0.196	0.0306
Testis	NA	NA	NA	NA	NA
Thymus	NA	NA	NA	NA	NA
Thyroid	NA	NA	NA	NA	NA
Urinary bladder	NA	NA	NA	NA	NA
Total					0.51942

NA, not applicable.

3. $R_{T\,(rat)} = AUC_{T\,(rat)}/AUC_{plasma\,(rat)}$, assuming comparable organ distribution patterns in rats and humans. The organs, whose exposure was not calculated using the transit model (Eq. 27.2), have been included in this calculation.

4. The equations given below have been defined in the recommendations of the International Commission on Radiological Protection and other references (Annals of the ICRP, 1987; Annals of the ICRP, 2007). The fundamental equation for calculation of the radiation dose H_T that has been taken up in a given organ or tissue is

$$H_T = k \times s \times AUC_{(0-\infty)(organ)}\,[mSv], \qquad (27.1)$$

where H_T is the radiation dose in any organ; k is the dose constant for ^{14}C $(=Ep \times f) = 25.6$

(mSv × g/μCi × d); s is the specific radioactivity of dose (20 μCi/mg); and $AUC_{(0-\infty)(organ)}$ is the area under the concentration–time curve $AUC_{(0-\infty)}$ of the ^{14}C-concentration in any organ.

5. The organs of the GI and renal tract were mostly exposed to radioactivity present in the luminal contents of the organs and irradiating the organ wall. The organ dose (i.e., the wall dose) was calculated by a modified equation (ICRP Publication 53, 1987), with the geometry factor 0.5, and with defined residence times (t) in intestinal tract segments and the renal tract, respectively (Table 27.4):

$$H_T = A \times F \times k \times t\ [mSv]/2m, \qquad (27.2)$$

where A is the administered radioactivity dose = 80 μCi (2.96 MBq); k is the dose constant for ^{14}C = 25.6 (mSv × g/μCi × d); t is the residence time (day); m is the organ contents (g); and F is the fraction of the administered dose passing the considered segment of the GI tract and renal tract. For the stomach: Since the test item was orally administered, 100% of the dose passed the stomach lumen. Thus, the radioactive dose in the stomach was calculated using Equation 27.2 with $F = 1.0$. For the small and large intestine: 80% of the administered dose passed the small and large intestine. Thus, the radioactive dose in the large intestine was calculated using Equation 27.2 with $F = 0.80$ for small intestine, and with F = 0.80 for the upper and lower parts of the large intestine. For the renal tract: 20% of the dose passed to the urine. Therefore, the radioactive dose calculation was performed using Equation 27.2 with $F = 0.20$.

In humans, the radiation dose in plasma was calculated based on Equation 27.1:

$$H_{plasma} = k \times s \times AUC_{(0-\infty)(plasma)} \text{ [mSv]}.$$

For individual human organs and tissues, the radiation doses H_T were calculated based on the radiation dose in plasma multiplied by the organ factor R_T, which was the ratio of AUC_T to AUC_{plasma} of ^{14}C in rat. The factor $R_{T (rat)}$ is determined from rat tissue distribution data and is used to calculate human exposure in comparable organs:

$$H_{T (man)} = H_{plasma (man)} \times R_{T (rat)} \text{ [mSv]}. \quad (27.3)$$

For example, the exposure of a human of salivary gland is calculated as follows:

AUC_T for rat salivary gland is 136.16 μg eq.·h/g; AUC_{plasma} for rat is 19.89 μg eq.·h/g; $R_{T (salivary gland)}$ = 136.16/19.89 = 6.85 ; $H_{plasma} = (K) (s) (AUC_{0-inf})/24$ h; K: dose constant for ^{14}C is 25.6 ((mSv) (g)/(μCi) (days)); s: specific activity of dose (1.6 Ci/mg); $H_{plasma} = 25.6 \times 1.6 \times$

0.005,668/24 = 0.0097 mSv; $H_{T (man)}$ for human salivary gland = 0.0097 mSv × 6.85 = 0.0664 mSv.

The whole-body burden Dwb (for stochastic effects), which was in accordance with ICRP publication 103 (2007) was determined from the individual organ and tissue doses (H_T) using Equation 27.3. In this study, only a standard list of mandatory organs and tissues with defined weighting factors was taken into account. This list included the five organs that had the highest W_T value and other most sensitive organs for a total of nine organs. All other organs were considered as remainder organs on average (Table 27.5).

Tables 27.6 and 27.7 summarize the predicted radiation exposure in the human GI and renal tract, individual organs, and whole body after oral administration of 80 μCi ^{14}C-drug. The limits of radiation exposure set by the FDA are 3 rems for whole-body and critical organs and 5 rems for other organs (Code of Federal Regulations 21 CFR: part 361.1.). In this case, the expected effective whole-body dose of the human volunteers weighing 70 kg and receiving a single oral administration of 80 μCi ^{14}C-drug is below the 3 rems (30 mSv) dose limit, and the predicted individual organs exposure is below the 3 or 5 rems (30 or 50 mSv) dose limit. Based on this prediction, it is safe to administer 80 μCi of radioactivity with drug B to human subjects.

27.5.5 Case Study 5: Placenta Transfer and Tissue Distribution in Pregnant Rats

Tissue distribution of [^{14}C]apixaban in pregnant rats—placenta distribution of radioactivity (Wang et al., 2010a,b).

27.5.5.1 Study Design The specific activity was 20.1 μCi/mg and radiochemical purity was >99.8% for [^{14}C]apixaban. The oral dosing solution was prepared on

TABLE 27.4. The Organ Weights and Organ Residence Times in Humans (ICRP Publication 53)

Organ	Organ Content (m) (g)	Organ Residence Time (t) (h)	Organ Residence Time (t) (day)	$k \times t/2m$ (mSv/μCi)
Stomach	250	1	0.0417	0.00213
Small intestine	1040	4	0.167	0.00205
Upper large intestine	220	13	0.542	0.03152
Lower large intestine	135	24	1.0	0.09481
Urinary tract	1400	24	1.0	0.00914

TABLE 27.5. Mandatory Organs and Tissues and Weighting Factors from ICRP Publication 103

Tissues	W_T	ΣW_T
Bone marrow (red), colon, lung, stomach, breast, remainder tissues[a]	0.12	0.72
Gonads	0.08	0.08
Bladder, esophagus, liver, thyroid	0.04	0.16
Bone surface, brain, salivary gland, skin	0.01	0.04
Total		1.00

[a] Remainder tissues: adrenals, extrathoracic region, gall bladder, heart, kidneys, lymphatic nodes, muscle, oral mucosa, pancreas, prostate, small intestine, spleen, thymus. $D_{wb} = \Sigma W_T \times H_T$ [mSv] (Eq. 27.4); H_T: dose taken up in an individual organ or tissue; W_T: weighting factor for that organ or tissue.

TABLE 27.6. Predicted Gastrointestinal and Renal Tract Exposure in Humans Given a Single PO Dose of [^{14}C]Drug B (2.96 MBq or 80 µCi)

Organ	A (µCi)	F (% of Dose)	k (Dose Constant)	t (Residence Time; Days)	m (Organ Weight; g)	H (mSv)
Stomach	80	1	25.6	0.0417	250	0.171
Upper large intestine	80	0.80	25.6	0.542	220	2.018
Lower large intestine	80	0.80	25.6	1	135	6.068
Small intestine	80	0.80	25.6	0.167	1040	0.132
Urinary tract	80	0.20	25.6	1	1400	0.146

Human GI and renal tract tissue exposure.
$H = (A)(F)(k \times t)/2(m)$.
A = administered human radioactive dose (µCi).
F = fraction of dose passing thru organ (0.8—GI tract; 0.2—renal tract).
k = dose constant for ^{14}C 25.6 [(mSv)(g)/(µCi)(days)].
t = ICRP residence time (stomach = 0.0417 d; small intestine = 0.167 d; upper large intestine = 0.542 d; lower large intestine = 1.0 d; renal = 1.0 d).
m = ICRP organ weight (stomach = 250 g; small intestine = 1040 g; upper large intestine = 220 g; lower large intestine = 135 g; renal = 1400 g).

TABLE 27.7. Estimated Organ/Tissue Radiation Doses and Whole-Body Dose in Humans Given a Single PO Dose of [^{14}C] Drug B (2.96 MBq or 80 µCi)

ICRP Tissue/Organs	AUC (µg eq.·h/g)	Rat RT Values	Organ Dose HT (mSv)	Equation Used	Weighting Factor	Human Exposure (mSv)
Stomach	NA		0.1708	2	0.12	0.02050
Lung	26.57	1.34	0.0130	3	0.12	0.00156
Lower large intestine (colon)			6.0681	2	0.12	0.72818
Bone marrow	20.87	1.05	0.0010	3	0.12	0.00122
Breast	18.9	0.95	0.0092	3	0.12	0.00111
Remainder tissue[a]				3	0.12	0.00353
Gonads (testis)	18.77	0.94	0.0091	3	0.08	0.00073
Esophagus	30.61	1.54	0.0146	3	0.04	0.00060
Thyroid	28.50	1.43	0.0014	3	0.04	0.00055
Urinary tract (bladder)	26.67		0.1463	2	0.04	0.00585
Liver	101.40	5.10	0.0495	3	0.04	0.00198
Bone surface	8.89	0.45	0.0044	3	0.01	0.00004
Skin (pigmented)	29.81	1.05	0.0102	3	0.01	0.00010
Brain	0	0	0	3	0.01	0
Salivary gland	136.16	6.68	0.0648	3	0.01	0.00065
Total					1.00	0.76660

[a] Remainder tissues included: adrenals, extrathoracic region, gall bladder, heart, kidneys, lymphatic nodes, muscle, oral mucosa, pancreas, prostate, small intestine, spleen, thymus.

the day of dosing by the addition of [^{14}C]apixaban to a 0.5% Tween 80 in a Labrfil® (w/w) vehicle.

Six pregnant female Sprague Dawley rats (body weight from 270 to 319 g) were used for whole-body autoradiography study. Each rat received a single oral dose of [^{14}C]apixaban at 5 mg/kg (100 µCi/kg). After dosing, rats were returned to their home cages and cage-side observations were performed on the day of dosing and at least daily for the remainder of the study. One pregnant female Sprague Dawley rat was euthanized at 0.5, 1, 4, 8, 24, and 48 h postdose.

27.5.5.2 Results   After single oral doses of [^{14}C]apixaban (5 mg/kg) were administered to pregnant female

TABLE 27.8. Tissue Distribution of Radioactivity in Pregnant Rats Following a Single Oral Dose of [^{14}C]Apixaban at 5 mg/kg

Tissue	μg eq. of Apixaban/g					
	0.5 h	1 h	4 h	8 h	24 h	48 h
Adrenal gland	1.49	1.90	2.30	0.18	NS	NS
Blood	1.43	1.44	1.30	0.09	ND	ND
Bone marrow	0.50	0.33	0.46	0.06	ND	ND
Brain	0.05	0.06	BLQ	BLQ	ND	ND
Brown fat	1.08	0.95	0.87	0.06	ND	ND
Harderian gland	1.00	1.09	1.1	0.19	ND	ND
Heart	0.68	0.93	0.83	0.06	ND	ND
Kidney (cortex)	2.11	2.03	1.93	0.21	0.05	ND
Liver	6.77	7.38	6.13	0.65	0.12	ND
Lung	1.65	1.56	1.33	0.15	BLQ	ND
Muscle	0.37	0.46	0.50	0.05	ND	ND
Pancreas	0.89	0.75	0.99	0.09	ND	ND
Pituitary	0.72	0.93	0.75	0.19	NS	NS
Salivary gland	0.73	0.95	1.03	0.08	ND	ND
Spleen	0.46	0.51	0.47	0.05	BLQ	ND
Thymus	0.38	0.58	0.54	0.05	ND	ND
Placenta	1.02	0.74	1.06	0.10	BLQ	ND
Fetal blood	0.27	0.31	0.51	0.07	ND	ND
Fetal brain	BLQ	0.05	0.06	BLQ	ND	ND

ND, not detectable; BLQ, below limit of quantitation (0.037 μg eq./g tissue); NS, not sampled, since not visualized on autoradiograph, considered as BLQ.

rats, the tissue distribution of radioactivity in the rats was determined using QWBA and the results are listed in Table 27.8. As shown in results, the distribution of [^{14}C]apixaban-derived radioactivity was extensive in the tissues of pregnant rats and only low levels of radioactivity were detected in the tissues of the fetuses (Figure 27.5). Fetal brains showed the lowest level of radioactivity. The C_{max} of radioactivity in fetal tissues was reached at 4 h postdose, and then the radioactivity concentrations in the tissues of fetuses declined quickly and were below the quantitation limit at 24 h postdose. At 48 h postdose, the overall radioactivity in maternal rat and fetuses was below the quantitation limit. The results also showed that radioactivity in the amnionic membrane was higher than in other fetal tissue at all time points from 0.5 to 24 h. The example demonstrates applications of QWBA for determining tissue-selective distribution and use for compound regulatory filing.

In summary, QWBA has been demonstrated to be an effective technique to study the distribution of radiolabeled compounds in animals. Combination with MALDI-MS, which is described in Chapter 26 of this book, will provide additional identification data for drugs and metabolites in tissues of interest at specific time points.

27.6 LIMITATIONS OF QWBA

QWBA has many different applications in drug discovery and development and can answer some specific questions related to the drug tissue distribution or accumulation. However, QWBA only provides a total radioactivity concentration and it cannot differentiate between parent drugs and metabolites; it is difficult to apply the QWBA technique to short half-life isotopes such as ^{90}Y ($t_{1/2}$: 2.67 days) and ^{18}F ($t_{1/2}$: 109.7 min); specific instrumentation is needed for whole-body sections; and QWBA can not provide the information about a radiolabeled drug at the cellular level (Solon et al., 2010).

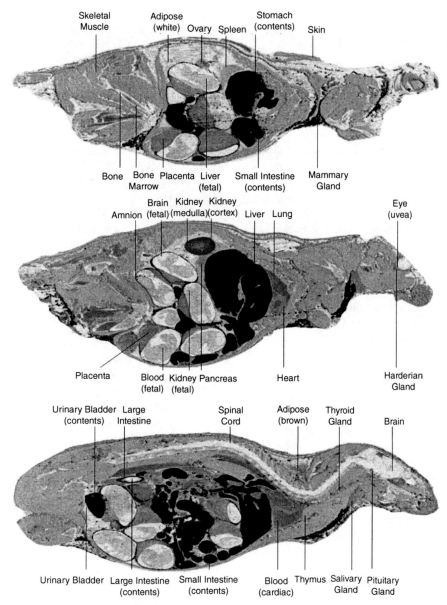

FIGURE 27.5. Representative whole-body autoradiogram of radioactivity distribution in a pregnant female rat at 1 h after a single oral administration of [^{14}C]apixaban at 5 mg/kg.

REFERENCES

21 CFR: Part 361.1 (2009) Radioactive drugs for certain research uses.

21 CRF: Part 361.1 (2009) United States food and drug administration code of federal regulations.

Annals of the ICRP (1987) Radiation dose to patients from radiopharmaceuticals. Appendix A3, Gastrointestinal tract model. ICRP Publication 53. Pergamon, Oxford, 1987. 18:16–18.

Annals of the ICRP (2007) The 2007 recommendations of the international commission on radiological protection. ICRP Publication 103, by Valentin J., ed., Elsevier Ltd.

Christopher LJ, Cui D, 1 Li W, Barros A, Arora VK, Zhang H, Wang L, Zhang D, Manning JA, He K, Fletcher AM, Ogan M, Lago M, Bonacorsi SL, Humphreys WG, Iyer R (2008) Biotransformation of [^{14}C]dasatinib: *In vitro* studies in rat, monkey, and human and disposition after administration to rats and monkeys. *Drug Metab Dispos* 36: 1341–1356.

Coro C, Robert P, Lancelot E, Martinell A, Santus R (2005) Distribution of [^{153}Gd]gadomelitol in a human breast tumor model in mice. *MAGMA* 18:138–143.

Dain JG, Collins JM, Robinson WT (1994) A regulatory and industrial perspective of the use of carbon-14 and tritium isotopes in human ADME studies. *Pharm Res* 11(6): 925–928.

Gan J, Wang L, Shen H, Jian W, Menard K, Hong Y, Tian Y, Zhang D, Zhang D, Bonacorsi SJ, Lee FY (2009) Tissue distribution and tumor uptake of novel folate receptor targeted epothilone, BMS-753493 in CD2F1 mice after systemic administration. *AACR Annual Meeting*.

He K, Lago MW, Iyer RA, Shyu W-C, Humphreys WG, Christopher LJ (2008) Lacteal secretion, fetal and maternal tissue distribution of dasatinib in rats. *Drug Metab Dispos* 36:2564–2570.

Igari Y, Sugiyama Y, Sawada Y, Iga T, Hanano M (1982) Tissue distribution of ^{14}C-diazepam and its metabolites in rats. *Drug Metab Dispos* 10:676–679.

Nair RS, Li AP, Shiotsuka RN, Wilson AGE (1992) Toxicity of calcium sodium metaphosphate fiber: I. *In vitro* and *in vivo* degradation and clearance studies. *Toxicol Sci* 19:69–78.

Oitate M, Hirota T, Koyama K, Inoue S, Kawa K, Ikeda T (2006) Covalent binding of radioactivity from [^{14}C]rofecoxib, but not [^{14}C]celecoxib or [^{14}C]CS-706, to the arterial elastin of rats. *Drug Metab Dispos* 34:1417–1422.

Osaka G, Carey K, Cuthbertson A, Godowski P, Patapoff T, Ryan A, Gadek T, Mordenti J (1996) Pharmacokinetics, tissue distribution, and expression efficiency of plasmid [^{33}P]DNA following intravenous administration of DNA/cationic lipid complexes in mice: Use of a novel radionuclide approach. *J Pharm Sci* 85(6):612–618.

Potchoiba MJ, Nocerini MR (2004) Utility of whole-body autoradioluminography in drug discovery for the quantification of tritium-labeled drug candidates. *Drug Metab Dispos* 32:1190–1198.

Riccobene TA, Miceli RC, Lincoln C, Knight Y, Meadows J, Stabin MG, Sung C (2003) Rapid and specific targeting of ^{125}I-labeled B lymphocyte stimulator to lymphoid tissues and B cell tumors in mice. *J Nucl Med* 44(3): 422–433.

Shaffer CL, Gunduz M, Thornburgh BA, Fate GD (2006) Using a tritiated compound to elucidate its preclinical metabolic and excretory pathways *in vivo*: Exploring tritium exchange risk. *Drug Metab Dispos* 34:1615–1623.

Shigematsu A, Aihara M, Motoji N, Hatori Y, Hamai Y, Asaumi M, Iwai S, Ogawa M, Miura K (1999) Proposition for assessment of quantitative whole-body autoradiography. *Exp Mol Pathol* 67:75–90.

Skotland T, Hustvedt SO, Oulie I, Jacobsen PB, Friisk GA, Langøy AS, Uran S, Sandosham J, Cuthbertson A, Toft KG (2006) NC100668, a new tracer for imaging of venous thromboembolism: Disposition and metabolism in rats. *Drug Metab Dispos* 34:111–120.

Solon E, Kraus L (2002b) Quantitative whole-body autoradiography in the pharmaceutical industry Survey results on study design, methods, and regulatory compliance. *J Pharmacol Toxicol Methods* 46:73–81.

Solon E, Lee F (2002c) Methods determining phosphor imaging limits of quantitation in whole-body autoradiography rodent tissue distribution studies affect predictions of ^{14}C human dosimetry. *J Pharmacol Toxicol Methods* 46:83–91.

Solon E, Schweitzer A, Stoeckli M, Prideaux B (2010) Autoradiography, MALDI-MS, and SIMS-MS imaging in pharmaceutical discovery and development. *AAPS J* 12(1):11–25.

Solon EG, Balani SK, Luo G, Yang TJ, Haines PJ, Wang L, Demond T, Diamond S, Christ D, Gan L-S, Lee FW (2002a) Interaction of ritonavir on tissue distribution of a [^{14}C] L-valinamide, a potent human immunodeficiency virus-1 protease inhibitor, in rats using quantitative whole-body autoradiography. *Drug Metab Dispos* 30:1164–1169.

Wang L, Zhang D, Raghavan N, Yao M, Ma L, Frost CA, Maxwell BD, Chen SY, He K, Goosen TC, Humphreys WG, Grossman SJ (2010a) *In vitro* assessment of metabolic drug-drug interaction potential of apixaban through cytochrome P450 phenotyping, inhibition, and induction studies. *Drug Dispos* 38:448–458.

Wang L, Maxwell B, Humphreys WG, Zhang D (2010b) QWBA analysis of tissue distribution in rats following oral administration of [^{14}C]apixaban. *MDO Meeting, Beijing, China*.

Wang L, Zhang D, Swaminathan A, Xue Y, Cheng PT, Wu S, Mosqueda-Garcia R, Aurang C, Everett DW, Humphreys WG (2006) Glucuronidation as a major metabolic clearance pathway of ^{14}C-labeled muraglitazar in humans: Metabolic profiles in subjects with or without bile collection. *Drug Metab Dispos* 34:427–439.

Xiang H, Nguyen CB, Kelley SK, Dybdal N, Escandón E (2004) Tissue distribution, stability, and pharmacokinetics of apo2 ligand/tumor necrosis factor-related apoptosis-inducing ligand in human colon carcinoma colo205 tumor-bearing nude mice. *Drug Metab Dispos* 32:1230–1238.

Yu RZ, Kim T-W, An Hong A, Watanabe TA, Gaus HJ, Geary RS (2007) Cross-species pharmacokinetic comparison from mouse to man of a second-generation antisense oligonucleotide, ISIS 301012, targeting human apolipoprotein B-100. *Drug Metab Dispos* 35:460–468.

Zhang D, He K, Raghavan N, Wang L, Mitroka J, Maxwell BD, Knabb RM, Frost CA, Schuster A, Hao F, Zheming Gu Z, Humphreys WG, Grossman SJ (2009) Comparative metabolism of ^{14}C-labeled apixaban in mice, rats, rabbits, dogs, and humans. *Drug Metab Dispos* 37:1738–1748.

Zhu M, Wang L, Gedamke R, Zhao W, Zhang D, Gozo S (2003) Determine radioactivity distribution and metabolite profiles in rat tissues using a combination of quantitative whole-body autoradiography (QWBA), microplate scintillation counting (MSC) and capillary LC/ion trap MS. *ISSX Meeting*.

PART D

NEW AND RELATED TECHNOLOGIES

28

GENETICALLY MODIFIED MOUSE MODELS IN ADME STUDIES

Xi-Ling Jiang and Ai-Ming Yu

28.1 INTRODUCTION

Drug discovery and development is a labor-intensive and time-consuming process, which involves thorough investigation of the chemical, pharmacological, and toxicological properties of new drug candidates. Understanding the pharmacokinetics of a drug when it enters the body is essential for preclinical and clinical research. Indeed, xenobiotic drugs may be subjected to diverse pathways, including absorption, distribution, metabolism, and excretion (ADME). In particular, drug metabolism or biotransformation catalyzed by Phase I oxidative enzymes (e.g., cytochrome P450 [P450 or CYP]) or Phase II conjugative enzymes (e.g., uridine 5′-diphospho-glucuronosyltransferase [UGT]) occurs mainly in the liver and small intestine (Guengerich, 2006). Intestinal absorption, tissue distribution, and renal and biliary elimination of drugs may be limited or facilitated by a variety of efflux (e.g., adenosine triphosphate [ATP]-binding cassette [ABC] family) or influx (e.g., solute carrier [SLC] family) membrane transporters (Beringer and Slaughter, 2005). The interplay between drugs and enzymes/transporters inevitably affects drug ADME profiles, which may consequently alter drug responses. In addition, different degrees of transcriptional regulation of enzyme and transporter expression, mediated by xenobiotic receptors (e.g., aryl hydrocarbon receptor [AHR], pregnane X receptor [PXR], and constitutive androstane receptor [CAR]), may cause considerable variations in drug metabolism and disposition (Tirona and Kim, 2005). Therefore,

understanding the functions of these enzymes, transporters, and xenobiotic receptors is essential for prediction of pharmacokinetics, achievement of desired therapeutic drug effects, and prevention of unwanted adverse drug reactions.

Many models have been developed and utilized for preclinical ADME research. Animal models such as rodent, dog, rabbit, and monkey are commonly used to predict pharmacokinetics in humans because these mammals share many similarities in anatomy, physiology, and genetics with humans (Martignoni et al., 2006; Mahmood, 2007). Mouse models are of particular value. Their small size is associated with relatively low husbandry and maintenance costs. Their short life span, large litter size, and fast breeding rate allow for growth of a large number of animals in a relatively short period of time, and therefore many studies can be conducted in a relatively shorter duration. In addition, approximately 99% of human genes have orthologs within the mouse genome (Guenet, 2005). With the development of molecular biological techniques, a target gene may be disrupted and a new gene may be inserted into mouse genome (Henderson and Wolf, 2003; Manis, 2007). Indeed, use of the resulting gene knockout and transgenic mouse models has advanced our understanding of the roles of specific drug metabolizing enzymes, transporters, or xenobiotic receptors in drug metabolism and disposition at the systemic level.

However, prediction of ADME in humans from animal data is not always straightforward due to complex differences in the ways that drugs are processed that are

ADME-Enabling Technologies in Drug Design and Development, First Edition. Edited by Donglu Zhang and Sekhar Surapaneni.

particular to a species (Martignoni et al., 2006; Muruganandan and Sinal, 2008). For instance, metabolic elimination of a drug may be high in mice (and other animal models), whereas it is actually low in humans. Metabolic drug–drug interactions (DDIs) may be minimal in mice, whereas it is alarmingly high in humans or vice versa. There are also human-specific metabolites that cannot be produced from the same drug or generated to similar levels in wild-type mouse models, whose pharmacological and toxicological properties need be understood. Species differences in ADME may be caused by the difference in composition, expression, and/or function of drug metabolizing enzymes, transporters or xenobiotic receptors (Dai and Wan, 2005; Xia et al., 2007; Cheung and Gonzalez, 2008). One approach to overcome species differences is to develop and use humanized transgenic mouse models, which may be achieved by breeding a knockout mouse with a transgenic mouse, both of which are created separately, or directly replacing a target mouse gene with the corresponding human gene or cDNA ("knockin") (Henderson and Wolf, 2003; Liggett, 2004). Studies with these humanized mouse models may not only provide mechanistic understanding of species difference in ADME but also improve prediction of pharmacokinetics in humans.

This chapter is to provide a brief review of drug metabolizing enzyme, transporter and xenobiotic receptor gene knockout, and transgenic and humanized mouse models that have been developed and utilized in ADME studies. More information may be found in many excellent review papers published recently (Henderson and Wolf, 2003; Dai and Wan, 2005; Gonzalez and Yu, 2006; Xia et al., 2007; Cheung and Gonzalez, 2008; Lin, 2008; Muruganandan and Sinal, 2008; Lagas et al., 2009).

28.2 DRUG METABOLIZING ENZYME GENETICALLY MODIFIED MOUSE MODELS

28.2.1 CYP1A1/CYP1A2

Human CYP1A1 and CYP1A2 are involved in the biotransformation of a variety of drugs and toxicants (Ma and Lu, 2007). Due to a high degree of interspecies conservation in CYP1A1/1A2 expression, function, and gene regulation, Cyp1a1(-/-) and Cyp1a2(-/-) mouse lines have been generated and used to study the function of human CYP1A orthologs (Pineau et al., 1995; Liang et al., 1996; Dalton et al., 2000). The Cyp1a2(-/-) mouse developed in Dr. Gonzalez's laboratory (Pineau et al., 1995) is the first reported P450 knockout mouse model (Table 28.1). The Cyp1a2(-/-) mice were created by disruption of exon 2 of Cyp1a2 on a mixed 129 × C57BL/6 mouse background, and exhibited

neonatal lethality associated with severe respiratory distress. Another Cyp1a2(-/-) mouse model was created in Dr. Nebert's laboratory by targeted disruption of exons 2–5 of Cyp1a2 on a mixed 129 × CF-1 background (Liang et al., 1996), and the mice did not show any apparent respiratory defects or other obvious phenotype. Such difference might be attributed to the difference in mouse background or targeting strategy (Henderson and Wolf, 2003). Nevertheless, human and rodent CYP1A isozymes do exhibit some difference in substrate selectivity and metabolic capacity (Bogaards et al., 2000), and extrapolation of rodent data to humans is prone to error. Therefore, CYP1A1_1A2 humanized mice were developed by crossing the Cyp1a1(-/-), Cyp1a2(-/-), or Cyp1a1/1a2(-/-) mice with the CYP1A1_1A2(+/+) mice, the latter of which were generated by incorporating a bacterial artificial chromosome (BAC) containing human CYP1A1 and CYP1A2 into the mouse genome (Jiang et al., 2005; Dragin et al., 2007). The Cyp1a1(-/-), Cyp1a2(-/-), and CYP1A1_1A2 humanized mouse models have been used for studying the role of CYP1A1/1A2 in the pharmacokinetics of a number of drugs, including caffeine and theophylline (Buters et al., 1996; Derkenne et al., 2005), and in metabolic detoxification or activation of many environmental carcinogens, including polycyclic aromatic hydrocarbons and heterocyclic aromatic amines/amides (Uno et al., 2004, 2006; Cheung et al., 2005a). Accordingly, these genetically modified mice are unique animal models for assessing the role of CYP1A1/1A2 in ADME and the risks of xenobiotics.

28.2.2 CYP2A6/Cyp2a5

Human CYP2A6 is involved in the biotransformation of several important therapeutic drugs and procarcinogens (Ingelman-Sundberg et al., 2007). It shares many common substrates (e.g., coumarin, nicotine, and cotinine) with Cyp2a5, a mouse ortholog (Su and Ding, 2004; Wong et al., 2005; Zhou et al., 2009). In addition, CYP2A6 expression is highly variable in humans, and CYP2A6 genetic variations have been shown to be associated with risks of smoking- and tobacco-related cancer (Rossini et al., 2008). To assess the role of CYP2A6 in drug metabolism in vivo, a CYP2A6(+/+) mouse model with liver-specific expression of human CYP2A6 has been generated using a full-length CYP2A6 cDNA, driven by the mouse liver-specific transthyretin promoter/enhancer (Zhang et al., 2005) (Table 28.1). More recently, a Cyp2a5(-/-) mouse model, in which exon 9 of Cyp2a5 was disrupted with a tk-neo-bpA cassette, has been developed on the C57BL/6J background (Zhou et al., 2009). Both CYP2A6 transgenic and Cyp2a5 knockout mice exhibited normal viability and fertility with no

TABLE 28.1. Drug Metabolizing Enzyme Genetically Modified Mouse Models

Gene	Category	Example Application	Reference
Cyp1a1	Knockout	Gene expression and function	Dalton et al. (2000)
Cyp1a2	Knockout	Potential physiological role of Cyp1a2	Pineau et al. (1995)
		Metabolism of xenobiotics	Liang et al. (1996)
Cyp1a1/1a2	Knockout	Assess the risk of environmental toxicant	Dragin et al. (2007)
CYP1A1/1A2	Humanized		
Cyp1b1	Knockout	Evaluate the role of Cyp1b1 in DMBA carcinogenesis	Buters et al. (1999)
Cyp2a5	Knockout	Role of Cyp2a5 in clearance of nicotine and cotinine	Zhou et al. (2009)
CYP2A6	Transgenic	Function of CYP2A6 in coumarin metabolism	Zhang et al. (2005)
CYP2D6	Transgenic/humanized	Debrisoquine metabolism and disposition	Corchero et al. (2001)
		Indolealkylamine metabolic pharmacogenetics	Yu et al. (2003b)
Cyp2e1	Knockout	Acetaminophen toxicity	Lee et al. (1996)
CYP2E1	Humanized		Cheung et al. (2005b)
CYP3A4	Transgenic	Intestinal metabolism of midazolam, drug–drug interaction, and physiological regulation	Granvil et al. (2003) Yu et al. (2005)
Cyp3a	Knockout	Docetaxel metabolism and pharmacokinetics	van Herwaarden et al. (2007)
CYP3A4	Humanized		
Cpr	Knockout	Drug metabolism and toxicity	Henderson et al. (2003) Gu et al. (2003); Wu et al. (2003)
Gstp	Knockout	DMBA carcinogenesis	Henderson et al. (1998)
		Acetaminophen toxicity	Henderson et al. (2000)
Sult1e1	Knockout	Physiological and pathological role of Sult1e1	Qian et al. (2001); Tong et al. (2005)
UGT1	Transgenic	Tissue distribution, induction and hormonal regulation	Chen et al. (2005)
Ugt1	Knockout	Physiological role in bilirubin homeostasis	Nguyen et al. (2008)

apparent developmental defects. These mouse models have been used to successfully delineate the role of CYP2A6/Cyp2a5 in systematic clearance of CYP2A substrate drugs such as coumarin, nicotine, and cotinine (Zhang et al., 2005; Zhou et al., 2009).

28.2.3 CYP2C19

Human *CYP2C19* is an important polymorphic drug metabolizing enzyme involved in biotransformation of many drugs, including antidepressants and proton pump inhibitors (Ingelman-Sundberg et al., 2007). A *CYP2C18/2C19* transgenic mouse model has been developed by introducing a *CYP2C18/2C19*-containing BAC clone into the C57BL/6 mouse genome (Lofgren et al., 2008, 2009). Sexually dimorphic expression and regulation of CYP2C18/C19 were observed at the

mRNA level in transgenic mouse livers and kidneys, while only CYP2C19 protein was detectable by immunoblot in heterozygous transgenic (*CYP2C18/19(+/-)*) mice. Studies also showed that liver microsomes from *CYP2C18/19(+/−)* mice exhibited higher catalytic capacity in *R*-omeprazole and *S*-mephenytoin metabolism than wild-type mice. These findings suggest that this transgenic mouse model may be employed for investigation of sex-dependent regulation of *CYP2C* and evaluation of CYP2C19-mediated drug metabolism *in vivo* (Lofgren et al., 2008, 2009).

28.2.4 CYP2D6

Human CYP2D6, one of the most important polymorphic P450 enzymes, participates in the biotransformation of many xenobiotic drugs and some endogenous neuroregulators (Zanger et al., 2004; Gonzalez and Yu, 2006; Yu, 2008). Considerable species difference is known for CYP2D6-mediated metabolism between mice and humans. *CYP2D6* is the only functional *CYP2D* gene in humans, whereas none of the nine murine *Cyp2d* genes appears to encode proteins with the same enzymatic activity as CYP2D6 (Bogaards et al., 2000). Hence, a humanized *CYP2D6(+/+)* mouse model was generated on an FVB/N background using a λ phage genomic clone that carried the whole wild-type *CYP2D6* gene (Corchero et al., 2001) (Table 28.1). Functional CYP2D6 protein was found in the livers, intestines, and kidneys of the *CYP2D6(+/+)* mice, leading to an elevated metabolism and elimination of debrisoquine. Urinary debrisoquine metabolic ratio (UMR) profiles (debrisoquine/4-hydroxydebrisoquine) in *CYP2D6(+/+)* mice and wild-type mice closely resembled those in human CYP2D6 extensive metabolizers and poor metabolizers, respectively. This finding suggests that *CYP2D6(+/+)* mice could be an unparalleled animal model for evaluation of CYP2D6 pharmacogenetics. Indeed, the *CYP2D6(+/+)* mice have been used to successfully define the role of CYP2D6 in pinoline metabolism (Jiang et al., 2009), as well as the effects of CYP2D6 status on harmaline pharmacokinetics and pharmacodynamics (Wu et al., 2009). On the other hand, when desipramine (DMI), another well-known CYP2D6 substrate, was administered to *CYP2D6*-humanized and wild-type control mice, the UMRs of DMI did not differ much between the two genotyped mice (Shen and Yu, 2009). The results indicate that mouse enzymes contribute notably to DMI 2-hydroxylation, which is primarily mediated by CYP2D6 in humans (Spina et al., 1984). In addition, this mouse model was used to explore endogenous substrates for CYP2D6 (Yu et al., 2003a,b,c), indicating its utility in evaluation of CYP2D6-catalyzed biotransformation in the whole body system.

28.2.5 CYP2E1

Human CYP2E1, an ethanol-inducible P450, is of particular interest due to its important role in the biotransformation of many toxicologically important organic solvents (e.g., ethanol, benzene, and chloroform), procarcinogens (e.g., *N,N*-dimethylnitrosamine), and therapeutic drugs (e.g., acetaminophen [APAP] and chlorzoxanzone) (Cederbaum, 2006; Gonzalez and Yu, 2006). Owing to a high degree of conservation of substrate specificity between mouse and human CYP2E1 (Martignoni et al., 2006; Muruganandan and Sinal, 2008), a *Cyp2e1(-/-)* mouse model was generated by deleting exon 2 of *Cyp2e1* on a mixed 129/SV × C57BL/6N mouse background (Lee et al., 1996) (Table 28.1). The *Cyp2e1(-/-)* mice displayed normal fertility and viability,

and were less sensitive to the hepatotoxic effects of APAP than wild-type mice. This finding supports the critical role of Cyp2e1 in the biotransformation of APAP into an active hepatotoxic metabolite. Nevertheless, species differences in CYP2E1 regulation and catalytic activity were also documented between mice and humans (Bogaards et al., 2000). Hence, a *CYP2E1*-humanized mouse model was developed by introducing a *CYP2E1*-containing BAC clone into the mouse genome and cross-breeding the *CYP2E1* transgenic mice with *Cyp2e1(-/-)* mice (Cheung et al., 2005b). *CYP2E1* transgene was inducible by acetone in the humanized mouse model. A difference in APAP toxicity was also shown between *CYP2E1*-humanized and wild-type mice. The *CYP2E1*-humanized mice may be useful to assess drug metabolism as well as pharmacological and toxicological effects of CYP2E1 substrates (Cheung et al., 2005b).

28.2.6 CYP3A4

Human CYP3A4 metabolizes a large set of drugs that includes benzodiazepines, human immunodeficiency virus (HIV) antivirals, immunomodulators, and steroids (Gonzalez and Yu, 2006). In addition, CYP3A4 is the most abundant P450 enzyme in the human liver and gut, and contributes to the first-pass metabolism of a number of drugs such as midazolam, nifedipine, and verapamil. Although mouse and male rat liver microsomes show similar testosterone 6β-hydroxylation activity as human liver microsomes (Bogaards et al., 2000), intestinal midazolam and nisoldipine oxidation rates are much lower in rats than those in humans (Komura and Iwaki, 2008). In addition, DDI between ketoconazole and midazolam is much stronger in humans than it is in rats (Kotegawa et al., 2002). To develop a *CYP3A4*-transgenic mouse model, a BAC clone containing a complete *CYP3A4* gene was used (Granvil et al., 2003) (Table 28.1). Interestingly, CYP3A4 expression in the transgenic mouse livers was subject to developmental and hormonal regulation (Yu et al., 2005; Cheung et al., 2006). In contrast, CYP3A4 was constitutively expressed in the small intestine of the transgenic mice. Comparative pharmacokinetic studies using adult male mice with CYP3A4 expression only in the gut support the role of CYP3A4 in first-pass metabolism of midazolam (Granvil et al., 2003). Furthermore, the impact of ketoconazole on midazolam pharmacokinetics in transgenic mouse models seems to be closer to that in humans (Table 28.2). These findings suggest that this *CYP3A4*-transgenic mouse model may offer a better assessment of intestinal CYP3A4-mediated drug metabolism and DDI. Nevertheless, the expression of some mouse P450 enzymes (e.g., Cyp3a and Cyp2c) was altered in *CYP3A4*-

TABLE 28.2. Effects of Ketoconazole on Midazolam Pharmacokinetics (AUC$_{(ketoconazole)}$/AUC$_{(control)}$) in Humans and Animal Models

	Intravenous Administration	Oral Administration	Reference
Humans	5	16	Tsunoda et al. (1999)
Rats	2	5	Kotegawa et al. (2002)
Wild-type mice	not available	3	Granvil et al. (2003)
Tg-CYP3A4 mice	not available	8	Granvil et al. (2003)

transgenic mouse livers, which might complicate interpretation of the data obtained from these transgenic mouse models (Yu et al., 2005; Felmlee et al., 2008).

Another *CYP3A4*-transgenic mouse model was generated in Dr. Schinkel's laboratory using an ApoE promoter-HCR1-driven expression cassette that contained *CYP3A4* cDNA (van Herwaarden et al., 2005) (Table 28.1). CYP3A4 was specifically expressed in transgenic mouse livers, which reduced the exposure to intravenously administered midazolam and cyclosporine A. Likewise, a villin-*CYP3A4*-transgenic mouse model was developed in which CYP3A4 was specifically expressed in the small intestines (van Herwaarden et al., 2007). Furthermore, *Cyp3a(-/-)* mice lacking all functional murine *Cyp3a* genes were created and crossed with the ApoE- and villin-*CYP3A4*-transgenic mice to produce liver (Cyp3a-/-A) and intestine (Cyp3a-/-V) *CYP3A4*-humanized mice, respectively (van Herwaarden et al., 2007). Extensive pharmacokinetic studies using these *CYP3A4*-genetically modified mouse models revealed that Cyp3a determined docetaxel pharmacokinetics, and consequently affected docetaxel toxicity (van Herwaarden et al., 2007). In addition, studies using these mouse models showed that intestinal CYP3A4 predominantly affected the systemic exposure to orally administered triazolam, and DDI between ketoconazole and triazolam depended on CYP3A4 status (van Waterschoot et al., 2009). These genetically modified mouse lines are useful models for better understanding of Cyp3a-mediated drug metabolism and DDIs. Rather, significant change in mouse Cyp2c expression was observed in the Cyp3a knockout mice (van Waterschoot et al., 2008). Since Cyp2c enzymes contribute to the metabolism of many Cyp3a substrates, it is necessary to be cautious when extrapolating data obtained from genetically modified mouse models.

28.2.7 Cytochrome P450 Reductase (CPR)

Human CPR, a NADPH : hemoprotein oxidoreductase, is the sole electron donor for microsomal P450s in catalyzing biotransformation of xenobiotic drugs. Disrup-

tion of CPR, therefore, would cause a global inactivation of reactions mediated by microsomal P450s. Since the ubiquitous deletion of *Cpr* led to embryonic lethality (Shen et al., 2002; Otto et al., 2003), a *Cre-loxP* approach was employed to create conditional knockout mice lacking Cpr expression specifically in liver tissues (Henderson et al., 2003; Wu et al., 2003). Other *Cpr(-/-)* mouse models were also generated, in which expression of the *Cpr* gene was controlled either by tissue-specific deletion, such as intestine (Finn et al., 2007; Zhang et al., 2009), lung (Weng et al., 2007), and cardiomyocytes (Fang et al., 2008b), or by global down-regulation, such as *Cpr*-low mice (Wu et al., 2005; Gu et al., 2007). These *Cpr* knockout or knockdown mouse models have been successfully utilized to investigate the physiological role of Cpr (Gu et al., 2003; Wu et al., 2005; Mutch et al., 2006) and the importance of Cpr in metabolic clearance and/or detoxification of xenobiotic drugs (Pass et al., 2005; Henderson et al., 2006; Weng et al., 2007; Fang et al., 2008a,b; Zhang et al., 2009).

28.2.8 Glutathione *S*-Transferase pi (GSTP)

Human GSTP is one of the most important glutathione *S*-transferase (GST) enzymes that catalyze the conjugation of glutathione (GSH) with electrophilic compounds, and are responsible for cellular defense against endobiotic and xenobiotic toxins (Meiers et al., 2007). Furthermore, GSTP is involved in carcinogenesis, and affects susceptibility to cancerous diseases in humans (Meiers et al., 2007; Simic et al., 2009). To evaluate the *in vivo* functions of GSTP, a *GstP1/P2(-/-)* mouse model was generated by disruption of the murine *Gstp* gene cluster (Henderson et al., 1998). Ablation of *GstP* did not cause obvious signs of distress or illness to the mice, but some *GstP*-null mice had a higher body weight. Well-controlled studies using the *GstP* knockout mouse models have demonstrated that GstP may be an important determinant of 7,12-dimethylbenz anthracene (DMBA)-induced carcinogenesis (Henderson et al., 1998), APAP-induced hepatotoxicity (Henderson et al., 2000), and tobacco-induced endothelial dysfunction (Conklin et al., 2009).

28.2.9 Sulfotransferase 1E1 (SULT1E1)

Estrogen sulfotransferase or SULT1E1 belongs to a group of SULT enzymes that mediate conjugation of sulfate with many hormones, neurotransmitters, and xenobiotic drugs (Lindsay et al., 2008). In particular, SULT1E1 catalyzes the sulfation of estrogens at the 3-hydroxyl position and contributes to estrogen homeostasis. A *Sult1e1(-/-)* mouse model was generated to examine the physiological role of Sult1e1 (Qian et al., 2001). As the null mice had no Sult1e1 expression in their testes, males developed age-dependent Leydig cell hypertrophy/hyperplasia and seminiferous tubule damage, which could be recapitulated by estradiol supplementation. These findings suggest that Sult1e1 plays an important role in estrogen metabolism and in intracrine and paracrine estrogen regulation (Qian et al., 2001). Further studies in *Sult1e1(-/-)* female mice demonstrate that Sult1e1 is a critical estrogen modulator in the placenta and affects estrogen levels and placental thrombosis (Tong et al., 2005). Nevertheless, this unique mouse model has not been reported for any ADME studies.

28.2.10 Uridine 5'-Diphospho-Glucuronosyltransferase 1 (UGT1)

UGTs are a superfamily of enzymes that convert many endogenous compounds (e.g., steroids, bile acids, hormones, and bilirubin) and numerous xenobiotic agents (e.g., medications, environmental toxicants, and carcinogens), as well as their oxidative metabolites, into hydrophilic glucuronides that are readily excreted from the body. Of particular note, enzymes in the UGT1 subfamily are more favored in the metabolism of xenobiotic drugs (Kiang et al., 2005). A BAC clone containing the entire human UGT1 locus was employed for the development of a *UGT1* transgenic mouse model (Chen et al., 2005). UGT1A proteins were differentially expressed in the livers and gastrointestinal tracts of *UGT1(+/+)* mice, and subjected to hormonal regulation. The *UGT1* transgene was also inducible by several ligands of xenobiotic receptors, including AHR, PXR, and peroxisome proliferator-activated receptor α (PPARα) (Chen et al., 2005; Bonzo et al., 2007; Senekeo-Effenberger et al., 2007). In addition, disruption of the *Ugt1* gene cluster resulted in severe hyperbilirubinemia in the newborn *Ugt1(-/-)* mice that resembles Crigler–Najjar type I disease in humans as well as an altered expression of some drug metabolizing enzymes (e.g., *Cyp2b9*) (Nguyen et al., 2008). The UGT1 transgenic and knockout mice would be useful models to investigate the regulation and function of glucuronidation as well as the physiological and pathological roles of UGT1.

28.3 DRUG TRANSPORTER GENETICALLY MODIFIED MOUSE MODELS

28.3.1 P-Glycoprotein (Pgp/MDR1/ABCB1)

Human Pgp, encoded by the *MDR1/ABCB1* gene, is expressed at high levels on the apical membrane of enterocytes, the biliary surface of hepatocytes, and the luminal (apical) side of kidney proximal tubule cells. In addition, MDR1 is abundantly expressed on the luminal membrane of brain microvessel endothelial cells. As numerous xenobiotic drugs and metabolites are MDR1 substrates, expression of MDR1 limits the entry or enhances the removal of these agents from particular cells/tissues (Borst et al., 1999). Thus, MDR1 not only contributes to multidrug resistance in chemotherapy but also affects drug absorption, distribution, and excretion (Zhou, 2008). In mice, Mdr1a and Mdr1b together provide a multidrug resistance function that is similar to that provided by human MDR1 (Borst et al., 1999). To define the physiological role of Pgp, mice deficient in *Mdr1a* (*Mdr1a(-/-)*) (Schinkel et al., 1994) or both *Mdr1a* and *Mdr1b* (*Mdr1a/b(-/-)*) (Schinkel et al., 1997) were created (Table 28.3). The knockout mice were viable and fertile and showed no physiological abnormalities. Studies in wild-type versus knockout mice demonstrated that absence of Mdr1a or Mdr1a/1b resulted in significant changes in brain penetration, pharmacokinetics, and toxicity of many drugs, such as ivermectin, vinblastine, dexamethasone, digoxin, cyclosporine A, ondansetron, loperamide, and rhodamine (Schinkel et al., 1994, 1995, 1996, 1997). These findings indicate that Mdr1 knockout mice are useful animal models to assess the pharmacological and toxicological roles of Pgp transporters. Rather, an obvious increase in Mdr1b expression in *Mdr1a(-/-)* mice (Schinkel et al., 1994) reminds one to use caution when extrapolating data from transgenic mice to humans.

28.3.2 Multidrug Resistance-Associated Proteins (MRP/ABCC)

Human MRP/ABCC membrane transporters share some substrate specificity with MDR1/ABCB1 (Lagas et al., 2009). Disruption of individual *Mrp1-4* did not affect the fertility of knockout mice (Lorico et al., 1997; Wijnholds et al., 1997; Leggas et al., 2004; Belinsky et al., 2005; Chu et al., 2006; Vlaming et al., 2006; Zelcer et al., 2006). These knockout mice did not show any physiological abnormalities, except that *Mrp2(-/-)* mice displayed mild hyperbilirubinemia due to an impaired biliary excretion of bilirubin monoglucuronides (Vlaming et al., 2006). These mouse models have been utilized to investigate the function of specific MRP transporters in drug disposition.

TABLE 28.3. Drug Transporter Genetically Modified Mouse Models

Gene	Category	Example Application	Reference
Bcrp1 (Abcg2)	Knockout	Drug absorption, distribution, and elimination, including the secretion of compounds into milk	van Herwaarden et al. (2003); Jonker et al. (2005)
Bsep (Abcb11)	Knockout	Bile acid distribution and excretion	Wang et al. (2001, 2003)
Mate1 (Slc47a1)	Knockout	Renal secretion of metformin	Tsuda et al. (2009)
Mdr1a (Abcb1a)	Knockout	Drug absorption, distribution, and elimination, including brain penetration	Schinkel et al. (1994, 1995, 1996)
Mdr1b (Abcb1b) Mdr1a/1b (Abcb1a/1b)	Knockout	Drug distribution	Schinkel et al. (1997)
Mrp1 (Abcc1)	Knockout	Disposition of xenobiotic drugs and endogenous leukotrienes	Wijnholds et al. (1997); Lorico et al. (1997)
Mrp2 (Abcc2)	Knockout	Glutathione excretion, bile flow, and drug disposition	Vlaming et al. (2006); Chu et al. (2006)
Mrp3 (Abcc3)	Knockout	Acetaminophen disposition and biliary excretion	Belinsky et al. (2005); Zelcer et al. (2006)
Mrp4 (Abcc4)	Knockout	Drug resistance and brain penetration	Leggas et al. (2004)
Oat1 (Slc22a6)	Knockout	Excretion of organic anions and drugs	Eraly et al. (2006)
Oat3 (Slc22a8)	Knockout	Uptake of organic anions and drugs	Sweet et al. (2002); Sykes et al. (2004)
OATP1B1 (SLCO1B1)	Transgenic	Hepatic uptake of mitoxantrone	van de Steeg et al. (2009)
Oatp1b2 (Slco1b2)	Knockout	Drug disposition and hepatic uptake	Zaher et al. (2008); Lu et al. (2008); Chen et al. (2008)
Oct1 (Slc22a1)	Knockout	Hepatic uptake and intestinal disposition of organic cation drugs	Jonker et al. (2001)
Oct2 (Slc22a2) Oct1/2 (Slc22a1/2)	Knockout	Renal excretion of drugs	Jonker et al. (2003)
Oct3 (Slc22a3)	Knockout	Role of Oct3 in drug distribution	Zwart et al. (2001)
Ostα	Knockout	Bile acid disposition	Rao et al. (2008)
Pept2 (Slc15a2)	Knockout	Role of Pept2 in drug distribution and excretion	Rubio-Aliaga et al. (2003); Shen et al. (2003, 2007)

MRP1/ABCC1 is ubiquitously expressed in normal tissues and has an important role in elimination of endobiotic and xenobiotic compounds (Borst and Elferink, 2002). Two different *Mrp1(-/-)* mouse lines have been generated by targeted disruption of either the exons encoding the first ATP-binding domain of mouse *Mrp1* gene (Wijnholds et al., 1997) or part of the second putative ATP-binding domain (Lorico et al., 1997). Increased sensitivity to anticancer drugs (e.g., etoposide and vincristine) was observed in both mouse lines (Lorico et al., 1997; Wijnholds et al., 1997), illustrating the critical role of Mrp1 in the defense against xenobiotics.

MRP2/ABCC2, localized mainly on the apical surface of epithelial cells in the liver, small intestine, and kidney,

mediates the active transport of many endogenous metabolites (e.g., glucuronosyl-bilirubin) and exogenous compounds (e.g., SN-38 and methotrexate). Overexpression of MRP2 in tumor cells may be associated with resistance to anticancer drugs (Borst and Elferink, 2002). Two *Mrp2(-/-)* mouse lines were generated and characterized (Chu et al., 2006; Vlaming et al., 2006). Deletion of *Mrp2* led to a reduction in biliary excretion of Mrp2 substrates and a significant increase in systemic exposure to several drugs and carcinogens after oral administration, demonstrating the important role of Mrp2 in the disposition of xenobiotics.

MRP3/ABCC3, expressed on the basolateral membrane of the liver, small intestine, and kidney, transports many anticancer drugs (e.g., etoposide, teniposide, and methotrexate) and glucuronidated compounds (e.g., etoposide glucuronide) (Borst et al., 2007). To examine the *in vivo* function of MRP3, two *Mrp3(-/-)* mouse models have been generated by targeted disruption of exons 6–8 with a *pgk-neo* cassette (Belinsky et al., 2005) or exons 2–8 with a *Hygromycin* cassette (Zelcer et al., 2006). The *Mrp3* knockout mice were used to evaluate the function of Mrp3 in transporting endogenous bile acids and bilirubin glucuronide (Belinsky et al., 2005; Zelcer et al., 2006). Furthermore, mice lacking Mrp3 had a significantly altered disposition of morphine, morphine-6-glucuronide, and APAP (Manautou et al., 2005; Zelcer et al., 2005).

MRP4/ABCC4 is localized either apically or basolaterally in the polarized cells of several tissues, including the kidney, the liver, and the brain, and is capable of transporting a variety of compounds (e.g., irinotecan and SN-38) (Borst et al., 2007). An *Mrp4(-/-)* mouse model developed by replacing exon 27 with a neomycin resistance cassette was employed to demonstrate the critical role of Mrp4 in renal elimination and brain penetration of anticancer drug topotecan (Leggas et al., 2004). The *Mrp4* knockout mouse model was also used to identify the contribution of Mrp4 to renal elimination of antiviral drugs (e.g., adefovir and tenofovir) (Imaoka et al., 2007), and cytotoxicity and tissue distribution of 9-(2-(phosphonomethoxy)ethyl)-adenine (Takenaka et al., 2007).

28.3.3 Breast Cancer Resistance Protein (BCRP/ABCG2)

BCRP/ABCG2, an ABC membrane transporter, is expressed ubiquitously in human tissues. ABCG2 plays an important role in the disposition of many xenobiotic drugs (e.g., mitoxantrone, doxorubicin, and topotecan) and metabolites (e.g., estrone-3-sulfate and estradiol-17-β-D-glucuronide) (Robey et al., 2009). In addition, overexpression of ABCG2 may confer multidrug resistance in cancer cells and represent a critical component of tumorigenic stem cells. The mouse homologue Bcrp1/Abcg2 has been constitutively knocked out to investigate the physiological and pharmacological functions of BCRP/ABCG2. One *Bcrp1(-/-)* mouse line was created by replacing exons 3–4 with a *pgk-neo* cassette on a mixed genetic background of 129/Ola × C57BL/6 (Zhou et al., 2002). Another *Bcrp1(-/-)* mouse model was developed by replacing exons 3–6 with a *pgk-hygro* cassette on an FVB or a mixed 129/Ola × FVB background (Jonker et al., 2002). Both stains of *Bcrp1(-/-)* mice were fertile and showed no abnormalities. Studies using *Bcrp1(-/-)* mouse models have revealed the critical role of Bcrp1 expression in protection of hematopoietic stem cells from toxic drugs (Zhou et al., 2002), and in

prevention from diet-dependent phototoxicity (Jonker et al., 2002). In addition, the *Bcrp1(-/-)* mice have been used to demonstrate the significant impact of Bcrp1 on absorption, distribution, and elimination of dietary carcinogens (e.g., 2-amino-1-methyl-6-phenylimidazo [4,5-b] pyridine) and therapeutic drugs (e.g., nitrofurantoin) (van Herwaarden et al., 2003; Breedveld et al., 2005; Jonker et al., 2005; Merino et al., 2005, 2006). Therefore, the *Bcrp1(-/-)* mice may serve as unique animal models for understanding *in vivo* ADME properties of drugs transported by BCRP.

28.3.4 Bile Salt Export Pump (BSEP/ABCB11)

Human BSEP/ABCB11, the sister of Pgp (sPgp) membrane transporter, is predominantly expressed in the canalicular membrane of the liver and profoundly contributes to canalicular secretion of monovalent bile salts (Stieger et al., 2007). A *Bsep(-/-)* mouse line (Table 28.3) has been generated by targeted inactivation of the murine homologue *Bsep* gene, in particular, replacement of a 1.4-kb fragment containing the coding region of Walker A in the *N*-terminal ATP-binding domain of *Bsep* with a neomycin-resistant cassette (Wang et al., 2001). The *Bsep(-/-)* mice were viable and fertile but showed growth retardation. In addition, the *Bsep(-/-)* mice exhibited various phenotypic characteristics, including intrahepatic cholestasis, reduced secretion of cholic acid and output of bile salt, liver necrosis, high mortality, and a number of changes in the liver genes, supporting the critical role of Bsep in the transport of bile acid and lipid homeostasis (Wang et al., 2001, 2003). Nevertheless, no study has been reported to examine drug disposition in this *Bsep(-/-)* mouse model.

28.3.5 Peptide Transporter 2 (PEPT2/SLC15A2)

Human PEPT2 is expressed specifically on the apical membrane of renal tubular cells and choroid plexus epithelial cells, and mediates cellular uptake of various di- or tripeptides and peptide-like drugs (e.g., cefadroxil, enalapril, betatin, and valacyclovir) (Kamal et al., 2008). To investigate the physiological and pharmacological functions of PEPT2, two *Pept2(-/-)* mouse models were created (Rubio-Aliaga et al., 2003; Shen et al., 2003). One constitutive *Pept2*-deficient mouse line was developed on a mixed C57BL/6 × 129/SvJ/R1 genetic background by targeted deletion of exons 1–3 (Shen et al., 2003). The other conditional *Pept2* knockout mouse line was generated on a mixed C57BL/6 × 129/Sv background using the Cre-loxP system (Rubio-Aliaga et al., 2003). The *Pept2(-/-)* mice were fertile and showed no obvious abnormal phenotypes. Well-controlled studies in knockout mouse models demonstrated the important role of PEPT2 in uptake of dipeptides (Rubio-Aliaga et al., 2003; Shen et al., 2003). Further studies revealed that renal PEPT2 was entirely responsible for reabsorption of cefadroxil in the kidneys, and choroid plexus PEPT2 limited brain penetration of cefadroxil (Shen et al., 2007). These findings suggest that Pept2 knockout mouse models may be useful in the development of peptidomimetic drugs.

28.3.6   Organic Cation Transporters (OCT/SLC22A)

The OCT/SLC22A proteins are mainly expressed in the liver and intestine (OCT1/SLC22A1), kidney (OCT1/SLC22A1 and OCT2/SLC22A2), and skeletal muscle and heart (OCT3/SLC22A3). OCTs translocate many endogenous compounds (e.g., adrenaline and dopamine) and xenobiotic agents (e.g., 1-methyl-4-phenylpyridinium, metformin, and cimetidine) that are positively charged at physiological pH (Koepsell et al., 2007). A number of *Oct* knockout mouse models (Table 28.3) were developed to examine the physiological and pharmacological functions of OCT transporters, which were all variable and fertile, with no obvious pheynotypic abnormalities (Jonker et al., 2001, 2003; Zwart et al., 2001). In particular, the *Oct1(-/-)* mice, in which exon 7 was deleted, were used to demonstrate the important role of Oct1 in distribution and excretion of tetraethylammonium, and hepatic accumulation of neurotoxin 1-methyl-4-phenylpyridium and the anticancer drug metaiodobenzylguanidine (Jonker et al., 2001). Further studies in *Oct2* single knockout and *Oct1/2* double knockout mice, developed by disrupting the exon 1 of *Slc22a2* on a wild-type and a *Oct1*[+/-] background, respectively, showed that Oct1 and Oct2 together were essential components for renal secretion of organic cations (Jonker et al., 2003). The mouse homologue of *OCT3*, namely *Orct3/Slc22a3*, was also disrupted to produce an *Orct3(-/-)* mouse line, which has been used to reveal the principle role of Orct3 in the accumulation of 1-methyl-4-phenylpyridium in the heart (Zwart et al., 2001).

28.3.7   Multidrug and Toxin Extrusion 1 (MATE1/SLC47A1)

Human MATE1/SLC47A1, a proton-dependent transporter predominantly found on the luminal side of renal proximal tubules and bile canaliculi, mediates the final excretion of many organic cations of exogenous or endogenous origins (Otsuka et al., 2005). Targeted disruption of murine *Mate1/Slc47a1* by replacing exon 1 with a *PGKp-neo[r]* cassette did not affect fertility or viability of the resulting knockout mice (Tsuda et al., 2009). Pharmacokinetic study revealed that loss of Mate1 led to a sharp decrease in renal clearance of

metformin and a significant increase in systemic exposure to metformin, suggesting an important role for Mate1 in metformin pharmacokinetics (Tsuda et al., 2009).

28.3.8   Organic Anion Transporters (OAT/SLC22A)

OAT/SLC22 transporters are located on the barrier epithelia of diverse tissues, including the proximal tubule of kidney and choroid plexus of the brain, where they mediate the absorption and excretion of many exogenous and endogenous organic anions (e.g., β-lactam antibiotics) (Ahn and Nigam, 2009). Disruption of murine *Oat1/Slc22a6* or *Oat3/Slc22a8* did not affect the fertility of knockout mice (Sweet et al., 2002; Eraly et al., 2006) (Table 28.3). *In vivo* and *ex vivo* studies with *Oat1(-/-)* mice, which were generated on a C57BL/6J mouse background by replacing exon 1 of *Oat1* with a LacZ-Neo cassette, indicated that Oat1 played an important role in renal secretion of organic anions and drugs (Eraly et al., 2006). *Oat3(-/-)* mice, developed by replacing exon 3 with an inverted neomycin cassette, have been utilized to demonstrate the critical contribution of Oat3 to renal and brain uptake of a number of substrates, such as *para*-aminohippurate and estrone sulfate taurocholate (Sweet et al., 2002; Sykes et al., 2004).

28.3.9   Organic Anion Transporting Polypeptides (OATP/SLCO)

Human liver-specific OATP/SLCO transporters, including OATP1B1, 1B3, and 2B1, are mainly expressed at the basolateral (sinusoidal) membrane of hepatocytes and contribute to the active hepatic uptake of a series of xenobiotics (Zair et al., 2008). Of note, the mouse liver-specific Oatp1b2/Slco1b2 is orthologous to human OATP1B1 and 1B3. To understand the role of Oatp1b2 in the hepatic uptake of xenobiotics, *Oatp1b2(-/-)* mouse models were developed by targeted disruption of *Oatp1b2/Slco1b2*, either by replacing exons 10–12 with a neomycin resistance gene cassette (Zaher et al., 2008) or by disrupting exon 3 with a *neo* cassette (Lu et al., 2008). Studies using *Oatp1b2(-/-)* mice demonstrated the critical role of Oatp1b2 in hepatic uptake, systemic disposition, and toxicity of several xenobiotic agents (e.g., phalloidin, microcystin-LR, pravastatin, rifampicin, and lovastatin) (Chen et al., 2008; Lu et al., 2008; Zaher et al., 2008). More recently, a transgenic mouse model with liver-specific expression of human OATP1B1 has been created using an ApoE promoter-HCR1-driven expression cassette consisting of *OATP1B1* cDNA (van de Steeg et al., 2009). The *OATP1B1(+/+)* mice, which were fertile and viable, showed significantly higher hepatic accumulation of methotrexate when compared with wild-type mice (van de Steeg et al., 2009). These genetically modified mice are useful models for defining the role of OATP in pharmacokinetics.

28.3.10   Organic Solute Transporter α (OSTα)

OSTα and OSTβ, which are coexpressed on the intestinal basolateral membrane, form a functional heteromer that transports many organic compounds, such as taurocholate, digoxin, and prostaglandin E1 (Ballatori, 2005). To delineate the role of Ostα-Ostβ heteromeric transporter in intestinal bile acid reabsorption, an *Ostα* knockout mouse model has been created by replacing the proximal promoter and exons 1 and 2 with a neomycin resistance gene. The *Ostα(-/-)* mice were viable and fertile, and indistinguishable from wild-type mice. A transileal transport study using the everted gut sac derived from *Ostα(-/-)* mice indicated that Ostα-Ostβ is critical for intestinal transport of bile acids (Rao et al., 2008).

28.4   XENOBIOTIC RECEPTOR GENETICALLY MODIFIED MOUSE MODELS

28.4.1   Aryl Hydrocarbon Receptor (AHR)

AHR is a member of the bHLH/PAS (basic Helix-Loop-Helix/Per-Arnt-Sim) family. Activation of AHR by xenobiotic agents (e.g., dioxins) leads to induction of P450 and UGT enzymes that may largely alter drug metabolism (Nebert et al., 2004). To date, three different *Ahr(-/-)* mouse lines (Table 28.4) have been created independently either by replacing exon 1 with a neomycin resistance cassette (Fernandez-Salguero et al., 1995) or a LacZ cassette (Mimura et al., 1997), or by replacing exon 2 with a neomycin resistance gene (Schmidt et al., 1996). These *Ahr*-null mouse strains exhibited some physiological changes such as lower growth rate, decreased fertility, and/or hepatic defects. Comparative studies in *Ahr(-/-)* versus wild-type mice demonstrated the essential role of Ahr in P450 induction, dioxin-induced toxicity, and carcinogenesis. Furthermore, an *AHR*-humanized mouse model in which murine *Ahr* was replaced by human *AHR* via homologous recombination was developed to investigate species difference in susceptibility to dioxin toxicity (Moriguchi et al., 2003). The *AHR*-humanized mice, without observed abnormalities, displayed different induction profiles toward the AHR ligands when compared with the wild-type *Ahr(+/+)* mice. These mouse models could be helpful tools for evaluating the risk of xenobiotics in humans.

28.4.2   Pregnane X Receptor (PXR/NR1I2)

Upon activation by endobiotics (e.g., steroids and bile acids) or xenobiotics (e.g., rifampicin, phenobarbital, and Saint-John's-wort), PXR up-regulates the expression of many important drug metabolizing enzymes (e.g., CYP3A4 and UGT1A) and transporters (e.g., ABCB1) that may alter the ADME of coadministered drugs (Ma et al., 2008). To investigate its physiological and pharmacological role, *Pxr(-/-)* mouse models were created by targeted deletion of exons 2–3 or exon 1 (Xie et al., 2000; Staudinger et al., 2001) (Table 28.4). Interestingly, deletion of Pxr led to around a fourfold increase in the basal level of hepatic Cyp3a mRNA (Staudinger et al., 2001). Loss of Pxr abolished the induction of various drug metabolizing enzymes and transporters by Pxr ligands (Xie et al., 2000; Staudinger et al., 2001). In addition, PXR-transgenic mice have been developed using a *PXR* cDNA driven by a liver-specific albumin promoter, a VP16 coactivator, or a BAC clone containing the complete *PXR* gene, which were then cross-bred with *Pxr*-null mice to generate *PXR*-humanized mouse models (Xie et al., 2000; Ma et al., 2007). Selective activation of human PXR by rifampicin in *PXR*-humanized mice demonstrated the functional difference in regulation of ADME by human PXR and mouse Pxr. Together, these findings indicate that *PXR* genetically modified mice are unique *in vivo* models to study PXR ligands and PXR-mediated DDIs.

28.4.3   Constitutive Androstane Receptor (CAR/NR1I3)

CAR is another important xenobiotic receptor that regulates the expression of various enzymes (e.g., CYP2B6, UGT1A1, and GSTA1) and transporters (e.g., ABCC2), and thus modulates the capacity of drug metabolism and disposition (Timsit and Negishi, 2007). To study the function of CAR *in vivo*, a *Car(-/-)* mouse model that showed no phenotypic abnormalities was created by replacing a segment of exons 1–2 with the β-*gal* and *neo* resistance genes. (Wei et al., 2000) (Table 28.4). The absence of Cyp2b10 induction by potent *Car* activators (e.g., phenobarbital) in *Car(-/-)* mice demonstrated the essential role of Car in regulation of Cyp2b10 expression (Wei et al., 2000). Because mouse and human CAR displayed significant difference in ligand specificity, a transgenic mouse line (VP-*CAR(+/+)*) with human CAR expression was developed (Saini et al., 2004). Compared with wild-type mice, the VP-*CAR(+/+)* mice exhibited remarkably higher hepatic sulfation activity toward lithocholic acid (LCA) and were more resistant to LCA-induced hepatotoxicity (Saini et al., 2004). In addition, a *CAR*-humanized mouse model was developed by breeding the *Car(-/-)* mice (Wei et al., 2000)

TABLE 28.4. Xenobiotic Receptor Genetically Modified Mouse Models

Gene	Category	Example Application	Reference
Ahr	Knockout	Role of Ahr in gene regulation and liver function	Fernandez-Salguero et al. (1995); Schmidt et al. (1996); Mimura et al. (1997)
AHR	Humanized	Species difference in AHR-mediated gene regulation	Moriguchi et al. (2003)
Car (Nr1i3)	Knockout	Role of Car in gene regulation and liver toxicity	Wei et al. (2000)
CAR (NR1I3)	Humanized	Species differences in CAR-mediated gene regulation	Zhang et al. (2002); Huang et al. (2004)
	Transgenic	Gene regulation and bile acid detoxification	Saini et al. (2004)
Pparα (Nr1c1)	Knockout	Gene regulation and peroxisome proliferation	Lee et al. (1995)
PPARα (NR1C1)	Humanized	Species difference in PPARα-mediated gene regulation	Cheung et al. (2004)
Pxr (Nr1i2)	Knockout	Role of Pxr in regulation of drug metabolism and disposition	Xie et al. (2000); Staudinger et al. (2001)
PXR (NR1I2)	Humanized	Species difference in PXR-mediated gene regulation	Xie et al. (2000); Ma et al. (2007)
Rxrα (Nr2b1)	Knockout	Physiological role of Rxrα (embryonic lethality)	Kastner et al. (1994); Sucov et al. (1994)
Rxrα (hepatocyte)	Knockout	Regulation of endobiotics and xenobiotics metabolism	Wan et al. (2000)

with *CAR(+/+)* mice, the latter of which were produced using human *CAR* cDNA driven by a liver-specific albumin promoter (Zhang et al., 2002). Studies using wild-type, *Car(-/-)*, and *CAR*-humanized mice revealed that meclizine is an agonist for mouse Car, whereas it is an inverse agonist for human CAR (Huang et al., 2004), indicating the potential application of *CAR*-humanized mice to preclinical assessment of ADME regulation.

28.4.4 Peroxisome Proliferator-Activated Receptor α (PPARα/NR1C1)

Within the PPAR subfamily, PPARα is of particular interest as PPARα not only regulates CYP4A expression but also mediates hepatocarcinogenesis in rodents (Gonzalez and Shah, 2008). PPARα ligands include naturally occurring steroids and lipids, herbicides, pesticides, and some widely used lipid-lowering fibrate drugs. By targeted disruption of exon 8 of murine *Pparα*, a *Pparα(-/-)* mouse line was generated to delineate the physiological and toxicological role of Pparα (Lee et al., 1995) (Table 28.4). *Pparα(-/-)* mice were viable and fertile, without phenotypic abnormalities. The findings that pleiotropic effects of peroxisome proliferators (e.g., clofibrate and Wy-14,643) and induction of Cyp4a were abolished in *Pparα(-/-)* mice demonstrated the critical role of Pparα in peroxisome proliferation and transcrip-

tional activation of target genes (Lee et al., 1995). To investigate the species difference in PPARα functions, a *PPARα*-humanized mouse model was created using human *PPARα* cDNA under the control of a tetracycline responsive regulatory system and being bred into the *Pparα*-null mouse background (Cheung et al., 2004). Induction of peroxisomal and mitochondrial fatty acid metabolizing enzymes by Wy-14643 was shown in both Pparα wild-type and *PPARα*-humanized mice. In sharp contrast, hepatocellular proliferation and hepatomegaly was only observed in wild-type mice, indicating the species difference in target gene activation by mouse Pparα and human PPARα (Cheung et al., 2004). These findings suggest that PPARα-humanized mice could be a better animal model to evaluate risk of peroxisome proliferators (e.g., lipid-lowering fibrate drugs) in humans.

28.4.5 Retinoid X Receptor α (RXRα/NR2B1)

The retinoid X receptors, consisting of three members coding for RXRα, β, and γ, are unique among the nuclear receptors because they dimerize with other nuclear receptors such as PXR, CAR, and PPARα to participate in the regulation of liver functions (Germain et al., 2006). Disruption of exon 3 or exon 4 of *RXRα/NR2B1* with a PGK-*neo* cassette led to embryonic lethality in these conventional *Rxrα(-/-)* mice, probably due to cardiac or placental developmental defects (Kastner et al., 1994; Sucov et al., 1994). Therefore, a Cre-mediated conditional disruption of *Rxrα* specifically in hepatocytes was used to generate a *Rxrα(-/-)ₕₑₚ* mouse model for investigation of its physiological functions (Wan et al., 2000) (Table 28.4). The *Rxrα(-/-)ₕₑₚ* mice were viable and exhibited no liver impairments. However, a range of hepatic metabolic pathways mediated by heteromerization between Rxrα and xenobiotic receptors (e.g., Pparα, Car, and Pxr) were altered in the absence of Rxrα (Wan et al., 2000; Wu et al., 2004; Dai et al., 2005). The *Rxrα(-/-)ₕₑₚ* mice are unique animal models used to define the regulatory role of Rxrα in homeostasis of cholesterol, fatty acids, bile acids, and steroids, as well as in xenobiotic metabolism and disposition.

28.5 CONCLUSIONS

Genetically modified mice are unparalleled and powerful animal models for delineating the physiological, pharmacological, and toxicological role of particular gene(s) in a whole body system. Studies in mice with lost (knockout) or gained (transgenic) functions have remarkably improved our understanding of the *in vivo* roles of drug metabolizing enzymes, transporters, and xenobiotic receptors in drug ADME, and the consequent effects on drug efficacy or toxicity. The use of humanized mouse models, in which the murine ortholog is replaced by a human gene, may overcome species differences in ADME, thus leading to a better understanding or prediction of drug metabolism and disposition in humans. Nevertheless, there is now accumulating evidence that disruption or introduction of one gene within the mouse genome may result in significant change in the expression of many other genes, which might complicate data interpretation. It is also difficult to determine what levels of gene expression are comparable between mice and humans. Thus, it is important to be aware of the limitations when using genetically modified mouse models to investigate the ADME properties of a drug.

REFERENCES

Ahn SY, Nigam SK (2009) Toward a systems level understanding of organic anion and other multispecific drug transporters: A remote sensing and signaling hypothesis. *Mol Pharmacol* 76:481–490.

Ballatori N (2005) Biology of a novel organic solute and steroid transporter, OSTalpha-OSTbeta. *Exp Biol Med (Maywood)* 230:689–698.

Belinsky MG, Dawson PA, Shchaveleva I, Bain LJ, Wang R, Ling V, Chen ZS, Grinberg A, Westphal H, Klein-Szanto A, Lerro A, Kruh GD (2005) Analysis of the *in vivo* functions of Mrp3. *Mol Pharmacol* 68:160–168.

Beringer PM, Slaughter RL (2005) Transporters and their impact on drug disposition. *Ann Pharmacother* 39:1097–1108.

Bogaards JJ, Bertrand M, Jackson P, Oudshoorn MJ, Weaver RJ, van Bladeren PJ, Walther B (2000) Determining the best animal model for human cytochrome P450 activities: A comparison of mouse, rat, rabbit, dog, micropig, monkey and man. *Xenobiotica* 30:1131–1152.

Bonzo JA, Belanger A, Tukey RH (2007) The role of chrysin and the ah receptor in induction of the human UGT1A1 gene *in vitro* and in transgenic UGT1 mice. *Hepatology* 45:349–360.

Borst P, Elferink RO (2002) Mammalian ABC transporters in health and disease. *Annu Rev Biochem* 71:537–592.

Borst P, Evers R, Kool M, Wijnholds J (1999) The multidrug resistance protein family. *Biochim Biophys Acta* 1461: 347–357.

Borst P, de Wolf C, van de Wetering K (2007) Multidrug resistance-associated proteins 3, 4, and 5. *Pflugers Arch* 453:661–673.

Breedveld P, Pluim D, Cipriani G, Wielinga P, van Tellingen O, Schinkel AH, Schellens JH (2005) The effect of Bcrp1 (Abcg2) on the *in vivo* pharmacokinetics and brain penetration of imatinib mesylate (Gleevec): Implications for the use of breast cancer resistance protein and P-glycoprotein inhibitors to enable the brain penetration of imatinib in patients. *Cancer Res* 65:2577–2582.

Buters JT, Tang BK, Pineau T, Gelboin HV, Kimura S, Gonzalez FJ (1996) Role of CYP1A2 in caffeine pharmacokinetics and metabolism: Studies using mice deficient in CYP1A2. *Pharmacogenetics* 6:291–296.

Buters JT, Sakai S, Richter T, Pineau T, Alexander DL, Savas U, Doehmer J, Ward JM, Jefcoate CR, Gonzalez FJ (1999) Cytochrome P450 CYP1B1 determines susceptibility to 7,12-dimethylbenz[a]anthracene-induced lymphomas. *Proc Natl Acad Sci U S A* 96:1977–1982.

Cederbaum AI (2006) CYP2E1—Biochemical and toxicological aspects and role in alcohol-induced liver injury. *Mt Sinai J Med* 73:657–672.

Chen C, Stock JL, Liu X, Shi J, Van Deusen JW, DiMattia DA, Dullea RG, de Morais SM (2008) Utility of a novel Oatp1b2 knockout mouse model for evaluating the role of Oatp1b2 in the hepatic uptake of model compounds. *Drug Metab Dispos* 36:1840–1845.

Chen S, Beaton D, Nguyen N, Senekeo-Effenberger K, Brace-Sinnokrak E, Argikar U, Remmel RP, Trottier J, Barbier O, Ritter JK, Tukey RH (2005) Tissue-specific, inducible, and hormonal control of the human UDP-glucuronosyltransferase-1 (UGT1) locus. *J Biol Chem* 280:37547–37557.

Cheung C, Gonzalez FJ (2008) Humanized mouse lines and their application for prediction of human drug metabolism and toxicological risk assessment. *J Pharmacol Exp Ther* 327:288–299.

Cheung C, Akiyama TE, Ward JM, Nicol CJ, Feigenbaum L, Vinson C, Gonzalez FJ (2004) Diminished hepatocellular proliferation in mice humanized for the nuclear receptor peroxisome proliferator-activated receptor alpha. *Cancer Res* 64:3849–3854.

Cheung C, Ma X, Krausz KW, Kimura S, Feigenbaum L, Dalton TP, Nebert DW, Idle JR, Gonzalez FJ (2005a) Differential metabolism of 2-amino-1-methyl-6-phenylimidazo[4,5-b]pyridine (PhIP) in mice humanized for CYP1A1 and CYP1A2. *Chem Res Toxicol* 18: 1471–1478.

Cheung C, Yu AM, Ward JM, Krausz KW, Akiyama TE, Feigenbaum L, Gonzalez FJ (2005b) The cyp2e1-humanized transgenic mouse: Role of cyp2e1 in acetaminophen hepatotoxicity. *Drug Metab Dispos* 33:449–457.

Cheung C, Yu AM, Chen CS, Krausz KW, Byrd LG, Feigenbaum L, Edwards RJ, Waxman DJ, Gonzalez FJ (2006) Growth hormone determines sexual dimorphism of hepatic cytochrome P450 3A4 expression in transgenic mice. *J Pharmacol Exp Ther* 316:1328–1334.

Chu XY, Strauss JR, Mariano MA, Li J, Newton DJ, Cai X, Wang RW, Yabut J, Hartley DP, Evans DC, Evers R (2006) Characterization of mice lacking the multidrug resistance protein MRP2 (ABCC2). *J Pharmacol Exp Ther* 317: 579–589.

Conklin DJ, Haberzettl P, Prough RA, Bhatnagar A (2009) Glutathione-*S*-transferase P protects against endothelial dysfunction induced by exposure to tobacco smoke. *Am J Physiol Heart Circ Physiol* 296:H1586–H1597.

Corchero J, Granvil CP, Akiyama TE, Hayhurst GP, Pimprale S, Feigenbaum L, Idle JR, Gonzalez FJ (2001) The CYP2D6 humanized mouse: Effect of the human CYP2D6 transgene and HNF4alpha on the disposition of debrisoquine in the mouse. *Mol Pharmacol* 60:1260–1267.

Dai G, Wan YJ (2005) Animal models of xenobiotic receptors. *Curr Drug Metab* 6:341–355.

Dai G, Chou N, He L, Gyamfi MA, Mendy AJ, Slitt AL, Klaassen CD, Wan YJ (2005) Retinoid X receptor alpha regulates the expression of glutathione *S*-transferase genes and modulates acetaminophen-glutathione conjugation in mouse liver. *Mol Pharmacol* 68:1590–1596.

Dalton TP, Dieter MZ, Matlib RS, Childs NL, Shertzer HG, Genter MB, Nebert DW (2000) Targeted knockout of Cyp1a1 gene does not alter hepatic constitutive expression of other genes in the mouse [Ah] battery. *Biochem Biophys Res Commun* 267:184–189.

Derkenne S, Curran CP, Shertzer HG, Dalton TP, Dragin N, Nebert DW (2005) Theophylline pharmacokinetics: Comparison of Cyp1a1(-/-) and Cyp1a2(-/-) knockout mice, humanized hCYP1A1_1A2 knock-in mice lacking either the mouse Cyp1a1 or Cyp1a2 gene, and Cyp1(+/+) wild-type mice. *Pharmacogenet Genomics* 15:503–511.

Dragin N, Uno S, Wang B, Dalton TP, Nebert DW (2007) Generation of "humanized" hCYP1A1_1A2_Cyp1a1/

1a2(-/-) mouse line. *Biochem Biophys Res Commun* 359:635–642.

Eraly SA, Vallon V, Vaughn DA, Gangoiti JA, Richter K, Nagle M, Monte JC, Rieg T, Truong DM, Long JM, Barshop BA, Kaler G, Nigam SK (2006) Decreased renal organic anion secretion and plasma accumulation of endogenous organic anions in OAT1 knock-out mice. *J Biol Chem* 281:5072–5083.

Fang C, Behr M, Xie F, Lu S, Doret M, Luo H, Yang W, Aldous K, Ding X, Gu J (2008a) Mechanism of chloroform-induced renal toxicity: Non-involvement of hepatic cytochrome P450-dependent metabolism. *Toxicol Appl Pharmacol* 227:48–55.

Fang C, Gu J, Xie F, Behr M, Yang W, Abel ED, Ding X (2008b) Deletion of the NADPH-cytochrome P450 reductase gene in cardiomyocytes does not protect mice against doxorubicin-mediated acute cardiac toxicity. *Drug Metab Dispos* 36:1722–1728.

Felmlee MA, Lon HK, Gonzalez FJ, Yu AM (2008) Cytochrome P450 expression and regulation in CYP3A4/CYP2D6 double transgenic humanized mice. *Drug Metab Dispos* 36:435–441.

Fernandez-Salguero P, Pineau T, Hilbert DM, McPhail T, Lee SS, Kimura S, Nebert DW, Rudikoff S, Ward JM, Gonzalez FJ (1995) Immune system impairment and hepatic fibrosis in mice lacking the dioxin-binding Ah receptor. *Science* 268:722–726.

Finn RD, McLaren AW, Carrie D, Henderson CJ, Wolf CR (2007) Conditional deletion of cytochrome P450 oxidoreductase in the liver and gastrointestinal tract: A new model for studying the functions of the P450 system. *J Pharmacol Exp Ther* 322:40–47.

Germain P, Chambon P, Eichele G, Evans RM, Lazar MA, Leid M, De Lera AR, Lotan R, Mangelsdorf DJ, Gronemeyer H (2006) International union of pharmacology. LXIII. Retinoid X receptors. *Pharmacol Rev* 58:760–772.

Gonzalez FJ, Shah YM (2008) PPARalpha: Mechanism of species differences and hepatocarcinogenesis of peroxisome proliferators. *Toxicology* 246:2–8.

Gonzalez FJ, Yu AM (2006) Cytochrome P450 and xenobiotic receptor humanized mice. *Annu Rev Pharmacol Toxicol* 46:41–64.

Granvil CP, Yu AM, Elizondo G, Akiyama TE, Cheung C, Feigenbaum L, Krausz KW, Gonzalez FJ (2003) Expression of the human CYP3A4 gene in the small intestine of transgenic mice: *In vitro* metabolism and pharmacokinetics of midazolam. *Drug Metab Dispos* 31:548–558.

Gu J, Weng Y, Zhang QY, Cui H, Behr M, Wu L, Yang W, Zhang L, Ding X (2003) Liver-specific deletion of the NADPH-cytochrome P450 reductase gene: Impact on plasma cholesterol homeostasis and the function and regulation of microsomal cytochrome P450 and heme oxygenase. *J Biol Chem* 278:25895–25901.

Gu J, Chen CS, Wei Y, Fang C, Xie F, Kannan K, Yang W, Waxman DJ, Ding X (2007) A mouse model with liver-specific deletion and global suppression of the NADPH-cytochrome P450 reductase gene: Characterization and utility for *in vivo* studies of cyclophosphamide disposition. *J Pharmacol Exp Ther* 321:9–17.

Guenet JL (2005) The mouse genome. *Genome Res* 15:1729–1740.

Guengerich FP (2006) Cytochrome P450s and other enzymes in drug metabolism and toxicity. *AAPS J* 8:E101–E111.

Henderson CJ, Wolf CR (2003) Transgenic analysis of human drug-metabolizing enzymes: Preclinical drug development and toxicology. *Mol Interv* 3:331–343.

Henderson CJ, Smith AG, Ure J, Brown K, Bacon EJ, Wolf CR (1998) Increased skin tumorigenesis in mice lacking pi class glutathione *S*-transferases. *Proc Natl Acad Sci U S A* 95:5275–5280.

Henderson CJ, Wolf CR, Kitteringham N, Powell H, Otto D, Park BK (2000) Increased resistance to acetaminophen hepatotoxicity in mice lacking glutathione *S*-transferase pi. *Proc Natl Acad Sci U S A* 97:12741–12745.

Henderson CJ, Otto DM, Carrie D, Magnuson MA, McLaren AW, Rosewell I, Wolf CR (2003) Inactivation of the hepatic cytochrome P450 system by conditional deletion of hepatic cytochrome P450 reductase. *J Biol Chem* 278:13480–13486.

Henderson CJ, Pass GJ, Wolf CR (2006) The hepatic cytochrome P450 reductase null mouse as a tool to identify a successful candidate entity. *Toxicol Lett* 162:111–117.

van Herwaarden AE, Jonker JW, Wagenaar E, Brinkhuis RF, Schellens JH, Beijnen JH, Schinkel AH (2003) The breast cancer resistance protein (Bcrp1/Abcg2) restricts exposure to the dietary carcinogen 2-amino-1-methyl-6-phenylimidazo[4,5-b]pyridine. *Cancer Res* 63:6447–6452.

van Herwaarden AE, Smit JW, Sparidans RW, Wagenaar E, van der Kruijssen CM, Schellens JH, Beijnen JH, Schinkel AH (2005) Midazolam and cyclosporin a metabolism in transgenic mice with liver-specific expression of human CYP3A4. *Drug Metab Dispos* 33:892–895.

van Herwaarden AE, Wagenaar E, van der Kruijssen CM, van Waterschoot RA, Smit JW, Song JY, van der Valk MA, van Tellingen O, van der Hoorn JW, Rosing H, Beijnen JH, Schinkel AH (2007) Knockout of cytochrome P450 3A yields new mouse models for understanding xenobiotic metabolism. *J Clin Invest* 117:3583–3592.

Huang W, Zhang J, Wei P, Schrader WT, Moore DD (2004) Meclizine is an agonist ligand for mouse constitutive androstane receptor (CAR) and an inverse agonist for human CAR. *Mol Endocrinol* 18:2402–2408.

Imaoka T, Kusuhara H, Adachi M, Schuetz JD, Takeuchi K, Sugiyama Y (2007) Functional involvement of multidrug resistance-associated protein 4 (MRP4/ABCC4) in the renal elimination of the antiviral drugs adefovir and tenofovir. *Mol Pharmacol* 71:619–627.

Ingelman-Sundberg M, Sim SC, Gomez A, Rodriguez-Antona C (2007) Influence of cytochrome P450 polymorphisms on drug therapies: Pharmacogenetic, pharmacoepigenetic and clinical aspects. *Pharmacol Ther* 116:496–526.

Jiang XL, Shen HW, Yu AM (2009) Pinoline may be used as a probe for CYP2D6 activity. *Drug Metab Dispos* 37:443–446.

Jiang Z, Dalton TP, Jin L, Wang B, Tsuneoka Y, Shertzer HG, Deka R, Nebert DW (2005) Toward the evaluation of function in genetic variability: Characterizing human SNP frequencies and establishing BAC-transgenic mice carrying the human CYP1A1_CYP1A2 locus. *Hum Mutat* 25:196–206.

Jonker JW, Wagenaar E, Mol CA, Buitelaar M, Koepsell H, Smit JW, Schinkel AH (2001) Reduced hepatic uptake and intestinal excretion of organic cations in mice with a targeted disruption of the organic cation transporter 1 (Oct1 [Slc22a1]) gene. *Mol Cell Biol* 21:5471–5477.

Jonker JW, Buitelaar M, Wagenaar E, Van Der Valk MA, Scheffer GL, Scheper RJ, Plosch T, Kuipers F, Elferink RP, Rosing H, Beijnen JH, Schinkel AH (2002) The breast cancer resistance protein protects against a major chlorophyll-derived dietary phototoxin and protoporphyria. *Proc Natl Acad Sci U S A* 99:15649–15654.

Jonker JW, Wagenaar E, Van Eijl S, Schinkel AH (2003) Deficiency in the organic cation transporters 1 and 2 (Oct1/Oct2 [Slc22a1/Slc22a2]) in mice abolishes renal secretion of organic cations. *Mol Cell Biol* 23:7902–7908.

Jonker JW, Merino G, Musters S, van Herwaarden AE, Bolscher E, Wagenaar E, Mesman E, Dale TC, Schinkel AH (2005) The breast cancer resistance protein BCRP (ABCG2) concentrates drugs and carcinogenic xenotoxins into milk. *Nat Med* 11:127–129.

Kamal MA, Keep RF, Smith DE (2008) Role and relevance of PEPT2 in drug disposition, dynamics, and toxicity. *Drug Metab Pharmacokinet* 23:236–242.

Kastner P, Grondona JM, Mark M, Gansmuller A, LeMeur M, Decimo D, Vonesch JL, Dolle P, Chambon P (1994) Genetic analysis of RXR alpha developmental function: Convergence of RXR and RAR signaling pathways in heart and eye morphogenesis. *Cell* 78:987–1003.

Kiang TK, Ensom MH, Chang TK (2005) UDP-glucuronosyltransferases and clinical drug-drug interactions. *Pharmacol Ther* 106:97–132.

Koepsell H, Lips K, Volk C (2007) Polyspecific organic cation transporters: Structure, function, physiological roles, and biopharmaceutical implications. *Pharm Res* 24:1227–1251.

Komura H, Iwaki M (2008) Species differences in *in vitro* and *in vivo* small intestinal metabolism of CYP3A substrates. *J Pharm Sci* 97:1775–1800.

Kotegawa T, Laurijssens BE, Von Moltke LL, Cotreau MM, Perloff MD, Venkatakrishnan K, Warrington JS, Granda BW, Harmatz JS, Greenblatt DJ (2002) *In vitro*, pharmacokinetic, and pharmacodynamic interactions of ketoconazole and midazolam in the rat. *J Pharmacol Exp Ther* 302:1228–1237.

Lagas JS, Vlaming ML, Schinkel AH (2009) Pharmacokinetic assessment of multiple ATP-binding cassette transporters: The power of combination knockout mice. *Mol Interv* 9:136–145.

Lee SS, Pineau T, Drago J, Lee EJ, Owens JW, Kroetz DL, Fernandez-Salguero PM, Westphal H, Gonzalez FJ (1995) Targeted disruption of the alpha isoform of the peroxisome proliferator-activated receptor gene in mice results in abolishment of the pleiotropic effects of peroxisome proliferators. *Mol Cell Biol* 15:3012–3022.

Lee SS, Buters JT, Pineau T, Fernandez-Salguero P, Gonzalez FJ (1996) Role of CYP2E1 in the hepatotoxicity of acetaminophen. *J Biol Chem* 271:12063–12067.

Leggas M, Adachi M, Scheffer GL, Sun D, Wielinga P, Du G, Mercer KE, Zhuang Y, Panetta JC, Johnston B, Scheper RJ, Stewart CF, Schuetz JD (2004) Mrp4 confers resistance to topotecan and protects the brain from chemotherapy. *Mol Cell Biol* 24:7612–7621.

Liang HC, Li H, McKinnon RA, Duffy JJ, Potter SS, Puga A, Nebert DW (1996) Cyp1a2(-/-) null mutant mice develop normally but show deficient drug metabolism. *Proc Natl Acad Sci U S A* 93:1671–1676.

Liggett SB (2004) Genetically modified mouse models for pharmacogenomic research. *Nat Rev Genet* 5:657–663.

Lin JH (2008) Applications and limitations of genetically modified mouse models in drug discovery and development. *Curr Drug Metab* 9:419–438.

Lindsay J, Wang LL, Li Y, Zhou SF (2008) Structure, function and polymorphism of human cytosolic sulfotransferases. *Curr Drug Metab* 9:99–105.

Lofgren S, Baldwin RM, Hiratsuka M, Lindqvist A, Carlberg A, Sim SC, Schulke M, Snait M, Edenro A, Fransson-Steen R, Terelius Y, Ingelman-Sundberg M (2008) Generation of mice transgenic for human CYP2C18 and CYP2C19: Characterization of the sexually dimorphic gene and enzyme expression. *Drug Metab Dispos* 36:955–962.

Lofgren S, Baldwin RM, Carleros M, Terelius Y, Fransson-Steen R, Mwinyi J, Waxman DJ, Ingelman-Sundberg M (2009) Regulation of human CYP2C18 and CYP2C19 in transgenic mice: Influence of castration, testosterone, and growth hormone. *Drug Metab Dispos* 37:1505–1512.

Lorico A, Rappa G, Finch RA, Yang D, Flavell RA, Sartorelli AC (1997) Disruption of the murine MRP (multidrug resistance protein) gene leads to increased sensitivity to etoposide (VP-16) and increased levels of glutathione. *Cancer Res* 57:5238–5242.

Lu H, Choudhuri S, Ogura K, Csanaky IL, Lei X, Cheng X, Song PZ, Klaassen CD (2008) Characterization of organic anion transporting polypeptide 1b2-null mice: Essential role in hepatic uptake/toxicity of phalloidin and microcystin-LR. *Toxicol Sci* 103:35–45.

Ma Q, Lu AY (2007) CYP1A induction and human risk assessment: An evolving tale of *in vitro* and *in vivo* studies. *Drug Metab Dispos* 35:1009–1016.

Ma X, Shah Y, Cheung C, Guo GL, Feigenbaum L, Krausz KW, Idle JR, Gonzalez FJ (2007) The PREgnane X receptor gene-humanized mouse: A model for investigating drug-drug interactions mediated by cytochromes P450 3A. *Drug Metab Dispos* 35:194–200.

Ma X, Idle JR, Gonzalez FJ (2008) The pregnane X receptor: From bench to bedside. *Expert Opin Drug Metab Toxicol* 4:895–908.

Mahmood I (2007) Application of allometric principles for the prediction of pharmacokinetics in human and veterinary drug development. *Adv Drug Deliv Rev* 59:1177–1192.

Manautou JE, de Waart DR, Kunne C, Zelcer N, Goedken M, Borst P, Elferink RO (2005) Altered disposition of acetaminophen in mice with a disruption of the Mrp3 gene. *Hepatology* 42:1091–1098.

Manis JP (2007) Knock out, knock in, knock down–genetically manipulated mice and the Nobel Prize. *N Engl J Med* 357:2426–2429.

Martignoni M, Groothuis GM, de Kanter R (2006) Species differences between mouse, rat, dog, monkey and human CYP-mediated drug metabolism, inhibition and induction. *Expert Opin Drug Metab Toxicol* 2:875–894.

Meiers I, Shanks JH, Bostwick DG (2007) Glutathione S-transferase pi (GSTP1) hypermethylation in prostate cancer: Review 2007. *Pathology* 39:299–304.

Merino G, Jonker JW, Wagenaar E, van Herwaarden AE, Schinkel AH (2005) The breast cancer resistance protein (BCRP/ABCG2) affects pharmacokinetics, hepatobiliary excretion, and milk secretion of the antibiotic nitrofurantoin. *Mol Pharmacol* 67:1758–1764.

Merino G, Alvarez AI, Pulido MM, Molina AJ, Schinkel AH, Prieto JG (2006) Breast cancer resistance protein (BCRP/ABCG2) transports fluoroquinolone antibiotics and affects their oral availability, pharmacokinetics, and milk secretion. *Drug Metab Dispos* 34:690–695.

Mimura J, Yamashita K, Nakamura K, Morita M, Takagi TN, Nakao K, Ema M, Sogawa K, Yasuda M, Katsuki M, Fujii-Kuriyama Y (1997) Loss of teratogenic response to 2,3,7,8-tetrachlorodibenzo-*p*-dioxin (TCDD) in mice lacking the Ah (dioxin) receptor. *Genes Cells* 2:645–654.

Moriguchi T, Motohashi H, Hosoya T, Nakajima O, Takahashi S, Ohsako S, Aoki Y, Nishimura N, Tohyama C, Fujii-Kuriyama Y, Yamamoto M (2003) Distinct response to dioxin in an aryl hydrocarbon receptor (AHR)-humanized mouse. *Proc Natl Acad Sci U S A* 100:5652–5657.

Muruganandan S, Sinal CJ (2008) Mice as clinically relevant models for the study of cytochrome P450-dependent metabolism. *Clin Pharmacol Ther* 83:818–828.

Mutch DM, Crespy V, Clough J, Henderson CJ, Lariani S, Mansourian R, Moulin J, Wolf CR, Williamson G (2006) Hepatic cytochrome P-450 reductase-null mice show reduced transcriptional response to quercetin and reveal physiological homeostasis between jejunum and liver. *Am J Physiol Gastrointest Liver Physiol* 291:G63–G72.

Nebert DW, Dalton TP, Okey AB, Gonzalez FJ (2004) Role of aryl hydrocarbon receptor-mediated induction of the CYP1 enzymes in environmental toxicity and cancer. *J Biol Chem* 279:23847–23850.

Nguyen N, Bonzo JA, Chen S, Chouinard S, Kelner MJ, Hardiman G, Belanger A, Tukey RH (2008) Disruption of the ugt1 locus in mice resembles human Crigler-Najjar type I disease. *J Biol Chem* 283:7901–7911.

Otsuka M, Matsumoto T, Morimoto R, Arioka S, Omote H, Moriyama Y (2005) A human transporter protein that mediates the final excretion step for toxic organic cations. *Proc Natl Acad Sci U S A* 102:17923–17928.

Otto DM, Henderson CJ, Carrie D, Davey M, Gundersen TE, Blomhoff R, Adams RH, Tickle C, Wolf CR (2003)

Identification of novel roles of the cytochrome p450 system in early embryogenesis: Effects on vasculogenesis and retinoic Acid homeostasis. *Mol Cell Biol* 23:6103–6116.

Pass GJ, Carrie D, Boylan M, Lorimore S, Wright E, Houston B, Henderson CJ, Wolf CR (2005) Role of hepatic cytochrome p450s in the pharmacokinetics and toxicity of cyclophosphamide: Studies with the hepatic cytochrome p450 reductase null mouse. *Cancer Res* 65:4211–4217.

Pineau T, Fernandez-Salguero P, Lee SS, McPhail T, Ward JM, Gonzalez FJ (1995) Neonatal lethality associated with respiratory distress in mice lacking cytochrome P450 1A2. *Proc Natl Acad Sci U S A* 92:5134–5138.

Qian YM, Sun XJ, Tong MH, Li XP, Richa J, Song WC (2001) Targeted disruption of the mouse estrogen sulfotransferase gene reveals a role of estrogen metabolism in intracrine and paracrine estrogen regulation. *Endocrinology* 142:5342–5350.

Rao A, Haywood J, Craddock AL, Belinsky MG, Kruh GD, Dawson PA (2008) The organic solute transporter alpha-beta, Ostalpha-Ostbeta, is essential for intestinal bile acid transport and homeostasis. *Proc Natl Acad Sci U S A* 105:3891–3896.

Robey RW, To KK, Polgar O, Dohse M, Fetsch P, Dean M, Bates SE (2009) ABCG2: A perspective. *Adv Drug Deliv Rev* 61:3–13.

Rossini A, de Almeida Simao T, Albano RM, Pinto LF (2008) CYP2A6 polymorphisms and risk for tobacco-related cancers. *Pharmacogenomics* 9:1737–1752.

Rubio-Aliaga I, Frey I, Boll M, Groneberg DA, Eichinger HM, Balling R, Daniel H (2003) Targeted disruption of the peptide transporter Pept2 gene in mice defines its physiological role in the kidney. *Mol Cell Biol* 23:3247–3252.

Saini SP, Sonoda J, Xu L, Toma D, Uppal H, Mu Y, Ren S, Moore DD, Evans RM, Xie W (2004) A novel constitutive androstane receptor-mediated and CYP3A-independent pathway of bile acid detoxification. *Mol Pharmacol* 65:292–300.

Schinkel AH, Smit JJ, van Tellingen O, Beijnen JH, Wagenaar E, van Deemter L, Mol CA, van der Valk MA, Robanus-Maandag EC, te Riele HP, Berns AJM, Borst P (1994) Disruption of the mouse mdr1a P-glycoprotein gene leads to a deficiency in the blood-brain barrier and to increased sensitivity to drugs. *Cell* 77:491–502.

Schinkel AH, Wagenaar E, van Deemter L, Mol CA, Borst P (1995) Absence of the mdr1a P-Glycoprotein in mice affects tissue distribution and pharmacokinetics of dexamethasone, digoxin, and cyclosporin A. *J Clin Invest* 96:1698–1705.

Schinkel AH, Wagenaar E, Mol CA, van Deemter L (1996) P-glycoprotein in the blood-brain barrier of mice influences the brain penetration and pharmacological activity of many drugs. *J Clin Invest* 97:2517–2524.

Schinkel AH, Mayer U, Wagenaar E, Mol CA, van Deemter L, Smit JJ, van der Valk MA, Voordouw AC, Spits H, van Tellingen O, Zijlmans JM, Fibbe WE, Borst P (1997) Normal viability and altered pharmacokinetics in mice lacking mdr1-type (drug-transporting) P-glycoproteins. *Proc Natl Acad Sci U S A* 94:4028–4033.

Schmidt JV, Su GH, Reddy JK, Simon MC, Bradfield CA (1996) Characterization of a murine Ahr null allele: Involvement of the Ah receptor in hepatic growth and development. *Proc Natl Acad Sci U S A* 93:6731–6736.

Senekeo-Effenberger K, Chen S, Brace-Sinnokrak E, Bonzo JA, Yueh MF, Argikar U, Kaeding J, Trottier J, Remmel RP, Ritter JK, Barbier O, Tukey RH (2007) Expression of the human UGT1 locus in transgenic mice by 4-chloro-6-(2,3-xylidino)-2-pyrimidinylthioacetic acid (WY-14643) and implications on drug metabolism through peroxisome proliferator-activated receptor alpha activation. *Drug Metab Dispos* 35:419–427.

Shen AL, O'Leary KA, Kasper CB (2002) Association of multiple developmental defects and embryonic lethality with loss of microsomal NADPH-cytochrome P450 oxidoreductase. *J Biol Chem* 277:6536–6541.

Shen H, Smith DE, Keep RF, Xiang J, Brosius FC, 3rd (2003) Targeted disruption of the PEPT2 gene markedly reduces dipeptide uptake in choroid plexus. *J Biol Chem* 278:4786–4791.

Shen H, Ocheltree SM, Hu Y, Keep RF, Smith DE (2007) Impact of genetic knockout of PEPT2 on cefadroxil pharmacokinetics, renal tubular reabsorption, and brain penetration in mice. *Drug Metab Dispos* 35:1209–1216.

Shen HW, Yu AM (2009) Difference in desipramine metabolic profile between wild-type and CYP2D6-humanized mice. *Drug Metab Lett* 3:234–241.

Simic T, Savic-Radojevic A, Pljesa-Ercegovac M, Matic M, Mimic-Oka J (2009) Glutathione *S*-transferases in kidney and urinary bladder tumors. *Nat Rev Urol* 6:281–289.

Spina E, Birgersson C, von Bahr C, Ericsson O, Mellstrom B, Steiner E, Sjoqvist F (1984) Phenotypic consistency in hydroxylation of desmethylimipramine and debrisoquine in healthy subjects and in human liver microsomes. *Clin Pharmacol Ther* 36:677–682.

Staudinger JL, Goodwin B, Jones SA, Hawkins-Brown D, MacKenzie KI, LaTour A, Liu Y, Klaassen CD, Brown KK, Reinhard J, Willson TM, Koller BH, Kliewer SA (2001) The nuclear receptor PXR is a lithocholic acid sensor that protects against liver toxicity. *Proc Natl Acad Sci USA* 98:3369–3374.

van de Steeg E, van der Kruijssen CM, Wagenaar E, Burggraaff JE, Mesman E, Kenworthy KE, Schinkel AH (2009) Methotrexate pharmacokinetics in transgenic mice with liver-specific expression of human organic anion-transporting polypeptide 1B1 (SLCO1B1). *Drug Metab Dispos* 37:277–281.

Stieger B, Meier Y, Meier PJ (2007) The bile salt export pump. *Pflugers Arch* 453:611–620.

Su T, Ding X (2004) Regulation of the cytochrome P450 2A genes. *Toxicol Appl Pharmacol* 199:285–294.

Sucov HM, Dyson E, Gumeringer CL, Price J, Chien KR, Evans RM (1994) RXR alpha mutant mice establish a genetic basis for vitamin A signaling in heart morphogenesis. *Genes Dev* 8:1007–1018.

Sweet DH, Miller DS, Pritchard JB, Fujiwara Y, Beier DR, Nigam SK (2002) Impaired organic anion transport in kidney and choroid plexus of organic anion transporter 3 (Oat3 [Slc22a8]) knockout mice. *J Biol Chem* 277: 26934–26943.

Sykes D, Sweet DH, Lowes S, Nigam SK, Pritchard JB, Miller DS (2004) Organic anion transport in choroid plexus from wild-type and organic anion transporter 3 (Slc22a8)-null mice. *Am J Physiol Renal Physiol* 286:F972–F978.

Takenaka K, Morgan JA, Scheffer GL, Adachi M, Stewart CF, Sun D, Leggas M, Ejendal KF, Hrycyna CA, Schuetz JD (2007) Substrate overlap between Mrp4 and Abcg2/Bcrp affects purine analogue drug cytotoxicity and tissue distribution. *Cancer Res* 67:6965–6972.

Timsit YE, Negishi M (2007) CAR and PXR: The xenobiotic-sensing receptors. *Steroids* 72:231–246.

Tirona RG, Kim RB (2005) Nuclear receptors and drug disposition gene regulation. *J Pharm Sci* 94:1169–1186.

Tong MH, Jiang H, Liu P, Lawson JA, Brass LF, Song WC (2005) Spontaneous fetal loss caused by placental thrombosis in estrogen sulfotransferase-deficient mice. *Nat Med* 11:153–159.

Tsuda M, Terada T, Mizuno T, Katsura T, Shimakura J, Inui K (2009) Targeted disruption of the multidrug and toxin extrusion 1 (mate1) gene in mice reduces renal secretion of metformin. *Mol Pharmacol* 75:1280–1286.

Tsunoda SM, Velez RL, von Moltke LL, Greenblatt DJ (1999) Differentiation of intestinal and hepatic cytochrome P450 3A activity with use of midazolam as an *in vivo* probe: Effect of ketoconazole. *Clin Pharmacol Ther* 66:461–471.

Uno S, Dalton TP, Sinclair PR, Gorman N, Wang B, Smith AG, Miller ML, Shertzer HG, Nebert DW (2004) Cyp1a1(-/-) male mice: Protection against high-dose TCDD-induced lethality and wasting syndrome, and resistance to intrahepatocyte lipid accumulation and uroporphyria. *Toxicol Appl Pharmacol* 196:410–421.

Uno S, Dalton TP, Dragin N, Curran CP, Derkenne S, Miller ML, Shertzer HG, Gonzalez FJ, Nebert DW (2006) Oral benzo[a]pyrene in Cyp1 knockout mouse lines: CYP1A1 important in detoxication, CYP1B1 metabolism required for immune damage independent of total-body burden and clearance rate. *Mol Pharmacol* 69:1103–1114.

Vlaming ML, Mohrmann K, Wagenaar E, de Waart DR, Elferink RP, Lagas JS, van Tellingen O, Vainchtein LD, Rosing H, Beijnen JH, Schellens JH, Schinkel AH (2006) Carcinogen and anticancer drug transport by Mrp2 *in vivo*: Studies using Mrp2 (Abcc2) knockout mice. *J Pharmacol Exp Ther* 318:319–327.

Wan YJ, An D, Cai Y, Repa JJ, Hung-Po Chen T, Flores M, Postic C, Magnuson MA, Chen J, Chien KR, French S, Mangelsdorf DJ, Sucov HM (2000) Hepatocyte-specific mutation establishes retinoid X receptor alpha as a heterodimeric integrator of multiple physiological processes in the liver. *Mol Cell Biol* 20:4436–4444.

Wang R, Salem M, Yousef IM, Tuchweber B, Lam P, Childs SJ, Helgason CD, Ackerley C, Phillips MJ, Ling V (2001) Targeted inactivation of sister of P-glycoprotein gene (spgp) in mice results in nonprogressive but persistent intrahepatic cholestasis. *Proc Natl Acad Sci U S A* 98:2011–2016.

Wang R, Lam P, Liu L, Forrest D, Yousef IM, Mignault D, Phillips MJ, Ling V (2003) Severe cholestasis induced by cholic acid feeding in knockout mice of sister of P-glycoprotein. *Hepatology* 38:1489–1499.

van Waterschoot RA, van Herwaarden AE, Lagas JS, Sparidans RW, Wagenaar E, van der Kruijssen CM, Goldstein JA, Zeldin DC, Beijnen JH, Schinkel AH (2008) Midazolam metabolism in cytochrome P450 3A knockout mice can be attributed to up-regulated CYP2C enzymes. *Mol Pharmacol* 73:1029–1036.

van Waterschoot RA, Rooswinkel RW, Sparidans RW, van Herwaarden AE, Beijnen JH, Schinkel AH (2009) Inhibition and stimulation of intestinal and hepatic CYP3A activity: Studies in humanized CYP3A4 transgenic mice using triazolam. *Drug Metab Dispos* 37:2305–2313.

Wei P, Zhang J, Egan-Hafley M, Liang S, Moore DD (2000) The nuclear receptor CAR mediates specific xenobiotic induction of drug metabolism. *Nature* 407:920–923.

Weng Y, Fang C, Turesky RJ, Behr M, Kaminsky LS, Ding X (2007) Determination of the role of target tissue metabolism in lung carcinogenesis using conditional cytochrome P450 reductase-null mice. *Cancer Res* 67:7825–7832.

Wijnholds J, Evers R, van Leusden MR, Mol CA, Zaman GJ, Mayer U, Beijnen JH, van der Valk M, Krimpenfort P, Borst P (1997) Increased sensitivity to anticancer drugs and decreased inflammatory response in mice lacking the multidrug resistance-associated protein. *Nat Med* 3:1275–1279.

Wong HL, Murphy SE, Hecht SS (2005) Cytochrome P450 2A-catalyzed metabolic activation of structurally similar carcinogenic nitrosamines: N'-nitrosonornicotine enantiomers, N-nitrosopiperidine, and N-nitrosopyrrolidine. *Chem Res Toxicol* 18:61–69.

Wu C, Jiang XL, Shen HW, Yu AM (2009) Effects of CYP2D6 status on harmaline metabolism, pharmacokinetics and pharmacodynamics, and a pharmacogenetics-based pharmacokinetic model. *Biochem Pharmacol* 78:617–624.

Wu L, Gu J, Weng Y, Kluetzman K, Swiatek P, Behr M, Zhang QY, Zhuo X, Xie Q, Ding X (2003) Conditional knockout of the mouse NADPH-cytochrome p450 reductase gene. *Genesis* 36:177–181.

Wu L, Gu J, Cui H, Zhang QY, Behr M, Fang C, Weng Y, Kluetzman K, Swiatek PJ, Yang W, Kaminsky L, Ding X (2005) Transgenic mice with a hypomorphic NADPH-cytochrome P450 reductase gene: Effects on development, reproduction, and microsomal cytochrome P450. *J Pharmacol Exp Ther* 312:35–43.

Wu Y, Zhang X, Bardag-Gorce F, Robel RC, Aguilo J, Chen L, Zeng Y, Hwang K, French SW, Lu SC, Wan YJ (2004) Retinoid X receptor alpha regulates glutathione homeostasis and xenobiotic detoxification processes in mouse liver. *Mol Pharmacol* 65:550–557.

Xia CQ, Milton MN, Gan LS (2007) Evaluation of drug-transporter interactions using *in vitro* and *in vivo* models. *Curr Drug Metab* 8:341–363.

Xie W, Barwick JL, Downes M, Blumberg B, Simon CM, Nelson MC, Neuschwander-Tetri BA, Brunt EM, Guzelian PS, Evans RM (2000) Humanized xenobiotic response in mice expressing nuclear receptor SXR. *Nature* 406: 435–439.

Yu AM (2008) Indolealkylamines: Biotransformations and potential drug-drug interactions. *AAPS J* 10:242–253.

Yu AM, Idle JR, Byrd LG, Krausz KW, Kupfer A, Gonzalez FJ (2003a) Regeneration of serotonin from 5-methoxytryptamine by polymorphic human CYP2D6. *Pharmacogenetics* 13:173–181.

Yu AM, Idle JR, Herraiz T, Kupfer A, Gonzalez FJ (2003b) Screening for endogenous substrates reveals that CYP2D6 is a 5-methoxyindolethylamine O-demethylase. *Pharmacogenetics* 13:307–319.

Yu AM, Idle JR, Krausz KW, Kupfer A, Gonzalez FJ (2003c) Contribution of individual cytochrome P450 isozymes to the O-demethylation of the psychotropic beta-carboline alkaloids harmaline and harmine. *J Pharmacol Exp Ther* 305:315–322.

Yu AM, Fukamachi K, Krausz KW, Cheung C, Gonzalez FJ (2005) Potential role for human cytochrome P450 3A4 in estradiol homeostasis. *Endocrinology* 146:2911–2919.

Zaher H, zu Schwabedissen HE, Tirona RG, Cox ML, Obert LA, Agrawal N, Palandra J, Stock JL, Kim RB, Ware JA (2008) Targeted disruption of murine organic anion-transporting polypeptide 1b2 (Oatp1b2/Slco1b2) significantly alters disposition of prototypical drug substrates pravastatin and rifampin. *Mol Pharmacol* 74:320–329.

Zair ZM, Eloranta JJ, Stieger B, Kullak-Ublick GA (2008) Pharmacogenetics of OATP (SLC21/SLCO), OAT and OCT (SLC22) and PEPT (SLC15) transporters in the intestine, liver and kidney. *Pharmacogenomics* 9:597–624.

Zanger UM, Raimundo S, Eichelbaum M (2004) Cytochrome P450 2D6: Overview and update on pharmacology, genetics, biochemistry. *Naunyn Schmiedebergs Arch Pharmacol* 369:23–37.

Zelcer N, van de Wetering K, Hillebrand M, Sarton E, Kuil A, Wielinga PR, Tephly T, Dahan A, Beijnen JH, Borst P (2005) Mice lacking multidrug resistance protein 3 show altered morphine pharmacokinetics and morphine-6-glucuronide antinociception. *Proc Natl Acad Sci U S A* 102:7274–7279.

Zelcer N, van de Wetering K, de Waart R, Scheffer GL, Marschall HU, Wielinga PR, Kuil A, Kunne C, Smith A, van der Valk M, Wijnholds J, Elferink RO, Borst P (2006) Mice lacking Mrp3 (Abcc3) have normal bile salt transport, but altered hepatic transport of endogenous glucuronides. *J Hepatol* 44:768–775.

Zhang J, Huang W, Chua SS, Wei P, Moore DD (2002) Modulation of acetaminophen-induced hepatotoxicity by the xenobiotic receptor CAR. *Science* 298:422–424.

Zhang QY, Gu J, Su T, Cui H, Zhang X, D'Agostino J, Zhuo X, Yang W, Swiatek PJ, Ding X (2005) Generation and characterization of a transgenic mouse model with hepatic expression of human CYP2A6. *Biochem Biophys Res Commun* 338:318–324.

Zhang QY, Fang C, Zhang J, Dunbar D, Kaminsky L, Ding X (2009) An intestinal epithelium-specific cytochrome P450 (P450) reductase-knockout mouse model: Direct evidence for a role of intestinal p450s in first-pass clearance of oral nifedipine. *Drug Metab Dispos* 37:651–657.

Zhou S, Morris JJ, Barnes Y, Lan L, Schuetz JD, Sorrentino BP (2002) Bcrp1 gene expression is required for normal numbers of side population stem cells in mice, and confers relative protection to mitoxantrone in hematopoietic cells *in vivo*. *Proc Natl Acad Sci U S A* 99:12339–12344.

Zhou SF (2008) Structure, function and regulation of P-glycoprotein and its clinical relevance in drug disposition. *Xenobiotica* 38:802–832.

Zhou X, Zhuo X, Xie F, Kluetzman K, Shu YZ, Humphreys WG, Ding X (2009) Role of CYP2A5 in the clearance of nicotine and cotinine: Insights from studies on a Cyp2a5-null mouse model. *J Pharmacol Exp Ther* 332:578–587.

Zwart R, Verhaagh S, Buitelaar M, Popp-Snijders C, Barlow DP (2001) Impaired activity of the extraneuronal monoamine transporter system known as uptake-2 in Orct3/Slc22a3-deficient mice. *Mol Cell Biol* 21:4188–4196.

29

PLURIPOTENT STEM CELL MODELS IN HUMAN DRUG DEVELOPMENT

David C. Hay

29.1 INTRODUCTION

Primary human hepatocytes are a scarce resource and exhibit variable function that diminishes with time in culture. As a consequence, their use in drug development is restricted. Other hepatocyte models exist but suffer from serious limitations often resulting in non-predictive outcomes. Human embryonic stem cells (hESCs) and induced pluripotent stem cells (iPSCs) are scalable resources that offer an unlimited supply of uniform hepatocyte-like cells (HLCs) for drug discovery from the desired genetic backgrounds (Hay, 2011). The enabling potential of stem cell-derived HLCs in combination with tissue engineering and large-scale manufacture will be discussed in this chapter.

29.2 HUMAN DRUG METABOLISM AND COMPOUND ATTRITION

Human drug development is a very costly process heavily determined by the rate of compound attrition. Approximately 5000–10,000 compounds are tested pre-clinically for each drug that reaches the market, demonstrating the requirement for deploying high-fidelity toxicity models early on in the human drug discovery process (PhRMA, 2005). Liver toxicity is one of the major causes of compound failure and therefore a serious concern. Hepatocytes constitute about two-thirds of the liver parenchyma and perform numerous

essential functions. One of their major roles is to metabolize foreign compounds and process them for excretion from the body. The pharmaceutical industry therefore has an interest in drug metabolizing enzyme function and the role that hepatocytes play in compound metabolism and excretion.

Drug metabolism can be broadly divided into three phases. The Phase I enzymes, such as cytochrome P450s (CYP P450s), catalyze the introduction of reactive groups onto xenobiotics. Phase I metabolites are then conjugated to polar compounds in Phase II (Guengerich, 2001). The creation of polar metabolites in Phase II prevents the diffusion of intermediates across the cell membrane, requiring active transport for their removal (Jakoby and Ziegler, 1990). Active transport is provided in Phase III by efflux transporters in an adenosine triphosphate (ATP)-dependent manner (Commandeur et al., 1995).

One of the major reasons for human drug attrition is the inability to accurately model human CYP P450 activity in both *in vitro* and *in vivo* models (Bachmann and Lewis, 2005). The human genome consists of 57 CYP P450s (Nelson, 2003) where as animals have as many or more CYP P450 genes than humans; for example, the murine genome consists of 101 P450 genes. Due to limitations with human hepatocytes, animal models are frequently used in drug discovery and toxicology testing. The relative abundance and diversity of CYP P450 function in animal models therefore only provides limited extrapolation to the human liver and can be nonpredictive.

ADME-Enabling Technologies in Drug Design and Development, First Edition. Edited by Donglu Zhang and Sekhar Surapaneni.
© 2012 John Wiley & Sons, Inc. Published 2012 by John Wiley & Sons, Inc.

29.3 HUMAN HEPATOCYTE SUPPLY

Primary adult human hepatocytes are the current gold standard for predictive toxicology testing. However, these cells are scarce, have limited life span, and exhibit variable and dramatic loss of function upon isolation. As a result, there has been much interest in identifying novel and readily available sources of hepatocytes that can be maintained long term in culture. Human liver cancer and immortalized cell lines have been established, but exhibit restricted drug metabolism (Dalgetty et al., 2009). Over the years, there have been significant advances in the identification and understanding of human stem cell populations that demonstrate potential. Stem cell populations exhibit the ability to self-renew; that is, produce exact copies of themselves, while retaining developmental plasticity offering, in theory, an inexhaustible supply of derivative hepatocytes (Dalgetty et al., 2009). There are a number of stem cell options one can consider when wishing to generate parenchymal hepatocytes. There are resident stem cell populations in both the developing and adult human liver. A number of studies have proposed the use of fetal liver stem cells (hepatoblasts) and adult liver stem cells (oval cells) to generate continuous supply of HLCs. Therefore, a pragmatic approach would be to expand the isolated hepatic progenitor cell populations and differentiate them to HLCs upon requirement. However, due to problems with large-scale isolation, purity, and limited *ex vivo* culture, hepatic progenitor cells are currently not suitable for large-scale expansion (Czyz et al., 2003). Therefore, in recent years, there has been a focus on deriving HLCs from other sources, in particular hESCs (Hay, 2010) and more recently human iPSCs (Medine et al., 2010). hESCs are highly primitive cells derived from the inner cell mass of blastocyst-stage embryos (Thomson et al., 1998). iPSCs, on the other hand, are derived from human somatic cells by the introduction of a defined set of transcription factors, first demonstrated by Shinya Yamanaka's group (Takahashi and Yamanaka, 2006; Takahashi et al., 2007).

29.4 hESCs

hESCs are derived from the inner cell mass of preimplantation embryos that are deemed unsuitable for human use. hESCs possess the ability to self-renew and demonstrate pluripotency, *in vitro* and *in vivo* (Thomson et al., 1998). These attributes allow the scalable production of hESCs capable of differentiation down all three germ layers (for a review, see Hannoun et al., 2010b). As such, hESCs have the potential to provide an unlim-

ited supply of the appropriate derivative cell types and offer significant advantages over their adult stem cell counterparts (Czyz et al., 2003).

29.5 hESC HLC DIFFERENTIATION

hESC differentiation to functional HLCs has been achieved either by spontaneous differentiation through the formation of multicellular aggregates, termed embryoid bodies (EBs), or by direct differentiation. Spontaneous differentiation of hESCs using EBs results in the production of a mixed cell population representative of all three germ layers (Itskovitz-Eldor et al., 2000). The differentiating cells within the EB structures are capable of further differentiation into HLCs (Imamura et al., 2004; Lavon and Benvenisty, 2005; Baharvand et al., 2006; Basma et al., 2009), albeit with limited efficiency and therefore require additional purification if homogeneous cultures are required; whereas, direct differentiation of hESCs to HLCs has proved faster and more efficient *in vitro* (Cai et al., 2007; Duan et al., 2007; Hay et al., 2007, 2008a,b; Fletcher et al., 2008; Agarwal et al., 2008; Hannoun et al., 2010a; Greenhough et al., 2010; Medine et al., 2011). Over the years, this has been an evolving process that has seen HLC yields and function drastically improve as human developmental parameters have been applied in *in vitro* models. In 2007, we described the differentiation of HLCs from hESCs with an efficiency of approximately 10% (Hay et al., 2007), which has been improved to approximately 90% (Hay et al., 2008a; Hannoun et al., 2010a). Further work in the laboratory also demonstrated that we could not only improve upon this yield but dramatically improve cell function and scalability by employing signaling parameters in line with human development (Hay et al., 2008a) and synthetic substrates (Hay et al., 2011).

29.6 iPSCs

The generation of human iPSCs from somatic cells has revolutionized the stem cell field. Human iPSCs share similar scalable and pluripotent characteristics to hESCs. The reprogramming of somatic cells was first demonstrated by Shinya Yamanaka's laboratory using both mouse and human fibroblasts and retroviral insertion of a defined set of transcription factors (Takahashi and Yamanaka, 2006; Takahashi et al., 2007). Inducing pluripotency in somatic cells has since been employed by a number of groups who have employed a variety of starting materials. Moreover, a variety of functional cell types have been derived from iPSCs, demonstrating that they are an important resource with which to model human biology (for a review, see Dalgetty et al., 2009). Most recently, we and others have demonstrated the translatable nature of model systems designed in hESC to iPSC cells, therefore making it possible to generate HLCs in sufficient quantity for downstream applications (Rashid et al., 2010; Si-Tayeb et al., 2010; Sullivan et al., 2010).

A major attraction of using human iPSCs in the drug discovery and toxicology testing process is the generation of somatic cells from the desired ethnic background and/or disease state. As such, iPSC-derived resources have the potential to revolutionize the manner in which we model human drug metabolism *in vitro* taking into account vital polymorphic variations in xenobiotic metabolism between individuals (e.g., CYP2C9/ warfarin) and ability to model disease genotypes (e.g., alpha 1-antitrypsin; Rashid et al., 2010). The reprogramming of the desired human genotype and differentiation to somatic cells displaying different metabolic features will facilitate the creation of a valuable human library.

29.7 CYP P450 EXPRESSION IN STEM CELL-DERIVED HLCs

Major progress has been made in the stem cell field with regard to Phase I CYP P450 expression and function from pluripotent stem cell-based models that approximates to that of adult human hepatocyte function. To date CYP 1A1, 1A2, 1B1, 2A6, 2A7, 2B6, 2C8, 2C9, 2C19, 2D6, 3A4, 3A7, and 7A1 P450 expression and/or function have been demonstrated *in vitro* using pluripotent stem cells (for a comprehensive review, see Asgari et al., 2010). While these studies are highly promising, it must not be overlooked that drug metabolism is an integrated process requiring the coordination of 3 phases. Therefore, future studies are needed to characterize the expression and function of Phase II and Phase III components in stem cell based systems.

29.8 TISSUE CULTURE MICROENVIRONMENT

Stem cell-derived HLC utility in cell-based assays depends on their functional repertoire. Cell populations generally have limited function without specialized tissue microenvironments. These key microenvironmental factors include signaling molecules, cell–cell association, and cell-matrix adhesions, which regulate homeostasis (Scadden, 2006; Discher et al., 2009) and maintain cellular phenotype (Ohno et al., 2009). An additional complexity of the cell niche *in vivo* is its three-dimensional (3D) architecture. The 3D structure is essential for physiological tissue function, of which cell polarity is known to be a vital component (Lee et al., 2008). *In vivo*, hepatocytes exhibit polarity with the division of the plasma membrane into three functional domains: basolateral, canalicular, and lateral. Upon isolation from the liver, hepatocytes lose their polarity and, as a consequence, some of their liver-specific function. Therefore, the ability to re-create hepatic polarity *in vitro* is important in maintaining and developing a broad range of high-fidelity liver models. Moreover, the in-depth study of the 3D environment may provide an insight into cellular physiology. (For a review, see Sharma et al., 2010.)

29.9 CULTURE DEFINITION FOR DERIVING HLCs FROM STEM CELLS

As stem cell-derived cell types move closer to application in the diagnostic arena, there is need for cell culture standardization. Essential to this process are the development of defined culture conditions that permit the cost-effective expansion and differentiation of stem cells. At present, hESCs and iPSCs are routinely maintained on extracellular matrices (ECMs) in medium that contain animal products. The presence of xenogenic products in stem cell culture creates a number of problems with respect to cell culture, including scale-up complications and data interpretation. We and several other groups have begun this process of culture definition using serum-free culture conditions for stem cell line maintenance and differentiation (Ludwig et al., 2006; Wang et al., 2007; Hannoun et al., 2010a). The next stage in the incremental process of culture definition will be the identification of defined and animal free ECMs that are able sustain pluripotent and differentiated cell populations. (Hay et al., 2011)

29.10 CONCLUSION

Recent progress in the production of stem cell-derived HLCs has provided an enormous opportunity to generate high-fidelity and predictive *in vitro* models of human liver. Although reproducible technology has been developed, there is still a clear requirement to broaden the repertoire stem cell-derived HLC function to that of the current gold standard, freshly isolated primary human hepatocytes. The elucidation of key developmental factors and novel tissue culture microenvironments, coupled with advances in stem cell technology, is essential.

REFERENCES

Agarwal S, Holton KL, Lanza R (2008) Efficient differentiation of functional hepatocytes from human embryonic stem cells. *Stem Cells* 26(5):1117–1127.

Asgari S, Pournasr B, Salekdeh GH, Ghodsizadeh A, Ott M, Baharvand H (2010) Induced pluripotent stem cells: A new era for hepatology. *J Hepatol* 53(4):738–751.

Bachmann KA, Lewis JD (2005) Predicting inhibitory drug-drug interactions and evaluating drug interaction reports using inhibition constants. *Ann Pharmacother* 39(6):1064–1072.

Baharvand H, Hashemi SM, Kazemi Ashtiani S, Farrokhi A (2006) Differentiation of human embryonic stem cells into hepatocytes in 2D and 3D culture systems *in vitro*. *Int J Dev Biol* 50(7):645–652.

Basma H, Soto-Gutiérrez A, Yannam GR, Liu L, Ito R, Yamamoto T, Ellis E, Carson SD, Sato S, Chen Y, Muirhead D, Navarro-Alvarez N, Wong RJ, Roy-Chowdhury J, Platt JL, Mercer DF, Miller JD, Strom SC, Kobayashi N, Fox IJ (2009) Differentiation and transplantation of human embryonic stem cell-derived hepatocytes. *Gastroenterology* 136(3):990–999.

Cai J, Zhao Y, Liu Y, Ye F, Song Z, Qin H, Meng S, Chen Y, Zhou R, Song X, Guo Y, Ding M, Deng H (2007) Directed differentiation of human embryonic stem cells into functional hepatic cells. *Hepatology* 45(5):1229–1239.

Commandeur JN, Stijntjes GJ, Vermeulen NP (1995) Enzymes and transport systems involved in the formation and disposition of glutathione S-conjugates. Role in bioactivation and detoxication mechanisms of xenobiotics. *Pharmacol Rev* 47(2):271–330.

Czyz J, Wiese C, Rolletschek A, Blyszczuk P, Cross M, Wobus AM (2003) Potential of embryonic and adult stem cells *in vitro*. *Biol Chem* 384:1391–1409.

Dalgetty DM, Medine C, Iredale JP, Hay DC (2009) Progress and future challenges in stem cell-derived liver technologies. *Am J Physiol Gastrointest Liver Physiol* 297(2):G241–G248.

Discher DE, Mooney DJ, Zandstra PW (2009) Growth factors, matrices, and forces combine and control stem cells. *Science* 324(5935):1673–1677.

Duan Y, Catana A, Meng Y, Yamamoto N, He S, Gupta S, Gambhir SS, Zern MA (2007) Differentiation and enrichment of hepatocyte-like cells from human embryonic stem cells *in vitro* and *in vivo*. *Stem Cells* 25(12):3058–3068.

Fletcher J, Cui W, Samuel K, Black JR, Hannoun Z, Currie IS, Terrace JD, Payne C, Filippi C, Newsome P, Forbes SJ, Ross JA, Iredale JP, Hay DC (2008) The inhibitory role of stromal cell mesenchyme on human embryonic stem cell hepatocyte differentiation is overcome by Wnt3a treatment. *Cloning Stem Cells* 10(3):331–339.

Greenhough S, Medine CN, Hay DC (2010) Pluripotent stem cell derived hepatocyte like cells and their potential in toxicity screening. *Toxicology* [Epub ahead of print].

Guengerich FP (2001) Common and uncommon cytochrome P450 reactions related to metabolism and chemical toxicity. *Chem Res Toxicol* 14(6):611–650.

Hannoun Z, Fletcher J, Greenhough S, Medine CN, Samuel K, Sharma R, Pryde A, Black JR, Ross JA, Wilmut I, Iredale JP, Hay DC (2010a) The comparison between conditioned media and serum free media in human embryonic stem cell culture and differentiation. *Cell Reprogram* 12(2):APR 12(2):133–140.

Hannoun Z, Filippi C, Sullivan G, Hay DC, Iredale JP (2010b) Hepatic endoderm differentiation from hESCs. *Curr Stem Cell Res Ther* 5(3):233–244.

Hay DC (2010) Cadaveric hepatocytes repopulate diseased livers: Life after death. *Gastroenterology* 139(3):729–731.

Hay DC (2011) The importance of stem cells and bioengineering in regenerative medicine. *Biochip Tissue Chip* DOI: 10.4172/2153-0777.1000e101.

Hay DC, Zhao D, Ross A, Mandalam R, Lebkowski J, Cui W (2007) Direct differentiation of human embryonic stem cells to hepatocyte-like cells exhibiting functional activities. *Cloning Stem Cells* 9(1):51–62.

Hay DC, Fletcher J, Payne C, Terrace JD, Gallagher RC, Snoeys J, Black JR, Wojtacha D, Samuel K, Hannoun Z, Pryde A, Filippi C, Currie IS, Forbes SJ, Ross JA, Newsome PN, Iredale JP (2008a) Highly efficient differentiation of hESCs to functional hepatic endoderm requires ActivinA and Wnt3a signaling. *Proc Natl Acad Sci U S A* 105(34):12301–12306.

Hay DC, Zhao D, Fletcher J, Hewitt ZA, McLean D, Urruticoechea-Uriguen A, Black JR, Elcombe C, Ross JA, Wolf R, Cui W (2008b) Efficient differentiation of hepatocytes from human embryonic stem cells exhibiting markers recapitulating liver development *in vivo*. *Stem Cells* 26(4):894–902.

Hay DC, Pernagallo S, Diaz-Mochon JJ, Medine CN, Greenhough S, Hannoun Z, Schrader J, Black JR, Fletcher J, Dalgetty D, Thomas AI, Newsome PN, Forbes SJ, Ross JA, Bradley M, Iredale JP (2011) Unbiased screening of polymer libraries to define novel substrates for functional hepatocytes with inducible drug metabolism. *Stem Cell Research* 6(2):92–102.

Imamura T, Cui L, Teng R, Johkura K, Okouchi Y, Asanuma K, Ogiwara N, Sasaki K (2004) Embryonic stem cell-derived embryoid bodies in three-dimensional culture system form hepatocyte-like cells *in vitro* and *in vivo*. *Tissue Eng* 10(11–12):1716–1724.

Itskovitz-Eldor J, Schuldiner M, Karsenti D, Eden A, Yanuka O, Amit M, Soreq H, Benvenisty N (2000) Differentiation of human embryonic stem cells into embryoid bodies compromising the three embryonic germ layers. *Mol Med* 6(2):88–95.

Jakoby WB, Ziegler DM (1990) The enzymes of detoxication. *Biol Chem* 265(34):20715–20718.

Lavon N, Benvenisty N (2005) Study of hepatocyte differentiation using embryonic stem cells. *J Cell Biochem* 96(6):1193–1202.

Lee J, Cuddihy MJ, Kotov NA (2008) Three-dimensional cell culture matrices: State of the art. *Tissue Eng Part B Rev* 14(1):61–86.

Ludwig ET, Levenstein ME, Jones JM, Berggren WT, Mitchen ER, Frane JL, Crandall LJ, Daigh CA, Conard KR, Piekarczyk MS, Llanas RA, Thomson JA (2006) Derivation of hESC in defined conditions. *Nat Biotechnol* 24(2):185–187.

Medine CN, Greenhough S, Hay DC (2010) Role of stem-cell-derived hepatic endoderm in human drug discovery. *Biochem Soc Trans* 38(4):1033–1036.

Medine CN, Lucendo-Villarin B, Zhou W, West OC, Hay DC (2011) Robust generation of hepatocyte-like cells from human embryonic stem cell populations. *J Vis Exp* Oct 26; (56), pii 2969. DOI: 10.3791/2969.

Nelson DR (2003) Comparison of P450s from human and fugu: 420 million years of vertebrate P450 evolution. *Arch Biochem Biophys* 409(1):18–24.

Ohno M, Motojima K, Okano T, Taniguchi A (2009) Maturation of the extracellular matrix and cell adhesion molecules in layered co-cultures of HepG2 and endothelial cells. *J Biochem* 145(5):591–597.

Pharmaceutical Research and Manufacturers of America (PhRMA) (2005) Pharmaceutical industry profile.

Rashid ST, Corbineau S, Hannan N, Marciniak SJ, Miranda E, Alexander G, Huang-Doran I, Griffin J, Ahrlund-Richter L, Skepper J, Semple R, Weber A, Lomas DA, Vallier L (2010) Modeling inherited metabolic disorders of the liver using human induced pluripotent stem cells. *J Clin Invest* 120(9):3127–3136.

Scadden DT (2006) The stem-cell niche as an entity of action. *Nature* 441(7097):1075–1079.

Sharma R, Greenhough S, Medine CN, Hay DC (2010) Three-dimensional culture of hESC derived hepatic endoderm and its role in bioartificial liver construction. *J Biomen Biotechnol* 2010:236–247.

Si-Tayeb K, Noto FK, Nagaoka M, Li J, Battle MA, Duris C, North PE, Dalton S, Duncan SA (2010) Highly efficient generation of human hepatocyte-like cells from induced pluripotent stem cells. *Hepatology* 51(1):297–305.

Sullivan GJ, Hay DC, Park IH, Fletcher J, Hannoun Z, Payne CM, Dalgetty D, Black JR, Ross JA, Samuel K, Wang G, Daley GQ, Lee JH, Church GM, Forbes SJ, Iredale JP, Wilmut I (2010) Generation of functional human hepatic endoderm from human iPS cells. *Hepatology* 51(1):329–335.

Takahashi K, Yamanaka S (2006) Induction of pluripotent stem cells from mouse embryonic and adult fibroblast cultures by defined factors. *Cell* 126:663–676.

Takahashi K, Tanabe K, Ohnuki M, Narita M, Ichisaka T, Tomoda K, Yamanaka S (2007) Induction of pluripotent stem cells from adult human fibroblasts by defined factors. *Cell* 131:861–872.

Thomson JA, Itskovitz-Eldor J, Shapiro SS, Waknitz MA, Swiergiel JJ, Marshall VS, Jones JM (1998) Embryonic stem cell lines derived from human blastocysts. *Science* 282: 1145–1147.

Wang L, Schulz TC, Sherrer ES, Dauphin DS, Shin S, Nelson AM, Ware CB, Zhan M, Song CZ, Chen X, Brimble SN, McLean A, Galeano MJ, Uhl EW, D'Amour KA, Chesnut JD, Rao MS, Blau CA, Robins AJ (2007) Self-renewal of human embryonic stem cells requires insulin-like growth factor-1 receptor and ERBB2 receptor signaling. *Blood* 110(12):4111–4119.

30

RADIOSYNTHESIS FOR ADME STUDIES

Brad D. Maxwell and Charles S. Elmore

30.1 BACKGROUND AND GENERAL REQUIREMENTS

The radiosynthesis of active pharmaceutical ingredients (API) is a highly specialized operation involving uniquely trained chemists operating in specialized laboratories. Most large pharmaceutical companies employ a staff of radiosynthetic chemists located within discovery chemistry, biotransformation, or pharmaceutical development departments, with responsibilities to synthesize, purify, analyze, and dispense radiolabeled and stable-labeled products for use in preclinical, clinical, and postmarketing studies. Small pharmaceutical companies typically pay for these services at contract research organizations when radiolabeled products are required.

There are many good references on the subject of radiochemistry. The most current and complete information can be found in *Preparation of Compounds Labeled with Tritium and Carbon-14* (Voges et al., 2009); however, this work does not include the radiosynthesis of API for human absorption, distribution, metabolism, and excretion (ADME) studies. The International Isotope Society is another good source of information on the synthesis of radiolabeled materials. In 2003, it held the first workshop on human ADME radiolabeled syntheses during the 8th International Symposium on the Synthesis and Applications of Isotopes and Isotopically Labelled Compounds. One of the presentations made at this conference included the various requirements for conducting radiosyntheses under current good manufacturing practice (cGMP) conditions (Lloyd

et al., 2004). There have also been publications detailing some of the aspects of conducting radiosyntheses of API under cGMP conditions for use in human ADME (Roberts, 2009/2010) and their use (Marathe et al., 2004), but none have provided a comprehensive description of this topic, thus necessitating this chapter.

30.1.1 Food and Drug Administration (FDA) Guidance

The FDA regulations governing the synthesis of materials for radiolabeled human ADME studies are found in the International Conference on Harmonization of Technical Requirements for Registration of Pharmaceuticals for Human Use (ICH) Q7A, Good Manufacturing Guidance for Active Pharmaceutical Ingredients, Section XIX, APIs for Use in Clinical Trials. The exact interpretation of how to comply with this guidance varies considerably from company to company as the latitude given in the documentation can be interpreted in vastly different manners. Typically, an agreement between quality assurance (QA) and radiochemistry is made on exactly how the API destined for a human ADME study will be handled and processed. Human ADME studies are governed by the Code of Federal Regulations Title 21 part 361 (21 CFR 361) (FDA Website, 2010), which requires that a Radioactive Drug Research Committee be established to oversee the use of radioactivity in humans. This committee ensures that a proper radioactive and pharmacological dose is administered and that the study's objectives justify the exposure of the subjects to radioactivity (Dain et al., 1994).

ADME-Enabling Technologies in Drug Design and Development, First Edition. Edited by Donglu Zhang and Sekhar Surapaneni.
© 2012 John Wiley & Sons, Inc. Published 2012 by John Wiley & Sons, Inc.

30.1.2 Third Clinical Study after Single Ascending Dose (SAD) and Multiple Ascending Dose (MAD) Studies

The timing of the radiolabeled human ADME study varies depending on the needs of the project, but it is typically after the SAD and MAD studies in late Phase I or early Phase II; although it can be delayed into late Phase II or even Phase III. The determination of the timing of the study is based on the requirements of the project. For projects that have toxicological or metabolite concerns, the study may occur early, but a project with lower concerns may wait until Phase IIb.

30.1.3 Formation of the ADME Team

An ADME team will guide the clinical trial and should be formed 6 to 9 months prior to the first dose of the study. The composition of the ADME team can vary, but generally consists of a representative from isotope or radiochemistry, DMPK (metabolism expert), analytical development, QA, regulatory affairs, pharmaceutics or formulations research, and the clinical study department. This team will also include a study director who may also fulfill one of the other roles. This team works together to plan the study including the location of the study, the dose to be given to the subjects, and the manner in which samples will be collected and processed, as well as whether bile will be collected as a part of the study. They also set the purity criteria for the API and the analyses that will be conducted to ensure the API is of appropriate purity and potency for the study. In most cases, the purity and potency minimum values will already have been established in the corporate standard operating procedure (SOP) for radiolabeled API used in human ADME clinical studies and is typically greater than 98%, with no single impurity being greater than 0.5%.

30.1.4 Human Dosimetry Projection

FDA regulations (21 CFR 361) set limits on the amount of radioactive dose that a subject can receive as a whole body dose and as a dose to single organs. To insure compliance with this regulation and for the safety of the subject, dose predictions are conducted by correlating the doses observed in animals to those projected in humans. Two general methods are used to determine the whole-body dose and organ dose in preclinical species: mass balance or quantitative whole-body autoradiography (QWBA). In both studies, several pigmented animals are dosed with the radiolabeled compound and the distribution of the radioactivity in the animal is followed until most of the radioactivity has been recovered. A mass balance experiment consists of the animal

being dissected, its organs homogenized, and the radioactive content of each organ determined by liquid scintillation counting (LSC). In QWBA, after dosing, the animal is sectioned and each section is imaged using a phosphor imager or other comparable imaging technology and referenced against a standard. The amount of radioactivity in each organ can then be determined. This value is then scaled to humans, and estimated doses for humans and individual organs can be estimated (Solon et al., 2002; Solon, 2010).

30.1.5 cGMP Synthesis Conditions

The purpose of cGMP is to insure the quality of the drug by controlling the manufacturing and testing of the API. cGMP is a set of guidelines and regulations that should be followed during the production of an API. There are a few key principles that define cGMP as applied to drug syntheses:

1. A well-documented protocol is written and validated.
2. Technicians are well trained to carry out their assigned job function.
3. The procedure is well defined and controls are in place to insure the results are as expected.
4. Any deviations from the protocol during manufacturing are noted and the impact investigated.
5. A detailed record of the process is kept that demonstrates that the protocol was followed.
6. The product is carefully labeled and controlled.

cGMP guidelines previously differed among Europe, Japan, and the United States. As a result, the ICH was formed in 1990 to draft a single guidance to satisfy all three regions. This effort has been quite fruitful in providing guidance for cGMP manufacturing.

30.1.6 Formation of One Covalent Bond

It is impractical for a preparation of C-14-labeled materials for human ADME studies to be conducted in a truly cGMP-compliant manner. Often the synthesis will be the second preparation of the labeled compound, which precludes an extensive batch history for the compound. Furthermore, since the laboratories that handle radiolabeled syntheses are specialized, they frequently prepare material for preclinical studies as well and may not have separate cGMP facilities. The validated route to the product using the process research scheme may not be appropriate for radiolabeling. Therefore, questions arise as to how to comply with the spirit of cGMP regulations in small-scale preparations of a single batch of the compound. Compromises are made that still insure the quality of the API and protect the subject receiving the drug while making the synthesis feasible in the radiochemistry laboratory.

One of the first issues to arise during a discussion of the preparation of API for a human ADME study is at what point to initiate the strictest of controls, the most cGMP-like. Some organizations require the entire synthesis be conducted in the spirit of cGMP compliance, while others conduct only the final purification in that spirit. An acceptable compromise in most organizations is to treat the entire synthesis with more controls than normal, but to institute the tightest of controls over the final bond-forming step—trying to comply with cGMP guidelines as closely as possible. This allows for good control over the environment leading up to the penultimate intermediate in the synthesis and strict, near-cGMP compliance for the final step. Since the final step, purification and analyses will be performed to nearly cGMP standards, the quality of the API is controlled, and the subject of the study protected, which are the ultimate goals of cGMP guidance.

30.2 RADIOSYNTHESIS STRATEGIES AND GOALS

There are several obstacles the radiosynthetic chemist must conquer to successfully prepare the radiolabeled API that is suitable for a human ADME study. Most of the strategies and goals for efficiently incorporating radiolabels into molecules for use in human ADME studies are the same as those for incorporating radiolabels into molecules for use in late-stage discovery metabolic profiling studies. Described below are some of the most important options and decisions that need to be made by the radiosynthetic chemist and ADME project team.

30.2.1 Determination of the Most Suitable Radioisotope for the Human ADME Study

Typical isotopes used for radiolabeled ADME studies are tritium (^{3}H, H-3, or T) or carbon-14 (C-14 or ^{14}C) since every API will have hydrogens and carbons present in their structure. In the majority of the cases, C-14 is preferred over tritium, as losses of the radiolabel due to hydroxylations and exchange more readily occur with tritiated API than with C-14. However, there are instances when C-14-labeled API is not suitable for use in a human ADME study. For example, if the API has high potency at low doses, the specific activity (SA) of the API will need to be higher than what can be achieved with C-14 as described in the synthesis of [^{3}H]apadenoson shown in Figure 30.1 (Hong et al., 2008). [1,2-^{3}H] Ethylamine hydrochloride was prepared by the tritium reduction of N-vinylphthalimide followed by a Gabriel synthesis to release the labeled product. [1,2-^{3}H]Ethylamine hydrochloride was coupled with the unlabeled acid to form [^{3}H]apadenoson in three short steps. The final SA of [^{3}H]apadenoson used in the human ADME was 48.07 Ci/mmol. This SA is much higher than is possible for C-14 labeling, even if every single carbon in apadenoson were labeled with C-14.

In other cases, tritium-labeled API is preferred over C-14 due to the difficulty or cost associated with synthesizing the C-14-labeled API compared with the ease of preparation of the tritiated API (Rotert et al., 2006; Moenius et al., 2001). In general though, most radiosyntheses of API for human ADME studies are completed with C-14.

30.2.2 Synthesize the API with the Radiolabel in the Most Metabolically Stable Position

The site of label incorporation should be directed by preclinical studies; it should be located such that it is not

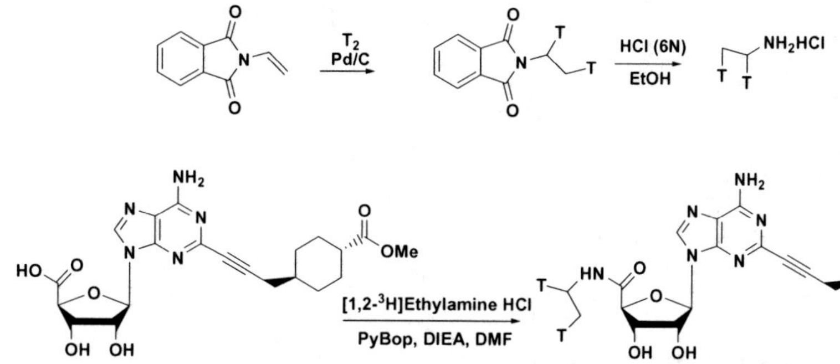

Figure 30.1. Synthesis of [^{3}H]apadenoson.

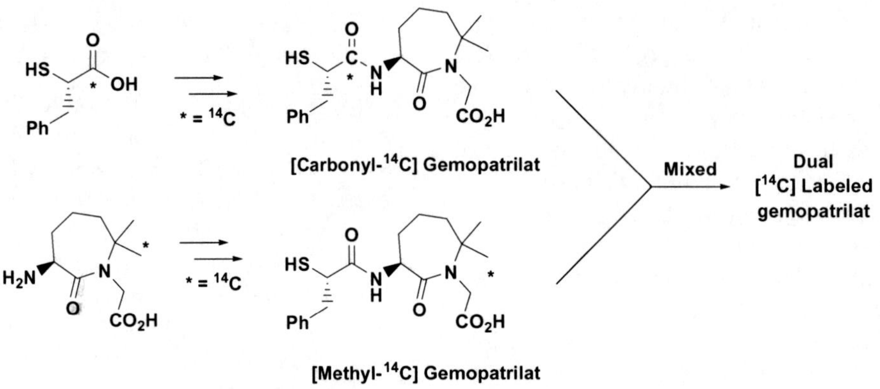

Figure 30.2. Synthesis of dual labeled [^{14}C]gemopatrilat.

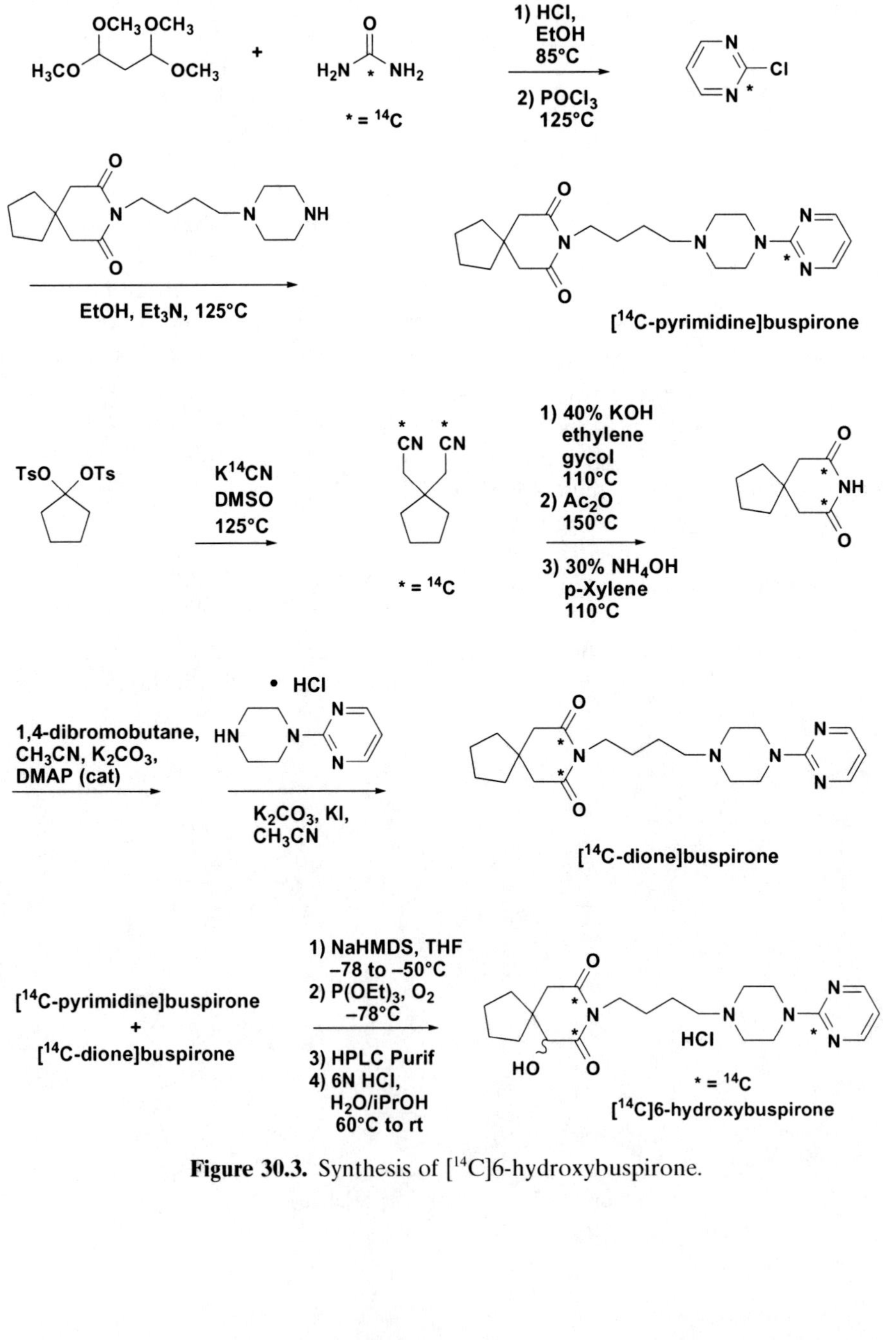

Figure 30.3. Synthesis of [^{14}C]6-hydroxybuspirone.

lost due to exchange or respiration and should be contained within a large portion of the metabolite(s). It is preferred to have the radiolabel in the largest metabolic fragment. If the molecule breaks into two large portions, a mixture of radiolabeled compounds containing the radiolabel in each half should be considered. For example, a major metabolic pathway for gemopatrilat involves peptide cleavage to give two large fragments. The human ADME team solved this by dosing a mixture of two C-14-labeled isotopomers of gemopatrilat with the two C-14 labels in opposite portions of the molecule as shown in Figure 30.2 (Wait et al., 2006).

Another example of dual labeling is in the synthesis of [^{14}C]6-hydroxybuspirone by Bristol-Myers Squibb radiochemists as shown in Figure 30.3 (Bonacorsi et al., 2007). [3-^{14}C]2-Chloropyrimidine was prepared from readily available [^{14}C]urea in two steps. The resulting product was coupled with unlabeled clinical-grade piperazinyl dione to form the penultimate compound, [^{14}C-pyrimidine]buspirone. [^{14}C-dione]Buspirone was prepared from readily available potassium [^{14}C]cyanide to form the spirocyclic dione that was coupled in two steps with 1,4-dibromobutane and the unlabeled piperazinylpyrimidine to form the second labeled penultimate compound. The two C-14-labeled buspirone products were mixed and converted to a mixture of [^{14}C]6-hydroxybuspirone allowing for both halves of the molecule to be followed in the ADME study.

30.2.3 Incorporate the Radiolabel as Late in the Synthesis as Possible

The radiosynthetic chemist would like to incorporate the radiolabel into the API as late in the synthesis as possible to minimize the number of radiochemical steps and thereby increase the yield of the API while minimizing the amount of radioactive waste generated. If the radiosynthetic route and clinical-grade process synthesis route are the same, then valuable advanced synthetic intermediates synthesized under cGMP conditions will be available for transformation into the radiolabeled API, thereby saving time and effort by the radiosynthetic chemist. These unlabeled advanced intermediates will include appropriate batch and release records, which will also simplify the preparation of the batch record. Unfortunately, it is not always possible to introduce the radiolabel late in the synthesis, especially if the radiolabel needs to be buried deep in the interior of the molecule to avoid metabolic or chemical soft spots. This becomes a trade-off between incorporating the radiolabel in the interior of the molecule earlier in the synthesis or labeling a side chain late in the synthesis and risking the possibility that it may get metabolized

and lost. A good example of this is the synthesis of [^{14}C] CP-529414 or [^{14}C]torcetrapib. Early in development, Pfizer radiochemists synthesized [^{14}C]torcetrapib with two separate C-14 labels in the bistrifluoromethylbenzyl and acetyl groups late in the synthesis. Instead of using one of these isotopomers, they chose to radiolabel the API for the human ADME study in nine steps as shown in Figure 30.4 (Kelley et al., 2004). The C-14 label was introduced by adding sodium [^{14}C]cyanide to commercially available (R)-2-amino-1-butanol, followed by a deprotection step with trifluoromethane sulfonic acid. Buchwald coupling and nitrile hydrolysis generated the labeled amide that was subsequently acetylated. The amide was regioselectively reduced and the product was stereoselectively cyclized with a catalytic amount of acid. The penultimate material was prepared by acetylation with ethylchloroformate and [^{14}C]CP-529414 was produced by alkylation of the penultimate amide with (bis)trifluoromethyl benzylbromide.

In contrast, Novartis radiochemists were able to place the radiolabel in the side chain of the API, thus avoiding a more difficult and lengthy preparation of the labeled adamantyl system by preparing [^{14}C]LAF237 ([^{14}C]vildagliptin) in five short steps from readily available [2-^{14}C]bromoacetyl bromide as shown in Figure 30.5 (Ciszewska et al., 2007). They stated that they also prepared [^{14}C]vildagliptin from [1-^{14}C]bromoacetyl bromide using the same synthetic route, but did not indicate which isotopomer was used in the human ADME study.

30.2.4 Use the Radiolabeled Reagent as the Limiting Reagent

Radiolabeled reagents are more expensive than their unlabeled counterparts. As such, it is desirable to incorporate the radiolabel into the radiolabeled API efficiently by using it as the limiting reagent to avoid purchasing unnecessary quantities of radiolabeled reagents and to decrease the amount of radioactive waste generated. This is impossible to accomplish for situations where the radiolabeled reagent is used in excess. At times, this can become a balancing act between using the radiolabeled reagent in excess to drive a reaction to completion and create excess radioactive waste or investing resources to redesign the synthesis to use the radiolabeled reagent as the limiting reagent. This can produce a new problem of unreacted excess unlabeled starting materials being present in the reaction mixture, which could complicate purifications. This can be especially challenging if the radiolabel synthetic route is the same as the unlabeled cGMP clinical synthesis route where the synthetic steps and purifications

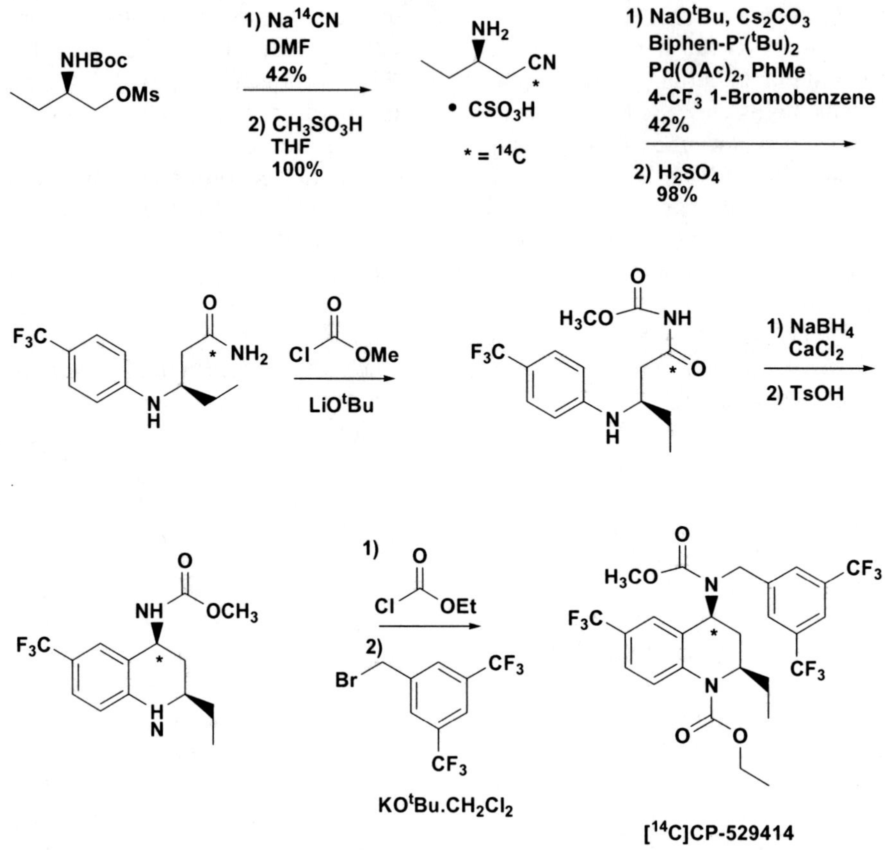

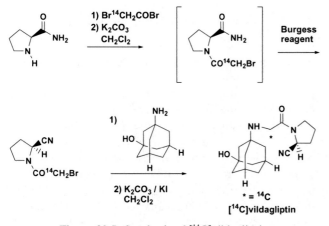

Figure 30.4. Synthesis of [^{14}C]CP-529414.

Figure 30.5. Synthesis of [^{14}C]vildagliptin.

are already highly optimized. This issue is somewhat resolved by the inherent small scale in which the radio-synthesis is completed, allowing for the use of flash chromatography or preparative high performance liquid chromatography (HPLC), which are usually avoided in large-scale syntheses. Thus, a balance must be struck

between the amount radiolabel reagents used versus the yield of the desired labeled product.

30.2.5 Consider Alternative Labeled Reagents and Strategies

All C-14-labeled APIs can trace their origin to the isolation of barium [^{14}C]carbonate in the production of C-14 (Voges et al., 2009). Barium [^{14}C]carbonate is then converted to more advanced C-14-labeled products by a series of well-known procedures by suppliers of radio-labeled products. While some pharmaceutical companies have the ability and equipment to convert barium [^{14}C]carbonate to the radiolabeled API, other companies contract the entire radiosynthesis of C-14-labeled API to the suppliers of radioactive materials. Most large pharmaceutical companies operate between these two extremes and purchase a suitable C-14-labeled intermediate and convert it to the C-14-labeled API. Generally, the more synthetic steps it takes to convert barium [^{14}C] carbonate to the C-14-labeled reagent purchased, the more expensive it is and less likely it will be readily available as a catalog item. Although not used in a

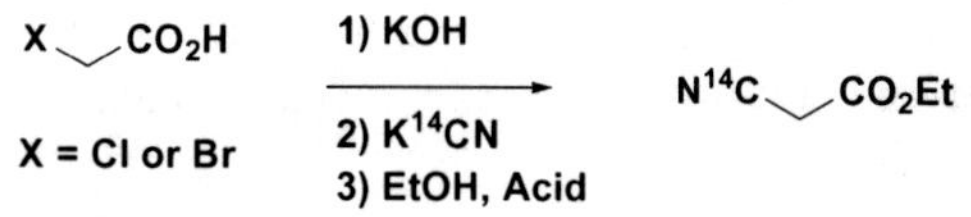

Figure 30.6. Synthesis of [^{14}C]cyanoethyl acetate.

cGMP synthesis, the preparation and use of [3-^{14}C]cyanoethyl acetate from halogenated acetic acid and potassium [^{14}C]cyanide (Maxwell, 2004; Coelho and Schildknegt, 2007) shown in Figure 30.6 illustrates some of the issues that need to be considered to determine when to start the C-14 synthesis and which C-14-labeled material to purchase. Potassium [^{14}C]cyanide is much less expensive and is typically more readily available than [3-^{14}C]cyanoethyl acetate. If the radiosynthetic chemist has the time and has practiced the preparation and purification of [3-^{14}C]cyanoethyl acetate with good results, then it may be best to purchase potassium [^{14}C]cyanide and prepare [3-^{14}C]cyanoethyl acetate themselves, especially if [3-^{14}C]cyanoethyl acetate is needed in large quantities. However, if small quantities of [3-^{14}C]cyanoethyl acetate are required and the radiosynthetic chemist cannot devote time to prepare and purify the [3-^{14}C]cyanoethyl acetate themselves, it may be best to purchase it from a supplier.

It is also wise to develop alternate labeling schemes and contingency strategies in the event a problem arises since the timelines for delivery of the radiolabeled API are usually tight and unforgiving. This is also why it is extremely important to fully complete practice reactions, purifications, and analyses on unlabeled materials, possibly even with small trace amounts of the radiolabeled materials present, to use test reagents and to be able to estimate the time needed to complete the entire synthesis. It is also critically important to complete the radiolabeled synthesis in exactly the same manner as the unlabeled practice chemistry. Altering the radiolabeled synthesis without practicing it first usually results in some degree of failure.

30.2.6 Develop One-Pot Reactions and Minimize the Number of Purification Steps

The advantage of completing multiple synthetic steps in a single reaction flask or pot is to increase yields of the radiolabeled API. Transferring products from one flask to another flask results in the loss of product and increases the potential for spillage. It is also important to minimize the purification of intermediates since significant amounts of products are lost during these processes. However, if the success of a synthetic step is dependent on the purity of the product from the previous step, a purification of the product will need

to be completed regardless of the amount lost during the purification.

30.2.7 Safety Considerations

Although neither C-14 nor tritium produce radiation that can penetrate the skin, it is important for radiosynthetic chemists to practice the concept of ALARA (as low as reasonably achievable) and avoid unnecessary exposure to radioactive material. This is best achieved by practicing the synthesis, developing short synthetic procedures, incorporating the radiolabel late in the synthesis, using the radiolabel as the limiting reagent, optimizing yields to reduce the amount of radioactive material used, and minimizing the number of product transfers to decrease the potential for exposure to radioactive material.

30.3 PREPARATION AND SYNTHESIS

30.3.1 Designated cGMP-Like Area

The area in which the material for human administration is to be synthesized should minimize the potential for contamination and cross-contamination (ICH 19.3). Some companies have taken a conservative approach to this concept by designating certain areas of a radiochemical lab or entire radiochemical labs as being for cGMP-like preparations of API only. This is a major commitment of resources and can only be done if there are many preparations of material for human ADME studies on going. Most companies designate the work area as cGMP-like only for the duration of the synthesis and limit the space to only one synthesis during that time to reduce the potential of cross-contamination. Some common cGMP-like areas can be shared for analytical instrumentation, but preparative HPLC and flash chromatography equipment must be dedicated to a single synthesis. When the instrument is to be converted for use from one compound to the next, the instrument must be certified as clean of any potential products to cross-contaminate the batch. Generally, this is accomplished by cleaning out the tubing of the instrument and using a new preparative HPLC column or fresh silica gel.

30.3.2 Cleaning

Prior to initiating a synthesis, the hood and bench areas to be used must be certified as being clean. The certification may come from either the chemist performing the work or from a QA scientist depending on the SOP. Generally, the hood is scrubbed to remove any potential contamination and swipe tests are performed to verify the efficacy of the cleaning. This also includes cleaning rotary evaporators and any other surfaces that may come in contact with radiolabeled intermediates or the API. Typically, new inert atmosphere gas lines and any in-line drying agent cartridges are also replaced. Further assays for specific compounds can be performed if highly potent, dangerous compounds or compounds that are known to cause allergic reactions in humans, such as beta-lactam antibiotics, were previously used in the hood. As with all steps of the cGMP-like process, the cleaning of the work area must be documented.

30.3.3 Glassware

All glassware must be certified to be clean. This can vary from a visual inspection to the purchase of new glassware, stirbars, spatulas, and all other equipment that may come in contact with the radiolabeled intermediates or API. This must also be documented.

30.3.4 Equipment and Calibration of Analytical Instruments

All equipment must be documented to provide the level of accuracy deemed necessary to support the study. Balance calibrations and cleaning should be conducted to insure that the balances used are functioning properly. All equipment should have well-documented and current maintenance records. The analytical instruments used for the final characterization must have a higher level of certification and must be verified and documented prior to use.

30.3.5 Reagents and Substrates

All reagents need to be evaluated prior to use to confirm identity and purity. A certificate of analysis (CoA) provided by the vendor coupled with either a test to confirm the identity of the compound, such as nuclear magnetic resonance (NMR), or a use test in practice reactions prior to the radiolabeled synthesis is usually required. The reagents need to be controlled so that there is no cross-contamination with other chemicals. The easiest method to insure this is to buy fresh reagents and to segregate them until the synthesis is completed. All reagents and solvents need to be assessed for the potential to transmit bovine spongiform encephalopathies/transmissible spongiform encephalopathies (BSE/TSE), and a certificate of origin should be obtained from the vendor indicating the compound has not been in contact with material of animal origin during its preparation. Since many vendors are reluctant or unable to supply this data, it is crucial to collect this prior to initiating the synthesis so that alternative reagents can be used or another supplier of the reagent can be found, if necessary.

30.3.6 Practice Reactions

Practice reactions are required prior to the actual radiosynthesis to provide a use test of the chemicals and to demonstrate the viability of the synthetic route. These probe reactions allow the chemist to generate the protocol that is to be followed during the preparation of the radiolabeled API. If a process chemistry route is being followed, which is the ideal situation, conditions will likely need to be altered to reflect a much smaller reaction scale and use of the radiolabeled reagent as the limiting reagent. Typically, more solvent and higher dilution is required to allow easier manipulation of the sample. These alterations to the process route need to be investigated to insure that the reactions work appropriately. Second, practice reactions allow the chemist to perform a use test on all the materials to be used in the synthesis. This satisfies the need to test the identity of the reagents and insures that the chemicals are of appropriate purity. Third, the practice reactions help the radiochemist project how much radiolabeled API will be prepared, which allows the radiochemist to adjust the scale of the synthesis appropriately to guarantee that sufficient radiolabeled API will be synthesized for the human ADME study with enough excess for release testing, stability samples, and for additional doses in the event of a misadministration.

30.3.7 Actual Radiolabel Synthesis

The actual production of the radiolabeled API can be approached in several ways. The material can be prepared in a single batch or in several smaller batches that are pooled after the completion of each step. The latter is especially attractive as it provides a record of several reactions to serve as batch documentation and reduces the risk of invalidating the batch if a single large-scale reaction fails. Typically, the radiosynthesis of C-14-labeled API is completed with material where the SA is relatively high or at an intermediate level until the API is isolated and analyzed. This material can be designated as the high specific activity API. Subsequently, the high specific activity API is mixed or cut with unlabeled clinical-grade API to the final target specific activity API for the human ADME study. The SA is reduced in stages so that adjustments can be made in the amounts of unlabeled API that are added to get as close as possible to the target SA for the study. Tritium-labeled API can also be cut to a desired SA with unlabeled API, but most times it is left at a higher value since this is typically the justification for using tritium instead of C-14. Everything must be documented in the notebook or on a batch document, and typically, a colleague must confirm that the work was conducted as described by countersigning the notebook or batch record. With the increase in the use of electronic notebooks, electronic signatures in place of hard copy signatures may be acceptable.

30.4 ANALYSIS AND PRODUCT RELEASE

30.4.1 Validated HPLC Analysis

The cGMP radiolabeled synthesis of the API must be completed under controlled conditions. One way to insure that this criterion is met is by conducting in process analysis of reactions and final products with validated HPLC methods and by comparison with validated standards of each intermediate and the API. One advantage to using the same synthetic route for the radiosynthesis of the API as that used in the preparation of clinical-grade API is the availability of validated HPLC methods. However, if a different synthetic route is used, it will be necessary for the radiosynthetic chemist to establish appropriate HPLC methods in order to follow the progress of the reactions and determine the purity and identity of all intermediates, the penultimate synthetic intermediate, the high specific activity API, and the low specific activity API. Typically, the radiosynthetic chemist will conduct their own HPLC analysis of intermediates, the penultimate compound, and the API first. If these are acceptable, independent HPLC analysis by a member of the analytical research department will be completed on the penultimate compound and the API.

30.4.2 Orthogonal HPLC Method

Some corporate SOPs for release testing of the radiolabeled API require a separate orthogonal HPLC analysis. For most radiolabeled APIs, reversed-phase HPLC analysis is the primary method of determining radiochemical and chemical purity. An orthogonal HPLC method typically involves the use of an HPLC column with a different bonded phase, different mobile phases, or the use of significantly different mobile phase modifiers.

30.4.3 Liquid Chromatography-Mass Spectrometry (LC-MS) Analysis

All intermediate products, the penultimate synthetic intermediate, and the radiolabeled API should be analyzed by LC-MS and the results compared against those

of unlabeled standards, especially if known impurities from the synthesis are expected. In some cases, new unknown impurities may be formed and detected due to the unstable nature of the API due to radiolysis (Jones et al., 2004). The use of different synthetic routes or different purification methods from those used in the process clinical synthesis may also lead to previously unknown impurities. For high specific activity singly labeled tritium and C-14 products, the enhanced M+2 peak should be present. For high specific activity uniformly labeled C-14 aromatic ring products, it may be possible to observe a range of mass spectral peaks from M+0 through M+12 peaks. Comparing the LC-MS results for the radiolabeled API with those of the unlabeled standard can be a useful secondary check of the SA of the high specific activity product (Lehmann and Kaspersen, 1984; Mayer et al., 2008). For low specific activity API at single-digit microcurie-per-milligram levels, the results may be very similar to the MS pattern of the unlabeled API and it may be impossible to determine the SA from the MS data.

30.4.4 Proton and Carbon-13 NMR

At a minimum, the penultimate synthetic intermediate and radiolabeled API should be analyzed by ^{1}H-NMR and the results compared against those of the corresponding unlabeled standards. ^{1}H-NMR analysis can also be used to measure residual solvent content. For some penultimate compounds and radiolabeled API, it may be necessary to complete ^{13}C-NMR analysis depending on corporate SOP requirements.

30.4.5 Determination of the SA of the High Specific Activity API

Typically, C-14-labeled starting materials containing one C-14 have specific activities ranging from 40 to 60 mCi/mmol with a maximum value of 62.4 mCi/mmol for a singly C-14-labeled API. The radiosynthetic chemist may choose to cut the SA during the synthesis with unlabeled material to generate the high specific activity API. After the high specific activity API passes the radiochemical purity criteria and other analytical tests, the SA is measured. Generally, the SA is measured gravimetrically by weighing a known amount into a volumetric flask, diluting with an appropriate solvent to thoroughly dissolve the API and then aliquots are removed and counted using LSC (Mayer et al., 2008). For some high specific activity API, there may also need to be a dilution step before aliquots can be counted by LSC. Based on the weight of the sample and the molecular weight of the API, the number of micromoles can be calculated. The results from LSC provide the number of microcuries present in the original volumetric flask and the SA in microcuries per micromole are calculated.

For single tritium-labeled products, the SA will typically be in the range of 10–25 Ci/mmol, with a maximum value of 28.8 Ci/mmol, which precludes the use of gravimetric analysis due to the need to weigh extremely small amounts of material. Instead, the SA can be measured by comparing the MS of the API with that of the unlabeled standard (Lehmann and Kaspersen, 1984; Mayer et al., 2008). Alternatively, the SA can be determined by HPLC using a standard curve. For this procedure to be accurate, the product needs to have a good UV chromophore and be of high purity. Solutions of the unlabeled standard are prepared in concentrations spanning at least three orders of magnitude. Varying amounts of these solutions are injected onto an HPLC and the area under the peak generated by the standard are measured. A plot of the area versus micrograms of material injected is generated to determine a straight line based on a minimum of three points. A solution of a known concentration of the high specific activity API is prepared between the lowest and highest concentration samples used for the unlabeled standard. A sample of the high specific activity API is injected onto the HPLC and the area under the peak from the UV detector is determined. Once the API has passed through the UV detector, the effluent corresponding to the UV peak of the API is collected in a volumetric flask. The contents in the volumetric flask are diluted with solvent, mixed thoroughly, and aliquots of this solution are analyzed by LSC. Based on the area of the peak generated by the API and using the equation of the line from the standard curve, one can calculate the micromoles present in the injection. The results from the LSC provide the number of microcuries. Taken together, the SA of the sample in millicuries per micromole is determined.

The accurate determination of the SA of the high specific activity API is critical since this is used to determine the weights of radiolabeled and unlabeled clinical-grade API to be mixed to prepare the low specific activity API that will be used in the human ADME study.

30.4.6   Mixing of the High Specific Activity API with Unlabeled Clinical-Grade API

Based on the SA of the high specific activity API, the appropriate amount of C-14-labeled API and unlabeled clinical-grade API are weighed together and dissolved in an appropriate solvent to form uniformly mixed API. If possible, solvents safe for human consumption should be used in the event residual solvent remains in the API. It is typical to make two to four dilutions of the high specific activity API with unlabeled API with SA deter-

minations at each cut in order to make more than one adjustment in the specific activities and lessen the risk of lowering the SA of the final API to a value too low to be used.

30.4.7   Determination of the SA of the Low Specific Activity API

The SA of the low specific activity API is typically determined gravimetrically as it was determined for the high specific activity API. Larger quantities of radiolabeled API will need to be weighed and longer counting times during LSC analysis to achieve statistically meaningful results will be needed. The determination of the SA of this material by MS is unlikely to be of use due to the small quantity of radiolabeled material present.

30.4.8   Other Potential Analyses

Depending on the corporate SOP or unique physical or chemical properties of the API, it may be necessary to complete additional analyses on the API or on one of the intermediate specific activity samples. Some potential analyses include the following:

1. Karl Fischer water analysis, especially if the API forms hydrates.
2. Gas chromatography and ^{1}H-NMR for residual solvent determination (the solvent to be measured should not be used for the ^{1}H-NMR analysis).
3. Heavy metal determination especially for synthetic sequences involving lead, tin, palladium, platinum, or iron. It is usually best to avoid these in the synthesis, if possible, since they can be difficult to remove completely.
4. Chiral purity assay of the final API (if chiral).
5. ^{19}F-NMR or ^{31}P-NMR, if appropriate.
6. Sterility, pyrogen, and bacterial endotoxin tests, if the radiolabeled API is to be administered as an injectable.
7. Acid or base titrations for acidic or basic compounds.
8. Weight percent analysis.

There may be additional routine tests that are required by the corporate SOP for clinical-grade API that may not be able to be completed on radioactive API such as elemental analysis.

30.4.9   Establishment of Use Date and Use Date Extensions

Upon completion of the analyses of the low specific activity API, a CoA is issued and a use date is assigned. The use date is typically based on stability studies. Stability studies involve storing the radiolabeled API with an

SA near to or above that being used in the human ADME study at a prescribed temperature for as long as storage data is needed. Samples of the radiolabeled API are removed on a regular basis, and are analyzed by HPLC with a radioactive detector to measure the radiochemical and chemical purities of the API over time. If the radiochemical and chemical purities stay above the radiochemical and chemical purity requirements for the human ADME, the material is deemed stable for the time from the original analysis to that time point at those storage conditions. If the radiochemical or chemical purity drops below the acceptable requirements, the API is determined to be unstable and a use date less than the duration of the stability study must be assigned. If stability studies have not been completed on the radiolabeled API prepared for the human ADME study, then a reasonable approximation is made based on the stability of the unlabeled API and any additional data from previous syntheses of radiolabeled API. Use dates typically range from 2 to 6 months. If the radiolabeled API is not used before the use date, it must be reanalyzed to determine the radiochemical purity and to establish identity. If the material still meets the purity and potency criteria for the radiolabeled API, a new use date is assigned and the material is deemed acceptable for use in the human ADME study. If the material does not meet the criteria, it may be repurified and reanalyzed until it does. The repurified API will be assigned a new batch number.

30.4.10   Analysis and Release of the Radiolabeled Drug Product

Once the radiolabeled drug product is manufactured, it is analyzed using whatever analytical tests are necessary and is released for use. Most human ADME studies are dosed using a drug-in–a-bottle approach. The drug-in-a-bottle approach avoids the preparation of capsules or tablets and the potential for contamination of tablet machines. Also, drug-in-a-bottle avoids the need for the determination of the crystal form and particle size prepared since the API is reconstituted in solution.

30.5   DOCUMENTATION

Proper documentation for the entire cGMP synthesis and analysis is required and is typically described by the corporate SOP. At a minimum, the entire experimental description of the synthesis with in-process analytical results should be captured in a hard copy or electronic notebook. This should include documentation describing the cleaning of the workspace and equipment. The experiments should also be properly witnessed either electronically or by signing hard copies of the notebook pages. The batch record should include the synthetic scheme; a list of reagents used including sources, purity, expiration dates, and CoAs; a list of instruments with model numbers, serial numbers, and most recent calibration date; a signed copy of the CoA of the API; a signed copy of a TSE/BSE assessment of the reagents and intermediates used in the synthesis; and signed notebook pages.

30.5.1   QA Oversight

The entire synthetic route must be reviewed and approved by the QA representative on the human ADME team. This is to insure that the procedure will meet the requirements necessary for use in the human ADME study. Once the synthetic route is approved and the API is prepared, analyzed, and released, the QA representative will review the results, the research notebook detailing the synthesis, and all of the documentation for the batch record. After these documents have been approved, the QA representative will sign off on the synthesis, which allows for the manufacture of the drug product.

30.5.2   TSE and BSE Assessment

All reagents, solvents, and materials that come into contact with the API must be reviewed to determine whether they have been derived from human or animal sources capable of transmitting BSE/TSE. A statement indicating that all reagents, solvents, and materials that have come into contact with the API and are not from animal or human sources should be included in the batch record.

30.6   SUMMARY

In summary, the synthesis of radiolabeled API for human ADME studies can be challenging, requiring significant planning, coordination, communication, and the ability to manage multiple details simultaneously. It is critical to fully understand the corporate SOP requirements and be able to translate them into practice, have contingency plans well worked out, be able to solve problems quickly to keep the synthesis on schedule, and fully document all aspects of the synthesis. Hopefully, this chapter has disseminated the most important issues that must be considered for completion of the syntheses of radiolabeled API for human ADME studies.

REFERENCES

Bonacorsi SJ Jr., Burrell RC, Luke GM, Depue JS, Rinehart JK, Balasubramanian B, Christopher LJ, Iyer R. (2007) Synthesis of the anxiolytic agent [^{14}C]6-hydroxybuspirone for use in a human ADME study. *J Labelled Comp Radiopharm* 50:65–71.

Ciszewska G, Allentoff A, Jones L, Wu A, Ray T (2007) Synthesis of radio- and stable-labelled LAF237 (Galvus, Vildagliptin). *J Labelled Comp Radiopharm* 50:593–594.

Coelho RV Jr., Schildknegt K (2007) An efficient large-scale synthesis of 1H-indazole-[3-^{14}C]carboxylic acid. *J Labelled Comp Radiopharm* 50:675–678.

Dain JG, Collins JM, Robinson WT (1994) A regulatory and industrial perspective on the use of carbon-14 and tritium isotopes in human ADME studies. *Pharm Res* 11:925–928.

FDA website In *Food and Drug Administration Website Describing Radioactive Drugs, 21 CFR 361*, http://www.accessdata.fda.gov/scripts/cdrh/cfdocs/cfcfr/CFRSearch.cfm?CFRPart=361&showFR=1, 2012.

Hong Y, Bonacorsi JSJ, Tian Y, Gong S, Zhang D, Humphreys WG, Balasubramanian B, Cheesman EH, Zhang Z, Castner JF, Crane PD (2008) Synthesis of [1,2-^{3}H]ethylamine hydrochloride and [^{3}H]-labeled apadenoson for a human ADME study. *J Labelled Comp Radiopharm* 51:113–117.

Jones AN, Braun M, Dean D, Elmore C, Jakubowski Y, Jenkins H, Melillo D, Miller R, Staskiewicz S, Wallace M (2004) The use of LC/MS to identify unknown radioactive impurities in clinical tracers. In *8th International Symposium on the Synthesis and Applications of Isotopes and Isotopically Labelled Compounds*, Dean DC, Filer CN, McCarthy KE, eds., pp. 289–292. John Wiley and Sons, Chichester, U.K.

Kelley RM, McCarthy KE, Miller SA, Nesler MJ, Schildknegt K, Wager CB, Zandi KS (2004) The synthesis of isotopically labelled CETP inhibitors. In *8th International Symposium on the Synthesis and Applications of Isotopically Labelled Compounds*, Dean DC, Filer CN, McCarthy KE, eds., pp. 11–14. John Wiley and Sons, Chichester, U.K.

Lehmann WD, Kaspersen FM (1984) Specific radioactivity determinations of ionic organic compounds of high specific activity by fast atom bombardment and field desorption mass spectrometry. *J Labelled Comp Radiopharm* 21:455–469.

Lloyd C, Potwin J, Wright C (2004) Radiosynthesis conducted under cGMP compliance. In *8th International Symposium on the Synthesis and Applications of Isotopes and Isotopically Labelled Compounds*, Dean DC, Filer CN, McCarthy KE, eds., pp. 251–254. John Wiley and Sons, Chichester, U.K.

Marathe PH, Shyu WC, Humphreys WG (2004) The use of radiolabeled compounds for ADME studies in discovery and exploratory development. *Curr Pharm Des* 10:2991–3008.

Maxwell BD (2004) The synthesis of 14C labeled silthiofam: A novel fungicide for wheat take-all disease. In *8th International Symposium on the Synthesis and Applications of Isotopes and Isotopically Labelled Compounds*, Dean DC, Filer CN, McCarthy KE, eds., pp. 161–164. John Wiley and Sons, Chichester, U.K.

Mayer P, Lacy L, Maniscalco M (2008) Labeling drug safety. *Drug Discov Devel* 11:31–33.

Moenius T, Baumann K, Bulusu M, Schweitzer A, Voges R (2001) Labeling of pharmacologically active macrocycles. In *7th International Symposium on the Synthesis and Applications of Isotopes and Isotopically Labelled Compounds*, Pleiss U, Voges R, eds., pp. 424–429. John Wiley and Sons, Chichester, U.K.

Roberts D (2009/2010) Custom carbon-14 radiolabelling investing to meet new challenges. *Drug Discov World* Winter 2009/10:53–60.

Rotert G, Fan L, Shevchenko VP, Nagaev IY, Myasoedov NF (2006) Synthesis of radiolabeled ABT-578 for ADME studies. *J Labelled Comp Radiopharm* 49:849–856.

Solon E (2010) Predicting human radiation dosimetry for clinical radiolabeled drug studies: A comparison of methods. In Maxwell BD, Wheeler WJ, eds. *Proceedings from the 10th International Symposium on the Synthesis and Applications of Isotopes and Isotopically Labelled Compounds. J Label Compd Radiopharm* 53:283–287.

Solon EG, Balani SK, Lee FW (2002) Whole-body autoradiography in drug discovery. *Curr Drug Metab* 3:451–462.

Voges R, Heys JR, Moenius T (2009) *Preparation of Compounds Labeled with Tritium and Carbon-14*. John Wiley and Sons, Chichester, U.K.

Wait JCM, Vaccharajani N, Mitroka J, Jemal M, Khan S, Bonacorsi JSJ, Rinehart K, Iyer RA (2006) Metabolism of [^{14}C] gemopatrilat after oral administration to rats, dogs and humans. *Drug Metab Dispos* 34:961–970.

31

FORMULATION DEVELOPMENT FOR PRECLINICAL *IN VIVO* STUDIES

Yuan-Hon Kiang, Darren L. Reid, and Janan Jona

31.1 INTRODUCTION

During drug discovery stages, one key area of pharmaceutical support is to identify acceptable formulations for *in vivo* studies. In general, there are three things that need to be considered for formulation development: (1) the routes of administration; (2) the endpoint (biomarker) that the *in vivo* assay measures for pharmacodynamic studies; and (3) the physicochemical properties of the active pharmaceutical ingredient (API). The routes of administration are of importance because they in turn affect the dose volume, the acceptable pH range, and viscosity of the formulation vehicle, as well as the types of formulation. Table 31.1 summarizes dose volumes in different species via common routes of administration that are considered good practice [1]. In clinical development and postlaunch, the route of administration is primarily determined by the therapeutic objectives. However, at the preclinical stage, the route of administration of a compound is usually determined based on the objective of the *in vivo* study. Depending on the pharmacokinetic and pharmaceutical properties of the compound, common routes used during drug discovery include intravenous, intraperitoneal, subcutaneous, and oral routes. Knowing the endpoints of the *in vivo* assay is important because some excipients used in the formulation may exhibit pharmacological effects and interfere with the measured endpoints. For example, surfactants used as wetting agents such as poloxamers may increase the blood glucose level in the diabetic rat model and therefore should be avoided or used with caution for metabolic disorder studies. Dimethylacetamide (DMAc), a well-known solubilizing agent cannot be used for oncology studies (xenograph studies) as it has an antitumor effect itself [2]. Basic knowledge of the physicochemical properties is also critical. Information about the pK_a, LogP, or LogD will help define the most promising formulation strategies and the limits of what can be achieved for a given approach. These intrinsic independent properties can often be easily determined or calculated. Other physicochemical properties such as, particle size distribution (PSD), morphology, solubility, and chemical stability have a complex relationship to the API, the physical form, and formulation. Knowledge of these dependent properties often develops as the API progresses through drug discovery. Once the routes of administration, the endpoint for pharmacodynamic studies, and the basic physicochemical properties of the API are defined, the formulation strategy can be developed.

31.2 FORMULATION CONSIDERATION FOR THE INTRAVENOUS ROUTE

The intravenous route is used in pharmacokinetic studies to obtain key parameters such as clearance and volume of distribution. Unlike other routes, only solutions or nanosuspensions can be administered intravenously [3]. Developing nanosuspension formulations requires significantly more resources and a longer

ADME-Enabling Technologies in Drug Design and Development, First Edition. Edited by Donglu Zhang and Sekhar Surapaneni.
© 2012 John Wiley & Sons, Inc. Published 2012 by John Wiley & Sons, Inc.

TABLE 31.1. Administration Volumes Considered Good Practice

Species	Route and Volumes (mL//kg)			
	Oral	Subcutaneous	Intraperitoneal	Intravenous (Bolus)
Mouse	10	10	20	5
Rat	10	5	10	5
Rabbit	10	1	5	2
Dog	5	1	1	2.5
Macaque	5	2	—	2
Marmoset	10	2	—	2.5
Minipig	10	1	1	2.5

timeline. For preclinical pharmacokinetic studies, intravenous administration is predominantly achieved using solution formulations. The vehicle selection for intravenous route is therefore limited by solubility. At the early discovery stage, in general, all the compounds that pass the *in vitro* criteria will be dosed intravenously to rats and due to resource and time constraints, it is not practical to attempt identifying aqueous-based formulation for every compound that goes into pharmacokinetic studies. Organic solvent, such as dimethyl sulfoxide (DMSO), is therefore the intravenous vehicle of choice for pharmacokinetic studies in rodent species. Although DMSO has excellent solubilizing power, toxicity, in particular hemolysis, is limiting the use of DMSO in higher species [4]. It is acceptable to dose pure DMSO in rats and dogs intravenously at the dose volume of 0.5 mL per dog and 0.5 mL/kg for rat. When DMSO cannot be used for reasons such as toxicity or interference with physiological endpoint, and in rare occasions where DMSO does not provide adequate solubility, a solubility screen in selected aqueous and nonaqueous systems needs to be conducted in order to identify an appropriate intravenous formulation vehicle.

31.3 FORMULATION CONSIDERATION FOR THE ORAL, SUBCUTANEOUS, AND INTRAPERITONEAL ROUTES

For oral, subcutaneous, and intraperitoneal routes, in addition to solution, suspensions can be administered. Suspension formulations are coarse dispersions in which insoluble solid particles distributed throughout a dispersion medium. The particle size of pharmaceutical suspensions is greater than 0.1 μm in diameter [5]. The dispersion medium, namely the formulation vehicle, usually contains polymers as the suspending agents and surfactants as wetting agents. In preclinical suspension formulations, hydroxypropyl methylcellulose (HPMC), methylcellulose (MC), and carboxymethylcellulose (CMC) are commonly used polymers. Polysorbate 80, Polysorbate 20,

TABLE 31.2. Common Suspension Formulation Vehicles

Suspending Agent		Surfactant
Name	Viscosity (mPa·s)	
2% HPMC	40–60[a]	1% Polysorbate 80
2% HPMC	40–60[a]	1% Pluronic F68
0.5% MC	1500[a]	
1% MC	1500[a]	1% Polysorbate 80
1% CMC	50–200[b]	1% Polysorbate 80
Ora-Plus		1% Polysorbate 80

[a] In 2% aqueous solution.
[b] In 4% aqueous solution.

and poloxamers are commonly used surfactants. The suspended particles should not settle rapidly and the settled particles should be redispersed into a homogenous mixture when the container is shaken. The viscosity of the vehicle should be sufficient to suspend the particles but not too high to prevent the vehicle to flow through an oral gavage tube. Since the viscosity of the formulation increases with the concentration of the compound, precaution should be taken to make sure formulations are dosable when formulating at high dose levels. A balance between settling time and viscosity should be obtained by adjusting the amount of polymer and surfactant in the vehicle.

Suspension formulations are generally more feasible because solubility is no longer a limiting factor. Moreover, most drugs are developed as a solid dosage form for oral administration. In the drug discovery phase, dosing with suspension formulation provides an early opportunity to evaluate the probability of success in the solid dosage form development for future clinical studies. Exposure and bioavailability obtained in preclinical pharmacokinetic studies are not only a critical part of candidate selection but also valuable to the clinical formulation development. A few commonly used formulation vehicles for oral dosing are summarized in Table 31.2. The surfactant level determines whether it is used purely as a wetting agent or also a solubilizer: If the surfactant level is below its critical micelle concentration, it acts as a wetting agent; when the level is above its critical micelle concentration, the surfactant not only wets the particles but also solubilizes the compound through micelle formation.

31.4   SPECIAL CONSIDERATION FOR THE INTRAPERITONEAL ROUTE

A few other routes are also used for compound administration in preclinical species.

Intraperitoneal route of administration is usually chosen in the early discovery stage to avoid the intestinal barrier. Although suspensions can be dosed intraperitoneally, solution is generally preferred as there is limited fluid supply to the peritoneal cavity to solubilize particles in the suspension. Due to resource constraints, it may not be practical to develop a solution formulation for every compound that is screened in the *in vivo* assays. It is therefore adventitious to take a generic formulation approach, using a vehicle containing solubilizing agent such as cyclodextrin and suspending agent and surfactant such as HPMC and poloxamer, for example, 20% hydroxypropyl-beta-cyclodextrin, 1% HPMC, and 1% Pluronic F68. The solubilizing agent in the vehicle will enhance solubility and increase the concentration of the active compound in solution. In case the concentration of the formulation still remains above its solubility in the vehicle containing solubilizing agent, a dosable suspension can be formed with the suspending agent and surfactant in the vehicle. Some ternary formulation vehicles containing cyclodextrin, suspending agent, and surfactant may be phase separated and appear to be cloudy. The cloudiness adds difficulty to screening for a solution formulation and therefore should be avoided. It is noteworthy that because the compound dosed via intraperitoneal route of administration ends up primarily in the portal vein, it is subject to hepatic first-pass metabolism [6]. To bypass the hepatic first-pass metabolism and intestinal barrier, subcutaneous route of administration is often used over oral route in early discovery stage. The subcutaneous space is proliferated with capillaries and absorption is rapid. Suspension formulations can be used for subcutaneous route. However, the absorption rate for a suspension formulation is determined by the dissolution rate of the particles in the extracellular fluid. The particle size is therefore important and needs to be characterized appropriately.

31.5   SOLUBILITY ENHANCEMENT

One of the most common requests a discovery pharmaceutical scientist receives from discovery team colleagues is to help improve poor oral bioavailability or achieve desired exposure through formulation approach. Oral bioavailability of a drug is mainly a function of its dissolution, permeability, metabolic stability and chemical stability in the gastrointestinal fluids. Figure 31.1 shows a schematic of the oral drug absorption.

Most drugs are absorbed via passive diffusion, and only drug in solution can get across the intestinal lumen and be absorbed into systemic circulation. In Figure 31.1, there are therefore two possible limiting steps for drug absorption: dissolution rate and permeability. Permeability is largely an intrinsic property of the chemical structure and is not commonly manipulated through formulation, although for P-glycoprotein (Pgp) substrates the permeability could be enhanced through adding Pgp inhibitors such as Polysorbate 80 to the formulation vehicle [7, 8]. The dissolution rate is well established and is described by the equation modified from the work of Noyes and Whitney [9]:

$$\frac{dm}{dt} = \frac{DA}{h}(C_s - C), \tag{31.1}$$

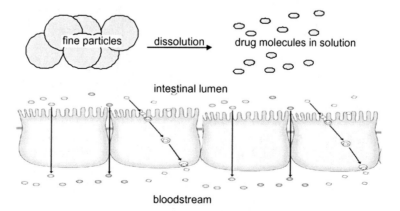

FIGURE 31.1. Schematic of the oral drug absorption.

where D is the diffusion constant, A is the surface area, C_s is the solubility, C is the concentration of the compound in the bulk solution, and h is the thickness of the diffusion layer. In order to increase the dissolution rate, one can reduce the particle size, and therefore increase the surface area, or increase the solubility of the solid. Common methods of increasing solubility in the preclinical stage include ionization/pH adjustment, cosolvency, amorphous solid dispersion, and complexation.

31.6   pH MANIPULATION

Many pharmacologically active compounds are weak bases or acids that can be ionized. In water, a weak base will be protonated at lower pH, depending on the pKa, and form an *in situ* salt that consists of the charged active compound and its counterion. Conversely, when the pH is high enough, a weak acid will be deprotonated to form the *in situ* salt. Figure 31.2 illustrates the pH-dependent ionization of a free base. The pH-dependent solubility of basic and acidic compounds is discussed below:

Basic compounds:

For any basic compound, the total solubility at any given pH is expressed as

$$S_{total} = S_o + S_{BH^+},   (31.2)$$

where S_{total} is the total solubility, S_o is the intrinsic solubility, and S_{BH^+} is the concentration of the salt.

When a basic compound dissolves in water, the general reaction is expressed as

$$BH \rightleftharpoons B + H^+.   (31.3)$$

The equilibrium expression of the above reaction is

$$K_a = \frac{[B][H^+]}{[BH^+]},   (31.4)$$

where K_a is the acid dissociation constant. Since

$$S_o = [B]   (31.5)$$

and

$$S_{BH^+} = [BH^+],   (31.6)$$

Equation 31.2 can be expressed as

$$S_{total} = [B] + [BH^+].   (31.7)$$

In the above equation, substituting $[BH^+]$ with $\dfrac{[B][H^+]}{K_a}$ (from Eq. 31.4), and expressing $[B]$ as S_o, it is obtained:

$$S_{total} = S_o\left(1 + \frac{[H^+]}{K_a}\right).   (31.8)$$

Substituting K_a and H^+ with pK_a and pH, Equation 31.8 can be rewritten as

$$S_{total} = S_o\left(1 + 10^{(pK_a - pH)}\right).   (31.9)$$

Equation 31.9 indicates the following:

1. At pH $\gg$ pK_a (more than 2 units), the total solubility of the base will be a constant and it is the intrinsic solubility of the base.
2. At pH close and lower than pK_a, the total solubility of a base increases exponentially as the pH decreases.

Although based on Equation 31.9, as the pH decreases, the total solubility increases exponentially, in reality the solubility will not increase indefinitely—when the pH decreases to a certain value, the solubility will reach its maximum. Therefore, the pH solubility profile of a base, shown in Figure 31.3, can be divided into three areas. The first area is when the pH is more

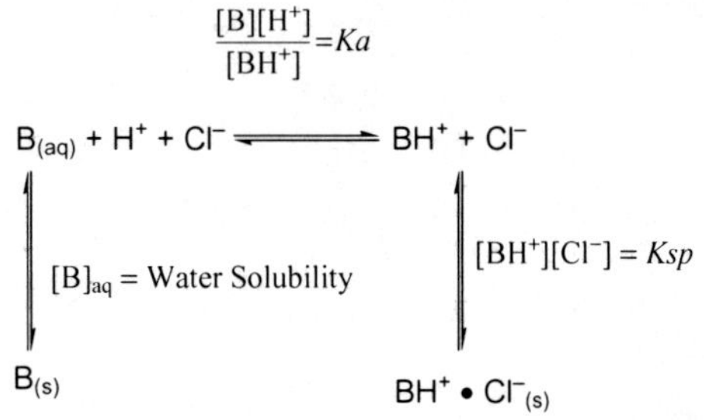

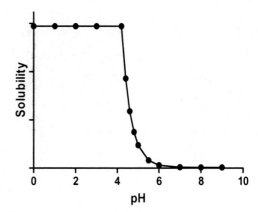

FIGURE 31.3. pH solubility profile of a free base with a pK_a value of 6.5.

$$\frac{[B][H^+]}{[BH^+]} = Ka$$

$$B_{(aq)} + H^+ + Cl^- \rightleftharpoons BH^+ + Cl^-$$

$[B]_{aq}$ = Water Solubility

$[BH^+][Cl^-] = Ksp$

$B_{(s)}$ $BH^+ \cdot Cl^-_{(s)}$

FIGURE 31.2. pH-dependent ionization of a free base.

than 2 units greater than the pK_a (pH > 8.5 in Figure 31.3), where the total solubility is close to a constant and is equal to the intrinsic solubility of the free base. In the second area, where the pH is closer and below the pK_a (pH < 8.5 in Figure 31.3), the solubility increases exponentially as the pH decreases. Both in these two areas, solubility is pH dependent following Equation 31.9. In Figure 31.3, the third area is in the pH range that is below 4, where the solubility is a constant. In this area, the solubility represents the saturated solubility of the salt and this is expressed as the solubility product K_{sp}. Using the system in Figure 31.2 as an example, where the pH is adjusted using hydrochloric acid, the K_{sp} is expressed as

$$K_{sp} = [BH^+][Cl^-]. \qquad (31.10)$$

When the product of the concentration of the protonated free base, BH^+ and the chloride Cl^- exceeds the solubility product constant, K_{sp}, the salt BH^+Cl^- will precipitate as a solid. The water solubility of the salt is therefore $\sqrt{K_{sp}}$. If the concentration of the counterion, Cl^-, increases, to keep the K_{sp} constant, the solubility of the salt will decrease. This phenomenon is called the common ion effect.

Acidic compounds:

When an acidic compound dissolves in water, the general reaction is expressed as

$$HA \rightleftharpoons A^- + H^+. \qquad (31.11)$$

The equilibrium expression of the above reaction is

$$K_a = \frac{[A^-][H^+]}{[HA]}. \qquad (31.12)$$

By replacing S_{BH^+} with S_{A^-} in Equation 31.2, the total solubility of an acid at any given pH is expressed as

$$S_{total} = S_o + S_{A^-}. \qquad (31.13)$$

Here

$$S_{A^-} = [A^-] \qquad (31.14)$$

and

$$S_o = [HA]. \qquad (31.15)$$

With Equations 31.12, 31.14, and 31.15, Equation 31.13 can be expressed as

$$S_{total} = S_o \left(\frac{K_a}{[H^+]} + 1 \right) \qquad (31.16)$$

or

$$S_{total} = S_o \left(10^{(pH-pK_a)} + 1 \right). \qquad (31.17)$$

The solubility of the salt could be expressed as solubility product:

$$K_{sp} = [A^-][N_a^+], \qquad (31.18)$$

where $[N_a^+]$ is the concentration of the counterion of the salt formed. The solubility of the salt will be $\sqrt{K_{sp}}$.

Following Equations 31.9 and 31.17, when the pH is at the same value of the compound's pK_a, the total solubility will be twice as much as the intrinsic solubility of the free base; when the pH is two units away from the pK_a, the total solubility is about 100-fold higher than the intrinsic solubility. The pH range of oral dosing is usually between pH 2 and 10. Thus, a free base with a pK_a above 4 and a free acid with a pK_a below 8 is a good candidate for *in situ* salt formulation by pH adjustment. In general, there are more basic pharmacologically active compounds than acidic ones. Moreover, the gastric fluids of human and preclinical species are acidic, with pH values ranging from 1.6 to 4 when fasted. Thus, it is not uncommon to acidify the oral formulation vehicle to pH around 2 for basic compounds. Since adjusting pH is essentially salt formation *in situ*, selection of acid/base that is used for pH adjustment may affect the total solubility. To avoid the common ion effect, the pH should not be adjusted for a compound that is already a salt using an acid or base of the same counter ion. Supersaturation in some cases can be utilized to enhance solubility-limited exposure. Instead of gradually adjusting pH from neutral condition to the desired pH, equivalent amount of acid/base of choice should be added to the formulation at once to form *in situ* salt. Once the *in situ* salt is formed, the final pH is adjusted to the acceptable range by strong base/acid (usually NaOH/HCl). Depending on the pH solubility profile shown in Figure 31.3, the *in situ* salt may disproportionate after the final pH adjustment. However, sometimes, the solution concentration of the *in situ* salt may remain for a period of time after the pH is adjusted to the range where salt disproportionation should occur. If this window of supersaturation is sufficient for absorption, bioavailability will be improved. In order to sustain supersaturation, inhibition of precipitation may be required through the use of pharmaceutical excipients or other components that interfere with nucleation and crystal growth.

31.7 COSOLVENTS UTILIZATION

If pH adjustment does not lead to the desired solubility, a cosolvency approach may be considered. Commonly used water-miscible cosolvents include *N,N*-dimethylacetamide (DMAc), *N*-methyl-2-pyrrolidone (NMP), ethanol, polyethylene glycol 400 (PEG 400), and propylene glycol (PG). With the cosolvency approach, the first step of formulation development is to conduct a quick solubility screen in a few water-miscible cosolvents that are acceptable for the *in vivo* study. The total API required is 5–10 mg, and visual solubility is sufficient for this step. The target formulation concentration is determined by the dose level as well as by the dose volume. Here in this chapter, let us use 10 mg/mL as the target formulation concentration and DMAc, ethanol, PEG 400, and propylene glycol as the acceptable cosolvents. Since DMAc usually has the highest solubility among the four cosolvents, we first weigh 1.0 mg of the compound in the vial and add 0.01 mL of DMAc. After sonication, if this forms a clear solution, then the solubility is >100 mg/mL. If the testing sample remains cloudy, then add another 0.01 mL of DMAc and sonicate. Continue adding DMAc at a 0.01-mL increment until a clear solution is formed to obtain a visual solubility. Visual solubility of other cosolvents can be obtained using a similar approach. However, since they will probably have lower solubility compared with DMAc, the initial weight of the API is lowered to 0.5 mg. Once the solubility of the compound in the selected cosolvents is obtained, the other key information needed to develop a solution formulation is the maximally allowable volume of the cosolvents. Table 31.3 summarizes this information. For cosolvents with limited information on the maximally allowable volume, a rule of thumb is that 10% of the LD_{50} may be used for one-time administration. In Figure 31.4, we provide an example using the cosolvency approach to make a solution formulation for dosing. In this example, the target concentration is 10 mg/mL. The solubility of the compound, Compound A, in the selected cosolvents is listed in Table 31.4.

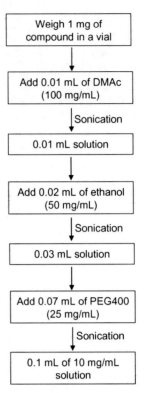

FIGURE 31.4. An example process using the cosolvency approach to make a solution formulation for dosing.

TABLE 31.4. Solubility of Compound A in Selected Cosolvents

Vehicle	Solubility
DMAc	>100 mg/mL
Ethanol	50–100 mg/mL
PEG 400	25–50 mg/mL
PG	<1 mg/mL

TABLE 31.3. Typical Maximum Use Levels of Common Cosolvents

	Rats				Mice				Dogs				Monkeys			
	Oral		IV		Oral		IV		Oral		IV		Oral		IV	
	% w/w	ml/ kg	% w/w	ml/ kg	% w/w	ml/ kg	% w/w	ml/ kg	% w/w	ml/ kg	% w/w	ml/ kg	% w/w	ml/ kg	% w/w	ml/ kg
DMSO	10	0.5	100	1	10	0.5	100	1	—	—	100	0.05	—	—	—	—
PEG400	100	5	80	0.8	100	5	80	0.8	100	2	80	0.8	100	2	80	0.8
Propylene glycol	100	2.5	80	1.3	80	2.5	80	0.83	80	2	80	2	50	2.5	10	0.12
DMAc	10	0.5	10	0.5	10	0.5	10	0.5	10	0.5	10	0.5	10	0.5	10	0.5
Ethanol	50	1	20	0.5	50	1	20	0.5	50	2	20	0.5	25	2.5	20	0.6
Oils	100	5	15	0.75	100	5	15	0.75	100	2	15	1.5	100	2	15	1.5
Polysorbate 80	10	1	5	0.25	10	1	5	0.58	1	0.05	—	—	1	0.15	1	0.05

31.8 COMPLEXATION

Complexation with cyclodextrins is another common approach to enhance exposure for poorly water-soluble compounds. Cyclodextrins are cyclic oligosaccharides containing a hydrophobic central cavity and a hydrophilic surface. Cyclodextrins can increase the equilibrium solubility of water-insoluble compounds by forming a noncovalent inclusion complex with the lipophilic molecule binding to the hydrophobic central cavity. The hydrophilic outer surface of the cyclodextrin will bring the complex into water. Sulfobutyl ether-β-cyclodextrin and hydroxypropyl-β-cyclodextrin are the two most commonly used water-soluble cyclodextrins. They are both modified β-cyclodextrin derivatives with seven glucopyranose units and a cavity diameter of 6.0–6.5 Å. The modification is important as it not only increases the intrinsic water solubility but also avoids the renal toxicity that renders β-cyclodextrin not suitable for use in parenteral delivery. A combination approach such as cyclodextrin, suspension agent, wetting agent, and pH adjustment can be the most effective default vehicle for the screening pharmacokinetic and pharmacodynamic studies at early stage for all common routes of administration except intravenous delivery.

For compounds of high lipophilicity (with LogP > 4), lipid-base vehicles are often used to increase solubility and therefore bioavailability. Since the body fluids are aqueous based, it is important that the active compound will not precipitate out of the lipid-based formulation and crystallize in the gastrointestinal tract. In this regard, mono- and diglycirides may work better than triglycirides because of their amphiphilic character. When the mono- and diglycirides-based formulation is dosed, the vehicle tends to form micelles with the compound incorporated, and therefore not precipitating in the gastrointestinal tract. Among the well-established lipid-based formulation approaches are emulsions, self-emulsifying drug delivery systems, and solid dispersions.

31.9 AMORPHOUS FORM APPROACH

Amorphous solid dispersion has recently gained increased popularity among discovery pharmaceutical scientists to enhance bioavailability for poorly water-soluble compounds. Amorphous solids are disordered in nature and therefore thermodynamically unstable. It exhibits a high degree of solubility due to the high-energy state. Although the high solubility makes amorphous material attractive from a bioavailability point of view, it is often difficult to develop an amorphous formulation with sufficient stability so that the amorphous solid does not crystallize before dosing. In the amorphous solid dispersion system, the amorphous compound is stabilized by a matrix carrier, which is usually a hydrophilic polymer such as various cellulose derivatives. Ideally, the compound is dispersed molecularly in the polymer matrix. Upon exposure to aqueous media, the molecularly dispersed compound is released from the matrix and immediately dissolves in the media. The high dissolution rate resulted from the molecular dispersion of the compound is mainly responsible for the improved bioavailability. For preclinical studies in rodents, it is necessary to formulate the amorphous solid dispersion into a suspension for dosing. It is important to monitor the physical stability of the solid dispersion in suspension. To improve physical stability of the amorphous dispersion in suspension, one can either increase the viscosity of the vehicle or decrease the drug load in the solid dispersion. The drawback is the increased amount of polymer that will be dosed along with the active compound.

31.10 IMPROVING THE DISSOLUTION RATE

As the Noyes–Whitney equation indicates, dissolution rate is proportional to the surface area. Particle surface area increases as particle size decreases. Thus, particle size reduction is commonly used to increase dissolution rate, and hence to improve oral absorption for poorly soluble compounds. During drug discovery phase, small-scale particle size reduction with 25–50 mg of material can be accomplished with a mortar and pestle or ball mill prior to preparing a suspension formulation. When a few hundred milligrams of material is available, particle size reduction can be achieved through either ball milling or jet milling. Homogenization with a tissue grinder sometimes also helps to break large aggregates in a suspension formulation.

31.11 FORMULATION FOR TOXICOLOGY STUDIES

Toxicology studies are aimed to characterize the potential adverse effects of the active compound. They are usually the final step of the discovery phase before the compound moves into preclinical development and approval to be administered in human is filed. At this stage, the pharmacokinetics of the compound in the relevant animal species is well understood. However, formulation development for toxicology studies is often challenging. Toxicology studies in general cover a wide range of dose levels, with the highest desired dose level typically 100-fold over the target dose, and in case the compound does not exhibit adverse effects, the dose is increased to a Food and Drug Administration (FDA)-recommended maximum of 1.5 g/kg [10]. The goal of such a high dose is to either establish a maximum tolerated dose when adverse effects are observed or a maximum feasible dose when adverse effects are not observed.

It is difficult to achieve dose proportional increase in exposure with high dose levels because of solubility-limited absorption for poorly soluble compounds. Suspension formulations often result in a plateau beyond a certain dose level with no further increase of exposure. In addition, high dose level suspension formulations tend to be viscous and prone to aggregation, which will lead to difficulty in administration and inhomogeneous distribution of the compound. In order to increase exposure and provide the ease of administration, solubility enhancement is often desired. Early in the chapter, we have mentioned a few common solubilizing approaches. Among them, cosolvency and complexation involve using excipients as solubilizers. Amorphous solid dispersion, although based on solubility and dissolution rate increase of the amorphous phase, also requires excipient (polymer) to increase physical stability. Unfortunately, the use of excipients is limited in toxicology formulations as any potential adverse effects from the vehicle may compromise a costly and time-consuming toxicology study. Usually, a preferred toxicology formulation vehicle would contain only low-level celluloses as suspending agent and a low level of surfactant as wetting agent. MC, CMC, and HPMC of 0.5–2% are commonly used suspending agents with Polysorbate 80 as the wetting agent. Salt formation through pH adjustment is an acceptable method to enhance solubility for toxicology formulation and should be tried first provided the compound has a suitable ionizable site. For oral route of administration, pH 2–10 is the acceptable range for nonbuffered vehicle, and for buffered vehicle pH 4–8 is acceptable. Combining the pH adjustment with various other solubilization approaches such as cyclodextrin and surfactant can be helpful. Substituted beta-cyclodextrins such as sulfobutyl ether-β-cyclodextrin and hydroxyl-beta-cyclodextrin at a level of 25% have been used to enhance solubility in toxicology formulations. However, use of hydroxypropyl-beta-cyclodextrin in the vehicle for carcinogenicity studies has raised some concerns due to the findings of increased tumors in rat in the pancreas and intestines after chronic dosing of hydroxypropyl-beta-cyclodextrin [11]. Cosolvents such as PEG 400, propylene glycol, lipids, and other organic solvents should be avoided due to either known toxicity at high levels or insufficient data on long-term safety pharmacology. Amorphous solid dispersion has recently seen increased use in toxicology studies.

A few pharmaceutically acceptable polymers such as hydroxypropylmethylcellulose acetate succinate (HPMC-AS), polyvinylpyrrolidone (PVP) and HPMC can be used to make amorphous solid dispersions that are stable under ambient conditions for a few months or even longer. However, for studies in rodents, formulations are dosed through an oral gavage tube and the amorphous solid dispersion (ASD) needs to be formulated in a suspension vehicle for dosing. It is therefore important that the ASD needs to be physically stable not only as is, but also as a formulated suspension. For a toxicology study longer than 4 days, it is desirable to establish physical stability of the ASD in the formulated form for a few days to reduce frequency of formulation preparation.

Another consideration during the toxicology formulation development is the chemical stability of the compound in the formulation over the period of time the formulation is stored and dosed. Since the toxicology findings result from not only the active compound of interest but also all impurities included in the formulation, impurity profile of the compound as well as degradants formed in the formulation should be well characterized. In order to minimize batch-to-batch variation in exposure, it is important to identify key solid-state properties such as crystalline form and particle size. For all batches used in toxicology studies, it is recommended to have a record of analysis including solid-state characterization to keep track of potential changes. When it is found that the properties of a new batch differs from the early batches, a bridging pharmacokinetic study is required to accurately assess the impact of the changes in the pharmaceutical properties and appropriately adjust the doses and design of the toxicology study.

31.12 TIMING AND ASSESSMENT OF PHYSICOCHEMICAL PROPERTIES

It is clear, based on the discussion for formulation approaches, that a basic knowledge of the physicochemical properties is critical. The independent properties such as pK_a, LogP, or LogD are solely determined by the chemical structure of the API in question and can often be easily calculated by widely available programs or quickly determine using automated high-throughput platforms. In addition, they are of immediate interest to discovery teams as they may correlate to pharmacokinetics and drug metabolism (PKDM). As a result, they are among the first properties evaluated for new APIs by discovery teams and are readily available to the pharmaceutical scientist developing a formulation strategy as described above. Typically, confirmatory low-throughput evaluation of these properties is only completed as a molecule progresses from discovery to clinical development. The second category of properties that can acutely affect formulation performance may be considered to be dependent properties. These properties include PSD, solubility, form stability, and chemical stability, and can depend on the final API form, manufacturing parameters, and the formulation of the compound. Since these properties depend on many unknown parameters, a complete assessment is not possible during early drug discovery. The strategies employed to assess these different properties vary widely and in general are related to the performance of the formulations themselves and the ability to control these properties during discovery.

31.13 CRITICAL ISSUES WITH SOLUBILITY AND STABILITY

Critical issues with solubility and stability are often first identified during early discovery.

31.13.1 Solubility

In the case of solubility, poor exposure that cannot be explained by PKDM parameters will lead to questions about solubility and dissolution resulting in targeted formulation screens designed to improve these properties. These studies may involve visual solubility screens as described above or quantitative assessments. In addition, solubility in physiologically relevant fluids (simulated intestinal fluid [SIF], simulated gastric fluid [SGF], 0.01 N HCl) may also be determined, although this typically occurs as an API approaches development. Assessment of chemical stability in discovery is similar in many ways to solubility determination.

31.13.2 Chemical Stability Assessment

In the same way that solubility assessment begins with determination and understanding of the pK_a, LogP, and LogD, stability assessment begins with an evaluation of the API's chemical structure for chemical liabilities that may result in degradation in the proposed formulation. For example, a compound susceptible to acid-catalyzed hydrolysis should not be formulated at pH 2 and stored at room temperature (RT) for an extended period. Initially, this evaluation may simply involve a quick conversation with the medicinal chemist that produced the compound or an expert in pharmaceutical degradation. Despite best efforts, chemical degradation is often identified due to poor performance of the formulation early in the discovery phase or during stability assessments

on actual formulations designed to support longer dosing protocols for pharmacodynamic assays or, more commonly, toxicology studies. Once a potential stability liability is detected, the need to determine the proper intervention will impart a strong desire to understand the basic chemistry. For example, will pH adjustment be necessary to control hydrolysis, or should antioxidants be added instead to prevent oxidation [12]. Liquid chromatography-mass spectrometry (LC-MS) is typically the first technique used to help identify degradants and provides a wealth of information [13]. To interpret these data, pharmaceutical scientists can employ a number of resources. Again, the medicinal chemist should be consulted, as well as the PKDM representative since drug stability often mirrors drug metabolism [14]. Significant developments have also been made in the area of degradation databases and expert knowledge systems for pharmaceutical degradation. Recent years have seen the development of the Pharma Drug Degradation Database (Pharma D3) [15]. This publicly accessible resource allows searches by compound name (common, generic, trade), compound number, condition, and most importantly, by functional group and change in molecular weight. Computer-aided prediction programs or expert knowledge bases are an additional resource used by some organizations. These programs generally predict chemical degradation pathways and degradants based on the identification of functional groups in the new chemical entity (NCE) and the known chemistry of these groups kept in a knowledge base [16–18] [19]. These programs have often been developed by an individual organization or have limited technical support; however, an expert knowledge base, Zeneth, is currently under development by Lhasa Ltd. and a consortium of pharmaceutical companies, with the first public release expected at the end of 2010 [20]. The program is built on the well-known Derek metabolite prediction platform and is being programmed based on data from public sources. These computer-aided resources can also be used to help assess compounds for chemical liabilities in preparation of formulation development. Note that some organizations take a very proactive approach to stability assessment. For example, standardized solution stability studies may be employed during lead identification in an attempt to determine the "intrinsic" stability liabilities. Excellent guidance now exists in the literature to help select reaction conditions that are more likely to give degradation products relevant to development [17, 18, 21–27]. Based on one strategy, advocated by Valentino Stella, stress testing can be initiated on a small number of molecules representing the synthetic scaffold of interest to identify critical liabilities influencing lead selection and formulation development [28].

31.13.3 Monitoring of the Physical and Chemical Stability

A number of the other dependent properties, such as PSD, morphology, and form stability, are often evaluated during early discovery to generate a basic lot history, and not necessarily to control the property of interest. This information represents an important opportunity for the pharmaceutical scientist to study the performance of the API and has the benefit of improving preclinical formulation support and development [29, 30]. Early on, the history can consist of some very basic information that includes polarized microscopy of the API and the formulation, a record of the highest dose at which different formulations were solutions, and the color of the API and formulations. This information can inform future questions about sudden changes in formulation performance due to changes in form or PSD, suitability of a given formulation for high-dose toxicological studies, and the chemical stability of the formulations. Often compound or formulation liabilities are identified based on these simple observations in combination with results from the corresponding pharmacokinetic and pharmacodynamic assays. This strategy is slowly expanded and employed on into clinical development. Likewise, as a molecule progresses, polarized micrographs will be supplemented with other techniques to more tightly control or monitor PSD (dynamic light scattering) or form (X-ray powder diffraction [XRD], differential scanning calorimetry [DSC], thermogravimetric analysis [TGA]), especially if the historical record indicates that these properties play an important role in performance.

31.14 GENERAL AND QUICK APPROACH FOR FORMULATION IDENTIFICATION AT THE EARLY DISCOVERY STAGES

For intravenous formulation for both rat and canine, use DMSO solution. This does not need any formulation preparation as the stock of the compound at this early stage is stored as DMSO solution. The only thing that needs to be done is to verify the concentration before dosing. If the DMSO is not acceptable or not practical for any reason, then the cosolvent approach will be the second choice. The formulation identified utilizing the cosolvent approach, as described under cosolvent utilization, could be the default formulation for all the compounds in the series and could be modified to accommodate other series. It is always advised that for this cosolvent-based formulation, a dilution test is conducted to ensure that no precipitation takes place at the site of the injection or when the compound enters the bloodstream. This could be as simple as dilution in 1:1 or 1:2 with phosphate buffered saline (PBS) buffer pH 7.0. Also, a slow injection is used during administration

For the oral formulation, start with basic vehicle, for example, 2% HPMC with 1% Polysorbate 80, or 0.5% CMC with 0.5% Polysorbate 80. This will be the default formulation for the project. Calculating the pK_a of the molecule is the next step. If the compound is basic and the pK_a is above 3, then adjust the pH of the formulation to pH 2–2.2 using, for example, hydrochloric acid. On the other hand, if the compound is acidic and the pK_a is below 8, then adjust the pH to pH 10 with sodium hydroxide solution. Make sure that there are no chemical stability issues as discussed in 31.13.3. Always plan on decreasing the particle size at this stage. If the required exposure is not achieved or higher exposure is needed, then utilize complexing agents such as cyclodextrins. Addition of 15–20% of hydroxypropyl-beta-cyclodextrin or sulfobutyl ether-β-cyclodextrin to enhance the solubility, and hence the exposure, will be next step. If the exposure is acceptable, then the default formulation for the project is identified and could be modified for certain compounds in the series.

REFERENCES

1. Diehl K-H, et al. (2001) A good practice guide to the administration of substances and removal of blood, including routes and volumes. *Journal of Applied Toxicology* 21:15–23.

2. Weiss AJ, et al. (1962) A phase I study of dimethylacetamide. *Cancer Chemotherapy Reports* 16:447–485.

3. Constantinides PP, Chaubal MV, Shorr R (2008) Advances in lipid nanodispersions for parenteral drug delivery and targeting. *Advanced Drug Delivery Reviews* 60:757–767.

4. Mottu F, et al. (2001) Comparative hemolytic activity of undiluted organic water-miscible solvents for intravenous and intra-arterial injection. *PDA Journal of Pharmaceutical Science and Technology* 55(1):16–21.

5. González-Caballero F, López-Durán JdDG Suspension formulation. In *Pharmaceutical Emulsions and Suspensions*, Nielloud F, Marti-Mestres G, eds. New York: Marcel Dekker, 2000:127–189.

6. Lukas G, Brindel SD, Greengard P (1971) The route of absorption of intraperitoneally administered compounds. *The Journal of Pharmacology and Experimental Therapeutics* 178(3):562–566.

7. Gough WB, et al. (1982) Hypertensive action of commercial intravenous amiodarone and polysorbate 80 in dogs. *Journal of Cardiovascular Pharmacology* 4:375–380.

8. Batrakova E, et al. (1999) Pluronic P85 increases permeability of a broad spectrum of drugs in polarized BBMEC and Caco-2 cell monolayers. *Pharmaceutical Research* 16:1366–1372.

9. Noyes AA, Whitney WR (1897) The degree of the solution of solid substances in their solutions. [machine transla-

tion]. *Journal of the American Chemical Society* 19: 930–934.

10. *Guidance for Industry: S1C(R2) Dose Selection for Carcinogenicity Studies*, Food and Drug Administration, ICH, Rockville, MD, September, 2008, Revision 1.

11. Gould S, Scott RC (2005) 2-Hydroxypropyl-b-cyclodextrin (HP-b-CD): A toxicology review. *Food and Chemical Toxicology* 43:1451–1459.

12. Reid DL, et al. (2004) Early prediction of pharmaceutical oxidation pathways by computational chemistry and forced degradation. *Pharmaceutical Research* 21(9): 1708–1717.

13. Rosenberg E (2003) The potential of organic (electrospray- and atmospheric pressure chemical ionisation) mass spectrometric techniques coupled to liquid-phase separation for speciation analysis. *Journal of Chromatography. A* 1000:841–889.

14. Kerns EH, Di L Accelerated stability profiling in drug discovery. In *Pharmacokinetic Profiling in Drug Research: Biological, Physicochemical, and Computational Strategies*, Testa B, et al., ed. Zurich, Switzerland: Wiley-VCH, 2006:281–306.

15. Pharma Drug Degradation Database, CambridgeSoft Corporation, 2004, Santafianos D, Alsante, K, Baertschi S, Jansen P, Reid DL, eds. http://d3.cambridgesoft.com.

16. Jorgensen WL, et al. (1990) CAMEO: A program for the logical prediction of the products of organic reactions. *Pure and Applied Chemistry* 62(10):1921–1932.

17. Alsante KM, et al. (2007) The role of degradant profiling in active pharmaceutical ingredients and drug products. *Advanced Drug Delivery Reviews* 59:29–37.

18. Baertschi SW (2006) Analytical methodologies for discovering and profiling degradation-related impurities. *Trends in Analytical Chemistry* 25(8):758–767.

19. Pole DL, Ando HY, Murphy ST (2007) Prediction of drug degradants using DELPHI: An expert system for focusing knowledge. *Molecular Pharmaceutics* 4(4): 539–549.

20. Lhasa, Ltd., Leeds, UK, http://www.lhasalimited.org.

21. Hovorka SW, Schöneich C (2001) Oxidative degradation of pharmaceuticals: Theory, mechanisms and inhibition. *Journal of Pharmaceutical Sciences* 90(3):253–269.

22. Jackson RA (2007) A forced degradation strategy: How Johnson & Johnson PRD approaches forced degradation. *American Pharmaceutical Review* 10(5):59, 60, 62, 63, 79.

23. Joshi BK, Reynolds DW (2005) Regulatory and experimental aspects of oxidative stress testing of pharmaceuticals. *American Pharmaceutical Review* 8(5):105–107.

24. Reynolds DW (2004) Forced degradation of pharmaceuticals. *American Pharmaceutical Review* 7(3):56–61.

25. Reynolds DW, et al. (2002) Available guidance and best practices for conducting forced degradation studies. *Pharmaceutical Technology* February:48–56.

26. Waterman KC, et al. (2002) Hydrolysis in pharmaceutical formulations. *Pharmaceutical Development and Technology* 7(2):113–146.

27. Waterman KC, et al. (2002) Stabilization of pharmaceuticals to oxidative degradation. *Pharmaceutical Development and Technology* 7(1):1–32.

28. Stella VJ (2005) Deciphering the chemical degradation profile of drug substances. *AAPS Annual Meeting and Exposition*. Nashville, TN.

29. Pudipeddi M, Serajuddin ATM (2005) Trends in solubility of polymorphs. *Journal of Pharmaceutical Sciences* 94:929–939.

30. Center for Drug Evaluation and Research (CDER) and Center for Biologics Evaluation and Research (CBER). *Guidance for Industry: Content and Format of INDs for Phase 1 Studies of Drugs, Including Well-Characterized, Therapeutic, Biotechnology-Derived Products*. Rockville, MD, November, 1995.

32

IN VITRO TESTING OF PROARRHYTHMIC TOXICITY

Haoyu Zeng and Jiesheng Kang

32.1 OBJECTIVES, RATIONALE, AND REGULATORY COMPLIANCE

Many non-antiarrhythmic prescription drugs have been withdrawn from markets or have received a black box warning due to safety issues, and ~50% of withdrawals/black box warnings since 1998 were caused by drug-induced (acquired) QT prolongation and ventricular arrhythmia, torsades de pointes. A list of these drugs with QT liability can be found in drug safety websites such as Arizona CERT (http://www.azcert.org). These withdrawals and black box warnings, along with delays in approvals, have brought huge liability and financial burdens to pharmaceutical companies. For these reasons and, most importantly, for the safety of patients, proarrhythmic toxicity has become one of the major safety concerns in drug discovery and development by pharmaceutical companies and for regulatory agencies worldwide.

After years of mechanistic research, it is clear that most, if not all, of drug-induced proarrhythmic adverse events, from QT interval prolongation to the rare yet fatal ventricular arrhythmia, torsades de pointes, are associated with cardiac ion channels that conduct action potentials and other electrical activities of cardiomyocytes. Thanks to technical feasibility, *in vitro* testing focusing on the effects of drug candidates on cardiac ion channels has been widely adopted and accepted by the pharmaceutical industry and regulatory agencies to predict potential proarrhythmic risks of the candidates and to avoid failure in later-stage drug development.

Because of its unique structure and biophysical properties (see a recent review by Sanguinetti and Tristani-Firouzi, 2006), the human ether-a-go-go-related gene (hERG) potassium channel that conducts I_{Kr} current in cardiomyocytes to restore resting membrane potential during Phase 3 of the cardiac action potential can be blocked by many pharmaceutical agents from almost all therapeutic classes with diversified structure scaffolds. Therefore, the hERG channel is the major target leading to drug-induced cardiac adverse events. In fact, almost all withdrawn, non-antiarrhythmic drugs associated with adverse electrocardiographic events are hERG blockers. For this reason, the International Conference on Harmonization of Technical Requirements for Registration of Pharmaceuticals for Human Use (ICH), an international organization founded in 1990 to harmonize drug discovery- and development-related issues among regulatory agencies and pharmaceutical associations of Europe, Japan, and the United States, issued the S7B guidelines to recommend *in vitro* hERG testing of all pharmaceutical compounds that are candidates for clinical trials in humans (www.ich.org). This guideline was accepted and adopted by the Food and Drug Administration (FDA) in 2005.

In addition to the hERG channel (or Kv11.1, coded by the KCNH2 gene), other cardiac ion channels, including KvLQT1/mink (or IKs, coded by KCNQ1/KCNE1), Kv4.3 (by KCND3), NaV1.5 (by SCN5A), Kir2.1 (by KCNJ2), and cardiac CaV1.2 (by CACNA1C), because of their individual yet irreplaceable roles in cardiac action potentials, are also potential targets for adverse cardiac events, and therefore are routinely counterscreened with *in vitro* testing to further reduce the potential risk of drug-induced proarrhythmic liability.

ADME-Enabling Technologies in Drug Design and Development, First Edition. Edited by Donglu Zhang and Sekhar Surapaneni.
© 2012 John Wiley & Sons, Inc. Published 2012 by John Wiley & Sons, Inc.

Due to advantages of higher throughput, homogeneity, easy availability, high reproducibility, and low cost, individual *in vitro* cardiac ion channel testing is conducted routinely using ion channels subcloned and overexpressed in heterologous expression systems (e.g., HEK293 or CHO cells). Other *in vitro* or *ex vivo* methods are also used on an as-needed basis for specific purposes. For example, individual ion channels from native cardiomyocytes are used to address involvement of indirect modulatory effects. Action potentials can be recorded from isolated cardiomyocytes to examine mixed ion channel effects (i.e., the combined effect of inward and outward ionic currents). Action potential duration (APD) is measured from isolated Purkinje fibers or papillary muscle to assess mixed ion channel effects and effects on electrical coupling and conductance with the involvement of the drug's tissue accessibility. Isolated working hearts (e.g., Langendorff preparation) can be used to examine drug effects at the whole-organ level. However, compared with assays using cultured cells, all of these *ex vivo* methods are labor-intensive with lower throughput, and the corresponding data interpretation is complicated and difficult, with significant data variation per animal and experimental condition. Therefore, they are usually used as second-tier methods or for investigational studies and so are only discussed briefly in the following sections.

32.2 STUDY SYSTEM AND DESIGN

Various *in vitro* testing systems commonly used by the pharmaceutical industry to assess proarrhythmic toxicity will be detailed in the following sections.

32.2.1 The Gold Standard Manual Patch Clamp System

All major cardiac ion channels that underlie the electrocardiographic activities of the heart belong to the voltage-gated ion channel superfamily, which is composed of pore-forming, multiple transmembrane proteins that can adjust the conformation of their pores in response to transmembrane potential change to pass or block ions across the membrane. A manual patch clamp technique that can control membrane potentials and measure passed currents, or vice versa, is widely used to study biophysiological properties of ion channels and cellular excitability. Because of the advantages of this Nobel Prize-wining technique, for example, precise control and measurement, various configurations, and wide-spanning recording capacity, the patch clamp technique is considered the gold standard for ion

channel studies and for *in vitro* cardiac ion channel safety testing.

Readers are referred to the book *Single-Channel Recording*, by Nobel laureates Bert Sakmann and Erwin Neher (Sakmann and Neher, 1995), for a detailed description of the technique. Briefly, a manual patch clamp system to conduct *in vitro* cardiac ion channel testing includes an inverted microscope with 20–60× magnification power to view, locate, and identify cells; a micromanipulator to stably hold and finely position a glass electrode; a perfusion system to handle drug/compound application and washout; an antivibration table to support all of these components; a Faraday cage to shield outside electrical noise; a temperature control unit (optional); a signal amplifier; and a computer system with software for data acquisition and analysis.

A general experimental procedure of manual patch clamp recording includes the following major steps: (1) manipulate a glass pipette electrode to the cell surface to archive a tight seal between the cell membrane and the tip of the glass electrode; (2) break the membrane by suction to obtain a whole-cell configuration with high seal resistance (usually with >1 GΩ electrical resistance, called gigaseal); (3) apply an appropriate voltage protocol to excite the desired ionic current from the cell that is perfused with vehicle control solution; (4) upon obtaining a stable baseline, challenge the ion channels with compounds to examine *in vitro* proarrhythmic toxicity. The following is an example of a well-accepted manual testing procedure using the hERG channel overexpressed in CHO cells:

1. Dissociate cells stably expressing hERG channels with 0.05% Trypsin-ethylenediaminetetraacetic acid (EDTA).
2. Resuspend cells in culture medium.
3. Load cells into the recording chamber, wait a few minutes to allow cells to settle, and then perfuse the chamber with extracellular solution.
4. Fabricate electrodes with 2–4 MΩ resistance from borosilicate glass using a pipette puller.
5. Zero the liquid junction potential with a compensation circuit of the amplifier.
6. Manipulate the pipette electrode filled with internal recording solution to the cell surface to form a gigaseal, then gently rupture the membrane with fine suction to generate a whole-cell configuration.
7. Perfuse the cell with vehicle control solution.
8. Elicit hERG currents with desired voltage protocols (such as a 4-S depolarizing step from a holding potential of −80 to +20 mV and then a 4-S repolarizing step to −50 mV) to elicit the tail

current for measurement. Repeat the voltage protocol at a desired interpulse interval.

9. Record current traces using data acquisition software with, usually, 60–80% compensation of series resistance.

10. Monitor hERG tail current amplitude for stability during control conditions before administration of test compounds at sequentially increasing concentrations, followed by a washout with vehicle control solution to determine reversibility.

11. Normalize peak tail current amplitudes to control to obtain the inhibitory effect of a test compound at a given concentration.

12. Calculate the half maximal inhibitory concentration with a Hill equation.

32.2.2 Semiautomated System

Although manual patch clamp is the gold standard, it does require extensive training to master the required techniques. Even to a well-trained veteran in electrophysiology, the success rate of data acquisition varies from day to day and is affected by many factors (e.g., leaky suction system, drifting manipulator, air table vibration, electrode glass properties, unstable pipette puller, electrical noise) that may lead to problems requiring years of experience to identify and resolve. Because of all of these uncertainties, it is not uncommon to get no data for many hours spent sitting in front of a recoding rig. Therefore, engineers and researchers have been trying to develop automated technologies that are user-friendly, less labor-intensive, require minimal training, are able to achieve higher throughput than the manual method, and that keep the advantages of manual patch clamp without compromising data quality. So far, quite a few semiautomated and automated systems have been commercialized.

The computer-aided manipulator (ezPATCH 100A, developed by Neo Biosystems, San Jose, CA, USA) uses a feedback system of electrode access resistance to automatically guide the electrode to the cell surface without user involvement. An associated pressure unit (ez-gSEAL 100) controlled by software is able to precisely generate positive or negative pressure. Incorporation of these two units into a conventional manual patch clamp system creates a personal semiautomated patch clamp system: ezPATCH replaces the regular micromanipulator to handle the pipette electrode manipulation with minimal user input, and ez-gSEAL then makes a seal, ruptures the cell membrane, and obtains whole-cell configurations by preset pressure steps to complete the most difficult and demanding part of manual patch clamp operation. This system can be viewed as a computer-aided manual patch clamp station as it keeps all of the advantages of the manual patch clamp while relieving users from the burden of making seals and obtaining whole-cell configuration, and it enables users to maintain full control of the system, much like a regular manual patch clamp system. The disadvantage of this system is that it requires users to manually handle test compound administration and data acquisition. Therefore, it is only a semiautomated system, and the throughput will not be much higher than that of a manual system.

32.2.3 Automated System

A fully automated recording system, Port-a-Patch developed by Nanion Technologies, Munich, Germany, is a miniaturized, self-contained, stand-alone unit requiring no air table, microscope, or Faraday cage. Only an amplifier and a computer are required for control and data acquisition. After suspended cells are manually added into its glass chip (NPC-1), the remainder of the study, including solution handling, is controlled by its data acquisition software. A unique feature of this system is that it can handle internal solution exchange, enabling test articles to challenge ion channels directly from the cytoplasmic side. This feature is excellent for lipidphobic or charged compounds that cannot easily pass through cell membrane by concentration gradient diffusion, and for modulation studies using molecules and/or proteins that cannot passively pass across membrane. It is very difficult, if not impossible, to realize internal solution exchange in other automated systems. Even with manual patch clamp, it is a very difficult task and requires special glass electrodes to conduct this type of solution exchange via passive (and time-consuming) diffusion. Since the whole-cell configuration is automatically formed inside its glass chip in the Port-a-Patch, the gigaseal success rate and data quality rely heavily on the quality of the chip. In addition, as it uses one cell at a time, the throughput is clearly less.

Most of the widely deployed automated electrophysiology systems in the pharmaceutical industry are parallel (mostly planar) patch clamp systems that can simultaneously acquire data from multiple (e.g., 16, 48, or more) cells with a proportional increase in throughput. Systems such as Dynaflow HT (Cellectricon, Mölndal, Sweden), FlyScreen (Flyion, Tübingen, Germany), IonWorks (Danaher, Sunnyvale, CA, USA), Patchliner (Nanion, Münich, Germany), PatchXpress (Danaher, Sunnyvale, CA, USA), and QPatch (Sophion, Ballerup, Denmark), with their individual pros and cons, are all self-contained, stand-alone units with fully automated cell additions, chip positions, whole-cell configurations, fluid handling, and data acquisition. The general concept underlying these automated machines is similar.

Suspended cells flow through or around the hole built inside every recording chamber on a chip. Ideally, one cell per chamber will be captured randomly and sealed with its aperture. The cell membrane will then be ruptured to form the whole-cell configuration by a preset automated procedure (mainly utilizing fine pressure/suction steps). Data are acquired using preset protocols with compound additions and washouts accomplished with built-in liquid handling machinery and fluidics. The procedure is controlled individually per chamber, and the parallel processes from multiple chambers greatly improve throughput. As the entire process mimics the manual patch technique with the ability to form gigaseals, the data quality is compatible with that of the manual patch clamp. In addition to their much higher throughput, these automated systems (represented by PatchXpress and QPatch, the two most popular automated systems) have consistent success rates and generate data of high quality due to their preset, fine, precise, and repeatable pressure/suction controls (assuming the use of high-quality chips). An example procedure using PatchXpress to run *in vitro* hERG testing is presented in Section 32.4 as a case study, and some practical issues are discussed in detail.

Practically, not all holes will always be occupied by cells, not all attached cells will be sealed and then ruptured, and not all whole-cell configurations have gigaseals and will last the entire recording procedure. The realistic success rate (defined as the number of completed and acceptable recordings over the total recording chambers on a chip) and the consequential throughput is far from optimal. From the user end, many optimization steps such as cell preparation modification, pressure/suction procedure refinement, and internal and external recording solution optimization have been utilized to improve or resolve these issues and increase the success rate. From the designer/manufacturer end, a technique called population patch clamp has been integrated into some of the latest models (e.g., QPatch HTX by Sophion) to help minimize those issues.

In population recording mode, multiple apertures are built on each recording chamber to capture multiple cells in every chamber, increasing the chance of cell capture per chamber (i.e., reducing or eliminating no-cell chambers) and the success rate of each of the steps that follow. Signals from all available cells within one chamber are recorded and combined as a "sum-cell" per chamber. Therefore, the population patch clamp technique does not increase the maximal throughput (i.e., the throughput with acceptable data from every chamber); it only improves throughput merely by reducing the number of "no-data" chambers but at the expense of compromising data quality (as tight and leaky signals, voltage errors, and noises are all summed together).

Among automated systems, IonWorks uses an exceptional principle referred to as perforated patch clamp technique. This method uses amphotericin B, an antibiotic, to generate small holes in the cell membrane and gain intracellular access, instead of breaking the cell membrane by suction as all other automated systems do. This technique significantly advances IonWorks over all other automated systems in terms of throughput. Enabled by the perforated patch clamp technique, IonWorks is able to use a 384-well plate as its platform, recording up to a theoretical maximum of 384 cells simultaneously. With the integration of the population patch clamp technique (i.e., not one, but 64 holes are on every one of the 384 wells, and up to 64 cells can be recorded together as "one" signal per well), which greatly reduces the number of empty/no-cell wells, the newest version of IonWorks, IonWorks Quattro, is able to generate data from most, if not all, of its 384 wells, and has a throughput close to nonelectrophysiological, surrogate assays (see below). However, the perforated patch clamp technique used in IonWorks has the disadvantage of low data quality since the system cannot form gigaohm seals (the seals formed are usually less than 200 MΩ, leading to substantial leak currents and voltage errors). For this reason, IonWorks is seldom used for *in vitro* ion channel testing for proarrhymic toxicity (with one exceptional application—see Section 32.5).

With the use of 384-well or even miniaturized 1536-well plates, nonelectrophysiological, automated assays for ion channels have much higher throughput than even automated electrophysiology systems. Commonly used, nonelectrophysiological cardiac ion channel assays include radioactive-labeled ligand displacement (e.g., ^{3}H-labeled dofetilide or astemizole binding assay), radioactive or nonradioactive ionic (e.g., Ca^{2+}, Rb^{+}, and Tl^{+}) influx or efflux, and atomic absorption spectroscopy using automated instruments such as VIPR (Aurora, San Diego, CA, USA), FLIPR (Danaher, Sunnyvale, CA, USA), ICPE-9000 (Shimadzu, Kyoto, Japan) and FDSS (Hammamatsu Photonics, Hamamatsu City, Japan). External potassium concentration can be adjusted to roughly manipulate intracellular membrane potentials in these assays, if needed. However, without a voltage clamp to lock measured voltage-gate ion channels into a uniform state, signals in these assays are a sum of multiple states that have or could have different responses to certain test compounds (e.g., closed channels will not respond to open channel blockers), resulting in compromised, inferior data. For this reason, these nonelectrophysiological assays are rarely used in cardiac ion channel testing for proarrhymic toxicity unless suitable electrophysiological methods are not available.

32.2.4 Comparison between Isolated Cardiomyocytes and Stably Transfected Cell Lines

In most, if not all, pharmaceutical companies, initial *in vitro* counterscreening of cardiac ion channels is conducted using ion channels overexpressed in heterologous expression systems (e.g., CHO or HEK293 cells) for their easy availability, homogeneous response, and, most importantly, their suitability for use in automated electrophysiology systems. (To the best of our knowledge, there have been no reports of successful assays using primary cardiomyocytes in any automated electrophysiology systems.) However, due to the nonnative environment, heterologously expressed cardiac ion channels have different responses or even a lack of response to modulation by native regulators. For example, L-type Cav1.2 expressed in HEK293 cells had different responses to modulation, and required cotransfection of a certain factor to be activated by Forskolin, a modulator that can enhance inward Ca^{2+} current conducted by L-type Cav1.2 in native cardiomyocytes (Gao et al., 1997). In an investigational study or on an as-needed basis (e.g., a candidate drug that targets a regulatory pathway of cardiac ion channels), freshly isolated cardiomyocytes will be required. Native cardiomyocyte isolation is a labor-intensive procedure, and the quality of isolated cardiomyocytes varies from day to day even when conducted by highly skilled biologists due to factors such as animal health, animal age and size, and variation in enzyme batches. In addition, because of their relatively larger size and nonspherical shape, it is extremely difficult, if not impossible, to seal isolated cardiomyocytes in automated systems, further limiting their utilization in *in vitro* testing of proarrhythmic toxicity.

32.3 GOOD LABORATORY PRACTICE (GLP)-hERG STUDY

The ICH S7B guideline recommends that an *in vitro* hERG assay should be conducted in compliance with GLP for all pharmaceutical compounds that are targeted for humans. Since the guideline was endorsed by the FDA in 2005, most, if not all, investigational new drug (IND) applications of new small-molecule entities submitted to the FDA include an *in vitro* hERG study conducted in compliance with GLP.

GLP hERG assays are conducted using a validated manual patch clamp system (to the best of our knowledge, no GLP hERG study has been conducted with an automated system). Per GLP, a study system must be validated prior to being used for any GLP studies. Like all other validations for GLP compliance, the validation procedure for a manual patch clamp system should

include installation qualification (i.e., the system installation should be executed and documented per vendor's specification), operation qualification (i.e., the system operation should be tested and documented per vendor's specification), and performance qualification (i.e., the system performance should be examined and documented per preset user's specification). A detailed explanation of GLP compliance is beyond the scope of this chapter, and readers are referred to 21 CFR (Code of Federal Regulations) Part 58—Good Laboratory Practice for Nonclinical Laboratory Studies for further information. Here we will describe the GLP hERG study from a more practical standpoint.

The GLP hERG study is not much different from other GLP toxicology studies in terms of GLP compliance, and is similar to a non-GLP manual hERG study in terms of electrophysiology (the same methods are used to seal and to get whole-cell configuration, and currents are recorded with the same voltage command and protocol). Besides a validated manual patch clamp system, other GLP-compliance-related elements include the study biologist(s), independent quality assurance personnel, a study director, the use of a preapproved study protocol, a GLP-compliant environment (e.g., controlled and documented temperature for all rooms, freezers, and refrigerators used), and predetermined data analysis procedure and exclusion criteria (e.g., acceptable limits for background leak current, abnormal activation and/or inactivation kinetics, excessive current rundown). These predefined criteria and procedures should be clearly described in the standard operating procedures (SOPs) and be strictly followed in data acquisition and analysis. Any deviations and/or unanticipated study events should be communicated to the study director in a timely fashion, evaluated for their impact on study integrity, and promptly documented. Failure to do so will severely affect the study integrity and the reliability of the results.

Since test article concentration is very important in pharmacology, determination of the actual test article concentrations that challenge the hERG channel in the recording chamber is very critical in GLP hERG testing. Nominal concentrations, obtained by weight, calculation, and dilutions, are prone to inaccuracy due to well-known factors such as entry/human error, weighing error, volume/pipette error, and compound stability in aqueous recording buffer. Actual test article concentrations are usually measured by analytical methods (e.g., HPLC) in a GLP-compliant analytical laboratory. The critical action in the biology laboratory is to faithfully collect solution samples from the recording chamber where the hERG channel meets the test articles, not just to submit the final solutions in a beaker or graduated cylinder for analysis. Due to certain biophysical properties, many compounds will stick to plastic or even glass surfaces. It is not unusual to have a significant concentration drop in solutions collected from the recording chamber as compared with those from the reservoir due to adsorption of the test article by the plastic perfusion tubing. Maximal use of glassware and Teflon-coated tubing can reduce but will not eliminate adsorption.

Theoretically, the true test article concentration should be the test article concentration on the cytoplasmic side. Clearly, it is not practical to collect samples inside the cell. In most cases, the test article concentration surrounding the cell (i.e., samples from the recording chamber) is a good estimate of the true test article concentration that is challenging the channel, but for lipidphobic or charged compounds that cannot easily diffuse across the cell membrane, this may not be the case. For those compounds, a simple resolution is to conduct the study at physiological temperature to accelerate the diffusion process, and to extend the recording time to reach a steady state at any given concentration.

As a footnote, there has been discussion of conducting the GLP hERG study with a validated automated system to improve assay efficiency. However, one practical issue is the difficulty in conducting accurate concentration analyses on the microliter volumes handled in multiwell plate platforms. Another limitation is the difficulty in maintaining the validated status of automated systems that require extensive routine maintenance and part replacement. Lack of precise temperature control in automated systems is one more hurdle.

32.4 MEDIUM-THROUGHPUT ASSAYS USING PATCHXPRESS AS A CASE STUDY

PatchXpress, a parallel planar automated patch clamp system developed by Axon (acquired by Molecular Devices, now part of Danaher Corp.) and commercialized in the early twenty-first century, is one of the earliest automated systems that can form gigaseals and obtain high-quality data. It has been widely used by pharmaceutical companies in drug discovery programs and cardiac ion channel counterscreening (Tao et al., 2004; Guo and Guthrie, 2005; Ly et al., 2007; Trepakova et al., 2007; Zeng et al., 2008).

As a brief introduction, PatchXpress uses a sealchip (a narrow plastic device with 16 center-holed chambers to hold cells and form up to 16 whole-cell configurations), an electrode-plate containing 16 silver-coated cooper electrodes that match with the 16 holes in a sealchip, a liquid handling system with a single-pipette head for cell and compound additions, and a 16-channel ground electrode containing a fluidics system linked to the extracellular solution reservoir and suction pump for solution exchange.

Many PatchXpress experimental procedures have been described in the literature, such as the one by Zeng et al. (2008). Briefly, the first step is to flush and clean the liquid handling system and all tubing with a built-in start-up procedure using internal and external solutions (many solution recipes are available in the literature and are usually modified from the corresponding solution recipes used with manual patch clamp). The second step is to load the compound plate and the plate layout file (describing compound location and concentration on the compound plate), to adjust the setting file containing the pressure protocols to seal, break-in, and get whole-cell configurations (these protocols are keys to get high success rate, and should be tailored to individual system/assay by "trial and error"), and to set the procedure file, including voltage protocols, and the cell and compound addition schedules. The last user-involved step is to treat and load the cells. As mentioned in the previous section, like other automated systems, PatchXpress requires suspended cells with spherical shapes. The cell treatment procedure, using reagents such as Trypsin, Versene™ (DOW Chemical Company, Midland, MI, USA), or Accutase® (Sigma-Aldrich, St. Louis, MO, USA) to dissociate cells from the culture dish, is critical for the subsequent experimental steps. Undertreatment makes cells difficult to seal and break in, while overdigestion leads to membrane damage, large leak currents, and early loss of whole-cell configurations. As reagent type, concentration, and treatment duration vary per cell type, passage number, and cell density, "trial and error" is the best way to find an optimized cell dissociation method in individual laboratories. After cells are loaded into the machine, the remaining steps, including data acquisition, will be automatically performed by the system.

One alternative to a problematic cell dissociation procedure is to use frozen or cryopreserved cells. The biophysical properties of most cardiac ion channels are not affected by a quick-thaw procedure after months of storage in a frozen state in liquid nitrogen. The success rate of frozen cells, in our own experience, is compatible with freshly dissociated cells. Users can optimize the procedure and make a large batch of cryopreserved cells for use over many months, or can order frozen cells from commercial sources (e.g., instant cell, EZcell, ready cell).

To accommodate the relatively large amount of data generated, data analysis is also "automated" with the associated data analysis software, DataXpress (Danaher), which allows users to customize analysis macros. After visual inspection of individual data files to exclude low-quality or abnormal files, a click of a button to launch data analysis macros will complete data analysis and calculate half maximal effect values or other user-defined values, resulting in a great reduction in data analysis time from hours to minutes. However, data interpretation is still completely up to the users, and it is beyond the scope of this chapter. Our own experience shows that if the same voltage protocol is used, most results from PatchXpress fall within a threefold difference of data generated with manual patch clamp.

Clearly, it is beneficial to optimize PatchXpress procedures to maximize throughput while maintaining high data quality. Solution recipe modification to incorporate potassium fluoride into the internal solution for a hERG assay is a good example. This modification can improve throughput up to sixfold without compromising data quality (Zeng et al., 2008). Other modifications to address certain assay-specific issues and improve throughput, such as adjusting osmolarity, adding ATP, and lowering cell culture temperature, have been well discussed in scientific conferences and the literature.

32.5 NONFUNCTIONAL AND FUNCTIONAL ASSAYS FOR hERG TRAFFICKING

Besides interacting directly with cardiac ion channels (mostly hERG), drugs can also induce proarrhythmic toxicity through indirect interaction by interrupting channel trafficking or via other possible mechanisms such as accelerating channel internalization and altering channel phosphorylation profiles. In this section, nonfunctional and functional assays for compound screening will be presented to select drug candidates with less potential for inducing long QT by interrupting channel trafficking.

A surface Western blot-type assay was first developed to address drug-induced hERG-trafficking deficiency (Wible et al., 2005). Cells expressing the hERG channel were incubated with compounds overnight and fixed, and hERG protein expression was detected by using anti-hERG antibody. Fluorescence signals were corrected for cell numbers and normalized to controls (cells incubated without compounds) to examine drug effects on hERG trafficking. This is a plate-based assay and can be easily miniaturized to gain high throughput. However, it has many intrinsic issues that might lead to false-positive or false-negative results. For example, there is no control of the housekeeping protein expression level, so the assay cannot determine whether an effect is hERG-specific or not. More importantly, it is not a functional assay and cannot differentiate whether the hERG channels still properly function or not, which might lead to false-negative results. For example, if a

compound changes the hERG gating properties, for example, through modifying its phosphorylation profile, the number of surface channels may remain the same, yet the net function has been altered. This type of drug-induced effect will not be detected by this Western blot assay.

Clearly, a better assay would be a functional test using an automated system to assess the surface hERG activity in the presence of test compounds. As it is impossible to obtain control data from the same groups of cells as those treated, the assay will be a population type assay: comparing drug-treated cells with cells incubated with vehicle only. Because of cell-to-cell variation, a relatively large number (at least tens) of data points is required in each group to gain enough statistical power. Therefore, IonWorks may be used for this function assay. A simple procedure is discussed below:

1. Evenly split cells into two flasks and culture them to 40–60% confluence.
2. Incubate one flask with test compound, the other one with vehicle (e.g., dimethyl sulfoxide) as control, for 24 h.
3. Wash cells intensively (e.g., 10 min × 6 washes) to completely wash away any residual compound to eliminate any direct effect.
4. Measure current amplitude of drug-treated cells and of controls with IonWorks (which can get up to 384 "cells" per condition).
5. Compare the current amplitudes of two groups with statistical analysis: A statistically significant difference between treated and control cells indicates that the tested compound might have indirect effects on the hERG channel.

As a footnote, a functional assay service using IonWorks to evaluate indirect effects on hERG and other cardiac ion channels has been commercially available, although its procedure and experimental details are not disclosed (www.chantest.com).

32.6 CONCLUSIONS AND THE PATH FORWARD

In this chapter, we have discussed *in vitro* testing of proarrhythmic toxicity, focusing on electrophysiological counterscreening of cardiac ion channels with respect to rationale, regulatory compliance, different assay procedures, and practical issues. Preclinical proarrhythmic toxicity assessment of pharmaceutical candidates is an integrated process. Individual *in vitro* assays to evaluate potential proarrhythmic risk of drug candidates each has its own advantages and limitations, and may not be able to predict the net outcome. (Verapamil is a good example in this case: Its effect on hERG is balanced by its effect on L-type CaV1.2 channel, resulting in no electrocardiographic impact.) As a rule of thumb, higher-throughput assays should be used in a first-tier screening to select a small pool of candidates, followed by manual GLP or GLP-like studies to confirm and cherry-pick the candidates for *in vivo* studies (e.g., telemetry assays using conscious dogs or monkeys). Labor-intensive *ex vivo* studies (e.g., primary cardiomyocytes or papillary muscle APD assay, Langendorff or working heart preparation for monophase action potential and ECG measurement) may be used for mechanism-related investigational studies, and be deployed on an as-needed basis for programs with known issues (i.e., as follow-up studies defined by ICH S7B guidance).

Currently, *in vitro* testing is mainly used for early detection and prediction of proarrhythmic liabilities, and *in vivo* telemetry studies with animals are still needed to confirm or clarify issues before clinical trials. With ongoing research in proarrhythmic toxicity, *in vitro* techniques and assays have been greatly advanced, and are being developed to better predict proarrhythmic liability and obtain safer drug candidates.

REFERENCES

Gao T, Yatani A, Dell'Acqua ML, Sako H, Green SA, Dascal N, Scott JD, Hosey MM (1997) cAMP-dependent regulation of cardiac L-type Ca2+ channels requires membrane targeting of PKA and phosphorylation of channel subunits. *Neuron* 19:185–196.

Guo L, Guthrie H (2005) Automated electrophysiology in the preclinical evaluation of drugs for potential QT prolongation. *J Pharmacol Toxicol Methods* 52:123–135.

Ly JQ, Shyy G, Misner DL (2007) Assessing hERG channel inhibition using PatchXpress. *Clin Lab Med* 27:201–208.

Sakmann B, Neher E (1995) *Single-Channel Recording*, 2nd ed. Plenum Press, New York.

Sanguinetti MC, Tristani-Firouzi M (2006) hERG potassium channels and cardiac arrhythmia. *Nature* 440:463–469.

Tao H, Santa AD, Guia A, Huang M, Ligutti J, Walker G, Sithiphong K, Chan F, Guoliang T, Zozulya Z, Saya S, Phimmachack R, Sie C, Yuan J, Wu L, Xu J, Ghetti A (2004) Automated tight seal electrophysiology for assessing the potential hERG liability of pharmaceutical compounds. *Assay Drug Dev Technol* 2:497–506.

Trepakova ES, Malik MG, Imredy JP, Penniman JR, Dech SJ, Salata JJ (2007) Application of patchxpress planar patch clamp technology to the screening of new drug candidates for cardiac KCNQ1/KCNE1 (I [Ks]) activity. *Assay Drug Dev Technol* 5:617–628.

Wible BA, Hawryluk P, Ficker E, Kuryshev YA, Kirsch G, Brown AM (2005) HERG-Lite: A novel comprehensive high-throughput screen for drug-induced hERG risk. *J Pharmacol Toxicol Methods* 52:136–145.

Zeng H, Penniman JR, Kinose F, Kim D, Trepakova ES, Malik MG, Dech SJ, Balasubramanian B, Salata JJ (2008) Improved throughput of PatchXpress hERG assay using intracellular potassium fluoride. *Assay Drug Dev Technol* 6:235–241.

33

TARGET ENGAGEMENT FOR PK/PD MODELING AND TRANSLATIONAL IMAGING BIOMARKERS

Vanessa N. Barth, Elizabeth M. Joshi, and Matthew D. Silva

33.1 INTRODUCTION

A guiding principle of biomarker research is the minimally invasive and serial measurement of biological processes to monitor tissue state and treatment response (Biomarkers Definitions Working Group, 2001). Of the broad class of biomarkers, medical imaging is emerging as a set of technologies that impact both preclinical research and clinical trials (Frank and Hargreaves, 2003; Rudin and Weissleder, 2003; Hargreaves, 2008; Willmann, 2008a, b). In some cases, the need for biomarkers is driven by the move toward targeted therapies, wherein the success of the agent is dictated by the ability to distribute, engage, and modulate the target with biologically relevant concentration and duration. The use of imaging technologies in combination with other strategies can not only provide useful information regarding the activity of a drug, but also allows for the comparison of that drug with other agents as a means of differentiation. The ultimate desire is to identify and treat patients most likely to respond, which can enable the rapid development of new agents and provide a level of personalization of medical treatment, which can be enhanced by utilizing imaging techniques (Richter, 2006; Eckelman et al., 2008).

It is increasingly common practice in experimental *in vivo* pharmacology to relate the observed pharmacological response to the administered dose, negating such factors as nonlinear concentration–dose relationships, concentration-dependent plasma protein binding, the role of active metabolite(s), and differences between species

bioavailability. These relationships are further enhanced by inclusion of drug receptor occupancy (RO) testing. By combining target engagement data with pharmacokinetics/pharmacodynamics (PK/PD) modeling, one would anticipate an increased probability of selecting the right compounds and their doses for development.

As technological advancements have been made, there are now a variety of ways to directly assess target engagement. Here, the focus is on the ability to leverage translatable imaging techniques from preclinical rodents and nonhuman primates into man. Liquid chromatography coupled to tandem mass spectrometry (LC-MS/MS)-based target occupancy, also referred to as RO and enzyme occupancy (EO), approaches have provided a quantitative, high-throughput *in vivo* solution to this challenge. The LC-MS/MS technology platform also provides an efficient mechanism to discover potential tracers for clinical biomarker strategies by way of nuclear imaging (i.e., positron emission tomography [PET]). The ability to translate target engagement across species and into the patient provides a much needed bridge to stringently test mechanistic hypotheses linking novel targets to disease biology and treatment. To successfully move from rodent/dog/nonhuman primate into man requires an understanding of PK/PD modeling to determine dose. These hurdles are most difficult for central nervous system (CNS) targets (G-protein coupled receptors, ion channels, intracellular enzymes, allosteric sites) given the restrictive blood–brain barrier (BBB). However, these approaches are also proving useful when applied to peripheral targets.

ADME-Enabling Technologies in Drug Design and Development, First Edition. Edited by Donglu Zhang and Sekhar Surapaneni.

The goal of this chapter is to discuss how preclinical quantitative pharmacology and PK/PD integration can guide compound and dose selection for more labor intensive *in vivo* assays, and how the technology lends itself to translational imaging biomarkers for both target engagement and PD measures in the clinic.

33.2   APPLICATION OF LC-MS/MS TO ASSESS TARGET ENGAGEMENT

Over the past decade, there have been significant increases in the sensitivity of mass spectrometry. As such, the ability to measure smaller and smaller concentrations of analytes in tissue matrices has become routine. Preclinical target engagement, such as RO, can be measured either *ex vivo* or *in vivo*, through quantification of changes in the amount of a radiolabeled ligand by liquid scintillation spectroscopy following pretreatment with an unlabeled test compound (Stockmeier et al., 1993; Zhang and Bymaster, 1999; Wadenberg et al., 2000). *Ex vivo* assessment of RO results from application of the radiolabeled ligand to tissue(s) that have been removed from an animal, whereas *in vivo* assessment of RO results from intravenous administration of the ligand to a live animal. In the latter condition, the ligand probes the available target population in a live animal, and then tissues are collected (Kapur et al., 2001). Previously, both cases have required the use of a radiolabeled ligand that is quantified by scintillation spectroscopy. Further, it could be argued that the "*in vivo*" version just described is more an *ex vivo* assessment because the final measurements are made outside of the animal.

The major advantages of the LC-MS/MS approach are the ability to assess more rapidly target engagement in the preclinical setting, and secondly, the ability to search for novel tracers at unprecedented rates. Leveraging these advantages coupled with nuclear imaging techniques, the impact of PET imaging on the drug discovery and development process could be greatly enhanced by the creation of novel probes, addressing the paucity of available tracers in common use (Guo et al., 2009).

33.2.1   Advantages and Disadvantages of Technology and Study Designs

LC-MS/MS provides a new method for detecting and quantifying changes in ligand concentration as a function of compound pretreatment. This approach obviates the need for radiolabeled ligands and has a number of practical advantages. The RO assays supported by this methodology are shorter in duration. The time spent developing the analytical method, conducting the live-phase portion of the study, processing samples, and analyzing data results occurs within 36 h. The rapid turnaround for data permits timely decision making regarding compound selection and dose selection for additional follow-up. Secondly, in addition to understanding target engagement, exposure data on the compound administered is also available in both tissue and plasma, and even cerebral spinal fluid (CSF) if needed. An understanding of the exposure/response relationship can be made for individual compounds as well as within a chemical series. Additionally, multiple tracers can be coinjected into a single animal to assess target engagement across multiple sites within the same animal (Need et al., 2007). Thirdly, avoidance of the radiolabeled ligand reduces the overhead for disposal of radioactive waste, purchasing and tracking material, and laboratory certification and regulations around working with radioactive material. Instead of measuring radioactivity, which is not necessarily parent tracer, a mass-to-charge ratio (m/z) is used to identify the parent tracer. This avoids the issue of potential confounding metabolites—active or inactive—which would regularly be picked up if tracking radioactivity alone. Lastly, and most importantly, the mass spectrometry application of detecting microdoses of compounds provides an efficient mechanism to evaluate novel ligands to enable new RO assays. In doing so, the iterative cycle for identification of translatable ligands/tracers for PET imaging resulting in clinical target engagement and/or PD markers is reduced. Compounds that are difficult to label become easy to assess *in vivo* as tracers as a result of sidestepping the radiolabeling process. However, one of the draw backs to the LC-MS/MS-based technology is the costly capital investment of the mass spectrometer and chromatography components.

33.3   LC-MS/MS-BASED RO STUDY DESIGNS AND THEIR CALCULATIONS

There are two formats for RO assays, referred to as the ratio method and the positive control method. Depending on the target, the biology drives the decision regarding assay format (after identifying the ligand/tracer to enable the assay). The ratio method requires a null-region tissue (Wadenberg et al., 2000). This refers to a tissue that is devoid of, or has extremely low levels, of target expression. The null region signifies the *in vivo* nonspecific binding of the tracer. One assumption inherent in this model is that a total binding region, rich with target expression, and the null region are similar with respect to nonspecific binding. When comparing across brain regions this assumption is not difficult to justify,

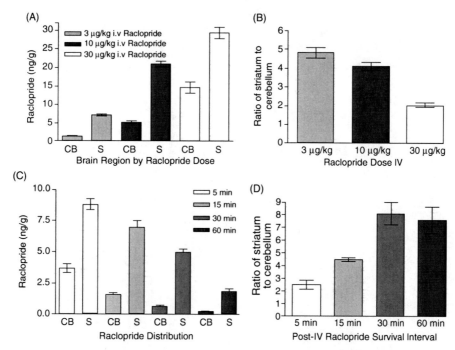

FIGURE 33.1. (A) Differential distribution of raclopride 15 min postdose in rat striatum (S) relative to cerebellum (CB) as measured by LC-MS/MS following several intravenous (IV) doses. (B) Ratio of raclopride levels in striatum to cerebellum for various intravenous tracer doses 15 min postdose. (C) Kinetics and distribution of raclopride in rat striatum and cerebellum following a 3 µg/kg dose at 5, 15, 30, and 60 min. (D). Ratio of striatal to cerebellar raclopride levels following a 3 µg/kg dose at various time points. $n = 4$ for all treatment groups; mean ± SEM.

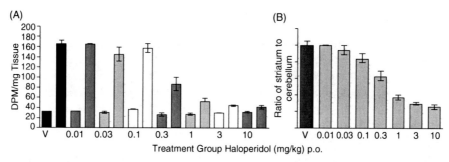

FIGURE 33.2. (A) Dose-dependent reduction in tritiated raclopride levels in striatal tissue (right bar) with increasing oral doses of haloperidol; cerebellar raclopride levels (left bar). (B) Dose-dependent reduction in striatal-to-cerebellar ratio of raclopride levels with increasing oral dose of haloperidol. V, vehicle; mean ± SEM.

but in the periphery it becomes more challenging. In the example of dopamine D_2 RO, the striatum is a receptor-rich brain region representing total binding (specific and nonspecific), and the cerebellum is the null region (Kohler et al., 1985). Following a 3 µg/kg dose of raclopride, a selective dopamine D_2/D_3 tracer employed in PET imaging and *in vitro* binding, it differentially distributes 4.5-fold in favor of striatum relative to cerebellum in Figure 33.1 (Barth et al., 2006).

With only vehicle pretreatment, this ratio is equivalent to 0% D_2 occupancy. As increasing doses of a dopamine D_2 antagonist, such as haloperidol, are administered 1 h before the raclopride tracer, raclopride levels in the

striatum are dose-dependently reduced while cerebellar levels remain constant. Eventually, the raclopride levels in the striatum will equal that of the cerebellum (Figure 33.2B). This results in a ratio of 1 and represents 100% dopamine D_2 occupancy whereby all specific binding of raclopride to dopamine D_2 receptors is blocked. Occupancy values can be calculated using the well-established ratio method (Wadenberg et al., 2000) using the following equation:

$$100\left[1-\left(\frac{Ratio_t - 1}{Ratio_c - 1}\right)\right] = \% \text{ Occupancy.}$$

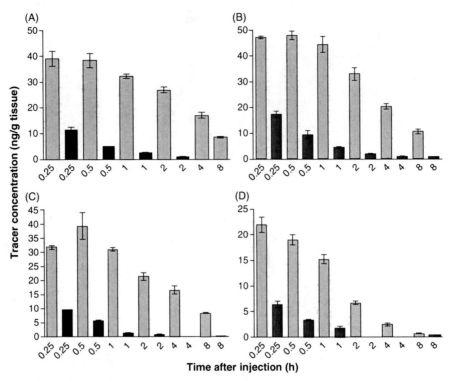

FIGURE 33.3. Time course evaluation in the frontal cerebral cortex of male Sprague Dawley rats after administration (30 μg/kg, intravenous) of 8 (A), 16 (B), 20 (C), and 22 (D) under baseline conditions and after pretreatment with 3 μg/kg (10 mg/kg, intravenous) 15 min before ligand injection: baseline (light stippling) and pretreatment (dark stippling).

Each "Ratio" refers to the ratio of tracer in a brain area rich in target receptor to the tracer detected in an area with little or no receptor density as detected by LC-MS/MS. "Ratiot" refers to animals treated with test compound, while "Ratioc" refers to the average ratio in vehicle-treated rats. In this case, the cerebellum was the null region, and the striatum was the total binding region per the expression pattern of dopamine D_2 receptors. The vehicle ratio of the tracer levels in striatum relative to cerebellum represents 0% occupancy. A ratio of 1 represents 100% occupancy. One-hundred percent occupancy is achieved when all specific binding to the D_2 receptor tracer is blocked. The intermediate ratios of striatal to cerebellar tracer from the pretreated groups were interpolated linearly between the ratio of tracer levels in the vehicle-treated animals (0% occupancy) and a ratio of 1 (100% occupancy) in order to determine the percent D_2 RO (Hume et al., 1998).

If a null region or tissue does not exist, then the alternate positive control method is employed. Here, a single dose of a selective blocking compound determines the residual/nonspecifically bound tracer remaining at 100% target occupancy (Donohue et al., 2008). To identify this dose of a compound, increasing doses are of compound are administered to separate groups of animals. As target occupancy increases, tracer levels decrease relative to the vehicle group. At some dose,

there is no further reduction in the tracer levels with increasing doses of blocker. Under this condition, the blocking compound achieves 100% occupancy and the residual tracer measured is nonspecific binding. Instead of using the ratio values to calculate % occupancy, the tissue levels of the tracer are employed. Thus, vehicle levels of the tracer in a single tissue represent 0% occupancy and the residual tracer levels described above in the positive control treatment group represent 100% occupancy (Figure 33.3).

33.3.1 Sample Analysis

The tissues collected are determined by the biology of the target being evaluated. When searching for a novel tracer, or reverse translating (moving a tracer from the clinic back into discovery research) an existing PET tracer, tissues are collected representing high and low target expression. Tissue samples are collected and kept on ice, following cervical dislocation of the animal. This is typically conducted within a 5-min time period Up to six samples, including blood, can be obtained under the above-mentioned conditions. Tissues are then extracted in an organic solvent, homogenized using an ultrasonic probe, and then centrifuged to remove protein (Chernet et al., 2005; Barth et al., 2006; Need et al., 2007). Supernatant is then diluted with sterile water and placed into an injection vial for subsequent LC-MS/MS analysis. Standards are prepared by adding known quantities of an analyte to tissue samples from nontreated rats and processed as described above.

33.3.2 Comparison and Validation versus Traditional Approaches

In order to validate the mass spectrometry approach to target engagement, we compared dopamine D_2 occupancy for a number of typical and atypical antipsychotics using both the LC-MS/MS and scintillation spectroscopy approach (Barth et al., 2006). Comparable D_2 occupancy curves and ED_{50} values were obtained regardless of the method of detection, and the percent of the receptor population labeled by the raclopride tracer dose was determined (Figure 33.4). ED_{50} values for the antipsychotics tested in rat were similar to those values producing changes in brain neurochemistry and behaviors (condition avoidance response) (data not published). Antipsychotic ED_{80} values were also approximate with doses producing effects in animal models of motoric side effects such as catalepsy (data not published). This resulted in a similar preclinical window of "opportunity" relative to the clinic; a desire to achieve approximately 60–80% dopamine D_2 occupancy to see efficacy without unwanted side effects (Kapur et al., 2000) (Nordström et al., 1993). The dopamine D_2

example demonstrates the ability to link target engagement (both preclinically and clinically) to neurochemical/behavioral measures. These data validated the mechanistic hypotheses, and provided an objective, quantitative foundation for the dose selection by PK/PD (where PD was the RO measurement) modeling for subsequent clinical studies.

33.4 LEVERAGING TARGET ENGAGEMENT DATA FOR DRUG DISCOVERY FROM AN ABSORPTION, DISTRIBUTION, METABOLISM, AND EXCRETION (ADME) PERSPECTIVE

Incorporating PK/PD principles along with target engagement data has allowed project teams to accelerate decision making around potential therapeutics, often narrowing the structure–activity relationship (SAR) and ultimately selecting a testable clinical candidate. Over the next few sections, a few key ADME-related principles will be discussed, with case examples included to illustrate the informed decisions that result from combined target engagement and PK/PD data.

33.4.1 Drug Exposure Measurement

Before any meaningful PK/PD correlations can be established to understand the relationship of exposure

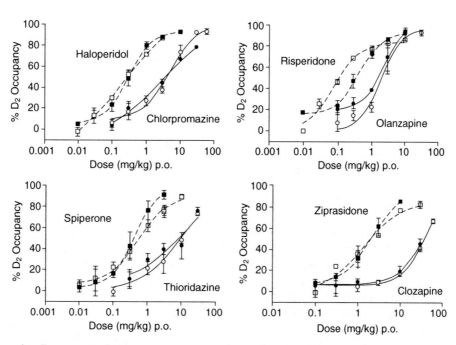

FIGURE 33.4. Dopamine D_2 receptor dose–occupancy curves for orally administered antipsychotic drugs were measured using raclopride as a tracer. Data were generated for each of the eight drugs using either radiolabeled (solid symbols) or unlabeled (open symbols) raclopride tracer. The x-axis represents the oral dose in milligrams per kilogram administered 1 h before the raclopride tracer. The y-axis represents the percent dopamine D_2 receptor occupancy. Two drugs are shown per panel.

to the extent of target engagement, one must ensure an accurate measurement of drug concentration. The bioanalytical assay employed must be reliable, specific, and include the appropriate level of quality controls. In drug discovery research, the use of LC-MS/MS has become the technique of choice for the quantification of analytes and has been the focus of a thorough review in *Drug Discovery Today* (Korfmacher, 2005).

33.4.2 Protein Binding and Unbound Concentrations

Protein binding and its effects on the pharmacological activity of drugs has been a source of debate within the scientific community (Sansom and Evans, 1995; Benet and Hoener, 2002). However, the "free drug hypothesis" is generally accepted in the PK field, which describes the role of free or unbound drug concentration as ultimately being responsible for the interaction with drug targets in the body (Wilkinson, 2001). An extension of this hypothesis has been made for CNS-targeted therapeutics, which assumes the unbound drug concentration found in the brain interstitial spaces is ultimately responsible for interaction with the drug receptors at the site of action, often referred to as the so-called biophase. Understanding the relationship between measured concentrations (total and unbound) and target engagement (RO or EO) data can be critical in understanding discrepancies, especially when studies involve multiple species, concomitant use of other drugs, and/or comparisons made between multiple compounds and genders (van Steeg et al., 2008). Additionally, the development of a reliable PK/PD relationship within a given platform or target can enable the acceleration of new drug development.

While a number of simple and efficient computational methods exist and have shown utility in predicting relative plasma and tissue-specific protein binding values, an accurate experimental determination is always the best means toward an accurate assessment (Kratochwil et al., 2002; Waterbeemd and Gifford, 2003; Emoto et al., 2009). A brief background on protein binding and its potential importance will be reviewed. Additionally, a case example where this parameter was utilized to effectively guide a discovery team toward elucidation of its PK/PD relationship will be discussed.

The total concentration (C) of drug can be determined by the ratio of its unbound concentration (C_u) measured in a given matrix (plasma or tissue of interest) to the measured unbound free fraction (f_u) as illustrated in the following equation:

$$C = \frac{C_u}{f_u}.$$

Using this equation, one can see that the total concentration of a drug can be affected by changes in protein binding or by the total protein concentration in the matrix of interest. While the unbound fraction (f_u) is often considered to be constant at pharmacological concentrations, any increase in C_u will translate into an increase in f_u as the unbound concentration approaches the concentration of binding sites in the matrix. At low unbound concentrations, binding is constant and f_u is then independent of C_u. However, at higher unbound concentrations, more binding sites become occupied and f_u can then increase as binding becomes saturated. Once saturation is achieved, f_u approaches unity, making C and C_u concentrations equivalent. Having an appreciation for this relationship becomes critical when the concentration–response relationship falls outside of the region of constant binding.

For compounds directed toward targets that reside within the CNS, the extent of drug transport across the BBB dictates the *in vivo* potency of centrally acting drugs. A number of *in situ*, *in vitro*, and *in vivo* methods exist for determining CNS distribution of a new chemical entity (NCE); however, the overall experimental complexity and labor intensity has limited their widespread use (Takasato et al., 1984; Kakee et al., 1996; Garberg, 1998; Dagenais et al., 2000) in the discovery setting. More recently, a brain homogenate method has been proposed as a surrogate approach to estimating unbound brain concentrations (Kalvass and Maurer, 2002); however, this approach can be limiting if the molecule of interest is known to be restricted to the vascular spaces in the brain (Friden et al., 2010).

Because the majority of preclinical target engagement (RO) has focused on interaction with targets in the CNS, a number of methods have been utilized to understand the distribution of molecules into the CNS. The brain/plasma partition coefficient, $K_{p,brain}$, is the most widely used *in vivo* parameter to assess the extent of CNS distribution and is known to be influenced by the binding affinity of a substrate for proteins in plasma versus various tissue proteins (Kurz and Fichtl, 1983). For compounds undergoing simple passive diffusion, the unbound concentration in tissue will be equal to the unbound concentration in plasma at equilibrium making the steady-state tissue partition coefficient explained using the following equation:

$$K_{p,tissue} = \frac{f_{u,plasma}}{f_{u,tissue}}.$$

Therefore, when plasma and brain unbound fractions are similar, then $K_{p,brain}$ would be equal to unity. Recently, a number of literature studies have utilized the $f_{u,plasma}/f_{u,brain}$ ratio to make predictions of $K_{p,brain}$ in order

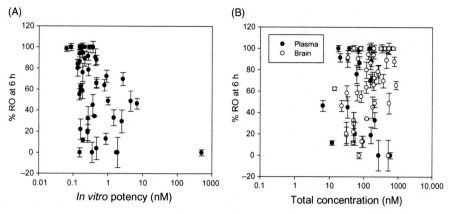

FIGURE 33.5. Relationship between % receptor occupancy (RO) and target potency at the neuropeptide receptor (A) or total concentration in either plasma or brain (B).

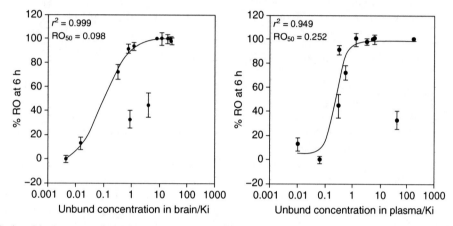

FIGURE 33.6. Relationship between % receptor occupancy (RO) and unbound brain concentration (A) or unbound plasma concentration (B), both normalized for *in vitro* target potency measured of the neuropeptide receptor.

to assess the distribution of CNS targeted compounds; however, issues with over- and underprediction can exist. More recently, it has been suggested that utilizing free fraction data and applying it to measured concentrations to result in an understanding of the unbound plasma concentration ($C_{plasma,u}$), unbound brain concentration, ([brain]$_u$) and relating these ratios ($C_{plasma,u}/C_{brain,u}$) may be a simpler model to assess the CNS distribution of drugs independent of the mechanism(s) involved in its transport across the BBB (Kalvass et al., 2007).

In some cases, *in vitro* binding and total exposure do not always predict the observed outcome during an *in vivo* RO assessment within a SAR (Figure 33.5). In the following case example, a discovery project team targeting a neuropeptide receptor in the brain did not observe any relationship between the *in vitro* potency measured at the target and the measured RO at 6 h, despite similar binding affinities across the molecules (unpublished data). Additionally, there appeared to be no correlation between the measured $C_{brain(total)}$ or $C_{plasma(total)}$ exposures

that might explain the wide range of RO values observed for this target. In an effort to elucidate a relationship between the target and its PK, 25 compounds best representing the SAR space being explored were selected and protein binding values in both rat plasma and whole brain homogenate were measured using equilibrium dialysis. As depicted in Figure 33.6, the free plasma and free brain concentrations, normalized for compound binding affinity to the target, were shown to give rise to a sigmoidal relationship allowing elucidation of the exposure multiples required to achieve an acceptable RO (deemed as >80%) at 6 h. While exploring this relationship, a few molecules were observed to fall outside this dose–response curve suggesting impaired CNS penetration. Further evaluation of these compounds showed that the $f_{u,plasma}/f_{u,brain}$ ratio was less than 0.1 suggesting limited CNS distribution possibly attributed to active efflux or poor passive permeability. While the latter was not experimentally observed *in vitro*, follow-up experiments probing the possible role of P-glycoprotein

(Pgp)-mediated efflux confirmed that these compounds were efficient substrates for the transport pump. These data effectively explained that poor CNS penetration resulted in low RO (Elsinga et al., 2004). While simple in its data interpretation, this example highlights the need to integrate measure brain and plasma levels, along with the target engagement data to understand the degree of CNS distribution, irrespective of the mechanism(s) involved. This data integration was essential to the forward progression of this SAR and ultimate selection of the clinical candidate for development.

While the previous example highlights the role of Pgp-mediated efflux from the CNS, there are a number of other mechanisms associated with limited CNS exposure. Poor passive permeability, non-Pgp-mediated efflux, metabolism, high nonspecific binding, and CSF bulk flow can also impair CNS exposure and should be considered for further evaluation within a given SAR.

33.4.3 Metabolism and Active Metabolites

Not only does one need to be concerned with the "active" component of an NCE under development, but it is also critical to be aware of its metabolic fate. The metabolism of an NCE may produce pharmacologically active metabolite(s) that may retain similar (or even increased) activity that can confound the PD measurements, or even elicit effects at another "site" possibly leading toward unwanted toxicity. Understanding the nature of the metabolic products within drug discovery project can expand the SAR scope, especially if the metabolite lends itself to increased potency. However, if the metabolic clearance is associated with a high, "first-pass" effect, it can limit the extent of NCE available to the systemic circulation. This is not only a potential problem for peripheral targets, but can severely limit the amount available for delivery to the brain, decreasing the opportunity to effectively reach any CNS-related targets.

Many of the currently marketed psychotropic drugs have been shown to form one or more active metabolites during metabolic clearance in humans and/or animals. In some cases, these metabolites are rapidly conjugated and excreted; however, in other cases, they have been shown to achieve plasma and/or brain concentrations within a similar range, and sometimes even higher than that of the parent compound. While it is often unknown whether the *in vitro* potency and selectivity will translate to similar *in vivo* potency, venlafaxine is one example where the major metabolite has been studied and is now currently being marketed at significantly lower prescribed doses. Venlafaxine (Effexor®, Pfizer, New York, New York, USA), an arylalkanol-amine, was first synthesized in the early 1980s and was

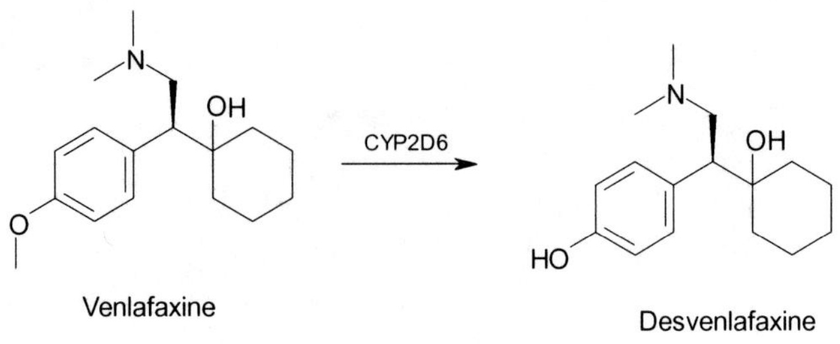

FIGURE 33.7. Major metabolic clearance pathway for venlafaxine.

found to be a serotonin and norpinephrine reuptake inhibitor (SNRI) effective in the treatment of depression. It received Food and Drug Administration (FDA) approval in 1993, with typical daily doses of 75–450 mg/day. During the course of drug development, oxidative O-demethylation of venlafaxine by cytochrome P450 (CYP) metabolizing enzyme 2D6 was found to lead to the production of desvenlafaxine (Figure 33.7). Clinical studies showed this metabolite to be a potent SNRI with similar activity to parent compound (Muth et al., 1991). Desvenlafaxine (Pristiq®, Pfizer, New York, New York, USA]) has recently been approved by the FDA as a treatment for major depressive disorder, with a recommended dose of up to 50 mg/day.

In addition to metabolism concerns within a drug candidate discovery project, additional consideration of metabolism should also be made when working to discover a suitable tracer ligand(s). Given that the properties of a drug candidate are not always identical with the desired properties of a tracer ligand, properties such as high target-site selectivity, high specific/nonspecific binding ratio, high sensitivity, and minimal metabolism (or a thorough understanding thereof) are critical to finding a useful tracer within any program. Additionally, those programs whose target resides within the CNS must also identify molecules with high passive permeability and minimal influence from efflux through transporters, such as Pgp, in order to achieve brain penetration. While a thorough understanding of metabolism is one component of successful tracer development, it is often the least understood due to the complexities of testing. Given this dilemma, a series of examples will be used to illustrate some common metabolic issues uncovered during the development of RO/PET ligands, along with a discussion on some of the common assumptions.

Metabolic stability is one of the hallmarks for an effective imaging agent, especially as one translates RO/EO ligands, initially developed for rodent, into PET tracers utilized for imaging in man. It is critical that radiolabeled metabolites are not present to any appreciable extent at the target tissue, as this will confound the resulting measurement. For CNS-related targets,

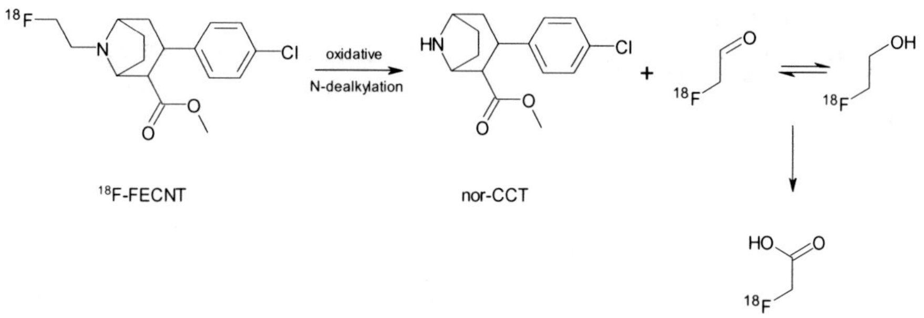

FIGURE 33.8. Speculated oxidative metabolic route for ^{18}F-FECNT N-dealkylation.

metabolites may still be able to cross the BBB, or could even be produced in the brain, thereby competing for the same site at the intended tracer (Dauchy et al., 2008). Additionally, the radiolabel should be placed in a nonlabile site such that a suitable signal can be obtained during the duration of an evaluation. During any discovery tracer effort, careful attention should be paid to species-related differences in metabolic clearance, as metabolic pathways are not always conserved. It is important to point out that an understanding of the metabolites generated is key to developing a successful PET tracer; however, this should not be confused with the desire to have a ligand that does not accumulate in target tissues due to slow clearance or high nonspecific binding of the compound.

During tracer discovery efforts, the tools available to interrogate the metabolism are often limited. Radiolabeled metabolism studies are not typically conducted until Phase I/II of clinical development, which is often too late to determine the liabilities associated with a novel tracer ligand. One can assess metabolism by way of *in vitro* methods (microsomal preparations, hepatocytes, brain homogenates); however, these studies are often performed with unlabeled compound and assume a reasonable *in vitro–in vivo* correlation. Additionally, due to sensitivity issues, they are often conducted at higher substrate concentrations (1–20 μM) that may not be relevant given the microdoses associated with tracers (Lappin and Garner, 2003, 2008). Depending on the chemistry associated with the ligand of interest, radiolabeled synthesis (^{14}C and/or ^{3}H) can be pulled forward in the discovery paradigm to increase confidence in the identification of metabolic pathways, but may not be cost-effective until later-stage discovery efforts are pursued.

A number of assumptions are often made within tracer discovery projects when thinking about metabolic concerns. It is often thought that metabolism of a PET tracer will not be a concern due to the fact that it is administered intravenously. However, despite direct systemic administration and avoidance of any "first-pass" effects, the compound will eventually run through the liver and could be metabolized to a compound that

retains the radiolabel. During studies assessing pathophysiology, disease progression, and effects of therapy in Parkinson's disease, noninvasive imaging of the dopamine transporter (DAT) has been successful in the brain using the ^{18}F-2β-carbomethoxy-3β-(4-chlorophenyl)-8-(2-fluorethyl)nortropane (^{18}F-FECNT) tracer (Goodman et al., 2000). In studies with rodents, an accumulating radiolabeled metabolite was observed and found to be present nonspecifically in various regions of brain resulting in an increase of total activity in the background (Zoghbi et al., 2006). N-dealkylation of ^{18}F-FECNT resulted in the formation of 2β-carbomethoxy-3β-(4-chlorophenyl)-8-nortropane (nor-CCT), which was confirmed from the metabolic characterization of rat brain homogenates following a single infusion of ^{18}F-FECNT. Presumably, the metabolic clearance of ^{18}F-FECNT also gave rise to an equivalent amount of the ^{18}F-2-fluoroacetaldehyde, which could be further oxidized to the acid or be found in equilibrium as the 2-carbon alcohol (Figure 33.8), and thought to be responsible for the accumulating metabolite(s) observed. Since ^{18}F-FECNT was found to be stable when incubated with rat brain homogenates, the data suggest the metabolism occurs in the periphery after intravenous administration followed by metabolite crossing the BBB.

While the nonspecific accumulation of radiolabeled metabolites can provide challenges to imaging data interpretation if using a reference tissue model, the cleavage of fluorinated radiolabels can also pose problems via potential bone uptake of liberated or free fluorine during imaging experiments. In the case of ^{18}F-FCWAY   (N-{2-[4-(2-methoxyphenyl)piperazino]}-N-(2-pyridinyl) *trans*-4-fluorocyclohexanecarboxamide), a radioligand utilized in PET imaging of brain 5-HT1a receptors, defluorination of the radiolabel was observed and found to accumulate in the skull over the course of 2 h (Ryu et al., 2007). This defluorination was determined to be mediated through CYP2E1 and could be inhibited by classic CYP2E1 inhibitors in man such as disulfiram. Cleavage of other radiolabels (e.g., ^{11}C, ^{13}N, or ^{15}O) is generally considered acceptable as long as suitable residence times can be achieved for imaging.

33.5 APPLICATION OF LC-MS/MS TO DISCOVERY NOVEL TRACERS

PET imaging are limited in their application by the existing tools, or tracers, available to assess target engagement as well as biological/disease processes (Burns et al., 2001; Fowler et al., 2003; Guo et al., 2009). The traditional approach for selecting compounds to then radiolabel and test *in vivo* is a lengthy and costly process with no guarantee of success. The application of mass spectrometry to assess tissue levels of compound after microdosing provides a unique high-throughput approach to triage compounds prior to investing in radiolabeling chemistry and *in vivo* scan time. Development of the LC-MS/MS method and *in vivo* assessment is approximately a day and a half (in comparison with at least 3–4 weeks for radiolabeling and *in vivo* scanning and data analysis) to attain data indicating if the compound is likely to enable a preclinical RO assay and potential a PET tracer. The latter is often a higher hurdle as additional parameters must be in place including little to no interference by radiometabolites and appropriate tissue uptake or the radiotracer. Often, as a result of these additional considerations, the tracer enabling the preclinical target engagement measures is identified first, and additional work is required to identify the translatable clinical PET tracer.

33.5.1 Characterization of the Dopamine D₂ PET Tracer Raclopride by LC-MS/MS

In vivo bioanalysis of raclopride will be used to exemplify the type of data that can be collected to assess the potential a compound has to function as a tracer. More often than not, our results under these circumstances suggest the majority of compounds examined under microdosing conditions do not perform as tracers. A number of physicochemical and inherent chemical properties converge such that a molecule can differentially distribute toward a target-rich tissue in a specific manner. These include molecular weight, target affinity, target selectivity (built into compound, or biology), lipophilicity, brain uptake/permeability, and nonspecific binding. While generalizations have been made regarding such properties, the structural features that confer suitable tracer-like properties are poorly understood (Wong and Pomper, 2003) (Waterhouse, 2003). All of these properties become even more important as the expression of the target decreases. Thus, more highly expressed targets are easier to approach and do not require an optimized tracer. The dopamine D₂ receptor is expressed in rat at approximately 515 fmol/mg protein in striatal tissue (Barth et al., 2006). LC-MS/MS assays have been established for targets both in the CNS and

in the periphery with expression levels as low as 5–10 fmol/mg protein. Application of this technique has been successful in identifying tracers for CNS and peripheral GPCRs, intracellular enzymes (as they related to EO), ion channels, and allosteric antagonists. Following intravenous microdosing of raclopride, and collection and processing of appropriate tissues as described above, the following parameters can be assessed that are analogous to PET parameters obtained in rodent scans: nonspecific binding *in vivo* (i.e., can the compound differentially distribute to match the target biology), tissue exposure or % standardized uptake value (SUV) (i.e., does a high enough proportion of the injected dose reach the tissue such that enough radioactivity would be available to make measurements under low mass dose conditions), and kinetics (i.e., depending on if a labeling site exists or would be introduced, does the uptake and washout from relevant tissues occur in the appropriate time frame relative to the desired radioisotope).

Assessing different microdoses of raclopride (3, 10, and 30 μg/kg) leads to a differential distribution in all cases (Figure 33.1A,B); however, with the goal of achieving the largest signal-to-noise and trying to label a small percentage of the receptor population as possible, the 3 μg/kg dose yields a ratio of about 4.5, or a binding potential of 3.5 (ratio-1) (Figure 33.1A,B). Eventually, a low-enough dose would be difficult to detect and accurately quantify, especially in the null region. Thus, the desire to have a large signal-to-noise is balanced by the need for reliable detection and quantification. In addition to dose, the survival interval, or the time between intravenous tracer injection and sacrifice, can be manipulated (Figure 33.1C) (Chernet et al., 2005; Barth et al., 2006). At early time points following intravenous administration of the tracer, free nonspecific tracer has not had enough time to clear from both the interstitial tissue space and plasma compartments to measure a differential distribution, or a robust differential distribution where the measured difference of the tracer accumulation 2.5- to 3-fold greater in the total binding region. With time, the tracer clears more rapidly from the null region relative to the total binding region revealing the potential of the compound to differentially distribute (Figure 33.1C) (Chernet et al., 2005). If a differential distribution matching the target biology is not attained within an hour following intravenous administration, then it is unlikely the compound will function as a tracer. When this situation occurs, it is most likely the result of higher-than-desired nonspecific binding and/or a lower-than-desired kinetics/tissue clearance. Raclopride already demonstrates a differential distribution at 5 min and the signal-to-noise continues to increase with increasing survival interval (time between intravenous tracer injection and sacrifice) until one of two things happens (Figure 33.1C,D): (1) the inability to reliably detect and quantify tracer levels in the null region; (2) collapse of the window (the difference between the total binging and null tissue tracer levels) as all tracer clears from the system. After establishing that a potential tracer dose differentially distribute, matching the target biology, there is a need to demonstrate that this results from specific binding to the intended target. Pharmacologically, validation of a tracer requires a convergence of evidence by demonstrating the ability of structurally distinct and selective compounds to dose-dependently reduce tracer levels in appropriate tissue regions and not in others (Figure 33.4). Additionally, if available, target knockout mice when compared with wild-type mice can demonstrate the abrogation of the differential distribution. This is a compelling experiment to conduct that demonstrates the tracer is indeed binding the assumed target.

By evaluating time points as early as 5 min and as long as 1 h, the kinetics under the microdosing paradigm suggest raclopride has appropriate brain kinetics for ^{11}C labeling given the 20-min half-life of ^{11}C. Additionally, at these time points, we are able to understand the proportion of the injected dose reaching the tissue. At 5 min postdose, excellent brain tissue uptake is seen with approximately 300% SUV (Figure 33.1C), which is calculated as the ratio of the tracer level in tissue (ng/g) to the injected tracer dose (μg/kg).

To summarize, the LC-MS/MS microdosing paradigm provides the following information: (1) ability to distribute toward target (low nonspecific binding suitable for target expression level); (2) attribute distribution to specific target binding via blockade studies and/or knockout mice; (3) appropriate kinetics (rapid uptake and washout on timescale that matches desired isotope); and (4) tissue uptake. This set of data can be collected within a week on a single compound.

33.5.2 Discovery of Novel Tracers

Using the LC-MS/MS approach, a number of novel PET tracers that also have enabled preclinical RO assays have been discovered and recently published including CB1 inverse agonist (Yasuno et al., 2008; Terry et al., 2009) and a novel kappa opioid receptor antagonist (in press). Figure 33.3 illustrates the LC-MS/MS analysis of four structurally distinct, candidate CB1 PET tracers under "baseline" and blocked conditions. Recognize that each time point for baseline/vehicle and blocked conditions is an individual group of animals. From this data set, it is apparent that all four potential candidates have appropriate kinetics in rat, with rapid uptake and washout occurring by 2 h posttracer dose. After 15 min

postdose, all potential tracers have reasonable uptake with 70% SUV (Figure 33.3D) to 160% SUV (Figure 33.3B). However, the tracer in Figure 33.3A,B appears to have more suitable SUV values (greater than 100% at earlier time points—in this case 15 min). All potential tracers demonstrated specific binding in rat cortex when comparing vehicle with treated/blocked animals. Compounds in panels A and B have both the highest uptake and larger signal windows (comparison of total with nonspecific binding). Panel C appears to have slightly slower kinetics, with peak uptake taking place near 30 min postdose, and panel D does not appear to have sufficient uptake. It would be useful to measure uptake at earlier time points such as 5 min. However, the data clearly demonstrated that these compounds have the potential to be translatable PET ligands—albeit without addressing radiometabolites or the feasibility of labeling chemistry.

LC-MS/MS assessment for potential tracers can extend beyond target engagement into PD markers. Animal models can be employed to screen for tracers that distribute and track biology. Examples of areas where this can be expanded, based on tracer availability and positive translational science, include tracers for apoptosis, metabolism, inflammation, and amyloid deposition. Such results can be validated and reconciled with techniques such as autoradiography and immunohistochemistry. Additionally, as PET ligands are discovered and clinically validated to assess such conditions, they can be reverse translated into preclinical LC-MS/MS based models—having higher throughput to triage future drugs and their doses to move forward into more labor-intensive and costly studies.

33.6 NONINVASIVE TRANSLATIONAL IMAGING

Thus far, the discussion of imaging has been focused on nuclear imaging technologies (i.e., PET or single-photon emission computerized tomography [SPECT]) for the monitoring of customized radioligands for the purpose of kinetic, microdosing studies of target distribution and engagement. While these modalities provide high sensitivity for target engagement and PK/PD modeling, the field of anatomical, functional, nuclear, and molecular imaging encompasses a broader collection of technologies. It is often useful when considering the use of imaging to understand some of the fundamental principles, strengths, and weakness of each technique. It is important to note that the physics governing the sensitivity and contrast (as it pertains to both exogenous and endogenous contrast mechanisms) remains the fundamental difference in the methods. Further, these differences in sensitivity and contrast mechanisms provides the basic justification for a multimodal imaging laboratory—in that the most appropriate imaging method can be deployed to answer the biological question of interest.

PET is an imaging technique that maps the location of positron-emitting radionuclides injected into living subjects. The distribution, as discussed throughout this chapter, is influenced by both physiological (e.g., blood flow, permeability, necrosis) and biological (e.g., target expression and accessibility) features of the tissue. The probe may be targeted (e.g., ^{11}C-raclopride) or related to a biological process (e.g., glucose metabolism by ^{18}F-fluorodeoxyglucose [FDG]). Regardless, the scan relies on the same physics, and images are reconstructed using common principles.

Among those common radionuclides used in PET imaging, the most widely used are ^{18}F and [11]C. The former, ^{18}F, balances medium energy with relatively short half-life, making it a good choice for an imaging agent. [18]F is used in the glucose metabolism probe [^{18}F] FDG and the proliferation probe [^{18}F]fluorothymidine (FLT). Carbon-11, on the other hand, can be substituted in small molecules without altering the chemical structure to enable short-term imaging of biodistribution, target engagement, and metabolite monitoring. It should be noted that for the successful implementation of PET into a drug discovery and development platform, there are significant costs and logistical challenges. Beyond instrument and facilities costs, the production of a radiolabeled ligand or small molecule requires significant radiochemistry resources. For that reason, to date, many of the uses of PET imaging have been for the monitoring of treatment response by the [^{18}F]FDG and [^{18}F]FLT (Gambhir, 2002; Kelloff et al., 2005a,b; Willmann et al., 2008a,b). A representative [^{18}F]FDG scan of a mouse bearing a flank tumor xenograft is shown in Figure 33.9. It should be noted that for [^{18}F]FDG there is no tumor "targeting" of the probe; rather, it is the tumor's reliance on glucose and glycolytic metabolism that results in high levels of [^{18}F]FDG uptake.

The PET scanning procedure does not detect the emitted positron; rather, following decay, the positron (which is an antielectron) traverses tissue, scatters, loses energy (primarily via coulomb interactions). Upon the loss of kinetic energy, the positron transiently interacts with an electron, ultimately resulting in an annihilation event, which produces a pair of coincident, 0.511-MeV photons (gamma rays). The PET scanner is designed as a ring of crystals that detects the photon pair. Coincidence refers to the fact that the annihilation photons are released and travel in opposite directions (i.e., $180° \pm 0.25°$), so that an "event" is defined as paired detection at opposing detectors in the ring within a

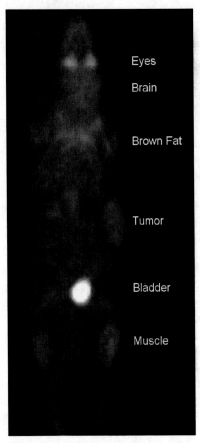

FIGURE 33.9. A representative whole-body FDG PET image of a mouse bearing a flank xenograft tumor. Example regions with high FDG uptake are noted.

predetermined timing window (typically <5 ns). The sensitivity of the PET scanner is then dictated by several factors, including geometry, detector efficiency, and detector/electronic dead-time. There is also a fundamental limit to the resolution of a PET scan in that the production of the coincident photon pair is the detected event; thus, spatial localization is not at the site of positron decay. The distance that a positron travels in tissue is primarily related to the energy of the positron and thus can be an important consideration in the selection of an isotope for labeling. Specifically, the image blurring associated with ^{18}F-probes results from the mean range of approximately 0.5 mm in tissue. Levin and Hoffman (1999) provide a detailed and thorough discussion of the implications for imaging; however, sufficed to say that the research must understand that most preclinical scanners have an imaging resolution of 1–2 mm and clinical scanners in the range of 4–8 mm, which is dependent on the scanner and the probe.

PET data are acquired either after a predetermine period of radiotracer distribution in the subject (commonly referred to as a static scan) or dynamically. The static scanning procedure carries the assumption that at

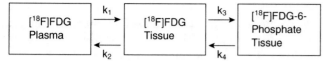

FIGURE 33.10. A compartmental model of $[^{18}F]$fluorodeoxy-glucose (FDG) *in vivo*. From the blood pool (plasma) the tracer is transported into tissue via glucose transporters (time constant k_1 and the reverse constant k_2) and then phosphory-lated via hexokinase to $[^{18}F]$FDG-6-phosphate (k_3). Often, due to low levels of glucose 6-phosphatase, the dephosphorylation of $[^{18}F]$FDG-6-phosphate (k_4) is considered negligible. The uptake of the $[^{18}F]$FDG is proportional to the metabolic demands of the tissue.

the time of the scan postinjection of the radiotracer, the blood pool signal is not the major contributor of signal (unless, of course, blood signal is desired for perfusion measurements). Dynamic scanning is performed during the distribution of the radiotracer; that is, the scanning is performed during the injection phase and continuously while the tracer distributes through the blood compartment to the tissue compartment. As an example, consider the basic compartmental distribution model for glucose metabolism as measured using $[^{18}F]$FDG shown in Figure 33.10. From the blood pool, the $[^{18}F]$FDG is transported into tissue/cells with a rate constant k_1 wherein it is phosphorylated by hexokinase (k_3) and trapped—further catabolism is not possible without the oxygen atom on the C-2 position. ^{18}F decays to ^{18}O with a half-life of 109.8 min and thereafter is metabolized. In the static imaging procedure, the phosphorylated form, $[^{18}F]$-6-phosphate, is thought to reflect the glucose metabolism within the tissue and static scan times are chosen (typically 45–75 min) such that this form is the predominate source of signal. It is important to note that specifically for $[^{18}F]$FDG, the dephosphorylation of $[^{18}F]$-6-phosphate (k_4 in Figure 33.10) is an insignificant contributor due to the low level of glucose-6-phosphatase in many cells, including tumors and brain, but excluding the liver and kidney cortex. Although more time-consuming, the true measure of glycolytic rate in tissue by $[^{18}F]$FDG PET requires dynamic scanning (Phelps et al., 1979; Reivich et al., 1979; Phelps et al., 1983). In cases wherein metabolism and excretion of the radiotracer may play a more significant role, one must more carefully consider the kinetics of tracer states; and static scanning may yield the combination of multiple biological states.

Dynamic scanning monitors changes in tracer uptake in tissue relative to changes in the blood compartment and potential other nontarget tissues, enabling estimation of the kinetic parameters (e.g., k_{1-4} in Figure 33.10). The situation is analogous for other radiolabeled probes or ligands, wherein the creation of a compartmental model may be necessary to fully

appreciate the kinetics and role of each of the biological constituents of the system. Of particular interest in imaging applications is that this analysis can be conducted nondestructively thereby enabling studying changes in kinetics over the time in disease or as a result of treatment. In addition, the imaging data allow for regional (whole organ/lesion or organ/lesion segments) or voxel-by-voxel analysis that can better characterize the heterogeneity of disease. This is useful in oncology imaging applications wherein inclusion of both viable and necrotic regions in the analysis can reduce sensitivity and affect interpretation. In other diseases, such as those of the CNS, region-specific analysis is critical as compound distribution as well as receptor expression will have spatial dependence.

For the duration of the PET scan, which can last 5–90 min or longer depending on experimental design and static versus dynamic acquisition mode, hundreds of millions of events are recorded. Following, the data are reconstructed to form images that map the events to a spatial location. Thus, the intensities of the image elements are proportional to the radioactive tracer concentration in that region of the subject. Clearly, dynamic scanning is performed to extract the time constants and compartmental concentrations. Static imaging analysis, on the other hand, is typically displayed and quantified as the SUV, which was introduced previously in this chapter (Zasadny and Wahl, 1993) or as the percent injected dose per unit tissue (% ID/g).

SPECT differs from PET in that the detected radiation is the gamma ray emitted during the radioactive decay of the radionuclide. In practice, the imaging is performed often by rotating a detector (or multiple detectors to increase count rate and decrease scan time) around the subject collecting the three-dimensional (3D) information necessary to reconstruct a volumetric image. Unlike PET, which utilizes the coincidence of gamma rays in opposing detectors, SPECT scanners must use physical collimation to reduce scatter. The collimation reduces the detection sensitivity, and thereby SPECT often requires injections of higher doses of radioactive tracers and longer scans to improve the signal-to-noise of the image. In favor of SPECT, however, is the availability of low-energy, long-half-life isotopes that would be required for the study of ligands or antibodies with long biological half-life *in vivo*. Ultimately, the decision to use either a PET or SPECT radiotracer in a preclinical or clinical study would be dictated by the biological process or compound of study. One can imagine situations wherein the use of a long-lived isotope would enable serial scanning over several days provided that the distribution to target tissue over background was adequate. On the other hand, there are situations wherein a short-lived isotope is not only relevant to the biological process but may also allow for more than one scan to be conducted in a short period of time. Saleem et al. (2003) report the use of ^{11}C-labeled temozolomide for two purposes: to study the mechanism of *in vivo* metabolic activation and to evaluate PK in tissue compartments (tumor, normal, and plasma). The study utilized a dual-labeling strategy where temozolomide was radiolabeled with ^{11}C separately in the 3-*N*-methyl and 4-carbonyl positions.

A recent optical imaging method has been introduced that utilizes the release of the high-energy positron in tissue and the subsequent production of visible light consistent with Cerenkov radiation (Robertson et al., 2009). Cerenkov radiation is produced when high-energy charged particles (>263 keV) travel through tissue at a speed that exceeds the speed of light in that tissue (i.e., depending on the refractive index of the medium). There are several radionuclides that satisfy the condition and would produce visible light. Radionuclides used for radiotherapy also qualify, such as iodine-131 and yttrium-90 (half-life of 64 h and energy of 682 keV). Despite the complexity of light absorption and scatter, emerging studies demonstrate that tomographic imaging is feasible (Li et al., 2010). Cerenkov luminescence imaging (CLI) and tomography (CLT) may provide in the future alternate detection methods for imaging radiotracers.

Beyond the nuclear imaging technologies, it is well regarded that magnetic resonance imaging (MRI) is perhaps the most versatile and provides exquisite anatomical images as well as functional data on tissue perfusion and diffusion. However, MRI is technically challenging, expensive, relatively insensitive for many molecular markers (10^{-3}–10^{-4} mol/L), and limited to nuclei with appropriate spin systems, such as hydrogen, fluorine, and phosphorous. That said, because of the versatility and various contrasts that can be achieved in images, MRI offers the advantage of possibility of acquiring various scans in a single session and has been a key technology in the drug discovery and development process (Evelhoch et al., 2000; Gillies et al., 2000). Predominantly, magnetic resonance image contrasts and weightings can be achieved by probing water molecules (hydrogen nuclei) in tissue, which are perturb from their equilibrium state by the application of radiofrequency energy to the system. During the return to equilibrium, the protons transfer energy to the local environment (the "lattice" in MRI terms) or to adjacent protons ("spins") measured as the T_1 and T_2 relaxation time constants, respectively. In addition, the signal of the image can be weighted to the amount of water in the tissue (M_0). There are a multitude of other contrasts that can be achieved in an MRI session, such as diffusion-weighted imaging (Le Bihan, 1991; Le Bihan

et al., 1991; Moseley et al., 1991), magnetization transfer (Balaban and Ceckler, 1992; Wolff and Balaban, 1994), functional MRI (fMRI) (Moonen et al., 1990; Matthews et al., 2006), and dynamic contrast-enhanced (DCE) MRI (Taylor et al., 1999; Padhani and Husband, 2001; Hayes et al., 2002). For techniques like DCE MRI or to alter the contrast in tissue, exogenous agents can be administered intravenously. Most commonly, MRI contrast agents such as gadolinium chelates are used, such as gadopentetic acid, gadodiamide, and gadoteridol.

Finally, in addition to the imaging applications, nuclear magnetic resonance (NMR) visible nuclei can be spectroscopically probed (magnetic resonance spectroscopy [MRS]). For example, a 1H MRS study can be used to study choline, creatine, myo-Inositol, *N*-acetylaspartate, and lipids; and a 31P MRS scan can be used to study phosphocreatine, inorganic phosphates, phosphodiesters, and adenosine triphosphate (ATP) (Dager et al., 2008; Soares and Law, 2009; van der Graaf, 2010). Further, compounds, probes, or ligands that contain an NMR-visible nuclei (e.g., fluorine-19) can be designed and used to monitor tissue distribution (Wolf et al., 1998). The complex chemical exchange of magnetization that occurs can also be leveraged to enable chemical exchange-sensitive techniques, such as chemical exchange saturation transfer (CEST) and paramagnetic CEST (PARACEST) (Zhang et al., 2003; Woods et al., 2006; Sherry and Woods, 2008).

Computed tomography (CT) and X-ray are cost-effective modes and effective technologies for anatomical imaging. The image contrast is generated by the tissue absorption properties of the transmitted X-rays during the scanner procedure. For this reason, unlike MRI, CT is limited in the available contrast mechanisms. Further, only iodinated intravenous contrast agents are available for CT, which are nonspecifically distributed to tissues to alter image properties. For tumor imaging, contrast-enhanced CT is routinely used to aid in diagnosis, localization, and monitoring of treatment effects. Of all the imaging modes, CT and X-ray are superb for the visualization of bone defects. However, as a molecular imaging tool, CT (and X-ray) remains limited due to insensitivity, even with the addition of iodinated contrast agents. In modern preclinical and clinical nuclear imaging scanners, CT is included in the same gantry to provide both anatomical registration as well as tissue attenuation correction in a single imaging session.

Ultrasound is a powerful and safe technology, and has applications in cardiovascular (e.g., blood flow kinetics in neck or extremities) and abdominal (e.g., OB/GYN) imaging applications. Ultrasound images are created following detection of reflected high-frequency sound waves off tissues. A range of frequencies (1–40 MHz) is used to improve image resolution and depth. To improve image quality and aid in the calculation of blood flow, microbubble contrast agents can be used to alter the echo properties tissue they distribute to (typically blood vessels or leaking into tumor tissue). Recently, some groups have investigate targeting the microbubbles by surface labeling (Willmann et al., 2008a,b, 2010) and size discrimination (Streeter et al., 2010).

All of the aforementioned technologies—MRI, PET/SPECT, CT/X-ray, and ultrasound—are employed clinically. Future advancements are primarily in the development of new imaging methods or probes/tracers or in the standardization of the most promising methods (e.g., DCE-MRI, FDG-PET). In addition, currently limited to use in preclinical research are the powerful optical imaging technologies, which include fluorescence and bioluminescence. These methods are on the cusp of clinical use and may provide new and innovative ways to monitor molecular targets *in vivo*. Both methods offer sensitive ($<10^{-15}$ mol/L) and rapid scanning with relatively low entry cost for new instrumentation. The main limitation of the method is that the absorption and scatter of light in tissue complicates the imaging of deep structures (millimeters); however, the expanded use of near-infrared (NIR) probes is greatly enhancing the imaging methods (Michalet et al., 2005; Hilderbrand and Weissleder, 2010), and clinical applications are also emerging (Kosaka et al., 2009; Te Velde et al., 2010). The light production methods of bioluminescence and fluorescence are different; however, the imaging scanners are based on charged coupled device (CCD) cameras, often cooled, such as the Xenogen scanners (Caliper Life Sciences, Alameda, CA).

The production of light for bioluminescence imaging is the result of a chemiluminescent reaction between an enzyme (e.g., firefly luciferase or sea pansy *Renilla*) and its substrate (e.g., D-luciferin for luciferase or coelenterazine for *Renilla*). *In vivo*, in the presence of oxygen, the enzyme-substrate reaction releases photons (peak emissions wavelength 580 nm and 480 nm, respectively for luciferase and *Renilla*) (Kalish et al., 2005). Molecular biology expertise is required to stably express the protein in engineered cell lines (e.g., cancer cells for xenograft studies). Early in the use of bioluminescence imaging for biological research, the primary goal was cell tracking: With no background signal (i.e., only expressing cells bioluminescence), a researcher can sensitively track and monitor cell viable *in vivo*. More recently, numerous examples of engineered probes for the monitoring of protein and gene function have emerged and will continue to advance bioluminescence imaging for both basic and translational research

(Hoffman and Gambhir, 2007; Massoud and Gambhir, 2007).

Fluorescence imaging is based on the absorption of energy from an external excitation source and the secondary release of light that is shifted to another wavelength, typically higher. The first fluorescing protein utilized for imaging applications was green fluorescent protein (GFP), a discovery that earned Drs. Chalfie, Shimomura, and Tsien the 2008 Nobel Prize in Chemistry. GFP is excited in the visible blue band with emission in the green (e.g., a mutant by Heim et al. [1995] noted excitation at 488 nm and emission at 509 nm). To that point, there are many variants to GFP, but the excitation and emission range is fairly similar. Green light undergoes substantial scatter and absorption *in vivo* and thus can be challenging for imaging at depths beyond a couple of millimeters. As a result, there have been mutations to optimize the emission wavelengths and optimization of assays to study disease (Gross and Piwnica-Worms, 2005; Weissleder and Pittet, 2008). Specifically for the purpose of *in vivo* imaging, there has been special interest in NIR probes, which have greater tissue penetration depth due to the longer emission wavelength enhancing imaging of deep structures (Michalet et al., 2005; Hilderbrand and Weissleder, 2010), and clinical applications are also emerging (Kosaka et al., 2009; Te Velde et al., 2010).

The detection schemes for fluorescence have been similar to bioluminescence, wherein the amount of light emitted is quantified. However, fluorescence imaging is burdened by the fact that most biological tissues also have autofluorescence. Simply put, when exposed to the excitation light, some tissue and cellular constituents fluorescence (e.g., mitochondria, lysosomes, and collagen), and the term "auto" implies that it is not the fluorophore signal but rather the endogenous structures that are fluorescing. This background autofluorescence complicates the imaging procedure, and spectral filters or mathematical models are often employed to remove unwanted signal. Optical imaging of fluorescence and bioluminescence remain primarily research tools; although there are emerging clinical examples. For example, one can envision translating LC-MS/MS discovered and optimized ligands that are both radio- and fluorolabeled for PK/PD modeling from animal studies into humans.

33.7 CONCLUSIONS AND THE PATH FORWARD

Target modulation is a fundamental aspect of *in vitro* drug discovery, which is carried into *in vivo* studies to confirm biological engagement that results in drug treatment effects. The ability to translate the study of target modulation, engagement, and dynamics across the discovery and development phases enables testing of the mechanistic linkage between disease biology and therapeutic intervention. Quantitative preclinical pharmacology, integrated with PK/PD testing and modeling, can guide compound and dose selection and, in some cases, may enable the development of translational imaging biomarkers for both preclinical and clinical research.

Imaging technologies continue to find new ways to influence the practice of medicine and the development of new therapeutics. Interestingly, just as the initial adoption of the technologies themselves was weighed carefully relative to the cost and benefit to the patient (Hillman et al., 1984), the emergence of novel imaging applications draws continued deliberation and protracted acceptance, especially as it pertains to the evaluation of drug therapies. Although there could be several reasons governing the inclusion of imaging assays in clinical trials, anecdotal evidence suggests that a few of the main reasons include lack of understanding of the imaging technologies themselves, general uncertainty in the value of biomarkers, and the historical comfort and tradition in the design and execution of clinical trials.

There can be little disagreement that ability to noninvasively see within the body is of transformative value for the practice of medicine as well as applied research. Further, the ability to visualize target engagement, modulation, and biological consequence can be of incalculable value to the pharmaceutical industry and, ultimately, to patients. However, the clinical trial process for the evaluation of novel drugs is governed by stringent guidelines and escalating costs (DiMasi et al., 2003, 2010; Frank and Hargreaves, 2003; DiMasi and Grabowski, 2007), and thereby, the role of imaging must be considered carefully. Ultimately, these technologies are not new to the medical community nor are they foreign to the drug development process; however, the deployment of novel biomarkers (including imaging) requires thorough characterization, founding in strong science, and a commitment that these monitoring techniques are fit to answer the posed questions.

REFERENCES

Balaban RS, Ceckler TL (1992) Magnetization transfer contrast in magnetic resonance imaging. *Magn Reson Q* 8:116–137.

Barth VN, Chernet E, Martin LJ, Need AB, Rash KS, Morin M, Phebus LA (2006) Comparison of rat dopamine D2 receptor occupancy for a series of antipsychotic drugs measured using radiolabeled or nonlabeled raclopride tracer. *Life Sci* 78:3007–3012.

Benet LZ, Hoener BA (2002) Changes in plasma protein binding have little clinical relevance. *Clin Pharmacol Ther* 71:115–121.

Biomarkers Definitions Working Group (2001) Biomarkers and surrogate endpoints: Preferred definitions and conceptual framework. *Clin Pharmacol Ther* 69:89–95.

Burns HD, Hamill TG, Eng W-S, Hargreaves R (2001) PET ligands for assessing receptor occupancy *in vivo*. *Annu Rep Med Chem* 36:267–276.

Chernet E, Martin LJ, Li D, Need AB, Barth VN, Rash KS, Phebus LA (2005) Use of LC/MS to assess brain tracer distribution in preclinical, *in vivo* receptor occupancy studies: Dopamine D2, serotonin 2A and NK-1 receptors as examples. *Life Sci* 78:340–346.

Dagenais C, Rousselle C, Pollack GM, Scherrmann JM (2000) Development of an *in situ* mouse brain perfusion model and its application to mdr1a P-glycoprotein-deficient mice. *J Cereb Blood Flow Metab* 20:381–386.

Dager SR, Corrigan NM, Richards TL, Posse S (2008) Research applications of magnetic resonance spectroscopy to investigate psychiatric disorders. *Top Magn Reson Imaging* 19:81–96.

Dauchy S, Dutheil F, Weaver R, Chassoux F, Daumas-Duport C, Courad P, Scherrmann J, De Waziers I, Decleves X (2008) ABC transporters, cytochromes P450 and their main transcription factors: Expression at the human blood–brain barrier. *J Neurochem* 107:1518–1528.

DiMasi JA, Grabowski HG (2007) Economics of new oncology drug development. *J Clin Oncol* 25:209–216.

DiMasi JA, Hansen RW, Grabowski HG (2003) The price of innovation: New estimates of drug development costs. *J Health Econ* 22:151–185.

DiMasi JA, Feldman L, Seckler A, Wilson A (2010) Trends in risks associated with new drug development: Success rates for investigational drugs. *Clin Pharmacol Ther* 87:272–277.

Donohue SR, Krushinski JH, Pike VW, Chernet E, Phebus LA, Chesterfield AK, Felder CC, Halldin C, Schaus JM (2008) Synthesis, *ex vivo* evaluation, and radiolabeling of potent 1,5-diphenylpyrrolidin-2-one cannabinoid subtype-1 receptor ligands as candidates for *in vivo* imaging. *J Med Chem* 51:5833–5842.

Eckelman WC, Reba RC, Kelloff GJ (2008) Targeted imaging: An important biomarker for understanding disease progression in the era of personalized medicine. *Drug Discov Today* 13:748–759.

Elsinga P, Hendrikse N, Joost B, Vaalburg W, Waarde A (2004) PET studies on P-glycoprotein function in the blood-brain barrier: How it affects uptake and binding of drugs within the CNS. *Curr Pharm Des* 10:1493–1503.

Emoto C, Murayama N, Rostami-Hodjegan A, Yamazaki H (2009) Utilization of estimated physicochemical properties as an integrated part of predicting hepatic clearance in the early drug-discovery stage: Impact of plasma and microsomal binding. *Xenobiotica* 39(3):227–235.

Evelhoch JL, Gillies RJ, Karczmar GS, Koutcher JA, Maxwell RJ, Nalcioglu O, Raghunand N, Ronen SM, Ross BD, Swartz HM (2000) Applications of magnetic resonance in model systems: Cancer therapeutics. *Neoplasia* 2:152–165.

Fowler JS, Ding YS, Volkow ND (2003) Radiotracers for positron emission tomography imaging. *Semin Nucl Med* 33:14–27.

Frank R, Hargreaves R (2003) Clinical biomarkers in drug discovery and development. *Nat Rev Drug Discov* 2:566–580.

Friden M, Ljungqvist H, Middleton B, Bredberg U, Hammarlund-Udenaes M (2010) Improved measurement of drug exposure in the brain using drug-specific correction for residual blood. *J Cereb Blood Flow Metab* 31:150–161.

Gambhir SS (2002) Molecular imaging of cancer with positron emission tomography. *Nat Rev* 2:683–693.

Garberg P (1998) In vitro models of the blood–brain barrier. *ATLA* 26:821–847.

Gillies RJ, Bhujwalla ZM, Evelhoch J, Garwood M, Neeman M, Robinson SP, Sotak CH, Van Der Sanden B (2000) Applications of magnetic resonance in model systems: Tumor biology and physiology. *Neoplasia* 2:139–151.

Goodman M, Kilts C, Keil R (2000) ^{18}F-Labeled FECNT: A selective radioligand for PET imaging of brain dopamine transporters. *Neuroimage* 27:1–12.

Gross S, Piwnica-Worms D (2005) Spying on cancer: Molecular imaging *in vivo* with genetically encoded reporters. *Cancer Cell* 7:5–15.

van der Graaf M (2010) *In vivo* magnetic resonance spectroscopy: Basic methodology and clinical applications. *Eur Biophys J* 39:527–540.

Guo Q, Brady M, Gunn RN (2009) A biomathematical modeling approach to central nervous system radioligand discovery and development. *J Nucl Med* 50:1715–1723.

Hargreaves R (2008) The role of molecular imaging in drug discovery and development. *Clin Pharmacol Ther* 83:349–353.

Hayes C, Padhani AR, Leach MO (2002) Assessing changes in tumour vascular function using dynamic contrast-enhanced magnetic resonance imaging. *NMR Biomed* 15:154–163.

Heim R, Cubitt A, Tsien R (1995) Improved green fluorescence. *Nature* 373:663–664.

Hilderbrand SA, Weissleder R (2010) Near-infrared fluorescence: Application to *in vivo* molecular imaging. *Curr Opin Chem Biol* 14:71–79.

Hillman BJ, Winkler JD, Phelps CE, Aroesty J, Williams AP (1984) Adoption and diffusion of a new imaging technology: A magnetic resonance imaging prospective. *AJR Am J Roentgenol* 143:913–917.

Hoffman JM, Gambhir SS (2007) Molecular imaging: The vision and opportunity for radiology in the future. *Radiology* 244:39–47.

Hume S, Gunn R, Jones T (1998) Pharmacological constraints associated with positron emission tomographic scanning of small laboratory animals. *Eur J Nucl Med* 25:173–176.

Kakee A, Terasaki T, Sugiyama Y (1996) Brain efflux index as a novel method of analyzing efflux transport at the blood–brain barrier. *J Pharmacol Exp Ther* 277:1550–1559.

Kalish F, Rice BW, Contag CH (2005) Emission spectra of bioluminescent reporters and interaction with mammalian tissue determine the sensitivity of detection *in vivo*. *J Biomed Opt* 10:41210.

Kalvass J, Maurer T (2002) Influence of nonspecific brain and plasma binding on CNS exposure: Implications for rational drug discovery. *Biopharm Drug Dispos* 23:327–338.

Kalvass J, Maurer T, Pollack G (2007) Use of plasma and brain unbound fractions to assess the extent of brain distribution of 34 drugs: Comparison of unbound concentration ratios to in vivo P-glycoprotein efflux ratios. *Drug Metab Dispos* 35:660–666.

Kapur S, Zipursky R, Jones C, Remington G, Houle S (2000) Relationship between dopamine D(2) occupancy, clinical response, and side effects: A double-blind PET study of first-episode schizophrenia. *Am J Psychiatry* 157:514–520.

Kapur S, Barlow K, Vanderspek SC, Javanamard M, Norbrega JN (2001) Drug-induced receptor occupancy: Substantial differences in measurements made in vivo vs ex vivo. *Psychopharmacology (Berl)* 157(2):168–171.

Kelloff GJ, Hoffman JM, Johnson B, Scher HI, Siegel BA, Cheng EY, Cheson BD, O'Shaughnessy J, Guyton KZ, Mankoff DA, Shankar L, Larson SM, Sigman CC, Schilsky RL, Sullivan DC (2005a) Progress and promise of FDG-PET imaging for cancer patient management and oncologic drug development. *Clin Cancer Res* 11:2785–2808.

Kelloff GJ, Krohn KA, Larson SM, Weissleder R, Mankoff DA, Hoffman JM, Link JM, Guyton KZ, Eckelman WC, Scher HI, O'Shaughnessy J, Cheson BD, Sigman CC, Tatum JL, Mills GQ, Sullivan DC, Woodcock J (2005b) The progress and promise of molecular imaging probes in oncology drug development. *Clin Cancer Res* 11:7967–7985.

Kohler C, Hall H, Ogren SO, Gawell L (1985) Specific *in vitro* and *in vivo* binding of 3H-raclopride. A potent substituted benzamide drug with high affinity for dopamine D-2 receptors in the rat brain. *Biochem Pharmacol* 34:2251–2259.

Korfmacher W (2005) Principles and applications of LC-MS in new drug discovery. *Drug Discov Today* 10:1357–1367.

Kosaka N, Ogawa M, Choyke PL, Kobayashi H (2009) Clinical implications of near-infrared fluorescence imaging in cancer. *Future Oncol* 5:1501–1511.

Kratochwil N, Huber W, Muller F, Kansy M, Gerber P (2002) Predicting plasma protein binding of drugs: A new approach. *Biochem Pharmacol* 64:1355–1374.

Kurz H, Fichtl B (1983) Binding of drugs to tissues. *Drug Metab Rev* 14:467–510.

Lappin G, Garner R (2003) Big physics, small doses: The use of AMS and PET in human microdosing of development drugs. *Nat Rev Drug Discov* 2:233–240.

Lappin G, Garner R (2008) The utility of microdosing over the past 5 years. *Expert Opin Drug Metab Toxicol* 4:1499–1506.

Le Bihan D (1991) Molecular diffusion nuclear magnetic resonance imaging. *Magn Reson Q* 7:1–30.

Le Bihan D, Turner R, Moonen CT, Pekar J (1991) Imaging of diffusion and microcirculation with gradient sensitization: Design, strategy, and significance. *J Magn Reson Imaging* 1:7–28.

Levin CS, Hoffman EJ (1999) Calculation of positron range and its effect on the fundamental limit of positron emission tomography system spatial resolution. *Phys Med Biol* 44:781–799.

Li C, Mitchell GS, Cherry SR (2010) Cerenkov luminescence tomography for small-animal imaging. *Opt Lett* 35: 1109–1111.

Massoud TF, Gambhir SS (2007) Integrating noninvasive molecular imaging into molecular medicine: An evolving paradigm. *Trends Mol Med* 13:183–191.

Matthews PM, Honey GD, Bullmore ET (2006) Applications of fMRI in translational medicine and clinical practice. *Nat Rev Neurosci* 7:732–744.

Michalet X, Pinaud FF, Bentolila LA, Tsay JM, Doose S, Li JJ, Sundaresan G, Wu AM, Gambhir SS, Weiss S (2005) Quantum dots for live cells, *in vivo* imaging, and diagnostics. *Science* 307:538–544.

Moonen CT, van Zijl PC, Frank JA, Le Bihan D, Becker ED (1990) Functional magnetic resonance imaging in medicine and physiology. *Science* 250:53–61.

Moseley ME, Wendland MF, Kucharczyk J (1991) Magnetic resonance imaging of diffusion and perfusion. *Top Magn Reson Imaging* 3:50–67.

Muth E, Moyer J, Haskins J, Andree TH, Husbands GEM (1991) Biochemical, neurophysiological, and behavioural effects of Wy-45,233 and other identified metabolites of the antidepressant venlafaxine. *Drug Dev Res* 23:191–199.

Need AB, McKinzie JH, Mitch CH, Statnick MA, Phebus LA (2007) *In vivo* rat brain opioid receptor binding of LY255582 assessed with a novel method using LC/MS/MS and the administration of three tracers simultaneously. *Life Sci* 81:1389–1396.

Nordström AL, Farde L, Wiesel FA, Forslund K, Pauli S, Halldin C, Uppfeldt G (1993) Central D2-dopamine receptor occupancy in relation to antipsychotic drug effects: A double-blind PET study of schizophrenic patients. *Biol Psychiatry* 33:227–235.

Padhani AR, Husband JE (2001) Dynamic contrast-enhanced MRI studies in oncology with an emphasis on quantification, validation and human studies. *Clin Radiol* 56:607–620.

Phelps ME, Huang S-C, Hoffman EJ, Selin C, Sokoloff L, Kuhl DE (1979) Tomographic measurement of local cerebral glucose metabolic rate in humans with [F-18]2-fluoro-2-deoxy-D-glucose: Validation of method. *Ann Neurol* 6: 371–388.

Phelps ME, Huang S-C, Mazziotta JC, Hawkins RA (1983) Alternate approach for examining stability of the deoxy-glucose model lumped constant. *J Cereb Blood Flow Metab* 3:S13–S14.

Reivich M, Kuhl D, Wolf A, Greenberg J, Phelps M, Ido T, Casella V, Fowler J, Hoffman E, Alavi A, Som P, Sokoloff L (1979) The [18F]fluorodeoxyglucose method for the measurement of local cerebral glucose utilization in man. *Circ Res* 44:127–137.

Richter WS (2006) Imaging biomarkers as surrogate endpoints for drug development. *Eur J Nucl Med Mol Imaging* 33:S6–S10.

Robertson R, Germanos MS, Li C, Mitchell GS, Cherry SR, Silva MD (2009) Optical imaging of Cerenkov light generation from positron-emitting radiotracers. *Phys Med Biol* 54:N355–N365.

Rudin M, Weissleder R (2003) Molecular imaging in drug discovery and development. *Nat Rev Drug Discov* 2:123–131.

Ryu YH, Liow J, Zoghbi S, Fujita M, Collins J, Tipre D, Sangare J, Hong J, Pike VW, Innis RB (2007) Disulfiram inhibits defluorination of ^{18}F-FCWAY, reduces bone radioactivity, and enhances visualization of radioligand binding to serotonin 5-HT$_{1A}$ receptors in human brain. *J Nucl Med* 48:1154–1161.

Saleem A, Brown GD, Brady F, Aboagye EO, Osman S, Luthra SK, Ranicar ASO, Brock CS, Stevens MFG, Newlands E, Jones T, Price P (2003) Metabolic activation of temozolomide measured *in vivo* using positron emission tomography. *Cancer Res* 63:2409–2415.

Sansom LN, Evans AM (1995) What is the true clinical significance of plasma protein binding displacement interactions? *Drug Saf* 12:227–233.

Sherry AD, Woods M (2008) Chemical exchange saturation transfer contrast agents for magnetic resonance imaging. *Annu Rev Biomed Eng* 10:391–411.

Soares DP, Law M (2009) Magnetic resonance spectroscopy of the brain: Review of metabolites and clinical applications. *Clin Radiol* 64:12–21.

Stockmeier CA, DiCarlo JJ, Zhang Y, Thompson P, Meltzer HY (1993) Characterization of typical and atypical antipsychotic drugs based on *in vivo* occupancy of serotonin2 and dopamine2 receptors. *J Pharmacol Exp Ther* 266:1374–1384.

Streeter JE, Gessner R, Miles I, Dayton PA (2010) Improving sensitivity in ultrasound molecular imaging by tailoring contrast agent size distribution: *In vivo* studies. *Mol Imaging* 9:87–95.

Takasato Y, Rapoport SI, Smith QR (1984) An in situ brain perfusion technique to study cerebrovascular transport in the rat. *Am J Physiol* 245:R303–R310.

Taylor JS, Tofts PS, Port R, Evelhoch JL, Knopp M, Reddick WE, Runge VM, Mayr N (1999) MR imaging of tumor microcirculation: Promise for the new millennium. *J Magn Reson Imaging* 10:903–907.

Te Velde EA, Veerman T, Subramaniam V, Ruers T (2010) The use of fluorescent dyes and probes in surgical oncology. *Eur J Surg Oncol* 36:6–15.

Terry GE, Liow JS, Zoghbi SS, Hirvonen J, Farris AG, Lerner A, Tauscher JT, Schaus JM, Phebus L, Felder CC, Morse CL, Hong JS, Pike VW, Halldin C, Innis RB (2009) Quantitation of cannabinoid CB1 receptors in healthy human brain using positron emission tomography and an inverse agonist radioligand. *Neuroimage* 48:362–370.

Van Steeg TJ, Boralli VB, Krekels EH, Slijkerman P, Freijer J, Danhof M, de Lange EC (2008) Influence of plasma protein binding on pharmacodynamics: Estimation of *in vivo* receptor affinities of beta blockers using a new mechanism based PK-PD modeling approach. *J Pharm Sci* 98:3816–3828.

Wadenberg M-L, Kapur S, Soliman A, Jones C, Vaccarino F (2000) Dopamine D2 receptor occupancy predicts catalepsy and the suppression of conditioned avoidance response behavior in rats. *Psychopharmacology* 150:422–429.

Waterbeemd H, Gifford E (2003) ADMET *in silico* modeling: Towards prediction paradise? *Nat Rev Drug Discov* 2:192–204.

Waterhouse RN (2003) Determination of lipophilicity and its use as a predictor of blood–brain barrier penetration of molecular imaging agents. *Mol Imaging Biol* 5:376–389.

Weissleder R, Pittet MJ (2008) Imaging in the era of molecular oncology. *Nature* 452:580–589.

Wilkinson GR (2001) Pharmacokinetics: The dynamics of drug absorption, distribution and elimination. In *Goodman & Gilman's The Pharmacological Basis of Therapeutics*, Hardman JG, Limbird LE, Gilman G, eds., pp. 3–30. McGraw-Hill, New York.

Willmann JK, Lutz AM, Paulmurugan R, Patel MR, Chu P, Rosenberg J, Gambhir SS (2008a) Dual-targeted contrast agent for US assessment of tumor angiogenesis *in vivo*. *Radiology* 248:936–944.

Willmann JK, Nicholas van Bruggen N, Dinkelborg LM, Gambhir SS (2008b) Molecular imaging in drug development. *Nat Rev Drug Discov* 7:591–607.

Willmann JK, Kimura RH, Deshpande N, Lutz AM, Cochran JR, Gambhir SS (2010) Targeted contrast-enhanced ultra-sound imaging of tumor angiogenesis with contrast microbubbles conjugated to integrin-binding knottin peptides. *J Nucl Med* 51:433–440.

Wolf W, Waluch V, Presant CA (1998) Non-invasive 19F-NMRS of 5-fluorouracil in pharmacokinetics and pharmacodynamic studies. *NMR Biomed* 11:380–387.

Wolff SD, Balaban RS (1994) Magnetization transfer imaging: Practical aspects and clinical applications. *Radiology* 192:593–599.

Wong DF, Pomper MG (2003) Predicting the success of a radiopharmaceutical for *in vivo* imaging of central nervous system neuroreceptor systems. *Mol Imaging Biol* 5:350–362.

Woods M, Woessner DE, Sherry AD (2006) Paramagnetic lanthanide complexes as PARACEST agents for medical imaging. *Chem Soc Rev* 35:500–511.

Yasuno F, Brown AK, Zoghbi SS, Krushinski JH, Chernet E, Tauscher J, Schaus JM, Phebus LA, Chesterfield AK, Felder CC, Gladding RL, Hong J, Halldin C, Pike VW, Innis RB (2008) The PET radioligand [11C]MePPEP binds reversibly and with high specific signal to cannabinoid CB1 receptors in nonhuman primate brain. *Neuropsychopharmacology* 33:259–269.

Zasadny KR, Wahl RL (1993) Standardized uptake values of normal tissues at PET with 2-[fluorine 18]-fluoro-2-deoxy-D-glucose: Variations with body weight and a method for correction. *Radiology* 189:847–850.

Zhang S, Merritt M, Woessner DE, Lenkinski RE, Sherry AD (2003) PARACEST agents: Modulating MRI contrast via water proton exchange. *Acc Chem Res* 36:783–790.

Zhang W, Bymaster FP (1999) The *in vivo* effects of olanzapine and other antipsychotic agents on receptor occupancy and antagonism of dopamine D1, D2, D3, 5HT2A and muscarinic receptors. *Psychopharmacology* 141:267–278.

Zoghbi S, Shetty H, Masanori I (2006) PET imaging of the dopamine transporter with ^{18}F-FECNT: A polar radiometabolite confounds brain radioligand measurements. *J Nucl Med* 47:520–527.

34

APPLICATIONS OF iRNA TECHNOLOGIES IN DRUG TRANSPORTERS AND DRUG METABOLIZING ENZYMES

Mingxiang Liao and Cindy Q. Xia

34.1 INTRODUCTION

Drug metabolizing enzymes and drug transporters play important roles in the absorption, distribution, metabolism, and elimination (ADME) of drugs administered to animals and humans. The inhibition and/or induction of drug metabolizing enzymes and transporters by some drugs may significantly alter the pharmacokinetic properties of the drugs and/or co-administered drugs, thus resulting in increased toxicity or decreased efficacy, known as drug–drug interaction (DDI). Drug transporters are generally categorized into the adenosine triphosphate (ATP)-binding cassette (ABC) superfamily and the solute carrier (SLC) family. SLC transporters may mediate the uptake of some drugs across the biological membranes into the cells. Once the drug molecules enter the cells, they may undergo Phase I metabolic reactions catalyzed mainly by cytochrome P450 (CYP) enzymes, and/or Phase II conjugating reactions. ABC transporters can efflux some drugs and/or their metabolites out of the cells in an ATP-dependent process.

Both drug metabolizing enzymes and transporters represent a superfamily of proteins. They are usually classified into families and subfamilies, with each family or subfamily consisting of multiple isoforms. The members of the same family or subfamily can share extensive homology in messenger RNA (mRNA) and protein sequences. For example, the human CYP3A subfamily contains three isoforms, namely CYP3A4,

3A5, and 3A7, which have over 80% identity in their gene and amino acid sequences (Stevens et al., 2003). There is also about 80% gene sequence homology in the three isoforms of hepatic SLC transporters OATP1B1, 1B3, and 2B1. As a result, there is substantial overlap in substrate and inhibitor specificities among the isoforms of drug metabolizing enzymes or transporters, and even between transporters and drug metabolizing enzymes. Additionally, some drugs serve as both the inhibitors and substrates for the individual drug metabolizing enzymes and transporters. For instance, verapamil, a drug used to treat high blood pressure and control angina, is a well-known substrate and also a potent inhibitor for P-glycoprotein (Pgp, a member of ABC transporters) (Pauli-Magnus et al., 2000).

Given the critical importance of drug metabolizing enzymes and transporters in drug discovery and development, there is an increasing need to prospectively define the roles played by those proteins in ADME. However, the complexity and multiplicity of drug metabolizing enzymes and transporters as well as the limitations of the currently available *in vitro* tools have frustrated efforts to understand the interplay between drug metabolizing enzymes and transporters in the pharmacokinetic behavior of drugs and DDI issues. In addition, in terms of drug transporters, the lack of selective probe substrates, chemical inhibitors, and inhibitory antibodies has precluded the elucidation of the contributions of individual drug transporters to drug

ADME-Enabling Technologies in Drug Design and Development, First Edition. Edited by Donglu Zhang and Sekhar Surapaneni.
© 2012 John Wiley & Sons, Inc. Published 2012 by John Wiley & Sons, Inc.

disposition and DDI using the traditional tools. An emerging technology, RNA interference (RNAi) has become an invaluable tool to characterize the function of specific genes in the *in vitro* and *in vivo* model systems (Hannon, 2002; Lee and Sinko, 2006; Lares et al., 2010). In 2003, RNAi was first reported to silence multidrug resistance (MDR) protein 1; since then, it has been extensively used in the characterization of the functions of drug transporters and drug metabolizing enzymes as well as drug resistance (Yu, 2007). Compared with traditional methods, RNAi provides a quick, reversible, efficient, and cost-effective gene silencing approach in a wide range of cells, tissues, and whole animals.

Since the discovery of RNAi in 1998 (Fire et al., 1998), the mechanisms of gene silencing by RNAi and the roles of RNAi have been investigated extensively around the world. As a highly conserved biological pathway, RNAi exists in almost all organisms, such as plants, yeasts, and humans, and involves several essential cellular processes, such as posttranscriptional regulation, cellular defense against viral invasion, and transposon expansion (Hannon, 2002; McManus and Sharp, 2002). The applications of RNAi as both valuable genetic tools and potential therapeutics were promptly identified and developed. In less than one decade following its discovery, RNAi rapidly demonstrated its value in biological research and medical applications, for which the 2006 Nobel Prize in Physiology or Medicine was awarded. Numerous attempts to elucidate the mechanism of RNAi revealed that RNAi is a double-stranded RNA (dsRNA)-dependent process that leads to sequence-specific mRNA degradation and target gene silencing (Hannon, 2002). RNAi involves several sequential steps controlled by RNA-induced silencing complex (RISC) and initiated by dsRNA, which shares a sequence-specific homology with target mRNA. There are typically two methods to generate dsRNA within cells, either through the direct delivery of exogenously synthesized small interfering RNA (siRNA) or the introduction of a DNA construct encoding short-hairpin RNA (shRNA) to be cleaved to siRNAs by an RNAse III-like enzyme known as Dicer in the cells. The functional siRNAs are small RNA duplexes of 21–23 nucleotides (nt). The siRNAs are subsequently incorporated into RISC. RISC experiences an ATP-dependent activation with the unwinding of the double stranded siRNAs into single-stranded RNA. With the aid of siRNA and nucleases in this complex, the activated RISC targets and cleaves the mRNAs complementary to the siRNA (Figure 34.1).

RNAi is a complicated technology relying on multiple molecular and cellular tools. The commercial availability of products from dozens of RNAi suppliers makes this approach practicable and able to be performed routinely. This book chapter provides the general procedures and recommendations for the application of RNAi to evaluate the functions of drug metabolizing enzymes, transporters, and their regulation factors. The advantages, limitations, and throughput modes of different approaches are also discussed.

34.2 EXPERIMENTAL DESIGNS

34.2.1 siRNA Design

The selection of siRNA sequences determines the efficiency and specificity of gene silencing. The typical siRNAs consist of a short (21 base pair [bp]) dsRNA with 2 bp overhangs at each 3′ end. General guidelines for the design of siRNAs have been developed (Elbashir et al., 2002). First, one should locate the coding sequence that starts 75–100 nt downstream from the AUG start codon of the target mRNA. Second, one should scan for all 21-nt sequences, beginning with an adenosine dinucleotide (AA) in that coding area. siRNA sequences with 3′ overhanging urasine dinucleotides (UU) have the highest silencing potency (Elbashir et al., 2001). If there is no suitable sequence identified, siRNA with other dinucleotide overhangs can also be selected; however, it is recommended to avoid guanosine (G) due to the potential for the RNase to cleave the siRNA at the singled-stranded G residues. Third, one should calculate the GC (C = cytosine) contents of the siRNA sequence. The acceptable GC content should be approximately 30–50%. Finally, one should check the specificity of a candidate siRNA by comparing the siRNA sequence with known expressed sequence tags (ESTs) in nucleotide databases, such as BLAST (http://www/ncbi.nlm.nih.gov/blast/). The comparison is used to confirm the uniqueness of a candidate siRNA and to minimize the off-target effects. The siRNA sequences with 15-nt or less homology to other mRNAs in the genome are proper choices. siRNAs that are designed based on the above guidelines have variable successful rates, and reports have shown that typically more than half of randomly designed siRNAs provide at least a 50% reduction in target mRNA levels, and approximately one in four siRNAs provide a 75–95% reduction (Li et al., 2004). Therefore, it is advised to begin with evaluating multiple siRNAs in order to obtain a potent siRNA. Generally, two to four siRNAs to the target mRNA are selected. In addition, using a mixture of multiple siRNAs that target different regions in the same target mRNA is also recommended. Such mixtures were found to increase the silencing potency and overcame the inefficiency caused by the polymorphic target sites without enhancing the off-target effects.

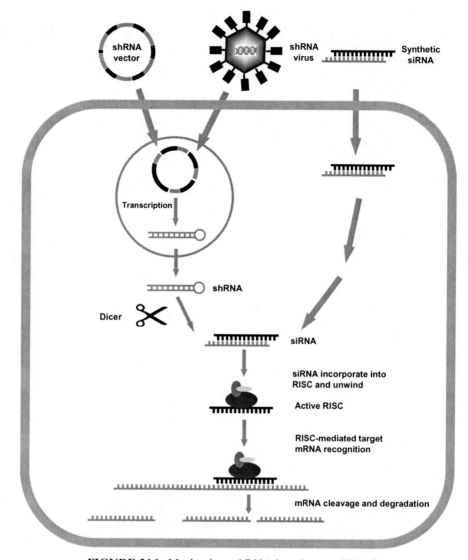

FIGURE 34.1. Mechanism of RNA interference (RNAi).

Currently, a number of siRNA design tools are provided by commercial siRNA suppliers, and public and private institutes, such as Dharmacon, Ambion, Invitrogen, Genelinks, Cold Spring Harbor Laboratories, and the Whitehead Institute. The tools are using siRNA design algorithms to assist researchers to design and select candidate sequences. Some of those design tools are available online and easy to use.

List of the representative websites for siRNA design:

Ambion: http://www.ambion.com/techlib/misc/siRNA_
finder.html

Dharmacon: http://www.dharmacon.com/seedlocator/
default.aspx

GenScript Corp: http://www.genscript.com/

shRNA Designer: http://shRNAdesigner.med.unc.edu

siSearch: http://sonnhammer.cgb.ki.se/siSearch/
siSearch_1.7.html

sFold: http://sfold.wadsworth.org/

34.2.2 Methods for siRNA Production

There are various approaches for siRNA production, including chemical synthesis, *in vitro* transcription, *in vivo* expression of siRNA duplexes, or shRNA using transfected vectors.

34.2.2.1 Chemical Synthesis Several commercial suppliers are providing ready-to-use chemically synthesized RNA oligonucleotides. Custom-synthesized siRNAs are also offered in different formats and amounts to meet the specific needs of researchers. The synthetic single-stranded sense and antisense of RNAs are annealed to form the siRNA duplexes. Chemically synthesized siRNA duplex has many advantages over the other commonly used siRNA production methods, such as (1) high fidelity and yields; (2) reproducibility; (3) availability for immediate use with minimal preparation steps; (4) easy storage and delivery; (5) high-throughput mode capability; and most importantly, (6) various chemical modifications can be introduced into siRNA duplex to increase siRNA stability, specificity, and potency, or to facilitate the detection of silencing effects by adding fluorescent labels.

The chemically synthesized siRNA duplex is readily delivered into most cultured cells, tissues, or animals using various delivery methods, including lipid-based transfection, electroporation, and microinjection. It is noteworthy that chemically synthesized siRNA functions as the most efficient reagent to knock down target genes, and the delivery of siRNA into cells through lipid-based transfection is one of the most attractive methods for current siRNA studies. However, the chemical synthesis of siRNAs is costly, and most of individual synthetic siRNA exhibits less than 80% knockdown efficiency for target genes.

34.2.2.2 In Vitro Transcription

siRNA generated by *in vitro* transcription needs a DNA template and an enzymatic transcription process. With the aid of T7 RNA polymerase, single-stranded sense and antisense RNAs are transcribed from a DNA template encoding a T7 promoter and the target DNA sequence, and then annealed to form siRNA duplexes (Donze and Picard, 2002). Alternatively, a DNA template can be designed to encode an shRNA in one strand that is recognized and processed by the RNAi machinery in cells where the annealing step can be skipped. The vectors and enzymes are commercially available for the generation of siRNA duplex, which employs multiple steps, including transcription and/or digestion.

In vitro transcribed siRNA is low cost and produced on a small scale; thus, this method is great for preliminary screening before the larger-scale production of the suitable duplex. However, as compared with the chemical synthesis method, *in vitro* transcription is labor-intensive, with relatively low yield. The current commercial kits generally produce several nanograms of siRNA duplexes, and thus *in vitro* transcription is not suitable for large-scale production of siRNAs and high-throughout screening. Another limitation is the low purity of the siRNA duplexes due to the early termination of transcription. Therefore, additional purification is required to acquire high-quality siRNA duplexes.

Alternatively, single-stranded sense and antisense RNAs can be produced using *in vitro* transcription with

the DNA template covering a large area of the target gene. The annealed long dsRNAs are cleaved into a mixture of several siRNA by recombinant prokaryotic RNase III or a eukaryotic Dicer (Myers et al., 2003). The mixture of 21-23-nt siRNA duplexes are purified and transfected into the cells for siRNA effect. Using this technique, the efforts for the functional siRNA selections, such as the designing, screening, and sequence verification can be obviated. However, the mixed 21–23-nt siRNA duplexes may induce the off-target effect, and jeopardize the effectiveness and specificity of gene silencing. Moreover, a thorough purification is pivotal for this approach due to the possible interferon response induced by long dsRNAs in mammalian cells (Schiffelers et al., 2004). Nevertheless, an *in vitro* transcription-digestion technique is good for evaluation of the siRNA effect in a fast and inexpensive manner, but not for specific gene function analysis.

34.2.2.3 Polymerase Chain Reaction (PCR) Amplification

The PCR is a powerful *in vitro* technique for rapid amplification of specific DNA from a DNA template. In this approach to producing siRNA, the DNA template contains an H1 or U6 RNA polymerase III promoter, the sequence encoding the sense, antisense, or hairpin for specific siRNA, and an RNA polymerase termination signal (Castanotto et al., 2002). After amplification, the PCR products are introduced into cells. The RNA polymerase III system incorporated in the PCR products transcribes the inserted dsDNA to the preferred siRNA or shRNA.

The PCR amplification approach is more inexpensive and time-efficient than other methods. In addition, the PCR cassettes containing the RNA polymerase III system and the desired dsDNA can also be incorporated into plasmids or viral vectors for shRNA production.

Scrutinous design is critical for the DNA template and functional cassette to generate efficient siRNA or shRNA. Similar to the *in vitro* transcription technique, generation of siRNA and shRNA by PCR amplification cannot introduce chemical modification to increase siRNA stability and specificity, and it is not compatible with a high throughput pattern.

34.2.2.4 Plasmid and Viral Vectors

Both plasmid and viral vectors produce shRNAs and provide important ways to achieve stable silencing effects in cells. The procedures for the generation of plasmids and viral vectors encoding desired shRNA are similar, including (1) identify the sequence of the target gene, (2) design and synthesize the DNA fragment encoding shRNA serving as the template for shRNA, (3) clone the DNA fragment into the appropriate vectors, and (4) verify DNA sequence of the construct by DNA sequencing.

There are various plasmid and viral vectors for shRNA production available commercially.

The plasmids constructed through the above procedures are ready to be transfected into the target cell lines. The resistance genes, such as neomycin or zeomycin resistance genes, incorporated into the plasmids allow for the selection of the transfected cells. Plasmids typically utilize an RNA polymerase III promoter system to produce shRNAs in cells. Thus, the plasmid-based shRNAs are comparable to siRNAs for long-term silencing in mammalian cells and can be employed in several types of cells. However, in contrast to the methods to generate siRNA directly, the creation of shRNA using plasmid vectors is time- and labor-intensive. Exacting design, comprehensive experimental manipulations, and meticulous validations are required.

Alternatively, in terms of viral vector techniques, additional viral amplification is required. The constructed vector containing the desired shRNA sequence and the helper plasmid encoding critical components of the virus, including the viral envelop proteins, are cotransfected into a virus packaging cell line, such as HEK293 cells. The viral genomes are produced and packaged in the packaging cells, and then the viral particles are released into the culture medium. The medium is collected, purified, and assessed for viral titer. Finally, the viral particles are used to transduce the target cells for gene silencing. There are several types of viral systems employed for shRNA expression, including retrovirus, adenovirus, and lentivirus (Kumara and Clarkeb, 2007). Retrovirus can be integrated into the host genome of dividing cells. Similarly, lentivirus can also be integrated into the host genome. The lentiviral system is more complex than the retroviral and is able to infect a broad range of host cells, including nondividing and terminally differentiated cells such as hepatocytes and macrophages. In contrast, the adenoviral system is a nonintegrating virus whose genome contains a single dsDNA fragment. Similar to the lentivirus, the adenovirus is also able to infect both dividing and nondividing cells. However, unlike lentivirus, the adenovirus is maintained as free episomes in the cells.

As a consequence, because of its highly efficient delivery, virus-mediated gene silencing is an appealing approach for gene knockdown in primary cells and other cells refractory to transfection. This very promising approach can result in a long-term suppression of target genes. However, caution should be given when working with viruses. A biosafety level of BL-2 is required for working with the viral systems. Furthermore, the viral vector method is an extremely time- and labor-intensive process that may span several weeks and months to generate effective siRNA. Well-designed optimizations are required for the procedures, including sequence design, virus amplification, and transduction.

In summary, the various methods for siRNA production each have their own advantages and limitations. Chemical synthesis produces the most accurate and specific siRNA, and adapts to high-throughput mode. In addition, chemical modification can be introduced to enhance the specificity and stability of siRNA. However, synthetic siRNA is expensive, and its predictability for knockdown efficiency is low. The *in vitro* transcription and digestion method is low cost but lacks accuracy. The shRNA expressed from plasmids and viral vectors is capable of silencing gene functions over a long term. However, challenges remain for the tedious and time-consuming procedures. To date, virus-mediated shRNA is the most attractive approach to silencing gene functions in nondividing and hard-to-transfect cells.

34.2.3 Controls and Delivery Methods Selection

34.2.3.1 Control Selections Owing to the complexity of siRNA, multiple appropriate controls should be included in a well-designed siRNA study. Several controls are recommended for siRNA studies in order to assist data interpretation and experimental troubleshooting. These controls (positive and/or negative) can be employed to monitor the delivery efficiency, the silencing efficiency, and nontargeting effect of siRNA.

Commonly used controls in siRNA studies are as follows:

Untreated Cells. Untreated cells, representing the normal cell population, should be included in every experiment as negative controls. The expression and function of target genes in the untreated cells can be used as the standard for comparisons with other samples, including other negative and positive controls. In order to monitor the difference between different samples, such as cell numbers, the expression levels of target genes can be normalized to housekeeping gene expression levels. For example, the target gene expression in treated cells is calculated as the ratio of the target mRNA : housekeeping gene mRNA ratio for the treated samples to that for the untreated cells (Eq. 34.1):

$$\text{Target Gene Expression (\% of Control)} =$$
$$(\text{Target:Housekeeping})_{\text{treaetd}} / \qquad (34.1)$$
$$(\text{Target:Housekeeping})_{\text{untreated}} \times 100\%.$$

Transfection without siRNA. Similar to the cells in the treated group, the cells in this control group are exposed to the transfection reagents and processes but in the absence of any siRNA. The transfection reagents and processes may cause significant cell toxicity, thus disturbing the normal functions of the treated cells. This control is included to evaluate the extent of the disturbance caused by the transfection.

Transfection with Nonfunctional/Nontargeting siRNA. The nonfunctional/nontargeting siRNAs are transfected into target cells as negative controls. One of the commonly used controls is scrambled siRNA, which is synthesized by scrambling the nucleotide sequence of the working siRNA. The sequence of the scrambled siRNA is compared to the gene database by performing a BLAST to ensure that it is deficient in homology to any genes in the target genome, and therefore the scrambled siRNA will not be able to knock down any gene expression in that species. RISC-free siRNA represents another type of control siRNA that loses its capacity to interact with RISC. RISC-free siRNAs may carry a fluorescent label through chemical modification to facilitate the knockdown detection process. In addition, a RISC-free control can serve as a simple control to assess delivery efficiency by monitoring the fluorescence signals using flow cytometry or confocal microscopy.

Delivery Efficiency Control siRNAs. Delivery efficiency is one of the major technical concerns in RNAi studies. Low transfection efficiency has always been the first suspect in an unsuccessful gene silencing. The appropriate controls to evaluate and optimize siRNA delivery efficiency in cells, especially in hard-to-transfect cells, are pivotal to a successful gene knockdown. As described previously, some delivery control siRNAs contain a fluorescently labeled fragment to facilitate the detection process. Another recently developed transfection control can also be used to evaluate the delivery efficiency by examining the extent of cell toxicity caused by the control siRNA. When entering the cells, the siRNA induces apoptosis, and the transfection efficiency may be assessed through the quantification of the levels of cell death, such as the cell viability. Occasionally, the nonfunctional/nontargeting siRNA control and positive silencing control can also serve as delivery efficiency controls.

Positive Silencing Control siRNAs. Positive controls are designed to knock down well-characterized genes in target cells. Because positive control siRNAs are subjected to all the processes of trans-

fection and silencing, they play important roles in the evaluation of siRNA silencing effect and optimization of the siRNA experimental conditions. Positive controls generally target housekeeping genes or ubiquitously expressed genes, including glyceraldehyde 3-phosphate dehydrogenase (GADPH), cyclophilin B, and lamin A/C. Chemical modifications, such as fluorescent modification, may be introduced to the siRNAs for easy detection.

A number of control siRNAs are currently available from siRNA manufacturers. These controls are carefully designed and validated by the suppliers, and would make the siRNA approach more feasible and easier to conduct.

34.2.3.2 siRNA Delivery into Cultured Mammalian Cells

A successful siRNA-mediated gene silencing relies mainly on superb transfection efficiency. Although the knockdown potencies for most of the siRNAs range from 70% to 95%, lower silencing efficacies (e.g., <50%) would be anticipated if only partial cells (e.g., <50%) are transfected with siRNA. To date, various state-of-the-art delivery techniques have been developed to introduce siRNA into a wide range of cultured cells, tissues, and whole animals.

Commonly used delivery techniques for the introduction of siRNA into mammalian cells include transfection, electroporation, microinjection, and virus transduction. Other approaches have been recently exploited and applied in siRNA delivery, exemplified by the nanoparticles and Accell® siRNA system developed by Dharmacon Inc. (Lafayette, CO, USA) (Donze, 2002; Weil et al., 2002; Venkatraman et al., 2010; Wang et al., 2010). In general, the choice of delivery approach depends on various factors, including the cell types (e.g., nondividing cells or dividing cells), the half-life of the target protein, and the desired duration of the silencing effect (e.g., short-term or long-term knockdown). Once the delivery method is determined, a meticulous and thorough optimization of transfection conditions, such as cell confluence, transfection reagent amount, and the ratio of reagent:siRNA, should be considered and performed in order to achieve an effective gene silencing.

34.2.3.2.1 Lipid-Mediated Transfection

Lipid-mediated transfection is currently the most attractive approach to introduce siRNA into mammalian cells (Elbashir et al., 2001; Weil, 2002). With lipid reagents, siRNA molecules are bound to liposomes to form an siRNA–lipid complex and taken up into cytoplasm via an endocytosis pathway. There are several suppliers providing lipid-based transfection reagents that share similar optimization protocols with differences in reagents, culture conditions, and siRNA concentrations, and so on. To obtain maximum transfection efficiency, optimization of experimental conditions by varying cell density, transfection reagents and siRNA amount, and the lipid:siRNA ratio must be performed for every assay.

34.2.3.2.1.1 CULTURE MEDIUM The culture medium is usually recommended by the manufacturer from whom the transfection reagent is purchased. The proteins in the serum can bind to the lipid to make it inaccessible to siRNA, thus significantly inhibiting the formation of siRNA–lipid complex. Additionally, increased concentrations of antibiotics during transfection become toxic to cells. Therefore, most transfection reagents act in the absence of serum and antibiotics to avoid the potential of decreased transfection efficiency. In contrast, there are studies showing that the presence of serum in the culture medium may increase the transfection efficiency for some lipids by decreasing the toxicity to cells.

34.2.3.2.1.2 CELL TYPE AND CONFLUENCY Optimization and transfection procedures are normally unique for individual cell types. Generally, target cells at low passages are recommended rather than at high passages due to possible changes in morphology and genetics for the high-passage cells. Furthermore, the target cells should be maintained in the exponential growth phase with active cell division occurring during transfection. This is critical to optimize the cell density of plating and the cell confluence of transfection. High-density cells may be more sensitive to transfection reagents, possibly leading to cytotoxicity. In contrast, a relatively lower percentage of cells is less accessible to lipid–siRNA complex and frustrates the transfection efficiency. Similarly, low density can also induce cytotoxicity owing to the excessive exposure of cells to transfection reagents and lipid–siRNA complex.

34.2.3.2.1.3 TRANSFECTION REAGENT The composition of a lipid has an important impact on lipid-mediated transfection approach. The selection of a suitable transfection reagent is the key to successful siRNA delivery. Literature and the instructions from manufacturers provide valuable information on the applications, advantages, and limitations of each transfection reagent for different cells, and should be used as the primary source of information as to how to start the assay. Different cell types may show various responses to transfection reagents. If the transfection of target cells using one certain transfection reagent leads to undesirable transfection efficiency, alternative reagents are highly recommended for transfection of the target cells.

34.2.3.2.1.4 AMOUNT OF TRANSFECTION REAGENT AND siRNA The total amount of lipids, siRNA, and the lipid:siRNA ratios are crucial parameters for the optimization of transfection. Each transfection reagent supplier commonly provides its own suggestions on these parameters, for instance, the concentrations of siRNA may range from 1 nM to 1 µM or higher. In general, higher concentrations of siRNA and lipids may cause cytotoxicity, whereas lower concentrations can result in insufficient transfection. The optimization of transfection by varying the amount of transfection reagents, siRNA, and the ratio of reagent:siRNA should be conducted in order. Generally speaking, the lowest concentration of siRNA that induces the highest knockdown effect with the least toxicity is considered to the optimal choice for transfection.

34.2.3.2.1.5 LENGTH OF TRANSFECTION The exposure of cells to lipid–siRNA complexes usually lasts for few hours to few days. The incubation period should be determined by evaluation of a time-dependent transfection effect in target cells. The length of transfection should be adjusted to obtain the optimal transfection efficiency. Occasionally, repeating transfection several times on the same cells over several days should be considered to obtain higher or long-term silencing efficiency.

In summary, a successful siRNA-mediated gene silencing depends heavily on the balance of transfection efficiency, cell toxicity, and gene silencing.

34.2.3.2.2 Electroporation Electroporation is a highly efficient method to introduce siRNA into almost all types of cells and various intact tissues (Jiang et al., 2003; Gilmore et al., 2006). It utilizes a strong electric pulse to temporarily permeate the cell plasma membrane to allow siRNA molecules enter the cells (Purves et al., 2001). The transfection efficiency of electroporation depends on several factors, including the duration and density of electric pulse, as well as the cell characteristics. The high delivery rate typically requires the cells grown at a mid-log phase with a density of 10^6–10^7 cells/mL. Electroporation on the late-phase cells may result in lower transfection efficiencies and survival rates. Considering the low survival rates of cells, electroporation normally requires a smaller amount of siRNA and a larger amount of cells than other methods.

34.2.3.2.3 Microinjection Microinjection is a process that involves the insertion of materials, such as nucleic acids, into various types of cells and model organisms using needles under microscopy (Donnelly et al., 2010). Microinjection is a time-consuming and laborious method that requires expensive equipment. Microinjection generally introduces materials into a single cell.

34.2.3.2.4 Viral Transduction Viral transduction refers to a process of delivery of DNA into target cells by a virus. There is considerable interest in the application of viral vectors in siRNA delivery, due to its ability to induce long-term knockdown and introduce siRNA into difficult-to-transfect cells, including primary cells (Couto and High, 2010). As described previously, the commonly used viruses include retrovirus, adenovirus, and lentivirus. The viral particles containing the target shRNA transduce the target cells, and only the cells transfected with the target shRNA that is generally incorporated with a drug resistance gene can survive in the drug selection. The clones of cells with desired production levels of shRNA are selected and screened for the target gene silencing. Extensive optimization and validation are required for the process of viral transduction, including virus selection, sequence design, vector construction, virus production, and transduction into the host cells. A number of manufacturers are providing various reagents and vectors for virus-mediated siRNA.

34.2.3.2.5 Other Methods The traditional delivery systems exhibit a limited capacity for introducing siRNA into particular types of cells, such as primary, nondividing, and differentiated cells. In addition, the unmodified siRNA is typically stable only for a short time in cells. Several alternative methods have been recently developed. In one method, nanoparticles made of polymers encapsulate siRNA to protect it from cleavage and efficiently deliver the siRNA into target cells (Venkatraman et al., 2010). A nanoparticle siRNA transfection system offers unique advantages over the other approaches with (1) great delivery efficiency, especially for cells difficult to be transfected, (2) less cell toxicity, (3) simple procedures, and (4) easy adaption to high-throughput mode.

The Accell siRNA delivery system developed by Dharmacon is another approach for providing high delivery efficiency of siRNA introduction into a wide range of cells. It is specially modified for use without a transfection reagent. Unlike the other delivery methods, the Accell siRNA delivery system is rather straightforward and easy to use. It is noteworthy that the tedious optimization steps can be completely skipped using this technique. The Accell siRNA delivery system can be used on a wide range of cell types, including nondividing cells, and also for long-term knockdown with minimal disruption of the morphology and function of cells.

34.2.4 Gene Silencing Effects Detection

Each step of gene expression can be measured for gene silencing effects, including (1) DNA-RNA transcriptional level for determining the changes of target mRNA expression, (2) RNA-protein translational level for quantifying the changes of target protein concentration, and (3) altered phenotypes affected by target proteins.

Generally, the scopes of research as well as the availability of instruments and experimental materials provide the basis for the selection of method(s) to be used for the detection of the extent of gene silencing. Although the mechanisms of siRNA are not fully understood, it is believed that siRNA represses gene expression through mRNA degradation caused by specifically binding to the target mRNA (Hannon, 2002). Proteins and phenotypes may be affected by some other mechanisms other than siRNA-induced mRNA degradation. Thus, the quantification of mRNA changes is primarily to determine the siRNA silencing effect and to exclude other mechanisms that lead to the changes of proteins and functions.

Frequently, used assays for the detection of mRNA levels include Northern blot analysis, quantitative reverse transcriptase polymerase chain reaction (qRT-PCR), and DNA microarray. The techniques for the evaluation of the levels of target protein, such as Western blot, generally require probes including specific antibodies. The functional assays consist of diverse approaches to appraise the altered phenotypes mediated by siRNA, such as methods to identify the alteration of drug uptake or transport, or the quantitative changes in drug metabolism during ADME.

34.2.4.1 mRNA Quantification The most acceptable methods to quantify mRNA levels include Northern blot analysis, qRT-PCR, and DNA microarray. The choice of mRNA detection assays depends on instrument availability, cost, and scalability.

34.2.4.1.1 Northern Blot Analysis Typical Northern blotting procedures include (1) extraction of total RNA from samples, (2) isolation of mRNA using polyT cellulose chromatography, (3) separation of RNA by gel electrophoresis, (4) transfer of RNA from a gel to a nitrocellulose or nylon membrane by passive diffusion or electroblotting, (5) hybridization of samples with radio- or chemiluminescent-labeled probes, and (6) quantification of the level of desired mRNA expression by autoradiography or phosphorimaging. Northern blot method requires specific probes that are composed of nucleic acids with a complementary sequence to all or part of the target mRNA. The probes are commonly labeled with radioactive isotopes (e.g., ^{32}P) or chemiluminescence. Northern blot analysis is relatively simple and inexpensive. The commercial availability of Northern blot products from multiple suppliers makes the procedure easy to conduct. It is less sensitive but more specific than RT-PCR, while more sensitive and more specific than DNA microarray. Note that Northern blot is designed to detect the expression of a single gene, and therefore it is incompatible with high-throughput screening.

34.2.4.1.2 qRT-PCR qRT-PCR is a new powerful approach for the measurement of mRNA expression levels. The procedure starts with the isolation of mRNA from samples followed by the generation of cDNA with mRNA as a template by reverse transcription. The cDNA is then used as template for the amplification of target genes by PCR using gene-specific primers. The quantification of mRNA expression is performed by measuring the changes in the fluorescent signal of a probe. Specific primers and/or probes are critical for using RT-PCR methods to evaluate the efficient siRNA effect. RT-PCR is more sensitive than other mRNA quantification methods. In addition, RT-PCR can be used to monitor gene expression at real time during the amplification process.

Two kinds of probes are routinely utilized for RT-PCR assays (Lutfalla and Uze, 2006). One assay uses the target-specific probe complementary to a short fragment of the cDNA sequence to be amplified. The probe is labeled with two conjugates of a reporter fluorophore on the 5′-terminus and a quencher on the 3′-terminus. When the probe is free, the close proximity of the quencher to the fluorophore reduces the fluorescence resonance energy transfer (FRET) (Clegg, 1995), whereas, during the process of amplification, the probe anneals to the complementary region of the amplified cDNA, and is cleaved by Taq DNA polymerase. The cleavage of the probe divides the reporter from the quencher and increases the fluorescence intensity. The measurement of fluorescence intensity is to accurately evaluate the mRNA expression. This technique offers high specificity by using both amplification primers and a detection probe specifically hybridizing with the target cDNA.

The other method employs SYBR Green I dye to assess the amount of the target PCR product. During the amplification process, SYBR Green I dye can bind to each copy of the double-stranded PCR product. The fluorescence intensity is proportional to the level of the target mRNA. SYBR Green I dye can be universally applied to any RT-PCR, which bypasses the meticulous process of probe design. This technique requires less time for experimental design, and is less costly than the other method using a specific probe. However, as the dye can bind to any dsDNA nonspecifically, this technique may result in more false-positives or an overestimate of the mRNA expression level.

34.2.4.1.3 DNA Microarray DNA microarray is a technique to measure the changes in multiple genes expressions in a single experiment (e.g., gene expression profiling) (Schena et al., 1995). It consists of an arrayed series of thousands of spots containing amounts of a specific DNA sequence, called probes. Each probe is a short fragment of a gene to hybridize DNA or RNA samples, and is typically attached to a solid surface, such as a glass or silicon chip, by covalent binding. Similar to other methods, DNA samples generated from the isolated mRNA are added to the chip. The hybridization occurs between the target DNA in the samples and the probe DNAs on the chip. Such hybridization is generally quantified by a fluorescent or chemiluminescent method. The microarray method can be used as a high throughput assay to accelerate the investigation by detecting the expression levels of thousands of genes. The microarray analysis has a unique advantage over other methods of identification of the siRNA-induced off-target effect in an experiment. The data obtained by microarrays are usually consistent with those from Northern blot. The major drawbacks of this approach are high cost and low sensitivity, as compared with the other two mRNA detection methods. The relatively low specificity is another disadvantage for microarray analysis.

34.2.4.2 Protein Quantification Protein quantification is usually accompanied by mRNA detection. The most commonly used protein detection methods include immunoblotting, enzyme-linked immunosorbent assay (ELISA), immunoprecipitation (IP), and the methods to quantify and localize target proteins, such as immunocytochemistry (ICC), immunohistochemistry (IHC), or immunofluorescence (IF). The method selected depends on the study proposal, as well as the availability of required instruments and antibodies. Generally, the alteration of proteins generally lags behind the mRNA change by 24 h or longer, which relies on the stability and abundance of the target proteins. The time-dependent changes of protein levels may be performed following siRNA delivery until observing the reduction of protein expression.

34.2.4.2.1 Western Blot Analysis The Western blot analysis is the most popular technique to quantify the levels of protein expressions. The Western blot procedures include (1) isolation of the total cellular proteins, (2) separation of proteins by polyacrylamide gel electrophoresis (PAGE), (3) transfer of the proteins to a membrane, such as nitrocellulose or specific polyvinylidene fluoride (PVDF) by electroblotting, (4) blockage of the nonspecific binding between the membrane and antibodies by blocking reagents, such as bovine serum albumin (BSA) or nonfat dry milk, (5) incubation of the membrane with unmodified or conjugated primary and secondary antibodies, (6) detection of proteins by the probes linked to the modified antibodies via colorimetric reaction, and (7) quantification of the levels of the target proteins by phosphorimaging. Western blot is a relatively easy assay to conduct, and is widely used to quantify protein expressions in many fields. There are currently several suppliers providing instruments, materials, and monoclonal and polyclonal antibodies against thousands of proteins. The choice of the specific primary antibody is the key to the successful recognition of the target protein. The application of the antibodies, which may also recognize similar epitopes other than the target proteins, may lead to misleading conclusions. Thus, the unavailability of appropriate antibodies may keep this technique from extensive application in gene silencing assays. In contrast, the conjugation of probes, such as biotin, alkaline phosphatase (AP), or horseradish peroxidase (HRP), with commercial secondary antibodies, along with signal amplification via a reaction with chemiluminescent or colorimetric substrate, make the recognition and quantification of target proteins easier and more accurate.

34.2.4.2.2 ELISA ELISA is an immunoassay technique to determine the presence and amount of proteins in samples. This assay is typically performed in 96- or 384-well plates, allowing high-throughput screening. The target proteins can be quantified by measuring the amount of either the target proteins or antibodies specifically binding to the target proteins. For example, when using ELISA to quantify the proteins through the determination of specific antibodies, certain amounts of cell lysates containing the proteins of interest may be absorbed to the surfaces of the plates. A primary antibody against a target protein is then applied on the surface so that it can bind to the antigen. After extensive washing to remove the unbound antibodies, the secondary antibody conjugated with AP or HRP against the primary antibody is added to the reactance. The proteins of interest are quantified by the addition of a colorimetric or chemiluminescent substrate for the conjugated enzymes. ELISA is an easy-to-use technique and has a wide linear dynamic range for protein quantification. In addition, it is commonly used for high-throughput screening to detect various proteins in a single experiment. However, the relatively high number of false-positive results obtained by ELISA may limit its application to the quantification of some proteins.

34.2.4.2.3 IP IP is used to determine protein qualitatively or quantitatively by extraction of a protein out of the cell lysates, which contain thousands of proteins, using an antibody specifically binding to this protein (Bonifacino et al., 2001). The process starts with the isolation of cell lysates, followed by a precleaning step to remove the proteins that may bind to the antibody nonspecifically. The protein of interest is then immunoprecipitated by passing the crude sample through a column modified with a specific antibody, or by incubating the sample with beads cross-linked with the appropriate antibody. Finally, the desired protein is eluted from the column or the beads by detergent or a solution with a high content of salts, and quantified by electrophoresis. The complicated procedures make IP a time- and labor-intensive assay. Protein quantification using IP is less accurate than quantification using Western blot or ELISA. A number of products for IP analysis are currently commercially available from several companies that greatly facilitate IP procedures.

34.2.4.2.4 ICC and IF The two methods employ antibodies to interact with specific protein antigens within the cells and can be used not only to measure the protein levels but also to determine the intracellular localization of a protein of interest (Ryan, 1986). Antibodies conjugated with fluorophore or other signal amplification reagents are generally required in both methods. Either adherent or suspended cells can be stained and observed by microscopy, such as with a confocal microscope. The two methods require tedious preparation steps and measure protein levels on a cell-by-cell basis, limiting their use for large populations of cells.

34.2.4.2.5 Flow Cytometry Flow cytometry is a powerful technique and is widely used to evaluate siRNA efficiency by quantification of the expression levels of mRNAs or proteins (Chan et al., 2006). The presence and amount of mRNAs are detected with fluorescence-labeled siRNAs, and proteins are evaluated via specific antibodies conjugated with fluorophores. Flow cytometry measures the fluorescent signals from the suspended cells in a cell-by-cell mode. Flow cytometry is an elegant and accurate method for the determination of the efficiency and specificity of siRNAs at both mRNA and protein levels. The method can be easily automated and used to screen for effective siRNAs and quantify the genes of interest at high throughput. However, the flow cytometry method usually requires expensive instruments and well-trained operators.

34.2.4.3 Functional Assays Compared with assays for evaluation of siRNA effect on mRNA and protein levels, functional assays are more diversified and used to assess more sophisticated changes of various phenotypes induced by siRNA. These assays can be categorized into several classes based on the functions of the proteins of interest, including growth and differentiation assays, cell viability assays, and apoptosis assays.

This section will examine ADME-related gene silencing mediated by siRNA.

34.2.4.3.1 Functional Assays of Drug Metabolizing Enzymes The reports regarding the application of siRNA on the functional investigation of drug-metabolizing enzymes are very limited. Functional assays detecting siRNA effects are mainly focused on the biotransformation of endo- and xenobiotics metabolized by specific enzymes. Considering the insufficient information regarding the effects of siRNA on Phase I and Phase II drug metabolizing enzymes, these functional assays are not discussed in detail in this section.

34.2.4.3.2 Transporter Assays siRNA technology has been employed to investigate the contribution of drug transporters to drug distribution and DDI. The commonly used functional assays in assessing the siRNA effects include transport assays, uptake assays, and biliary elimination assays in hepatocytes (Xia et al., 2007).

34.2.4.3.2.1 TRANSPORT ASSAYS siRNA effects may be measured by comparing the permeability of a probe substrate across cell monolayers in siRNA target and intact cells. In transport assays, the polarized cells, such as Caco-2 cells, are seeded onto Transwell™ (Grand Island, NY, USA) inserts and ready for use when the cells reach confluence. The assay is initiated by the addition of a probe substrate solution to either the apical (for apical-to-basolateral transport) or basolateral (for basolateral-to-apical transport) compartment (Figure 34.2). The apparent permeability coefficient (P_{app}) of unidirectional fluxes for the substrate is calculated as Equation 34.2:

$$P_{app} = (dQ/dt)/(A * C_0), \qquad (34.2)$$

where dQ/dt: total amount of drug present in the receiving chamber per unit time (e.g., nmol/s)

A: surface area (cm^2)

C_0: initial drug concentration in the donor chamber (e.g., nmol/mL)

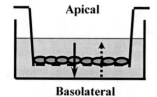

Apical

Basolateral

AB = Apical-to-Basolateral
BA = Basolateral-to-Apical
Ratio = BA/AB

FIGURE 34.2. Transwell system.

P_{app}: apparent permeability expressed as cm/s $\times 10^{-6}$

The efflux ratio is calculated as Equation 34.3:

$$\text{Efflux Ratio (B:A)} = P_{app,\,B\text{-to-A}} : P_{app,\,A\text{-to-B}}, \quad (34.3)$$

where $P_{app,\,A\text{-to-B}}$: apparent permeability when the substrate solution is added in the apical side

$P_{app,\,B\text{-to-A}}$: apparent permeability when the substrate solution is added in the basolateral side

The changes in P_{app} and efflux ratio are used to monitor the efficiency and specificity of siRNA on the knockdown of transporter genes. The transport assay is the most direct assay to evaluate transporter functions. However, it usually takes 3–4 weeks for the cells to reach confluence, thus the transport assay is not suitable for high-throughput applications. The overgrown cells may be more sensitive to cell toxicity and more resistant to siRNA molecules.

34.2.4.3.2.2 UPTAKE ASSAY The uptake assay determines the function of a given transporter by measuring the amount of a probe substrate accumulated in cells. The probe substrate is pumped into cells, such as Caco-2 cells or hepatocytes, and the substrate concentration within the cells may be quantified by several analytical tools, such as liquid chromatography-mass spectrometry (LC-MS), liquid scintillation, or fluorescence detection, based on the properties of the substrate. The flow charts of uptake assays in cultured hepatocytes are shown in Figure 34.3. The difference in the substrate concentrations within the cells in the presence or absence of siRNA reflects the knockdown efficiency of siRNA on the given transporter. The uptake assay is easy to perform, and the uses of fluorescent, colorimetric, or radiolabeled substrates can further simplify the analysis of substrates. In addition, the uptake assay is compatible with high-throughput testing. Several well-known fluorescent substrates are utilized in uptake assays, exemplified by daunorubicin and rhodamine 123 for Pgp, calcein for multidrug resistance protein (MRP), LysoTracker for breast cancer resistance protein (BCRP), and H2FDA and BODIPY for bile salt export pump (BSEP).

34.2.4.3.2.3 TRANSPORT ASSAYS IN HEPATOCYTES The ability of sandwich-cultured hepatocytes to form functional bile duct canalicular networks during culture makes them a valuable model for investigating the functions of both uptake and efflux transporters. The bile duct canalicular networks are maintained by tight

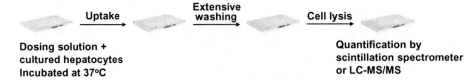

FIGURE 34.3. Flow chart of uptake assay in cultured hepatocytes.

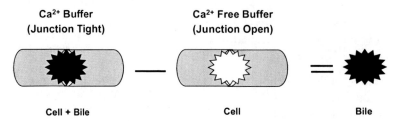

FIGURE 34.4. Sandwich-cultured hepatocytes model.

junctions between hepatocytes; for which the presence of calcium in the culture medium is required to maintain the integrity of these junctions. Incubation of cells in a calcium-free medium or buffer can cause opening of the tight junctions and release of the contents of the bile ducts (Liu et al., 1999a,b). Therefore, the biliary excretion index (BEI) can be calculated as the difference between the amounts of substrate accumulation in the presence and absence of intact canalicular networks, normalized for substrate accumulation in the presence of intact canalicular networks (Figure 34.4) (Eq. 34.4):

$$BEI\,(\%) = (Accumulation_{cells+bile} - Accumulation_{cells})/$$
$$Accumulation_{cells+bile} \times 100\%,$$

$$(34.4)$$

where $Accumulation_{cells+bile}$: the total amount of drug present in the hepatocytes and the bile duct canalicular networks per unit protein when hepatocytes are incubated in standard Hank's balanced salt solution (HBSS) (e.g., pmol/mg protein).

$Accumulation_{cells}$: the total amount of drug present in the hepatocytes per unit protein when hepatocytes are incubated in calcium-free HBSS (e.g., pmol/mg protein).

The efficacy and specificity of siRNA effects on efflux transporters can be evaluated by the changes in the BEI

values of the probe substrates in the intact hepatocytes and the siRNA knockdown hepatocytes. This approach is very powerful in measuring the function of efflux transporters *in vitro*. However, the results obtained from the hepatocytes of one donor may not be reflective of the whole human population because of possible transporter polymorphism.

34.2.4.3.2.4 CYTOTOXICTY ASSAYS Cytotoxicity assay is an indirect measurement of the accumulation of cytotoxic compounds in cells and is a frequently used surrogate assay to determine whether a compound is a substrate of a given transporter. Transporter substrates can be identified by comparing the IC_{50} (the concentration that inhibits the cell growth by 50%) in wild-type (naïve), transporter-expressing (drug resistant), or transporter knockdown cells, while the inhibition of transporter can be identified by the ability to potentiate (for efflux transporters) or attenuate (for uptake transporters) the cytotoxicity of a substrate in transporter-expressing or drug-resistant cells. The activity of the reversal agent is generally expressed as a fold reversion or multidrug resistance (MDR) ratio. The MDR ratio is the ratio of the IC_{50} of the cytotoxic drug alone to the IC_{50} of the cytotoxic drug in the presence of the transporter modulator.

Cytotoxicity measurements can be applied in high-throughput mode. However, this method is limited in its utility for identifying transporter substrates since the substrate must be an antiproliferative compound.

34.2.5 Challenges in siRNA

Since its discovery, siRNA has been quickly recognized as a powerful tool for investigating the functions of particular genes. This emerging technology offers quick, reversible, and efficient gene silencing in various types of cells, tissues, and whole animals. However, some challenges exist for this method, including the relatively low knockdown efficiency and lack of specificity. These challenges are related to a variety of technical obstacles related to the selection of effective siRNA sequences, efficient delivery methods, successful gene silencing effect assays, and the half-life of the protein of interest. Overcoming these obstacles needs the full understanding of siRNA mechanisms and innovation of the methodologies. For this reason, numerous academic and industrial researchers have been devoting their efforts to tackling the unresolved challenges presented by siRNA.

34.2.5.1 Challenge in Selection of Effective siRNA

It has been reported that only half of the randomly designed siRNAs can cause a 50% reduction, and approximately a quarter of siRNAs resulted in more than a 70% reduction in the mRNA levels of the target genes (Li et al., 2004). Selection of multiple randomly designed siRNAs against the different segments of a target gene was recommended to improve the efficacy of siRNA. However, a reduced silencing potency for some target genes was observed when multiple siRNA oligos were used, possibly due to the saturation of RISC with low or nonfunctional siRNAs that could compete with the effective siRNA for the RNAi apparatus (McManus et al., 2002). A number of useful *in silico* approaches incorporating thermodynamic and sequence-specific parameters have been developed to facilitate the screening and selection of functional siRNAs.

34.2.5.2 Challenge in Selection of Efficient Delivery Methods

The effectiveness of siRNA largely depends on the efficiency of delivery. It is quite challenging to achieve a successful siRNA delivery, especially for the difficult-to-transfect cells, such as primary cells, stem cells, and some obstinate cell lines. A number of experimental factors must be well considered, including cell growth properties and viability, delivery reagents and concentrations, and the duration of delivery. At present, a variety of reagents for delivery are commercially available. Detailed use protocols as well as the advantages and limitations of the reagents can be provided by their suppliers. However, a full validation is required for almost all delivery methods. Nowadays, lipid-mediated transfection appears to be the most popular method used by researchers. It is simple, cost-effective, and suitable for most cell lines. However, it is generally not compatible with nondividing cells and several hard-to-transfect cell lines, nor with long-term gene silencing. Alternatively, vector-based expression systems can be used to introduce genes into difficult-to-transfect cells and achieve extended gene suppression. In contrast to other delivery methods, vector-based expression systems introduce shRNA into cells in the form of plasmids or viral vectors. They generally produce relatively lower gene silencing for shRNA (75% reduction) than for siRNA (90% reduction). Furthermore, the preparation and validation of vector-based delivery is tremendously time- and labor-intensive.

34.2.5.3 Challenge in Minimizing Off-Target Effects

Off-target effects can complicate the interpretation of siRNA effects on gene silencing and lead to undesirable toxicities. siRNA can trigger the down-regulation of genes that contain less than 100% identity with the siRNA sense and/or antisense strand with incomplete complementary base pair interaction. Indeed, siRNAs are able to induce the off-target cleavage of genes with as few as 11 bases of contiguous identity. The unintended alterations of gene expression profiles caused by the siRNAs have been observed in multiple studies (Jackson et al., 2003; Semizarov et al., 2003). Because the isoforms within one family or subfamily of drug metabolizing enzymes and drug transporters typically have high homology in an mRNA sequence, the off-target effects may become an important concern when an isoform is knocked down. Furthermore, both of the unwound single-stranded sense and antisense siRNAs can be loaded into the RISC to recognize and cleave the complementary mRNA independently (Nykanen et al., 2001). The off-target effects arise when the sense siRNA directs the RISC to recognize and degrade the undesirable mRNA sequences.

In general, off-target effects can be minimized by several approaches. (1) Using well-designed methods and bioinformatic screening. As described previously, the deliberately designed siRNAs following the guidance (see Session 34.2.1, siRNA design) significantly reduced the likelihood of off-target effects. Further bioinformatic analysis, such as BLASTn, to compare the target siRNA to a nucleotide database of the genome can exclude the siRNAs with high homology to known genes and ensures the unique identity of a candidate siRNA with respect to the gene of interest. (2) Chemical modification provides a useful tool to diminish the off-target effect caused by sense siRNA. Since RISC may favor one strand of siRNA rather than another based on the thermodynamics of the siRNA (Khvorova et al., 2003), modification induces the sense siRNAs to lose their capability to be loaded to RISC. Consequently, the sufficient RISC preferentially interacts with the antisense siRNA, leading to higher siRNA potency and lower off-target effects.

In summary, the combined applications of rational siRNA design, systemic bioinformatic examination, and chemical modification can efficiently minimize the off-target effects. Notably, these techniques may only be used to partially address the off-target issues, while other technologies, including microarray assay, as well as appropriate positive and negative controls would be a useful addition to the verification of on-target effects and data interpretation.

34.2.5.4 Challenges in Performing Long-Term Gene Silencing

CYP and transporter proteins typically have long half-lives; as a result, a prolonged waiting time may be required before observing target protein suppressions and consequent functional changes. The introduction of an siRNA duplex can generally knock down gene expression transiently, especially in rapidly dividing cells, not only because the exogenously synthesized siRNA oligos have no capability of replication in mammalian cells but also because these synthetic siRNAs are generally not stable in cells. Note that chemical modification is able to increase the stability of siRNAs in cell culture systems. In most instances, the vector-based expression systems are the primary consideration when seeking to achieve long-term gene knockdown. The plasmid or viral vectors containing an RNA polymerase III promoter (e.g., U6 or H1) and the desirable siRNA sequences can be transcribed and processed to functional siRNAs by RNA polymerase III and Dicer in mammalian cells. However, these vector-based systems can be time-consuming and labor-intensive.

34.2.5.5 Challenges in in vivo siRNA

Although less extensively employed than *in vitro* siRNA, *in vivo* siRNA targeting gene expression has drawn enormous attention for its potential to be used therapeutically for various diseases (Lee and Sinko, 2006; Phalon et al., 2010; Walton et al., 2010). Delivering siRNA *in vivo* to animal tissues is complicated and involves using physical, chemical, or biological approaches, and in some cases a combination. Because the main goal of *in vivo* delivery is to have active siRNA oligos in the target cells, the stability of siRNA oligos in the extracellular and intracellular environments after systemic administration is the most challenging issue. The first hurdle is the size of the 21-nucleotide double-stranded siRNA oligos. These oligos are relatively small and thus are rapidly excreted through urine when administered into the bloodstream, even if siRNA molecules remain stable through chemical modifications. Secondly, the double-stranded siRNA oligos are relatively unstable in the serum environment and they can be degraded by RNase activity within a short period of time. Thirdly, when siRNA is administered systemically, the nonspecific distribution of these oligos throughout the body will significantly decrease the local concentration where the

disease occurs. In addition, the siRNA oligos need to overcome the blood vessel endothelial wall and multiple tissue barriers to reach the target cells. Finally, when siRNA reaches the target cells, cellular uptake of those oligos and intracellular RNAi activity require efficient endocytosis and intact double-stranded oligos, respectively (Xie et al., 2006).

Recent advances have been made in resolving these issues, such as using chemically modified siRNA or encapsulated siRNA, optimizing the administration route, employing viral siRNA vectors, and applying receptor-mediated siRNA to specifically bind to target cells. Such innovations are able to improve efficiency and reduce the toxicity of siRNA *in vivo* (Shim and Kwon, 2010). Chemical modifications, including changes in the oligo backbone, replacement of individual nucleotides with nucleotide analogs, and addition of conjugates to the oligo, increase its stability and the overall cellular uptake of siRNA, but this did not solve the problems of urine excretion and targeted delivery (Soutschek et al., 2004). Therefore, a delivery system capable of protecting siRNA oligos from urine excretion and RNase degradation, transporting siRNA oligos through the physical barriers to the target tissue, and enhancing cellular uptake of the siRNAs is key to the success of *in vivo* siRNA application. The accessibility of different tissue types, various delivery routes and a variety of pharmacological requirements makes it impossible to have a universal *in vivo* delivery system suitable to every scenario of siRNA delivery. In terms of *in vivo* delivery vehicles for siRNA, the "nonviral" carriers are the major type being investigated so far, but some physical and viral delivery approaches are also very effective. The routes of *in vivo* deliveries are commonly categorized as local or systemic. Some of the delivery vehicles and delivery routes are very efficient for target validation in animals but might not be useful for delivery of siRNA therapeutics in humans. A method that has been used in the preclinical setting with some success involves the hydrodynamic intravenous (IV) injection of naked siRNA or shRNA. This method enhances siRNA delivery and involves injecting a large volume (2 mL/mouse) of siRNA solution over a short period of time (<10 s). This method has been successfully used in mice. Thus, *in vivo* siRNA delivery carriers and methods can also be classified as clinically viable or nonclinically viable, according to their suitability in humans (Xie et al., 2006; Shim and Kwon, 2010).

Over the decade since its discovery, siRNA technology has been proven to be a powerful tool for investigating gene functions. Currently, a variety of new technologies have been developed to overcome the technical challenges in the design, specificity, stability, and potency of siRNAs. Continuing innovations should eventually mature siRNA technology as a discovery tool for new therapeutics for humans on a routine basis.

34.3 APPLICATIONS OF RNAi IN DRUG METABOLIZING ENZYMES AND TRANSPORTERS

34.3.1 Applications of Silencing Drug Transporters

The RNAi technique has been successfully launched as an experimental tool to identify new drug transporters, delineate the function of transporters, and provide a novel approach to circumventing MDR and predict DDI.

34.3.1.1 Applications in Identifying New Transporters

Combined with other molecular and cell biology techniques, RNAi has been used to identify new transporters. 3-Iodothyronamine (T_1AM) is a recently discovered endogenous metabolite of thyroid hormone with dramatic physiological actions when administered *in vivo* (Scanlan et al., 2004). Specific cellular uptake of T_1AM has been observed in a variety of cultured cell lines, suggesting a ubiquitous transport mechanism consistent with the widespread tissue accumulation of T_1AM and its wide range of actions, including hypothermia, bradycardia, hyperglycemia, and general behavioral inactivity (Jansen et al., 2005; Sonders et al., 2005; Braulke et al., 2008 and Friesema et al., 2005). In an attempt to identify plasma membrane transporters responsible for the uptake of T_1AM, Ianculescu et al. (2009) developed a high-throughput RNAi screening method in which a library of siRNAs targeting all of the SLC series of membrane transporters was transfected into HeLa cells, and the siRNAs producing the greatest degree of reduction of T_1AM uptake were identified (Figure 34.5). The transporters targeted by these siRNAs are likely to be involved in T_1AM uptake into cells. A total of 34 of 403 transporters were initially identified as facilitating T_1AM uptake in HeLa cells. The 34 included several heavy metal transporters and various inorganic and organic ion transporters. As would have been expected, none of the likely candidate transporters previously tested and ruled out as T_1AM transporters displayed reduced T_1AM uptake after siRNA transfection. After examining endogenous expression levels in HeLa cells and cellular localization of the 34 transporters, the authors obtained a list of eight transporters that were retested and consistently displayed decreased T_1AM uptake function when knocked down (Figures 34.5 and 34.6). Although direct testing of transporters identified by the RNAi screen was inconclusive, the RNAi screening method developed in this study for

T_1AM is a broadly applicable approach to potentially identify all transporters involved in the uptake of any particular compound in a particular cell type or tissue.

Folates are essential nutrients that are required for one-carbon biosynthetic and epigenetic processes (Stover, 2004). Mammals cannot synthesize folates; hence, dietary sources must meet metabolic needs, necessitating an efficient intestinal absorptive mechanism. Absorption of folates occurs primarily in the duodenum and upper jejunum and involves a carrier-mediated process with a low-pH optimum that operates efficiently within the acidic microclimate of the intestinal surface in this region (Selhub and Rosenberg, 1981; McEwan et al., 1990; Mason and Rosenberg, 1994). It is known that folates are absorbed in the acidic milieu of the upper small intestine, but the underlying absorption mechanism has not been defined. Qiu et al. (2006) have identified a human proton-coupled, high-affinity folate transporter (PCFT/HCP1, encoded by G21) that recapitulates properties of folate transport and absorption in the intestine and in various cell types at low pH. The function of this new transporter has been further recognized in Caco-2 cells by RNAi. Two small shRNA vectors, targeted to two different regions of the G21 transcript, were stably cotransfected into Caco-2 cells. This resulted in a 55% reduction in [^{3}H]folic acid influx at pH 5.5 and a similar (50%) decrease in G21 mRNA as quantified by RT-PCR. Wild-type Caco-2 cells were subjected to transient transfection with siRNA duplex using the Amaxa system, resulting in a 60% reduction in [^{3}H]MTX influx and a 50% decrease in G21 mRNA as compared with negative siRNA-transfected cells. When the stably transfected Caco-2 cells were also subjected to transient transfection with siRNA, there was a further suppression of G21 mRNA, resulting in an 80% decrease in [^{3}H]MTX influx at pH 5.5 and a 75% decrease in G21 mRNA as compared with vector control-transfected cells. Taken together, these studies demonstrate that G21 is the major, and possibly the only, low-pH folate transporter in Caco-2 cells. The identification of a molecular basis underlying folate transport mediated by a proton-coupled carrier offers a new dimension to the understanding of the physiology of folate transport, in particular, intestinal folate absorption and the mechanism of delivery of folates to peripheral tissues in which this activity is expressed.

Riboflavin, a water-soluble vitamin also known as vitamin B2, is essential for normal cellular functions. Its most important biologically active forms, flavin adenine dinucleotide (FAD) and flavin mononucleotide (FMN), act as intermediaries in the transfer of electrons in biological oxidation–reduction reactions. Under conditions of physiological and pathological stress, humans are susceptible to developing riboflavin deficiency. Such a

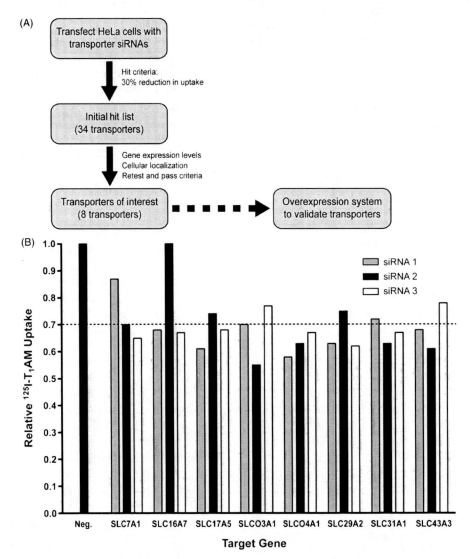

FIGURE 34.5. RNAi screen design and resulting candidates. (A) Experimental scheme for high-throughput screening of the transporter siRNA library to identify transporters involved in uptake of T_1AM. The siRNA library used for the screening consisted of 403 total transporter targets, including all SLC series transporters, with three distinct siRNAs per target. A total of 1209 unique siRNAs were individually transfected into HeLa cells using NeoFX transfection agent, and the transfected cells were assayed for uptake of T_1AM 48 h after transfection. A positive hit was defined as a target for which at least two of the three siRNAs produced a 30% or greater reduction in T_1AM uptake, when compared with cells transfected with negative control siRNA; these hits comprised the preliminary transporter candidates. After analysis of endogenous gene expression levels and cellular localization (plasma membrane vs. vesicular/mitochondrial membrane), the transporter siRNAs were retested, and the candidate transporters were reduced to a subset of eight transporters. Future studies are required to validate the transporter candidates through an approach such as overexpression in cell lines and direct measurement of T_1AM uptake. (B) Average levels of T_1AM uptake in HeLa cells transfected with siRNAs targeting the final eight transporter candidates identified in the screening, expressed relative to uptake in cells transfected with negative control (Neg.) siRNA. Each value represents the mean of triplicate determinations with variations of 1–6%.

deficiency in pregnancy and adolescence induces developmental abnormalities and has been implicated as a risk factor for anemia, cancer, cardiovascular disease, and neurodegeneration (Powers, 2003). Humans are unable to synthesize riboflavin and thus acquire it as a nutrient. Riboflavin transporters are thought to be

essential for the maintenance of riboflavin homeostasis in the intestine and kidney. The novel human and rat riboflavin transporters hRFT1 and rRFT1 were identified on the basis of the rat kidney mRNA expression database (Horiba et al., 2004). The function of the new transporter has been further confirmed by the RNAi

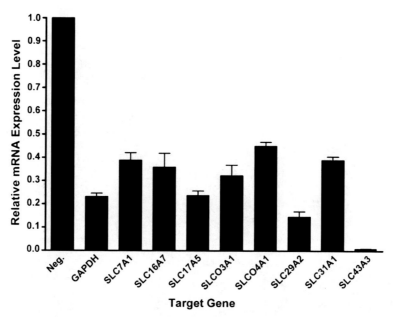

FIGURE 34.6. Expression of GAPDH and membrane transporters after siRNA transfection. The mRNA levels of GAPDH after positive control GAPDH siRNA transfection were determined by qRT-PCR and used to confirm successful transfection and gene knockdown for each set of transfections performed during transporter siRNA library screening. GAPDH expression was normalized to PGK1, and for each transfection experiment, at least 70% knockdown of GAPDH mRNA was observed. After obtaining the final eight transporter candidates, qRT-PCR was used to determine levels of gene knockdown after siRNA transfection for each of these specific transporters and was typically at least 50–60%, although knockdown efficiency varied among the different transporters. Transporter expression levels were normalized to GAPDH in cells transfected only with transporter siRNA, and the levels obtained after transfection with each of the three siRNAs targeting the particular transporter were averaged. All expression levels are shown relative to the expression of the corresponding target gene in cells transfected with negative control (Neg.) siRNA and were calculated using the comparative Ct method.

knockdown HEK293 and Caco-2 cells (Yonezawa et al., 2008). The transfection of siRNA targeting both hRFT1 and hRFT1sv significantly decreased the uptake of [³H]-riboflavin by HEK293 (Figure 34.7) and Caco-2 cells. Kinetic analyses of the results obtained from wild-type and hRFT1 knockdown cells demonstrated that the Michaelis–Menten constants for the uptake by HEK293 (Figure 34.7) and Caco-2 cells were 28.1 and 63.7 nM, respectively. The uptake of [³H]cimetidine, [³H]estrone sulfate, [³H]choline, and [³H]thiamine was unchanged in HEK293 cells transfected with hRFT1/hRFT1sv siRNA. These results suggested that hRFT1 could play, at least in part, a role in the absorption of riboflavin in the intestine as a high-affinity transporter.

34.3.1.2 Applications in Delineating the Transporter Function in Drug Absorption, Distribution, Elimination, and DDI
RNAi has been used to elucidate the transporter function in different cells lines and primary cultured cells. The efflux transporters, Pgp (encoded by MDR1), BCRP, and MRP2, are located in the brush-border membrane of intestinal absorptive cells and reduce the bioavailability of a wide range of orally administered drugs. Using transporter inhibitors in transport experiments in Caco-2 cell monolayers is widely accepted as an efficient way to estimate the contribution of Pgp to the intestinal absorption of drugs. However, there still remain some arguments that the inhibitors may affect the function of other proteins. Vector-based RNAi was used to establish a stable Caco-2 cell line with a persistent knockdown of efflux transporters (Celius et al., 2004; Watanabe et al., 2005; Zhang et al., 2009, Darnell et al., 2010; Graber-Maier et al., 2010). The effective sites of RNAi were selected using siRNA libraries and single siRNAs, and MDR1 stable knockdown Caco-2 cells were constructed using a tRNAval-shRNA expression vector (Watanabe et al., 2005). In siRNA stably expressed Caco-2 cells, the expression level of MDR1 was reduced at mRNA and protein levels. Transcellular transport studies using digoxin revealed that the Pgp function was suppressed completely, similar to that in verapamil-treated cells. Zhang et al. (2009) engineered and characterized Caco-2 cell clones with stable knockdown of BCRP expression using shRNA/BCRP lentiviral particles. In one of their stable BCRP knockdown Caco-2 cell clones, the expression of BCRP mRNA was knocked down with a maximum of 97% silencing. The expression of BCRP

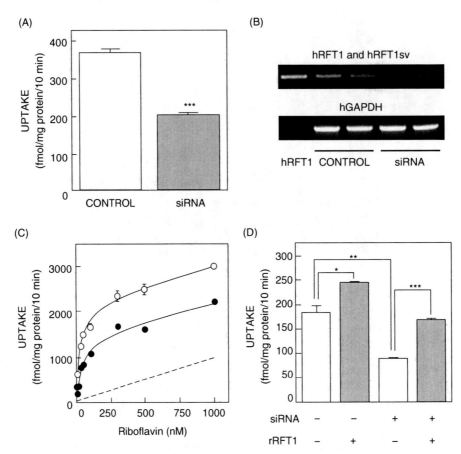

FIGURE 34.7. Effect of siRNA for hRFT1 on the [^{3}H]riboflavin uptake by HEK293 cells. (A) Uptake of [^{3}H]riboflavin by HEK293 cells transfected with control siRNA (control) or siRNA for hRFT1 and hRFT1sv (siRNA). (B) RT-PCR amplification of hRFT1 and hRFT1sv mRNAs in HEK293 cells transfected with hRFT1/hRFT1sv siRNA. The PCR product of 457 bp corresponds to hRFT1 and hRFT1sv. The quality of the RNA samples was also checked by RT-PCR of GAPDH as an internal control. Plasmid DNAs encoding hRFT1 and hRFT1sv were used as a positive control for hRFT1 and hRFT1sv. (C) concentration-dependent uptake of [^{3}H]riboflavin by HEK293 cells transfected with control siRNA (○) or hRFT1/hRFT1sv siRNA (●). The dashed line is $Kd[S]$. (D) uptake of [^{3}H]riboflavin by HEK293 cells transfected with control siRNA or hRFT1/hRFT1sv siRNA and empty vector or rRFT1.

protein was also reduced significantly. Functionally, BCRP knockdown was reflected in a significant reduction of the efflux ratio of estrone-3-sulfate (E3S) and pheophorbide A (PhA). Silencing of BCRP gene expression was maintained for at least 25 passages. MDR1, BCRP, or MRP2 single or double stable knockdown Caco-2 cells have been successfully constructed using RNAi technologies. This will consequently allow the development of a selection system for candidate drugs with improved absorption properties.

Transporter knockdown stable Caco-2 cells have been used to quantify the contributions of efflux transporter to their substrates. Pgp and BCRP play an important role in the transport of a wide variety of endogenous compounds, drugs, and toxins. For example, transport of tyrosine kinase inhibitor (TKI), imatinib, is influenced by both Pgp and BCRP. Stable single and double knock-

downs of Pgp and BCRP were obtained by RNAi in Caco-2 cells (Graber-Maier et al., 2010). Transporter expression was measured on RNA and protein levels using real-time RT-PCR and Western blot, respectively. Functional activity was quantified by the transport of specific substrates across Caco-2 cells. MDR1 and BCRP mRNA expression was reduced to 75% and 90% compared with wild-type control in single MDR1 and BCRP knockdown clones, respectively. In double knockdown clones, MDR1 expression decreased to 95% and BCRP expression to 80%. Functional activity of Pgp and BCRP was diminished as transport of the Pgp-specific substrate ^{3}H-digoxin and the BCRP-specific substrate ^{14}C-2-amino-1-methyl-6-phenylimidazo[4,5-b]pyridine (^{14}C-PhIP) was augmented in the opposite direction when the respective transporter was knocked down. Similar effects were observed by chemical inhibition of the respective transporter. Bidirectional transport studies with ^{14}C-imatinib revealed an abrogation of asymmetric transport when Pgp was knocked down, either in single or double knockdown clones compared with wild-type cells. This was not observed in single BCRP knockdown clones. The results indicated that the contribution of Pgp for imatinib transport seems to be more important compared with BCRP in this Caco-2 cell system. The Pgp or MRP2 single knockdown Caco-2 cells have been used to further describe the efflux properties of ximelagatran and its metabolites (Darnell et al., 2010). The data obtained from bidirectional transport studies in these cells indicate a clear involvement of Pgp but not of MRP2 in the transport of ximelagatran, hydroxymelagatran, and melagatran across the apical cell membrane. The results suggest that inhibition of hepatic Pgp but not MRP2 is involved in the erythromycin–ximelagatran interaction seen in clinical studies. Statins have been reported as Pgp, BCRP, and MRP2 substrates. Pgp, BCRP, or MRP2 single knockdown Caco-2 cells have been used to determine the efflux pathways of several statin drugs (Volpe et al., 2010). Atorvastatin in acid form was transported by all three efflux transporters with a larger reduction in basolateral-to-apical transport for Pgp than BCRP or MRP2. Fluvastatin and rosuvastatin in acid form exhibited reduced basolateral-to-apical permeability in all knockdown cell monolayers. Lovastatin and simvastatin dosed in lactone form showed no efflux in the vector control or knockdown cell monolayers. A substantial amount of the lactone form was converted to the acid form during the assay period, and the newly generated acid also showed no efflux (Table 34.1). The present studies have demonstrated that shRNA-mediated knockdown Caco-2 cells are a valuable tool to investigate the contribution of specific transporters in the transcellular transport of drug molecules and to predict potential sites of pharmacokinetic interactions.

Hepatobiliary transporters, together with drug metabolism enzymes, govern drug distribution in the liver and elimination in bile. Knockdown of drug transporters in sandwich-cultured hepatocytes by siRNA or shRNA has been a feasible *in vitro* approach to study the expression and function of hepatobiliary transport-

ers and predict DDI (Tian et al., 2004; Yue et al., 2009; Liao et al., 2010; Swift et al., 2010). Canalicular Mrp2 and basolateral Mrp3 mediate the excretion of organic anions, including conjugated and unconjugated xenobiotics and bile acids, from the liver. Specific knockdown of Mrp2 (approximately 50% decrease in expression) in rat hepatocytes resulted in an approximately 45% decrease in the BEI of carboxydichlorofluorescein (CDF) (9.3% vs. 16.5%), but did not affect Mrp3 or radixin expression (Tian et al., 2004). Specific Mrp3 knockdown (approximately 50% decrease in expression) in rat hepatocytes resulted in significantly higher accumulation of CDF in cells plus bile canaliculi (32.3 vs. 24.4 pmol/mg of protein/10 min), but no change in cellular accumulation (13.7 vs. 15.6 pmol/mg of protein/10 min), consistent with an approximately 60% increase in the BEI of CDF. The extent of protein knockdown was in good agreement with changes in CDF disposition. The contribution of BCRP to drug/metabolite biliary excretion has been evaluated in the sandwich-cultured rat hepatocytes with Bcrp knockdown by shRNA (Yue et al., 2009, 2011; and Swift et al., 2010). Adenoviral vectors expressing shRNA targeting Bcrp (Ad-si01Bcrp) or a nontarget control (Ad-siNT) were packaged and infected into sandwich-cultured rat hepatocytes. shRNA targeting Bcrp efficiently knocked down Bcrp in sandwich-cultured rat hepatocytes, while levels of other transport proteins (Pgp, Mrp2, Bsep, Mrp4, and Oatp1a1) were unaffected. BEI and *in vitro* biliary clearance ($CL_{biliary}$) of nitrofurantoin (BCRP substrates) and digoxin (Pgp substrate) were compared among noninfected, Ad-siNT-, and Ad-si01Bcrp-infected sandwich-cultured rat hepatocytes (Yue et al., 2009). In sandwich-cultured rat hepatocytes exhibiting Bcrp knockdown, cellular accumulation of nitrofurantoin was increased markedly, and nitrofurantoin BEI and *in vitro* $CL_{biliary}$ were decreased to 11% and 14% of control, respectively. Digoxin values were unaffected by knockdown of Bcrp. The BEI and *in vitro* $CL_{biliary}$ of pitavastatin were decreased significantly to approximately 58% and 52% of control, respectively, in Bcrp knockdown sandwich-cultured rat hepatocytes (Yue et al., 2011), indicating that Bcrp plays a role in pitavastatin biliary excretion. The Bcrp/BCRP knockdown rat or human

TABLE 34.1. B-to-A P_{app} ($\times 10^{-6}$; cm/s; $n = 3$) of Statin Drugs across the Monlayers of Knockdown Cells

BL-to-AP P_{app} ($\times 10^{-6}$; cm/s)	Vector Control	Pgp Knockdown Cells	BCRP Knockdown Cells	MRP2 Knockdown Cells
Atorvastatin	28.1	8.6	20.3	21.5
Fluvastatin	96.1	25.3	57.3	48.3
Rosuvastatin	14.1	4.6	5.8	6.6
Lovastatin	22.6	26.4	35.4	27.3
Simvastatin	10.0	10.5	16.1	12.3

hepatocytes also demonstrated that Bcrp/BCRP was not involved in the biliary elimination of ^{99}mTc-mebrofenin (MEB) and ^{99}mTc-sestamibi (MIBI) (Swift et al., 2010).

The OATP-mediated DDI was evaluated by using RNAi in cultured human hepatocytes (Liao et al., 2010). In the human liver, OATP1B1, 1B3, and 2B1 are located at the basolateral membrane of hepatocytes and are involved in hepatic drug uptake and biliary elimination. Clinically significant DDIs mediated by hepatic OATPs have drawn great attention from clinical practitioners and researchers. However, there are considerable challenges to prospectively understanding the extent of OATP-mediated DDIs because of the lack of specific OATP inhibitors or substrates and the limitations of *in vitro* tools. Quantitative PCR, microarray, and immunoblotting analyses along with uptake assays illustrated that the expression and transport activity of hepatic OATPs were reduced by siRNA efficiently and specifically in this model. Although OATP siRNA decreased only 20–30% of the total uptake of cerivastatin into human hepatocytes, it caused a 50% reduction in cerivastatin metabolism, which was observed by monitoring the formation of the two major metabolites of cerivastatin. The results suggest that coadministration of a drug that is a hepatic OATP inhibitor could significantly alter the pharmacokinetic profile of cerivastatin in clinical studies. Further studies with this novel siRNA hepatocyte model demonstrated that OATP and CYP have a synergistic effect on cerivastatin–gemfibrozil interactions. The siRNA knockdown sandwich-cultured human hepatocytes may provide a new powerful model for evaluating DDIs.

Uptake transporters have been examined using the human alveolar epithelial-derived cell line A549 as a model (Seki et al., 2009) by RNAi. siRNAs targeted to the OATP2B1 gene decreased the OATP2B1 mRNA expression level in A549 cells up to approximately 50% and reduced the uptake of amiodarone up to approximately 40%. These results indicate that amiodarone uptake mediated by carriers, including OATP2B1, might lead to accumulation of amiodarone in the lung and AMD-induced pulmonary toxicity (AIPT).

Monocarboxylate transporters (MCTs) are proton-linked membrane carriers involved in the transport of monocarboxylate such as lactate, pyruvate, and ketone bodies, and some therapeutic drugs such as β-lactam antibiotics. The tissue distribution of MCT isoforms is quite different. MCT1 is ubiquitously distributed, being present in the intestine, colon, muscle, heart, brain, kidney, and red blood cells (Garcia et al., 1994; Lin et al., 1998; Price et al., 1998). The tissue distribution of MCT2 is more restricted, with MCT2 being found mainly in the testis, brain, skeletal muscle, heart, and kidney (Lin et al., 1998). MCT4 is present in the testis, small intestine, lung, brain, heart, kidney, and spleen (Dimmer et al., 2000). γ-Hydroxybutyric acid (GHB) is present in the mammalian brain as a metabolite of γ-aminobutyric acid (Bessman and Fishbein, 1963); studies have revealed that GHB is a neurotransmitter or neuromodulator with its own specific receptors (Snead, 2000). GHB has been detected in various tissues other than brain tissue, including tissues of the heart, kidney, liver, lung, muscle, and gastrointestinal tract (Nelson et al., 1981; Tedeschi et al., 2003). GHB is approved in the United States for treatment of the sleep disorder narcolepsy (Mamelak et al., 1986) and is used in the treatment of alcohol dependence in Europe (Gallimberti et al., 2000). The abuse of GHB by bodybuilders as a popular steroid alternative due to its growth hormone-releasing effects (Okun et al., 2001), and by drug abusers as a recreational drug at nightclubs and rave parties for its euphoric effects, has resulted in serious adverse effects, including coma, seizure, and death (Mason and Kerns, 2002). The distribution of GHB to various tissues, including the brain, is also dependent on specific transporters since at physiological pH, more than 99% of GHB is ionized and cannot readily diffuse across cellular membranes. Wang et al. (Wang et al., 2006; 2007a; Wang and Morris, 2007b) have employed a transient siRNA method to selectively knock down the expression of MCT1, MCT2, and MCT4, respectively, in human kidney HK-2 cells. Silencing MCT1 resulted in a 50% decrease in GHB transport, suggesting that MCT1 may be a major carrier for renal reabsorption of GHB, and modulation of MCT1 activity represents a potential detoxification strategy for treating GHB overdoses. Silencing MCT2 or MCT4 in mammalian MDA-MB231 cells significantly decreased their protein expression and the uptake of GHB; however, the decrease in GHB uptake with MCT2 inhibition was smaller than that for MCT4. This investigation demonstrated that GHB is a substrate for both MCT2 and MCT4. MCT4 represents the major transporter for GHB in these cells, with MCT2 responsible for a minor component of the uptake of GHB. These transporters may be important in both the bioavailability and renal clearance of GHB, as well as influencing its pharmacological activity by influencing its brain uptake and distribution.

34.3.1.3 Applications in Delineating the Transporter Function in Drug Efficacy and Toxicity

Fulfilling their role in detoxification, several ABC transporters have been found to be overexpressed in cancer cell lines cultured under selective pressure. Tissue culture studies have consistently shown that the major mechanism of MDR in most cultured cancer cells involves Pgp, MRP, or BCRP. RNAi has provided a new approach to further understand the role of these efflux pumps in tumor cells and to identify the importance of individual efflux pumps to chemotherapy. It has also presented an alternative way to overcome drug resistance.

RNAi has been widely used to reveal that MDR protein is one of the factors to cause drug resistance in chemotherapy and provide an alternative way to circumvent MDR. By using RNAi in tumor cells, it has been further discovered that MRP1 conferred drug resistance to MTX in HL60 cells (Golalipour et al., 2006), MRP4 conferred drug resistance to cisplatin in SGC7901, a gastric cancer cell line (Zhang et al., 2010), MRP3 conferred drug resistance to paclitaxel and monomethyl-auristatin-E (MMAE) in breast cancer cells (O'Brien et al., 2008), MRP5 conferred drug resistance to 5-florouracil in pancreatic carcinoma cells (Hagmann et al., 2009), BCRP conferred drug resistance to mitoxantrone (Lv et al., 2007) and 5-florouracil (Yuan et al., 2009) in breast cancer cells, and to mitoxantrone and topotecan in human choriocarcinoma BeWo cells (Ee et al., 2004), and Pgp but not MRP1 conferred drug resistance to etoposide, doxorubicin, and vincristine in human small-cell lung carcinoma (SCLC) cells (Sadava and Hamman, 2007).

Studies on the cellular disposition of targeted anticancer TKIs have mostly focused on imatinib while the functional importance of Pgp remains controversial for more recent TKIs. By using RNAi-mediated knockdown of MDR1, Haouala et al. (2010) have investigated and compared the specific functional consequence of Pgp on the cellular disposition of the major clinical in-use TKIs imatinib, dasatinib, nilotinib, sunitinib, and sorafenib. siRNA-mediated knockdown in K562/Dox cell lines provides a unique opportunity to dissect the specific contribution of Pgp to TKIs intracellular disposition. In these conditions, abrogating specifically Pgp-mediated efflux *in vitro* revealed the remarkable and statistically significant cellular accumulation of imatinib, dasatinib, sunitinib, and sorafenib, confirming that these TKIs are all substrates of Pgp. By contrast, no statistically significant difference in cellular disposition of nilotinib was observed as a result of MDR1 expression silencing, indicating that differential expression and/or function of Pgp is unlikely to affect nilotinib cellular disposition. This study enables for the first time a direct estimation of the specific contribution of one transporter among the various efflux and influx carriers involved in the cellular trafficking of these major TKIs *in vitro*. Knowledge on the distinct functional consequence of Pgp expression for the cellular distribution of these various TKIs is necessary to better appreciate the efficacy, toxicity, and potential DDIs of TKIs with other classes of therapeutic agents at the systemic, tissue, and cellular levels.

Although MDR proteins have been demonstrated to play important roles in chemotherapy resistance, less is

known about the function of MDR proteins in tumor cell growth. Several laboratories have used RNAi to evaluate the roles of MDR proteins in tumor cells. Katoh et al. (2008) demonstrated that knockdown of the MDR1 gene suppressed tumor cell proliferation *in vitro* and induced the passage of the cell cycle into the G1/G0 phase. They further showed that Mdr1 knockdown of tumor cells inhibited tumor expansion in a mouse xenograft tumor formation assay. These results suggest that MDR1 plays a role in regulation of tumor cells proliferation. Tagami et al. (2010) reported that RNAi-mediated MRP4 knockdown in human retinal microvascular endothelial cells (HRECs) did not affect cell proliferation but enhanced cell migration. Moreover, cell apoptosis induced by serum starvation was less prominent in MRP4 siRNA-treated HRECs as compared with control siRNA-treated HRECs. In a Matrigel-based tube-formation assay, although MRP4 knockdown did not lead to a significant change in the total tube length, MRP4 siRNA-treated HRECs assembled and aggregated into a massive tube-like structure, which was not observed in control siRNA-treated HRECs. These results suggest that MRP4 is uniquely involved in retinal angiogenesis. Dependence on glycolysis is a hallmark of malignant tumors. As a consequence, these tumors generate more lactate, which is effluxed from cells by MCTs. MCT 1 and 2 were the primary isoforms expressed in human glioblastoma multiforme and glioma-derived cell lines. In contrast, MCT 3 was the predominantly expressed isoform in normal brain. siRNA specific for MCT 1 and 2 reduced expression of these isoforms in U-87 MG cells (a glioma-derived cell line) to barely detectable levels and reduced lactate efflux by 30% individually and 85% in combination, with a concomitant decrease of intracellular pH by 0.6 units (a fourfold increase in intracellular H(+)). Prolonged silencing of both MCTs reduced viability by 75% individually and 92% in combination, as measured by both phenotypic and flow cytometric analyses. MCT targeting significantly reduced the viability of U-87 MG cells mediated by both apoptosis and necrosis. This indicates that this strategy may be a useful therapeutic avenue for treatment of patients with malignant glioma (Mathupala et al., 2004).

The usefulness and efficacy of cisplatin, a chemotherapeutic drug, are limited by its toxicity to normal tissues and organs, including the kidneys. The uptake of cisplatin in renal tubular cells is high, leading to cisplatin accumulation and tubular cell injury and death, culminating in acute renal failure. Extensive investigations have been focused on the signaling pathways of cisplatin nephrotoxicity, but much less is known about the mechanism of cisplatin uptake by renal cells and tissues. The copper transporter Ctr1 is highly expressed in the renal tubular cells; however, its role in cisplatin nephrotoxicity is not known. Pabla et al. (2009) demonstrated that Ctr1 was mainly expressed in both proximal and distal tubular cells in mouse kidneys and was mainly localized on the basolateral side of these cells, a proposed site for cisplatin uptake. Down-regulation of Ctr1 by siRNA or copper pretreatment resulted in decreased cisplatin uptake. Consistently, down-regulation of Ctr1 suppressed cisplatin toxicity, including cell death by both apoptosis and necrosis. These results are the first evidence of a role for Ctr1 in cisplatin uptake and nephrotoxicity.

34.3.2 Applications of Silencing Drug Metabolizing Enzymes

Compared with the number of studies using siRNA to evaluate drug transporters, there are relatively fewer studies that have utilized siRNA in evaluating the functions of human DMEs, presumably due to easy access to selective chemical inhibitors, inhibitory antibodies, and recombinant enzymes. The CYP function in the efficacy of cytotoxic agents has been evaluated by RNAi in cancer cells. To knock down *CYP3A4* gene expression, three different constructs have been designed and prepared to produce shRNA molecules (Chen et al., 2006a). A scrambled sequence was used as a negative control. The efficiency of individual constructs in silencing the *CYP3A4* gene was investigated in Chinese hamster cells that stably express CYP3A4. Interestingly, only a shRNA targeting 3′-UTR of CYP3A4 mRNA was found to knock down about 70–80% of CYP3A4 mRNA and protein levels, whereas other constructs targeting CYP3A4 coding regions did not. Suppression of CYP3A4 expression persisted until day 3, and recovery to regular expression levels was observed on day 5. Although it knocked down CYP3A4 expression, this shRNA had no impact on CYP3A5 expression, whose amino acid sequence is 84% identical to CYP3A4. As expected, silencing CYP3A4 expression by siRNA led to a sharp decrease in nifedipine oxidation. Furthermore, repressed CYP3A4 expression resulted in a reduced sensitivity to the anticancer drugs cyclophosphamide and ifosfamide, which are both bioactivated by CYP3A4. Compared with wild-type cells, CYP3A4-expressing cells were very sensitive to oxazaphosphrine drugs with an IC_{50} of 210 µM for cyclophosphamide and 55 µM for ifosfamide. When CYP3A4-expressing cells were transfected with CYP3A4 siRNA, cells became considerably resistant to both drugs with a half maximal inhibitory concentration (IC50) $IC_{50} > 1000$ µM. This represents one successful example of employing RNAi to investigate posttranscriptional gene silencing of human CYP drug metabolizing enzymes and the subsequent effects on drug metabolism.

RNAi has been applied to further confirm the CYP contributions to the biotansformation pathway. In order to clarify the relative role of CYP1A1 and CYP1B1 in bioactivation of aromatic hydrocarbons (polycyclic aromatic hydrocarbon [PAH]) in the human lung, Uppstad et al. (2010) performed RNA-interference studies with human lung cell lines *in vitro*. The capacity to bioactivate benzo[*a*]pyrene (B[*a*]P) in lung cells transfected with siRNA against CYP1A1, CYP1B1, or both was compared with control siRNA transfected cells. Subsequent fluorescence high-performance liquid chromatography (HPLC) analysis revealed that formation of B[*a*]P-tetrol-I-1 (hydrolyzed form of its ultimate carcinogenic metabolite B[*a*]P-diol-epoxide-I-1 [BPDE I-1] form) was dependent primarily on the expression of CYP1A1 in both cell lines. The precursors of the B[*a*] P-diol-epoxides (BPDEs), the B[*a*]P-*cis*- and *trans*-7,8-dihydrodiol isomers (B[*a*]P-DHDs) were, however, readily formed in cells expressing high levels of either CYP1A1 or CYP1B1. Simultaneous down-regulation of CYP1A1 and CYP1B1 mRNA resulted in low formation of metabolites overall, and residual unmetabolized B[*a*]P levels followed the expression of CYP1A1 in an inverse manner. This study shows that CYP1A1 is the most important enzyme in the bioactivation of B[*a*]P to the ultimate carcinogenic B[*a*]P-7,8-diol-9,10-epoxide metabolites in human lung cell lines *in vitro*. On the other hand, CYP1B1 may play an equal role compared with CYP1A1 in the formation of B[*a*]P-7,8-DHDs. Overall metabolism of B[*a*]P appears to be primarily dependent on CYP1A1.

There are a growing number of studies utilizing RNAi to define the role of nonhuman drug metabolizing enzymes in biotransformation of endobiotics and xenobiotics (Menzel et al., 2005; Kulas et al., 2008; Benenati et al., 2009; Schaefer et al., 2009).

34.3.3 Applications of Silencing Nuclear Receptors (NRs)

NRs regulate the expression of Phase I and II drug metabolizing enzymes and drug transporters and thus have a major impact on the metabolism and transport of xenobiotics or endobiotics. In some cases, altered expression of drug metabolizing enzymes and drug transporters via the activation of NRs may result in unwanted DDIs in clinical drug therapy. Silencing an NR with RNAi will not only aid in understanding the role of this NR in controlling its target gene expression but will also provide insight into potential effects on drug metabolism and transport.

Pregnane X receptor (PXR, NR1I2) also called the steroid X receptor (SXR), activates the expression of CYPs, glutathione *S*-transferases (GSTs), sulfotransferases (SULTs), uridine 5'-diphospho-glucuronosyltransferases (UDP-glucuronosyltransferases, UGTs) and ABC transporters that are important in the metabolism and transport of a large number of endogenous steroids and clinically used drugs. To delineate the mechanisms underlying PXR-mediated suppression of bile acid biosynthesis, Bhalla et al. (2004) examined the functional cross-talk between human PXR and HNF-4, a key hepatic activator of genes involved in bile acid biosynthesis including the cholesterol 7-α hydroxylase (CYP7A1) and sterol 12-α hydroxylase (CYP8B1) genes. Treatment with rifampicin resulted in repression of endogenous human CYP7A1 expression in HepG2 cells that was reversed by PXR siRNA.

Farnesoid X receptor (FXR, NR1H4) has been characterized as a bile acid NR and plays a pivotal role in maintaining bile salt and lipid homeostasis by regulating the expression of several key genes involved in bile acid synthesis, metabolism, and transport in the liver. However, as a permissive NR, FXR can also be activated through the ligand of its obligated heterodimeric partner, the retinoid X receptor (RXR, NR2B). Promoter reporter assays demonstrated that all-*trans* retinoic acid (atRA) specifically activated FXR/RXR. However, detailed molecular analyses indicated that this activation was through RXR, whose ligand is 9-*cis* RA. Knocking down of FXR or RXRα by siRNA in human hepatocytes increased CYP7A1 basal expression (Cai et al., 2010), but the repressive effect of atRA persisted, suggesting there are also FXR/RXR-independent mechanisms mediating atRA repression of CYP7A1 expression. The study demonstrates that atRA specifically represses human CYP7A1 expression. This suppression is mediated through both FXR/RXR-dependent and FXR/RXR-independent mechanisms. Neither atRA nor its metabolites are ligands for FXR. Rather, atRA and other RAs that are capable of conversion to 9cRA then permissively activate FXR/RXR. These findings may provide a potential explanation for hyperlipidemia in patients treated with high doses of RAs.

The CYP family/subfamily, as well as related transcription factor(s), is responsible for the estrogen-dependent synthesis of epoxyeicosatrienoic acids (EETs) to initiate shear stress-induced vasodilation. Knockdown of RXRgamma in isolated mesenteric arteries/arterioles of female endothelial nitric oxide synthase-knockout mice significantly inhibited the production of EETs, parallel to reduced vasodilation. RXRgamma siRNA not only silenced vascular RXR-gamma expression but also synchronously down-regulated CYP2C29 expression, leading to a reduced EET synthesis. These data provide the first evidence for a specific signaling cascade by which estrogen potentially activates the CYP2C29 gene in the absence of nitric oxide to synthesize EETs in response to shear stress via an RXRgamma-related regulatory mechanism (Sun et al., 2010).

Constitutively activated receptor (CAR, NR1I3) has specific and overlapping functions in regulating the expression of CYP2B, CYP3A, GSTs, SULTs, UGTs, and ABC transporters. These genes consist of many functional response elements for both PXR and CAR. Furthermore, both PXR and CAR may be activated by the same ligand, such as chlorpromazine or phenobarbital. Studies on the selectivity of CAR- and PXR-mediated gene regulation could be confounded by such "cross-talk" between PXR and CAR. Specific silencing of CAR, in combination with other techniques, would be helpful in understanding the role of CAR in drug metabolizing enzyme and drug transporter gene regulation. Indeed, Chen et al. (2006b) have employed both reporter assay and RNAi technique in delineating SULT2A1 gene regulation by CAR.

Aryl hydrocarbon receptor (AhR) is a member of the Per-Arnt-Sim, beta helix-loop-helix superfamily, which activates the expression of CYP1A, CYP1B, GSTs, UGTs, and other drug metabolizing enzymes. AhR mediates the expression of CYP1A1 induced by PAHs and is thought to be involved in the regulation of cell growth and differentiation. Human umbilical vein endothelial cells (ECs) were exposed to laminar shear stress (SS) and thereafter collected to evaluate the expression, activity, and transcription of CYP1A1 and the expression of AhR and cell cycle-related proteins (Han et al., 2008). A physiological level of laminar SS (15 dynes/cm^2) markedly increased the expression level and enzymatic activity of CYP1A1. An siRNA for AhR significantly suppressed laminar SS-induced CYP1A1 expression. The siRNA also abolished SS-induced cell cycle arrest, the expression of the cyclin-dependent kinase inhibitor p21(Cip1), and dephosphorylation of retinoblastoma protein. The results indicate that laminar SS stimulated the transcription of CYP1A1 through the activation of AhR in a way that is similar to the effects of PAH and AhR is involved in cell cycle arrest induced by SS.

34.3.4   Applications in *in vivo*

Contrary to the great interest and progress of *in vitro* siRNA transfection, there are few reports of *in vivo* use of siRNA due to the critical issue of siRNA delivery. As discussed in Section 34.2.5.5, at present, the *in vivo* siRNA systems are still facing extensive challenges, including low stability in blood, nonspecific binding in tissues, poor cellular uptake, low efficiency, and rapid excretion in urine. To study the biodistribution of siRNA, rats were injected IV with radiolabeled siRNA or radiolabel alone (control), and scintigraphic images

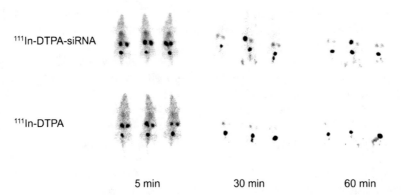

FIGURE 34.8. Scintigraphic images of the uptake of radiolabeled siRNA. Rats ($n = 3$ for each treatment group) were injected intravenously with either ^{111}In-DTPA (control) or ^{111}In-DTPA-siRNAMrp4, and the biodistribution of radiolabel was followed on a gamma camera for 1 h.

were acquired at different time intervals postinjection. The siRNA preferentially accumulated in the kidneys and was excreted in the urine. One hour after injection, the amount of siRNA present in both kidneys was on average 40 times higher than in other tissues (liver, brain, intestine, muscle, lung, spleen, and blood) (Figure 34.8) (van de Water et al., 2006). Besides the biodistribution, the effect of siRNA on Mrp2/Abcc2 (siRNAMrp2) in renal proximal tubules was investigated. Mrp2 function was assessed by measuring the excretion of its fluorescent substrate calcein in the isolated perfused rat kidney. Four days after administration, siRNAMrp2 reduced the urinary calcein excretion rate significantly (35% inhibition over the period 80–150 min of perfusion). This down-regulation was specific because another siRNA sequence directed against a different transporter in the proximal tubule, Mrp4 (Abcc4, siRNAMrp4), did not alter the Mrp2-mediated excretion of calcein. These studies indicate that siRNA accumulates spontaneously in the kidney after IV injection, where it selectively suppresses gene function in the proximal tubules. Therefore, intravenously administered siRNA may provide a novel experimental and potential therapeutic tool for gene silencing in the kidney.

Yumi et al. (2005) employed a hydrodynamics-based procedure to repeatedly treat mice with synthetic siRNAs or siRNA-expressing plasmid DNAs in naked form intravenously via a large-volume and high-speed injection. The amounts of targeted mRNA and Pgp in the liver were determined by real-time PCR and Western blot analysis, respectively. Following administration of synthetic siRNAs or siRNA-expressing plasmid DNAs (pDNAs) directed against mdr1a, the mRNA level in the liver was significantly reduced to approximately 50–60% of that in control mice. Furthermore, a slight reduction was observed at the protein level. Similar results were obtained in the experiments using siRNA-expressing

pDNA directed against mdr1a/1b. These results demonstrate that sequence-specific suppression of mdr1 gene expression is possible at the mRNA level as well as the protein level in mice following IV delivery of siRNA effectors. Hino et al. (2006) suppressed gene expression in mouse brain capillary endothelial cells (BCECs) by intravenous injection of a large dose of OAT3 siRNA using with the hydrodynamic technique. The OAT3 siRNA could be delivered to the BCECs and efficiently inhibited the endogenously expressed protein of the BCECs. The suppressive effect of siRNA on OAT3 is sufficient to reduce the brain-to-blood transport of OAT3 substrate, benzylpenicillin at the blood brain barrier (BBB). The *in vivo* siRNA-silencing method with hydrodynamic technique may be useful for the study of BBB function and gene therapy targeting BCECs.

To date, most of the *in vivo* RNAi technology applied to transporters has been used to confirm their roles in MDR. It is important to restrict the delivery of RNAi to targeted cells. Use of antigen-modified retroviral- and adenoviral-based as well as other vector systems may minimize problems associated with targeted delivery of siRNAs. Using a MDR1-FLuc fusion construct, *in vivo* down-regulation of Pgp protein levels has been directly and noninvasively visualized in living animals, following injection of D-luciferin, the FLuc substrate (Pichler et al., 2005). *In vivo* delivery of retroviral-mediated shRNAi targeting MDR1-FLuc stably expressed in mouse livers showed a specific fourfold reduction in FLuc-mediated bioluminescence (Figure 34.9). Scrambled shRNAi sequences or an empty vector plasmid had no effect on light output. Similar experiments were done with hydrodynamic liver injections of pMDR1-eGFP combined with the same shRNAi constructs. Using fluorescence microscopy of liver sections, down-regulation of the MDR1-eGFP fusion was only detected in mice treated with shRNAi against MDR1 and not in

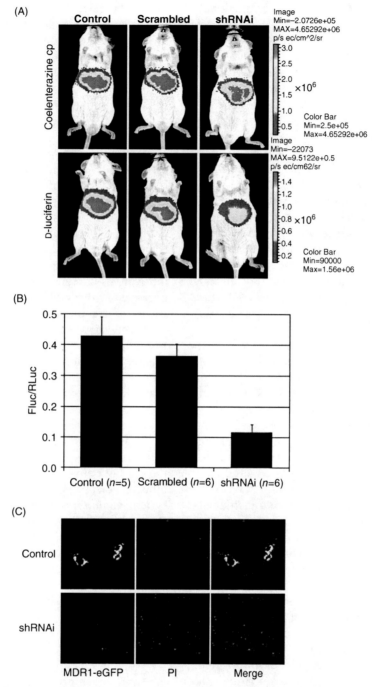

FIGURE 34.9. Down-regulation of P-glycoprotein *in vivo*: shRNAi-mediated reduction in P-glycoprotein protein levels. (A) Representative bioluminescence images of RLuc expression with coelenterazine cp (top) and P-glycoprotein-FLuc expression with D-luciferin (bottom). Mice were hydrodynamically transfected with pMDR1-FLuc (1 μg) combined with pRLuc-N3 (1 μg; as transfection control) and treated as indicated with a 10-fold excess of control (left), scrambled shRNAi (middle), or shRNAi against MDR1 (right). (B) Quantification of data obtained from animals treated with control ($n = 5$), scrambled shRNAi ($n = 6$), or shRNAi against MDR1 ($n = 6$). Columns, total flux ratio of FLuc/RLuc to correct for variation in injections of expression cassettes; bars, ±SE. $P < 0.002$ comparing control mice with shRNAi-treated mice; $P < 0.003$ comparing scrambled treated mice with shRNAi-treated mice. (C) Fluorescence confocal microscopy images (10×) from liver cryosections harvested from animals hydrodynamically transfected with pMDR1-eGFP (1 μg) and fivefold excess of control (top) or shRNAi (bottom). Tissue fluorescence emitted by excitation of eGFP (left), by excitation of propidium iodide (PI) (middle), and merged images (right).

control mice. Thus, targeted delivery and functional expression of shRNAi is feasible *in vivo* and can be used to down-regulate expression of MDR1 mRNA. shRNAi effectively inhibited MDR1 expression and function in cultured cells, tumor implants, and mammalian liver, documenting the feasibility of a knockdown approach to reversing MDR *in vivo*. Imaging Pgp transport activity and inhibition in a variety of tissues within the intact animal and direct documentation of shRNAi-targeted effects *in vivo* using bioluminescence provide further evidence to establish the potential of knockdown strategies for circumventing MDR.

Chen et al. (2008, 2009) explored the potential use of gene therapy with adenoviral-delivered RNAi against mdr1b as a method to sensitize refractory epilepsy to antiepileptic drugs (AEDs). They constructed replication-deficient recombinant adenovirus Adeno-mdr1b1 carrying shRNA targeting against mdr1b, and successfully infected the established Sprague Dawley rat astrocyte model of coriaria lactone-induced Pgp over-expression. The expression levels of mdr1b and Pgp and the rhodamine123 efflux ratio in trial groups were significantly lower than that of the blank control during the first 7 days postinfection, with the most inhibition at 48 h. The results suggest that knockdown of MDR using adenovirus not only avoided the toxicity and low rate of plasmid nucleofection but also overcame its poor efficiency of mdr1b silencing. Jiang et al. (2007) demonstrated knockdown Pgp and enhanced the therapeutic effect of cisplatin by using attenuated *Salmonella* Typhi as an *in vivo* delivery vector for MDR gene (MDR1) siRNA in a mouse model bearing human tongue squamous cell cancer. Pan et al. (2009) used the vector-based RNAi system for the knockdown of MDR1, restoring its sensitivity to adriamycin and doxorubicin in Pgp over expression tumor cells both *in vitro* and in mice. *In vivo*, after transfected with vector pSUPER-shRNA/mdr1 through the peritoneal cavity and *in situ*, carcinoma cells(HepG2 and HepG2/mdr1) measured by flow cytometry and IHC analysis showed that dsRNA/mdr1 was successfully and effectively brought into cells, and that Pgp expression in positive groups was significantly down-regulated compared with control groups (65.1% vs. 94.1%, $P < 0.05$). The tumor suppression rate for test groups was 57.8%. After chemotherapy, the growth rate of tumors for test groups was slower than that for control groups (700 vs. 1659, $P < 0.05$). Li et al. (2009) combined plasmid pSUPER-BCRP and lipofectamine to deliver BCRP siRNA by intraperitoneal injection in multidrug-resistant hepatocellular carcinoma-bearing nude mice. RNAi plasmid pSUPER-BCRP decreased the BCRP mRNA transcription and protein expression levels in hepatocellular carcinoma tissues of nude mice effectively and reversed

MDR at a certain degree. After adriamycin treatment, the sizes of carcinomas in mice in the RNAi injected group decreased significantly compared with control mouse group.

Although viral vectors have successfully delivered siRNA or shRNA to tumor cells in xenographed mouse or disease rat models, viral vectors do not have the dual capacity of shRNA, and drug delivery and successful targeting of tumor cells *in vivo* remains problematic. MacDiarmid et al. (2009) provided a strategy that involved dual sequential treatments. First, a known drug resistance mechanism (e.g., overexpression of the MDR Pgp MDR1) was knocked down by means of si/shRNA-containing minicells targeted to tumors via bispecific antibody (BsAb). After allowing for sufficient time to achieve substantial knockdown of the drug resistance-mediating protein, a second wave of therapy involving intravenous administration of BsAb-targeted minicells packaged with cytotoxic drug was followed. The two waves of treatment, involving minicells loaded with both types of payload enabled complete survival without toxicity in mice with tumor xenografts while involving several thousandfold less drug, siRNA, and antibody than needed for conventional systemic administration of cancer therapies.

While *in vivo* RNAi has been demonstrated in preclinical models, selection of a siRNA specific only to the target gene is critical to *in vivo* transfection. Delivery is still a problem especially in clinical applications since the current delivery methods such as hydrodynamic IV infusion, electroporation, or a vector-based delivery system cannot be applied to human subjects safely. Because transporters and enzymes are expressed in normal tissues as well, the need to silence a given transporter in all cells will likely be limited to preclinical and mechanistic research. Nevertheless, therapeutic use of siRNA in clinic will likely require an additional building block to limit exposure to target cells.

34.4 CONCLUSIONS

RNAi has been proven to be an invaluable addition to the existing tools for gene function analyses since the discovery of the RNAi pathway. The use of RNAi is spreading rapidly to nearly every aspect of biomedical research. The gene silencing capability of RNAi is being used to study the biological functions of individual genes and their roles in biochemical pathways. Down-regulation of specific genes with the RNAi approach helps to define the critical role of individual drug metabolism enzymes, drug transporters, or NRs in drug metabolism and disposition, gene regulation, and DDIs. However, off-target effects have been observed both *in vitro* and *in vivo*. While the specificity of siRNA molecules is not defined, one should be cautious when interpreting RNAi data because other mechanisms may be provoked. The efficacy of RNAi depends on efficient delivery of the intermediates of RNAi, siRNA, and shRNA oligonucleotides. The delivery challenge is even greater when the aim is to inhibit the expression of target genes in animal models. The potential use of siRNAs as a therapeutic agent is also exciting and holds great promise for the future. For the study of drug transporter functions in ADME and in the treatment of disease, RNAi presents an alternative way to assemble interpretable mechanistic data. In spite of a few promising clinical trials, effective delivery of siRNA *in vivo* remains a pivotal challenge in translating RNAi into a conventional treatment option in the clinic.

ACKNOWLEDGMENT

The authors would like to thank Ms. Heidi Correira for expert proofreading.

REFERENCES

Benenati G, Penkov S, Mueller-Reichert T, Entchev EV, Kurzchalia TV (2009) Two cytochrome P450s in *Caenorhabditis elegans* are essential for the organization of eggshell, correct execution of meiosis and the polarization of embryo. *Mech Dev* 126(5–6):382–393.

Bessman SP, Fishbein WN (1963) Gamma-Hydroxybutyrate, a normal brain metabolite. *Nature* 200:1207–1208.

Bhalla S, Ozalp C, Gang S, Xiang L, Kemper JK (2004) Ligand-activated pregnane X receptor interferes with HNF-4 signaling by targeting a common coactivator PGC-1α: Functional implications in hepatic cholesterol and glucose metabolism. *J Biol Chem* 279:45139–45147.

Bonifacino JS, Dell'Angelica EC, Springer TA (2001) Immunoprecipitation. In *Current Protocols in Molecular Biology*, 10.16.1–10.16.29. John Wiley & Sons, Inc.

Braulke LJ, Klingenspor M, DeBarber A, Tobias SC, Grandy DK, Scanlan TS, Heldmaier G (2008) 3-Iodothyronamine: A novel hormone controlling the balance between glucose and lipid utilisation. *J Comp Physiol* 178:167–177.

Cai SY, He H, Nguyen T, Mennone A, Boyer JL (2010) Retinoic acid represses CYP7A1 expression in human hepatocytes and HepG2 cells by FXR/RXR-dependent and independent mechanisms. *J Lipid Res* 51:2265–2274.

Castanotto D, Li H, Rossi JJ (2002) Functional siRNA expression from transfected PCR products. *RNA* 8(11): 1454–1460.

Celius T, Garberg P, Lundgren B (2004) Stable suppression of MDR1 gene expression and function by RNAi in Caco-2 cells. *Biochem Biophys Res Commun* 324(1):365–371.

Chan SM, Olson JA, Utz PJ (2006) Single-cell analysis of siRNA-mediated gene silencing using multiparameter flow cytometry. *Cytometry A* 69(2):59–65.

Chen J, Yang XX, Huang M, Hu ZP, He M, Duan W, Chan E, Sheu FS, Chen X, Zhou AF (2006a) Small interfering RNA-mediated silencing of cytochrome P450 3A4 gene. *Drug Metab Dispos* 34:1650–1657.

Chen L, Tian L, Yang T, Cheng X, Hermann S, Zhou D (2008) Reversal of Mdr1b-dependent multidrug resistance in a rat astrocyte model by adenoviral-delivered short hairpin RNA. *Cell Mol Neurobiol* 28(8):1057–1066.

Chen L, Cheng X, Tian L, Yang T, Herman S, Zhou D (2009) Inhibition of P-glycoprotein over-expression by shRNA-mdr1b in rat astrocytes. *Neurochem Res* 34(3):411–417.

Chen X, Zhang J, Baker SM, Chen GP (2006b) Human constitutive androstane receptor mediated methotrexate induction of human dehydroepiandrosterone sulfotransferase (hSULT2A1). *Toxicology* 231:224–233.

Clegg RM (1995) Fluorescence resonance energy transfer. *Curr Opin Biotechnol* 6(1):103–110.

Couto LB, High KA (2010) Viral vector-mediated RNA interference. *Curr Opin Pharmacol* 10(5):534–542.

Darnell M, Karlsson JE, Owen A, Hidalgo IJ, Li J, Zhang W, Andersson TB (2010) Investigation of the involvement of P-glycoprotein and multidrug resistance-associated protein 2 in the efflux of ximelagatran and its metabolites by using short hairpin RNA knockdown in Caco-2 cells. *Drug Metab Dispos* 38(3):491–497.

Dimmer KS, Friedrich B, Lang F, Deitmer JW, Broer S (2000) The low-affinity monocarboxylate transporter MCT4 is adapted to the export of lactate in highly glycolytic cells. *Biochem J* 350(Pt 1):219–227.

Donnelly RF, Raj Singh TR, Woolfson AD (2010) Microneedle-based drug delivery systems: Microfabrication, drug delivery, and safety. *Drug Deliv* 17(4):187–207.

Donze O, Picard D (2002) RNA interference in mammalian cells using siRNAs synthesized with T7 RNA polymerase. *Nucleic Acids Res* 30(10):e46.

Ee PLR, He X, Ross DD, Beck WT (2004) Modulation of breast cancer resistance protein (BCRP/ABCG2) gene expression using RNA interference. *Mol Cancer Ther* 3(12):1577–1584.

Elbashir SM, Harborth J, Lendeckel W, Yalcin A, Weber K, Tuschl T (2001) Duplexes of 21-nucleotide RNAs mediate RNA interference in cultured mammalian cells. *Nature* 411(6836):494–498.

Elbashir SM, Harborth J, Weber K, Tuschl T (2002) Analysis of gene function in somatic mammalian cells using small interfering RNAs. *Methods* 26(2):199–213.

Fire A, Xu S, Montgomery MK, Kostas SA, Driver SE, Mello CC (1998) Potent and specific genetic interference by double-stranded RNA in Caenorhabditis elegans. *Nature* 391(6669):806–811.

Friesema EC, Jansen J, Visser TJ (2005) Thyroid hormone transporters. *Biochem Soc Trans* 33:228–232.

Gallimberti L, Spella MR, Soncini CA, Gessa GL (2000) Gamma-hydroxybutyric acid in the treatment of alcohol and heroin dependence. *Alcohol* 20:257–262.

Garcia CK, Goldstein JL, Pathak RK, Anderson RG, Brown MS (1994) Molecular characterization of a membrane transporter for lactate, pyruvate, and other monocarboxylates: Implications for the Cori cycle. *Cell* 76:865–873.

Gilmore IR, Fox SP, Hollins AJ, Akhtar S (2006) Delivery strategies for siRNA-mediated gene silencing. *Curr Drug Deliv* 3(2):147–145.

Golalipour M, Mahjoubi F, Sanati MH (2006) RNAi induced inhibition of MRP1 expression and reversal of drug resistance in human promyelocytic HL60 cell line. *Iran J Biotechnol* 4(3):169–173.

Graber-Maier A, Gutmann H, Drewe J (2010) A new intestinal cell culture model to discriminate the relative contribution of P-gp and BCRP on transport of substrates such as imatinib. *Mol Pharm* 7(5):1618–1628.

Hagmann W, Jesnowski R, Faissner R, Guo C, Loehr JM (2009) ATP-binding cassette C transporters in human pancreatic carcinoma cell lines. *Pancreatology* 9(1–2):136–144.

Han Z, Miwa Y, Obikane H, Mitsumata M, Takahashi-Yanaga F, Morimoto S, Sasaguri T (2008) Aryl hydrocarbon receptor mediates laminar fluid shear stress-induced CYP1A1 activation and cell cycle arrest in vascular endothelial cells. *Cardiovasc Res* 77(4):809–818.

Hannon GJ (2002) RNA interference. *Nature* 418:244–251.

Haouala A, Rumpold H, Untergasser G, Buclin T, Ris HB, Widmer N, Decosterd LA (2010) siRNA-mediated knockdown of P-glycoprotein expression reveals distinct cellular disposition of anticancer tyrosine kinases inhibitors. *Drug Metab Lett* 4(2):114–119.

Hino T, Yokota T, Ito S, Nishina K, Kang YS, Mori S, Hori S, Kanda T, Terasaki T, Mizusawa H (2006) *In vivo* delivery of small interfering RNA targeting brain capillary endothelial cells. *Biochem Biophys Res Commun* 340(1):263–267.

Horiba N, Masuda S, Takeuchi A, Saito H, Okuda M, Inui K (2004) Gene expression variance based on random sequencing in rat remnant kidney. *Kidney Int* 66:29–45.

Ianculescu AG, Giacomini KM, Scanlan TS (2009) Identification and characterization of 3-iodothyronamine intracellular transport. *Endocrinology* 150(4):1991–1999.

Jackson AL, Bartz SR, Schelter J, Kobayashi SV, Burchard J, Mao M, Li B, Cavet G, Linsley PS (2003) Expression profiling reveals off-target gene regulation by RNAi. *Nat Biotechnol* 21(6):635–637.

Jansen J, Friesema EC, Milici C, Visser TJ (2005) Thyroid hormone transporters in health and disease. *Thyroid* 15:757–768.

Jiang Z, Zhao P, Zhou Z, Liu J, Qin L, Wang H (2007) Using attenuated salmonella typhi as tumor targeting vector for MDR1 siRNA delivery an experimental study. *Cancer Biol Ther* 6(4):555–560.

Jiang ZY, Zhou QL, Coleman KA, Chouinard M, Boese Q, Czech MP (2003) Insulin signaling through Akt/protein kinase B analyzed by small interfering RNA-mediated gene silencing. *Proc Natl Acad Sci U S A* 100(13): 7569–7574.

Katoh SY, Ueno M, Takakura N (2008) Involvement of MDR1 function in proliferation of tumor cells. *J Biochem* 143(4):517–524.

Khvorova A, Reynolds A, Jayasena S (2003) Functional siRNAs and miRNAs exhibit strand bias. *Cell* 115(1): 209–216.

Kulas J, Schmidt C, Rothe M, Schunck WH, Menzel R (2008) Cytochrome P450-dependent metabolism of eicosapentaenoic acid in the nematode Caenorhabditis elegans. *Arch Biochem Biophys* 472(1):65–75.

Kumara LD, Clarkeb AR (2007) Opportunities and challenges for therapeutic gene silencing using RNAi and microRNA technologies. *Adv Drug Deliv Rev* 59:87–100.

Lares MR, Rossi JJ, Ouellet DL (2010) RNAi and small interfering RNAs in human disease therapeutic applications. *Trends Biotechnol* 28(11):570–579.

Lee SH, Sinko PJ (2006) siRNA—Getting the message out. *Eur J Pharm Sci* 27:401–410.

Li T, Chang CY, Jin DY, Lin PJ, Khvorova A, Stafford DW (2004) Identification of the gene for vitamin K epoxide reductase. *Nature* 427(6974):541–544.

Li T, Zeng G, Chen Y, Yu S, Wang X, Gong J (2009) *In vivo* experimental study on reversion of multidrug-resistance of hepatocellular carcinoma by suppression of breast cancer resistance protein gene. *Zhongguo Shengwu Zhipinxue Zazhi* 22(12):1161–1168.

Liao M, Raczynski AR, Chen M, Chuang BC, Zhu Q, Shipman R, Morrison J, Lee D, Lee FW, Balani SK, Xia CQ (2010) Inhibition of hepatic organic anion-transporting polypeptide by RNA interference in sandwich-cultured human hepatocytes: An *in vitro* model to assess transporter-mediated drug-drug interactions. *Drug Metab Dispos* 38(9):1612–1622.

Lin RY, Vera JC, Chaganti RS, Golde DW (1998) Human monocarboxylate transporter 2 (MCT2) is a high affinity pyruvate transporter. *J Biol Chem* 273:28959–28965.

Liu X, LeCluyse EL, Brouwer KR, Gan LS, Lemasters JJ, Stieger B, Meier PJ, Brouwer KL (1999a) Biliary excretion in primary rat hepatocytes cultured in a collagen-sandwich configuration. *Am J Physiol* 277(40):G12–G21.

Liu X, LeCluyse EL, Brouwer KR, Lightfoot RM, Lee JI, Brouwer KL (1999b) Use of Ca2+ modulation to evaluate biliary excretion in sandwich-cultured rat hepatocytes. *J Pharmacol Exp Ther* 289(3):1592–1599.

Lutfalla G, Uze G (2006) Performing quantitative reverse-transcribed polymerase chain reaction experiments. *Methods Enzymol* 410:386–400.

Lv H, He Z, Liu X, Yuan J, Yu Y, Chen Z (2007) Reversal of BCRP-mediated multidrug resistance by stable expression of small interfering RNAs. *J Cell Biochem* 102(1):75–81.

MacDiarmid JA, Amaro-Mugridge NB, Madrid-Weiss J, Sedliarou I, Wetzel S, Kochar K, Brahmbhatt VN, Phillips L, Pattison ST, Petti C, Stillman B, Graham RM, Brahmbhatt H (2009) Sequential treatment of drug-resistant tumors with targeted minicells containing siRNA or a cytotoxic drug. *Nat Biotechnol* 27(7):643–651.

Mamelak M, Scharf MB, Woods M (1986) Treatment of narcolepsy with gammahydroxybutyrate. A review of clinical and sleep laboratory findings. *Sleep* 9:285–289.

Mason B, Rosenberg IH (1994) Intestinal absorption of folate. In *Physiology of the Gastrointestinal Tract*, Johnson LR, ed., pp. 1979–1995. Raven Press, New York.

Mason PE, Kerns WP (2002) Gamma hydroxybutyric acid (GHB) intoxication. *Acad Emerg Med* 9:730–739.

Mathupala SP, Parajuli P, Sloan AE (2004) Silencing of monocarboxylate transporters via small interfering ribonucleic acid inhibits glycolysis and induces cell death in malignant glioma: An *in vitro* study. *Neurosurgery* 55(6):1410–1419.

McEwan GT, Lucas ML, Denvir M, Raj M, McColl KE, Russell RI, Mathan VI (1990) A combined TDDA-PVC pH and reference electrode for use in the upper small intestine. *J Med Eng Technol* 14:16–20.

McManus MT, Sharp PA (2002) Gene silencing in mammals by small interfering RNAs. *Nat Rev Genet* 3(10):737–747.

McManus MT, Haines BB, Dillon CP, Whitehurst CE, van Parijs L, Chen J, Sharp PA (2002) Small interfering RNA-mediated gene silencing in T lymphocytes. *J Immunol* 169(10):5754–5760.

Menzel R, Roedel M, Kulas J, Steinberg CEW (2005) CYP35: Xenobiotically induced gene expression in the nematode *Caenorhabditis elegans*. *Arch Biochem Biophys* 438(1):93–102.

Myers JW, Jones JT, Meyer T, Ferrell JE Jr. (2003) Recombinant Dicer efficiently converts large dsRNAs into siRNAs suitable for gene silencing. *Nat Biotechnol* 21(3):324–328.

Nelson T, Kaufman E, Kline J, Sokoloff L (1981) The extraneural distribution of gammahydroxybutyrate. *J Neurochem* 37:1345–1348.

Nykanen A, Haley B, Zamore PD (2001) ATP requirements and small interfering RNA structure in the RNA interference pathway. *Cell* 107(3):309–321.

O'Brien C, Cavet G, Pandita A, Hu X, Haydu L, Mohan S, Toy K, Rivers CS, Modrusan Z, Amler LC, Lackner MR (2008) Functional genomics identifies ABCC3 as a mediator of taxane resistance in HER2-amplified breast cancer. *Cancer Res* 68(13):5380–5389.

Okun MS, Boothby LA, Bartfield RB, Doering PL (2001) GHB: An important pharmacologic and clinical update. *J Pharm Pharm Sci* 4:167–175.

Pabla N, Murphy RF, Liu K, Dong Z (2009) The copper transporter Ctr1 contributes to cisplatin uptake by renal tubular cells during cisplatin nephrotoxicity. *Am J Physiol Renal Physiol* 296(3):F505–F511.

Pan GD, Yang JQ, Lv NY, Chu GP, Xiao Y, Lin Y (2009) Reversal of multi-drug resistance by pSUPER-shRNA-mdr1 *in vivo* and *in vitro*. *World J Gastroenterol* 15(4):431–440.

Pauli-Magnus C, von Richter O, Burk O, Ziegler A, Mettang T, Eichelbaum M, Fromm MF (2000) Characterization of the major metabolites of verapamil as substrates and inhibitors of P-glycoprotein. *J Pharmacol Exp Ther* 293(2):376–382.

Phalon C, Rao DD, Nemunaitis J (2010) Potential use of RNA interference in cancer therapy. *Expert Rev Mol Med* 12:e26.

Pichler A, Zelcer N, Pior JL, Kuil AJ, Piwnica-Worms D (2005) *In vivo* RNA interference-mediated ablation of MDR1 P-glycoprotein. *Clin Cancer Res* 11:4487–4494.

Powers HJ (2003) Riboflavin (vitamin B2) health. *Am J Clin Nutr* 77:1352–1360.

Price NT, Jackson VN, Halestrap AP (1998) Cloning and sequencing of four new mammalian monocarboxylate transporter (MCT) homologues confirms the existence of a transporter family with an ancient past. *Biochem J* 329(Pt 2):321–328.

Purves D, Lotto RB, Williams SM, Nundy S, Yang Z (2001) Why we see things the way we do: Evidence for a wholly empirical strategy of vision. *Philos Trans R Soc Lond B Biol Sci* 356(1407):285–297.

Qiu A, Jansen M, Sakaris A, Min SH, Chattopadhyay S, Tsai E, Sandoval C, Zhao R, Akabas MH, Goldman ID (2006) Identification of an intestinal folate transporter and the molecular basis for hereditary folate malabsorption. *Cell* 127(5):917–928.

Ryan US (1986) Immunofluorescence and immunocytochemistry of endothelial surface antigens. *Methods Cell Sci* 10(1):27–30.

Sadava D, Hamman W (2007) RNA interference against MDR1 but not MRP1 reverses drug resistance in human small-cell lung cancer cells. *J Cancer Mol* 3(3):91–94.

Scanlan TS, Suchland KL, Hart ME, Chiellini G, Huang Y, Kruzich PJ, Frascarelli S, Crossley DA, Bunzow JR, Ronca-Testoni S, Lin ET, Hatton D, Zucchi R, Grandy DK (2004) 3-Iodothyronamine is an endogenous and rapid-acting derivative of thyroid hormone. *Nat Med* 10:638–642.

Schaefer P, Mueller M, Krueger A, Steinberg CEW, Menzel R (2009) Cytochrome P450-dependent metabolism of PCB52 in the nematode *Caenorhabditis elegans*. *Arch Biochem Biophys* 488(1):60–68.

Schena M, Shalon D, Davis RW, Brown PO (1995) Quantitative monitoring of gene expression patterns with a complementary DNA microarray. *Science* 270(5235):467–470.

Schiffelers RM, Woodle MC, Scaria P (2004) Pharmaceutical prospects for RNA interference. *Pharm Res* 21(1):1–7.

Seki S, Kobayashi M, Itagaki S, Hirano T, Iseki K (2009) Contribution of organic anion transporting polypeptide OATP2B1 to amiodarone accumulation in lung epithelial cells. *Biochim Biochim Biophys Acta* 1788(5):911–917.

Selhub J, Rosenberg IH (1981) Folate transport in isolated brush border membrane vesicles from rat intestine. *J Biol Chem* 256:4489–4493.

Semizarov D, Frost L, Sarthy A, Kroeger P, Halbert DN, Fesik SW (2003) Specificity of short interfering RNA determined through gene expression signatures. *Proc Natl Acad Sci U S A* 100(11):6347–6352.

Shim MS, Kwon YJ (2010) Efficient and targeted delivery of siRNA *in vivo*. *FEBS J* 277(23):4814–4827.

Snead OC 3rd (2000) Evidence for a G protein-coupled gamma-hydroxybutyric acid receptor. *J Neurochem* 75: 1986–1996.

Sonders MS, Quick M, Javitch JA (2005) How did the neurotransmitter cross the bilayer? A closer view. *Curr Opin Neurobiol* 15:296–304.

Soutschek J, et al. (2004) Therapeutic silencing of an endogenous gene by systemic administration of modified siRNAs. *Nature* 432:173–178.

Stevens JC, Hines RN, Gu C, Koukouritaki SB, Manro JR, Tandler PJ, Zaya MJ (2003) Developmental expression of the major human hepatic CYP3A enzymes. *J Pharmacol Exp Ther* 307(2):573–582.

Stover PJ (2004) Physiology of folate and vitamin B12 in health and disease. *Nutr Rev* 62:S3–S12.

Sun D, Yang YM, Jiang H, Wu H, Ojaimi C, Kaley G, Huang A (2010) Roles of CYP2C29 and RXR gamma in vascular EET synthesis of female mice. *Am J Physiol Regul Integr Comp Physiol* 298(4):R862–R869.

Swift B, Yue W, Brouwer KLR (2010) Evaluation of (^{99}m)technetium-mebrofenin and (^{99}m)technetium-sestamibi as specific probes for hepatic transport protein function in rat and human hepatocytes. *Pharm Res* 27(9):1987–1998.

Tagami M, Kusuhara S, Imai H, Uemura A, Honda S, Tsukahara Y, Negi A (2010) MRP4 knockdown enhances migration, suppresses apoptosis, and produces aggregated morphology in human retinal vascular endothelial cells. *Biochem Biophys Res Commun* 400(4):593–598.

Tedeschi L, Carai MA, Frison G, Favretto D, Colombo G, Ferrara SD, Gessa GL (2003) Endogenous gamma-hydroxybutyric acid is in the rat, mouse and human gastrointestinal tract. *Life Sci* 72:2481–2488.

Tian X, Zamek-Gliszczynski MJ, Zhang P, Brouwer KLR (2004) Modulation of multidrug resistance-associated protein 2 (Mrp2) and Mrp3 expression and function with small interfering RNA in sandwich-cultured rat hepatocytes. *Mol Pharmacol* 66(4):1004–1010.

Uppstad H, Ovrebo S, Haugen A, Mollerup S (2010) Importance of CYP1A1 and CYP1B1 in bioactivation of benzo[a]pyrene in human lung cell lines. *Toxicol Lett* 192(2):221–228.

van de Water FM, Boerman OC, Wouterse AC, Peters JGP, Russel FGM, Masereeuw R (2006) Intravenously administered short interfering RNA accumulates in the kidney and selectively suppresses gene function in renal proximal tubules. *Drug Metab Dispos* 34(8):1393–1397.

Venkatraman SS, Ma LL, Natarajan JV, Chattopadhay S (2010) Polymer- and liposome-based nanoparticles in targeted drug delivery. *Front Biosci* 2:801–814.

Volpe DA, Li J, Wang Y, Zhang W, Bode C, Owen A, Hidalgo IJ (2010) Use of caco-2 knockdown cells to investigate transporter-mediated efflux of statin drugs. *AAPS. Abstract* T3371.

Walton SP, Wu M, Gredell JA, Chan C (2010) Designing highly active siRNAs for therapeutic applications. *FEBS J* 277(23):4806–4813.

Wang M, Orwar O, Olofsson J, Weber SG (2010) Single-cell electroporation. *Anal Bioanal Chem* 397(8):3235–3248.

Wang Q, Morris ME (2007b) The role of monocarboxylate transporter 2 and 4 in the transport of gamma-hydroxybutyric acid in mammalian cells. *Drug Metab Dispos* 35(8):1393–1399.

Wang Q, Lu Y, Yuan M, Darling IM, Repasky EA, Morris ME (2006) Characterization of monocarboxylate transport in human kidney HK-2 cells. *Mol Pharm* 3:675–685.

Wang Q, Lu Y, Morris ME (2007a) Monocarboxylate transporter (MCT) mediates the transport of gamma-hydroxybutyrate in human kidney HK-2 cells. *Pharm Res* 24:1067–1078.

Watanabe T, Onuki R, Yamashita S, Taira K, Sugiyama Y (2005) Construction of a functional transporter analysis system using MDR1 knockdown caco-2 cells. *Pharm Res* 22(8):1287–1293.

Weil D, Garçon L, Harper M, Duménil D, Dautry F, Kress M (2002) Targeting the kinesin Eg5 to monitor siRNA transfection in mammalian cells. *Biotechniques* 33(6): 1244–1248.

Xia CQ, Milton MN, Gan LS (2007) Evaluation of drug-transporter interactions using *in vitro* and *in vivo* models. *Curr Drug Metab* 8(4):341–363.

Xie FY, Woodle MC, Lu PY (2006) Harnessing *in vivo* siRNA delivery for drug discovery and therapeutic development. *Drug Discov Today* 11:67–73.

Yonezawa A, Masuda S, Katsura T, Ken-ichi I (2008) Identification and functional characterization of a novel human and rat riboflavin transporter, RFT1. *Am J Physiol Cell Physiol* 295(3):C632–C641.

Yu AM (2007) Small interfering RNA in drug metabolism and transport. *Curr Drug Metab* 8:700–708.

Yuan J, Lv H, Peng B, Wang C, Yu Y, He Z (2009) Role of BCRP as a biomarker for predicting resistance to 5-fluorouracil in breast cancer. *Cancer Chemother Pharmacol* 63(6):1103–1110.

Yue W, Abe K, Brouwer KLR (2009) Knocking down breast cancer resistance protein (Bcrp) by adenoviral vector-mediated RNA interference (RNAi) in sandwich-cultured rat hepatocytes: A novel tool to assess the contribution of Bcrp to drug biliary excretion. *Mol Pharm* 6(1):134–143.

Yue W, Lee JK, Abe K, Sugiyama Y, Brouwer KLR (2011) Decreased hepatic breast cancer resistance protein expression and function in multidrug resistance-associated protein 2-deficient (TR-) rats. *Drug Metab Dispos* 39(3): 441–447.

Yumi M, Naoki K, Makiya N, Yoshinobu T (2005) Sequence-specific suppression of mdr1a/1b expression in mice via RNA interference. *Pharm Res* 22(12):2091–2098.

Zhang YH, Wu Q, Xiao XY, Li DW, Wang XP (2010) Silencing MRP4 by small interfering RNA reverses acquired DDP resistance of gastric cancer cell. *Cancer Lett* 291(1): 76–82.

Zhang W, Li J, Allen SM, Weiskircher EA, Huang Y, George RA, Fong RG, Owen A, Hidalgo IJ (2009) Silencing the breast cancer resistance protein expression and function in caco-2 cells using lentiviral vector-based short hairpin RNA. *Drug Metab Dispos* 37:737–744.

APPENDIX

DRUG METABOLIZING ENZYMES AND BIOTRANSFORMATION REACTIONS[1]

Natalia Penner, Caroline Woodward, and Chandra Prakash

A.1 INTRODUCTION

Drug metabolizing enzymes (DMEs) are a diverse group of proteins that are responsible for metabolizing a vast array of xenobiotic chemicals, including drugs, carcinogens, pesticides, pollutants, and food toxicants, as well as endogenous compounds, such as steroids, prostaglandins, and bile acids (Coon, 2005; Brown et al., 2008; Rendic and Guengerich, 2010). Metabolic biotransformation of chemicals by DMEs form more hydrophilic (water soluble), polar entities, which not only enhance their elimination from the body but also leads to compounds that are generally pharmacologically inactive and relatively nontoxic. As a result, the chemical and functional homeostasis of the cell is maintained in the face of chemical challenges. However, metabolic biotransformation of chemicals by DMEs at times can lead to the formation of metabolites with pharmacological activity (Fura et al., 2004) and/or toxicity (Baillie, 2003, Kalgutkar et al., 2005, Zhou et al., 2005).

Xenobiotics, including drugs and environmental chemicals, are metabolized by four different kinds of reactions: oxidation, reduction, hydrolysis, and conjugation (Parkinson and Ogilvie, 2008). The first three reactions (oxidation, reduction, and hydrolysis) are often grouped together and called functionalization (Phase I) reactions, and the conjugation reactions are called Phase II reactions (Table A.1). Recently, a third

phase of metabolism has been proposed (Phase III), in recognition of the role of membrane transporters on the biliary excretion of drugs and their metabolites, as well as the efflux of these compounds across the hepatocellular membrane.

Phase I reactions introduce or unmask a functional group (e.g., -OH, -CO₂H, -NH₂, or -SH) within a molecule to enhance its hydrophilicity. It can occur through direct introduction of the functional group (e.g., aromatic and aliphatic hydroxylation) or by modifying existing functionalities (e.g., reduction of the ketones and aldehydes to alcohols; oxidation of alcohols to acids; hydrolysis of ester and amides; reduction of azo and nitro compounds; oxidative N-, O-, and S-dealkylation) (Parkinson and Ogilvie, 2008). Oxidative Phase I DMEs include cytochrome P450s (CYPs or P450s), flavin-containing monooxygenases (FMOs), monoamine oxidases (MAOs), and xanthine oxidase/aldehyde oxidase (XO/AO). Phase II biotransformation reactions include glucuronidation, sulfonation, methylation, acetylation, amino acids (such as glycine, glutamic acid, and taurine), and glutathione (GSH) conjugation (Table A.1). The cofactors of these reactions react with functional groups that are either present on the xenobiotics or are introduced during Phase I biotransformation. Conjugative Phase II DMEs include uridine 5'-diphospho (UDP)-glucuronosyltransferases (UGTs), sulfotransferases (SULTs), glutathione S-transferases (GSTs), N-acetyltransferases (NATs), and methyl

---

[1] This chapter arrived late and normally would have been placed after Chapter 2 in Part A. Because the Editors feel that this is important information to which the reader would like to have access, we have decided to include this contribution in the book even though it is not in the ideal place.

ADME-Enabling Technologies in Drug Design and Development, First Edition. Edited by Donglu Zhang and Sekhar Surapaneni.
© 2012 John Wiley & Sons, Inc. Published 2012 by John Wiley & Sons, Inc.

Table A.1. Common Biotransformation Reactions, DME Enzymes, Major Liver Isoforms, and Their Cellular Localization

Reaction	DME Involved	Major Liver Isoform	Cellular Localization	Cofactor Requirement
Oxidation	Cytochrome P450	CYP3A4, 2D6, 2C, 1A2, 2E1	Microsomes	O_2, NADPH
	Flavin-containing monooxygenase	FMO3, FMO4, FMO5	Microsomes	O_2, NADPH
	Peroxidase			
	Monoamine oxidase	MAO-A, MAO-B	Mitochondrial outer membrane	O_2, H_2O
	Alchohol dehydrogenase	ADH1A, 1B, 1C	Cytosol	NAD+
	Aldehyde dehydrogenase	ALDH1, ALDH2	Mitochondria, cytosol	NAD(P)+
	Aldehyde oxidase	AO	Cytosol	O_2, H_2O
	Xanthine oxidase	XO	Cytosol	O_2, H_2O
	Prostaglandin H synthase	PHS-1, PHS-2	Microsomes	O_2
Reduction	Nitro-reductase	P450, non-P450 enzymes	Microsomes, microflora	NADPH
	Azo-reductase	P450, non-P450 enzymes	Microsomes, microflora	NADPH
	Aldo-keto reductase	AKR1A1, 1B1, 1C1-4, 1D1	Cytosol, microsomes	NADPH, NADH
	Quinone reductase	NQO1, P450 reductase	Cytosol, microsomes	NAD(P)H, NADPH
Hydrolysis	Epoxide hydrolase	EPHX1 (mEH), EPHX2 (sEH)	Microsomes, cytoplasm	H_2O
	Esterase	hCE1, hCE2	Microsomes, cytosol, lysosomes	H_2O
	Peptidase	Aminopeptidase, carboxypeptidase, endopeptidase	Lysosomes	H_2O
	Alkaline phosphatase		Plasma membrane	H_2O
Conjugation	Uridine diphospho-glucuronosyltransferase	UGT1A1, 1A3, 1A4, 1A6, 1A9, 2B	Microsomes	UDPGA
	Sulfotransferase	SULT1A1, 1B1, 1E1, 2A1	Cytosol	PAPS
	Methyltransferase	COMT, PNMT, TPMT, and so on	Cytosol, microsomes	SAM
	N-acetyltransferase	NAT1, NAT2	Cytosol, mitochondria	Acetyl CoA
	Amino acid conjugation enzyme	Acyl-CoA synthetase, acyl-CoA: amino acid N-acyltransferase, and so on	Cytosol, microsomes, mitochondria	ATP, acetyl CoA, amino acids
	Glutathione S-transferase	GST A1-1, M1-1, P1-1	Cytoplasm	GSH

(N-methyl-, thiomethyl-, and thiopurinemethyl-) transferases. Conjugated metabolites are relatively more polar and hence are readily excreted (urine or bile depending on the molecular weight [MW]) from the body. Indeed, majority (>75%) of the top 200 marketed drugs are eliminated by metabolism. Of the DMEs involved in the metabolism of drugs, the dominant players are P450 enzymes, followed by UGTs and esterases. Together, these reactions account for ~95% of the drug metabolism. The other enzymes together metabolize only ~5% of the marketed drugs (Williams et al., 2004).

Metabolism is considered one of the main reasons for the poor bioavailability (first-pass effect), interindivdual variability in drug response, and/or drug–drug interactions via inhibition or induction of the DMEs (Rock et al., 2008). Currently, data from *in vivo* preclinical and *in vitro* human tissue studies are used in predicting the safety and human pharmacokinetics, and assessing the potential of a new chemical entity (NCE) as a successful human drug candidate. However, >50% of drugs are failed in the Phase I clinical trial due to toxicity or poor pharmacokinetics. Even after a drug is marketed there is a possibility that the drug is either withdrawn from the market or acquires a warning label (black box) due to some adverse drug reactions, which were not seen in earlier clinical trials. This is primarily due to species-related differences in the expression, activity, inhibition, induction, pharmacogenetics, and regulation of DMEs (Guengerich, 2003; Zhou et al., 2009; Crettol et al., 2010).

Many of the DMEs exhibit genetic polymorphism and can be inhibited or induced by the coadministered drugs and or/diet, which can alter their catalytic activity or levels of expression. Nuclear receptors such as aryl hydrocarbon receptor (AhR), pregnane X receptor (PXR), and constitutive androstane receptor (CAR) together are believed to play a critical role in the regulation of the catalytic activity of DMEs (Crettol et al., 2010). In addition, the expression and activity of DMEs in humans are markedly influenced by various other factors such as chronic disease conditions, age, and hormonal variations during pregnancy and environment. Tremendous progress has been made in the last six decades in the characterization, expression, function, and regulation of the DMEs in animals and humans (Coon, 2005; Rendic and Guengerich, 2010). Each enzyme exhibits distinctive substrate selectivity in the drugs that they metabolize, although there is some overlap. In some cases, several enzymes within a group or even enzymes from different groups can catalyze the same reaction. In this review, we summarize the most recent advances in our knowledge and application of these enzymes in drug discovery and development with particular emphasis on their involvement in the metabolism of drugs. In addition, most common and uncommon biotransformation reactions mediated by DMEs will also be discussed.

A.2 OXIDATIVE ENZYMES

Oxidative reactions are catalyzed by P450s, FMOs, MAOs, molybdenum hydroxylases (AO/XO), alcohol dehydrogenases (ADHs), aldehyde dehydrogenases (ALDHs), and prostaglandin H synthase (PHS). These enzymes are responsible for the metabolism of the majority (>75%) of the marketed drugs (Guengerich and Rendic, 2010).

A.2.1 P450

The term cytochrome P450 refers to a large group of enzymes located in the endoplasmic reticulum membrane and belongs to a superfamily of hemoproteins. P450s are found in most living systems, from bacteria to humans, with more than 11,500 genes reported to date. The substrates of P450s range from small (ethylene, MW = 28) to large (cyclosporine, MW = 1201) molecules (Testa and Kramer, 2007). The catalytic cycle of P450 oxidation is a complex multistep process that involves CYP enzyme, NADPH (nicotinamide adenine dinucleotide phosphate) as the electron donor, and flavin adenine dinucleotide (FAD)-containing P450 reductase as the electron transfer bridge. P450s, acting

as the monooxygenases, activate molecular oxygen with electrons from NADPH via NADPH-P450 reductase, and insert one atom of molecular oxygen into the substrate while reducing the other atom of oxygen to water (Eqs. A.1 and A.2). The presence and adequate functionality of the reductase is determinant for effective drug oxidation by the P450s. P450-mediated reactions are inhibited by carbon monoxide (CO) because ferrous iron preferentially binds CO just like hemoglobin:

$$XH + O_2 + NADPH + H^+ \rightarrow XOH + H_2O + NAD(P)^+ \tag{A.1}$$

$$X + O_2 + NADPH + H^+ \rightarrow XO + H_2O + NAD(P)^+. \tag{A.2}$$

P450 enzymes are grouped in families and subfamilies according to their amino acid sequence homology. The drug metabolizing P450s are confined to subfamilies 1, 2, 3, and 4. These subfamilies are further divided into isoforms (Prakash and Vaz, 2009). There are approximately 57 human P450 genes and 58 pseudogenes exhibiting major differences with respect to their catalytic specificity and patterns of tissue expression. In the human liver, there are at least 18 distinct P450s, while only 10 from families 1, 2, and 3 (CYP1A2, CYP2A6, CYP2B6, CYP2C8, CYP2C9, CYP2C19, CYP2D6, CYP2E1, CYP3A4, and CYP3A5) are responsible for the hepatic metabolism of most of the marketed drugs. The list of clinically relevant substrates, inhibitors, and inducers of major P450s is constantly updated and can be found in Parkinson and Ogilvie (2008).

CYP1A2 is mainly confined to the liver and expressed in low levels in extrahepatic tissue. In human liver, it accounts for ~10–15% of the total CYP content. CYP1A2 substrates are planner polyaromatic/heterocyclic amines and amides with one putative H-bond donor site. CYP1A2 is involved in the metabolism of ~4% of the marketed drugs including acetaminophen, phenacetin, tacrine, ropinirole, riluzole, theophylline, propafenone, tamoxifene, and caffeine. It also plays an important role in the metabolic activation of polycyclic aromatic hydrocarbons (PAHs), aromatic amines, and heterocyclic amines.

CYP2A6 is expressed in the liver and accounts for ~4% of total hepatic CYP contents. It is the principal and perhaps the sole catalyst for human liver microsomal coumarin 7-hydroxylation. CYP2A6 substrates are nonplanner molecules of relatively low molecular weight with two H-bonds acceptors 2–3 Å apart and 5–7 Å from the site of metabolism. CYP2A6 is involved in the oxidation of ~1% of drugs such as nicotine, cyclophosphamide, ifosfamide, zidovudine, and fadrozole. In addition, CYP2A6 also plays an important role in the activation of several procarcinogens and promutagens, especially the nitrosamines.

CYP2B6 is expressed in the liver and in some extrahepatic tissues and accounts for ~1% of total hepatic CYP content. CYP2B6 is involved in the metabolism of ~2% of the marketed drugs, such as the anticancer drug: cyclophosphamide and tamoxifene, and the anesthetics ketamine and propofol. Significant interindividual differences from 25- to 250-fold have been reported in hepatic CYP2B6 expression.

CYP2C8, 2C9, and 2C19 are the members of the human CYP2C subfamily, and all together account for ~20–25% of the total CYP in the human liver. CYP2C8 and 2C9 are the major isoforms, accounting for 35% and 60%, respectively, of the total human CYP2C, whereas 2C19 (2%) is the minor expressed CYP2C isoforms. CYP2C subfamily is involved in the metabolism of ~20–25% of the marketed drugs. CYP2C8 is involved in the metabolism of retinol and retinoic acid, arachidonic acid, benzo[a]pyrene, and the anticancer drug paclitaxel. CYP2C9 substrates are neutral and acidic molecules with lipophilic site of oxidation at 5–8 Å from one or two H-bond donor/acceptors. CYP2C9 plays an important role in the metabolism of a number of clinically significant drugs, including tolbutamide, phenytoin, *S*-warfarin, ibuprofen, diclofenac, piroxicam, tenoxicam, mefenamic acid, losartan, glipizide, and torasemide. CYP2C19 substrates are neutral or weakly basic, with two or three H-bond donor/acceptors 4–5 Å apart and 5–8 Å from the site of metabolism. CYP2C19 metabolizes (*S*)-mephenytoin, omeprazole, imipramine diazepam, some barbiturates, and proguanil.

CYP2D6 enzyme is expressed in various tissues including the liver, kidney, placenta, brain, breast, lungs, and small intestine. Although CYP2D6 is expressed at a low level in human liver accounting for only ~3% of total CYP protein, this enzyme is responsible for the metabolism of numerous therapeutically used drugs such as amitriptyline, bufuralol, codeine, debrisoquine, dextromethorphan, fentanyl, morphine, paroxetine, propafenone, sparteine, and tamoxifen. A common feature of drug metabolized by CYP2D6 is the presence of at least one basic nitrogen atom at a distance of 5–7 A° from the site of oxidation.

CYP2E1 is expressed in many tissues, such as the nose, lung, and the liver and accounts for ~7% of total CYP content in the human liver. CYP2E1 substrates are neutral, small, relatively planner with one or two H-bond donor/acceptors at 4–6 Å from the site of metabolism. CYP2E1 is involved in the metabolism of only 2–3% of the drugs, such as acetaminophen, caffeine, and chlorzoxazone, the latter being considered a marker of CYP2E1 activity. CYP2E1 is the most active CYP enzyme in forming reactive oxygen intermediates, such as superoxide radical, causing tissue injury.

The CYP3A subfamily of P450 in humans is composed of several enzymes and accounts for ~28–40% of total hepatic P450 content. The human CYP3A family is clinically very important because it has been shown to catalyze the metabolism of an amazingly large number of structurally diverse molecules from almost every therapeutic class of drugs. It is estimated that CYP3A subfamily participates in the metabolism of 35–50% of all marketed drugs. CYP3A4 is the major human liver enzyme of CYP3A subfamily, whereas CYP3A5 is presented only in~20% of human liver. CYP3A4 and 3A5 are also expressed in the stomach, lungs, small intestine, and renal tissue. CYP3A4 substrates range from small (ethylene, MW = 28) to very large (cyclosporine, MW = 1201) molecules. Most of the CYP3A4 substrates are also metabolized by CYP3A5. Some examples of drugs metabolized by CYP3A are terfenadine, midazolam, triazolam, quinidine, lidocaine, carbamazepine, nifedipine, tacrolimus, dapsone, erythromycin, and dextromethorphan.

A.2.2 FMOs

FMOs are NADPH and oxygen-dependent microsomal flavoenzymes located in endoplasmic reticulum, similar to P450s. FMOs oxygenate a number of drugs and xenobiotics that contain a "soft-nucleophile" heteroatom such as nitrogen, sulfur, and phosphorus. The mammalian FMO gene family comprises five enzymes designated as FMO1, FMO2, FMO3, FMO4, and FMO5, which are differentially expressed in tissues—liver, lung, intestine, kidney, and brain (Cashman and Zhang, 2006):

- FMO1—primarily located in kidney and fetal liver, absent in adult liver, high specificity toward tertiary amines and sulfides
- FMO2—located in lung and kidney
- FMO3—hepatic, polymorphic with several allelic variants, trimethylamine oxygenase—variants with diminished activity responsible for "fish odor syndrome"
- FMO4—more broadly distributed in liver, kidney, small intestine, and lung
- FMO5—expressed in human liver, lung, small intestine, and kidney

Of all the FMO isozymes, FMO3 has a wide substrate specificity, including the physiologically and plant-derived tertiary amine, trimethylamine, tyramine, and nicotine; commonly used drugs including cimetidine, ranitidine, clozapine methimazole, itopride, ketoconazole, tamoxifen, and sulindac sulfide; and agrichemicals, such as organophosphates and carbamates. A typical reaction scheme of FMO mediated oxidation of sulfide

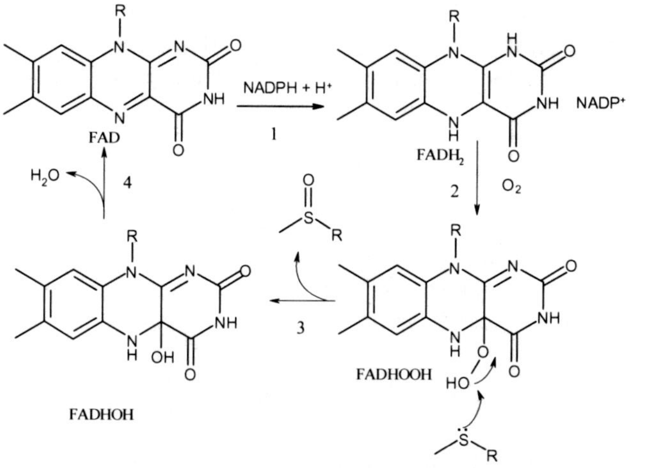

FIGURE A.1. The catalytic cycle of FMO.

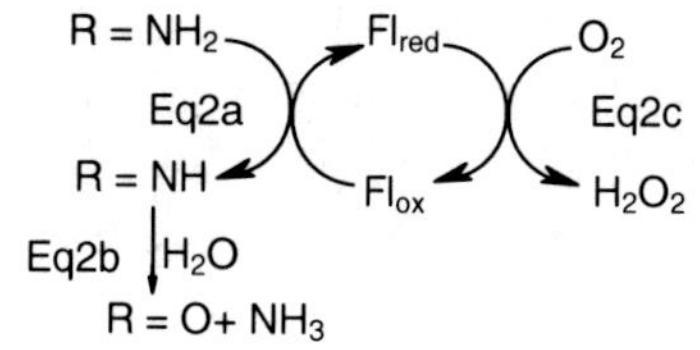

FIGURE A.2. The reaction cycle of MAO.

is shown in Figure A.1. The reaction starts with the reduction of FAD by NADPH (1), $FADH_2$ is then oxidized by O_2 (2), the resulting reactive complex FADHOOH reacts with a substrate (3), and the enzyme converts to the initial state after a loss of water molecule (4).

FMO-catalyzed oxidations are insensitive to CO inactivation and can be distinguished from P450 using CO inactivation assay. In contrast to P450, FMOs are thermally labile enzymes that can be denatured by incubating the reaction mixture in the absence of NADPH at 50°C.

A.2.3 MAOs

MAO enzymes belong to a family of flavoproteins—flavin-containing amine oxidoreductases—and contain one covalently bound FAD per polypeptide chain. These enzymes are found in the outer membrane of mitochondria in most cell types in the body but can be also be present in microsomal suspensions prepared from frozen tissues. MAOs catalyze the oxidative deamination and dehydrogenation of structurally diverse amines including neurotransmitters dopamine, norepinephrine, serotonin, tyramine, and 2-phenylethylamine, and some drugs and xenobiotics that contain cyclic and acyclic alkylamine functional amines. Two enzymes, MAO-A and MAO-B are identified, which are differentially expressed in tissues: brain, liver, lung, intestine, kidney, and blood platelets. Inhibitors of MAOs are used in psychiatry for the treatment of depressive disorders and in neurology for the treatment of Parkinson's disease. MAO-A and MAO-B play a critical role in the bioactivation of the neurotoxin 1-methyl-4-phenyl-1,2,3,6-tetrahydropyridine (MPTP) to a toxic metabolite that induces Parkinson's-like effects.

The reaction cycle of MAO is shown in Figure A.2. A two-electron oxidation results in the formation of an imine and reduced protein-bound FAD (Eq. A.2a). The imine is then nonenzymatically hydrolyzed to the carbonyl compound (Eq. A.2b). In the second half reaction, the reduced FAD ($FADH_2$) is reoxidized by molecular oxygen producing hydrogen peroxide (Eq. A.2).

Oxygen is required for the reaction with MAOs since it is involved in the regeneration of the enzyme. However, in contrast to P450, the oxygen atom incorporated into the final product is not from molecular oxygen; rather it is from water. This can be used for differentiation of CYP and non-CYP-mediated reactions by conducting the experiment in ^{18}O-labeled water. Additionally, unlike P450, MAO-mediated reactions are not inhibited by CO.

The presence of an alpha-hydrogen is necessary for the above reaction to proceed. Sterically hindered amines would not be able to bind to the flavin; therefore, they are not oxidized by MAOs. Some amine drugs such as fluoxetine (Prozac [Eli Lilly, Indianapolis, IN, USA]), methamphetamine, nortryptyline, tamoxifen, and propranolol are not oxidized by MAOs. The oxidation by MAOs is regioselective and depends on the acidity of alpha-hydrogen since an intermediate is a carboanion stabilized by resonance. Acidity of the benzylic and allylic hydrogen is much higher than the alkyl hydrogen, so the former two will preferentially be abstracted. MAOs are specifically and irreversibly inactivated by acetylenic-amine derivatives such as clorgyline (MAO-A), deprenyl, and pargyline (MAO-B), so these compounds can be used to distinguish between two enzymes.

A.2.4 Molybdenum Hydroxylases (AO and XO)

Molybdenum hydroxylases, which include AO and XO, are enzymes with an MW of about 300,000 Da comprising two subunits of equal size. Each of these contains one FAD molecule and one atom of molybdenum in a core in a form of a molybdopterin cofactor. These enzymes are expressed in most of the tissues such as heart, brain, liver, lung epithelial cells, kidney, small intestine, and placenta. Major differences have been observed in the tissue distribution between AO and XO. High XO expression and activities are consistently reported in the proximal intestine and lactating mammary gland, whereas AO activity is generally expressed at high levels in liver, lung, kidney, and brain. There is considerable variability of AO activity in the liver cytosol of mammals. Humans show the highest activity, rats and mice show low activity, and dogs have no detectable AO activity (Garattini et al., 2008; Prakash and Vaz, 2009).

Molybdenum hydroxylases catalyze both oxidation and reduction reactions. Unlike P450, oxidation catalyzed by the molybdenum hydroxylases generates the reducing equivalents and, although both enzymes use molecular oxygen as a cosubstrate, oxygen atom from water, not from the molecular oxygen, is inserted into the substrate. The reaction catalyzed by these enzymes obeys the general Equation A.3:

$$RH + H_2O \rightarrow ROH + 2e^- + 2H^+. \quad (A.3)$$

AO and XO are cytosolic enzymes and are closely related. However, they differ in their substrate/inhibitor specificities. AO catalyzes the oxidation of a wide range of aldehydes to their corresponding carboxylic acids. In addition, AO is involved in the metabolism of several clinically significant drugs such as famciclovir, zaleplon, zonisamide, and ziprasidone. XO has narrower substrate specificity than AO and is mainly active toward purines and pyrimidines. XO plays a role in the oxidation of several chemotherapeutic agents and has been implicated in the bioactivation of mitomycin B. In general, oxidation involves nucleophilic attack at an electron deficient carbon to form either a cyclic lactam or a carboxylic acid from aromatic N-heterocyclic and aldehydes, respectively. In addition, AO and XO also catalyze the reduction of N- and S-functional groups such as the *N*-oxide, sulfoxides, hydroxamic acids, and reductive ring cleavage of the thiazole and isothiazole moieties (Benedetti et al., 2006; Prakash and Vaz, 2009). AO and XO can be distinguished by using menadione and allopurinol, respectively, *in vitro*.

A.2.5 ADHs

ADHs are a family of zinc-containing enzymes that facilitate the reversible oxidation of alcohols to aldehydes or ketones using NAD$^+$/NADH as a cofactor (Eq. A.4):

$$CH_3CH_2OH + NAD^+ \rightarrow CH_3CHO + NADH^+ + H^+.$$
$$(A.4)$$

ADHs are located almost exclusively in the cytoplasm of cells and are grouped into seven classes. The ADH classes share <70% amino acid sequence identity within the same organism, and ADH isozymes share >80% sequence identities within a single class. Human ADH is a dimeric protein consisting of two 40-kDa subunits and is involved in the metabolism of several xenobiotics and drugs such as ethanol, ethambutol, hydoxyzine, celecoxib, avacavir, and felbamate. Pyrazole and its 4-alkyl-substituted derivatives are potent inhibitors of many ADHs (Benedetti et al., 2006).

A.2.6 ALDHs

ALDHs are a superfamily of NAD(P)$^+$-dependent enzymes that catalyze the oxidation of a wide range of aldehydes to their corresponding carboxylic acids (Eq. A.5):

$$RCHO + NAD + H_2O \rightarrow RCOOH + NADH_2. \quad (A.5)$$

There are approximately 17 human ALDH genes that are classified into 10 families and 13 subfamilies according to their amino acid sequence homology. Similar to P450s, proteins sharing ≥40% homology are assigned to a particular family designated by an Arabic number, whereas sharing ≥60% identity are classified in the same subfamily designated by a letter. In addition to endobiotics, human ALDHs are involved in the metabolism of xenobiotics and drugs such as ethanol, ethambutol, hydoxyzine cyclophosphamide, and nitroglycerine (Benedetti et al., 2006).

A.3 REDUCTIVE ENZYMES

Reductive reactions are catalyzed by aldo-keto reductases (AKRs), azoreductases (AZRs) nitroreductases (NTRs), quinone reductases (QR s), P450s, ADH, and NADPH-P450 reductase.

A.3.1 AKRs

AKRs are soluble NADPH-dependent oxidoreductases that are capable of reducing aldehydes and ketones to alcohols. The known human AKR enzymes can convert a vast range of substrates, which can lead to either their bioactivation or detoxification (Jin and Penning, 2007; Barski et al., 2008). The nomenclature of AKRs includes a number to identify the family (AKR1), a letter to designate the subfamily (AKR1A), and a second number to assign the unique protein (AKR1A1) (e.g., human aldehyde reductase). Mammalian AKR are found in the AKR1, AKR6, and AKR7 families, with AKR1 being the largest of the 15 families. A complete list of AKR members can be found at http://www.med.upenn.edu/akr (Jin and Penning, 2007).

The majority of the human AKRs belong to the AKR1 family including the human homologues of aldehyde reductase (AKR1A1), aldose reductases (AKR1B1 and AKR1B10), hydroxysteroid dehydrogenases (HSDs) (AKR1C1–AKR1C4), and steroid 5β-reductase (AKR1D1). Other human AKRs include the human homologues of aflatoxin aldehyde reductases (AKR7A2 and AKR7A3) (Jin and Penning, 2007; Barski et al., 2008).

The cofactor (NADPH) and a substrate bind to the different sites of the enzyme and converge at the active site. The hydride transfer from NADPH to the substrate is generally stereospecific. Human AKRs are involved in metabolism of synthetic hormones (Kang and Kim, 2008), chemotherapeutic agents (Jin and Penning, 2007; Novotna et al., 2008), and CNS drugs (Jin and Penning, 2007). Natural substrates for AKRs include sugar and lipid aldehydes, retinals, steroids, and prostaglandins (Jin and Penning, 2007; Barski et al., 2008). AKRs also participate in detoxification of nicotine-derived carcinogens (Jin and Penning, 2007).

A.3.2 AZRs and NTRs

AZRs reduce the azo bond (N = N) in azo dyes to produce corresponding amines (Eq. A.6):

$$Ar–N=N–Ar' + 2\,NAD(P)H \rightarrow Ar–NH_2 + NH_2–Ar' + 2\,NAD(P)^+. \tag{A.6}$$

Most AZR isoenzymes can reduce methyl red, but are not able to reduce sulfonated azo dyes. Crude cell extracts from *Enterococcus faecalis* have been shown to utilize both NADH and NADPH as electron donors for azo dye reduction (Macwana et al., 2010).

The NTR family is a group of flavin mononucleotide (FMN) or FAD-dependent and NADPH-dependent enzymes, which metabolize nitrosubstituted compounds in a wide range of substrates to produce the corresponding hydroxylamines (Figure A.3).

AZR and NTR enzymes are normally associated with bacteria and are absent from most eukaryotes except for trypanosomes (Hall et al., 2010). The predominant AZR and NTR producing bacteria belong to *Clostridium* and *Eubacterium*. The AZR and NTR have wide substrate specificities with different species of bacteria producing various amounts of the enzymes. Both AZR and NTR are extracellular and oxygen-sensitive, and produced constitutively. The activity of both enzymes increases in the presence of added FAD. The bacteria with NTR activity decrease the mutagenicity of nitroaromatic compounds in the Ames assay, whereas AZRs convert some azo dyes to potentially genotoxic

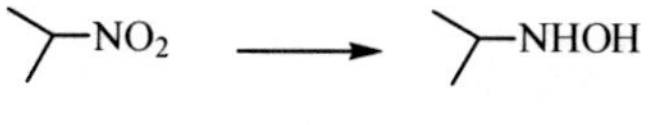

FIGURE A.3. Reduction of nitrosubstituted compounds.

compounds (Rafii and Cerniglia, 1995; Oppermann and Maser, 2000).

Although the role of NTRs *in vivo* is unclear, they have been identified as useful in the metabolism of a number of prodrugs in anticancer gene therapy. NTR NfsB from *Escherichia coli* is used for activation of the prodrug CB1954 to a potent bifunctional alkylating agent (Vass et al., 2009).

A.3.3 QRs

The QRs (QR1 and QR2) are enzymes that catalyze the two-electron reduction of quinones into hydroquinones utilizing NAD(P)H as an electron donor (Kepa et al., 1997; Long and Jaiswal, 2000; Oppermann and Maser, 2000; Ross, 2004). QRs function via a "ping-pong" mechanism where NAD(P)H binds to QR, reduces the FAD cofactor, and is then released, allowing the quinone substrate to bind the enzyme and to be reduced. QR1 is known to catalyze the reduction of a broad range of reactive substrates including quinones, quinone-imines, and azo compounds and protect cells from redox cycling, oxidative stress, and neoplasia (Kepa et al., 1997). QR's antioxidant function of the cell is maintaining vitamin E and ubiquinones (coenzyme Q) in their reduced and active states. NRH:quinone oxidoreductase 2 (NQO2) is resistant to typical inhibitors of NQO1, such as dicoumarol, cibacron blue, and phenindone. Flavones, including quercetin and benzo(a)pyrene, are known inhibitors of NQO2 (Long and Jaiswal, 2000).

A.3.4 ADH, P450, and NADPH-P450 Reductase

The reaction with ADH is simply a $2e^-$ reduction of an aldehyde or imine (Eq. A.7):

$$RCHO + NAD(P)H + H^+ \rightarrow RCH_2OH + NAD^+. \tag{A.7}$$

Reductive reactions with P450 and NADPH-P450 reductase are not common, but reductions of nitro, nitroso, hydroxyl amines, and *N*-oxides involving these enzymes have been documented (Guengerich, 2001).

A.4 HYDROLYTIC ENZYMES

A.4.1 Epoxide Hydrolases (EHs)

The EHs belong to a subcategory of a broad group of hydrolytic enzymes that include esterases, proteases, dehalogenases, and lipases (Beetham et al., 1995). EHs are a class of proteins that catalyze the hydration of chemically reactive epoxides to their corresponding 1,2-diol products. In mammalian species, there are at least five EH forms: microsomal cholesterol 5,6-oxide hydrolase, hepoxilin A hydrolase, leukotriene A hydrolase, soluble epoxide hydrolase (sEH), and microsomal epoxide hydrolase (mEH). Each of these enzymes is distinct chemically and immunologically (Fretland and Omiecinski, 2000).

sEH and mEH have been the most studied EHs over the past 30 years. Two-step hydration mechanism with the formation of enzyme-substrate intermediate is generally accepted. The last step, dissociation of the complex, is the rate-limiting. The single turnover experiment (excess of enzyme) in $H_2^{18}O$ showed that the ^{18}O was not incorporated in the formed glycol but rather in the protein (Lacourciere and Armstrong, 1993, 1994), which supports the mechanism as shown in Figure A.4.

Further evidence was gained through the isolation of the covalent intermediates for the sEH and mEH (Lacourciere and Armstrong, 1993, 1994). Chemical characterization of the enzyme-product intermediate indicated a structure consistent with an α-hydroxyl alkyl-enzyme complex. Mammalian mEH are highly expressed in the liver and show broad substrate selectivity. The mEH are capable of hydrating a large number of structurally different, highly reactive epoxides such as epoxide of benzopyrene and androstane.

Taken together, EHs play a central role in the detoxification of genotoxic epoxides and have an important function in the regulation of physiological processes by the control of signaling molecules with an epoxide structure (Decker et al., 2009). sEH participate in arachidonic acid metabolism, which is important for regulation of vascular, renal, and cardiac functions (Morisseau and Hammock, 2005).

A.4.2   Esterases and Amidases

Esterases and amidases catalyze the addition of a water molecule to an ester, thio-ester, or an amide resulting in the formation of the corresponding acids and alcohols or amines. Like EHs, esterases are members of the α/β-hydrolase fold family of enzymes. They share a common structure and catalytic mechanism involving the formation and hydrolysis of a covalent intermediate. Both enzymes add water to the substrate with no additional cofactors needed (Figure A.5).

While the EH's reaction is irreversible, the action of esterases and amidases is reversible and in some conditions these enzyme could be used for the synthesis of the esters or amides. Furthermore, the esterases and amidases use two substrates (ester and water) to form two products (acid and alcohol or amine). Esterases and amidases play an important role in maintaining normal physiology and metabolism, detoxifying various drugs and environmental toxicants in living systems and are increasingly important for chemical synthesis.

The mammalian carboxylesterases (CEs) belong to a family of proteins encoded by multiple genes, and the isozymes are classified into four main CE groups (CE1-CE4) and several subgroups according to the homology of the amino acid sequence. The two major human isozymes hCE-1 and hCE-2 belong to classes CE1 and CE2, respectively, and these two enzymes exhibit 48% sequence identity. Human hCE-1 is highly expressed in the liver but low in the gastrointestinal tract. On the other hand, human CE-2 is present in the small

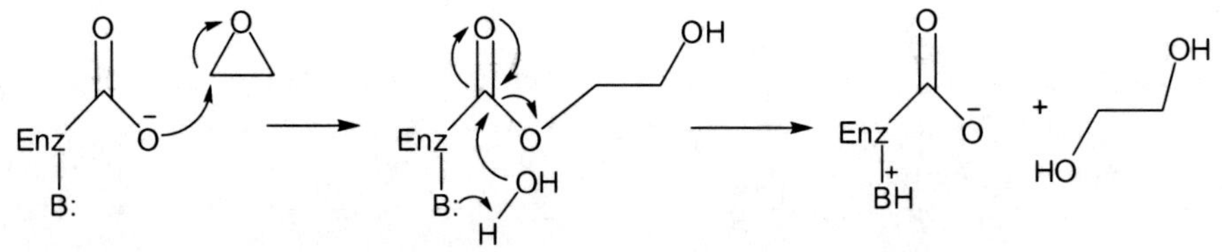

FIGURE A.4. Mechanism for hydration of epoxides.

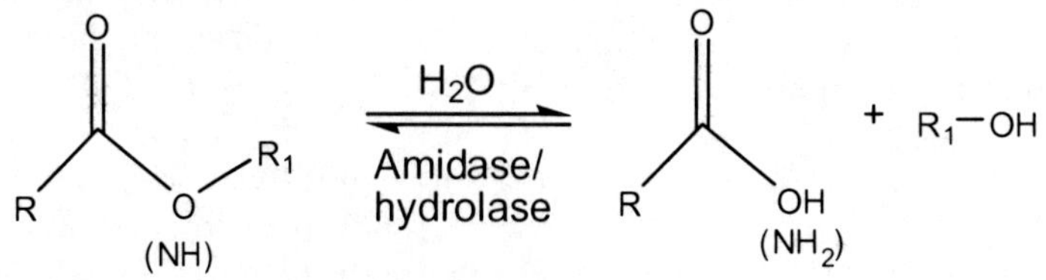

FIGURE A.5. Hydrolysis of esters and amides.

intestine, colon, kidney, liver heart, brain, and testis. hCE-1 preferentially hydrolyzed esters with a small alcohol group, while hCE-2 catalyzes the hydrolysis of compounds with a small acyl group and large alcohol group. A new class of human CEs, hCE-3, is expressed in the liver and gastrointestinal tract at an extremely low level compared with hCE-1 and hCE-2.

A.5 CONJUGATIVE (PHASE II) DMEs

Conjugative enzymes include Phase II enzymes such as UGTs, SULTs, GSTs, NATs, and methyl (N-methyl-, thiomethyl-, and thiopurinemethyl-) transferases. GSH conjugates are further metabolized to cysteine and N-acetylcysteine adducts. Most Phase II reactions result in a compound's concomitant increase in hydrophilicity and decrease in volume of distribution (VD_{ss}), which together greatly facilitate its excretion from the body.

A.5.1 UGTs

Glucuronidation reaction is catalyzed by UGTs located predominantly in the endoplasmic reticulum (ER) of liver, kidney, intestine, lungs, prostate, mammary glands, skin, brain, spleen, and nasal mucosa. Glucuronidation requires the cofactor uridine-5-diphospho-α-D-glucuronic acid (UDPGA). The active site of UGTs faces the lumen of the ER. Nonpolar substrates can diffuse through the ER membrane and can be conjugated in the ER lumen. However, UDPGA must be transported into the ER and the product formed must be transported out of the ER (Bossyut and Blanckeart, 1994a,b). There is an evidence for multiple transporters located in ER membrane, which are responsible for the transport of glucuronidated products from the lumen to the cytoplasm (Csala et al., 2004). *In vitro* assays for glucuronide screening require the use of a detergent or a pore-forming antibiotic such as alamethecin to disrupt the membrane barrier and enhance access for both substrate and cofactor to the active site.

The UGT family of enzymes is subdivided into two subfamilies, UGT1 (1A1, 1A3, 1A4, 1A5 1A6, 1A7, 1A8, 1A9, and 1A10) and UGT2 (2A1, 2B4, 2B7, 2B10, 2B11, 2B15, 2B17, and 2B28), on the basis of sequence homology (http://www.flinders.edu.au/medicine/sites/clinical-pharmacology/ugt-homepage.cfm). All classes of drugs containing a wide range of acceptor groups including phenols, alcohols, aliphatic and aromatic amines, thiols, and carboxylic acids are substrates for UGTs and this pathway has been estimated to account for ~15% of all drugs metabolized by DMEs. The sites for glucuronidation reaction are electron-rich nucleophiles such as O, N, or S heteroatoms. Unusual cases such as bisglucuronides, where two different functional groups are glucuronidated (e.g., bilirubin and morphine); diglucuronides, where two glucuronides are attached in tandem to a single site; N-carbamoyl glucuronides (e.g., sertraline and varenceline), where the carbonate is incorporated in the glucuronide; and glycosylation with UDP-sugars (e.g., glucosidation of barbiturates) have been observed *in vivo*. Glucuronidation is observed in all mammalian species except for the cat family (Tukey and Strassburg, 2000).

A.5.2 SULTs

Sulfonation reaction is catalyzed by SULTs, a large multigene family, found primarily in the liver, kidney, intestine, lung, platelets, and brain. There are two classes of SULTs found in mammals: (1) membrane-bound SULTs in Golgi apparatus are important for many biological processes such as cell adhesion, axon function, T-cell response, cell proliferation, and modulation of viral and bacterial infection (Grunwell and Bertozzi, 2002; Grunwell et al., 2002); and (2) soluble SULTs in the cytoplasm (Gamage et al., 2006) are known for sulfonation of various drugs.

Sulfonation requires the cofactor 3′-phosphoadenosine-5′-phosphosulfate (PAPS) and involves the transfer of a sulfonate (SO_3^-) and not a sulfate (SO_4^-) from PAPS to the xenobiotics and is catalyzed by the SULT enzymes. Conjugation can occur at -C-OH, -N-OH, and -NH side chains yielding O-sulfates and N-sulfates. The cofactor PAPS is synthesized from inorganic sulfate and ATP by the enzymes sulfuryl and adenosine 5′-phosphusulate kinase in prokaryotes and a bifunctional enzyme PAPS synthetase (PAPSS) in higher organisms including humans.

To date, 11 SULT isoforms have been identified in humans and are divided into two subfamilies, phenolic SULTS (SULT1) and hydroxysteroid SULTS (SULT2) (Blanchard et al., 2004). SULT1 consists of at least eight isoforms and are responsible for conjugation of small phenolic compounds, estrogens, catecholamines, and many therapeutic drugs (Meloche and Falany, 2001). SULT1A isoform has been reported to be responsible for the metabolism of several therapeutic drugs. Although sulfonation can lead to decreased pharmacological and toxicological activity, there are some cases where toxicity has been linked to toxicity (Wang and James, 2006). Other drugs must be converted to a sulfonate conjugate in order to produce the desired pharmacological effect (Wang and James, 2006).

A.5.3 Methyltransferases (MTs)

MTs are involved in the transfer of a methyl group from S-adenosylmethionine (SAM) to xenobiotics or endogenous substrates that contain, -C, -O, -N, or -S functional groups. Unlike other conjugation reactions, methylation of xenobiotics results in more hydrophobic metabolites, except for cases of N-methylation of pyridine-containing xenobiotics such as nicotine and S-methylation of thioethers. The methyl group bound to sulfonium ion in SAM has the characteristic of a carbonium ion and is transferred to xenobiotics or endogenous substrates by nucleophilic attack from an electron-rich heteroatom. There are a number of MTs distributed in the body. The major ones contributing to xenobiotic methylation are nicotinamide *N*-methyltransferase (NNMT), thiopurine methyltransferase (TPMT), thiol methyltransferase (TMT), catechol-*O*-methyltransferase (COMT), and histamine *N*-methyltransferase (HNMT).

NNMT methylates compounds containing pyridine rings such as nicotinamide or an indole ring such as tryptophan and serotonin. S-methylation is catalyzed by two enzymes, TPMT and TMT. TPMT is found in the cytosol and preferentially methylates aromatic and heterocyclic compounds such as thiopurine drugs. TMT, on the other hand, is a microsomal enzyme that preferentially methylates aliphatic sulfhydryl compounds.

TPMT and TMT have distinct inhibitor sensitivity, making it easy to distinguish their relative contribution in methylation of xenobiotics. 2,3-Dichloro-α-methylbenzylamine (DCMB) is an inhibitor of TMT while *p*-anisic acid is an inhibitor of TPMT. Although these two enzymes are independently regulated, their expression is determined by genetic factors. Most notably, TPMT genetic polymorphism is responsible for clinically significant interindividual variations in the toxicity and therapeutic efficacy of thiopurine drugs (Weinshilboum et al., 1999). COMT was the first MT to be characterized biochemically. COMT is responsible for the O-methylation of both endogenous and xenobiotic catechol compounds. The substrates of COMT are neurotransmitters such as dopamine, norepinephrine, and epinephrine, and catechol drugs such as the anti-Parkinson's disease agent L-dopa and antihypertensive methyldopa.

A.5.4 NATs

NATs catalyze the acetyl coenzyme A (CoA)-dependent N-acetylation of arylamines and arylhydrazines, and O-acetylation of the *N*-hydroxyarylamines. NATs are cytosolic enzymes and found in liver and many other tissues of most mammals. Humans express two distinct isozymes, designated NAT1 and NAT2, which showed 75–90% sequence homology. NAT1 expressed in many tissues, whereas NAT2 is primarily in the gut and liver. Overlapping substrate specificity and species differences in N-acetylation (human vs. rat) have been characterized. In most cases, this reaction is generally considered to result in the detoxification of potentially toxic exogenous compounds. However, NATs are also involved in bioactivation reactions via O-acetylation of *N*-hydroxyarylamines to unstable acetoxy esters that decompose to highly reactive mutagens that form adducts with cellular macromolecules.

N-acetylation reaction occurs in two sequential steps. In the first step, the acetyl group from acetyl-CoA is transferred to a cysteine residue in the NAT active site with release of the CoA (Figure A.6). In the second step, the acetyl group is transferred from the acetylated enzyme to the amino group of the substrate with regeneration of the enzyme. The basicity of the amine determines the rate of N-acetylation. For basic amines, the rate of N-acetylation is determined by the first step, whereas for weak amines, the rate of N-acetylation is determined by the second step.

A.5.5 GSTs

The GST family of enzymes catalyzes the nucleophilic attack of the tripeptide (γ-glu-cys-gly) (GSH) on a wide variety of soft electrophiles, formed during Phase I oxidation of xenobiotics. GSTs play an important role not only in the detoxification of electrophilic xenobiotics but also in the inactivation of endogenous secondary metabolites formed during oxidative stress such as α-β-

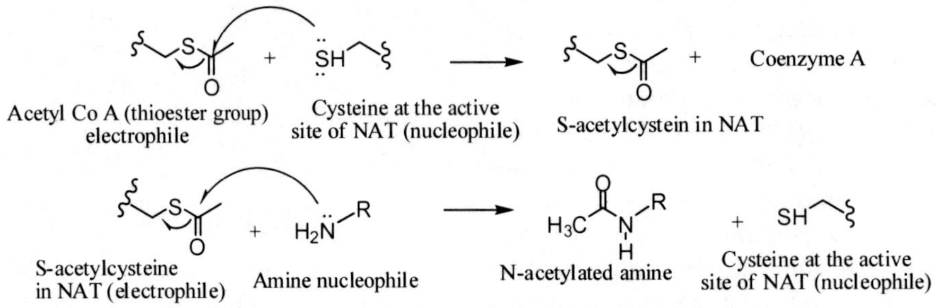

FIGURE A.6. Mechanism of N-acetylation.

unsaturated aldehydes, quinones, epoxides and, hydroperoxides. There are two GST superfamilies: (1) the membrane-bound GST isozymes and leukotriene C_4 synthetase; and (2) the cytosolic soluble enzymes, each of which displays different intracellular distribution and distinct catalytic, as well as noncatalytic binding properties. Thirteen different human GST subunits, GSTA1 through GSTA4, GSTM1 through GSTM5, GSTP1, GSTT1, and GSTT2, and GSTZ1 have been identified belonging to seven distinct classes: alpha (α), mu (μ), omega (ω), pi (π), sigma (σ), theta (θ), and zeta (ζ) (Hayes and Pulford, 1995). GSTs appear to be ubiquitously distributed in human tissues. Some examples of clinically significant drugs that form GSH conjugates include acetaminophen, sulfonamides, irinotecan, carbamazepine, rotonavir, clozapine, procainamide, hydralazine, cyclosporine A, diclofenac, estrogens, and tamoxifen (Zhou et al., 2005).

Although there is a strong evidence of GSH conjugation being a detoxification pathway, there are many examples where it has been implicated in the formation of metabolites that maybe cytotoxic, mutagenic, or carcinogenic (Parkinson and Ogilvie, 2008).

A.5.6 Amino Acid Conjugation

Amino acid conjugation is observed in substrates containing carboxylic acid or aromatic hydroxylamines. Amino acid conjugation with carboxylic acids involves three steps (Figure A.7): (1) activation of carboxylic acid with ATP to generate an acyl adenylate and pyrophosphate; (2) reaction of acyl adenylate with CoA

yielding a reactive acyl-CoA; and (3) linkage of the activated acyl group to the amino group of amino acid. Aromatic hydroxylamines conjugate with the carboxylic acid of the amino acid such as serine and proline. The amino acid is activated by aminoacyl-tRNA, which reacts with an aromatic hydroxylamine to form a reactive N-ester (Kato and Yamazoe, 1994).

Amino acid conjugation of xenobiotics depends on other groups around the aromatic ring system or on steric hindrance that may be found around the carboxylic acid. Where conjugation of amino acid to carboxylic groups is a detoxification pathway, conjugation of glucuronide to carboxylic acid may lead to toxicity. However, amino acid conjugation of hydroxylamines produces N-esters that upon degradation can form electrophilic nitrenium and carbonium ions (Parkinson and Ogilvie, 2008). Examples of compounds that undergo amino acid conjugation are benzoic acid (glycine), valporic acid (glutamine, glutamate, and glycine), and ibuprofen (taurine).

A.6 FACTORS AFFECTING DME ACTIVITIES

Clearance and plasma half-life are the two important determinants of drug disposition since these parameters are used to establish the doses and dosing regimen of drugs. The majority (>75%) of marketed drugs are eliminated from the body, at least in part, by metabolism, and therefore, metabolism plays a major influential role in interindividual variability in drug clearance in humans. Several factors such as disease state, age,

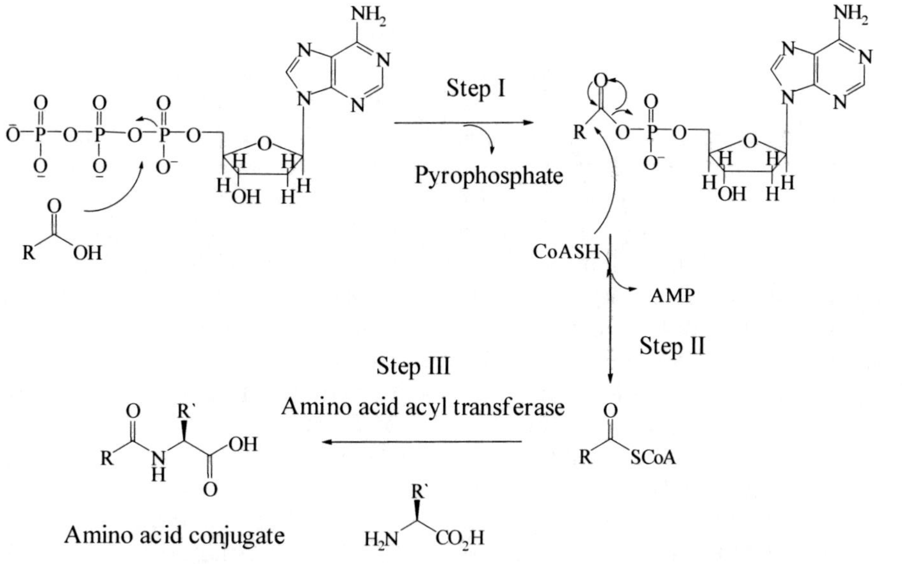

FIGURE A.7. Mechanism of amino acid conjugation.

gender, pharmacogenetics, and comedications may alter the DME activities, resulting in variable drug clearance in humans.

A.6.1   Species and Gender

DMEs in mammals have common ancestral roots. However, species- and gender-related differences in drug metabolism are well established and have led to the characterization of DMEs involved in the metabolism and their differences in expression and catalytic activities across several species including humans. Interspecies specificity of some DMEs can differ significantly (Martignoni et al., 2006). The species-specific isoforms of CYP1A, 2C, 2D, and 3A families show appreciable interspecies differences in the catalytic activity and caution should be exercised when extrapolating metabolism data from animals to humans. Similarly, when AO is a major enzyme involved in metabolic clearance, species selection for pharmacokinetics and toxicology testing is critical. AO activity is high in monkey and human, moderate in rodent, and not detected in dog liver. In addition, some gender-related differences in drug metabolism have also been reported. CYP1A2, CYP2D6, CYP2E1, TMT, GT, and COMT substrates exhibit higher clearance in men, while CYP3A4 and XO substrates exhibit higher clearance in women. No gender-related differences were observed in the clearance of CYP2C9, 2C19, XO, and NAT substrates (Franconi et al., 2007). Thus, in addition to identifying metabolites, the enzymes involved for the formation of major metabolites of an NCE at early stages in the drug development process can be critical in the selection of the species for pharmacokinetic and toxicological testing.

A.6.2   Polymorphism of DMEs

Genetic variability in DMEs is a significant contributor to the variability in human drug pharmacokinetics. CYP2C9, CYP2C19, CYP2D6, UGT1A1, NAT, and TMT are particularly good examples of the importance of knowing polymorphisms in DMEs. CYP2D6 polymorphisms are of the major concern as many of its substrates have narrow therapeutic margin especially in psychiatry, where approximately 50% of the patients use at least one drug primarily metabolized by CYP2D6, which can be a reason for adverse drug reactions and poor response to treatment. There are three categories of individuals (poor metabolizers [PM], extensive metabolizers [EM], and ultrarapid metabolizers) depending on the nature of CYP2D6 polymorphism. Approximately 7–10% of Caucasians, 0–5% of Africans, and 0–1% of Asians lack CYP2D6 activity due to the presence of one or several mutant alleles at the CYP2D6 gene locus, and these individuals are called PMs. They do not respond to codeine therapy due to their inability

of formation of the active metabolite, morphine, catalyzed by CYP2D6. Compared with normal or EMs, PMs demonstrate markedly greater area-under-the-curve (AUC) values for parent drugs that are metabolized by CYP2D6, and therefore require lower doses to achieve therapeutic effects. Other polymorphic P450 enzymes are CYP2C9 and CYP2C19. Approximately 20% of Asians and 3% Caucasians are PMs of CYP2C19. The patients carrying allelic variants CYP 2C9*6 or CYP2C19*2 are shown to be susceptible to neutropenia on the treatment of anticancer drug indisulam, which is metabolized primarily by CYP2C9 and 2C19.

The polymorphic Phase II enzymes, which have received attention, are UGT1A1 and NAT2. Sixty-two allelic variants have been reported for the UGT1A1 gene and 16 allelic variants of UGT1A1 result in the less lethal, type II Crigler-Najjar (CN) syndrome. The active metabolite of irinotecan, SN38, and the cancer drug etoposide are drugs that are cleared primarily by UGT1A1 and have narrow therapeutic index (TI). To maintain drug concentrations within the therapeutic window, it is critical to genotype patients to be treated with such drugs. Genetic polymorphism for N-acetylation has been well documented in humans, hamsters, rabbits, and mice. Slow acetylator phenotype is high in Middle Easterners (~92%), intermediate in Caucasian and African populations, and low in Asian populations (Sirot et al., 2006). Understanding the expression pattern of an enzyme is therefore important as the polymorphism has been reported to result in altered enzyme activity, which in turn has an effect on drug clearance (Salman et al., 2009).

A.6.3   Comedication and Diet

There is a wealth of information on the alteration of DMEs catalytic activity by diet or coadministered drugs. Many of the DMEs are inducible and can be inhibited. Therefore, the inhibition and/or induction of DMEs by one drug could have a significant effect ranging from loss of efficacy to serious adverse drug reactions in the disposition of the other drug. The risk is more profound if the victim drug has high intrinsic clearance or is selectively cleared via only one major pathway.

Significant decrease of exposures of several drugs such as midazolam, amitriptyline, cyclosporine, digoxin, indinavir, irinotecan, warfarin, alprazolam, simvastatin, and ethinylestradiol have been reported in the presence of Saint-John's-wort, which has been shown to be a potent inducer of both CYP3A4. Induction of Phase I and Phase II DMEs is regulated primarily by a large family of nuclear receptors such as AhR, PXR, and CAR. CYP1A2, 2B6, and 3A4 induction are suggested to be indicative of activation of AhR, CAR, and PXR receptors, respectively.

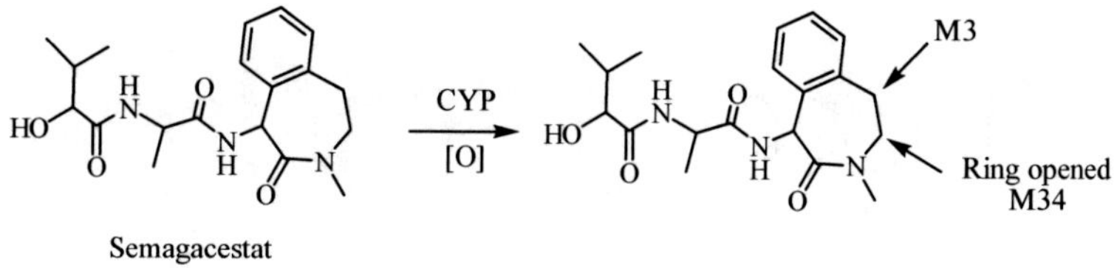

FIGURE A.8. Oxidation of semagacestat (Yi et al., 2010).

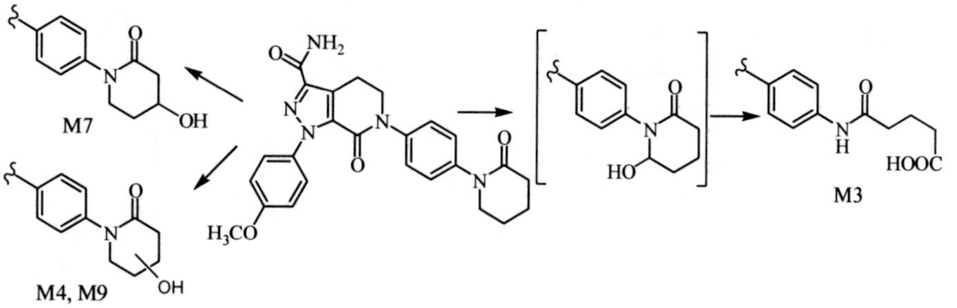

FIGURE A.9. Aliphatic hydroxylation of apixaban.

A.7 BIOTRANSFORMATION REACTIONS

A.7.1 Oxidation

Oxidation is the most common Phase I reaction and can be mediated by multiple DMEs, both CYPs and non-CYPs. A variety of substrates and functional groups can be susceptible to oxidation. Examples of different types of DME-mediated oxidation are listed below.

Aliphatic hydroxylation is a CYP-mediated reaction that often can result in multiple hydroxylated products; a product forms preferentially from the most stable radical (resonance-stabilized).

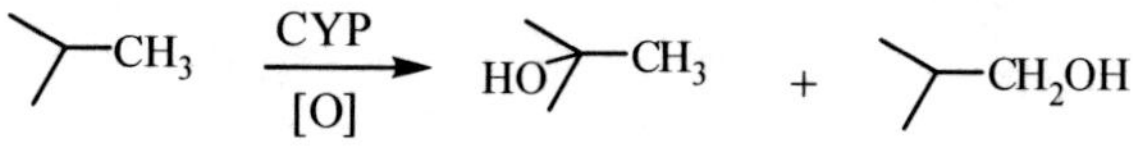

Oxidation of the lactam ring of semagacestat (Figure A.8) results in the formation of two metabolites, M3 and M34. M3 is a product of hydroxylation of the benzylic carbon, while M34 is formed after ring opening followed by hydroxylation of carbon atom next to nitrogen and addition of a water molecule (Yi et al., 2010).

Similarly, oxidation of the lactam moiety of apixaban results in the formation of three isomeric hydroxylated products (M4, M7, M9), as well as the ring opening metabolite (M3) as shown in Figure A.9 (Wang et al., 2010).

Aromatic hydroxylation can be mediated by CYPs and cytosolic enzymes such as AO/XO. CYP-mediated reactions often result in the formation of isomeric hydroxylated products. Due to resonance-stabilization, for monosubstituted phenyl groups, the rate of formation of hydroxylated metabolites is usually as follows: para > ortho > meta.

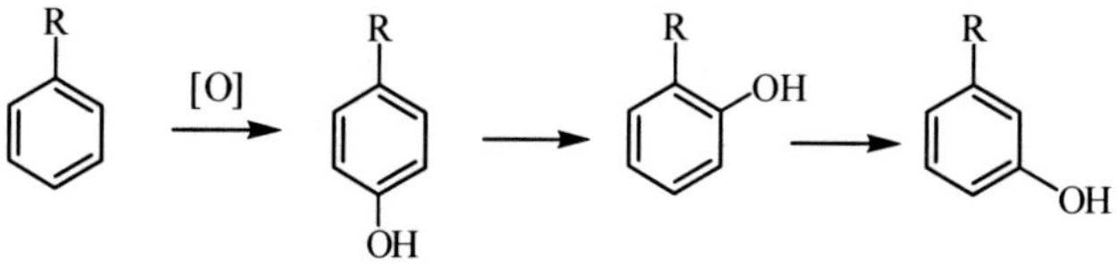

Oxidation of lasofoxifene is primarily catalyzed by CYP3A4/3A5 and CYP2D6 and leads to the formation of isomeric phenols followed by conjugation (Figure A.10) (Prakash et al., 2008).

Aromatic hydroxylation of heterocycles can lead to formation of the carbonyl group-containing metabolites. For example, CYP-mediated oxidation of clopidogrel (Kazui et al., 2010) formed 2-oxo-clopidogrel (thioester metabolite), and AO-mediated oxidation of zoniporide resulted in the formation of a lactam (Dalvie et al., 2010) (Figure A.11).

Epoxidation. Compounds containing double, triple bonds, and aromatic groups can be subjected to CYP-mediated epoxidation.

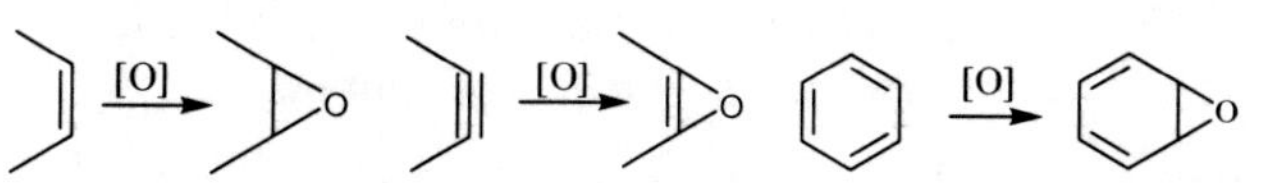

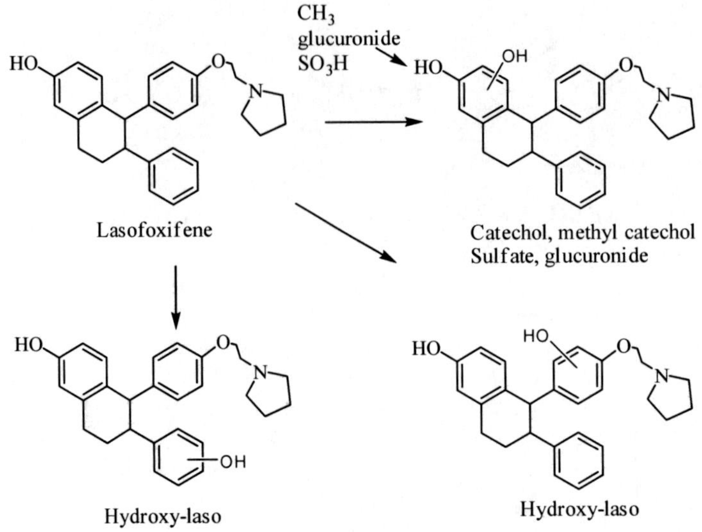

FIGURE A.10. Aromatic hydroxylation followed conjugation of lasofoxifene. (Prakash et al., 2008).

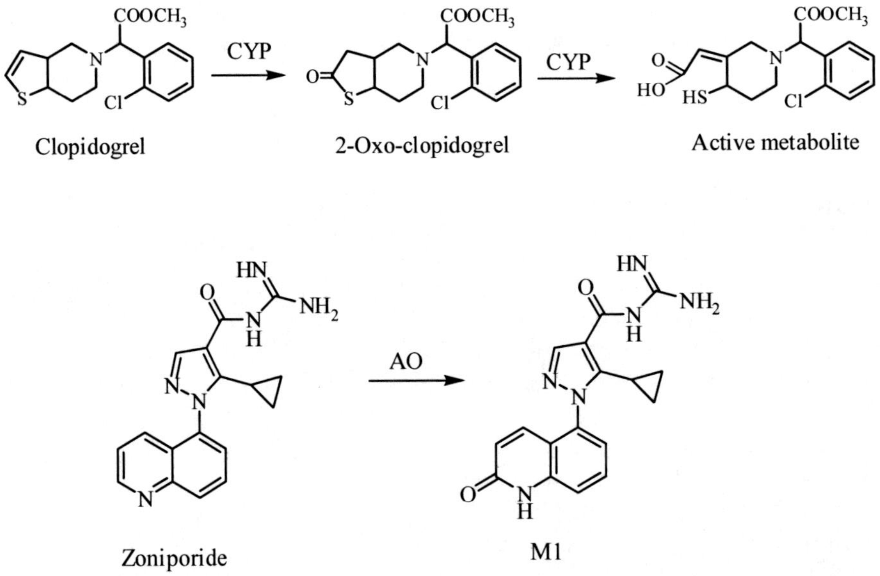

FIGURE A.11. Oxidation of clopidogrel and zoniporide.

Epoxidation results in the formation of unstable products, which hydrolyze by EHs to form diols or react with nucleophilic groups in macromolecules to initiate toxicological effects. (Guengerich, 2001). Epoxides can also be further biotransformed to stable metabolites such as in case of formation of a carboxylic acid metabolite of erlotinib (Figure A.12) (Ling et al., 2006).

Oxidation of Amines. Secondary, tertiary, and aromatic amines are subjected to N-oxidation, which is mediated by a large spectrum of enzymes including CYPs and FMOs. This reaction can result in the formation of either *N*-oxides or hydroxylamines.

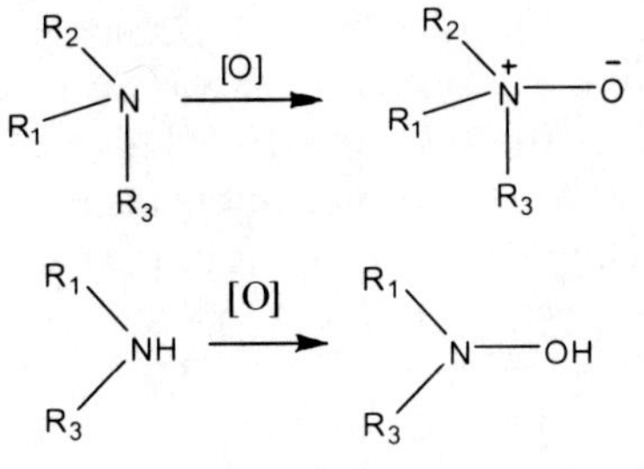

Multiple enzymes can be responsible for the oxidation at the same position as shown for voriconazole

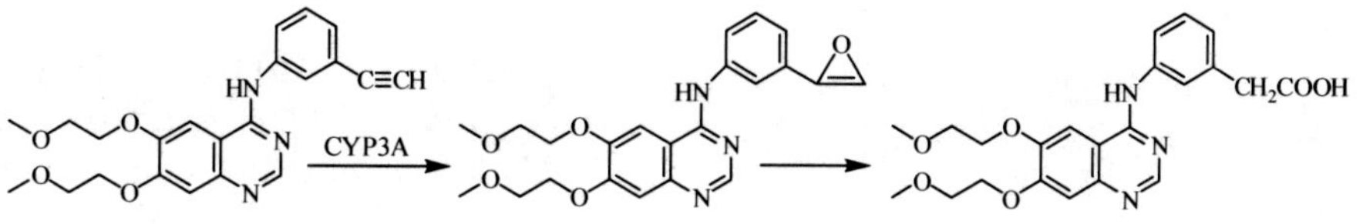

FIGURE A.12. Metabolism of erlotinib to carboxylic acid metabolite.

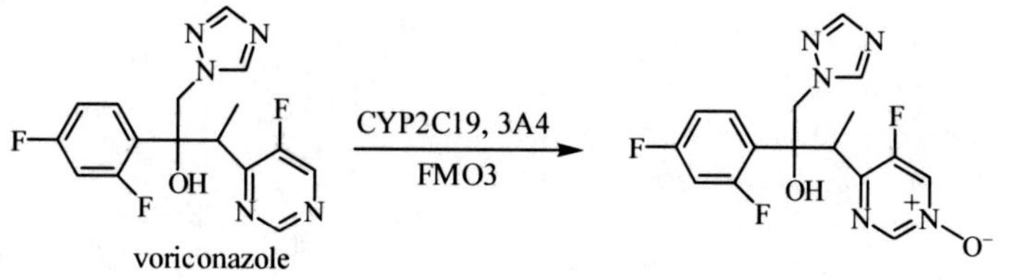

FIGURE A.13. N-oxidation of voriconazole by CYP2C19, 3A4, and FMO3.

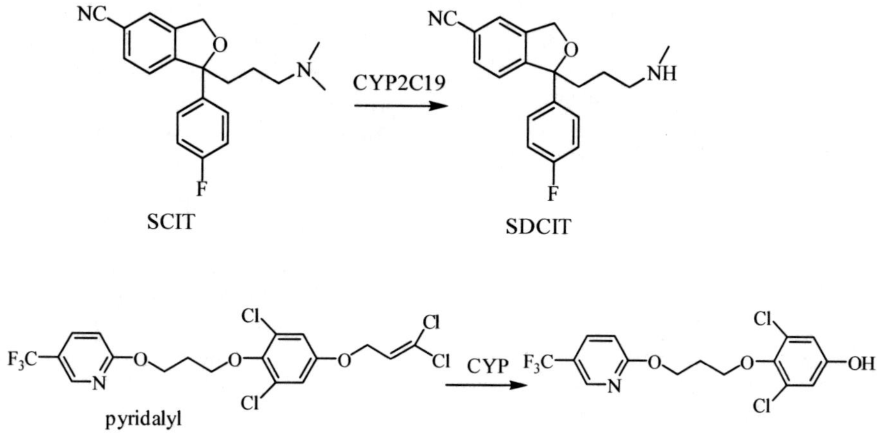

FIGURE A.14. N-demethylation of SCIT and O-demethylation of pyridalyl.

where the N-oxidation is mediated by CYPs and FMO3 (Figure A.13) (Yanni et al., 2010).

N- and O(S)-dealkylation is a common reaction involving drugs containing a secondary or tertiary amine, alkoxy group, or an alkyl-substituted thiol. The reaction is mediated by CYPs and proceeds in two steps: oxidation of a carbon atom alpha to N, O, or S followed by the decomposition of an unstable intermediate.

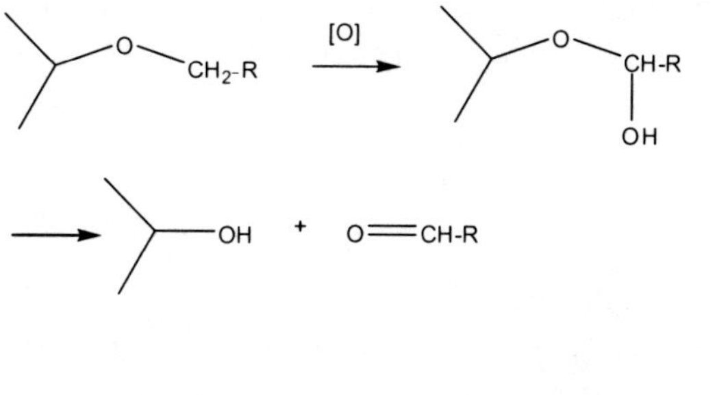

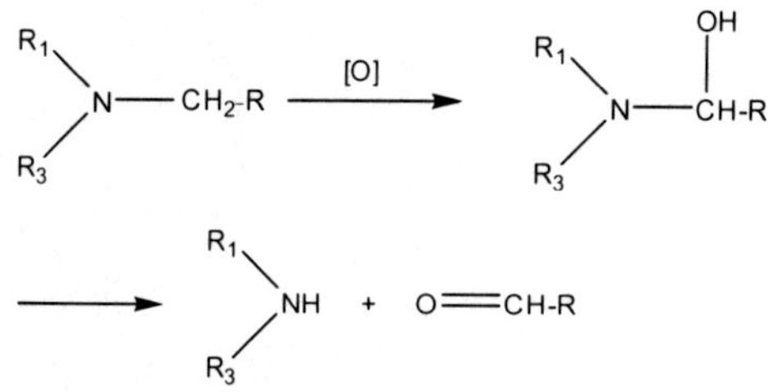

N-demethylation of *S*-citalopram (SCIT) (Rudberg et al., 2009) and O-dealkylation of pyridalyl (Nagahori et al., 2009) are both mediated by CYPs (Figure A.14).

Oxidative ester and amide hydrolysis is another common reaction that involves a multistep process: oxidation, dissociation of an unstable intermediate, and decarboxylation.

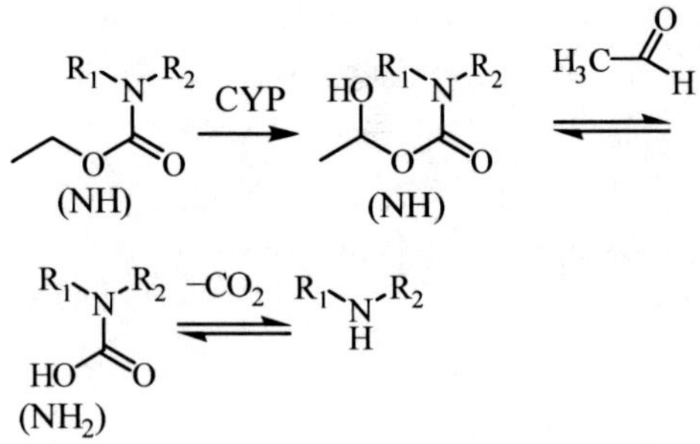

Hydrolysis can also be mediated by non-CYP enzymes such as hydrolases through nonoxidative process. The distinction between the oxidative reaction and nonoxidative hydrolysis is demonstrated by the dependence on NADPH-P450 reductase and NADPH (Guengerich, 2001). Oxidative hydrolysis mediated by CYP 3A4 and 2C19 was a major metabolic pathway for the loratadine to its active metabolite desloratadine (Figure A.15) (Ramanathan et al., 2007; Ghosal et al., 2009).

On the contrary, hydrolysis of semagacestat is mediated by non-CYP enzymes (Figure A.16) (Yi et al., 2010), and this process does not include the oxidation step.

Oxidative deamination is mediated by both CYPs and MAOs. For example, deamination of sumatriptan is mediated by MAO, while SCIT is metabolized by CYP2C19 (Figure A.17) (Dixon et al., 1994; Rudberg et al., 2009).

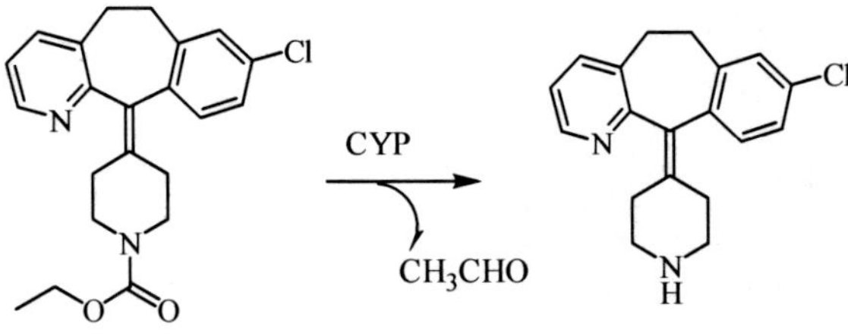

Loratadine Desloratadine

FIGURE A.15. Oxidative hydrolysis of loratadine to its active metabolite desloratadine.

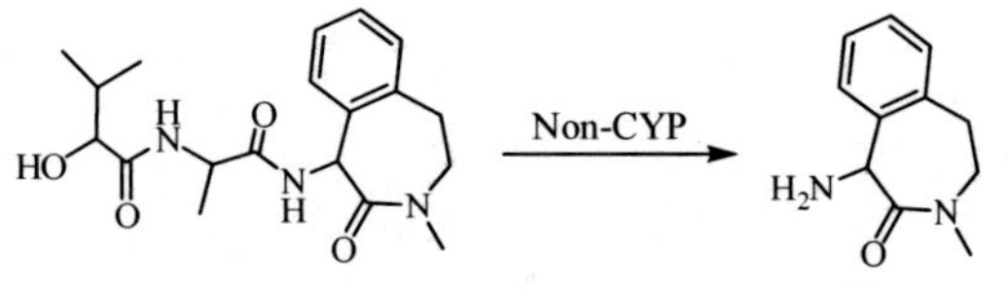

Semagacestat

FIGURE A.16. Non-CYP-mediated hydrolysis of semagacestat.

A.7.2 Reduction

Reduction can be catalyzed by a variety of enzymes. Flavoproteins are most often responsible for reducing endogenous compounds. CYPs are also able to mediate reductive reactions due to the reactivity of ferrous form (Isin and Guengerich, 2007). Drugs containing nitro groups are sequentially reduced to hydroxyl amines (Figure A.18) (Gu et al., 2010; Guise et al., 2010).

$$RNO_2 \rightarrow RN{=}O \rightarrow RNHOH \rightarrow RNH_2.$$

N-hydroxyl amines and *N*-oxides are reduced by P450s and other enzymes such as AO. The reduction reactions catalyzed by CYPs seem to involve the transfer of an electron from the iron to the substrate.

AKRs are also able to reduce ketones to form the corresponding alcohols.

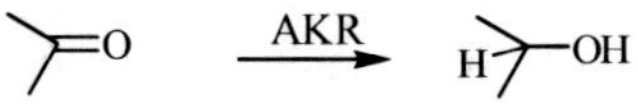

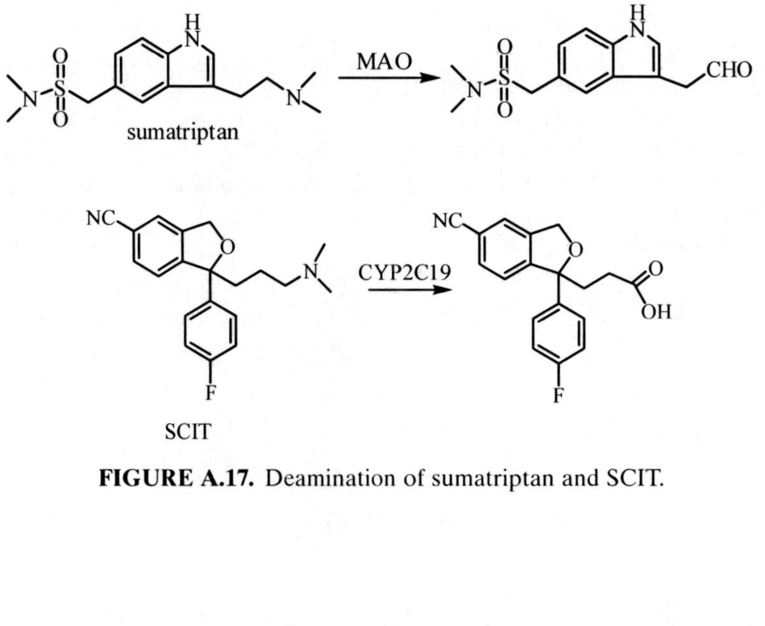

FIGURE A.17. Deamination of sumatriptan and SCIT.

Reduction of tibolone to hydroxytibolone can be carried through by multiple isoforms of AKR (Figure A.19) (Kang and Kim, 2008).

A.7.3   Conjugation Reactions

Conjugation reactions or *Phase II reactions* include glucuronidation, sulfonation, GSH, and amino acid conjugation. Carboxyl, hydroxyl, amino, and sulfhydryl groups are subjected to conjugation.

The same functional group can be a target of multiple conjugation reactions as shown for lasofoxifene (Figure A.20) (Prakash et al., 2008).

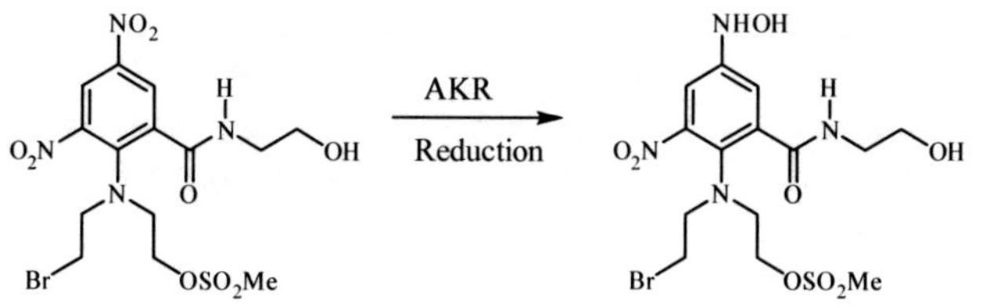

FIGURE A.18. The reduction of nitro compounds.

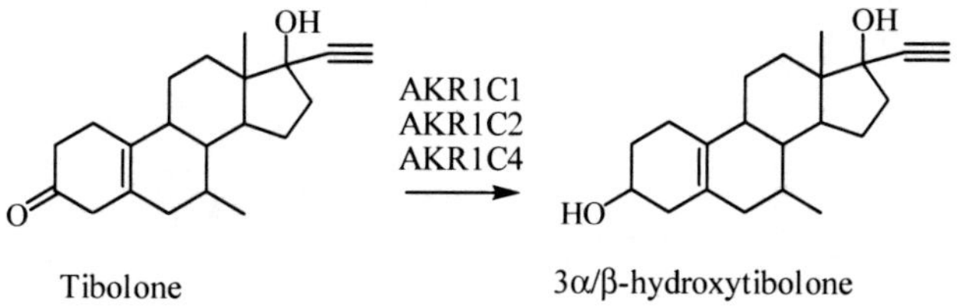

FIGURE A.19. Reduction of tibolone.

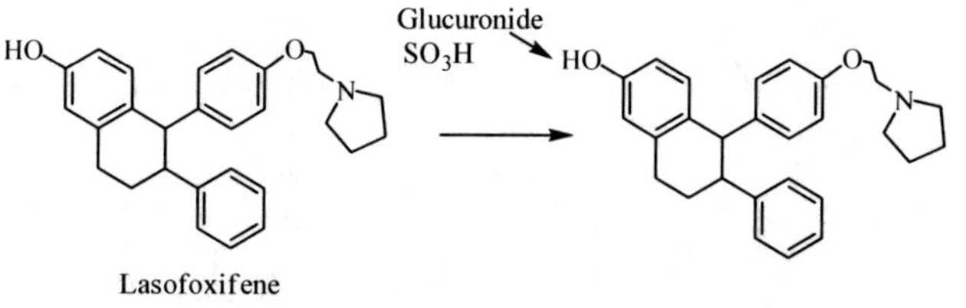

FIGURE A.20. Conjugation of lasofoxifene.

Conjugation in most cases is preceded by Phase I reactions. Flupirtine is first hydrolyzed by esterases to form a diaminopyridine metabolite, which is then converted to D13223 by NAT (Figure A.21) (Methling et al., 2009).

Drugs containing a carboxylic acid group can be metabolized via conjugation with glucuronic acid to from the acyl glucuronides or an amino acid such as glycine, alanine, glutamine, and taurine. For example, amino acid conjugation is the predominant route of metabolism of salicylic acid, with salicyluric acid (the glycine conjugate) accounting for 75% of aspirin's excretion in urine. Carboxylic acid groups are the most common target for conjugation, but there are cases when other groups such as ketones are involved in conjugative reactions (Figure A.22) (Herebian et al., 2010).

GSH conjugation is the major pathway of detoxification of reactive electrophiles. The biotransformation reactions with GSH include either the direct addition of GSH to the molecule (Michael acceptors), or conjugation following bioactivation such as the formation of reactive epoxides or quinone/quinone imine intermediates as demonstrated for erlotinib (Figure A.23) (Li et al., 2010).

A.8   SUMMARY

The field of drug metabolism has come a long way since the first recorded study conducted by Alexander Ure in 1841. Now it is well established that knowledge of DMEs has become one of key importance in the drug development process, which has several steps, from identifying a targeted disease model to choosing a series of compounds that are selective to their target. Role of DMEs is only one variable among many that define a successful path of an NCE from discovery to its clinical efficacy. The use of subcellular fractions has been the most popular and invaluable tool for predicting drug disposition in humans and for choosing the appropriate toxicology species. Furthermore, there has been an explosion in technological advancements of tools available that allow scientists to look at various pathways involved in the fate of xenobiotics.

A majority of the human enzymes involved in drug metabolism such as the CYPs, UGTs, MAOs, FMOs,

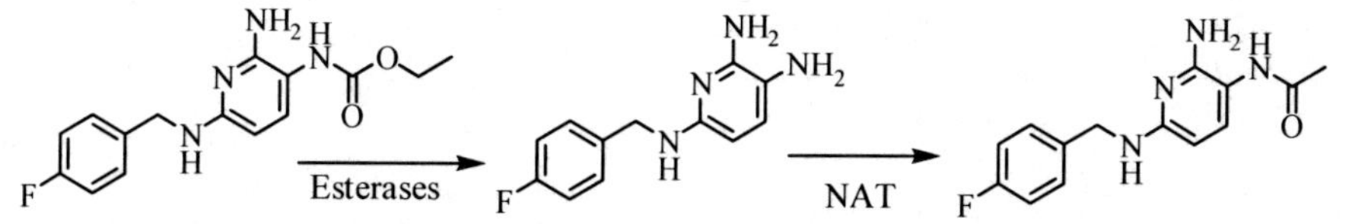

FIGURE A.21. N-acetylation of flupirtine.

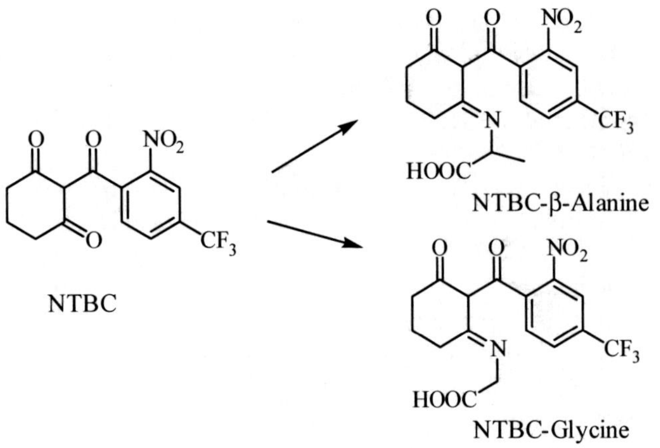

FIGURE A.22. Glycine and β-alanine conjugation of NTBC.

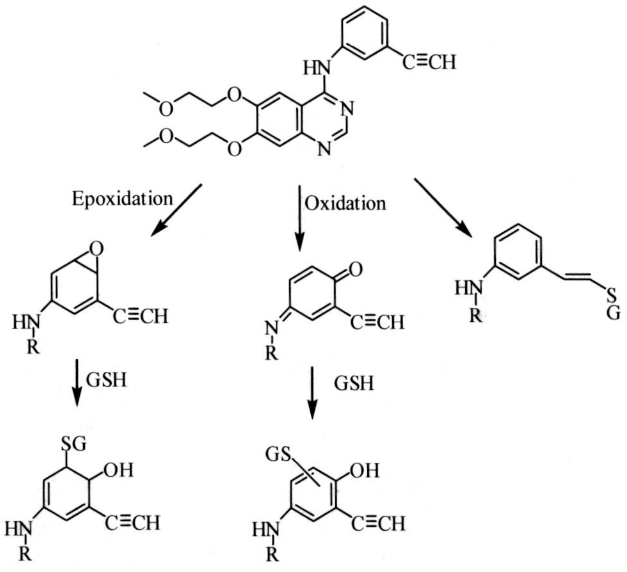

FIGURE A.23. Bioactivation and detoxification of erlotinib (Li et al., 2010).

crystal structures allows for *in silico* predictive models to be constructed.

Knowledge of non-CYP metabolic pathways has also expanded. Tools are available to assess if enzymes such as MAOs, UGTs, SULTs, AOs, and XOs contribute to the clearance. As medicinal chemists become more effective in limiting the CYP-mediated metabolism of NCEs, non-CYP metabolic pathways will become increasingly involved in drug clearance, and consequently, better tools are necessary to evaluate the role such enzymes may play in the metabolism of NCEs early in the drug development process.

Technological advancement in tools available to identify metabolites (e.g., the introduction of ultrapressure liquid chromatography coupled with high-resolution mass spectrometry) allows for the rapid characterization of metabolites and identification of metabolic hot spots and bioactivation pathways (Prakash et al., 2007). When done in an iterative manner, this can aid in designing compounds with better pharmacokinetic properties.

ACKNOWLEDGMENT

The authors would like to thank Dr. Lewis Klunk for reviewing this article and providing insightful comments.

REFERENCES

Baillie TA (2003) Drug metabolizing enzymes: Cytochrome P450 and other enzymes. In *Drug Discovery and Development*, Lee JS, Obach RS, Fisher MB, eds., pp. 147–154. Marcel Dekker, New York.

Barski OA, Tipparaju SM, Bhatnagar A (2008) The aldo-keto reductase super family and its role in drug metabolism and detoxification. *Drug Metab Rev* 40:553–624.

Beetham JK, Grant D, Arand M, Garbarino J, Kiyosue T, Pinot F, Oesch F, Belknap WR, Shinozaki K, Hammock BD (1995) Gene evolution of epoxide hydrolases and recommended nomenclature. *DNA Cell Biol* 14:61–71.

Benedetti M, Womsley R, Baltes E (2006) Involvement of enzymes other than CYPs in the oxidative metabolism of xenobiotics. *Expert Opin Drug Metab Toxicol* 2:895–921.

Blanchard RL, Freimuth RR, Buck J, Weinshilboum RM, Coughtrie MW (2004) A proposed nomenclature system for the cytosolic sulfotransferase (SULT) super family. *Pharmacogenetics* 14:199–211.

Bossyut X, Blanckeart N (1994a) Carrier-mediated transport of intact UDP-glucuronic acid into the lumen of endoplasmic-reticulum-derived vesicles from rat liver. *Biochem J* 302:261–269.

UGTs, and some of the lesser encountered DMEs have been isolated or heterologously expressed and characterized, and more recently, some crystal structures have been solved, providing a better understanding of the structural basis for their broad specificity. Several CYP isoform-specific inhibitors and substrates are known, which allow for the elucidation of some primary metabolic pathways and an evaluation of the potential for drug–drug interactions. Progress in molecular genetics of drug metabolizing and related enzyme systems has enabled some understanding of the molecular basis for inherited traits such as poor and extensive metabolizers. Expansion of our knowledge of the reaction mechanisms of CYP enzymes, their substrate specificities, and

Bossyut X, Blanckeart N (1994b) Functional characterization of carrier-mediated transport of uridine diphosphate N-acetylgylucosamine across the endoplasmic reticulum membrane. *Eur J Biochem* 223:981–988.

Brown CM, Reisfeld B, Mayeno AN (2008) Cytochromes P450: A structure-based summary of biotransformations using representative substrates. *Drug Metab Rev* 40:1–100.

Cashman JR, Zhang J (2006) Human flavin-containing mono-oxygenases. *Annu Rev Pharmacol Toxicol* 46:65–100.

Coon MJ (2005) Cytochrome P450: Nature's most versatile biological catalyst. *Annu Rev Pharmacol Toxicol* 45:1–25.

Crettol S, Petrovic N, Murray M (2010) Pharmacogenetics of phase I and phase II drug metabolism. *Curr Pharm Des* 16:204–219.

Csala M, Staines AG, Bangheyi G, Mandl J, Coughtrie M, Burchell B (2004) Evidence for multiple glucuronide transporters in rat liver microsomes. *Biochem Pharmacol* 68:1353–1362.

Dalvie D, Zhang C, Chen W, Smolarek T, Obach RS, Loi C-M (2010) Cross-species comparison of the metabolism and excretion of zoniporide: Contribution of aldehyde oxidase to interspecies differences. *Drug Metab Dispos* 38:641–654.

Decker M, Arand M, Cronin A (2009) Mammalian epoxide hydrolases in xenobiotic metabolism and signalling. *Arch Toxicol* 83:297–318.

Dixon CM, Partk GR, Tarbit MH (1994) Characterization of the enzyme responsible for the metabolism of sumatriptan in human liver. *Biochem Pharmacol* 47:1253–1257.

Franconi F, Brunelleschi S, Steardo L, Cuomo V (2007) Gender differences in drug responses. *Pharmacol Res* 55:81–95.

Fretland AJ, Omiecinski CJ (2000) Epoxide hydrolases: Biochemistry and molecular biology. *Chem Biol Interact* 129:41–59.

Fura A, Shu YZ, Zhu M, Hanson RL, Roongta V, Humphreys WG (2004) Discovering drugs through biological transformation: Role of pharmacologically active metabolites in drug discovery. *J Med Chem* 47:4339–4351.

Gamage N, Barnett A, Hempel N, Duggleby RG, Windmill KF, Martin JL, McManus ME (2006) Human sulfotransferases and their role in chemical metabolism. *Toxicol Sci* 90:5–22.

Garattini E, Fratelli M, Terao M (2008) Mammalian aldehyde oxidases: Genetics, evolution and biochemistry. *Cell Mol Life Sci* 65:1019–1048.

Ghosal A, Gupta S, Ramanathan R, Yuan Y, Lu X, Su AD, Alvarez N, Zbaida S, Chowdhury SK, Alton KB (2009) Metabolism of loratadine and further characterization of its *in vitro* metabolites. *Drug Metab Lett* 3:162–170.

Grunwell JR, Bertozzi CR (2002) Carbohydrate sulfotransferases of the GalNAc/Gal/GlcNAc6ST family. *Biochemistry* 41:13117–13126.

Grunwell JR, Rath VL, Rasmussen J, Cabrilo Z, Bertozzi CR (2002) Characterization and mutagenesis of Gal/GlcNAc-6-O-sulfotransferases. *Biochemistry* 41:15590–15600.

Gu Y, Atwell GJ, Wilson WR (2010) Metabolism and excretion of the novel bioreductive prodrug PR-104 in mice, rats, dogs, and humans. *Drug Metab Dispos* 38:498–508.

Guengerich FP (2001) Common and uncommon cytochrome P450 reactions related to metabolism and chemical toxicity. *Chem Res Toxicol* 14:611–650.

Guengerich FP (2003) Cytochromes P450, drugs, and diseases. *Mol Interv* 3:194–204.

Guengerich FP, Rendic S (2010) Update information on drug metabolism systems-2009, Part I. *Curr Drug Metab* 11:1–3.

Guise CP, Abbattista MR, Singleton RS, Holford SD, Connolly J, Dachs GU, Fox SB, Pollock R, Harvey J, Guilford P, Donate F, Wilson WR, Patterson AV (2010) The bioreductive prodrug PR-104A is activated under aerobic conditions by human aldo-keto reductase 1C3. *Cancer Res* 70:1573–1584.

Hall BS, Wu X, Hu L, Wilkinson SR (2010) Exploiting the drug-activating properties of a novel trypanosomal nitroreductase. *Antimicrob Agents Chemother* 54:1193–1199.

Hayes JD, Pulford DJ (1995) The glutathione S-transferase supergene family: Regulation of GST and the contribution of the isoenzymes to cancer chemoprotection and drug resistance. *Crit Rev Biochem Mol Biol* 30:445–600.

Herebian D, Lamshoeft M, Mayatepek E, Spiekerkoetter U (2010) Identification of NTBC metabolites in urine from patients with hereditary tyrosinemia type 1 using two different mass spectrometric platforms: Triple stage quadrupole and LTQ-Orbitrap. *Rapid Commun Mass Spectrom* 24:791–800.

Isin EM, Guengerich FP (2007) Complex reactions catalyzed by cytochrome P 450 enzymes. *Biochim Biophys Acta* 1770:314–329.

Jin Y, Penning TM (2007) Aldo-keto reductases and bioactivation/detoxication. *Annu Rev Pharmacol Toxicol* 47:263–292.

Kalgutkar AS, Gardner I, Obach RS, Shaffer CL, Callegari E, Henne KR, Mutlib AE, Dalvie DK, Lee JS, Nakai Y, O'Donnell JP, Boer J, Harriman SP (2005) A comprehensive listing of bioactivation pathways of organic functional groups. *Curr Drug Metab* 6:161–225.

Kang KW, Kim YG (2008) Bioequivalence studies of tibolone in premenopausal women and effects on expression of the tibolone-metabolizing enzyme AKR1C (aldo-keto reductase) family caused by estradiol. *J Clin Pharmacol* 48:1430–1437.

Kato R, Yamazoe Y (1994) Metabolic activation of N-hydroxylated metabolites of carcinogenic and mutagenic arylamines and arylamides by esterification. *Drug Metab Rev* 26:413–429.

Kazui M, Nishiya Y, Ishizuka T, Hagihara K, Farid NA, Okazaki O, Ikeda T, Kurihara A (2010) Identification of the human cytochrome P450 enzymes involved in the two oxidative steps in the bioactivation of clopidogrel to its pharmacologically active metabolite. *Drug Metab Dispos* 38:92–99.

Kepa JK, Traver RD, Siegel D, Winski SL, Ross D (1997) Regulation and function of NAD(P)H:quinone oxidoreductase (NQO1). *Rev Toxicol (Amsterdam)* 1:53–73.

Lacourciere GM, Armstrong RN (1993) The catalytic mechanism of microsomal epoxide hydrolase involves an ester intermediate. *J Am Chem Soc* 115:10466–10467.

Lacourciere GM, Armstrong RN (1994) Microsomal and soluble epoxide hydrolases are members of the same family of C-X bond hydrolase enzymes. *Chem Res Toxicol* 7:121–124.

Li X, Kamenecka TM, Cameron MD (2010) Cytochrome P450-mediated bioactivation of the EGFR inhibitor erlotinib to a reactive electrophile. *Drug Metab Dispos* 38:1238–1245. [Epub April 9, 2010]

Ling J, Johnson KA, Miao Z, Rakhit A, Pantze MP, Hamilton M, Lum BL, Prakash C (2006) Metabolism and excretion of erlotinib, a small molecule inhibitor of epidermal growth factor receptor tyrosine kinase, in healthy male volunteers. *Drug Metab Dispos* 34:420–426.

Long DJ, Jaiswal AK (2000) NRH:quinone oxidoreductase2 (NQO2). *Chem Biol Interact* 129:99–112.

Macwana SR, Punj S, Cooper J, Schwenk E, John GH (2010) Identification and isolation of an azoreductase from *Enterococcus faecium*. *Curr Issues Mol Biol* 12:43–48.

Martignoni M, Groothuis GM, de Kanter R (2006) Species differences between mouse, rat, dog, monkey and human CYP-mediated drug metabolism, inhibition and induction. *Expert Opin Drug Metab Toxicol* 2:875–894.

Meloche CA, Falany CN (2001) Expression and characterization of the human 3β-hydroxysteroid sulfotransferases (SULT2B1a and SULT2B1b). *J Steroid Biochem Mol Biol* 77(4–5):261–269.

Methling K, Reszka P, Lalk M, Vrana O, Scheuch E, Siegmund W, Terhaag B, Bednarski PJ (2009) Investigation of the *in vitro* metabolism of the analgesic flupirtine. *Drug Metab Dispos* 37:479–493.

Morisseau C, Hammock BD (2005) Epoxide hydrolases: Mechanisms, inhibitor designs, and biological roles. *Annu Rev Pharmacol Toxicol* 45:311–333. 313 plates.

Nagahori H, Saito K, Tomigahara Y, Isobe N, Kaneko H (2009) Metabolism of pyridalyl in rats. *Drug Metab Dispos* 37:2284–2289.

Novotna R, Wsol V, Xiong G, Maser E (2008) Inactivation of the anticancer drugs doxorubicin and oracin by aldo-keto reductase (AKR) 1C3. *Toxicol Lett* 181:1–6.

Oppermann UCT, Maser E (2000) Molecular and structural aspects of xenobiotic carbonyl metabolizing enzymes. Role of reductases and dehydrogenases in xenobiotic phase I reactions. *Toxicology* 144:71–81.

Parkinson A, Ogilvie B (2008) Biotransformation of xenobiotics. In *Casarett & Doull's Toxicology: The Basic Science of Poisons*, 7th ed., Klaassen CD, ed., pp. 161–304. McGraw-Hill, New York.

Prakash C, Vaz ADN (2009) Drug metabolism: Significance and challenges. In *Nuclear Receptors in Drug Metabolism*, Wen X, ed., pp. 1–41. John Wiley & Sons, Hoboken, NJ.

Prakash C, Shaffer Christopher L, Nedderman A (2007) Analytical strategies for identifying drug metabolites. *Mass Spectrom Rev* 26:340–369.

Prakash C, Johnson KA, Gardner MJ (2008) Disposition of lasofoxifene, a next-generation selective estrogen receptor modulator, in healthy male subjects. *Drug Metab Dispos* 36:1218–1226.

Rafii F, Cerniglia CE (1995) Role of reductive enzymes from human intestinal bacteria in the metabolism of azo dyes and nitro-polycyclic aromatic hydrocarbons. *Mikroökologie Ther* 23:111–123.

Ramanathan R, Reyderman L, Kulmatycki K, Su AD, Alvarez N, Chowdhury SK, Alton KB, Wirth MA, Clement RP, Statkevich P, Patrick JE (2007) Disposition of loratadine in healthy volunteers. *Xenobiotica* 37:753–769.

Rendic S, Guengerich FP (2010) Update information on drug metabolism systems—2009, part II: Summary of information on the effects of diseases and environmental factors on human cytochrome P450 (CYP) enzymes and transporters. *Curr Drug Metab* 11:4–84.

Rock D, Wahlstrom J, Wienkers L (2008) Cytochrome P450s: Drug-drug interactions. *Methods Princ Med Chem* 38:197–246.

Ross D (2004) Quinone reductases multitasking in the metabolic world. *Drug Metab Rev* 36:639–654.

Rudberg I, Reubsaet JLE, Hermann M, Refsum H, Molden E (2009) Identification of a novel CYP2C19-mediated metabolic pathway of S-citalopram *in vitro*. *Drug Metab Dispos* 37:2340–2348.

Salman ED, Kadlubar SA, Falany CN (2009) Expression and localization of cytosolic sulfotransferase (SULT) 1A1 and SULT1A3 in normal human brain. *Drug Metab Dispos* 37:706–709.

Sirot EJ, van der Velden JW, Rentsch K, Eap CB, Baumann P (2006) Therapeutic drug monitoring and pharmacogenetic tests as tools in pharmacovigilance. *Drug Saf* 29:735–768.

Testa B, Kramer SD (2007) The biochemistry of drug metabolism—an introduction: Part 2. Redox reactions and their enzymes. *Chem Biodivers* 4:257–405.

Tukey RH, Strassburg CP (2000) Human UDP-glucuronosyltransferases: Metabolism, expression, and disease. *Annu Rev Pharmacol Toxicol* 40:581–616.

Vass SO, Jarrom D, Wilson WR, Hyde EI, Searle PF (2009) *E. coli* NfsA: An alternative nitroreductase for prodrug activation gene therapy in combination with CB1954. *Br J Cancer* 100:1903–1911.

Wang L, Zhang D, Raghavan N, Yao M, Ma L, Frost CA, Maxwell BD, Chen SY, He K, Goosen TC, Humphreys WG, Grossman SJ (2010) *In vitro* assessment of metabolic drug-drug interaction potential of apixaban through cytochrome P450 phenotyping, inhibition, and induction studies. *Drug Metab Dispos* 38:448–458.

Wang L-Q, James MO (2006) Inhibition of sulfotransferases by xenobiotics. *Curr Drug Metab* 7:83–104.

Weinshilboum RM, Otterness DM, Szumlanski CL (1999) Methylation pharmacogenetics: Catechol O-methyltransferase, thiopurine methyltransferase, and histamine N-methyltransferase. *Annu Rev Pharmacol Toxicol* 39:19–52.

Williams JA, Hyland R, Jones BC, Smith DA, Hurst S, Goosen TC, Peterkin V, Koup JR, Ball SE (2004) Drug-drug interactions for UDP-glucuronosyltransferase substrates: A pharmacokinetic explanation for typically observed low exposure (AUCI/AUC) ratios. *Drug Metab Dispos* 32:1201–1208.

Yanni SB, Annaert PP, Augustijns P, Ibrahim JG, Benjamin DK, Jr., Thakker DR (2010) *In vitro* hepatic metabolism explains higher clearance of voriconazole in children versus adults: Role of CYP2C19 and flavin-containing monooxygenase 3. *Drug Metab Dispos* 38:25–31.

Yi P, Hadden C, Kulanthaivel P, Calvert N, Annes W, Brown T, Barbuch RJ, Chaudhary A, Ayan-Oshodi MA, Ring BJ (2010) Disposition and metabolism of semagacestat, a {gamma}-secretase inhibitor, in humans. *Drug Metab Dispos* 38:554–565.

Zhou S, Chan E, Duan W, Huang M, Chen YZ (2005) Drug bioactivation, covalent binding to target proteins and toxicity relevance. *Drug Metab Rev* 37:41–213.

Zhou SF, Liu JP, Chowbay B (2009) Polymorphism of human cytochrome P450 enzymes and its clinical impact. *Drug Metab Rev* 41:89–295.

INDEX

ADME-Enabling Technologies in Drug Design and Development, First Edition. Edited by Donglu Zhang and Sekhar Surapaneni.
© 2012 John Wiley & Sons, Inc. Published 2012 by John Wiley & Sons, Inc.